Financial Mathematics
A Comprehensive Treatment in Continuous Time

The book has been tested and refined through years of classroom teaching experience. With an abundance of examples, problems, and fully worked-out solutions, the text introduces the financial theory and relevant mathematical methods in a mathematically rigorous yet engaging way.

This textbook provides complete coverage of continuous-time financial models that form the cornerstones of financial derivative pricing theory. Unlike similar texts in the field, this one presents multiple problem-solving approaches, linking related comprehensive techniques for pricing different types of financial derivatives.

Key features:

- In-depth coverage of continuous-time theory and methodology.
- Numerous, fully worked out examples and exercises in every chapter.
- Mathematically rigorous and consistent yet bridging various basic and more advanced concepts.
- Judicious balance of financial theory and mathematical methods.
- Guide to material.

This revision contains:

- Almost 150 pages worth of new material in all chapters.
- An expanded set of solved problems and additional exercises.
- Answers to all exercises.

This book is a comprehensive, self-contained, and unified treatment of the main theory and application of mathematical methods behind modern-day financial mathematics.

The text complements ***Financial Mathematics: A Comprehensive Treatment in Discrete Time,*** by the same authors, also published by CRC Press.

Chapman & Hall/CRC Financial Mathematics Series

Series Editors

M.A.H. Dempster
Centre for Financial Research
Department of Pure Mathematics and Statistics
University of Cambridge, UK

Dilip B. Madan
Robert H. Smith School of Business
University of Maryland, USA

Rama Cont
Department of Mathematics
Imperial College, UK

Robert A. Jarrow
Lynch Professor of Investment Management
Johnson Graduate School of Management
Cornell University, USA

Recently Published Titles

Stochastic Modelling of Big Data in Finance
Anatoliy Swishchuk

Introduction to Stochastic Finance with Market Examples, Second Edition
Nicolas Privault

Commodities: Fundamental Theory of Futures, Forwards, and Derivatives Pricing, Second Edition
Edited by M.A.H. Dempster, Ke Tang

Introducing Financial Mathematics: Theory, Binomial Models, and Applications
Mladen Victor Wickerhauser

Financial Mathematics: From Discrete to Continuous Time
Kevin J. Hastings

Financial Mathematics: A Comprehensive Treatment in Discrete Time
Giuseppe Campolieti and Roman N. Makarov

Introduction to Financial Derivatives with Python
Elisa Alòs, Raúl Merino

The Handbook of Price Impact Modeling
Dr. Kevin Thomas Webster

For more information about this series please visit: https://www.crcpress.com/Chapman-and-HallCRC Financial-Mathematics-Series/book series/CHFINANCMTH

Financial Mathematics
A Comprehensive Treatment in Continuous Time
Volume II

Giuseppe Campolieti
Roman N. Makarov

CRC Press
Taylor & Francis Group
Boca Raton London New York

CRC Press is an imprint of the
Taylor & Francis Group, an **informa** business
A CHAPMAN & HALL BOOK

First edition published 2023
by CRC Press
6000 Broken Sound Parkway NW, Suite 300, Boca Raton, FL 33487-2742

and by CRC Press
4 Park Square, Milton Park, Abingdon, Oxon, OX14 4RN

CRC Press is an imprint of Taylor & Francis Group, LLC

Library of Congress Cataloging-in-Publication Data

Names: Campolieti, Giuseppe (Mathematics professor), author. | Makarov, Roman N., author.
Title: Financial mathematics Volume II : a comprehensive treatment in continuous time / Giuseppe Campolieti and Roman N. Makarov.
Description: First Edition. | Boca Raton, FL : CRC Press, an imprint of Taylor and Francis, 2023. | Series: Textbooks in mathematics | Includes bibliographical references and index.
Identifiers: LCCN 2022032096 (print) | LCCN 2022032097 (ebook) | ISBN 9781138603639 (hardback) | ISBN 9781032392592 (paperback) | ISBN 9780429468889 (ebook)
Subjects: LCSH: Finance--Mathematical models.
Classification: LCC HG106 .C35 2023 (print) | LCC HG106 (ebook) | DDC 332.01/5195--dc23/eng/20221104
LC record available at https://lccn.loc.gov/2022032096
LC ebook record available at https://lccn.loc.gov/2022032097

ISBN: 978-1-138-60363-9 (hbk)
ISBN: 978-1-032-39259-2 (pbk)
ISBN: 978-0-429-46888-9 (ebk)

DOI: 10.1201/9780429468889

Typeset in CMR10 font
by KnowledgeWorks Global Ltd.

Publisher's note: This book has been prepared from camera-ready copy provided by the authors.

To our students and colleagues

Contents

List of Figures

Preface

Objectives and Audience

This book has evolved from financial mathematics courses that the authors have developed and taught mainly within the bachelor's and master's programs in financial mathematics at Wilfrid Laurier University over the past 15+ years. The material has been tested and refined through years of classroom teaching experience. The book is a continuation of the first volume, "Financial Mathematics: A Comprehensive Treatment in Discrete Time," published in 2021. Both volumes revise and extend the first edition, "Financial Mathematics: A Comprehensive Treatment," published in 2014. The first edition combines discrete-time and continuous-time asset pricing theory, with various models and applications into a single volume consisting of Parts I-IV. This book is a revision and extension of Part III of the single-volume edition.

Part III of the first edition constitutes material for senior undergraduate and master's level graduate courses in continuous-time asset pricing theory. It provides a comprehensive coverage of stochastic (Itô) calculus for Brownian motion in one and many dimensions, no-arbitrage pricing in the Black–Scholes–Merton framework for single and multi-asset European options, cross-currency options, American options, as well as other path-dependent options. Various option pricing formulae are derived using different techniques. Part III presents risk-neutral asset pricing theory within a mathematically rigorous framework that incorporates equivalent martingale measures and change of numéraire methods for pricing with various option pricing applications. Part III also tackles interest-rate modelling and pricing of fixed-income derivatives, as well as alternative asset pricing models, including the local volatility model and solvable state-dependent volatility (e.g., the CEV diffusion) models; stochastic volatility models; jump-diffusion and pure jump processes and variance gamma models.

In comparison with Part III of the first edition, this volume contains new material in all chapters of Parts I and II (former Part III) and an expanded set of problems throughout each chapter, including answers to all problems in Appendix C. In total, this volume has almost 150 pages worth of new material. As the title suggests, this book is a comprehensive, self-contained, and unified treatment of the main theory and application of mathematical methods behind modern-day financial mathematics in the continuous time setting. In writing this book, the authors have really strived to create a single source that can be used as a complete standard university textbook for several interrelated courses in financial mathematics at the undergraduate as well as graduate levels. As such, the authors have aimed to introduce both the financial theory and the relevant mathematical methods in a mathematically rigorous yet student-friendly and engaging style that includes an abundance of examples, problem exercises, and fully worked-out solutions.

In contrast to most published manuscripts on the subject of financial mathematics, this book presents multiple problem-solving approaches. It hence bridges together related comprehensive techniques for pricing different types of financial derivatives. The book contains a rather complete and in-depth comprehensive coverage of continuous-time financial models that form the foundation of financial derivative pricing theory. This book also provides a self-contained introduction to stochastic calculus and martingale theory, which are important cornerstones in quantitative finance. The material in many of the chapters is presented at a level that is mainly accessible to undergraduate students of mathematics, finance, actuarial science, economics, and other related quantitative fields. The textbook covers a breadth of material, from beginner to more advanced levels, that is required, i.e., absolutely essential, in the core curriculum courses on financial mathematics currently taught at the second, third, and senior year undergraduate levels at many universities across the globe. As well, a significant portion of the more advanced material in the textbook is meant to be used in courses at the master's graduate level. These courses include formal derivative pricing theory and stochastic calculus. The combination of analytical methods for solving various derivative pricing problems can also be a useful reference for researchers and practitioners in quantitative finance.

The book has the following key features:

- comprehensive treatment covering a complete undergraduate program in financial mathematics as well as some master's level courses in financial mathematics;
- student-friendly presentation with numerous fully worked out examples and exercise problems in every chapter;
- in-depth coverage of continuous-time theory and methodology;
- mathematically rigorous and consistent, yet simple, style that bridges various basic and more advanced concepts and techniques;
- judicious balance of financial theory, and mathematical methods.

Guide to Material

This book is divided into two main parts with each part consisting of several chapters. There are a total of seven chapters and one appendix on general probability theory. A second appendix contains several useful expectation (integral) identities. A third appendix contains solutions (and in some cases hints or partial steps) to all exercises. Each chapter ends with a comprehensive and exhaustive set of exercises of varying difficulty.

Parts I and II (i.e., Chapters 1 to 7) can be considered as a complete text in continuous-time financial mathematics for senior undergraduates and master's level students.

Chapter 1 lays down the foundation for standard Brownian motion. Chapter 2 is a comprehensive coverage of stochastic (Itô) calculus that is required for a large portion of the material in the rest of the book chapters. Chapters 3 and 4 are the main chapters on continuous-time derivative pricing theory, which also include the Black–Scholes–Merton theory of European option pricing. The central concepts of dynamic hedging and replication are presented. Chapter 3 deals with derivative pricing and hedging in the Black–Scholes–Merton model with a single risky asset. Chapter 3 also covers path-dependent derivative pricing within the Black–Scholes–Merton framework. Chapter 4 extends the theory and methodology to derivative pricing and hedging with multiple underlying assets as well as the valuation of cross-currency options. The chapter combines different techniques for pricing multi-asset financial derivatives. Various option pricing formulae are then derived. Chapter 4 also presents risk-neutral asset pricing theory within a mathematically rigorous framework that incorporates equivalent martingale measures and change of numéraire methods for

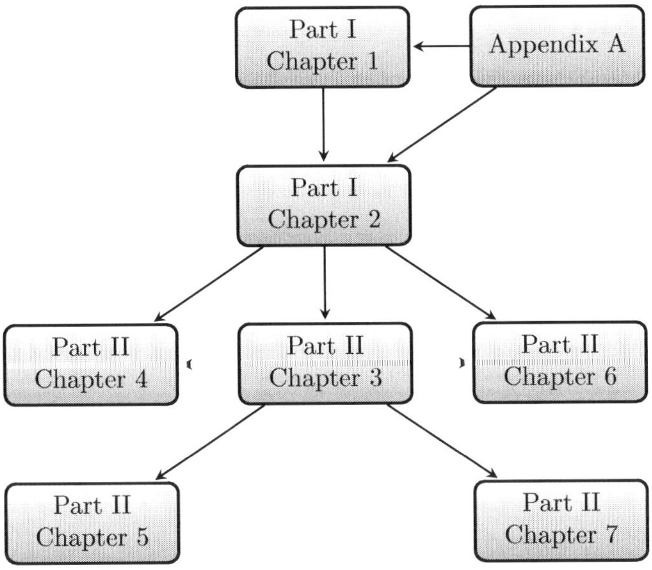

FIGURE: Guide to material.

pricing. Chapter 5 is devoted to pricing American options on a single asset. Chapter 6 covers interest-rate modelling and derivative pricing for fixed-income products. Chapter 7 introduces some alternative asset price models, including the local volatility model and solvable state-dependent volatility (e.g., the CEV diffusion) models; stochastic volatility models (e.g., the Heston model); models based on jump-diffusion and pure jump processes (e.g., the Merton jump-diffusion model and the variance gamma model).

Lastly, Appendix A summarizes the main theoretical concepts in formal probability theory as it relates to measure theory. This also provides some mathematical foundation for later chapters that deal with continuous time modelling and stochastic calculus.

The inter-relationship among the different chapters is summarized in the figure, which represents a flowchart of the material in the textbook. Each arrow indicates a connection between the material in the respective chapters; i.e., when a chapter is viewed as a prerequisite for the other. Finally, the table below is a reference guide for instructors. It displays two different courses for which this book can be adopted as a required textbook at both the undergraduate and graduate levels. The relevant chapters for each course and the prerequisites are indicated in the table.

Course	**Chapters**	**Prerequisites**
Stochastic Calculus with Brownian Motion	1, 2, Appendix A	Analysis, Probability Theory
Continuous-Time Derivative Pricing	3–7	Analysis, Probability Theory, Differential Equations

Acknowledgements

We would like to thank all our past and current undergraduate and graduate students for their valuable comments, feedback, and advice. We are also grateful for the feedback we have received from our colleagues in the Mathematics Department at Wilfrid Laurier University.

Giuseppe Campolieti and Roman N. Makarov

Waterloo, Ontario
April 2022

Authors

Giuseppe Campolieti is Professor of Mathematics at Wilfrid Laurier University in Waterloo, Canada. He has been Natural Sciences and Engineering Research Council postdoctoral research fellow and university research fellow at the University of Toronto. In 1998, he joined the Masters in Mathematical Finance as an instructor and later as an adjunct professor in financial mathematics until 2002. Dr. Campolieti also founded a financial software and consulting company in 1998. He joined Laurier in 2002 as Associate Professor of Mathematics and as SHARCNET Chair in Financial Mathematics.

Roman N. Makarov is Professor of Mathematics at Wilfrid Laurier University. Prior to joining Laurier in 2003, he was an Assistant Professor of Mathematics at Siberian State University of Telecommunications and Informatics and a senior research fellow at the Laboratory of Monte Carlo Methods at the Institute of Computational Mathematics and Mathematical Geophysics in Novosibirsk, Russia.

Part I

Stochastic Calculus with Brownian Motion

1

One-Dimensional Brownian Motion and Related Processes

Brownian motion is a keystone in the foundation of mathematical finance. Brownian motion is a continuous-time stochastic process that can be constructed as a limiting case of symmetric random walks. Recall that a similar approach can be used to obtain the log-normal price model as a limiting case of binomial models. Since the probability distribution of Brownian motion is normal, it is reasonable to begin with reviewing some of the important basic facts about the multivariate normal distribution.

1.1 Multivariate Normal Distributions

1.1.1 Multivariate Normal Distribution

The n-variate normal distribution is determined by an $n \times 1$ mean vector $\boldsymbol{\mu}$ and an $n \times n$ positive definite covariance matrix $\mathbf{C} = \mathbf{C}^\top$:

$$\boldsymbol{\mu} = \left[\mu_1, \mu_2, \ldots, \mu_n\right]^\top, \quad \mathbf{C} = [C_{ij}]_{i,j=1,\ldots,n}.$$

Note: throughout we use superscript $\top$ to denote the transpose. We denote the n-variate normal distribution with mean vector $\boldsymbol{\mu}$ and covariance matrix $\mathbf{C}$ by $Norm_n(\boldsymbol{\mu}, \mathbf{C})$. Assuming the nondegenerate case with a positive definite covariance matrix, the joint probability density function (PDF) is an n-variate real-valued function

$$f_{\mathbf{X}}(\mathbf{x}) = \frac{1}{(2\pi)^{n/2}\,[\det \mathbf{C}]^{n/2}} \mathrm{e}^{-\frac{1}{2}(\mathbf{x}-\boldsymbol{\mu})^\top \mathbf{C}^{-1}(\mathbf{x}-\boldsymbol{\mu})}, \quad \mathbf{x} \in \mathbb{R}^n. \tag{1.1}$$

This PDF is that of random $n \times 1$ vector $\mathbf{X} = \left[X_1, X_2, \ldots, X_n\right]^\top \sim Norm_n(\boldsymbol{\mu}, \mathbf{C})$, i.e., $\mathbf{X}$ has n-variate normal distribution with mean $\boldsymbol{\mu} = \mathrm{E}[\mathbf{X}]$ and covariance $\mathbf{C} = \mathrm{E}[\mathbf{X}\mathbf{X}^\top] - \boldsymbol{\mu}\boldsymbol{\mu}^\top$. As components, $\mu_i = \mathrm{E}[X_i]$ and $C_{ij} = \mathrm{Cov}(X_i, X_j) = \mathrm{E}[X_i X_j] - \mu_i\mu_j, i, j = 1, 2\ldots, n$. It is a well-known fact that normal random variables are independent iff they are uncorrelated with zero covariances $C_{ij} = 0$ for all $i \neq j$. Therefore, if the covariance is a diagonal matrix, $\mathbf{C} = \mathrm{diag}(\sigma_1^2, \sigma_2^2, \ldots, \sigma_n^2)$, where $\sigma_i^2 = \mathrm{E}[(X_i - \mu_i)^2]$, then the components of $\mathbf{X}$ are jointly independent normally distributed random variables. In this case, the joint PDF is a product of univariate (one-dimensional) marginal densities:

$$f_{\mathbf{X}}(\mathbf{x}) = \prod_{i=1}^{n} f_{X_i}(x_i) = \prod_{i=1}^{n} \frac{1}{\sigma_i} n\left(\frac{x_i - \mu_i}{\sigma_i}\right) = \prod_{i=1}^{n} \frac{1}{\sqrt{2\pi}\sigma_i} \mathrm{e}^{-\frac{(x_i-\mu_i)^2}{2\sigma_i^2}}.$$

The joint CDF (cummulative distribution function) is given by $F_{\mathbf{X}}(\mathbf{x}) = \prod_{i=1}^{n} \mathcal{N}\left(\frac{x_i - \mu_i}{\sigma_i}\right)$.

DOI: 10.1201/9780429468889-1

Throughout the text, we use the notation: $n(z) = \mathcal{N}'(z) = \frac{e^{-z^2/2}}{\sqrt{2\pi}}$ for the PDF and $\mathcal{N}(z) := \int_{-\infty}^{z} n(x)\,dx$ for the CDF of the standard univariate normal random variable Z. In terms of probabilities: $\mathbb{P}(Z \leqslant z) = \mathcal{N}(z)$ and $\mathbb{P}(Z \in dz) = n(z)\,dz$, where $Z \sim Norm(0,1)$.

Another well-known fact is that a sum of normal random variables is again normally distributed:

$$X_i \sim Norm\left(\mu_i, \sigma_i^2\right),\ i = 1,2 \implies X_1 + X_2 \sim Norm\left(\mu_1 + \mu_2, \operatorname{Var}\left(X_1 + X_2\right)\right), \tag{1.2}$$

where $\operatorname{Var}(X_1 + X_2) = \sigma_1^2 + \sigma_2^2 + 2\operatorname{Cov}(X_1, X_2)$. This property extends to the multivariate case as follows. Let $\mathbf{a}$ and $\mathbf{B}$ be an $m \times 1$ constant vector and an $m \times n$ constant matrix, respectively, with $1 \leqslant m \leqslant n$. Suppose that $\mathbf{X} \sim Norm_n(\boldsymbol{\mu}, \mathbf{C})$, then the m-variate random vector $\mathbf{a} + \mathbf{B}\mathbf{X}$ is normally distributed with mean vector $\mathbf{a} + \mathbf{B}\boldsymbol{\mu}$. Its $m \times m$ covariance matrix is given by $\mathrm{E}[\mathbf{B}(\mathbf{X}-\boldsymbol{\mu})(\mathbf{B}(\mathbf{X}-\boldsymbol{\mu}))^\top] = \mathbf{B}\,\mathrm{E}[(\mathbf{X}-\boldsymbol{\mu})(\mathbf{X}-\boldsymbol{\mu})^\top]\,\mathbf{B}^\top = \mathbf{B}\mathbf{C}\mathbf{B}^\top$. Hence, provided that $\det(\mathbf{B}\mathbf{C}\mathbf{B}^\top) \neq 0$, we have

$$\mathbf{X} \sim Norm_n(\boldsymbol{\mu}, \mathbf{C}) \implies \mathbf{a} + \mathbf{B}\mathbf{X} \sim Norm_m(\mathbf{a} + \mathbf{B}\boldsymbol{\mu}, \mathbf{B}\mathbf{C}\mathbf{B}^\top). \tag{1.3}$$

For example, applying (1.3) by setting $\mathbf{B} = \mathbf{b} = [b_1, \ldots, b_n]$ ($1 \times n$ constant vector) gives a univariate (scalar) normal random variable

$$Y := \mathbf{b}\mathbf{X} \equiv \sum_{i=1}^{n} b_i X_i \sim Norm(\mu_Y, \sigma_Y^2)$$

with mean $\mu_Y = \sum_{i=1}^n b_i \mu_i$ and variance $\sigma_Y^2 = \sum_{i=1}^n b_i^2 \sigma_i^2 + 2\sum\sum_{1\leqslant i<j\leqslant n} b_i b_j C_{ij}$. Note that the double summation term is zero in case the X_i are jointly independent where $C_{ij} = 0$.

The identity in (1.3) allows us to express an arbitrary multivariate normal random vector in terms of jointly independent standard normal random variables. Since the covariance matrix $\mathbf{C}$ is positive definite, it admits the Cholesky factorization: $\mathbf{C} = \mathbf{L}\mathbf{L}^\top$ with a lower-triangular $n \times n$ matrix $\mathbf{L}$. Let $Z_1, Z_2, \ldots, Z_n$ be i.i.d. standard normal random variables: $Z_i \sim Norm(0,1)$, $\operatorname{Cov}(Z_i, Z_j) = \delta_{ij}$. The vector $\mathbf{Z} = \left[Z_1, Z_2, \ldots, Z_n\right]^\top$ is then an n-variate normal with $n \times 1$ zero mean vector and having the $n \times n$ identity covariance matrix $\mathbf{I}$. Applying (1.3), while setting $\mathbf{X} \equiv \mathbf{Z} \sim Norm_n(\mathbf{0}, \mathbf{I})$ and $\mathbf{a} \equiv \boldsymbol{\mu}$, gives $\boldsymbol{\mu} + \mathbf{L}\mathbf{Z} \sim Norm_n(\boldsymbol{\mu}, \mathbf{C})$. That is, the random $n \times 1$ vector defined by $\mathbf{Y} := \boldsymbol{\mu} + \mathbf{L}\mathbf{Z}$, where $\boldsymbol{\mu}$ is *any* constant $n \times 1$ vector, is a normal $n \times 1$ random vector with mean $\mathrm{E}[\mathbf{Y}] = \boldsymbol{\mu} + \mathbf{L}\mathbf{0} = \boldsymbol{\mu}$ and covariance matrix $\mathbf{C} = \mathbf{L}\mathbf{I}\mathbf{L}^\top = \mathbf{L}\mathbf{L}^\top$. Hence, this gives us a simple method for generating a normal $n \times 1$ random vector $\mathbf{Y}$ with a prescribed mean vector $\boldsymbol{\mu}$ and covariance matrix $\mathbf{C}$ by employing the Cholesky factorization and generating a vector $\mathbf{Z}$ of i.i.d. standard normal random variables.

1.1.2 Conditional Normal Distributions

Suppose that $\mathbf{X} = \left[X_1, X_2, \ldots, X_n\right]^\top \sim Norm_n(\boldsymbol{\mu}, \mathbf{C})$, $n \geqslant 2$. Let us split (partition) the vector $\mathbf{X}$ into two parts:

$$\mathbf{X} = \begin{bmatrix} \mathbf{X}_1 \\ \mathbf{X}_2 \end{bmatrix}, \text{ where } \mathbf{X}_1 \in \mathbb{R}^m \text{ and } \mathbf{X}_2 \in \mathbb{R}^{n-m}$$

for some m with $1 < m < n$. Correspondingly, we split the $n \times 1$ mean vector $\boldsymbol{\mu}$ and $n \times n$ covariance matrix $\mathbf{C}$ to represent them in block form:

$$\boldsymbol{\mu} = \begin{bmatrix} \boldsymbol{\mu}_1 \\ \boldsymbol{\mu}_2 \end{bmatrix} \text{ and } \mathbf{C} = \begin{bmatrix} \mathbf{C}_{11} & \mathbf{C}_{12} \\ \mathbf{C}_{21} & \mathbf{C}_{22} \end{bmatrix},$$

where $\boldsymbol{\mu}_1 \in \mathbb{R}^m$ and $\boldsymbol{\mu}_2 \in \mathbb{R}^{n-m}$ are the respective mean vectors of $\mathbf{X}_1$ and $\mathbf{X}_2$; $\mathbf{C}_{11} = \mathrm{E}[\mathbf{X}_1\mathbf{X}_1^\top] - \boldsymbol{\mu}_1\boldsymbol{\mu}_1^\top$ is $m \times m$, $\mathbf{C}_{12} = \mathrm{E}[\mathbf{X}_1\mathbf{X}_2^\top] - \boldsymbol{\mu}_1\boldsymbol{\mu}_2^\top$ is $m \times (n-m)$, $\mathbf{C}_{21} = \mathbf{C}_{12}^\top$ is $(n-m) \times m$, and $\mathbf{C}_{22} = \mathrm{E}[\mathbf{X}_2\mathbf{X}_2^\top] - \boldsymbol{\mu}_2\boldsymbol{\mu}_2^\top$ is $(n-m) \times (n-m)$. Then, the *conditional distribution of* $\mathbf{X}_1$ *given the value of* $\mathbf{X}_2 = \mathbf{x}_2$ *is normal:*

$$\mathbf{X}_1|\{\mathbf{X}_2 = \mathbf{x}_2\} \sim Norm_m\left(\boldsymbol{\mu}_1 + \mathbf{C}_{12}\,\mathbf{C}_{22}^{-1}(\mathbf{x}_2 - \boldsymbol{\mu}_2), \mathbf{C}_{11} - \mathbf{C}_{12}\,\mathbf{C}_{22}^{-1}\mathbf{C}_{21}\right). \tag{1.4}$$

The two arguments in (1.4) represent the respective conditional mean vector and conditional covariance matrix of $\mathbf{X}_1$ given $\{\mathbf{X}_2 = \mathbf{x}_2\}$. We remark that the partitioned vectors $\mathbf{X}_1$ and $\mathbf{X}_2$ can involve any permutation (re-ordering) of the original components of $\mathbf{X}$. In particular, we also have

$$\mathbf{X}_2|\{\mathbf{X}_1 = \mathbf{x}_1\} \sim Norm_{n-m}\left(\boldsymbol{\mu}_2 + \mathbf{C}_{21}\,\mathbf{C}_{11}^{-1}(\mathbf{x}_1 - \boldsymbol{\mu}_1), \mathbf{C}_{22} - \mathbf{C}_{21}\,\mathbf{C}_{11}^{-1}\mathbf{C}_{12}\right). \tag{1.5}$$

For example, consider a two-dimensional (bivariate) normal random vector

$$\mathbf{X} = \begin{bmatrix} X_1 \\ X_2 \end{bmatrix} \sim Norm_2\left(\begin{bmatrix} \mu_1 \\ \mu_2 \end{bmatrix}, \begin{bmatrix} \sigma_1^2 & \rho\sigma_1\sigma_2 \\ \rho\sigma_1\sigma_2 & \sigma_2^2 \end{bmatrix}\right).$$

In conformity with (1.4), X_1 conditional on X_2 (and X_2 conditional on X_1) is normally distributed. In this case, $m = n - m = 1$ so the block matrices are simply scalars: $\mathbf{C}_{11} = \sigma_1^2$, $\mathbf{C}_{12} = \mathbf{C}_{21} = \rho\sigma_1\sigma_2$, $\mathbf{C}_{22} = \sigma_2^2$, and $\mathbf{x}_2 - \boldsymbol{\mu}_2 = x_2 - \mu_2$. In particular,

$$X_1 \mid \{X_2 = x_2\} \sim Norm\left(\mu_1 + \rho\frac{\sigma_1}{\sigma_2}(x_2 - \mu_2), \sigma_1^2(1-\rho^2)\right) \tag{1.6}$$

and $X_2 \mid \{X_1 = x_1\}$ is given by simply interchanging indicies $1 \leftrightarrow 2$ in (1.6). In the special case of standard normal random variables $X = Z_1, Y = Z_2$, where $\mu_1 = \mu_2 = 0, \sigma_1 = \sigma_2 = 1, \mathrm{Cov}(Z_1, Z_2) = \rho$, we have $Z_1|\{Z_2 = y\} \stackrel{d}{=} Norm(\rho y, 1-\rho^2)$ and $Z_2|\{Z_1 = x\} \stackrel{d}{=} Norm(\rho x, 1-\rho^2)$.

1.2 Standard Brownian Motion

1.2.1 One-Dimensional Symmetric Random Walk

Consider the binomial sample space $\Omega \equiv \Omega_\infty$:

$$\Omega_\infty = \prod_{i=1}^{\infty}\{\mathsf{D}, \mathsf{U}\} = \{\omega_1\omega_2\ldots \mid \omega_i \in \{\mathsf{D}, \mathsf{U}\}, \text{ for all } i \geqslant 1\}$$

generated by a Bernoulli experiment using any number of repeated up (U) or down (D) moves. We assume a probability measure $\mathbb{P}$ such that the up and down moves are equally probable. That is, each outcome $\omega = \omega_1\omega_2\ldots \in \Omega_\infty$ is a sequence of U's and D's where $\mathbb{P}(\omega_i = \mathsf{D}) = \mathbb{P}(\omega_i = \mathsf{U}) = \frac{1}{2}$ for all $i \geqslant 1$. In contrast to the binomial model where we have a finite number of moves $N \geqslant 1$ with $\Omega = \Omega_N$, the number of moves is now infinite and the sample space Ω_∞ is *uncountable*. For a thorough discussion of the finite period binomial model, we refer the reader to Chapter 6 of Volume I. A short review is also provided in Section 3.1 of this volume where random variables $\mathsf{U}_n, \mathsf{D}_n$, etc., are defined. The sample space now also admits uncountable partitions. Obviously, we still have all the countable partitions in the same manner as discussed for Ω_N. For example, the event that the first move is up is the atom $A_{\mathsf{U}} := \{\omega_1 = \mathsf{U}\} = \{\mathsf{U}\omega_2\omega_3 \ldots \mid \omega_i \in \{\mathsf{D},\mathsf{U}\}, \text{ for all } i \geqslant 2\}$ and its complement is $A_{\mathsf{D}} := \{\omega_1 = \mathsf{D}\} = \{\mathsf{D}\omega_2\omega_3 \ldots \mid \omega_i \in \{\mathsf{D},\mathsf{U}\}, \text{ for all } i \geqslant 2\}$ where $\Omega_\infty = A_{\mathsf{U}} \cup A_{\mathsf{D}}$. In particular, the atoms that correspond to fixing (resolving) the first $n \geqslant 1$ moves are given similarly,

$$A_{\omega_1^* \ldots \omega_n^*} := \{\omega_1 = \omega_1^*, \ldots, \omega_n = \omega_n^*\} = \{\omega_1^* \ldots \omega_n^* \omega_{n+1} \ldots \mid \omega_i \in \{\mathsf{D},\mathsf{U}\},\ i \geqslant n+1\}.$$

The union of these 2^n atoms is a partition of Ω_∞, for every choice of $n \geqslant 1$. We denote by $\mathcal{F}_\infty$ the σ-algebra generated by all the above atoms corresponding to any number of moves $n \geqslant 1$. Hence, we have a sequence of power sets $\mathcal{F}_n = 2^{\Omega_n}$ and $\mathcal{F}_\infty = 2^{\Omega_\infty}$. Recall that for every $N \geqslant 1$, $\{\mathcal{F}_n\}_{0 \leqslant n \leqslant N}$ is the natural filtration generated by the moves up to time N. We saw that the natural filtration can be equivalently generated by sets of random variables that contain the information on all the moves, e.g., $\mathcal{F}_0 = \{\emptyset, \Omega\}$, $\mathcal{F}_n = \sigma(\omega_1, \ldots, \omega_n) = \sigma(\{\mathsf{U}_k : 1 \leqslant k \leqslant n\})$. For every N, $\{\mathcal{F}_n\}_{0 \leqslant n \leqslant N}$ is the natural filtration for the binomial model up to time N, where $\mathcal{F}_n \subset \mathcal{F}_{n+1}$ for all $n \geqslant 0$. In the limiting case of $N \to \infty$, we have $\mathcal{F}_\infty$ containing all the possible events in the binomial model with an arbitrary number of moves for which we can compute probabilities of occurrence. For every given $N \geqslant 1$, $(\Omega_N, \mathcal{F}_N, \mathbb{P})$ is a probability space on which we can define any random variable that may depend on up to N numbers of moves (i.e., random variables that are $\mathcal{F}_n$-measurable for $1 \leqslant n \leqslant N$). Passing to the limiting case, the triplet $(\Omega_\infty, \mathcal{F}_\infty, \mathbb{P})$ is then a probability space for any random variable that depends on any number of moves (i.e., random variables that are $\mathcal{F}_n$-measurable for all $n \geqslant 1$), i.e., $(\Omega_\infty, \mathcal{F}_\infty, \mathbb{P}, \{\mathcal{F}_n\}_{n \geqslant 1})$ is a filtered probability space.

Recall that the symmetric random walk $\{M_n\}_{n \geqslant 0}$ is defined by $M_0 = 0$ and

$$M_n(\omega) = \sum_{k=1}^{n} X_k(\omega), \text{ where } X_k(\omega) = \begin{cases} 1 & \text{if } \omega_k = \mathsf{U}, \\ -1 & \text{if } \omega_k = \mathsf{D}, \end{cases}$$

for $n, k \geqslant 1$. Note that all market moves are assumed mutually independent, i.e., all X_k's are independent and identically distributed (i.i.d.). Let us summarize the properties of the symmetric random walk.

1. $\{M_n\}_{n \geqslant 0}$ is a martingale[1] w.r.t. $\{\mathcal{F}_n\}_{n \geqslant 0}$ with zero mean, $\mathrm{E}[M_n] \equiv 0$ for all times $n \geqslant 0$.

   Proof. (i) $M_0 = 0$ is $\mathcal{F}_0$-measurable and for each $n \geqslant 1$, M_n is $\mathcal{F}_n$-measurable since $\mathcal{F}_n = \sigma(\omega_1, \ldots, \omega_n) = \sigma(M_0, M_1, \ldots, M_n)$, i.e., the process is adapted to its natural filtration. (ii) Each M_n is integrable since $\mathrm{E}[|M_n|] < \sum_{k=1}^{n} \mathrm{E}[|X_k|] = n < \infty$. (iii) Writing $M_{n+1} = M_n + X_{n+1}$ and using the fact that M_n is $\mathcal{F}_n$-measurable and X_{n+1} is independent of $\mathcal{F}_n$ gives

   $$\mathrm{E}[M_{n+1}|\mathcal{F}_n] = \mathrm{E}[M_n|\mathcal{F}_n] + \mathrm{E}[X_{n+1}|\mathcal{F}_n] = M_n + \mathrm{E}[X_{n+1}] = M_n, \quad n \geqslant 0.$$

   Note: for every $k \geqslant 1$, $\mathrm{E}[X_k] = p(1) + (1-p)(-1) = 0$ since $p = \mathbb{P}(\mathsf{U}) = 1/2$. Based on (i)-(iii), we conclude the martingale property. Since the mean value of a martingale is constant over time, we have $\mathrm{E}[M_n] \equiv \mathrm{E}[M_0] = 0$, for $n \geqslant 0$. □

2. $\{M_n\}_{n \geqslant 0}$ is a square-integrable process, where $\mathrm{E}[M_n^2] = \mathrm{Var}(M_n) = n$ and $\mathrm{Cov}(M_n, M_k) = n \wedge k \equiv \min(n, k)$, for all $n, k \geqslant 0$.

[1] We recall the definition of a discrete-time martingale process from Chapter 6 of Volume I. Given a probability measure $\mathbb{P}$ with expectation E under this measure, a process $\{X_n\}_{n \geqslant 0}$ is a martingale w.r.t. a filtration $\{\mathcal{F}_n\}_{n \geqslant 0}$ if: (i) it is adapted to the filtration, i.e., X_n is $\mathcal{F}_n$-measurable for all $n \geqslant 0$; (ii) it is integrable, i.e., $\mathrm{E}[|X_n|] < \infty$ for all $n \geqslant 0$; (iii) it satisfies the conditional martingale property, i.e., $\mathrm{E}[X_m|\mathcal{F}_n] = X_n$, $0 \leqslant n \leqslant m$. Due to the tower property, this conditional expectation property is equivalent to the single-time-step property: $\mathrm{E}[X_{n+1}|\mathcal{F}_n] = X_n$, for all $n \geqslant 0$.

Proof. Computing the variance, where $\text{Var}(X_k) = 1$ and $\{X_k\}_{k\geqslant 1}$ are i.i.d., gives

$$\text{Var}(M_n) = \sum_{k=1}^{n} \text{Var}(X_k) = n.$$

Let $0 \leqslant k \leqslant n$. Since $M_n - M_k$ is independent of M_k we have

$$\text{Cov}(M_n, M_k) = \text{Cov}(M_n - M_k, M_k) + \text{Cov}(M_k, M_k) = 0 + \text{Var}(M_k) = k.$$

Hence, for any $n, k \geqslant 0$ we have $\text{Cov}(M_n, M_k) = n \wedge k$. Since the mean is zero, $\text{E}[M_n^2] = \text{Var}(M_n) = n$, for $n \geqslant 0$. □

It is useful to recall here the general property $\text{Cov}(M_n, M_k) = \text{Var}(M_k)$, for $0 \leqslant k \leqslant n$, where $\{M_n\}_{n\geqslant 0}$ is *any discrete-time martingale process with zero mean.* Indeed, applying the tower property in reverse, while using the martingale property, $\text{E}[M_n|\mathcal{F}_k] = M_k$, and the fact that M_k is $\mathcal{F}_k$-measurable gives

$$\text{Cov}(M_n, M_k) = \text{E}[M_n M_k] = \text{E}[\text{E}[M_n M_k|\mathcal{F}_k]] = \text{E}[M_k \text{E}[M_n|\mathcal{F}_k]] = \text{E}[M_k^2] = \text{Var}(M_k)$$

Later, we also make use of this property for *any continuous-time zero-mean martingale process* $\{X_t\}_{t\geqslant 0}$, i.e., $\text{Cov}(X_s, X_t) = \text{Var}(X_t)$, for real $0 \leqslant t \leqslant s$.

3. It is a process with independent (nonoverlapping in time) increments and stationary increments such that $\text{E}[M_n - M_k] = 0$ and $\text{Var}(M_n - M_k) = |n - k|$, for $n, k \geqslant 0$.

   Proof. The independence and stationarity follows directly from the fact that $\{X_k\}_{k\geq 1}$ are i.i.d. random variables. In particular, all single-step increments $M_{n+1} - M_n = X_{n+1}$ for different n value are jointly independent. Hence, all increments nonoverlapping in time are independent. Moreover, for any $0 \leqslant k < n$ and $\ell \geqslant 0$, $M_n - M_k \stackrel{d}{=} M_{n+\ell} - M_{k+\ell} \stackrel{d}{=} \sum_{i=1}^{n-k} X_i$. Hence, the distribution of the increments is the same for equal time step, i.e., the process satisfies stationarity. The expected value and variance are

$$\begin{aligned} \text{E}[M_n - M_k] &= \text{E}[M_n] - \text{E}[M_k] = 0 - 0 = 0, \\ \text{Var}(M_n - M_k) &= \text{Var}(M_n) - 2\,\text{Cov}(M_n, M_k) + \text{Var}(M_k) \\ &= n - 2(n \wedge k) + k = n \vee k - n \wedge k = |n - k|. \end{aligned}$$ □

   Note: $n \vee k \equiv \max(n, k)$, $n \wedge k \equiv \min(n, k)$ and $n + k = n \wedge k + n \vee k$.

We hence see that the variance of any increment is equal to the length of the time step.

Using the above symmetric random walk as a backbone, we now construct a *scaled symmetric random walk* in continuous time. For each *fixed* $n \in \mathbb{N}$, we denote this process by $W^{(n)} \equiv \{W^{(n)}(t)\}_{t\geqslant 0}$. For $t \geqslant 0$, define $W^{(n)}(t) := \frac{1}{\sqrt{n}} M_{nt}$ provided that nt is an integer. Suppose that nt is noninteger. Find s such that ns is an integer and $ns < nt < ns+1$ holds. Since ns is the largest integer less than or equal to nt, it is equal to $\lfloor nt \rfloor$. Note that $\lfloor x \rfloor$ denotes the floor of x (i.e., integer part of x). We hence define $W^{(n)}(t) := W^{(n)}(s) = \frac{1}{\sqrt{n}} M_{ns}$ to obtain the following general formula of the scaled symmetric random walk:

$$W^{(n)}(t) = \frac{1}{\sqrt{n}} M_{\lfloor nt \rfloor}, \quad t \geqslant 0. \tag{1.7}$$

As is seen from (1.7), $W^{(n)}$ is a *continuous-time* stochastic process with piecewise-constant right-continuous sample paths all starting at zero. At every time value $t = \frac{k}{n}$, $k = 1, 2, \ldots,$

the process has a jump of size $\frac{X_k}{\sqrt{n}} = \pm\frac{1}{\sqrt{n}}$. The process is constant in between jump times, i.e.,

$$W^{(n)}(t) = \frac{1}{\sqrt{n}} M_k\,, \quad \text{for}\, t \in \left[\frac{k}{n}, \frac{k+1}{n}\right)\,, \quad k = 0, 1, \ldots.$$

For a given realization (outcome) $\omega = \omega_1\omega_2\ldots$, the path of the process is given by

$$W^{(n)}(t,\omega) = \frac{1}{\sqrt{n}} M_k(\omega_1\omega_2\ldots\omega_k) = \frac{1}{\sqrt{n}}(X_1(\omega_1) + \ldots + X_k(\omega_k))\,,\ t \in \left[\frac{k}{n}, \frac{k+1}{n}\right),$$

for $t \geqslant 1/n$ and $W^{(n)}(t,\omega) = 0$ for $t \in [0, 1/n)$. For any real time $t \geqslant 0$, we can equivalently represent the realized path as:

$$W^{(n)}(t,\omega) = \frac{1}{\sqrt{n}} M_{\lfloor nt \rfloor}(\omega) = \frac{1}{\sqrt{n}} \sum_{k=1}^{\infty} \mathbb{I}_{\{\frac{k}{n} \leqslant t < \frac{k+1}{n}\}} M_k(\omega_1\omega_2\ldots\omega_k),$$

where $M_k(\omega_1\omega_2\ldots\omega_k) = X_1(\omega_1) + \ldots + X_k(\omega_k)$. Figure 1.1 depicts a sample path of this process for $n = 100$.

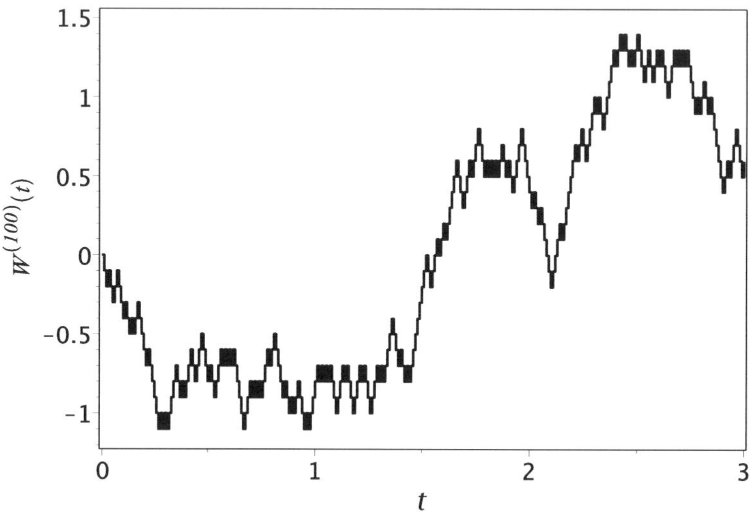

FIGURE 1.1: A sample path of $W^{(100)}$.

Hence, the time-t realization $W^{(n)}(t)$ depends on the first $k = \lfloor nt \rfloor$ sequence of up or down moves $\omega_1, \ldots, \omega_k$. Thus, for all $t \geqslant 0$, $W^{(n)}(t)$ is measurable w.r.t. $\mathcal{F}_{\lfloor nt \rfloor} = \sigma(X_1, \ldots, X_{\lfloor nt \rfloor}) = \sigma(M_1, \ldots, M_{\lfloor nt \rfloor}) = \sigma(W^{(n)}(u) : 0 \leqslant u \leqslant t)$. Hence, the filtration defined by $\mathbb{F}^{(n)} := \{\mathcal{F}^{(n)}(t) := \mathcal{F}_{\lfloor nt \rfloor};\, t \geqslant 0\}$ is in fact the natural filtration for the process $\{W^{(n)}(t)\}_{t \geqslant 0}$. The process is integrable and adapted to $\mathbb{F}^{(n)}$. Since the scaled symmetric random walk is obtained by re-scaling (and constant interpolation in between jumps) of the symmetric random walk, it inherits its main properties as follows.

1. The process $\{W^{(n)}(t)\}_{t \geqslant 0}$ is a martingale[2] w.r.t. its natural filtration and $\mathrm{E}[W^{(n)}(t)] \equiv 0$.

[2]We recall the definition of a continuous-time martingale process from Chapter 6 of Volume I. Given a probability measure $\mathbb{P}$ with expectation E under this measure, a process $\{X(t)\}_{0 \leqslant t \leqslant T}$, over real-time values where $T > 0$ can be finite or $T = \infty$, is a martingale w.r.t. a filtration $\{\mathcal{F}_t\}_{0 \leqslant t \leqslant T}$ if: (i) it is adapted to the filtration, i.e., $X(t)$ is $\mathcal{F}_t$-measurable for $0 \leqslant t \leqslant T$; (ii) it is integrable, i.e., $\mathrm{E}[|X(t)|] < \infty$ for $0 \leqslant t \leqslant T$; (iii) it satisfies the conditional martingale property, i.e., $\mathrm{E}[X(t)|\mathcal{F}_s] = X(s)$, $0 \leqslant s \leqslant t \leqslant T$.

2. For all $s, t \geqslant 0$, $\mathrm{Cov}(W^{(n)}(t), W^{(n)}(s)) = \frac{\lfloor n(s \wedge t) \rfloor}{n}$. Hence, $\mathrm{Var}(W^{(n)}(t) - W^{(n)}(s)) = \frac{|\lfloor nt \rfloor - \lfloor ns \rfloor|}{n}$ and $\mathrm{Var}(W^{(n)}(t)) = \frac{\lfloor nt \rfloor}{n}$.

3. Let $0 \leqslant t_0 < t_1 < \cdots < t_m < \infty$, for any $m \geqslant 1$, then the increments $W^{(n)}(t_1) - W^{(n)}(t_0), \ldots, W^{(n)}(t_m) - W^{(n)}(t_{m-1})$ are jointly independent and stationary.

The proof of these properties is left as an exercise for the reader (see Exercise 1.5).

Consider the limit $n \to \infty$, i.e., as the time between jumps $1/n \to 0$. For real x, $x - 1 < \lfloor x \rfloor \leqslant x$, i.e., $nt - 1 < \lfloor nt \rfloor \leqslant nt \implies t - \frac{1}{n} < \frac{\lfloor nt \rfloor}{n} \leqslant t$. Therefore, $\frac{\lfloor nt \rfloor}{n} \to t$, as $n \to \infty$. Hence, as $n \to \infty$, the above variance and covariance expressions simplify to

$$\mathrm{Cov}(W^{(n)}(t), W^{(n)}(s)) \to s \wedge t, \quad \mathrm{Var}(W^{(n)}(t) - W^{(n)}(s)) \to |t - s|,$$

for all $s, t \geqslant 0$.

An important characteristic of a stochastic process is the so-called *quadratic variation.* This quantity is a random variable that characterizes the variability of a stochastic process. We further explicitly define and discuss this in Section 1.2.5. For $t \geqslant 0$, the quadratic variation of the scaled random walk $W^{(n)}$ on $[0, t]$ is denoted by $[W^{(n)}, W^{(n)}](t)$. It is calculated for any given path $\omega \in \Omega$ as the limit $m \to \infty$ of the sum of squared differences of the process over m adjacent time intervals of a partition, $0 = t_0 < t_1 < \cdots < t_m = t$, where the length of all adjacent time steps in the partition approach zero in the limit $m \to \infty$. Since the process $W^{(n)}$ is constant in between the jump times $\frac{k}{n}, k \geqslant 1$, the quadratic variation is then simply the sum of squared differences (square of the jump size) of the process over all the jump times within the time interval $[0, t]$:

$$\begin{aligned} [W^{(n)}, W^{(n)}](t) &= \sum_{k=1}^{\lfloor nt \rfloor} \left[W^{(n)}\left(\frac{k}{n}\right) - W^{(n)}\left(\frac{k-1}{n}\right) \right]^2 \\ &= \sum_{k=1}^{\lfloor nt \rfloor} \left(\frac{1}{\sqrt{n}} X_k \right)^2 = \sum_{k=1}^{\lfloor nt \rfloor} \frac{1}{n} = \frac{\lfloor nt \rfloor}{n}. \end{aligned} \tag{1.8}$$

Note that $X_k^2 = 1, \forall k \geqslant 1$, i.e., $X_k^2(\omega) = 1$ for all $\omega \in \Omega$, $k \geqslant 1$. It should be observed here that the quadratic variation turns out to be a constant random variable, i.e., for any realized path $\omega \in \Omega$ we have $[W^{(n)}, W^{(n)}](t, \omega) = \frac{\lfloor nt \rfloor}{n}$. Moreover, the quadratic variation has the same value as the variance of the process for all $t \geqslant 0$: $[W^{(n)}, W^{(n)}](t) = \mathrm{Var}(W^{(n)}(t)) = \frac{\lfloor nt \rfloor}{n}$. At values of time t such that $\lfloor nt \rfloor = nt \in \mathbb{N}$, $[W^{(n)}, W^{(n)}](t) = \mathrm{Var}(W^{(n)}(t)) = t$. In the limit $n \to \infty$, we have both quantities equal to time t: $\mathrm{Var}(W^{(n)}(t)) \to t$ and $[W^{(n)}, W^{(n)}](t) \to t$ for all real $t \geqslant 0$.

Let $m \equiv \lfloor nt \rfloor$, $m = 0, 1, 2, \ldots$. The symmetric random walk value at time m is given by $M_m = 2\,\mathsf{U}_m - m$, where U_m is the number of up moves up to time m. Hence, $W^{(n)}(t) = \frac{1}{\sqrt{n}} M_m = \frac{1}{\sqrt{n}}(2\,\mathsf{U}_m - m)$ is a shifted and scaled binomial random variable where $\mathsf{U}_m \sim Bin(m, \frac{1}{2})$. By the equivalence of events, $\{W^{(n)}(t) = \frac{2k-m}{\sqrt{n}}\} \equiv \{\mathsf{U}_m = k\}$, we have the probability distribution (PMF) of $W^{(n)}(t)$.

$$\mathbb{P}\left(W^{(n)}(t) = \frac{2k - m}{\sqrt{n}} \right) = \mathbb{P}\left(\mathsf{U}_m = k\right) = \frac{m!}{k!\,(m-k)!} \left(\frac{1}{2} \right)^m, \quad k = 0, 1, \ldots, m \equiv \lfloor nt \rfloor.$$

The range of $W^{(n)}(t)$ is given by the set $\{\frac{-m}{\sqrt{n}}, \frac{-m+2}{\sqrt{n}}, \ldots, \frac{m-2}{\sqrt{n}}, \frac{m}{\sqrt{n}}\}$, $m \equiv \lfloor nt \rfloor$. For example, $W^{(100)}(0.1)$ (with $n = 100, t = 0.1, m = \lfloor nt \rfloor = 10, \sqrt{n} = 10$) takes its value on the set

$$\{-1, -0.8, -0.6, -0.4, -0.2, 0, 0.2, 0.4, 0.6, 0.8, 1\}.$$

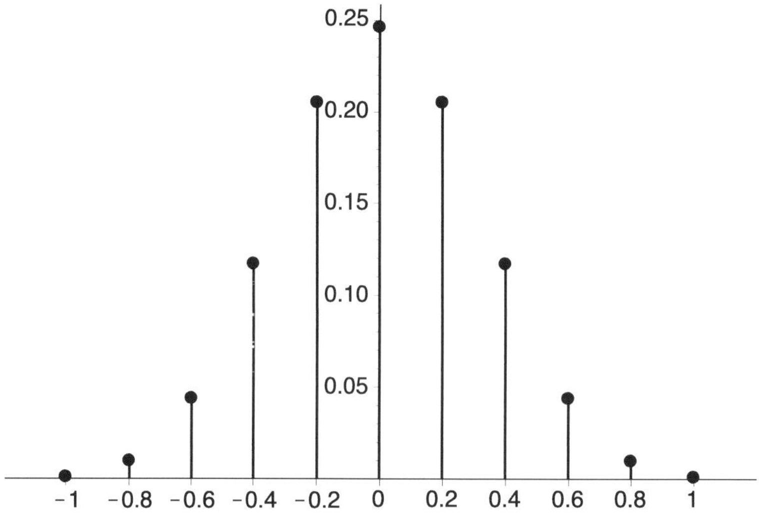

FIGURE 1.2: The probability mass function of $W^{(100)}(0.1)$.

The PMF is plotted in Figure 1.2.

Let us determine the limiting distribution of $W^{(n)}(t)$, as $n \to \infty$, for fixed $t > 0$. Since $W^{(n)}(t)$ is equal to a scaled sum of i.i.d. random variables X_k, $k \geqslant 1$, with common mean $\mathrm{E}[X_1] = 0$ and variance $\mathrm{Var}(X_1) = 1$, the Central Limit Theorem can be applied to yield

$$W^{(n)}(t) = \frac{1}{\sqrt{n}} \sum_{k=1}^{\lfloor nt \rfloor} X_k = \sqrt{\frac{\lfloor nt \rfloor}{n}} \left(\frac{1}{\sqrt{\lfloor nt \rfloor}} \sum_{k=1}^{\lfloor nt \rfloor} X_k \right) \xrightarrow{d} \sqrt{t}\, Z, \text{ as } n \to \infty,$$

where $Z \sim \mathit{Norm}(0, 1)$. In the above derivation, we used the limit $\lim_{n\to\infty} \frac{\lfloor nt \rfloor}{n} = t$. Thus,

$$W^{(n)}(t) \xrightarrow{d} \mathit{Norm}(0, t), \text{ as } n \to \infty.$$

To demonstrate this convergence of the probability distribution of $W^{(n)}(t)$ to the normal distribution, we plot and compare a histogram for $W^{(n)}(t)$ and a normal density curve for $\mathit{Norm}(0, t)$ in Figure 1.3. The histogram is constructed for $W^{(100)}(0.1)$ by replacing each mass probability in Figure 1.2 by a bar with width 0.2 (since the distance between mass points equals 0.2) and height chosen so that the area of the bar is equal to the respective mass probability. For example, $\mathbb{P}(W^{(100)}(0.1) = 0.2) = \mathbb{P}(\mathsf{U}_{10} = 6) \cong 0.20508$, hence we draw a histogram bar centred at 0.2 with width 0.2 and height $\frac{0.20508}{0.2} \cong 1.02539$. As a result, the total area of the histogram is one (as well as the area under the normal curve). Note the close agreement between the histogram and the normal density in Figure 1.3.

The limit of scaled random walks $\{W^{(n)}(t)\}_{t\geqslant 0}$ taken simultaneously, as $n \to \infty$, for all $t \in [0, T]$, $T > 0$, gives us a construction of a continuous-time stochastic process who's probability law is that of *Brownian motion*. Brownian motion, which is also referred to as the Wiener process, is more formally introduced in the next section. It is denoted by $\{W(t)\}_{t\geqslant 0}$. Brownian motion can in a sense be viewed as a scaled random walk with infinitesimally small steps so that the process moves equally likely upward or downward by $\sqrt{\mathrm{d}t}$ in each infinitesimal time interval $\mathrm{d}t$. To govern the behaviour of Brownian motion, the number of (up or down) moves becomes infinite in any time interval. Equivalently, if each move is thought of as a coin toss, then the coin needs to be tossed "infinitely fast." Thus, $W(t)$,

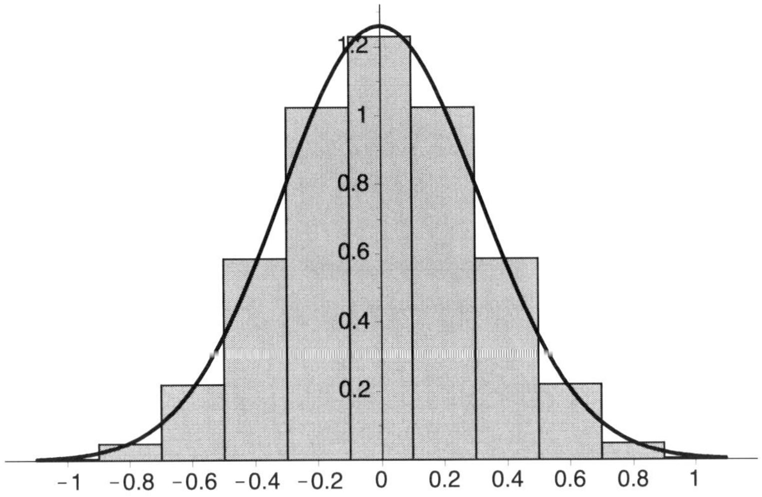

FIGURE 1.3: Comparison of the histogram constructed for $W^{(100)}(0.1)$ and the density curve for $Norm(0, 0.1)$.

for all $t > 0$, can be viewed as a function of an uncountably infinite sequence of elementary moves (or coin tosses). Different outcomes produce different sample paths. A particular outcome $\omega \in \Omega_\infty$ produces a particular path $\{W(t, \omega)\}_{t \geqslant 0}$. Although sample paths of $W^{(n)}$ are piecewise-constant functions of t that are discontinuous at times $t = k/n$, $k \geqslant 1$, the size of jumps at the points of discontinuity goes to zero, as $n \to \infty$. Moreover, the length of intervals where $W^{(n)}$ is constant goes to zero as $n \to \infty$, as well. In the limiting case, the paths become continuous everywhere and nonconstant (in fact, nonmonotonic) on any interval no matter how small. The properties of continuity and nonmonotonicity of Brownian paths will be later discussed more precisely in Section 1.2.5.

Let us summarize the differences between a scaled random walk $\{W^{(n)}(t)\}_{t \in [0,T]}$, for any fixed finite n, and Brownian motion $\{W(t)\}_{t \in [0,T]}$, in the following table.

	$W^{(n)}(t)$	$W(t)$
RANGE:	discrete and bounded	continuous on $(-\infty, \infty)$
DISTRIBUTION:	approximately normal	normal
SAMPLE PATHS:	piecewise-constant	continuous and nonconstant on any time interval

1.2.2 Formal Definition and Basic Properties of Brownian Motion

There are various equivalent ways to define standard Brownian motion. One formal definition is as follows.

Definition 1.1. A real-valued continuous-time stochastic process $\{W(t)\}_{t \geqslant 0} \in \mathbb{R}$ defined on a probability space $(\Omega, \mathcal{F}, \mathbb{P})$ is called a *standard Brownian motion* if it satisfies the following.

1. (Almost every path starts at the origin.) $W(0) = 0$ with probability one.

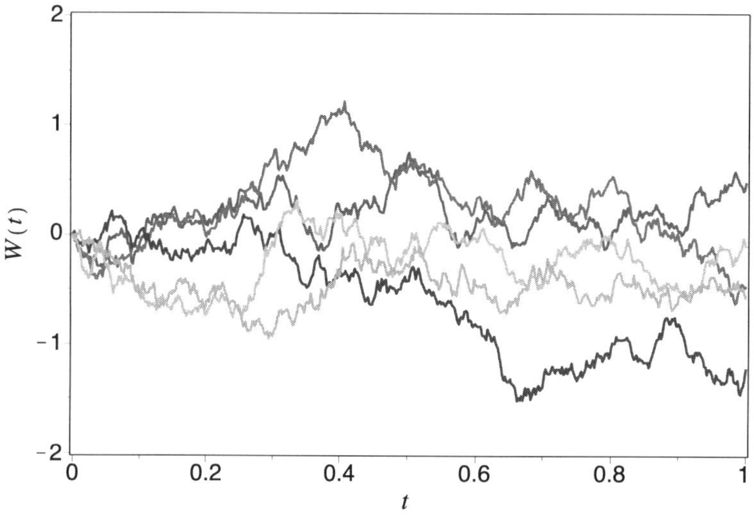

FIGURE 1.4: Sample paths of a standard Brownian motion.

2. (Independence of nonoverlapping increments) For all $n \in \mathbb{N}$ and every choice of time partition $0 = t_0 < t_1 < \cdots < t_n < \infty$, the increments

$$W(t_1) \equiv W(t_1) - W(t_0), W(t_2) - W(t_1), \ldots, W(t_n) - W(t_{n-1})$$

are jointly independent.

3. (Normality and stationarity of increments) For all $0 \leqslant s < t$, the increment $W(t) - W(s)$ is normally distributed with mean 0 and variance $t - s$:

$$\begin{aligned} &\mathrm{E}[W(t) - W(s)] = 0, \quad \mathrm{Var}(W(t) - W(s)) = t - s, \\ &W(t) - W(s) \sim Norm(0, t - s). \end{aligned} \tag{1.9}$$

4. (Continuity of paths) For almost all $\omega \in \Omega$ (i.e., with probability one), the sample path $W(t, \omega)$ is a continuous function of time $t \geqslant 0$.

We remark that property #2 can also be stated more simply as: $W(t) - W(s)$ and $W(v) - W(u)$ are independent for any $0 \leqslant s < t < u < v < \infty$. Property #3 implies that $W(t) \sim Norm(0, t)$, i.e., $W(t) \stackrel{d}{=} \sqrt{t}Z$, $Z \sim Norm(0, 1)$. Moreover, $W(t) - W(s) \sim Norm(0, |t - s|)$ for any $s, t \geqslant 0$.

Brownian motion (which we also abbreviate as BM) is named after the botanist Robert Brown who was the first to observe and describe the motion of a pollen particle suspended in fluid as an irregular random motion. In 1900, Louis Bachelier used Brownian motion as a model for movements of stock prices in his mathematical theory of speculations. In 1905, Albert Einstein obtained differential equations for the distribution function of Brownian motion. He argued that the random movement observed by Robert Brown is due to bombardment of the particle by molecules of the fluid. However, it was Norbert Wiener who constructed the mathematical foundation of BM as a stochastic process in 1931. Sometimes, this process is also called the Wiener process and this fact explains the notation used to denote Brownian motion by the letter W.

Although the processes $W^{(n)}$ and W have different probability distributions and sample paths, some of the properties of a scaled random walk are inherited by Brownian motion. For example, using the independence of increments for adjacent (nonoverlapping) time intervals $W(t) - W(s)$ and $W(s) - W(0) \equiv W(s)$, and writing $W(t) = W(t) - W(s) + W(s)$, the covariance of $W(s)$ and $W(t)$, for $0 \leqslant s \leqslant t$, is given by

$$\begin{aligned}\text{Cov}(W(s), W(t)) &= \text{Cov}(W(s), W(t) - W(s) + W(s)) \\ &= \text{Cov}(W(s), W(t) - W(s)) + \text{Var}(W(s)) = 0 + s = s.\end{aligned}$$

In general, we have

$$\text{Cov}(W(s), W(t)) = s \wedge t, \text{ for all } s, t \geqslant 0. \tag{1.10}$$

Another feature that BM and symmetric random walks have in common is the martingale property. The *natural filtration for Brownian motion* is the collection of σ-algebras generated by the Brownian motion observed up to time $t \geqslant 0$:

$$\mathcal{F}_t^W = \sigma\big(\{W(s) : 0 \leqslant s \leqslant t\}\big).$$

BM is hence automatically adapted to the filtration $\mathbb{F}^W \equiv \{\mathcal{F}_t^W\}_{t\geqslant 0}$, i.e., $W(t)$ is obviously $\mathcal{F}_t^W$-measurable for every $t \geqslant 0$. In what follows we will assume any appropriate filtration $\mathbb{F} = \{\mathcal{F}_t\}_{t\geqslant 0}$ for BM (of which $\mathbb{F}^W$ is one such filtration) such that all of the defining properties of Brownian motion hold. For all $0 \leqslant u \leqslant s < t$, $W(t) - W(s)$ is independent of $W(u)$ and hence it follows that the increment $W(t) - W(s)$ is independent of the σ-algebra $\mathcal{F}_s^W$ generated by all Brownian paths up to time s. So, for any appropriate filtration $\mathbb{F}$ for BM, we must have that $W(t) - W(s)$ is independent of $\mathcal{F}_s$. In what follows, we shall also sometimes use the shorthand notation for the conditional expectation of a random variable X w.r.t. a given σ-algebra $\mathcal{F}_t$, i.e., we shall often interchangeably write $\text{E}[X \mid \mathcal{F}_t] \equiv \text{E}_t[X]$ (for short) whenever convenient.

Below we consider some examples of processes that are martingales w.r.t. any assumed filtration $\{\mathcal{F}_t\}_{t\geqslant 0}$ for BM[3]. One such process is standard Brownian motion. For any finite time $t \geqslant 0$, the process is obviously integrable since, by Jensen's inequality, $\text{E}[|W(t)|] = \text{E}[\sqrt{W^2(t)}] \leqslant \sqrt{\text{E}[W^2(t)]} = \sqrt{t} < \infty$. In fact, this expectation can also be computed exactly since $W(t) \stackrel{d}{=} \sqrt{t}Z$, where $Z \sim Norm(0,1)$, giving $\text{E}[|W(t)|] = \sqrt{t}\text{E}[|Z|] = 2\sqrt{t}\int_0^\infty z n(z)\,\text{d}z = -2\sqrt{t}\int_0^\infty \text{d}n(z) = 2\sqrt{t}n(0) = \sqrt{\frac{2t}{\pi}}$.

Theorem 1.1. *Standard Brownian motion $\{W(t)\}_{t\geqslant 0}$ is a martingale w.r.t. $\{\mathcal{F}_t\}_{t\geqslant 0}$.*

Proof. BM is adapted to its natural filtration and hence adapted to $\{\mathcal{F}_t\}_{t\geqslant 0}$. Since we already showed that $W(t)$ is integrable, we need only show the martingale conditional expectation property, i.e., $\text{E}[W(t) \mid \mathcal{F}_s] \equiv \text{E}_s[W(t)] = W(s)$, for $0 \leqslant s \leqslant t$. Writing $W(t) = W(t) - W(s) + W(s)$ and using the fact that $W(s)$ is $\mathcal{F}_s$-measurable and $W(t) - W(s)$ is independent of $\mathcal{F}_s$:

$$\begin{aligned}\text{E}_s[W(t)] &= \text{E}_s[(W(t) - W(s)) + W(s)] = \text{E}_s[W(t) - W(s)] + \text{E}_s[W(s)] \\ &= \text{E}[W(t) - W(s)] + W(s) = 0 + W(s) = W(s).\end{aligned}$$

□

The next example shows that squaring BM and subtracting by the time gives a martingale.

Example 1.1. Prove that $\{X(t) := W^2(t) - t\}_{t\geqslant 0}$ is a martingale w.r.t. any filtration $\{\mathcal{F}_t\}_{t\geqslant 0}$ for BM.

---

[3]Note: to avoid excessive use of brackets, throughout this volume we write $W^\alpha(t) \equiv (W(t))^\alpha$ for any real exponent α. For example, $W^2(t) \equiv (W(t))^2$ is the square of standard BM at time t.

Solution. Since $W(t)$ is $\mathcal{F}_t$-measurable, then $X(t) = f(t, W(t))$ is $\mathcal{F}_t$-measurable, where $f(t,x) := x^2 - t$ is a Borel function on $\mathbb{R} \times [0,\infty)$, i.e., $\sigma(X(t)) \subset \sigma(W(t)) \subset \mathcal{F}_t$. Moreover, by the triangular inequality, $\mathrm{E}[|X(t)|] \leqslant \mathrm{E}[W^2(t)] + t = 2t < \infty$. It remains to show that $\mathrm{E}_s[X(t)] = X(s)$, for $0 \leqslant s \leqslant t < \infty$. Writing $W(t) = W(t) - W(s) + W(s)$, we now calculate the conditional time-s expectation of $W^2(t)$ for $s \leqslant t$, by using the fact that $W(s)$ and $W^2(s)$ are $\mathcal{F}_s$-measurable and that $W(t) - W(s)$ and $(W(t) - W(s))^2$ are independent of $\mathcal{F}_s$:

$$\begin{aligned}
\mathrm{E}_s\left[W^2(t)\right] &= \mathrm{E}_s\left[(W(t) - W(s) + W(s))^2\right] \\
&= \mathrm{E}_s\left[(W(t) - W(s))^2 + 2\,W(s)(W(t) - W(s)) + W^2(s)\right] \\
&= \mathrm{E}_s\left[(W(t) - W(s))^2\right] + 2\mathrm{E}_s\left[W(s)(W(t) - W(s))\right] + \mathrm{E}_s\left[W^2(s)\right] \\
&= \mathrm{E}\left[(W(t) - W(s))^2\right] + 2W(s)\,\mathrm{E}[W(t) - W(s)] + W^2(s) \\
&= \mathrm{Var}(W(t) - W(s)) + 2W(s)\cdot 0 + W^2(s) = t - s + W^2(s).
\end{aligned}$$

Therefore, $\mathrm{E}_s[X(t)] \equiv \mathrm{E}_s\left[W^2(t) - t\right] = t - s + W^2(s) - t = W^2(s) - s \equiv X(s)$. □

We note that an alternative way to compute $\mathrm{E}_s\left[W^2(t)\right] \equiv \mathrm{E}\left[W^2(t) \mid \mathcal{F}_s\right]$ is to use the fact that $W(s)$ is $\mathcal{F}_s$-measurable and $Y \equiv W(t) - W(s)$ is independent of $\mathcal{F}_s$. Hence, $\mathrm{E}\left[W^2(t) \mid \mathcal{F}_s\right] = \mathrm{E}\left[(Y + W(s))^2 \mid \mathcal{F}_s\right] = g(W(s)) = t - s + W^2(s)$, where Proposition A.6 gives $g(x) = \mathrm{E}\left[(Y + x)^2\right] = \mathrm{E}[Y^2] + 2x\mathrm{E}[Y] + x^2 = t - s + 2x \cdot 0 + x^2 = t - s + x^2$. [Note that in Proposition A.6, we have set $\mathcal{G} \equiv \mathcal{F}_s$, $X \equiv W(s)$, $Y \equiv W(t) - W(s)$.]

The following is an example of a so-called *exponential martingale*. This type of martingale will turn out to be important when pricing options. Recall (e.g., see Appendix B) $\mathrm{E}[\mathrm{e}^{\alpha X}] = \mathrm{e}^{\alpha\mu + \alpha^2\sigma^2/2}$ is the moment generating function (MGF) of a normal random variable $X \sim Norm(\mu, \sigma^2)$. Hence, the MGF of $W(t)$ is given by $\mathrm{E}[\mathrm{e}^{\alpha W(t)}] = \mathrm{e}^{\alpha^2 t/2}$ since $W(t) \sim Norm(0,t)$.

Example 1.2. Prove that $\{X(t) := \mathrm{e}^{\alpha W(t) - \alpha^2 t/2}\}_{t \geqslant 0}$, for any $\alpha \in \mathbb{R}$, is a martingale w.r.t. any filtration $\{\mathcal{F}_t\}_{t\geqslant 0}$ for BM.

Solution. Since $W(t)$ is $\mathcal{F}_t$-measurable, then $X(t) = f(t, W(t))$ is $\mathcal{F}_t$-measurable, where $f(t,x) := \mathrm{e}^{\alpha x - \alpha^2 t/2}$ is a Borel function on $\mathbb{R} \times [0,\infty)$. The process is integrable since $\mathrm{E}[|X(t)|] = \mathrm{e}^{-\alpha^2 t/2}\mathrm{E}[\mathrm{e}^{\alpha W(t)}] = 1$. We now show that $\mathrm{E}_s[X(t)] = X(s)$, for $0 \leqslant s \leqslant t$. Indeed, since $\mathrm{e}^{\alpha W(s)}$ is $\mathcal{F}_s$-measurable and $W(t) - W(s) \sim Norm(0, t - s)$ is independent of $\mathcal{F}_s$:

$$\mathrm{E}_s\left[\mathrm{e}^{\alpha W(t)}\right] = \mathrm{E}_s\left[\mathrm{e}^{\alpha(W(t) - W(s))}\mathrm{e}^{\alpha W(s)}\right] = \mathrm{e}^{\alpha W(s)}\mathrm{E}\left[\mathrm{e}^{\alpha(W(t) - W(s))}\right] = \mathrm{e}^{\alpha W(s)}\mathrm{e}^{\alpha^2(t-s)/2}.$$

Therefore, $\mathrm{E}_s[X(t)] \equiv \mathrm{E}[\mathrm{e}^{\alpha W(t) - \alpha^2 t/2} \mid \mathcal{F}_s] = \mathrm{e}^{\alpha W(s) - \alpha^2 s/2} \equiv X(s)$. □

Again, we note that the above conditional expectation is readly evaluated by applying Proposition A.6 with $\mathcal{G} \equiv \mathcal{F}_s$, $X \equiv W(s)$, $Y \equiv W(t) - W(s)$. In particular, $\mathrm{E}_s[\mathrm{e}^{\alpha W(t)}] = \mathrm{E}_s[\mathrm{e}^{\alpha W(s)}\mathrm{e}^{\alpha Y}] = g(W(s))$, where $g(x) = \mathrm{e}^{\alpha x}\mathrm{E}[\mathrm{e}^{\alpha Y}] = \mathrm{e}^{\alpha x}\mathrm{e}^{\alpha^2(t-s)/2}$.

1.2.3 Multivariate Distribution of Brownian Motion

From the formal definition of standard BM, the random variables $W(t_1), W(t_2), \ldots, W(t_n)$ are jointly normally distributed for all time points $0 < t_1 < t_2 < \cdots < t_n < \infty$. Their joint n-variate distribution is determined by the mean vector and covariance matrix. In accordance with (1.9) and (1.10), we have

$$\mu_i = \mathrm{E}[W(t_i)] = 0 \text{ and } C_{ij} = \mathrm{Cov}\left(W(t_i), W(t_j)\right) = \mathrm{E}\left[W(t_i)\,W(t_j)\right] = t_i \wedge t_j,$$

for $1 \leqslant i, j \leqslant n$. Therefore, $\mathbf{W} := \left[W(t_1), W(t_2), \cdots, W(t_n)\right]^\top$, where each component random variable corresponds to the Brownian path at the respective time points and has the n-variate normal distribution with mean vector zero and covariance matrix

$$\begin{bmatrix} \mathrm{E}[W^2(t_1)] & \mathrm{E}[W(t_1)\,W(t_2)] & \cdots & \mathrm{E}[W(t_1)\,W(t_n)] \\ \mathrm{E}[W(t_2)\,W(t_1)] & \mathrm{E}[W^2(t_2)] & \cdots & \mathrm{E}[W(t_2)\,W(t_n)] \\ \vdots & \vdots & \ddots & \vdots \\ \mathrm{E}[W(t_n)\,W(t_1)] & \mathrm{E}[W(t_n)\,W(t_2)] & \cdots & \mathrm{E}[W^2(t_n)] \end{bmatrix} = \begin{bmatrix} t_1 & t_1 & \cdots & t_1 \\ t_1 & t_2 & \cdots & t_2 \\ \vdots & \vdots & \ddots & \vdots \\ t_1 & t_2 & \cdots & t_n \end{bmatrix} \tag{1.11}$$

Thus, the joint PDF is given by (1.1) with $\boldsymbol{\mu} = \mathbf{0}$ and covaraince matrix $\mathbf{C}$ in (1.11):

$$f_{\mathbf{W}}(\mathbf{x}) = \frac{1}{(2\pi)^{n/2}\,[\det \mathbf{C}]^{n/2}} e^{-\frac{1}{2}\mathbf{x}^\top \mathbf{C}^{-1}\mathbf{x}}, \quad \mathbf{x} \in \mathbb{R}^n. \tag{1.12}$$

Example 1.3. Let $0 < t_1 < t_2 < t_3$. Determine the probability distribution of

$$W(t_1) - 2W(t_2) + 3W(t_3).$$

Solution. The vector $\mathbf{W} = \left[W(t_1), W(t_2), W(t_3)\right]^\top$ has the trivariate normal distribution:

$$\mathbf{W} \sim Norm_3\left(\mathbf{0} \equiv \begin{bmatrix} 0 \\ 0 \\ 0 \end{bmatrix}, \mathbf{C} = \begin{bmatrix} t_1 & t_1 & t_1 \\ t_1 & t_2 & t_2 \\ t_1 & t_2 & t_3 \end{bmatrix}\right).$$

Set $\mathbf{b} = \left[1, -2, 3\right]^\top$. Then, in accordance with (1.3), $X := W(t_1) - 2W(t_2) + 3W(t_3) = \mathbf{b}^\top \mathbf{W}$ is a scalar normal random variable with mean $\mathrm{E}[X] = \mathbf{b}^\top E[\mathbf{W}] = \mathbf{b}^\top \mathbf{0} \equiv 0$ and variance

$$\mathrm{Var}(X) = \mathbf{b}^\top \mathbf{C}\,\mathbf{b} = \left[1, -2, 3\right] \begin{bmatrix} t_1 & t_1 & t_1 \\ t_1 & t_2 & t_2 \\ t_1 & t_2 & t_3 \end{bmatrix} \begin{bmatrix} 1 \\ -2 \\ 3 \end{bmatrix} = 3t_1 - 8t_2 + 9t_3,$$

i.e., $W(t_1) - 2W(t_2) + 3W(t_3) \sim Norm(0, 3t_1 - 8t_2 + 9t_3)$.

Alternative method #1: By using the property that a linear combination of normal random variables is a univariate normal random variable, we have that $X := W(t_1) - 2W(t_2) + 3W(t_3)$ is normally distributed with mean

$$\mathrm{E}[X] = \mathrm{E}[W(t_1)] - 2\mathrm{E}[W(t_2)] + 3\mathrm{E}[W(t_3)] = 0$$

and variance

$$\begin{aligned} \mathrm{Var}(X) &= \mathrm{Var}(W(t_1)) + \mathrm{Var}(-2W(t_2)) + \mathrm{Var}(3W(t_3)) \\ &\quad + 2\,\mathrm{Cov}(W(t_1), -2W(t_2)) + 2\,\mathrm{Cov}(-2W(t_2), 3W(t_3)) + 2\,\mathrm{Cov}(W(t_1), 3W(t_3)) \\ &= t_1 + 4t_2 + 9t_3 - 4t_1 - 12t_2 + 6t_1 = 3t_1 - 8t_2 + 9t_3, \end{aligned}$$

i.e., $X \sim Norm(0, 3t_1 - 8t_2 + 9t_3)$.

Alternative method #2: Set $X_1 = W(t_1) - W(0) \equiv W(t_1), X_2 = W(t_2) - W(t_1), X_3 = W(t_3) - W(t_2)$: $W(t_1) = X_1, W(t_2) = X_1 + X_2, W(t_3) = X_1 + X_2 + X_3$. Hence, $X := W(t_1) - 2W(t_2) + 3W(t_3) = 2X_1 + X_2 + 3X_3$ is a linear combination of X_i's which are independent (nonoverlapping Brownian increment) normal random variables. The variance is given by

$$\mathrm{Var}(X) = 2^2 \cdot \mathrm{Var}(X_1) + \mathrm{Var}(X_2) + 3^2 \cdot \mathrm{Var}(X_3) = 4t_1 + (t_2 - t_1) + 9(t_3 - t_2) = 3t_1 - 8t_2 + 9t_3$$

and the mean $\mathrm{E}[X] = 0$. We again conclude that $X \sim Norm(0, 3t_1 - 8t_2 + 9t_3)$. □

Probabilities of elementary events associated with Brownian paths can be computed in terms of standard normal CDFs. For Brownian motion at a single time $t > 0$, we write $W(t) \stackrel{d}{=} \sqrt{t}Z$, where $Z := \frac{W(t)}{\sqrt{t}} \sim Norm(0,1)$. For Brownian motion at multiple times $0 < t_1 < t_2 < \ldots < t_n$, we have $W(t_i) \stackrel{d}{=} \sqrt{t_i}Z_i$, where $Z_i := \frac{W(t_i)}{\sqrt{t_i}} \sim Norm(0,1)$ with correlation $\rho_{ij} \equiv \operatorname{Corr}(Z_i, Z_j) = \operatorname{Cov}(Z_i, Z_j)$, for $t_i < t_j$, $1 \leqslant i < j \leqslant n$:

$$\rho_{ij} = \frac{1}{\sqrt{t_i t_j}} \operatorname{Cov}(W(t_i), W(t_j)) = \frac{t_i \wedge t_j}{\sqrt{t_i t_j}} = \sqrt{\frac{t_i}{t_j}},$$

$\rho_{ji} = \rho_{ij}$, $\rho_{ii} = 1$.

Example 1.4. Derive expressions for the following probabilities:

(a) $\mathbb{P}(a < W(t) < b)$ for real $a < b$, $t > 0$;

(b) $\mathbb{P}(W(s) \leqslant a, W(t) \leqslant b)$ for $a, b \in \mathbb{R}$, $0 < s < t$;

(c) $\mathbb{P}(W(t) > W(s) > a)$ for $a \in \mathbb{R}$, $0 < s < t$.

Solution.
(a) Using $Z := \frac{W(t)}{\sqrt{t}} \sim Norm(0,1)$:

$$\mathbb{P}(a < W(t) < b) = \mathbb{P}\left(\frac{a}{\sqrt{t}} < Z < \frac{b}{\sqrt{t}}\right) = \mathcal{N}\left(\frac{b}{\sqrt{t}}\right) - \mathcal{N}\left(\frac{a}{\sqrt{t}}\right).$$

(b) Let $Z_1 := \frac{W(s)}{\sqrt{s}} \sim Norm(0,1)$, $Z_2 := \frac{W(t)}{\sqrt{t}} \sim Norm(0,1)$, where $\operatorname{Cov}(Z_1, Z_2) = \sqrt{\frac{s}{t}}$. Hence,

$$\mathbb{P}(W(s) \leqslant a, W(t) \leqslant b) = \mathbb{P}\left(Z_1 \leqslant \frac{a}{\sqrt{s}}, Z_2 \leqslant \frac{b}{\sqrt{t}}\right) = \mathcal{N}_2\left(\frac{a}{\sqrt{s}}, \frac{b}{\sqrt{t}}; \sqrt{\frac{s}{t}}\right).$$

$\mathcal{N}_2$ is the bivariate standard normal CDF (see Appendix B for the definition and basic properties).

(c) Writing the joint event as $\{W(t) > W(s) > a\} = \{W(t) - W(s) > 0, W(s) > a\}$ and by independence of the BM increments over nonoverlapping time intervals:

$$\begin{aligned}
\mathbb{P}(W(t) > W(s) > a) &= \mathbb{P}(W(t) - W(s) > 0, W(s) > a) \\
&= \mathbb{P}(W(t) - W(s) > 0) \cdot \mathbb{P}(W(s) > a) \\
&= \mathbb{P}(\sqrt{t-s}Z > 0) \cdot \mathbb{P}(Z > a/\sqrt{s}) \\
&= \frac{1}{2}\left[1 - \mathcal{N}\big(\frac{a}{\sqrt{s}}\big)\right] \equiv \frac{1}{2}\mathcal{N}\big(-\frac{a}{\sqrt{s}}\big).
\end{aligned}$$

□

Example 1.5. Calculate the following probabilities:

(a) $\mathbb{P}(W(t) \leqslant 0)$ for $t > 0$;

(b) $\mathbb{P}(W(1) \leqslant 0, W(2) \leqslant 0)$.

Solution.
(a) Since for all $t > 0$, $W(t) \stackrel{d}{=} \sqrt{t}Z$, where $Z \sim Norm(0,1)$, we have

$$\mathbb{P}(W(t) \leqslant 0) = \mathbb{P}(Z \leqslant 0) = \mathcal{N}(0) = \frac{1}{2}.$$

Note: this result also follows from the expression derived in part (a) of Example 1.4 with $a \to -\infty$, $b = 0$, i.e., $\mathbb{P}(W(t) < 0) = \mathbb{P}(-\infty < W(t) < 0) = \mathcal{N}(0) - \mathcal{N}(-\infty) = 1/2 - 0 = 1/2$.
(b) The increments $Z_1 := W(1) - W(0) \equiv W(1)$ and $Z_2 := W(2) - W(1)$ are independent standard normal random variables with joint PDF $f_{Z_1,Z_2}(z_1,z_2) = f_{Z_1}(z_1)f_{Z_2}(z_2) = n(z_1)\, n(z_2)$. Hence,

$$\begin{aligned}
\mathbb{P}(W(1) \leqslant 0, W(2) \leqslant 0) &= \mathbb{P}(W(1) \leqslant 0, W(1) + (W(2) - W(1)) \leqslant 0) \\
&= \mathbb{P}(Z_1 \leqslant 0, Z_1 + Z_2 \leqslant 0) \\
&= \iint_D n(z_1)\, n(z_2)\, \mathrm{d}z_1\, \mathrm{d}z_2 = \int_{-\infty}^{0} \left(\int_{-\infty}^{-z_1} n(z_2)\, \mathrm{d}z_2 \right) n(z_1)\, \mathrm{d}z_1 \\
&\big(\text{where } D = \{(z_1, z_2) \in \mathbb{R}^2 \,:\, z_1 + z_2 \leqslant 0 \text{ and } z_1 \leqslant 0\}\big) \\
&= \int_{-\infty}^{0} \mathcal{N}(-z_1)\, n(z_1)\, \mathrm{d}z_1 = \int_{0}^{\infty} \mathcal{N}(x)\, n(x)\, \mathrm{d}x \\
&\big(\text{by change of variables } x = -z_1 \text{ and where } n(-x) = n(x)\big) \\
&= \int_{0}^{\infty} \mathcal{N}(x)\mathcal{N}'(x)\, \mathrm{d}x = \frac{1}{2}\mathcal{N}^2(x)\bigg|_0^\infty = \frac{1}{2}(1^2 - (1/2)^2) = \frac{3}{8}.
\end{aligned}$$

Note: as an alternative derivation to the above first steps, we can simply use the expression derived in part (b) of Example 1.4, where we set $s = 1, t = 2$, $\rho = \sqrt{s/t} = 1/\sqrt{2}$, $a = b = 0$, and (B.23) in Appendix B:

$$\mathbb{P}(W(1) \leqslant 0, W(2) \leqslant 0) = \mathcal{N}_2(0, 0; 1/\sqrt{2}) = \int_{-\infty}^{0} n(x)\mathcal{N}(-x)\, \mathrm{d}x\,.$$

The remaining steps are the same as above.
Not surprisingly, this example verifies that $W(1)$ and $W(2)$ are dependent random variables since the joint probability

$$\mathbb{P}(W(1) \leqslant 0, W(2) \leqslant 0) = \frac{3}{8} \neq \frac{1}{4} = \mathbb{P}(W(1) \leqslant 0)\,\mathbb{P}(W(2) \leqslant 0).$$ □

1.2.4 The Markov Property and the Transition PDF

Our goal is the derivation of the transition probability distribution (i.e., transition PDF) of standard Brownian motion. In particular, given a filtration $\{\mathcal{F}_t\}_{t \geqslant 0}$ for standard BM, we wish to derive a formula for calculating conditional probabilities of the form

$$\mathbb{P}(W(t) \in A \mid \mathcal{F}_s) \text{ for } 0 \leqslant s \leqslant t \text{ and Borel set } A \in \mathcal{B}(\mathbb{R}).$$

In doing so, we shall also prove the important Markov property which allows us to readily compute such conditional probabilities as probabilities conditional on $W(s)$, the BM at time s. That is, the expected value of any (Borel) function of $W(t)$ conditional on the complete history of the BM paths (contained in $\mathcal{F}_s$) up to any time $s \leqslant t$ is equivalent to the expected value conditional only on $W(s)$ (i.e., the σ-algebra $\sigma(W(s))$). We recall that this defines the Markov property for any continuous-time stochastic process, where conditioning on the natural filtration of the process up to time s is equivalent to conditioning on the process value at time s.

Two other properties of Brownian motion are time and space homogeneity. The process defined by $X(t) := x + W(t)$, $x \in \mathbb{R}$, is a Brownian motion started at x. Standard BM and this process have identical increments: $X(t+s) - X(t) = W(t+s) - W(t)$, $s > 0, t \geqslant 0$. Hence, $\{X(t)\}_{t \geqslant 0}$ is a continuous-time stochastic process with the same independent normal increments as W and continuous sample paths. The only difference w.r.t. standard BM is that this process starts at any given real value x: $X(0) = x$. The property that $x + W(t)$ is a Brownian motion for all x is called *space homogeneity*. Another property of BM is its *time homogeneity*, meaning that the process defined by $\widehat{W}(t) := W(t+T) - W(T)$, $t \geqslant 0$, is a standard Brownian motion for any fixed $T \geqslant 0$. In other words, translation of a Brownian sample path along both space and time axes gives us again a Brownian path with the same transition PDF. As shown further below, the transition PDF is a function of the time difference and spatial difference.

Theorem 1.2. *Brownian motion is a Markov process.*

Proof. Fix any (Borel) function $h\colon \mathbb{R} \to \mathbb{R}$ and take $\{\mathcal{F}_t\}_{t \geqslant 0}$ as a filtration for BM. We need to show that there exists a (Borel) function $g\colon \mathbb{R} \to \mathbb{R}$ such that

$$\mathrm{E}[h(W(t)) \mid \mathcal{F}_s] \equiv \mathrm{E}_s[h(W(t))] = g(W(s)), \text{ for all } 0 \leqslant s \leqslant t,$$

i.e., this means that $\mathrm{E}[h(W(t)) \mid \mathcal{F}_s] = \mathrm{E}[h(W(t)) | W(s)]$ for all $0 \leqslant s \leqslant t$. Indeed, since $W(s)$ is $\mathcal{F}_s$-measurable and $Y := W(t) - W(s)$ is independent of $\mathcal{F}_s$, we can apply Proposition A.6, giving

$$\mathrm{E}_s[h(W(t))] = \mathrm{E}_s[h(Y + W(s))] = g(W(s)),$$

where $g(x) = \mathrm{E}[h(Y + x)] = \mathrm{E}[h(\sqrt{t-s}\, Z + x)]$, $Z \sim \mathit{Norm}(0,1)$. In fact, using the density of Z we have $g(x) = \int_{\mathbb{R}} h(\sqrt{t-s}\, z + x) \mathfrak{n}(z)\, dz$. □

We remark that, for $s = t$, this result simply gives $g(x) = h(x)$ and recovers the trivial case that $\mathrm{E}_s[h(W(s))] = h(W(s))$, i.e., $W(s)$, and hence $h(W(s))$, is $\mathcal{F}_s$-measurable. For $s < t$, we can also re-express $g(x)$ by using a linear change of integration variables $z \to y = \sqrt{t-s}\, z + x$, $z = (y - x)/\sqrt{t-s}$, giving:

$$g(x) = \int_{\mathbb{R}} h(y) p(s, t; x, y)\, \mathrm{d}y \tag{1.13}$$

where $p(s,t;x,y) = \frac{1}{\sqrt{t-s}} \mathfrak{n}\left(\frac{y-x}{\sqrt{t-s}}\right) = \frac{\mathrm{e}^{-\frac{(y-x)^2}{2(t-s)}}}{\sqrt{2\pi(t-s)}}$. Below, we shall see that this function has a very special role and is the so-called transition PDF of BM. The above Markov property means that when taking an expectation of a function of BM at a future time t conditional on all the information (path history) of the BM up to an earlier time $s < t$, it is the same as only conditioning on knowledge of the BM at time s. In particular,

$$\mathrm{E}_s[h(W(t))] = \mathrm{E}[h(W(t)) \mid W(s)] = \int_{\mathbb{R}} h(y) p(s, t; W(s), y)\, \mathrm{d}y \equiv g(W(s)).$$

Based on this formula, we can then compute the expected value of $h(W(t))$ conditional on any real value $W(s) = x$ of BM at time $s < t$ as

$$\mathrm{E}[h(W(t)) \mid W(s) = x] = \int_{\mathbb{R}} h(y) p(s, t; x, y)\, \mathrm{d}y. \tag{1.14}$$

By this formula it is then clear that $p(s,t;x,y)$ is in fact the conditional PDF of random variable $W(t)$, given $W(s)$, i.e.,

$$\mathrm{E}[h(W(t)) \mid W(s)=x] = \int_{\mathbb{R}} h(y) f_{W(t)|W(s)}(y|x)\,\mathrm{d}y \implies f_{W(t)|W(s)}(y|x) = p(s,t;x,y).$$

By the Markov property of W we have, for all $0 \leqslant s \leqslant t$ and Borel set A,

$$\mathbb{P}(W(t) \in A \mid \mathcal{F}_s) = \mathrm{E}_s[\mathbb{I}_{\{W(t)\in A\}}] = \mathrm{E}[\mathbb{I}_{\{W(t)\in A\}} \mid W(s)] = \mathbb{P}(W(t) \in A \mid W(s)), \quad (1.15)$$

i.e., this probability is a function of $W(s)$ and A. Hence, the evaluation of probabilities of the form (1.15) and other expectations of functions of BM reduces to calculation of integrals involving $p(s,t;x,y)$.

From the defining properties of Brownian motion, we also have the following Markov property

$$\mathrm{E}[h(W(t)) \mid W(t_1),\dots,W(t_n)] = \mathrm{E}[h(W(t)) \mid W(t_n)] \quad (1.16)$$

or equivalently

$$\mathbb{P}(W(t) \leqslant y \mid W(t_1),\dots,W(t_n)) = \mathbb{P}(W(t) \leqslant y \mid W(t_n)), \quad (1.17)$$

for any $t_1 < \ldots < t_n < t$ and $y \in \mathbb{R}$. Note that (1.17) implies the equivalence of conditional densities $f_{W(t)|W(t_1),\dots,W(t_n)}(y \mid x_1,\dots,x_n) = f_{W(t)|W(t_n)}(y \mid x_n)$.

Equation (1.16) or (1.17) is implied by Theorem 1.2. Indeed, Theorem 1.2 asserts that $\mathrm{E}[h(W(t)) \mid \mathcal{F}_{t_n}] = \mathrm{E}[h(W(t))|W(t_n)]$. Using this relation and applying nested conditioning via the tower property, where the σ-algebra $\sigma(W(t_1),\dots,W(t_n)) \subset \mathcal{F}_{t_n}$:

$$\begin{aligned}\mathrm{E}[h(W(t)) \mid W(t_1),\dots,W(t_n)] &= \mathrm{E}[\mathrm{E}[h(W(t)) \mid \mathcal{F}_{t_n}] \mid W(t_1),\dots,W(t_n)]\\ &= \mathrm{E}[\mathrm{E}[h(W(t)) \mid W(t_n)] \mid W(t_1),\dots,W(t_n)]\\ &= \mathrm{E}[h(W(t)) \mid W(t_n)]\,.\end{aligned}$$

The last line follows since $\mathrm{E}[h(W(t)) \mid W(t_n)]$ is $\sigma(W(t_n))$–measurable and hence it is $\sigma(W(t_1),\dots,W(t_n))$–measurable so that it may be pulled out of the outer expectation. An alternative direct proof of (1.16) also follows by applying Proposition A.6 where $Y := W(t) - W(t_n)$ is independent of $\mathcal{G} \equiv \sigma(W(t_1),\dots,W(t_n))$ and $W(t_n)$ is $\mathcal{G}$–measurable, giving $\mathrm{E}[h(W(t)) \mid W(t_1),\dots,W(t_n)] \equiv \mathrm{E}[h(W(t)) \mid \mathcal{G}]$:

$$\mathrm{E}[h(W(t)) \mid \mathcal{G}] = \mathrm{E}[h(Y + W(t_n)) \mid \mathcal{G}] = g(W(t_n)),\ g(x) = \mathrm{E}[h(Y+x)]\,,$$

i.e., $\mathrm{E}[h(W(t)) \mid \mathcal{G}] = \mathrm{E}[h(W(t)) \mid \sigma(W(t_n))] \equiv \mathrm{E}[h(W(t)) \mid W(t_n)]$.

The transition PDF, and its associated transition probability function, plays a pivotal role in the theory of continuous-time Markov processes.

Definition 1.2. A *transition probability function* for a continuous-time Markov process $\{X(t)\}_{t\geqslant 0} \in \mathbb{R}$ is a real-valued function P given by the conditional probability[4]

$$P(s,t;x,y) := \mathbb{P}(X(t) \leqslant y \mid X(s) = x), \quad 0 \leqslant s \leqslant t, \quad x,y \in \mathbb{R}.$$

We also use the shorter notation $\mathbb{P}_{s,x}(X(t) \leqslant y) \equiv \mathbb{P}(X(t) \leqslant y \mid X(s) = x)$. Suppose that P is absolutely continuous (w.r.t. the Lebesgue measure), i.e., there exists a real-valued nonnegative function p such that, for all $s < t$,

$$P(s,t;x,y) = \int_{-\infty}^{y} p(s,t;x,z)\,\mathrm{d}z. \quad (1.18)$$

[4]Note that for $s = t$ we simply have $P(t,t;x,y) = \mathbb{P}(X(t) \leqslant y \mid X(t) = x) = \mathbb{I}_{\{x\leqslant y\}}$.

p is called a *transition PDF* of the process X. Differentiating (1.18) gives

$$p(s,t;x,y) = \frac{\partial}{\partial y} P(s,t;x,y). \tag{1.19}$$

By definition, the transition probability function P (also called transition CDF) and transition PDF p are, respectively, the conditional CDF and PDF of random variable $X(t)$ given a value for $X(s)$:

$$F_{X(t)|X(s)}(y \mid x) = P(s,t;x,y) \text{ and } f_{X(t)|X(s)}(y \mid x) = p(s,t;x,y). \tag{1.20}$$

The transition CDF $P(s,t;x,y)$ represents the probability that the value of the process at time t, $X(t)$, is at most y given that it has a known value $X(s) = x$ at an earlier time $s < t$. In other words, in terms of a path-wise description, it represents the probability of the collection of events $\{\omega \in \Omega \mid X(t,\omega) \leqslant y,\ X(s,\omega) = x,\ 0 \leqslant s < t\}$. The transition PDF represents the density of such paths in an infinitesimal interval $\mathrm{d}y$ about the value y, i.e., we formally write

$$\mathbb{P}_{s,x}\big(X(t) \in \mathrm{d}y\big) \equiv \mathbb{P}\big(X(t) \in (y, y+\mathrm{d}y] \mid X(s) = x\big) = p(s,t;x,y)\,\mathrm{d}y.$$

We recall that a process $\{X(t)\}_{t\geqslant 0}$ with a PDF p is *conservative* on its state space (here simply assumed as $\mathbb{R}$) if $P(s,t;x,\infty) = 1$, i.e., for $s < t$, $x \in \mathbb{R}$:

$$\mathbb{P}_{s,x}\big(X(t) \in \mathbb{R}\big) \equiv \int_{\mathbb{R}} p(s,t;x,y)\,\mathrm{d}y = 1.$$

[Note: for $s = t$ it is trivial that $P(t,t;x,\infty) = \mathbb{I}_{\{x \leqslant \infty\}} = 1$.]

An important class of processes occurring in many models of finance and other fields is one in which the transition CDF (and PDF) can be written as a function of the difference $t - s$ of the time variables s and t. This stationarity property is stated formally below.

Definition 1.3. A stochastic process $\{X(t)\}_{t\geqslant 0}$ is called *time-homogeneous or stationary* if $P(s,t;x,y) \equiv P(t-s;x,y)$; i.e., the transition probability is a function of the time difference $t - s$.

Hence, for all $s < t$, $t' \geqslant 0$, $x \in \mathbb{R}$, and $A \in \mathcal{B}(\mathbb{R})$,

$$\mathbb{P}(X(t+t') \in A \mid X(s+t') = x) = \mathbb{P}(X(t) \in A \mid X(s) = x). \tag{1.21}$$

Taking $A = (-\infty, y]$ in (1.21) gives

$$\mathbb{P}(X(t+t') \leqslant y \mid X(s+t') = x) = \mathbb{P}(X(t) \leqslant y \mid X(s) = x).$$

Hence, by (1.19) the transition PDF is also a function of the difference:

$$p(s,t;x,y) = p(t-s;x,y)\,.$$

Hence, in the time-homogeneous case we can write both transition CDF and PDF as functions of one time variable and two spatial variables. The above expressions relate the functions expressed in terms of two and one time argument and are equally expressed as

$$P(t;x,y) = P(s,s+t;x,y) \text{ and } p(t;x,y) = p(s,s+t;x,y) \text{ for } s \geqslant 0,\ t > 0,\ x,y \in \mathbb{R},$$

where transition probabilities of the process are calculated by integrating the time-homogeneous transition PDF $p(t;x,y)$:

$$\mathbb{P}(X(s+t) \in A \mid X(s) = x) = \int_A p(t;x,y)\,\mathrm{d}y, \text{ for } s \geqslant 0,\ t > 0,\ A \in \mathcal{B}(\mathbb{R}).$$

Brownian motion W is a Markov process, hence it is possible to define its transition probability function. In fact, since h is any Borel function in (1.13), we have already found the transition PDF of W as part of the proof of Theorem 1.2. We now derive the transition CDF and PDF based simply on Definition 1.2. Writing $W(t) = W(s)+Y, Y := W(t)-W(s)$, where $Y \stackrel{d}{=} \sqrt{t-s}Z, Z \sim Norm(0,1)$, is independent of $W(s)$ allows us to express the transition CDF of W as an unconditional probability which is evaluated explicitly:

$$P(s,t;x,y) := \mathbb{P}(W(t) \leqslant y \mid W(s) = x) = \mathbb{P}(W(s) + Y \leqslant y \mid W(s) = x) = \mathbb{P}(x + Y \leqslant y)$$
$$= \mathbb{P}\left(Z \leqslant \frac{y-x}{\sqrt{t-s}}\right) = \mathcal{N}\left(\frac{y-x}{\sqrt{t-s}}\right), \quad s < t, \ x, y \in \mathbb{R}. \tag{1.22}$$

By (1.19), the transition PDF for W is hence:

$$p(s,t;x,y) = \frac{\partial}{\partial y}\mathcal{N}\left(\frac{y-x}{\sqrt{t-s}}\right) = \frac{1}{\sqrt{t-s}}n\left(\frac{y-x}{\sqrt{t-s}}\right) = \frac{\mathrm{e}^{-\frac{(y-x)^2}{2(t-s)}}}{\sqrt{2\pi(t-s)}}, \quad s < t, \ x, y \in \mathbb{R}. \tag{1.23}$$

Note that this is exactly the expression we obtained above.
Remark: As an alternative derivation, the transition CDF in (1.22) also follows by a straightforward application of (1.6), where we set $x_2 \equiv x$, $X_1 \equiv W(t)$, $X_2 \equiv W(s)$, $\mu_1 = \mu_2 = \mathrm{E}[W(t)] = \mathrm{E}[W(s)] = 0$, $\sigma_1^2 = \mathrm{Var}(W(t)) = t$, $\sigma_2^2 = \mathrm{Var}(W(s)) = s$, $\rho = \mathrm{Cov}(W(t), W(s))/\sqrt{st} = \sqrt{s/t}$, and hence

$$W(t) \mid \{W(s) = x\} \sim Norm(x, t-s)\,, \ s < t, \ x, y \in \mathbb{R}.$$

Hence, the conditional CDF $F_{W(t)|W(s)}(y,x) = \mathcal{N}\left(\frac{y-x}{\sqrt{t-s}}\right)$ is the expression in (1.22).

As is seen from (1.23), the transition PDF of standard Brownian motion is a function of the spatial point difference $y - x$ and time difference $t - s$; i.e., this shows that standard Brownian motion has the time and space homogeneity property. Moreover, the PDF is invariant to the interchange of x and y. Throughout we let $p_0(t;z)$ denote the PDF of $W(t)$, i.e.,

$$p_0(t;z) := \frac{\partial}{\partial z}\mathbb{P}(W(t) \leqslant z) = \frac{\partial}{\partial z}\mathcal{N}\left(\frac{z}{\sqrt{t}}\right) = \frac{1}{\sqrt{t}}n\left(\frac{z}{\sqrt{t}}\right) = \frac{\mathrm{e}^{-\frac{z^2}{2t}}}{\sqrt{2\pi t}}, \ t > 0, \ z \in \mathbb{R}. \tag{1.24}$$

Then, the transition PDF of W in (1.23) is equivalently written as

$$p(s,t;x,y) = p_0(t-s; y-x), \quad s < t, \quad x, y, \in \mathbb{R}. \tag{1.25}$$

We observe that the PDF is an even function of the spatial variable, $p_0(t-s; y-x) = p_0(t-s; x-y)$, i.e., $p_0(t;z) = p_0(t;-z)$. Standard BM is clearly conservative on $\mathbb{R}$:

$$\mathbb{P}_{s,x}(W(t) \in \mathbb{R}) = \int_{\mathbb{R}} p_0(t-s; y-x)\,\mathrm{d}y = \int_{\mathbb{R}} p_0(t-s; z)\,\mathrm{d}z = 1, \ \ s < t, \, x \in \mathbb{R}.$$

This is also seen directly from the transition CDF in (1.22) in the limit $y \to \infty$: $P(s,t;x,\infty) = \mathcal{N}(\infty) = 1$.

In the following simple examples we demonstrate the direct use of the transition PDF to compute conditional expectations. Note that we already computed these conditional expectations without the need to employ the transition PDF and integration (i.e., we employed simple known properties of BM increments and independence via Proposition A.6).

Example 1.6. Evaluate each conditional expectation with the use of the transition PDF, where $\{\mathcal{F}_t\}_{t \geqslant 0}$ is a filtration for BM.

(a) $\mathrm{E}[W(t) \mid \mathcal{F}_s]$, $0 < s < t$;

(b) $\mathrm{E}[W^2(t) \mid \mathcal{F}_s]$, $0 < s < t$;

Solution.
(a) From the Markov property we have $\mathrm{E}[W(t) \mid \mathcal{F}_s] = \mathrm{E}[W(t) \mid W(s)] \equiv g(W(s))$, with $g(x) = \mathrm{E}[h(W(t))|W(s) = x]$ computed using (1.14) where $h(y) = y$. Using the transition PDF $p(s,t;x,y) = p_0(t-s; y-x)$ in (1.24) and by changing integration variables $z := y - x, y = z + x$:

$$\begin{aligned}\mathrm{E}[W(t) \mid W(s) = x] &= \int_{\mathbb{R}} y\, p_0(t-s; y-x)\, \mathrm{d}y \\ &= \int_{\mathbb{R}} z\, p_0(t-s; z)\, \mathrm{d}z + x \int_{\mathbb{R}} p_0(t-s; z)\, \mathrm{d}z = x,\end{aligned}$$

i.e., $g(x) = x, x \in \mathbb{R}$. Hence, $\mathrm{E}[W(t) \mid \mathcal{F}_s] = W(s)$.
Note: $p_0(t-s;z)$ is a function that is even in z, i.e., $z\, p_0(t-s;z)$ is odd in z so that $\int_{\mathbb{R}} z\, p_0(t-s;z)\, \mathrm{d}z = 0$. Moreover, $p_0(t-s;z)$ is a PDF on $\mathbb{R}$ that integrates to unity, $\int_{\mathbb{R}} p_0(t-s;z) dz = 1$.

(b) Using (1.14), where $h(y) = y^2$, and letting $z := y - x, y = z + x$:

$$\begin{aligned}\mathrm{E}[W^2(t) \mid W(s) = x] &= \int_{\mathbb{R}} y^2\, p_0(t-s; y-x)\, \mathrm{d}y = \int_{\mathbb{R}} (z+x)^2\, p_0(t-s; z)\, \mathrm{d}z \\ &= \int_{\mathbb{R}} z^2\, p_0(t-s; z)\, \mathrm{d}z + 2x \int_{\mathbb{R}} z\, p_0(t-s; z)\, \mathrm{d}z + x^2 \int_{\mathbb{R}} p_0(t-s; z) dz \\ &= t - s + 2x \cdot 0 + x^2 \cdot 1 = t - s + x^2.\end{aligned}$$

Hence, $\mathrm{E}[W^2(t) \mid \mathcal{F}_s] = \mathrm{E}[W^2(t) \mid W(s)] = t - s + W^2(s)$. □

In the previous section, we obtained the joint PDF of the path skeleton, $W(t_1), \ldots, W(t_n)$, of Brownian motion evaluated at n arbitrary time points $0 = t_0 < t_1 < t_2 < \cdots < t_n < \infty$. The resulting n-variate Gaussian distribution formula in (1.12) looks somewhat complicated since it involves the evaluation of the inverse of the covariance matrix $\mathbf{C}$ given in (1.11). However, by using the Markov property of Brownian motion, we can derive a simpler expression for the joint density. As is well known, a joint PDF of a random vector can be expressed as a product of marginal and conditional univariate densities by successively conditioning:

$$\begin{aligned} f_{\mathbf{W}}(\mathbf{x}) &\equiv f_{W(t_1),\ldots,W(t_n)}(x_1, \ldots, x_n) \\ &= f_{W(t_1)}(x_1)\, f_{W(t_2)|W(t_1)}(x_2 \mid x_1) \times \cdots \times f_{W(t_n)|W(t_1),\ldots,W(t_{n-1})}(x_n \mid x_1, \ldots, x_{n-1}).\end{aligned}$$

From the Markov property in (1.17), we have the equivalences in distribution:

$$W(t_k) \mid \{W(t_1), \ldots, W(t_{k-1})\} \stackrel{d}{=} W(t_k) \mid W(t_{k-1}), \quad 2 \leqslant k \leqslant n,$$

i.e., $f_{W(t_k)|W(t_1),\ldots,W(t_{k-1})}(x_k \mid x_1, \ldots, x_{k-1}) = f_{W(t_k)|W(t_{k-1})}(x_k \mid x_{k-1})$. Thus, we obtain

$$\begin{aligned} f_{\mathbf{W}}(\mathbf{x}) &= \prod_{k=1}^{n} f_{W(t_k)|W(t_{k-1})}(x_k \mid x_{k-1}) = \prod_{k=1}^{n} p(t_{k-1}, t_k; x_{k-1}, x_k) \\ &= \prod_{k=1}^{n} p_0(t_k - t_{k-1}; x_k - x_{k-1}) = \prod_{k=1}^{n} \frac{1}{\sqrt{t_k - t_{k-1}}} \cdot n\left(\frac{x_k - x_{k-1}}{\sqrt{t_k - t_{k-1}}}\right) \\ &= \frac{1}{(2\pi)^{n/2}\sqrt{(t_1 - t_0)\cdots(t_n - t_{n-1})}} \cdot \exp\left(-\frac{1}{2}\sum_{k=1}^{n} \frac{(x_k - x_{k-1})^2}{t_k - t_{k-1}}\right), \end{aligned} \tag{1.26}$$

where $f_{W(t_1)|W(t_0)}(x_1 \mid x_0) \equiv f_{W(t_1)}(x_1) = p_0(t_1; x_1)$ since $W(t_0) \equiv W(0) = x_0 \equiv 0$ for a standard Brownian motion. So now we have two formulas, (1.12) and (1.26), of the joint PDF $f_{\mathbf{W}}$ of Brownian motion on n time points. However, one can show that they are equivalent (see Exercise 1.19). Equation (1.26) clearly shows that the joint PDF of a Brownian path along any discrete set of time points is the product of the transition PDF for each time step.

Hence, for any integrable Borel function $h : \mathbb{R}^n \to \mathbb{R}$, we can express the expected value of $h(\mathbf{W}) = h(W(t_1), \dots, W(t_n))$ as an n-dimensional integral

$$\begin{aligned} \mathrm{E}[h(\mathbf{W})] &= \int_{\mathbb{R}^n} h(x_1, \dots, x_n) f_{W(t_1),\dots,W(t_n)}(x_1, \dots, x_n)\, \mathrm{d}x_1 \dots, \mathrm{d}x_n \\ &= \int_{\mathbb{R}^n} h(x_1, \dots, x_n) \prod_{k=1}^{n} p_0(t_k - t_{k-1}; x_k - x_{k-1})\, \mathrm{d}x_1 \dots, \mathrm{d}x_n, \end{aligned} \tag{1.27}$$

where $0 \equiv t_0 < t_1 < t_2 < \cdots < t_n < \infty$ and $x_0 = 0$. From the Markov property we also readily have the joint density of $W(t_1), \dots, W(t_n)$ conditional on $W(t_0) = x_0$, where $0 < t_0 < t_1 < \cdots < t_n < \infty$, $x_0 \in \mathbb{R}$:

$$f_{W(t_1),\dots,W(t_n)|W(t_0)}(x_1, \dots, x_n | x_0) = \prod_{k=1}^{n} p_0(t_k - t_{k-1}; x_k - x_{k-1}).$$

Note that here the initial time $t_0 > 0$ and the BM at time t_0 takes on any given real value $W(t_0) = x_0$. Hence,

$$\begin{aligned} &\mathrm{E}[h(W(t_1), \dots, W(t_n)) | W(t_0) = x_0] \\ &= \int_{\mathbb{R}^n} h(x_1, \dots, x_n) f_{W(t_1),\dots,W(t_n)|W(t_0)}(x_1, \dots, x_n | x_0)\, \mathrm{d}x_1 \dots, \mathrm{d}x_n \\ &= \int_{\mathbb{R}^n} h(x_1, \dots, x_n) \prod_{k=1}^{n} p_0(t_k - t_{k-1}; x_k - x_{k-1})\, \mathrm{d}x_1 \dots, \mathrm{d}x_n. \end{aligned} \tag{1.28}$$

The Markov property proven in Theorem 1.2 extends to the evaluation of expectations of functions of BM at future joint times. In particular, given $0 < s < t_1 < \cdots < t_n < \infty$ and any filtration $\{\mathcal{F}_t\}_{t \geqslant 0}$ for BM,

$$\mathrm{E}[h(W(t_1), \dots, W(t_n)) | \mathcal{F}_s] = \mathrm{E}[h(W(t_1), \dots, W(t_n)) | W(s)]. \tag{1.29}$$

The r.h.s. expectation is conditional only on the BM at time s (i.e., on $\sigma(W(s))$). This is readily proven. In particular, for each $k = 1, \dots, n$, write $W(t_k) = Y_k + W(s)$, where each $Y_k := W(t_k) - W(s)$ is independent of $\mathcal{F}_s$ and $W(s)$ is $\mathcal{F}_s$-measurable. By Proposition A.7 we then have $\mathrm{E}[h(W(t_1), \dots, W(t_n)) | \mathcal{F}_s] = g(W(s))$, where $g(x) := \mathrm{E}[h(Y_1 + x, \dots, Y_n + x)$. This therefore proves (1.29). Hence, the random variable on the l.h.s. of (1.29) can be derived by evaluating the conditional expectation in (1.28), where $t_0 \equiv s$, as function of dummy variable x_0 and finally replacing x_0 by the random variable $W(s)$.

We remark that (as shown in previous simple examples) the evaluation of the conditional expectation in (1.28) can be performed in whatever manner is convenient, which may or may not involve the transition PDF. The solution to the next relatively simple example demonstrates the use of different methods for computing conditional expectations. The first two are based on basic properties of BM while the third method uses the transition PDF.

Example 1.7. Evaluate $\mathrm{E}[W(t)W(u) \mid \mathcal{F}_s]$, $0 < s < t < u$, where $\{\mathcal{F}_t\}_{t \geqslant 0}$ is a filtration for BM.

Solution.
Method #1. We can show that $\mathrm{E}[W(t)W(u) \mid \mathcal{F}_s] = \mathrm{E}[W^2(t) \mid \mathcal{F}_s]$. One way is to use the tower property in reverse, where $\mathcal{F}_s \subset \mathcal{F}_t$ and $W(t)$ is $\mathcal{F}_t$-measurable, and the martingale property:

$$\begin{aligned}\mathrm{E}[W(t)W(u) \,|\, \mathcal{F}_s] &= \mathrm{E}[\mathrm{E}[W(t)W(u) \,|\, \mathcal{F}_t] \mid \mathcal{F}_s] \\ &= \mathrm{E}[W(t) \underbrace{\mathrm{E}[W(u) \mid \mathcal{F}_t]}_{=W(t)} \mid \mathcal{F}_s] = \mathrm{E}[W^2(t) \,|\, \mathcal{F}_s] = t - s + W^2(s).\end{aligned}$$

The last expression follows from the solution to Example 1.1.

Method #2. Write $W(t) = Y_1 + W(s), Y_1 := W(t) - W(s)$ and $W(u) = Y_2 + W(s), Y_2 := W(u) - W(s)$. Since Y_1, Y_2 are independent of $\mathcal{F}_s$ and $W(s)$ is $\mathcal{F}_s$-measurable, we apply Proposition A.7 to give

$$\mathrm{E}[W(t)W(u) \,|\, \mathcal{F}_s] = \mathrm{E}[(Y_1 + W(s))(Y_2 + W(s)) \,|\, \mathcal{F}_s] = g(W(s)),$$

where

$$g(x) = \mathrm{E}[(Y_1 + x)(Y_2 + x)] = \mathrm{E}[Y_1 Y_2] + x^2 = \mathrm{Cov}(Y_1, Y_2) + x^2 = t - s + x^2 .$$

Hence, $\mathrm{E}[W(t)W(u) \mid \mathcal{F}_s] = t - s + W^2(s)$. Note: $\mathrm{E}[Y_1] = \mathrm{E}[Y_2] = 0$, $\mathrm{Cov}(Y_1, Y_2) = \mathrm{Cov}(W(t) - W(s), W(u) - W(s)) = \mathrm{Var}(W(t) - W(s)) = t - s$.

Method #3. Using (1.28) for $n = 2$, $h(y, z) = yz$, where $t_0 = s, t_1 = t, t_2 = u$, gives

$$\begin{aligned}\mathrm{E}[W(u)W(t)|W(s) = x] &= \int_{\mathbb{R}}\int_{\mathbb{R}} f_{W(t),W(u)|W(s)}(y, z|x) yz \, \mathrm{d}y \, \mathrm{d}z \\ &= \int_{\mathbb{R}} \left(\int_{\mathbb{R}} p_0(u - t; z - y) \, z dz \right) p_0(t - s; y - x) y \, \mathrm{d}y \\ &= \int_{\mathbb{R}} y^2 \, p_0(t - s; y - x) dy = t - s + x^2 .\end{aligned}$$

By the Markov property in (1.29), $\mathrm{E}[W(t)W(u) \,|\, \mathcal{F}_s] = \mathrm{E}[W(t)W(u) \mid W(s)] = t - s + W^2(s)$. Note: the inner integral in z equals y (see the solution to part (a) of Example 1.6). This also corresponds to the martingale property of standard BM. The last integral was already evaluated in the solution to part (b) of Example 1.6. □

In closing this section, we note that the transition PDF also satisfies an important general identity due to the Markov property. Let $s < t' < t$, $x, y \in \mathbb{R}$. The transition probability density, $p(s, t; x, y)$, for all paths $(s, x) \to (t, y)$ originating at time-space point (s, x) and ending at time-space point (t, y) is equivalent to the integral, over all intermediate values $x' \in \mathbb{R}$, of the product of transition probability densities, $p(s, t'; x, x')p(t', t; x', y)$, corresponding to the joint density of paths $(s, x) \to (t', y') \to (t, y)$. In precise mathematical terms we have:

$$
\begin{aligned}
p(s,t;x,y) = f_{W(t)|W(s)}(y|x) &= \frac{f_{W(s),W(t)}(x,y)}{f_{W(s)}(x)} \\
&= \int_{\mathbb{R}} \frac{f_{W(s),W(t'),W(t)}(x,x',y)}{f_{W(s)}(x)}\,\mathrm{d}x' \\
&= \int_{\mathbb{R}} f_{W(t')|W(s)}(x'|x)\, f_{W(t)|W(t')}(y|x')\,\mathrm{d}x' \\
&= \int_{\mathbb{R}} p(s,t';x,x')\,p(t',t;x',y)\,\mathrm{d}x' \\
&= \int_{\mathbb{R}} p_0(t'-s;z-x)\,p_0(t-t';y-z)\,\mathrm{d}z\,. \qquad (1.30)
\end{aligned}
$$

This is the so-called Chapman-Kolmogorov equation and is derived again in Section 2.7.1 of Chapter 2 for any diffusion process. The relation in (1.30) can be directly verified by using (1.24) and a bit of tedious algebraic manipulation in completing the square of the combined exponent as a function of z and thereby reducing the integral to a standard gaussian.

1.2.5 Quadratic Variation and Nondifferentiability of Paths

Although sample paths of Brownian motion $W(t,\omega)$, $\omega \in \Omega$, are almost surely, i.e., with probability one, continuous functions of time index t, they do not look like regular functions that appear in a textbook on calculus. First, Brownian sample paths are fractals, meaning that any zoomed-in part of a Brownian path looks very much the same as the original trajectory. Second, Brownian sample paths are (almost surely) not monotone on any finite interval $(t, t+\delta t)$ and are not differentiable w.r.t. t at any point. A formal proof of the second statement is based on the concept of the variation of a function, discussed below. However, let us first provide a simple and instructive probabilistic argument against the differentiability of $W(t)$. Consider a finite difference of the process over a time $\delta t > 0$. Then, in distribution we can write

$$\frac{W(t+\delta t) - W(t)}{\delta t} \stackrel{d}{=} \frac{\sqrt{\delta t}\,Z}{\delta t} = \frac{Z}{\sqrt{\delta t}}, \quad Z \sim Norm(0,1). \qquad (1.31)$$

Clearly, for any constant $\kappa > 0$,

$$\mathbb{P}\left(\left|\frac{Z}{\sqrt{\delta t}}\right| > \kappa\right) = \mathbb{P}(|Z| > \kappa\sqrt{\delta t}) \to \mathbb{P}(|Z| > 0) = 1, \text{ as } \delta t \to 0.$$

Therefore, the ratio $\frac{W(t+\delta t)-W(t)}{\delta t}$ is unbounded, as $\delta t \to 0$, with probability one. So the ratio in (1.31) cannot converge to some finite limit. Recall that for a differentiable function we have the convergence $\frac{f(t+\delta t)-f(t)}{\delta t} \to f'(t)$, as $\delta t \to 0$.

Definition 1.4. The nonnegative quantity defined by

$$V^{(p)}_{[a,b]}(f) = \limsup_{\delta t^{(n)} \to 0} \sum_{i=1}^{n} |f(t_i) - f(t_{i-1})|^p,$$

where the limit is taken over all possible partitions $a = t_0 < t_1 < \cdots < t_n = b$ of $[a,b]$ shrinking as $n \to \infty$ and $\delta t^{(n)} := \max_{i=1}^{n}(t_i - t_{i-1}) \to 0$, is called the *p-variation* (for $p > 0$) of a function $f\colon \mathbb{R} \to \mathbb{R}$ on $[a,b]$, $a < b$. If $V^{(p)}_{[a,b]}(f) < \infty$, then f is said to be a *function of bounded p-variation* on $[a,b]$.

Of particular interest are the first ($p = 1$) and quadratic (second) ($p = 2$) variations. Let us consider regular functions such as monotone and differentiable functions and let us find their variations on an interval $[a, b]$.

Proposition 1.3.

1. *A bounded monotone function has bounded first variation.*
2. *A function with bounded derivative has bounded first variation.*
3. *A function with bounded derivative has zero quadratic variation.*

Proof.

1. Consider a function f that is monotone and defined on $[a, b]$. Assume that f is nondecreasing. Then, for any partition $a = t_0 < t_1 < \cdots < t_n = b$,

$$\sum_{i=1}^{n} |f(t_i) - f(t_{i-1})| = \sum_{i=1}^{n} (f(t_i) - f(t_{i-1})) = f(b) - f(a),$$

since $f(t_i) \geqslant f(t_{i-1})$. Therefore, the first variation of a nondecreasing (increasing) function f is equal to $f(b) - f(a)$. For a nonincreasing (decreasing) function, $V^{(1)}_{[a,b]}(f) = f(a) - f(b)$. By combining these two cases, we obtain $V^{(1)}_{[a,b]}(f) = |f(b) - f(a)| < \infty$.

2. Now, consider a differentiable function $f : [a, b] \to \mathbb{R}$, where $|f'(t)| \leqslant M < \infty$, for all $t \in [a, b]$. By the mean value theorem, for $u, v \in [a, b]$ with $u < v$, there exists $t^* \in [u, v]$ such that $\frac{f(v)-f(u)}{v-u} = f'(t^*)$. Therefore, for every partition of $[a, b]$, we obtain

$$\sum_{i=1}^{n} |f(t_i) - f(t_{i-1})| = \sum_{i=1}^{n} |f'(t_i^*)| \, (t_i - t_{i-1}) \leqslant M \sum_{i=1}^{n} (t_i - t_{i-1}) = M\,(b - a),$$

where $t_i^* \in [t_{i-1}, t_i]$, $1 \leqslant i \leqslant n$. Therefore,

$$V^{(1)}_{[a,b]}(f) \leqslant M\,(b - a) < \infty.$$

Moreover, since $|f'|$ is bounded on $[a, b]$, we have the first variation given by the Riemann integral of $|f'|$ on $[a, b]$:

$$V^{(1)}_{[a,b]}(f) = \lim_{n\to\infty} \sum_{i=1}^{n} |f'(t_i^*)| \, (t_i - t_{i-1}) = \int_a^b |f'(t)| \, \mathrm{d}t.$$

3. First, let us find an upper bound for the sum of squared increments by using the mean value theorem and the fact that $(t_i - t_{i-1})^2 \leqslant \delta t^{(n)} (t_i - t_{i-1})$ and $|f'(t)| \leqslant M < \infty$:

$$\begin{aligned} \sum_{i=1}^{n} (f(t_i) - f(t_{i-1}))^2 &= \sum_{i=1}^{n} (f'(t_i^*))^2 \, (t_i - t_{i-1})^2 \\ &\leqslant \delta t^{(n)} \, M^2 \sum_{i=1}^{n} (t_i - t_{i-1}) = \delta t^{(n)} \, M^2 \, (b - a). \end{aligned}$$

Since $\delta t^{(n)} \to 0$ as $n \to \infty$, the above upper bound converges to zero. Hence, the quadratic variation $V^{(2)}_{[a,b]}(f) = 0$. □

Now, we turn our attention to Brownian motion. In the theorem that follows, we prove that the first variation of BM is infinite and the second variation is nonzero almost surely (a.s.). Therefore, almost surely, a Brownian path cannot be monotone on any time interval since otherwise its first variation on that interval would be finite. Moreover, Brownian sample paths are (a.s.) not differentiable w.r.t. t, since otherwise a differentiable path would have zero quadratic variation and a finite first variation. Since Brownian motion is time homogeneous, it is sufficient to consider the interval $[0,t]$. Note that it is customary to denote the quadratic variation of the standard BM process W on $[0,t]$ by $[W,W](t)$, i.e., $V^{(2)}_{[0,t]}(W) \equiv [W,W](t)$.

Theorem 1.4. *$V^{(1)}_{[0,t]}(W) = \infty$ and $[W,W](t) = t$ for all $t > 0$.*

Proof. Let us first prove that the quadratic variation of BM on $[0,t]$ is finite and equals t. Consider a partition $0 = t_0 < t_1 < \cdots < t_n = t$ of $[0,t]$. The expected value and variance of $V_n = \sum_{i=1}^n (W(t_i) - W(t_{i-1}))^2$ are

$$\mathrm{E}[V_n] = \sum_{i=1}^n \mathrm{E}\left[(W(t_i) - W(t_{i-1}))^2\right] = \sum_{i=1}^n (t_i - t_{i-1}) = t$$

$$\mathrm{Var}(V_n) = \sum_{i=1}^n \mathrm{Var}\left((W(t_i) - W(t_{i-1}))^2\right) = 2\sum_{i=1}^n (t_i - t_{i-1})^2 \leqslant 2\,t\,\delta t^{(n)},$$

$\delta t^{(n)} := \max_{i=1}^n (t_i - t_{i-1})$. Here, we used the independence of Brownian increments on nonoverlapping time intervals, i.e., $W(t_i) - W(t_{i-1}) \stackrel{d}{=} \sqrt{t_i - t_{i-1}}\, Z_i$, where $\{Z_i\}_{n\geqslant 1}$ are i.i.d. $Norm(0,1)$. Hence,

$$\begin{aligned}\mathrm{Var}\left((W(t_i) - W(t_{i-1}))^2\right) = (t_i - t_{i-1})^2\ \mathrm{Var}(Z_i^2) &= (t_i - t_{i-1})^2\ \left\{\mathrm{E}[Z_i^4] - (\mathrm{E}[Z_i^2])^2\right\}\\ &= (t_i - t_{i-1})^2\,(3 - 1^2) = 2(t_i - t_{i-1})^2.\end{aligned}$$

Therefore, $\mathrm{Var}(V_n) \to 0$, as $n \to \infty$ (i.e., $\delta t^{(n)} \to 0$). Since $\mathrm{Var}(V_n) = \mathrm{E}[(V_n - t)^2]$, we have shown $\mathrm{E}[(V_n - t)^2] \to 0$, as $n \to \infty$, i.e., this proves convergence of the sequence, $V_n \to t$, in $L^2(\Omega)$. This implies convergence of the sequence in probability. Hence, there exists a subsequence converging almost surely to t. Thus, $[W,W](t) = t$ (a.s.).

Now, consider the first variation of BM. We find a lower bound of the partial sum of the absolute value of Brownian increments:

$$s_n := \sum_{i=1}^n |W(t_i) - W(t_{i-1})| = \sum_{i=1}^n \frac{(W(t_i) - W(t_{i-1}))^2}{|W(t_i) - W(t_{i-1})|} \geqslant \frac{\sum_{i=1}^n (W(t_i) - W(t_{i-1}))^2}{\sup_{1\leqslant i\leqslant n} |W(t_i) - W(t_{i-1})|}.$$

In the limiting case, $\sum_{i=1}^n (W(t_i) - W(t_{i-1}))^2 \to [W,W](t) = t < \infty$, as $n \to \infty$. It can be shown that almost every path of BM is uniformly continuous, therefore (a.s.)

$$\sup_{1\leqslant i\leqslant n} |W(t_i) - W(t_{i-1})| \to 0, \text{ as } n \to \infty.$$

[Note that we are not proving the known uniform continuity of Brownian paths.] Thus, the first variation of BM, given by $\lim_{n\to\infty} s_n$, is infinite since s_n is bounded from below by a ratio converging to ∞ as $n \to \infty$. □

The results of Theorem 1.4 agree with similar properties of the scaled random walk. In Section 1.2.1 we proved that $[W^{(n)}, W^{(n)}](t) = \frac{\lfloor nt \rfloor}{n}$, which is given by time t as $n \to \infty$. So the quadratic variations of the processes W and $W^{(n)}$ are the same in the limit $n \to \infty$. Moreover, one can show that the first variation of $W^{(n)}$ on $[0,t]$ is equal to $\frac{\lfloor nt \rfloor}{\sqrt{n}}$ ($= \sqrt{n}t$ if nt is an integer). As $n \to \infty$, we showed how scaled random walks $W^{(n)}$ converge to Brownian motion and now we also see that $V^{(1)}_{[0,t]}\left(W^{(n)}\right) = \frac{\lfloor nt \rfloor}{\sqrt{n}} \to \infty = V^{(1)}_{[0,t]}(W)$.

1.3 Some Processes Derived from Brownian Motion

1.3.1 Drifted Brownian Motion

Let μ and $\sigma > 0$ be real constants. A scaled Brownian motion with linear drift, denoted by $W^{(\mu,\sigma)}$, is defined by

$$W^{(\mu,\sigma)}(t) := \mu t + \sigma W(t). \tag{1.32}$$

We call this process (and its extensions described below) *drifted Brownian motion.* Note: $W(t) \equiv W^{(0,1)}(t)$. The expected value and variance of $W^{(\mu,\sigma)}(t)$ are

$$\mathrm{E}\left[W^{(\mu,\sigma)}(t)\right] = \mathrm{E}[\mu t + \sigma W(t)] = \mu t + \sigma \mathrm{E}[W(t)] = \mu t,$$
$$\mathrm{Var}\left(W^{(\mu,\sigma)}(t)\right) = \mathrm{Var}(\mu t + \sigma W(t)) = \sigma^2 \,\mathrm{Var}(W(t)) = \sigma^2 t.$$

Since $W(t)$ is normally distributed, the sum $\mu t+\sigma W(t)$ is a normal random variable as well. Moreover, any increment of drifted Brownian motion, $W^{(\mu,\sigma)}(t) - W^{(\mu,\sigma)}(s)$, with $0 \leqslant s < t$, is normally distributed:

$$\begin{aligned} W^{(\mu,\sigma)}(t) - W^{(\mu,\sigma)}(s) &= \mu\,(t-s) + \sigma\,(W(t) - W(s)) \\ &\stackrel{d}{=} \mu\,(t-s) + \sigma\,\sqrt{t-s}\,Z \sim Norm(\mu\,(t-s), \sigma^2\,(t-s)), \end{aligned}$$

where $Z \sim Norm(0,1)$.

The transition CDF of drifted BM is easily derived using the same steps as in the derivation of (1.22). Writing

$$W^{(\mu,\sigma)}(t) = W^{(\mu,\sigma)}(t) - W^{(\mu,\sigma)}(s) + W^{(\mu,\sigma)}(s) = \mu(t-s) + \sigma Y + W^{(\mu,\sigma)}(s)$$

where $Y := W(t) - W(s) \stackrel{d}{=} \sqrt{t-s}Z$ is independent of $W^{(\mu,\sigma)}(s)$, gives

$$\begin{aligned} P(s,t;x,y) &:= \mathbb{P}\left(W^{(\mu,\sigma)}(t) \leqslant y \mid W^{(\mu,\sigma)}(s) = x\right) \\ &= \mathbb{P}\left(\mu(t-s) + \sigma Y + W^{(\mu,\sigma)}(s) \leqslant y \mid W^{(\mu,\sigma)}(s) = x\right) \\ &= \mathbb{P}\left(Z \leqslant \frac{y-x-\mu(t-s)}{\sigma\sqrt{t-s}}\right) = \mathcal{N}\left(\frac{y-x-\mu(t-s)}{\sigma\sqrt{t-s}}\right), \quad s<t,\ x,y \in \mathbb{R}. \end{aligned} \tag{1.33}$$

Differentiating w.r.t. y gives the transition PDF of drifted BM:

$$p(s,t;x,y) = \frac{1}{\sigma\sqrt{t-s}}\, n\left(\frac{y-x-\mu(t-s)}{\sigma\sqrt{t-s}}\right) = \frac{\mathrm{e}^{-\frac{(y-x-\mu(t-s))^2}{2\sigma^2(t-s)}}}{\sigma\sqrt{2\pi(t-s)}}, \quad s<t,\ x,y\in\mathbb{R}. \tag{1.34}$$

We hence see that $W^{(\mu,\sigma)}$ is time and space homogeneous. The transition PDF/CDF are functions of $y-x$ and $t-s$, and hence are invariant to a shift in either space or time origins. However, the transition PDF is not invariant to the interchange of x and y unless the drift parameter $\mu = 0$.

The probability distribution of $W^{(\mu,\sigma)}(t)$ is

$$\mathbb{P}\left(W^{(\mu,\sigma)}(t) \leqslant z\right) = \mathbb{P}\left(\mu t + \sigma\sqrt{t}\,Z \leqslant z\right) = \mathcal{N}\left(\frac{z-\mu t}{\sigma\sqrt{t}}\right), \quad t>0,\ z\in\mathbb{R}, \tag{1.35}$$

with corresponding PDF, which we denote by $p_0^{(\mu,\sigma)}(t;z)$:

$$p_0^{(\mu,\sigma)}(t;z) := \frac{\partial}{\partial z}\mathbb{P}\left(W^{(\mu,\sigma)}(t) \leqslant z\right) = \frac{1}{\sigma\sqrt{t}}\, n\left(\frac{z-\mu t}{\sigma\sqrt{t}}\right) = \frac{e^{-\frac{(z-\mu t)^2}{2\sigma^2 t}}}{\sigma\sqrt{2\pi t}},\ t > 0,\ z \in \mathbb{R}. \quad (1.36)$$

The transition PDF in (1.34) is equivalently written as

$$p(s,t;x,y) = p_0^{(\mu,\sigma)}(t-s;y-x). \quad (1.37)$$

$W^{(\mu,\sigma)}$ is clearly conservative on $\mathbb{R}$:

$$\mathbb{P}_{s,x}\left(W^{(\mu,\sigma)}(t) \in \mathbb{R}\right) = \int_{\mathbb{R}} p_0^{(\mu,\sigma)}(t-s;y-x)\,\mathrm{d}y = \int_{\mathbb{R}} p_0^{(\mu,\sigma)}(t-s;z)\,\mathrm{d}z = 1,\quad s < t,\ x \in \mathbb{R}.$$

This is also seen directly from (1.33) where taking $y \to \infty$ gives $P(s,t;x,\infty) = \mathcal{N}(\infty) = 1$.

Drifted BM, $\{X(t), t \geqslant 0\}$ with constant drift and scale parameters, starting at an arbitrary point $x_0 \in \mathbb{R}$ is defined by

$$X(t) := x_0 + W^{(\mu,\sigma)}(t) = x_0 + \mu t + \sigma W(t). \quad (1.38)$$

Clearly, the transition CDF and PDF of this process are the same as that of $W^{(\mu,\sigma)}$ above for $0 < s < t$. This is easily shown by the definition of the transition CDF and is in fact the spatial homogeneity property of $W^{(\mu,\sigma)}$. Note that the CDF of $X(t)$ and its corresponding PDF are given by (1.35)–(1.36) with the replacement $z \to z - x_0$.

Let us now consider any process, $\{X(t), t \geqslant 0\} \in \mathbb{R}$, having independent increments over nonoverlapping times; i.e., assume $X(t) - X(s)$ and $X(u) - X(v)$ are an independent pair of random variables for any $0 \leqslant s < t < u < v$. For such a process, $Y := X(t) - X(s)$ is independent of $X(s)$ for any $s < t$. Hence, for all $s < t$, $x, y \in \mathbb{R}$, the transition CDF is reduced to

$$P(s,t;x,y) = \mathbb{P}(Y + X(s) \leqslant y \mid X(s) = x) = \mathbb{P}(Y \leqslant y - x) \equiv F_Y(y-x). \quad (1.39)$$

The process is therefore spatially homogeneous, where F_Y is the CDF of Y, and it has transition PDF $p(s,t;x,y) = F_Y'(y-x) = f_Y(y-x)$, where f_Y is the PDF of Y. The transition CDF and PDF hence follow directly from the distribution of an increment of the process. An example of such a process is the drifted BM defined in (1.38) and, of course, standard BM itself. For instance, the formula in (1.33) also follows as a simple application of (1.39) where $Y \sim Norm(\mu(t-s), \sigma^2(t-s))$.

A simple extension of drifted BM is the case where the drift and scale parameters are generally (nonrandom) time-dependent functions, $\mu = \mu(t)$ and $\sigma = \sigma(t) > 0$, with process starting at $X(0) = x_0$:

$$X(t) := x_0 + \mu(t)t + \sigma(t)W(t). \quad (1.40)$$

Unless $\sigma(t)$ is constant in time, this process does not have independent increments and the process is not spatially homogeneous; i.e., (1.39) is not applicable. However, the expressions for the transition CDF and PDF are readily derived (see Exercise 1.13).

1.3.2 Geometric Brownian Motion

Geometric Brownian motion (GBM) is the most well-known stochastic model for modelling positive asset price processes and pricing contingent claims. It is a keystone of the Black–Scholes–Merton theory of option pricing. However, from the mathematical point of view, the GBM is just an exponential function of a drifted Brownian motion. For constant drift μ and volatility parameter σ, (standard) GBM is defined by

$$S(t) := \mathrm{e}^{x_0+W^{(\mu,\sigma)}(t)} = \mathrm{e}^{x_0+\mu t+\sigma W(t)} = S_0\,\mathrm{e}^{\mu t+\sigma W(t)}, \quad t \geqslant 0, \quad S_0 > 0. \tag{1.41}$$

The process starts at $S_0 = \mathrm{e}^{x_0}$, since $W^{(\mu,\sigma)}(0) = W(0) = 0$. This process is hence strictly positive for all $t \geqslant 0$. Since $S(t)$ is an exponential function of a normal variable, the probability distribution of the time-t realization of GBM is log-normal. The CDF of $S(t)$ is

$$\mathbb{P}(S(t) \leqslant y) = \mathbb{P}\left(S_0\,\mathrm{e}^{\mu t+\sigma W(t)} \leqslant y\right) = \mathbb{P}\left(W(t) \leqslant \frac{\ln(y/S_0)-\mu t}{\sigma}\right) = \mathcal{N}\left(\frac{\ln(y/S_0)-\mu t}{\sigma\sqrt{t}}\right),$$

for $y > 0$, $t > 0$, and zero for $y \leqslant 0$. The PDF of $S(t)$ follows by differentiating the CDF w.r.t. y:

$$f_{S(t)}(y) = \frac{1}{y\sigma\sqrt{t}}\,n\left(\frac{\ln(y/S_0)-\mu t}{\sigma\sqrt{t}}\right) = \frac{1}{y\sigma\sqrt{2\pi t}}\mathrm{e}^{-\frac{(\ln(y/S_0)-\mu t)^2}{2\sigma^2 t}}, y > 0, t > 0.$$

By combining the expression (1.41) written for $S(t_1)$ and $S(t_0)$ with $0 \leqslant t_0 < t_1$, we can express $S(t_1)$ in terms of $S(t_0)$ as follows:

$$S(t_1) = S(t_0)\,\mathrm{e}^{\mu\,(t_1-t_0)+\sigma\,(W(t_1)-W(t_0))} \stackrel{d}{=} S(t_0)\,\mathrm{e}^{\mu\,(t_1-t_0)+\sigma\sqrt{t_1-t_0}\,Z},$$

where $Z \sim Norm(0,1)$. We observe that the standard GBM process is time-homogeneous and its logarithm

$$X(t) := \ln S(t) = x_0 + \mu t + \sigma W(t) \tag{1.42}$$

is a drifted BM. Hence, the ratios $\frac{S(t)}{S(s)}$ and $\frac{S(v)}{S(u)}$ are independent random variables for $0 \leqslant s < t \leqslant u < v$. In particular,

$$\ln\frac{S(t)}{S(s)} = X(t) - X(s) = \mu(t-s) + \sigma(W(t) - W(s))$$

and

$$\ln\frac{S(v)}{S(u)} = X(v) - X(u) = \mu(v-u) + \sigma(W(v) - W(u))$$

are jointly independent.

The transition CDF of the GBM process is readily derived from its mapping to the drifted BM via the logarithm: [5]

$$\begin{aligned} P(s,t;x,y) &= \mathbb{P}\left(S(t) \leqslant y\,|S(s) = x\right) = \mathbb{P}\left(X(t) \leqslant \ln y\,|X(s) = \ln x\right) \\ &= \mathbb{P}(X(t) - X(s) \leqslant \ln y - \ln x) \\ &= \mathcal{N}\left(\frac{\ln(y/x)-\mu(t-s)}{\sigma\sqrt{t-s}}\right), \quad s < t,\ x,y > 0. \end{aligned} \tag{1.43}$$

Note that the second line above follows from independence of the X-increments; i.e., (1.39)

[5] Alternatively, by writing $S(t) = \frac{S(t)}{S(s)}S(s)$ and using the independence of $\frac{S(t)}{S(s)}$ and $S(s)$, we have: $P(s,t;x,y) = \mathbb{P}\left(\frac{S(t)}{S(s)}S(s) \leqslant y\,|S(s) = x\right) = \mathbb{P}\left(\frac{S(t)}{S(s)}x \leqslant y\right) = \mathbb{P}\left(\frac{S(t)}{S(s)} \leqslant \frac{y}{x}\right) = \mathbb{P}\left(\ln\frac{S(t)}{S(s)} \leqslant \ln\frac{y}{x}\right) = \mathbb{P}\left(\mu(t-s) + \sigma(W(t) - W(s)) \leqslant \ln\frac{y}{x}\right) = \mathcal{N}\left(\frac{\ln(y/x)-\mu(t-s)}{\sigma\sqrt{t-s}}\right)$.

applies directly in this case where $X(t) - X(s) \sim Norm(\mu(t-s), \sigma^2(t-s))$. Differentiating the above CDF w.r.t. y gives the transition PDF of standard GBM:

$$p(s,t;x,y) = \frac{1}{y\sigma\sqrt{2\pi(t-s)}} \exp\left(-\frac{[\ln(y/x) - \mu(t-s)]^2}{2\sigma^2(t-s)}\right), \quad s < t,\ x, y > 0. \tag{1.44}$$

GBM is clearly conservative on its state space $(0,\infty)$ since integrating the transition PDF in (1.44) over $y \in (0,\infty)$ gives unity, i.e., taking $y \to \infty$ in (1.43) gives $\mathbb{P}_{s,x}(S(t) \in (0,\infty)) \equiv P(s,t;x,\infty) = \mathcal{N}(\infty) = 1$. Both X and S processes are conservative on their respective state spaces and are related by a monotonic mapping (exponential or logarithm).

As discussed further in Section 1.3.3 below, the transition CDF/PDF in (1.43)–(1.44) are also readily derived from the corresponding CDF/PDF in (1.33)–(1.34) of the drifted BM in (1.38) since they are connected by a strictly increasing mapping $S(t) = F(X(t))$, with $F : \mathbb{R} \to (0,\infty)$ defined by $F(x) := e^x$. A simple extension of standard GBM is to consider the exponential of the drifted BM in (1.40). Of course, the process in this case still has a lognormal distribution. The transitions PDF/CDF are again readily derived from the corresponding PDF/CDF of the process in (1.40) where $S(t) = F(X(t))$, $F(x) := e^x$.

1.3.3 Processes related by a monotonic mapping

Let $X(t) \in \mathcal{I}_X$ be any continuous-time process with interval state space $\mathcal{I}_X \subset \mathbb{R}$ and assume $F : \mathcal{I}_X \to \mathcal{I}_Y \subset \mathbb{R}$ is a strictly increasing mapping. The case for strictly decreasing mapping follows in similar fashion. Consider the process defined via this mapping: $Y(t) := F(X(t))$. The mapping has a unique inverse $F^{-1} : \mathcal{I}_Y \to \mathcal{I}_X$. Hence, $X(t) = F^{-1}(Y(t))$. Let us denote the transition CDF of the respective X and Y processes by P^X and P^Y. By applying the strictly increasing mapping F^{-1} within the definition of the transition CDF, we have a relationship between the two transition CDFs:

$$\begin{aligned} P^Y(s,t;x,y) &:= \mathbb{P}(Y(t) \leqslant y \,|Y(s) = x) = \mathbb{P}\left(X(t) \leqslant F^{-1}(y) \,|X(s) = F^{-1}(x)\right) \\ &= P^X(s,t;\hat{x},\hat{y}) \end{aligned} \tag{1.45}$$

where $\hat{x} \equiv F^{-1}(x), \hat{y} \equiv F^{-1}(y)$, $x, y \in \mathcal{I}_Y$, $s \leqslant t$. Equivalently, we also have

$$P^X(s,t;x,y) = P^Y(s,t;F(x),F(y)), \quad x, y \in \mathcal{I}_X,\ s \leqslant t. \tag{1.46}$$

If we further assume that F has a continuous derivative, $F' > 0$, then the corresponding transition PDFs are related as follows, upon differentiating (1.45) w.r.t. y:

$$p^Y(s,t;x,y) = (F^{-1})'(y) \cdot p^X(s,t;F^{-1}(x),F^{-1}(y)) \equiv \frac{p^X(s,t;\hat{x},\hat{y})}{F'(\hat{y})}. \tag{1.47}$$

Equivalently, $p^X(s,t;x,y) = F'(y) \cdot p^Y(s,t;F(x),F(y))$. The reader may recognize these relations as simply change-of-variable formulas for conditional distributions.

For example, let us specialize (1.45) and (1.47) to the case where $X(t)$ is the drifted BM in (1.42) with $\mathcal{I}_X \equiv \mathbb{R}$ and P^X given by (1.33), i.e., $P^X(s,t;\hat{x},\hat{y}) = \mathcal{N}(\frac{\hat{y}-\hat{x}-\mu(t-s)}{\sigma\sqrt{t-s}})$. Hence, the transition CDF, and PDF, of the process $Y(t) := F(X(t))$ have the form

$$P^Y(s,t;x,y) = \mathcal{N}\left(\frac{F^{-1}(y) - F^{-1}(x) - \mu(t-s)}{\sigma\sqrt{t-s}}\right), \tag{1.48}$$

$$p^Y(s,t;x,y) = \frac{1}{\sigma\sqrt{t-s}F'(F^{-1}(y))} \cdot n\left(\frac{F^{-1}(y) - F^{-1}(x) - \mu(t-s)}{\sigma\sqrt{t-s}}\right) \tag{1.49}$$

$x, y \in \mathcal{I}_Y$, for any continuously differentiable strictly increasing function $F : \mathbb{R} \to \mathcal{I}_Y$. Setting $\sigma = 1, \mu = 0$ in these expressions gives the transition CDF and PDF of $Y(t) := F(W(t))$.

As a simple example, let $F(x) := \mathrm{e}^x$. Then, $F'(x) = F(x)$ with inverse $F^{-1}(y) = \ln y$ and $F'(F^{-1}(y)) = F(F^{-1}(y)) = y$, $y \in (0, \infty)$. The process $Y(t) := F(X(t)) = \mathrm{e}^{X(t)}$ is the GBM in (1.41) and the application of (1.48)–(1.49) recovers the expressions in (1.43)–(1.44).

1.3.4 Brownian Bridge

A bridge process is obtained by conditioning on the value of the original process at some future time. For example, consider a standard Brownian motion *pinned at the origin* at some time $T > 0$; i.e., all paths ω such that $W(0) = W(T, \omega) = 0$. As a result, we obtain a continuous time process defined for time $t \in [0, T]$ such that its probability distribution is the distribution of a standard BM *conditional on* $W(T) = 0$. This process is called a *standard Brownian bridge*. We shall denote it by $B^{(0,0)}_{[0,T]}$ or simply B. Figure 1.5 depicts some sample paths of this process for $T = 1$. The process can be expressed in terms of a standard BM as follows:

$$B(t) = W(t) - \frac{t}{T} W(T), \quad 0 \leqslant t \leqslant T. \tag{1.50}$$

First, we show that it satisfies the boundary condition $B(0) = B(T) = 0$:

$$B(0) = W(0) - \frac{0}{T} W(T) = 0, \quad B(T) = W(T) - \frac{T}{T} W(T) = W(T) - W(T) = 0.$$

Second, we show that realizations of the Brownian bridge have the same normal probability distribution as those of Brownian motion $W(t)$ conditional on $W(T) = 0$, for all $t \in (0, T)$. $B(t)$ is normally distributed as a linear combination of normal random variables. The mean and variance are

$$\begin{aligned}
\mathrm{E}[B(t)] &= \mathrm{E}[W(t)] - \frac{t}{T} \mathrm{E}[W(T)] = 0, \\
\mathrm{Var}(B(t)) &= \mathrm{Var}(W(t)) - 2\frac{t}{T} \mathrm{Cov}(W(t), W(T)) + \frac{t^2}{T^2} \mathrm{Var}(W(T)) \\
&= t - 2\frac{t}{T} t + \frac{t^2}{T^2} T = t - \frac{t^2}{T} = \frac{t(T-t)}{T}.
\end{aligned}$$

Let us now find the conditional distribution of $W(t)$ given $W(T)$. Since $[W(t), W(T)]$ is jointly normally distributed with zero mean vector and covariance matrix $\begin{bmatrix} t & t \\ t & T \end{bmatrix}$, the conditional distribution is again normal. Applying (1.6) gives $W(t)$, conditional on $W(T) = y$, for any real y, as normally distributed with mean $y\frac{t}{T}$ and variance $\frac{t(T-t)}{T}$, i.e.,

$$W(t) | \{W(T) = y\} \sim \mathit{Norm}\left(y\frac{t}{T}, \frac{t(T-t)}{T}\right), \quad 0 < t < T.$$

Note that this corresponds to the distribution of the BM bridge process started at zero and pinned at value y at time T: $B^{(0,y)}_{[0,T]}(t) = y\frac{t}{T} + B^{(0,0)}_{[0,T]}(t) = W(t) + \frac{t}{T}(y - W(T))$. Therefore, for the standard Brownian bridge pinned at zero at time T we set $y = 0$:

$$B(t) \sim \mathit{Norm}\left(0, \frac{t(T-t)}{T}\right), \quad 0 < t < T. \tag{1.51}$$

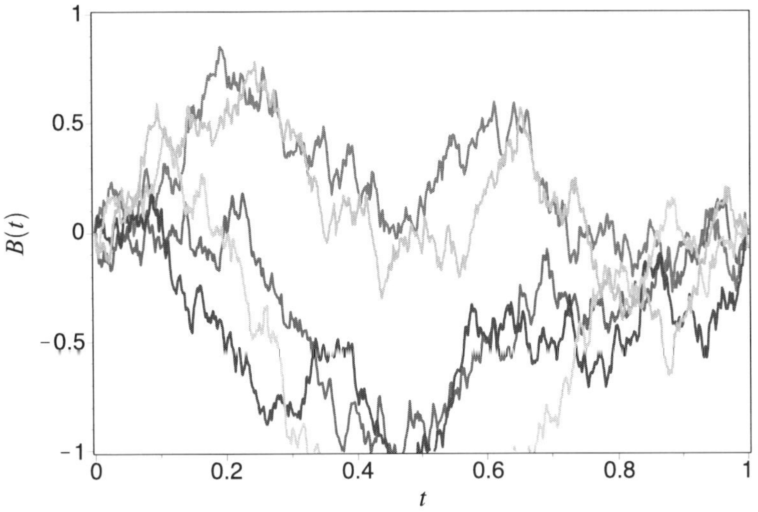

FIGURE 1.5: Sample paths of a standard Brownian bridge for $t \in [0,1]$. Note that all paths begin and end at value zero.

An alternative way to realize this is to re-write the conditioning in terms of transition PDFs. In particular, the conditional PDF of $W(t)$, given $W(T) = y$, can be expressed in terms of transition PDFs of BM in (1.25) as follows:

$$\begin{aligned} f_{W(t)|W(T)}(x \mid y) &= \frac{f_{W(t),W(T)}(x,y)}{f_{W(T)}(y)} = \frac{f_{W(t)}(x)\, f_{W(T)|W(t)}(y \mid x)}{f_{W(T)}(y)} \\ &= \frac{p(0,t;0,x)\, p_0(t,T;x,y)}{p(0,T;0,y)} = \frac{p_0(t;x)\, p_0(T-t;y-x)}{p_0(T;y)} \\ &= \frac{\frac{1}{\sqrt{2\pi t}} \exp\left(-\frac{x^2}{2t}\right) \frac{1}{\sqrt{2\pi(T-t)}} \exp\left(-\frac{(y-x)^2}{2(T-t)}\right)}{\frac{1}{\sqrt{2\pi T}} \exp\left(-\frac{y^2}{2T}\right)} \\ &= \frac{1}{\sqrt{2\pi \frac{t(T-t)}{T}}} \exp\left(-\frac{1}{2}\left(\frac{x^2}{t} + \frac{(y-x)^2}{T-t} - \frac{y^2}{T}\right)\right). \end{aligned} \tag{1.52}$$

Simplifying the exponent in the above expression gives

$$\begin{aligned} \frac{x^2}{t} + \frac{(y-x)^2}{T-t} - \frac{y^2}{T} &= \frac{x^2T(T-t) + (y-x)^2tT - y^2t(T-t)}{tT(T-t)} \\ &= \frac{x^2T^2 - 2xytT + y^2t^2}{tT(T-t)} = \frac{(xT-yt)^2}{tT(T-t)} = \frac{(x-yt/T)^2}{t(T-t)/T}. \end{aligned}$$

Finally, we obtain

$$f_{W(t)|W(T)}(x \mid y) = \frac{1}{\sqrt{2\pi \frac{t(T-t)}{T}}} \exp\left(-\frac{1}{2}\frac{\left(x - y\frac{t}{T}\right)^2}{\frac{t(T-t)}{T}}\right) = \frac{1}{\sqrt{\frac{t(T-t)}{T}}}\, n\left(\frac{x - y\frac{t}{T}}{\sqrt{\frac{t(T-t)}{T}}}\right). \tag{1.53}$$

Again, we conclude that $B^{(0,y)}_{[0,T]}(t) \stackrel{d}{=} W(t) \mid \{W(T) = y\} \sim Norm\left(y\frac{t}{T}, \frac{t(T-t)}{T}\right)$, $0 < t < T$.

A simple extension of the above is to consider a Brownian bridge from a to b on $[0, T]$, denoted by $B^{(a,b)}_{[0,T]}(t)$. For any real constants a, b, this process is constructed by adding a linear drift function to the standard Brownian bridge:

$$B^{(a,b)}_{[0,T]}(t) = a + \frac{(b-a)t}{T} + B^{(0,0)}_{[0,T]}(t) = a + \frac{(b-a)t}{T} + W(t) - \frac{t}{T}W(T), \quad t \in [0, T]. \tag{1.54}$$

Since adding a nonrandom function to a normal random variable again gives a normal random variable, the realizations of the Brownian bridge are normally distributed:

$$B^{(a,b)}_{[0,T]}(t) \sim Norm\left(a + \frac{(b-a)t}{T}, \frac{t(T-t)}{T}\right), \quad t \in (0, T).$$

Note that the mean is a linear function of time having values a and b at the respective endpoint times $t = 0$ and $t = T$ where the process is pinned.

A further extension is to define a Brownian bridge having values a and b on the respective endpoints of any future time interval $[t_0, T]$, $0 \leqslant t_0 \leqslant t \leqslant T$:

$$B^{(a,b)}_{[t_0,T]}(t) := a + (b-a)\frac{t-t_0}{T-t_0} + B^{(0,0)}_{[t_0,T]}(t), \tag{1.55}$$

where $B^{(0,0)}_{[t_0,T]}(t) = W(t) - W(t_0) - \frac{t-t_0}{T-t_0}(W(T) - W(t_0))$ is a standard Brownian bridge on $[t_0, T]$. Note that this gives (1.50) and (1.55) gives (1.54) when $t_0 = 0$. The mean and variance of $B^{(a,b)}_{[t_0,T]}(t)$ are readily computed using the same steps as employed above for the case of the standard bridge, giving

$$B^{(a,b)}_{[t_0,T]}(t) \sim Norm\left(a + (b-a)\frac{t-t_0}{T-t_0}, \frac{(t-t_0)(T-t)}{T-t_0}\right). \tag{1.56}$$

As in (1.52), the distribution is readily expressed in terms of transition PDFs. We have

$$B^{(a,b)}_{[t_0,T]}(t) \stackrel{d}{=} W(t)|\{W(t_0) = a, W(T) = b\},$$

where the PDF of $B^{(a,b)}_{[t_0,T]}(t)$, $f_{B^{(a,b)}_{[t_0,T]}(t)}(x) = f_{W(t)|W(t_0),W(T)}(x|a,b)$, $a, b, x \in \mathbb{R}$, $t_0 < t < T$, is given by

$$\begin{aligned} f_{W(t)|W(t_0),W(T)}(x|a,b) &= \frac{f_{W(t_0),W(t),W(T)}(a,x,b)}{f_{W(t_0),W(T)}(a,b)} \\ &= \frac{f_{W(t_0)}(a) \cdot f_{W(t)|W(t_0)}(x|a) \cdot f_{W(T)|W(t)}(b|x)}{f_{W(t_0)}(a) \cdot f_{W(T)|W(t_0)}(b|a)} \\ &= \frac{p(t_0, t; a, x)\, p(t, T; x, b)}{p(t_0, T; a, b)} \\ &= \frac{p_0(t-t_0; x-a)\, p_0(T-t; b-x)}{p_0(T-t_0; b-a)}. \end{aligned} \tag{1.57}$$

Note: setting $t_0 = 0, a = 0, b = y$ recovers the density in (1.53). The expression in (1.57) can be written as a single Gaussian density that corresponds to the normal distribution in (1.56). This can also be verified by an appropriate application of (1.4) or (1.5) where the scalar $W(t)$ is conditioned on the vector $[W(t_0), W(T)] = [a, b]$.

Let us now consider any Markov process $\{X(t)\}_{t \geqslant 0}$ with state space interval $\mathcal{I}_X \subset \mathbb{R}$ and transition PDF $p(s, t; x, y)$, $s < t$, $x, y \in \mathcal{I}_X$. A bridge process is then realized where the process is pinned at the endpoint times of an interval $[t_0, T]$, $t_0 \leqslant t \leqslant T$, i.e., $X(t_0) = a, X(T) = b$, for any $a, b \in \mathcal{I}_X$. Denoting this bridge process by $B^{(a,b)}_{[t_0,T]}(t)$, we have

$$B^{(a,b)}_{[t_0,T]}(t) \stackrel{d}{=} X(t) | \{X(t_0) = a, X(T) = b\} .$$

The PDF of this process, $f_{B^{(a,b)}_{[t_0,T]}(t)}(x) = f_{X(t)|X(t_0),X(T)}(x|a,b)$, $a, b, x \in \mathcal{I}_X$, $t_0 < t < T$, is hence given by simply applying the Markov property (as in (1.57)):

$$f_{X(t)|X(t_0),X(T)}(x|a,b) = \frac{f_{X(t)|X(t_0)}(x|a) \cdot f_{X(T)|X(t)}(b|x)}{f_{X(T)|X(t_0)}(b|a)} = \frac{p(t_0, t; a, x)\, p(t, T; x, b)}{p(t_0, T; a, b)} . \tag{1.58}$$

From the general Chapman-Kolmogorov relation, we see that integrating this PDF of $B^{(a,b)}_{[t_0,T]}(t)$ over $x \in \mathcal{I}_X$ gives unity. We also remark that the PDF for the standard Brownian bridge is obviously recovered in the special case that $X(t) = W(t)$.

1.3.5 Gaussian Processes

Most of the continuous-time processes we have considered so far belong to the class of Gaussian processes. The most well-known representative of such a class is Brownian motion. Other examples of Gaussian processes include BM with drift and the Brownian bridge process. Gaussian processes are so named because their realizations have a normal (Gaussian) probability distribution.

Definition 1.5. A continuous-time process $\{X(t)\}_{t \geqslant 0}$ is called a *Gaussian process*, if for every partition $0 \leqslant t_1 < t_2 < \cdots < t_n$, the random variables $X(t_1), X(t_2), \ldots, X(t_n)$ are jointly normally distributed.

The probability distribution of a normal vector $\mathbf{X} = \left[X(t_1), X(t_2), \ldots, X(t_n)\right]^\top$ is determined by the mean vector and covariance matrix. Thus, a Gaussian process is determined by the *mean value function* defined by $m_X(t) := \mathrm{E}[X(t)]$ and *covariance function* defined by $c_X(t,s) := \mathrm{Cov}(X(t), X(s)) = \mathrm{E}[(X(t) - m_X(t))(X(s) - m_X(s))]$, for $t, s \geqslant 0$. The probability distribution of the time-t realization is normal: $X(t) \sim Norm\,(m_X(t), c_X(t,t))$, where $c_X(t,t) = \mathrm{Var}(X(t))$. The probability distribution of the vector $\mathbf{X}$ is $Norm_n(\boldsymbol{\mu}, \mathbf{C})$ with

$$\boldsymbol{\mu} = \left[m_X(t_1), m_X(t_2), \ldots, m_X(t_n)\right]^\top \text{ and } \mathbf{C}_{ij} = c_X(t_i, t_j),\ 1 \leqslant i, j \leqslant n.$$

The function $m_X(t)$ defines a curve on the time-space plane where the sample paths of $\{X(t)\}$ concentrate around it.

Two of the simplest examples of Gaussian processes are as follows.

1. A constant distribution process $X(t) \equiv Z$, where $Z \sim Norm(a, b^2)$, with $m_X(t) = a$ and $c_X(t,s) = b^2$.

2. A piecewise constant process $Y(t) = Z_{\lfloor t \rfloor}$, where $Z_k \sim Norm(a_k, b_k^2)$, $k \in \mathbb{N}_0$, are independent random variables, with $m_Y(t) = a_{\lfloor t \rfloor}$ and $c_Y(t,s) = b_{\lfloor t \rfloor} b_{\lfloor s \rfloor} \mathbb{I}_{\{\lfloor t \rfloor = \lfloor s \rfloor\}}$, i.e., the covariance function is zero unless $\lfloor s \rfloor = \lfloor t \rfloor$, in which case $c_Y(t,s) = b^2_{\lfloor t \rfloor}$.

Example 1.8. Let $\{Z_k\}_{k \in \mathbb{N}}$ be i.i.d. standard normal random variables and define $X(t) = \sum_{k=1}^{\lfloor t \rfloor} Z_k$, $t \geqslant 0$. Show that $\{X(t)\}_{t \geqslant 0}$ is a Gaussian process. Find its mean function and covariance function.

Solution. For any time $t \geqslant 0$, $X(t)$ is a sum of standard normals, hence $X(t)$ is normal. Therefore, any finite-dimensional distribution of the process X is multivariate normal as well. The expected value and variance of $X(t)$ are

$$\mathrm{E}[X(t)] = \sum_{k=1}^{\lfloor t \rfloor} \mathrm{E}[Z_k] = 0 \text{ and } \mathrm{Var}(X(t)) = \sum_{k=1}^{\lfloor t \rfloor} \mathrm{Var}(Z_k) = \lfloor t \rfloor.$$

For $0 \leqslant s < t$ we have

$$X(t) = X(s) + \sum_{k=\lfloor s \rfloor+1}^{\lfloor t \rfloor} Z_k.$$

Since $X(t) - X(s)$ and $X(s)$ are independent, we have

$$\mathrm{Cov}(X(s), X(t)) = \mathrm{Cov}(X(s), X(s)) + \mathrm{Cov}(X(s), X(t) - X(s)) = \mathrm{Var}(X(s)) + 0 = \lfloor s \rfloor.$$

The above argument applies similarly if we assume $s \geqslant t$, giving $\mathrm{Cov}(X(s), X(t)) = \lfloor t \rfloor$. Thus, $m_X(t) = 0$ and $c_X(s,t) = \lfloor s \wedge t \rfloor$. □

Clearly, Brownian motion and some other processes derived from it are Gaussian processes.

- Standard Brownian motion is a Gaussian process with zero mean function $m(t) = \mathrm{E}[W(t)] = 0$ and covariance function $c(t,s) = \mathrm{E}[W(s)W(t)] = s \wedge t$, $s,t \geqslant 0$. In fact, a Gaussian process having such covariance and mean functions and continuous sample paths is a standard Brownian motion (see Exercise 1.20).

- Brownian motion starting at $x \in \mathbb{R}$ is a Gaussian process with $m(t) = x$ and $c(s,t) = s \wedge t$, $s,t \geqslant 0$.

- Clearly, adding a nonrandom function to a Gaussian process produces another Gaussian process. Therefore, the drifted Brownian motion in (1.32) is a Gaussian process with $m(t) = x + \mu t$ and $c(s,t) = \sigma^2 (s \wedge t)$, $s,t \geqslant 0$. The drifted BM defined in (1.40) is also a Gaussian process with $m(t) = x + \mu(t)t$ and $c(s,t) = \sigma(s)\sigma(t)\,(s \wedge t)$.

- The standard Brownian bridge $B(t) \equiv B^{(0,0)}_{[0,T]}(t)$, is a Gaussian process with mean $m(t) = 0$ and $c(s,t) = s \wedge t - \frac{st}{T}$, $s,t \in [0,T]$.

  Proof. The standard Brownian bridge is defined in (1.50). Clearly $B(t)$ is normally distributed and $B(t_1), \ldots, B(t_n)$ are jointly normally distributed for $0 \leqslant t_1 < \ldots < t_n \leqslant T$. We already calculated its mean $m(t) = 0$. For $s,t \in [0,T]$, we simply calculate the covariance function using the identity $\mathrm{E}[W(s)\,W(t)] = s \wedge t$ (note: $s \wedge T = s, t \wedge T = t$):

$$\begin{aligned} c(s,t) &= \mathrm{E}\left[\left(W(s) - \frac{s}{T}W(T)\right)\left(W(t) - \frac{t}{T}W(T)\right)\right] \\ &= \mathrm{E}[W(s)\,W(t)] - \frac{t}{T}\mathrm{E}[W(s)\,W(T)] - \frac{s}{T}\mathrm{E}[W(t)\,W(T)] + \frac{st}{T^2}\mathrm{E}[W^2(T)] \\ &= s \wedge t - \frac{st}{T} - \frac{st}{T} + \frac{stT}{T^2} = s \wedge t - \frac{st}{T}. \end{aligned}$$ □

- The general Brownian bridge process defined in (1.55) is a Gaussian process. The proof is left as an exercise for the reader (see Exercise 1.21).

- Geometric Brownian motion is clearly not a Gaussian process, but the logarithm of GBM is. Nevertheless, we can derive the mean and covariance functions for a GBM (see Exercise 1.22).

1.4 First Hitting Times and Maximum and Minimum of Brownian Motion

From Chapter 6 of Volume I, we recall the discussion and basic definitions of first passage times for a stochastic process. We are now interested in developing various formulae for the distribution of first passage times, the distribution of the sampled maximum (or minimum) up to a given time $t > 0$, as well as the joint distribution of the sampled maximum (or minimum) and the process value at a given time $t > 0$ for standard BM as well as translated and scaled BM with drift.

1.4.1 The Reflection Principle: Standard Brownian Motion

Consider a standard BM, $W = \{W(t)\}_{t\geqslant 0}$, under a given (fixed) measure $\mathbb{P}$. The first hitting time, to a given level $m \in \mathbb{R}$ is defined by

$$\mathcal{T}_m := \inf\{t \geqslant 0 : W(t) = m\}. \tag{1.59}$$

We recall that either terminology, first hitting time or first passage time, is equivalent since the paths of W are continuous (a.s.) functions of time. In what follows we can equally say that $\mathcal{T}_m$ is a first hitting time or a first passage time to level m. The sampled maximum and minimum of W, from time 0 to time $t \geqslant 0$, are, respectively, denoted by[6]

$$M(t) := \sup_{0\leqslant u\leqslant t} W(u) \tag{1.60}$$

and

$$m(t) := \inf_{0\leqslant u\leqslant t} W(u). \tag{1.61}$$

At time zero we have $W(0) = M(0) = m(0) = 0$. Clearly, $M(t) \geqslant 0$ and $M(t)$ is increasing in t and $m(t) \leqslant 0$ and $m(t)$ is decreasing in t. If $m > 0$, then $\mathcal{T}_m$ is the same as the first hitting time *up* to level m. If $m < 0$, then $\mathcal{T}_m$ is the same as the first hitting time *down* to level m. In particular, we have the equivalence of events:

$$\{\mathcal{T}_m \leqslant t\} = \{M(t) \geqslant m\}, \text{ if } m > 0; \quad \{\mathcal{T}_m \leqslant t\} = \{m(t) \leqslant m\}, \text{ if } m < 0. \tag{1.62}$$

The path symmetries and Markov (memoryless) properties of standard BM are key ingredients in what follows. We already know that BM is time and space homogeneous. In particular, $\{W(t+s) - W(s)\}_{t\geqslant 0}$ is a standard BM for any fixed time $s \geqslant 0$. A more general version of this is the property that $\{W(t+\mathcal{T}) - W(\mathcal{T})\}_{t\geqslant 0}$ is a standard BM, independent of $\{W(u) : 0 \leqslant u \leqslant \mathcal{T}\}$, where $\mathcal{T}$ is a stopping time w.r.t. a filtration for BM. We don't prove this latter property but it is a consequence of what is known as the *strong Markov property* of BM. Based on this property, the distribution of $\mathcal{T}_m$ is now easily derived.

Proposition 1.5 (First Hitting Time Distribution for Standard BM). *The cumulative distribution function of the first hitting time in (1.59), for any real level $m \neq 0$, is given by*

$$\mathbb{P}(\mathcal{T}_m \leqslant t) = 2\,\mathbb{P}(W(t) \geqslant |m|) = 2\left[1 - \mathcal{N}\left(\frac{|m|}{\sqrt{t}}\right)\right] \equiv 2\mathcal{N}\left(-\frac{|m|}{\sqrt{t}}\right), \quad 0 < t < \infty, \tag{1.63}$$

and zero for all $t \leqslant 0$.

[6]Note: throughout the text, the random variable $m(t)$ should not be confused with the constant m. We also recall that $M(t)$ and $m(t)$ are $\mathcal{F}_t$–measurable, where $\{\mathcal{F}_t\}_{t\geqslant 0}$ is any filtration for Brownian motion.

Remark: for $m = 0$ the formula holds trivially for any $t > 0$ where $\mathbb{P}(\mathcal{T}_0 \leqslant t) = 2\mathcal{N}(0) = 2(1/2) = 1$. In fact, $\mathbb{P}(\mathcal{T}_0 = 0) = 1$ since BM starts at zero; $\mathbb{P}(\mathcal{T}_m = 0) = 0$ for $m \neq 0$ as also given by the limit $t \to 0$ in (1.63).

Proof. We prove the result for $|m| = m > 0$, as the formula follows for $m < 0$ ($-m = |m|$) by simple reflection symmetry of BM where $-W$ is also a standard BM. If a BM path lies above m at time t, this implies that the path has already hit level m by time t. By the above strong Markov property of BM, $\{W(t + \mathcal{T}_m) - W(\mathcal{T}_m)\}_{t\geqslant 0}$ is a standard BM. By spatial symmetry of BM paths, given $t > \mathcal{T}_m$, then $W(t) - W(\mathcal{T}_m) \sim Norm(0, t - \mathcal{T}_m)$; i.e., we have the conditional probability $\mathbb{P}(W(t) - W(\mathcal{T}_m) > 0 \mid \mathcal{T}_m < t) = \frac{1}{2}$. Now, note that the event that BM is greater than level m at time t is equivalent to the joint event that BM already hit level m by time t and BM is above $W(\mathcal{T}_m) = m$; i.e.,

$$\{W(t) > m\} = \{W(t) > m, \mathcal{T}_m \leqslant t\} \cup \{W(t) > m, \mathcal{T}_m > t\} = \{W(t) > m, \mathcal{T}_m \leqslant t\}$$

since $\{W(t) > m, \mathcal{T}_m > t\} \equiv \{W(t) > m, M(t) < m\} = \emptyset$. So, we have the equivalence of probabilities:

$$\begin{aligned}\mathbb{P}(W(t) > m) &= \mathbb{P}(W(t) - W(\mathcal{T}_m) > 0, \mathcal{T}_m \leqslant t) \\ &= \mathbb{P}(W(t) - W(\mathcal{T}_m) > 0 \mid \mathcal{T}_m \leqslant t)\,\mathbb{P}(\mathcal{T}_m \leqslant t) \\ &= \frac{1}{2}\mathbb{P}(\mathcal{T}_m \leqslant t)\,.\end{aligned}$$

This gives (1.63) for $m > 0$ since $\mathbb{P}(W(t) > m) = \mathcal{N}(-m/\sqrt{t})$. □

From (1.63) we see that, given an arbitrary long (infinite) time, BM will eventually hit any finite fixed level with probability one, i.e.,

$$\mathbb{P}(\mathcal{T}_m < \infty) = \lim_{t\to\infty} \mathbb{P}(\mathcal{T}_m \leqslant t) = 2\lim_{t\to\infty} \mathcal{N}\left(-\frac{|m|}{\sqrt{t}}\right) = 2\mathcal{N}(0) = 2 \cdot \frac{1}{2} = 1\,.$$

In the opposing limit, given any $m \neq 0$,

$$\mathbb{P}(\mathcal{T}_m = 0) = \lim_{t\searrow 0} \mathbb{P}(\mathcal{T}_m \leqslant t) = 2\lim_{t\to\infty} \mathcal{N}\left(-\frac{|m|}{\sqrt{t}}\right) = 2\mathcal{N}(-\infty) = 2 \cdot 0 = 0.$$

Moreover, $\mathbb{P}(\mathcal{T}_m < 0) = 0$. Hence, the function defined by $F_{\mathcal{T}_m}(t) := \mathbb{P}(\mathcal{T}_m \leqslant t)$ is a proper CDF. The PDF (density) of the first hitting time is given by differentiating (1.63) w.r.t. t, $f_{\mathcal{T}_m}(t) := \frac{\partial}{\partial t}F_{\mathcal{T}_m}(t) \equiv \frac{\partial}{\partial t}\mathbb{P}(\mathcal{T}_m \leqslant t)$:

$$f_{\mathcal{T}_m}(t) = \frac{|m|}{t\sqrt{t}}\, n\left(\frac{|m|}{\sqrt{t}}\right) = \frac{|m|}{t\sqrt{2\pi t}}\mathrm{e}^{-m^2/2t}\,, \quad 0 < t < \infty. \tag{1.64}$$

An important symmetry, which we state without any proof, is called the *reflection principle* for standard BM. This states that, given a stopping time $\mathcal{T}$ (w.r.t. a filtration for BM), the process $\mathcal{W} = \{\mathcal{W}(t)\}_{t\geqslant 0}$ defined by

$$\mathcal{W}(t) := \begin{cases} W(t) & \text{for } t \leqslant \mathcal{T}, \\ 2W(\mathcal{T}) - W(t) & \text{for } t > \mathcal{T}, \end{cases}$$

is also a standard BM. In particular, we set $\mathcal{T} \equiv \mathcal{T}_m$. Hence, $W(\mathcal{T}) = W(\mathcal{T}_m) = m$ and every path of $\mathcal{W}$ is a path of W up to the first hitting time $\mathcal{T}_m$ to level m and the reflection of a path of W for time $t > \mathcal{T}_m$. Since $\mathcal{W}$ and W are equivalent realizations of standard BM,

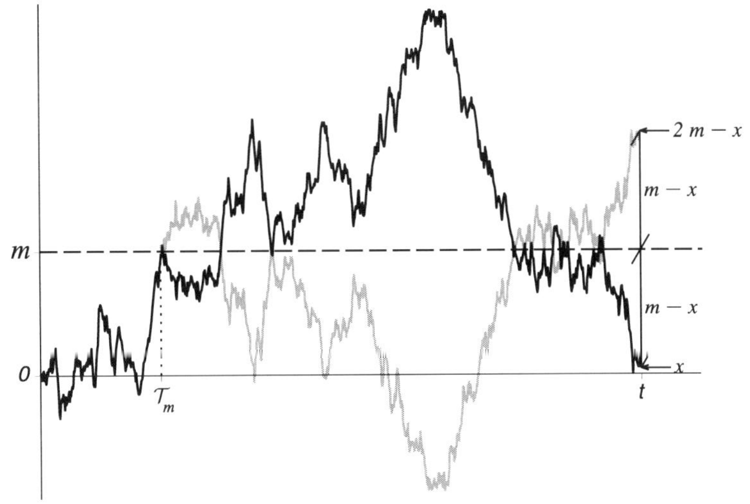

FIGURE 1.6: A sample path of standard BM reaching level $m > 0$ at the first hitting time $\mathcal{T}_m$ and its reflected (gray) path about level m for times past $\mathcal{T}_m$.

this means that every path of a standard Brownian has a reflected path. This is depicted in Figure 1.6.

Based on the reflection principle, we can now obtain the probability of the joint event that BM at time t, $W(t)$, is below a value $x \in \mathbb{R}$ and the sampled maximum of BM in the interval $[0, t]$, $M(t)$, is above level $m > 0$, where $x \leqslant m$. This then leads to the joint distribution for the pair of random variables $M(t), W(t)$, as follows.

Proposition 1.6 (Joint Distribution of $M(t), W(t)$)**.** *For all* $0 < m < \infty$, $-\infty < x \leqslant m$, $t \in (0, \infty)$, *we have*

$$\mathbb{P}(M(t) \geqslant m, W(t) \leqslant x) = \mathbb{P}(W(t) \geqslant 2m - x) = \mathcal{N}\left(\frac{x - 2m}{\sqrt{t}}\right). \tag{1.65}$$

Hence, for all $t \in (0, \infty)$, *the joint PDF of* $(M(t), W(t))$ *is*

$$f_{M(t),W(t)}(m, x) = \frac{2(2m - x)}{t\sqrt{2\pi t}} \mathrm{e}^{-(x-2m)^2/2t}, \quad \text{for } x \leqslant m, \ m > 0, \tag{1.66}$$

and zero otherwise.

Proof. As depicted in Figure 1.6, observe that for every Brownian path that hits level m before $t > 0$ and has value $x \leqslant m$ at time t there is a (reflected) Brownian path that has level $2m - x$. Indeed, by the reflection principle (taking $\mathcal{T} = \mathcal{T}_m$, so $\mathcal{W}(t) = 2m - W(t)$ for $t \geqslant \mathcal{T}_m$) and the equivalence in (1.63) for $m > 0$:

$$\begin{aligned}
\mathbb{P}(M(t) \geqslant m, W(t) \leqslant x) &= \mathbb{P}(\mathcal{T}_m \leqslant t, W(t) \leqslant x) \\
&= \mathbb{P}(\mathcal{T}_m \leqslant t, 2m - W(t) \geqslant 2m - x) \\
&= \mathbb{P}(\mathcal{T}_m \leqslant t, \mathcal{W}(t) \geqslant 2m - x) \\
&= \mathbb{P}(\mathcal{W}(t) \geqslant 2m - x).
\end{aligned}$$

The last equality obtains as follows. Since $x \leqslant m$, then $2m - x \geqslant 2m - m = m$, which implies

$\mathcal{W}(t) \geqslant m$; i.e., the reflected BM and hence the BM has already reached level m by time t. This implies that the joint probability is just the probability of the event $\{\mathcal{W}(t) \geqslant 2m - x\}$ and so (1.65) obtains since $\mathcal{W}$ and W are both standard BM. The formula in (1.66) follows by the standard definition of the joint PDF as the second mixed partial derivative of the joint CDF. Note the minus sign here since we are directly differentiating the joint probability in (1.65), which involves $\{M(t) \geqslant m\}$ instead of $\{M(t) \leqslant m\}$:

$$\begin{aligned} f_{M(t),W(t)}(m,x) &= -\frac{\partial^2}{\partial m \partial x}\mathbb{P}(M(t) \geqslant m, W(t) \leqslant x) \\ &= \frac{\partial}{\partial m}\frac{\partial}{\partial x}\mathcal{N}\left(\frac{2m-x}{\sqrt{t}}\right) \\ &= \frac{-1}{\sqrt{t}}\frac{\partial}{\partial m}n\left(\frac{2m-x}{\sqrt{t}}\right) \\ &= \frac{-1}{\sqrt{t}}\left[\frac{-2(2m-x)}{t}\right]n\left(\frac{2m-x}{\sqrt{t}}\right), \end{aligned}$$

$n(-z) = n(z) \equiv \frac{e^{-z^2/2}}{\sqrt{2\pi}}$, which is the right-hand side of (1.66). □

The joint CDF of $(M(t), W(t))$ now follows by writing the event $\{W(t) \leqslant x\}$ as a union of two mutually exclusive events,

$$\{W(t) \leqslant x\} = \{M(t) > m, W(t) \leqslant x\} \cup \{M(t) \leqslant m, W(t) \leqslant x\},$$

and equating probabilities on both sides,

$$\mathbb{P}(W(t) \leqslant x) = \mathbb{P}(M(t) > m, W(t) \leqslant x) + \mathbb{P}(M(t) \leqslant m, W(t) \leqslant x).$$

Isolating the second term on the right-hand side and using (1.65) for the first term on the right-hand side (where $\mathbb{P}(M(t) > m, W(t) \leqslant x) = \mathbb{P}(M(t) \geqslant m, W(t) \leqslant x)$ since $M(t)$ is continuous) gives the joint CDF for $m > 0, x \leqslant m$:

$$\begin{aligned} F_{M(t),W(t)}(m,x) := \mathbb{P}(M(t) \leqslant m, W(t) \leqslant x) &= \mathbb{P}(W(t) \leqslant x) - \mathbb{P}(W(t) \geqslant 2m - x) \\ &= \mathcal{N}\left(\frac{x}{\sqrt{t}}\right) - \mathcal{N}\left(\frac{x-2m}{\sqrt{t}}\right). \end{aligned} \tag{1.67}$$

For $x > m$, we recover the marginal CDF of $M(t)$. Indeed, applying the law of total probability and the fact that $\{M(t) \leqslant m, m < W(t) \leqslant x\} = \emptyset$:

$$\begin{aligned} \mathbb{P}(M(t) \leqslant m, W(t) \leqslant x) &= \mathbb{P}(M(t) \leqslant m, m < W(t) \leqslant x) + \mathbb{P}(M(t) \leqslant m, W(t) \leqslant m) \\ &= \mathbb{P}(M(t) \leqslant m, W(t) \leqslant m) \equiv F_{M(t),W(t)}(m,m). \end{aligned}$$

This holds for any $x \geqslant m$. Taking $x \to \infty$ gives $F_{M(t),W(t)}(m,m) = F_{M(t),W(t)}(m,\infty) = \mathbb{P}(M(t) \leqslant m) = F_{M(t)}(m)$. The reader may also show that the above steps are equivalent to expressing the joint probability as a double integral and noting the domain in which the joint PDF in (1.66) is nonzero. In the trivial case when $m \leqslant 0$, we have $F_{M(t),W(t)}(m,x) \equiv 0$ for all $t \in (0,\infty)$.

Hence, the (marginal) CDF of $M(t)$ follows by setting $x = m$ in (1.67),

$$F_{M(t)}(m) = F_{M(t),W(t)}(m,m) = \mathcal{N}\left(\frac{m}{\sqrt{t}}\right) - \mathcal{N}\left(-\frac{m}{\sqrt{t}}\right) = 2\mathcal{N}\left(\frac{m}{\sqrt{t}}\right) - 1, \tag{1.68}$$

for $m > 0$ and identically zero for $m \leqslant 0$. Note that this is a CDF for the continuous random variable $M(t)$ where $F_{M(t)}$ is continuous on $\mathbb{R}$, monotonically increasing for positive argument, with $F_{M(t)}(\infty) = 1$ and $F_{M(t)}(m) \equiv 0$ for $m \leqslant 0$. For $m > 0$, $\{\mathcal{T}_m \leqslant t\} \equiv \{M(t) \geqslant m\}$,

$$F_{\mathcal{T}_m}(t) = \mathbb{P}(M(t) \geqslant m) = 1 - \mathbb{P}(M(t) \leqslant m) = 1 - F_{M(t)}(m) = 2 - 2\mathcal{N}\left(\frac{m}{\sqrt{t}}\right), \quad t > 0.$$

This therefore recovers the CDF of $\mathcal{T}_m$ in (1.63) in the case that $m > 0$.

The complete formula for the above joint CDF of $(M(t), W(t))$ is summarized as follows:

$$F_{M(t),W(t)}(m,x) = \begin{cases} \mathcal{N}\left(\frac{x}{\sqrt{t}}\right) - \mathcal{N}\left(\frac{x-2m}{\sqrt{t}}\right) & , m > 0,\ x \leqslant m, \\ 2\mathcal{N}\left(\frac{m}{\sqrt{t}}\right) - 1 & , m > 0,\ x \geqslant m, \\ 0 & , m \leqslant 0,\ x \in \mathbb{R}. \end{cases} \tag{1.69}$$

The next proposition follows simply from Proposition 1.6 and leads to the joint distribution of the pair $(m(t), W(t))$; i.e., the sampled minimum of BM in the interval $[0, t]$, $m(t)$, and the value of standard BM at time t.

Proposition 1.7 (Joint Distribution of $m(t), W(t)$)**.** *For all* $-\infty < m < 0$, $m \leqslant x < \infty$, $t \in (0, \infty)$, *we have*

$$\mathbb{P}(m(t) \leqslant m, W(t) \geqslant x) = \mathbb{P}(W(t) \geqslant x - 2m) = \mathcal{N}\left(\frac{2m - x}{\sqrt{t}}\right). \tag{1.70}$$

Hence, for all $t \in (0, \infty)$, *the joint PDF of* $(m(t), W(t))$ *is*

$$f_{m(t),W(t)}(m,x) = \frac{2(x - 2m)}{t\sqrt{2\pi t}} \mathrm{e}^{-(x-2m)^2/2t}, \quad \text{for } x \geqslant m,\ m < 0, \tag{1.71}$$

and zero otherwise.

Proof. One way to prove the result is to apply the same steps as in the proof of Proposition 1.6, but with arguments applying to a picture that is the reflection (about the vertical axis) of the picture in Figure 1.6. Here we simply make use of the fact that reflected BM defined by $\{B(t) := -W(t)\}_{t \geqslant 0}$ is also a standard BM. Let $M_B(t) := \sup_{0 \leqslant u \leqslant t} B(u)$. Note that $-M_B(t) = -\sup_{0 \leqslant u \leqslant t}(-W(u)) = \inf_{0 \leqslant u \leqslant t} W(u) \equiv m(t)$. Since $\{B(t)\}_{t \geqslant 0}$ is a standard BM, by Proposition 1.6, with values $m' > 0$, $x' \leqslant m'$, $t > 0$, we have

$$\mathbb{P}(M_B(t) \geqslant m', B(t) \leqslant x') = \mathbb{P}(B(t) \geqslant 2m' - x') = \mathcal{N}\left(\frac{x' - 2m'}{\sqrt{t}}\right).$$

The left-hand side is re-expressed as:

$$\begin{aligned} \mathbb{P}(M_B(t) \geqslant m', B(t) \leqslant x') &= \mathbb{P}(-M_B(t) \leqslant -m', -W(t) \leqslant x') \\ &= \mathbb{P}(m(t) \leqslant -m', W(t) \geqslant -x') \end{aligned}$$

Substituting $m' = -m$, $x' = -x$, where $m < 0$, $x \geqslant m$, gives the result in (1.70). The joint density in (1.71) follows by differentiating as in the above proof of (1.66). □

The (marginal) CDF of $m(t)$, $F_{m(t)}$, follows by expressing $\{W(t) \geqslant m\}$ as a union of two mutually exclusive events (note: $\{m(t) \geqslant m\} = \{m(t) \geqslant m, W(t) \geqslant m\}$ since $\{m(t) \geqslant m, W(t) < m\} = \emptyset$):

$$\begin{aligned}\{W(t) \geqslant m\} &= \{m(t) < m, W(t) \geqslant m\} \cup \{m(t) \geqslant m, W(t) \geqslant m\} \\ &= \{m(t) < m, W(t) \geqslant m\} \cup \{m(t) \geqslant m\}.\end{aligned}$$

Computing probabilities and using (1.70) for $x = m$ (note: $\mathbb{P}(m(t) > m) = \mathbb{P}(m(t) \geqslant m)$):

$$\begin{aligned}\mathbb{P}(m(t) > m) &= \mathbb{P}(W(t) \geqslant m) - \mathbb{P}(m(t) \leqslant m, W(t) \geqslant m) \\ &= \mathcal{N}\left(-\frac{m}{\sqrt{t}}\right) - \mathcal{N}\left(\frac{m}{\sqrt{t}}\right) = 1 - 2\mathcal{N}\left(\frac{m}{\sqrt{t}}\right)\end{aligned}$$

where we used $\mathcal{N}(-z) = 1 - \mathcal{N}(z)$. Hence,

$$F_{m(t)}(m) := \mathbb{P}(m(t) \leqslant m) = 1 - \mathbb{P}(m(t) > m) = 2\mathcal{N}\left(\frac{m}{\sqrt{t}}\right) \tag{1.72}$$

for $-\infty < m < 0$ and $F_{m(t)}(m) \equiv 1$ for $m \geqslant 0$. The reader can easily check that this is a proper CDF for the continuous random variable $m(t)$. Moreover, this also recovers the CDF of the first hitting time $\mathcal{T}_m$ in (1.63) for $m = -|m| < 0$, i.e., we have the equivalence of the CDFs, $F_{m(t)}(m) = F_{\mathcal{T}_m}(t)$, for all $t > 0, m < 0$.

By writing $\{W(t) \geqslant x\} = \{m(t) \leqslant m, W(t) \geqslant x\} \cup \{m(t) > m, W(t) \geqslant x\}$ and taking probabilities of this disjoint union, while using (1.70), gives another useful relation:

$$\begin{aligned}\mathbb{P}(m(t) > m, W(t) \geqslant x) &= \mathbb{P}(W(t) \geqslant x) - \mathbb{P}(W(t) \geqslant x - 2m) \\ &= \mathcal{N}\left(-\frac{x}{\sqrt{t}}\right) - \mathcal{N}\left(\frac{2m - x}{\sqrt{t}}\right) = \mathcal{N}\left(\frac{x - 2m}{\sqrt{t}}\right) - \mathcal{N}\left(\frac{x}{\sqrt{t}}\right)\end{aligned} \tag{1.73}$$

for $m < 0$, $m \leqslant x < \infty$, $t > 0$. For any given $t > 0$, the joint CDF of $m(t), W(t)$ follows by using (1.70) and (1.72):

$$\begin{aligned}F_{m(t),W(t)}(m, x) := \mathbb{P}(m(t) \leqslant m, W(t) \leqslant x) &= \mathbb{P}(m(t) \leqslant m) - \mathbb{P}(m(t) \leqslant m, W(t) \geqslant x) \\ &= 2\mathcal{N}\left(\frac{m}{\sqrt{t}}\right) - \mathcal{N}\left(\frac{2m - x}{\sqrt{t}}\right),\end{aligned} \tag{1.74}$$

for $m < 0$, $m \leqslant x < \infty$. Note that the marginal CDF in (1.72) for values of $m < 0$ is indeed recovered by taking $x \to \infty$ in (1.74): $F_{m(t),W(t)}(m, \infty) = 2\mathcal{N}\left(\frac{m}{\sqrt{t}}\right) - \mathcal{N}(-\infty) = 2\mathcal{N}\left(\frac{m}{\sqrt{t}}\right)$.

We can derive the joint CDF for all other cases. In particular, consider the case where $x \leqslant m < 0$. By total law of probabilities and the fact that $\mathbb{P}(m(t) > m, W(t) \leqslant x) = 0$:

$$\mathbb{P}(W(t) \leqslant x) = \mathbb{P}(m(t) \leqslant m, W(t) \leqslant x) + \mathbb{P}(m(t) > m, W(t) \leqslant x) = \mathbb{P}(m(t) \leqslant m, W(t) \leqslant x).$$

Hence, $F_{m(t),W(t)}(m, x) = \mathbb{P}(W(t) \leqslant x) = \mathcal{N}\left(\frac{x}{\sqrt{t}}\right)$ when $x \leqslant m < 0$. The only other case is when $m \geqslant 0$, for all $x \in \mathbb{R}$. Since $m(t) \leqslant 0$ (a.s.), i.e., for $m \geqslant 0$ we have $\{m(t) \leqslant m\} = \{m(t) \leqslant 0\} = \Omega$, this case gives joint CDF

$$\mathbb{P}(m(t) \leqslant m, W(t) \leqslant x) = \mathbb{P}(W(t) \leqslant x) = \mathcal{N}\left(\frac{x}{\sqrt{t}}\right).$$

Analogous to (1.69), the complete formula for the joint CDF of $(m(t), W(t))$ is summarized as follows:

$$F_{m(t),W(t)}(m,x) = \begin{cases} 2\mathcal{N}\left(\frac{m}{\sqrt{t}}\right) - \mathcal{N}\left(\frac{2m-x}{\sqrt{t}}\right) & , m < 0,\ x \geqslant m, \\ \mathcal{N}\left(\frac{x}{\sqrt{t}}\right) & , m < 0,\ x \leqslant m, \\ \mathcal{N}\left(\frac{x}{\sqrt{t}}\right) & , m \geqslant 0,\ x \in \mathbb{R}. \end{cases} \tag{1.75}$$

Before closing this section, we make another important connection to BM that is killed at a given level m. Given $m \neq 0$, standard *BM killed at level* m is standard BM up to the first hitting time $\mathcal{T}_m$ at which time the process is "killed and sent to the so-called cemetery state $\partial^\dagger$." [Remark: we can also define a similar BM that is "frozen or absorbed" at the level m.] Let $\{W_{(m)}(t)\}_{t\geqslant 0}$ denote the standard BM killed at a given level m, then

$$W_{(m)}(t) := \begin{cases} W(t) & \text{for } t < \mathcal{T}_m, \\ \partial^\dagger & \text{for } t \geqslant \mathcal{T}_m. \end{cases} \tag{1.76}$$

For $m > 0$, according to (1.62), $\{t < \mathcal{T}_m\} = \{M(t) < m\}$. The event that the process $W_{(m)}$ has not been killed and hence lies below or at any $x < m$ at time t is the joint event that W lies below or at $x < m$ at time t and the maximum of W up to time t is less than m:

$$\begin{aligned} \mathbb{P}(W_{(m)}(t) \leqslant x) &= \mathbb{P}(M(t) < m, W(t) \leqslant x) = F_{M(t),W(t)}(m,x) \\ &= \mathcal{N}\left(\frac{2m-x}{\sqrt{t}}\right) - \mathcal{N}\left(-\frac{x}{\sqrt{t}}\right). \end{aligned}$$

Here, we used (1.67). Differentiating this expression w.r.t. x gives the density $p^{W_{(m)}}$ for killed standard BM (starting at $W_{(m)}(0) = 0$) on its state space $x \in (-\infty, m)$, $m > 0$:

$$p^{W_{(m)}}(t; 0, x)\,\mathrm{d}x \equiv \mathbb{P}(W_{(m)}(t) \in \mathrm{d}x) = \mathbb{P}(M(t) \leqslant m, W(t) \in \mathrm{d}x) \tag{1.77}$$

where

$$\begin{aligned} p^{W_{(m)}}(t; 0, x) &= \frac{\partial}{\partial x}\mathbb{P}(W_{(m)}(t) \leqslant x) = \frac{\partial}{\partial x}F_{M(t),W(t)}(m,x) \\ &= \frac{1}{\sqrt{t}}\left[n\left(\frac{x}{\sqrt{t}}\right) - n\left(\frac{2m-x}{\sqrt{t}}\right)\right] \\ &= p_0(t; x) - p_0(t; 2m-x), \end{aligned} \tag{1.78}$$

for $x < m$, and $p^{W_{(m)}}(t; 0, x) \equiv 0$ for $x \geqslant m$. [Note: $n(z) = n(-z)$.] Here $p_0(t; x)$, defined in (1.24), is the Gaussian density for standard BM. Observe that the density $p^{W_{(m)}}(t; 0, x)$ is indeed strictly positive for $x < m$ since $(2m - x)^2 > x^2$, i.e., $p_0(t; x) > p_0(t; 2m - x)$.

Note that we also readily obtain the transition PDF for the killed BM by time and space homogeneity (see (1.25)). In particular, the density of paths starting at $x < m$ at time $s > 0$ and transitioning to level $y < m$ at time $t > s$, with killing at level m, is the same as the density of paths starting at 0 at time 0 and transitioning to the level given by the difference $y - x$ at time $t - s$, with killing at level $m - x$: $p^{W_{(m)}}(s, t; x, y) = p^{W_{(m-x)}}(t - s; 0, y - x)$. Hence, (1.78) gives the transtion PDF

$$\begin{aligned} p^{W_{(m)}}(s, t; x, y) &= p_0(t - s; y - x) - p_0(t - s; 2(m - x) - (y - x)) \\ &= p_0(t - s; y - x) - p_0(t - s; 2m - x - y), \end{aligned} \tag{1.79}$$

which is strictly positive for $s < t$, $-\infty < x, y < m$, and identically zero for $x, y \geqslant m$. This density holds for any real upper level m.

For $m < 0$, $\{t < \mathcal{T}_m\} = \{m(t) > m\}$. The event that $W_{(m)}$ has not been killed and hence lies at or above a value $x > m$ at time t is the joint event that $W(t)$ lies above or at $x > m$ and the minimum of W up to time t is greater than m:

$$\begin{aligned}\mathbb{P}(W_{(m)}(t) \geqslant x) &= \mathbb{P}(m(t) > m, W(t) \geqslant x)\\ &= \mathcal{N}\left(\frac{x-2m}{\sqrt{t}}\right) - \mathcal{N}\left(\frac{x}{\sqrt{t}}\right),\end{aligned}$$

where we used (1.73). Following the same steps as above gives the density $p^{W_{(m)}}$ for killed standard BM (starting at $W_{(m)}(0) = 0$) on its state space $x \in (m, \infty)$, $m < 0$:

$$p^{W_{(m)}}(t; 0, x)\,\mathrm{d}x \equiv \mathbb{P}(W_{(m)}(t) \in \mathrm{d}x) = \mathbb{P}(m(t) > m, W(t) \in \mathrm{d}x) \tag{1.80}$$

where

$$p^{W_{(m)}}(t; 0, x) = -\frac{\partial}{\partial x}\mathbb{P}(W_{(m)}(t) \geqslant x) = p_0(t; x) - p_0(t; x - 2m), \tag{1.81}$$

for $x > m$ and $p^{W_{(m)}}(t; 0, x) \equiv 0$ for $x \leqslant m$. Note that, since $p_0(t; z) = p_0(t; -z)$, this expression is the same as that in (1.78), but now it holds for $x > m$, $m < 0$. The transition PDF is given by the expression in (1.79) for $s < t$, $x, y > m$ and is identically zero for $x, y \leqslant m$.

The densities in (1.78) and (1.81) are strictly positive on their respective domains of definition. However, they do not integrate to unity, as is clear from their definitions. In fact, they can be used to obtain the distribution of $M(t)$ and $m(t)$, respectively, and hence thereby also obtain the distribution of the first hitting time $\mathcal{T}_m$. These distributions were already derived above. However, it is instructive to see how they can be derived based on the density for the killed BM process. For example, let's take the case with $m > 0$. Then, the probability of event $\{\mathcal{T}_m \leqslant t\} = \{M(t) \geqslant m\}$ is the probability that the process has been killed and is hence in the state $\partial^\dagger$ at time $t > 0$. Hence, the CDF of the first hitting time up to level $m > 0$ for standard BM is equivalently given by:

$$\begin{aligned}F_{\mathcal{T}_m}(t) = 1 - \mathbb{P}(M(t) < m) &= 1 - \int_{-\infty}^{m} \mathbb{P}(M(t) < m, W(t) \in \mathrm{d}x)\\ &= 1 - \int_{-\infty}^{m} p^{W_{(m)}}(t; 0, x)\,\mathrm{d}x\\ &= 1 - \int_{-\infty}^{m} p_0(t; x)\,\mathrm{d}x + \int_{-\infty}^{m} p_0(t; x - 2m)\,\mathrm{d}x\\ &= 1 - \int_{-\infty}^{m} \frac{1}{\sqrt{t}}n\left(\frac{x}{\sqrt{t}}\right)\mathrm{d}x + \int_{-\infty}^{m} \frac{1}{\sqrt{t}}n\left(\frac{x-2m}{\sqrt{t}}\right)\mathrm{d}x\\ &= 1 - \int_{-\infty}^{\frac{m}{\sqrt{t}}} n(z)\,\mathrm{d}z + \int_{-\infty}^{-\frac{m}{\sqrt{t}}} n(z)\,\mathrm{d}z\\ &= 1 - \mathcal{N}\left(\frac{m}{\sqrt{t}}\right) + \mathcal{N}\left(-\frac{m}{\sqrt{t}}\right) = 2\mathcal{N}\left(-\frac{m}{\sqrt{t}}\right).\end{aligned} \tag{1.82}$$

This recovers our previously derived formula in (1.63) when $m > 0$. We leave it to the reader to verify that the CDF in (1.63) for the first hitting time down to level $m < 0$ for standard BM is also recovered by integrating the density in (1.81):

$$\begin{aligned}F_{\mathcal{T}_m}(t) = 1 - \mathbb{P}(m(t) > m) &= 1 - \int_{m}^{\infty} \mathbb{P}(m(t) > m, W(t) \in \mathrm{d}x)\\ &= 1 - \int_{m}^{\infty} p^{W_{(m)}}(t; 0, x)\,\mathrm{d}x = 2\mathcal{N}\left(\frac{m}{\sqrt{t}}\right).\end{aligned} \tag{1.83}$$

Note that the above integrals of $p^{W_{(m)}}(t;0,x)$ over the respective state spaces $(-\infty, m)$ (or (m,∞)) represent the survival probability (up to time t) of the killed BM process $W_{(m)}$.

1.4.2 Translated and Scaled Driftless Brownian Motion

Let us now consider the process $X = \{X(t)\}_{t\geqslant 0}$ as a scaled and translated BM defined as in (1.38), $x_0 \in \mathbb{R}, \sigma > 0$, but with *zero drift* $\mu = 0$:

$$X(t) := x_0 + W^{(0,\sigma)}(t) = x_0 + \sigma W(t)\,. \tag{1.84}$$

Hence, all paths of X start at the point $X(0) = x_0 \in \mathbb{R}$. The relationship between $X(t)$ and $W(t)$ is simply a monotonically increasing mapping, i.e., $X(t) = h(W(t))$ where $h:\mathbb{R}\to\mathbb{R}$ is defined by $h(w) := x_0 + \sigma w$, with inverse $h^{-1}(x) := (x - x_0)/\sigma$. As a consequence, all formulae derived in Section 1.4.1 for the CDF (PDF) of the first hitting time, maximum and minimum, readily follow as shown just below.

The first hitting time of the X process to a level $m \in \mathbb{R}$, denoted by $\mathcal{T}_m^X$, is given by the first hitting time of standard BM, W, to the level $h^{-1}(m) = (m - x_0)/\sigma$:

$$\begin{aligned}\mathcal{T}_m^X &:= \inf\{t \geqslant 0 \,:\, X(t) = m\}\\ &= \inf\{t \geqslant 0 \,:\, W(t) = (m - x_0)/\sigma\} \equiv \mathcal{T}_{(m-x_0)/\sigma}\,.\end{aligned} \tag{1.85}$$

That is, sending $m \to (m - x_0)/\sigma$ in $\mathcal{T}_m$ (hitting time for W) gives $\mathcal{T}_m^X$. Similarly, for the sampled maximum and minimum of X, from time 0 to time $t \geqslant 0$, we have the simple relations

$$\begin{aligned}M^X(t) := \sup_{0\leqslant u\leqslant t} X(u) &= \sup_{0\leqslant u\leqslant t}[x_0 + \sigma W(u)]\\ &= x_0 + \sigma \sup_{0\leqslant u\leqslant t} W(u) = x_0 + \sigma M(t)\end{aligned} \tag{1.86}$$

and similarly

$$m^X(t) := \inf_{0\leqslant u\leqslant t} X(u) = x_0 + \sigma\, m(t)\,. \tag{1.87}$$

Note: $x_0 \equiv X(0) = M^X(0) = m^X(0)$, $m^X(t) \leqslant X(t) \leqslant M^X(t)$, where $M^X(t)$ is increasing and $m^X(t)$ is decreasing in t.

Based on (1.85), the formula in (1.63) directly gives us the CDF of $\mathcal{T}_m^X$, for all $m \neq x_0 \in \mathbb{R}$:

$$\mathbb{P}(\mathcal{T}_m^X \leqslant t) = \mathbb{P}(\mathcal{T}_{\frac{(m-x_0)}{\sigma}} \leqslant t) = 2\mathcal{N}\left(-\frac{|m - x_0|}{\sigma\sqrt{t}}\right), \quad 0 < t < \infty. \tag{1.88}$$

Remark: for $m = x_0$ the formula holds trivially for any $t > 0$ where $\mathbb{P}(\mathcal{T}_{x_0}^X \leqslant t) = 1$. In fact, $\mathbb{P}(\mathcal{T}_{x_0}^X = 0) = 1$ since the process starts at $X(0) = x_0$. Also, $\mathbb{P}(\mathcal{T}_m^X = 0) = 0$ for $m \neq x_0$ as given by the limit $t \to 0$ of the CDF in (1.88). Clearly, for all real x, x_0, m, we have the equivalence of events:

$$\begin{aligned}\{M^X(t) \geqslant m\} &= \{M(t) \geqslant (m - x_0)/\sigma\},\\ \{m^X(t) \leqslant m\} &= \{m(t) \leqslant (m - x_0)/\sigma\},\\ \{X(t) \leqslant x\} &= \{W(t) \leqslant (x - x_0)/\sigma\}.\end{aligned}$$

Based on these relations, all relevant formulae for the marginal and joint CDFs immediately follow from those for standard BM upon *sending* $m \to (m - x_0)/\sigma$ and $x \to (x - x_0)/\sigma$ in the CDF formulae of Section 1.4.1. For example, using (1.65), for $m > x_0$, $x \leqslant m$:

$$\begin{aligned}\mathbb{P}(M^X(t) \geqslant m, X(t) \leqslant x) &= \mathbb{P}(M(t) \geqslant (m - x_0)/\sigma, W(t) \leqslant (x - x_0)/\sigma) \\ &= \mathcal{N}\left(\frac{x + x_0 - 2m}{\sigma\sqrt{t}}\right). \end{aligned} \tag{1.89}$$

For all $t > 0$, the joint CDF of $M^X(t), X(t)$ is obtained by sending $m \to (m - x_0)/\sigma$ and $x \to (x - x_0)/\sigma$ in (1.67):

$$F_{M^X(t),X(t)}(m,x) = \mathcal{N}\left(\frac{x - x_0}{\sigma\sqrt{t}}\right) - \mathcal{N}\left(\frac{x + x_0 - 2m}{\sigma\sqrt{t}}\right), \quad m > x_0, m \geqslant x. \tag{1.90}$$

Moreover, $F_{M^X(t),X(t)}(m,x) \equiv 0$ for $m \leqslant x_0$ and the (marginal) CDF of $M^X(t)$ is recovered for $x > m > x_0$: $F_{M^X(t),X(t)}(m,x) = F_{M^X(t)}(m)$. Differentiating gives the joint PDF,

$$f_{M^X(t),X(t)}(m,x) = \frac{2(2m - (x + x_0))}{\sigma^3 t\sqrt{2\pi t}} \mathrm{e}^{-(2m-(x+x_0))^2/2\sigma^2 t}, \tag{1.91}$$

for all real values $x < m, x_0 < m$ and zero otherwise.

The (marginal) CDF of $M^X(t)$, $t > 0$, follows simply from (1.68) upon sending $m \to (m - x_0)/\sigma$, or from (1.90) with $x = m$, since $F_{M^X(t),X(t)}(m,\infty) = F_{M^X(t),X(t)}(m,m) = F_{M^X(t)}(m)$:

$$F_{M^X(t)}(m) := \mathbb{P}(M^X(t) \leqslant m) = 2\mathcal{N}\left(\frac{m - x_0}{\sigma\sqrt{t}}\right) - 1, \tag{1.92}$$

for $m > x_0$ and $F_{M^X(t)}(m) \equiv 0$ for $m \leqslant x_0$. The complete joint CDF of $M^X(t), X(t)$ is given by (1.69) upon sending $m \to (m - x_0)/\sigma$ and $x \to (x - x_0)/\sigma$:

$$F_{M^X(t),X(t)}(m,x) = \begin{cases} \mathcal{N}\left(\frac{x - x_0}{\sigma\sqrt{t}}\right) - \mathcal{N}\left(\frac{x + x_0 - 2m}{\sigma\sqrt{t}}\right) & , m > x_0,\ x \leqslant m, \\ 2\mathcal{N}\left(\frac{m - x_0}{\sigma\sqrt{t}}\right) - 1 & , m > x_0,\ x \geqslant m, \\ 0 & , m \leqslant x_0,\ x \in \mathbb{R}. \end{cases} \tag{1.93}$$

Similarly, for $t > 0$, the joint variables $m^X(t), X(t)$ satisfy (sending $m \to (m - x_0)/\sigma$, $x \to (x - x_0)/\sigma$ in (1.70)):

$$\mathbb{P}(m^X(t) \leqslant m, X(t) \geqslant x) = \mathcal{N}\left(\frac{2m - x - x_0}{\sigma\sqrt{t}}\right) \tag{1.94}$$

for $x_0 > m, x \geqslant m$ and with joint PDF

$$f_{m^X(t),X(t)}(m,x) = \frac{2(x + x_0 - 2m)}{\sigma^3 t\sqrt{2\pi t}} \mathrm{e}^{-(2m-(x+x_0))^2/2\sigma^2 t}, \tag{1.95}$$

for all real $x > m, x_0 > m$ and zero otherwise.

Upon sending $m \to (m - x_0)/\sigma$ and $x \to (x - x_0)/\sigma$, the relation in (1.73) gives

$$\mathbb{P}(m^X(t) > m, X(t) \geqslant x) = \mathcal{N}\left(\frac{x + x_0 - 2m}{\sigma\sqrt{t}}\right) - \mathcal{N}\left(\frac{x - x_0}{\sigma\sqrt{t}}\right) \tag{1.96}$$

for all real values $m < x_0$, $m \leqslant x < \infty$, $t > 0$. The complete joint CDF of $m^X(t), X(t)$ is given by (1.75) upon sending $m \to (m - x_0)/\sigma$ and $x \to (x - x_0)/\sigma$:

$$F_{m^X(t),X(t)}(m,x) = \begin{cases} 2\mathcal{N}\left(\frac{m-x_0}{\sigma\sqrt{t}}\right) - \mathcal{N}\left(\frac{2m-x-x_0}{\sigma\sqrt{t}}\right) & , m < x_0,\ x \geqslant m, \\ \mathcal{N}\left(\frac{x-x_0}{\sigma\sqrt{t}}\right) & , m < x_0,\ x \leqslant m, \\ \mathcal{N}\left(\frac{x-x_0}{\sigma\sqrt{t}}\right) & , m \geqslant x_0,\ x \in \mathbb{R}. \end{cases} \tag{1.97}$$

Formulae for the transition PDF of driftless BM killed at level m and started at $x_0 < m$, or $x_0 > m$, are given by (1.78) or (1.81) upon making the substitution $m \to (m - x_0)/\sigma$ and $x \to (x - x_0)/\sigma$ and dividing the expressions by σ.

1.4.3 Brownian Motion with Drift

We now consider BM started at zero and with a constant drift parameter $\mu \in \mathbb{R}$; i.e., let the process $\{X(t)\}_{t \geqslant 0}$ be defined by (1.32) for $\sigma = 1$:

$$X(t) := W^{(\mu,1)}(t) \equiv \mu t + W(t). \tag{1.98}$$

We remark that it suffices to consider this case, as it also leads directly to the respective formulae for the process defined in (1.38), where the BM is started at an arbitrary point x_0 and has arbitrary volatility parameter $\sigma > 0$. This is due to the relation

$$W^{(\mu,\sigma)}(t) \equiv \mu t + \sigma W(t) = \sigma[\frac{\mu}{\sigma}t + W(t)] = \sigma W^{(\frac{\mu}{\sigma},1)}(t)\,. \tag{1.99}$$

So, the process in (1.38) is obtained by scaling with σ and translating by x_0 the process X in (1.98), as in the previous section, and now by applying the additional scaling of the drift, i.e., send $\mu \to \frac{\mu}{\sigma}$. For clarity, we shall discuss this scaling and translation in Section 1.4.3.1.

The sampled maximum and minimum of the process $X \equiv W^{(\mu,1)}$ are defined by

$$M^X(t) := \sup_{0 \leqslant s \leqslant t} X(s) = \sup_{0 \leqslant s \leqslant t} [\mu s + W(s)] \tag{1.100}$$

and

$$m^X(t) := \inf_{0 \leqslant s \leqslant t} X(s) = \inf_{0 \leqslant s \leqslant t} [\mu s + W(s)]\,. \tag{1.101}$$

In contrast to our previous expressions for driftless or standard BM, the drift term is a *time-dependent function* that is added to the standard (zero drift) BM. Hence, we cannot use the reflection principle in this case. In the next chapter, we shall see how to explicitly implement methods in probability measure changes in order to calculate expectations of random variables that are functionals of standard Brownian motion. A very powerful tool is Girsanov's Theorem. Section 2.8.1 contains some applications of Girsanov's Theorem. Here, we simply borrow the key results derived in Section 2.8.1 that allow us to derive formulae for the CDF of the first hitting time of drifted BM and other joint event probabilities, PDFs and CDFs for the above sampled maximum and minimum random variables.

Our two main formulae are given by (2.139) and (2.140). The joint CDFs in (2.141) and (2.142) are important cases. These are used in Section 2.8.1 to derive the joint PDF of $M^X(t), X(t)$ and of $m^X(t), X(t)$, given in (2.144) and (2.145). The joint CDF in (2.141) also leads to the useful relation

$$\mathbb{P}(M^X(t) \leqslant m, X(t) \in \mathrm{d}x) = \mathrm{e}^{-\frac{1}{2}\mu^2 t + \mu x}\,\mathbb{P}(M(t) \leqslant m, W(t) \in \mathrm{d}x)\,, \tag{1.102}$$

for $x < m, m > 0$.

Note that the probabilities in (1.102) represent the probability for $X(t)$ (or respectively $W(t)$) to take on a value in an infinitesimal interval around the point x, assuming that the sampled maximum $M^X(t)$ (or respectively $M(t)$) is less than m; i.e., the respective probabilities correspond to a density in x (obtained by differentiating the joint CDF w.r.t. x) times the infinitesimal $\mathrm{d}x$,

$$\mathbb{P}(M^X(t) \leqslant m, X(t) \in \mathrm{d}x) = \frac{\partial}{\partial x} F_{M^X(t),X(t)}(m,x)\,\mathrm{d}x \tag{1.103}$$

and

$$\mathbb{P}(M(t) \leqslant m, W(t) \in \mathrm{d}x) = \frac{\partial}{\partial x} F_{M(t),W(t)}(m,x)\,\mathrm{d}x\,. \tag{1.104}$$

Hence, the relation in (1.102) takes the equivalent form:

$$\frac{\partial}{\partial x} F_{M^X(t),X(t)}(m,x) = \mathrm{e}^{-\frac{1}{2}\mu^2 t+\mu x}\frac{\partial}{\partial x} F_{M(t),W(t)}(m,x). \tag{1.105}$$

This relation follows simply by differentiating w.r.t. x on both sides of (2.143):

$$\begin{aligned}\frac{\partial}{\partial x} F_{M^X(t),X(t)}(m,x) &= \mathrm{e}^{-\frac{1}{2}\mu^2 t}\frac{\partial}{\partial x}\int_{-\infty}^{x} \mathrm{e}^{\mu y}\left[\int_0^m f_{M(t),W(t)}(w,y)\,\mathrm{d}w\right]\mathrm{d}y\\ &= \mathrm{e}^{-\frac{1}{2}\mu^2 t+\mu x}\int_0^m f_{M(t),W(t)}(w,x)\,\mathrm{d}w\\ &= \mathrm{e}^{-\frac{1}{2}\mu^2 t+\mu x}\frac{\partial}{\partial x} F_{M(t),W(t)}(m,x)\,.\end{aligned}$$

We have already seen that the quantity in (1.104) is related to the density for standard BM killed at $m > 0$. This is given by (1.77) and (1.78). The quantity in (1.103) is similarly related to the density for drifted BM killed at $m > 0$. For any $m \neq 0$, *drifted BM killed at level* m is defined as

$$X_{(m)}(t) := \begin{cases} X(t) & \text{for } t < \mathcal{T}_m^X,\\ \partial^\dagger & \text{for } t \geqslant \mathcal{T}_m^X,\end{cases} \tag{1.106}$$

where $\mathcal{T}_m^X$ is the first hitting time of the drifted BM, $X = W^{(\mu,1)}$, to level m. For $m > 0$, the process $X_{(m)}$ has state space $(-\infty, m)$ and starts at zero, as does $W_{(m)}$. Hence, taking any value $x < m$, the joint CDF $F_{M^X(t),X(t)}(m,x)$ is equivalent to the probability that the drifted BM with killing at m lies below x, i.e.,

$$\mathbb{P}(X_{(m)}(t) \leqslant x) = \mathbb{P}(M^X(t) < m, X(t) \leqslant x) = F_{M^X(t),X(t)}(m,x) \tag{1.107}$$

and the analogue of (1.77) is

$$p^{X_{(m)}}(t;0,x)\,\mathrm{d}x \equiv \mathbb{P}(X_{(m)}(t) \in \mathrm{d}x) = \mathbb{P}(M^X(t) \leqslant m, X(t) \in \mathrm{d}x)\,. \tag{1.108}$$

Hence, (1.102) is equivalent to

$$\mathbb{P}(X_{(m)}(t) \in \mathrm{d}x) = \mathrm{e}^{-\frac{1}{2}\mu^2 t+\mu x}\,\mathbb{P}(W_{(m)}(t) \in \mathrm{d}x)\,. \tag{1.109}$$

In particular, the density $p^{X_{(m)}}$ for the drifted killed BM, started at $X_{(m)}(0) = 0$, is related to the density $p^{W_{(m)}}$ in (1.78) by

$$\begin{aligned}p^{X_{(m)}}(t;0,x) &= \mathrm{e}^{-\frac{1}{2}\mu^2 t+\mu x} p^{W_{(m)}}(t;0,x)\\ &= \mathrm{e}^{-\frac{1}{2}\mu^2 t+\mu x}\left[p_0(t;x) - p_0(t;2m-x)\right],\end{aligned} \tag{1.110}$$

for $x < m$ and zero otherwise. We can rewrite this density in a more convenient form by multiplying and completing the squares in the two exponents,

$$
\begin{aligned}
p^{X_{(m)}}(t;0,x) &= \mathrm{e}^{-\frac{1}{2}\mu^2 t+\mu x}\left(\frac{\mathrm{e}^{-x^2/2t}}{\sqrt{2\pi t}} - \frac{\mathrm{e}^{-(2m-x)^2/2t}}{\sqrt{2\pi t}}\right) \\
&= \frac{\mathrm{e}^{-(x-\mu t)^2/2t}}{\sqrt{2\pi t}} - \mathrm{e}^{2\mu m}\,\frac{\mathrm{e}^{-(x-\mu t-2m)^2/2t}}{\sqrt{2\pi t}} \\
&= p_0(t;x-\mu t) - \mathrm{e}^{2\mu m}\,p_0(t;x-\mu t-2m)\,,
\end{aligned}
\tag{1.111}
$$

for $x < m$ and zero otherwise. This expression involves a linear combination of densities for standard BM at any point $x \in (-\infty, m)$. Note that setting $\mu = 0$ recovers $p^{W_{(m)}}$ in (1.78). In the limit $m \to \infty$, this density recovers the density $p^X(t;0,x) \equiv p_0(t;x-\mu t)$ for drifted BM, $X \equiv W^{(\mu,1)} \in \mathbb{R}$, with no killing.

The joint CDF of $M^X(t), X(t)$ follows by evaluating a double integral of the joint PDF. Alternatively, since we now have the density $p^{X_{(m)}}$ at hand, we can simply use it to compute a single integral to obtain the joint CDF. Indeed, by (1.108) and (1.111):

$$
\begin{aligned}
F_{M^X(t),X(t)}(m,x) &= \int_{-\infty}^{x} \mathbb{P}(M^X(t) \leqslant m, X(t) \in \mathrm{d}y) = \int_{-\infty}^{x} \mathbb{P}(X_{(m)}(t) \in \mathrm{d}y) \\
&= \int_{-\infty}^{x} p^{X_{(m)}}(t;0,y)\,\mathrm{d}y \\
&= \int_{-\infty}^{x} p_0(t;y-\mu t)\,\mathrm{d}y - \mathrm{e}^{2\mu m}\int_{-\infty}^{x} p_0(t;y-\mu t-2m)\,\mathrm{d}y \\
&= \int_{-\infty}^{x-\mu t} p_0(t;y)\,\mathrm{d}y - \mathrm{e}^{2\mu m}\int_{-\infty}^{x-\mu t-2m} p_0(t;y)\,\mathrm{d}y \\
&= \mathbb{P}(W(t) \leqslant x-\mu t) - \mathrm{e}^{2\mu m}\,\mathbb{P}(W(t) \leqslant x-\mu t-2m) \\
&= \mathcal{N}\left(\frac{x-\mu t}{\sqrt{t}}\right) - \mathrm{e}^{2\mu m}\,\mathcal{N}\left(\frac{x-\mu t-2m}{\sqrt{t}}\right),
\end{aligned}
\tag{1.112}
$$

for $x \leqslant m, m > 0$. For $x \geqslant m > 0$, $F_{M^X(t),X(t)}(m,x) = F_{M^X(t),X(t)}(m,m) = F_{M^X(t)}(m)$ and $F_{M^X(t),X(t)}(m,x) \equiv 0$ for $m \leqslant 0$.

The (marginal) CDF of $M^X(t)$ is given by setting $x = m$ in (1.112),

$$
\begin{aligned}
F_{M^X(t)}(m) &= F_{M^X(t),X(t)}(m,m) = \mathbb{P}(X_{(m)}(t) \leqslant m) \\
&= \mathcal{N}\left(\frac{m-\mu t}{\sqrt{t}}\right) - \mathrm{e}^{2\mu m}\,\mathcal{N}\left(\frac{-m-\mu t}{\sqrt{t}}\right),\quad m > 0,
\end{aligned}
\tag{1.113}
$$

and $F_{M^X(t)}(m) \equiv 0$ for $m \leqslant 0$. The first hitting time up to level $m > 0$ has CDF

$$
F_{\mathcal{T}_m^X}(t) = 1 - F_{M^X(t)}(m) = \mathcal{N}\left(\frac{-m+\mu t}{\sqrt{t}}\right) + \mathrm{e}^{2\mu m}\,\mathcal{N}\left(\frac{-m-\mu t}{\sqrt{t}}\right),
\tag{1.114}
$$

for $t > 0$, and zero otherwise. In the trivial case where $m \leqslant 0$, we have $F_{M^X(t),X(t)}(m,x) \equiv 0$. Hence, the complete joint CDF of $M^X(t), X(t)$ is given by

$$
F_{M^X(t),X(t)}(m,x) = \begin{cases}
\mathcal{N}\left(\frac{x-\mu t}{\sqrt{t}}\right) - \mathrm{e}^{2\mu m}\,\mathcal{N}\left(\frac{x-\mu t-2m}{\sqrt{t}}\right) & , m > 0,\ x \leqslant m, \\
\mathcal{N}\left(\frac{m-\mu t}{\sqrt{t}}\right) - \mathrm{e}^{2\mu m}\,\mathcal{N}\left(\frac{-m-\mu t}{\sqrt{t}}\right) & , m > 0,\ x \geqslant m, \\
0 & , m \leqslant 0,\ x \in \mathbb{R}.
\end{cases}
\tag{1.115}
$$

We now consider the case where $m < 0$. The analogue of (1.102) for the joint pair $m^X(t), X(t)$ is

$$\mathbb{P}(m^X(t) \geqslant m, X(t) \in \mathrm{d}x) = \mathrm{e}^{-\frac{1}{2}\mu^2 t + \mu x}\, \mathbb{P}(m(t) \geqslant m, W(t) \in \mathrm{d}x)\,, \tag{1.116}$$

for $x > m, m < 0$. The derivation is very similar to that given above for (1.102). We can relate this to the first hitting time $\mathcal{T}_m^X$, which is now the first hitting time of the drifted BM down to level $m < 0$. The process $X_{(m)}$ has state space (m, ∞). The identity in (1.109) is now valid for $x > m$, $m < 0$. The analogues of (1.107) and (1.108) are

$$\mathbb{P}(X_{(m)}(t) \geqslant x) = \mathbb{P}(m^X(t) > m, X(t) \geqslant x) \tag{1.117}$$

and

$$p^{X_{(m)}}(t; 0, x)\, \mathrm{d}x \equiv \mathbb{P}(X_{(m)}(t) \in \mathrm{d}x) = \mathbb{P}(m^X(t) \geqslant m, X(t) \in \mathrm{d}x)\,, \tag{1.118}$$

$x > m, m < 0$. By (1.109), the density $p^{X_{(m)}}(t; 0, x)$ for the process $X_{(m)} \in (m, \infty)$ is still given by the expression in (1.110), or equivalently in (1.111), but now for $x > m$, $m < 0$. The density $p^{X_{(m)}}(t; 0, x)$ is identically zero for $x \leqslant m$.

In what follows it is useful to compute the following joint probability (using similar steps as in (1.112)):

$$\begin{aligned}\mathbb{P}(m^X(t) > m, X(t) \geqslant x) &= \int_x^\infty p^{X_{(m)}}(t; 0, y)\, \mathrm{d}y \\ &= \mathbb{P}(W(t) \geqslant x - \mu t) - \mathrm{e}^{2\mu m}\, \mathbb{P}(W(t) \geqslant x - \mu t - 2m) \\ &= \mathcal{N}\left(\frac{-x + \mu t}{\sqrt{t}}\right) - \mathrm{e}^{2\mu m}\, \mathcal{N}\left(\frac{-x + \mu t + 2m}{\sqrt{t}}\right), \end{aligned} \tag{1.119}$$

for $x \geqslant m, m \leqslant 0$. The CDF of the first hitting time down to level $m < 0$ follows from this expression for $x = m$:

$$\begin{aligned} F_{\mathcal{T}_m^X}(t) &= 1 - \mathbb{P}(m^X(t) \geqslant m) = 1 - \mathbb{P}(m^X(t) \geqslant m, X(t) \geqslant m) \\ &= 1 - \int_m^\infty p^{X_{(m)}}(t; 0, y)\, \mathrm{d}y \\ &= 1 - \mathcal{N}\left(\frac{-m + \mu t}{\sqrt{t}}\right) + \mathrm{e}^{2\mu m}\, \mathcal{N}\left(\frac{-m + \mu t + 2m}{\sqrt{t}}\right) \\ &= \mathcal{N}\left(\frac{m - \mu t}{\sqrt{t}}\right) + \mathrm{e}^{2\mu m}\, \mathcal{N}\left(\frac{m + \mu t}{\sqrt{t}}\right) \end{aligned} \tag{1.120}$$

for $t > 0$, and zero otherwise. For given $t > 0$, this expression is also the CDF of $m^X(t)$, $F_{m^X(t)}(m) = F_{\mathcal{T}_m}(t)$ for $-\infty < m \leqslant 0$, and $F_{m^X(t)}(m) \equiv 1$ for $m \geqslant 0$. Note the trivial case where $F_{\mathcal{T}_m^X}(t) \equiv 1$ for $m = 0$, $t \geqslant 0$; i.e., the process starts at zero.

We now derive the joint CDF of $m^X(t), X(t)$. Using (1.119) and the total law of probabilities,

$$\begin{aligned}\mathbb{P}(m^X(t) \leqslant m, X(t) \geqslant x) &= \mathbb{P}(X(t) \geqslant x) - \mathbb{P}(m^X(t) > m, X(t) \geqslant x) \\ &= \mathrm{e}^{2\mu m}\, \mathcal{N}\left(\frac{2m - x + \mu t}{\sqrt{t}}\right), \end{aligned}$$

and hence the joint CDF for $x \geqslant m, m < 0$ is given by

$$\begin{aligned}\mathbb{P}(m^X(t) \leqslant m, X(t) \leqslant x) &= \mathbb{P}(m^X(t) \leqslant m) - \mathbb{P}(m^X(t) \leqslant m, X(t) \geqslant x) \\ &= F_{m^X(t)}(m) - \mathbb{P}(m^X(t) \leqslant m, X(t) \geqslant x) \\ &= \mathcal{N}\left(\frac{m - \mu t}{\sqrt{t}}\right) + \mathrm{e}^{2\mu m}\left[\mathcal{N}\left(\frac{m + \mu t}{\sqrt{t}}\right) - \mathcal{N}\left(\frac{2m - x + \mu t}{\sqrt{t}}\right)\right]. \end{aligned}$$

In the case that $x \leqslant m < 0$, we have $\mathbb{P}(x < m^X(t) \leqslant m, X(t) \leqslant x) = 0$ and hence

$$\begin{aligned}\mathbb{P}(m^X(t) \leqslant m, X(t) \leqslant x) &= \mathbb{P}(m^X(t) \leqslant x, X(t) \leqslant x) + \mathbb{P}(x < m^X(t) \leqslant m, X(t) \leqslant x) \\ &= \mathbb{P}(m^X(t) \leqslant x, X(t) \leqslant x) = \mathcal{N}\left(\frac{x - \mu t}{\sqrt{t}}\right).\end{aligned}$$

The last line follows by putting $m = x$ in the previous formula for the CDF. In the remaining case where $m \geqslant 0$, $x \in \mathbb{R}$, we have $\{m^X(t) \leqslant m\} = \{m^X(t) \leqslant 0\} = \Omega$, i.e.,

$$\mathbb{P}(m^X(t) \leqslant m, X(t) \leqslant x) = \mathbb{P}(X(t) \leqslant x) = \mathcal{N}\left(\frac{x - \mu t}{\sqrt{t}}\right).$$

Summarizing the above formulas into one gives the complete joint CDF of $m^X(t), X(t)$:

$$F_{m^X(t),X(t)}(m,x) = \begin{cases} \mathcal{N}\left(\frac{m-\mu t}{\sqrt{t}}\right) + \mathrm{e}^{2\mu m}\left[\mathcal{N}\left(\frac{m+\mu t}{\sqrt{t}}\right) - \mathcal{N}\left(\frac{2m-x+\mu t}{\sqrt{t}}\right)\right] & , m < 0,\ x \geqslant m, \\ \mathcal{N}\left(\frac{x-\mu t}{\sqrt{t}}\right) & , m < 0,\ x \leqslant m, \\ \mathcal{N}\left(\frac{x-\mu t}{\sqrt{t}}\right) & , m \geqslant 0,\ x \in \mathbb{R}. \end{cases} \tag{1.121}$$

Other probabilities related to (1.102) and (1.116) also follow. For example, we can compute the probability that drifted BM process X is within an infinitesimal interval containing $x \in \mathbb{R}$ jointly with the condition that the process has already hit level $m > 0$ (i.e., its sampled maximum $M^X(t) \geqslant m$). This probability is given by the total law

$$\begin{aligned}\mathbb{P}(M^X(t) \geqslant m, X(t) \in \mathrm{d}x) &= \mathbb{P}(X(t) \in \mathrm{d}x) - \mathbb{P}(M^X(t) \leqslant m, X(t) \in \mathrm{d}x) \\ &= \mathbb{P}(X(t) \in \mathrm{d}x) - \mathbb{P}(X_{(m)}(t) \in \mathrm{d}x) \\ &= [p^X(t;0,x) - p^{X_{(m)}}(t;0,x)]\,\mathrm{d}x,\end{aligned} \tag{1.122}$$

where we used (1.108) and (1.109). We can write down the explicit expression by combining $p^X(t;0,x) = p_0(t;x-\mu t)$ with the density in (1.111), i.e., for $m > 0$:

$$\begin{aligned}&\mathbb{P}(M^X(t) \geqslant m, X(t) \in \mathrm{d}x) \\ &= \{p_0(t;x-\mu t) - [p_0(t;x-\mu t) - \mathrm{e}^{2\mu m}\, p_0(t;x-\mu t-2m)]\mathbb{I}_{\{x<m\}}\}\,\mathrm{d}x \\ &= \begin{cases} \mathrm{e}^{2\mu m}\, p_0(t;x-\mu t-2m)\,\mathrm{d}x & , x \in (-\infty, m) \\ p_0(t;x-\mu t)\,\mathrm{d}x & , x \in [m,\infty). \end{cases}\end{aligned} \tag{1.123}$$

For $m < 0$, a similar derivation gives the probability that the drifted BM process is within an infinitesimal interval jointly with the condition that the process has already hit level $m < 0$ (i.e., its sampled minimum $m^X(t) \leqslant m$):

$$\mathbb{P}(m^X(t) \leqslant m, X(t) \in \mathrm{d}x) = \begin{cases} \mathrm{e}^{2\mu m}\, p_0(t;x-\mu t-2m)\,\mathrm{d}x & , x \in (m,\infty), \\ p_0(t;x-\mu t)\,\mathrm{d}x & , x \in (-\infty, m]. \end{cases} \tag{1.124}$$

1.4.3.1 Translated and Scaled Brownian Motion with Drift

All the formulae derived for the process $W^{(\mu,1)}$ in the previous section are trivially extended to generate all the corresponding formulae for drifted BM starting at arbitrary $x_0 \in \mathbb{R}$ and with any volatility parameter $\sigma > 0$. To see this, consider the process defined by (1.38). According to (1.99),

$$X(t) = x_0 + \mu t + \sigma W(t) = x_0 + \sigma W^{(\frac{\mu}{\sigma},1)}(t). \tag{1.125}$$

The sampled maximum of this process is given by

$$M^X(t) := \sup_{0\leqslant u\leqslant t} X(u) = x_0 + \sigma \sup_{0\leqslant u\leqslant t} W^{(\frac{\mu}{\sigma},1)}(u) \equiv x_0 + \sigma M^{(\frac{\mu}{\sigma},1)}(t)\,. \tag{1.126}$$

Here, we are using $M^{(\mu,\sigma)}(t)$ to denote the sampled maximum of process $W^{(\mu,\sigma)}$.

Computing the joint CDF of $M^X(t), X(t)$,

$$\begin{aligned} F_{M^X(t),X(t)}(m,x) &= \mathbb{P}(M^X(t) \leqslant m, X(t) \leqslant x) \\ &= \mathbb{P}\left(x_0 + \sigma M^{(\frac{\mu}{\sigma},1)}(t) \leqslant m,\, x_0 + \sigma W^{(\frac{\mu}{\sigma},1)}(t) \leqslant x\right) \\ &= \mathbb{P}\left(M^{(\frac{\mu}{\sigma},1)}(t) \leqslant \frac{m-x_0}{\sigma},\, W^{(\frac{\mu}{\sigma},1)}(t) \leqslant \frac{x-x_0}{\sigma}\right) \\ &= \mathbb{P}\left(M^{(\mu',1)}(t) \leqslant m',\, W^{(\mu',1)}(t) \leqslant x'\right), \end{aligned} \tag{1.127}$$

where $\mu' = \frac{\mu}{\sigma}$, $m' = \frac{m-x_0}{\sigma}$, and $x' = \frac{x-x_0}{\sigma}$. Hence, this is the joint CDF of the random variable pair $M^{(\mu',1)}(t), W^{(\mu',1)}(t)$, where the function is evaluated at arguments m', x'. By definition, this is given by the formula for the joint CDF in (1.112) with the *variable replacements*:

$$m \to \frac{m-x_0}{\sigma}, \quad x \to \frac{x-x_0}{\sigma}, \quad \mu \to \frac{\mu}{\sigma}. \tag{1.128}$$

These are precisely the same variable replacements that we saw in Section 1.4.2 for driftless BM. In the case of drifted BM we see that the volatility parameter σ also enters into the adjusted drift. By simply making the above variable replacements in (1.115), we obtain the complete joint CDF of the pair $M^X(t), X(t)$ for any $m \in \mathbb{R}$:

$$F_{M^X(t),X(t)}(m,x) = \begin{cases} \mathcal{N}\left(\frac{x-x_0-\mu t}{\sigma\sqrt{t}}\right) - \mathrm{e}^{\frac{2\mu}{\sigma^2}(m-x_0)}\mathcal{N}\left(\frac{x+x_0-2m-\mu t}{\sigma\sqrt{t}}\right) & , x \leqslant m,\ x_0 \leqslant m, \\ \mathcal{N}\left(\frac{m-x_0-\mu t}{\sigma\sqrt{t}}\right) - \mathrm{e}^{\frac{2\mu}{\sigma^2}(m-x_0)}\mathcal{N}\left(\frac{x_0-m-\mu t}{\sigma\sqrt{t}}\right) & , x \geqslant m,\ x_0 \leqslant m, \\ 0 & , x_0 \geqslant m,\ x \in \mathbb{R}. \end{cases} \tag{1.129}$$

For $m \in [x_0, \infty)$, the (marginal) CDF of $M^X(t)$ is given by the above middle expression:

$$F_{M^X(t)}(m) = \mathcal{N}\left(\frac{m-x_0-\mu t}{\sigma\sqrt{t}}\right) - \mathrm{e}^{\frac{2\mu}{\sigma^2}(m-x_0)}\mathcal{N}\left(\frac{x_0-m-\mu t}{\sigma\sqrt{t}}\right) \tag{1.130}$$

and $F_{M^X(t)}(m) \equiv 0$ for $m \in (-\infty, x_0)$. Hence, the CDF of the first hitting time to an upper level is given by $F_{\mathcal{T}^X_m}(t) = 1 - F_{M^X(t)}(m)$:

$$F_{\mathcal{T}^X_m}(t) = \mathcal{N}\left(\frac{x_0-m+\mu t}{\sigma\sqrt{t}}\right) + \mathrm{e}^{\frac{2\mu}{\sigma^2}(m-x_0)}\mathcal{N}\left(\frac{x_0-m-\mu t}{\sigma\sqrt{t}}\right), \quad m > x_0, t > 0. \tag{1.131}$$

The sampled minimum is given by

$$m^X(t) := \inf_{0\leqslant u\leqslant t} X(u) = x_0 + \sigma \inf_{0\leqslant u\leqslant t} W^{(\frac{\mu}{\sigma},1)}(u) \equiv x_0 + \sigma m^{(\frac{\mu}{\sigma},1)}(t), \tag{1.132}$$

where $m^{(\mu,\sigma)}(t)$ denotes the sampled minimum of $W^{(\mu,\sigma)}$. Analoguous to (1.127) we have the joint CDF of $m^X(t), X(t)$:

$$F_{m^X(t),X(t)}(m,x) = \mathbb{P}\left(m^{(\mu',1)}(t) \leqslant m',\, W^{(\mu',1)}(t) \leqslant x'\right). \tag{1.133}$$

Hence, from the joint CDF in (1.121) and applying the parameter replacements in (1.128), we have the complete joint CDF of the pair $m^X(t), X(t)$ for any $m \in \mathbb{R}$:

$$F_{m^X(t),X(t)}(m,x) = \begin{cases} \mathcal{N}\left(\frac{m-x_0-\mu t}{\sigma\sqrt{t}}\right) \\ +\mathrm{e}^{\frac{2\mu}{\sigma^2}(m-x_0)}\left[\mathcal{N}\left(\frac{m-x_0+\mu t}{\sigma\sqrt{t}}\right) - \mathcal{N}\left(\frac{2m-x-x_0+\mu t}{\sigma\sqrt{t}}\right)\right] & , x \geqslant m,\ x_0 \geqslant m, \\ \mathcal{N}\left(\frac{x-x_0-\mu t}{\sigma\sqrt{t}}\right) & , x \leqslant m,\ x_0 \geqslant m, \\ \mathcal{N}\left(\frac{x-x_0-\mu t}{\sigma\sqrt{t}}\right) & , x_0 \leqslant m,\ x \in \mathbb{R}. \end{cases} \tag{1.134}$$

The (marginal) CDF of $m^X(t)$ follows from (1.120) upon using (1.128):

$$F_{m^X(t)}(m) = \mathcal{N}\left(\frac{m - x_0 - \mu t}{\sigma\sqrt{t}}\right) + \mathrm{e}^{\frac{2\mu}{\sigma^2}(m-x_0)}\mathcal{N}\left(\frac{m - x_0 + \mu t}{\sigma\sqrt{t}}\right) \tag{1.135}$$

for $m \in (-\infty, x_0)$ and $F_{m^X(t)}(m) \equiv 1$ for $m \in [x_0, \infty)$. Hence, the CDF of the first hitting time to a lower level is given by the expression in (1.135); i.e., $F_{\mathcal{T}_m^X}(t) = F_{m^X(t)}(m)$, where $m < x_0, t > 0$. Note the trivial case where $F_{\mathcal{T}_m^X}(t) \equiv 1$ for $m = x_0$, $t \geqslant 0$; i.e., the process starts at x_0.

By the above analysis we showed that all formulae for event probabilities, e.g., marginal CDFs or joint CDFs of the random variables $M^X(t), m^X(t), X(t)$, as well as the CDF of the first hitting time, derived in the previous section for the process $X(t) = W^{(\mu,1)}(t)$, give rise to the respective formulae for the process defined by (1.125) upon making the variable replacements in (1.128). Note that for every differentiation of a (joint) CDF w.r.t. variables x, m, or t, there is an extra factor of $\frac{1}{\sigma}$ that multiplies the expression for the resulting PDF. For example, let $X_{(m)}$ be the process X in (1.125) killed at an upper level $m > x_0$. The density for $X_{(m)}(t)$ is then given by applying (1.128) to (1.111) and multiplying by $\frac{1}{\sigma}$:

$$p^{X_{(m)}}(t; x_0, x) = \frac{\mathrm{e}^{-(x-x_0-\mu t)^2/2\sigma^2 t}}{\sigma\sqrt{2\pi t}} - \mathrm{e}^{\frac{2\mu}{\sigma^2}(m-x_0)}\frac{\mathrm{e}^{-(x+x_0-\mu t-2m)^2/2\sigma^2 t}}{\sigma\sqrt{2\pi t}} \tag{1.136}$$

for $x_0, x < m$, and zero otherwise. The same expression for the density is also valid for the domain $x_0, x > m$, i.e., where m is a lower killing level. Similarly, the corresponding expressions for the probability densities in (1.123) and (1.124) follow upon making the variable replacements in (1.128) and multiplying by $\frac{1}{\sigma}$.

All of the above formulae obviously recover those derived in the previous sections when $x_0 = 0$ or $\mu = 0$ or $\sigma = 1$ (e.g., the corresponding joint and marginal CDFs and PDFs for standard BM are recovered when $x_0 = 0, \mu = 0$ and $\sigma = 1$).

We conclude this chapter by noting that the above formulae are readily extended to processes that are related to drifted BM by a monotonic mapping. We recall our discussion in Section 1.3.3. Let $X(t)$ be drifted BM defined in (1.125) and let $F : \mathbb{R} \to \mathcal{I}_Y \subset \mathbb{R}$ be a strictly increasing mapping with process $Y(t) := F(X(t)), t \geqslant 0$. The mapping has a unique inverse $F^{-1} : \mathcal{I}_Y \to \mathbb{R}$, where $X(t) = F^{-1}(Y(t))$. Since F is strictly increasing, the sampled maximum and minimum of the Y-process is simply given in terms of the sampled maximum and minimum of the drifted BM:

$$\begin{aligned} M^Y(t) &:= \sup_{0 \leqslant u \leqslant t} Y(u) = F\left(\sup_{0 \leqslant u \leqslant t} X(u)\right) = F(M^X(t)) \\ m^Y(t) &:= \inf_{0 \leqslant u \leqslant t} Y(u) = F\left(\inf_{0 \leqslant u \leqslant t} X(u)\right) = F(m^X(t)) \,. \end{aligned} \tag{1.137}$$

By the equivalence of events, we clearly have the equivalence of joint CDFs

$$\mathbb{P}(M^Y(t) \leqslant B, Y(t) \leqslant y) = \mathbb{P}(M^X(t) \leqslant F^{-1}(B), X(t) \leqslant F^{-1}(y)),$$
$$\mathbb{P}(m^Y(t) \leqslant B, Y(t) \leqslant y) = \mathbb{P}(m^X(t) \leqslant F^{-1}(B), X(t) \leqslant F^{-1}(y)), \tag{1.138}$$

and marginal CDFs

$$\mathbb{P}(Y(t) \leqslant y) = \mathbb{P}(X(t) \leqslant F^{-1}(y)),$$
$$\mathbb{P}(M^Y(t) \leqslant B) = \mathbb{P}(M^X(t) \leqslant F^{-1}(B)),\ \mathbb{P}(m^Y(t) \leqslant B) = \mathbb{P}(m^X(t) \leqslant F^{-1}(B)), \tag{1.139}$$

for all $B, y \in \mathcal{I}_Y$.

An important example is the exponential mapping $F(x) = \mathrm{e}^x, x \in \mathbb{R}$ with unique inverse $F^{-1}(y) = \ln y, y \in (0, \infty)$. Based on this mapping and the above relations, probabilities of (joint) events associated with the sampled maximum, minimum, and first hitting times of a GBM process are readily derived (see Exercise 1.23).

1.5 Exercises

Exercise 1.1. Let $X_1 \sim \mathit{Norm}(\mu_1, \sigma_1^2)$ and $X_2 \sim \mathit{Norm}(\mu_2, \sigma_2^2)$ with correlation coefficient $\mathrm{Corr}(X_1, X_2) = \rho$.

(a) Determine the joint distribution of $[X, Y]^\top := [X_1 + X_2, X_1 - X_2]^\top$.

(b) Determine the conditional distributions of $X|\{Y = y\}$ and $Y|\{X = x\}$ for any $x, y \in \mathbb{R}$.

Exercise 1.2. Consider a log-normal random variable $S = \mathrm{e}^{a+bZ}$ with $Z \sim \mathit{Norm}(0, 1)$. Derive formulae for the following expected values:

(a) $\mathrm{E}[\,\mathbb{I}_{\{S>K\}}\,]$, where K is a real positive constant;

(b) $\mathrm{E}[\,S\,\mathbb{I}_{\{S>K\}}\,]$;

(c) $\mathrm{E}[\,\max(S, K)\,]$, where you may use the property

$$\max(S, K) = S\,\mathbb{I}_{\{S>K\}} + K\,\mathbb{I}_{\{K \geqslant S\}} = S\,\mathbb{I}_{\{S>K\}} - K\,\mathbb{I}_{\{S>K\}} + K.$$

[Note: for (b-c), (B.1) in Appendix B can be used, e.g., $\mathrm{E}[e^{BZ}\mathbb{I}_{\{Z>A\}}] = e^{B^2/2}\mathcal{N}(B - A)$, where $Z \sim \mathit{Norm}(0, 1)$.] Express your answers in terms of the standard normal CDF function $\mathcal{N}$ and constants a, b, and K. [Note: This calculation is connected to the Black–Scholes–Merton theory of pricing vanilla European style options—a topic covered in depth in Chapter 3.]

Exercise 1.3. Suppose $\mathbf{X} = [X_1, X_2, X_3]^\top$ is a three-dimensional Gaussian (normal) random vector with mean vector zero and covariance matrix

$$\mathbf{C} = \mathrm{E}\left[(\mathbf{X} - \mathrm{E}[\mathbf{X}])\,(\mathbf{X} - \mathrm{E}[\mathbf{X}])^\top\right] = \mathrm{E}\left[\mathbf{X}\,\mathbf{X}^\top\right] = \begin{bmatrix} 4 & 1 & 2 \\ 1 & 2 & 1 \\ 2 & 1 & 3 \end{bmatrix}.$$

Set $Y = 1 + X_1 - 2X_2 + X_3$ and $Z = X_1 - 2X_3$.

(a) Determine the probability distribution of Y.

(b) Determine the probability distribution of the vector $[Y, Z]^\top$.

(c) Find a linear combination $W = aY + bZ$ that is independent of X_1.

Exercise 1.4. Take $X_0 = 0$ and define $X_k = X_{k-1} + Z_k$, for $k = 1, 2, \ldots, n$, where Z_k are i.i.d. standard normals. So we have

$$X_k = \sum_{j=1}^{k} Z_j \, . \tag{1.140}$$

Let $\mathbf{X} = [X_1, X_2, \ldots, X_n]^\top$.

(a) Show that $\mathbf{X}$ is a multivariate normal.

(b) Use the formula (1.140) to calculate the covariances $\mathrm{Cov}(X_k, X_j)$, for $1 \leqslant k, j \leqslant n$.

(c) Use the answers to part (b) to write a formula for the elements of $\mathbf{C} = \mathrm{Cov}(\mathbf{X}, \mathbf{X}) = \mathrm{E}\left[(\mathbf{X} - \mathrm{E}[\mathbf{X}])\,(\mathbf{X} - \mathrm{E}[\mathbf{X}])^\top\right]$.

(d) Write the joint PDF of $\mathbf{X}$.

Exercise 1.5. Prove the main properties 1–3 of the scaled symmetric random walk $\{W^{(n)}(t)\}_{\geqslant 0}$, as stated in Section 1.2.1.

Exercise 1.6. Show that each of the following processes are standard Brownian motions.

(a) $X(t) := -W(t)$.

(b) $X(t) := W(T + t) - W(T)$, where $T \in (0, \infty)$.

(c) $X(t) := cW(t/c^2)$, where $c \neq 0$ is a real constant.

(d) $X(t) := tW(1/t)$, $t > 0$, and $X(0) := 0$.

(e) $X(t) := \frac{1}{\sqrt{a}}[W(at + b) - W(b)]$, $t \geqslant 0$, with constants $a > 0$, $b \geqslant 0$.

Exercise 1.7. Let $\{B(t)\}_{t\geqslant 0}$ and $\{W(t)\}_{t\geqslant 0}$ be two independent standard Brownian motions. Show that $X(t) = (B(t) + W(t))/\sqrt{2}$, $t \geqslant 0$, is also a standard Brownian motion. Determine the coefficient of correlation between $B(t)$ and $X(t)$.

Exercise 1.8. Determine the distribution of $W(1) + W(2) + \cdots + W(n)$ for $n = 1, 2, \ldots$

Exercise 1.9. Let $a_1, a_2, \ldots, a_n \in \mathbb{R}$ and $0 < t_1 < t_2 < \cdots < t_n$. Determine the distribution of $\sum_{k=1}^{n} a_k\, W(t_k)$. Note that the choice with $a_k = \frac{1}{n}$, $1 \leqslant k \leqslant n$, leads to Asian options.

Exercise 1.10. Evaluate each of the following for $0 \leqslant s \leqslant t$.

(a) $\mathrm{E}[W^3(t) | \mathcal{F}_s^W]$, with $\{\mathcal{F}_t^W\}_{t\geqslant 0}$ as the natural filtration for Brownian motion;

(b) $\mathrm{E}[(W(t) - t)^2 | W(s)]$;

(c) $\mathrm{Cov}(W(s), W^2(t))$;

(d) $\mathrm{Cov}(W(t) + W(s), W(t) - W(s))$;

(e) $\mathrm{Cov}\big(X(s), X(t)\big)$, where $X(t) := W^2(t) - t$;

(f) $\mathrm{E}[e^{W(t)}\mathbb{I}_{\{W(t)<b\}} \mid W(s) = x]$, where $x, b \in \mathbb{R}$.

[Note: (B.1) in Appendix B can be used, e.g., $\mathrm{E}[e^{BZ}\mathbb{I}_{\{Z>A\}}] = e^{B^2/2}\mathcal{N}(B - A)$, where $Z \sim Norm(0, 1)$.]

Exercise 1.11. Show that each process is a martingale w.r.t. any filtration for Brownian motion.

(a) $X(t) := W^3(t) - 3tW(t), t \geqslant 0$;

(b) $X(t) := W^4(t) - 6tW^2(t) + 3t^2, t \geqslant 0$.

[Note: $\mathrm{E}[Z^{2n}] = (2n - 1)(2n - 3) \cdots 3 \cdot 1$, for $n = 1, 2, \ldots$, $Z \sim Norm(0, 1)$.]

Exercise 1.12. Suppose that the processes $\{X(t)\}_{t\geqslant 0}$ and $\{Y(t)\}_{t\geqslant 0}$ are, respectively, given by $X(t) = x_0 + \mu_x t + \sigma_x W(t)$ and $Y(t) = y_0 + \mu_y t + \sigma_y W(t)$, where x_0, y_0, μ_x, μ_y, $\sigma_x > 0$, and $\sigma_y > 0$ are real constants. Determine the covariance $\mathrm{Cov}(X(t), Y(s))$ for $s, t \geqslant 0$.

Exercise 1.13. Consider the process defined by (1.40).

(a) Compute $\mathrm{Cov}(X(s), X(t))$, for $s < t$, and thereby show that

$$X(t)|\{X(s) = x\} \sim Norm\left(x_0 + \mu(t)t + \frac{\sigma(t)}{\sigma(s)}(x - x_0 - \mu(s)s),\ \sigma^2(t)(t - s)\right).$$

Hence, determine the expressions for the transition CDF, $P(s, t; x, y)$, and transition PDF, $p(s, t; x, y)$, of the process.

(b) Show that the transition CDF of the process is given by

$$P(s, t; x, y) = P^W(s, t; x', y'),\ \ x' \equiv \frac{x - x_0 - \mu(s)s}{\sigma(s)},\ y' \equiv \frac{y - x_0 - \mu(t)t}{\sigma(t)},$$

where P^W is the transition CDF of standard BM. Hence, recover the formula for $P(s, t; x, y)$ in part (a).

Exercise 1.14. Consider the process $X(t) := x_0 + \mu t + \sigma W(t)$, $t \geqslant 0$, where x_0, μ, and $\sigma > 0$ are real constants. Show that

$$\mathrm{E}[\max(X(t) - K, 0)] - (x_0 + \mu t - K)\mathcal{N}\left(\frac{x_0 + \mu t - K}{\sigma\sqrt{t}}\right) + \sigma\sqrt{t}\, n\left(\frac{x_0 + \mu t - K}{\sigma\sqrt{t}}\right).$$

Exercise 1.15. By directly calculating partial derivatives, verify that the transition PDF

$$p_0(t; x) = \frac{1}{\sqrt{2\pi t}} e^{-\frac{x^2}{2t}}$$

of standard Brownian motion satisfies the *heat equation*, also called the *diffusion equation*:

$$\frac{\partial u(t, x)}{\partial t} = \frac{1}{2}\frac{\partial^2 u(t, x)}{\partial x^2}.$$

Exercise 1.16. Consider $X(t) := \mu t + \sigma W(t)$, with constants $\mu \in \mathbb{R}$, $\sigma > 0$, and where $\{W(t)\}_{t\geqslant 0}$ is standard Brownian motion. Define the Y-process: $Y(t) := y_0 e^{\alpha X(t)}, t \geqslant 0$, with constant $Y(0) = y_0 > 0$ and $\alpha > 0$.

(a) Derive the transition CDF, $P(s, t; x, y)$, and transition PDF, $p(s, t; x, y)$, *of the Y-process* for times $0 \leqslant s < t$ and $x, y > 0$.

(b) Derive an expression for the conditional probability $\mathbb{P}(Y(t) \in (0, K_1) \cup (K_2, \infty) | Y(s) = x)$ for times $0 \leqslant s < t$, any $x > 0$, and where $0 < K_1 < K_2$.

(c) Using (a), provide an expression for the Y*-bridge density* of the process at $Y(t) = x$, given $Y(T) = y$, for any $x, y > 0$, $0 < t < T$. Moreover, derive a formula for the conditional probability: $\mathbb{P}(Y(t) < b | Y(T) = y)$ for any $b > 0$. [Hint: relate this to a Brownian bridge.]

Exercise 1.17. Derive the transition PDF of $\{S^n(t)\}_{t \geqslant 0}$, where $S(t) = S_0 \, \mathrm{e}^{\mu t + \sigma W(t)}, t \geqslant 0$, $S_0 > 0$, for $n = 1, 2, \ldots$.

Exercise 1.18. Derive the transition probability function and transition PDF of the process $X(t) := |W(t)|$, $t \geqslant 0$. [Hint: make use of $\{|W(t)| < x\} = \{-x < W(t) < x\}$.] We note that this process corresponds to nonnegative standard BM reflected at the origin.

Exercise 1.19. Show that formulae (1.12) and (1.26) are equivalent by considering successive Brownian increments and applying (1.3).

Exercise 1.20. Prove that a Gaussian process with the covariance function $c(s, t) = s \wedge t$, mean function $m(t) \equiv 0$, and continuous sample paths is a standard Brownian motion.

Exercise 1.21. Show that the Brownian bridge defined in (1.55) is a Gaussian process and derive its mean and covariance functions.

Exercise 1.22. Derive expressions for the mean and covariance functions of the GBM process defined by (1.41).

Exercise 1.23. Consider the GBM process $S(t) = S_0 \, \mathrm{e}^{\mu t + \sigma W(t)}, t \geqslant 0$, $S_0 > 0$. The respective sampled maximum and minimum of this process are defined by

$$M^S(t) := \sup_{0 \leqslant u \leqslant t} S(u) \quad \text{and} \quad m^S(t) := \inf_{0 \leqslant u \leqslant t} S(u),$$

and the first hitting time to a level $B > 0$ is defined by $\mathcal{T}_B^S = \inf\{t \geqslant 0 : S(t) = B\}$. Derive expressions for the following:

(a) $F_{M^S(t), S(t)}(y, x) := \mathbb{P}(M^S(t) \leqslant y, S(t) \leqslant x)$, for all $t > 0$, $0 < x \leqslant y < \infty$, $S_0 \leqslant y$;

(b) $F_{M^S(t)}(y) := \mathbb{P}(M^S(t) \leqslant y)$, for all $t > 0$, $S_0 \leqslant y < \infty$;

(c) $F_{\mathcal{T}_B^S}(t) := \mathbb{P}(\mathcal{T}_B^S \leqslant t)$, for all $t > 0$, $S_0 < B$;

(d) $\mathbb{P}(M^S(t) \leqslant y, S(t) \in \mathrm{d}x)$, for all $t > 0$, $x \leqslant y < \infty$, $S_0 \leqslant y$;

(e) $\mathbb{P}(m^S(t) \geqslant y, S(t) \geqslant x)$, for all $t > 0$, $0 < y \leqslant x < \infty$, $S_0 \geqslant y$;

(f) $F_{\mathcal{T}_B^S}(t) := \mathbb{P}(\mathcal{T}_B^S \leqslant t)$, for all $t > 0$, $S_0 > B$;

(g) $\mathbb{P}(m^S(t) \geqslant y, S(t) \in \mathrm{d}x)$, for all $t > 0$, $0 < y \leqslant x < \infty$, $S_0 > y$.

Exercise 1.24. Consider the drifted BM defined by $X(t) := \mu t + W(t)$, $t \geqslant 0$, with $M^X(t)$ and $m^X(t)$ defined by (1.100) and (1.101). For any $T > 0$, show that

$$\mathbb{P}\left(M^X(T) \leqslant m \,|\, X(T) = x\right) = 1 - \mathrm{e}^{-2m(m-x)/T}, \quad x \leqslant m, \, m \geqslant 0,$$

and

$$\mathbb{P}\left(m^X(T) \geqslant m \,|\, X(T) = x\right) = 1 - \mathrm{e}^{-2m(m-x)/T}, \quad x \geqslant m, \, m \leqslant 0.$$

Exercise 1.25. Consider the geometric Brownian motion as in Exercise 1.23. For any time $T > 0$, and levels $0 < y < B < \infty$, derive a formula for the conditional probability:

$$\mathbb{P}\left(M^S(T) \leqslant B \,|\, S(T) = y\right).$$

Hint: You may relate this to Exercise 1.24 and use the hint given for Exercise 1.24 in Appendix C.

Exercise 1.26. Let $\{X(t)\}_{t\geqslant 0} \in \mathbb{R}$ be *any process with continuous paths* where we define $M^X(t) := \sup_{0\leqslant u\leqslant t} X(u)$ and $m^X(t) := \inf_{0\leqslant u\leqslant t} X(u)$. Express each of the following probabilities in terms of appropriate combinations of joint CDFs, $F_{M^X(t),X(t)}$ or $F_{m^X(t),X(t)}$, and/or marginal CDFs, $F_{M^X(t)}$, $F_{m^X(t)}$ or $F_{X(t)}$. For example, $\mathbb{P}(M^X(t) \leqslant y, X(t) \leqslant x) = F_{M^X(t),X(t)}(y, x)$, $\mathbb{P}(X(t) \leqslant x) = F_{X(t)}(x)$, $\mathbb{P}(M^X(t) \leqslant y) = F_{M^X(t)}(y)$ follow simply by definition.

(a) $\mathbb{P}(M^X(t) < m, X(t) > x)$, for $x < m$;

(b) $\mathbb{P}(M^X(t) < m, X(t) \in (a, b))$, for $a < b < m$;

(c) $\mathbb{P}(M^X(t) \geqslant m, X(t) \in (a, b))$, for $a < b < m$;

(d) $\mathbb{P}(m^X(t) \leqslant m, X(t) \in (a, b))$, for $b > a > m$;

(e) $\mathbb{P}(m^X(t) > m, X(t) > x)$, for $x > m$.

Exercise 1.27. Consider geometric Brownian motion as in Exercise 1.23. Derive a formula for the probability that this process will eventually hit level B; i.e., $\mathbb{P}(\mathcal{T}_B^S < \infty)$ in the two separate cases where $S_0 < B$ and where $S_0 > B$. Hint: Consider the limit $t \to \infty$ of $F_{\mathcal{T}_B^S}(t)$.

Exercise 1.28. Consider the drifted Brownian bridge process defined by

$$B^{(\mu)}(t) := a + \frac{(b-a)t}{T} + X(t) - \frac{t}{T}X(T),\ 0 \leqslant t \leqslant T,$$

where $X(t) := \mu t + W(t)$ and a, b are real constants. Note: for $\mu = 0$ this corresponds to the process $B_{[0,T]}^{(a,b)}(t)$ defined in (1.54).

(a) Show that $\{B^{(\mu)}(t)\}_{0\leqslant t\leqslant T}$ and $\{B_{[0,T]}^{(a,b)}(t)\}_{0\leqslant t\leqslant T}$ are the same Gaussian processes. This shows that the drift parameter μ vanishes in the Brownian bridge X-process.

(b) Determine the conditional distribution $B^{(\mu)}(t)| \{B^{(\mu)}(s) = x\}$ for $0 < s < t < T$, $x \in \mathbb{R}$. Hence, provide the expression for the transition density $p^{B^{(\mu)}}(s, t, x, y)$ for the bridge process $\{B^{(\mu)}(t)\}_{0\leqslant t\leqslant T}$.

Exercise 1.29. Let $X(t) \in \mathcal{I}_X \subset \mathbb{R}$ be any continuous-time process and assume $F : \mathcal{I}_X \to \mathcal{I}_Y \subset \mathbb{R}$ is a strictly increasing continuously differentiable mapping. Define $Y(t) := F(X(t))$ and consider the Y-bridge process, denoted by $Y_{[t_0,T]}^{(a,b)}(t)$, pinned at $Y(t_0) = a, Y(T) = b$, $a, b \in \mathcal{I}_Y$, $t_0 < t < T$.

(a) Show that the density at $x \in \mathcal{I}_Y$ of the Y-bridge process is given in terms of the density of the X-bridge process according to the relation:

$$f_{Y_{[t_0,T]}^{(a,b)}(t)}(x) = \frac{1}{F'(\hat{x})} f_{X_{[t_0,T]}^{(\hat{a},\hat{b})}(t)}(\hat{x}),$$

$\hat{a} = F^{-1}(a), \hat{b} = F^{-1}(b), \hat{x} = F^{-1}(x)$.

(b) Determine the PDF for the geometric Brownian bridge process defined in Exercise 1.23 pinned at $S(t_0) = a$ and $S(T) = b$, $a, b \in (0, \infty)$, $t_0 < t < T$.

2

Introduction to Continuous-Time Stochastic Calculus

2.1 The Riemann Integral of Brownian Motion

2.1.1 The Riemann Integral

Let f be a real-valued function defined on $[0, T]$. We now recall the precise definition of the Riemann integral of f on $[0, T]$ as follows.

- For $n \in \mathbb{N}$, consider a partition P_n of the interval $[0, T]$:
$$P_n = \{t_0, t_1, \ldots, t_n\}, \quad 0 = t_0 < t_1 < \cdots < t_n = T.$$
Define $\Delta t_i = t_i - t_{i-1}$, $i = 1, 2, \ldots, n$.
- Introduce an intermediate partition Q_n for the partition P_n:
$$Q_n = \{s_1, s_2, \ldots, s_n\}, \quad t_{i-1} \leqslant s_i \leqslant t_i, \quad i = 1, 2, \ldots, n.$$
- Define the Riemann (nth partial) sum as a weighted average of the values of f:
$$S_n = S_n(f, P_n, Q_n) = \sum_{i=1}^{n} f(s_i)\Delta t_i\,.$$
- Suppose that the mesh size $\delta t^{(n)} := \max_{1 \leqslant i \leqslant n} \Delta t_i$ goes to zero as $n \to \infty$. If the limit $\lim_{n\to\infty} S_n$ exists and does not depend on the choice of partitions P_n and Q_n, then this limit is called the *Riemann integral* of f on $[0, T]$, denoted as usual by $\int_0^T f(t)\,dt$. The function f is called the *integrand*. If the Riemann integral exists, then f is said to be *Riemann integrable* on $[0, T]$. For instance, if f is continuous on $[0, T]$ (or the set of discontinuities of f is finite), then f is Riemann integrable on $[0, T]$. In fact, the Riemann integral exists if f is m–a.e. (i.e., Lebesgue almost everywhere) continuous on $[0, T]$.

2.1.2 The Integral of a Brownian Path

Our goal is now to consider computing the Riemann integral of a Brownian sample path w.r.t. time $t \in [0, T]$, i.e., for an outcome $\omega \in \Omega$:

$$I(T, \omega) := \int_0^T W(t, \omega)\,\mathrm{d}t\,. \tag{2.1}$$

Recall that, with probability one (i.e., for almost all $\omega \in \Omega$), Brownian paths are continuous functions of time t. Hence, almost any sample path $(t, W(t, \omega))$, $0 \leqslant t \leqslant T$, is continuous. The Riemann integral (2.1) of such a sample path hence exists and is given by

$$\int_0^T W(t, \omega)\,\mathrm{d}t = \lim_{\delta t^{(n)} \to 0} \sum_{i=1}^{n} W(s_i, \omega)(t_i - t_{i-1})\,.$$

DOI: 10.1201/9780429468889-2

It hence suffices to consider a uniform partition P_n of $[0,T]$ with step size $\Delta t = \frac{T}{n}$ and time points $t_i = i\,\Delta t$ for $0 \leqslant i \leqslant n$. Let Q_n be chosen so that $s_i = t_i$, $1 \leqslant i \leqslant n$. We then have the nth Riemann sum $S_n(\omega) := \Delta t \cdot \sum_{i=1}^{n} W(i\Delta t, \omega)$ and its limit converging to the Riemann integral of the Brownian path:

$$I(T,\omega) = \lim_{n\to\infty} S_n(\omega) = \lim_{n\to\infty} \frac{T}{n} \sum_{i=1}^{n} W\big(i\frac{T}{n}, \omega\big)$$

for almost all $\omega \in \Omega$. Hence, as a random variable, the Riemann integral of Brownian motion is (a.s.) uniquely given by

$$I(T) \equiv \int_0^T W(t)\,\mathrm{d}t = \lim_{n\to\infty} S_n = \lim_{n\to\infty} \frac{T}{n} \sum_{i=1}^{n} W\big(i\frac{T}{n}\big).$$

We now show that the nth Riemann sum S_n is a normally distributed random variable.

Proposition 2.1. *The Riemann sum S_n is normally distributed with mean and variance*

$$\mathrm{E}[S_n] = 0 \quad \textit{and} \quad \mathrm{E}[(S_n)^2] = \frac{T(T+\Delta t)(T+\Delta t/2)}{3}.$$

Proof. Since S_n is a linear combination of jointly normal random variables, it is normally distributed. The expected value is

$$\mathrm{E}[S_n] = \mathrm{E}\left[\Delta t \cdot \sum_{i=1}^{n} W(t_i)\right] = \Delta t \sum_{i=1}^{n} \mathrm{E}[W(t_i)] = 0.$$

Then, $\mathrm{Var}(S_n) = \mathrm{E}[(S_n)^2]$ is given by [note: $t_i \wedge t_j = (\Delta t)(i \wedge j)$]

$$\begin{aligned}
\mathrm{E}[(S_n)^2] = \mathrm{E}\left[\left(\Delta t \cdot \sum_{i=1}^{n} W(t_i)\right)^2\right] &= (\Delta t)^2\, \mathrm{E}\left[\left(\sum_{i=1}^{n} W(t_i)\right)\left(\sum_{j=1}^{n} W(t_j)\right)\right] \\
&= (\Delta t)^2 \sum_{i=1}^{n}\sum_{j=1}^{n} \mathrm{E}[W(t_i)W(t_j)] = (\Delta t)^2 \sum_{i=1}^{n}\sum_{j=1}^{n} t_i \wedge t_j \\
&= (\Delta t)^3 \sum_{i=1}^{n}\sum_{j=1}^{n} i \wedge j = (\Delta t)^3 \sum_{k=1}^{n} k^2 = (\Delta t)^3\, \frac{n(n+1)(2n+1)}{6} \\
&= \frac{(\Delta t \cdot n)(\Delta t \cdot n + \Delta t)(\Delta t \cdot n + \Delta t/2)}{3} = \frac{T(T+\Delta t)(T+\Delta t/2)}{3}
\end{aligned}$$

since $\Delta t \cdot n = T$. □

The Reimann sums $\{S_n\}_{n\geqslant 1}$ hence form a sequence of normally distributed random variables. The limit of such a sequence is hence normally distributed, and this implies that $I(T) = \lim_{n\to\infty} S_n$ is a normal random variable. This is true for all time values $T > 0$. As $n \to \infty$ (and hence $\Delta t = T/n \to 0$), we obtain the mean and variance of $I(T)$:

$$\mathrm{E}[S_n] \to \mathrm{E}[I(T)] = 0, \quad \mathrm{E}[S_n^2] \to \mathrm{E}[I^2(T)] = \frac{T^3}{3},$$

i.e., $S_n \to \mathit{Norm}(0, T^3/3)$. Thus, the stochastic process $\{I(t)\}_{t\geqslant 0}$ is a Gaussian process where $I(t) \sim \mathit{Norm}(0, t^3/3)$.

We can obtain the covariance function of this Gaussian process by applying the Fubini Theorem that allows for changing the order of the time integral and expectation integral. For $0 \leqslant s \leqslant t$:

$$
\begin{aligned}
\mathrm{E}[I(s)I(t)] &= \mathrm{E}\left[\left(\int_0^s W(u)\,\mathrm{d}u\right)\left(\int_0^t W(v)\,\mathrm{d}v\right)\right] = \int_0^t \int_0^s \mathrm{E}[W(u)W(v)]\,\mathrm{d}u\,\mathrm{d}v \\
&= \int_0^t \int_0^s \min\{u,v\}\,\mathrm{d}u\,\mathrm{d}v = \int_0^s \int_0^s \min\{u,v\}\,\mathrm{d}u\,\mathrm{d}v + \int_s^t \left(\int_0^s \min\{u,v\}\,\mathrm{d}u\right)\mathrm{d}v \\
&= \int_0^s \int_0^s \min\{u,v\}\,\mathrm{d}u\,\mathrm{d}v + \int_s^t \left(\int_0^s u\,\mathrm{d}u\right)\mathrm{d}v \\
&= \frac{s^3}{3} + (t-s)\frac{s^2}{2}.
\end{aligned}
$$

The last line follows by computing each integral separately. The second integral follows trivially. The first integral was readily computed by writing $\min\{u,v\} = \min\{u,v\}\mathbb{I}_{u\leqslant v} + \min\{u,v\}\mathbb{I}_{v\leqslant u}$ and by symmetry:

$$
\begin{aligned}
\int_0^s \int_0^s \min\{u,v\}\,\mathrm{d}u\,\mathrm{d}v &= \int_0^s \int_0^s \min\{u,v\}\mathbb{I}_{u\leqslant v}\,\mathrm{d}u\,\mathrm{d}v + \int_0^s \int_0^s \min\{u,v\}\mathbb{I}_{v\leqslant u}\,\mathrm{d}v\,\mathrm{d}u \\
&= \int_0^s \left(\int_0^v u\,\mathrm{d}u\right)\mathrm{d}v + \int_0^s \left(\int_0^u v\,\mathrm{d}v\right)\mathrm{d}u \\
&= 2\int_0^s \left(\int_0^v u\,\mathrm{d}u\right)\mathrm{d}v = \int_0^s v^2\,\mathrm{d}v = s^3/3.
\end{aligned}
$$

Hence, applying the above formula for $s = t$ gives the variance $\mathrm{E}[I^2(t)] = \mathrm{E}[I(t)I(t)] = \frac{t^3}{3}$. The mean function of the integral process is zero, $m_I(t) = \mathrm{E}[I(t)] = 0$, and the covariance function is

$$
c_I(s,t) = \mathrm{Cov}(I(s), I(t)) = \mathrm{E}[I(s)I(t)] = \frac{(s\wedge t)^3}{3} + |t-s|\frac{(s\wedge t)^2}{2}, \quad s,t \geqslant 0.
$$

Example 2.1. Show that $Y(t) := W^3(t) - 3\int_0^t W(u)\,\mathrm{d}u$, $t \geqslant 0$, is a martingale w.r.t. any filtration $\{\mathcal{F}_t\}_{t\geqslant 0}$ for Brownian motion.

Solution. First we note that $Y(t) := W^3(t) - 3I(t)$, where the integral $I(t) \equiv \int_0^t W(u)\,\mathrm{d}u$ is a function of the history of the Brownian motion up to time t and is hence $\mathcal{F}_t$-measurable. That is, the integral process $\{I(t)\}_{t\geqslant 0}$ is adapted to the filtration and hence so is the process $\{Y(t)\}_{t\geqslant 0}$. The process is also integrable since $\mathrm{E}[|Y(t)|] \leqslant \mathrm{E}[|W^3(t)|] + 3\mathrm{E}[|I(t)|] \leqslant \mathrm{E}[|W^3(t)|] + 3\int_0^t \mathrm{E}[|W(u)|]\,\mathrm{d}u < \infty$. Now, for times $s, t \geqslant 0$ we consider $\mathrm{E}[I(t+s) \mid \mathcal{F}_t] \equiv \mathrm{E}_t[I(t+s)]$ (using shorthand notation):

$$
\begin{aligned}
\mathrm{E}_t[I(t+s)] &= \mathrm{E}_t\left[I(t) + \int_t^{t+s} W(u)\,\mathrm{d}u\right] = I(t) + \mathrm{E}_t\left[\int_t^{t+s} W(u)\,\mathrm{d}u\right] \\
&= I(t) + \int_t^{t+s} \mathrm{E}_t[W(u)]\,\mathrm{d}u = I(t) + \int_t^{t+s} W(t)\,\mathrm{d}u \\
&= I(t) + W(t)\int_t^{t+s} \mathrm{d}u = I(t) + sW(t),
\end{aligned}
$$

where we used the martingale property of Brownian motion and Fubini's theorem in one of the terms. Note that in explicit integral form we have shown

$$
\mathrm{E}_t\left[\int_0^{t+s} W(u)\,\mathrm{d}u\right] = \int_0^t W(u)\,\mathrm{d}u + sW(t).
$$

Using the fact that $\{W^3(t) - 3tW(t)\}_{t\geqslant 0}$ and $\{W(t)\}_{t\geqslant 0}$ are martingales, we obtain

$$\begin{aligned}\mathrm{E}_t\left[W^3(t+s)\right] &= \mathrm{E}_t\left[W^3(t+s) - 3(t+s)W(t+s)\right] + \mathrm{E}_t\left[3(t+s)W(t+s)\right] \\ &= W^3(t) - 3tW(t) + 3(t+s)W(t) = W^3(t) + 3sW(t).\end{aligned}$$

Therefore,

$$\begin{aligned}\mathrm{E}_t[Y(t+s)] &= \mathrm{E}_t\left[W^3(t+s)\right] - 3\mathrm{E}_t[I(t+s)] \\ &= W^3(t) + 3sW(t) - 3I(t) - 3sW(t) = W^3(t) - 3I(t) = Y(t).\end{aligned}$$ □

2.2 The Riemann–Stieltjes Integral of Brownian Motion

Since it is possible to integrate Brownian paths (and functions of Brownian motion) w.r.t. time, it is interesting to find out what other integrals can be calculated for Brownian motion. The Riemann–Stieltjes integral generalizes the Riemann integral. It provides an integral of one function w.r.t. another appropriate one. In what follows, our goal is to define the integral of one stochastic process w.r.t. another one, in particular, w.r.t. Brownian motion.

2.2.1 The Riemann–Stieltjes Integral

The construction of the Riemann–Stieltjes integral goes as follows. Let f be a bounded function and g be a monotonically increasing function, both defined on $[0, T]$.

- For $n \in \mathbb{N}$, introduce partitions P_n and Q_n in the same manner as is done for the Riemann integral.
- If the limit of the partial sum over any (shrinking) partition,

$$\lim_{\substack{n\to\infty \\ \delta t^{(n)}\to 0}} \sum_{i=1}^{n} f(s_i)(g(t_i) - g(t_{i-1})),$$

  exists and is independent of the choice of P_n and Q_n, then it is called the *Riemann–Stieltjes integral* of f w.r.t. g on $[0, T]$ and is denoted by

$$\int_0^T f(t)\,\mathrm{d}g(t).$$

  The function g is called the *integrator*.
- By taking $g(x) = x$, the Riemann integral is simply seen to be a special case of the Riemann–Stieltjes integral.

Let us take a look at some important examples, as follows.

(1) Consider the indicator function

$$\mathbb{I}_{\{x\geqslant 0\}} \equiv \mathbb{I}_{[0,\infty)}(x) = \begin{cases} 0 & \text{if } x < 0, \\ 1 & \text{if } x \geqslant 0. \end{cases}$$

Let f be continuous at an interior point $s \in (0,T)$, c be a nonnegative constant, and let $g(x) = c\,\mathbb{I}_{[0,\infty)}(x-s) \equiv c\,\mathbb{I}_{\{x \geqslant s\}}$. Then,

$$\int_0^T f(t)\,\mathrm{d}g(t) = c\,f(s).$$

Hence, when integrator g is a simple step function, the integral simply picks out one value of the continuous function f and this value corresponds to the value of f at the point of discontinuity of g. This is a *sifting property* of the step function integrator g.

(2) The first example extends into the more general case of a step function $g(x)$ assumed as a mixture of step functions: $g(x) = \sum_{n=1}^{\infty} c_n \mathbb{I}_{[0,\infty)}(x - s_n) \equiv \sum_{n=1}^{\infty} c_n \mathbb{I}_{\{x \geqslant s_n\}}$, where $c_n \geqslant 0$ for $n = 1, 2, 3, \ldots$ are chosen such that $\sum_{n=1}^{\infty} c_n$ converges and $\{s_n\}_{n \geqslant 1}$ is a sequence of distinct points in $(0,T)$. If f is continuous on $[0,T]$, then

$$\int_0^T f(t)\,\mathrm{d}g(t) = \sum_{n=1}^{\infty} c_n f(s_n).$$

We see that the integral is a sum over f evaluated at all points of discontinuity of g within the integration interval $[0,T]$. This extends the sifting property in the above first example.

(3) Suppose that f and g' are Riemann integrable on $[0,T]$. In that case

$$\int_0^T f(t)\,\mathrm{d}g(t) = \int_0^T f(t) g'(t)\,\mathrm{d}t.$$

Hence, when the integrator is differentiable, the Riemann–Stieltjes integral is simply the Riemann integral of fg'; i.e., we formally have the differential $\mathrm{d}g(t) = g'(t)\,\mathrm{d}t$.

(4) Consider a CDF F that is a mixture of a discrete CDF F_1 and a continuous CDF F_2:

$$F(x) = w_1 F_1(x) + w_2 F_2(x), \quad F_1(x) = \sum_{n=1}^{\infty} p_n\, \mathbb{I}_{\{x \geqslant x_n\}}, \quad F_2(x) = \int_{-\infty}^{x} p(t)\,\mathrm{d}t,$$

where w_1 and w_2 are nonnegative weights summing to one, $\{p_n\}_{n \geqslant 1}$ and $\{x_n\}_{n \geqslant 1}$ are, respectively, the mass probabilities and mass points of the discrete distribution, and $p(x) = F_2'(x)$ is the PDF of the continuous distribution. Then, for a bounded f:

$$\begin{aligned}\int_{-\infty}^{\infty} f(x)\,\mathrm{d}F(x) &= w_1 \int_{-\infty}^{\infty} f(x)\,\mathrm{d}F_1(x) + w_2 \int_{-\infty}^{\infty} f(x)\,\mathrm{d}F_2(x)\\ &= w_1 \sum_{n=1}^{\infty} p_n f(x_n) + w_2 \int_{-\infty}^{\infty} f(x)\,p(x)\,\mathrm{d}x.\end{aligned}$$

In the first integral, with F_1 as integrator, we used the result in example (2). This is an example in which the Riemann–Stieltjes integral gives us the expected value of a function $f(X)$ of a random variable X having a mixture distribution given by F_1 and F_2 with respective mixture probabilities (weights) w_1 and w_2. In particular, the above equation can be read as $\mathrm{E}[f(X)] = w_1 \mathrm{E}^{(1)}[f(X)] + w_2 \mathrm{E}^{(2)}[f(X)]$.

The Riemann–Stieltjes integral $\int_0^T f(t)\,\mathrm{d}g(t)$ can be extended on a larger class of functions. Recall that the p-variation of $f\colon [0,T] \to \mathbb{R}$ is

$$V_{[0,T]}^{(p)}(f) = \limsup_{\delta t^{(n)} \to 0} \sum_{i=1}^{n} |f(t_i) - f(t_{i-1})|^p,$$

where the limit is taken over all possible partitions $0 = t_0 < t_1 < \cdots < t_n = T$, shrinking as $n \to \infty$. The following result (stated without proof) shows that we can consider the Riemann–Stieltjes integral on a fairly extensive combination of functions f and integrator g whose combined variational properties satisfy a certain condition.

Proposition 2.2. *Assume that f and g do not have discontinuities at the same points within the integration interval $[0, T]$. Let the p-variation of f and the q-variation of g be finite for some $p, q > 0$, such that $\frac{1}{p} + \frac{1}{q} > 1$. Then, the Riemann–Stieltjes integral $\int_0^T f(t)\,\mathrm{d}g(t)$ exists and is finite.*

For example, if the integrator g is a function of bounded variation on $[0, T]$, and f is a continuous function, then both functions have finite (first) variation. Hence, we may use $p = q = 1$ ($\frac{1}{p} + \frac{1}{q} = 2 > 1$) in the above proposition and this confirms that the Riemann–Stieltjes integral of f w.r.t. g is defined.

2.2.2 Integrals w.r.t. Brownian Motion: Preliminary Discussion

It is known that (a.s.) the p-variation of a Brownian sample path on $[0, T]$ is finite for $p \geqslant 2$ and infinite for $p < 2$. In particular, we proved that the quadratic variation of Brownian motion is bounded but the first variation is unbounded. Hence, if we apply Proposition 2.2 for $q = 2$ to the Riemann–Stieltjes integral $\int_0^T f(t)\,\mathrm{d}W(t)$ we see that such an integral w.r.t. Brownian motion (taken as the integrator) is well-defined if the p-variation of f is finite for some $p \in (0, 2)$. For example, the integral exists if f is a function of bounded variation ($p = 1$) such as a monotone function or a continuously differentiable function. Thus, for example, the integrals

$$\int_0^T \mathrm{e}^t\,\mathrm{d}W(t), \quad \int_0^T t^\alpha\,\mathrm{d}W(t) \ \ (\alpha \geqslant 1)$$

exist as Riemann–Stieltjes integrals. However, the integral

$$\int_0^T W(t)\,\mathrm{d}W(t) \tag{2.2}$$

does *not* (a.s.) exist as a Riemann–Stieltjes integral. First, note that Proposition 2.2 is not applicable to the integral in (2.2). Indeed, for given ω, consider $\int_0^T W(t,\omega)\,\mathrm{d}W(t,\omega)$ as a Riemann–Stieltjes integral with integrand and integrator $f(t) = g(t) = W(t,\omega)$. We recall that $V^{(p)}_{[0,T]}(W)$ is finite (a.s.) iff $p \geqslant 2$. Since $f(t) = g(t)$ we have $p = q$, where $\frac{1}{p} + \frac{1}{p} \leqslant \frac{1}{2} + \frac{1}{2} = 1$. Hence, Proposition 2.2 cannot guarantee the existence of the Riemann–Stieltjes integral.

Let us now evaluate the integral in (2.2) for different intermediate partitions Q_n. Consider the Riemann–Stieltjes sum

$$S_n = \sum_{i=1}^{n} W(t_{i-1})(W(t_i) - W(t_{i-1})) \tag{2.3}$$

with the intermediate (left endpoint) nodes $s_i = t_{i-1}$ for $i = 1, 2, \ldots, n$, $0 = t_0 < \ldots < t_n = T$. Rewriting S_n, upon using the algebraic identity $a(b-a) = -\frac{1}{2}(b-a)^2 + \frac{1}{2}(b^2 - a^2)$ with

$a \equiv W(t_{i-1}), b \equiv W(t_i)$:

$$S_n = -\frac{1}{2}\sum_{i=1}^{n}\left\{(W(t_i) - W(t_{i-1}))^2 - (W^2(t_i) - W^2(t_{i-1}))\right\}$$
$$= -\frac{1}{2}\sum_{i=1}^{n}(W(t_i) - W(t_{i-1}))^2 + \frac{1}{2}\underbrace{\sum_{i=1}^{n}(W^2(t_i) - W^2(t_{i-1}))}_{=W^2(t_n)-W^2(t_0)=W^2(T)-W^2(0)}.$$

As $n \to \infty$ and $\delta t^{(n)} \to 0$, we have $\sum_{i=1}^{n}(W(t_i) - W(t_{i-1}))^2 \to [W,W](T) = T$, i.e., the quadratic variation of Brownian motion on $[0, T]$. Therefore,

$$\lim_{n\to\infty} S_n = \frac{1}{2}\left(W^2(T) - W^2(0)\right) - \frac{T}{2}.$$

We shall see later that this limit corresponds to the *Itô integral* of BM:

$$\int_0^T W(t)\,\mathrm{d}W(t) = \frac{1}{2}\left(W^2(T) - W^2(0)\right) - \frac{T}{2} = \frac{1}{2}(W^2(T) - T).$$

Recall that for a differentiable function f we would have obtained (the ordinary calculus result)

$$\int_0^T f(t)\,\mathrm{d}f(t) = \frac{f^2(T) - f^2(0)}{2}.$$

Consider now another choice of the intermediate partition with (right endpoint) $s_i = t_i$ for every $i = 1, 2, \ldots, n$. The Riemann–Stieltjes sum is then, using $a(a-b) = \frac{1}{2}(a-b)^2 + \frac{1}{2}(a^2 - b^2)$ with $a \equiv W(t_i), b \equiv W(t_{i-1})$:

$$S_n^* = \sum_{i=1}^{n} W(t_i)(W(t_i) - W(t_{i-1}))$$
$$= \underbrace{\frac{1}{2}\sum_{i=1}^{n}(W(t_i) - W(t_{i-1}))^2}_{\to [W,W](T)=T, \text{ as } n\to\infty} + \underbrace{\frac{1}{2}\sum_{i=1}^{n}(W^2(t_i) - W^2(t_{i-1}))}_{=W^2(T)-W^2(0)}$$
$$\to \frac{1}{2}\left(W^2(T) - W^2(0)\right) + \frac{T}{2} = \frac{1}{2}\left(W^2(T) + T\right), \text{ as } n \to \infty.$$

For $0 \leqslant \alpha \leqslant 1$, we can consider a weighted average of S_n and S_n^*:

$$\alpha S_n + (1-\alpha)S_n^* = \sum_{i=1}^{n}(\alpha W(t_{i-1}) + (1-\alpha)W(t_i))(W(t_i) - W(t_{i-1}))$$
$$\to \frac{1}{2}\left(W^2(T) - W^2(0)\right) + \frac{T}{2} - \alpha T = \frac{1}{2}W^2(T) + \left(\frac{1}{2} - \alpha\right)T, \text{ as } n \to \infty.$$

An interesting case is when the midpoint is used, i.e., $\alpha = \frac{1}{2}$. The respective limit is called the *Stratonovich integral* of BM:

$$\int_0^T W(t) \circ \mathrm{d}W(t) = \frac{1}{2}\left(W^2(T) - W^2(0)\right) = \frac{1}{2}W^2(T).$$

The Stratonovich integral satisfies the usual rules of ordinary calculus, such as the chain rule and integration by parts. The two types of stochastic integrals are related as

$$\int_0^T W(t) \circ \mathrm{d}W(t) = \int_0^T W(t)\,\mathrm{d}W(t) + \frac{T}{2}.$$

For a continuously differentiable function $f\colon \mathbb{R} \to \mathbb{R}$, it can be shown that the following conversion formula applies:

$$\int_0^T f(W(t)) \circ \mathrm{d}W(t) = \int_0^T f(W(t))\,\mathrm{d}W(t) + \frac{1}{2}\int_0^T f'(W(t))\,\mathrm{d}t,$$

where the respective integrals correspond to the Stratonovich and Itô integrals of a differentiable function $f(W(t))$ of BM. We note, however, that we have yet to give a precise general definition of such stochastic integrals. Of importance to us is the general definition and properties of the stochastic Itô integral. This is the topic of the next section.

2.3 The Itô Integral and Its Basic Properties

2.3.1 The Itô Integral for Simple Processes

Our goal is to give a construction of the Itô stochastic integral w.r.t. standard Brownian motion. Throughout, we shall assume that the integrand is a stochastic process that is adapted to any chosen filtration $\mathbb{F} = \{\mathcal{F}_t\}_{t\geqslant 0}$ for Brownian motion. In particular, recall that $W(t)$ is $\mathcal{F}_t$-measurable and $W(t) - W(s)$ is independent of $\mathcal{F}_s$, for all $0 \leqslant s \leqslant t < \infty$. In what follows, we shall consider all processes defined on some time interval $[0, T]$, for some $T \in (0, \infty)$. We begin by considering integrands that are simple processes with piecewise-constant paths.

Definition 2.1. A continuous-time stochastic process $X \equiv \{X(t)\}_{0\leqslant t\leqslant T}$ defined on the filtered probability space $(\Omega, \mathcal{F}, \mathbb{P}, \mathbb{F})$ is said to be a square-integrable $\mathbb{F}$-adapted *simple* (or *step-stochastic*) *process* if there exists a time partition $P_n = \{t_0, t_1, \ldots, t_n\}$, where $t_0 = 0$ and $t_n = T$, such that [1]

1. $X(t) = \sum_{i=0}^{n-1} \xi_i \mathbb{I}_{[t_i,t_{i+1})}(t)$, i.e., $X(t) = \xi_i$ for $t \in [t_i, t_{i+1})$;
2. ξ_i is $\mathcal{F}_{t_i}$-measurable and $\mathrm{E}[\xi_i^2] < \infty$, for all $i = 0, 1, \ldots, n-1$.

From the first condition we observe that an ω-path of the process is fixed at $X(t,\omega) = \xi_i(\omega)$ for $t \in [t_i, t_{i+1})$. The step process generally has right-continuous paths. The second condition is equivalent to stating that the process is $\mathbb{F}$-adapted, i.e., $X(t)$ is $\mathcal{F}_t$-measurable for all $t \in [0, T]$, and *square-integrable* on $[0, T]$, i.e.,

$$\int_0^T \mathrm{E}\big[X^2(t)\big]\,\mathrm{d}t < \infty. \tag{2.4}$$

Note that $t \in [t_k, t_{k+1})$ for some $k \in \{0, 1, \ldots, n-1\}$, where $X(t) = \xi_k$ is $\mathcal{F}_{t_k}$-measurable. Since $t_k \leqslant t$ we have $\mathcal{F}_{t_k} \subset \mathcal{F}_t$, implying $X(t)$ is $\mathcal{F}_t$-measurable; i.e., X is $\mathbb{F}$-adapted. The

[1] The process is also equivalently written as $X(t) = \sum_{i=1}^{n} \xi_{i-1}\mathbb{I}_{[t_{i-1},t_i)}(t)$ and we may also use the equivalent indicator notation: $\mathbb{I}_{[t_{i-1},t_i)}(t) \equiv \mathbb{I}_{\{t_{i-1}\leqslant t<t_i\}}$.

square-integrability condition in (2.4) is equivalent to $\mathrm{E}[\xi_i^2] < \infty$ for all $i = 0, 1, \ldots, n-1$. To see this, we use the identity $(\mathbb{I}_{[t_i,t_{i+1})}(t))^2 = \mathbb{I}_{[t_i,t_{i+1})}(t)$ and $\mathbb{I}_{[t_i,t_{i+1})}(t)\mathbb{I}_{[t_j,t_{j+1})}(t) = 0$ if $i \neq j$. Hence,

$$X^2(t) = \sum_{i=0}^{n-1}\sum_{j=0}^{n-1} \xi_i \xi_j \mathbb{I}_{[t_i,t_{i+1})}(t)\mathbb{I}_{[t_j,t_{j+1})}(t) = \sum_{i=0}^{n-1} \xi_i^2 \, \mathbb{I}_{[t_i,t_{i+1})}(t) \tag{2.5}$$

i.e., $X^2(t) = \xi_i^2$ for $t \in [t_i, t_{i+1})$. The time integral in (2.4) is then given explicitly as

$$\begin{aligned}\int_0^T \mathrm{E}\big[X^2(t)\big]\,\mathrm{d}t = \sum_{i=0}^{n-1}\mathrm{E}[\xi_i^2]\int_0^T \mathbb{I}_{[t_i,t_{i+1})}(t)\,\mathrm{d}t &= \sum_{i=0}^{n-1}\mathrm{E}[\xi_i^2]\int_{t_i}^{t_{i+1}}\mathrm{d}t \\ &= \sum_{i=0}^{n-1}\mathrm{E}[\xi_i^2](t_{i+1}-t_i). \end{aligned} \tag{2.6}$$

Clearly, this is bounded iff each $\mathrm{E}[\xi_i^2]$ is bounded.

We remark that Definition 2.1 may also include square-integrable $\mathbb{F}$-adapted simple processes defined for $t \in [0, \infty)$, i.e., with $T = \infty$. We also note that a nonrandom process where $X(t)$ is an ordinary peicewise constant function of time (with all $\xi_i \equiv a_i$ as nonrandom) is automatically a square-integrable $\mathbb{F}$-adapted simple process. In this special case, we simply have $\mathrm{E}[\xi_i^2] = a_i^2$ in (2.6). Simple processes satisfy some basic properties. For instance, the reader can verify that any linear combination of square-integrable adapted simple processes is again a square-integrable adapted simple process. More generally, Borel functions of adapted simple processes are also adapted simple processes, although not necessarily square-integrable unless the corresponding square-integrability condition holds.

As a primary example, consider a piecewise-constant approximation of Brownian motion defined by $X(t) = \xi_i \equiv W(t_i)$ for $t \in [t_i, t_{i+1})$, $0 \leqslant t_0 < t_1 < \ldots$. For all $t \in [0, \infty)$, this is a simple process that can be represented as

$$X(t) = \sum_{i=0}^{\infty} W(t_i)\,\mathbb{I}_{[t_i,t_{i+1})}(t).$$

Note that $X(t)$ on the interval $t \in [t_i, t_{i+1})$ is given by BM at time t_i, where $X(t) = W(t_i) \stackrel{d}{=} Norm(0, t_i)$. For any path ω, the graph of $X(t, \omega)$ as function of time $t \in [0, T]$ is a piecewise constant (step function) with fixed value $X(t, \omega) = W(t_i, \omega)$ on every time interval $t \in [t_i, t_{i+1})$. Clearly, $X(t)$ is $\mathcal{F}_t$–measurable (every $\xi_i = W(t_i)$ is $\mathcal{F}_{t_i}$–measurable). Moreover, $\mathrm{E}[\xi_i^2] = \mathrm{E}[W^2(t_i)] = t_i < \infty$. Hence, X is a square-integrable $\mathbb{F}$-adapted simple process on any finite interval $[0, T]$. Figure 2.1 depicts a sample path of $W(t)$ and an approximation by $X(t)$ for a path on the interval $[0, 1]$ where $t_0 = 0, t_1 = 0.2, t_2 = 0.4, t_3 = 0.6, t_4 = 0.8, t_5 = 1.0$.

The Itô integral of a simple process can be defined as a Riemann–Stieltjes sum evaluated at the *left endpoint* of the subintervals. The simplest nonzero (trivial) case of an Itô integral is that of a nonrandom single indicator function $X(t) = \mathbb{I}_{[a,b]}(t)$, $0 \leqslant a < b \leqslant T$. As a very special case within the definition given below, the Itô integral is the Riemann–Stieltjes integral w.r.t. W:

$$\int_0^T \mathbb{I}_{[a,b]}(t)\,\mathrm{d}W(t) = \int_a^b \mathrm{d}W(t) = W(b) - W(a).$$

The Riemann–Stieltjes sum evaluated at the *left endpoint* of a step function gives us the working definition of the Itô integral of a simple process as follows.

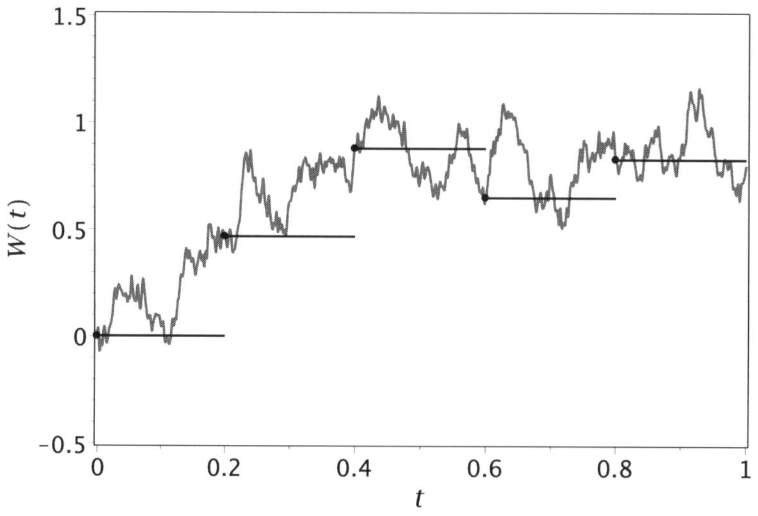

FIGURE 2.1: A Brownian sample path and its approximation by a simple process.

Definition 2.2. (Itô integral of a simple process)
Let $X(t) = \sum_{i=0}^{n-1} \xi_i \mathbb{I}_{[t_i,t_{i+1})}(t)$, where $a = t_0 < t_1 < \ldots < t_n = b$, $[a,b] \in [0,\infty)$, and $\{\xi_i\}_{i\geqslant 0}$ satisfying the conditions in Definition 2.1. The Itô integral of this process on $[a,b]$ is defined by

$$\int_a^b X(t)\,\mathrm{d}W(t) := \sum_{i=0}^{n-1} \xi_i (W(t_{i+1}) - W(t_i)). \tag{2.7}$$

We remark that, throughout the text, the Itô integral is defined only for square-integrable adapted processes.[2] Based on this definition, the Itô integral on any interval $[0,t]$, $0 \leqslant t \leqslant T$, of a simple process $X(t) = \sum_{i=0}^{n-1} \xi_i \mathbb{I}_{[t_i,t_{i+1})}(t)$, $0 = t_0 < t_1 < \ldots < t_n = T$, is given by [3]

$$\begin{aligned} I(t) \equiv \int_0^t X(s)\,\mathrm{d}W(s) &= \int_0^T \mathbb{I}_{[0,t)}(s) X(s)\,\mathrm{d}W(s) \\ &= \sum_{i=0}^{k-1} \xi_i (W(t_{i+1}) - W(t_i)) + \xi_k (W(t) - W(t_k)), \end{aligned} \tag{2.8}$$

for $t_k \leqslant t \leqslant t_{k+1}$, $k = 0, 1, \ldots, n-1$. Within (2.8) we used the simple process

$$\mathbb{I}_{[0,t)}(s) X(s) = \sum_{i=0}^{n-1} \xi_i \mathbb{I}_{[0,t)\cap[t_i,t_{i+1})}(s) = \sum_{i=0}^{k-1} \xi_i \mathbb{I}_{[t_i,t_{i+1})}(s) + \xi_k \mathbb{I}_{[t_k,t)}(s)\,.$$

Setting $t = t_n = T$, (2.8) gives the Itô integral of X on $[0,T]$,

$$I(T) = \int_0^T X(s)\,\mathrm{d}W(s) = \sum_{i=0}^{n-1} \xi_i (W(t_{i+1}) - W(t_i)).$$

The Itô integral in (2.8) represents a continuous-time stochastic process $\{I(t)\}_{0\leqslant t\leqslant T}$.

[2] An alternate definition of the Itô integral can also be given for a larger class of functions where the square-integrability condition $\int_a^b \mathrm{E}[X^2(t)]\mathrm{d}t < \infty$ is replaced by $\int_a^b X^2(t)\mathrm{d}t < \infty$ (a.s.).

[3] Note that in this case the "dummy integration variable" s is used since the Itô integral is a function of the endpoint time variable t.

A trivial property that follows directly from Definition 2.2 is additivity of Itô integrals over integration intervals. That is, for $a < b < c$ we have $X(t)\mathbb{I}_{[a,c)}(t) = X(t)\mathbb{I}_{[a,b)}(t) + X(t)\mathbb{I}_{[b,c)}(t)$ and hence

$$\int_a^c X(t)\,\mathrm{d}W(t) = \int_a^b X(t)\,\mathrm{d}W(t) + \int_b^c X(t)\,\mathrm{d}W(t).$$

Before developing the main important properties of the Itô integral we consider an instructive example.

Example 2.2. Let $X(t) = W(n)$, for $n \leqslant t < n+1$ and all integers $n = 0, 1, 2, \ldots$.

(a) Write $X(t)$ and $X^2(t)$ as simple (step) stochastic processes for any real $t \in [0, \infty)$.

(b) Express the Itô integral $\int_0^t X(s)dW(s)$ in terms of Brownian increments for any real $t \in [0, \infty)$.

(c) Compute the mean $\mathrm{E}\big[\int_0^t X(s)dW(s)\big]$ and variance $\mathrm{Var}\left(\int_0^t X(s)dW(s)\right)$ for any real $t \in [0, \infty)$.

Solution.

(a) By definition, $X(t)$ is given by BM at integer time i, $W(i)$, on every time interval $t \in [t_i, t_{i+1}) \equiv [i, i+1)$, for all integers $i \geqslant 0$. Hence,

$$X(t) = \sum_{i=0}^{\infty} W(i)\mathbb{I}_{[i,i+1)}(t) = \sum_{i=1}^{\infty} W(i)\mathbb{I}_{[i,i+1)}(t).$$

The last equality is due to $W(0) = 0$. [Note: we can alternatively use the fact that $\sum_{i=0}^{\infty} \mathbb{I}_{[i,i+1)}(t) \equiv 1$ for all $t \in [0, \infty)$; i.e., $X(t) = X(t)\sum_{i=0}^{\infty} \mathbb{I}_{[i,i+1)}(t) = \sum_{i=0}^{\infty} X(t)\mathbb{I}_{[i,i+1)}(t)$ which is the same sum as above since $X(t)\mathbb{I}_{[i,i+1)}(t) = W(i)\mathbb{I}_{[i,i+1)}(t)$.]

Squaring the above sum (see also (2.5)) while using $\mathbb{I}_{[i,i+1)}(t)\mathbb{I}_{[j,j+1)}(t) = \mathbb{I}_{[i,i+1)}(t)$ if $i = j$, and zero if $i \neq j$:

$$X^2(t) = \sum_{i=0}^{\infty} W^2(i)\mathbb{I}_{[i,i+1)}(t) = \sum_{i=1}^{\infty} W^2(i)\mathbb{I}_{[i,i+1)}(t).$$

(b) $\int_0^t X(s)dW(s) = 0$ for $t \in [0, 1]$. For time $t > 1$, we apply (2.8) with $t_i = i$, $\xi_i = W(i)$:

$$\int_0^t X(s)dW(s) = \int_1^t X(s)dW(s) = \sum_{i=1}^{k-1} W(i)(W(i+1) - W(i)) + W(k)(W(t) - W(k)),$$

$t \in [k, k+1)$, $k \geqslant 1$. We can write this equivalently in terms of the integer part of t, since $t \in [\lfloor t \rfloor, \lfloor t \rfloor + 1)$ (where $k \equiv \lfloor t \rfloor$):

$$\int_0^t X(s)dW(s) = \sum_{i=1}^{\lfloor t \rfloor - 1} W(i)(W(i+1) - W(i)) + W(\lfloor t \rfloor)(W(t) - W(\lfloor t \rfloor)).$$

Note that this expression is valid for all $t \in [0, \infty)$ since it vanishes for $t \in [0, 1]$.

(c) [**Remark**: further below we state and prove the Itô isometry formula that allows a simple and direct calculation of the variance of an Itô integral. We also learn that any Itô integral has zero mean. In this solution, we shall assume no such knowledge.]

For every i, $W(i)$ and $W(i+1) - W(i)$ are independent and $W(\lfloor t \rfloor)$ is independent of $W(t) - W(\lfloor t \rfloor)$. Hence, $\mathrm{E}[W(i)(W(i+1) - W(i))] = \mathrm{E}[W(i)]\mathrm{E}[W(i+1) - W(i)] = 0$ and $\mathrm{E}[W(\lfloor t \rfloor)(W(t) - W(\lfloor t \rfloor))] = \mathrm{E}[W(\lfloor t \rfloor)]\mathrm{E}[W(t) - W(\lfloor t \rfloor)] = 0$. Hence, the expectation of the above sum is zero, $\mathrm{E}\big[\int_0^t X(s)dW(s)\big] = 0$.

Since the mean is zero, $\mathrm{Var}\left(\int_0^t X(s)dW(s)\right) = \mathrm{E}\left[\left(\int_0^t X(s)dW(s)\right)^2\right]$. For $t \in [0,1]$ this is zero. For $t > 1$, where $k = \lfloor t \rfloor$, we square the above Itô integral sum and take expectations (to simplify notation we define standard normal i.i.d. $Z_i := W(i+1) - W(i)$):

$$\begin{aligned}\mathrm{E}\left[\left(\int_0^t X(s)dW(s)\right)^2\right] &= \sum_{i=1}^{k-1} \mathrm{E}[W^2(i)Z_i^2] + 2 \sum\sum_{0 \leqslant i < j \leqslant k-1} \mathrm{E}\big[W(i)W(j)Z_iZ_j\big] \\ &+2\sum_{i=1}^{k-1} \mathrm{E}\big[W(i)Z_iW(k)(W(t) - W(k))\big] + \mathrm{E}[W^2(k)(W(t) - W(k))^2].\end{aligned}$$

For $i < j$, the Brownian increment $Z_j = W(j+1) - W(j)$ is independent of BM at times i and j, $W(i)$ and $W(j)$, and also independent of the Brownian increment Z_i (i.e., Brownian increments on nonoverlapping time intervals are independent), i.e., Z_j is independent of the product $W(i)W(j)Z_i$. Hence, $\mathrm{E}\big[W(i)W(j)Z_iZ_j\big] = \mathrm{E}\big[W(i)W(j)Z_i\big] \cdot \mathrm{E}[Z_j] = 0$ since $\mathrm{E}[Z_j] = 0$. The above double sum is hence zero. Similarly, the third summation is zero since $W(t) - W(k)$ $(t \geqslant k)$ is independent of $W(i)Z_iW(k)$, for $i \leqslant k-1$, i.e., $\mathrm{E}[W(i)Z_iW(k)(W(t) - W(k))] = \mathrm{E}[W(i)Z_iW(k)] \cdot \mathrm{E}[(W(t) - W(k))] = 0$. For the first and last terms, we use the independence of $W^2(i)$ and Z_i^2, and independence of $W^2(k)$ and $(W(t) - W(k))^2$, i.e., $\mathrm{E}[W^2(i)Z_i^2] = \mathrm{E}[W^2(i)] \cdot \mathrm{E}[Z_i^2] = i \cdot 1 = i$ and $\mathrm{E}[W^2(k)(W(t) - W(k))^2] = \mathrm{E}[W^2(k)] \cdot \mathrm{E}[(W(t) - W(k))^2] = k(t-k)$. Finally, the variance is

$$\begin{aligned}\mathrm{Var}\left(\int_0^t X(s)dW(s)\right) &= \sum_{i=1}^{k-1} i + k(t-k) \\ &= \frac{k}{2}(k-1) + k(t-k) = \frac{\lfloor t \rfloor}{2}(\lfloor t \rfloor - 1) + \lfloor t \rfloor(t - \lfloor t \rfloor).\end{aligned}$$

Note that this expression holds for all $t \in [0, \infty)$.

□

As seen in (2.7) or (2.8), the Itô integral is a random variable constructed as a sum of terms, where each term is a product of a BM increment $W(t_{i+1}) - W(t_i)$, which is independent of $\mathcal{F}_{t_i}$, and a $\mathcal{F}_{t_i}$-measurable random variable ξ_i; i.e., ξ_i is independent of $W(t_{i+1}) - W(t_i)$. The independence allowed us to explicitly compute the mean and variance of the Itô integral in Example 2.2. We are now ready to state and prove the main properties of any Itô integral for a simple process.

Theorem 2.3. Properties of the Itô Integral
The Itô integral has the following properties.

(1) ***Continuity.*** *Sample paths* $\{I(t,\omega)\}_{0 \leqslant t \leqslant T}$ *of* $I(t)$ *given by (2.8) are (a.s.) continuous functions of time* t.

(2) ***Adaptivity.*** *The process* $\{I(t)\}_{0\leqslant t\leqslant T}$ *with* $I(t)$ *given by (2.8) is* $\mathbb{F}$*-adapted i.e.,* $I(t)$ *is* $\mathcal{F}_t$*-measurable for all* $t\in[0,T]$.

(3) ***Linearity.*** *Let* $I_1(t)=\int_0^t X_1(s)\,\mathrm{d}W(s)$ *and* $I_2(t)=\int_0^t X_2(s)\,\mathrm{d}W(s)$, *where* X_1 *and* X_2 *are simple processes on* $[0,T]$ *in accordance with Definition 2.1. Then,*

$$c_1I_1(t)+c_2I_2(t)=\int_0^t (c_1X_1(s)+c_2X_2(s))\,\mathrm{d}W(s)\,,\quad \textit{for any constants } c_1,c_2\in\mathbb{R}.$$

(4) ***Martingale.*** *The process* $\{I(t)\}_{0\leqslant t\leqslant T}$, *with* $I(t)$ *given by (2.8), is a martingale w.r.t. filtration* $\mathbb{F}$.

(5) ***Zero mean.*** $\mathrm{E}[I(t)]\equiv\mathrm{E}\left[\int_0^t X(t)\,\mathrm{d}W(s)\right]=0$, *for all* $0\leqslant t\leqslant T$ *with* $I(t)$ *given by (2.8).*

(6) ***Itô isometry.*** *The Itô integral in (2.8) has variance*

$$\begin{aligned}\mathrm{Var}(I(t))=\mathrm{E}[I^2(t)]&\equiv\mathrm{E}\left[\left(\int_0^t X(s)\,\mathrm{d}W(s)\right)^2\right]\\ &=\int_0^t \mathrm{E}[X^2(s)]\,\mathrm{d}s=\sum_{i=0}^{k-1}\mathrm{E}[\xi_i^2](t_{i+1}-t_i)+\mathrm{E}[\xi_k^2](t-t_k)\,,\end{aligned}\tag{2.9}$$

i.e., the variance of an Itô integral on $[0,t]$ *is equal to a Riemann integral (w.r.t. time variable* s*) of the second moment of the integrand process as a function of time* $s\in[0,t]$.

(7) ***Covariance and Independence.*** *For all* $0\leqslant t\leqslant u\leqslant T$,

$$\mathrm{Cov}\left(I(t),I(u)\right)=\mathrm{Var}\left(I(t)\right)\ \ \textit{i.e.,}\ \mathrm{Cov}\left(\int_0^t X(s)\,\mathrm{d}W(s),\int_t^u X(s)\,\mathrm{d}W(s)\right)=0.$$

Proof. (1) Fix $\omega\in\Omega$. Then $I(t,\omega)$ is a Riemann–Stieltjes integral of a piecewise-constant step function $X(s,\omega)$ with respect to the continuous integrator function $W(s,\omega)$ on the interval $s\in[0,t]$. Such an integral is a continuous function of the upper limit t. Alternatively, continuity follows from (2.8) which is a function of t as a Borel function of $W(t)$, which is (a.s.) continuous and hence $I(t)$ is (a.s.) continuous in t.

(2) It is clear from (2.8) that $I(t)$ is $\mathcal{F}_t$-measurable since it is a function of only Brownian motions up to time t and all $\xi_i, 0\leqslant i\leqslant k$, are assumed $\mathcal{F}_{t_i}$-measurable and hence $\mathcal{F}_t$-measurable since all $\mathcal{F}_{t_i}\subset\mathcal{F}_t$.

(3) Any two simple processes X_1 and X_2 on $[0,T]$ can be expressed as simple processes on a common time partition. In particular, for any $t\in[0,T]$, there is a partition $P_m=\{0=t_0<t_1<\ldots<t_m=t\}$, for some $m\geqslant 1$ with

$$X_j(s)=\sum_{i=0}^{m-1}\xi_i^{(j)}\mathbb{I}_{[t_i,t_{i+1})}(s),\quad j=1,2,$$

$s\in[0,t]$, with each $\xi_i^{(j)}$ as $\mathcal{F}_{t_i}$-measurable. Then, the linear combination $X(s):=c_1X_1(s)+c_2X_2(s)=\sum_{i=0}^{m-1}(c_1\xi_i^{(1)}+c_2\xi_i^{(2)})\mathbb{I}_{[t_i,t_{i+1})}(s)$ is a simple process on $[0,t]$. By definition, the Itô integral of X on $[0,t]$ is

$$\int_0^t (c_1 X_1(s) + c_2 X_2(s))\,\mathrm{d}W(s) = \sum_{i=0}^{m-1} (c_1\xi_i^{(1)} + c_2\xi_i^{(2)})(W(t_{i+1}) - W(t_i))$$
$$= c_1 \sum_{i=0}^{m-1} \xi_i^{(1)}(W(t_{i+1}) - W(t_i)) + c_2 \sum_{i=0}^{m-1} \xi_i^{(2)}(W(t_{i+1}) - W(t_i))$$
$$= c_1 \int_0^t X_1(s)\,\mathrm{d}W(s) + c_2 \int_0^t X_2(s)\,\mathrm{d}W(s).$$

(4) Let us show that $\mathrm{E}[I(t) \mid \mathcal{F}_s] = I(s)$, $0 \leqslant s \leqslant t \leqslant T$. Suppose that $s \in [t_m, t_{m+1}]$ and $t \in [t_k, t_{k+1}]$ for some $0 \leqslant m \leqslant k \leqslant n-1$. We can then represent the expression in (2.8) as

$$I(t) = \underbrace{\sum_{i=0}^{m-1} \xi_i(W(t_{i+1}) - W(t_i)) + \xi_m(W(s) - W(t_m))}_{=I(s) \text{ is } \mathcal{F}_s\text{-measurable}}$$
$$+ \xi_m(W(t_{m+1}) - W(s)) + \sum_{j=m+1}^{k-1} \xi_j(W(t_{j+1}) - W(t_j)) + \xi_k(W(t) - W(t_k)).$$

Taking conditional expectations on both sides of the above equation (with shorthand $\mathrm{E}[\cdot \mid \mathcal{F}_s] \equiv \mathrm{E}_s[\cdot]$) while using the fact that $I(s)$ is $\mathcal{F}_s$–measurable (from property (2)) and pulling out what is known in the second term (i.e., $\mathcal{F}_{t_{m-1}} \subset \mathcal{F}_s$) gives

$$\mathrm{E}_s[I(t)] = I(s) + \underbrace{\xi_m \mathrm{E}_s[W(t_{m+1}) - W(s)]}_{\left(\text{since } \xi_m \text{ is } \mathcal{F}_s\text{-measurable}\right)}$$
$$+ \sum_{j=m+1}^{k-1} \mathrm{E}_s\left[\xi_j(W(t_{j+1}) - W(t_j))\right] + \mathrm{E}_s\left[\xi_k(W(t) - W(t_k))\right].$$

Since $W(t_{m+1}) - W(s)$ is independent of $\mathcal{F}_s$, we have

$$\mathrm{E}_s[W(t_{m+1}) - W(s)] = \mathrm{E}[W(t_{m+1}) - W(s)] = 0.$$

Applying the tower property, $\mathrm{E}_s[\,\cdot\,] = \mathrm{E}_s[\mathrm{E}_{t_j}[\,\cdot\,]]$ for $s < t_j$, where ξ_j is $\mathcal{F}_{t_j}$-measurable and $W(t_{j+1}) - W(t_j)$ is independent of $\mathcal{F}_{t_j}$, gives

$$\begin{aligned}\mathrm{E}_s\left[\xi_j\,(W(t_{j+1}) - W(t_j))\right] &= \mathrm{E}_s\left[\mathrm{E}_{t_j}\left[\xi_j\, W(t_{j+1}) - W(t_j)\right]\right] \\ &= \mathrm{E}_s\left[\xi_j\,\mathrm{E}_{t_j}\left[W(t_{j+1}) - W(t_j)\right]\right] \\ &= \mathrm{E}_s\left[\xi_j\,\mathrm{E}\left[W(t_{j+1}) - W(t_j)\right]\right] = 0\end{aligned}$$

for each j in the above sum. A similar step can be applied to the last expectation $\mathrm{E}_s\left[\xi_k(W(t) - W(t_k))\right] = \mathrm{E}_s\left[\xi_k\,\mathrm{E}_{t_k}[W(t) - W(t_k)]\right] = \mathrm{E}_s\left[\xi_k\,\mathrm{E}[W(t) - W(t_k)]\right] = 0$. Therefore, we have $\mathrm{E}_s[I(t)] = I(s)$. From property (2), the Itô integral is $\mathbb{F}$-adapted. Moreover, since the process is square-integrable it is integrable: $\mathrm{E}[|I(t)|] = \mathrm{E}[\sqrt{I^2(t)}] \leqslant \sqrt{\mathrm{E}[I^2(t)]} < \infty$. [Note: this equivalently follows from $\mathrm{E}[|\xi_i|] \leqslant \sqrt{\mathrm{E}[\xi_i^2]} < \infty$, for all $i \geqslant 0$.] Hence, $\{I(t)\}_{0 \leqslant t \leqslant T}$ is a martingale w.r.t. $\mathbb{F}$.

(5) One way to prove this follows the same steps as in Example 2.2. That is, in each summation term in (2.8), ξ_i is independent of $W(t_{i+1}) - W(t_i)$ and ξ_k is independent of $W(t) - W(t_k)$. Hence, $\mathrm{E}\left[\xi_i(W(t_{i+1}) - W(t_i))\right] = \mathrm{E}\left[\xi_i\right]\mathrm{E}\left[W(t_{i+1}) - W(t_i)\right] = 0$ and $\mathrm{E}\left[\xi_k(W(t) - W(t_k))\right] = \mathrm{E}\left[\xi_k\right]\mathrm{E}\left[W(t) - W(t_k)\right] = 0$ and linearity of expectations gives $\mathrm{E}[I(t)] = 0$.

Alternatively: the zero mean property follows directly from the martingale property since the expected value $\mathrm{E}[I(t)]$ is constant in time and equals $\mathrm{E}[I(0)] = 0$ for all $0 \leqslant t \leqslant T$. The martingale property also gives us the following identity for $0 \leqslant s \leqslant t \leqslant T$:

$$\mathrm{E}\left[\int_s^t X(u)\,\mathrm{d}W(u) \,\middle|\, \mathcal{F}_s\right] = \mathrm{E}[I(t) - I(s) \mid \mathcal{F}_s] = \mathrm{E}[I(t) \mid \mathcal{F}_s] - I(s) = I(s) - I(s) = 0.$$

(6) For ease of presentation, we define $Y_i := W(t_{i+1}) - W(t_i) \stackrel{d}{=} Norm(0, t_{i+1} - t_i), i = 0, \ldots, k-1$ and $Y_k := W(t) - W(t_k) \stackrel{d}{=} Norm(0, t - t_k)$. By (2.8), we have

$$I(t) = \sum_{i=0}^{k} \xi_i\, Y_i.$$

Squaring $I(t)$ and taking its expectation gives

$$\mathrm{E}\big[I^2(t)\big] = \sum_{i=0}^{k} \mathrm{E}\big[\xi_i^2 Y_i^2\big] + 2 \sum\sum_{0 \leqslant i < j \leqslant k} \mathrm{E}\big[\xi_i \xi_j Y_i Y_j\big].$$

The second summation involves expectations of products with $i < j$ ($j \geqslant i+1$), and hence ξ_i, ξ_j and Y_i are $\mathcal{F}_{t_j}$-measurable. Hence, applying the tower property in reverse by conditioning on $\mathcal{F}_{t_j}$ gives

$$\begin{aligned} \mathrm{E}\big[\xi_i \xi_j Y_i Y_j\big] &= \mathrm{E}\big[\mathrm{E}[\xi_i \xi_j Y_i Y_j \mid \mathcal{F}_{t_j}]\big] \quad (\xi_i \xi_j Y_i \text{ is } \mathcal{F}_{t_j}\text{-measurable}) \\ &= \mathrm{E}\big[\xi_i \xi_j Y_i\, \mathrm{E}[Y_j \mid \mathcal{F}_{t_j}]\big] = \mathrm{E}\big[\xi_i \xi_j Y_i\, \mathrm{E}[Y_j]\big] = 0. \end{aligned}$$

The last line is due to the fact that Y_j is independent of $\mathcal{F}_{t_j}$ and $\mathrm{E}[Y_j] = 0$. [Note that the above zero expectation is alternatively derived by simply using the independence of Y_j and $\xi_i \xi_j Y_i$, i.e., $\mathrm{E}[\xi_i \xi_j Y_i Y_j] = \mathrm{E}[\xi_i \xi_j Y_i] \cdot \mathrm{E}[Y_j] = 0$.] Therefore, the above double summation vanishes and we have from the first sum (upon using the independence of each pair of ξ_i and Y_i):

$$\begin{aligned} \mathrm{E}[I^2(t)] = \sum_{i=0}^{k} \mathrm{E}\big[\xi_i^2 Y_i^2\big] = \sum_{i=0}^{k} \mathrm{E}\big[\xi_i^2\big]\mathrm{E}\big[Y_i^2\big] &= \sum_{i=0}^{k-1} \mathrm{E}\big[\xi_i^2\big](t_{i+1} - t_i) + \mathrm{E}\big[\xi_k^2\big](t - t_k) \\ &= \int_0^t \mathrm{E}[X^2(s)]\,\mathrm{d}s, \end{aligned}$$

The last equality follows from the (nonrandom) step function of time s: $\mathrm{E}[X^2(s)] = \sum_{i=0}^{k-1} \mathrm{E}[\xi_i^2]\, \mathbb{I}_{[t_i, t_{i+1})}(s) + \mathrm{E}[\xi_k^2]\, \mathbb{I}_{[t_k, t)}(s)$, $0 \leqslant s < t$.

(7) Employing the tower property, in combination with properties (2), (4), and (5),

$$\operatorname{Cov}(I(t), I(u)) = \mathrm{E}\left[I(t) I(u)\right] = \mathrm{E}\left[I(t)\mathrm{E}[I(u)|\mathcal{F}_t]\right] = \mathrm{E}\left[I^2(t)\right] = \operatorname{Var}(I(t)).$$

Using this identity and the fact that $\int_t^u X(s)\,\mathrm{d}W(s) = I(u) - I(t)$, then

$$\operatorname{Cov}\left(\int_0^t X(s)\,\mathrm{d}W(s), \int_t^u X(s)\,\mathrm{d}W(s)\right) = \operatorname{Cov}\left(I(t), I(u) - I(t)\right)$$
$$= \operatorname{Cov}\left(I(t), I(u)\right) - \operatorname{Var}(I(t)) = 0.$$

□

Remark: The reader should realize that analogous statements to the above theorem hold for any Itô integral of the form in (2.7) with interval $[0, t]$ replaced by $[a, b]$ and any simple process defined accordingly on $[a, b]$. For instance, $I(t) := \int_a^t X(s)\,\mathrm{d}W(s)$, $a \leqslant t \leqslant b$, is a zero mean, $\mathbb{F}$–adapted martingale having continuous paths and variance given by the isometry formula, i.e., $\operatorname{Var}\left(\int_a^t X(s)\,\mathrm{d}W(s)\right) = \int_a^t \mathrm{E}[X^2(s)]\,\mathrm{d}s$. By property (7) we also have $\operatorname{Cov}\left(\int_a^b X(s)\,\mathrm{d}W(s), \int_c^d X(s)\,\mathrm{d}W(s)\right) = 0$ for $a \leqslant b \leqslant c \leqslant d$. By the isometry formula, or property (7), we observe that the variance of an Itô integral is additive over a time interval.

The isometry formula in (2.9) is very useful for computing the variance of an Itô integral. For instance, we can now compute the variance in Example 2.2(c) by simply applying (2.9). In this case we have $\mathrm{E}[\xi_i^2] = \mathrm{E}[W^2(i)] = i$, with $t_i = i$, $t_{i+1} - t_i = 1$. Hence, for any $t \in [0, \infty)$, we have $t \in [\lfloor t \rfloor, \lfloor t \rfloor + 1)$, i.e., $t_k = k \equiv \lfloor t \rfloor$. Using (2.9) gives

$$\operatorname{Var}(I(t)) = \sum_{i=1}^{k-1} i + k(t-k) = \frac{\lfloor t \rfloor}{2}(\lfloor t \rfloor - 1) + \lfloor t \rfloor (t - \lfloor t \rfloor).$$

As required, this is the same expression we derived in the solution to Example 2.2(c).

2.3.2 The Itô Integral for General Processes

Based on Definition 2.2 of the Itô integral for simple processes, we now consider defining the Itô integral of a more general (not necessarily simple) integrand process $X \equiv \{X(t)\}_{0 \leqslant t \leqslant T}$ that is square-integrable and adapted to a filtration $\mathbb{F}$ for BM. The main idea is that the process X can be approximated arbitrarily accurately in a mean-square convergence sense (made precise below) as the limit of a sequence of simple processes $\{X^{(n)}(t)\}_{0 \leqslant t \leqslant T}, n \geqslant 1$. The Itô integral, on a given time interval, for each simple process $X^{(n)}$ is given according to Definition 2.2, i.e. (2.8). The Itô integral of X on the given time interval is then given as a limit of a sequence of the Itô integrals of the simple processes that approximate X.

We shall consider the definition of the Itô integral only for a square-integrable continuous-time process $\{X(t)\}_{0 \leqslant t \leqslant T}$ adapted to $\mathbb{F}$; i.e., assume the square-integrability condition (2.4) holds and that $X(t)$ is $\mathcal{F}_t$-measurable for $0 \leqslant t \leqslant T$. The square-integrability condition (2.4) is also denoted by writing $X \in L^2([0, T], \Omega)$. Throughout we assume that the integrand process X is measurable. That is, for every Borel set $B \in \mathcal{B}(\mathbb{R})$, the sets $\{(t, \omega) : X(t, \omega) \in B\} \in \mathcal{B}(\mathbb{R}_+) \times \mathcal{F}$. By Fubini's Theorem, assuming $\mathrm{E}[X^2(t)] < \infty$ for all $t \in [0, T]$, then this expectation is a Lebesgue-measurable function of time t and we may interchange the expectation integral with the time integral:

$$\mathrm{E}\left[\int_0^T X^2(t)\,\mathrm{d}t\right] = \int_0^T \mathrm{E}\left[X^2(t)\right]\,\mathrm{d}t\,.$$

We already showed that this condition is trivially satisfied by all simple processes in Definition 2.1. The following result establishes the existence of a sequence of $\mathbb{F}$–adapted square-integrable simple processes converging to a continuous-time $\mathbb{F}$–adapted square-integrable process.

Lemma 2.4. *Let $\{X(t)\}_{0 \leqslant t \leqslant T}$ be a process adapted to a filtration $\mathbb{F}$ for Brownian motion and assume the square-integrability condition (2.4) holds. Then, there exists a sequence of $\mathbb{F}$-adapted square-integrable simple processes $\{X^{(n)}(t)\}_{0 \leqslant t \leqslant T}, n \geqslant 1$, such that*

$$\lim_{n \to \infty} \int_0^T \mathrm{E}\big[\big(X(t) - X^{(n)}(t)\big)^2\big]\,\mathrm{d}t = 0. \tag{2.10}$$

Remark: We also say that $X^{(n)} \to X$ in $L^2([0,T],\Omega)$, as $n \to \infty$.

Proof. Consider the class of processes whereby $h(s,t) := \mathrm{E}[X(s)X(t)]$ is a continuous (a.e.) function of $(s,t) \in [0,T]^2$. We now show that the sequence of simple processes defined by $\xi_i = X(t_i)$, i.e., for $0 \leqslant t \leqslant T$,

$$X^{(n)}(t) = \sum_{i=0}^{n-1} X(t_i)\mathbb{I}_{[t_i,t_{i+1})}(t), \tag{2.11}$$

with $0 = t_0 < t_1 < \ldots < t_n = T$, satisfies (2.10) and where each $X^{(n)}$ is $\mathbb{F}$-adapted and square-integrable. [Note: the time points in the sequence of partitions depend upon on n, i.e., $t_i = t_i^{(n)}$, but we simply write t_i. Also, the maximum time subinterval $\delta t^{(n)} \to 0$, as $n \to \infty$.]

By assumption, $X(t)$ is $\mathcal{F}_t$–measurable for any $t \in [0,T]$. Hence, for each n, $X^{(n)}(t) = X(t_i)$, where $t \in [t_i, t_{i+1})$, for some $i \geqslant 0$, is $\mathcal{F}_{t_i}$-measurable. It follows that each $X^{(n)}(t)$ is $\mathcal{F}_t$–measurable since $\mathcal{F}_{t_i} \subset \mathcal{F}_t$. Each $X^{(n)}$ is square-integrable since $\mathrm{E}[\xi_i^2] \equiv \mathrm{E}[X^2(t_i)] < \infty$, for all $i = 0, 1, \ldots, n$. Indeed, $\mathrm{E}[X^2(t_i)] = h(t_i, t_i)$, $t_i \in [0,T]$, is bounded by the assumption of continuity of h on $[0,T]^2$.

The continuity of h implies $\mathrm{E}\big[\big(X(t) - X^{(n)}(t)\big)^2\big] \to 0$, as $n \to \infty$, for every $t \in [0,T)$, as follows. First note that, upon substituting for $X^{(n)}(t)$ and $(X^{(n)}(t))^2$, and writing $h(t,t) = \sum_{i=0}^{n-1} \mathbb{I}_{[t_i,t_{i+1})}(t) h(t,t)$ (note: $\sum_{i=0}^{n-1} \mathbb{I}_{[t_i,t_{i+1})}(t) \equiv \mathbb{I}_{[0,T)}(t)$, which equals one for all $t \in [0,T)$), we can write (for all $t \in [0,T)$):

$$\begin{aligned}
\mathrm{E}\big[\big(X(t) - X^{(n)}(t)\big)^2\big] &= \mathrm{E}[X^2(t)] - 2\mathrm{E}[X(t)X^{(n)}(t)] + \mathrm{E}[(X^{(n)}(t))^2] \\
&= \mathrm{E}[X^2(t)] - 2\sum_{i=0}^{n-1} \mathbb{I}_{[t_i,t_{i+1})}(t)\mathrm{E}[X(t_i)X(t)] + \sum_{i=0}^{n-1} \mathbb{I}_{[t_i,t_{i+1})}(t)\mathrm{E}[X^2(t_i)] \\
&= h(t,t) - 2\sum_{i=0}^{n-1} \mathbb{I}_{[t_i,t_{i+1})}(t) h(t_i,t) + \sum_{i=0}^{n-1} \mathbb{I}_{[t_i,t_{i+1})}(t) h(t_i,t_i) \\
&= \sum_{i=0}^{n-1} \left[(h(t,t) - h(t_i,t)) + (h(t_i,t_i) - h(t_i,t))\right] \mathbb{I}_{[t_i,t_{i+1})}(t).
\end{aligned}$$

Given a value of t and fixed n, the above sum has only one nonzero ith term where $t \in [t_i, t_{i+1}) \equiv [t_i^{(n)}, t_{i+1}^{(n)})$. As $n \to \infty$, every $t_{i+1}^{(n)} - t_i^{(n)} \to 0$ and $t - t_i \equiv t - t_i^{(n)} \to 0$. By continuity of h, every $h(t,t) - h(t_i,t) \to 0$ and $h(t_i,t_i) - h(t_i,t) \to 0$. Hence, $\mathrm{E}\big[\big(X(t) - X^{(n)}(t)\big)^2\big] \to 0$, as $n \to \infty$.

The sequence of nonnegative functions defined by $g_n(t) := \mathrm{E}\big[\big(X(t) - X^{(n)}(t)\big)^2\big]$, $n \geqslant 1$, hence converges pointwise to zero for every $t \in [0,T)$. Moreover, each function $g_n(t)$ is bounded on $[0,T]$ by the continuity of h. [This boundedness also follows by noting that $\mathrm{E}\big[\big(X(t) - X^{(n)}(t)\big)^2\big] \leqslant 2(\mathrm{E}[X^2(t)] + \mathrm{E}[(X^{(n)}(t))^2]) \leqslant 4 \sup_{0 \leqslant t \leqslant T} \mathrm{E}[X^2(t)] < \infty$.] Hence, we can apply the well-known Lebesgue's Dominated Convergence Theorem giving

$\lim_{n\to\infty} \int_0^T g_n(t)\,\mathrm{d}t = 0$, which is the condition in (2.10). The proof of this Lemma can also be extended to the most general case, where we only assume condition (2.4). An alternative sequence to that in (2.11) can be found such that (2.10) holds. For ease of presentation, we shall omit the technical details here. □

[Remark: It should be clear that the above Lemma holds for processes defined on any time interval $[a,b] \subset [0,\infty)$ where the time interval $[0,T]$ in (2.10) is replaced by $[a,b]$.]

It follows that for any two sequences of simple processes, $\{X_1^{(n)}(t), n \geqslant 1\}$ and $\{X_2^{(n)}(t), n \geqslant 1\}$, satisfying (2.10) we have

$$\int_0^T \mathrm{E}\big[\big(X_1^{(n)}(t) - X_2^{(m)}(t)\big)^2\big]\,\mathrm{d}t \to 0\,, \quad \text{as } n,m \to \infty. \tag{2.12}$$

This is readily shown by writing $X_1^{(n)}(t) - X_2^{(m)}(t) = X(t) - X_2^{(m)}(t) - (X(t) - X_1^{(n)}(t))$. Then, using the algebraic inequality $(x-y)^2 \leqslant 2(x^2+y^2)$, where $x \equiv X(t) - X_2^{(m)}(t)$, $y \equiv X(t) - X_1^{(n)}(t)$, gives $(X_1^{(n)}(t) - X_2^{(m)}(t))^2 \leqslant 2[(X(t) - X_2^{(m)}(t))^2 + (X(t) - X_1^{(n)}(t))^2]$, i.e.,

$$\int_0^T \mathrm{E}\big[\big(X_1^{(n)}(t) - X_2^{(m)}(t)\big)^2\big]\mathrm{d}t \leqslant 2\int_0^T \mathrm{E}\big[\big(X(t) - X_2^{(m)}(t)\big)^2\big]\mathrm{d}t + 2\int_0^T \mathrm{E}\big[\big(X(t) - X_1^{(n)}(t)\big)^2\big]\mathrm{d}t.$$

Since both sequences satisfy (2.10), the limits $n \to \infty\,, m \to \infty$ give (2.12). In particular, the above implies

$$\int_0^T \mathrm{E}\big[\big(X^{(n)}(t) - X^{(m)}(t)\big)^2\big]\,\mathrm{d}t \to 0\,, \quad \text{as } n,m \to \infty. \tag{2.13}$$

Based on the above unique convergence of simple processes in $L^2([0,T],\Omega)$, we are now ready to define the stochastic Itô integral for any square-integrable adapted process. The important connection between the convergence in $L^2([0,T],\Omega)$ of a sequence of simple processes with the convergence in $L^2(\Omega)$ of the corresponding sequence of Itô integrals is provided by the isometry property #6 in Theorem 2.3.

For any given sequence of $\mathbb{F}$-adapted square-integrable simple processes $X^{(n)}(t) = \sum_{i=0}^{n-1} \xi_i^{(n)} \mathbb{I}_{[t_i^{(n)}, t_{i+1}^{(n)})}(t)$, $0 \leqslant t \leqslant T$, $n \geqslant 1$, as in Lemma 2.4, we have the corresponding sequence of Itô integrals $\{I^{X^{(n)}}(T) \equiv I^{(n)}(T),\ n \geqslant 1\}$, given by Definition 2.2 where, for each $n \geqslant 1$, we denote the Itô integral of $X^{(n)}$ on $[0,T]$ by $I^{(n)}(T)$. Hence, by employing the linearity property #3 (where $I^{X^{(n)}-X^{(m)}}(T)$ denotes the Itô integral of the difference process $X^{(n)} - X^{(m)}$ on $[0,T]$) and the isometry property #6 in Theorem 2.3, we have

$$\begin{aligned}\mathrm{E}\big[\big(I^{(n)}(T) - I^{(m)}(T)\big)^2\big] &= \mathrm{E}\big[\big(I^{X^{(n)}-X^{(m)}}(T)\big)^2\big] \\ &= \int_0^T \mathrm{E}\big[\big(X^{(n)}(t) - X^{(m)}(t)\big)^2\big]\,\mathrm{d}t \to 0\,, \quad \text{as } n,m \to \infty.\end{aligned} \tag{2.14}$$

This therefore shows that the sequence of Itô integrals $\{I^{(n)}(T),\ n \geqslant 1\}$ is a Cauchy sequence converging in $L^2(\Omega)$.

In summary, for any square-integrable $\mathbb{F}$-adapted process X, Lemma 2.4 assures the existence of a sequence of square-integrable $\mathbb{F}$-adapted simple processes satisfying (2.10), and hence (2.13), i.e., convergence in $L^2([0,T],\Omega)$. In turn, (2.14) assures the unique convergence of the corresponding sequence of Itô integrals in $L^2(\Omega)$. We see this also from (2.12) and the isometry property which assures us that any two sequences of simple processes that converge to X in $L^2([0,T],\Omega)$ must have corresponding sequences of Itô integrals that converge to a common random variable $I(T)$ in $L^2(\Omega)$, i.e., $I^{(n)}(T) \to I(T)$ in $L^2(\Omega)$. This clearly applies to any time subinterval $[0,t]$, $0 \leqslant t \leqslant T$, or $[a,b] \in [0,T]$. We hence have the following definition.

Definition 2.3. **(Itô Integral)** The *Itô integral* on $[0,t]$, denoted by $I(t) \equiv \int_0^t X(s)\,\mathrm{d}W(s)$, of a square-integrable $\mathbb{F}$-adapted process $X = \{X(t)\}_{0\leqslant t\leqslant T}$ is defined as the limit in $L^2(\Omega)$ of a sequence of Itô integrals, $I^{(n)}(t) \equiv \int_0^t X^{(n)}(s)\,\mathrm{d}W(s)$, $n \geqslant 1$, where $\{X^{(n)}, n \geqslant 1\}$ is a sequence of square-integrable $\mathbb{F}$-adapted simple processes satisfying (2.10). We write:

$$I(t) \equiv \int_0^t X(s)\,\mathrm{d}W(s) := \lim_{n\to\infty} \int_0^t X^{(n)}(s)\,\mathrm{d}W(s) \equiv \lim_{n\to\infty} I^{(n)}(t) \tag{2.15}$$

where $I^{(n)}(t) \to I(t)$ in $L^2(\Omega)$, as $n \to \infty$, for every $t \in [0,T]$.

Note that this definition also applies to processes on a time interval $[a,b]$ where $\int_a^b X(t)\,\mathrm{d}W(t) = \lim_{n\to\infty} \int_a^b X^{(n)}(t)\,\mathrm{d}W(t)$ in $L^2(\Omega)$, given $X^{(n)} \to X$ in $L^2([a,b],\Omega)$.

It follows that all the properties we have shown to hold true for the Itô integral of a simple process also hold true for the Itô integral of any generally continuous-time (square-integrable $\mathbb{F}$-adapted) process. We state this important result as a theorem.

Theorem 2.5. *Properties (1)-(7) in Theorem 2.3 hold true for all processes satisfying the conditions of Lemma 2.4 and where the Itô integral is defined in Definition 2.3.*

This theorem can be proven by making use of the properties already established for simple processes and employing convergence arguments. We shall not present the proof of all the properties. However, it is instructive to show how the Itô isometry formula and martingale property are proven. Consider the case where the sequence $\{X^{(n)}, n \geqslant 1\}$ is given by (2.11), i.e., with continuous $h(s,t) \equiv \mathrm{E}[X(s)X(t)]$. Then, by applying (2.9) for each n, where $I^{(n)}(T) \to I(T)$ in $L^2(\Omega)$:

$$\mathrm{E}[I^2(T)] = \lim_{n\to\infty} \mathrm{E}\big[\big(I^{(n)}(T)\big)^2\big] = \lim_{n\to\infty} \sum_{i=0}^{n-1} \mathrm{E}[X^2(t_i)](t_{i+1} - t_i) = \int_0^T \mathrm{E}[X^2(t)]\,\mathrm{d}t.$$

Here, we used the fact that $h(t,t) = \mathrm{E}[X^2(t))]$ is a continuous function of $t \in [0,T]$ with convergent Riemann time integral. This is the isometry formula on an interval $[0,T]$ and it can also be proven for more general square-integrable processes.

To prove the martingale property of the Itô integral process $\{I(t)\}_{0\leqslant t\leqslant T}$, first note that property (2) in Theorem 2.5 and the isometry formula assure us that the process is $\mathbb{F}$-adapted and integrable, i.e., for every $t \in [0,T]$, $I(t)$ is $\mathcal{F}_t$–measurable and $\mathrm{E}[|I(t)|] \leqslant \sqrt{\mathrm{E}[I^2(t)]} = \sqrt{\int_0^t \mathrm{E}[X^2(u)]\,\mathrm{d}u} < \infty$. It remains to show the martingale conditional expectation property: $\mathrm{E}[I(t)|\mathcal{F}_s] = I(s)$ (a.s.), for $0 \leqslant s \leqslant t \leqslant T$. Since $I(s)$ is $\mathcal{F}_s$–measurable, this conditional expectation property is satisfied iff $\mathrm{E}[I(t) - I(s)|\mathcal{F}_s] = 0$ (a.s.). From Theorem 2.3, each simple process $I^{(n)}$ is a martingale w.r.t. $\mathbb{F}$. To show that $\mathrm{E}[I(t) - I(s)|\mathcal{F}_s] = 0$ in $L^2(\Omega)$, i.e., $\mathrm{E}\big[\big(E[(I(t) - I(s)|\mathcal{F}_s]\big)^2\big] = 0$, we first write

$$I(t) - I(s) = \big(I(t) - I^{(n)}(t)\big) + \big(I^{(n)}(t) - I^{(n)}(s)\big) + I^{(n)}(s) - I(s).$$

Taking expectations conditional on $\mathcal{F}_s$, while using the martingale property of each simple process i.e., $\mathrm{E}[I^{(n)}(t) - I^{(n)}(s)|\mathcal{F}_s] = 0$, and the fact that $I^{(n)}(s) - I(s)$ is $\mathcal{F}_s$–measurable,

$$\mathrm{E}[I(t) - I(s)|\mathcal{F}_s] = \mathrm{E}[I(t) - I^{(n)}(t)|\mathcal{F}_s] - \big(I(s) - I^{(n)}(s)\big).$$

By squaring both sides and taking expectations, we have

$$\mathrm{E}\big[\big(E[I(t)-I(s)|\mathcal{F}_s]\big)^2\big] \leqslant 2\,\mathrm{E}\big[\big(\mathrm{E}[I(t)-I^{(n)}(t)|\mathcal{F}_s]\big)^2\big] + 2\,\mathrm{E}[(I(s)-I^{(n)}(s))^2].$$

By applying the conditional Jensen's inequality (see Appendix A for a convex function $\phi(x) := x^2$ and σ-algebra $\mathcal{G} \equiv \mathcal{F}_s$) to the inner part of the nested expectation on the r.h.s., followed by the tower property, gives

$$\mathrm{E}\big[\big(\mathrm{E}[I(t)-I^{(n)}(t)|\mathcal{F}_s]\big)^2\big] \leqslant \mathrm{E}\big[\mathrm{E}[\big(I(t)-I^{(n)}(t)\big)^2|\mathcal{F}_s]\big] = \mathrm{E}\big[\big(I(t)-I^{(n)}(t)\big)^2\big].$$

Hence,

$$\mathrm{E}\big[\big(E[I(t)-I(s)|\mathcal{F}_s]\big)^2\big] \leqslant 2\,\mathrm{E}\big[\big(I(t)-I^{(n)}(t)\big)^2\big] + 2\,\mathrm{E}[(I(s)-I^{(n)}(s))^2],$$

for every $n \geqslant 1$. Taking $n \to \infty$ gives zero since $I^{(n)}(t) \to I(t)$ in $L^2(\Omega)$, for all $t \in [0,T]$, according to (2.15). This implies convergence to zero in probability and the existence of a subsequence with almost sure convergence, i.e., $\mathrm{E}[I(t)|\mathcal{F}_s] = I(s)$ (a.s.).

Consider $\mathbb{F}$-adapted processes $\{X(t)\}_{0\leqslant t\leqslant T}$ whereby $\mathrm{E}[X(s)X(t)]$ is assumed to be a continuous function of $(s,t) \in [0,T]^2$. As seen later, this is an important and quite common class of processes arising in several applications. We showed in the proof of Lemma 2.4 that (2.10) holds with $X^{(n)}(t)$ given by (2.11). The Itô integral of each simple process $X^{(n)}$ on $[0,T]$ is given by

$$I^{(n)}(T) \equiv \int_0^T X^{(n)}(t)\,\mathrm{d}W(t) = \sum_{i=0}^{n-1} X(t_i)(W(t_{i+1}) - W(t_i)), \tag{2.16}$$

where $0 = t_0 < t_1 < \ldots < t_n = T$. By the isometry formula we have $I^{(n)}(T) \to I(T)$ in $L^2(\Omega)$. Hence, according to (2.15), we have just proven the following useful Proposition where the Itô integral is the limit of a Riemann–Stieltjes sum with integrand process evaluated at the left endpoint of every time subinterval.

Proposition 2.6. *Let $\{X(t)\}_{0\leqslant t\leqslant T}$ be adapted to a filtration for Brownian motion (i.e., $\mathbb{F}$-adapted) with $\mathrm{E}[X(s)X(t)]$ assumed to be a continuous (a.e.) function of $(s,t) \in [0,T]^2$. Then, in $L^2(\Omega)$,*

$$I(T) \equiv \int_0^T X(t)\,\mathrm{d}W(t) = \lim_{n\to\infty} \sum_{i=0}^{n-1} X(t_i)(W(t_{i+1}) - W(t_i)), \tag{2.17}$$

where $0 = t_0 < \ldots < t_n = T$ and $\delta t^{(n)} := \max_{0\leqslant i\leqslant n-1}(t_{i+1} - t_i) \to 0$, as $n \to \infty$.

We remark that, if $\mathrm{E}[X(s)X(t)]$ is continuous for $(s,t) \in [a,b]^2$ with $\int_a^b \mathrm{E}[X^2(t)]\,\mathrm{d}t < \infty$, then $\int_a^b X(t)\,\mathrm{d}W(t)$ is given by (2.17) where $a = t_0 < \ldots < t_n = b$. For $[a,b] \subset [0,T]$, we clearly have $\int_a^b X(t)\,\mathrm{d}W(t) - \int_0^b X(t)\,\mathrm{d}W(t) - \int_0^a X(t)\,\mathrm{d}W(t) \equiv I(b) - I(a)$.

A common application of (2.17) is when the integrand is a function of standard BM, i.e., $X(t) = f(W(t))$, such that $\mathrm{E}[f(W(s))f(W(t))]$ is continuous in s and t. A simple example of this is $f : \mathbb{R} \to \mathbb{R}$, where $f(x) := x$, i.e., $X(t) := W(t)$. The Itô integral can be explicitly evaluated in this case, as covered in the following example. Proposition 2.6 now justifies our previous evaluation of this Itô integral in Section 2.2.2.

Example 2.3. Evaluate $\int_0^T W(t)\,\mathrm{d}W(t)$ and compute its variance for any $T > 0$.

Solution. The integrand $X(t) \equiv W(t)$, where $\mathrm{E}[X(s)X(t)] \equiv \mathrm{E}[W(s)W(t)] = s \wedge t \equiv \min\{s,t\}$ is a continous function for all values of s and t, i.e., continuous for $(s,t) \in [0,T]^2$ for any $T > 0$. Hence, we can use (2.17), giving the limit of the partial sum S_n in (2.3) which we already evaluated:

$$\int_0^T W(t)\,\mathrm{d}W(t) = \lim_{n\to\infty} \sum_{i=0}^{n-1} W(t_i)(W(t_{i+1}) - W(t_i)) = \frac{1}{2}(W^2(T) - T)\,.$$

By the isometry formula,

$$\mathrm{Var}\left(\int_0^T W(t)\,\mathrm{d}W(t)\right) = \int_0^T \mathrm{E}[W^2(t)]\,\mathrm{d}t = \int_0^T t\,\mathrm{d}t = \frac{T^2}{2}\,.$$

Note: the variance can also be equivalently computed from the explicit r.h.s. expression for the Itô integral (where $\mathrm{E}[W^2(T)] = T, \mathrm{E}[W^4(T)] = 3T^2$):

$$\mathrm{Var}\left(\int_0^T W(t)\,\mathrm{d}W(t)\right) = \mathrm{Var}\left(\frac{1}{2}W^2(T)\right) = \frac{1}{4}\left[\mathrm{E}[W^4(T)] - (\mathrm{E}[W^2(T)])^2\right] = \frac{T^2}{2}\,.$$

□

From the above example, we also trivially have

$$\int_a^b W(t)\,\mathrm{d}W(t) = \int_0^b W(t)\,\mathrm{d}W(t) - \int_0^a W(t)\,\mathrm{d}W(t) = \frac{1}{2}\left[W^2(b) - W^2(a) - (b-a)\right],$$

$0 \leqslant a \leqslant b < \infty$. Moreover, $I(t) := \int_0^t W(s)\,\mathrm{d}W(s) = \frac{1}{2}(W^2(t) - t)$, $t \geqslant 0$, defines a process that is a (square-integrable) martingale w.r.t. $\mathbb{F}$. From the above proof of the martingale property in Theorem 2.5, we know this must be true for any Itô integral process. This is consistent with our proof of the martingale property of $\{W^2(t) - t\}_{t\geqslant 0}$ (see Example 1.1 in Chapter 1) where we made no use of any Itô integral. We also make note here of the fact that multiplying a martingale process by a constant (e.g., $\frac{1}{2}$) still gives a martingale.

The above example points to a special case where the Itô integral can be evaluated in terms of a known explicit function of standard BM at the endpoints of integration. Generally, and in fact in most situations that we encounter, this is not the case. However, in many cases, we can re-express the Itô integral as a sum of a function of standard BM, evaluated at the endpoints of integration, and a Riemann integral of some function of standard BM. Later in this chapter, we will develop the main tools of stochastic calculus that will allow us to much more easily perform such manipulations. In particular, the various versions of the so-called Itô formula and corresponding Itô "antiderivative" formulas will be powerful tools for such purposes. At this point we do not yet use such tools. We now consider our first example of an Itô integral that cannot be evaluated in terms of only a known explicit function of standard BM, but which necessarily also involves a Riemann integral over BM paths. It is instructive to see how we can make use of similar manipulations of partial sums as in Section 2.2.2 and employ $L^2(\Omega)$-convergence arguments to arrive at the final expression.

Example 2.4. Express $I(T) := \int_0^T W^2(t)\,\mathrm{d}W(t)$ in terms of $W(T)$ and a Riemann integral in $W(t)$ over $t \in [0,T]$.

Solution. The integrand $X(t) \equiv W^2(t)$, where $\mathrm{E}[W^2(s)W^2(t)] = s(2s+t)$ is continuous for $0 \leqslant s \leqslant t \leqslant T$, $T > 0$. Hence, according to (2.17) we have, in $L^2(\Omega)$,

$$I(T) = \lim_{n\to\infty} \sum_{i=0}^{n-1} W^2(t_i)(W(t_{i+1}) - W(t_i))\,,$$

where $\delta t^{(n)} \to 0$, as $n \to \infty$. The key is to re-express each term in this partial sum in terms of Brownian increments squared upon using the algebraic identity: $(a-b)^3 = a^3 - b^3 - 3a^2b + 3ab^2 \implies a^2(b-a) = \frac{1}{3}(b^3 - a^3) - \frac{1}{3}(b-a)^3 - a(b-a)^2$. Setting $a \equiv W(t_i)$, $b \equiv W(t_{i+1})$ gives

$$\begin{aligned}\sum_{i=0}^{n-1} W^2(t_i)(W(t_{i+1}) - W(t_i)) &= \frac{1}{3}\sum_{i=0}^{n-1}(W^3(t_{i+1}) - W^3(t_i)) - \frac{1}{3}\sum_{i=0}^{n-1}(W(t_{i+1}) - W(t_i))^3 \\ &\quad - \sum_{i=0}^{n-1} W(t_i)(W(t_{i+1}) - W(t_i))^2 \\ &\equiv \frac{1}{3}W^3(T) - \frac{1}{3}R_n - S_n\,.\end{aligned}$$

Note that the first sum is telescopic with value $W^3(t_n) - W^3(t_0) = W^3(T) - W^3(0) = W^3(T)$. The second sum converges to zero in $L^2(\Omega)$, i.e., $\lim_{n\to\infty} \mathrm{E}[(R_n - 0)^2] \equiv \lim_{n\to\infty} \mathrm{E}[R_n^2] = 0$. This follows by taking the expected value of R_n^2 while using the independence of BM increments for nonoverlapping time intervals; i.e., $Y_i := W(t_{i+1}) - W(t_i) \stackrel{d}{=} \sqrt{t_{i+1} - t_i}Z_i$, where $\{Z_i\}_{n\geqslant 1}$ are i.i.d. $Norm(0,1)$. In particular, for $i \neq j$: $\mathrm{E}[Y_i^3 Y_j^3] = \mathrm{E}[Y_i^3]\mathrm{E}[Y_j^3] = 0$ since $\mathrm{E}[Z_i^3] = 0$ for all $i \geqslant 1$. Hence,

$$\begin{aligned}\mathrm{E}[R_n^2] = \mathrm{E}\left[\left(\sum_{i=0}^{n-1} Y_i^3\right)^2\right] &= \sum_{i=0}^{n-1}\sum_{j=0}^{n-1} \mathrm{E}\big[Y_i^3 Y_j^3\big] = \sum_{i=0}^{n-1} \mathrm{E}\big[Y_i^6\big] \\ &= 15\sum_{i=0}^{n-1}(t_{i+1} - t_i)^3 \leqslant 15(\delta t^{(n)})^2 \sum_{i=0}^{n-1}(t_{i+1} - t_i) = 15(\delta t^{(n)})^2 T\,.\end{aligned}$$

Here, we used the sixth moment of the standard normal, $\mathrm{E}\big[Z_i^6\big] = 15$. Since $\delta t^{(n)} \to 0$, as $n \to \infty$, we have $\mathrm{E}[R_n^2] \to 0$, as $n \to \infty$.

For the second sum, S_n, we consider $\lim_{n\to\infty} \mathrm{E}[(S_n - A_n)^2] = 0$ where we define the partial sum $A_n := \sum_{i=0}^{n-1} W(t_i)(t_{i+1} - t_i)$. Recall that the sequence $\{A_n\}_{n\geqslant 1}$ converges (a.s. and in $L^2(\Omega)$) to the Riemann integral, i.e., $A_n \to \int_0^T W(t)\,\mathrm{d}t$, in $L^2(\Omega)$, as $n \to \infty$. Hence, $S_n \to \int_0^T W(t)\,\mathrm{d}t$, in $L^2(\Omega)$, if we show that $\mathrm{E}[(S_n - A_n)^2] \to 0$, as $n \to \infty$, as follows. To simplify notation, we denote $\delta_i \equiv t_{i+1} - t_i$ and use the above defined random variables:

$$\begin{aligned}\mathrm{E}[(S_n - A_n)^2] &= \mathrm{E}\left[\left(\sum_{i=0}^{n-1} W(t_i)(Y_i^2 - \delta_i)\right)^2\right] \\ &= \sum_{i=0}^{n-1} \mathrm{E}\big[W^2(t_i)(Y_i^2 - \delta_i)^2\big] + 2\sum_{j=0}^{n-1}\sum_{i=0}^{j-1} \mathrm{E}\big[W(t_i)W(t_j)(Y_i^2 - \delta_i)(Y_j^2 - \delta_j)\big]\,.\end{aligned}$$

Since each $W^2(t_i)$ is independent of Y_i^2, the terms in the first sum are given by (note: $\mathrm{E}[Z_i^4] = 3, \mathrm{E}[Z_i^2] = 1$)

$$\mathrm{E}\big[W^2(t_i)(Y_i^2 - \delta_i)^2\big] = \mathrm{E}\big[W^2(t_i)\big]\mathrm{E}\big[(Y_i^2 - \delta_i)^2\big] = t_i\big(\delta_i^2\mathrm{E}[Z_i^4] - 2\delta_i^2\mathrm{E}[Z_i^2] + \delta_i^2\big) = 2t_i\delta_i^2\,.$$

Summing these terms, noting that $t_i \leqslant T, \delta_i \leqslant \delta t^{(n)}$, gives the inequality:

$$\sum_{i=0}^{n-1} \mathrm{E}\big[W^2(t_i)(Y_i^2 - \delta_i)^2\big] = 2\sum_{i=0}^{n-1} t_i\delta_i^2 \leqslant 2T\delta t^{(n)} \sum_{i=0}^{n-1} \delta_i = 2T^2\delta t^{(n)}\,.$$

Since $\delta t^{(n)} \to 0$, this sum goes to zero as $n \to \infty$. The double sum (with $i < j$) is identically zero. This follows by the tower property in reverse by conditioning on $\mathcal{F}_{t_j}$ in each term, where $W(t_i)W(t_j)(Y_i^2 - \delta_i)$ is $\mathcal{F}_{t_j}$-measurable, or more simply from the fact that $(Y_j^2 - \delta_j)$ is independent of $W(t_i)W(t_j)(Y_i^2 - \delta_i)$. Hence, using $\mathrm{E}[Y_j^2] = \delta_j \mathrm{E}[Z_j^2] = \delta_j$,

$$\mathrm{E}\big[W(t_i)W(t_j)(Y_i^2 - \delta_i)(Y_j^2 - \delta_j)\big] = \mathrm{E}\big[W(t_i)W(t_j)(Y_i^2 - \delta_i)\big]\mathrm{E}\big[Y_j^2 - \delta_j\big] = 0\,.$$

In summary, we have shown that $S_n \to \int_0^T W(t)\,\mathrm{d}t$ and $R_n \to 0$, in $L^2(\Omega)$. Hence, in $L^2(\Omega)$,

$$\int_0^T W^2(t)\,\mathrm{d}W(t) = \frac{1}{3}W^3(T) - \int_0^T W(t)\,\mathrm{d}t\,.$$

□

By the martingale property of any Itô integral process, we automatically know that $I(t) = W^3(t) - 3\int_0^t W(u)\,du$, $t \geqslant 0$, is a (square-integrable) martingale w.r.t. $\mathbb{F}$. Again, this is consistent with the martingale property we showed holds explicitly in Example 2.1 where we made no use of Itô integrals. By the isometry formula (where $\mathrm{E}[W^4(t)] = t^2\mathrm{E}[Z^4] = 3t^2$):

$$\mathrm{Var}\,\big(I(T)\big) \equiv \mathrm{Var}\left(\int_0^T W^2(t)\,\mathrm{d}W(t)\right) = \int_0^T \mathrm{E}[W^4(t)]\,\mathrm{d}t = 3\int_0^T t^2\,\mathrm{d}t = T^3\,.$$

The isometry formula generally gives us the simplest direct way to compute the variance and is indeed the method of choice! As a side note, the same variance can be computed (in a much less direct and cumbersome fashion) by using only the r.h.s. expression (and the identity $\mathrm{Var}(X - Y) = \mathrm{Var}(X) + \mathrm{Var}(Y) - 2\,\mathrm{Cov}(X, Y)$):

$$\begin{aligned}
\mathrm{Var}\,\big(I(T)\big) &= \mathrm{Var}\left(\frac{1}{3}W^3(T)\right) + \mathrm{Var}\left(\int_0^T W(t)\,\mathrm{d}t\right) - 2\,\mathrm{Cov}\left(\frac{1}{3}W^3(T), \int_0^T W(t)\,\mathrm{d}t\right)\\
&= \frac{1}{9}\mathrm{E}[W^6(T)] + \frac{T^3}{3} - \frac{2}{3}\int_0^T \mathrm{Cov}(W^3(T), W(t))\,\mathrm{d}t\\
&= \frac{5}{3}T^3 + \frac{T^3}{3} - T\int_0^T 2t\,\mathrm{d}t = 2T^3 - T^3 = T^3\,.
\end{aligned}$$

Here, we needed to compute $\mathrm{Cov}(W^3(T), W(t)) = 3tT$, $\mathrm{E}[W^6(T)] = T^3\mathrm{E}[Z^6] = 15T^3$ and also make use of the variance $\mathrm{Var}(\int_0^T W(t)\,\mathrm{d}t) = T^3/3$ (which was derived in Section 2.1.2).

It should not be a surprise to the reader at this point that not all properties of Riemann integrals necessarily hold for Itô integrals. For example, suppose that two processes X and Y satisfy $X(t) \leqslant Y(t)$ (a.s.), i.e., $\mathbb{P}(X(t) \leqslant Y(t)) = 1$, for $0 \leqslant t \leqslant T$. Then, it is true that $\int_0^t X(s)\,\mathrm{d}s \leqslant \int_0^t Y(s)\,\mathrm{d}s$ (a.s.) for $0 \leqslant t \leqslant T$. However, this type of integral inequality is *not* generally valid for Itô integrals $I_X(t) = \int_0^t X(s)\,\mathrm{d}W(s)$ and $I_Y(t) = \int_0^t Y(s)\,\mathrm{d}W(s)$. For example, consider the trivial case of constant processes $X(t) \equiv 0$ and $Y(t) \equiv 1$. Clearly, $\mathbb{P}(X(t) \leqslant Y(t)) = \mathbb{P}(0 \leqslant 1) = 1$. However, $I_X(t) \equiv 0$ and $I_Y(t) = \int_0^t 1 \cdot \mathrm{d}W(s) = W(t)$ so that $\mathbb{P}(I_X(t) \leqslant I_Y(t)) = \mathbb{P}(0 \leqslant W(t)) = 1/2 \neq 1$.

Given an integral expression w.r.t. BM, i.e., $\int_a^b X(t)\,\mathrm{d}W(t)$, a basic question is whether or not it is well-defined as an Itô integral. According to Definition 2.3, it is a well-defined Itô integral iff the following two conditions are satisfied:

(i) the integrand is $\mathbb{F}$-adapted, i.e., $X(t)$ is $\mathcal{F}_t$-measurable for all t in the interval of integration $[a, b]$;

(ii) the square-integrability condition is satisfied on the interval of integration, i.e., $\int_a^b \mathrm{E}[X^2(t)]\,\mathrm{d}t < \infty$ (finite variance).

Example 2.5. Show whether or not each expression is a well-defined Itô integral.

(a) $\int_0^T e^{W(t)}\,dW(t)$, $T \geqslant 0$.

(b) $\int_0^T W(t+1)\,dW(t)$, $T \geqslant 0$.

(c) $\int_0^t e^{W^2(s)}\,dW(s)$ for $t \geqslant 0$.

(d) $\int_0^1 (1-t)^{-a}\,dW(t)$ for $a \in \mathbb{R}$.

Solution.

(a) For $t \geqslant 0$, the integrand $X(t) := e^{W(t)}$ is a Borel function of $W(t)$ and hence is $\mathcal{F}_t$-measurable, i.e., condition (i) holds. Now, we check condition (ii) for square-integrability (using the m.g.f. of $Z \sim Norm(0,1)$, where $E[e^{2W(t)}] = E[e^{2\sqrt{t}Z}] = e^{(2\sqrt{t})^2/2} = e^{2t}$):

$$\int_0^T E[X^2(t)]\,dt \equiv \int_0^T E[e^{2W(t)}]\,dt = \int_0^T e^{2t}\,dt = \frac{1}{2}(e^{2T}-1).$$

This is bounded iff T is finite. Therefore, the expression is a well-defined Itô integral iff $T \in [0,\infty)$.

(b) Note, in this case the integrand $X(t) := W(t+1)$ is $\mathcal{F}_{t+1}$-measurable, but not $\mathcal{F}_t$-measurable. Hence, the expression is not defined as an Itô integral for any $T \geqslant 0$.

(c) $X(s) := e^{W^2(s)}$ is a Borel function of $W(s)$ and hence is $\mathcal{F}_s$-measurable for all $s \in [0,t]$, i.e., condition (i) holds. To check condition (ii), we first compute the second moment of the integrand (note: $W^2(s) \stackrel{d}{=} sZ^2$, $Z \sim Norm(0,1)$ with PDF $n(z)$):

$$E\big[X^2(s)\big] \equiv E\big[e^{2W^2(s)}\big] = \int_{-\infty}^{\infty} e^{2sz^2}\,n(z)\,dz = \int_{-\infty}^{\infty} \frac{1}{\sqrt{2\pi}} e^{-\frac{1}{2}(1-4s)z^2}\,dz.$$

This has finite value $\frac{1}{\sqrt{1-4s}}$ iff $s < \frac{1}{4}$. Hence, $\int_0^t E\big[X^2(s)\big]\,ds = \int_0^t \frac{ds}{\sqrt{1-4s}}$ is finite iff $0 \leqslant t \leqslant \frac{1}{4}$, where $s = 1/4$ is an integrable singularity. So the expression is well-defined as an Itô integral only for values of $t \in [0, \frac{1}{4}]$.

(d) Note that the integrand $X(t) = (1-t)^{-a}$ is just an ordinary (nonrandom) function of time t, i.e., $\mathcal{F}_0$-measurable and hence $\mathcal{F}_t$-measurable for all $t \geqslant 0$. For condition (ii), note that $E[X^2(t)] = X^2(t)$. Hence, $\int_0^1 E[X^2(t)]\,dt = \int_0^1 (1-t)^{-2a}\,dt < \infty$ iff $a < \frac{1}{2}$, i.e., the point $t = 1$ is an integrable singularity iff $-2a > -1$. So the Itô integral is defined iff $a < \frac{1}{2}$.

□

Before discussing further properties of the Itô integral, we now present a useful formula for computing the covariance between two Itô integrals as follows by Itô isometry. In particular, let X and Y be two $\mathbb{F}$-adapted processes such that each satisfies the square integrability condition, i.e., assume $\int_0^t E[X^2(s)]\,ds < \infty$ and $\int_0^t E[Y^2(s)]\,ds < \infty$. Then, $I_X(t) := \int_0^t X(s)\,dW(s)$ and $I_Y(t) := \int_0^t Y(s)\,dW(s)$ have zero mean, $E\big[I_X(t)\big] = E\big[I_Y(t)\big] = 0$, and covariance $\mathrm{Cov}\big(I_X(t), I_Y(t)\big) = E[I_X(t)I_Y(t)]$:

$$E[I_X(t)I_Y(t)] \equiv E\left[\int_0^t X(s)\,dW(s) \int_0^t Y(s)\,dW(s)\right] = \int_0^t E[X(s)Y(s)]\,ds. \tag{2.18}$$

We remark that the Itô isometry formula is recovered as a special case of (2.18) when $X \equiv Y$. The formula in (2.18) is readily proven using the algebraic identity $xy = \frac{1}{2}(x+y)^2 - \frac{x^2}{2} - \frac{y^2}{2}$:

$$I_X I_Y = \frac{1}{2}(I_X + I_Y)^2 - \frac{1}{2}I_X^2 - \frac{1}{2}I_Y^2 = \frac{1}{2}I_{X+Y}^2 - \frac{1}{2}I_X^2 - \frac{1}{2}I_Y^2$$

where the linearity property $I_X + I_Y = I_{X+Y}$ is used, i.e., the sum of two Itô integrals is the Itô integral of the sum of their integrands. Now, using linearity of expectations, followed by applying Itô isometry on each expectation, proves the result:

$$\begin{aligned} \mathrm{E}[I_X(t)I_Y(t)] &= \frac{1}{2}\left(\mathrm{E}[I_{X+Y}^2(t)] - \mathrm{E}[I_X^2(t)] - \mathrm{E}[I_Y^2(t)]\right) \\ &= \frac{1}{2}\left(\int_0^t \mathrm{E}[(X(s)+Y(s))^2]\,\mathrm{d}s - \int_0^t \mathrm{E}[X^2(s)]\,\mathrm{d}s - \int_0^t \mathrm{E}[Y^2(s)]\,\mathrm{d}s\right) \\ &= \int_0^t \mathrm{E}\left[\frac{1}{2}(X(s)+Y(s))^2 - \frac{1}{2}X^2(s) - \frac{1}{2}Y^2(s)\right]\mathrm{d}s = \int_0^t \mathrm{E}[X(s)Y(s)]\,\mathrm{d}s. \end{aligned}$$

In the third equation line, we combined the Riemann integrals into one integral where the integrand is cast as a single expectation which is simplified by the above algebraic identity.

The result in (2.18) also leads to a formula for the covariance between two Itô integrals for $0 \leqslant t \leqslant u$:

$$\mathrm{Cov}\left(I_X(t), I_Y(u)\right) \equiv \mathrm{Cov}\left(\int_0^t X(s)\,\mathrm{d}W(s), \int_0^u Y(s)\,\mathrm{d}W(s)\right) = \int_0^t \mathrm{E}[X(s)Y(s)]\,\mathrm{d}s. \tag{2.19}$$

Note that $\mathrm{Cov}(I_X(t), I_Y(u)) = \mathrm{E}[I_X(t)I_Y(u)]$. The formula in (2.19) follows directly from the martingale property of an Itô integral process and by conditioning on $\mathcal{F}_t$, with $I_X(t)$ as $\mathcal{F}_t$-measurable, while using the tower property:

$$\mathrm{E}[I_X(t)I_Y(u)] = \mathrm{E}\big[\mathrm{E}[I_X(t)I_Y(u) \mid \mathcal{F}_t]\big] = \mathrm{E}\big[I_X(t)\,\mathrm{E}[I_Y(u) \mid \mathcal{F}_t]\big] = \mathrm{E}[I_X(t)I_Y(t)]$$

which gives the Riemann integral in (2.19) from (2.18). Note that $\mathrm{Cov}\left(I_X(t), I_Y(u)\right) = \mathrm{Cov}\left(I_X(t), I_Y(u) - I_Y(t)\right) + \mathrm{Cov}\left(I_X(t), I_Y(t)\right)$. According to (2.19), we also have $\mathrm{Cov}\left(I_X(t), I_Y(u)\right) = \mathrm{Cov}\left(I_X(t), I_Y(t)\right)$. Hence, $\mathrm{Cov}\left(I_X(t), I_Y(u) - I_Y(t)\right) = 0$, i.e.,

$$\mathrm{Cov}\left(\int_0^t X(s)\,\mathrm{d}W(s), \int_t^u Y(s)\,\mathrm{d}W(s)\right) = 0$$

for $0 \leqslant t \leqslant u$. In case $X \equiv Y$, this formula recovers property (7) stated in Theorem 2.3.

2.4 Itô Processes and Their Properties

2.4.1 Gaussian Processes Generated by Itô Integrals

The Itô integral of a nonrandom (ordinary) differentiable function f can be considered as a Riemann–Stieltjes integral with any path of Brownian motion acting as the integrator function w.r.t. time. Thus it can be reduced to a Riemann integral by using the integration by parts formula:

$$I(t) = \int_0^t f(s)\,\mathrm{d}W(s) = f(t)W(t) - f(0)W(0) - \int_0^t f'(s)W(s)\,\mathrm{d}s.$$

We showed that the Riemann integral of Brownian motion in time is a Gaussian process. In the same way, by considering the limit of a Riemann sum, we can prove that the integral $\int_0^t f'(s)W(s)\,\mathrm{d}s$ is also a Gaussian process as a function of time t. Thus, $I(t)$ above is a Gaussian process as well. The following theorem states that $I(t)$ is a Gaussian process in the general case where f is square-integrable and not necessarily differentiable.

Theorem 2.7. *Let f be a nonrandom real function such that $\int_0^T f^2(t)\,\mathrm{d}t < \infty$ for some $T > 0$. Then, the Itô integral process $I(t) = \int_0^t f(u)\,\mathrm{d}W(u)$, $0 \leqslant t \leqslant T$, is a Gaussian process with mean zero and covariance function given by*

$$c_I(t,s) := \mathrm{Cov}(I(t), I(s)) = \int_0^{s\wedge t} f^2(u)\,\mathrm{d}u, \quad 0 \leqslant s,t \leqslant T. \tag{2.20}$$

Proof. First, let us show that $I(T) \stackrel{d}{=} Norm(0,\sigma^2)$, where $\sigma^2 := \int_0^T f^2(t)\,\mathrm{d}t$. Since the integrand $X(t) \equiv f(t)$ is nonrandom, the square-integrability condition holds by assumption: $\int_0^T \mathrm{E}[f^2(t)]\,\mathrm{d}t \equiv \int_0^T f^2(t)\,\mathrm{d}t < \infty$. Moreover, $\mathrm{E}[X(s)X(t)] \equiv f(s)f(t)$ is continuous (a.e.) in s and t, $0 \leqslant s,t \leqslant T$. Hence, using (2.17) we have $L^2(\Omega)$-convergence:

$$I(T) \equiv \int_0^T f(t)\,\mathrm{d}W(t) = \lim_{n\to\infty} I^{(n)} \equiv \lim_{n\to\infty} \sum_{i=0}^{n-1} f(t_i)(W(t_{i+1}) - W(t_i)),$$

where $0 = t_0 < \ldots < t_n = T$ and $\delta t^{(n)} \to 0$, as $n \to \infty$. Hence, we have a sequence of normal random variables $\{I^{(n)}\}_{n\geqslant 1}$ where $I^{(n)} \stackrel{d}{=} Norm(0,\sigma_n^2)$, $\sigma_n^2 := \sum_{i=0}^{n-1} f^2(t_i)(t_{i+1}-t_i)$, converging to $I(T)$ in $L^2(\Omega)$. It follows that $I(T) \stackrel{d}{=} Norm(0,\sigma^2)$, where

$$\sigma^2 = \lim_{n\to\infty} \sigma_n^2 = \lim_{n\to\infty} \sum_{i=0}^{n-1} f^2(t_i)(t_{i+1} - t_i) = \int_0^T f^2(t)\,\mathrm{d}t.$$

It follows that, for any $t \in [0,T]$, $I(t) \stackrel{d}{=} Norm\big(0, \int_0^t f^2(u)\,\mathrm{d}u\big)$. The covariance formula in (2.20) follows simply from property (7) in Theorem 2.3 where $\mathrm{E}[X^2(u)] \equiv f^2(u)$, i.e., $\mathrm{Cov}(I(t), I(s)) = \mathrm{Var}(I(s\wedge t)) = \int_0^{s\wedge t} f^2(u)\,\mathrm{d}u$, $s \wedge t \equiv \min\{s,t\}$. □

It should be remarked that the above Gaussian property of $I(t)$ can be stated as

$$\int_0^t f(u)\,\mathrm{d}W(u) \stackrel{d}{=} Norm\left(0, \int_0^t f^2(u)\,\mathrm{d}u\right) \stackrel{d}{=} W(g(t))$$

where $g(t) := \int_0^t f^2(s)\,\mathrm{d}s$. Note that $g(t)$ is a nondecreasing function of time t. That is, the Itô integral of the ordinary function f on $[0,t]$ has the same distribution as standard Brownian motion at a time given by $g(t)$. This is a simple type of *time-changed Brownian motion* where in this case the time change, $t \to g(t)$, is an ordinary function of t.

Example 2.6. The process $X(t) = \int_0^t s\,\mathrm{d}W(s)$ is a Gaussian process with mean zero and variance $\mathrm{Var}(X(t)) = \int_0^t s^2\,\mathrm{d}s = t^3/3$, i.e., $X(t) \stackrel{d}{=} W(g(t))$ where $g(t) = t^3/3$.

2.4.2 Itô Processes

The sum of an Itô integral of a stochastic process and an ordinary (Riemann) integral generates another stochastic process called an Itô process.

Definition 2.4. Let $\{\mu(t)\}_{t\geqslant 0}$ and $\{\sigma(t)\}_{t\geqslant 0}$ be adapted to a filtration $\{\mathcal{F}_t\}_{t\geqslant 0}$ for standard Brownian motion and satisfying

$$\int_0^T \mathrm{E}[|\mu(t)|]\,\mathrm{d}t < \infty \text{ and } \int_0^T \mathrm{E}[\sigma^2(t)]\,\mathrm{d}t < \infty.$$

Then, the process

$$X(t) = X_0 + \int_0^t \mu(s)\,\mathrm{d}s + \int_0^t \sigma(s)\,\mathrm{d}W(s) \tag{2.21}$$

is well-defined for $0 \leqslant t \leqslant T$. It is called an *Itô process*. The processes $\{\mu(t)\}_{t\geqslant 0}$ and $\{\sigma(t)\}_{t\geqslant 0}$ are, respectively, called the *drift coefficient* process and the *diffusion or volatility coefficient* process.

As discussed further below, the Itô process X can also be described by its so-called *stochastic differential equation (SDE)* which is obtained by "formally differentiating" (2.21) w.r.t. the time parameter t:

$$\mathrm{d}X(t) = \mu(t)\,\mathrm{d}t + \sigma(t)\,\mathrm{d}W(t). \tag{2.22}$$

We note that this SDE, along with the initial condition $X(0) = X_0$, is a shorthand way of writing the stochastic integral equation in (2.21). We interpret (2.22) through (2.21), where the latter has proper mathematical meaning as a sum of a Riemann integral and an Itô stochastic integral. That is, the Itô process $X \equiv \{X(t)\}_{t\geqslant 0}$ can be viewed as a solution to the SDE in (2.22) with the initial condition $X(0) = X_0$. The differential representation in (2.22) only has rigorous mathematical meaning by way of the respective integral representations in (2.21).

Some examples of Itô processes are as follows.

(a) Let $X(0) = x_0$, $\mu(t) \equiv \mu$ and $\sigma(t) \equiv \sigma$ be constants. Then, we obtain a drifted BM (i.e., BM with constant drift μ):

$$X(t) = x_0 + \int_0^t \mu\,\mathrm{d}s + \int_0^t \sigma\,\mathrm{d}W(s) = x_0 + \mu t + \sigma W(t).$$

(b) Let $\mu(t,x)$ and $\sigma(t,x)$ be given (ordinary) functions of both time and spatial variables t, x. The Itô process implicitly defined by the stochastic integral equation

$$X(t) = X_0 + \int_0^t \mu(s, X(s))\,\mathrm{d}s + \int_0^t \sigma(s, X(s))\,\mathrm{d}W(s)$$

is called a *diffusion process*.

(c) Let $\mu(t)$ and $\sigma(t)$ be nonrandom (ordinary) functions of time t. Then,

$$X(t) = X_0 + \int_0^t \mu(u)\,\mathrm{d}u + \int_0^t \sigma(u)\,\mathrm{d}W(u)\,, \quad t \geqslant 0,$$

with constant $X_0 \in \mathbb{R}$, is a Gaussian process with mean function $m_X(t) := \mathrm{E}[X(t)]$ and covariance $c_X(t,s) := \mathrm{Cov}(X(s), X(t))$ given by

$$m_X(t) = X_0 + \int_0^t \mu(u)\,\mathrm{d}u \quad \text{and} \quad c_X(t,s) = \int_0^{t\wedge s} \sigma^2(u)\,\mathrm{d}u.$$

The mean follows by taking expectations where the drift portion, $X_0 + \int_0^t \mu(u)\,\mathrm{d}u$, is nonrandom and $\mathrm{E}\left[\int_0^t \sigma(u)\,\mathrm{d}W(u)\right] = 0$. The covariance function arises simply from (2.20) where $\mathrm{Cov}(X(s), X(t)) = \mathrm{Cov}\left(\int_0^s \sigma(u)\,\mathrm{d}W(u), \int_0^t \sigma(u)\,\mathrm{d}W(u)\right)$.

The Itô process defined in (2.21) is given by a sum of a Riemann time-integral of the drift coefficient and an Itô integral of the volatility coefficient. Both integrals being considered as functions of the upper limit t have (a.s.) continuous sample paths. Therefore, an Itô process of the general form in (2.21) has continuous sample paths as well.

So far, we have defined the Itô process as a stochastic integral w.r.t. Brownian motion. More generally, we can extend this to also define a stochastic integral w.r.t. an Itô process. Let the process $\{Y(t)\}_{t\geqslant 0}$ be adapted to a filtration for BM. We define the stochastic integral of Y *w.r.t. the Itô process* X, defined in (2.21), as follows:

$$\int_0^t Y(s)\,\mathrm{d}X(s) := \int_0^t Y(s)\mu(s)\,\mathrm{d}s + \int_0^t Y(s)\sigma(s)\,\mathrm{d}W(s), \quad t \geqslant 0.$$

Note that this is like substituting the stochastic differential $\mathrm{d}X(s) = \mu(s)\,\mathrm{d}s + \sigma(s)\,\mathrm{d}W(s)$ (given by (2.22) with time parameter t replaced by s) into the left-hand integral and writing it as a sum of a Riemann integral and an Itô integral w.r.t. Brownian motion W. Note that in case the process is standard Brownian motion, i.e., $X(t) \equiv W(t)$ with $\mu \equiv 0, \sigma \equiv 1$ in (2.21), we simply recover $\int_0^t Y(s)\,\mathrm{d}W(s)$, i.e., the Itô integral of Y w.r.t. standard Brownian motion W.

2.4.3 Quadratic and Co-Variation of Itô Processes

An important characteristic of a stochastic process is the *quadratic variation* that measures the accumulated variability of the process along its path. The quadratic variation is a path-dependent quantity. Recall that for Brownian motion we derived its quadratic variation on a time interval $[0, t]$ as $[W, W](t) = t$. So Brownian motion *accumulates quadratic variation at rate one per unit time.* This gives us a simple stochastic differential "rule" for the square of an infinitesimal Brownian increment:

$$\mathrm{d}[W, W](t) \equiv \mathrm{d}W(t)\,\mathrm{d}W(t) \equiv (\mathrm{d}W(t))^2 = \mathrm{d}t.$$

A way of thinking about this result is to say that a Brownian increment is of order $\mathcal{O}((\mathrm{d}t)^{1/2})$ as $\mathrm{d}t \to 0$. We essentially already used this fact in showing the non-differentiability of Brownian paths. We also saw that, formally, the quadratic variation of a continuously differentiable function f is zero. This fact is also realized by noting that $\mathrm{d}[f, f](t) = (\mathrm{d}f(t))^2 = (f'(t))^2(\mathrm{d}t)^2 = \mathcal{O}((\mathrm{d}t)^2)$ is negligible as $\mathrm{d}t \to 0$, i.e., we write $(\mathrm{d}t)^2 \equiv 0$.

For an $\mathbb{F}$-adapted square integrable process $\{X(t)\}_{0\leqslant t\leqslant T}$, the quadratic variation of the process $I(t) := \int_0^t X(s)\,\mathrm{d}W(s)$ is given by

$$[I, I](t) = \int_0^t X^2(s)\,\mathrm{d}s, \quad 0 \leqslant t \leqslant T. \tag{2.23}$$

It is instructive to give a condensed proof of this, as follows. Consider the sequence $\{I^{(n)}(t)\}_{n\geqslant 1}$ defined by $I^{(n)}(t) := \sum_{i=0}^{n-1} X(t_i)(W(t_{i+1}) - W(t_i))$, as in (2.17), where we now set $0 = t_0 < \ldots < t_n = t$. We seek to compute the quadratic variation of $[I^{(n)}, I^{(n)}](t)$ over the time interval $[0, t]$ by considering the sum of quadratic variations over each adjacent time subinterval $[t_i, t_{i+1}]$: $[I^{(n)}, I^{(n)}](t) = \sum_{i=0}^{n-1} \left\{[I^{(n)}, I^{(n)}](t_{i+1}) - [I^{(n)}, I^{(n)}](t_i)\right\}$. For each fixed subinterval $[t_i, t_{i+1}]$, we introduce another time partition $t_i = \tilde{t}_0 < \tilde{t}_1 < \ldots < \tilde{t}_m = t_{i+1}$ such that $\max_{0\leqslant j\leqslant m-1}(\tilde{t}_{j+1} - \tilde{t}_j) \to 0$, as $m \to \infty$. Hence, using the fact that the difference on each subinterval $[t_i, t_{i+1}]$ is $I^{(n)}(\tilde{t}_{j+1}) - I^{(n)}(\tilde{t}_j) = X(t_i)(W(\tilde{t}_{j+1}) - W(\tilde{t}_j))$, with $X(t_i)$ having no dependence on j, the quadratic variation over each subinterval $[t_i, t_{i+1}]$

is computed as:

$$
\begin{aligned}
[I^{(n)}, I^{(n)}](t_{i+1}) - [I^{(n)}, I^{(n)}](t_i) &= \lim_{m\to\infty} \sum_{j=0}^{m-1} \left(I^{(n)}(\tilde{t}_{j+1}) - I^{(n)}(\tilde{t}_j)\right)^2 \\
&= X^2(t_i) \lim_{m\to\infty} \sum_{j=0}^{m-1} \left(W(\tilde{t}_{j+1}) - W(\tilde{t}_j)\right)^2 \\
&= X^2(t_i)\big([W,W](t_{i+1}) - [W,W](t_i)\big) = X^2(t_i)(t_{i+1} - t_i)\,.
\end{aligned}
$$

In the last equality, we simply used the quadratic variation of BM on $[t_i, t_{i+1}]$. Hence, adding up the quadratic variations over each time interval gives

$$
[I^{(n)}, I^{(n)}](t) = \sum_{i=0}^{n-1} X^2(t_i)(t_{i+1} - t_i).
$$

Taking the limit $n \to \infty$, $\delta t^{(n)} \to 0$, of this partial sum gives the Riemann integral in (2.23) where $I^{(n)}(t) \to I(t)$, i.e., $[I^{(n)}, I^{(n)}](t) \to [I, I](t)$.

So, according to (2.23), the Itô integral $I(t)$ *accumulates quadratic variation at the (generally random) rate of the square of the integrand,* $X^2(t)$*, per unit time.* That is, the differential form of (2.23) gives us the "rule" for the square of the stochastic differential of an Itô integral:

$$
\mathrm{d}[I, I](t) \equiv \mathrm{d}I(t)\,\mathrm{d}I(t) \equiv (\,\mathrm{d}I(t))^2 = X^2(t)\,\mathrm{d}t. \tag{2.24}
$$

Similarly, we can define the *quadratic covariation* of two processes, X and Y, on a time interval $[0, t]$:

$$
[X, Y](t) = \lim_{n\to\infty} \sum_{i=1}^{n} (X(t_i) - X(t_{i-1}))(Y(t_i) - Y(t_{i-1})), \tag{2.25}
$$

where $0 = t_0 < \ldots < t_n = t$, $\delta t^{(n)} \to 0$, as $n \to \infty$. Clearly, the quadratic covariation is a bilinear functional. Let us consider two main cases.

1. Let $X(t)$ be a continuously differentiable ($C^1(\mathbb{R})$) function that satisfies $\mathrm{d}X(t) = \mu_X(t)\,\mathrm{d}t$ and let $Y(t)$ be an Itô process (and hence a.s. continuous). Then, their covariation is identically zero $[X, Y](t) = 0$ for $t \geqslant 0$. In differential form, this fact reads as $\mathrm{d}X(t)\,\mathrm{d}Y(t) = 0$. Since Brownian motion is itself an Itô process and the function $X(t) = t$ belongs to $C^1(\mathbb{R})$, we have $[t, W](t) = 0$. This last fact is recorded in stochastic differential form as the multiplication "rule":

$$
\mathrm{d}t\,\mathrm{d}W(t) \equiv \mathrm{d}W(t)\,\mathrm{d}t = 0\,,
$$

i.e., a differential of time multiplying an infinitesimal Brownian increment is identically zero.

Proof. For $t \geqslant 0$,

$$
\begin{aligned}
\big|[X, Y](t)\big| &\leqslant \lim_{n\to\infty} \left| \sum_{i=1}^{n} (X(t_i) - X(t_{i-1}))(Y(t_i) - Y(t_{i-1})) \right| \\
&\leqslant \underbrace{\lim_{n\to\infty} \max_{1\leqslant i\leqslant n} \big|Y(t_i) - Y(t_{i-1})\big|}_{=0 \text{ a.s.}} \cdot \underbrace{\lim_{n\to\infty} \sum_{i=1}^{n} \big|X(t_i) - X(t_{i-1})\big|}_{=V_X^{(1)}(t) < \infty} = 0
\end{aligned}
$$

Here, we applied the Heine–Cantor theorem, which states that a continuous function (in this case Y is a.s. continuous) on a finite interval is uniformly continuous and the fact that sample paths of X have finite first variation. $\square$

2. The covariation of two Itô processes X and Y defined by

$$X(t) = X(0) + \int_0^t \mu_X(s)\,\mathrm{d}s + \int_0^t \sigma_X(s)\,\mathrm{d}W(s)$$

and

$$Y(t) = Y(0) + \int_0^t \mu_Y(s)\,\mathrm{d}s + \int_0^t \sigma_Y(s)\,\mathrm{d}W(s)$$

is given by

$$[X,Y](t) = \int_0^t \sigma_X(s)\sigma_Y(s)\,\mathrm{d}s. \tag{2.26}$$

Proof. Let us denote the respective Riemann and Itô integrals by $R_X(t) := \int_0^t \mu_X(s)\,\mathrm{d}s$, $R_Y(t) := \int_0^t \sigma_Y(s)$, $I_X(t) := \int_0^t \sigma_X(s)\,\mathrm{d}W(s)$, $I_Y(t) := \int_0^t \sigma_Y(s)\,\mathrm{d}W(s)$, where

$$X(t) = X(0) + R_X(t) + I_X(t), \quad Y(t) = Y(0) + R_Y(t) + I_Y(t).$$

Since $R_X(t)$ and $R_Y(t)$ are continuously differentiable and $I_X(t)$ and $I_Y(t)$ are continuous (a.s.) in t, we have $[I_X, R_Y](t) = [R_X, I_Y](t) = [R_X, R_Y](t) = 0$. Hence, using the bilinearity property (where $X(0)$ and $Y(0)$ have zero contribution to the covariation):

$$\begin{aligned}[X,Y](t) &= [R_X + I_X, R_Y + I_Y](t) \\ &= \underbrace{[R_X, R_Y](t) + [R_X, I_Y](t) + [I_X, R_Y](t)}_{=0} + [I_X, I_Y](t) = [I_X, I_Y](t).\end{aligned}$$

That is, the covariation of two Itô processes is given by the covariation of their Itô integral component processes. The covariation of the two Itô integral component processes is given by

$$[I_X, I_Y](t) = \int_0^t \sigma_X(s)\sigma_Y(s)\,\mathrm{d}s \tag{2.27}$$

which completes the proof of (2.26). Note: The proof of (2.27) follows similar steps as in the above proof of (2.23) so we do not present it here. $\square$

Note: $[X,X](t) = [R_X + I_X, R_X + I_X](t) = [I_X, I_X](t)$ and $[Y,Y](t) = [R_Y + I_Y, R_Y + I_Y](t) = [I_Y, I_Y](t)$. Hence, applying (2.23) to I_X and I_Y, the Itô processes X and Y have the respective quadratic variation

$$[X,X](t) = \int_0^t \sigma_X^2(s)\,\mathrm{d}s \quad \text{and} \quad [Y,Y](t) = \int_0^t \sigma_Y^2(s)\,\mathrm{d}s. \tag{2.28}$$

Differentiating the Riemann integrals in (2.26) and (2.28), gives the differential form of the covariation and quadratic variation of the Itô processes:

$$\mathrm{d}[X,Y](t) \equiv \mathrm{d}X(t)\,\mathrm{d}Y(t) = \sigma_X(t)\sigma_Y(t)\,\mathrm{d}t \tag{2.29}$$

and

$$\mathrm{d}[X,X](t) \equiv \big(\mathrm{d}X(t)\big)^2 = \sigma_X^2(t)\,\mathrm{d}t, \quad \mathrm{d}[Y,Y](t) \equiv \big(\mathrm{d}Y(t)\big)^2 = \sigma_Y^2(t)\,\mathrm{d}t. \tag{2.30}$$

A simple way to realize (2.26)–(2.30) is to make recourse to the above "stochastic differential multiplication rules." We begin by writing the above Itô processes X and Y in their respective stochastic differential forms:

$$\mathrm{d}X(t) = \mu_X(t)\,\mathrm{d}t + \sigma_X(t)\,\mathrm{d}W(t) \quad \text{and} \quad \mathrm{d}Y(t) = \mu_Y(t)\,\mathrm{d}t + \sigma_Y(t)\,\mathrm{d}W(t).$$

By multiplying out all terms in the two differentials and collecting terms in $(\mathrm{d}t)^2$, $\mathrm{d}t\,\mathrm{d}W(t)$ and $(\mathrm{d}W(t))^2$, we have

$$\begin{aligned}\mathrm{d}X(t)\,\mathrm{d}Y(t) &= \{\mu_X(t)\,\mathrm{d}t + \sigma_X(t)\,\mathrm{d}W(t)\}\{\mu_Y(t)\,\mathrm{d}t + \sigma_Y(t)\,\mathrm{d}W(t)\} \\ &= \mu_X(t)\mu_Y(t)(\mathrm{d}t)^2 + (\mu_X(t)\sigma_Y(t) + \mu_Y(t)\sigma_X(t))\,\mathrm{d}t\,\mathrm{d}W(t) \\ &\quad + \sigma_X(t)\sigma_Y(t)(\mathrm{d}W(t))^2.\end{aligned}$$

Now, using the rules: $(\mathrm{d}t)^2 \equiv 0$, $\mathrm{d}t\,\mathrm{d}W(t) \equiv 0$ and $(\mathrm{d}W(t))^2 = dt$ gives the differential form in (2.29) and whose integral form is (2.26). Similarly, by squaring the stochastic differentials and applying the same multiplication rules gives

$$\begin{aligned}\big(\mathrm{d}X(t)\big)^2 &= \big(\mu_X(t)\,\mathrm{d}t + \sigma_X(t)\,\mathrm{d}W(t)\big)^2 = \sigma_X^2(t)\,\mathrm{d}t, \\ \big(\mathrm{d}Y(t)\big)^2 &= \big(\mu_Y(t)\,\mathrm{d}t + \sigma_Y(t)\,\mathrm{d}W(t)\big)^2 = \sigma_Y^2(t)\,\mathrm{d}t,\end{aligned}$$

which are the relations in (2.30). The integral form of these expressions recover (2.28).

An important application of quadratic covariation is the integration by parts formula. This is derived by rewriting the partial sum in (2.25) as follows:

$$\begin{aligned}\sum_{i=1}^{n}(X(t_i) - X(t_{i-1}))(Y(t_i) - Y(t_{i-1})) &= \underbrace{\sum_{i=1}^{n}\big(X(t_i)Y(t_i) - X(t_{i-1})Y(t_{i-1})\big)}_{=X(t)Y(t)-X(0)Y(0)} \\ &- \underbrace{\sum_{i=1}^{n}X(t_{i-1})\big(Y(t_i) - Y(t_{i-1})\big)}_{\to\int_0^t X(s)\,\mathrm{d}Y(s),\text{ as } n\to\infty} - \underbrace{\sum_{i=1}^{n}Y(t_{i-1})\big(X(t_i) - X(t_{i-1})\big)}_{\to\int_0^t Y(s)\,\mathrm{d}X(s),\text{ as } n\to\infty}.\end{aligned}$$

Thus, in the limit $n \to \infty$, this gives the quadratic covariation of two Itô processes as

$$[X,Y](t) = X(t)Y(t) - X(0)Y(0) - \int_0^t X(s)\,\mathrm{d}Y(s) - \int_0^t Y(s)\,\mathrm{d}X(s).$$

Alternatively, we write

$$X(t)Y(t) - X(0)Y(0) = \int_0^t X(s)\,\mathrm{d}Y(s) + \int_0^t Y(s)\,\mathrm{d}X(s) + [X,Y](t). \tag{2.31}$$

The r.h.s. involves an Itô integral of process X w.r.t. Y and an Itô integral of process Y w.r.t. X. In differential form, this gives us the important *Itô product rule*:

$$\mathrm{d}(X(t)Y(t)) = X(t)\,\mathrm{d}Y(t) + Y(t)\,\mathrm{d}X(t) + \mathrm{d}X(t)\,\mathrm{d}Y(t). \tag{2.32}$$

The reader should observe that the stochastic differential of a product of two processes *does not* obey the same differential product rule as in ordinary calculus. The extra term $\mathrm{d}X(t)\,\mathrm{d}Y(t) \equiv \mathrm{d}[X,Y](t)$ is the product of two stochastic differentials, which is generally nonzero. In particular, if both processes X and Y are driven by a Brownian increment $\mathrm{d}W(t)$ then their paths are nondifferentiable and hence the quadratic covariation $[X,Y](t)$ is nonzero. Later we shall see that the above product rule also follows as a special case of the Itô formula derived for smooth functions of two processes.

2.5 Itô's Formula for Functions of BM and Itô Processes

2.5.1 Itô's Formula for Functions of BM

The Itô formula is a stochastic chain rule that allows us to find stochastic differentials of functions of Brownian motion as well as functions of an Itô process. The ordinary chain rule written for two differentiable functions f and g is as follows:

- $\frac{\mathrm{d}}{\mathrm{d}t} f(g(t)) = f'(g(t))\, g'(t)$ (derivative form);
- $\mathrm{d}f(g(t)) = f'(g(t))\, \mathrm{d}g(t)$ (differential form);
- $f(g(t)) - f(g(0)) = \int_0^t f'(g(s))\, \mathrm{d}g(s)$ (integral form).

However, we cannot apply this rule to $f(W(t))$ since Brownian motion W has (a.s.) non-differentiable sample paths! Assume that $f : \mathbb{R} \to \mathbb{R}$ has continuous derivatives of first, second, and higher orders. Consider the Taylor series expansion for a smooth function f about the value $W(t)$:

$$f(W(t+\delta t)) - f(W(t)) = f'(W(t)) \underbrace{\left(W(t+\delta t) - W(t)\right)}_{\text{of order } (\delta t)^{\frac{1}{2}}} + \frac{1}{2} f''(W(t)) \underbrace{\left(W(t+\delta t) - W(t)\right)^2}_{\text{of order } \delta t}$$
$$+ \frac{1}{6} f'''(W(t)) \underbrace{\left(W(t+\delta t) - W(t)\right)^3}_{\text{of order } (\delta t)^{\frac{3}{2}}} + \cdots,$$

where δt is a small time increment. A heuristic argument that leads us to the simplest version of the Itô formula goes as follows. In the infinitesimal limit, we take $\delta t \to \mathrm{d}t$ and $W(t+\delta t) - W(t) \to \mathrm{d}W(t)$, and we neglect all terms of order $(\delta t)^{3/2}$ and smaller (of higher power than $3/2$ in δt) to obtain

$$\mathrm{d}f(W(t)) = f'(W(t))\, \mathrm{d}W(t) + \frac{1}{2} f''(W(t)) (\mathrm{d}W(t))^2.$$

By applying the rule $(\mathrm{d}W(t))^2 = \mathrm{d}t$, we obtain the *Itô formula* for $f(W(t))$, which can be stated in the respective differential and integral forms:

$$\mathrm{d}f(W(t)) = \frac{1}{2} f''(W(t))\, \mathrm{d}t + f'(W(t))\, \mathrm{d}W(t), \tag{2.33}$$

$$\int_0^t \mathrm{d}f(W(s)) := f(W(t)) - f(W(0)) = \frac{1}{2} \int_0^t f''(W(s))\, \mathrm{d}s + \int_0^t f'(W(s))\, \mathrm{d}W(s). \tag{2.34}$$

This formula holds for any twice continuously differentiable function $f \in C^2(\mathbb{R})$. Throughout we assume the Itô integral in (2.34) is well-defined for all $t \in [0, T]$ and some $T > 0$, i.e., the square-integrability condition in (2.4), with integrand $X(t) \equiv f'(W(t))$, is assumed to hold, i.e., $\int_0^T \mathrm{E}\left[\left(f'(W(t))\right)^2\right] \mathrm{d}t < \infty$. [4] Note that $f'(W(t))$ is a continuous function of $W(t)$ and hence is $\mathcal{F}_t$–measurable for all $t \geqslant 0$. The Riemann integral in (2.34) is well-defined

[4]Technically, the Itô formula in (2.34) applies without the need for the square-integrability condition when adopting a definition of the Itô integral for a larger class of $\mathbb{F}$-adapted integrands s.t. $\int_0^T X^2(t)\mathrm{d}t < \infty$ (a.s.). Within this class, the Itô integral in (2.34) is well-defined where $\int_0^T \left(f'(W(t))\right)^2 \mathrm{d}t < \infty$ (a.s.) since $f'(W(t))$ is a continuous function of $W(t)$ which is bounded (a.s.) for all $0 \leqslant t \leqslant T$.

since the integrand $f''(W(t))$ is a composite function that is continuous in t. The expression in (2.34) tells us that $F(t) := f(W(t)), 0 \leqslant t \leqslant T$, is an Itô process with initial value $F(0) = f(W(0)) = f(0)$.

Only the skeleton of a proof of the Itô formula (2.34) is outlined below:

Proof. Let $0 = t_0 < \ldots < t_n = t$ be a partition of $[0, t]$, with $\delta t^{(n)} := \max_{1 \leqslant i \leqslant n} (t_i - t_{i-1}) \to 0$, as $n \to \infty$. Write $f(W(t)) - f(W(0))$ as a telescopic sum

$$f(W(t)) - f(W(0)) = \sum_{i=1}^{n} \big(f(W(t_i)) - f(W(t_{i-1}))\big).$$

Now, apply Taylor's expansion formula to each term of the above sum:

$$f(W(t_i)) - f(W(t_{i-1})) = f'(W(t_{i-1}))\big(W(t_i) - W(t_{i-1})\big) + \frac{1}{2} f''(\theta_i)\big(W(t_i) - W(t_{i-1})\big)^2,$$

where θ_i lies between $W(t_{i-1})$ and $W(t_i)$ for $i = 1, 2, \ldots, n$. By taking the limit $n \to \infty$, the partial sums converge (in $L^2(\Omega)$) to the respective integrals:

$$\sum_{i=1}^{n} f'(W(t_{i-1}))\big(W(t_i) - W(t_{i-1})\big) \to \int_0^t f'(W(s))\,\mathrm{d}W(s) \quad \text{(an Itô integral)},$$

$$\sum_{i=1}^{n} f''(\theta_i)\big(W(t_i) - W(t_{i-1})\big)^2 \to \int_0^t f''(W(s))\,\mathrm{d}s \quad \text{(a Riemann integral)}. \qquad \square$$

Example 2.7. Find the stochastic differential $\mathrm{d}f(W(t))$ for functions:

(a) $f(x) = x^n$, $n \in \mathbb{N}$;

(b) $f(x) = \mathrm{e}^{\alpha x}$, $\alpha \in \mathbb{R}$.

Solution.
(a) Differentiating gives $f'(x) = nx^{n-1}$, $f''(x) = n(n-1)x^{n-2}$. Thus, (2.34) with $f(W(t)) = (W(t))^n \equiv W^n(t), f(0) = 0$ reads

$$W^n(t) = \frac{n(n-1)}{2} \int_0^t W^{n-2}(s)\,\mathrm{d}s + n \int_0^t W^{n-1}(s)\,\mathrm{d}W(s).$$

The differential form of the above representation is

$$\mathrm{d}W^n(t) = \frac{n(n-1)}{2} W^{n-2}(t)\,\mathrm{d}t + nW^{n-1}(t)\,\mathrm{d}W(t).$$

Setting $n = 2$ recovers the well-known formula for the Itô integral of Brownian motion that we laboriously derived previously (note: $W^0(t) \equiv (W(t))^0 = 1$):

$$W^2(t) = \int_0^t \mathrm{d}s + 2\int_0^t W(s)\,\mathrm{d}W(s) \implies \int_0^t W(s)\,\mathrm{d}W(s) = \frac{1}{2}W^2(t) - \frac{t}{2}.$$

Setting $n = 3$ gives

$$W^3(t) = 3\int_0^t W(s)\,\mathrm{d}s + 3\int_0^t W^2(s)\,\mathrm{d}W(s).$$

which is the same relation we (quite laboriously) derived in Example 2.3.2 before we had any knowledge of Itô's formula. As we know by now, the Itô integral $\int_0^t W^2(s)\,\mathrm{d}W(s)$ is a (square-integrable) martingale; i.e., the process $Y(t) := W^3(t) - 3\int_0^t W(s)\,\mathrm{d}s$, $t \geqslant 0$, is a martingale w.r.t. any Brownian filtration. We have also proven this fact earlier. However, now it follows more simply by applying the Itô formula for $f(x) = x^3$.

(b) Differentiating gives $f'(x) = \alpha f(x)$ and $f''(x) = \alpha^2 f(x)$. Denote $X(t) := f(W(t)) = \mathrm{e}^{\alpha W(t)}$. Recall that X is a geometric Brownian motion (GBM). Now, by applying the Itô formula in (2.33), we have the stochastic differential of this GBM:

$$\mathrm{d}X(t) = \frac{\alpha^2}{2}X(t)\,\mathrm{d}t + \alpha X(t)\,\mathrm{d}W(t).$$

In integral form, we have (note: $X(0) = f(0) = \mathrm{e}^0 = 1$)

$$X(t) = 1 + \frac{\alpha^2}{2}\int_0^t X(u)\,\mathrm{d}u + \alpha\int_0^t X(u)\,\mathrm{d}W(u). \qquad \square$$

There are various important extensions of the Itô formula. In particular, consider the case of a stochastic process defined by $X(t) := f(t, W(t))$, $0 \leqslant t \leqslant T$, where the function $f(t,x) \in C^{1,2}$ i.e., we assume that the functions $f_t(t,x) := \frac{\partial f}{\partial t}(t,x)$, $f_x(t,x) := \frac{\partial f}{\partial x}(t,x)$, $f_{tx}(t,x) := \frac{\partial^2 f}{\partial t\partial x}(t,x)$, and $f_{xx}(t,x) := \frac{\partial^2 f}{\partial x^2}(t,x)$ are continuous. Let us heuristically apply a Taylor expansion to the differential $\mathrm{d}f(t,W(t)) = f(t+\mathrm{d}t, W(t)+\mathrm{d}W(t)) - f(t,W(t))$ and keep only terms up to second order in the Brownian increment $\mathrm{d}W(t)$ and first order in the time increment $\mathrm{d}t$:

$$\begin{aligned}\mathrm{d}f(t,W(t)) &= f_t(t,W(t))\,\mathrm{d}t + f_x(t,W(t))\,\mathrm{d}W(t)\\ &\quad + f_{tx}(t,W(t))\,\mathrm{d}t\,\mathrm{d}W(t) + \frac{1}{2}f_{xx}(t,W(t))(\,\mathrm{d}W(t))^2 + \cdots\end{aligned}$$

By the differential multiplication rules we set

$$(\,\mathrm{d}W(t))^2 = \,\mathrm{d}t,\ (\,\mathrm{d}t)^2 = 0,\ \,\mathrm{d}t\,\mathrm{d}W(t) = 0.$$

Collecting the coefficient terms in $\mathrm{d}t$ and $\mathrm{d}W(t)$, the differential and integral forms of the *Itô formula* for $f(t, W(t))$ are then, respectively, given by

$$\mathrm{d}f(t,W(t)) = \left(f_t(t,W(t)) + \frac{1}{2}f_{xx}(t,W(t))\right)\mathrm{d}t + f_x(t,W(t))\,\mathrm{d}W(t), \quad (2.35)$$

$$\begin{aligned}f(t,W(t)) - f(0,W(0)) &= \int_0^t \left(f_u(u,W(u)) + \frac{1}{2}f_{xx}(u,W(u))\right)\mathrm{d}u\\ &\quad + \int_0^t f_x(u,W(u))\,\mathrm{d}W(u), \qquad (2.36)\end{aligned}$$

for all $0 \leqslant t \leqslant T$. We do not give a proof of this result. The main steps in the proof are very similar to those for (2.34). From (2.36) we see that $F(t) := f(t,W(t)), 0 \leqslant t \leqslant T$, is an Itô process with initial value $F(0) = f(0,W(0)) = f(0,0)$.

We assume the Itô integral in (2.36) is well-defined, i.e., based on our definition of the Itô integral we assume the square-integrability condition $\int_0^T \mathrm{E}\big[\big(f_x(t,W(t))\big)^2\big]\mathrm{d}t < \infty$ holds.[5] Note that the integrand $f_x(t,W(t))$ is $\mathcal{F}_t$-measurable for all $0 \leqslant t \leqslant T$. The Riemann

[5] As in our previous technical footnote about the Itô integral within (2.34), the Itô formula (2.36) is applicable without the square-integrability condition if one is adopting a definition of the Itô integral for a larger class of $\mathbb{F}$-adapted integrands s.t. $\int_0^T X^2(t)\mathrm{d}t < \infty$ (a.s.). Within this class, the Itô integral in (2.36) is well-defined where $\int_0^T \big(f_x(t,W(t))\big)^2\mathrm{d}t < \infty$ (a.s.) since $f_x(t,W(t))$ is a continuous function of $W(t)$ which is bounded (a.s.) for all $0 \leqslant t \leqslant T$.

integral in (2.36) is well-defined since the integrand $f_t(t, W(t)) + \frac{1}{2} f_{xx}(t, W(t))$ is (a.s.) continuous for all $0 \leqslant t \leqslant T$.

Example 2.8. Find the stochastic differential of the GBM process $S(t) = S_0 e^{\alpha t + \sigma W(t)}$, $t \geqslant 0$, with constants $S_0 > 0$, $\alpha, \sigma \in \mathbb{R}$.

Solution. We represent $S(t) = f(t, W(t))$, where $f(t, x) := S_0 e^{\alpha t + \sigma x}$. Hence,

$$f_t(t, x) = \alpha f(t, x), \quad f_x(t, x) = \sigma f(t, x), \quad f_{xx}(t, x) = \sigma^2 f(t, x).$$

Substituting these partial derivatives into the Itô formula (2.35) gives

$$\begin{aligned} dS(t) &= \left(\alpha f(t, W(t)) + \frac{\sigma^2}{2} f(t, W(t)) \right) dt + \sigma f(t, W(t))\, dW(t) \\ &= (\alpha + \frac{\sigma^2}{2}) S(t)\, dt + \sigma S(t)\, dW(t). \end{aligned}$$

□

Note that, in the above example, if we put $\alpha = \mu - \sigma^2/2$, with parameter $\mu \in \mathbb{R}$, then $S(t) = S_0 e^{(\mu - \sigma^2/2)t + \sigma W(t)}$ is a GBM satisfying the SDE

$$dS(t) = \mu S(t)\, dt + \sigma S(t)\, dW(t)$$

with initial condition $S(0) = S_0$. In integral form, the GBM process satisfies the stochastic integral equation:

$$S(t) = S_0 + \mu \int_0^t S(u)\, du + \sigma \int_0^t S(u)\, dW(u). \tag{2.37}$$

Hence, $S(t) = S_0 e^{(\mu - \sigma^2/2)t + \sigma W(t)}$, for $t \geqslant 0$, is an explicit solution to the above stochastic integral (or differential) equation whereby $S(t)$ is explicitly given as (an exponential) function of $W(t)$. It is a martingale iff the drift coefficient $\mu = 0$. In fact, in the previous chapter, we already directly proved that $M(t) := e^{-\frac{\sigma^2}{2} t + \sigma W(t)}$ is a martingale. It follows that the discounted process given by $e^{-\mu t} S(t) = S_0 M(t), t \geqslant 0$, is a martingale w.r.t. any filtration for BM. In fact, the GBM process is a square-integrable martingale since $E[S^2(t)] < \infty$. Note that the Itô integral in (2.37) is well-defined since $S(u)$ is $\mathcal{F}_u$-measurable and the square-integrability condition holds: $\int_0^t E[S^2(u)]\, du = S_0^2 \int_0^t e^{(2\mu + \sigma^2)u}\, du < \infty$ for all $t \geqslant 0$.

2.5.2 An "Antiderivative" Formula for Evaluating Itô Integrals

The Itô formula in (2.36) for time-dependent functions, and its corresponding reduced form in (2.34) for time-independent functions, can be simply recast in a manner that expresses an Itô integral of a given explicit integrand function in terms of an explicit function of Brownian motion and a Riemann time-integral. To see this, consider a function $F(t, x) \in C^{1,2}$ and let $F_x(t, x) \equiv \frac{\partial}{\partial x} F(t, x) := f(t, x)$. Hence, $F(t, x)$ is an antiderivative of f in x,

$$F(t, x) = \int f(t, x)\, dx + h(t),$$

where $h(t)$ is an arbitrary function of only t. Since $F_t(t, x) \equiv \frac{\partial}{\partial t} F(t, x)$ is continuous, the derivative $h'(t)$ is continuous in t. Below we shall see that we can simply set $h(t) = 0$. Now, applying the Itô formula in (2.36) to the process $F(t, W(t))$, i.e., with f replaced by F, while using the fact that $F_x(t, x) = f(t, x)$, $F_{xx}(t, x) \equiv \frac{\partial}{\partial x}(\frac{\partial}{\partial x} F(t, x)) = \frac{\partial}{\partial x} f(t, x)$, gives

$$\begin{aligned} F(t, W(t)) - F(0, 0) = &\int_0^t \left[\frac{\partial F}{\partial u}(u, W(u)) + \frac{1}{2} \frac{\partial f}{\partial x}(u, W(u)) \right] du \\ &+ \int_0^t f(u, W(u))\, dW(u). \end{aligned}$$

Isolating the Itô integral gives

$$\int_0^t f(u,W(u))\,\mathrm{d}W(u) = F(t,W(t)) - F(0,0) - \int_0^t \left[\frac{\partial F}{\partial u}(u,W(u)) + \frac{1}{2}\frac{\partial f}{\partial x}(u,W(u))\right]\mathrm{d}u. \tag{2.38}$$

We remark that this formula is valid for continuous functions $f, \frac{\partial f}{\partial x}, \frac{\partial F}{\partial t}$. Also, based on our definition of a well-defined Itô integral, we require the usual square-integrability condition to hold (as in all our Itô formulas). This condition is not required if we define Itô integrals within the more general class which exist by continuity of the integrand (see our previous footnote pertaining to the Itô formula). Note that F is any antiderivative of f in x. We see from the Riemann integral of $\frac{\partial F}{\partial u}(u,W(u))$ that adding any function $h(t)$ (where $h'(t)$ is continuous) to an antiderivative will cancel out from the term $F(t,W(t)) - F(0,0)$.

We refer to the formula in (2.38) as the "antiderivative" formula for the Itô integral of a given function f of time and BM. That is, the r.h.s. involves two terms. The first is the antiderivative, F, in the spatial variable x of the integrand function f evaluated at the two time endpoints of integration. The second term is a Riemann time-integral with integrand given equivalently as $\left[\frac{\partial}{\partial u}F(u,x) + \frac{1}{2}\frac{\partial}{\partial x}f(u,x)\right]_{x=W(u)} \equiv \left[\frac{\partial}{\partial u}F(u,x) + \frac{1}{2}\frac{\partial^2}{\partial x^2}F(u,x)\right]_{x=W(u)}$. If $F(t,x)$ satisfies $\frac{\partial}{\partial t}F(t,x) + \frac{1}{2}\frac{\partial^2}{\partial x^2}F(t,x) = 0$, for all t,x, then the Riemann integral in (2.38) is identically zero. In such special cases, the Itô integral is given purely in terms of the antiderivative term $F(t,W(t)) - F(0,0)$. For any time interval $(a,b) \subset [0,\infty)$, where the above assumptions are satisfied, the formula in (2.38) is equivalent to

$$\int_a^b f(t,W(t))\,\mathrm{d}W(t) = F(b,W(b)) - F(a,W(a)) - \int_a^b \left[\frac{\partial F}{\partial t}(t,W(t)) + \frac{1}{2}\frac{\partial f}{\partial x}(t,W(t))\right]\mathrm{d}t. \tag{2.39}$$

If we consider functions $f(t,x) \equiv f(x)$ that are time-independent, then the above formulas simplify where partial derivatives in x are now ordinary derivatives, i.e., $\frac{\partial f}{\partial x} = f'(x)$, and where $\frac{\partial}{\partial t}F(t,x) = \frac{\partial}{\partial t}F(x) \equiv 0$ with antiderivative $F(x) = \int f(x)\,\mathrm{d}x$ (plus an irrelevant constant), with $F'(x) = f(x)$. This is also seen from the Itô formula in (2.34). Hence, (2.38)–(2.39) simplify to give

$$\int_0^t f(W(u))\,\mathrm{d}W(u) = F(W(t)) - F(0) - \frac{1}{2}\int_0^t f'(W(u))\,\mathrm{d}u\,, \tag{2.40}$$

$$\int_a^b f(W(t))\,\mathrm{d}W(t) = F(W(b)) - F(W(a)) - \frac{1}{2}\int_a^b f'(W(t))\,\mathrm{d}t\,. \tag{2.41}$$

Example 2.9. Evaluate each integral as an expression involving *no Itô integral*, i.e., only in terms of a Riemann integral and possibly other terms.

(a) $\int_0^t W(u)\,\mathrm{d}W(u)$.

(b) $\int_0^t (W^2(u) - u)\,\mathrm{d}W(u)$.

(c) $\int_0^T W(t)\cos(W^2(t) + t)\,\mathrm{d}W(t)$.

Solution.
(a) Let $f(x) := x$. We have $f'(x) \equiv 1$ and $F(x) = \int x\,\mathrm{d}x = \frac{x^2}{2}$ (plus an irrelevant constant). Applying (2.40) gives (note: $F(0) = 0$):

$$\int_0^t W(u)\,\mathrm{d}W(u) = \frac{1}{2}W^2(t) - \frac{1}{2}\int_0^t 1\cdot\mathrm{d}u = \frac{1}{2}(W^2(t) - t).$$

This is the well-known result we derived earlier, but now with much less effort thanks to the antiderivative formula!
(b) Let $f(t,x) := x^2 - t$. We have $\frac{\partial f}{\partial x} = 2x$, $F(t,x) = \int (x^2 - t)\,\mathrm{d}x = \frac{x^3}{3} - tx$ and $\frac{\partial F}{\partial t} = -x$. Applying (2.38) gives (note: $F(0,0) = 0$ and $\frac{\partial F}{\partial t} + \frac{1}{2}\frac{\partial f}{\partial x} \equiv 0$):

$$\int_0^t (W^2(u) - u)\,\mathrm{d}W(u) = \frac{1}{3}W^3(t) - tW(t)\,.$$

Note: this is a case where the Riemann integral portion is identically zero and hence the Itô integral is evaluated using only the antiderivative term.
(c) Let $f(t,x) := x\cos(x^2 + t)$. Then, $\frac{\partial f}{\partial x} = \cos(x^2+t) - 2x^2\sin(x^2+t)$, $F(t,x) = \int x\cos(x^2+t)\,\mathrm{d}x = \frac{1}{2}\sin(x^2+t)$ and $\frac{\partial F}{\partial t} = \frac{1}{2}\cos(x^2+t)$. Hence, $F(0,0) = \frac{1}{2}\sin(0) = 0$ and $\frac{\partial F}{\partial t} + \frac{1}{2}\frac{\partial f}{\partial x} = \cos(x^2+t) - x^2\sin(x^2+t)$. Substituting these terms in (2.39) with $a = 0, b = T$ (or (2.38) with upper endpoint T) gives

$$\int_0^T W(t)\cos(W^2(t)+t)\,\mathrm{d}W(t) = \frac{1}{2}\sin(W^2(T)+T)$$
$$- \int_0^T \left[\cos(W^2(t)+t) - W^2(t)\sin(W^2(t)+t)\right]\mathrm{d}t$$

□

2.5.3 Itô's Formula for Itô Processes

We are now ready to extend the Itô formula to the case of a process defined in terms of a smooth function of an Itô process and time t. Consider an Itô process $\{X(t)\}_{0 \leqslant t \leqslant T}$ with the stochastic differential

$$\mathrm{d}X(t) = \mu(t)\,\mathrm{d}t + \sigma(t)\,\mathrm{d}W(t),$$

where $\mu(t)$ and $\sigma(t)$ are $\mathcal{F}_t$-measurable coefficients processes. As in the previous version of the Itô formula obtained above, assume that $f(t,x) \in C^{1,2}$, i.e., $f_t(t,x)$, $f_x(t,x)$, $f_{tx}(t,x)$, and $f_{xx}(t,x)$ are continuous functions. Then, $Y(t) := f(t,X(t))$, $0 \leqslant t \leqslant T$, is also an Itô process. To obtain its stochastic differential, we apply a Taylor expansion and keep only terms up to second order in the increment $\mathrm{d}X(t)$ and first order in the time increment $\mathrm{d}t$:

$$\mathrm{d}Y(t) = f_t(t,X(t))\,\mathrm{d}t + f_x(t,X(t))\,\mathrm{d}X(t) + \frac{1}{2}f_{xx}(t,X(t))(\mathrm{d}X(t))^2. \tag{2.42}$$

Note that the mixed partial derivative term $f_{tx}(t,X(t))\,\mathrm{d}t\,\mathrm{d}X(t) \equiv 0$ since $\mathrm{d}t\,\mathrm{d}X(t) = \mu(t)(\mathrm{d}t)^2 + \sigma(t)\,\mathrm{d}t\,\mathrm{d}W(t) \equiv 0$ upon using the rules: $(\mathrm{d}t)^2 \equiv 0$, $\mathrm{d}t\,\mathrm{d}W(t) \equiv 0$. Also, $(\mathrm{d}X(t))^2 \equiv \mathrm{d}[X,X](t) = \sigma^2(t)(\mathrm{d}W(t))^2 = \sigma^2(t)\,\mathrm{d}t$, and inserting the differential for $\mathrm{d}X(t)$ into the above equation and combining all coefficients multiplying $\mathrm{d}t$ and $\mathrm{d}W(t)$ finally gives us the *Itô formula* in differential form:

$$\mathrm{d}Y(t) \equiv \mathrm{d}f(t,X(t)) = \left(f_t(t,X(t)) + \mu(t)f_x(t,X(t)) + \frac{1}{2}\sigma^2(t)f_{xx}(t,X(t))\right)\mathrm{d}t$$
$$+ \sigma(t)f_x(t,X(t))\,\mathrm{d}W(t). \tag{2.43}$$

The integral form of this is

$$f(t,X(t)) - f(0,X(0)) = \int_0^t \left(f_s(s,X(s)) + \mu(s)f_x(s,X(s)) + \frac{1}{2}\sigma^2(s)f_{xx}(s,X(s))\right)\mathrm{d}s$$
$$+ \int_0^t \sigma(s)f_x(s,X(s))\,\mathrm{d}W(s), \tag{2.44}$$

for $0 \leqslant t \leqslant T$. As in the previous Itô formulas, we are assuming that the respective Riemann and Itô integrals in (2.44) are well-defined.

Note that in case $Y(t) = f(X(t))$, i.e., $f(t,x) \equiv f(x)$ is not an explicit function of the time variable, then $f_t(t,x) \equiv 0$ and all partial derivatives are simply ordinary derivatives: $f_x(t,x) = f'(x), f_{xx}(t,x) = f''(x)$. The differential form of the Itô formula is

$$\mathrm{d}f(X(t)) = f'(X(t))\,\mathrm{d}X(t) + \frac{1}{2}f''(X(t))(\mathrm{d}X(t))^2\,, \tag{2.45}$$

i.e.,

$$\mathrm{d}f(X(t)) = \left(\mu(t)f'(X(t)) + \frac{1}{2}\sigma^2(t)f''(X(t))\right)\mathrm{d}t + \sigma(t)f'(X(t))\,\mathrm{d}W(t) \tag{2.46}$$

and in integral form

$$\begin{aligned} f(X(t)) - f(X(0)) &= \int_0^t \left(\mu(s)f'(X(s)) + \frac{1}{2}\sigma^2(s)f''(X(s))\right)\mathrm{d}s \\ &\quad + \int_0^t \sigma(s)f'(X(s))\,\mathrm{d}W(s). \end{aligned} \tag{2.47}$$

Observe that (2.33) and (2.34) are recovered by (2.46) and (2.47) in the special case where the Itô process is Brownian motion: $X = W$ where $\mu \equiv 0$, $\sigma \equiv 1$. Similarly, (2.35) and (2.36) are special cases of (2.43) and (2.44).

Example 2.10. Let $Y(t) := \ln X(t)$, $t \geqslant 0$, where $\{X(t)\}_{t\geqslant 0}$ is an Itô process with stochastic differential

$$\mathrm{d}X(t) = aX(t)\,\mathrm{d}t + bX(t)\,\mathrm{d}W(t).$$

Find the SDE for the process Y and then find explicit representations for $Y(t)$ and $X(t)$ in terms of $W(t)$.

Solution. In this case, we define $f(x) := \ln x$, where $Y(t) = f(X(t))$. Differentiating gives $f'(x) = \frac{1}{x}, f''(x) = -\frac{1}{x^2}$. Applying (2.46) with $\mu(t) = aX(t), \sigma(t) = bX(t)$ gives

$$\begin{aligned} \mathrm{d}Y(t) &= \left(aX(t)\frac{1}{X(t)} + \frac{1}{2}(bX(t))^2\left(\frac{-1}{X^2(t)}\right)\right)\mathrm{d}t + bX(t)\frac{1}{X(t)}\,\mathrm{d}W(t) \\ &= \left(a - \frac{b^2}{2}\right)\mathrm{d}t + b\,\mathrm{d}W(t). \end{aligned}$$

As an alternative derivation, we note that this equation follows directly from (2.45) which also provides a useful compact form for the stochastic differential of the logarithm of any Itô process X:

$$\mathrm{d}\ln X(t) = \frac{1}{X(t)}\,\mathrm{d}X(t) - \frac{1}{2}\frac{1}{X^2(t)}(\mathrm{d}X(t))^2 \equiv \frac{\mathrm{d}X(t)}{X(t)} - \frac{1}{2}\left(\frac{\mathrm{d}X(t)}{X(t)}\right)^2. \tag{2.48}$$

Now, writing the original SDE of $X(t)$ as $\frac{\mathrm{d}X(t)}{X(t)} = a\,\mathrm{d}t + b\,\mathrm{d}W(t)$ and then substituting this, and $\left(\frac{\mathrm{d}X(t)}{X(t)}\right)^2 = b^2\,\mathrm{d}t$, into (2.48) gives the same SDE as above:

$$\mathrm{d}Y(t) \equiv \mathrm{d}\ln X(t) = a\,\mathrm{d}t + b\,\mathrm{d}W(t) - \frac{1}{2}b^2\,\mathrm{d}t = \left(a - \frac{b^2}{2}\right)\mathrm{d}t + b\,\mathrm{d}W(t).$$

Integrating this equation therefore shows that the process Y is a drifted Brownian motion starting at $Y(0) = \ln X(0)$:

$$Y(t) = \ln X(0) + \left(a - \frac{b^2}{2}\right)t + bW(t).$$

By inverting the transformation, we find the original process $X(t)$ as a closed-form expression in $W(t)$:

$$X(t) = \mathrm{e}^{Y(t)} = \mathrm{e}^{\ln X(0) + \left(a - \frac{b^2}{2}\right)t + bW(t)} = X(0)\mathrm{e}^{\left(a - \frac{b^2}{2}\right)t + bW(t)}. \qquad \square$$

We recall the basic fact that an Itô integral process, say $I(t) = \int_0^t \sigma(s)\,\mathrm{d}W(s)$, $t \geqslant 0$, is a martingale (provided that the stochastic integral is well-defined). However, a Riemann time-integral $\int_0^t \mu(s)\,\mathrm{d}s$ is generally not a martingale. The Itô formula can be used to verify whether or not a stochastic process that is a function of an Itô process is a martingale. This is done by recasting the original expression of a given process into the form of an Itô process upon taking stochastic differentials and employing an appropriate Itô formula. The following example illustrates this technique.

Example 2.11. Verify whether or not the following processes are martingales w.r.t. a filtration for BM:

(a) $X(t) = Z^2(t) - \int_0^t f^2(u)\,\mathrm{d}u$, where $Z(t) = \int_0^t f(u)\,\mathrm{d}W(u)$ and f is an ordinary (nonrandom) continuous function for $t \geqslant 0$;

(b) $Y(t) = V^2(t) - \frac{t^2}{2}$, where $V(t) = \int_0^t W(u)\,\mathrm{d}W(u)$.

Solution.
(a) The process Z is an $\mathbb{F}$-adapted Gaussian process with the stochastic differential $\mathrm{d}Z(t) = f(t)\,\mathrm{d}W(t) = \mu(t)\,\mathrm{d}t + \sigma(t)\,\mathrm{d}W(t)$ where $\mu(t) \equiv 0$ and $\sigma(t) \equiv f(t)$. The process X is given by $X(t) = g(t, Z(t))$ with $g(t,x) := x^2 - \int_0^t f^2(u)\,\mathrm{d}u$. Taking derivatives of g:

$$g_t(t,x) = -f^2(t), \quad g_x(t,x) = 2x, \quad g_{xx}(t,x) \equiv 2.$$

Applying (2.43) gives a stochastic differential with zero drift,

$$\begin{aligned}
\mathrm{d}X(t) &= \left[g_t(t, Z(t)) + 0 \cdot g_x(t, Z(t)) + \frac{1}{2}f^2(t)g_{xx}(t, Z(t))\right]\mathrm{d}t + f(t)g_x(t, Z(t))\,\mathrm{d}W(t) \\
&= \left(-f^2(t) + \frac{1}{2}(2f^2(t))\right)\mathrm{d}t + 2f(t)Z(t)\,\mathrm{d}W(t) = 2f(t)Z(t)\,\mathrm{d}W(t).
\end{aligned}$$

An alternative direct way to derive this relation is to add the stochastic differentials while employing (2.45) on $Z^2(t)$:

$$\begin{aligned}
\mathrm{d}X(t) = \mathrm{d}Z^2(t) - f^2(t)\,\mathrm{d}t &= 2Z(t)\,\mathrm{d}Z(t) + (\mathrm{d}Z(t))^2 - f^2(t)\,\mathrm{d}t \\
&= 2Z(t)\,\mathrm{d}Z(t) = 2Z(t)f(t)\,\mathrm{d}W(t).
\end{aligned}$$

Note: in the last equation we used $\mathrm{d}Z(t) = f(t)\,\mathrm{d}W(t)$ and $(\mathrm{d}Z(t))^2 = f^2(t)(\mathrm{d}W(t))^2 = f^2(t)\,\mathrm{d}t$.

In integral form, $X(t) = 2\int_0^t f(u)Z(u)\,\mathrm{d}W(u)$ since $X(0) = 0$. Thus, $X(t)$ is an Itô integral process with no Riemann integral (drift) term. Note that the Itô integral is well-defined as it satisfies the square-integrability condition:

$$\mathrm{E}\left[\int_0^t (f(u)Z(u))^2\,\mathrm{d}u\right] = \int_0^t f^2(u)\mathrm{E}[Z^2(u)]\,\mathrm{d}u < \infty$$

since $\mathrm{E}[Z^2(u)] = \mathrm{Var}\left(\int_0^u f(s)\,\mathrm{d}W(s)\right) = \int_0^u f^2(s)\,\mathrm{d}s$ is a continuous function of $u \geqslant 0$. Hence the process X is a martingale.

(b) First, compute the stochastic differential of $Y(t)$:

$$\mathrm{d}Y(t) = 2V(t)\,\mathrm{d}V(t) + (\,\mathrm{d}V(t))^2 - t\,\mathrm{d}t = (W^2(t) - t)\,\mathrm{d}t + 2V(t)W(t)\,\mathrm{d}W(t).$$

Note that $\mathrm{d}V(t) = W(t)\,\mathrm{d}W(t)$ and $(\,\mathrm{d}V(t))^2 = W^2(t)(\,\mathrm{d}W(t))^2 = W^2(t)\,\mathrm{d}t$. Thus, $Y(t)$ is a sum of an Itô integral (which the reader can verify as being well-defined and thereby a martingale) and a Riemann integral of a function of Brownian motion:

$$Y(t) = 2\int_0^t V(u)W(u)\,\mathrm{d}W(u) + \int_0^t (W^2(u) - u)\,\mathrm{d}u.$$

Note that $Y(0) = 0$. Let us show that the Riemann integral above is not a martingale. As a first simple check, we can try to verify whether the expected value of $I(t) := \int_0^t (W^2(u) - u)\,\mathrm{d}u$ is nonconstant over time:

$$\mathrm{E}[I(t)] = \int_0^t (\underbrace{\mathrm{E}\left[W^2(u)\right]}_{u} - u)\,\mathrm{d}u = 0$$

for $t \geqslant 0$. So the expectation is constant and we cannot yet conclude whether or not the process is a martingale. We hence need to necessarily calculate the conditional expectation to verify whether the process satisfies the martingale property. For $t, s > 0$, we have, upon using the martingale property of $\{W^2(t) - t\}_{t \geqslant 0}$:

$$\begin{aligned}
\mathrm{E}_t\left[I(t+s)\right] &= I(t) + \mathrm{E}_t\left[\int_t^{t+s} (W^2(u) - u)\,\mathrm{d}u\right] = I(t) + \int_t^{t+s} \mathrm{E}_t\left[(W^2(u) - u)\right]\,\mathrm{d}u \\
&= I(t) + \int_t^{t+s} (W^2(t) - t)\,\mathrm{d}u \\
&= I(t) + (W^2(t) - t)\int_t^{t+s} \mathrm{d}u = I(t) + s(W^2(t) - t) \neq I(t).
\end{aligned}$$

In conclusion, $\mathrm{E}_t\left[Y(t+s)\right] \neq Y(t)$ and hence the process Y is not a martingale. □

2.6 Stochastic Differential Equations

An equation of the form of an Itô stochastic differential

$$\mathrm{d}X(t) = \mu(t, X(t))\,\mathrm{d}t + \sigma(t, X(t))\,\mathrm{d}W(t) \tag{2.49}$$

where the drift coefficient $\mu(t, x)$ and volatility $\sigma(t, x)$ are given (known) functions and $X(t)$ is the unknown process is called a *stochastic differential equation (SDE)*. Equations of this form are of great importance in financial modelling. In practice, (2.49) is subject to an initial condition $X(0) = X_0$ where X_0 is either a random variable or simply a constant $X_0 = x \in \mathbb{R}$. As was mentioned in a previous section, an SDE of the type in (2.49), with constant X_0, is also called a diffusion. We will study diffusions in some depth a little later in the text.

A process X is a solution, or so-called *strong solution to the SDE* in (2.49) if, for all $t \geqslant 0$ (or $t \in [0, T]$ if time is restricted to some finite interval $[0, T]$), the process satisfies

$$X(t) = X(0) + \int_0^t \mu(s, X(s))\,\mathrm{d}s + \int_0^t \sigma(s, X(s))\,\mathrm{d}W(s)$$

almost surely, where both integrals are assumed to exist. The randomness is completely driven by the underlying Brownian motion. So, in case $\sigma \equiv 0$ the equation is simply an ordinary first order ODE. It is important to note that a solution $X(t)$ is an adapted process that is some *representation or functional* written in terms of the Brownian motion up to time t, i.e., $X(t) = F(t, \{W(s); 0 \leqslant s \leqslant t\})$. We have in fact already seen some cases (see Examples 2.8 and 2.10), where the solution $X(t) = F(t, W(t))$ is just a function of the Brownian motion at the endpoint time t. A strong solution hence also gives a path-wise representation of the process $\{X(t)\}_{t \geqslant 0}$. In most cases, strong solutions to SDEs cannot be found explicitly, although we can still compute a number of important properties of the process. An alternative and important type of solution is a so-called *weak solution*, which is a solution in distribution. We now turn our attention to so-called linear SDEs, as these form the simplest class of SDEs that have some applications in finance and for which a unique strong solution can be found explicitly.

2.6.1 Solutions to Linear SDEs

A linear SDE is an equation of the form

$$\mathrm{d}X(t) = (\alpha(t) + \beta(t)X(t))\,\mathrm{d}t + (\gamma(t) + \delta(t)X(t))\,\mathrm{d}W(t)\,, \tag{2.50}$$

subject to an initial value $X(0)$, where $\alpha(t), \beta(t), \gamma(t), \delta(t)$ are given $\mathbb{F}$-adapted processes. These are assumed to be continuous (or piecewise continuous) in time t. We note that they can simply be ordinary (nonrandom) functions of t or may also be random but not functions of the process $X(t)$. The key point is that the coefficients in $\mathrm{d}t$ and $\mathrm{d}W(t)$ in (2.50) are linear functions of $X(t)$. A special class of linear SDEs is when $\alpha(t), \beta(t), \gamma(t), \delta(t)$ are non-random functions of time. In this case, the process is a diffusion with linear SDE of the form $\mathrm{d}X(t) = a(t, X(t))\,\mathrm{d}t + b(t, X(t))\,\mathrm{d}W(t)$, with both coefficient functions being linear in the state variable x: $a(t, x) = \alpha(t) + \beta(t)x$ and $b(t, x) = \gamma(t) + \delta(t)x$. The stochastic differential equations considered in Examples 2.8 and 2.10 are simple linear SDEs. The nice thing about an SDE of the form (2.50) is that we have explicit solutions, as we now derive.

Equation (2.50) is readily solved by first considering the simpler case when $\alpha(t) \equiv \gamma(t) \equiv 0$. Denoting the simpler process by U, the SDE in (2.50) takes the form

$$\mathrm{d}U(t) = \beta(t)U(t)\,\mathrm{d}t + \delta(t)U(t)\,\mathrm{d}W(t). \tag{2.51}$$

This SDE is now solved by considering the logarithm of the process, $Y(t) := \ln U(t)$, and applying Itô's formula (see Example 2.10):

$$\mathrm{d}Y(t) = \mathrm{d}\ln U(t) = \frac{\mathrm{d}U(t)}{U(t)} - \frac{1}{2}\left(\frac{\mathrm{d}U(t)}{U(t)}\right)^2 = \left(\beta(t) - \frac{1}{2}\delta^2(t)\right)\mathrm{d}t + \delta(t)\,\mathrm{d}W(t).$$

Putting this SDE in integral form and using $U(t) = \mathrm{e}^{Y(t)}$ gives

$$U(t) = U(0)\exp\left[\int_0^t \left(\beta(s) - \frac{1}{2}\delta^2(s)\right)\mathrm{d}s + \int_0^t \delta(s)\,\mathrm{d}W(s)\right]. \tag{2.52}$$

This solution is compactly written as a product: $U(t) = U(0)\mathrm{e}^{\int_0^t \beta(s)\,\mathrm{d}s}\mathcal{E}_t(\delta \cdot W)$, where we denote the *stochastic exponential* of an adapted process $\{\delta(s), 0 \leqslant s \leqslant t\}$, w.r.t. BM on the time interval $[0, t]$, by

$$\mathcal{E}_t(\delta \cdot W) := \exp\left[-\frac{1}{2}\int_0^t \delta^2(s)\,\mathrm{d}s + \int_0^t \delta(s)\,\mathrm{d}W(s)\right]. \tag{2.53}$$

Note that, by setting $\beta(t) \equiv 0$ in (2.51), the solution to the SDE

$$\mathrm{d}U(t) = \delta(t)U(t)\,\mathrm{d}W(t) \quad \text{subject to } U(0) = 1$$

is the stochastic exponential in (2.53). This type of process plays an important role in derivative pricing theory so we shall revisit it in Section 2.8, as well as its multidimensional version in Section 2.10. Note that when the coefficients $\beta(t)$ and $\delta(t)$ are nonrandom functions of time, and $U(0)$ is taken as a positive constant, the Itô integral $\int_0^t \delta(s)\,\mathrm{d}W(s) \sim Norm(0, \int_0^t \delta^2(s)\,\mathrm{d}s)$; i.e., it is a Gaussian process, where $\ln \frac{U(t)}{U(0)} \sim Norm(\mu_t, v_t)$ with mean $\mu_t = \int_0^t \left(\beta(s) - \frac{1}{2}\delta^2(s)\right)\mathrm{d}s$ and variance $v_t = \int_0^t \delta^2(s)\,\mathrm{d}s$. That is, $U(t)$ is a lognormal random variable and hence $\{U(t)\}_{t\geqslant 0}$ is a GBM process. In particular, for the case of constant coefficients we recover a GBM process as in Example 2.10. In more complicated general cases where $\delta(t)$ and $\beta(t)$ are random variables (for example, functionals of BM up to time t), then the exponent in (2.52) is not a normal random variable and hence the process $\{U(t)\}_{t\geqslant 0}$ is not a GBM.

The solution $X(t)$ to the general linear SDE in (2.50) is now readily derived based on the solution to (2.51). The key step is to seek a solution as a product, i.e., let $X(t) = U(t)V(t)$, where $U(t)$ is given in (2.52) and satisfies (2.51), and where $V(t)$ is an Itô process solving the SDE

$$\mathrm{d}V(t) = \mu_V(t)\,\mathrm{d}t + \sigma_V(t)\,\mathrm{d}W(t). \tag{2.54}$$

The $\mathbb{F}$-adapted coefficients $\mu_V(t)$ and $\sigma_V(t)$ are determined as follows. We set $U(0) = 1$ and $V(0) = X(0)$, thereby matching any given initial condition $U(0)V(0) = X(0)$. Applying the Itô product formula (2.32) on $X(t) = U(t)V(t)$ and substituting for $\mathrm{d}V(t)$ using (2.54) and $\mathrm{d}U(t)$ using (2.51), multiplying out all terms, where $(\mathrm{d}t)^2 = \mathrm{d}t\,\mathrm{d}W(t) = 0$, $(\mathrm{d}W(t))^2 = \mathrm{d}t$, and then collecting coefficients in $\mathrm{d}t$ and $\mathrm{d}W(t)$, gives

$$\begin{aligned}
\mathrm{d}X(t) &= U(t)\,\mathrm{d}V(t) + V(t)\,\mathrm{d}U(t) + \mathrm{d}U(t)\,\mathrm{d}V(t)\\
&= U(t)\big[\mu_V(t)\,\mathrm{d}t + \sigma_V(t)\,\mathrm{d}W(t)\big] + V(t)\big[\beta(t)U(t)\,\mathrm{d}t + \delta(t)U(t)\,\mathrm{d}W(t)\big]\\
&\quad + \sigma_V(t)\delta(t)U(t)\,\mathrm{d}t\\
&= \big[\mu_V(t) + \beta(t)V(t) + \sigma_V(t)\delta(t)\big]U(t)\,\mathrm{d}t + \big[\sigma_V(t) + \delta(t)V(t)\big]U(t)\,\mathrm{d}W(t)\\
&= \big(\mu_V(t)U(t) + \sigma_V(t)\delta(t)U(t) + \beta(t)X(t)\big)\,\mathrm{d}t + \big(\sigma_V(t)U(t) + \delta(t)X(t)\big)\,\mathrm{d}W(t).
\end{aligned}$$

The r.h.s. of this SDE must be equivalent to the r.h.s. in (2.50). Hence, equating the terms in $\mathrm{d}t$ and $\mathrm{d}W(t)$ gives two equations in the two unknown coefficients $\mu_V(t)$ and $\sigma_V(t)$ which are readily solved due to canceling terms:

$$\gamma(t) + \cancel{\delta(t)X(t)} = \sigma_V(t)U(t) + \cancel{\delta(t)X(t)} \implies \sigma_V(t) = \frac{\gamma(t)}{U(t)}$$

$$\alpha(t) + \cancel{\beta(t)X(t)} = \mu_V(t)U(t) + \sigma_V(t)\delta(t)U(t) + \cancel{\beta(t)X(t)} \implies \mu_V(t) = \frac{\alpha(t) - \delta(t)\gamma(t)}{U(t)}$$

Hence, substituting these into (2.54) gives the SDE for $V(t)$:

$$\mathrm{d}V(t) = \left[\frac{\alpha(t) - \gamma(t)\delta(t)}{U(t)}\right]\mathrm{d}t + \frac{\gamma(t)}{U(t)}\,\mathrm{d}W(t) \tag{2.55}$$

s.t. $V(0) = X(0)$, where all coefficients are known functions of t. Hence, an explicit solution for $V(t)$ is obtained simply from the integral form of (2.55):

$$V(t) = X(0) + \int_0^t \frac{\alpha(s) - \gamma(s)\delta(s)}{U(s)}\,\mathrm{d}s + \int_0^t \frac{\gamma(s)}{U(s)}\,\mathrm{d}W(s). \tag{2.56}$$

Finally, the unique solution to the general linear SDE (2.50) is given by $X(t) = U(t)V(t)$ where $U(t)$ and $V(t)$ are, respectively, given by (2.52) and (2.56) with $U(0) = 1$. That is,

$$X(t) = U(t)\left(X(0) + \int_0^t [\alpha(s) - \gamma(s)\delta(s)]U^{-1}(s)\,\mathrm{d}s + \int_0^t \gamma(s)U^{-1}(s)\,\mathrm{d}W(s)\right) \tag{2.57}$$

where $U(t) = \mathrm{e}^{\int_0^t \left(\beta(u) - \frac{1}{2}\delta^2(u)\right)\mathrm{d}u + \int_0^t \delta(u)\,\mathrm{d}W(u)} = \mathrm{e}^{\int_0^t \beta(u)\,\mathrm{d}u}\mathcal{E}_t(\delta\cdot W)$ and $U^{-1}(s) \equiv 1/U(s) = \mathrm{e}^{-\int_0^s \left(\beta(u) - \frac{1}{2}\delta^2(u)\right)\mathrm{d}u - \int_0^s \delta(u)\,\mathrm{d}W(u)}$.

The unique solution to (2.50) subject to an initial condition $X(t_0)$, $0 \leqslant t_0 < t$, is derived in exactly the same fashion as above with the initial condition placed at time t_0. By repeating the above steps, where the lower integration point in (2.52) and (2.56) is t_0, with $U(t_0) = 1$ replacing $U(0) = 1$ and $V(t_0) = X(t_0)$ replacing $V(0) = X(0)$, we have

$$X(t) = U(t)\left(X(t_0) + \int_{t_0}^t [\alpha(s) - \gamma(s)\delta(s)]U^{-1}(s)\,\mathrm{d}s + \int_{t_0}^t \gamma(s)U^{-1}(s)\,\mathrm{d}W(s)\right) \tag{2.58}$$

with $U(t) = \mathrm{e}^{\int_{t_0}^t \left(\beta(u) - \frac{1}{2}\delta^2(u)\right)\mathrm{d}u + \int_{t_0}^t \delta(u)\,\mathrm{d}W(u)}$, $U^{-1}(s) = \mathrm{e}^{-\int_{t_0}^s \left(\beta(u) - \frac{1}{2}\delta^2(u)\right)\mathrm{d}u - \int_{t_0}^s \delta(u)\,\mathrm{d}W(u)}$.

It should be noted that in the special case when $\gamma(t) \equiv \delta(t) \equiv 0$, with arbitrary initial condition $X(t_0) = x_0 \in \mathbb{R}$, the solution in (2.58) recovers the well-known solution to the initial value problem for a first-order linear ODE (ordinary differential equation). In this case, the random (Brownian) component vanishes and the SDE in (2.50) is simply the ODE:

$$\frac{\mathrm{d}}{\mathrm{d}t}X(t) - \beta(t)X(t) = \alpha(t)$$

subject to $X(t_0) = x_0$. Hence, (2.58) recovers the well-known unique solution

$$X(t) = U(t)\left(x_0 + \int_{t_0}^t \alpha(s)U^{-1}(s)\,\mathrm{d}s\right) = \mathrm{e}^{\int_{t_0}^t \beta(u)\,\mathrm{d}u}\left(x_0 + \int_{t_0}^t \alpha(s)\mathrm{e}^{-\int_{t_0}^s \beta(u)\,\mathrm{d}u}\,\mathrm{d}s\right).$$

This solution is an ordinary (deterministic) funtion of t, i.e. $X(t,\omega) \equiv X(t)$, $\forall\,\omega \in \Omega$.

Example 2.12. Solve the SDE

$$\mathrm{d}X(t) = (\alpha - \beta X(t))\,\mathrm{d}t + \sigma\,\mathrm{d}W(t)$$

for all $t \geqslant 0$, subject to $X(0) = x$ with constants $x, \alpha, \beta, \sigma \in \mathbb{R}$.

Solution. Note that the SDE is of the form in (2.50) with constant coefficients $\alpha(t) = \alpha$, $\beta(t) = -\beta$, $\gamma(t) = \sigma$, $\delta(t) \equiv 0$. The expression in (2.53) simplifies to $\mathcal{E}_t(\delta\cdot W) = \mathcal{E}_t(0) = 1$. Hence, $U(t) = \mathrm{e}^{-\beta t}$ and $U^{-1}(s) = \mathrm{e}^{\beta s}$. Substituting into (2.57) gives the solution

$$\begin{aligned} X(t) &= \mathrm{e}^{-\beta t}\left(x + \alpha\int_0^t \mathrm{e}^{\beta s}\,\mathrm{d}s + \sigma\int_0^t \mathrm{e}^{\beta s}\,\mathrm{d}W(s)\right) \\ &= \mathrm{e}^{-\beta t}x + \frac{\alpha}{\beta}(1 - \mathrm{e}^{-\beta t}) + \sigma\mathrm{e}^{-\beta t}\int_0^t \mathrm{e}^{\beta s}\,\mathrm{d}W(s). \end{aligned} \tag{2.59}$$

□

Note: the reader can verify that this solves the SDE. One way is to take the stochastic differential of the r.h.s. of (2.59) while using the Itô product rule. Alternatively, we can set $X(t) := f(t, Y(t))$, $Y(t) = \int_0^t \mathrm{e}^{\beta s}\,\mathrm{d}W(s)$, where $f(t,y) := \mathrm{e}^{-\beta t}x + \frac{\alpha}{\beta}(1-\mathrm{e}^{-\beta t}) + \sigma \mathrm{e}^{-\beta t}y$, and apply the appropriate Itô formula.

We note that a simple alternative method to arrive at this solution in (2.59) (without using (2.57)) is to define $\tilde{X}(t) := \mathrm{e}^{\beta t}X(t)$. This has the effect of eliminating the state variable dependence in the SDE. Indeed, by the Itô product rule, we have $\mathrm{d}\tilde{X}(t) = \alpha\mathrm{e}^{\beta t}\,\mathrm{d}t + \sigma\mathrm{e}^{\beta t}\,\mathrm{d}W(t)$, with coefficients independent of $\tilde{X}(t)$. Integrating, with $\tilde{X}(0) = X(0) = x$, gives the solution to $\tilde{X}(t)$:

$$\tilde{X}(t) = x + \frac{\alpha}{\beta}(\mathrm{e}^{\beta t} - 1) + \sigma \int_0^t \mathrm{e}^{\beta s}\,\mathrm{d}W(s). \tag{2.60}$$

The solution in (2.59) follows by $X(t) = \mathrm{e}^{-\beta t}\tilde{X}(t)$.

In Example 2.12, the solution represented in (2.59) is a Gaussian process involving an Itô integral that is a normal random variable, i.e., $X(t) = m_X(t) + \sigma\mathrm{e}^{-\beta t}Y(t)$ with mean function $m_X(t) := \mathrm{E}[X(t)] = \mathrm{e}^{-\beta t}x + \frac{\alpha}{\beta}(1-\mathrm{e}^{-\beta t})$ and $Y(t) := \int_0^t \mathrm{e}^{\beta s}\,\mathrm{d}W(s)$ is a zero-mean normal random variable:

$$Y(t) \sim Norm(0, \mathrm{Var}(Y(t)) = Norm\left(0, \int_0^t \mathrm{e}^{2\beta s}\,\mathrm{d}s\right) = Norm\left(0, \frac{1}{2\beta}(\mathrm{e}^{2\beta t}-1)\right).$$

Hence, $X(t) \stackrel{d}{=} Norm(m_X(t), \sigma^2\mathrm{e}^{-2\beta t}\,\mathrm{Var}(Y(t)))$, i.e.,

$$X(t) \stackrel{d}{=} Norm\left(\mathrm{e}^{-\beta t}x + \frac{\alpha}{\beta}(1-\mathrm{e}^{-\beta t}), \frac{\sigma^2}{2\beta}(1-\mathrm{e}^{-2\beta t})\right).$$

Applying the formulae in (2.19) to (2.59) also gives us the covariance function $c_X(t,v) := \mathrm{Cov}(X(t), X(v))$, for all $0 \leqslant t \leqslant v$:

$$\begin{aligned} c_X(t,v) &= \sigma^2\mathrm{e}^{-\beta(t+v)}\,\mathrm{Cov}\left(Y(t), Y(v)\right) = \sigma^2\mathrm{e}^{-\beta(t+v)}\,\mathrm{Var}\left(Y(t)\right) \\ &= \frac{\sigma^2}{2\beta}\mathrm{e}^{-\beta(t+v)}(\mathrm{e}^{2\beta t}-1) = \frac{\sigma^2}{2\beta}(\mathrm{e}^{-\beta(v-t)} - \mathrm{e}^{-\beta(v+t)}). \end{aligned}$$

We can also represent the solution as a functional of the Brownian motion up to time t by a trivial use of the antiderivative formula in (2.38). Here, we use the Itô product rule: $\mathrm{d}(\mathrm{e}^{\beta t}W(t)) = \beta\mathrm{e}^{\beta t}W(t)\,\mathrm{d}t + \mathrm{e}^{\beta t}\,\mathrm{d}W(t) \implies \mathrm{e}^{\beta t}W(t) = \beta\int_0^t \mathrm{e}^{\beta s}W(s)\,\mathrm{d}s + \int_0^t \mathrm{e}^{\beta s}\,\mathrm{d}W(s) \implies \int_0^t \mathrm{e}^{\beta s}\,\mathrm{d}W(s) = \mathrm{e}^{\beta t}W(t) - \beta\int_0^t \mathrm{e}^{\beta s}W(s)\,\mathrm{d}s$. Putting this last expression into (2.59) gives $X(t) = F(t, \{W(s); 0 \leqslant s \leqslant t\})$ with functional F of the Brownian path defined by

$$F(t, \{W(s); 0 \leqslant s \leqslant t\}) := \mathrm{e}^{-\beta t}x + \frac{\alpha}{\beta}(1-\mathrm{e}^{-\beta t}) + \sigma W(t) - \sigma\beta\int_0^t \mathrm{e}^{-\beta(t-s)}W(s)\,\mathrm{d}s. \tag{2.61}$$

In the chapter on interest rate modelling, we shall see that the above process, referred to as the Vasicek model for $\alpha, \beta, \sigma > 0$, is among the simplest one used to model the instantaneous (short) interest rate. A natural extension of this model is to allow the coefficients to be time-dependent functions. The resulting linear SDE is explicitly solved in the following example.

Example 2.13. Solve the SDE

$$\mathrm{d}X(t) = (a(t) - b(t)X(t))\,\mathrm{d}t + \sigma(t)\,\mathrm{d}W(t)$$

subject to $X(0) = x \in \mathbb{R}$, where $a(t), b(t), \sigma(t)$ are nonrandom continuous functions of time $t \geqslant 0$.

Solution. The SDE is of the form in (2.50) with coefficient functions $\alpha(t) \equiv a(t)$, $\beta(t) \equiv -b(t)$, $\gamma(t) \equiv \sigma(t)$, $\delta(t) \equiv 0$. Hence, $\mathcal{E}_t(\delta \cdot W) = \mathcal{E}_t(0) = 1$ and $U(t) = \mathrm{e}^{-\int_0^t b(s)\,\mathrm{d}s}, U^{-1}(s) = \mathrm{e}^{\int_0^s b(u)\,\mathrm{d}u}$. Substituting into (2.57) gives us the unique solution

$$\begin{aligned} X(t) &= \mathrm{e}^{-\int_0^t b(u)\,\mathrm{d}u}\left(x + \int_0^t \mathrm{e}^{\int_0^s b(u)\,\mathrm{d}u} a(s)\,\mathrm{d}s + \int_0^t \mathrm{e}^{\int_0^s b(u)\,\mathrm{d}u}\sigma(s)\,\mathrm{d}W(s)\right) \\ &= x\mathrm{e}^{-\int_0^t b(u)\,\mathrm{d}u} + \int_0^t \mathrm{e}^{-\int_s^t b(u)\,\mathrm{d}u} a(s)\,\mathrm{d}s + \int_0^t \mathrm{e}^{-\int_s^t b(u)\,\mathrm{d}u}\sigma(s)\,\mathrm{d}W(s). \qquad (2.62)\end{aligned}$$

□

[Note: the exponential functions of t can be moved inside the integrals in $\mathrm{d}s$ and $\mathrm{d}W(s)$.]

The above process is Gaussian since the integrand in the Itô integral is an ordinary (nonrandom) function of time. By isometry, and the zero-mean property of the Itô integral in (2.62), we have that $X(t)$ is normally distributed with mean and variance:

$$m_X(t) := \mathrm{E}[X(t)] = x\mathrm{e}^{-\int_0^t b(u)\,\mathrm{d}u} + \int_0^t \mathrm{e}^{-\int_s^t b(u)\,\mathrm{d}u} a(s)\,\mathrm{d}s\,, \qquad (2.63)$$

$$\mathrm{Var}(X(t)) = \mathrm{Var}\left(\int_0^t \mathrm{e}^{-\int_s^t b(u)\,\mathrm{d}u}\sigma(s)\,\mathrm{d}W(s)\right) = \int_0^t \mathrm{e}^{-2\int_s^t b(u)\,\mathrm{d}u}\sigma^2(s)\,\mathrm{d}s\,. \qquad (2.64)$$

The covariance function for the Gaussian process in (2.62) is computed by using (2.20) for $t \leqslant v$:

$$\begin{aligned} c_X(t,v) &= \mathrm{e}^{-\int_0^t b(u)\,\mathrm{d}u}\mathrm{e}^{-\int_0^v b(u)\,\mathrm{d}u}\,\mathrm{Cov}\left(\int_0^t \mathrm{e}^{\int_0^s b(u)\,\mathrm{d}u}\sigma(s)\,\mathrm{d}W(s), \int_0^v \mathrm{e}^{\int_0^s b(u)\,\mathrm{d}u}\sigma(s)\,\mathrm{d}W(s)\right) \\ &= \mathrm{e}^{-2\int_0^t b(u)\,\mathrm{d}u}\mathrm{e}^{-\int_t^v b(u)\,\mathrm{d}u}\int_0^t \mathrm{e}^{2\int_0^s b(u)\,\mathrm{d}u}\sigma^2(s)\,\mathrm{d}s \\ &= \mathrm{e}^{-\int_t^v b(u)\,\mathrm{d}u}\int_0^t \mathrm{e}^{-2\int_s^t b(u)\,\mathrm{d}u}\sigma^2(s)\,\mathrm{d}s. \end{aligned}$$

Example 2.14. Solve the SDE

$$\mathrm{d}X(t) = (\mathrm{e}^t - X(t))\,\mathrm{d}t + \mathrm{e}^{W(t)}\,\mathrm{d}W(t)\,,\ \ X(0) = x,$$

and compute the mean and variance of $X(t)$ for all $t \geqslant 0$.

Solution. This SDE is of the form in (2.50) with coefficients $\alpha(t) = \mathrm{e}^t, \beta(t) \equiv -1, \gamma(t) = \mathrm{e}^{W(t)}, \delta(t) \equiv 0$. Hence, $\mathcal{E}_t(\delta \cdot W) = \mathcal{E}_t(0) = 1$ and $U(t) = \mathrm{e}^{-t}, U^{-1}(s) = \mathrm{e}^s$. Substituting into (2.57) gives the unique solution

$$X(t) = \mathrm{e}^{-t}\left[x + \int_0^t \mathrm{e}^{2s}\,\mathrm{d}s + \int_0^t \mathrm{e}^{W(s)}\mathrm{e}^s\,\mathrm{d}W(s)\right] = x\mathrm{e}^{-t} + \sinh t + \int_0^t \mathrm{e}^{s-t}\mathrm{e}^{W(s)}\,\mathrm{d}W(s).$$

Note that the Itô integral is well-defined (as we compute the finite variance below). Hence, by the zero-mean property of the Itô, which is the only random component, we simply have

$$\mathrm{E}[X(t)] = x\mathrm{e}^{-t} + \sinh t \equiv (x - 1/2)\mathrm{e}^{-t} + \mathrm{e}^t/2\,.$$

Using Itô isometry, where $\mathrm{E}[\mathrm{e}^{2W(s)}] = \mathrm{e}^{2s}$,

$$\operatorname{Var}(X(t)) = \operatorname{Var}\left(\mathrm{e}^{-t}\int_0^t \mathrm{e}^{W(s)}\mathrm{e}^{s}\,\mathrm{d}W(s)\right) = \mathrm{e}^{-2t}\int_0^t \mathrm{e}^{2s}\mathrm{E}[\mathrm{e}^{2W(s)}]\,\mathrm{d}s$$
$$= \mathrm{e}^{-2t}\int_0^t \mathrm{e}^{4s}\,\mathrm{d}s = \frac{1}{2}\sinh 2t.$$

We note that this process is not a diffusion and is not Gaussian. We leave it as a simple exercise for the reader to compute the covariance function $c_X(t, v)$, $0 \leqslant t \leqslant v$. □

We close this section by describing a method for computing the moments; e.g., the mean, the second moment or variance, etc., of a process that satisfies a given SDE without the need to find a solution to the SDE. If the objective is to only compute such moments, then this method can be useful when the SDE is not readily solved or does not admit an explicit strong solution. It can also be used as an alternative to the approach that was used above where the SDE has an explicit solution which is firstly found, and then the moments (e.g., mean and variance) of the process are computed. As best demonstrated by example, the main idea is to take expectations on both sides of the given SDE in the process $X(t)$, and thereby arrive at an initial value ODE problem in time t for the mean function $m(t) := \mathrm{E}[X(t)]$. This ODE is then solved explicitly. To compute the second moment we apply Itô's formula, while using the given SDE in $X(t)$, to obtain a stochastic differential for $X^2(t)$. In integral form we then have $X^2(t)$ as an Itô process. By again taking expectations, we arrive at another initial value ODE problem in time t for the second moment $v(t) := \mathrm{E}[X^2(t)]$. Solving this explicitly gives $v(t)$ and hence the variance $\operatorname{Var}(X(t)) = v(t) - (m(t))^2$.

As a first simple example we consider the GBM process $\{S(t)\}_{t\geqslant 0}$ satisfying the SDE in integral form given in (2.37). Let $m(t) := \mathrm{E}[S(t)]$ be the mean function. Taking expectations on both sides of (2.37), while using Fubini's theorem where we move the expectation on the inside of the Riemann integral,

$$m(t) = S_0 + \mu\int_0^t \mathrm{E}[S(u)]\,\mathrm{d}u + \sigma\mathrm{E}\left[\int_0^t S(u)\,\mathrm{d}W(u)\right] = S_0 + \mu\int_0^t m(u)\,\mathrm{d}u.$$

Note that here we used the fact that $\mathrm{E}[S(u)] = m(u)$ and the zero-mean property of the Itô integral which we assume is well-defined. Differentiating the above integral equation on both sides w.r.t. t gives a linear first-order ODE: $m'(t) = \mu m(t)$ with intial condition $m(0) = S_0$. This is trivially solved to give $m(t) = S_0\mathrm{e}^{\mu t}$.

To compute the second moment function $v(t) := \mathrm{E}[S^2(t)]$, we consider the stochastic differential of $S^2(t)$. Using Itô's formula and the given SDE $\mathrm{d}S(t) = \mu S(t)\,\mathrm{d}t + \sigma S(t)\,\mathrm{d}W(t)$ gives

$$\mathrm{d}S^2(t) = 2S(t)\,\mathrm{d}S(t) + \big(\mathrm{d}S(t)\big)^2 = 2S(t)[\mu S(t)\,\mathrm{d}t + \sigma S(t)\,\mathrm{d}W(t)] + \sigma^2 S^2(t)\,\mathrm{d}t$$
$$= (2\mu + \sigma^2)S^2(t)\,\mathrm{d}t + 2\sigma S^2(t)\,\mathrm{d}W(t)\,.$$

In integral form we have (note: $S(0) = S_0$)

$$S^2(t) = S_0^2 + (2\mu + \sigma^2)\int_0^t S^2(u)\,\mathrm{d}u + 2\sigma\int_0^t S^2(u)\,\mathrm{d}W(u)\,.$$

[Note: this equation can also be derived directly from (2.47) where $f(x) := x^2$, $X \equiv S$, with volatility and drift coefficient processes $\sigma(t) \equiv \sigma S(t), \mu(t) \equiv \mu S(t)$.] Taking expectations on both sides of this equation, where $\mathrm{E}[S^2(u)] = v(u)$ and the assumed well-defined Itô integral has zero mean, gives

$$v(t) = S_0^2 + (2\mu + \sigma^2)\int_0^t \mathrm{E}[S^2(u)]\,\mathrm{d}u + 2\sigma\mathrm{E}\Big[\int_0^t S^2(u)\,\mathrm{d}W(u)\Big]$$
$$= S_0^2 + (2\mu + \sigma^2)\int_0^t v(u)\,\mathrm{d}u\,.$$

Differentiating on both sides w.r.t. t gives a linear first-order ODE: $v'(t) = (2\mu + \sigma^2)v(t)$, with intial condition $v(0) = S_0^2$. This is trivially solved to give $v(t) = S_0^2\mathrm{e}^{(2\mu+\sigma^2)t}$. The variance is hence given by

$$\mathrm{Var}(S(t)) = v(t) - (m(t))^2 = S_0^2\mathrm{e}^{(2\mu+\sigma^2)t} - S_0^2\mathrm{e}^{2\mu t} = S_0^2\mathrm{e}^{2\mu t}[\mathrm{e}^{\sigma^2 t} - 1].$$

In this simple example, $m(t), v(t), \mathrm{Var}(S(t))$ are more easily derived in the usual approach were we use our knowledge of the explicit solution for $S(t)$. In fact, we easily obtain the expectation of any power of the process simply from the m.g.f. of $W(t)$, i.e.,

$$\mathrm{E}[S^\alpha(t)] = S_0^\alpha\mathrm{e}^{\alpha(\mu-\sigma^2/2)t}\mathrm{E}[\mathrm{e}^{\alpha\sigma W(t)}] = S_0^\alpha\mathrm{e}^{\alpha(\mu-\sigma^2/2)t}\mathrm{e}^{\alpha^2\sigma^2 t/2} = S_0^\alpha\mathrm{e}^{\alpha\mu t}\mathrm{e}^{\alpha(\alpha-1)\sigma^2 t/2},\quad \alpha \in \mathbb{R}.$$

The same method can be applied to the SDE in Examples 2.12–2.14. For instance, in Example 2.12, the mean function $m(t) := \mathrm{E}[X(t)]$ satisfies

$$m'(t) + \beta m(t) = \alpha\,,\ \ m(0) = x\,,$$

with solution $m(t) \equiv m_X(t) = \big(x - \frac{\alpha}{\beta}\big)\mathrm{e}^{-\beta t} + \frac{\alpha}{\beta}$. Applying Itô's formula for $X^2(t)$ then gives

$$X^2(t) = X^2(0) + \int_0^t (2\alpha X(s) - 2\beta X^2(s) + \sigma^2)\,\mathrm{d}s + 2\sigma\int_0^t X(s)\,\mathrm{d}W(s)\,.$$

Taking expectations on both sides of this equation, with the Itô integral having zero mean, and then differentiating the resulting ordinary integral equation w.r.t. t gives a linear ODE for the second moment $v(t) := \mathrm{E}[X^2(t)]$:

$$v'(t) + 2\beta v(t) = 2\alpha m(t) + \sigma^2\,,\ \ v(0) = x^2.$$

Solving and upon simplifying gives $v(t) = (m(t))^2 + \frac{\sigma^2}{2\beta}(1 - \mathrm{e}^{-2\beta t})$. Hence, the variance is given by $\mathrm{Var}(X(t)) = \frac{\sigma^2}{2\beta}(1 - \mathrm{e}^{-2\beta t})$, which is the same expression we derived above where we solved the SDE and then applied Itô isometry.

As our next example we consider the SDE

$$\mathrm{d}X(t) = (\alpha - \beta X(t))\,\mathrm{d}t + \sigma\sqrt{X(t)}\,\mathrm{d}W(t)\,,\ \ X(0) = x > 0,$$

which in integral form is

$$X(t) = x + \int_0^t (\alpha - \beta X(s))\,\mathrm{d}s + \sigma\int_0^t \sqrt{X(s)}\,\mathrm{d}W(s).$$

It is important to remark that this SDE is **not linear** since the volatility function involves the square root of $X(t)$. As briefly discussed in a later chapter, this diffusion process is the so-called CIR process and can be used to model interest rates when α, β, σ are positive constants. Although beyond the scope of our discussion, it suffices to mention that this SDE does not admit a unique strong solution. However, the transition PDF for this strictly positive process on $(0, \infty)$ is known analytically for each type of boundary specification of the left endpoint (i.e., the origin). Without the use of any transition PDF, we can still employ the above ODE method for computing the mean and variance of the process. The validity of the method rests upon assumptions on the Itô integrals, as stated in the steps below.

We begin by taking expectations on both sides of the above equation with the square-integrability assumption that $\int_0^t \mathrm{E}\big[X(s)\big]\,\mathrm{d}s < \infty$ (hence $\mathrm{E}\big[\int_0^t \sqrt{X(s)}\,\mathrm{d}W(s)\big] = 0$) followed by differentiating the resulting ordinary integral equation w.r.t. t. This gives the same linear ODE as above with solution $m(t) \equiv m_X(t) = \big(x - \frac{\alpha}{\beta}\big)\mathrm{e}^{-\beta t} + \frac{\alpha}{\beta}$ as in Example 2.12. Assuming we can apply Itô's formula for $X^2(t)$ gives

$$X^2(t) = x^2 + \int_0^t [(2\alpha + \sigma^2)X(s) - 2\beta X^2(s)]\,\mathrm{d}s + 2\sigma \int_0^t X^{\frac{3}{2}}(s)\,\mathrm{d}W(s)\,.$$

We now take expectations on both sides of this equation and further assume that $\int_0^t \mathrm{E}\big[X^3(s)\big]\,\mathrm{d}s < \infty$ (hence $\mathrm{E}\big[\int_0^t X^{\frac{3}{2}}(s)\,\mathrm{d}W(s)\big] = 0$). This gives the ordinary integral equation in $v(t) := \mathrm{E}[X^2(t)]$,

$$v(t) = x^2 + \int_0^t [(2\alpha + \sigma^2)m(s) - 2\beta v(s)]\,\mathrm{d}s\,.$$

This is equivalent to: $v'(t) + 2\beta v(t) = (2\alpha + \sigma^2)m(t)$, $v(0) = x^2$, with $m(t)$ given above. The solution for $v(t)$ is then given by

$$\begin{aligned}
v(t) &= \mathrm{e}^{-2\beta t}x^2 + (2\alpha + \sigma^2)\mathrm{e}^{-2\beta t}\int_0^t \mathrm{e}^{2\beta s} m(s)\,\mathrm{d}s \\
&= \mathrm{e}^{-2\beta t}x^2 + (2\alpha + \sigma^2)\mathrm{e}^{-2\beta t}\int_0^t \mathrm{e}^{2\beta s}\left[\big(x - \frac{\alpha}{\beta}\big)\mathrm{e}^{-\beta s} + \frac{\alpha}{\beta}\right]\mathrm{d}s \\
&= \mathrm{e}^{-2\beta t}x^2 + (2\alpha + \sigma^2)\left[\big(x - \frac{\alpha}{\beta}\big)\Big(\frac{\mathrm{e}^{-\beta t} - \mathrm{e}^{-2\beta t}}{\beta}\Big) + \frac{\alpha}{\beta}\Big(\frac{1 - \mathrm{e}^{-2\beta t}}{2\beta}\Big)\right].
\end{aligned}$$

Finally, $\mathrm{Var}(X(t)) = v(t) - (m(t))^2$.

2.6.2 Existence and Uniqueness of a Strong Solution to an SDE

An important question when finding a strong solution to an SDE of the form given by (2.49) is whether such a solution exists and, if so, whether the solution is unique. The following theorem gives sufficient conditions for the existence and uniqueness of a strong solution to the SDE in (2.49). We omit the proof of this theorem as the technical details can be found in other more specialized textbooks on stochastic analysis. The conditions in the theorem are not necessary, but are rather mild sufficient conditions that guarantee the existence a unique strong solution, i.e., that there is a unique process $\{X(t)\}_{t \geqslant 0}$ satisfying (2.49).

Theorem 2.8. *Assume the following conditions are satisfied:*

- *The coefficient functions $\mu(t, x)$ and $\sigma(t, x)$ are locally Lipschitz in x, uniformly in t. That is, for arbitrary positive constants T and N, there exists a constant K depending possibly only on T and N such that*

$$|\mu(t, x) - \mu(t, y)| + |\sigma(t, x) - \sigma(t, y)| < K|x - y|,$$

whenever $|x|, |y| \leqslant N$ and $0 \leqslant t \leqslant T$.

- *The coefficient functions $\mu(t,x)$ and $\sigma(t,x)$ satisfy the linear growth condition in the variable x, i.e.,*
$$|\mu(t,x)| + |\sigma(t,x)| \leqslant K(1+|x|).$$

- *The initial value $X(0)$ is independent of the Brownian motion up to arbitrary time T and has a finite second moment, i.e., $X(0)$ is independent of $\mathcal{F}_T^W \equiv \sigma(W(t); 0 \leqslant t \leqslant T)$ and $\mathrm{E}[X^2(0)] < \infty$.*

Then, the SDE in (2.49) has a unique strong solution $\{X(t)\}_{t\geqslant 0}$ with continuous paths $X(t,\omega), t \geqslant 0$.

For a given SDE, the conditions in the above theorem can be readily checked. For example, the first (Lipschitz) condition holds if the coefficient functions have continuous first partial derivatives $\frac{\partial \mu}{\partial x}(t,x)$ and $\frac{\partial \sigma}{\partial x}(t,x)$. The second condition is satisfied when the coefficient functions $\mu(t,x)$ and $\sigma(t,x)$ have at most a linear growth in x for large values of x and are also bounded for arbitrarily small values of x. In most cases, the SDE is subject to a constant initial condition $X(0) \in \mathbb{R}$ so that the above third condition is automatically satisfied. In the case of a general linear SDE, we have already shown that the solution is given by (2.57). This includes cases when the coefficients $\alpha(t), \beta(t), \gamma(t), \delta(t)$ are bounded nonrandom functions of time. In these cases, $\mu(t,x)$ and $\sigma(t,x)$ are linear functions of x, for all t, and hence the conditions in the above theorem are indeed satisfied and so a unique strong solution exists. In Examples 2.12 and 2.13, we have solved the SDE and found the unique strong solution as given by (2.59) and (2.62), respectively. We note that unique strong solutions also exist when the coefficient functions in the SDE satisfy milder conditions than those listed in the above theorem. For instance, for a time-homogeneous SDE with (time-independent) drift and volatility coefficients $\mu(x)$ and $\sigma(x)$, it can be shown that there exists a unique strong solution when $\mu(x)$ satisfies a Lipschitz condition, $|\mu(x)-\mu(y)| < K|x-y|$, and $\sigma(x)$ satisfies a Hölder condition, $|\sigma(x)-\sigma(y)| < K|x-y|^\alpha$, with order $\alpha \geqslant 1/2$, for some constant K.

2.7 The Markov Property, Martingales, Feynman–Kac Formulae, and Transition CDFs and PDFs

We have already shown in Chapter 1 that Brownian motion is a Markov process. Generally, a process $\{X(t)\}_{t\geqslant 0}$ has the Markov property if the probability of the event $\{X(t) \leqslant y\}$, $y \in \mathbb{R}$, conditional on all the past information $\mathcal{F}_s$ (i.e., the information about the complete history of the process up to a prior time s) is the same as its probability conditional on only knowing the process endpoint value $X(s)$, for all $0 \leqslant s \leqslant t$. That is, the Markov property can be formally stated equivalently as

$$\mathbb{P}(X(t) \leqslant y \mid \mathcal{F}_s) = \mathbb{P}(X(t) \leqslant y \mid X(s)) \tag{2.65}$$

or, for any Borel function $h : \mathbb{R} \to \mathbb{R}$,

$$\mathrm{E}[h(X(t)) \mid \mathcal{F}_s] = \mathrm{E}[h(X(t)) \mid X(s)], \tag{2.66}$$

for all $0 \leqslant s \leqslant t$. Note that the specific choice of $h(x) := \mathbb{I}_{x\leqslant y}$ recovers (2.65). Hence, when the Markov property holds, conditioning on natural filtration $\mathcal{F}_s = \sigma(X(u); 0 \leqslant u \leqslant s)$ is equal to conditioning on $\sigma(X(s))$. In particular, for the case of a discrete-time process

$X(t_0), X(t_1), \ldots, X(t_{n-1}), X(t_n), \ldots$, with times $t_0 < t_1 < \ldots < t_{n-1} < t_n < \ldots$, the above property takes the form that is familiar in the theory of discrete-time Markov chains:

$$\mathbb{P}(X(t_m) \leqslant y \mid X(t_0), X(t_1), \ldots, X(t_{n-1}), X(t_n)) = \mathbb{P}(X(t_m) \leqslant y \mid X(t_n)) \tag{2.67}$$

for all $m \geqslant n$, i.e., when conditioning on the values of the process for a set $\{t_i\}_{0 \leqslant i \leqslant n}$ of previous times, the only conditioning that is relevant is the value of the process at the most recent time t_n. We note that this follows from (2.65) by setting $s = t_n, t = t_m$, i.e., $\mathcal{F}_s \equiv \mathcal{F}_{t_n} = \sigma(X(t_0), X(t_1), \ldots, X(t_{n-1}), X(t_n))$ and $\sigma(X(s)) = \sigma(X(t_n))$, and then using the usual shorthand notation $\mathbb{P}(A \mid \sigma(Y_1, \ldots, Y_n)) \equiv \mathbb{P}(A \mid Y_1, \ldots, Y_n)$ for expressing the probability of an event A conditional on a σ-algebra generated by a set of random variables $Y_1, \ldots, Y_n$.

We are interested in computing conditional expectations involving functions of a process $X = \{X(t)\}_{t \geqslant 0} \in \mathbb{R}$ that solves a given SDE as in (2.49) subject to some initial condition $X(0) = x$. In particular, we will need to compute expectations as in (2.66) for the case that the process X is Markov. Let us now fix some time $T \geqslant 0$. Then, we shall denote the conditional expectation of a function $h(X(T))$, conditioned on the process having a given value $X(t) = x \in \mathbb{R}$ at a time $t \leqslant T$, by

$$\mathrm{E}_{t,x}[h(X(T))] := \mathrm{E}[h(X(T)) \mid X(t) = x].$$

So the subscript t, x is shorthand notation for conditioning on a given value $X(t) = x$ of the process at time t. It should be clear that this conditional expectation is an ordinary (nonrandom) function of the ordinary variables t and x, i.e., $\mathrm{E}_{t,x}[h(X(T))] = g(t,x)$ for any fixed T. Therefore, the Markov property in (2.66) is expressible as $\mathrm{E}[h(X(T)) \mid \mathcal{F}_t] = \mathrm{E}_{t,X(t)}[h(X(T))] = g(t, X(t))$, for all $0 \leqslant t \leqslant T$. This expectation is now the *random variable* given by the function $g(t, X(t))$ of the random variable $X(t)$ and evaluates to $g(t,x)$ upon setting $X(t) = x$. Hence, if we know the conditional probability distribution of random variable $X(T)$, given $X(t) = x$, then we could compute $g(t,x)$. That is, assume the conditional probability density function (PDF) of $X(T)$, given $X(t)$, exists. Recall from the previous chapter that this PDF is the transition PDF for the process X. From Definition 1.2, we have

$$\mathrm{E}_{t,x}[h(X(T))] = \int_{\mathbb{R}} h(y) f_{X(T)|X(t)}(y \mid x)\, \mathrm{d}y = \int_{\mathbb{R}} h(y) p(t, T; x, y)\, \mathrm{d}y. \tag{2.68}$$

In some cases, this integral can be computed analytically. Otherwise, we need to employ a numerical method. For example, if the expression for the PDF is known, then we can compute the integral using an appropriate numerical quadrature algorithm. Monte Carlo methods can generally be used to compute the above integral by sampling (i.e., simulating) the paths of the process at time T given their fixed value x at time t. Different simulation approaches may be applied. One approach is to use a time-stepping algorithm for simulating the paths according to the SDE. Alternatively, if the transition PDF is known, then we can sample the (path endpoint) value $X(T)$ according to its distribution.

The next theorem tells us that solutions to an SDE are Markov processes.

Theorem 2.9. *Let $\{X(t)\}_{t \geqslant 0}$ be a solution to the SDE in (2.49) with some given initial condition. Then,*

$$\mathrm{E}[h(X(T)) \mid \mathcal{F}_t] = g(t, X(t)) \tag{2.69}$$

where $g(t,x) = \mathrm{E}_{t,x}[h(X(T))]$, for all $0 \leqslant t \leqslant T$ and Borel function h.

A rigorous proof of this Markov property is beyond the scope of this text. The important content of this result is that the expectation of any function of the process at a future time $T \geqslant t$, conditional on the filtration (or path history) at time t, is given simply by its expectation conditional only on the path value at time t. In practice, we can apply this theorem by first computing $\mathrm{E}_{t,x}[h(X(T))]$, and then putting $X(t)$ in the place of variable x.

A simple nonrigorous, yet instructive, argument that leads us to the fact that the solution to an SDE has the Markov property is to let $T = t + \Delta t$, for a small time step $\Delta t \approx 0$. Then, by the integral form of the SDE:

$$X(t+\Delta t) = X(t) + \int_t^{t+\Delta t} \mu(s, X(s))\,\mathrm{d}s + \int_t^{t+\Delta t} \sigma(s, X(s))\,\mathrm{d}W(s).$$

This expresses the value of the process at future time $t + \Delta t$ in terms of its value at any current time t plus an ordinary integral and an Itô integral. For small Δt, the integrals are well approximated by holding the integrand coefficient functions constant and evaluated at the left endpoint $s = t$ of the time interval $[t, t + \Delta t]$. This gives us the approximation $X(t + \Delta t) \approx X(t) + \mu(t, X(t))\Delta t + \sigma(t, X(t))\Delta W(t)$, where $\Delta W(t) \equiv W(t + \Delta t) - W(t)$. Hence, the left-hand side of (2.65), for $T = t + \Delta t$, is approximated by

$$\mathbb{P}(X(t+\Delta t) \leqslant y \mid \mathcal{F}_t) \approx \mathbb{P}(X(t) + \mu(t, X(t))\Delta t + \sigma(t, X(t))\Delta W(t) \leqslant y \mid \mathcal{F}_t) = g(X(t))$$

where $g(x) = \mathbb{P}(x+\mu(t,x)\Delta t+\sigma(t,x)\Delta W(t) \leqslant y)$. Here, we used the independence proposition, where the Brownian increment $\Delta W(t)$ is independent of $\mathcal{F}_t$ and $X(t)$ is $\mathcal{F}_t$-measurable. We can now put back the conditioning on $X(t)$ in the unconditional probability since $\Delta W(t)$ is independent of $X(t)$, so the function $g(x)$ is equally given by the conditional probability: $g(x) = \mathbb{P}(x + \mu(t,x)\Delta t + \sigma(t,x)\Delta W(t) \leqslant y \mid X(t)) \approx \mathbb{P}(X(t + \Delta t) \leqslant y \mid X(t))$. Hence, we recover (approximately) the Markov property in (2.65), $\mathbb{P}(X(t+\Delta t) \leqslant y \mid \mathcal{F}_t) \approx \mathbb{P}(X(t + \Delta t) \leqslant y \mid X(t))$. In the above theorem, this relation holds exactly.

By the Markov property of the process X, we also have the following martingale property for a process defined via an expectation of some function of $X(T)$ conditional on $X(t)$.

Proposition 2.10. *Let $\{X(t)\}_{t\geqslant 0}$ satisfy the SDE in (2.49) subject to some initial condition. Let $\phi : \mathbb{R} \to \mathbb{R}$ be a Borel function and define $f(t,x) := \mathrm{E}_{t,x}[\phi(X(T))]$ for fixed $T > 0$, assuming $\mathrm{E}_{t,x}[|\phi(X(T))|] < \infty$. Then, the stochastic process*

$$Y(t) := f(t, X(t)),\ 0 \leqslant t \leqslant T,$$

is a martingale w.r.t. any filtration $\{\mathcal{F}_t\}_{t\geqslant 0}$ for Brownian motion.

Proof. This process is an example of a Doob-Lévy martingale. Since $Y(t)$ is a Borel function of $X(t)$, where $X(t)$ is $\mathcal{F}_t$-measurable, then $Y(t)$ is $\mathcal{F}_t$-measurable for all $0 \leqslant t \leqslant T$. Also, $\mathrm{E}[|Y(t)|] = \mathrm{E}[|f(t, X(t))|] = \mathrm{E}[|\mathrm{E}[\phi(X(T))|X(t)]|] \leqslant \mathrm{E}[\mathrm{E}[|\phi(X(T))||X(t)]] < \infty$ by assumption.

Let $0 \leqslant s \leqslant t \leqslant T$. Note that, based on Theorem 2.9,

$$\mathrm{E}[\phi(X(T)) \mid \mathcal{F}_t] = \mathrm{E}_{t,X(t)}[\phi(X(T))] = f(t, X(t)) = Y(t)$$

for any $0 \leqslant t \leqslant T$. Using this relation and applying the tower property gives the martingale expectation property:

$$\mathrm{E}[Y(t) \mid \mathcal{F}_s] = \mathrm{E}[\,\mathrm{E}[\phi(X(T)) \mid \mathcal{F}_t] \mid \mathcal{F}_s] = \mathrm{E}[\phi(X(T)) \mid \mathcal{F}_s] = f(s, X(s)) = Y(s). \qquad \square$$

Based on the Markov property of any solution to an SDE, we are now ready to discuss the very important connection that exists between an SDE and a PDE. In what follows we will find it very convenient to make use of the differential operator $\mathcal{G}$ defined by

$$\mathcal{G}_{t,x} f(t,x) := \frac{1}{2}\sigma^2(t,x)\frac{\partial^2 f}{\partial x^2}(t,x) + \mu(t,x)\frac{\partial f}{\partial x}(t,x). \tag{2.70}$$

This differential operator in the variables (t, x) is the so-called *generator* for the process X and it acts on all functions $f \in C^{1,2}$, i.e., having continuous partial derivatives $\frac{\partial f}{\partial t}$, $\frac{\partial f}{\partial x}$ and $\frac{\partial^2 f}{\partial x^2}$. Using this operator we can now rewrite the differential and integral forms of the Itô formula in (2.43) and (2.44) in compact form as [6]

$$\mathrm{d}f(t, X(t)) = \left(\frac{\partial}{\partial t} + \mathcal{G}_{t,x}\right) f(t, X(t))\,\mathrm{d}t + \sigma(t, X(t))\frac{\partial f}{\partial x}(t, X(t))\,\mathrm{d}W(t). \tag{2.71}$$

and

$$f(t, X(t)) = f(0, X(0)) + \int_0^t \left(\frac{\partial}{\partial s} + \mathcal{G}_{s,x}\right) f(s, X(s))\,\mathrm{d}s + \int_0^t \sigma(s, X(s))\frac{\partial f}{\partial x}(s, X(s))\,\mathrm{d}W(s). \tag{2.72}$$

Here, we are assuming that the integrals are well-defined for all $0 \leqslant t \leqslant T$, for some fixed time $T > 0$. The Itô integral process in (2.72) is well-defined whereby the square integrability condition holds, i.e.,

$$\int_0^T \mathrm{E}\left[\left(\sigma(s, X(s))\frac{\partial f}{\partial x}(s, X(s))\right)^2\right] \mathrm{d}s < \infty\,. \tag{2.73}$$

This leads directly to a useful representation of a martingale Markov process. In particular, we see that the process $\{M_f(t), 0 \leqslant t \leqslant T\}$ defined by

$$M_f(t) := f(t, X(t)) - \int_0^t \left(\frac{\partial}{\partial s} + \mathcal{G}_{s,x}\right) f(s, X(s))\,\mathrm{d}s \tag{2.74}$$

is a martingale (w.r.t. a filtration for Brownian motion) since, according to (2.72),

$$M_f(t) = M_f(0) + \int_0^t \sigma(s, X(s))\frac{\partial f}{\partial x}(s, X(s))\,\mathrm{d}W(s), \tag{2.75}$$

where $M_f(0) = f(0, X(0))$. This is an Itô integral process and therefore satisfies the martingale property:

$$\mathrm{E}[M_f(T) \mid \mathcal{F}_t] = M_f(t)\,,\ 0 \leqslant t \leqslant T. \tag{2.76}$$

One nice application of (2.74) is that it offers us a method of constructing a martingale (w.r.t. a filtration for Brownian motion) via a smooth function of a given process that solves an SDE. We also have a way of verifying whether or not such a process is a martingale. We summarize this in the following points.

1. Take any given smooth function $f \in C^{1,2}$ where (2.73) holds. Then, the process (2.74) is a martingale w.r.t. any filtration for Brownian motion. In particular, given an X-process with solution $X(t)$ to the SDE in (2.49), we have a martingale process by subtracting $f(t, X(t))$ and the Riemann integral in (2.74). The Riemann integrand is computed by operating with the corresponding generator in (2.70) for the X-process.

2. Take any given smooth function $f \in C^{1,2}$ where (2.73) holds and where f is a solution to the PDE: $\frac{\partial}{\partial t} f(t, x) + \mathcal{G}_{t,x} f(t, x) = 0$ for all x in the domain of f, $0 < t < T$. Then, the Riemann integral in (2.74) is identically zero and the process $M_f(t) := f(t, X(t))$, $0 \leqslant t \leqslant T$, where $X(t)$ is a solution to the SDE in (2.49), is a martingale w.r.t. any filtration for Brownian motion. This process has zero drift component in (2.71)–(2.72) and hence also has the representation in (2.75).

[6]Note the shorthand notation: $\left(\frac{\partial}{\partial t} + \mathcal{G}_{t,x}\right) f(t, X(t)) \equiv \left[\frac{\partial}{\partial t} f(t, x) + \mathcal{G}_{t,x} f(t, x)\right]_{x=X(t)}$ and similarly within the integrals with t replaced by s.

Let's specialize the above analysis to the case of standard BM where we take $X(t) \equiv W(t)$ (or $W(t)$ plus a constant). The SDE in (2.49) is simply $\mathrm{d}X(t) = \mathrm{d}W(t)$ with drift $\mu(t,x) \equiv 0$ and volatility $\sigma(t,x) \equiv 1$. From (2.70) we see that the generator $\mathcal{G}_{t,x} \equiv \mathcal{G}_x^W$ for standard BM operates only on the x variable:

$$\mathcal{G}_x^W f(t,x) := \frac{1}{2}\frac{\partial^2}{\partial x^2} f(t,x). \tag{2.77}$$

Hence, according to (2.74) and (2.75), we have the martingale process (given $f \in C^{1,2}$)

$$M_f(t) := f(t, W(t)) - \int_0^t \left(\frac{\partial}{\partial s} + \frac{1}{2}\frac{\partial^2}{\partial x^2} \right) f(s, W(s))\,\mathrm{d}s \tag{2.78}$$

$$= f(0,0) + \int_0^t \frac{\partial f}{\partial x}(s, W(s))\,\mathrm{d}W(s). \tag{2.79}$$

Note the shorthand: $\left(\frac{\partial}{\partial s} + \frac{1}{2}\frac{\partial^2}{\partial x^2}\right) f(s, W(s)) \equiv \left[\frac{\partial}{\partial s} f(s,x) + \frac{1}{2}\frac{\partial^2}{\partial x^2} f(s,x)\right]_{x=W(s)}$.
In particular, $M_f(t) := f(t, W(t))$, $0 \leqslant t \leqslant T$, is a martingale w.r.t. any filtration for Brownian motion if f is a solution to the PDE: $\frac{\partial}{\partial t} f(t,x) + \frac{1}{2}\frac{\partial^2}{\partial x^2} f(t,x) = 0$, for all x in the domain of f, $0 < t < T$, and where the Itô integral in (2.79) is assumed well-defined with square-integrability condition $\int_0^T \mathrm{E}\left[\left(\frac{\partial f}{\partial x}(s, W(s))\right)^2\right] \mathrm{d}s < \infty$.

Example 2.15. Consider $f(t,x) := x$, $x \in \mathbb{R}$, $t \geqslant 0$. It is trivial to check that f is a solution to $\frac{\partial f}{\partial t} + \frac{1}{2}\frac{\partial^2 f}{\partial x^2} = 0$. Hence, the process defined by $M_f(t) := f(t, W(t)) = W(t)$, $t \geqslant 0$, is a martingale w.r.t. any filtration for Brownian motion. We also see this from (2.79), where $f(0,0) = 0$ and $\frac{\partial f}{\partial x} = 1$, giving

$$M_f(t) = 0 + \int_0^t 1 \cdot \mathrm{d}W(s) = W(t).$$

Of course, this is consistent with the martingale property of standard Brownian motion proven in Chapter 1.

Example 2.16. Consider $f(t,x) := x^2 - t$, $x \in \mathbb{R}$, $t \geqslant 0$. It is again trivial that $\frac{\partial f}{\partial t} + \frac{1}{2}\frac{\partial^2 f}{\partial x^2} = -1 + \frac{1}{2}(2) = 0$. Hence, $M_f(t) := f(t, W(t)) = W^2(t) - t$, $t \geqslant 0$, is a martingale w.r.t. any filtration for Brownian motion. We also see this from (2.79), where $f(0,0) = 0$ and $\frac{\partial f}{\partial x} = 2x$, giving

$$M_f(t) = 0 + \int_0^t 2W(s)\,\mathrm{d}W(s) = W^2(t) - t.$$

Again, this is consistent with the martingale property proven in Chapter 1.

Example 2.17. Consider $f(t,x) := \mathrm{e}^{-\frac{1}{2}\alpha^2 t + \alpha x}$, $x \in \mathbb{R}$, $t \geqslant 0$. We have $\frac{\partial f}{\partial t} + \frac{1}{2}\frac{\partial^2 f}{\partial x^2} = -\frac{1}{2}\alpha^2 + \frac{1}{2}\alpha^2 = 0$. Hence, $M_f(t) := f(t, W(t)) = \mathrm{e}^{-\frac{1}{2}\alpha^2 t + \alpha W(t)} \equiv \mathcal{E}_t(\alpha \cdot W)$, $t \geqslant 0$, is a martingale w.r.t. a filtration for Brownian motion. We recall that this process (also referred to as an exponential martingale) was already proven to be a martingale in Chapter 1. Using (2.79), where $f(0,0) = 1$ and $\frac{\partial f}{\partial x} = \alpha f$, we also have the representation:

$$M_f(t) = 1 + \alpha \int_0^t \mathrm{e}^{-\frac{1}{2}\alpha^2 s + \alpha W(s)}\,\mathrm{d}W(s).$$

Example 2.18. Let $f(t,x) := tx$, $x \in \mathbb{R}$, $t \geqslant 0$. We have $\frac{\partial f}{\partial t} = x$, $\frac{\partial f}{\partial x} = t$, $\frac{\partial^2 f}{\partial x^2} = 0$. Hence, $\frac{\partial f}{\partial t} + \frac{1}{2}\frac{\partial^2 f}{\partial x^2} = x \not\equiv 0$ and $f(t, W(t)) = tW(t)$ is not a martingale (as is also clear since

$\mathrm{E}[tW(t)|\mathcal{F}_s] = tW(s) \neq sW(s)$). However, (2.78) and (2.79) give us equivalent representations of a martingale w.r.t. a filtration for Brownian motion:

$$M_f(t) := tW(t) - \int_0^t W(s)\,\mathrm{d}s = \int_0^t s\,\mathrm{d}W(s)\,.$$

Consider now the case of GBM with constants $\mu \in \mathbb{R}, \sigma > 0$, i.e., with SDE

$$\mathrm{d}X(t) = \mu X(t)\,\mathrm{d}t + \sigma X(t)\,\mathrm{d}W(t)\,, \quad X(0) = x_0 > 0,\ t \geqslant 0.$$

The unique solution to this SDE is given by

$$X(t) = x_0 \mathrm{e}^{\mu t} \mathcal{E}_t(\sigma \cdot W) \equiv x_0 \mathrm{e}^{\mu t} \mathrm{e}^{-\frac{1}{2}\alpha^2 t + \alpha W(t)}. \tag{2.80}$$

The generator $\mathcal{G}_{t,x} \equiv \mathcal{G}_x$ for this GBM process (with respective drift and volatility functions $\mu(t,x) \equiv \mu x$ and $\sigma(t,x) \equiv \sigma x$) is independent of t:

$$\mathcal{G}_x f(t,x) := \frac{1}{2}\sigma^2 x^2 \frac{\partial^2}{\partial x^2} f(t,x) + \mu x \frac{\partial}{\partial x} f(t,x). \tag{2.81}$$

From (2.74) and (2.75) we have the martingale process (given $f \in C^{1,2}$)

$$M_f(t) := f(t, X(t)) - \int_0^t \left(\frac{\partial}{\partial s} + \frac{1}{2}\sigma^2 x^2 \frac{\partial^2}{\partial x^2} + \mu x \frac{\partial}{\partial x} \right) f(s, X(s))\,\mathrm{d}s \tag{2.82}$$

$$= x_0 + \sigma \int_0^t X(s) \frac{\partial f}{\partial x}(s, X(s))\,\mathrm{d}W(s), \tag{2.83}$$

where $f(0, X(0)) = f(0, x_0) = x_0$. Hence, $M_f(t) := f(t, X(t))$, $0 \leqslant t \leqslant T$, with $X(t)$ in (2.80), is a martingale w.r.t. any filtration for Brownian motion if f satisfies $\frac{\partial}{\partial t} f(t,x) + \frac{1}{2}\sigma^2 x^2 \frac{\partial^2}{\partial x^2} f(t,x) + \mu x \frac{\partial}{\partial x} f(t,x) = 0$, for all x in the domain of f, $0 < t < T$, and assuming the square-integrability condition $\int_0^T \mathrm{E}\big[X^2(s)\big(\frac{\partial f}{\partial x}(s, X(s))\big)^2\big]\,\mathrm{d}s < \infty$ holds.

Example 2.19. Consider $f(t,x) := x$, $x > 0$, $t \geqslant 0$. Then, $\frac{\partial}{\partial t} f(t,x) + \frac{1}{2}\sigma^2 x^2 \frac{\partial^2}{\partial x^2} f(t,x) + \mu x \frac{\partial}{\partial x} f(t,x) = \mu x$, which is zero for all $x > 0$ iff $\mu = 0$. Hence, the process $M_f(t) := f(t, X(t)) = X(t) = x_0 \mathrm{e}^{\mu t} \mathrm{e}^{-\frac{1}{2}\alpha^2 t + \alpha W(t)}$, $t \geqslant 0$, is a martingale w.r.t. any filtration for Brownian motion iff $\mu = 0$. Of course, this is consistent with what we have already shown previously. We also see this directly in the above SDE which has a martingale solution only when the drift term is identically zero in which case $x_0 \mathrm{e}^{-\frac{1}{2}\alpha^2 t + \alpha W(t)}$ solves the zero-drift SDE $\mathrm{d}X(t) = \sigma X(t)\,\mathrm{d}W(t)$. However, we can construct a martingale with the above choice of $f(t,x) = x$ and for any $\mu \in \mathbb{R}$ if we make use of (2.83), where $\frac{\partial f}{\partial x} = 1$:

$$M_f(t) := x_0 + \sigma \int_0^t X(s)\,\mathrm{d}W(s) = x_0 \left[1 + \sigma \int_0^t \mathrm{e}^{(\mu - \frac{1}{2}\alpha^2)s + \alpha W(s)}\,\mathrm{d}W(s) \right].$$

We can also express this in terms of a Riemann integral via (2.82). Note that this is a well-defined Itô integral process with square-integrability condition satisfied: $\int_0^t \mathrm{e}^{(2\mu - \alpha^2)s} \mathrm{E}\big[\mathrm{e}^{2\alpha W(s)}\big]\,\mathrm{d}s = \int_0^t \mathrm{e}^{(2\mu + \alpha^2)s} < \infty$, for all $t \in [0, \infty)$.

A very important consequence of the martingale property in (2.76) is the following theorem, which shows that a solution to certain parabolic PDEs can be represented as a conditional expectation.

Theorem 2.11 (Feynman–Kac). *Given a fixed $T > 0$, let $\{X(t)\}_{t\geqslant 0}$ satisfy the SDE in (2.49) and let $\phi : \mathbb{R} \to \mathbb{R}$ be a Borel function. Moreover, assume the square integrability condition (2.73) holds. Let $f(t,x)$ be a $C^{1,2}$ function solving the PDE $\frac{\partial f}{\partial t} + \mathcal{G}_{t,x} f = 0$, i.e.,*

$$\frac{\partial f}{\partial t}(t,x) + \frac{1}{2}\sigma^2(t,x)\frac{\partial^2 f}{\partial x^2}(t,x) + \mu(t,x)\frac{\partial f}{\partial x}(t,x) = 0\,, \tag{2.84}$$

for all x, $0 < t < T$, subject to the terminal condition $f(T,x) = \phi(x)$. Then, assuming that $\mathrm{E}_{t,x}[|\phi(X(T))|] < \infty$, $f(t,x)$ has the representation

$$f(t,x) = \mathrm{E}_{t,x}[\phi(X(T))] \equiv \mathrm{E}[\phi(X(T)) \mid X(t) = x] \tag{2.85}$$

for all x, $0 \leqslant t \leqslant T$.

We note that if $\phi(x)$ is continuous, then $f(T-,x) \equiv \lim_{t\nearrow T} f(t,x) = f(T,x) = \phi(x)$, i.e., we have continuity of the solution at $t = T$, for all x.

Proof. Assuming the square integrability condition (2.73), then according to the above discussion we have that the process defined in (2.74) satisfies the martingale property in (2.76). Now, let f satisfy the PDE in (2.84). This implies that the integral in (2.74) vanishes since the integrand function is identically zero, i.e., $\frac{\partial}{\partial s} f(s,x) + \mathcal{G}_{s,x} f(s,x) = 0$, for all $s > 0$ and all values of x. Hence, the process $M_f(t) := f(t, X(t)), 0 \leqslant t \leqslant T$, is a martingale. Combining this with the Markov property of the process, and finally substituting the terminal condition for the random variable $f(T, X(T)) = \phi(X(T))$, we have

$$f(t,X(t)) = \mathrm{E}\big[f(T,X(T)) \,\big|\, \mathcal{F}_t\big] = \mathrm{E}_{t,X(t)}\big[f(T,X(T))\big] = \mathrm{E}_{t,X(t)}\big[\phi(X(T))\big]$$

so that $f(t,x) = \mathrm{E}_{t,x}\big[\phi(X(T))\big]$ for all $x, 0 \leqslant t \leqslant T$. □

This theorem hence shows that the solution at current time t, given by (2.85), is in the form of a conditional expectation of the random variable $\phi(X(T))$, where ϕ is the given boundary value function and $X(T)$ is the random variable corresponding to the endpoint value of the process at future (terminal) time T, where the process solves the SDE in (2.49) subject to it having current time-t value $X(t) = x$. From our previous discussion surrounding (2.68), we see that this theorem gives us a probabilistic representation of the solution to the parabolic PDE in (2.84) subject to the (terminal time) boundary value function $f(T,x) = \phi(x)$. Alternatively, the theorem can be used in the opposite sense; that is, it provides a PDE approach for evaluating a conditional expectation of the form in (2.85).

We now consider the simplest example of how this theorem is used in the case where the underlying Itô process is just standard Brownian motion. In particular, we solve the simple heat equation on the real line and thereby obtain a probabilistic representation of the solution as an expectation involving the endpoint value of Brownian motion.

Example 2.20. Solve the boundary value problem:

$$\frac{\partial f}{\partial t} + \frac{1}{2}\frac{\partial^2 f}{\partial x^2} = 0\,,$$

for $x \in \mathbb{R}, 0 \leqslant t \leqslant T$, subject to $f(x,T) = \phi(x)$ where ϕ is an arbitrary function. Give the explicit solution for $\phi(x) = x^2$.

Solution. Observe that this PDE is of the same form as in (2.84) with coefficient functions $\sigma(t,x) \equiv 1, \mu(t,x) \equiv 0$. The corresponding SDE in (2.49) is then

$$\mathrm{d}X(t) = \mathrm{d}W(t)\,.$$

This trivial linear SDE has the solution $X(t) = x_0 + W(t)$. As seen below, the end result does not depend on x_0 since $X(T) - X(t) = W(T) - W(t)$. Using (2.85) and the fact that $W(T) - W(t)$ and $W(t)$ are independent, the solution takes the equivalent forms

$$\begin{aligned} f(t,x) &= \mathrm{E}[\phi(X(T)) \mid X(t) = x] = \mathrm{E}[\phi(W(T) - W(t) + X(t)) \mid X(t) = x] \\ &= \mathrm{E}[\phi(W(T) - W(t) + x)] \\ &= \mathrm{E}[\phi(W(T)) \mid W(t) = x] \\ &= \int_{-\infty}^{\infty} \phi(y) p(t,T;x,y)\, dy\,, \quad \text{for } t < T, \end{aligned}$$

where $p(t,T;x,y) := \frac{\mathrm{e}^{-\frac{(y-x)^2}{2(T-t)}}}{\sqrt{2\pi(T-t)}}$ is the transition PDF of standard Brownian motion (BM). The last line follows immediately from (1.14). Note that, for $t = T$, the boundary value condition is satisfied where $f(T,x) = \mathrm{E}[\phi(W(T) - W(T) + x)] = \mathrm{E}[\phi(x)] = \phi(x)$.

For $\phi(x) = x^2$, we simply have

$$\begin{aligned} f(t,x) &= \mathrm{E}[(W(T) - W(t) + x)^2] \\ &= \mathrm{E}[(W(T) - W(t))^2] + 2x\mathrm{E}[W(T) - W(t)] + x^2 = (T-t) + x^2\,, \end{aligned}$$

for all $0 \leqslant t \leqslant T$. Note that the terminal condition is satisfied, $f(T,x) = x^2$, and this function satisfies the above PDE since $\frac{\partial f}{\partial t} + \frac{1}{2}\frac{\partial^2 f}{\partial x^2} = -1 + \frac{1}{2}(2) = 0$. □

We note that the square integrability condition in Theorem 2.11 can be shown to hold in the above example. Moreover, $f(t,x)$ is a $C^{1,2}$ function since $p(t,T;x,y)$ is a $C^{1,2}$ function in the (t,x) variables for $t < T$. In the case $\phi(x) = x^2$, we see that this follows trivially. More generally, assuming an arbitrary ϕ function such that the above y-integral exists, we can verify that the above integral represents a solution to the PDE. Indeed, the corresponding linear differential operator in (2.70) is now $\mathcal{G}_{t,x} f := \frac{1}{2}\frac{\partial^2 f}{\partial x^2}$, and reversing the order of differentiation and integration (w.r.t. the y variable) we have, for all $t < T$:

$$\left(\frac{\partial}{\partial t} + \mathcal{G}_{t,x}\right) f(t,x) = \int_{-\infty}^{\infty} \phi(y) \left(\frac{\partial}{\partial t} + \mathcal{G}_{t,x}\right) p(t,T;x,y)\, dy = 0\,,$$

since (as can be verified explicitly and directly) the above PDF $p = p(t,T;x,y)$ solves the PDE $\frac{\partial p}{\partial t} + \mathcal{G}_{t,x} p = 0$. For continuous $\phi(x)$, the solution is also continuous w.r.t. time $t \in [0,T]$ where $f(T-,x) = f(T,x) = \phi(x)$. This is the case as the transition PDF approaches the *Dirac delta function* centred at zero, denoted by $\delta(\cdot)$, as $t \nearrow T$, i.e.,

$$p(T-,T;x,y) \equiv \lim_{t \nearrow T} p(t,T;x,y) = \lim_{\tau \searrow 0} \frac{\mathrm{e}^{-\frac{(y-x)^2}{2\tau}}}{\sqrt{2\pi\tau}} = \delta(x-y).$$

Here, we have used one representation of the Dirac delta function as the limit of an infinitesimally narrow Gaussian PDF. The Dirac delta function is even, $\delta(x-y) = \delta(y-x)$, and has the defining (sifting) property:

$$f(T-,x) = \int_{\mathbb{R}} p(T-,T;x,y)\phi(y)\, dy = \int_{\mathbb{R}} \delta(x-y)\phi(y)\, dy = \phi(x)\,,$$

for any function ϕ that is continuous at x. The above delta function terminal condition is a general property of any transition PDF, as shown in Proposition 2.12 below. The above sifting property arises naturally by the Dirac measure defined in (A.29). Viewed as a distribution over $\mathbb{R}$, the only outcome that occurs with probability one is the single point x.

The Dirac delta function can then be related to the Dirac (singular) measure $\delta_x(y)$ for a given point $x \in \mathbb{R}$, $\mathrm{d}\delta_x(y) = \delta(y-x)\,\mathrm{d}y$. Formally, the Dirac delta function $\delta(x)$ is also related to the Heaviside unit step function $H(x)$, where $H(x)$ equals 1 for $x > 0$, equals 0 for $x < 0$ and equals $1/2$ at $x = 0$. In particular, its derivative is the delta function, $H'(x) = \delta(x)$. As a function of y, we write the differential $\mathrm{d}H(y-x) = H'(y-x)\,\mathrm{d}y = \delta(y-x)\,\mathrm{d}y$. Hence, when considered as a Riemann–Stieltjes integral with integrator $H(y-x)$, a function $\phi(y)$ that is continuous at the point $y = x$ will exhibit the sifting property:

$$\int_I \phi(y)\delta(y-x)\,\mathrm{d}y \equiv \int_I \phi(y)\,\mathrm{d}H(y-x) = \phi(x)$$

if x is in any interval $I \subset \mathbb{R}$ and the integral is zero if $x \not\subset I$. Note that when $I = \mathbb{R}$ the integral equals $\phi(x)$. The reader will note that exactly the same property is satisfied if we use the unit indicator function $\mathbb{I}_{\{y \geqslant x\}}$ as integrator in the place of $H(y-x)$. The two are equivalent for all $y \neq x$ and they are used as alternate definitions of the unit step function.

We can now use Theorem 2.11 to obtain the *backward Kolmogorov PDE* (in the so-called backward-time variables t, x) that is solved by any transition CDF, $P(t,T;x,y) := \mathbb{P}(X(T) \leqslant y \mid X(t) = x)$, and hence, its corresponding transition PDF for a diffusion process with SDE in (2.49).

Proposition 2.12. *Assume the square-integrability condition in Theorem 2.11 holds. Then, a transition PDF, $p = p(t,T;x,y)$, for the process $\{X(t)\}_{t\geqslant 0}$ with the generator in (2.70) solves the backward Kolmogorov PDE:*

$$\left(\frac{\partial}{\partial t} + \mathcal{G}_{t,x}\right) p = 0\,, \tag{2.86}$$

where $\lim_{t \nearrow T} p(t,T;x,y) \equiv p(T-,T;x,y) = \delta(x-y)$.

Proof. We begin by writing the transition probability function as a conditional expectation:

$$P(t,T;x,y) = \mathbb{P}(X(T) \leqslant y \mid X(t) = x) = \mathrm{E}[\mathbb{I}_{\{X(T)\leqslant y\}} \mid X(t) = x]\,.$$

By the above Feynman–Kac theorem, then P (for fixed T, y) solves the PDE

$$\left(\frac{\partial}{\partial t} + \mathcal{G}_{t,x}\right) P(t,T;x,y) = 0\,,$$

with terminal condition $P(T,T;x,y) = \mathbb{P}(y \geqslant x) = \mathbb{I}_{\{y\geqslant x\}} \equiv \phi(x)$. Taking partial derivatives w.r.t. y on both sides of the above PDE, and using the fact that the order of the differential operators $\left(\frac{\partial}{\partial t} + \mathcal{G}_{t,x}\right)$ and $\frac{\partial}{\partial y}$ can be reversed, gives

$$\left(\frac{\partial}{\partial t} + \mathcal{G}_{t,x}\right) \frac{\partial}{\partial y} P(t,T;x,y) = 0.$$

This is exactly (2.86) since the transition PDF $p(t,T;x,y) := \frac{\partial}{\partial y} P(t,T;x,y)$. The delta function terminal condition is seen to arise as follows, since the transition function approaches the unit step function as $t \nearrow T$,

$$\begin{aligned}\mathrm{d}P(T-,T;x,y) &= \mathrm{d}H(y-x) = \delta(y-x)\,\mathrm{d}y \\ &= p(T-,T;x,y)\,\mathrm{d}y\,.\end{aligned}$$

□

We remark that any function p (or P) that is a solution to the backward Kolmogorov PDE and is a conditional density (or distribution) function of some Markov process, as a diffusion or Itô process, is a transition PDF (or CDF). In fact, a transition PDF $p = p(t,T;x,y)$ is called a *fundamental solution* to the Kolmogorov PDE in (2.86) and its defining properties are that: (i) p is nonnegative, jointly continuous in the variables $t,T;x,y$, twice continuously differentiable in the spatial variables and continuously differentiable in the time variables; (ii) for any bounded Borel function ϕ then the function defined by $u(t,x) := \int_{\mathbb{R}} \phi(y)p(t,T;x,y)\,\mathrm{d}y$ is bounded and also satisfies the same Kolmogorov PDE; (iii) for continuous ϕ, $\lim_{t\nearrow T} u(t,x) \equiv u(T-,x) = \phi(x)$ for all x. Property (iii) is equivalent to the Dirac delta function limit, $\lim_{t\nearrow T} p(t,T;x,y) = p(T-,T;x,y) = \delta(x-y)$. Hence, generally, if given a transition PDF p, the conditional expectation in (2.85), i.e.,

$$f(t,x) = \int_{\mathbb{R}} \phi(y)\, p(t,T;x,y)dy\,, \tag{2.87}$$

solves the backward Kolmogorov PDE in (2.84) with terminal condition $f(T,x) = \phi(x)$. We showed this specifically for the simple case of BM in the above example.

Example 2.21. Consider a GBM process $\{S(t)\}_{t\geqslant 0} \in \mathbb{R}_+$ with SDE

$$\mathrm{d}S(t) = \mu S(t)\,\mathrm{d}t + \sigma S(t)\,\mathrm{d}W(t)\,,$$

where $\mu, \sigma > 0$ are constants.

(a) Provide the corresponding backward Kolmogorov PDE and obtain the transition CDF and PDF.

(b) Solve the PDE
$$\frac{\partial f}{\partial t} + \frac{1}{2}\sigma^2 x^2 \frac{\partial^2 f}{\partial x^2} + \mu x \frac{\partial f}{\partial x} = 0\,,$$
for $x > 0, t \leqslant T$, subject to $f(T,x) = \phi(x)$ where ϕ is an arbitrary function. Give the explicit solution for $\phi(x) = x\mathbb{I}_{\{x>a\}}$, with constant $a > 0$.

Solution.
(a) The drift and diffusion coefficient functions are time independent linear functions: $\mu(t,x) = \mu x$ and $\sigma(t,x) = \sigma x$. According to (2.70), the generator $\mathcal{G}_{t,x} \equiv \mathcal{G}_x$ is the differential operator

$$\mathcal{G}_x := \frac{1}{2}\sigma^2 x^2 \frac{\partial^2}{\partial x^2} + \mu x \frac{\partial}{\partial x}\,.$$

The transition CDF $P = P(t,T;x,y)$ hence solves the PDE in (2.86):

$$\frac{\partial P}{\partial t} + \frac{1}{2}\sigma^2 x^2 \frac{\partial^2 P}{\partial x^2} + \mu x \frac{\partial P}{\partial x} = 0\,,$$

for all $t < T, x, y > 0$ with $P(T,T;x,y) = \mathbb{I}_{\{x\leqslant y\}}$. The transition CDF is given by the conditional expectation:

$$P(t,T;x,y) = \mathbb{P}(S(T) \leqslant y \mid S(t) = x) = \mathrm{E}[\mathbb{I}_{\{S(T)\leqslant y\}} \mid S(t) = x]\,.$$

We have already computed this in the previous chapter by substituting the strong solution for GBM in the form

$$S(T) = S(t)\mathrm{e}^{(\mu - \frac{1}{2}\sigma^2)(T-t) + \sigma(W(T)-W(t))}\,,$$

giving

$$\begin{aligned} P(t,T;x,y) &= \mathrm{E}[\mathbb{I}_{\{\ln(S(T)/y)\leqslant 0\}} \mid S(t)=x] \\ &= \mathbb{P}\left(\frac{W(T)-W(t)}{\sqrt{T-t}} \leqslant -\frac{\ln(x/y)+(\mu-\frac{1}{2}\sigma^2)(T-t)}{\sigma\sqrt{T-t}}\right) \\ &= \mathcal{N}\left(\frac{\ln(y/x)-(\mu-\frac{1}{2}\sigma^2)(T-t)}{\sigma\sqrt{T-t}}\right). \end{aligned} \tag{2.88}$$

Here, we used the fact that $W(T)-W(t)$ is independent of $W(t)$, and hence independent of $S(t)$, where $W(T)-W(t) \stackrel{d}{=} \sqrt{T-t}Z$, $Z \sim Norm(0,1)$. Differentiating the above CDF with respect to y gives the known lognormal density (see (1.44) with the drift replacement $\mu \to \mu - \frac{1}{2}\sigma^2$):

$$p(t,T;x,y) = \frac{1}{y\sigma\sqrt{2\pi(T-t)}} \exp\left(-\frac{[\ln(y/x)-(\mu-\frac{1}{2}\sigma^2)(T-t)]^2}{2\sigma^2(T-t)}\right), \tag{2.89}$$

for all $x, y > 0, t < T$, and zero otherwise. The reader can verify that the transition CDF in (2.88) has limit $P(T-,T;x,y) = H(y-x)$.

(b) The solution $f(t,x)$ can be obtained by the Feynman–Kac Theorem 2.11 or, alternatively, directly from (2.87). Let's solve for $f(t,x)$ using both equivalent approaches. In the first approach, we use the above strong solution to the SDE. By (2.85), and the independence of $W(T)-W(t)$ and $S(t)$, we have

$$\begin{aligned} f(t,x) &= \mathrm{E}[\phi(S(T)) \mid S(t)=x] = \mathrm{E}[\phi(S(t)\mathrm{e}^{(\mu-\frac{1}{2}\sigma^2)(T-t)+\sigma(W(T)-W(t))}) \mid S(t)=x] \\ &= \mathrm{E}[\phi(x\mathrm{e}^{(\mu-\frac{1}{2}\sigma^2)(T-t)+\sigma(W(T)-W(t))})] \\ &= \mathrm{E}[\phi(x\mathrm{e}^{(\mu-\frac{1}{2}\sigma^2)(T-t)+\sigma\sqrt{T-t}Z})] \\ &= \int_{-\infty}^{\infty} \phi(x\mathrm{e}^{(\mu-\frac{1}{2}\sigma^2)(T-t)+\sigma\sqrt{T-t}z})n(z)\,\mathrm{d}z\,. \end{aligned}$$

This integral, assuming it exists, represents the solution to the PDE for arbitrary function ϕ. In particular, for $\phi(y) = y\mathbb{I}_{\{y>a\}} = y\mathbb{I}_{\{\ln(y/a)>0\}}$ we have

$$\begin{aligned} f(t,x) &= x\mathrm{e}^{(\mu-\frac{1}{2}\sigma^2)(T-t)}\mathrm{E}\left[\mathrm{e}^{\sigma\sqrt{T-t}Z}\mathbb{I}_{\{\ln(x/a)+(\mu-\frac{1}{2}\sigma^2)(T-t)+\sigma\sqrt{T-t}Z>0\}}\right] \\ &= x\mathrm{e}^{(\mu-\frac{1}{2}\sigma^2)(T-t)}\mathrm{E}\left[\mathrm{e}^{\sigma\sqrt{T-t}Z}\mathbb{I}_{\{Z>A\}}\right] \end{aligned}$$

with constant $A \equiv -\frac{\ln(x/a)+(\mu-\frac{1}{2}\sigma^2)(T-t)}{\sigma\sqrt{T-t}}$, for all $t < T$. For $t = T$, we simply have $f(T,x) = x\mathbb{I}_{\{x>a\}}$. This expectation is evaluated (see identity (B.1)) using $\mathrm{E}[\mathrm{e}^{BZ}\mathbb{I}_{\{Z>A\}}] = \mathrm{e}^{B^2/2}\mathcal{N}(B-A)$, with constant $B \equiv \sigma\sqrt{T-t}$, giving

$$\begin{aligned} f(t,x) &= x\mathrm{e}^{(\mu-\frac{1}{2}\sigma^2)(T-t)} \cdot \mathrm{e}^{\frac{1}{2}\sigma^2(T-t)}\mathcal{N}\left(\sigma\sqrt{T-t} + \frac{\ln(x/a)+(\mu-\frac{1}{2}\sigma^2)(T-t)}{\sigma\sqrt{T-t}}\right) \\ &= x\mathrm{e}^{\mu(T-t)}\mathcal{N}\left(\frac{\ln(x/a)+(\mu+\frac{1}{2}\sigma^2)(T-t)}{\sigma\sqrt{T-t}}\right). \end{aligned}$$

The reader can check that this expression solves the above PDE by computing the partial derivatives $\frac{\partial f}{\partial t}, \frac{\partial^2 f}{\partial x^2}, \frac{\partial f}{\partial x}$. Moreover, in the limit $t \nearrow T$ (defining $\tau = T-t$):

$$f(T-,x) = \lim_{\tau\searrow 0} x\mathrm{e}^{\mu\tau}\mathcal{N}\left(\frac{\ln(x/a)+(\mu-\frac{1}{2}\sigma^2)\tau}{\sigma\sqrt{\tau}}\right) = \lim_{\tau\searrow 0} x\,\mathcal{N}\left(\frac{\ln(x/a)}{\sigma\sqrt{\tau}}\right) = x\,H(x-a)\,.$$

[Note that this equals $f(T,x) = \phi(x) = x\,\mathbb{I}_{\{x>a\}}$ for all x, except at the point of discontinuity $x = a$ of $\phi(x)$, i.e., $\phi(a) = 0$ and $f(T-,a) = aH(0) = a/2$.]

In the second approach, we use (2.87) and insert the above transition PDF to obtain

$$
\begin{aligned}
f(t,x) &= \int_0^\infty \phi(y)\,p(t,T;x,y)\,\mathrm{d}y = \frac{1}{\sigma\sqrt{2\pi(T-t)}}\int_0^\infty \phi(y)\mathrm{e}^{-\frac{1}{2}\left[\frac{\ln(y/x)-\mu(T-t)}{\sigma\sqrt{T-t}}\right]^2}\frac{\mathrm{d}y}{y} \\
&\left(\text{let } z = \frac{\ln(y/x)-\mu(T-t)}{\sigma\sqrt{T-t}},\ y = x\mathrm{e}^{(\mu-\frac{1}{2}\sigma^2)(T-t)+\sigma\sqrt{T-t}z},\ \frac{\mathrm{d}y}{y} = \sigma\sqrt{T-t}\,\mathrm{d}z\right) \\
&= \int_0^\infty \phi(x\mathrm{e}^{(\mu-\frac{1}{2}\sigma^2)(T-t)+\sigma\sqrt{T-t}z})\,\frac{\mathrm{e}^{-\frac{1}{2}z^2}}{\sqrt{2\pi}}\,\mathrm{d}z\,.
\end{aligned}
$$

As required, this produces exactly the same solution as we have above by the first approach. Of course, we should not be surprised by this fact since, by definition, $p(t,T;x,y)$ is the conditional density of $S(T)$ at y, given $S(t) = x$, and hence $f(t,x) = \mathrm{E}[\phi(S(T)) \mid S(t) = x] = \int_0^\infty \phi(y)\,p(t,T;x,y)\,\mathrm{d}y$. □

In the next example, we obtain the transition CDF/PDF for the GBM process with time-dependent drift and diffusion coefficients.

Example 2.22. Consider a GBM process $\{S(t)\}_{t\geqslant 0} \in \mathbb{R}_+$ with SDE

$$\mathrm{d}S(t) = \mu(t)S(t)\,\mathrm{d}t + \sigma(t)S(t)\,\mathrm{d}W(t)\,, \tag{2.90}$$

where $\mu(t), \sigma(t) > 0$ are continuous (ordinary) functions of time $t \geqslant 0$. State the corresponding backward Kolmogorov PDE and obtain the transition CDF and PDF.

Solution. We have a linear SDE with coefficient functions $\mu(t,x) = \mu(t)x$ and $\sigma(t,x) = \sigma(t)x$. The corresponding generator is the differential operator

$$\mathcal{G}_{t,x} := \frac{1}{2}\sigma^2(t)x^2\frac{\partial^2}{\partial x^2} + \mu(t)x\frac{\partial}{\partial x}.$$

The transition CDF $P = P(t,T;x,y)$ solves the PDE in (2.86):

$$\frac{\partial P}{\partial t} + \frac{1}{2}\sigma^2(t)x^2\frac{\partial^2 P}{\partial x^2} + \mu(t)x\frac{\partial P}{\partial x} = 0\,,$$

for all $t < T, x, y > 0$ with $P(T,T;x,y) = \mathbb{I}_{\{x\leqslant y\}}$. By the Feynman–Kac Theorem 2.11, the transition CDF is obtained by evaluating the conditional expectation

$$P(t,T;x,y) = \mathbb{P}(S(T) \leqslant y \mid S(t) = x) = \mathbb{P}(X(T) \leqslant \ln y \mid X(t) = \ln x)\,,$$

where we define the process $X(t) := \ln S(t), t \geqslant 0$. The SDE (2.90) has unique strong solution given by (2.57) with $\alpha(t) \equiv \gamma(t) \equiv 0$, $\beta(t) \equiv \mu(t)$, $\delta(t) \equiv \sigma(t)$, i.e.,

$$S(t) = S(0)\mathrm{e}^{\int_0^t \mu(s)\,\mathrm{d}s - \frac{1}{2}\int_0^t \sigma^2(s)\,\mathrm{d}s + \int_0^t \sigma(s)\,\mathrm{d}W(s)},$$

hence

$$S(T) = S(t)\,\mathrm{e}^{\int_t^T \mu(s)\,\mathrm{d}s - \frac{1}{2}\int_t^T \sigma^2(s)\,\mathrm{d}s + \int_t^T \sigma(s)\,\mathrm{d}W(s)}$$

and

$$X(T) = X(t) + \int_t^T \mu(s)\,\mathrm{d}s - \frac{1}{2}\int_t^T \sigma^2(s)\,\mathrm{d}s + \int_t^T \sigma(s)\,\mathrm{d}W(s)\,.$$

It is convenient to define the *time-averaged* drift and volatility functions:

$$\bar{\mu}(t,T) := \frac{1}{T-t}\int_t^T \mu(s)\,\mathrm{d}s\,, \quad \bar{\sigma}(t,T) := \sqrt{\frac{1}{T-t}\int_t^T \sigma^2(s)\,\mathrm{d}s}\,. \tag{2.91}$$

Since $\int_t^T \sigma(s)\,\mathrm{d}W(s) \stackrel{d}{=} W\left(\bar{\sigma}^2(t,T)(T-t)\right) \stackrel{d}{=} \bar{\sigma}(t,T)\sqrt{T-t}Z$, $Z \sim Norm(0,1)$,

$$X(T) \stackrel{d}{=} X(t) + [\bar{\mu}(t,T) - \frac{1}{2}\bar{\sigma}^2(t,T)](T-t) + \bar{\sigma}(t,T)\sqrt{T-t}Z\,,$$

where $X(T) - X(t)$, and hence Z, is independent of $X(t)$. Combining these facts into the above gives:

$$\begin{aligned}
P(t,T;x,y) &= \mathbb{P}\left(X(T) - X(t) \leqslant \ln y - X(t) \mid X(t) = \ln x\right) = \mathbb{P}\left(X(T) - X(t) \leqslant \ln(y/x)\right) \\
&= \mathbb{P}\left([\bar{\mu}(t,T) - \frac{1}{2}\bar{\sigma}^2(t,T)](T-t) + \bar{\sigma}(t,T)\sqrt{T-t}Z \leqslant \ln(y/x)\right) \\
&= \mathcal{N}\left(\frac{\ln(y/x) - [\bar{\mu}(t,T) - \frac{1}{2}\bar{\sigma}^2(t,T)](T-t)}{\bar{\sigma}(t,T)\sqrt{T-t}}\right).
\end{aligned} \tag{2.92}$$

We note that this is the form of the transition CDF for standard GBM in Example 2.21, wherein the drift and volatility coefficients in (2.88) are now replaced by the time-averaged ones: $\mu \to \bar{\mu}(t,T)$ and $\sigma \to \bar{\sigma}(t,T)$. Observe, however, that the CDF in (2.92) is not a function of only $T-t$, i.e., the GBM process with time-dependent coefficients is an example of a time-inhomogeneous process (i.e., not time-homogeneous as in Example 2.21). Differentiating (2.92) with respect to y gives the lognormal density (analogous to 2.89):

$$p(t,T;x,y) = \frac{1}{y\bar{\sigma}\sqrt{2\pi(T-t)}}\exp\left(-\frac{[\ln(y/x) - (\bar{\mu} - \frac{1}{2}\bar{\sigma}^2)(T-t)]^2}{2\bar{\sigma}^2(T-t)}\right), \tag{2.93}$$

for all $x, y > 0, t < T$, and zero otherwise, where $\bar{\mu} \equiv \bar{\mu}(t,T)$, $\bar{\sigma} \equiv \bar{\sigma}(t,T)$. □

Example 2.23. Consider the process $\{X(t)\}_{t\geqslant 0} \in \mathbb{R}$ in Example 2.12, i.e.,

$$\mathrm{d}X(t) = (\alpha - \beta X(t))\,\mathrm{d}t + \sigma\,\mathrm{d}W(t)\,. \tag{2.94}$$

For $\alpha = 0$, this process is specifically called the *Ornstein–Uhlenbeck process* or OU process for short. Derive the corresponding transition CDF and PDF.

Solution. From the analysis in Example 2.12, we have the strong solution for $X(T)$ in terms of $X(t) = x$, which we can write in equivalent forms using time-changed BM:

$$\begin{aligned}
X(T) &= \mathrm{e}^{-\beta(T-t)}x + \frac{\alpha}{\beta}(1 - \mathrm{e}^{-\beta(T-t)}) + \sigma\int_t^T \mathrm{e}^{-\beta(T-s)}\,\mathrm{d}W(s) \\
&\stackrel{d}{=} \mathrm{e}^{-\beta(T-t)}x + \frac{\alpha}{\beta}(1 - \mathrm{e}^{-\beta(T-t)}) + \sigma W\left(\frac{1 - \mathrm{e}^{-2\beta(T-t)}}{2\beta}\right) \\
&\stackrel{d}{=} \mathrm{e}^{-\beta(T-t)}x + \frac{\alpha}{\beta}(1 - \mathrm{e}^{-\beta(T-t)}) + \sigma\mathrm{e}^{-\beta(T-t)}W\left(\frac{\mathrm{e}^{2\beta(T-t)} - 1}{2\beta}\right) \\
&\stackrel{d}{=} \mathrm{e}^{-\beta(T-t)}x + \frac{\alpha}{\beta}(1 - \mathrm{e}^{-\beta(T-t)}) + \sigma\sqrt{\frac{1 - \mathrm{e}^{-2\beta(T-t)}}{2\beta}}\,Z\,.
\end{aligned}$$

The last line displays $X(T)$ as a normal random variable, where $Z \sim Norm(0,1)$ and independent of $X(t)$. Hence, the transition CDF is a normal CDF:

$$P(t,T;x,y) = \mathbb{P}\left(X(T) \leqslant y \mid X(t) = x\right) = \mathcal{N}\left(\frac{y - [\mathrm{e}^{-\beta(T-t)}x + \frac{\alpha}{\beta}(1 - \mathrm{e}^{-\beta(T-t)})]}{\sigma\sqrt{(1 - \mathrm{e}^{-2\beta(T-t)})/2\beta}}\right), \tag{2.95}$$

and the transition PDF is the Gaussian function

$$p(t,T;x,y) = \frac{1}{\sigma}\sqrt{\frac{2\beta}{1 - \mathrm{e}^{-2\beta(T-t)}}}\; n\left(\frac{y - [\mathrm{e}^{-\beta(T-t)}x + \frac{\alpha}{\beta}(1 - \mathrm{e}^{-\beta(T-t)})]}{\sigma\sqrt{(1 - \mathrm{e}^{-2\beta(T-t)})/2\beta}}\right), \tag{2.96}$$

for all $x, y \in \mathbb{R}$, $t < T$. □

Note that a transition PDF p (or CDF P) solving a given Kolmogorov PDE as in (2.86), subject to $p(T-,T;x,y) = \delta(x-y)$, is in general cases not necessarily a unique solution. This is the case even if we require p (or P) to be a PDF (or CDF). If a diffusion has one or both of its endpoints (left or right endpoint) as a regular boundary, then the behaviour of the process at the endpoint can be specified differently. An example of this is the specification of a regular reflecting boundary versus a regular killing (absorbing) boundary as in the case of BM that is either reflected or killed at an upper or lower finite boundary point. The known transition PDFs for both respective cases are of course different, yet both solve the same Kolmogorov PDE for BM and have limit $p(T-,T;x,y) = \delta(x-y)$. The key point is that the Kolmogorov PDE and terminal time condition make no mention of the boundary conditions imposed on the solution as a function of the spatial variable x. To obtain a unique fundamental solution that corresponds to a transition PDF (assuming of course that such a solution exists) one generally needs to also specify the spatial boundary conditions at both endpoints of the process. In some cases, such as for BM or drifted BM on $\mathbb{R}$, both endpoints $\pm\infty$ of the process are natural boundaries (not regular) and there is then a unique transition PDF on $\mathbb{R}$, i.e., the Gaussian PDF we have already derived. Similarly, for GBM the two endpoints of the state space $(0,\infty)$ are natural boundaries and hence the process has a unique transition PDF on $\mathbb{R}_+$, i.e., the known lognormal PDF.

The following result extends Theorem 2.11 and, as we shall see in later chapters, is used for pricing (single-asset) financial derivatives via a PDE based approach.

Theorem 2.13 ("Discounted" Feynman–Kac). *Fix $T > 0$ and let $\{X(t)\}_{t\geqslant 0}$ satisfy the SDE in (2.49). Let the same assumptions stated in Theorem (2.11) hold and assume $r(t,x)$: $[0,T] \times \mathbb{R} \to \mathbb{R}$ is a lower-bounded continuous function. Then, the function defined by the conditional expectation*

$$f(t,x) := \mathrm{E}_{t,x}[\mathrm{e}^{-\int_t^T r(u,X(u))\,\mathrm{d}u}\phi(X(T))] \equiv \mathrm{E}[\mathrm{e}^{-\int_t^T r(u,X(u))\,\mathrm{d}u}\phi(X(T)) \mid X(t) = x] \tag{2.97}$$

solves the PDE $\frac{\partial f}{\partial t} + \mathcal{G}_{t,x}f - r(t,x)f = 0$, *i.e.,*

$$\frac{\partial f}{\partial t}(t,x) + \frac{1}{2}\sigma^2(t,x)\frac{\partial^2 f}{\partial x^2}(t,x) + \mu(t,x)\frac{\partial f}{\partial x}(t,x) - r(t,x)f(t,x) = 0\,, \tag{2.98}$$

for all x, $0 < t < T$, subject to the terminal condition $f(T,x) = \phi(x)$.

Proof. This result follows by first rewriting the exponential factor as

$$\mathrm{e}^{-\int_t^T r(u,X(u))\,\mathrm{d}u} = \mathrm{e}^{-\int_0^T r(u,X(u))\,\mathrm{d}u} \cdot \mathrm{e}^{\int_0^t r(u,X(u))\,\mathrm{d}u}\,.$$

The process defined by $g_t := \mathrm{e}^{-\int_0^t r(u,X(u))\,\mathrm{d}u} f(t, X(t))$ is a martingale since $g_t = \mathrm{E}[g_T \mid \mathcal{F}_t]$, where $g_T = \mathrm{e}^{-\int_0^T r(u,X(u))\,\mathrm{d}u} f(T, X(T)) = \mathrm{e}^{-\int_0^T r(u,X(u))\,\mathrm{d}u}\phi(X(T))$:

$$\begin{aligned} g_t = \mathrm{e}^{-\int_0^t r(u,X(u))\,\mathrm{d}u} f(t, X(t)) &= \mathrm{E}_{t,X(t)}[\mathrm{e}^{-\int_0^T r(u,X(u))\,\mathrm{d}u}\phi(X(T))] \\ &= \mathrm{E}[\mathrm{e}^{-\int_0^T r(u,X(u))\,\mathrm{d}u}\phi(X(T)) \mid \mathcal{F}_t]\,. \end{aligned}$$

Note that g_T is $\mathcal{F}_T$-measurable and assumed integrable. The last step consists of computing the stochastic differential of g_t via the Itô product formula. To do so, define $I(t) := \int_0^t r(u, X(u))\,\mathrm{d}u$ giving $\mathrm{d}I(t) = r(t, X(t))\,\mathrm{d}t$, $(\,\mathrm{d}I(t))^2 \equiv 0$ and

$$\mathrm{d}[\mathrm{e}^{-\int_0^t r(u,X(u))\,\mathrm{d}u}] = \mathrm{d}\mathrm{e}^{-I(t)} = -\mathrm{e}^{-I(t)}\,\mathrm{d}I(t) = -\mathrm{e}^{-I(t)} r(t, X(t))\,\mathrm{d}t\,.$$

Hence, using this and (2.71) within the Itô product formula, where $g_t = \mathrm{e}^{-I(t)} f(t, X(t))$, gives (note: $\mathrm{d}t \cdot \mathrm{d}f(t, X(t)) = 0$)

$$\begin{aligned} \mathrm{d}g_t &= \mathrm{e}^{-I(t)} \cdot \mathrm{d}f(t, X(t)) + f(t, X(t)) \cdot \mathrm{d}\mathrm{e}^{-I(t)} + \mathrm{d}\mathrm{e}^{-I(t)} \cdot \mathrm{d}f(t, X(t)) \\ &= \mathrm{e}^{-I(t)} \cdot \mathrm{d}f(t, X(t)) + f(t, X(t)) \cdot \mathrm{d}\mathrm{e}^{-I(t)} \\ &= \mathrm{e}^{-I(t)}\left[\mathrm{d}f(t, X(t)) - f(t, X(t)\,\mathrm{d}I(t)\right] \\ &= \mathrm{e}^{-I(t)}\bigg[\left(\frac{\partial}{\partial t} f(t, X(t)) + \mathcal{G}_{t,x} f(t, X(t)) - r(t, X(t)) f(t, X(t))\right)\mathrm{d}t \\ &\qquad + \sigma(t, X(t))\frac{\partial f}{\partial x}(t, X(t))\,\mathrm{d}W(t)\bigg]\,. \end{aligned}$$

By the martingale condition, the drift coefficient (i.e., the expression multiplying $\mathrm{d}t$) must vanish for all values $X(t) = x$ and time t; namely, $\left(\frac{\partial}{\partial t} + \mathcal{G}_{t,x} - r(t,x)\right) f(t,x) = 0$. This is precisely the PDE in (2.98). Finally, the terminal condition follows trivially from (2.97) for $t = T$: $f(T, x) = \mathrm{E}[\mathrm{e}^{-\int_T^T r(u,X(u))\,\mathrm{d}u}\phi(X(T)) \mid X(T) = x] = \mathrm{E}[\phi(X(T)) \mid X(T) = x] = \phi(x)$. ☐

An important special case is when $r(t, x) = r$ is a constant. Then, the function defined by the conditional expectation, $f(t, x) := \mathrm{e}^{-r(T-t)}\mathrm{E}_{t,x}[\phi(X(T))]$, solves

$$\frac{\partial f}{\partial t}(t,x) + \frac{1}{2}\sigma^2(t,x)\frac{\partial^2 f}{\partial x^2}(t,x) + \mu(t,x)\frac{\partial f}{\partial x}(t,x) - r f(t,x) = 0\,, \tag{2.99}$$

with terminal condition $f(T, x) = \phi(x)$. This is also readily seen by setting $f(t, x) = \mathrm{e}^{-r(T-t)} g(t, x)$, where $g(t, x) := \mathrm{E}_{t,x}[\phi(X(T))]$ satisfies (2.84), i.e., $\frac{\partial g}{\partial t} + \mathcal{G}_{t,x} g = 0$. Since $\frac{\partial f}{\partial t} = r f + \mathrm{e}^{-r(T-t)}\frac{\partial g}{\partial t}$ and $\mathcal{G}_{t,x} f = \mathrm{e}^{-r(T-t)}\mathcal{G}_{t,x} g$, we have

$$\frac{\partial f}{\partial t} + \mathcal{G}_{t,x} f - r f = r f + \mathrm{e}^{-r(T-t)}\frac{\partial g}{\partial t} + \mathrm{e}^{-r(T-t)}\mathcal{G}_{t,x} g - r f = \mathrm{e}^{-r(T-t)}\left(\frac{\partial g}{\partial t} + \mathcal{G}_{t,x} g\right) = 0.$$

2.7.1 Forward Kolmogorov PDE

Proposition 2.12 states that a transition PDF solves the Kolmogorov PDE (2.86) in the *backward variables* (t, x). We can also define the differential operator $\tilde{\mathcal{G}} \equiv \tilde{\mathcal{G}}_{T,y}$ acting on the *forward variables* (T, y):

$$\tilde{\mathcal{G}} f(T, y) := \frac{1}{2}\frac{\partial^2}{\partial y^2}\left(\sigma^2(T, y) f(T, y)\right) - \frac{\partial}{\partial y}\left(\mu(T, y) f(T, y)\right). \tag{2.100}$$

This is also referred to as the differential adjoint to the generator $\mathcal{G}$. It can be shown that under fairly general conditions the transition PDF $p = p(t, T; x, y)$ (considered as a function of T, y, for any fixed t, x) satisfies the so-called *forward Kolmogorov or Fokker–Planck* PDE

$$\frac{\partial p}{\partial T} = \tilde{\mathcal{G}}\, p\,, \tag{2.101}$$

with $\lim_{T\searrow t} p(t, T; x, y) = \delta(y - x)$. The name forward derives from the fact that y refers to the value of the process at future time $T > t$.

The formal proof of (2.101), and under what conditions it holds true, requires a rather technical discussion that is beyond our scope. However, it is instructive to see how (2.101) arises from the backward PDE. Let the interval $\mathcal{I}$ denote the state space of process X. For example, $\mathcal{I} = \mathbb{R}$ for standard BM, $\mathcal{I} = \mathbb{R}_+$ for GBM, $\mathcal{I} = (L, \infty)$ for GBM killed at a lower level $L > 0$, etc. In our heuristic justification of (2.101), we shall now make use of the Chapman–Kolmogorov relation:

$$p(t, T; x, y) = \int_{\mathcal{I}} p(t, t'; x, x') p(t', T; x', y)\, \mathrm{d}x' \tag{2.102}$$

for any $t < t' < T$, $x, y \in \mathcal{I}$. Equation (2.102) is an important general property that follows from the Markov property. To derive this relation, consider the joint PDF of the triplet $(X(T), X(t'), X(t))$ and applying conditioning gives

$$f_{X(T),X(t'),X(t)}(y, x', x) = f_{X(t)}(x) \cdot f_{X(t')|X(t)}(x'|x) \cdot f_{X(T)|X(t')}(y|x')$$

where $f_{X(T)|X(t'),X(t)}(y|x', x) = f_{X(T)|X(t')}(y|x')$ by the Markov property. Dividing both sides by the PDF of $X(t)$, $f_{X(t)}(x)$, and using the definition of the transition PDF in (1.20), gives the joint PDF of the pair $X(T), X(t')$ conditional on $X(t) = x$:

$$f_{X(T),X(t')|X(t)}(y, x'|x) \equiv \frac{f_{X(T),X(t'),X(t)}(y, x', x)}{f_{X(t)}(x)} = p(t, t'; x, x') p(t', T; x', y)\,.$$

Integrating out the x' variable gives the PDF of $X(T)$ conditional on $X(t) = x$:

$$f_{X(T)|X(t)}(y|x) = \int_{\mathcal{I}} f_{X(T),X(t')|X(t)}(y, x'|x)\, \mathrm{d}x' = \int_{\mathcal{I}} p(t, t'; x, x') p(t', T; x', y)\, \mathrm{d}x'\,.$$

By definition, $f_{X(T)|X(t)}(y|x) = p(t, T; x, y)$ and therefore we obtain (2.102).

To arrive at (2.101) we begin by differentiating both sides of (2.102) w.r.t. t' and note that $\frac{\partial}{\partial t'} p(t, T; x, y) \equiv 0$, giving

$$\int_{\mathcal{I}} \left[p(t', T; x', y) \frac{\partial}{\partial t'} p(t, t'; x, x') + p(t, t'; x, x') \frac{\partial}{\partial t'} p(t', T; x', y) \right] \mathrm{d}x' \equiv 0\,. \tag{2.103}$$

We leave the first integral term as is, but re-express the second part of the integral by using the backward PDE, $\frac{\partial}{\partial t'} p(t', T; x', y) = -\mathcal{G}_{t',x'} p(t', T; x', y)$, to obtain

$$\int_{\mathcal{I}} p(t, t'; x, x') \frac{\partial}{\partial t'} p(t', T; x', y)\, \mathrm{d}x' = -\int_{\mathcal{I}} p(t, t'; x, x')\, \mathcal{G}_{t',x'} p(t', T; x', y)\, \mathrm{d}x'\,.$$

The next step consists of using the differential operator $\mathcal{G}_{t',x'}$, applying integration by parts on the above right-hand integral and assuming that contributions from the boundaries of $\mathcal{I}$ vanish (see Exercise 2.36) to obtain

$$\int_{\mathcal{I}} p(t, t'; x, x')\, \mathcal{G}_{t',x'} p(t', T; x', y)\, \mathrm{d}x' = \int_{\mathcal{I}} p(t', T; x', y)\, \tilde{\mathcal{G}}_{t',x'} p(t, t'; x, x')\, \mathrm{d}x'\,. \tag{2.104}$$

This shows that $\tilde{\mathcal{G}}$ indeed acts as the corresponding adjoint operator to $\mathcal{G}$. Using this relation into the second term in the integrand of (2.103) gives

$$\int_{\mathcal{I}} p(t', T; x', y) \left[\frac{\partial}{\partial t'} p(t, t'; x, x') - \tilde{\mathcal{G}}_{t',x'} p(t, t'; x, x') \right] \mathrm{d}x' \equiv 0 . \tag{2.105}$$

Since this integral is identically zero for arbitrary given values $t' > t$, $x' \in \mathcal{I}$, then (assuming a large enough family of positive transition PDFs $p(t', T; x', y)$ as functions of x') the integrand must be zero for all $x' \in \mathcal{I}$. This implies that the term in brackets in the integrand must equal zero; i.e., for fixed backward variables t, x we have the forward Kolmogorov PDE, $\frac{\partial p}{\partial t'} = \tilde{\mathcal{G}}_{t',x'} p$, in the forward variables t', x' for an arbitrary transition PDF $p \quad p(t, t'; x, x')$.

2.7.2 Transition CDF/PDF for Time-Homogeneous Diffusions

In many applications, including derivative pricing, the stochastic process is assumed to be time-homogeneous. We recall the definition of a time-homogeneous process from the previous chapter, i.e., the relation in (1.21). For a time-homogeneous diffusion process, this means that the drift and diffusion coefficient functions are only functions of the "spatial variable" and are not functions of time t: $\mu(x, t) = \mu(x)$ and $\sigma(x, t) = \sigma(x)$. The generator $\mathcal{G}_{t,x} \equiv \mathcal{G}_x$ for such a process is then of the form

$$\mathcal{G}_x := \frac{1}{2}\sigma^2(x)\frac{\partial^2}{\partial x^2} + \mu(x)\frac{\partial}{\partial x} . \tag{2.106}$$

Since the transition PDF (or CDF) satisfies a *time-homogeneous PDE*, it is then a *function of the time difference*: $\tau \equiv T - t$ i.e., we write it as $p(\tau; x, y)$ and the transition CDF as $P(\tau; x, y)$. This time dependence on $\tau = T - t$ can also be realized from the conditional expectation definition of the transition CDF. Indeed, the defining relation in (1.21) implies

$$\begin{aligned} P(t, T; x, y) \equiv \mathbb{P}(X(T) \leqslant y \mid X(t) = x) &= \mathbb{P}(X(t+\tau) \leqslant y \mid X(t) = x) \\ &= \mathbb{P}(X(\tau) \leqslant y \mid X(0) = x) \\ &= P(0, \tau; x, y) \equiv P(\tau; x, y) , \end{aligned}$$

and $p(\tau; x, y) = \frac{\partial}{\partial y} P(\tau; x, y)$. Writing $p(t, T; x, y) = p(\tau; x, y)$ and using the fact that $\frac{\partial \tau}{\partial T} = 1$ and $\frac{\partial \tau}{\partial t} = -1$ gives

$$\frac{\partial p(t, T; x, y)}{\partial T} = \frac{\partial p(\tau; x, y)}{\partial \tau} \quad \text{and} \quad \frac{\partial p(t, T; x, y)}{\partial t} = -\frac{\partial p(\tau; x, y)}{\partial \tau} .$$

The backward and forward Kolmogorov PDEs in the respective τ, x and τ, y variables are then given by

$$\frac{\partial p}{\partial \tau} = \frac{1}{2}\sigma^2(x)\frac{\partial^2 p}{\partial x^2} + \mu(x)\frac{\partial p}{\partial x} \qquad \text{(backward)} \tag{2.107}$$

$$\frac{\partial p}{\partial \tau} = \frac{1}{2}\frac{\partial^2}{\partial y^2}\left(\sigma^2(y)p\right) - \frac{\partial}{\partial y}\left(\mu(y)p\right) \qquad \text{(forward)} \tag{2.108}$$

for a transition PDF $p = p(\tau; x, y)$. The previous terminal condition is now an *initial condition* where

$$\lim_{\tau \searrow 0} p(\tau, x, y) \equiv p(0+, x, y) = \delta(x - y) \text{ and } P(0, x, y) = \mathbb{I}_{\{x \leqslant y\}} . \tag{2.109}$$

Note that, by time homogeneity, the conditional expectation in (2.85) in the above Feynman–Kac Theorem gives the conditional expectation:

$$\mathrm{E}[\phi(X(T)) \mid X(t) = x] = \mathrm{E}[\phi(X(t+\tau)) \mid X(t) = x] = \mathrm{E}[\phi(X(\tau)) \mid X(0) = x] := f(\tau, x)\,.$$

That is, (2.87) now reads

$$f(\tau, x) = \int_{\mathbb{R}} \phi(y)\, p(\tau; x, y) dy\,, \tag{2.110}$$

where f solves the backward Kolmogorov PDE in the variables τ, x,

$$\frac{\partial f}{\partial \tau} = \frac{1}{2}\sigma^2(x)\frac{\partial^2 f}{\partial x^2} + \mu(x)\frac{\partial f}{\partial x} \tag{2.111}$$

with initial condition $f(0, x) = \phi(x)$. Note: $f(0+, x) \equiv f(0, x)$ for continuous $\phi(x)$.

Assuming a constant discount function $r(t, x) = r$, we observe that the discounted expectation is also a function of variables τ, x, i.e., we have the function

$$v(\tau, x) = \mathrm{e}^{-r(T-t)}\mathrm{E}_{t,x}[\phi(X(T))] = \mathrm{e}^{-r\tau} f(\tau, x) = \mathrm{e}^{-r\tau}\int_{\mathbb{R}} \phi(y)\, p(\tau; x, y) dy$$

satisfying the PDE

$$\frac{\partial v}{\partial \tau} = \frac{1}{2}\sigma^2(x)\frac{\partial^2 v}{\partial x^2} + \mu(x)\frac{\partial v}{\partial x} - rv \tag{2.112}$$

with initial condition $v(0, x) = \phi(x)$. This is the time-homogeneous version of (2.99) in the variables τ, x.

We have already seen several specific examples of time-homogeneous processes such as standard BM, GBM in Example 2.21, and the OU process in Example 2.23. In Example 2.21, the GBM process is time homogeneous with coefficient functions $\mu(x) = \mu x$ and $\sigma(x) = \sigma x$, and having respective transition CDF and PDF:

$$P(\tau; x, y) = \mathcal{N}\left(\frac{\ln(y/x) - (\mu - \frac{1}{2}\sigma^2)\tau}{\sigma\sqrt{\tau}}\right) \tag{2.113}$$

and

$$p(\tau; x, y) = \frac{1}{y\sigma\sqrt{2\pi\tau}} \exp\left(-\frac{[\ln(y/x) - (\mu - \frac{1}{2}\sigma^2)\tau]^2}{2\sigma^2\tau}\right), \tag{2.114}$$

$x, y > 0, \tau > 0$. The reader can verify by direct differentiation that both functions satisfy (2.107) and (2.108) with the appropriate initial condition in (2.109). This is also the case for the time-homogeneous OU process, where setting $\tau = T - t$ in (2.95) and (2.96) gives the transition CDF and PDF that satisfy the above time-homogeneous Kolmogorov PDEs with $\mu(x) = \alpha - \beta x$, $\sigma(x) = \sigma$. In contrast, for nonconstant $\mu(t)$ and (or) nonconstant $\sigma(t)$, the GBM process in Example 2.21 is time inhomogeneous; i.e., the transition functions in (2.92) and (2.93) cannot be written as functions of only $\tau = T - t$ in the time variables, but rather depend on both t and T, separately, via the time-averaged quantities in (2.91).

2.8 Radon–Nikodym Derivative Process and Girsanov's Theorem

Our main goal in this section is to use and build upon the basic tools and ideas developed in Section A.5 of Appendix A in order to understand how to construct a certain type of probability measure change which introduces a drift in the BM. In particular, we are interested in a measure change, say $\mathbb{P} \to \widehat{\mathbb{P}}$, whereby we begin with $W := \{W(t)\}_{t\geqslant 0}$ as a standard BM under measure $\mathbb{P}$ and then define a new process, which we denote by $\widehat{W} := \{\widehat{W}(t)\}_{t\geqslant 0}$, such that $\widehat{W}$ is a standard BM under the new measure $\widehat{\mathbb{P}}$. We will see that the measure change from $\mathbb{P} \to \widehat{\mathbb{P}}$ is constructed by using a positive random variable that is an exponential $\mathbb{P}$-martingale and that there is a precise relationship between the two Brownian motions W and $\widehat{W}$ which will differ only by a drift component. This is the essence of Girsanov's Theorem, whose statement and proof are given later. The change of measure has many useful applications and will also allow us to compute conditional expectations of processes or random variables that are functionals of Brownian motion under two different probability measures $\mathbb{P}$ and $\widehat{\mathbb{P}}$. These two measures will be equivalent in the sense that (as we recall from our previous discussion on equivalent probability measures) all events having zero probability under one measure also have zero probability under the other measure.

Let us begin by fixing a filtered probability space $(\Omega, \mathcal{F}, \mathbb{P}, \mathbb{F})$, where $\mathbb{F} = \{\mathcal{F}_t\}_{t\geqslant 0}$ is any filtration for standard Brownian motion and recall Definition 1.1 of a $(\mathbb{P}, \mathbb{F})$-BM. This is shorthand for a standard Brownian motion (BM) $W := \{W(t)\}_{t\geqslant 0}$ w.r.t. a given filtration $\mathbb{F}$ and a measure $\mathbb{P}$. That is, W has (a.s.) continuous paths started at $W(0) = 0$ and, under given measure $\mathbb{P}$, it has normally distributed increments $W(t) - W(s) \sim Norm(0, t-s)$ that are independent of $\mathcal{F}_s$, for all $0 \leqslant s < t$. From these original defining properties we then showed that W has quadratic variation $[W, W](t) = t$ and that it is a $(\mathbb{P}, \mathbb{F})$-martingale (shorthand for a martingale w.r.t. filtration $\mathbb{F}$ and measure $\mathbb{P}$). Now, let's assume that we have a process that we know is a continuous martingale started at zero and with the same quadratic variation formula as standard BM. The question is whether or not this process is a standard BM. It turns out that the answer is yes, the process is a standard BM as stated in the following theorem, originally due to Lévy. In what follows, this will give us a useful way to recognize when a martingale process is in fact a standard BM. The characterization makes no assumption of the normality and independence of increments! Rather, these properties are implied. Besides the martingale property, the requirement of continuity of all paths and the fact that they must start at zero, the recognition that we have a BM follows from the assumption of the above quadratic variation formula.

[*Technical Remark*: We note that the proof of Theorem 2.14 below makes use of a general version of the Itô formula in (2.35). Although we do not prove it, it turns out that we have the same Itô formula as in (2.35) if W is replaced by a continuous martingale process $M := \{M(t)\}_{t\geqslant 0}$, that starts at zero and has quadratic variation $[M, M](t) = t$, i.e., $\mathrm{d}M(t)\,\mathrm{d}M(t) = \mathrm{d}[M, M](t) = \mathrm{d}t$:

$$\mathrm{d}f(t, M(t)) = \left(f_t(t, M(t)) + \frac{1}{2} f_{xx}(t, M(t))\right) \mathrm{d}t + f_x(t, M(t))\,\mathrm{d}M(t). \tag{2.115}$$

Essentially one can think of this as the Itô formula in (2.43) where $X \equiv M$ with zero drift $\mu \equiv 0$ and unit diffusion function $\sigma \equiv 1$. The integral form of (2.115) is (see (2.36))

$$f(t, M(t)) = f(0, M(0)) + \int_0^t \left(f_u(u, M(u)) + \frac{1}{2} f_{xx}(u, M(u))\right)\mathrm{d}u + \int_0^t f_x(u, M(u))\,\mathrm{d}M(u). \tag{2.116}$$

Assuming the usual square-integrability condition as we did for any Itô integral w.r.t. BM, the above stochastic integral w.r.t. the increment $\mathrm{d}M(u)$ is defined in a similar fashion and is a martingale having zero expected value. Note that if $f(t, x)$ is a $C^{1,2}$ function satisfying the PDE $f_t(t, x) + \frac{1}{2} f_{xx}(t, x) = 0$, then the process defined by $Y(t) := f(t, M(t))$ is a martingale w.r.t. any filtration for BM.]

Theorem 2.14 (Lévy's characterization of standard BM). *Let the process* $\{M(t)\}_{t\geqslant 0}$ *be a continuous* $(\mathbb{P},\mathbb{F})$*-martingale started at* $M(0) = 0$ *(a.s.) and with quadratic variation* $[M,M](t) = t$ *for all* $t \geqslant 0$*. Then,* $\{M(t)\}_{t\geqslant 0}$ *is a standard* $(\mathbb{P},\mathbb{F})$*-BM.*

Proof. Since we have already assumed that M has continuous paths all starting at zero, from the definition of a $(\mathbb{P},\mathbb{F})$-BM we have left to show that $M(t) - M(s) \sim Norm(0, t-s)$ and that these increments are independent of $\mathcal{F}_s$ for all $0 \leqslant s < t$. For this purpose, consider the function $f(t,x) = \mathrm{e}^{-\frac{1}{2}\theta^2 t+\theta x}$ with arbitrary real parameter θ. Since $f(t,x)$ satisfies the PDE $f_t(t,x) + \frac{1}{2}f_{xx}(t,x) = 0$, from the above discussion we have that the process $f(t,M(t)) = \mathrm{e}^{-\frac{1}{2}\theta^2 t+\theta M(t)}, t \geqslant 0$, is a martingale. In fact, we recognize this as an example of an exponential martingale. Taking the expectation of the process at time t, conditional on filtration $\mathcal{F}_s$, $s \leqslant t$, and using the martingale property gives the *conditional* moment-generating function (m.g.f.) of $M(t) - M(s)$ as a function of θ:

$$\mathrm{E}\left[\mathrm{e}^{-\frac{1}{2}\theta^2 t+\theta M(t)} \mid \mathcal{F}_s\right] = \mathrm{e}^{-\frac{1}{2}\theta^2 s+\theta M(s)} \implies \mathrm{E}\left[\mathrm{e}^{\theta(M(t)-M(s))} \mid \mathcal{F}_s\right] = \mathrm{e}^{\frac{1}{2}\theta^2(t-s)}.$$

This is equivalent to the m.g.f. of $M(t) - M(s)$, which is the m.g.f. of a $Norm(0, t-s)$ random variable (by the tower property):

$$\mathrm{E}\left[\mathrm{e}^{\theta(M(t)-M(s))}\right] = \mathrm{E}\left[\mathrm{E}\left[\mathrm{e}^{\theta(M(t)-M(s))} \mid \mathcal{F}_s\right]\right] = \mathrm{e}^{\frac{1}{2}\theta^2(t-s)}.$$

Hence, as function of θ, the m.g.f. and the m.g.f. conditional on $\mathcal{F}_s$ are the same and correspond to that of a $Norm(0, t-s)$ random variable; i.e., $M(t) - M(s) \sim Norm(0, t-s)$ and $M(t) - M(s)$ is independent of $\mathcal{F}_s$, for all $s \leqslant t$. □

[*Remark*: In what follows, we will only distinguish between different probability measures while *fixing a filtration* $\mathbb{F}$ *for BM.* Hence, we shall also write $\mathbb{P}$-BM to mean standard $(\mathbb{P},\mathbb{F})$-BM, i.e., standard BM w.r.t. filtration $\mathbb{F}$ and under measure $\mathbb{P}$. Equivalently, we shall also say that W is a BM under the measure $\mathbb{P}$. We shall also sometimes loosely say BM (or Brownian motion) where we clearly really mean *standard* BM. Also, we simply say $\mathbb{P}$-martingale to mean a $(\mathbb{P},\mathbb{F})$-martingale and $\widehat{\mathbb{P}}$-martingale to mean a $(\widehat{\mathbb{P}},\mathbb{F})$-martingale when $\mathbb{F}$ is fixed.]

In what follows we let the probability measure $\widehat{\mathbb{P}} \equiv \widehat{\mathbb{P}}^{(\varrho)}$ be defined by (A.112) of Section A.5, i.e., $\widehat{\mathbb{P}}(A) := \int_A \varrho(\omega)\,\mathrm{d}\mathbb{P}(\omega)$, $A \in \mathcal{F}$, with Radon–Nikodym random variable $\varrho \equiv \frac{\mathrm{d}\widehat{\mathbb{P}}}{\mathrm{d}\mathbb{P}}$ assumed positive (almost surely) with unit expectation under measure $\mathbb{P}$, $\mathrm{E}[\varrho] = 1$. We recall how ϱ is used in (A.113) and (A.114) for computing the expectation of any integrable random variable under measures $\widehat{\mathbb{P}}$ and $\mathbb{P}$, respectively. Shortly we shall explicitly specify this random variable and, in fact, its precise specification is a key ingredient in Girsanov's Theorem. However, for the moment we can keep our assumptions on ϱ as is (which are as general as possible). In preparation for our main result, we will need to define and discuss some basic properties of a so-called *Radon–Nikodym derivative process of* $\widehat{\mathbb{P}}$ *w.r.t.* $\mathbb{P}$. In Volume I on discrete-time financial models, we defined a similar process but in a discrete-time stochastic setting. In continuous time we shall fix some terminal time $T > 0$ and define the Radon Nikodym derivative process $\{\varrho_t\}_{0\leqslant t\leqslant T}$ (of measure $\widehat{\mathbb{P}} \equiv \widehat{\mathbb{P}}^{(\varrho)}$ w.r.t. measure $\mathbb{P}$ for a given filtration $\mathbb{F}$) by

$$\varrho_t := \mathrm{E}[\,\varrho \mid \mathcal{F}_t],\ \ 0 \leqslant t \leqslant T. \tag{2.117}$$

We remark that it is customary to also use the following *more explicit equivalent notations* for the random variable ϱ_t:

$$\varrho_t \equiv \left(\frac{\mathrm{d}\widehat{\mathbb{P}}}{\mathrm{d}\mathbb{P}}\right)_t \overset{\text{or}}{\equiv} \left(\frac{\mathrm{d}\widehat{\mathbb{P}}^{(\varrho)}}{\mathrm{d}\mathbb{P}}\right)_t \overset{\text{or}}{\equiv} \left(\frac{\mathrm{d}\widehat{\mathbb{P}}^{(\varrho)}}{\mathrm{d}\mathbb{P}}\right)_{\mathcal{F}_t}.$$

Hence (2.117) is also written as $\left(\frac{d\widehat{\mathbb{P}}}{d\mathbb{P}}\right)_t := \mathrm{E}[\frac{d\widehat{\mathbb{P}}}{d\mathbb{P}} \mid \mathcal{F}_t]$. These notations really spell out the definition in (2.117) and also visually remind us of the "direction of the measure change," e.g., $\mathbb{P} \to \widehat{\mathbb{P}}$. In what follows we shall try to keep our notation less cumbersome as long as there is no ambiguity.

Clearly ϱ_t is $\mathcal{F}_t$-measurable and integrable, $\mathrm{E}[|\varrho_t|] \leqslant \mathrm{E}[|\varrho|] = \mathrm{E}[\varrho] = 1 < \infty$, for all $t \in [0, T]$. By the tower property and the definition in (2.117), we immediately we see that the process $\{\varrho_t\}_{0\leqslant t\leqslant T}$ is a $\mathbb{P}$-martingale (recall the Doob-Lévy martingale):

$$\mathrm{E}[\,\varrho_t \mid \mathcal{F}_s] = \mathrm{E}\big[\mathrm{E}[\,\varrho \mid \mathcal{F}_t] \mid \mathcal{F}_s\big] = \mathrm{E}[\,\varrho \mid \mathcal{F}_s] = \varrho_s\,, \quad 0 \leqslant s \leqslant t \leqslant T.$$

By definition, the process also starts with unit value: $\varrho_0 = \mathrm{E}[\,\varrho \mid \mathcal{F}_0] \equiv \mathrm{E}[\,\varrho\,] = 1$. Hence, by the martingale property, the process has unit expectation, $\mathrm{E}[\varrho_t] = \varrho_0 = 1$, for all $t \in [0, T]$.

The next proposition gives a useful formula for computing the $\widehat{\mathbb{P}}$-measure expectation of an $\mathcal{F}_t$-measurable random variable X, conditional on information up to a time s prior to time t, as a $\mathbb{P}$-measure conditional expectation of $X \cdot (\varrho_t/\varrho_s)$. The ratio ϱ_t/ϱ_s of the Radon–Nikodym derivative process at times s and t adjusts for the change of measure in the conditional expectation.

Proposition 2.15. *Let $\widehat{\mathbb{P}}$ be defined by $\widehat{\mathbb{P}}(A) := \int_A \varrho(\omega)\,d\mathbb{P}(\omega)$, $A \in \mathcal{F}$, with process $\varrho_t := \mathrm{E}[\,\varrho \mid \mathcal{F}_t], 0 \leqslant t \leqslant T$. Assume the random variable X is integrable w.r.t. $\widehat{\mathbb{P}}$ and $\mathcal{F}_t$-measurable for a given time $t \in [0, T]$. Then, for all $0 \leqslant s \leqslant t$,*

$$\widehat{\mathrm{E}}\big[X \mid \mathcal{F}_s\big] = \varrho_s^{-1}\mathrm{E}\big[\varrho_t\, X \mid \mathcal{F}_s\big]\,. \tag{2.118}$$

Proof. This result follows as a simple application of Theorem A.9 where we set $\mathcal{G} \equiv \mathcal{F}_s$, and $\mathcal{F}_s \subset \mathcal{F}_t \subset \mathcal{F}$ implies $\mathcal{G} \subset \mathcal{F}$. Then, upon using the definition in (2.117) for time s, the formula in (A.116) gives

$$\widehat{\mathrm{E}}\big[X \mid \mathcal{F}_s\big] = \frac{\mathrm{E}[\,\varrho X \mid \mathcal{F}_s]}{\mathrm{E}[\,\varrho \mid \mathcal{F}_s]} = \varrho_s^{-1}\mathrm{E}[\,\varrho X \mid \mathcal{F}_s]\,.$$

The last expectation on the right is now recast by reversing the tower property, by conditioning on $\mathcal{F}_t$, and using the fact that X is $\mathcal{F}_t$-measurable (so it is pulled out of the inner expectation conditional on $\mathcal{F}_t$ below):

$$\mathrm{E}[\,\varrho X \mid \mathcal{F}_s] = \mathrm{E}\big[\mathrm{E}[\,\varrho X \mid \mathcal{F}_t] \mid \mathcal{F}_s\big] = \mathrm{E}\big[\,X\,\mathrm{E}[\,\varrho \mid \mathcal{F}_t] \mid \mathcal{F}_s\big] = \mathrm{E}\big[\,X\,\varrho_t \mid \mathcal{F}_s\big]\,.$$

In the last step, we used the definition $\mathrm{E}[\,\varrho \mid \mathcal{F}_t] = \varrho_t$. □

Note that a special case of (2.118) is when $s = 0$. Since $\varrho_0 = 1$, $\widehat{\mathrm{E}}\big[X \mid \mathcal{F}_0\big] = \widehat{\mathrm{E}}\big[X\big]$ and $\mathrm{E}[\,\varrho_t\, X \mid \mathcal{F}_0] = \mathrm{E}[\varrho_t\, X]$, we have

$$\widehat{\mathrm{E}}[X] = \mathrm{E}[\,\varrho_t\, X] \text{ for } \mathcal{F}_t\text{-measurable } X. \tag{2.119}$$

Consider a continuous-time stochastic process $\{X(t)\}_{t\geqslant 0}$ adapted to the filtration $\mathbb{F}$. Since $X(t)$ is $\mathcal{F}_t$-measurable for every $t \geqslant 0$, we may put $X = X(t)$ in (2.118) to obtain

$$\widehat{\mathrm{E}}\big[X(t) \mid \mathcal{F}_s\big] = \varrho_s^{-1}\mathrm{E}[\,\varrho_t\, X(t) \mid \mathcal{F}_s]\,, \;\; 0 \leqslant s \leqslant t \leqslant T. \tag{2.120}$$

As a consequence of this property, we have the following result.

Proposition 2.16. *A continuous-time adapted stochastic process $\{M(t)\}_{0\leqslant t\leqslant T}$ is a $\widehat{\mathbb{P}}$-martingale if and only if $\{\varrho_t M(t)\}_{0\leqslant t\leqslant T}$ is a $\mathbb{P}$-martingale.*

Proof. Assume $\{M(t)\}_{0\leqslant t\leqslant T}$ is a $\widehat{\mathbb{P}}$-martingale. Then, using (2.120) with $X(t) \equiv M(t)$,

$$M(s) = \widehat{\mathrm{E}}[M(t) \mid \mathcal{F}_s] = \varrho_s^{-1}\mathrm{E}[\,\varrho_t\, M(t) \mid \mathcal{F}_s] \implies \varrho_s\, M(s) = \mathrm{E}[\,\varrho_t\, M(t) \mid \mathcal{F}_s]$$

for $0 \leqslant s \leqslant t \leqslant T$, where the last relation is the $\mathbb{P}$-martingale property of $\{\varrho_t M(t)\}_{0\leqslant t\leqslant T}$. The converse follows since all the above steps may be reversed. Moreover, $\{M(t)\}_{0\leqslant t\leqslant T}$ is adapted to $\mathbb{F}$ and integrable w.r.t. $\widehat{\mathbb{P}}$ if and only if $\{\varrho_t M(t)\}_{0\leqslant t\leqslant T}$ is adapted to $\mathbb{F}$ and integrable w.r.t. $\mathbb{P}$. □

We are now finally ready to state and prove Girsanov's Theorem for the case of standard Brownian motion.

Theorem 2.17 (Girsanov's Theorem for BM)**.** *Let $\{W(t)\}_{0\leqslant t\leqslant T}$ be a standard $\mathbb{P}$-BM w.r.t. a filtration $\mathbb{F} = \{\mathcal{F}_t\}_{0\leqslant t\leqslant T}$ and assume the process $\{\gamma(t)\}_{0\leqslant t\leqslant T}$ is adapted to $\mathbb{F}$, for a given $T > 0$. Define*

$$\varrho_t := \exp\left(-\frac{1}{2}\int_0^t \gamma^2(s)\,\mathrm{d}s + \int_0^t \gamma(s)\,\mathrm{d}W(s)\right), 0 \leqslant t \leqslant T, \tag{2.121}$$

and the probability measure $\widehat{\mathbb{P}} \equiv \widehat{\mathbb{P}}^{(\varrho)}$ by the Radon–Nikodym derivative $\frac{\mathrm{d}\widehat{\mathbb{P}}}{\mathrm{d}\mathbb{P}} = \left(\frac{\mathrm{d}\widehat{\mathbb{P}}}{\mathrm{d}\mathbb{P}}\right)_T \equiv \varrho_T$. Furthermore, assume the square-integrability condition holds:

$$\mathrm{E}\left[\int_0^T \varrho_s^2\,\gamma^2(s)\,\mathrm{d}s\right] < \infty\,. \tag{2.122}$$

Then, the process $\{\widehat{W}(t)\}_{0\leqslant t\leqslant T}$ defined by

$$\widehat{W}(t) := W(t) - \int_0^t \gamma(s)\,\mathrm{d}s \tag{2.123}$$

is a standard $\widehat{\mathbb{P}}$-BM w.r.t. filtration $\mathbb{F}$.

Some clarifying remarks on Theorem 2.17 before its proof:

1. The condition in (2.122) is required to ensure that $\{\varrho_t\}_{0\leqslant t\leqslant T}$ is a $\mathbb{P}$-martingale with $\mathrm{E}[\varrho_t] = 1$, i.e., this corresponds to the Itô process $\int_0^t \varrho_s\,\gamma(s)\,\mathrm{d}W(s), 0 \leqslant t \leqslant T$, being a martingale. An equivalent and more practically verified condition that guarantees the process $\{\varrho_t\}_{0\leqslant t\leqslant T}$ is a $\mathbb{P}$-martingale is the so-called *Novikov condition*:

$$\mathrm{E}\left[\exp\left(\frac{1}{2}\int_0^T \gamma^2(s)\,\mathrm{d}s\right)\right] < \infty\,. \tag{2.124}$$

2. The differential increments of the two Brownian motions are simply related: $\mathrm{d}W(t) = \mathrm{d}\widehat{W}(t) + \gamma(t)\,\mathrm{d}t$ and $\mathrm{d}\widehat{W}(t) = \mathrm{d}W(t) - \gamma(t)\,\mathrm{d}t$.

3. Pay attention to the consistent and correct use of the ± signs. In this regard, we note that the Radon–Nikodym derivative random variable in (2.121) can *equivalently* be written as

$$\varrho_t = \exp\left(-\frac{1}{2}\int_0^t \theta^2(s)\,\mathrm{d}s - \int_0^t \theta(s)\,\mathrm{d}W(s)\right).$$

Note the minus sign instead of the plus sign in front of the Itô integral. Then, (2.123) is replaced by $\widehat{W}(t) := W(t) + \int_0^t \theta(s)\,\mathrm{d}s$, i.e., $\mathrm{d}\widehat{W}(t) = \mathrm{d}W(t) + \theta(t)\,\mathrm{d}t$. This is obtained simply by setting $\gamma(t) = -\theta(t)$ in the original definition where $\gamma^2(t) = \theta^2(t)$.

4. In general, γ is an adapted process so that ϱ_t is a functional of BM from time 0 to t. In particular, the Radon–Nikodym derivative process has the form of an exponential $\mathbb{P}$-martingale in the process γ w.r.t. the $\mathbb{P}$-BM; i.e., by the definition in (2.53) we have $\varrho_t \equiv \varrho_t^{(\gamma)} = \mathcal{E}_t(\gamma \cdot W)$. Dividing the process value at any two times $0 \leqslant s < t \leqslant T$ gives

$$\frac{\varrho_t}{\varrho_s} \equiv \frac{(\frac{d\widehat{\mathbb{P}}}{d\mathbb{P}})_t}{(\frac{d\widehat{\mathbb{P}}}{d\mathbb{P}})_s} = \frac{\mathcal{E}_t(\gamma \cdot W)}{\mathcal{E}_s(\gamma \cdot W)} = \exp\left(-\frac{1}{2}\int_s^t \gamma^2(u)\, du + \int_s^t \gamma(u)\, dW(u)\right).$$

5. Note that $\mathbb{F}$ is any filtration for BM. It can, but need not be, the natural filtration $\mathbb{F}^W$ generated by W.

6. In the simplest case, we can choose a constant process, $\gamma(t) = \gamma =$ constant, where

$$\varrho_t \equiv \varrho_t^{(\gamma)} = e^{-\frac{1}{2}\gamma^2 t + \gamma W(t)} \tag{2.125}$$

and $\widehat{W}(t) \equiv \widehat{W}^{(\gamma)}(t) := W(t) - \gamma t, 0 \leqslant t \leqslant T$, is a $\widehat{\mathbb{P}}$-BM.

Proof. First let us verify that $\{\varrho_t\}_{0 \leqslant t \leqslant T}$ is a Radon–Nikodym derivative process. By the assumption in (2.122) (or the Novikov condition) we have that $\{\varrho_t\}_{0 \leqslant t \leqslant T}$ is a $\mathbb{P}$-martingale; in fact it is an exponential $\mathbb{P}$-martingale. This can be seen by applying Itô's formula to the stochastic exponential in (2.121), giving

$$d\varrho_t = \varrho_t \gamma(t)\, dW(t) \implies \varrho_t = \varrho_0 + \int_0^t \varrho_s \gamma(s)\, dW(s)$$

where the Itô integral is a martingale (under measure $\mathbb{P}$) by the condition in (2.122). Because of the $\mathbb{P}$-martingale property, $E[\varrho_t] = \varrho_0 = e^0 = 1$, $0 \leqslant t \leqslant T$. In particular, $E[\varrho_T] = 1$ and ϱ_T is also nonnegative. Hence, $\varrho \equiv \frac{d\widehat{\mathbb{P}}}{d\mathbb{P}} = \varrho_T$ is a proper Radon–Nikodym derivative and by the $\mathbb{P}$-martingale property the process in (2.121) satisfies the definition in (2.117), i.e., it is indeed a Radon–Nikodym derivative process.

We now show that the process $\widehat{W}$ defined by (2.123) is a standard $\widehat{\mathbb{P}}$-BM by verifying all the defining properties in Theorem 2.14 with measure $\widehat{\mathbb{P}}$ (filtration $\mathbb{F}$ fixed):

(i) The process starts at zero, $\widehat{W}(0) = W(0) = 0$, and is continuous in time since $\widehat{W}(t) := W(t) - \int_0^t \gamma(s)\, ds$ where $W(t)$ and the integral $\int_0^t \gamma(s)\, ds$ are both continuous in $t \geqslant 0$.

(ii) $d[\widehat{W}, \widehat{W}](t) = d\widehat{W}(t)\, d\widehat{W}(t) = (dW(t) - \gamma(t)\, dt)(dW(t) - \gamma(t)\, dt) = dW(t)\, dW(t) = dt$, i.e., the process has quadratic variation $[\widehat{W}, \widehat{W}](t) = t$.

(iii) $\{\widehat{W}(t)\}_{0 \leqslant t \leqslant T}$ is a $\widehat{\mathbb{P}}$-martingale. By Proposition 2.16, this follows if we can show that the process $\{\varrho_t \widehat{W}(t)\}_{0 \leqslant t \leqslant T}$ is a $\mathbb{P}$-martingale. To show the latter, we compute the stochastic differential by Itô's product rule (using $d\varrho_t = \varrho_t \gamma(t)\, dW(t)$ and $d\widehat{W}(t) = dW(t) - \gamma(t)\, dt$ and setting $dW(t)\, dW(t) = dt$, $dW(t)\, dt = 0$):

$$\begin{aligned} d\big(\varrho_t \widehat{W}(t)\big) &= \varrho_t\, d\widehat{W}(t) + \widehat{W}(t)\, d\varrho_t + d\varrho_t\, d\widehat{W}(t) \\ &= \varrho_t[dW(t) - \gamma(t)\, dt] + \varrho_t \gamma(t) \widehat{W}(t)\, dW(t) + \varrho_t \gamma(t)\, dW(t)[dW(t) - \gamma(t)\, dt] \\ &= \varrho_t[1 + \gamma(t)\widehat{W}(t)]\, dW(t). \end{aligned}$$

This is a stochastic differential with a zero drift term (i.e., the coefficient in dt is zero). In integral form, where $\varrho_0 \widehat{W}(0) = 0$, we have

$$\varrho_t \widehat{W}(t) = \int_0^t \varrho_s\, [1 + \gamma(s)\widehat{W}(s)]\, dW(s), \quad 0 \leqslant t \leqslant T.$$

By the assumed boundedness of $\int_0^t \gamma(s)\,\mathrm{d}s$, and the fact that the BM $W(t)$ is bounded (a.s.), $\widehat{W}(t)$ is bounded (a.s.) for all $0 \leqslant t \leqslant T$. Combining this fact with the square-integrability condition (2.122), it follows that the above Itô integral is defined as it satisfies the square-integrability condition, $\mathrm{E}\left[\int_0^T \varrho_s^2\,[1+\gamma(s)\widehat{W}(s)]^2\,\mathrm{d}s\right] < \infty$, and is hence a $\mathbb{P}$-martingale, i.e., $\{\varrho_t \widehat{W}(t)\}_{0\leqslant t\leqslant T}$ is a $\mathbb{P}$-martingale. □

2.8.1 Some Applications of Girsanov's Theorem

Let's begin by considering a simple example of how Girsanov's Theorem can be applied to change probability measures so as to eliminate the drift in a drifted Brownian process.

Example 2.24. Let $X(t) \equiv W^{(\mu,\sigma)}(t)$ be a drifted BM process (recall (1.32))

$$X(t) := \mu t + \sigma W(t),$$

where $\{W(t)\}_{t\geqslant 0}$ is a standard $\mathbb{P}$-BM. Find a measure under which $\{X(t)\}_{0\leqslant t\leqslant T}$, for any $T > 0$, is a scaled BM with zero drift.

Solution. We note that the drift μ and volatility parameter $\sigma > 0$ are constants. Hence, by using Girsanov's Theorem we define a measure $\widehat{\mathbb{P}}$, $\frac{\mathrm{d}\widehat{\mathbb{P}}}{\mathrm{d}\mathbb{P}} = \varrho_T$, where ϱ_t is given by (2.125). Now, $\widehat{W}(t) := W(t) - \gamma t$ is a standard $\widehat{\mathbb{P}}$-BM and writing $X(t)$ in terms of $\widehat{W}(t)$ gives

$$X(t) = \mu t + \sigma W(t) = \mu t + \sigma(\widehat{W}(t) + \gamma t) = (\mu + \sigma\gamma)t + \sigma\widehat{W}(t).$$

So the drift coefficient of $X(t)$ is now $\mu+\sigma\gamma$, while the volatility parameter multiplying the standard $\widehat{\mathbb{P}}$-BM is still σ. Note that we can also see this in stochastic differential form:

$$\mathrm{d}X(t) = \mu\,\mathrm{d}t + \sigma\,\mathrm{d}W(t) = \mu\,\mathrm{d}t + \sigma(\,\mathrm{d}\widehat{W}(t) + \gamma\,\mathrm{d}t) = (\mu + \sigma\gamma)\,\mathrm{d}t + \sigma\,\mathrm{d}\widehat{W}(t).$$

Hence, choosing $\gamma = -\mu/\sigma$ gives zero drift, $\mu + \sigma\gamma = 0$, and the process $X(t) = \sigma\widehat{W}(t)$ is a zero-drift scaled BM under measure $\widehat{\mathbb{P}}$. The measure change $\mathbb{P} \to \widehat{\mathbb{P}}$, $\frac{\mathrm{d}\widehat{\mathbb{P}}}{\mathrm{d}\mathbb{P}} = \varrho_T$, is defined explicitly by the Radon–Nikodym derivative process

$$\varrho_t \equiv \left(\frac{\mathrm{d}\widehat{\mathbb{P}}}{\mathrm{d}\mathbb{P}}\right)_t = \mathcal{E}_t(\gamma\cdot W) = \exp\left(-\frac{\mu^2 t}{2\sigma^2} - \frac{\mu}{\sigma}W(t)\right),\quad 0 \leqslant t \leqslant T. \tag{2.126}$$

□

For the above example, we can also find the CDF of $X(t)$ in the $\widehat{\mathbb{P}}$-measure, denoted by $\widehat{F}_{X(t)}$. It is instructive to see the two ways to obtain this CDF. One way is to simply use $X(t) = \sigma\widehat{W}(t) \stackrel{d}{=} \sigma\sqrt{t}\widehat{Z}$, $\widehat{Z} \sim Norm(0,1)$ under measure $\widehat{\mathbb{P}}$:

$$\widehat{F}_{X(t)}(x) \equiv \widehat{\mathbb{P}}(X(t) \leqslant x) = \widehat{\mathbb{P}}\left(\widehat{Z} \leqslant \frac{x}{\sigma\sqrt{t}}\right) = \mathcal{N}\left(\frac{x}{\sigma\sqrt{t}}\right).$$

The other way is to compute a $\mathbb{P}$-measure expectation using ϱ_t and apply the identity in (2.119) since $\mathbb{I}_{\{X(t)\leqslant x\}} = \mathbb{I}_{\{\mu t+\sigma W(t)\leqslant x\}}$ is an $\mathcal{F}_t$-measurable random variable:

$$\begin{aligned}
\widehat{F}_{X(t)}(x) = \widehat{\mathrm{E}}\big[\mathbb{I}_{\{X(t)\leqslant x\}}\big] = \mathrm{E}\big[\varrho_t \mathbb{I}_{\{X(t)\leqslant x\}}\big] &= \mathrm{E}\big[\mathrm{e}^{-\frac{1}{2}\gamma^2 t+\gamma W(t)}\mathbb{I}_{\{\mu t+\sigma W(t)\leqslant x\}}\big] \\
&= \mathrm{e}^{-\frac{1}{2}\gamma^2 t}\mathrm{E}\big[\mathrm{e}^{\gamma W(t)}\mathbb{I}_{\{W(t)\leqslant (x-\mu t)/\sigma\}}\big] \\
&= \mathrm{e}^{-\frac{1}{2}\gamma^2 t}\cdot \mathrm{e}^{\frac{1}{2}\gamma^2 t}\mathcal{N}\left(\frac{x-\mu t}{\sigma\sqrt{t}} - \gamma\sqrt{t}\right) \\
&= \mathcal{N}\left(\frac{x-(\mu+\sigma\gamma)t}{\sigma\sqrt{t}}\right) = \mathcal{N}\left(\frac{x}{\sigma\sqrt{t}}\right)
\end{aligned}$$

where $\mu + \sigma\gamma = 0$ since $\gamma = -\mu/\sigma$. Note that here we used the expectation identity (B.2) in Appendix B, where $W(t) \sim Norm(0, t)$ under measure $\mathbb{P}$.

The CDF of $X(t)$ in the $\mathbb{P}$-measure was already computed in Section 1.3.1, i.e.,

$$F_{X(t)}(x) \equiv \mathbb{P}(X(t) \leqslant x) \equiv \mathbb{P}(W^{(\mu,\sigma)}(t) \leqslant x) = \mathcal{N}\left(\frac{x - \mu t}{\sigma\sqrt{t}}\right).$$

We therefore see from the above two expressions for the CDF of the process at time t (in the two different measures) that the measure change $\mathbb{P} \to \widehat{\mathbb{P}}$ eliminates the drift μt when $\gamma = -\mu/\sigma$. Observe that $X(t) \sim Norm(0, \sigma^2 t)$ under the $\widehat{\mathbb{P}}$-measure:

$$\widehat{\mathrm{E}}[X(t)] = \sigma\widehat{\mathrm{E}}[\widehat{W}(t)] = 0, \quad \widehat{\mathrm{E}}[X^2(t)] = \sigma^2\widehat{\mathrm{E}}[\widehat{W}^2(t)] = \sigma^2 t.$$

In contrast, $X(t) \sim Norm(\mu t, \sigma^2 t)$ under the $\mathbb{P}$-measure.

In previous chapters of Volume I we saw how measure changes are employed in discrete-time asset price models such as the binomial model. In particular, we discussed various risk-neutral measures. By using Girsanov's Theorem, we can now consider our first example of how to construct a risk-neutral measure for a single stock GBM price process in *continuous time.*

Example 2.25. (Changing the drift in GBM) Assume a non-dividend-paying stock price process with SDE

$$\mathrm{d}S(t) = S(t)\big[\mu\,\mathrm{d}t + \sigma\,\mathrm{d}W(t)\big],$$

where $\{W(t)\}_{t\geqslant 0}$ is a standard BM under the physical (real-world) measure $\mathbb{P}$, μ is a constant physical (i.e., historical) growth rate, and $\sigma > 0$ is a constant volatility. Find the risk-neutral probability measure $\widetilde{\mathbb{P}}$ defined such that the discounted stock price process $\{\overline{S}(t) := \mathrm{e}^{-rt}S(t)\}_{0\leqslant t\leqslant T}$, for any $T > 0$, is a $\widetilde{\mathbb{P}}$-martingale, where r is a constant interest rate.

Solution. By the strong solution of the SDE

$$S(t) = S(0)\,\mathrm{e}^{(\mu-\sigma^2/2)t+\sigma W(t)} \implies \overline{S}(t) = \mathrm{e}^{-rt}S(t) = S(0)\,\mathrm{e}^{(\mu-r)t}\cdot\mathrm{e}^{-\sigma^2 t/2+\sigma W(t)}$$
$$\equiv S(0)\,\mathrm{e}^{(\mu-r)t}\cdot\mathcal{E}_t(\sigma\cdot W).$$

We recognize $\{\mathcal{E}_t(\sigma\cdot W) := \mathrm{e}^{-\sigma^2 t/2+\sigma W(t)}\}_{t\geqslant 0}$ as a (exponential) $\mathbb{P}$-martingale with unit expectation, $\mathrm{E}[\mathcal{E}_t(\sigma\cdot W)] = 1$. So we now proceed to eliminate the drift $\mu - r$ by expressing W in terms of a new BM, $\widetilde{W}$, in the new measure $\widetilde{\mathbb{P}}$. Since $\mu - r$ and σ are constants, we can accomplish this by employing a measure change as in the above example:

$$\varrho_t \equiv \left(\frac{\mathrm{d}\widetilde{\mathbb{P}}}{\mathrm{d}\mathbb{P}}\right)_t = \mathrm{e}^{-\frac{1}{2}\gamma^2 t+\gamma W(t)} \equiv \mathcal{E}_t(\gamma\cdot W), \tag{2.127}$$

where $\frac{\mathrm{d}\widetilde{\mathbb{P}}}{\mathrm{d}\mathbb{P}} = \varrho_T$ and $\widetilde{W}(t) = W(t) - \gamma t$ is a standard $\widetilde{\mathbb{P}}$-BM. Substituting $W(t) = \widetilde{W}(t) + \gamma t$ into the above exponential expression gives

$$\overline{S}(t) = S(0)\,\mathrm{e}^{(\mu-r)t}\cdot\mathrm{e}^{-\sigma^2 t/2+\sigma(\widetilde{W}(t)+\gamma t)} = \overline{S}(0)\,\mathrm{e}^{(\mu-r+\sigma\gamma)t}\cdot\mathcal{E}_t(\sigma\cdot\widetilde{W}) \tag{2.128}$$

where $\overline{S}(0) = S(0)$. Note that $\{\mathcal{E}_t(\sigma\cdot\widetilde{W}) := \mathrm{e}^{-\sigma^2 t/2+\sigma\widetilde{W}(t)}\}_{t\geqslant 0}$ is a $\widetilde{\mathbb{P}}$-martingale where:

$$\widetilde{\mathrm{E}}[\,\mathcal{E}_t(\sigma\cdot\widetilde{W}) \mid \mathcal{F}_u] = \mathcal{E}_u(\sigma\cdot\widetilde{W}),\ u \leqslant t.$$

Clearly, by setting $\gamma = (r-\mu)/\sigma$, we have $\mu - r + \sigma\gamma = 0$ and this gives the unique measure change for eliminating the drift in (2.128), giving the discounted stock price process as a $\widetilde{\mathbb{P}}$-martingale, i.e.,

$$\overline{S}(t) = \overline{S}(0) \cdot \mathcal{E}_t(\sigma \cdot \widetilde{W}), \ \ 0 \leqslant t \leqslant T, \tag{2.129}$$

where

$$\widetilde{\mathrm{E}}[\overline{S}(t) \mid \mathcal{F}_u] = \overline{S}(u), \ \ 0 \leqslant u \leqslant t \leqslant T. \tag{2.130}$$

In summary, the risk-neutral measure is the unique measure obtained with the Radon–Nikodym derivative process and measure change defined by (2.127) with $\gamma = (r - \mu)/\sigma$:

$$\left(\frac{\mathrm{d}\widetilde{\mathbb{P}}}{\mathrm{d}\mathbb{P}}\right)_t = \mathcal{E}_t\left(\frac{(r-\mu)}{\sigma} \cdot W\right), \ 0 \leqslant t \leqslant T; \quad \frac{\mathrm{d}\widetilde{\mathbb{P}}}{\mathrm{d}\mathbb{P}} = \mathcal{E}_T\left(\frac{(r-\mu)}{\sigma} \cdot W\right). \tag{2.131}$$

□

Note that the measure $\widetilde{\mathbb{P}}$ is uniquely specified by (2.131), where $\gamma = (r - \mu)/\sigma$ always exists since $\sigma > 0$. We can also see directly how to choose the above measure change by working with the SDE where the Brownian increment $\mathrm{d}W(t) = \mathrm{d}\widetilde{W}(t) + \gamma\,\mathrm{d}t$ is used within the original SDE:

$$\mathrm{d}S(t) = S(t)\big[\mu\,\mathrm{d}t + \sigma(\,\mathrm{d}\widetilde{W}(t) + \gamma\,\mathrm{d}t\,)\big] = S(t)\big[(\mu + \sigma\gamma)\,\mathrm{d}t + \sigma\,\mathrm{d}\widetilde{W}(t)\big]. \tag{2.132}$$

Taking the stochastic differential of $\overline{S}(t) \equiv \mathrm{e}^{-rt}S(t)$ and using the above $\mathrm{d}S(t)$ term:

$$\begin{aligned} \mathrm{d}\overline{S}(t) = \mathrm{d}(\mathrm{e}^{-rt}S(t)) &= \mathrm{e}^{-rt}\big[\mathrm{d}S(t) - rS(t)\,\mathrm{d}t\big] \\ &= \mathrm{e}^{-rt}S(t)\big[(\mu - r + \sigma\gamma)\,\mathrm{d}t + \sigma\,\mathrm{d}\widetilde{W}(t)\big] \\ &= \overline{S}(t)\big[(\mu - r + \sigma\gamma)\,\mathrm{d}t + \sigma\,\mathrm{d}\widetilde{W}(t)\big] \qquad (2.133) \\ \implies \mathrm{d}\overline{S}(t) &= \sigma\overline{S}(t)\,\mathrm{d}\widetilde{W}(t) \qquad (2.134) \end{aligned}$$

where the last expression with zero drift is obtained by choosing $\gamma = (r - \mu)/\sigma$, i.e., by employing the measure change defined in (2.131). Note that the SDE in (2.134) with initial condition $\overline{S}(0)$ is equivalent to (2.129), which is its unique solution. For an arbitrary choice of γ the SDE with drift in (2.133) subject to initial condition $\overline{S}(0)$ is equivalent to (2.128), which is its unique solution. Finally, note that choosing $\gamma = (r - \mu)/\sigma$ in (2.132) gives the stock price *drifting at the risk-free rate within the risk-neutral measure*:

$$\mathrm{d}S(t) = S(t)\big[r\,\mathrm{d}t + \sigma\,\mathrm{d}\widetilde{W}(t)\big] \tag{2.135}$$

with unique solution

$$S(t) = S(0)\,\mathrm{e}^{rt} \cdot \mathcal{E}_t(\sigma \cdot \widetilde{W}) = S(0)\,\mathrm{e}^{(r - \sigma^2/2)t + \sigma\widetilde{W}(t)} \tag{2.136}$$

equivalent to (2.129). The $\widetilde{\mathbb{P}}$-martingale property in (2.130) is equivalently expressed as

$$\widetilde{\mathrm{E}}[\,S(t) \mid \mathcal{F}_u] = \mathrm{e}^{r(t-u)}S(u), \ \ 0 \leqslant u \leqslant t \leqslant T. \tag{2.137}$$

In Example 2.24 we used Girsanov's Theorem to obtain a new measure $\widehat{\mathbb{P}}$, defined by the Radon–Nikodym process in (2.126), such that the process $X(t) \equiv W^{(\mu,\sigma)}(t)$ is a scaled standard $\widehat{\mathbb{P}}$-BM. We now employ the same measure change and thereby compute expectations and joint probabilities of events associated with the sampled maximum or minimum of BM with drift. In particular, let's simply set $\sigma = 1$ and consider the process defined by (1.98) in Section 1.4.3, i.e.,

$$X(t) = \mu t + W(t) = \widehat{W}(t).$$

[The reader can think of this as $\widehat{W}(t) = W(t) - \gamma t$ with $\gamma \equiv -\mu$.] The expression in (2.126), for $\sigma = 1$, gives the Radon–Nikodym derivative for the change of measure $\mathbb{P} \to \hat{\mathbb{P}}$, $\varrho_t = \left(\frac{\mathrm{d}\hat{\mathbb{P}}}{\mathrm{d}\mathbb{P}}\right)_t = \mathrm{e}^{-\frac{1}{2}\mu^2 t - \mu W(t)}$. Hence, the Radon–Nikodym derivative for the change of measure $\hat{\mathbb{P}} \to \mathbb{P}$ is expressed in terms of the $\hat{\mathbb{P}}$-BM, $\widehat{W}(t) = W(t) + \mu t$, as

$$\frac{1}{\varrho_t} = \left(\frac{\mathrm{d}\mathbb{P}}{\mathrm{d}\hat{\mathbb{P}}}\right)_t = \mathrm{e}^{\frac{1}{2}\mu^2 t + \mu W(t)} = \mathrm{e}^{-\frac{1}{2}\mu^2 t + \mu \widehat{W}(t)}. \tag{2.138}$$

Let A, B be any two Borel sets in $\mathbb{R}$ and consider the $\mathcal{F}_t$-measurable indicator random variables $\mathbb{I}_{\{M^X(t) \in A, X(t) \in B\}}$ and $\mathbb{I}_{\{m^X(t) \in A, X(t) \in B\}}$ where the respective sampled maximum, $M^X(t)$, and minimum, $m^X(t)$, of the drifted BM process X are defined in (1.100) and (1.101). That is,

$$M^X(t) = \sup_{0 \leqslant u \leqslant t} X(u) = \sup_{0 \leqslant u \leqslant t} \widehat{W}(u) \equiv M^{\widehat{W}}(t)$$

and

$$m^X(t) = \inf_{0 \leqslant u \leqslant t} X(u) = \inf_{0 \leqslant u \leqslant t} \widehat{W}(u) \equiv m^{\widehat{W}}(t).$$

The sampled maximum $M(t) \equiv M^W(t)$ and minimum $m(t) \equiv m^W(t)$ of the standard $\mathbb{P}$-BM, W, are defined in (1.60) and (1.61). Applying the change of measure while using (2.138) within (2.119) gives

$$\begin{aligned}
\mathbb{P}(M^X(t) \in A, X(t) \in B) &\equiv \mathrm{E}\big[\mathbb{I}_{\{M^X(t) \in A, X(t) \in B\}}\big] \\
&= \widehat{\mathrm{E}}\big[\varrho_t^{-1} \mathbb{I}_{\{M^X(t) \in A, X(t) \in B\}}\big] \\
&= \mathrm{e}^{-\frac{1}{2}\mu^2 t} \widehat{\mathrm{E}}\big[\mathrm{e}^{\mu \widehat{W}(t)} \mathbb{I}_{\{M^{\widehat{W}}(t) \in A, \widehat{W}(t) \in B\}}\big] \\
&= \mathrm{e}^{-\frac{1}{2}\mu^2 t} \mathrm{E}\big[\mathrm{e}^{\mu W(t)} \mathbb{I}_{\{M(t) \in A, W(t) \in B\}}\big].
\end{aligned} \tag{2.139}$$

In the last equation line we simply removed all "hats" since the random variables $M^{\widehat{W}}(t)$ and $\widehat{W}(t)$ under measure $\widehat{\mathbb{P}}$ are the same as $M(t)$ and $W(t)$ under measure $\mathbb{P}$. By the same steps as in (2.139), we have

$$\mathbb{P}(m^X(t) \in A, X(t) \in B) = \mathrm{e}^{-\frac{1}{2}\mu^2 t} \mathrm{E}\big[\mathrm{e}^{\mu W(t)} \mathbb{I}_{\{m(t) \in A, W(t) \in B\}}\big]. \tag{2.140}$$

Equations (2.139) and (2.140) can be used to compute the probability of any joint event involving either pair $M^X(t), X(t)$ or $m^X(t), X(t)$. For example, taking intervals $A = (-\infty, m]$, $B = (-\infty, x]$ gives the respective joint CDFs

$$\begin{aligned}
F_{M^X(t), X(t)}(m, x) &:= \mathbb{P}(M^X(t) \leqslant m, X(t) \leqslant x) \\
&= \mathrm{e}^{-\frac{1}{2}\mu^2 t} \mathrm{E}\big[\mathrm{e}^{\mu W(t)} \mathbb{I}_{\{M(t) \leqslant m, W(t) \leqslant x\}}\big]
\end{aligned} \tag{2.141}$$

and

$$\begin{aligned}
F_{m^X(t), X(t)}(m, x) &:= \mathbb{P}(m^X(t) \leqslant m, X(t) \leqslant x) \\
&= \mathrm{e}^{-\frac{1}{2}\mu^2 t} \mathrm{E}\big[\mathrm{e}^{\mu W(t)} \mathbb{I}_{\{m(t) \leqslant m, W(t) \leqslant x\}}\big].
\end{aligned} \tag{2.142}$$

Expressing the expectation in (2.141) as an integral over the joint density of $M(t), W(t)$:

$$F_{M^X(t),X(t)}(m,x) = \mathrm{e}^{-\frac{1}{2}\mu^2 t} \int_0^m \int_{-\infty}^x \mathrm{e}^{\mu y} f_{M(t),W(t)}(w,y)\,\mathrm{d}y\,\mathrm{d}w\,. \tag{2.143}$$

Differentiating, and making use of the known joint PDF of $M(t), W(t)$ in (1.66), gives the joint PDF of $M^X(t), X(t)$

$$\begin{aligned} f_{M^X(t),X(t)}(m,x) &= \mathrm{e}^{-\frac{1}{2}\mu^2 t+\mu x} f_{M(t),W(t)}(m,x) \\ &= \frac{2(2m-x)}{t\sqrt{2\pi t}} \mathrm{e}^{-\frac{1}{2}\mu^2 t+\mu x-(2m-x)^2/2t}, \end{aligned} \tag{2.144}$$

for $x \leqslant m, m > 0$ and zero otherwise. Similarly, the joint PDF of $m^X(t), X(t)$ follows from (2.142) and the joint PDF in (1.71),

$$\begin{aligned} f_{m^X(t),X(t)}(m,x) &= \mathrm{e}^{-\frac{1}{2}\mu^2 t+\mu x} f_{m(t),W(t)}(m,x) \\ &= \frac{2(x-2m)}{t\sqrt{2\pi t}} \mathrm{e}^{-\frac{1}{2}\mu^2 t+\mu x-(x-2m)^2/2t}, \end{aligned} \tag{2.145}$$

for $x \geqslant m, m < 0$, and zero otherwise. Other applications of (2.139) and (2.140) are given in Section 1.4.3.

2.9 Brownian Martingale Representation Theorem

Before moving on to the next section on multidimensional (vector) BM, we state a result that we will later see has some theoretical importance in replication (hedging) and pricing derivative contracts within a continuous-time financial model driven by a single BM. We have already learned that, given an adapted process $\{X(t)\}_{0\leqslant t\leqslant T}$ with $\int_0^T \mathrm{E}[X^2(t)]\,\mathrm{d}t < \infty$, the Itô process $\{I(t) := \int_0^t X(s)\,\mathrm{d}W(s)\}_{0\leqslant t\leqslant T}$ is a $(\mathbb{P},\mathbb{F})$-martingale where $\{W(t)\}_{t\geqslant 0}$ is a $(\mathbb{P},\mathbb{F})$-BM. A question that one may ask is: Are all $(\mathbb{P},\mathbb{F})$-martingales expressible as an Itô process? It turns out that this is the case if we consider martingales that are *square integrable* and we also restrict the filtration to be the *natural filtration generated by the* BM, i.e., if $\mathbb{F} = \mathbb{F}^W = \{F_t^W\}_{t\geqslant 0} := \{\sigma(W(s) : 0 \leqslant s \leqslant t)\}_{t\geqslant 0}$. We summarize this in the following known theorem without proof.

Theorem 2.18 (Brownian Martingale Representation Theorem). *Assume $\{M(t)\}_{0\leqslant t\leqslant T}$ is a $(\mathbb{P},\mathbb{F}^W)$-martingale and that it is square integrable, i.e.,*

$$\mathrm{E}[M^2(t)] < \infty, \quad \textit{for every } t \in [0,T].$$

Then, there exists an $\mathbb{F}^W$-adapted process $\{\theta(t)\}_{0\leqslant t\leqslant T}$ such that (a.s.)

$$M(t) = M(0) + \int_0^t \theta(u)\,\mathrm{d}W(u)\,. \tag{2.146}$$

This theorem tells us that if a process is a square-integrable martingale, w.r.t. a given measure $\mathbb{P}$ and natural filtration $\mathbb{F}^W$ generated by the standard $\mathbb{P}$-BM W, then it can be expressed as a sum of its initial value and an Itô integral in W. The integrand of the Itô integral is a process that is adapted to $\mathbb{F}^W$. Note that the Itô integral itself is a square-integrable $(\mathbb{P}, \mathbb{F}^W)$-martingale and also continuous in time. So the martingale having this representation is also continuous in time (i.e., the process has no jumps).

We are now ready to state a closely related result that is a consequence of the above theorem and will later be applicable to our discussion of derivative replication in Chapter 3. Let us consider what happens when we change measures $\mathbb{P} \to \widehat{\mathbb{P}}$ as defined in Girsanov's Theorem 2.17. As we already noted, $\mathbb{F}$ could be any filtration for the $\mathbb{P}$-BM, W. Now set $\mathbb{F} = \mathbb{F}^W$ where $\{\gamma(t)\}_{0 \leqslant t \leqslant T}$ is assumed to be $\mathbb{F}^W$-adapted and clearly the time integral of this process occurring in (2.123) is $\mathbb{F}^W$-adapted. In particular, the σ-algebra $\sigma\left(\int_0^t \gamma(s)\, ds\right) \subset \mathcal{F}_t^W$ for every $t \geqslant 0$. Then, by the definition in (2.123), the σ-algebra $\mathcal{F}_t^{\widehat{W}} := \sigma(\widehat{W}(u) : 0 \leqslant u \leqslant t) = \mathcal{F}_t^W$. Hence, if $\{\gamma(t)\}_{0 \leqslant t \leqslant T}$ is chosen as an $\mathbb{F}^W$-adapted process, then the natural filtration $\mathbb{F}^{\widehat{W}} = \{\mathcal{F}_t^{\widehat{W}}\}_{0 \leqslant t \leqslant T}$, generated by $\widehat{W}$ in (2.123), is equal to the natural filtration $\mathbb{F}^W$, generated by W, i.e., $\mathbb{F}^W = \mathbb{F}^{\widehat{W}}$. In summary, by combining these facts with Theorem 2.18 we have the result below. This states that, if the change of measure $\mathbb{P} \to \widehat{\mathbb{P}}$ is defined via Girsanov's Theorem with an $\mathbb{F}^W$-adapted process, then we can always express a square-integrable $(\widehat{\mathbb{P}}, \mathbb{F}^W)$-martingale as its initial value plus an Itô integral in the $\widehat{\mathbb{P}}$-BM.

Proposition 2.19. *Let the measure $\widehat{\mathbb{P}}$ be defined as in Girsanov's Theorem 2.17 with the assumption that the process $\{\gamma(t)\}_{0 \leqslant t \leqslant T}$ is $\mathbb{F}^W$-adapted. If $\{M(t)\}_{0 \leqslant t \leqslant T}$ is a square-integrable $(\widehat{\mathbb{P}}, \mathbb{F}^W)$-martingale, then there exists an adapted process, say $\{\widehat{\theta}(t)\}_{0 \leqslant t \leqslant T}$, such that (a.s.)*

$$M(t) = M(0) + \int_0^t \widehat{\theta}(u)\, d\widehat{W}(u)\,. \tag{2.147}$$

Proof. By the above argument we have $\mathbb{F}^W = \mathbb{F}^{\widehat{W}}$. Hence, $\{M(t)\}_{0 \leqslant t \leqslant T}$ is a square-integrable $(\widehat{\mathbb{P}}, \mathbb{F}^{\widehat{W}})$-martingale where $\widehat{\mathrm{E}}[M^2(t)] = \mathrm{E}[\varrho_t M^2(t)] \leqslant \mathrm{E}[M^2(t)] < \infty$. It now follows trivially by Theorem 2.18 that there exists an $\mathbb{F}^{\widehat{W}}$-adapted (and hence $\mathbb{F}^W$-adapted) process $\{\widehat{\theta}(t)\}_{0 \leqslant t \leqslant T}$ such that (2.147) holds (a.s.). □

2.10 Stochastic Calculus for Multidimensional BM

2.10.1 The Itô Integral and Itô's Formula for Multiple Processes on Multidimensional BM

We now extend the definition of one-dimensional standard BM $\{W(t)\}_{t \geqslant 0}$ into d dimensions for any finite integer $d \geqslant 1$. As seen below, the extension to multiple dimensions is fairly straightforward as we take each component as an independent one-dimensional standard BM. Notation needs to be introduced to precisely denote each component BM and boldface is used for a vector BM.

Definition 2.5. A standard BM in $\mathbb{R}^d$ (or standard d-dimensional BM) is a vector process

$$\mathbf{W}(t) \equiv (W_1(t), W_2(t), \ldots, W_d(t))\,,\ t \geqslant 0,$$

where each component process $\{W_i(t)\}_{t \geqslant 0}$, $1 \leqslant i \leqslant d$, is an *independent one-dimensional standard BM* in $\mathbb{R}$.

Hence, each component is *i.i.d.* where $W_i(t) \sim Norm(0,t)$ and $W_i(t) - W_i(s) \sim Norm(0, t-s)$, $1 \leqslant i \leqslant d$, $0 \leqslant s \leqslant t$. We call this a standard vector BM since, by construction, each component is an identical and independent copy of a one-dimensional standard BM. That is, $W_i(t)$ and $W_j(t)$ are independent if $i \neq j$. A filtration $\mathbb{F} = \{\mathcal{F}_t\}_{t \geqslant 0}$ is a filtration for standard d-dimensional BM if it is a filtration for each component BM, $\{W_i(t)\}_{t\geqslant 0}$. The natural filtration for $\{\mathbf{W}(t)\}_{t\geqslant 0}$, denoted by $\mathbb{F}^{\mathbf{W}}$, is the filtration generated by all components of the standard d-dimensional BM. Given any filtration $\mathbb{F}$ for $\{\mathbf{W}(t)\}_{t\geqslant 0}$, we must have that $\{\mathbf{W}(t)\}_{t\geqslant 0}$ is $\mathbb{F}$-adapted, i.e., $\mathbf{W}(t)$ is $\mathcal{F}_t$-measurable, and that each Brownian *vector increment* $\mathbf{W}(t+s) - \mathbf{W}(t)$ is independent of $\mathcal{F}_t$ for $s, t \geqslant 0$.

Since each component is a standard BM, then we have the usual properties such as the quadratic variation formula for each $1 \leqslant i \leqslant d$:

$$[W_i, W_i](t) = t \implies \mathrm{d}[W_i, W_i](t) \equiv \mathrm{d}W_i(t)\,\mathrm{d}W_i(t) \equiv \big(\mathrm{d}W_i(t)\big)^2 = \mathrm{d}t\,. \tag{2.148}$$

Moreover, $[f, W_i](t)$ has zero covariation for any continuously differentiable function $f(t)$. In particular, for each $1 \leqslant i \leqslant d$,

$$[t, W_i](t) = 0 \implies \mathrm{d}W_i(t)\,\mathrm{d}t = 0\,. \tag{2.149}$$

The covariation of two independent Brownian motions is zero,

$$[W_i, W_j](t) = 0 \implies \mathrm{d}[W_i, W_j](t) \equiv \mathrm{d}W_i(t)\,\mathrm{d}W_j(t) = 0\,, \quad \text{for } i \neq j. \tag{2.150}$$

It is simple to see how this arises by considering a time partition $\{0 = t_0 < t_1 < \ldots < t_n = t\}$ and forming the partial sum of products of individual Brownian increments:

$$Q_n^{i,j}(t) := \sum_{k=1}^{n} \left(W_i(t_k) - W_i(t_{k-1})\right)\left(W_j(t_k) - W_j(t_{k-1})\right)$$

for $i \neq j$. Using the fact that the increments are all mutually independent with mean zero, $\mathrm{E}[W_i(t_k) - W_i(t_{k-1})] = 0$, for every k, then $\mathrm{E}[Q_n^{i,j}(t)] = 0$. Since all n terms in the sum are mutually independent, the variance of the sum is the sum of the individual variances. Using the independence of the product terms, where $\mathrm{E}[(W_i(t_k) - W_i(t_{k-1}))^2] = \mathrm{E}[(W_j(t_k) - W_j(t_{k-1}))^2] = t_k - t_{k-1}$, gives

$$\begin{aligned}
\mathrm{Var}\left(Q_n^{i,j}(t)\right) &= \sum_{k=1}^{n} \mathrm{E}[(W_i(t_k) - W_i(t_{k-1}))^2]\,\mathrm{E}[(W_j(t_k) - W_j(t_{k-1}))^2] \\
&= \sum_{k=1}^{n} (t_k - t_{k-1})^2 \\
&\leqslant \Delta_n \sum_{k=1}^{n} (t_k - t_{k-1}) = \Delta_n (t_n - t_0) = \Delta_n t,
\end{aligned}$$

where $\Delta_n := \max_{k=1,\ldots n} (t_k - t_{k-1})$ is the maximum time increment over the partition. Clearly, $\mathrm{Var}(Q_n^{i,j}(t)) \to 0$ as $n \to \infty$, $\Delta_n \to 0$, i.e., this implies that, for all $t \geqslant 0$, $Q_n^{i,j}(t)$ converges to its expected value $\mathrm{E}[Q_n^{i,j}(t)] = 0$ as $n \to \infty$, $\Delta_n \to 0$. Hence, the co-variation $[W_i, W_j](t) := \lim_{n\to\infty, \Delta_n \to 0} Q_n^{i,j}(t)$, for $i \neq j$, converges to zero in $L^2(\Omega)$. This implies convergence to zero in probability. Hence, there exists a subsequence converging almost surely to zero, i.e., $[W_i, W_j](t) = 0$ (a.s.).

For convenience we summarize the above "basic rules" for the multiplication of stochastic increments as follows:

$$\mathrm{d}W_i(t)\,\mathrm{d}W_j(t) = \delta_{ij}\,\mathrm{d}t\,, \quad \mathrm{d}W_i(t)\,\mathrm{d}t = 0\,, \quad (\mathrm{d}t)^2 = 0, \tag{2.151}$$

with Kronecker delta $\delta_{ij} = 1$ if $i = j$, and 0 if $i \neq j$.

As in the case of standard BM in one dimension, there is a similar useful characterization of a standard d-dimensional BM due to Lévy which we state in the following Theorem. The result can be proven based on multidimensional extensions of the Itô formula.

Theorem 2.20 (Lévy's Characterization of a Standard Multidimensional BM). *Consider the vector-valued process* $\{\mathbf{M}(t) := M_1(t), \ldots, M_d(t)\}_{t \geqslant 0}$ *where each component process* $\{M_i(t)\}_{t \geqslant 0}$, $1 \leqslant i \leqslant d$, *is a continuous* $(\mathbb{P}, \mathbb{F})$*-martingale starting at* $M_i(0) = 0$ *(a.s.) and having quadratic variation* $[M_i, M_i](t) = t$, *for all* $t \geqslant 0$. *Also, assume* $[M_i, M_j](t) = 0$ *for* $i \neq j$. *Then,* $\{\mathbf{M}(t)\}_{t \geqslant 0}$ *is a standard d-multidimensional* $(\mathbb{P}, \mathbb{F})$*-BM.*

According to this result, a vector process is a standard vector BM (in a given measure $\mathbb{P}$ and filtration $\mathbb{F}$) if we can verify that *every component process* is a martingale with continuous paths starting at zero, has the same quadratic variation as a standard BM, and all covariations among different components are zero. Basically, this means that each component is an i.i.d. standard one-dimensional BM.

Let us fix a filtration $\mathbb{F} = \{\mathcal{F}_t\}_{t \geqslant 0}$ for BM in $\mathbb{R}^d$ for a given integer $d \geqslant 1$. The formulae and concepts we developed in previous sections on the Itô integral, Itô's formula for a function of an Itô process and SDEs can be generalized to a multiple (vector) BM and multiple Itô processes that are driven by the vector BM in $\mathbb{R}^d$. Let us first discuss this extension for the case of BM in $\mathbb{R}^2$, i.e., $d = 2$ where $\mathbf{W}(t) = (W_1(t), W_2(t))$. We can have any number of Itô processes that can be represented as an Itô integral w.r.t. $\mathbf{W}(t)$ plus a drift term which is a Riemann integral. Consider two Itô processes $X \equiv \{X(t)\}_{t \geqslant 0}$ and $Y \equiv \{Y(t)\}_{t \geqslant 0}$ which form a vector process, $(X(t), Y(t))_{t \geqslant 0}$. Let $\{\mu_X(t)\}_{t \geqslant 0}$ and $\{\mu_Y(t)\}_{t \geqslant 0}$ be $\mathbb{F}$-adapted drift coefficients for processes X and Y, respectively. The diffusion or volatility coefficient vectors are $\mathbb{F}$-adapted vectors in $\mathbb{R}^2$ denoted by

$$\boldsymbol{\sigma}_X(t) = (\sigma_{X,1}(t), \sigma_{X,2}(t)) \quad \text{and} \quad \boldsymbol{\sigma}_Y(t) = (\sigma_{Y,1}(t), \sigma_{Y,2}(t))$$

for processes X and Y, respectively. For each $i = 1, 2$, $\sigma_{X,i}(t)$ is the time-t volatility of process X w.r.t. the ith BM and $\sigma_{Y,i}(t)$ is the time-t volatility of process Y w.r.t. the ith BM. The two processes have the representations:

$$X(t) = X(0) + \int_0^t \mu_X(u)\,\mathrm{d}u + \int_0^t \boldsymbol{\sigma}_X(u) \cdot \mathrm{d}\mathbf{W}(u), \tag{2.152}$$

$$Y(t) = Y(0) + \int_0^t \mu_Y(u)\,\mathrm{d}u + \int_0^t \boldsymbol{\sigma}_Y(u) \cdot \mathrm{d}\mathbf{W}(u). \tag{2.153}$$

In each case, the first (Riemann) time integral is the drift term and the second integral is a sum of two Itô integrals; one is taken w.r.t. the first component of the volatility vector and the first BM and the second is w.r.t. the second component of the volatility vector and the second BM. That is, we define

$$\int_0^t \boldsymbol{\sigma}_X(u) \cdot \mathrm{d}\mathbf{W}(u) := \int_0^t \sigma_{X,1}(u)\,\mathrm{d}W_1(u) + \int_0^t \sigma_{X,2}(u)\,\mathrm{d}W_2(u), \tag{2.154}$$

$$\int_0^t \boldsymbol{\sigma}_Y(u) \cdot \mathrm{d}\mathbf{W}(u) := \int_0^t \sigma_{Y,1}(u)\,\mathrm{d}W_1(u) + \int_0^t \sigma_{Y,2}(u)\,\mathrm{d}W_2(u). \tag{2.155}$$

Given a time $T > 0$, throughout we shall assume the square-integrability condition holds for any d-dimensional $\mathbb{F}$-adapted vector integrand process $\boldsymbol{\sigma}(t) = (\sigma_1(t), \ldots, \sigma_d(t))$,

$$\int_0^T \mathrm{E}\left[\|\boldsymbol{\sigma}(t)\|^2\right]\,\mathrm{d}t < \infty, \tag{2.156}$$

where $\|\boldsymbol{\sigma}(t)\|^2 \equiv \sum_{i=1}^d \sigma_i^2(t)$ is the square magnitude of the volatility vector. We note the obvious extension of the definition for the d-dimensional Itô integral (for any $d \geqslant 1$) which is stated later in (2.186). For $d = 2$ we have $\|\boldsymbol{\sigma}(t)\|^2 = \sigma_1^2(t) + \sigma_2^2(t)$. This condition is equivalent to requiring $\int_0^T \mathrm{E}\left[\sigma_i^2(t)\right]\, dt < \infty$, for every component i. For example, in (2.154)–(2.155), we assume $\int_0^T \mathrm{E}\left[\sigma_{X,i}^2(t)\right]\, dt < \infty$ and $\int_0^T \mathrm{E}\left[\sigma_{Y,i}^2(t)\right]\, dt < \infty$ for $i = 1, 2$.

Based on the martingale property that we have proven for any one-dimensional Itô integral process, and the fact that a d-dimensional Itô integral is a sum of one-dimensional Itô integrals, it follows that the condition in (2.156) guarantees the martingale property of any d-dimensional Itô integral process $I(t) := \int_0^t \boldsymbol{\sigma}(u) \cdot \mathrm{d}\mathbf{W}(u)$, where $\mathrm{E}[I(t)|\mathcal{F}_s] = I(s)$, $0 \leqslant s \leqslant t \leqslant T$, i.e.,

$$\mathrm{E}\left[\int_0^t \boldsymbol{\sigma}(u) \cdot \mathrm{d}\mathbf{W}(u) \,\middle|\, \mathcal{F}_s\right] = \int_0^s \boldsymbol{\sigma}(u) \cdot \mathrm{d}\mathbf{W}(u). \tag{2.157}$$

Any d-dimensional Itô integral has zero mean (as its a sum of zero-mean one-dimensional Itô integrals). We have the Itô isometry formula for a d-dimensional Itô integral:

$$\begin{aligned}\mathrm{E}\left[\left(\int_0^t \boldsymbol{\sigma}(u) \cdot \mathrm{d}\mathbf{W}(u)\right)^2\right] = \mathrm{Var}\left(\int_0^t \boldsymbol{\sigma}(u) \cdot \mathrm{d}\mathbf{W}(u)\right) &= \int_0^t \mathrm{E}\left[\|\boldsymbol{\sigma}(u)\|^2\right] \mathrm{d}u \\ &= \sum_{i=1}^d \int_0^t \mathrm{E}\left[\sigma_i^2(u)\right] \mathrm{d}u\,. \end{aligned} \tag{2.158}$$

This is a special case of the covariance formula:

$$\mathrm{Cov}\left(\int_0^s \boldsymbol{\sigma}(u) \cdot \mathrm{d}\mathbf{W}(u), \int_0^t \boldsymbol{\gamma}(u) \cdot \mathrm{d}\mathbf{W}(u)\right) = \int_0^s \mathrm{E}\left[\boldsymbol{\sigma}(u) \cdot \boldsymbol{\gamma}(u)\right] \mathrm{d}u\,, \tag{2.159}$$

for all $0 \leqslant s \leqslant t \leqslant T$, where $\{\boldsymbol{\sigma}(t)\}_{t \geqslant 0}$ and $\{\boldsymbol{\gamma}(t)\}_{t \geqslant 0}$ are $\mathbb{F}$-adapted d-dimensional vectors. The formula in (2.159) is readily derived by writing out the two Itô integrals as sums of (one-dimensional) Itô integrals, and then using the known covariance relation for each pair of one-dimensional Itô integrals. Let us denote $I_{\sigma_i}(s) \equiv \int_0^s \sigma_i(u) \cdot \mathrm{d}W_i(u)$ and $I_{\gamma_i}(t) \equiv \int_0^t \gamma_i(u) \cdot \mathrm{d}W_i(u)$. Using the independence of Itô integrals w.r.t. different Brownian motions, i.e., $\mathrm{Cov}\left(I_{\sigma_i}(s), I_{\gamma_j}(t)\right) = 0$ for $i \neq j$, we have

$$\begin{aligned}\mathrm{Cov}\left(\int_0^s \boldsymbol{\sigma}(u) \cdot \mathrm{d}\mathbf{W}(u), \int_0^t \boldsymbol{\gamma}(u) \cdot \mathrm{d}\mathbf{W}(u)\right) &= \sum_{i=1}^d \sum_{j=1}^d \mathrm{Cov}\left(I_{\sigma_i}(s), I_{\gamma_j}(t)\right) \\ = \sum_{i=1}^d \mathrm{Cov}\left(I_{\sigma_i}(s), I_{\gamma_i}(t)\right) = \sum_{i=1}^d \int_0^s \mathrm{E}\left[\sigma_i(u)\gamma_i(u)\right] \mathrm{d}u &= \int_0^s \mathrm{E}\left[\boldsymbol{\sigma}(u) \cdot \boldsymbol{\gamma}(u)\right] \mathrm{d}u\,.\end{aligned}$$

In the second equation line, we used the covariance formula (2.19) for each ith pair of Itô integrals w.r.t. a single standard BM W_i. The isometry formula in (2.158) is recovered if we set $s = t$ and the two vector processes as the same, $\boldsymbol{\sigma} \equiv \boldsymbol{\gamma}$.

Also, throughout we assume that any drift coefficient $\mu(t)$ in the Riemann integral portion of an Itô process is integrable,

$$\int_0^T \mathrm{E}\left[|\mu(t)|\right] \mathrm{d}t < \infty\,. \tag{2.160}$$

The Itô integrals in (2.154)–(2.155) are the one-dimensional Itô integrals w.r.t. a single standard BM which is taken as either W_1 or W_2. The Riemann integrals in (2.152) and (2.153) are continuous functions of time and therefore have zero quadratic variation. To obtain the quadratic variation of the X process, note that the quadratic variation of each Itô integral in (2.154),

$$I_{X,1}(t) := \int_0^t \sigma_{X,1}(u)\,\mathrm{d}W_1(u) \quad \text{and} \quad I_{X,2}(t) := \int_0^t \sigma_{X,2}(u)\,\mathrm{d}W_2(u),$$

is computed according to (2.23) (where W_1 and W_2 individually act as W):

$$[I_{X,1}, I_{X,1}](t) \equiv \int_0^t \sigma_{X,1}^2(u)\,\mathrm{d}u \quad \text{and} \quad [I_{X,2}, I_{X,2}](t) \equiv \int_0^t \sigma_{X,2}^2(u)\,\mathrm{d}u\,. \tag{2.161}$$

Since $[W_1, W_2](t) = 0$, i.e., $\mathrm{d}W_1(t)\,\mathrm{d}W_2(t) = 0$, the covariation of the two integrals is zero: $[I_{X,1}, I_{X,2}](t) = 0$. Hence, the quadratic variation of the X process in (2.152) is the quadratic variation of the Itô integral in (2.154), which, in turn, is the sum of the two quadratic variations in (2.161):

$$\begin{aligned}[X, X](t) = [I_{X,1}, I_{X,1}](t) + [I_{X,2}, I_{X,2}](t) &= \int_0^t \left(\sigma_{X,1}^2(u) + \sigma_{X,2}^2(u)\right)\,\mathrm{d}u \\ &= \int_0^t \|\boldsymbol{\sigma}_X(u)\|^2\,\mathrm{d}u\,.\end{aligned} \tag{2.162}$$

Similarly, the Y process has quadratic variation

$$[Y, Y](t) = \int_0^t \left(\sigma_{Y,1}^2(u) + \sigma_{Y,2}^2(u)\right)\,\mathrm{d}u = \int_0^t \|\boldsymbol{\sigma}_Y(u)\|^2\,\mathrm{d}u\,. \tag{2.163}$$

The stochastic differential forms of (2.162) and (2.163) are

$$\mathrm{d}[X, X](t) = \mathrm{d}X(t)\,\mathrm{d}X(t) = \left(\sigma_{X,1}^2(t) + \sigma_{X,2}^2(t)\right)\,\mathrm{d}t = \|\boldsymbol{\sigma}_X(t)\|^2\,\mathrm{d}t\,, \tag{2.164}$$

$$\mathrm{d}[Y, Y](t) = \mathrm{d}Y(t)\,\mathrm{d}Y(t) = \left(\sigma_{Y,1}^2(t) + \sigma_{Y,2}^2(t)\right)\,\mathrm{d}t = \|\boldsymbol{\sigma}_Y(t)\|^2\,\mathrm{d}t\,. \tag{2.165}$$

It is a simple matter to derive (2.164) and (2.165) by working directly with the stochastic differential forms of (2.152) and (2.153),

$$\begin{aligned}\mathrm{d}X(t) &= \mu_X(t)\,\mathrm{d}t + \boldsymbol{\sigma}_X(t)\cdot\mathrm{d}\mathbf{W}(t) \equiv \mu_X(t)\,\mathrm{d}t + \sigma_{X,1}(t)\,\mathrm{d}W_1(t) + \sigma_{X,2}(t)\,\mathrm{d}W_2(t)\,,\\ \mathrm{d}Y(t) &= \mu_Y(t)\,\mathrm{d}t + \boldsymbol{\sigma}_Y(t)\cdot\mathrm{d}\mathbf{W}(t) \equiv \mu_Y(t)\,\mathrm{d}t + \sigma_{Y,1}(t)\,\mathrm{d}W_1(t) + \sigma_{Y,2}(t)\,\mathrm{d}W_2(t)\,,\end{aligned}$$

and then applying the elementary rules in (2.151). For example, by squaring the differential $\mathrm{d}X(t)$ and setting the terms $\mathrm{d}t\,\mathrm{d}W_1(t) = \mathrm{d}t\,\mathrm{d}W_2(t) = 0$, $\mathrm{d}W_1(t)\,\mathrm{d}W_2(t) = 0$, $(\mathrm{d}t)^2 = 0$, and $(\mathrm{d}W_1(t))^2 = (\mathrm{d}W_2(t))^2 = \mathrm{d}t$, we obtain

$$\begin{aligned}\mathrm{d}X(t)\,\mathrm{d}X(t) \equiv (\mathrm{d}X(t))^2 &= (\mu_X(t)\,\mathrm{d}t + \boldsymbol{\sigma}_X(t)\cdot\mathrm{d}\mathbf{W}(t))^2 \\ &= \boldsymbol{\sigma}_X(t)\cdot\boldsymbol{\sigma}_X(t)\,\mathrm{d}t = \|\boldsymbol{\sigma}_X(t)\|^2\,\mathrm{d}t\,.\end{aligned}$$

This recovers the result in (2.164). A similar derivation based on squaring $\mathrm{d}Y(t)$ gives (2.165). The covariation is also simpler to compute based on this differential approach. By multiplying the two stochastic differentials and applying the simple rules in (2.151),

$$\begin{aligned}\mathrm{d}[X, Y](t) = \mathrm{d}X(t)\,\mathrm{d}Y(t) &= (\mu_X(t)\,\mathrm{d}t + \boldsymbol{\sigma}_X(t)\cdot\mathrm{d}\mathbf{W}(t))(\mu_Y(t)\,\mathrm{d}t + \boldsymbol{\sigma}_Y(t)\cdot\mathrm{d}\mathbf{W}(t)) \\ &= (\boldsymbol{\sigma}_X(t)\cdot\mathrm{d}\mathbf{W}(t))(\boldsymbol{\sigma}_Y(t)\cdot\mathrm{d}\mathbf{W}(t)) \\ &= \boldsymbol{\sigma}_X(t)\cdot\boldsymbol{\sigma}_Y(t)\,\mathrm{d}t\,.\end{aligned} \tag{2.166}$$

The last equation line is obtained as follows:

$$
\begin{aligned}
(\boldsymbol{\sigma}_X(t)\cdot \mathrm{d}\mathbf{W}(t))\,(\boldsymbol{\sigma}_Y(t)\cdot \mathrm{d}\mathbf{W}(t)) &= \sum_{i=1}^{d}\sum_{j=1}^{d}\sigma_{X,i}(t)\sigma_{Y,j}(t)\,\underbrace{\mathrm{d}W_i(t)\,\mathrm{d}W_j(t)}_{=\delta_{ij}\,\mathrm{d}t} \\
&= \left(\sum_{i=1}^{d}\sigma_{X,i}(t)\sigma_{Y,i}(t)\right)\mathrm{d}t = \boldsymbol{\sigma}_X(t)\cdot\boldsymbol{\sigma}_Y(t)\,\mathrm{d}t\,.
\end{aligned}
$$

The integral form of (2.166) gives the covariation of the two Itô processes,

$$
[X,Y](t) = \int_0^t \boldsymbol{\sigma}_X(u)\cdot\boldsymbol{\sigma}_Y(u)\,\mathrm{d}u = \int_0^t (\sigma_{X,1}(u)\sigma_{Y,1}(u)+\sigma_{X,2}(u)\sigma_{Y,2}(u))\,\mathrm{d}u\,.
$$

The Itô formula in (2.43) and (2.44) for a function of one Itô process, and time t, extends further to the slightly more general case of a *function of two Itô processes and time* t. We simply state this important result as a lemma (without proof). The main idea, and a simple way to remember the formula in (2.167), is to Taylor expand $f(t,x,y)$ up to terms of order $\mathrm{d}t$, $(\mathrm{d}x)^2, (\mathrm{d}y)^2$ and then replace ordinary variables $x \to X(t)$, $y \to Y(t)$ and ordinary differentials by their respective stochastic differentials: $\mathrm{d}x \to \mathrm{d}X(t)$, $\mathrm{d}y \to \mathrm{d}Y(t)$, $(\mathrm{d}x)^2 \to (\mathrm{d}X(t))^2 \equiv \mathrm{d}[X,X](t)$, $(\mathrm{d}y)^2 \to (\mathrm{d}Y(t))^2 \equiv \mathrm{d}[Y,Y](t)$, and $\mathrm{d}x\,\mathrm{d}y \to \mathrm{d}X(t)\,\mathrm{d}Y(t) \equiv \mathrm{d}[X,Y](t)$.

Lemma 2.21 (Itô Formula for a Function of Two Processes)**.** *Assume $f(t,x,y)$ is a $C^{1,2,2}$ function on $\mathbb{R}_+\times\mathbb{R}^2$, i.e., having continuous derivatives $f_t \equiv \frac{\partial f}{\partial t}$, $f_x \equiv \frac{\partial f}{\partial x}$, $f_y \equiv \frac{\partial f}{\partial y}$, $f_{xx} \equiv \frac{\partial^2 f}{\partial x^2}$, $f_{xy} \equiv \frac{\partial^2 f}{\partial x \partial y}$ and $f_{yy} \equiv \frac{\partial^2 f}{\partial y^2}$. Let the processes X and Y be Itô processes as given in (2.152) and (2.153). Then, the process defined by $F(t) := f(t,X(t),Y(t)), t \geqslant 0$, has stochastic differential $\mathrm{d}F(t) \equiv \mathrm{d}f(t,X(t),Y(t))$ given by*

$$
\begin{aligned}
\mathrm{d}f(t,X(t),Y(t)) = f_t(t,X(t),Y(t))\,\mathrm{d}t + f_x(t,X(t),Y(t))\,\mathrm{d}X(t) + f_y(t,X(t),Y(t))\,\mathrm{d}Y(t) \\
+\frac{1}{2}f_{xx}(t,X(t),Y(t))\;\mathrm{d}[X,X](t) + \frac{1}{2}f_{yy}(t,X(t),Y(t))\;\mathrm{d}[Y,Y](t) \\
+ f_{xy}(t,X(t),Y(t))\;\mathrm{d}[X,Y](t)\,. \qquad (2.167)
\end{aligned}
$$

In integral form,

$$
\begin{aligned}
f(t,X(t),Y(t)) &= f(0,X(0),Y(0)) \\
&+ \int_0^t \Bigg[f_u(u,X(u),Y(u)) + \frac{1}{2}\|\boldsymbol{\sigma}_X(u)\|^2 f_{xx}(u,X(u),Y(u)) \\
&\qquad + \frac{1}{2}\|\boldsymbol{\sigma}_Y(u)\|^2 f_{yy}(u,X(u),Y(u)) + \boldsymbol{\sigma}_X(u)\cdot\boldsymbol{\sigma}_Y(u)\, f_{xy}(u,X(u),Y(u))\Bigg]\,\mathrm{d}u \\
&+ \int_0^t f_x(u,X(u),Y(u))\,\mathrm{d}X(u) + \int_0^t f_y(u,X(u),Y(u))\,\mathrm{d}Y(u)\,. \qquad (2.168)
\end{aligned}
$$

It should be remarked (and we shall see later when we present the general form of the Itô formula for functions of multiple processes driven by multiple Brownian motions) that this lemma is generally valid for any number $d \geqslant 1$ of underlying Brownian motions, although we have taken $d = 2$ as the base case in our examples. For $d \geqslant 2$ the volatilities are d-dimensional vectors and the standard BM is a d-dimensional vector (standard) BM. For the case that $d = 1$ we simply have the vectors becoming scalars, e.g., $\boldsymbol{\sigma}_X(t) \to \sigma_X(t)$, $\boldsymbol{\sigma}_Y(t) \to \sigma_Y(t)$, and $\mathbf{W}(t) \to W(t)$.

Observe that the first integral in (2.168) is a Riemann integral on the time interval $[0, t]$, whereas the second and third integrals are stochastic integrals w.r.t. the Itô processes X in (2.152) and Y in (2.153). The representation of $\mathrm{d}f(t, X(t), Y(t))$ in (2.167) and its corresponding integral form in (2.168) is written in terms of the stochastic differentials of X and Y. The Itô formula is also equivalently rewritten by substituting the above stochastic differentials for $\mathrm{d}X(t)$ and $\mathrm{d}Y(t)$. Then, (2.167) takes the form

$$\begin{aligned}\mathrm{d}f &= \left(f_t + \mu_X(t) f_x + \mu_Y(t) f_y + \frac{1}{2}\|\boldsymbol{\sigma}_X(t)\|^2 f_{xx} + \frac{1}{2}\|\boldsymbol{\sigma}_Y(t)\|^2 f_{yy} + \boldsymbol{\sigma}_X(t)\cdot\boldsymbol{\sigma}_Y(t) f_{xy}\right)\mathrm{d}t \\ &\quad + (f_x\boldsymbol{\sigma}_X(t) + f_y\boldsymbol{\sigma}_Y(t))\cdot\mathrm{d}\mathbf{W}(t) \\ &\equiv \mu_f(t)\,\mathrm{d}t + \boldsymbol{\sigma}_f(t)\cdot\mathrm{d}\mathbf{W}(t), \end{aligned} \tag{2.169}$$

where $f \equiv f(t, X(t), Y(t))$, $f_x \equiv f_x(t, X(t), Y(t))$, etc., is used to compact the expressions. In the second equation line we simply identified the drift $\mu_f(t)$ and volatility vector $\boldsymbol{\sigma}_f(t)$ for the process $\{f(t, X(t), Y(t))\}_{t\geqslant 0}$. We see that $\mu_f(t)$ and $\boldsymbol{\sigma}_f(t)$ are adapted processes defined explicitly as functions of $f(t, X(t), Y(t))$ and its partial derivatives, as well as functions of linear combinations of the drift and volatility vector coefficients of processes X and Y. In particular, for $d = 2$, the volatility vector $\boldsymbol{\sigma}_f(t) := f_x\,\boldsymbol{\sigma}_X(t) + f_y\,\boldsymbol{\sigma}_Y(t) = (\sigma_{f,1}(t), \sigma_{f,2}(t))$ has components

$$\sigma_{f,1}(t) = f_x\sigma_{X,1}(t) + f_y\sigma_{Y,1}(t)\,, \quad \sigma_{f,2}(t) = f_x\sigma_{X,2}(t) + f_y\sigma_{Y,2}(t)\,. \tag{2.170}$$

Hence, $\{F(t)\}_{t\geqslant 0} \equiv \{f(t, X(t), Y(t))\}_{t\geqslant 0}$ is an Itô process satisfying the stochastic integral equation

$$\begin{aligned} F(t) &= F(0) + \int_0^t \mu_f(u)\,\mathrm{d}u + \int_0^t \boldsymbol{\sigma}_f(u)\cdot\mathrm{d}\mathbf{W}(u) \\ &\equiv F(0) + \int_0^t \mu_f(u)\,\mathrm{d}u + \int_0^t \sigma_{f,1}(u)\,\mathrm{d}W_1(u) + \int_0^t \sigma_{f,2}(u)\,\mathrm{d}W_2(u)\,, \end{aligned} \tag{2.171}$$

where the second line is the explicit form for dimension $d = 2$.

The following example shows that the Itô product rule, derived previously, follows simply by applying the Itô formula in (2.167).

Example 2.26. Let $\{X(t)\}_{t\geqslant 0}$ and $\{Y(t)\}_{t\geqslant 0}$ be Itô processes. Derive the stochastic differential of their product.

Solution. Defining the function $f(t, x, y) := xy$ gives the product $F(t) := f(t, X(t), Y(t)) = X(t)Y(t)$ as an Itô process whose stochastic differential is given according to (2.167). In this case the function is independent of t and has derivatives:

$$f_t = 0,\ f_x = y,\ f_y = x,\ f_{xy} = 1,\ f_{xx} = f_{yy} = 0\,.$$

Substituting these terms into (2.167) (with $x = X(t), y = Y(t)$) gives

$$\begin{aligned} \mathrm{d}(X(t)Y(t)) &\equiv \mathrm{d}f(t, X(t), Y(t)) = 0\cdot\mathrm{d}t + Y(t)\,\mathrm{d}X(t) + X(t)\,\mathrm{d}Y(t) + \frac{1}{2}\cdot 0\cdot\mathrm{d}X(t)\,\mathrm{d}X(t) \\ &\quad + \frac{1}{2}\cdot 0\cdot\mathrm{d}Y(t)\,\mathrm{d}Y(t) + 1\cdot\mathrm{d}X(t)\,\mathrm{d}Y(t) \\ &= Y(t)\,\mathrm{d}X(t) + X(t)\,\mathrm{d}Y(t) + \mathrm{d}X(t)\,\mathrm{d}Y(t)\,. \end{aligned} \tag{2.172}$$

Assuming $X(t)Y(t) \neq 0$, we note that a useful way to represent this is to divide by $X(t)Y(t)$ (i.e., factor out the product), giving the relative differential

$$\frac{\mathrm{d}F(t)}{F(t)} \equiv \frac{\mathrm{d}(X(t)Y(t))}{X(t)Y(t)} = \frac{\mathrm{d}X(t)}{X(t)} + \frac{\mathrm{d}Y(t)}{Y(t)} + \frac{\mathrm{d}X(t)}{X(t)}\frac{\mathrm{d}Y(t)}{Y(t)}\,. \tag{2.173}$$

We can also write this in the form of (2.169):

$$\frac{\mathrm{d}(X(t)Y(t))}{X(t)Y(t)} = \left(\frac{\mu_X(t)}{X(t)} + \frac{\mu_Y(t)}{Y(t)} + \frac{\boldsymbol{\sigma}_X(t)}{X(t)} \cdot \frac{\boldsymbol{\sigma}_Y(t)}{Y(t)}\right)\mathrm{d}t + \left(\frac{\boldsymbol{\sigma}_X(t)}{X(t)} + \frac{\boldsymbol{\sigma}_Y(t)}{Y(t)}\right) \cdot \mathrm{d}\mathbf{W}(t)$$
$$\equiv \mu_{XY}(t)\,\mathrm{d}t + \boldsymbol{\sigma}_{XY}(t) \cdot \mathrm{d}\mathbf{W}(t)\,. \tag{2.174}$$

This shows how the drift $\mu_{XY}(t)$ and volatility vector $\boldsymbol{\sigma}_{XY}(t)$ for the product process $F = XY$ are related to the drifts and volatility vectors of the processes X and Y. □

Another important example is the *Quotient Rule* for the stochastic differential of a ratio of two Itô processes. This rule is useful when pricing derivatives when we need to compute the drift and volatility of a process defined by a ratio of two asset price processes.

Example 2.27. Let $\{X(t)\}_{t\geqslant 0}$ and $\{Y(t)\}_{t\geqslant 0} \neq 0$ be Itô processes. Derive the stochastic differential of their ratio $F(t) := \frac{X(t)}{Y(t)}$.

Solution. Let $f(t,x,y) := x/y$, i.e., $f(t,X(t),Y(t)) = X(t)/Y(t)$ is an Itô process with its stochastic differential given by (2.167). The relevant partial derivatives are

$$f_t = 0,\;\; f_x = \frac{1}{y},\;\; f_y = -\frac{x}{y^2},\;\; f_{xy} = -\frac{1}{y^2},\;\; f_{yy} = \frac{2x}{y^3},\;\; f_{xx} = 0\,.$$

Substituting these terms into (2.167) (with $x = X(t), y = Y(t)$) gives

$$\mathrm{d}\left(\frac{X(t)}{Y(t)}\right) \equiv \mathrm{d}F(t) \equiv \mathrm{d}f(t,X(t),Y(t))$$
$$= \frac{1}{Y(t)}\,\mathrm{d}X(t) - \frac{X(t)}{Y^2(t)}\,\mathrm{d}Y(t) + \frac{X(t)}{Y^3(t)}(\,\mathrm{d}Y(t))^2 - \frac{1}{Y^2(t)}\,\mathrm{d}X(t)\,\mathrm{d}Y(t)$$
$$= \left(\frac{\mathrm{d}X(t)}{Y(t)} - \frac{X(t)}{Y(t)}\frac{\mathrm{d}Y(t)}{Y(t)}\right)\left(1 - \frac{\mathrm{d}Y(t)}{Y(t)}\right). \tag{2.175}$$

This is written in a more convenient form (for later use) by dividing through by $X(t)/Y(t)$:

$$\frac{\mathrm{d}F(t)}{F(t)} \equiv \frac{\mathrm{d}\frac{X(t)}{Y(t)}}{\frac{X(t)}{Y(t)}} = \left(\frac{\mathrm{d}X(t)}{X(t)} - \frac{\mathrm{d}Y(t)}{Y(t)}\right)\left(1 - \frac{\mathrm{d}Y(t)}{Y(t)}\right). \tag{2.176}$$

Substituting the expressions for $\mathrm{d}X(t)$ and $\mathrm{d}Y(t)$, applying the basic rules, and combining terms in $\mathrm{d}t$ and $\mathrm{d}\mathbf{W}(t)$ gives the form in (2.169) as

$$\frac{\mathrm{d}\frac{X(t)}{Y(t)}}{\frac{X(t)}{Y(t)}} = \left(\frac{\mu_X(t)}{X(t)} - \frac{\mu_Y(t)}{Y(t)} + \frac{\boldsymbol{\sigma}_Y(t)}{Y(t)}\cdot\left(\frac{\boldsymbol{\sigma}_Y(t)}{Y(t)} - \frac{\boldsymbol{\sigma}_X(t)}{X(t)}\right)\right)\mathrm{d}t + \left(\frac{\boldsymbol{\sigma}_X(t)}{X(t)} - \frac{\boldsymbol{\sigma}_Y(t)}{Y(t)}\right)\cdot \mathrm{d}\mathbf{W}(t)$$
$$\equiv \mu_{\frac{X}{Y}}(t)\,\mathrm{d}t + \boldsymbol{\sigma}_{\frac{X}{Y}}(t) \cdot \mathrm{d}\mathbf{W}(t)\,. \tag{2.177}$$

This gives the drift $\mu_{\frac{X}{Y}}(t)$ and volatility vector $\boldsymbol{\sigma}_{\frac{X}{Y}}(t)$ for the quotient process $F = \frac{X}{Y}$ in terms of the drifts and volatility vectors of the individual processes X and Y. □

The Itô product and quotient rules in (2.174) and (2.177) take on a more compact form if the processes X and Y are represented in terms of the so-called *log-drifts and log-volatility vectors* (sometimes also referred to as local drift and local volatility). It will turn out to be particularly convenient when we later model the asset (e.g., stock) price processes. That is, assume processes X and Y satisfy the SDEs

$$\frac{\mathrm{d}X(t)}{X(t)} = \mu_X(t)\,\mathrm{d}t + \boldsymbol{\sigma}_X(t)\cdot \mathrm{d}\mathbf{W}(t)\,, \tag{2.178}$$

$$\frac{\mathrm{d}Y(t)}{Y(t)} = \mu_Y(t)\,\mathrm{d}t + \boldsymbol{\sigma}_Y(t)\cdot \mathrm{d}\mathbf{W}(t)\,, \tag{2.179}$$

where $\mu_X(t), \mu_Y(t), \boldsymbol{\sigma}_X(t), \boldsymbol{\sigma}_Y(t)$ are $\mathcal{F}_t$-adapted log-drifts and log-volatility vectors. Note that these SDEs are quite general. The difference is that the previous coefficients are related to these "log-coefficients" by sending the previous coefficients $\mu_X(t) \to \mu_X(t)X(t)$, $\mu_Y(t) \to \mu_Y(t)Y(t)$, $\boldsymbol{\sigma}_X(t) \to \boldsymbol{\sigma}_X(t)X(t)$, $\boldsymbol{\sigma}_Y(t) \to \boldsymbol{\sigma}_Y(t)Y(t)$. The Itô formula applied to (2.178) and (2.179) still gives (2.173) and (2.175). However, now the terms occurring in (2.174) and (2.177) simplify, where we replace the previous ratios $\frac{\mu_X(t)}{X(t)} \to \mu_X(t)$, $\frac{\mu_Y(t)}{Y(t)} \to \mu_Y(t)$, $\frac{\boldsymbol{\sigma}_X(t)}{X(t)} \to \boldsymbol{\sigma}_X(t)$, $\frac{\boldsymbol{\sigma}_Y(t)}{Y(t)} \to \boldsymbol{\sigma}_Y(t)$, giving

$$\begin{aligned}\frac{\mathrm{d}(X(t)Y(t))}{X(t)Y(t)} &= \big(\mu_X(t) + \mu_Y(t) + \boldsymbol{\sigma}_X(t)\cdot\boldsymbol{\sigma}_Y(t)\big)\,\mathrm{d}t + \big(\boldsymbol{\sigma}_X(t) + \boldsymbol{\sigma}_Y(t)\big)\cdot \mathrm{d}\mathbf{W}(t) \\ &\equiv \mu_{XY}(t)\,\mathrm{d}t + \boldsymbol{\sigma}_{XY}(t)\cdot \mathrm{d}\mathbf{W}(t)\,.\end{aligned} \tag{2.180}$$

Here, $\mu_{XY}(t)$ and $\boldsymbol{\sigma}_{XY}(t)$ denote the log-drift and log-volatility vector of process XY and

$$\begin{aligned}\frac{\mathrm{d}\frac{X(t)}{Y(t)}}{\frac{X(t)}{Y(t)}} &= \big(\mu_X(t) - \mu_Y(t) + \boldsymbol{\sigma}_Y(t)\cdot(\boldsymbol{\sigma}_Y(t) - \boldsymbol{\sigma}_X(t))\big)\,\mathrm{d}t + \big(\boldsymbol{\sigma}_X(t) - \boldsymbol{\sigma}_Y(t)\big)\cdot \mathrm{d}\mathbf{W}(t) \\ &\equiv \mu_{\frac{X}{Y}}(t)\,\mathrm{d}t + \boldsymbol{\sigma}_{\frac{X}{Y}}(t)\cdot \mathrm{d}\mathbf{W}(t),\end{aligned} \tag{2.181}$$

where $\mu_{\frac{X}{Y}}(t)$ and $\boldsymbol{\sigma}_{\frac{X}{Y}}(t)$ denote the log-drift and log-volatility vector of process $\frac{X}{Y}$.

For $d = 1$, the SDEs in (2.178)–(2.181) are all of the form in (2.51) where all volatility coefficients are scalars. The processes can therefore be represented as in (2.52). Given initial values $X(0)$ and $Y(0)$:

$$\begin{aligned}X(t) &= X(0)\exp\left[\int_0^t (\mu_X(s) - \frac{1}{2}\sigma_X^2(s))\,\mathrm{d}s + \int_0^t \sigma_X(s)\,\mathrm{d}W(s)\right], \\ Y(t) &= Y(0)\exp\left[\int_0^t (\mu_Y(s) - \frac{1}{2}\sigma_Y^2(s))\,\mathrm{d}s + \int_0^t \sigma_Y(s)\,\mathrm{d}W(s)\right], \\ X(t)Y(t) &= X(0)Y(0)\exp\left[\int_0^t (\mu_{XY}(s) - \frac{1}{2}\sigma_{XY}^2(s))\,\mathrm{d}s + \int_0^t \sigma_{XY}(s)\,\mathrm{d}W(s)\right], \\ \frac{X(t)}{Y(t)} &= \frac{X(0)}{Y(0)}\exp\left[\int_0^t (\mu_{\frac{X}{Y}}(s) - \frac{1}{2}\sigma_{\frac{X}{Y}}^2(s))\,\mathrm{d}s + \int_0^t \sigma_{\frac{X}{Y}}(s)\,\mathrm{d}W(s)\right].\end{aligned}$$

The reader can verify that the above third equation obtains by multiplying the expressions in the first and second equations, while the fourth equation obtains by dividing the expressions in the first and second equations. In the special case that the log-drift and log-volatility vectors are nonrandom (constants or ordinary functions of time t), the above processes are all GBM processes.

The above representations readily extend to the general vector case of $d \geqslant 1$. Consider the X process. Its natural logarithm has SDE:

$$\begin{aligned}
\mathrm{d}\ln X(t) &= \frac{\mathrm{d}X(t)}{X(t)} - \frac{1}{2}\left(\frac{\mathrm{d}X(t)}{X(t)}\right)^2 \\
&= \mu_X(t)\,\mathrm{d}t + \boldsymbol{\sigma}_X(t)\cdot \mathrm{d}\mathbf{W}(t) - \frac{1}{2}\left(\boldsymbol{\sigma}_X(t)\cdot \mathrm{d}\mathbf{W}(t)\right)^2 \\
&= \left(\mu_X(t) - \frac{1}{2}\|\boldsymbol{\sigma}_X(t)\|^2\right)\mathrm{d}t + \boldsymbol{\sigma}_X(t)\cdot \mathrm{d}\mathbf{W}(t)\,.
\end{aligned}$$

In integral form,

$$\ln\frac{X(t)}{X(0)} = \int_0^t \left(\mu_X(s) - \frac{1}{2}\|\boldsymbol{\sigma}_X(s)\|^2\right)\mathrm{d}s + \int_0^t \boldsymbol{\sigma}_X(s)\cdot \mathrm{d}\mathbf{W}(s)\,.$$

By exponentiating, $\frac{X(t)}{X(0)} = \exp(\ln\frac{X(t)}{X(0)})$,

$$X(t) = X(0)\exp\left[\int_0^t \left(\mu_X(s) - \frac{1}{2}\|\boldsymbol{\sigma}_X(s)\|^2\right)\mathrm{d}s + \int_0^t \boldsymbol{\sigma}_X(s)\cdot \mathrm{d}\mathbf{W}(s)\right]. \tag{2.182}$$

This expresses $X(t)$ in the general case of d-dimensional BM and reduces to the above expression in case $d = 1$. Similar expressions hold for the other processes. In fact, given an adapted drift $\mu(t)$ and volatility vector $\boldsymbol{\sigma}(t)$ (satisfying the usual integrability assumptions), the SDE

$$\frac{\mathrm{d}U(t)}{U(t)} = \mu(t)\,\mathrm{d}t + \boldsymbol{\sigma}(t)\cdot \mathrm{d}\mathbf{W}(t) \tag{2.183}$$

with initial value $U(0)$ is equivalent to the representation

$$\begin{aligned}
U(t) &= U(0)\exp\left[\int_0^t \left(\mu(s) - \frac{1}{2}\|\boldsymbol{\sigma}(s)\|^2\right)\mathrm{d}s + \int_0^t \boldsymbol{\sigma}(s)\cdot \mathrm{d}\mathbf{W}(s)\right] \\
&= U(0)\mathrm{e}^{\int_0^t \mu(s)\,\mathrm{d}s}\cdot \mathcal{E}_t(\boldsymbol{\sigma}\cdot\mathbf{W}),
\end{aligned} \tag{2.184}$$

where the vector BM version of the stochastic exponential in (2.53) is defined by

$$\mathcal{E}_t(\boldsymbol{\sigma}\cdot\mathbf{W}) := \exp\left[-\frac{1}{2}\int_0^t \|\boldsymbol{\sigma}(s)\|^2\,\mathrm{d}s + \int_0^t \boldsymbol{\sigma}(s)\cdot \mathrm{d}\mathbf{W}(s)\right]. \tag{2.185}$$

Hence, each process satisfying (2.178)–(2.181) has an equivalent representation as in (2.184). For the X process we see that (2.182) has precisely the form in (2.184). For processes Y, XY, and X/Y the same form obtains in the obvious manner where the corresponding drifts and volatility vectors, for the respective processes, are substituted into (2.184).

We now further extend our discussion to arbitrary dimensions $d \geqslant 1$. As already noted, all the formulae presented so far are valid for $d \geqslant 1$. Given a d-dimensional $\mathbb{F}$-adapted vector $\boldsymbol{\gamma}(t) = (\gamma_1(t), \ldots, \gamma_d(t))$, the Itô integral w.r.t. d-dimensional BM is defined by the sum of one-dimensional Itô integrals,

$$\int_0^t \boldsymbol{\gamma}(s)\cdot \mathrm{d}\mathbf{W}(s) := \sum_{i=1}^d \int_0^t \gamma_i(s)\,\mathrm{d}W_i(s)\,. \tag{2.186}$$

Throughout we shall assume that all such Itô integrals are square-integrable martingales for all times $0 \leqslant t \leqslant T$, given some $T > 0$, and where (2.160) holds for every drift coefficient function.

Let $\{\mathbf{X}(t) \equiv (X_1(t), X_2(t), \ldots, X_n(t))\}_{t \geqslant 0} \in \mathbb{R}^n$, $n \geqslant 1$, be an n-dimensional Itô vector process where each component is a real-valued Itô process driven by a d-dimensional BM:

$$\begin{aligned} \mathrm{d}X_i(t) &= \mu_i(t)\,\mathrm{d}t + \sum_{j=1}^{d} \sigma_{ij}(t)\,\mathrm{d}W_j(t) \\ &\equiv \mu_i(t)\,\mathrm{d}t + \boldsymbol{\sigma}_i(t) \cdot \mathrm{d}\mathbf{W}(t) \end{aligned} \tag{2.187}$$

with corresponding integral form

$$\begin{aligned} X_i(t) &= \int_0^t \mu_i(s)\,\mathrm{d}s + \sum_{j=1}^{d} \int_0^t \sigma_{ij}(s)\,\mathrm{d}W_j(s) \\ &\equiv \int_0^t \mu_i(s)\,\mathrm{d}s + \int_0^t \boldsymbol{\sigma}_i(s) \cdot \mathrm{d}\mathbf{W}(s) \end{aligned} \tag{2.188}$$

for $i = 1, \ldots, n$. Each $\{\mu_i(t)\}_{t \geqslant 0}$ is an integrable $\mathbb{F}$-adapted process. The coefficients $\sigma_{ij}(t)$ are $\mathbb{F}$-adapted and satisfy the square-integrability condition, $\int_0^T \mathrm{E}[\sigma_{ij}^2(s)]\,\mathrm{d}s < \infty$. The $n \times d$ matrix of coefficients, which we denote by $\boldsymbol{\sigma}(t) := [\sigma_{ij}(t)]_{i=1,\ldots,n;\, j=1,\ldots,d}$, is the *matrix of volatilities where the ith row gives the volatility vector* $\boldsymbol{\sigma}_i(t)$ *of the ith process* X_i:

$$\boldsymbol{\sigma}_i(t) = (\sigma_{i1}(t), \sigma_{i2}(t), \ldots, \sigma_{id}(t)) \ , \quad i = 1, \ldots, n \, .$$

The jth component of $\boldsymbol{\sigma}_i(t)$ is the volatility coefficient $\sigma_{ij}(t)$.

Being $\mathbb{F}$-adapted, the coefficients $\mu_i(t)$ and $\sigma_{ij}(t)$ can generally depend on the entire path of the vector process $\mathbf{X}$ up to time t, e.g., they can be functionals of the Brownian path $\{\mathbf{W}(s) : 0 \leqslant s \leqslant t\}$. Generally this can be an intractable situation. However, an important case (and one that leads to some tractable models) is when these coefficients are known (defined) functions of the endpoint value of the process: $\mu_i(t) := \mu_i(t, \mathbf{X}(t))$ and $\boldsymbol{\sigma}_i(t) := \boldsymbol{\sigma}_i(t, \mathbf{X}(t))$. In this case, we say that the coefficients are *state-dependent* and the vector process $\{\mathbf{X}(t)\}_{t \geqslant 0}$ is a *vector-valued diffusion process* with each component process solving an SDE of the form

$$\mathrm{d}X_i(t) = \mu_i(t, \mathbf{X}(t))\,\mathrm{d}t + \boldsymbol{\sigma}_i(t, \mathbf{X}(t)) \cdot \mathrm{d}\mathbf{W}(t) \, . \tag{2.189}$$

We will return to this case later.

We now turn to the more general multidimensional version of the Itô formula by extending Lemma 2.21 to the case of a function of time and any $n \geqslant 1$ processes. In preparation, we already computed the quadratic variation and the covariation of two Itô processes driven by a vector BM (see (2.162), (2.163), and (2.166)). In particular, any component process X_i has quadratic variation

$$[X_i, X_i](t) = \int_0^t \|\boldsymbol{\sigma}_i(u)\|^2\,\mathrm{d}u = \sum_{j=1}^{d} \int_0^t \sigma_{ij}^2(u)\,\mathrm{d}u \tag{2.190}$$

which is written equivalently in differential form as

$$\mathrm{d}[X_i, X_i](t) \equiv \mathrm{d}X_i(t)\,\mathrm{d}X_i(t) \equiv (\,\mathrm{d}X_i(t))^2 = \|\boldsymbol{\sigma}_i(t)\|^2\,\mathrm{d}t \, . \tag{2.191}$$

Any pair of processes X_i, X_j has covariation

$$[X_i, X_j](t) = \int_0^t \boldsymbol{\sigma}_i(u) \cdot \boldsymbol{\sigma}_j(u)\, \mathrm{d}u\,, \tag{2.192}$$

or in differential form,

$$\mathrm{d}[X_i, X_j](t) \equiv \mathrm{d}X_i(t)\, \mathrm{d}X_j(t) = \boldsymbol{\sigma}_i(t) \cdot \boldsymbol{\sigma}_j(t)\, \mathrm{d}t\,. \tag{2.193}$$

It is also useful to write the above as $\mathrm{d}[X_i, X_j](t) = C_{ij}(t)\, \mathrm{d}t$ by defining the coefficients

$$C_{ij}(t) := \boldsymbol{\sigma}_i(t) \cdot \boldsymbol{\sigma}_j(t) = \sum_{k=1}^{n} \sigma_{ik}(t)\, \sigma_{jk}(t)\,. \tag{2.194}$$

These are the elements of an $n \times n$ matrix $\mathbf{C}(t) := [C_{ij}(t)]_{i,j=1,\ldots,n}$, where $C_{ij}(t)$ are related to the *instantaneous covariances* between the differential increments of the two processes X_i and X_j. In terms of the matrix $\boldsymbol{\sigma}(t)$, the instantaneous covariances are given by $C_{ij}(t) = (\boldsymbol{\sigma}(t)\, \boldsymbol{\sigma}(t)^\top)_{ij}$ where $\boldsymbol{\sigma}(t)^\top$ is the $d \times n$ transpose of the matrix $\boldsymbol{\sigma}(t)$. Given the instantaneous covariances, we also define the *instantaneous correlations*

$$\rho_{ij}(t) := \frac{\boldsymbol{\sigma}_i(t) \cdot \boldsymbol{\sigma}_j(t)}{\|\boldsymbol{\sigma}_i(t)\|\, \|\boldsymbol{\sigma}_j(t)\|} = \frac{C_{ij}(t)}{\|\boldsymbol{\sigma}_i(t)\|\, \|\boldsymbol{\sigma}_j(t)\|}\,. \tag{2.195}$$

[Remark: We can see, at least heuristically, that these coefficients are a measure of the instantaneous correlations between the increments of a pair of processes X_i and X_j when we divide the differential of the covariation with the square root of the product of the differentials of the variations, $\frac{\mathrm{d}[X_i, X_j](t)}{\sqrt{\mathrm{d}[X_i, X_i](t) \cdot \mathrm{d}[X_j, X_j](t)}} = \frac{C_{ij}(t)}{\|\boldsymbol{\sigma}_i(t)\|\, \|\boldsymbol{\sigma}_j(t)\|} = \rho_{ij}(t)$.]

As in the case of Lemma 2.21, the main idea, which also gives us a simple way to remember the Itô formula for smooth functions $f(t, x_1, \ldots, x_n)$ of n variables and time t, is to apply a Taylor expansion of f up to terms of first order in the time increment $\mathrm{d}t$, and up to second order in the increments $\mathrm{d}x_1, \ldots, \mathrm{d}x_n$. The stochastic differential form of the Itô formula is then obtained upon replacing the ordinary variables $x_i \to X_i(t)$, and replacing all ordinary differentials by the respective stochastic differentials, $\mathrm{d}x_i \to \mathrm{d}X_i(t)$, $\mathrm{d}x_i\, \mathrm{d}x_j \to \mathrm{d}[X_i, X_j](t)$, $i, j = 1, \ldots, n$. We then also write the formula more explicitly by applying the basic rules in (2.151). The formal proof (which we omit) follows in the same manner as the proof for the two variable case in Lemma 2.21. Here, we simply state this important result as a lemma.

Lemma 2.22 (Itô Formula for a Function of Several Processes). *Let the vector-valued process $\{\mathbf{X}(t) \equiv (X_1(t), X_2(t), \ldots, X_n(t))\}_{t \geqslant 0}$ satisfy the SDE in (2.187) and assume $f(t, \mathbf{x}) \equiv f(t, x_1, \ldots, x_n)$ is a real-valued function on $\mathbb{R}_+ \times \mathbb{R}^n$ that is continuously differentiable with respect to time t and twice continuously differentiable with respect to the n variables $x_1, \ldots, x_n$. Then, the process defined by $F(t) := f(t, \mathbf{X}(t)) \equiv f(t, X_1(t), \ldots, X_n(t))$, $t \geqslant 0$, is an Itô process with stochastic differential $\mathrm{d}F(t) \equiv \mathrm{d}f(t, \mathbf{X}(t))$ given by*

$$\begin{aligned} \mathrm{d}F(t) = {} & \frac{\partial f}{\partial t}(t, \mathbf{X}(t))\, \mathrm{d}t + \sum_{i=1}^{n} \frac{\partial f}{\partial x_i}(t, \mathbf{X}(t))\, \mathrm{d}X_i(t) \\ & + \frac{1}{2} \sum_{i=1}^{n} \sum_{j=1}^{n} \frac{\partial^2 f}{\partial x_i \partial x_j}(t, \mathbf{X}(t))\, \mathrm{d}[X_i, X_j](t) \end{aligned} \tag{2.196}$$

Note that the formula in (2.167) is recovered in the special case that $n = 2$ by setting

$X_1(t) \equiv X(t), X_2(t) \equiv Y(t)$. The stochastic differential in (2.196) has meaning when written as an Itô process in integral form. Using (2.193), (2.196) takes the form

$$\mathrm{d}f(t, \mathbf{X}(t)) = \left[\frac{\partial f}{\partial t}(t, \mathbf{X}(t)) + \frac{1}{2}\sum_{i=1}^{n}\sum_{j=1}^{n} C_{ij}(t)\frac{\partial^2 f}{\partial x_i \partial x_j}(t, \mathbf{X}(t))\right] \mathrm{d}t$$
$$+ \sum_{i=1}^{n} \frac{\partial f}{\partial x_i}(t, \mathbf{X}(t))\, \mathrm{d}X_i(t) . \tag{2.197}$$

This expression involves the time differential and a linear combination of stochastic differentials in the component processes. By further making use of (2.187), we obtain Itô's formula, extending (2.160), where the stochastic differential is written in terms of the vector BM increment:

$$\mathrm{d}f(t, \mathbf{X}(t)) = \left[\frac{\partial f}{\partial t}(t, \mathbf{X}(t)) + \frac{1}{2}\sum_{i=1}^{n}\sum_{j=1}^{n} C_{ij}(t)\frac{\partial^2 f}{\partial x_i \partial x_j}(t, \mathbf{X}(t)) + \sum_{i=1}^{n} \mu_i(t)\frac{\partial f}{\partial x_i}(t, \mathbf{X}(t))\right] \mathrm{d}t$$
$$+ \sum_{i=1}^{n} \frac{\partial f}{\partial x_i}(t, \mathbf{X}(t))\, \boldsymbol{\sigma}_i(t) \cdot \mathrm{d}\mathbf{W}(t)$$
$$\equiv \mu_f(t)\, \mathrm{d}t + \boldsymbol{\sigma}_f(t) \cdot \mathrm{d}\mathbf{W}(t) . \tag{2.198}$$

In the second equation line, we identified the drift and volatility vector of the process $\{f(t, \mathbf{X}(t))\}_{t \geqslant 0}$.

2.10.2 Multidimensional SDEs, Feynman–Kac Formulae, and Transition CDFs and PDFs

The main concepts, theorems, and formulae that we established in Section 2.7 for the case of a single process driven by a one-dimensional BM also carry over into the multidimensional case with appropriate assumptions in place. Here, we only give a very brief account of some of the relevant results. Our main starting point is the n-dimensional diffusion process solving the system of SDEs in (2.189), i.e.,

$$\mathrm{d}X_i(t) = \mu_i(t, \mathbf{X}(t))\, \mathrm{d}t + \sum_{j=1}^{d} \sigma_{ij}(t, \mathbf{X}(t))\, \mathrm{d}W_j(t) , \quad i = 1, \ldots, n . \tag{2.199}$$

In integral form,

$$X_i(t) = X_i(0) + \int_0^t \mu_i(s, \mathbf{X}(s))\, \mathrm{d}s + \sum_{j=1}^{d} \int_0^t \sigma_{ij}(s, \mathbf{X}(s))\, \mathrm{d}W_j(s) , \quad i = 1, \ldots, n . \tag{2.200}$$

The drift and volatility coefficients, $\mu_i(t, \mathbf{x})$ and $\sigma_{ij}(t, \mathbf{x})$, are given functions of time and variables $\mathbf{x} = (x_1, \ldots, x_n)$.

Theorem 2.8, which provides sufficient conditions on the existence and uniqueness of a strong solution to the one-dimensional SDE (2.49), extends in a similar manner to the above system of SDEs. The absolute values for the Lipschitz condition and the linear growth condition on the coefficients are now replaced by appropriate vector and matrix norms. We denote the drift vector by $\boldsymbol{\mu}(t, \mathbf{x}) = (\mu_1(t, \mathbf{x}), \ldots, \mu_n(t, \mathbf{x}))$. The norm of a vector $\mathbf{v} \in \mathbb{R}^n$ is $\|\mathbf{v}\| := \sqrt{\sum_{i=1}^n v_i^2}$ and the norm of a matrix A with elements a_{ij} is defined similarly as $\|A\| := \sqrt{\sum_{i,j} a_{ij}^2}$. If there is a constant $K > 0$ such that the Lipschitz condition

$$\|\boldsymbol{\mu}(t, \mathbf{x}) - \boldsymbol{\mu}(t, \mathbf{y})\| + \|\boldsymbol{\sigma}(t, \mathbf{x}) - \boldsymbol{\sigma}(t, \mathbf{y})\| \leqslant K\|\mathbf{x} - \mathbf{y}\|$$

and the linear growth condition

$$\|\boldsymbol{\mu}(t,\mathbf{x})\| + \|\boldsymbol{\sigma}(t,\mathbf{x})\| \leqslant K(1+\|\mathbf{x}\|)$$

are both satisfied, for $\mathbf{x},\mathbf{y} \in \mathbb{R}^n$, then this ensures that there is a unique vector process $\{\mathbf{X}(t)\}_{t\geqslant 0}$ solving (2.200) and that the paths of the vector process are continuous in time. As in the one-dimensional case, these conditions are not necessary, but are sufficient conditions to guarantee the existence of a unique strong solution.

The solution to (2.200) is also a vector Markov process, i.e.,

$$\mathbb{P}(\mathbf{X}(t) \leqslant \mathbf{y} \mid \mathcal{F}_s) = \mathbb{P}(\mathbf{X}(t) \leqslant \mathbf{y} \mid \mathbf{X}(s)) \tag{2.201}$$

for all $\mathbf{y} \in \mathbb{R}^n$ or, for Borel function $h : \mathbb{R}^n \to \mathbb{R}$,

$$\mathrm{E}[h(\mathbf{X}(t)) \mid \mathcal{F}_s] = \mathrm{E}[h(\mathbf{X}(t)) \mid \mathbf{X}(s)], \quad 0 \leqslant s \leqslant t. \tag{2.202}$$

The conditional expectation of $h(\mathbf{X}(T))$, given the vector value $\mathbf{X}(t) = \mathbf{x} \in \mathbb{R}^n$ at a time $t \leqslant T$, is denoted by

$$\mathrm{E}_{t,\mathbf{x}}[h(\mathbf{X}(T))] := \mathrm{E}[h(\mathbf{X}(T)) \mid \mathbf{X}(t) = \mathbf{x}].$$

As in the scalar case, the subscripts $t, \mathbf{x}$ are shorthand for conditioning on a given vector value $\mathbf{X}(t) = \mathbf{x}$ at time t. The conditional expectation is a function of the ordinary variables $\mathbf{x}$ and t, i.e., $\mathrm{E}_{t,\mathbf{x}}[h(\mathbf{X}(T))] = g(t,\mathbf{x})$ for fixed T. The Markov property is expressible as $\mathrm{E}[h(\mathbf{X}(T)) \mid \mathcal{F}_t] = \mathrm{E}_{t,\mathbf{X}(t)}[h(\mathbf{X}(T))] = g(t,\mathbf{X}(t))$, for all $0 \leqslant t \leqslant T$. Hence, if we know the conditional probability distribution of the random vector $\mathbf{X}(T)$, given $\mathbf{X}(t) = x$, then we can compute the function $g(t,\mathbf{x})$.

As in the scalar case, the conditional PDF of $\mathbf{X}(T)$, given $\mathbf{X}(t)$ is the (joint) transition PDF, $p(t,T;\mathbf{x},\mathbf{y}) \equiv p(t,T;x_1,\ldots,x_n,y_1,\ldots,y_n)$, for the vector process $\mathbf{X}$ obtained by differentiating the corresponding (joint) transition CDF, $P(t,T;\mathbf{x},\mathbf{y})$:

$$p(t,T;\mathbf{x},\mathbf{y}) = \frac{\partial^n}{\partial y_1 \ldots \partial y_n} P(t,T;\mathbf{x},\mathbf{y}), \tag{2.203}$$

$$\begin{aligned} P(t,T;\mathbf{x},\mathbf{y}) &:= \mathbb{P}(\mathbf{X}(T) \leqslant \mathbf{y} \mid \mathbf{X}(t) = \mathbf{x}) \\ &= \mathbb{P}(X_1(T) \leqslant y_1, \ldots, X_n(T) \leqslant y_n \mid X_1(t) = x_1, \ldots, X_n(t) = x_n) \\ &= \int_{-\infty}^{y_1} \cdots \int_{-\infty}^{y_n} p(t,T;\mathbf{x},\mathbf{z})\, \mathrm{d}\mathbf{z}\,. \end{aligned} \tag{2.204}$$

As in the one-dimensional case, the Markov and tower property lead to the multidimensional version of the Chapman–Kolmogorov relation:

$$p(s,t;\mathbf{x},\mathbf{y}) = \int_{\mathbb{R}^n} p(s,u;\mathbf{x},\mathbf{z}) p(u,t;\mathbf{z},\mathbf{y})\, \mathrm{d}\mathbf{z}, \quad s < u < t. \tag{2.205}$$

Given any Borel set $B \in \mathcal{B}(\mathbb{R}^n)$, the probability that the time-t vector process has value in B, given that it has value $\mathbf{x} \in \mathbb{R}^n$ at some earlier time $s < t$, is given by integrating the transition PDF over B:

$$P(\mathbf{X}(t) \in B \mid \mathbf{X}(s) = \mathbf{x}) = \int_B p(s,t;\mathbf{x},\mathbf{y})\, \mathrm{d}\mathbf{y}\,. \tag{2.206}$$

The multidimensional analogue of (2.68) for computing a conditional expectation is an integral over $\mathbb{R}^n$:

$$\mathrm{E}_{t,\mathbf{x}}[h(\mathbf{X}(T))] = \int_{\mathbb{R}^n} h(\mathbf{y}) p(t,T;\mathbf{x},\mathbf{y})\; \mathrm{d}\mathbf{y}, \quad t < T. \tag{2.207}$$

As in the case of a scalar diffusion on $\mathbb{R}$ with the generator in (2.70), the generator for the above vector diffusion process $\{\mathbf{X}(t)\}_{t\geqslant 0}$ on $\mathbb{R}^n$ is defined by the differential operator $\mathcal{G}_{t,\mathbf{x}}$ acting on a smooth function $f = f(t,\mathbf{x})$,

$$\mathcal{G}_{t,\mathbf{x}} f := \frac{1}{2}\sum_{i=1}^{n}\sum_{j=1}^{n} C_{ij}(t,\mathbf{x})\frac{\partial^2 f}{\partial x_i \partial x_j} + \sum_{i=1}^{n}\mu_i(t,\mathbf{x})\frac{\partial f}{\partial x_i} \tag{2.208}$$

where $[C_{ij}(t,\mathbf{x}) = \sum_{k=1}^{d}\sigma_{ik}(t,\mathbf{x})\sigma_{jk}(t,\mathbf{x})]_{i,j=1,\ldots,n}$ is the *diffusion matrix*. We shall assume that this matrix is positive definite where $\mathbf{v}^\top \mathbf{C}\mathbf{v} > 0$ for nonzero $\mathbf{v} \in \mathbb{R}^n$. The differential and integral forms of the Itô formula are now written compactly (extending (2.71) and (2.72) to the multidimensional case):

$$\mathrm{d}f(t,\mathbf{X}(t)) = \left(\frac{\partial}{\partial t} + \mathcal{G}_{t,\mathbf{x}}\right) f(t,\mathbf{X}(t))\,\mathrm{d}t + \sum_{j=1}^{d}\left(\sum_{i=1}^{n}\frac{\partial f}{\partial x_i}(t,\mathbf{X}(t))\,\sigma_{ij}(t,\mathbf{X}(t))\right)\mathrm{d}W_j(t) \tag{2.209}$$

and

$$\begin{aligned} f(t,\mathbf{X}(t)) = f(0,\mathbf{X}(0)) &+ \int_0^t \left(\frac{\partial}{\partial s} + \mathcal{G}_{s,\mathbf{x}}\right) f(s,\mathbf{X}(s))\,\mathrm{d}s \\ &+ \sum_{j=1}^{d}\int_0^t \left(\sum_{i=1}^{n}\frac{\partial f}{\partial x_i}(s,\mathbf{X}(s))\,\sigma_{ij}(s,\mathbf{X}(s))\right)\mathrm{d}W_j(s)\,. \end{aligned} \tag{2.210}$$

The analogues of (2.73)–(2.76) also follow if we fix a time $T > 0$ and assume the square-integrability condition,

$$\int_0^T \mathrm{E}\left[\left(\sum_{i=1}^{n}\frac{\partial f}{\partial x_i}(s,\mathbf{X}(s))\sigma_{ij}(s,\mathbf{X}(s))\right)^2\right]\mathrm{d}s < \infty\,,\quad j = 1,\ldots,n, \tag{2.211}$$

which ensures that all Itô integrals (w.r.t. each BM, W_j) in (2.210) are martingales. By using a similar argument as in the one-dimensional case, the process $\{M_f(t)\}_{0\leqslant t\leqslant T}$ defined by

$$M_f(t) := f(t,\mathbf{X}(t)) - \int_0^t \left(\frac{\partial}{\partial s} + \mathcal{G}_{s,\mathbf{x}}\right) f(s,\mathbf{X}(s))\,\mathrm{d}s \tag{2.212}$$

is a martingale, i.e., (2.76) holds. As a particular application of (2.212), we obtain a martingale defined by $M_f(t) := f(t,\mathbf{X}(t))$ if the function f solves the PDE: $\frac{\partial f}{\partial t} + \mathcal{G}_{t,\mathbf{x}} f = 0$.

The martingale property of the process defined in (2.212) allows us to extend Theorems 2.11 and 2.13 to the multidimensional case. Here, we simply state useful versions of the multidimensional extensions. Their proofs involve some similar steps as in the one-dimensional case.

Theorem 2.33 (Multidimensional Feynman–Kac). *Let $\{\mathbf{X}(t) := (X_1(t),\ldots,X_n(t))\}_{0\leqslant t\leqslant T}$ solve the system of SDEs in (2.199) and let $\phi : \mathbb{R}^n \to \mathbb{R}$ be a Borel function. Also, assume the square-integrability condition (2.211) holds. Suppose the function $f(t,\mathbf{x})$ is a solution to the backward Kolmogorov PDE $\frac{\partial f}{\partial t} + \mathcal{G}_{t,\mathbf{x}} f = 0$, i.e.,*

$$\frac{\partial f}{\partial t}(t,\mathbf{x}) + \frac{1}{2}\sum_{i=1}^{n}\sum_{j=1}^{n} C_{ij}(t,\mathbf{x})\frac{\partial^2 f}{\partial x_i \partial x_j}(t,\mathbf{x}) + \sum_{i=1}^{n}\mu_i(t,\mathbf{x})\frac{\partial f}{\partial x_i}(t,\mathbf{x}) = 0\,, \tag{2.213}$$

for all $\mathbf{x} \in \mathbb{R}^n$, $t < T$, *subject to the terminal condition* $f(T, \mathbf{x}) = \phi(\mathbf{x})$. *Then, assuming* $\mathrm{E}_{t,\mathbf{x}}[|\phi(\mathbf{X}(T))|] < \infty$, $f(t, \mathbf{x})$ *has the representation*

$$f(t, \mathbf{x}) = \mathrm{E}_{t,\mathbf{x}}[\phi(\mathbf{X}(T))] \equiv \mathrm{E}[\phi(\mathbf{X}(T)) \mid \mathbf{X}(t) = \mathbf{x}] \tag{2.214}$$

for all $\mathbf{x} \in \mathbb{R}^n$, $0 \leqslant t \leqslant T$.

The slightly more general result below includes an additional exponential discount factor via a discounting function $r(t, \mathbf{x})$. Theorem 2.23 is then a particular case of this theorem by simply setting $r(t, \mathbf{x}) \equiv 0$.

Theorem 2.24 ("Discounted" Feynman–Kac). *Let* $\{\mathbf{X}(t) := (X_1(t), \ldots, X_n(t))\}_{0 \leqslant t \leqslant T}$ *solve the system of SDEs in (2.199) and assume the square integrability condition (2.211) holds. Let* $\phi : \mathbb{R}^n \to \mathbb{R}$ *be a Borel function and* $r(t, \mathbf{x}) : [0, T] \times \mathbb{R}^n \to \mathbb{R}$ *be a lower-bounded continuous function. Then, the function defined by the conditional expectation*

$$f(t, \mathbf{x}) := \mathrm{E}_{t,\mathbf{x}}[\mathrm{e}^{-\int_t^T r(u, \mathbf{X}(u))\,\mathrm{d}u}\phi(\mathbf{X}(T))] \equiv \mathrm{E}[\mathrm{e}^{-\int_t^T r(u, \mathbf{X}(u))\,\mathrm{d}u}\phi(\mathbf{X}(T)) \mid \mathbf{X}(t) = x] \tag{2.215}$$

solves the PDE $\frac{\partial f}{\partial t} + \mathcal{G}_{t,\mathbf{x}} f - r(t, \mathbf{x}) f = 0$, *i.e.,*

$$\frac{\partial f}{\partial t}(t, \mathbf{x}) + \frac{1}{2}\sum_{i=1}^{n}\sum_{j=1}^{n} C_{ij}(t, \mathbf{x})\frac{\partial^2 f}{\partial x_i \partial x_j}(t, \mathbf{x}) + \sum_{i=1}^{n}\mu_i(t, \mathbf{x})\frac{\partial f}{\partial x_i}(t, \mathbf{x}) - r(t, \mathbf{x}) f(t, \mathbf{x}) = 0\,, \tag{2.216}$$

for all $\mathbf{x}$, $0 < t < T$, *subject to the terminal condition* $f(T, \mathbf{x}) = \phi(\mathbf{x})$.

In the special case that the discount function is a constant, $r(t, \mathbf{x}) \equiv r$, then (2.215) simplifies since the discount factor is simply $\mathrm{e}^{-r(T-t)}$ where $f(t, \mathbf{x}) = \mathrm{e}^{-r(T-t)}\mathrm{E}_{t,\mathbf{x}}[\phi(\mathbf{X}(T))]$.

The multidimensional version of Proposition 2.12 also follows, where both the transition PDF and CDF solve the backward Kolmogorov PDE in (2.213) in the (backward) variables $(t, \mathbf{x})$. In particular, fixing a time $T > 0$ and a vector $\mathbf{y} \in \mathbb{R}^n$, and setting $\phi(\mathbf{x}) = \mathbb{I}_{\{x_1 \leqslant y_1, \ldots, x_n \leqslant y_n\}}$ in Proposition 2.23 implies that the transition CDF

$$P(t, T; \mathbf{x}, \mathbf{y}) \equiv \mathrm{E}[\mathbb{I}_{\{X_1(T) \leqslant y_1, \ldots, X_n(T) \leqslant y_n\}} \mid X_1(t) = x_1, \ldots, X_n(t) = x_n]$$

solves the PDE in (2.213) with terminal condition as the indicator function, $P(T, T; \mathbf{x}, \mathbf{y}) = \mathbb{P}(x_1 \leqslant y_1, \ldots, x_n \leqslant y_n) = \mathbb{I}_{\{x_1 \leqslant y_1, \ldots, x_n \leqslant y_n\}}$. The transition PDF, $p = p(t, T; \mathbf{x}, \mathbf{y})$, is obtained from the CDF by differentiating in the $\mathbf{y}$ variables, according to (2.203). Hence, p also solves (2.213) and the terminal condition is given by

$$\lim_{t \nearrow T} p(t, T; \mathbf{x}, \mathbf{y}) = \delta(\mathbf{y} - \mathbf{x}),$$

where $\delta(\mathbf{y} - \mathbf{x}) = \delta(y_1 - x_1) \cdots \delta(y_n - x_n)$ is the n-dimensional Dirac delta function as a product of univariate delta functions.

The transition PDF $p(t, T; \mathbf{x}, \mathbf{y})$ is the conditional PDF of the random vector $\mathbf{X}(T)$ at $\mathbf{y}$, given $\mathbf{X}(t) = \mathbf{x}$. Hence, according to (2.214), the solution to the PDE problem takes the form of an integral of the product of the transition PDF and the function $\phi(\mathbf{y})$:

$$f(t, \mathbf{x}) = \mathrm{E}[\phi(\mathbf{X}(T)) \mid \mathbf{X}(t) = \mathbf{x}] = \int_{\mathbb{R}^n} \phi(\mathbf{y}) p(t, T; \mathbf{x}, \mathbf{y})\,\mathrm{d}\mathbf{y}\,. \tag{2.217}$$

That is, the transition PDF is the fundamental solution to the PDE problem stated in Theorem 2.23.

For many applications the vector diffusion process is time homogeneous where the drift and diffusion matrix are time independent, $\mu_i(t,\mathbf{x}) \equiv \mu_i(\mathbf{x})$ and $C_{ij}(t,\mathbf{x}) = C_{ij}(\mathbf{x}) = \sum_{k=1}^{d} \sigma_{ik}(\mathbf{x})\sigma_{jk}(\mathbf{x})$. The generator is then a differential operator only in the $\mathbf{x}$ variables,

$$\mathcal{G}_{t,\mathbf{x}} = \mathcal{G}_{\mathbf{x}} := \frac{1}{2}\sum_{i=1}^{n}\sum_{j=1}^{n} C_{ij}(\mathbf{x})\frac{\partial^2}{\partial x_i \partial x_j} + \sum_{i=1}^{n} \mu_i(\mathbf{x})\frac{\partial}{\partial x_i}.$$

Defining $\tau := T - t$, the solution in (2.214) is a function of $(\tau, \mathbf{x})$, i.e., $f = f(\tau, \mathbf{x})$, and the backward PDE in (2.213) takes the form

$$\frac{\partial f}{\partial \tau} = \frac{1}{2}\sum_{i=1}^{n}\sum_{j=1}^{n} C_{ij}(\mathbf{x})\frac{\partial^2 f}{\partial x_i \partial x_j} + \sum_{i=1}^{n} \mu_i(\mathbf{x})\frac{\partial f}{\partial x_i} \tag{2.218}$$

subject to the initial condition $f(0,\mathbf{x}) = \phi(\mathbf{x})$. Moreover, if the discount function in Theorem 2.24 is time independent, $r(t,\mathbf{x}) \equiv r(\mathbf{x})$, then the operator $\mathcal{G}_{t,\mathbf{x}} - r(t,\mathbf{x}) \equiv \mathcal{G}_{\mathbf{x}} - r(\mathbf{x})$ is time independent; i.e., the PDE in (2.216) is time-homogeneous:

$$\frac{\partial f}{\partial \tau} = \frac{1}{2}\sum_{i=1}^{n}\sum_{j=1}^{n} C_{ij}(\mathbf{x})\frac{\partial^2 f}{\partial x_i \partial x_j} + \sum_{i=1}^{n} \mu_i(\mathbf{x})\frac{\partial f}{\partial x_i} - r(\mathbf{x})f\,, \tag{2.219}$$

with the solution represented in (2.215) as a function $f = f(\tau,\mathbf{x})$ having initial condition $f(0,\mathbf{x}) = \phi(\mathbf{x})$.

For the time-homogeneous case, we hence have the transition CDF and PDF as functions of $\tau, \mathbf{x}, \mathbf{y}$ where we equivalently write $P(t,T;\mathbf{x},\mathbf{y})$ as $P(\tau;\mathbf{x},\mathbf{y})$ and $p(t,T;\mathbf{x},\mathbf{y})$ as $p(\tau;\mathbf{x},\mathbf{y})$. Both $P(\tau;\mathbf{x},\mathbf{y})$ and $p(\tau;\mathbf{x},\mathbf{y})$ solve the PDE in (2.218) where $p(0+;\mathbf{x},\mathbf{y}) = \delta(\mathbf{x}-\mathbf{y})$ and $P(0;\mathbf{x},\mathbf{y})$ given by the n-dimensional unit step function $\mathbb{I}_{\{\mathbf{y}\geqslant\mathbf{x}\}} = \mathbb{I}_{\{y_1\geqslant x_1,\ldots,y_n\geqslant x_n\}}$.

As a first example of how Theorem 2.23 can be applied in practice, consider a simple two-dimensional process ($n = 2$) driven by a two-dimensional BM ($d = 2$). That is, let $\{\mathbf{X}(t) = [X_1(t), X_2(t)]^\top\}_{t\geqslant 0} \in \mathbb{R}^2$ be two scaled and drifted Brownian motions satisfying the system of SDEs:

$$\begin{aligned} \mathrm{d}X_1(t) &= \mu_1\,\mathrm{d}t + \sigma_1\,\mathrm{d}W_1(t) \equiv \mu_1\,\mathrm{d}t + \boldsymbol{\sigma}_1\cdot\mathrm{d}\mathbf{W}(t)\,, \\ \mathrm{d}X_2(t) &= \mu_2\,\mathrm{d}t + \sigma_2\rho\,\mathrm{d}W_1(t) + \sigma_2\sqrt{1-\rho^2}\,\mathrm{d}W_2(t) \equiv \mu_2\,\mathrm{d}t + \boldsymbol{\sigma}_2\cdot\mathrm{d}\mathbf{W}(t)\,, \end{aligned} \tag{2.220}$$

where $\rho \in (-1,1)$ is a constant correlation coefficient, $\boldsymbol{\sigma}_1 = [\sigma_1, 0]$ and $\boldsymbol{\sigma}_2 = [\sigma_2\rho, \sigma_2\sqrt{1-\rho^2}]$ are volatility vectors with magnitudes $\|\boldsymbol{\sigma}_1\| = \sigma_1$, $\|\boldsymbol{\sigma}_2\| = \sigma_2$. Note that $\boldsymbol{\sigma}_1\cdot\boldsymbol{\sigma}_2 = \rho\sigma_1\sigma_2$. We can also represent this system of SDEs in vector-matrix notation, $\mathrm{d}\mathbf{X}(t) = \boldsymbol{\mu}\,\mathrm{d}t + \boldsymbol{\sigma}\,\mathrm{d}\mathbf{W}(t)$:

$$\begin{bmatrix} \mathrm{d}X_1(t) \\ \mathrm{d}X_2(t) \end{bmatrix} = \begin{bmatrix} \mu_1 \\ \mu_2 \end{bmatrix} \mathrm{d}t + \begin{bmatrix} \sigma_1 & 0 \\ \sigma_2\rho & \sigma_2\sqrt{1-\rho^2} \end{bmatrix} \begin{bmatrix} \mathrm{d}W_1(t) \\ \mathrm{d}W_2(t) \end{bmatrix}.$$

The above 2×2 is the $\boldsymbol{\sigma}$-matrix whose rows correspond to the volatility vectors $\boldsymbol{\sigma}_1$ and $\boldsymbol{\sigma}_2$. The diffusion matrix $\mathbf{C} = \boldsymbol{\sigma}\boldsymbol{\sigma}^\top$ is then

$$\mathbf{C} = \begin{bmatrix} \sigma_1 & 0 \\ \sigma_2\rho & \sigma_2\sqrt{1-\rho^2} \end{bmatrix} \begin{bmatrix} \sigma_1 & \sigma_2\rho \\ 0 & \sigma_2\sqrt{1-\rho^2} \end{bmatrix} = \begin{bmatrix} \sigma_1^2 & \rho\sigma_1\sigma_2 \\ \rho\sigma_1\sigma_2 & \sigma_2^2 \end{bmatrix},$$

where the 2×2 matrix $\boldsymbol{\sigma}$ is the lower Cholesky factorization of $\mathbf{C}$.

The SDEs in (2.220), subject to arbitrary initial conditions $X_1(t) = x_1, X_2(t) = x_2$, are solved by simply integrating from time t to T:

$$\begin{aligned} X_1(T) &= x_1 + \mu_1(T-t) + \boldsymbol{\sigma}_1\cdot(\mathbf{W}(T) - \mathbf{W}(t))\,, \\ X_2(T) &= x_2 + \mu_2(T-t) + \boldsymbol{\sigma}_2\cdot(\mathbf{W}(T) - \mathbf{W}(t))\,. \end{aligned} \tag{2.221}$$

We can express these random variables in terms of two standard normal random variables:

$$\begin{aligned} X_1(T) &= x_1 + \mu_1\tau + \sigma_1\sqrt{\tau}Z_1\,, \\ X_2(T) &= x_2 + \mu_2\tau + \sigma_2\sqrt{\tau}Z_2\,. \end{aligned} \tag{2.222}$$

where

$$Z_1 := \frac{\boldsymbol{\sigma}_1\cdot(\mathbf{W}(T) - \mathbf{W}(t))}{\sigma_1\sqrt{\tau}}, \quad Z_2 := \frac{\boldsymbol{\sigma}_2\cdot(\mathbf{W}(T) - \mathbf{W}(t))}{\sigma_2\sqrt{\tau}}, \tag{2.223}$$

$\tau := T - t$. The correlation (and covariance) between these two standard normals equals ρ. This follows from the fact that $W_i(T) - W_i(t), i = 1, 2$, are i.i.d. $Norm(0, \tau)$:

$$\begin{aligned} \text{Cov}(Z_1, Z_2) &= \frac{1}{\sigma_1\sigma_2\tau}\text{Cov}\left(\boldsymbol{\sigma}_1\cdot(\mathbf{W}(T) - \mathbf{W}(t)), \boldsymbol{\sigma}_2\cdot(\mathbf{W}(T) - \mathbf{W}(t))\right) \\ &= \frac{1}{\sigma_1\sigma_2\tau}\sum_{i=1}^{2}\sum_{j=1}^{2}\sigma_{1i}\sigma_{2j}\,\text{Cov}\left(W_i(T) - W_i(t), W_j(T) - W_j(t)\right) \\ &= \frac{1}{\sigma_1\sigma_2\tau}\sum_{i=1}^{2}(\sigma_{1i}\sigma_{2i})\tau = \frac{\boldsymbol{\sigma}_1\cdot\boldsymbol{\sigma}_2}{\sigma_1\sigma_2} = \rho\,. \end{aligned}$$

Hence, by (2.222), the matrix of correlations, ρ_{ij}, of the component processes is the 2×2 correlation matrix given by $\rho_{11} = \rho_{22} = 1$, $\rho_{12} = \rho_{21} = \rho$, i.e.,

$$\rho_{12} := \text{Corr}\left(X_1(T), X_2(T)\right) = \frac{\text{Cov}(X_1(T), X_2(T))}{\sqrt{\text{Var}(X_1(T))\,\text{Var}(X_2(T))}} = \text{Cov}\left(Z_1, Z_2\right) = \rho\,.$$

The covariance matrix of $\mathbf{X}(T)$ is given by $\text{Cov}(X_i(T), X_j(T)) = C_{ij}\tau = (\boldsymbol{\sigma}_i\cdot\boldsymbol{\sigma}_j)\tau = \rho_{ij}\sigma_i\sigma_j\tau$; $i, j = 1, 2$. The time-scaled solution vector $\frac{1}{\sqrt{\tau}}\mathbf{X}(T)$ is a bivariate normal,

$$\left[\frac{X_1(T)}{\sqrt{\tau}}, \frac{X_2(T)}{\sqrt{\tau}}\right]^\top \sim Norm_2\left(\left[\frac{x_1 + \mu_1\tau}{\sqrt{\tau}}, \frac{x_2 + \mu_2\tau}{\sqrt{\tau}}\right]^\top, \mathbf{C}\right).$$

The time-homogeneous transition CDF for the vector process is obtained by computing a joint conditional probability while using (2.221), or (2.222), and the fact that $\mathbf{W}(T) - \mathbf{W}(t)$ is independent of $\mathbf{X}(t)$, i.e., the pair Z_1, Z_2 is independent of the pair $X_1(t), X_2(t)$:

$$\begin{aligned} P(\tau; x_1, x_2, y_1, y_2) &:= \mathbb{P}\left(X_1(T) \leqslant y_1, X_2(T) \leqslant y_2 \mid X_1(t) = x_1, X_2(t) = x_2\right) \\ &= \mathbb{P}\left(x_1 + \mu_1\tau + \sigma_1\sqrt{\tau}Z_1 \leqslant y_1, x_2 + \mu_2\tau + \sigma_2\sqrt{\tau}Z_2 \leqslant y_2\right) \\ &= \mathbb{P}\left(Z_1 \leqslant \frac{y_1 - x_1 - \mu_1\tau}{\sigma_1\sqrt{\tau}}, Z_2 \leqslant \frac{y_2 - x_2 - \mu_2\tau}{\sigma_2\sqrt{\tau}}\right) \\ &= \mathcal{N}_2\left(\frac{y_1 - x_1 - \mu_1\tau}{\sigma_1\sqrt{\tau}}, \frac{y_2 - x_2 - \mu_2\tau}{\sigma_2\sqrt{\tau}}; \rho\right). \end{aligned} \tag{2.224}$$

This is a bivariate normal CDF and differentiating (using the chain rule) gives the transition PDF as a bivariate normal density,

$$\begin{aligned} p(\tau; x_1, x_2, y_1, y_2) &\equiv \frac{\partial^2}{\partial y_1\partial y_2}P(\tau; x_1, x_2, y_1, y_2) \\ &= \frac{1}{\sigma_1\sigma_2\tau}n_2\left(\frac{y_1 - x_1 - \mu_1\tau}{\sigma_1\sqrt{\tau}}, \frac{y_2 - x_2 - \mu_2\tau}{\sigma_2\sqrt{\tau}}; \rho\right) \\ &= \frac{1}{2\pi\tau\sigma_1\sigma_2\sqrt{1-\rho^2}}\exp\left(-\frac{z_1^2 + z_2^2 - 2\rho z_1 z_2}{2(1-\rho^2)}\right), \end{aligned} \tag{2.225}$$

where $z_1 = \frac{y_1 - x_1 - \mu_1 \tau}{\sigma_1 \sqrt{\tau}}$, $z_2 = \frac{y_2 - x_2 - \mu_2 \tau}{\sigma_2 \sqrt{\tau}}$. According to Theorem 2.23, both transition CDF and PDF, P and p, solve the time-homogeneous PDE in (2.218) in the variables τ, x_1, x_2, for fixed arbitrary real values of y_1, y_2. Using the above explicit constant expressions for $C_{11} = \sigma_1^2, C_{22} = \sigma_2^2, C_{12} = C_{21} = \rho\sigma_1\sigma_2$, and the constant drift coefficients μ_1 and μ_2, p and P solve the backward PDE, i.e.,

$$\frac{\partial p}{\partial \tau} = \frac{1}{2}\sigma_1^2 \frac{\partial^2 p}{\partial x_1^2} + \frac{1}{2}\sigma_2^2 \frac{\partial^2 p}{\partial x_2^2} + \rho\sigma_1\sigma_2 \frac{\partial^2 p}{\partial x_1 \partial x_2} + \mu_1 \frac{\partial p}{\partial x_1} + \mu_2 \frac{\partial p}{\partial x_2}$$

and similarly for P. We leave it as an exercise for the reader to show that the transition CDF in (2.224) has limit $P(0+; x_1, x_2, y_1, y_2) = \mathbb{I}_{\{y_1 \geqslant x_1, y_2 \geqslant x_2\}}$ for all $\mathbf{x} \neq \mathbf{y}$. The Dirac delta function initial condition for the transition PDF then follows by formally differentiating the step function to obtain $p(0+; x_1, x_2, y_1, y_2) = \delta(y_1 - x_1)\delta(y_2 - x_2)$. The reader can verify by direct differentiation that the transition PDF in (2.225) satisfies the above PDE and is hence the fundamental solution. As a density in the variables y_1, y_2, the transition PDF should also integrate to unity over $\mathbb{R}^2$. That is, the event $\{(X_1(T), X_2(T)) \in \mathbb{R}^2\}$, conditional on $X_1(t) = x_1, X_2(t) = x_2$, must have unit probability. This is directly verified as follows:

$$\begin{aligned}
&\mathbb{P}\left(X_1(T) < \infty, X_2(T) < \infty \mid X_1(t) = x_1, X_2(t) = x_2\right) \\
&= \lim_{y_1 \to \infty, y_2 \to \infty} P(\tau; x_1, x_2, y_1, y_2) \\
&= \lim_{y_1 \to \infty, y_2 \to \infty} \mathcal{N}_2\left(\frac{y_1 - x_1 - \mu_1 \tau}{\sigma_1 \sqrt{\tau}}, \frac{y_2 - x_2 - \mu_2 \tau}{\sigma_2 \sqrt{\tau}}; \rho\right) \\
&= \mathcal{N}_2(\infty, \infty; \rho) = 1\,.
\end{aligned}$$

From (2.204), this implies that the PDF integrates to unity for all $x_1, x_2 \in \mathbb{R}$,

$$\int_{-\infty}^{\infty} \int_{-\infty}^{\infty} p(\tau; x_1, x_2, y_1, y_2)\, \mathrm{d}y_1\, \mathrm{d}y_2 = 1\,.$$

Alternatively, this is easily shown by directly integrating the bivariate density in (2.225). Note that in the special case when $\rho = 0$, the two processes are independent (uncorrelated) drifted and scaled BM. The joint transition PDF and CDF are simply products of the one-dimensional PDFs and CDFs of the component processes. This is consistent with the fact that the above PDE is separable in the variables x_1 and x_2 and hence admits a solution as a product of individual functions of x_1 and x_2.

Let us now consider a *multidimensional GBM process*. This is an important example that arises in later chapters where we consider derivative pricing within a standard economic model containing multiple stocks whose price processes are correlated geometric Brownian motions. In particular, consider $n \geqslant 1$ strictly positive stock price processes $\mathbf{S}(t) := (S_1(t), \ldots, S_n(t))$, $t \geqslant 0$, that are driven by a standard $d \geqslant 1$ dimensional BM, $\mathbf{W}(t) = (W_1(t), \ldots, W_d(t))$:

$$\begin{aligned}
\mathrm{d}S_i(t) &= S_i(t)\left[\mu_i\, \mathrm{d}t + \sum_{j=1}^{d} \sigma_{ij}\, \mathrm{d}W_j(t)\right] \\
&\equiv S_i(t)\left[\mu_i\, \mathrm{d}t + \boldsymbol{\sigma}_i \cdot \mathrm{d}\mathbf{W}(t)\right], \quad i = 1, \ldots, n\,, \qquad (2.226)
\end{aligned}$$

where the log-drifts μ_i and log-volatilities σ_{ij} are assumed to be constant parameters. It is important to stress the distinction that these are *log-coefficients*, although by standard convention we are still using similar symbols for the drift and volatility coefficient functions!

To be precise, by identifying the SDE in (2.226) with that in (2.199) (where $\mathbf{X}(t) \to \mathbf{S}(t)$) we see that the drift and volatility *coefficient functions* for the ith stock price in (2.226) are state-dependent (time-independent) linear functions of the ith variable

$$\mu_i(t, \mathbf{x}) \equiv \mu_i(\mathbf{x}) = \mu_i x_i \quad \text{and} \quad \sigma_{ij}(t, \mathbf{x}) \equiv \sigma_{ij}(\mathbf{x}) = \sigma_{ij} x_i, \tag{2.227}$$

where on the right of each equality are the log-drift and log-volatility parameters μ_i and σ_{ij}. [We note that the symbols μ_i and σ_{ij} are constant parameters when denoted without arguments and they are the drift and volatility coefficient functions when denoted with arguments.] So the above SDEs are time homogeneous with linear functions $\mu_i(t, \mathbf{S}(t)) \equiv \mu_i(\mathbf{S}(t)) = \mu_i S_i(t)$ and $\sigma_{ij}(t, \mathbf{S}(t)) \equiv \sigma_{ij}(\mathbf{S}(t)) = S_i(t)\sigma_{ij}$.

The *log-volatility coefficient matrix* $\boldsymbol{\sigma} = [\sigma_{ij}]_{i=1,\dots,n; j=1,\dots,d}$ is an $n \times d$ constant matrix with the ith row being the $1 \times d$ volatility vector $\boldsymbol{\sigma}_i = [\sigma_{i1} \dots, \sigma_{id}]$ for the ith stock price process. The system of SDEs in (2.226) has matrix-vector form:

$$\begin{bmatrix} \frac{\mathrm{d}S_1(t)}{S_1(t)} \\ \vdots \\ \frac{\mathrm{d}S_n(t)}{S_n(t)} \end{bmatrix} = \begin{bmatrix} \mu_1 \\ \vdots \\ \mu_n \end{bmatrix} \mathrm{d}t + \begin{bmatrix} \sigma_{11} & \cdots & \sigma_{1d} \\ \vdots & & \vdots \\ \sigma_{n1} & \cdots & \sigma_{nd} \end{bmatrix} \begin{bmatrix} \mathrm{d}W_1(t) \\ \vdots \\ \mathrm{d}W_d(t) \end{bmatrix}. \tag{2.228}$$

As shown just below, the *log-diffusion matrix* $\mathbf{C} = \boldsymbol{\sigma}\boldsymbol{\sigma}^\top$ is proportional to the $n \times n$ matrix of covariances among the log-returns of the stocks. We assume that $\mathbf{C}$ is nonsingular. As usual, we define the $n \times n$ matrix of correlations, $\boldsymbol{\rho} := [\rho_{ij}]_{i,j=1,\dots,n}$ where $C_{ij} = \boldsymbol{\sigma}_i \cdot \boldsymbol{\sigma}_j = \rho_{ij}\sigma_i\sigma_j$, where the ith volatility vector has magnitude denoted by $\sigma_i > 0$,

$$C_{ii} = \sigma_i^2 \equiv \|\boldsymbol{\sigma}_i\|^2 = \sigma_{i1}^2 + \ldots + \sigma_{id}^2 .$$

The system in (2.226) is readily solved by considering the log-prices defined by $X_i(t) := \ln S_i(t)$. In fact, we have already solved this problem. See (2.183)–(2.185), where each SDE in (2.226) is of the form in (2.183) with solution of the form in (2.184). For the sake of clarity, we repeat the same steps here by using Itô's formula where (upon substituting the expression in (2.226)):

$$\begin{aligned} \mathrm{d}X_i(t) = \mathrm{d}\ln S_i(t) = \frac{\mathrm{d}S_i(t)}{S_i(t)} - \frac{1}{2}\left(\frac{\mathrm{d}S_i(t)}{S_i(t)}\right)^2 &= \left(\mu_i - \frac{1}{2}\|\boldsymbol{\sigma}_i\|^2\right)\mathrm{d}t + \boldsymbol{\sigma}_i \cdot \mathrm{d}\mathbf{W}(t) \\ &= \left(\mu_i - \frac{1}{2}\sigma_i^2\right)\mathrm{d}t + \boldsymbol{\sigma}_i \cdot \mathrm{d}\mathbf{W}(t) \end{aligned}$$

with initial condition $X_i(0) = \ln S_i(0)$, where $S_i(0), i = 1, \dots, n$, are the initial stock prices. Integrating and exponentiating gives the stock prices $S_i(t) = \mathrm{e}^{X_i(t)}$ for all $t \geqslant 0$:

$$S_i(t) = S_i(0)\,\mathrm{e}^{(\mu_i - \frac{1}{2}\sigma_i^2)t + \boldsymbol{\sigma}_i \cdot \mathbf{W}(t)} = S_i(0)\,\mathrm{e}^{\mu_i t}\mathcal{E}_t(\boldsymbol{\sigma}_i \cdot \mathbf{W}). \tag{2.229}$$

It is easy to verify by computing the stochastic differential of this expression, upon directly applying Itô's formula, that each $S_i(t)$ solves (2.226). The solution in (2.229) is in fact the unique strong solution subject to the initial price vector $[S_1(0), \dots, S_n(0)]^\top$.

The second expression in (2.229) involves an exponential $\mathbb{P}$-martingale,

$$\mathcal{E}_t(\boldsymbol{\sigma}_i \cdot \mathbf{W}) = \exp\left[-\frac{1}{2}\sigma_i^2 t + \boldsymbol{\sigma}_i \cdot \mathbf{W}(t)\right] = \exp\left[-\frac{1}{2}\sigma_i^2 t + \sum_{j=1}^{d}\sigma_{ij}W_j(t)\right].$$

To see that this is a $\mathbb{P}$-martingale with expectation one, note that $\boldsymbol{\sigma}_i \cdot \mathbf{W}(t)$ is a scalar normal random variable with mean $\mathrm{E}[\boldsymbol{\sigma}_i \cdot \mathbf{W}(t)] = \sum_{j=1}^{d} \sigma_{ij}\mathrm{E}[W_j(t)] = 0$ and variance

$$\mathrm{Var}\left(\boldsymbol{\sigma}_i \cdot \mathbf{W}(t)\right) = \mathrm{Var}\left(\sum_{j=1}^{d} \sigma_{ij} W_j(t)\right) = \sum_{j=1}^{d} \sigma_{ij}^2 \,\mathrm{Var}(W_j(t)) = \|\boldsymbol{\sigma}_i\|^2 t = \sigma_i^2 t\,.$$

Here, we used the fact that all $W_j(t)$ BMs are i.i.d. *Norm*$(0,t)$. Hence, by the expression for the m.g.f. of a normal random variable, $\mathrm{E}[\mathrm{e}^{\alpha\,\boldsymbol{\sigma}_i\cdot\mathbf{W}(t)}] = \mathrm{e}^{\frac{1}{2}\alpha^2\sigma_i^2 t}$ for any α, i.e., $\mathrm{E}[\mathrm{e}^{\boldsymbol{\sigma}_i\cdot\mathbf{W}(t)}] = \mathrm{e}^{\frac{1}{2}\sigma_i^2 t}$ and $\mathrm{E}[\mathcal{E}_t(\boldsymbol{\sigma}_i \cdot \mathbf{W})] = 1$, for all $t \geqslant 0$. So the mean of the price in (2.229) is

$$\mathrm{E}\left[S_i(t)\right] = S_i(0)\mathrm{e}^{\mu_i t}\mathrm{E}\left[\mathcal{E}_t(\boldsymbol{\sigma}_i \cdot \mathbf{W})\right] = S_i(0)\,\mathrm{e}^{\mu_i t}\,. \tag{2.230}$$

As in the one-dimensional GBM stock model, the log-normal drift parameter μ_i is therefore the (physical) growth rate of the ith price process in the (physical) measure $\mathbb{P}$.

From the strong solution in (2.229), we have

$$S_i(T) = S_i(t)\,\mathrm{e}^{(\mu_i - \frac{1}{2}\sigma_i^2)(T-t) + \boldsymbol{\sigma}_i\cdot(\mathbf{W}(T)-\mathbf{W}(t))}\,. \tag{2.231}$$

It is convenient to define the log-return random variables over a time interval $\tau := T - t$,

$$X_i := \ln\frac{S_i(T)}{S_i(t)} = \alpha_i + \sigma_i\sqrt{\tau}Z_i, \tag{2.232}$$

$\alpha_i \equiv (\mu_i - \frac{1}{2}\sigma_i^2)\tau$, where

$$Z_i := \frac{\boldsymbol{\sigma}_i \cdot (\mathbf{W}(T) - \mathbf{W}(t))}{\sigma_i\sqrt{\tau}} = \frac{1}{\sigma_i\sqrt{\tau}}\sum_{j=1}^{d}\sigma_{ij}(W_j(T) - W_j(t)),$$

$i = 1, \ldots, n$. The Z_i's are all *Norm*$(0,1)$ random variables. They are correlated (not i.i.d.). In fact, the vector $\mathbf{Z} = [Z_1, \ldots, Z_n]^\top$ has multivariate normal distribution:

$$[Z_1, \ldots, Z_n]^\top \sim \mathit{Norm}_n\left(\mathbf{0}, \boldsymbol{\rho}\right)\,. \tag{2.233}$$

To verify this, the covariances are computed using the same steps as in the calculation of the covariance of Z_1 and Z_2 in (2.223):

$$\begin{aligned}\mathrm{Cov}\left(Z_i, Z_j\right) &= \frac{1}{\sigma_i\sqrt{\tau}}\frac{1}{\sigma_j\sqrt{\tau}}\mathrm{Cov}\left(\boldsymbol{\sigma}_i \cdot (\mathbf{W}(T) - \mathbf{W}(t)), \boldsymbol{\sigma}_j \cdot (\mathbf{W}(T) - \mathbf{W}(t))\right) \\ &= \frac{1}{\sigma_i\sqrt{\tau}}\frac{1}{\sigma_j\sqrt{\tau}}\boldsymbol{\sigma}_i \cdot \boldsymbol{\sigma}_j\tau = \frac{\boldsymbol{\sigma}_i \cdot \boldsymbol{\sigma}_j}{\sigma_i\sigma_j} = \rho_{ij}\,.\end{aligned} \tag{2.234}$$

Hence, $\mathrm{Cov}(X_i, X_j) = \tau\sigma_i\sigma_j\,\mathrm{Cov}(Z_i, Z_j) = \tau\sigma_i\sigma_j\rho_{ij} = \tau C_{ij}$. The log-returns are therefore jointly normally distributed:

$$[X_1, \ldots, X_n]^\top \sim \mathit{Norm}_n\left([\alpha_1, \ldots, \alpha_n]^\top, \tau\mathbf{C}\right)\,. \tag{2.235}$$

In particular, the matrix $\boldsymbol{\rho}$ is in fact the *matrix of correlations of the stock price log-returns*:

$$\mathrm{Corr}\left(X_i, X_j\right) = \frac{\mathrm{Cov}\left(X_i, X_j\right)}{\sqrt{\mathrm{Var}(X_i)\,\mathrm{Var}(X_j)}} = \frac{\tau C_{ij}}{\sqrt{(\sigma_i^2\tau)(\sigma_j^2\tau)}} = \frac{C_{ij}}{\sigma_i\sigma_j} = \rho_{ij}\,. \tag{2.236}$$

The above GBM process is time homogeneous. Let $\mathbf{x} = (x_1, \ldots, x_n)$ and $\mathbf{y} = (y_1, \ldots, y_n)$ be strictly positive vectors in $\mathbb{R}_+^n$. The (joint) transition CDF of the stock price process $\mathbf{S}(t)$

is given by the conditional probability below which is calculated by using independence among all log-returns $\{X_i \equiv \frac{S_i(T)}{S_i(t)}\}_{i=1,\dots,n}$ and time-t stock prices $\{S_i(t)\}_{i=1,\dots,n}$:

$$\begin{aligned}
P(\tau; \mathbf{x}, \mathbf{y}) &:= \mathbb{P}\left(S_1(T) \leqslant y_1, \dots, S_n(T) \leqslant y_n \mid S_1(t) = x_1, \dots, S_n(t) = x_n\right) \\
&= \mathbb{P}\left(\ln \frac{S_1(T)}{S_1(t)} \leqslant \ln \frac{y_1}{x_1}, \dots, \ln \frac{S_n(T)}{S_n(t)} \leqslant \ln \frac{y_n}{x_n} \,\middle|\, S_1(t) = x_1, \dots, S_n(t) = x_n\right) \\
&= \mathbb{P}\left(X_1 \leqslant \ln \frac{y_1}{x_1}, \dots, X_n \leqslant \ln \frac{y_n}{x_n}\right) \\
&= \mathbb{P}\left(Z_1 \leqslant a_1, \dots, Z_n \leqslant a_n\right) \\
&= \mathcal{N}_n\left(a_1 \dots, a_n; \boldsymbol{\rho}\right)
\end{aligned} \tag{2.237}$$

where $a_i \equiv \frac{\ln \frac{y_i}{x_i} - \alpha_i}{\sigma_i \sqrt{\tau}} = \frac{\ln \frac{y_i}{x_i} - (\mu_i - \sigma_i^2/2)\tau}{\sigma_i \sqrt{\tau}}$. The function $\mathcal{N}_n(a_1, \dots, a_n; \boldsymbol{\rho})$ is the n-variate standard normal CDF of $\mathbf{Z}$ with given correlation matrix $\boldsymbol{\rho}$:

$$\mathcal{N}_n(a_1, \dots, a_n; \boldsymbol{\rho}) = \int_{-\infty}^{a_n} \cdots \int_{-\infty}^{a_1} n_n(z_1, \dots, z_n; \boldsymbol{\rho})\, \mathrm{d}z_1 \dots \mathrm{d}z_n .$$

The standard normal PDF of $\mathbf{Z}$, $n_n(z_1, \dots, z_n; \boldsymbol{\rho}) = \frac{\partial^n}{\partial z_1 \dots \partial z_n} \mathcal{N}_n\left(z_1 \dots, z_n; \boldsymbol{\rho}\right)$, is given by the n-variate Gaussian density

$$n_n(z_1, \dots, z_n; \boldsymbol{\rho}) = \frac{1}{\sqrt{(2\pi)^n \det \boldsymbol{\rho}}} \exp\left(-\frac{1}{2} \mathbf{z} \cdot \boldsymbol{\rho}^{-1} \cdot \mathbf{z}^\top\right), \tag{2.238}$$

$\mathbf{z} = [z_1, \dots, z_n] \in \mathbb{R}^n$. Differentiating according to (2.203), and applying the chain rule, gives the (joint) transition PDF for the time-homogeneous GBM stock price process:

$$\begin{aligned}
p(\tau; \mathbf{x}, \mathbf{y}) &= \frac{\partial^n}{\partial y_1 \dots \partial y_n} P(\tau; \mathbf{x}, \mathbf{y}) = \left(\prod_{i=1}^n \frac{\partial a_i}{\partial y_i}\right) \frac{\partial^n}{\partial a_1 \dots \partial a_n} \mathcal{N}_n\left(a_1 \dots, a_n; \boldsymbol{\rho}\right) \\
&= \left(\prod_{i=1}^n \frac{1}{y_i \sigma_i \sqrt{\tau}}\right) n_n\left(a_1 \dots, a_n; \boldsymbol{\rho}\right) \\
&= \frac{1}{y_1 \cdots y_n \sigma_1 \cdots \sigma_n \sqrt{(2\pi\tau)^n \det \boldsymbol{\rho}}} \exp\left(-\frac{1}{2} \mathbf{a} \cdot \boldsymbol{\rho}^{-1} \cdot \mathbf{a}^\top\right),
\end{aligned} \tag{2.239}$$

$\mathbf{a} = [a_1, \dots, a_n]$ and a_i's defined above. This is of the form of a *multivariate log-normal density*. Note that this density can also be written in terms of the covariance matrix $\mathbf{C} = \mathbf{D}\boldsymbol{\rho}\mathbf{D}$, $\mathbf{D} = \mathrm{diag}(\sigma_1, \dots, \sigma_n)$, $\boldsymbol{\rho}^{-1} = \mathbf{D}\mathbf{C}^{-1}\mathbf{D}$.

By (2.227), the diffusion matrix function has elements

$$C_{ij}(\mathbf{x}) = \sum_{k=1}^d \sigma_{ik}(\mathbf{x})\, \sigma_{jk}(\mathbf{x}) = \sum_{k=1}^d x_i \sigma_{ik}\, x_j \sigma_{jk} = x_i x_j \sum_{k=1}^d \sigma_{ik} \sigma_{jk} = x_i x_j C_{ij}$$

with constants $C_{ij} = \boldsymbol{\sigma}_i \cdot \boldsymbol{\sigma}_j = \rho_{ij} \sigma_i \sigma_j$. Hence, the time-homogeneous PDE in (2.218) takes the equivalent form:

$$\begin{aligned}
\frac{\partial f}{\partial \tau} &= \frac{1}{2} \sum_{i=1}^n \sum_{j=1}^n \rho_{ij} \sigma_i \sigma_j x_i x_j \frac{\partial^2 f}{\partial x_i \partial x_j} + \sum_{i=1}^n \mu_i x_i \frac{\partial f}{\partial x_i} \\
&= \frac{1}{2} \sum_{i=1}^n \sigma_i^2 x_i^2 \frac{\partial^2 f}{\partial x_i^2} + \sum_{j=1}^n \sum_{i<j} \rho_{ij} \sigma_i \sigma_j x_i x_j \frac{\partial^2 f}{\partial x_i \partial x_j} + \sum_{i=1}^n \mu_i x_i \frac{\partial f}{\partial x_i} .
\end{aligned} \tag{2.240}$$

The Feynman–Kac Theorem 2.23 assures us that the transition CDF in (2.237) and PDF in (2.239) both solve the PDE (2.240) in the variables $\tau > 0, \mathbf{x} \in \mathbb{R}_+^n$, for fixed $\mathbf{y} \in \mathbb{R}_+^n$. The initial condition $P(0; \mathbf{x}, \mathbf{y}) = \mathbb{I}_{\{\mathbf{x} \leqslant \mathbf{y}\}} \equiv \mathbb{I}_{\{x_1 \leqslant y_1, \ldots, x_n \leqslant y_n\}}$ follows from the basic limit properties of the multivariate normal CDF. We leave it to the reader to verify. Then, by multiple differentiation of the step functions, the initial condition $p(0+; \mathbf{x}, \mathbf{y}) = \delta(\mathbf{x} - \mathbf{y})$ is obtained for the transition PDF.

A common case is when $n = d = 2$. The above formulation simplifies since we have only one correlation coefficient ρ for the log-returns of stocks 1 and 2, where $\rho_{12} = \rho_{21} \equiv \rho$, $\rho_{11} = \rho_{22} = 1$. The log-diffusion matrix of covariances has elements $C_{11} = \sigma_1^2, C_{22} = \sigma_2^2, C_{12} = C_{21} = \rho\sigma_1\sigma_2$. The log-volatility vectors are $\boldsymbol{\sigma}_1 = [\sigma_{11}, \sigma_{12}] = [\sigma_1, 0]$ and $\boldsymbol{\sigma}_2 = [\sigma_{21}, \sigma_{22}] = [\rho\sigma_2, \sigma_2\sqrt{1-\rho^2}]$. In this case, the system of SDEs in (2.226) simplifies for two stock prices driven by two BMs:

$$\begin{aligned} \mathrm{d}S_1(t) &= S_1(t)[\mu_1\, \mathrm{d}t + \sigma_1\, \mathrm{d}W_1(t)] \\ \mathrm{d}S_2(t) &= S_2(t)[\mu_2\, \mathrm{d}t + \sigma_2\rho\, \mathrm{d}W_1(t) + \sigma_2\sqrt{1-\rho^2}\, \mathrm{d}W_2(t)]\,. \end{aligned} \tag{2.241}$$

The unique solution is

$$\begin{aligned} S_1(t) &= S_1(0)\, \mathrm{e}^{(\mu_1 - \frac{1}{2}\sigma_1^2)t + \sigma_1 W_1(t)}\,, \\ S_2(t) &= S_2(0)\, \mathrm{e}^{(\mu_2 - \frac{1}{2}\sigma_2^2)t + \sigma_2(\rho W_1(t) + \sqrt{1-\rho^2} W_2(t))}\,. \end{aligned}$$

The transition CDF and PDF obtain as a special case of (2.237) and (2.239) for $n = 2$:

$$P(\tau; x_1, x_2, y_1, y_2) = \mathcal{N}_2\left(\frac{\ln\frac{y_1}{x_1} - (\mu_1 - \frac{1}{2}\sigma_1^2)\tau}{\sigma_1\sqrt{\tau}}, \frac{\ln\frac{y_2}{x_2} - (\mu_2 - \frac{1}{2}\sigma_2^2)\tau}{\sigma_2\sqrt{\tau}}; \rho\right) \tag{2.242}$$

and

$$p(\tau; x_1, x_2, y_1, y_2) = \frac{1}{y_1 y_2 \sigma_1 \sigma_2 \tau} n_2\left(\frac{\ln\frac{y_1}{x_1} - (\mu_1 - \frac{1}{2}\sigma_1^2)\tau}{\sigma_1\sqrt{\tau}}, \frac{\ln\frac{y_2}{x_2} - (\mu_2 - \frac{1}{2}\sigma_2^2)\tau}{\sigma_2\sqrt{\tau}}; \rho\right). \tag{2.243}$$

By the Feynman–Kac Theorem 2.23, these functions solve the time-homogeneous PDE in (2.240) for $n = 2$:

$$\frac{\partial f}{\partial \tau} = \frac{1}{2}\sigma_1^2 x_1^2 \frac{\partial^2 f}{\partial x_1^2} + \frac{1}{2}\sigma_2^2 x_2^2 \frac{\partial^2 f}{\partial x_2^2} + \rho\sigma_1\sigma_2 x_1 x_2 \frac{\partial^2 f}{\partial x_1 \partial x_2} + \mu_1 x_1 \frac{\partial f}{\partial x_1} + \mu_2 x_2 \frac{\partial f}{\partial x_2}\,. \tag{2.244}$$

The reader can verify that the transition CDF has the limiting form $P(0+; x_1, x_2, y_1, y_2) = \mathbb{I}_{\{y_1 \geqslant x_1, y_2 \geqslant x_2\}}$, for all $\mathbf{x} \neq \mathbf{y}$, and $p(0+; x_1, x_2, y_1, y_2) = \delta(y_1 - x_1)\delta(y_2 - x_2)$.

2.10.3 Girsanov's Theorem for Multidimensional BM

We recall Girsanov's Theorem 2.17 where the measure change was constructed in terms of a Radon–Nikodym process which has the form of an exponential martingale involving a single standard BM in the original measure $\mathbb{P}$. Based on our knowledge of multidimensional BM and Itô integrals on multidimensional BM, we can now consider the multidimensional version of Girsanov's Theorem. The main ingredients are as in Theorem 2.17 where the single BM is now a multidimensional BM. As usual, we fix some filtered probability space $(\Omega, \mathcal{F}, \mathbb{P}, \mathbb{F})$, where $\mathbb{F} = \{\mathcal{F}_t\}_{t \geqslant 0}$ is any filtration for standard BM.

Theorem 2.25 (Girsanov's Theorem for Multidimensional BM). *Fix a time $T > 0$ and let $\mathbf{W}(t) = (W_1(t), \ldots, W_d(t)), 0 \leqslant t \leqslant T$, be a standard d-dimensional $\mathbb{P}$-BM with respect to a filtration $\mathbb{F} = \{\mathcal{F}_t\}_{0 \leqslant t \leqslant T}$. Assume the vector process $\boldsymbol{\gamma}(t) = (\gamma_1(t), \ldots, \gamma_d(t)), 0 \leqslant t \leqslant T$, is adapted to $\mathbb{F}$ such that*

$$\mathrm{E}\left[\exp\left(\frac{1}{2}\int_0^T \|\boldsymbol{\gamma}(s)\|^2\,\mathrm{d}s\right)\right] < \infty\,. \tag{2.245}$$

Define

$$\varrho_t := \exp\left(-\frac{1}{2}\int_0^t \|\boldsymbol{\gamma}(s)\|^2\,\mathrm{d}s + \int_0^t \boldsymbol{\gamma}(s)\cdot \mathrm{d}\mathbf{W}(s)\right), 0 \leqslant t \leqslant T, \tag{2.246}$$

and the probability measure $\widehat{\mathbb{P}} \equiv \widehat{\mathbb{P}}^{(\varrho)}$ by the Radon–Nikodym derivative $\frac{\mathrm{d}\widehat{\mathbb{P}}}{\mathrm{d}\mathbb{P}} = \left(\frac{\mathrm{d}\widehat{\mathbb{P}}}{\mathrm{d}\mathbb{P}}\right)_T \equiv \varrho_T$. Then, the process $\widehat{\mathbf{W}}(t) = (\widehat{W}_1(t), \ldots, \widehat{W}_d(t)), 0 \leqslant t \leqslant T$, defined by

$$\widehat{\mathbf{W}}(t) := \mathbf{W}(t) - \int_0^t \boldsymbol{\gamma}(s)\,\mathrm{d}s \tag{2.247}$$

is a standard d-dimensional $\widehat{\mathbb{P}}$-BM w.r.t. filtration $\mathbb{F}$.

We don't provide the proof of this result here. We leave it as an exercise where one can apply similar steps as in (i)–(iii) in the proof of Theorem 2.17. In the multidimensional case, we have d-dimensional BM and Lévy's characterization in Theorem 2.20 can be applied.

The same remarks as were stated for Theorem 2.17 also apply to Theorem 2.25, where the adapted process is now the vector $\boldsymbol{\gamma}$ rather than the scalar γ. Of course, this multidimensional version generalizes Theorem 2.17, which obtains in the simplest case with $d = 1$. The Radon–Nikodym derivative process that defines the change of measure $\mathbb{P} \to \widehat{\mathbb{P}}$, such that the new d-dimensional BM, $\widehat{\mathbf{W}}$, is a $\widehat{\mathbb{P}}$-BM, is given by the exponential $\mathbb{P}$-martingale in (2.246). It can be proven that the Novikov condition in (2.245) guarantees that the process $\{\varrho_t\}_{0 \leqslant t \leqslant T}$ is indeed a proper Radon–Nikodym derivative process i.e., it is a $\mathbb{P}$-martingale with constant unit expectation, $\mathrm{E}[\varrho_t] = 1$ for all $t \in [0, T]$. By Itô's formula, the stochastic exponential in (2.246) is equivalent to the stochastic differential (using $\varrho_0 = 1$):

$$\mathrm{d}\varrho_t = \varrho_t \boldsymbol{\gamma}(t)\cdot \mathrm{d}\mathbf{W}(t) \implies \varrho_t = 1 + \int_0^t \varrho_s \boldsymbol{\gamma}(s)\cdot \mathrm{d}\mathbf{W}(s)\,.$$

The Novikov condition assures us that the Itô integral satisfies the integrability condition as in (2.156) and is therefore a $\mathbb{P}$-martingale with zero expectation. By the definition in (2.185) we have $\varrho_t \equiv \varrho_t^{(\boldsymbol{\gamma})} = \mathcal{E}_t(\boldsymbol{\gamma}\cdot\mathbf{W})$. Dividing the process value at any two times $0 \leqslant s < t \leqslant T$ gives

$$\frac{\varrho_t}{\varrho_s} \equiv \frac{\left(\frac{\mathrm{d}\widehat{\mathbb{P}}}{\mathrm{d}\mathbb{P}}\right)_t}{\left(\frac{\mathrm{d}\widehat{\mathbb{P}}}{\mathrm{d}\mathbb{P}}\right)_s} = \frac{\mathcal{E}_t(\boldsymbol{\gamma}\cdot\mathbf{W})}{\mathcal{E}_s(\boldsymbol{\gamma}\cdot\mathbf{W})} = \exp\left(-\frac{1}{2}\int_s^t \|\boldsymbol{\gamma}(u)\|^2\,\mathrm{d}u + \int_s^t \boldsymbol{\gamma}(u)\cdot \mathrm{d}\mathbf{W}(u)\right).$$

In general, $\boldsymbol{\gamma}$ is an adapted vector process so that ϱ_t is a functional of d-dimensional BM from time 0 to t. By choosing a constant vector, $\boldsymbol{\gamma}(t) = \boldsymbol{\gamma}$, we have the simple and important special case where the Radon–Nikodym derivative process is also a GBM expressed equivalently as

$$\varrho_t \equiv \varrho_t^{(\boldsymbol{\gamma})} = \mathrm{e}^{-\frac{1}{2}\|\boldsymbol{\gamma}\|^2 t + \boldsymbol{\gamma}\cdot\mathbf{W}(t)} = \mathrm{e}^{\frac{1}{2}\|\boldsymbol{\gamma}\|^2 t + \boldsymbol{\gamma}\cdot\widehat{\mathbf{W}}(t)} \tag{2.248}$$

where $\widehat{\mathbf{W}}(t) \equiv \widehat{\mathbf{W}}^{(\boldsymbol{\gamma})}(t) := \mathbf{W}(t) - \boldsymbol{\gamma}t$, $0 \leqslant t \leqslant T$, is a $\widehat{\mathbb{P}}$-BM. We recall that the random variable $\boldsymbol{\gamma}\cdot\mathbf{W}(t) \sim Norm(0, \|\boldsymbol{\gamma}\|^2 t)$. Hence, from its m.g.f. (under measure $\mathbb{P}$) we have

$\mathrm{E}[\mathrm{e}^{\boldsymbol{\gamma}\cdot\mathbf{W}(t)}] = \mathrm{e}^{\frac{1}{2}\|\boldsymbol{\gamma}\|^2 t}$, i.e., $\mathrm{E}[\varrho_t] = 1$ for all $t \geqslant 0$. This also follows trivially from the fact that the process is easily verified to be a $\mathbb{P}$-martingale and therefore must have constant expectation under measure $\mathbb{P}$, i.e., $\mathrm{E}[\varrho_t] = \mathrm{E}[\varrho_0] = \mathrm{E}[1] = 1$.

The multidimensional version of Girsanov's Theorem 2.25 has many far-reaching applications. We now give one of its applications that is particularly important for financial derivative pricing theory. Namely, we shall use Girsanov's Theorem to find a risk-neutral measure such that all stock prices $S_1(t), \ldots, S_n(t)$ in the multidimensional GBM model have a *common drift rate equal to a constant* r. Here, r is again the fixed (continuously compounded) interest rate. This problem is equivalent to applying Girsanov's Theorem to construct an equivalent martingale measure $\widehat{\mathbb{P}} \equiv \widetilde{\mathbb{P}}$, such that all discounted stock price processes defined by $\{\overline{S}_i(t) := \mathrm{e}^{-rt} S_i(t)\}_{t\geqslant 0}$, $i = 1, \ldots, n$, are $\widetilde{\mathbb{P}}$-martingales. For the case of a single stock ($n = 1$) driven by one BM ($d = 1$), we have already solved this problem in Example 2.25.

For each ith stock, the (log-)drift μ_i and all components of the (log-)volatility vector $\boldsymbol{\sigma}_i$ in (2.226) are constants. It follows that the measure change that we will need to employ uses (2.248), i.e., with constant d-dimensional vector $\boldsymbol{\gamma} = [\gamma_1, \ldots, \gamma_d]$:

$$\varrho_t \equiv \left(\frac{\mathrm{d}\widetilde{\mathbb{P}}}{\mathrm{d}\mathbb{P}}\right)_t = \mathrm{e}^{-\frac{1}{2}\|\boldsymbol{\gamma}\|^2 t + \boldsymbol{\gamma}\cdot\mathbf{W}(t)} \tag{2.249}$$

with d-dimensional $\widetilde{\mathbb{P}}$-BM given by $\widetilde{\mathbf{W}}(t) \equiv \widehat{\mathbf{W}}^{(\gamma)}(t) := \mathbf{W}(t) - \boldsymbol{\gamma} t$. In terms of stochastic differentials, $\mathrm{d}\widetilde{\mathbf{W}}(t) = \mathrm{d}\mathbf{W}(t) - \boldsymbol{\gamma}\,\mathrm{d}t$. A quick method of arriving at the change of measure is to consider the SDE satisfied by the stock prices (with respect to the physical $\mathbb{P}$-BM) in (2.226) and write $\mathrm{d}\mathbf{W}(t) = \mathrm{d}\widetilde{\mathbf{W}}(t) + \boldsymbol{\gamma}\,\mathrm{d}t$:

$$\begin{aligned} \mathrm{d}S_i(t) &= S_i(t)\big[\mu_i\,\mathrm{d}t + \boldsymbol{\sigma}_i\cdot(\,\mathrm{d}\widetilde{\mathbf{W}}(t) + \boldsymbol{\gamma}\,\mathrm{d}t)\big] \\ &= S_i(t)\big[(\mu_i + \boldsymbol{\sigma}_i\cdot\boldsymbol{\gamma})\,\mathrm{d}t + \boldsymbol{\sigma}_i\cdot\mathrm{d}\widetilde{\mathbf{W}}(t)\big]\,,\ i = 1, \ldots, n. \end{aligned} \tag{2.250}$$

Hence, a risk-neutral measure exists if we can find a vector $\boldsymbol{\gamma}$ such that the log-drift coefficient equals r for every $i = 1, \ldots, n$. That is, we have

$$\mathrm{d}S_i(t) = S_i(t)\big[r\,\mathrm{d}t + \boldsymbol{\sigma}_i\cdot\mathrm{d}\widetilde{\mathbf{W}}(t)\big]\,,\ i = 1, \ldots, n, \tag{2.251}$$

which is equivalent to $\{\overline{S}_i(t)\}_{t\geqslant 0}$, $i = 1, \ldots, n$, being $\widetilde{\mathbb{P}}$-martingales, if and only if $\boldsymbol{\gamma}$ solves $\mu_i + \boldsymbol{\sigma}_i\cdot\boldsymbol{\gamma} = r$, for each $i = 1, \ldots, n$. This is a linear system of n-equations in d unknowns $\gamma_1, \ldots, \gamma_d$. Using the components of the (log-)volatility vectors, $\boldsymbol{\sigma}_i = [\sigma_{i1}, \ldots, \sigma_{id}]$, we then have an $n \times d$ linear system

$$\begin{bmatrix} \sigma_{11} & \cdots & \sigma_{1d} \\ \vdots & & \vdots \\ \sigma_{n1} & \cdots & \sigma_{nd} \end{bmatrix} \begin{bmatrix} \gamma_1 \\ \vdots \\ \gamma_d \end{bmatrix} = \begin{bmatrix} r - \mu_1 \\ \vdots \\ r - \mu_n \end{bmatrix}. \tag{2.252}$$

In compact notation, this reads $\boldsymbol{\sigma}\,\boldsymbol{\gamma}^\top = \mathbf{b}$, where $\boldsymbol{\sigma}$ is $n \times d$, $\boldsymbol{\gamma}^\top$ is $d \times 1$, and $\mathbf{b} := r\mathbf{1} - \boldsymbol{\mu}$ is $n \times 1$, where $\boldsymbol{\mu} = [\mu_1, \ldots, \mu_n]^\top$, $\mathbf{1} = [1, \ldots, 1]^\top$.

Hence, the question of the existence of a risk-neutral measure is answered quite simply by applying standard linear algebra. Generally a solution vector $\boldsymbol{\gamma}$ exists if and only if $\mathbf{b}$ is spanned by the d column vectors of $\boldsymbol{\sigma}$. Here we should point out that we are seeking a solution for *arbitrary physical drift vector* $\boldsymbol{\mu} \in \mathbb{R}^n$. A solution vector $\boldsymbol{\gamma}$ exists for any $\mathbf{b} \in \mathbb{R}^n$, and hence for any $\boldsymbol{\mu} \in \mathbb{R}^n$, if the d column vectors of $\boldsymbol{\sigma}$ span $\mathbb{R}^n$, i.e., if $\mathrm{rank}(\boldsymbol{\sigma}) = n$. In the case when $\mathrm{rank}(\boldsymbol{\sigma}) = n < d$, we have an infinite (continuum) number of solution vectors $\boldsymbol{\gamma}$

and each corresponds to a (different) risk-neutral measure $\widetilde{\mathbb{P}} \equiv \widetilde{\mathbb{P}}^{(\gamma)}$. This is therefore the case where the risk-neutral measure exists and is not unique. If $\text{rank}(\boldsymbol{\sigma}) = n = d$, which is the case where the *number of stocks equals the number of independent BMs and the* $n \times n$ *matrix* $\boldsymbol{\sigma}$ *has an inverse* $\boldsymbol{\sigma}^{-1}$, then the risk-neutral measure $\widetilde{\mathbb{P}}$ exists and is uniquely given by $\boldsymbol{\gamma}^\top = \boldsymbol{\sigma}^{-1}(r\mathbf{1} - \boldsymbol{\mu})$, i.e., $\gamma_j = \sum_{i=1}^n (\boldsymbol{\sigma}^{-1})_{ji}(r - \mu_i)$, $j = 1, \ldots, d$. Finally, if $d < n$, then $\text{rank}(\boldsymbol{\sigma}) \leqslant d < n$ (the d column vectors of $\boldsymbol{\sigma}$ do not span all of $\mathbb{R}^n$) and hence a solution vector $\boldsymbol{\gamma}$ exists only for $\boldsymbol{\mu}$ vectors such that $\mathbf{b}$ is in the span of the column vectors of $\boldsymbol{\sigma}$. In this case, there does not exist a risk-neutral measure $\widetilde{\mathbb{P}}$ for arbitrary $\boldsymbol{\mu}$.

2.10.4 Martingale Representation Theorem for Multidimensional BM

In closing this chapter, we simply state (without proof) the multidimensional version of Theorem 2.18. This is of importance when discussing hedging financial derivatives in an economy with multiple assets that are driven by a multidimensional BM. The theorem is quite similar and extends Theorem 2.18 in a rather obvious fashion whereby the Itô integrals, and hence the Itô processes, are defined with respect to a d-dimensional BM where $d \geqslant 1$. The theorem basically states that a square-integrable $(\mathbb{P}, \mathbb{F}^W)$-martingale is expressible as its initial value plus an Itô integral in the d-dimensional BM and some $\mathbb{F}^W$-adapted vector process as integrand. In the result below, we combine Theorem 2.18 and Proposition 2.19 into one Theorem for the more general case of multidimensional BM.

Theorem 2.26 (Multidimensional Brownian Martingale Representation Theorem). *Let* $\mathbf{W}(t) = (W_1(t), \ldots, W_d(t))$ *be a d-dimensional standard BM where* $\mathbb{F}^W$ *denotes its natural filtration. Assume* $\{M(t)\}_{0 \leqslant t \leqslant T}$ *is a square-integrable* $(\mathbb{P}, \mathbb{F}^W)$*-martingale. Then, there exists an* $\mathbb{F}^W$*-adapted d-dimensional process (a.s.)* $\boldsymbol{\theta}(t) = (\theta_1(t), \ldots, \theta_d(t)), 0 \leqslant t \leqslant T$, *such that*

$$M(t) = M(0) + \int_0^t \boldsymbol{\theta}(u) \cdot \mathrm{d}\mathbf{W}(u) \equiv M(0) + \sum_{j=1}^d \int_0^t \theta_j(u) \cdot \mathrm{d}W_j(u)\,, \tag{2.253}$$

for all $t \in [0, T]$. *Moreover, let* $\widehat{\mathbb{P}}$ *be a measure constructed using Girsanov's Theorem 2.25 with the assumption that the d-dimensional process* $\{\gamma(t)\}_{0 \leqslant t \leqslant T}$ *is* $\mathbb{F}^W$*-adapted. If the process* $\{\widehat{M}(t)\}_{0 \leqslant t \leqslant T}$ *is a square-integrable* $(\widehat{\mathbb{P}}, \mathbb{F}^W)$*-martingale, then there exists an adapted d-dimensional process* $\{\widehat{\boldsymbol{\theta}}(t) = (\widehat{\theta}_1(t), \ldots, \widehat{\theta}_d(t))\}_{0 \leqslant t \leqslant T}$, *such that (a.s.)*

$$\widehat{M}(t) = \widehat{M}(0) + \int_0^t \widehat{\boldsymbol{\theta}}(u) \cdot \mathrm{d}\widehat{\mathbf{W}}(u)\,. \tag{2.254}$$

Again we stress that the martingale having this representation is (a.s.) continuous in time (i.e., the process has no jumps) since it is an Itô process.

2.11 Exercises

Exercise 2.1. In each case, show whether or not the integral expression is well-defined as an Itô integral.

(a) $\int_0^1 (W^2(t) + W(t) + 1)\,\mathrm{d}W(t)$;

(b) $\int_0^1 |W(t)|^{-1/2}\,\mathrm{d}W(t)$;

(c) $\int_0^1 W(\frac{1}{t})\,\mathrm{d}W(t)$;

(d) $\int_0^1 |tW(t)|^{-1/4}\,\mathrm{d}W(t)$;

(e) $\int_0^1 W^a(t)\,\mathrm{d}W(t)$.

For part (e) find all values of the parameter $a \in \mathbb{R}$ for which the integral is well-defined. [Hint for singular integrands: Recall that if $t = 0$ is a singular point of an otherwise continuous function $f(t)$ for $t \in (0, T]$, where $f(t) = \mathcal{O}(|t|^p)$, as $t \to 0$, then $\int_0^T f(t)\,\mathrm{d}t < \infty$ iff $p > -1$.]

Exercise 2.2. In each case, show whether or not the integral expression is well-defined as an Itô integral.

(a) $\int_0^1 W(2t)\,\mathrm{d}W(t)$;

(b) $\int_0^1 W(\frac{t}{2})\,\mathrm{d}W(t)$;

(c) $\int_1^\infty W(\frac{1}{t})\,\mathrm{d}W(t)$;

(d) $\int_0^1 |W(t)|^{1/2}\,\mathrm{d}W(t)$.

Exercise 2.3. For any $\alpha \in (0, 1)$, define the stochastic integral

$$\int_0^T W(t) \diamond \mathrm{d}_\alpha W(t) := \lim_{\delta t^{(n)} \to 0} \sum_{i=1}^n [\alpha W(t_{i-1}) + (1-\alpha) W(t_i)](W(t_i) - W(t_{i-1})),$$

where $P_n = \{0 = t_0 < t_1 < \ldots < t_n = T\}$ is a finite partition of $[0, T]$, and $\delta(P_n)$ is the mesh size. Write $\int_0^T W(t) \diamond \mathrm{d}_\alpha W(t)$ as a linear combination of the Itô integral $\int_0^T W(t)\,\mathrm{d}W(t)$ and the Stratonovich integral $\int_0^T W(t) \circ \mathrm{d}W(t)$.

Exercise 2.4. Calculate

$$\lim_{n\to\infty,\ \delta(P_n)\to 0} \sum_{i=1}^n (2W(t_{i-1}) + W(t_i))(W(t_i) - W(t_{i-1})),$$

where $P_n = \{0 = t_0 < t_1 < t_2 < \ldots < t_n = T\}$ is a partition of $[0, T]$ with mesh size $\delta(P_n) = \max_{i=1,\ldots,n} |t_i - t_{i-1}|$.

Exercise 2.5. Evaluate the following repeated (double) stochastic integral:

$$\int_0^t \left(\int_0^s \mathrm{d}W(u) \right) \mathrm{d}W(s).$$

Exercise 2.6. Use an appropriate isometry formula and compute each of the following:

(a) $\mathrm{Var}\left(\int_0^t |W(s)|^{1/2}\,\mathrm{d}W(s) \right)$;

(b) $\mathrm{Var}\left(\int_0^t |W(s) + s|^2\,\mathrm{d}W(s) \right)$;

(c) $\mathrm{Var}\left(\int_0^t (W^2(s) + s)^{3/2}\,\mathrm{d}W(s) \right)$;

(d) $\mathrm{Var}\left(\int_0^t \mathrm{e}^{W(s)}\,\mathrm{d}W(s) \right)$;

(e) $\text{Var}\left(\int_a^b \sqrt{t}e^{W(t)}\,dW(t)\right)$, $0 \leqslant a < b < \infty$;

(f) $\text{Cov}\left(\int_0^t \sin(W^2(s))\,dW(s), \int_0^t W(s)\,dW(s)\right)$.

Exercise 2.7. Use Itô's formula to find the stochastic differentials for the following functions of Brownian motion:

(a) $e^{W(t)}$; (b) $W^k(t)$, $k \geqslant 0$; (c) $\cos(tW(t))$; (d) $e^{W^2(t)}$; (e) $\arctan(t + W(t))$.

Exercise 2.8. Assume $g(t)$ is continuously differentiable for $t \geqslant 0$. Determine the general form for $g(t)$ such that

$$\int_0^t g(u)e^{\alpha W(u)}\,dW(u) = \frac{1}{\alpha}g(u)e^{\alpha W(u)}\Big|_0^t \equiv \frac{1}{\alpha}\left[g(t)e^{\alpha W(t)} - g(0)\right].$$

Exercise 2.9. Use an appropriate Itô formula to show that each process is a martingale w.r.t. a filtration for Brownian motion.

(a) $X(t) := W^4(t) - 6\int_0^t W^2(u)\,du$, $t \geqslant 0$;

(b) $M(t) := e^{t/2}\sin(W(t))$, $t \geqslant 0$;

(c) $Y(t) := W^3(t) - 3tW(t)$, $t \geqslant 0$.

Exercise 2.10. Use an appropriate Itô formula to show that for any integer $k \geqslant 2$,

$$E[W^k(t)] = \frac{k(k-1)}{2}\int_0^t E[W^{k-2}(s)]\,ds,$$

and use this to derive a formula for all the moments of the standard normal distribution.

Exercise 2.11. Show that for any nonrandom, continuously differentiable function $f(t)$, the following formula of integration by parts is true:

$$\int_0^t f(s)\,dW(s) = f(t)W(t) - \int_0^t f'(s)W(s)\,ds.$$

Exercise 2.12. Evaluate each integral as an expression involving *no Itô integral.*

(a) $\int_0^T \frac{1}{1+W^2(t)}\,dW(t)$;

(b) $\int_0^T \cos(W^2(t) + t)W(t)\,dW(t)$;

(c) $\int_0^t W^2(u)e^{W(u)}\,dW(u)$;

(d) $\int_0^t W(u)e^{W(u)}\,dW(u)$;

(e) $\int_0^T W(t)(e^{W(t)} + t)\,dW(t)$.

Exercise 2.13. Let $X(t) := e^{W(t)} - 1 - \frac{1}{2}\int_0^t e^{W(s)}ds$.

(a) Express $X(t)$ as an Itô process.

(b) Compute the mean and variance of $X(t)$.

Exercise 2.14. Define $Z(t) = \exp(\sigma W(t))$. Use Itô's formula to write down a stochastic differential for $Z(t)$. Then, by taking the mathematical expectation, find an ordinary (deterministic) first order linear differential equation for $m(t) := \mathrm{E}[Z(t)]$ and solve it to show that

$$\mathrm{E}[\exp(\sigma W(t))] = \exp\left(\frac{\sigma^2}{2}t\right).$$

Exercise 2.15. Let $\mathcal{N}(x)$ be the standard normal CDF and consider the process

$$X(t) := \mathcal{N}\left(\frac{W(t)}{\sqrt{T-t}}\right), \quad 0 \leqslant t < T.$$

Express this process as an Itô process; i.e. you need to *determine the explicit expressions for the adapted drift* $\mu(t)$ *and diffusion* $\sigma(t)$ *and provide the explicit form for* $X(t)$ as:

$$X(t) = X(0) + \int_0^t \mu(s)\,ds + \int_0^t \sigma(s)\,dW(s).$$

Show that the process is a martingale w.r.t. any filtration for BM. Find the limiting value $X(T-) = \lim_{t \nearrow T} X(t)$. Determine the state space for the X process.

Exercise 2.16. Suppose that the processes $X := \{X(t)\}_{t \geqslant 0}$ and $Y := \{Y(t)\}_{t \geqslant 0}$ have the log-normal dynamics:

$$dX(t) = X(t)(\mu_X\,\mathrm{d}t + \sigma_X\,\mathrm{d}W(t))$$
$$dY(t) = Y(t)(\mu_Y\,\mathrm{d}t + \sigma_Y\,\mathrm{d}W(t)).$$

Show that the process $Z(t) := \frac{Y(t)}{X(t)}$ is also log-normal, with dynamics

$$dZ(t) = Z(t)(\mu_Z\,\mathrm{d}t + \sigma_Z\,\mathrm{d}W(t)),$$

and determine the coefficients μ_Z and σ_Z in terms of those of X and Y. Solve the same problem now assuming that X and Y are governed by two correlated Brownian motions W^X and W^Y, respectively, where $\mathrm{Corr}(W^X(t), W^Y(t)) = \rho t$, i.e., $\mathrm{d}W^X(t)\,\mathrm{d}W^Y(t) = \rho\,\mathrm{d}t$, for a given correlation coefficient $-1 \leqslant \rho \leqslant 1$.

Exercise 2.17. Let a time-homogeneous diffusion $X(t)$ have a stochastic differential with drift coefficient function $\mu(x) = 3x - 1$ and diffusion coefficient function $\sigma(x) = 2\sqrt{x}$. Assuming that $X(t) \geqslant 0$, find the stochastic differential for the process $Y(t) := \sqrt{X(t)}$. Find the generator $\mathcal{G}^Y$ for $Y(t)$.

Exercise 2.18. Let $X(t) = tW^2(t)$ and $Y(t) = e^{W(t)}$. Determine the stochastic differential of $Z(t) := \frac{X(t)}{Y(t)}$. Compute the mean and variance of $Z(t)$. [Hint: use the representation of $Z(t)$ as a function $f(t, W(t))$. For the mean and variance, you may use expectation identities in Appendix B.]

Exercise 2.19. Let $X(t)$ be a time-homogeneous diffusion process solving an SDE with drift and diffusion coefficient functions $\mu(x) = cx$ and $\sigma(x) = \sigma$, respectively, where c, σ are constants and with initial condition $X(0) = x \in \mathbb{R}$. Consider the process defined by $Y(t) := X^2(t) - 2c\int_0^t X^2(s)\,\mathrm{d}s - \sigma^2 t$, $t \geqslant 0$.

(a) Represent the Y process as an Itô process and show that is a martingale w.r.t. any filtration for Brownian motion.

(b) Compute the mean and variance of $Y(t)$ for all $t \geqslant 0$.

Exercise 2.20. Consider the linear SDE in (2.50) in the case where $\delta(t) \equiv 0$ and where $\alpha(t), \beta(t), \gamma(t)$ are continuous nonrandom functions of time $t \geqslant 0$. Assume a constant initial condition $X(0) = x_0$. Show that the process $\{X(t)\}_{t \geqslant 0}$ is a Gaussian process and compute its mean and covariance functions.

Exercise 2.21. Use the Itô formula to write down stochastic differentials for the following processes:

(a) $Y(t) = \exp\left(\sigma W(t) - \frac{1}{2}\sigma^2 t\right)$,

(b) $Z(t) = f(t)W(t)$ where f is a continuously differentiable function.

Exercise 2.22. A time-homogeneous diffusion process X has a stochastic differential with respective drift and diffusion coefficient functions $\mu(x) = 0$ and $\sigma(x) = x(1-x)$. Assuming $0 < X(t) < 1$, show that the process $Y(t) := \ln\left(\frac{X(t)}{1-X(t)}\right)$ has a constant diffusion coefficient.

Exercise 2.23. Let $X(t) := (1-t)\int_0^t \frac{\mathrm{d}W(s)}{1-s}$, where $0 \leqslant t < 1$. Provide the stochastic differential equation for $X(t)$ in the form $\mathrm{d}X(t) = a(t, X(t))\,\mathrm{d}t + b(t, X(t))\,\mathrm{d}W(t)$. Check your answer by solving the SDE obtained subject to the initial condition $X(0) = 0$.

Exercise 2.24. Solve the following linear SDEs:

(a) $\mathrm{d}X(t) = W(t)X(t)\,\mathrm{d}t + W(t)X(t)\,\mathrm{d}W(t)\,, \quad X(0) = 1;$

(b) $\mathrm{d}X(t) = \alpha(\theta - X(t))\,\mathrm{d}t + \sigma X(t)\,\mathrm{d}W(t)\,, \quad X(0) = x \in \mathbb{R};$

(c) $\mathrm{d}X(t) = a(t)X(t)\,\mathrm{d}t + \sigma X(t)\,\mathrm{d}W(t)\,, \quad X(0) = x \in \mathbb{R}.$

(d) $\mathrm{d}X(t) = X(t)\,\mathrm{d}t + X(t)\,\mathrm{d}W(t),\ X(0) = 1.$

(e) $\mathrm{d}X(t) = (2 - X(t))\,\mathrm{d}t + W(t)\,\mathrm{d}W(t),\ X(0) = 1.$

(f) $\mathrm{d}X(t) = W(t)\,\mathrm{d}t + W(t)\,\mathrm{d}W(t),\ X(0) = 1.$

(g) $\mathrm{d}X(t) = (e^{bt} - bX(t))\,\mathrm{d}t + t\,\mathrm{d}W(t),\ X(0) = 1.$

Assume α, θ, σ are positive constants in parts (b)-(c).

Exercise 2.25. Let $g(y)$ be a given function of y, and suppose that $y = f(x)$ is a solution of the ODE $\mathrm{d}y = g(y)\,\mathrm{d}x$, that is, $f'(x) = g(f(x))$. Show that $X(t) = f(W(t))$ is a solution of the SDE

$$\mathrm{d}X(t) = \frac{1}{2}g(X(t))g'(X(t))\,\mathrm{d}t + g(X(t))\,\mathrm{d}W(t).$$

Exercise 2.26. Using Exercise 2.25, solve the following nonlinear SDEs, subject to $X(0) = x_0 \in \mathbb{R}$. See the hint in the solution to this exercise.

(a) $\mathrm{d}X(t) = \frac{\sigma^2}{4}\,\mathrm{d}t + \sigma\sqrt{X(t)}\,\mathrm{d}W(t);$

(b) $\mathrm{d}X(t) = X^3(t)\,\mathrm{d}t + X^2(t)\,\mathrm{d}W(t);$

(c) $\mathrm{d}X(t) = \frac{1}{2}\mathrm{e}^{2X(t)}\,\mathrm{d}t + \mathrm{e}^{X(t)}\,\mathrm{d}W(t).$

In each case, determine the time interval for which the solution is bounded. For parts (b) and (c), the solution exists up to an "explosion time" when the solution becomes unbounded. Provide a precise description of the explosion time in terms of a first hitting time for standard Brownian motion.

Exercise 2.27. Show that for any $u \in \mathbb{R}$, the function $f(t,x) = \exp(ux - u^2t/2)$ solves the backward PDE for Brownian motion. Take the first, second, and third derivatives of $\exp(ux - u^2t/2)$ w.r.t. u, and set $u = 0$, to show that functions x, $x^2 - t$, $x^3 - 3tx$ also solve the backward equation for Brownian motion. Deduce that $W^2(t) - t$ and $W^3(t) - 3tW(t)$ are martingales.

Exercise 2.28. (a) Consider a time-homogeneous diffusion $\{X(t)\}_{0\leqslant t\leqslant T} \in \mathbb{R}$ having generator $\mathcal{G}^X f(x) := \frac{1}{2}\sigma^2(x)f''(x) + \mu(x)f'(x)$ acting on a twice continuously differentiable function $f(x)$, for all $x \in \mathbb{R}$. Derive a general expression for f such that $Y(t) := f(X(t)), 0 \leqslant t \leqslant T$, is a martingale w.r.t. any filtration for BM. State any square-integrability condition that must hold.
(b) Consider drifted BM, $X(t) = x_0 + \mu t + \sigma W(t)$, and let $Y(t) := f(X(t))$, $t \geqslant 0$. Derive a general expression for a twice continuously differentiable function $f(x)$, for all $x \in \mathbb{R}$, such that $\{Y(t)\}_{t\geqslant 0}$ is a martingale w.r.t. any filtration for BM.
(c) Consider a diffusion X having generator as in part (a) and a given state space in $\mathbb{R}$. Assume f is a twice continuously differentiable real-valued function with a unique inverse f^{-1}. By applying an Itô formula, show that $Y(t) := f(X(t)), 0 \leqslant t \leqslant T$ is a time-homogeneous diffusion with generator $\mathcal{G}^Y h(y) := \frac{1}{2}\sigma_Y^2(y)h''(y) + \mu_Y(y)h'(y)$, acting on any twice continuously differentiable function $h(y)$, for all y in the state space of the process Y, where $\mu_Y(y) = (\mathcal{G}^X f)(f^{-1}(y))$ and $\sigma_Y(y) = \sigma(f^{-1}(y)) \cdot f'(f^{-1}(y))$.

Exercise 2.29. Derive a system of diffusion-type SDEs for the coupled processes $X(t) = \cos(W(t))$ and $Y(t) = \sin(W(t))$.

Exercise 2.30. Consider the process defined by $X(t) = \sinh(C + t + W(t))$, $t \geqslant 0$, where $C = \sinh^{-1} x_0$ with initial condition $X(0) = x_0$. This process is a diffusion on $\mathbb{R}$ and it satisfies an SDE of the form

$$\mathrm{d}X(t) = \mu(X(t))dt + \sigma(X(t))\,\mathrm{d}W(t)\,.$$

(i) Determine the coefficient functions $\mu(x)$ and $\sigma(x)$.
(ii) Provide the backward Kolmogorov PDE and the terminal condition for the transition PDF $p(t,T;x,y)$.
(iii) Derive analytical expressions for the transition CDF and PDF of the process X.

Exercise 2.31. Give the probabilistic representation of the solution $f(t,x)$ of the PDE

$$\frac{\partial f}{\partial t} + \frac{x^2}{2}\frac{\partial^2 f}{\partial x^2} = 0,\ 0 \leqslant t \leqslant T, \quad f(T,x) = x^2\,.$$

Solve this PDE using the solution of the respective SDE.

Exercise 2.32. Consider the boundary value problem for the heat equation:

$$\frac{\partial V}{\partial t} + \frac{1}{2}\frac{\partial^2 V}{\partial x^2} = 0, \quad V(1,x) = f(x)$$

where f, the boundary value for time $t = 1$, is given, and where we are looking for a solution $V = V(t,x)$ defined for $0 \leqslant t \leqslant 1$ and $x \in \mathbb{R}$. Show that the solution is

$$V(t,x) = \int_{-\infty}^{\infty} f(y) e^{-\frac{(x-y)^2}{2(1-t)}} \frac{\mathrm{d}y}{\sqrt{2\pi(1-t)}}\,.$$

Can you think of a function f for which the solution formula would not make sense?

Exercise 2.33. Consider the boundary value problem for the heat equation with a drift term:

$$\frac{\partial V}{\partial t} + \frac{1}{2}\frac{\partial^2 V}{\partial x^2} + a\frac{\partial V}{\partial x} = 0, \quad V(1,x) = f(x)$$

where f, the boundary value for time $t = 1$, is given, and a is a real constant. Derive an explicit (integral) formula for the solution $V = V(t,x)$.

Exercise 2.34. Let $f(t,x)$ satisfy the PDE

$$\frac{\partial f}{\partial t} + \frac{1}{2}\sigma^2 x^2 \frac{\partial^2 f}{\partial x^2} + \mu x \frac{\partial f}{\partial x} = 0, \quad 0 \leqslant t \leqslant T,\ x \in \mathbb{R}_+,$$

for fixed $T > 0$, with real constants $\sigma > 0$, μ. Solve for $f(t,x)$ subject to the terminal condition $f(T,x) = \mathbb{I}_{\{K_1 < x < K_2\}}$, where $K_2 > K_1 > 0$ are constants.

Exercise 2.35. Let $f(t,x)$ satisfy the PDE

$$\frac{\partial f}{\partial t} + \frac{1}{2}\sigma^2 t \frac{\partial^2 f}{\partial x^2} + \mu \frac{\partial f}{\partial x} = 0, \quad 0 < t < T,\ x \in \mathbb{R},$$

for fixed $T > 0$, with real constants $\sigma > 0$, μ. By applying the Feynman-Kac Theorem, solve for $f(t,x)$ subject to the terminal condition:
(a) $f(T,x) = x\mathbb{I}_{\{x>y\}}$, for any real constant y;
(b) $f(T,x) = \mathbb{I}_{\{x \leqslant y\}}$, for any real constant y.

Exercise 2.36. To compact notation, we suppress all other variables except x' and denote $f(x') \equiv p(t,t';x,x')$ and $g(x') \equiv p(t',T;x',y)$. Using the definition of $\mathcal{G}_{t',x'}$, the left-hand integral in (2.104) becomes

$$\int_{\mathcal{I}} f(x')\, \mathcal{G}_{t',x'} g(x')\, \mathrm{d}x' = \frac{1}{2}\int_{\mathcal{I}} f(x')\sigma^2(t',x')\frac{\partial^2 g}{\partial x'^2}\, \mathrm{d}x' + \int_{\mathcal{I}} f(x')\mu(t',x')\frac{\partial g}{\partial x'}\, \mathrm{d}x' .$$

Now apply integration by parts to both integrals (in the first integral note that $\frac{\partial^2 g}{\partial x'^2} = \frac{\partial}{\partial x'}\left(\frac{\partial g}{\partial x'}\right)$). State appropriate assumptions that allow you to set the boundary terms to zero. Then, apply integration by parts again on the remaining integral containing $\sigma^2(t',x')$. Again, state appropriate assumptions that allow you to set the boundary terms to zero. In the end, obtain

$$\int_{\mathcal{I}} f(x')\, \mathcal{G}_{t',x'} g(x')\, \mathrm{d}x' = \int_{\mathcal{I}} g(x')\, \tilde{\mathcal{G}}_{t',x'} f(x')\, \mathrm{d}x'$$

where $\tilde{\mathcal{G}}_{t',x'} f = \frac{1}{2}\frac{\partial^2}{\partial x'^2}\left(\sigma^2(t',x')f\right) - \frac{\partial}{\partial x'}\left(\mu(t',x')f\right)$.

Exercise 2.37. Assume that a stock price process $\{S(t)\}_{t\geqslant 0}$ satisfies the SDE

$$\mathrm{d}S(t) = rS(t)\,\mathrm{d}t + \sigma S(t)\,\mathrm{d}\widetilde{W}(t)\,,$$

with constants $r, \sigma > 0$, and where $\{\widetilde{W}(t)\}_{t\geqslant 0}$ is a standard $\widetilde{\mathbb{P}}$-BM. By using Girsanov's Theorem, find the explicit expression for the Radon–Nikodym derivative process

$$\varrho_t := \left(\frac{\mathrm{d}\widehat{\mathbb{P}}}{\mathrm{d}\widetilde{\mathbb{P}}}\right)_t$$

such that the process defined by $\widehat{S}(t) := \frac{e^{rt}}{S(t)}$, $t \geqslant 0$, is a $\widehat{\mathbb{P}}$-martingale. Give the SDE satisfied by the stock price $S(t)$ w.r.t. the $\widehat{\mathbb{P}}$-BM.

Exercise 2.38. Consider a stock price process $\{S(t)\}_{t \geqslant 0}$ that obeys the SDE

$$\mathrm{d}S(t) = \mu S(t)dt + \sigma(S(t))^{1+\beta}\,\mathrm{d}W(t), \quad S(0) = S_0 > 0,$$

with constant parameters $\sigma \neq 0$, β, μ.

(a) Assume the process $\int_0^t (S(u))^{1+\beta}\,\mathrm{d}W(u)$ is a $\mathbb{P}$-martingale w.r.t. the filtration generated by the standard $\mathbb{P}$-BM $\{W(t)\}_{t \geqslant 0}$. Determine an exact expression for the mean of the process $\mathrm{E}[S(t)]$. [Hint: you may write $S(t)$ in terms of an exponential martingale by considering the process defined by $X(t) := \ln S(t)$.]

(b) Fix $T > 0$ and assume the existence of a Radon–Nikodym derivative process $\varrho_t := \left(\frac{\mathrm{d}\widetilde{\mathbb{P}}}{\mathrm{d}\mathbb{P}}\right)_t$, $0 \leqslant t \leqslant T$, such that $\{e^{-rt}S(t)\}_{0 \leqslant t \leqslant T}$ is a $\widetilde{\mathbb{P}}$-martingale. Give the form of ϱ_t as an exponential $\mathbb{P}$-martingale.

Exercise 2.39. Consider a one-dimensional general diffusion process $\{X(t)\}_{t \geqslant 0}$ having a transition PDF $p(s,t;x,y)$, $s < t$, w.r.t. a given probability measure $\mathbb{P}$, for all x, y in the state space of the process. Assume a change of measure $\mathbb{P} \to \widehat{\mathbb{P}}$ is defined by a Radon–Nikodym derivative process

$$\varrho_t := \left(\frac{\mathrm{d}\widehat{\mathbb{P}}}{\mathrm{d}\mathbb{P}}\right)_t = h(t, X(t))$$

for all $t \geqslant 0$ and where $h(t,x)$ is some given Borel function of t, x. Let $\widehat{p}(s,t;x,y)$ be the transition PDF w.r.t. the measure $\widehat{\mathbb{P}}$. Show that the two transition PDFs are related by

$$\widehat{p}(s,t;x,y) = \frac{h(t,y)}{h(s,x)}\,p(s,t;x,y)\,.$$

[Hint: consider the definition of the transition CDF (w.r.t. the measure $\widehat{\mathbb{P}}$):

$$\widehat{P}(s,t;x,y) = \widehat{\mathbb{P}}(X(t) \leqslant y \mid X(s) = x) = \widehat{\mathrm{E}}[\mathbb{I}_{\{X(t) \leqslant y\}} \mid X(s) = x]$$

and make use of the Markov property and the change of measure for computing expectations.]

Part II

Continuous-Time Modelling

3

Risk-Neutral Pricing in the (B, S) Economy: One Underlying Stock

At this point, we have the necessary tools in stochastic analysis for developing the theory of derivative pricing and hedging in continuous time in an economy where risky assets are modeled as Itô processes. In this chapter, we consider an economy with two securities: a single tradable risky asset, namely, a stock, and a money market (bank) account or a bond. We refer to this as a (B, S) economy with only two tradable assets: B stands for the bank account or bond and S stands for the stock. More specifically, this chapter is devoted to presenting the theoretical framework solely within the classical case of an economy where the interest rate is fixed and the stock price is modeled as a standard geometric Brownian motion (GBM) with constant growth rate and constant volatility. The case in which the stock pays a dividend yield is also included later. The only source of randomness driving the stock price is a single Brownian motion. This is also referred to as the standard Black–Scholes framework. This can be thought of as the continuous-time analogue of the standard binomial tree model which was formally discussed in great detail in Volume I. The analogues of the up and down market moves in discrete time are now random movements of the underlying (driving) Brownian motion (BM). The same important underlying concepts of self-financing, replication, hedging, arbitrage, and no-arbitrage pricing of derivative contracts in discrete time now carry over into the continuous-time setting.

All appropriately discounted tradable assets are martingales under an equivalent martingale (risk-neutral) measure. In particular, by using a self-financing replication strategy, we will arrive at the risk-neutral pricing formula that expresses the current price of any attainable European-style derivative, including contracts with path-dependent payoffs, as a conditional expectation (under the risk-neutral or equivalent martingale measure) of the discounted payoff. The class of derivative securities that are attainable is quite general. This chapter also includes a discussion of this and its relation to hedging and pricing. When pricing a non-path-dependent European-style derivative, the inherent Markov property reduces the risk-neutral pricing formula to a conditional expectation where the (discounted) Feynman-Kac formula allows us to arrive at the infamous Black–Scholes–Merton PDE for the current price of the derivative (option) contract. Recall that in Chapter 2, we made the important connection between the SDE of an Itô process, the (discounted) conditional expectation of a (payoff) function of the terminal value of the process, and the corresponding terminal (or initial) value PDE problem. This is the essence of the Feynman–Kac representation. Our discussion on the theoretical framework of the (B, S) economy culminates in the risk-neutral pricing formulation. We then apply this formulation to derivative pricing problems, where we explicitly derive pricing and hedging formulae for various European contracts, such as standard calls and puts, as well as more complex options such as compound options. We finally also apply the risk-neutral pricing formulation to path-dependent derivatives such as barrier options and lookback options.

DOI: 10.1201/9780429468889-3

3.1 From the CRR Model to the BSM Model

Before we start exploring continuous-time asset price models based on a GBM process, let us review the standard binomial tree model studied in great detail in Volume I. In this section, we also discuss the special case known as the Cox–Ross–Rubinstein (CRR) model. Our main interest is the derivation of the Black–Scholes–Merton (BSM) asset price equation as a limiting case of the CRR model. We show how to derive no-arbitrage pricing formulae for basic European-type options under the BSM model and the Black–Scholes PDE, when the length of one period goes to zero, while the number of periods goes to infinity.

Review of the Standard Binomial Tree Model

Consider a standard discrete-time binomial tree model with N periods. Each market scenario can be visualized as a path in an N-period recombining binomial tree, when the stock price goes up or down during each single period $(t-1, t]$ with $t \in \{1, 2, \ldots, N\}$. There are 2^N possible market scenarios. The *market move* ω_t equals D or U, if the stock price moves down or up at time t, respectively. The sample space is a collection of N-tuples:

$$\Omega \equiv \Omega_N = \left\{\omega_1\omega_2\cdots\omega_N \ : \ \omega_t \in \{\mathsf{D}, \mathsf{U}\}\right\}.$$

In the filtration, $\mathbb{F} = \{\mathcal{F}_t\}_{0 \leqslant t \leqslant N}$ with $\mathcal{F}_t \subset \mathcal{F}_{t+1}$ for all t, the sigma-algebra $\mathcal{F}_t$ is generated by the first t market moves $\omega_1, \omega_2, \ldots, \omega_t$. The event space $\mathcal{F} \equiv \mathcal{F}_N$ is the power set 2^Ω.

To define the probability function $\mathbb{P}$ on $\mathcal{F}$, we introduce two counting functions:

$$\begin{aligned}\mathsf{U}_t(\omega) &= \text{the number of } \mathsf{U}\text{'s in the first } t \text{ symbols of } \omega,\\ \mathsf{D}_t(\omega) &= \text{the number of } \mathsf{D}\text{'s in the first } t \text{ symbols of } \omega,\end{aligned}$$

where $t = 1, \ldots, N$, and $\mathsf{U}_0 \equiv \mathsf{D}_0 \equiv 0$. The $\mathcal{F}_t$-measurable random variables U_t and D_t give the number of times the stock price moves up and down, respectively, during the first t periods. Clearly, $\mathsf{U}_t(\omega) + \mathsf{D}_t(\omega) = t$, and thus we can express D_t in terms of U_t and vice versa: $\mathsf{D}_t(\omega) = t - \mathsf{U}_t(\omega)$. Let $p \in (0, 1)$ be the probability that the stock price moves up during a single period. That is, $\mathbb{P}(\omega_t = \mathsf{U}) = p$ and $\mathbb{P}(\omega_t = \mathsf{D}) = 1 - p$ hold for any $t \in \{1, 2, \ldots, N\}$. The probability $\mathbb{P}(\omega)$ of a single outcome $\omega \in \Omega$ is given by

$$\mathbb{P}(\omega) = p^{\mathsf{U}_N(\omega)}(1-p)^{\mathsf{D}_N(\omega)} = p^{\mathsf{U}_N(\omega)}(1-p)^{N-\mathsf{U}_N(\omega)}.$$

The function $\mathbb{P}$, defined on the event space $\mathcal{F}$, allows for computing the probability of an event as a sum of probabilities of all singeltons included in the event:

$$\mathbb{P}(E) = \sum_{\omega \in E} \mathbb{P}(\omega) \text{ for } E \in \mathcal{F}.$$

Clearly, $\mathbb{P}$ satisfies the Kolmogorov axioms. The quadruple $(\Omega, \mathcal{F}, \mathbb{P}, \mathbb{F})$ is a filtered probability space that forms a foundation for the binomial tree model.

Since Ω is a finite sample space, the (unconditional) mathematical expectation of random variable X is calculated as a sum of $X(\omega)\mathbb{P}(\omega)$ over all possible market scenarios $\omega \in \Omega$:

$$\mathrm{E}[X] = \sum_{\omega \in \Omega} X(\omega)\mathbb{P}(\omega).$$

Similarly, the expectation of X conditional on event $E \neq \emptyset$ is calculated as

$$\mathrm{E}[X \mid E] = \sum_{\omega \in \Omega} X(\omega)\mathbb{P}(\omega \mid E) = \frac{\mathrm{E}[X\,\mathbb{I}_E]}{\mathbb{P}(E)}.$$

The expectation of X conditional on the sigma-algebra $\mathcal{F}_t$, denoted $\mathrm{E}_t[X] \equiv \mathrm{E}[X \mid \mathcal{F}_t]$, is a function of $\omega_1, \omega_2, \ldots, \omega_t$ such that

$$\mathrm{E}_t[X](\omega'_1, \omega'_2, \ldots, \omega'_t) = \mathrm{E}[X \mid \mathcal{A}_{\omega'_1,\omega'_2,\ldots,\omega'_t}] = \frac{\mathrm{E}[X\,\mathbb{I}_{\mathcal{A}_{\omega'_1,\omega'_2,\ldots,\omega'_t}}]}{\mathbb{P}(\mathcal{A}_{\omega'_1,\omega'_2,\ldots,\omega'_t})}.$$

holds for all $\omega'_1, \omega'_2, \ldots, \omega'_t \in \{\mathsf{U}, \mathsf{D}\}$. Here, $\mathcal{A}_{\omega'_1,\omega'_2,\ldots,\omega'_t}$ denotes the atom of $\mathcal{F}_t$ generated by the first t market moves $\omega'_1, \omega'_2, \ldots, \omega'_t$. Hence, $\mathcal{F}_t$ is a collection of 2^{2^t} events generated by the 2^t atoms.

Let us recall some basic financial concepts. A *payoff* is a random variable defined on the probability space. That is, it is an $\mathcal{F}$-measurable function $X \colon \Omega \to \mathbb{R}$. Payoffs form a vector space, denoted $\mathrm{L}(\Omega)$, and thus we can add payoffs up and multiply them by a scalar to construct new ones. A *claim* is any nonnegative payoff. A *risk-free payoff* is constant on Ω, whereas for a *risky payoff* X there are at least two scenarios, ω' and ω'' so that $X(\omega') \neq X(\omega'')$.

A *security* is described by a nonnegative stochastic price process $\{A_t\}_{0 \leqslant t \leqslant N}$ adapted to the filtration $\mathbb{F}$. That is, the time-t price A_t is an $\mathcal{F}_t$-measurable nonnegative random variable for all $t \in \{0, 1, \ldots, N\}$. We call $A_0 \geqslant 0$ the initial price and $A_N \geqslant 0$ the terminal price (at time N). $\{\mathsf{U}_t\}_{0 \leqslant t \leqslant N}$ and $\{\mathsf{D}_t\}_{0 \leqslant t \leqslant N}$ are examples of risky price processes. For any fixed $\omega \in \Omega$, the paths $\mathsf{U}_t(\omega)$ and $\mathsf{D}_t(\omega)$ are nondecreasing functions of time t. Recall that both U_t and D_t are binomial random variables:

$$\mathsf{U}_t \sim \mathit{Bin}(t, p) \text{ and } \mathsf{D}_t \sim \mathit{Bin}(t, 1-p).$$

The standard binomial tree model includes two base securities: risky stock S and risk-free bank account B with respective time-t prices

$$B_t := (1+r)^t \text{ and } S_t(\omega) := S_0\, u^{\mathsf{U}_t(\omega)}\, d^{\mathsf{D}_t(\omega)} = S_0\, u^{\mathsf{U}_t(\omega)}\, d^{t-\mathsf{U}_t(\omega)} \text{ for } t = 0, 1, \ldots, N. \quad (3.1)$$

Here, $r \geqslant 0$ is the risk-free rate of return, d and u are two possible single-period returns on the stock such that $0 < d < u$. For simplicity, we assume here that the stock pays zero dividends. Since $S_t(\omega)$ is defined by the number of upward stock moves $\mathsf{U}_t(\omega)$, the dynamics of stock prices is described by a *recombining* binomial tree (see Figure 3.1). The fact that S_t is a function of the binomial random variable $\mathsf{U}_t \sim \mathit{Bin}(t, p)$ allows us to find the asset price probability distribution for any $t = 1, 2, \ldots, N$:

$$S_t = S_0\, u^k\, d^{t-k} \text{ with probability } \binom{t}{k} p^k\,(1-p)^{t-k}, \quad k = 0, 1, 2, \ldots, t.$$

For any n and t with $0 \leqslant t < n \leqslant N$, the stock return $\dfrac{S_n}{S_t}$ is a function of $\omega_{t+1}, \ldots, \omega_n$, and hence it is independent of $\mathcal{F}_t$. Moreover, we can represent $\dfrac{S_n}{S_t}$ as a product of $n - t$ i.i.d. single-period returns:

$$\frac{S_n}{S_t} = \frac{S_{t+1}}{S_t}\frac{S_{t+2}}{S_{t+1}} \cdots \frac{S_n}{S_{n-1}} = \prod_{k=t}^{n-1} \frac{S_{k+1}}{S_k} = \prod_{k=t}^{n-1} u^{\mathcal{B}_k} d^{1-\mathcal{B}_k},$$

where $\mathcal{B}_k = \mathsf{U}_{k+1} - \mathsf{U}_k \sim \mathrm{Be}(p)$, $k \geqslant 1$, are i.i.d. Bernoulli random variables. Hence, $\mathrm{E}[S_n/S_t] = \prod_{k=t}^{n-1} \mathrm{E}[u^{\mathcal{B}_k} d^{1-\mathcal{B}_k}] = \prod_{k=t}^{n-1}(pu + (1-p)d) = (pu + (1-p)d)^{n-t}$. Since S_t is $\mathcal{F}_t$-measurable, and S_n/S_t is independent of $\mathcal{F}_t$, the expectation of S_n conditional on $\mathcal{F}_t$ is given by

$$\mathrm{E}_t[S_n] = \mathrm{E}_t\left[S_t \frac{S_n}{S_t}\right] = S_t \mathrm{E}_t\left[\frac{S_n}{S_t}\right] = S_t \mathrm{E}\left[\frac{S_n}{S_t}\right] = S_t(pu + (1-p)d)^{n-t}.$$

Note that this also follows by applying Proposition A.6.

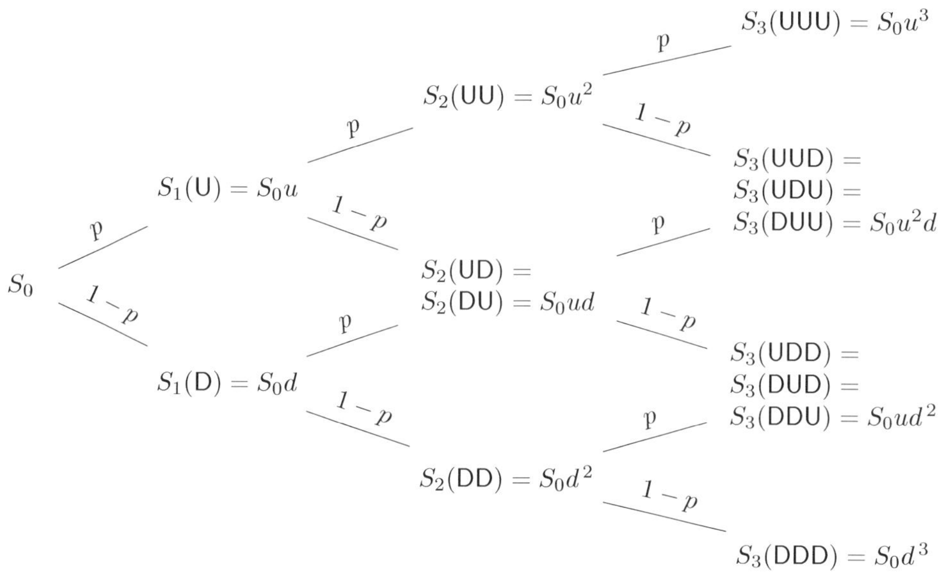

FIGURE 3.1: A schematic representation of the binomial tree model with three periods.

3.1.1 Portfolio Strategies in the Binomial Tree Model

Now, let us see how base assets can be used to create static portfolios and dynamic trading strategies in the binomial tree model. A *static portfolio* is a vector $(\beta, \Delta) \in \mathbb{R}^2$. Its time-$t$ value is $\Pi_t^{(\beta,\Delta)} := \beta\, B_t + \Delta\, S_t$. A *portfolio strategy* is a stochastic process, $\{(\beta_t, \Delta_t)\}_{t \geqslant 0}$ adapted to the filtration $\mathbb{F}$. That is, both β_t and Δ_t are $\mathcal{F}_t$-measurable.

A *self-financing strategy* does not allow for injecting or withdrawing funds. That is, for all $t = 1, 2, \ldots, N$, the acquisition value, $\Pi_t^{(\beta_t,\Delta_t)} \equiv \beta_t\, B_t + \Delta_t\, S_t$, is the same as the liquidation value, $\Pi_t^{(\beta_{t-1},\Delta_{t-1})} \equiv \beta_{t-1}\, B_t + \Delta_{t-1}\, S_t$:

$$\Pi_t^{(\beta_t,\Delta_t)} = \Pi_t^{(\beta_{t-1},\Delta_{t-1})} \equiv \Pi_t.$$

Equivalently, for a self-financing strategy, we have

$$\delta\Pi_t = \beta_t\, \delta B_t + \Delta_t\, \delta S_t, \quad t = 0, 1, \ldots, N-1, \tag{3.2}$$

where $\delta\Pi_t := \Pi_{t+1} - \Pi_t$, $\delta S_t := S_{t+1} - S_t$, and $\delta B_t := B_{t+1} - B_t$. A portfolio strategy with nonnegative values is called *admissible*.

In the binomial tree model, any self-financing strategy is fully defined by the Δ-process, $\{\Delta_t\}_{t \geqslant 0}$, and the initial wealth Π_0. The position in the bank account, β_t, is given by

$$\beta_t = \frac{\Pi_t - \Delta_t S_t}{B_t}.$$

The time-t portfolio value can be calculated iteratively using the wealth equation

$$\Pi_{t+1} = \Delta_t S_{t+1} + (\Pi_t - \Delta_t S_t)(1 + r), \quad t = 0, 1, \ldots, N-1.$$

The *cumulative gain* $\Pi_N - \Pi_0$ of a self-financing strategy $\{(\beta_t, \Delta_t)\}_{0 \leqslant t \leqslant N-1}$ is equal to a sum of weighted increments δS_t and δB_t:

$$\Pi_N - \Pi_0 = \sum_{t=0}^{N-1} \delta\Pi_t = \sum_{t=0}^{N-1} \beta_t \delta B_t + \sum_{t=0}^{N-1} \Delta_t \delta S_t. \tag{3.3}$$

In the next section, we will get to know continuous-time analogues of equations (3.2) and (3.3). The difference equation in (3.2) changes to the stochastic differential $\mathrm{d}\Pi_t = \beta_t \, \mathrm{d}B_t + \Delta_t \, \mathrm{d}S_t$, whereas the cumulative gain is expressed as a sum of two stochastic integrals.

A key concept of mathematical finance is *replication*. Suppose that for a given payoff $X \in \mathrm{L}(\Omega)$, we can find a self-financing portfolio strategy in base assets with the portfolio value process $\{\Pi_t\}_{t \geqslant 0}$ such that

$$\Pi_N(\omega) = X(\omega) \text{ holds for all scenarios } \omega \in \Omega.$$

Such a portfolio strategy is said to *replicate* the payoff X. It is also called a *hedge* for X. The payoff X is said to be *attainable*. If every payoff is attainable, the market model is said to be *complete*. Otherwise, if there exists a non-replicable payoff, the market model is *incomplete*.

Another crucial concept of mathematical finance is *arbitrage*, which is an admissible self-financing portfolio strategy $\varphi \equiv \{\varphi_t\}_{t \geqslant 0}$ such that its initial value Π_0^φ is zero, whereas its terminal value Π_N^φ is nonnegative for all market scenarios and strictly positive for some ω's:

$$\Pi_0^\varphi = 0, \quad \mathbb{P}(\Pi_N^\varphi \geqslant 0) = 1, \quad \text{and} \quad \mathbb{P}(\Pi_N^\varphi > 0) > 0.$$

We say that a market model is *arbitrage free*, if there are no arbitrage strategies. Our primary objective is risk-neutral pricing of financial contracts in arbitrage-free market models.

Consider some claim with terminal payoff V_N. Suppose that V_N is attainable and that the self-financing portfolio strategy φ replicates V_N. The fair initial price V_0 of the claim should be equal to Π_0^φ; otherwise, there is an arbitrage opportunity if the actual initial price of the claim differs from Π_0^φ. This method of no-arbitrage evaluation is called *pricing by replication.* Another approach described below is based on the use of an *equivalent martingale measure (EMM).*

A numéraire asset is any trading asset g with a strictly positive price process. For any base asset A (it is B or S for the standard binomial tree model), we define the discounted asset price process as $\{\overline{A}_t := A_t/g_t\}_{t \geqslant 0}$. An EMM relative to a non-dividend paying numéraire asset g, denoted $\widetilde{\mathbb{P}}^{(g)}$, is a probability measure defined such that (1) it is equivalent to the real-world probability measure $\mathbb{P}$, and (2) every discounted base-asset price process is a $\widetilde{\mathbb{P}}^{(g)}$-martingale. It can be proved that the value of any self-financing portfolio strategy in the binomial model, $\{\Pi_t\}_{t \geqslant 0}$, discounted by g is a $\widetilde{\mathbb{P}}^{(g)}$-martingale. That is, $\widetilde{\mathrm{E}}_t[\overline{\Pi}_n] = \overline{\Pi}_t$ for any t and n s.t. $0 \leqslant t \leqslant n \leqslant N$, where $\overline{\Pi}_t := \Pi_t/g_t$ for all $t \geqslant 0$. Here and below, $\widetilde{\mathrm{E}}[\,]$ denotes the mathematical expectation under the EMM $\widetilde{\mathbb{P}}^{(g)}$.

Typically, we use the risk-free bank account B as a numéraire asset. The first Fundamental Theorem of Asset Pricing (FTAP) states that a market model is arbitrage free iff there exists an EMM. If the EMM is unique (for the given numéraire), then, according to the second FTAP, the market model is also complete.

Consider a self-financing portfolio strategy replicating some claim V_N. Assuming that the EMM $\widetilde{\mathbb{P}} \equiv \widetilde{\mathbb{P}}^{(g)}$ (relative to some numéraire g) exists, the discounted portfolio value process $\{\overline{\Pi}_t\}_{t \geqslant 0}$ is a $\widetilde{\mathbb{P}}$-martingale. Therefore, the no-arbitrage initial value V_0, which is equal to Π_0, can be calculated as follows:

$$\frac{\Pi_0}{g_0} = \widetilde{\mathrm{E}}\left[\frac{\Pi_N}{g_N}\right] \implies \frac{V_0}{g_0} = \widetilde{\mathrm{E}}\left[\frac{V_N}{g_N}\right] \implies V_0 = \widetilde{\mathrm{E}}\left[\frac{g_0}{g_N} V_N\right]. \tag{3.4}$$

With the help of conditional expectations, we can generalize the no-arbitrage pricing formula (3.4) for any time $t = 0, 1, \ldots, N$. If a self-financing portfolio strategy with value process $\{\Pi_t\}_{t\geqslant 0}$ replicates claim V_N, then the time-t no-arbitrage value is given by $V_t = \Pi_t$. Therefore, thanks to the martingale property,

$$\frac{\Pi_t}{g_t} = \widetilde{\mathrm{E}}\left[\frac{\Pi_N}{g_N} \;\middle|\; \mathcal{F}_t\right] \implies V_t = \widetilde{\mathrm{E}}\left[\frac{g_t}{g_N} V_N \;\middle|\; \mathcal{F}_t\right]. \tag{3.5}$$

Particularly, if $g_t = B_t \equiv B_0(1+r)^t$, we have

$$V_t = (1+r)^{-(N-t)}\widetilde{\mathrm{E}}\left[V_N \mid \mathcal{F}_t\right]. \tag{3.6}$$

Let us summarize the main results about pricing and replicating derivatives under the standard binomial tree model from Volume I (where the stock pays zero dividend).

1. There exists a unique risk-neutral probability measure $\widetilde{\mathbb{P}}$ relative to the bank account B_t used as a numéraire asset if $d < 1 + r < u$. Indeed, the discounted stock price process, $\overline{S}_t = S_t/B_t$, $t \geqslant 0$, is a $\widetilde{\mathbb{P}}$-martingale, if

$$\begin{aligned}\widetilde{\mathrm{E}}_t[\overline{S}_{t+1}] = \overline{S}_t &\iff \widetilde{\mathrm{E}}_t[S_{t+1}] = (1+r)S_t \\ &\iff \tilde{p}\cdot u + (1-\tilde{p})\cdot d = 1 + r \iff \tilde{p} = \frac{1+r-d}{u-d}.\end{aligned}$$

   The risk-neutral probability $\tilde{p}$ is within $(0,1)$ if $d < 1 + r < u$ holds. The binomial tree model is arbitrage free in this case.

2. The binomial tree model is also complete if $d < 1 + r < u$. The self-financing portfolio strategy $\{\Delta_t\}_{0\leqslant t\leqslant N-1}$ replicating payoff V_N is constructed as follows:

$$\Delta_t(\omega_1\omega_2\ldots\omega_t) = \frac{V_{t+1}(\omega_1\omega_2\ldots\omega_t\mathsf{U}) - V_{t+1}(\omega_1\omega_2\ldots\omega_t\mathsf{D})}{S_{t+1}(\omega_1\omega_2\ldots\omega_t\mathsf{U}) - S_{t+1}(\omega_1\omega_2\ldots\omega_t\mathsf{D})}, \quad t = 0, 1, \ldots, N-1,$$

   where the derivative price process, $\{V_t\}_{0\leqslant t\leqslant N}$, can be obtained using the following backward recursion:

$$\begin{aligned}V_t(\omega_1\omega_2\ldots\omega_t) &= \frac{1}{1+r}\left[\tilde{p}V_{t+1}(\omega_1\omega_2\ldots\omega_t\mathsf{U}) + (1-\tilde{p})V_{t+1}(\omega_1\omega_2\ldots\omega_t\mathsf{D})\right] \\ &\equiv \frac{1}{1+r}\widetilde{\mathrm{E}}[V_{t+1} \mid \mathcal{F}_t](\omega_1\omega_2\ldots\omega_t), \quad t = 0, 1, \ldots, N-1,\end{aligned}$$

   with the risk-neutral probabilities $\tilde{p}$ and $1-\tilde{p}$. This strategy replicates the derivative price process at every time, i.e., $\Pi_t(\omega) = V_t(\omega)$ holds for all times $t = 0, 1, \ldots, N$ and all scenarios $\omega \in \Omega$.

3. Consider a European-style derivative with payoff $V_N = \Lambda(S_N)$, where $\Lambda(S)$ is a non-negative function of the stock price. The initial no-arbitrage price is given by

$$V_0 = (1+r)^{-N}\,\widetilde{\mathrm{E}}[\,\Lambda(S_N)\,]. \tag{3.7}$$

   Using the fact that $S_N \stackrel{d}{=} S_0\,u^{U_N}\,d^{N-U_N}$, where $U_N \sim Bin(N, \tilde{p})$ (under $\widetilde{\mathbb{P}}$), we have

$$\begin{aligned}V_0 \equiv V_0(S_0) &= (1+r)^{-N}\widetilde{\mathrm{E}}\left[\Lambda(S_0\,u^{U_N}\,d^{N-U_N})\right] \\ &= (1+r)^{-N}\sum_{k=0}^{N}\Lambda(S_0\,u^k\,d^{N-k})\widetilde{\mathbb{P}}(U_N = k) \\ &= (1+r)^{-N}\sum_{k=0}^{N}\Lambda(S_0\,u^k\,d^{N-k})\binom{N}{k}\tilde{p}^k(1-\tilde{p})^{N-k}.\end{aligned}$$

4. Consider the standard European call and put options with the respective payoff functions $C_N = (S_N - K)^+$ and $P_N = (K - S_N)^+$ at maturity N with a strike price $K > 0$. The binomial pricing formulae for the European call and put options are, respectively,

$$C_0(S_0) = S_0\left(1 - \mathcal{B}(m_N; N, \hat{p})\right) - \frac{K}{(1+r)^N}\left(1 - \mathcal{B}(m_N; N, \tilde{p})\right), \tag{3.8}$$

$$P_0(S_0) = \frac{K}{(1+r)^N}\mathcal{B}(m_N; N, \tilde{p}) - S_0\,\mathcal{B}(m_N; N, \hat{p}), \tag{3.9}$$

where $\tilde{p} = \dfrac{1+r-d}{u-d}$ and $\hat{p} = \dfrac{u}{1+r}\cdot\tilde{p}$ are probabilities, and $m_N \equiv m_N(S_0, K)$ is given by

$$m_N(S_0, K) = \max\left\{m \,:\, 0 \leqslant m \leqslant N;\ S_0\,u^m\,d^{N-m} \leqslant K\right\} = \left\lfloor \frac{\ln\left(\frac{K}{S_0}\right) - N\ln d}{\ln\left(\frac{u}{d}\right)} \right\rfloor.$$

Recall that $\tilde{p}$ and $\hat{p}$ are the probabilities of an upward move in a single period under the EMMs $\widetilde{\mathbb{P}}^{(B)}$ and $\widetilde{\mathbb{P}}^{(S)}$, respectively. Here, $\mathcal{B}(m; n, p)$ denotes the CDF of $\mathsf{X} \sim Bin(n, p)$; it is given by

$$\mathcal{B}(m; n, p) = \mathbb{P}(\mathsf{X} \leqslant m) = \sum_{k=0}^{m}\binom{n}{k}p^k\,(1-p)^{n-k}.$$

Note the option prices in (3.8) and (3.9) satisfy the *put-call parity* relation for zero stock dividends:

$$C_0(S_0) - P_0(S_0) = S_0 - \frac{K}{(1+r)^N}. \tag{3.10}$$

3.1.2 The Cox–Ross–Rubinstein Model and its Continuous-Time Limit

Our objective is to construct a continuous-time model for stock prices. Let $S(t)$ denote the time-t price of a risky stock. We fix the time horizon $T > 0$, the annual stock volatility σ, the annual expected rate of the stock log-return ν, and risk-free interest rate r compounded continuously. The parameters σ and ν are defined as follows:

$$\operatorname{Var}\left[\ln\left(\frac{S(T)}{S(0)}\right)\right] = \sigma^2 T, \quad \mathrm{E}\left[\ln\left(\frac{S(T)}{S(0)}\right)\right] = \nu T.$$

For any given $n \geqslant 1$, consider an N-period binomial tree model with $N = \lfloor nT \rfloor \geqslant 1$, where trading is allowed at times $t_k = k\delta_n$ with $k = 0, 1, 2, \ldots, N$. Here, $\delta_n = \frac{1}{n}$ is the length of one period. Clearly, $N\delta_n \to T$, as $n \to \infty$. The initial price is $S_0 \equiv S(0)$. We use the following parameters:

$$r_n = \mathrm{e}^{\delta_n r} - 1, \quad p_n = \frac{1}{2} + \frac{1}{2}\frac{\nu}{\sigma}\sqrt{\delta_n}, \quad u_n = \mathrm{e}^{\sigma\sqrt{\delta_n}}, \quad d_n = \mathrm{e}^{-\sigma\sqrt{\delta_n}}.$$

The binomial tree model under such a parametrization is called the Cox–Ross–Rubinstein (CRR) model.

Consider a sequence of CRR models parametrized by $n = 1, 2, \ldots$. For each n, the model has two base assets, denoted $B^{(n)}$ and $S^{(n)}$, respectively. At time $t_k = k\delta_n$ with $k = 0, 1, 2, \ldots, N$, the bank account value is $B_k^{(n)} = (1 + r_n)^k = \mathrm{e}^{\delta_n r k}$, and the stock price is $S_k^{(n)} \in \{S_0\,u_n^\ell\,d_n^{k-\ell} \mid \ell = 0, 1, \ldots, k\}$. Since $u_n d_n = 1$, we can write

$$S_k^{(n)} = S_0\,u_n^{2\mathsf{U}_k^{(n)} - k}, \text{ where } \mathsf{U}_k^{(n)} \sim Bin(k, p_n).$$

The process $M_k^{(n)} := 2\mathsf{U}_k^{(n)} - k$ with $k = 0, 1, 2, \ldots, N$ is known as a random walk process. It is asymmetric if $p_n \neq 1/2$. So, we can also write $S_k^{(n)} = S_0\, u_n^{M_k^{(n)}}$.

Let us consider asymptotic properties of the model under the real-world probability measure $\mathbb{P}$. The aggregate log-return equals $\ln\left(\frac{S_N^{(n)}}{S_0}\right)$. It has the following expected value and variance:

$$\mathrm{E}\left[\ln\left(\frac{S_N^{(n)}}{S_0}\right)\right] = N\delta_n \nu \to \nu T, \text{ as } n \to \infty,$$

$$\mathrm{Var}\left[\ln\left(\frac{S_N^{(n)}}{S_0}\right)\right] = N\delta_n \sigma^2 + \mathcal{O}(1/n) \to \sigma^2 T, \text{ as } n \to \infty.$$

Thus, binomial prices converge to log-normal prices thanks to the Central Limit Theorem:

$$\ln\left(\frac{S_N^{(n)}}{S_0}\right) \stackrel{d}{\to} \nu T + \sigma\sqrt{T}Z \implies S_N^{(n)} \stackrel{d}{\to} S_0\, \mathrm{e}^{\nu T + \sigma\sqrt{T}Z}, \; Z \sim \mathit{Norm}(0, 1). \tag{3.11}$$

Under the risk-neutral probability measure $\widetilde{\mathbb{P}}$, we have the following asymptotic behaviour of the mean and variance of the aggregate log-return, as $n \to \infty$:

$$\widetilde{\mathrm{E}}\left[\ln\left(\frac{S_N^{(n)}}{S_0}\right)\right] \to (r - \frac{\sigma^2}{2})T,$$

$$\widetilde{\mathrm{Var}}\left[\ln\left(\frac{S_N^{(n)}}{S_0}\right)\right] \to \sigma^2 T.$$

Thus, the risk-neutral distribution of binomial prices converges to the log-normal distribution with the mean parameter different than that in (3.11):

$$S_N^{(n)} \stackrel{d}{\to} S_0\, \mathrm{e}^{(r - \frac{\sigma^2}{2})T + \sigma\sqrt{T}\widetilde{Z}}, \; \widetilde{Z} \sim \mathit{Norm}(0, 1). \tag{3.12}$$

In general, for any $t \in [0, T]$, as $n \to \infty$, we have the following pathwise convergences:

$$B_{\lfloor nt \rfloor}^{(n)} \equiv (1 + r_n)^{\lfloor nt \rfloor} \to \mathrm{e}^{rt} \equiv B(t),$$

$$S_{\lfloor nt \rfloor}^{(n)} \equiv S_0\, u_n^{M_{\lfloor nt \rfloor}^{(n)}} \stackrel{d}{\to} S_0\, \mathrm{e}^{\nu t + \sigma\sqrt{t}Z_t} \equiv S(t) \text{ (under } \mathbb{P}),$$

$$S_{\lfloor nt \rfloor}^{(n)} \equiv S_0\, u_n^{M_{\lfloor nt \rfloor}^{(n)}} \stackrel{d}{\to} S_0\, \mathrm{e}^{(r - \frac{\sigma^2}{2})t + \sigma\sqrt{t}\widetilde{Z}_t} \equiv S(t) \text{ (under } \widetilde{\mathbb{P}}),$$

with $Z_t, \widetilde{Z}_t \sim \mathit{Norm}(0, 1)$. The sequence of normal random variables, $\{\sqrt{t}Z_t\}_{0 \leqslant t \leqslant T}$, has the same probability distribution as a standard BM, $\{W(t)\}_{0 \leqslant t \leqslant T}$. So, the continuous-time process for stock prices (with the initial value $S(0) \equiv S_0$) takes the following form:

$$S(t) = S(0)\, \mathrm{e}^{\nu t + \sigma W(t)} \text{ (under } \mathbb{P}) \text{ or } S(t) = S(0)\, \mathrm{e}^{(r - \frac{\sigma^2}{2})t + \sigma\widetilde{W}(t)} \text{ (under } \widetilde{\mathbb{P}}).$$

We call it the log-normal, or the geometric Brownian motion (GBM), or the Black–Scholes–Merton (BSM) model. In this chapter, we also refer to it as the standard (B, S) model implying that there are two base assets trading on this continuous-time market, namely, the risky stock S and the risk-free bank account B.

We can now find the limit of the CRR pricing formula in (3.7), as $n \to \infty$, by combining the following two facts:

1. Under $\widetilde{\mathbb{P}}$, $S_N^{(n)} \xrightarrow{d} S_0\,\mathrm{e}^{(r-\frac{\sigma^2}{2})T+\sigma\sqrt{T}\widetilde{Z}}$, as $n \to \infty$, where $\widetilde{Z} \sim Norm(0,1)$.

2. For a bounded continuous function f and a sequence of random variables $\{X_n\}_{n\geqslant 1}$ s.t. $X_n \xrightarrow{d} X$, we have that $\mathrm{E}[f(X_n)] \to \mathrm{E}[f(X)]$, as $n \to \infty$.

Consider a European-style derivative with a bounded payoff $\Lambda(S)$. For example, it can be a put option with the payoff $\Lambda(S) := (K - S)^+$. For the CRR model (with $N = \lfloor nT \rfloor$ periods), the initial no-arbitrage price of the derivative is given by the discounted risk-neutral expectation of the payoff function:

$$V_0^{(n)}(S_0) = (1+r_n)^{-N}\,\widetilde{\mathrm{E}}\left[\Lambda\left(S_N^{(n)}\right)\right].$$

As $n \to \infty$, we have that $(1+r_n)^{-N} \to \mathrm{e}^{-rT}$ and $\widetilde{\mathrm{E}}\left[\Lambda\left(S_N^{(n)}\right)\right] \to \widetilde{\mathrm{E}}[\Lambda(S(T))]$, where $S(T) \stackrel{d}{=} S_0\,\mathrm{e}^{(r-\sigma^2/2)T+\sigma\sqrt{T}\widetilde{Z}}$ with $\widetilde{Z} \sim Norm(0,1)$ under $\widetilde{\mathbb{P}}$. Therefore, the initial no-arbitrage value of a European derivative under the log-normal model is

$$V(0, S_0) = \mathrm{e}^{-rT}\,\widetilde{\mathrm{E}}[\Lambda(S(T)) \mid S(0) = S_0].$$

We can now derive the BSM pricing formulae for a European put option by computing a respective mathematical expectation. Using the put-call parity relation, we can then derive the no-arbitrage pricing formula for a European call option. This approach is discussed in detail in the next section of this chapter.

Alternatively, we can derive pricing formulae for standard European options under the log-normal model by proceeding to the limit in (3.8) and (3.9). Let us consider a European put option. For the CRR model with n periods per year, formula (3.9) takes the form:

$$P_0^{(n)}(S_0) = \frac{K}{(1+r_n)^N}\mathcal{B}\left(m_N^{(n)}; N, \tilde{p}_n\right) - S_0\,\mathcal{B}\left(m_N^{(n)}; N, \hat{p}_n\right), \tag{3.13}$$

where $\tilde{p}_n = \dfrac{1+r_n-d_n}{u_n-d_n}$, $\hat{p}_n = \dfrac{u_n}{1+r_n}\cdot\tilde{p}_n$, $N = \lfloor nT \rfloor$, and $m_N^{(n)} := \left\lfloor \dfrac{\ln\left(\frac{K}{S_0}\right) - N\ln d_n}{\ln\left(\frac{u_n}{d_n}\right)} \right\rfloor$.
Firstly, notice that both $\tilde{p}_n$ and $\hat{p}_n$ converge to $\frac{1}{2}$, as $n \to \infty$. Secondly, we can show that

$$\frac{m_N^{(n)} - N\tilde{p}_n}{\sqrt{N\tilde{p}_n(1-\tilde{p}_n)}} \to -d_-(S_0/K, T) \quad\text{and}\quad \frac{m_N^{(n)} - N\hat{p}_n}{\sqrt{N\hat{p}_n(1-\hat{p}_n)}} \to -d_+(S_0/K, T), \text{ as } n \to \infty,$$

where $d_\pm(x,\tau) := \dfrac{\ln x + (r \pm \sigma^2/2)\tau}{\sigma\sqrt{\tau}}$. Lastly, we apply the De Moivre–Laplace theorem stated below.

Theorem 3.1. *Consider a sequence $\{p_n \in (0,1)\}_{n\geqslant 1}$ that converges to $p \in (0,1)$, as $n \to \infty$. Let $\{Y_n \sim Bin(n, p_n)\}_{n\geqslant 1}$ be a sequence of binomial random variables. Then the sequence of standardized random variables*

$$Y_n^* := \frac{Y_n - \mathrm{E}[Y_n]}{\sqrt{\mathrm{Var}(Y_n)}} = \frac{Y_n - np_n}{\sqrt{np_n(1-p_n)}}$$

converges weakly (in distribution) to a standard normal random variable, as $n \to \infty$.

Therefore, we have that

$$\mathcal{B}(m_N; N, \tilde{p}) \to \mathcal{N}(-d_-(S_0/K, T)) \quad\text{and}\quad \mathcal{B}(m_N; N, \hat{p}) \to \mathcal{N}(-d_+(S_0/K, T)), \text{ as } n \to \infty.$$

So, in the limiting case, equation (3.13) takes the form

$$P(0, S_0) = K\mathrm{e}^{-rT}\mathcal{N}(-d_-(S_0/K, T)) - S_0\mathcal{N}(-d_+(S_0/K, T)).$$

We recognize this as the well-known Black–Scholes–Merton formula for the no-arbitrage price of a standard European put option.

Moreover, as discussed in Chapter 4 of Volume I, we can derive the Black–Scholes partial differential equation (PDE) by proceeding to a time limit in the single-period pricing equation for the CRR model. Let $V(t, S)$ denote the no-arbitrage pricing function at time t for spot S. On a small time interval $[t - \delta, t]$ with $\delta \equiv \delta_n = \frac{1}{n}$, where n is the number of periods per year, the continuous-time model can be approximated by a single-period binomial model with factors $u_n = \mathrm{e}^{\sigma\sqrt{\delta}}$ and $d_n = \mathrm{e}^{-\sigma\sqrt{\delta}}$, and rate $r_n = \mathrm{e}^{r\delta} - 1$. The single-period approximation of the option price (on the interval $[t - \delta, t]$) is

$$V(t - \delta, S) = \mathrm{e}^{-r\delta}\widetilde{\mathrm{E}}[V(t, S_t) \mid S_{t-\delta} = S].$$

In the binomial case, it can be written as follows:

$$\mathrm{e}^{r\delta}\, V(t - \delta, S) = \tilde{p}_n\, V\big(t, S\mathrm{e}^{\sigma\sqrt{\delta}}\big) + (1 - \tilde{p}_n)\, V\big(t, S\mathrm{e}^{-\sigma\sqrt{\delta}}\big),$$

where $\tilde{p}_n = \frac{1+r_n-d_n}{u_n-d_n}$. We can apply Taylor's formula to the option value function to obtain

$$\begin{aligned}
V(t - \delta, S) &= V - V_t\delta + \mathcal{O}(\delta^2)\,,\\
V(t, S\mathrm{e}^{\pm\sigma\sqrt{\delta}}) &= V + \left(\pm\sigma\sqrt{\delta} + \sigma^2\delta/2\right) SV_S + (\sigma^2\delta/2)S^2V_{SS} + \mathcal{O}(\delta^{3/2})\,,\\
\tilde{p}_n &= \frac{1}{2} + \left(r - \frac{\sigma^2}{2}\right)\frac{\sqrt{\delta}}{2\sigma} + \mathcal{O}(\delta^{3/2}),
\end{aligned}$$

where $V \equiv V(t, S)$, $V_t \equiv \frac{\partial V(t,S)}{\partial t}$, $V_S \equiv \frac{\partial V(t,S)}{\partial S}$, and $V_{SS} \equiv \frac{\partial^2 V(t,S)}{\partial S^2}$. Taking a limit as $\delta \to 0$ gives us the *Black–Scholes PDE*

$$\frac{\partial V}{\partial t} + \frac{\sigma^2}{2}S^2\frac{\partial^2 V}{\partial S^2} + rS\frac{\partial V}{\partial S} - rV = 0,$$

where $V = V(t, S)$ for $0 \leqslant t < T$ and $S > 0$, subject to the terminal condition

$$V(T, S) = \Lambda(S) \text{ for } S > 0.$$

3.2 Replication (Hedging) and Derivative Pricing in the Simplest Black–Scholes Economy

Following Section 2.8 of Chapter 2, we fix a filtered probability space $(\Omega, \mathcal{F}, \mathbb{P}, \mathbb{F})$, where $\mathbb{F} = \{\mathcal{F}_t\}_{0\leqslant t\leqslant T}$ is a filtration for standard Brownian motion, i.e., $\{W(t)\}_{t\geqslant 0}$ is a standard $(\mathbb{P}, \mathbb{F})$-BM where $\mathbb{P}$ is the physical (real-world) measure. The first base security, B, in this market is the bank account whose price process is denoted by $\{B(t)\}_{0\leqslant t\leqslant T}$. In this section we assume a constant interest rate r where $B(t) = \mathrm{e}^{rt}$, i.e., $B(0) = 1$ and $B(t)$ is one unit (dollar) of investment compounded continuously with fixed rate r over time $[0, t]$. The second base security is the stock whose price process is assumed to be a standard GBM satisfying the SDE

$$\mathrm{d}S(t) = \mu S(t)\,\mathrm{d}t + \sigma S(t)\,\mathrm{d}W(t)\,,$$

with constant drift μ and constant volatility $\sigma > 0$. We recall Example 2.25 in Section 2.8 of Chapter 2. There we showed, by a simple application of Girsanov's Theorem, that there is a unique risk-neutral measure $\widetilde{\mathbb{P}}$, defined by (2.131), where $\{\widetilde{W}(t) := W(t) + \frac{(\mu - r)}{\sigma} t\}_{t \geqslant 0}$ is a standard $\widetilde{\mathbb{P}}$-BM. The discounted stock price process is a $\widetilde{\mathbb{P}}$-martingale. For convenience, we repeat some of the important equations here. In particular, the stock satisfies the SDE

$$\mathrm{d}S(t) = rS(t)\,\mathrm{d}t + \sigma S(t)\,\mathrm{d}\widetilde{W}(t)\,, \tag{3.14}$$

with solution

$$S(t) = S(0)\,\mathrm{e}^{(r - \frac{1}{2}\sigma^2)t + \sigma \widetilde{W}(t)}\,. \tag{3.15}$$

The discounted stock price process, $\overline{S}(t) = \mathrm{e}^{-rt}S(t) = D(t)S(t)$ with discount factor $D(t) := 1/B(t) = \mathrm{e}^{-rt}$, is a $\widetilde{\mathbb{P}}$-martingale:

$$\mathrm{d}\overline{S}(t) = \sigma \overline{S}(t)\,\mathrm{d}\widetilde{W}(t)\,, \quad \text{i.e.,} \quad \overline{S}(t) = S(0) + \sigma \int_0^t \overline{S}(u)\,\mathrm{d}\widetilde{W}(u)\,. \tag{3.16}$$

Note also that $\mathrm{d}B(t) = rB(t)\,\mathrm{d}t$ has an identically zero coefficient in $\mathrm{d}\widetilde{W}(t)$ and that $\overline{B}(t) := B(t)/B(t) = 1$ is trivially a (constant) martingale under any measure. Hence, $\widetilde{\mathbb{P}}$ is an equivalent martingale measure (EMM) with the bank account as a numéraire asset.

As in the discrete-time models, we assume no arbitrage in the market and will, below, set up a *self-financing replicating portfolio strategy* in the two base securities such that the value of the portfolio replicates the value of the financial derivative at some future (maturity) time T. In the absence of arbitrage, the time-t value of the self-financing replicating portfolio must therefore equal the no-arbitrage price of the derivative. Let $\{V_t\}_{0 \leqslant t \leqslant T}$ denote the *price process of the derivative security* where V_t is the price of the derivative at time $t \leqslant T$. At maturity T, the derivative price is given by the payoff value V_T, which is an $\mathcal{F}_T$-measurable random variable. In the present model, V_T is generally some functional of the BM up to time T or equivalently some functional of the underlying stock price process $\{S(t)\}_{0 \leqslant t \leqslant T}$. This functional can be quite complex for a general path-dependent payoff, although for a non-path-dependent derivative (as, for example, a standard call or put) the payoff is only a function of the terminal stock value $S(T)$.

In this (B, S) economy any portfolio (trading) strategy is a continuous sequence of portfolios in the two base assets: (β_t, Δ_t), $0 \leqslant t \leqslant T$, with each process $\{\beta_t\}_{0 \leqslant t \leqslant T}$ and $\{\Delta_t\}_{0 \leqslant t \leqslant T}$ assumed to be adapted to the filtration $\mathbb{F}$. As in the binomial model, β_t represents the time-t position in the bank account where $\beta_t < 0$ corresponds to a loan and $\beta_t > 0$ is an investment. The hedge position Δ_t is the number of shares held in the stock at time t where $\Delta_t < 0$ corresponds to shorting the stock and $\Delta_t > 0$ is a long position. Since we are in continuous time, trading (i.e., portfolio re-balancing) is allowed at any moment in time. The investor begins with a given initial wealth Π_0, which completely finances the initial portfolio with positions (β_0, Δ_0) and subsequently trades at every time $t \in [0, T]$ while holding Δ_t shares in the stock and β_t units in the bank account; i.e., this is represented by the portfolio value process for $t \in [0, T]$:

$$\Pi_t = \Delta_t S(t) + \beta_t B(t). \tag{3.17}$$

In what follows, we will only consider *self-financing* portfolio strategies. In analogy with the binomial model, a self-financing portfolio strategy is one in which the differential change in portfolio value is due only to differential changes in the prices of the base assets. Essentially this means that the investor holds the positions (β_t, Δ_t) during the infinitesimal time window $[t, t + \mathrm{d}t)$. At time $t + \mathrm{d}t$ the BM will have changed by a differential amount $\mathrm{d}W(t)$, the stock will have changed its share price by a differential amount $\mathrm{d}S(t)$, and the investment in the bank account will have either accrued interest (if $\beta_t > 0$) or will have decreased in value if $\beta_t < 0$. Formally, a self-financing portfolio strategy is then a portfolio strategy $(\beta_t, \Delta_t), 0 \leqslant t \leqslant T$, such that the cumulative gain in portfolio value is given by

$$\Pi_t = \Pi_0 + \int_0^t \beta_u \,\mathrm{d}B(u) + \int_0^t \Delta_u \,\mathrm{d}S(u)\,, \quad \text{for all } t \in [0, T], \tag{3.18}$$

with probability one (a.s.). It is more convenient to work with the differential form of (3.18),

$$\mathrm{d}\Pi_t = \beta_t \,\mathrm{d}B(t) + \Delta_t \,\mathrm{d}S(t)\,.$$

From (3.17) we can express the bank account investment (or loan) as $\beta_t B(t) = \Pi_t - \Delta_t S(t)$ and substituting this into the above differential, where $\beta_t \,\mathrm{d}B(t) = r\beta_t B(t)\,\mathrm{d}t$, gives

$$\mathrm{d}\Pi_t = r\beta_t B(t)\,\mathrm{d}t + \Delta_t \,\mathrm{d}S(t) = r(\Pi_t - \Delta_t S(t))\,\mathrm{d}t + \Delta_t \,\mathrm{d}S(t)\,. \tag{3.19}$$

We can recognize this as a continuous-time analogue of the wealth equation we encountered in the standard binomial tree model.

As in the binomial model, the discounted self-financing portfolio value process, defined by $\{\overline{\Pi}_t := \mathrm{e}^{-rt}\Pi_t\}_{0\leqslant t\leqslant T}$, is a $\widetilde{\mathbb{P}}$-martingale. This is readily shown by applying the Itô product rule to the process $\mathrm{e}^{-rt}\Pi_t \equiv D(t)\Pi_t$, where $\mathrm{d}D(t)\,\mathrm{d}\Pi_t = -rD(t)\,\mathrm{d}t\,\mathrm{d}\Pi_t \equiv 0$, and using (3.19):

$$\begin{aligned}
\mathrm{d}\overline{\Pi}_t \equiv \mathrm{d}(D(t)\Pi_t) &= \Pi_t \,\mathrm{d}D(t) + D(t)\,\mathrm{d}\Pi_t \\
&= -rD(t)\Pi_t \,\mathrm{d}t + D(t)\left[\, r(\Pi_t - \Delta_t S(t))\,\mathrm{d}t + \Delta_t \,\mathrm{d}S(t)\,\right] \\
&= \Delta_t \, D(t)\left[\, -rS(t)\,\mathrm{d}t + \,\mathrm{d}S(t)\,\right] \\
&= \Delta_t \,\mathrm{d}\overline{S}(t) \\
&= \Delta_t \,\sigma\overline{S}(t)\,\mathrm{d}\widetilde{W}(t)\,.
\end{aligned} \tag{3.20}$$

In the last equation line, we made use of (3.16). This therefore shows that $\{\overline{\Pi}_t\}_{0\leqslant t\leqslant T}$ is a $\widetilde{\mathbb{P}}$-martingale and that changes in this discounted self-financing portfolio are due only to changes in the discounted stock price. In integral form, we have

$$\overline{\Pi}_t = \overline{\Pi}_0 + \int_0^t \Delta_u \,\mathrm{d}\overline{S}(u) = \Pi_0 + \int_0^t \sigma\overline{S}(u)\Delta_u \,\mathrm{d}\widetilde{W}(u)\,, \quad 0 \leqslant t \leqslant T. \tag{3.21}$$

Note: we are assuming that the above Itô integral is square integrable. This guarantees that the process defined by (3.21) is a $\widetilde{\mathbb{P}}$-martingale. This technical detail can be verified later once the option price, and hence the delta position, is obtained.

As in the binomial model, we wish to price derivative contracts that can be replicated by a self-financing portfolio strategy. Let us therefore give a definition of this for the above continuous-time model.

Definition 3.1. A self-financing strategy $(\beta_t, \Delta_t), 0 \leqslant t \leqslant T$, is said to *replicate* the $\mathcal{F}_T$-measurable payoff V_T at maturity T if $\Pi_T = V_T$ (a.s.), i.e., $\mathbb{P}(\Pi_T = V_T) = 1$. We also say that the payoff V_T is attainable.

Let us suppose that a self-financing portfolio strategy that replicates the derivative payoff V_T exists. Later we give some discussion on this existence. Then, the cost at any time $t \leqslant T$ to set up such a strategy, i.e., the portfolio value Π_t, must equal the *time-t price of the derivative* V_t in order for the investor to hedge (at time t) the short position in the derivative security that has the given attainable payoff V_T at future time T, and hence avoid arbitrage. Therefore, we set $V_t = \Pi_t$ for all $0 \leqslant t \leqslant T$ for any attainable payoff V_T.

The key step now is to make use of the $\widetilde{\mathbb{P}}$-martingale property in (3.21). In particular, we have

$$D(t)\Pi_t = \widetilde{\mathrm{E}}[\, D(T)\Pi_T \mid \mathcal{F}_t]$$

and, since $V_t = \Pi_t$, then $D(t)V_t = \widetilde{\mathrm{E}}[\, D(T)V_T \mid \mathcal{F}_t]$. The discounted derivative price process, $\{D(t)V_t \equiv \mathrm{e}^{-rt}V_t\}_{0\leqslant t\leqslant T}$, is therefore a $\widetilde{\mathbb{P}}$-martingale. We can combine the discount factors, where $D(T)/D(t) = B(t)/B(T) = \mathrm{e}^{-r(T-t)}$, giving

$$V_t = B(t)\,\widetilde{\mathrm{E}}\left[\frac{V_T}{B(T)}\,\middle|\,\mathcal{F}_t\right] = \mathrm{e}^{-r(T-t)}\widetilde{\mathrm{E}}[\, V_T \mid \mathcal{F}_t]\,, \quad 0 \leqslant t \leqslant T. \tag{3.22}$$

This is the *risk-neutral pricing formula* for the above (B, S) model with a constant interest rate. It is a continuous-time analogue of (3.6) for the binomial tree model studied in Chapter 7 of Volume I and reviewed in Section 3.1. We remark that (3.22) was derived for the simplest case where the stock is a standard GBM, but in Chapter 4, we will arrive at the same formula for more general continuous-time stock price processes as long as the payoff is attainable and there exists a risk-neutral measure $\widetilde{\mathbb{P}}$ where the discounted stock (base asset) price process is a $\widetilde{\mathbb{P}}$-martingale. In Chapter 4, we shall also define an arbitrage strategy in the multi-asset continuous-time framework where it is shown that the existence of a risk-neutral measure implies that no arbitrage strategies are possible within the market model.

The formula in (3.22) may seem very simple; however, its practical use rests upon our ability to calculate the expectation of the payoff conditional on the filtration at time t. The payoff is an $\mathcal{F}_T$-measurable random variable and it can, in some cases, be quite complex, as it may have a complicated dependence on the path of the stock price process. Hence, the conditional expectation can be quite challenging to compute for complex path-dependent payoffs. The main idea is to simplify the $\mathcal{F}_t$-conditional expectation to an expectation that can be readily computed. In most practical situations, the derivative has a payoff structure that is not too complex. In the binomial model, we have already used the risk-neutral pricing framework to value derivatives having commonly encountered payoffs, such as the standard European call and put options as well as path-dependent options such as lookback and Asian options. Our main tools for analytically pricing such options were the Markov property and the Independence Proposition A.6 and A.7 in Appendix A (see also Chapter 6 of Volume I). For continuous-time models, these tools will also be used within the risk-neutral framework. Moreover, we shall also have other tools at our disposal, such as the PDE approach that is a result of the Feynman–Kac theorems.

Let us now consider the case of a standard (non-path-dependent) European option with payoff $V_T = \Lambda(S(T))$ as a $\mathcal{F}_T$-measurable random variable that is a function of only the *terminal (maturity time T) value of the stock price*. The important simplification that now follows is due to the Markov property where the conditioning on $\mathcal{F}_t$ is replaced with a conditioning on $S(t)$. We remind the reader of our discussions in Section 2.7 on conditional expectations and the Markov property. In particular, recall that a solution to an SDE is a Markov process and hence (2.69) in Theorem 2.9 applies. The time t derivative value (expressed as a $\sigma(S(t))$-measurable random variable) then takes the form

$$\begin{aligned} V_t &= \mathrm{e}^{-r(T-t)}\widetilde{\mathrm{E}}[\, \Lambda(S(T)) \mid \mathcal{F}_t] \\ &= \mathrm{e}^{-r(T-t)}\widetilde{\mathrm{E}}[\, \Lambda(S(T)) \mid S(t)] = V(t, S(t))\,. \end{aligned} \tag{3.23}$$

The *derivative pricing function*, $V(t, S)$, is a function of calendar (actual) time t and the spot[1] (ordinary variable) $S > 0$ and is given by the discounted expectation of the payoff conditional on $S(t) = S$:

$$V(t, S) = \mathrm{e}^{-r(T-t)}\widetilde{\mathrm{E}}_{t,S}[\,\Lambda(S(T))] := \mathrm{e}^{-r(T-t)}\widetilde{\mathrm{E}}[\,\Lambda(S(T)) \mid S(t) = S\,]\,. \tag{3.24}$$

Given the share price for the stock at time $t \leqslant T$, which is the spot $S > 0$, then (3.24) gives us the no-arbitrage time-t price of a European option having non-path-dependent attainable payoff $\Lambda(S(T))$ at maturity T, where $\Lambda : \mathbb{R}_+ \to \mathbb{R}$ is the payoff function.

We can now generally express the pricing function in (3.24) as an integral over the standard normal density $n(z)$ or equivalently as an integral over the risk-neutral transition PDF. This has essentially already been done in Example 2.21 of Chapter 2 where related formulae were derived under measure $\mathbb{P}$ with drift μ (instead of $\widetilde{\mathbb{P}}$ with drift r). Using (3.15), the time-T stock price is given in terms of the time-t price and the $\widetilde{\mathbb{P}}$-BM increment:

$$S(T) = S(t)\,\mathrm{e}^{(r-\frac{1}{2}\sigma^2)(T-t)+\sigma(\widetilde{W}(T)-\widetilde{W}(t))} \equiv S(t)\,\mathrm{e}^{(r-\frac{1}{2}\sigma^2)\tau+\sigma\sqrt{\tau}\widetilde{Z}}\,. \tag{3.25}$$

Throughout, we conveniently define $\tau := T - t$ as the time to maturity and

$$\widetilde{Z} := \frac{\widetilde{W}(T) - \widetilde{W}(t)}{\sqrt{\tau}} \tag{3.26}$$

which is a standard normal random variable, i.e., $\widetilde{Z} \sim Norm(0, 1)$ under measure $\widetilde{\mathbb{P}}$. Note that $\widetilde{Z}$ (or $\frac{S(T)}{S(t)}$) is independent of $S(t)$ since $\widetilde{W}(T) - \widetilde{W}(t)$ is independent of $\widetilde{W}(t)$. Combining these facts into (3.24) gives

$$\begin{aligned} V(t, S) &= \mathrm{e}^{-r\tau}\widetilde{\mathrm{E}}\left[\Lambda\big(S(t)\,\mathrm{e}^{(r-\frac{1}{2}\sigma^2)\tau+\sigma\sqrt{\tau}\widetilde{Z}}\big) \mid S(t) = S\right] \\ &= \mathrm{e}^{-r\tau}\widetilde{\mathrm{E}}\left[\Lambda\big(S\,\mathrm{e}^{(r-\frac{1}{2}\sigma^2)\tau+\sigma\sqrt{\tau}\widetilde{Z}}\big)\right] \\ &= \mathrm{e}^{-r\tau}\int_{-\infty}^{\infty} \Lambda(S\,\mathrm{e}^{(r-\frac{1}{2}\sigma^2)\tau+\sigma\sqrt{\tau}z})\,n(z)\,\mathrm{d}z\,. \end{aligned} \tag{3.27}$$

By changing integration variables in (3.27), i.e., letting $y = S\,\mathrm{e}^{(r-\frac{1}{2}\sigma^2)\tau+\sigma\sqrt{\tau}z}$, gives the equivalent pricing formula

$$V(t, S) = \mathrm{e}^{-r\tau}\int_0^{\infty} \Lambda(y)\,\widetilde{p}(\tau; S, y)\,\mathrm{d}y\,, \tag{3.28}$$

where $\widetilde{p}(\tau; S, y)$ denotes the risk-neutral transition PDF of the stock price process in (3.15):

$$\widetilde{p}(\tau; S, y) = \frac{1}{y\sigma\sqrt{2\pi\tau}} \exp\left(-\frac{[\ln(y/S) - (r - \frac{1}{2}\sigma^2)\tau]^2}{2\sigma^2\tau}\right)\,;\quad S, y > 0,\ \tau > 0. \tag{3.29}$$

We recall that this is the time-homogeneous log-normal density in y given by (2.114) with drift parameter μ set to the risk-free rate r. Note that when the stock price process is time homogeneous, as in the present case, the pricing formula is a function of the time to maturity $\tau = T - t$ and we shall also denote it by $v(\tau, S) := V(t, S)$, i.e., $v(\tau, S) := V(T - \tau, S)$.

We see that (3.27) and (3.28) provide two equivalent expectation (integral) approaches for pricing standard European options on a stock. The transition PDF $\widetilde{p}(\tau; S, y)$ is the

[1] Assuming there is no confusion, when discussing pricing formulae we prefer to choose more appropriate letters for some of the ordinary variables. In this case, we denote the spot value by using the dummy variable S, instead of using some other letter like x, in the context of a pricing formula.

fundamental solution to the PDE in (2.107) with the spot value as a backward variable, $x \equiv S$, linear diffusion coefficient function $\sigma(x) \equiv \sigma(S) = \sigma S$, and linear drift coefficient function $\mu(x) \equiv \mu(S) = rS$. The discounted transition PDF, $\mathrm{e}^{-r\tau}\widetilde{p}(\tau; S, y)$, and hence the pricing function $v = v(\tau, S)$ in (3.28), satisfies the PDE in the variables (S, τ) (see (2.112) in Chapter 2):

$$\frac{\partial v}{\partial \tau} = \frac{1}{2}\sigma^2 S^2 \frac{\partial^2 v}{\partial S^2} + rS\frac{\partial v}{\partial S} - rv\,, \tag{3.30}$$

subject to the payoff condition $v(0+, S) = \Lambda(S)$, which is an initial condition in the time to maturity $\tau \searrow 0$. This is the Black–Scholes partial differential equation (BSPDE) for the pricing function *expressed as function of the spot and time to maturity variables* (S, τ). Since $\frac{\partial V}{\partial t} = \frac{\partial \tau}{\partial t} \cdot \frac{\partial v}{\partial \tau} = -\frac{\partial v}{\partial \tau}$, then (3.30) is equivalent to

$$\frac{\partial V}{\partial t} + \frac{1}{2}\sigma^2 S^2 \frac{\partial^2 V}{\partial S^2} + rS\frac{\partial V}{\partial S} - rV = 0\,, \tag{3.31}$$

subject to the payoff condition $V(T-, S) = \Lambda(S)$, which is a terminal condition in time $t \nearrow T$. This is the BSPDE satisfied by the pricing function $V = V(t, S)$ in the variables (t, S). In fact, the BSPDE in (3.31) arises by direct application of the discounted Feynman–Kac Theorem 2.13 to the conditional expectation in (3.24). [Note that Theorem 2.13 is the same when we replace $\mathbb{P}$, E, and W everywhere by $\widetilde{\mathbb{P}}$, $\widetilde{\mathrm{E}}$, and $\widetilde{W}$, respectively. All this means is that the probability measure is now the risk-neutral measure $\widetilde{\mathbb{P}}$. Here, we have the dummy variable $x \equiv S$ and the process $X(t) \equiv S(t)$ is GBM with coefficient functions defined by $\mu(t, S) := rS, \sigma(t, S) := \sigma S$ in the SDE (3.14). In particular, for constant interest rate r the pricing function $V(t, S)$ satisfies the PDE (2.99) in the variables t and S, which is the BSPDE in (3.31).]

Let us now turn our attention to the problem of replicating a derivative claim, i.e., *the hedging problem* in the simple (B, S) model. We first see how this problem is solved in the case of non-path-dependent payoffs where $V_T = \Lambda(S(T))$ and $V_t = V(t, S(t))$. By applying the Itô formula with the differential in (3.14) we obtain:

$$\begin{aligned}
\mathrm{d}[\mathrm{e}^{-rt}V(t, S(t))] &= \mathrm{e}^{-rt}\left[\mathrm{d}V(t, S(t)) - rV(t, S(t))\,\mathrm{d}t\right] \\
&= \mathrm{e}^{-rt}\left(\frac{\partial V}{\partial t}(t, S(t)) + \mathcal{G}\,V(t, S(t)) - rV(t, S(t))\right)\mathrm{d}t \\
&\quad + \sigma\mathrm{e}^{-rt}S(t)\frac{\partial V}{\partial S}(t, S(t))\,\mathrm{d}\widetilde{W}(t)\,,
\end{aligned}$$

where $\mathcal{G}\,V(t, S) := \frac{1}{2}\sigma^2 S^2 \frac{\partial^2}{\partial S^2}V(t, S) + rS\frac{\partial}{\partial S}V(t, S)$ is the differential generator corresponding to the GBM process with SDE in (3.14). Now using the fact that the pricing function $V(t, S)$ satisfies the BSPDE in (3.31), i.e., $\left(\frac{\partial}{\partial t} + \mathcal{G} - r\right)V(t, S) = 0$ for all (t, S), then

$$\mathrm{d}[\mathrm{e}^{-rt}V(t, S(t))] = \sigma\overline{S}(t)\frac{\partial V}{\partial S}(t, S(t))\,\mathrm{d}\widetilde{W}(t)\,.$$

In integral form:

$$\mathrm{e}^{-rt}V(t, S(t)) = V(0, S(0)) + \int_0^t \sigma\overline{S}(u)\frac{\partial V}{\partial S}(u, S(u))\,\mathrm{d}\widetilde{W}(u) \tag{3.32}$$

for all $t \in [0, T]$. Assuming the square-integrability condition on the integrand of this Itô integral, $\{\mathrm{e}^{-rt}V(t, S(t))\}_{0 \leqslant t \leqslant T}$ is a $\widetilde{\mathbb{P}}$-martingale. For portfolio replication, we require $\Pi_t = V(t, S(t))$, or $\overline{\Pi}_t = \mathrm{e}^{-rt}V(t, S(t))$, for all $t \in [0, T]$. We see from (3.21) that this can be achieved by setting the initial value of the portfolio $\Pi_0 = V_0 = V(0, S(0))$ and by choosing the delta position such that the respective Itô integrals in (3.21) and (3.32) are equal, i.e., the delta hedge is achieved by choosing

$$\Delta_t = \frac{\partial V}{\partial S}(t, S(t)) \equiv \frac{\partial V}{\partial S}(t, S)\bigg|_{S=S(t)}, \quad \text{for all } t \in [0, T). \tag{3.33}$$

Hence, (3.33) gives the *time-t (delta) position in the stock required to dynamically replicate the derivative claim in a self-financing portfolio strategy.* Clearly, Δ_t is $\mathcal{F}_t$-measurable (in fact it is $\sigma(S(t))$-measurable) and is given uniquely by the first derivative (w.r.t. the spot variable S) of the pricing function evaluated at $S = S(t)$. Defining the function $\Delta(t, S) := \frac{\partial V}{\partial S}(t, S)$, then $\Delta(t, S)$ gives the time-t position in the stock given the spot value S. Once the pricing function $V(t, S)$ is known (and hence its derivative computed) the self-financing replicating portfolio in (3.17) is given by $\Delta_t = \Delta(t, S(t))$ positions in the stock and $\beta_t = \mathrm{e}^{-rt}[V(t, S(t)) - S(t)\Delta(t, S(t))]$ units in the bank account, i.e., the value of the bank account portion of the portfolio at time t is $V(t, S(t)) - S(t)\Delta(t, S(t))$.

Consider the general case where V_T is any $\mathcal{F}_T$-measurable payoff (i.e., generally path dependent). Now, we generally have $V_t \neq V(t, S(t))$ (i.e., we do not generally have V_t as a function of t and $S(t)$) and hence the above BSPDE and Feynman–Kac results are not generally applicable. However, what is important is that the discounted derivative value process $\{D(t)V_t\}_{0 \leqslant t \leqslant T}$ is a $(\widetilde{\mathbb{P}}, \mathbb{F})$-martingale. We now show that we can guarantee that V_T is attainable if we make two general assumptions on the payoff:

1. V_T is square integrable, i.e., $\mathrm{E}[V_T^2] < \infty$.
2. V_T is $\mathcal{F}_T^W$-measurable.

[Note that these two conditions are already implicit in the above non-path-dependent case.] The discounted price process $\{D(t)V_t\}_{0 \leqslant t \leqslant T}$ is then a square-integrable $(\widetilde{\mathbb{P}}, \mathbb{F}^W)$-martingale. We can now make use of Theorem 2.18, in particular Proposition 2.19 of Chapter 2, where we identify $M(t) \equiv D(t)V_t$ and identify the measure $\widehat{\mathbb{P}} = \widetilde{\mathbb{P}}$, with constant $\gamma(t) = \gamma \equiv (r - \mu)/\sigma$ being $\mathcal{F}_t^W$-adapted. This implies[2] $\mathbb{F}^W = \mathbb{F}^{\widetilde{W}}$ and hence $\{D(t)V_t\}_{0 \leqslant t \leqslant T}$ is a square-integrable $(\widetilde{\mathbb{P}}, \mathbb{F}^{\widetilde{W}})$-martingale. Hence, there exists an adapted process, we now denote by $\{\widetilde{\theta}(t)\}_{0 \leqslant t \leqslant T}$, such that (note $D(0)V_0 = V_0$):

$$D(t)V_t = V_0 + \int_0^t \widetilde{\theta}(t)\, \mathrm{d}\widetilde{W}(u)\,, \quad 0 \leqslant t \leqslant T. \tag{3.34}$$

For portfolio replication, we require $\Pi_t = V_t$, for all $t \in [0, T]$. By (3.34) we see that this is achieved by setting $\Pi_0 = V_0$ and by choosing the delta position such that the integrands in (3.21) and (3.34) are equal, i.e., the delta hedge is achieved by setting [note $\overline{S}(t) = D(t)S(t)$]

$$\widetilde{\theta}(t) = \sigma D(t)S(t)\Delta_t\,, \quad \text{i.e.,} \quad \Delta_t = \frac{\widetilde{\theta}(t)}{\sigma D(t)S(t)} \tag{3.35}$$

for all $t \in [0, T]$. This solution for Δ_t exists, for every $\widetilde{\theta}(t)$, since we are assuming that the volatility parameter $\sigma > 0$. Note also that the stock price is a GBM with $S(0) > 0$ and therefore cannot hit zero in any finite time. Of course, implicit in this existence are also the above assumptions on the derivative payoff. Namely, V_T must be $\mathcal{F}_T^W$-measurable and this

[2]It is also trivial to see in this case that the natural filtration $\mathbb{F}^W = \{\mathcal{F}_t^W\}_{t \geqslant 0}$ generated by W ($\mathbb{P}$-measure BM) is the same as the natural filtration $\mathbb{F}^{\widetilde{W}} = \{\mathcal{F}_t^{\widetilde{W}}\}_{t \geqslant 0}$ generated by $\widetilde{W}$ ($\widetilde{\mathbb{P}}$-measure BM) since the two Brownian motions differ only by a linear term: $\widetilde{W}(t) = W(t) + \frac{(\mu - r)}{\sigma}t$, so $\mathcal{F}_t^{\widetilde{W}} = \mathcal{F}_t^W$ for every $t \geqslant 0$.

means that we can only guarantee replication of payoffs whose only source of randomness is the BM driving the stock itself.

We have therefore shown that the model can replicate (hedge) any complex arbitrary path-dependent derivatives with a payoff satisfying the above two assumptions. In this sense we can say that the (B, S) model is a *complete market model.* The formula in (3.35) asserts the existence of a hedging strategy and therefore justifies the use of the risk-neutral pricing formula in (3.22). We remark that generally (3.35) does not give a practical (or explicit) construction of the hedging strategy. However, there are important examples of common payoffs for which we do have an explicit formula for the delta hedge. For example, in the particular case of a non-path-dependent standard European option, we already showed that $\widetilde{\theta}(t)$ is given explicitly in terms of the derivative of the pricing function:

$$\widetilde{\theta}(t) = \sigma D(t) S(t) \frac{\partial V}{\partial S}(t, S(t))$$

with the delta position given by (3.33).

3.2.1 Pricing Standard European Calls and Puts

We now use the risk-neutral pricing formulation to derive the well-known Black–Scholes–Merton formulae for the prices of a standard call and put option in the simplest (B, S) model of Section 3.2, where the stock price is a GBM with constant volatility $\sigma > 0$ and the bank account has constant interest rate r. Here, we are assuming a zero dividend on the stock, but the inclusion of a stock dividend is quite simple, as shown later.

Consider a call option with payoff $V_T \equiv C_T = (S(T) - K)^+$. This is an example of a non-path-dependent option with payoff function $\Lambda(x) := (x - K)^+$. Let $S(t) = S > 0$ be the spot price of the stock at time $t \leqslant T$. Since the payoff depends only on the terminal stock price $S(T)$, the (current) time-t price of this call with strike $K > 0$ can be found by directly evaluating the integral in (3.27) using the identity (B.1) in Appendix B. We leave this as an exercise for the reader. We recall that in Chapter 4 of Volume I a brute force integration of (3.27) was also used in deriving the time-0 price of the put option with payoff $\Lambda(x) = (K - x)^+$. Here, we carry out the derivation in an explicit manner that is instructive since it clearly displays the use of the conditional expectation approach and the connection between the different random variables and independence. In particular, using (3.24), the time-t price of the call, $C(t, S)$, is the discounted risk-neutral expectation of the payoff at time $T > t$, conditional on $S(t) = S$:

$$\begin{aligned} C(t,S) &= \mathrm{e}^{-r(T-t)}\, \widetilde{\mathrm{E}}_{t,S}\big[(S(T)-K)^+\big] \\ &= \mathrm{e}^{-r\tau}\, \widetilde{\mathrm{E}}_{t,S}\big[(S(T)-K)\,\mathbb{I}_{\{S(T)>K\}}\big] \\ &= \mathrm{e}^{-r\tau}\, \widetilde{\mathrm{E}}_{t,S}\big[S(T)\,\mathbb{I}_{\{S(T)>K\}}\big] - \mathrm{e}^{-r\tau} K \widetilde{\mathrm{E}}_{t,S}\big[\mathbb{I}_{\{S(T)>K\}}\big] \\ &= \mathrm{e}^{-r\tau}\, \widetilde{\mathrm{E}}_{t,S}\big[S(T)\,\mathbb{I}_{\{S(T)>K\}}\big] - \mathrm{e}^{-r\tau} K\, \widetilde{\mathbb{P}}_{t,S}\big(S(T)>K\big)\,, \end{aligned} \tag{3.36}$$

with $S(T)$ given by (3.25). Two conditional expectations need to be computed (where the last term has been expressed as a conditional probability). In both cases, we use the fact that $\widetilde{Z}$ in (3.26) is independent of $S(t)$ and, by the Independence Proposition A.6, this allows us to remove the conditioning upon setting $S(t) = S$. First, let us rewrite the event $\{S(T) > K\}$ in terms of $\widetilde{Z}$ and $S(t)$ by simply dividing $S(T)$ in (3.25) by $S(t)$, dividing K by $S(t)$, and taking natural logarithms:

$$\{S(T) > K\} = \left\{ \ln \frac{S(T)}{S(t)} > \ln \frac{K}{S(t)} \right\} = \left\{ \widetilde{Z} > -d_-\left(\frac{S(t)}{K}, \tau\right) \right\}$$

where we define

$$d_-(x,\tau) := \frac{\ln x + (r - \frac{1}{2}\sigma^2)\tau}{\sigma\sqrt{\tau}}\,, \quad d_+(x,\tau) := \frac{\ln x + (r + \frac{1}{2}\sigma^2)\tau}{\sigma\sqrt{\tau}} = d_-(x,\tau) + \sigma\sqrt{\tau} \quad (3.37)$$

for all $x > 0$, $\tau := T - t > 0$. We use the above representation for the event and substitute the expression for $S(T)$, given in (3.25), into the first expectation in (3.36), which is now readily computed by setting $S(t) = S$, using the independence of $\widetilde{Z}$ and $S(t)$, and then applying the identity[3] in (B.1) of Appendix B to evaluate the (unconditional) expectation:

$$\begin{aligned}
\widetilde{\mathrm{E}}_{t,S}\big[S(T)\,\mathbb{I}_{\{S(T)>K\}}\big] &= \widetilde{\mathrm{E}}\big[S(t)\,\mathrm{e}^{(r-\frac{1}{2}\sigma^2)\tau+\sigma\sqrt{\tau}\widetilde{Z}}\,\mathbb{I}_{\{\widetilde{Z}>-d_-(\frac{S(t)}{K},\tau)\}}\,\big|\,S(t)=S\big] \\
&= S\,\mathrm{e}^{(r-\frac{1}{2}\sigma^2)\tau}\,\widetilde{\mathrm{E}}\big[\mathrm{e}^{\sigma\sqrt{\tau}\widetilde{Z}}\,\mathbb{I}_{\{\widetilde{Z}>-d_-(\frac{S}{K},\tau)\}}\big] \\
&= S\,\mathrm{e}^{(r-\frac{1}{2}\sigma^2)\tau}\,\mathrm{e}^{\frac{1}{2}(\sigma\sqrt{\tau})^2}\,\mathcal{N}\left(\sigma\sqrt{\tau} + d_-(\frac{S}{K},\tau)\right) \\
&= \mathrm{e}^{r\tau} S\,\mathcal{N}\left(d_+(\frac{S}{K},\tau)\right). \qquad (3.38)
\end{aligned}$$

The conditional probability in (3.36) now follows quite simply:

$$\begin{aligned}
\widetilde{\mathbb{P}}_{t,S}\big(S(T) > K\big) = \widetilde{\mathbb{P}}\left(\widetilde{Z} > -d_-(\frac{S(t)}{K},\tau)\,\middle|\,S(t) = S\right) &= \widetilde{\mathbb{P}}\left(\widetilde{Z} > -d_-(\frac{S}{K},\tau)\right) \\
&= \mathcal{N}\left(d_-(\frac{S}{K},\tau)\right). \qquad (3.39)
\end{aligned}$$

Note that this is also obtained directly from the transition CDF, $\widetilde{P}$, of the GBM process under measure $\widetilde{\mathbb{P}}$, where

$$\widetilde{P}(t,T;S,K) \equiv \widetilde{\mathbb{P}}_{t,S}\big(0 < S(T) \leqslant K\big) = 1 - \widetilde{\mathbb{P}}_{t,S}\big(S(T) > K\big) = \mathcal{N}\left(-d_-(\frac{S}{K},\tau)\right).$$

Substituting (3.38) and (3.39) into (3.36) completes our derivation of the pricing formula for the standard call:

$$C(t,S) = S\mathcal{N}\left(d_+(\frac{S}{K},\tau)\right) - \mathrm{e}^{-r\tau} K\mathcal{N}\left(d_-(\frac{S}{K},\tau)\right)\;;\quad \tau = T - t. \qquad (3.40)$$

Having priced the call, we can now easily derive the pricing formula for the put price, $P(t, S)$, by recalling the simple symmetry between the call and put payoffs, i.e., a portfolio in one long call and one short put is equivalent to a portfolio in one long forward contract:

$$(S(T) - K)^+ - (K - S(T))^+ = S(T) - K\,. \qquad (3.41)$$

Taking discounted risk-neutral expectations on both sides of (3.41) gives

$$\underbrace{\mathrm{e}^{-r\tau}\,\widetilde{\mathrm{E}}_{t,S}\big[(S(T)-K)^+\big]}_{=C(t,S)} - \underbrace{\mathrm{e}^{-r\tau}\,\widetilde{\mathrm{E}}_{t,S}\big[(K-S(T))^+\big]}_{=P(t,S)} = \underbrace{\mathrm{e}^{-r\tau}\,\widetilde{\mathrm{E}}_{t,S}\big[S(T)\big]}_{=S} - \mathrm{e}^{-r\tau} K \qquad (3.42)$$

[3]In applying the expectation identities in Appendix B we need to simply identify the parameters A, B and the mean μ and variance σ^2 of the appropriate normal random variable X in the given measure. Note that the parameter σ used in the above equations is obviously the symbol for the volatility of the stock. Of course, this σ is not the same σ used throughout Appendix B! For the expectation in (3.38), we employ the formula in (B.1) by identifying $X \equiv \widetilde{Z}$ as $Norm(\mu = 0, \sigma^2 = 1)$ and $A \equiv -d_-(\frac{S}{K},\tau)$, $B \equiv \sigma\sqrt{\tau}$.

where we recognize the left-hand side as the difference in the time-t price of the call and put, at strike K. The right-hand side is the time-t price of a forward contract with payoff $S(T)-K$. This is the *put-call parity relation* that we have already encountered in Chapter 4 of Volume I on derivative securities. It is important to note that this relation is valid for quite general models (i.e., beyond the GBM model). The only assumption is that the discounted stock price is a $\widetilde{\mathbb{P}}$-martingale. Substituting the call price in (3.40) into (3.42) gives the pricing formula for the standard put:

$$\begin{aligned} P(t,S) &= C(t,S) + \mathrm{e}^{-r\tau} K - S \\ &= \mathrm{e}^{-r\tau} K \left[1 - \mathcal{N}\left(d_-(\frac{S}{K},\tau)\right)\right] - S\left[1 - \mathcal{N}\left(d_+(\frac{S}{K},\tau)\right)\right] \\ &= \mathrm{e}^{-r\tau} K \mathcal{N}\left(-d_-(\frac{S}{K},\tau)\right) - S\mathcal{N}\left(-d_+(\frac{S}{K},\tau)\right)\, ; \quad \tau = T - t. \end{aligned} \tag{3.43}$$

Typical plots of the call and put pricing functions are given in Figures 3.2 and 3.3.

We leave it as an exercise for the reader to show by direct differentiation that the above call and put pricing functions in (3.40) and (3.43) solve the BSPDE. The initial conditions $\tau \searrow 0$ (equivalently $t \nearrow T$) on the above pricing functions are readily shown to be satisfied by working out the limiting forms (see also Example 2.21):

$$\begin{aligned} \lim_{\tau\searrow 0} \mathcal{N}\left(d_\pm(\frac{S}{K},\tau)\right) &= \lim_{\tau\searrow 0} \mathcal{N}\left(\frac{\ln(S/K)}{\sigma\sqrt{\tau}} + (r \pm \frac{1}{2}\sigma^2)\frac{\sqrt{\tau}}{\sigma}\right) = \lim_{\tau\searrow 0} \mathcal{N}\left(\frac{\ln(S/K)}{\sigma\sqrt{\tau}}\right) \\ &= \begin{cases} \mathcal{N}(\infty) = 1 & \text{if } S > K \\ \mathcal{N}(0) = 1/2 & \text{if } S = K \\ \mathcal{N}(-\infty) = 0 & \text{if } S < K \end{cases} \\ &= H(S-K)\,. \end{aligned} \tag{3.44}$$

Hence, both limits equal the unit step function centred at $S = K$. Using both these limits in (3.40) verifies the payoff condition for the call:

$$\begin{aligned} \lim_{t\nearrow T} C(t,S) &= S \lim_{\tau\searrow 0} \mathcal{N}\left(d_+(\frac{S}{K},\tau)\right) - K \lim_{\tau\searrow 0} \mathrm{e}^{-r\tau}\mathcal{N}\left(d_-(\frac{S}{K},\tau)\right) \\ &= (S-K)H(S-K) \\ &= (S-K)\mathbb{I}_{\{S>K\}} = (S-K)^+. \end{aligned}$$

By the same steps applied to (3.43), we verify the payoff condition:

$$\lim_{t\nearrow T} P(t,S) = (K-S)H(K-S) = (K-S)^+.$$

The asymptotic values for the call and put pricing functions for small and large spot values are readily obtained. We leave it as an exercise to show, by directly computing the limits, that

$$P(t,S) \sim \mathrm{e}^{-r\tau} K\,, \text{ as } S \searrow 0, \quad \text{and} \quad P(t,S) \sim 0\,, \text{ as } S \nearrow \infty\,, \tag{3.45}$$

and

$$C(t,S) \sim 0\,, \text{ as } S \searrow 0, \quad \text{and} \quad C(t,S) \sim S - \mathrm{e}^{-r\tau} K \sim \infty\,, \text{ as } S \nearrow \infty\,. \tag{3.46}$$

The financial interpretation of the above limits is clear. In the limit that the time-t stock price (spot S) is very close to zero, it will remain close to zero within a finite time to maturity $\tau = T - t$. This means that the call will certainly expire out of the money (hence is worthless) and the put will expire completely in the money with payoff K and time-t value $\mathrm{e}^{-r\tau} K$. In the limit of arbitrarily large spot value, the stock price will remain arbitrarily large; i.e., the put will certainly expire out of the money (hence is worthless) and the call will expire in the money with an arbitrarily large payoff.

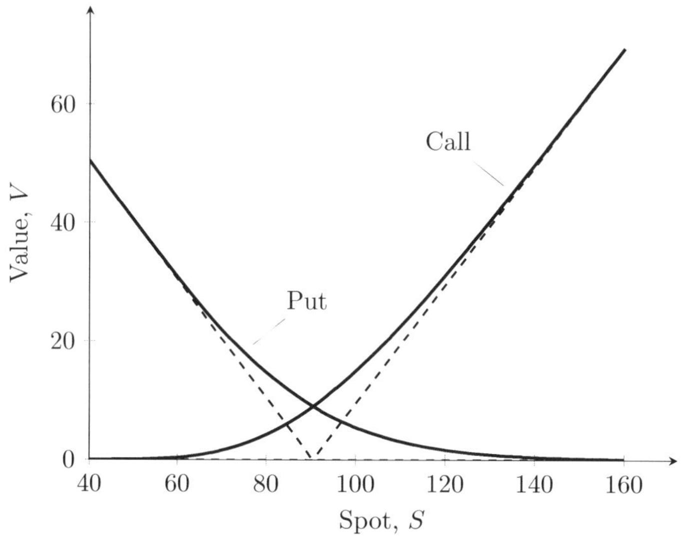

FIGURE 3.2: No-arbitrage values of standard European call and put options as functions of spot price S under the (B, S) model with $\tau = 1$, $K = 100$, $r = 10\%$, and $\sigma = 25\%$. The dashed lines represent the lower no-arbitrage bounds on the call and put options: $C \geqslant (S - \mathrm{e}^{-r\tau}K)^+$ and $P \geqslant (\mathrm{e}^{-r\tau}K - S)^+$.

3.2.2 Hedging Standard European Calls and Puts

In Chapter 4 of Volume I, we computed the "delta" of a standard call and put by differentiating the pricing functions. Let us denote the delta of the call at calendar time t by $\Delta_c(t, S) = \frac{\partial C}{\partial S}(t, S)$ and the delta of the put by $\Delta_p(t, S) = \frac{\partial P}{\partial S}(t, S)$. Note that Δ_c and Δ_p are also functions of (τ, S), since the pricing functions in (3.40) and (3.43) are expressible as functions of (τ, S). We recall a previous method to derive Δ_c by simply differentiating (3.40) w.r.t. S, where $\mathcal{N}'(x) = n(x) = \mathrm{e}^{-\frac{1}{2}x^2}/\sqrt{2\pi}$, while using $\frac{\partial}{\partial S} d_\pm\big(\frac{S}{K}, \tau\big) = \frac{1}{S\sigma\sqrt{\tau}}$,

$$\begin{aligned} \Delta_c(t, S) &= \mathcal{N}\left(d_+\big(\frac{S}{K}, \tau\big)\right) + \frac{K}{S\sigma\sqrt{2\pi\tau}}\left[\frac{S}{K}\,\mathrm{e}^{-\frac{1}{2}d_+^2(\frac{S}{K},\tau)} - \mathrm{e}^{-r\tau-\frac{1}{2}d_-^2(\frac{S}{K},\tau)}\right] \\ &= \mathcal{N}\left(d_+\big(\frac{S}{K}, \tau\big)\right). \end{aligned} \tag{3.47}$$

Here, we used the following identity with $x = S/K$ (which we leave as a somewhat tedious exercise in algebra for the reader to show):

$$x\,\mathrm{e}^{-\frac{1}{2}d_+^2(x,\tau)} \equiv \mathrm{e}^{\ln x - \frac{1}{2}d_+^2(x,\tau)} = \mathrm{e}^{-r\tau - \frac{1}{2}d_-^2(x,\tau)}. \tag{3.48}$$

The above derivation of Δ_c is correct but perhaps not the most instructive. We now

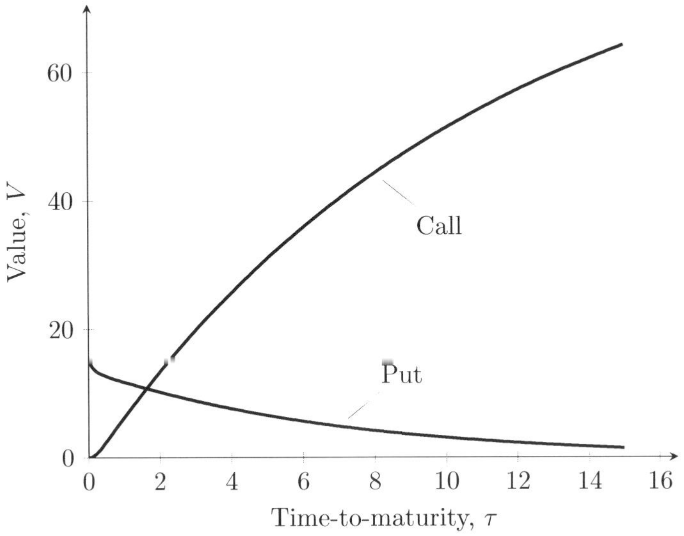

FIGURE 3.3: No-arbitrage values of standard European call and put options as functions of time to maturity τ under the (B, S) model with $S = 85$, $K = 100$, $r = 10\%$, and $\sigma = 25\%$.

give an alternate derivation of Δ_c by directly connecting it to the discounted risk-neutral conditional expectation in (3.38). From the risk-neutral pricing formula in (3.27),

$$C(t,S) = \mathrm{e}^{-r\tau}\widetilde{\mathrm{E}}\big[\big(S\,\mathrm{e}^{(r-\frac{1}{2}\sigma^2)\tau+\sigma\sqrt{\tau}\widetilde{Z}} - K\big)^+\big] = \mathrm{e}^{-r\tau}\int_{-\infty}^{\infty}\big(S\,\mathrm{e}^{(r-\frac{1}{2}\sigma^2)\tau+\sigma\sqrt{\tau}z} - K\big)^+ n(z)\,\mathrm{d}z\,.$$

We now differentiate w.r.t. S (as a parameter) inside the expectation or inside the integral and use the property $\frac{\partial}{\partial S}(aS-K)^+ = aH(aS-K) = a\mathbb{I}_{\{aS\geqslant K\}}$. In the integral we have $a \equiv a(z) \equiv \mathrm{e}^{(r-\frac{1}{2}\sigma^2)\tau+\sigma\sqrt{\tau}z}$ or in the expectation $a \equiv a(\widetilde{Z}) \equiv \mathrm{e}^{(r-\frac{1}{2}\sigma^2)\tau+\sigma\sqrt{\tau}\widetilde{Z}}$. We can write the steps compactly as follows by differentiating inside the expectation (note: $S(T) = Sa(\widetilde{Z})$),

$$\begin{aligned}\Delta_c(t,S) = \mathrm{e}^{-r\tau}\widetilde{\mathrm{E}}\left[\frac{\partial}{\partial S}\big(S\,a(\widetilde{Z}) - K\big)^+\right] &= \mathrm{e}^{-r\tau}\widetilde{\mathrm{E}}\big[a(\widetilde{Z})\,\mathbb{I}_{\{S\,a(\widetilde{Z})>K\}}\big]\\ &= \frac{1}{S}\mathrm{e}^{-r\tau}\widetilde{\mathrm{E}}_{t,S}\big[S(T)\,\mathbb{I}_{\{S(T)>K\}}\big]\\ &= \mathcal{N}\left(d_+\big(\frac{S}{K},\tau\big)\right).\end{aligned}$$

Here, we identified the conditional expectation in (3.38).

It is important to note that the delta of a call is strictly positive; i.e., $\Delta_c(t,S) > 0$ for all $t < T$ and $S > 0$, since the CDF $\mathcal{N}(x)$ is strictly positive for all $x \in \mathbb{R}$. Hence, according to (3.33), to replicate a call the writer is continuously re-balancing the self-financing portfolio while always maintaining a *delta positive (long) position in the stock* given by $\Delta_t \equiv \Delta_c(t, S(t)) = \mathcal{N}\left(d_+\big(\frac{S(t)}{K},\tau\big)\right)$ for $\tau = T - t > 0$. In particular, the time-t investment in the stock, given spot $S(t) = S$, is

$$\Delta_t S(t) = S\,\Delta_c(t,S) = S\mathcal{N}\left(d_+\big(\frac{S}{K},\tau\big)\right).$$

Since $\Pi_t = C(t, S(t)) = C(t, S)$, the time-$t$ value of the bank account (cash) portion of the self-financing portfolio is, upon using (3.40),

$$\beta_t B(t) = \Pi_t - \Delta_t S(t) = C(t,S) - S\,\Delta_c(t,S) = -\mathrm{e}^{-r\tau}\,K\mathcal{N}\left(d_-\left(\frac{S}{K},\tau\right)\right).$$

This quantity is negative for all $t < T$, $S > 0$. Hence, the writer of the call is always maintaining a *negative position (loan) in the bank account.*

The delta position at maturity $t = T$ is given by the limit $t \nearrow T$ of the function in (3.47). From (3.44) we see that, in the limit of zero time to maturity, Δ_c approaches the unit step function with discontinuity at $S = K$. Moreover, for any fixed $\tau > 0$, we observe that Δ_c is a strictly increasing function of S with limiting values of zero and unity:

$$\lim_{S\searrow 0} \Delta_c(t,S) = \mathcal{N}(-\infty) = 0 \quad \text{and} \quad \lim_{S\nearrow\infty} \Delta_c(t,S) \to \mathcal{N}(\infty) = 1\,.$$

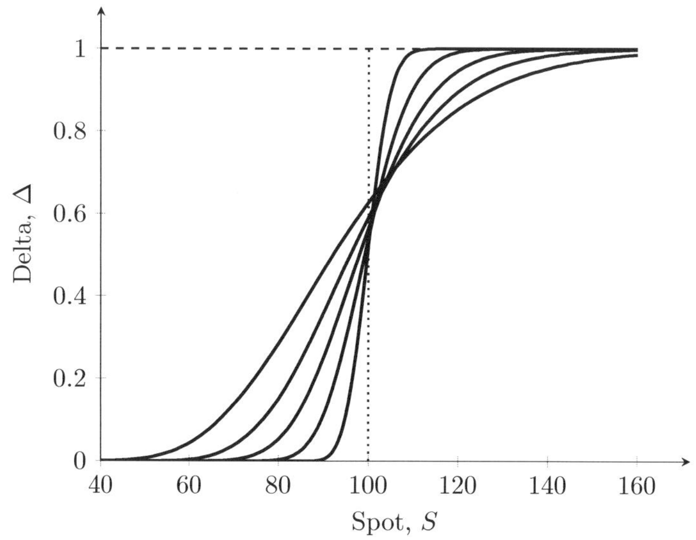

FIGURE 3.4: Plots of the call delta as a function of spot price S under the (B, S) model with $K = 100$, $r = 5\%$, and $\sigma = 25\%$. The time to maturity τ takes the following values: 1, 0.5, 0.25, 0.1, and 0.025. The graph becomes steeper, as τ approaches 0.

Figure 3.4 contains typical plots of Δ_c for different values of τ. We clearly see that Δ_c approaches the unit step function $H(S - K)$, as τ approaches 0. As $\tau \searrow 0$, (3.47) gives $\Delta_c \to 0$ for $S < K$, $\Delta_c \to 1$ for $S > K$, and $\Delta_c \to \frac{1}{2}$ for $S = K$. We see that Δ_c gets progressively steeper for smaller and smaller values of τ and eventually becomes the unit step function. At a time t just before expiry (i.e., $\tau \approx 0$), if the stock price lies above K, then $\Delta_c \approx 1$ and the hedging replicates the positive payoff of the call after paying off the loan in the amount of K (see (3.44) and the above expression for $\beta_t B(t)$). On the other hand, if the stock price is below K (out of the money), then $\Delta_c \approx 0$ and $\beta_t B(t) \approx 0$, so the hedge replicates the zero payoff of the call.

There are also scenarios where the stock price $S(t)$ just before expiry ($t \approx T$) stays very close to the strike K until time T. There is a fairly high probability (which is easily computed) that the stock price will fluctuate between values below K and above K. These are scenarios where the stock is said to be "pinning the strike." For values of τ close to zero, this would mean that the (hedge) position in the stock would have to be re-balanced, in a short time, between a very small long position ($\Delta_c \approx 0$), and a large long position where the replicating portfolio is almost all stock ($\Delta_c \approx 1$). This re-balancing requires either selling off a large portion of the underlying stock or buying up a large portion of stock so as to maintain the correct hedge. In particular, if the stock price suddenly moves just before expiry from one side of the strike to the other, then the writer or trader must rapidly trade enough of the underlying stock before expiration in order to hedge the loss against such a movement. In the (B, S) theoretical (idealized) model such transactions are assumed to occur instantaneously (as efficiently as is required!) and without any liquidity issues in trading; i.e., all scenarios are, in theory, hedged. In the real world, these kinds of scenarios cannot be hedged effectively in time since instantaneous re-balancing is obviously not possible. Moreover, there are transaction fees associated with each trade. The risk associated with options trading whereby the market price of the stock is pinning the strike is referred to as *pin risk*. Later we consider the pricing of a so-called soft-strike call. This contract differs from the standard call, as its payoff is everywhere differentiable and it has a continuous range of strikes. This type of option avoids the problem of pin risk when delta hedging.

The delta of a put option, Δ_p, follows trivially by put-call parity,

$$\Delta_p(t, S) = \frac{\partial}{\partial S}\left(C(t, S) - S + \mathrm{e}^{-r\tau}K\right) = \Delta_c(t, S) - 1 = \mathcal{N}\left(d_+\left(\frac{S}{K}, \tau\right)\right) - 1$$
$$= -\mathcal{N}\left(-d_+\left(\frac{S}{K}, \tau\right)\right). \quad (3.49)$$

The delta of a put is therefore strictly negative and is simply related to the call delta by $\Delta_p = \Delta_c - 1$. A put option is replicated by continuously re-balancing with a *delta negative (short) position in the stock* given by $\Delta_t \equiv \Delta_p(t, S(t)) = -\mathcal{N}\left(-d_+\left(\frac{S(t)}{K}, \tau\right)\right)$. Given spot $S(t) = S$, the time-t value of the stock portion of the replicating portfolio for the put is given by

$$\Delta_t S(t) = S\,\Delta_p(t, S) = -S\mathcal{N}\left(-d_+\left(\frac{S}{K}, \tau\right)\right).$$

The self-financing replicating portfolio value for the put is $\Pi_t = P(t, S(t)) = P(t, S)$, so the corresponding time-t investment in the bank account is

$$\beta_t B(t) = P(t, S) - S\,\Delta_p(t, S) = \mathrm{e}^{-r\tau}\,K\mathcal{N}\left(-d_-\left(\frac{S}{K}, \tau\right)\right).$$

In direct contrast to the call, the put is replicated by always maintaining a *positive investment in the bank account.*

At maturity $t = T$, $\Delta_p(T, S) = \Delta_c(T, S) - 1 = H(S - K) - 1$. For $\tau > 0$, Δ_p is a strictly increasing function of S with limiting values:

$$\lim_{S \searrow 0} \Delta_p(t, S) = -1 \quad \text{and} \quad \lim_{S \nearrow \infty} \Delta_p(t, S) \to 0\,.$$

The above discussion on hedging and pin risk associated with the call also applies in an obviously similar manner to the put with given strike K. We note that for the put delta the corresponding plots of Δ_p as a function of S are as in Figure 3.4, where the origin of the vertical axis is simply shifted up by unity.

The delta is one among other "*Greeks*" of an option that are of interest to practitioners. The Greeks are used to measure the sensitivity of an option portfolio or an individual derivative to small changes in underlying parameters. Here is a list of some commonly used Greeks.

Delta $\Delta = \frac{\partial V}{\partial S}$ is the rate of change in V w.r.t. the price of the underlying security.

Gamma $\Gamma = \frac{\partial^2 V}{\partial S^2}$ is the rate of change in Δ w.r.t. the price of the underlying security.

Theta $\theta = \frac{\partial V}{\partial t} = -\frac{\partial V}{\partial \tau}$ is the rate of change in V w.r.t. time.

Vega $\mathcal{V} = \frac{\partial V}{\partial \sigma}$ is the rate of change in V w.r.t. a change in volatility.

Rho $\rho = \frac{\partial V}{\partial r}$ is the rate of change in V w.r.t. the interest rate.

As demonstrated above for the delta, in the (B, S) model, the Greeks are obtained by differentiating the value function V with respect to the corresponding parameter. This chapter is focused on risk-neutral pricing and hedging of options. We leave it as an exercise for the reader to derive the corresponding formulae for the gamma, theta, vega, and rho of a standard call and put expressed as functions of S, τ, K, r, and σ.

The delta, gamma, and other Greeks are also useful in estimating values of portfolios in stock and options. Consider a portfolio whose value $\Pi \equiv \Pi(t, S)$ is a function of spot S and time t. Suppose that the stock price changes from S to $S + \delta S$ over a small time interval $[t, t + \delta t]$. According to Taylor's formula, the change in the portfolio value approximately equals $\frac{\partial \Pi}{\partial S} \delta S$. Thus, the portfolio value can be approximated using its delta:

$$\Pi(t + \delta t, S + \delta S) \cong \Pi(t, S) + \Delta_\Pi(t, S)\, \delta S.$$

More accurate approximations involve other Greeks such as theta and gamma:

$$\Pi(t + \delta t, S + \delta S) \cong \Pi(t, S) + \theta_\Pi(t, S)\delta t + \Delta_\Pi(t, S)\, \delta S,$$
$$\Pi(t + \delta t, S + \delta S) \cong \Pi(t, S) + \theta_\Pi(t, S)\delta t + \Delta_\Pi(t, S)\, \delta S + \Gamma_\Pi(t, S)\, \frac{(\delta S)^2}{2}.$$

Example 3.1. An investor writes and sells three-month call and put options with strike price $K = \$55$ on stock with initial price $S_0 = \$50$. Assume a log-normal price model with $r = 5\%$ and $\sigma = 30\%$.

(a) Find the no-arbitrage initial price, delta, theta, and gamma of each option.

(b) One month later, the stock price rises to \$55. Estimate the value of the portfolio with 100 call options using the Δ-, $\Delta\theta$-, and $\Delta\Gamma$-approximations and compare the estimates with the exact Black–Scholes price.

Solution. To price European call and put options, we first calculate the values of $d_\pm$, $\mathcal{N}(d_\pm)$, and $\mathcal{N}(-d_\pm) = 1 - \mathcal{N}(d_\pm)$:

$$\begin{aligned} &d_+ \cong -0.47706787, \quad &&\mathcal{N}(d_+) \cong 31.6657\%, \quad &&\mathcal{N}(-d_+) \cong 68.3343\%, \\ &d_- \cong -0.62706787, \quad &&\mathcal{N}(d_-) \cong 26.5307\%, \quad &&\mathcal{N}(-d_-) \cong 73.4693\%. \end{aligned}$$

(a) For single call and put options with $T = 3/12$ and $K = \$55$, we have

$$\begin{aligned} &C(0, 50) \cong \$1.42220, \quad &&P(0, 50) \cong \$5.73898, \\ &\Delta_c(0, 50) \cong 0.31666, \quad &&\Delta_p(0, 50) \cong -0.68334, \\ &\theta_c(0, 50) \cong -6.06101, \quad &&\theta_p(0, 50) \cong -3.34517, \\ &\Gamma_c(0, 50) \cong 0.04747, \quad &&\Gamma_p(0, 50) \cong 0.04747. \end{aligned}$$

The subscript (c for a call and p for a put) indicates the type of option for which the Greeks are calculated.

(b) At time $t = 1/12$, the exact value of the 100 call options with strike $K = 55$ and expiry $T = \frac{3}{12}$ is

$$V = 100 \cdot C({}^{1}\!/_{12}, 55) \cong 100 \cdot \$2.9089 = \$290.89.$$

Now, we find approximations for $V \equiv V({}^{1}\!/_{12}, 55)$ using the three approaches. Notice that $\delta S = 55 - 50 = 5$ and $\delta t = 1/12$.

Δ-approximation: $V \approx 100 \cdot C(0, 50) + 100 \cdot \Delta_c(0, 50) \cdot 5 \cong \$300.57.$

$\Delta\Gamma$-approximation: $V \approx 100 \cdot C(0, 50) + 100 \cdot \Delta_c(0, 50) \cdot 5 + \frac{1}{2} \cdot 100 \cdot \Gamma_c(0, 50) \cdot 5^2 \cong \$359.91.$

$\Delta\theta$-approximation: $V \approx 100 \cdot C(0, 50) + 100 \cdot \Delta_c(0, 50) \cdot 5 + 100 \cdot \theta_c(0, 50) \cdot \frac{1}{12} \cong \$250.06.$

□

The writer of a European option receives a cash premium upfront but has potential liabilities later on the option exercise date. The writer's profit or loss is the reverse of that for the purchaser of the option. For example, the profit of a call option writer is $e^{rT} C - (S(T) - K)^+$, where C is the no-arbitrage initial call value. If the option ends up deep in the money when $S(T) \gg K$, the writer is exposed to the risk of a large loss.

The writer of an option may reduce their risk over a small time interval by forming a suitable portfolio in the underlying security called a hedge or a hedging portfolio. In reality, it is impossible to hedge in a perfect way by designing a single (static) portfolio to be held for the whole period. The hedge has to be rebalanced dynamically to adapt it to changes in risk factors that affect the option value.

For small changes δt, δS, $\delta\sigma$, δr of the respective variables, the change in value of the portfolio value is approximated by Taylor's formula:

$$\begin{aligned}
\delta V &:= V(t + \delta t, S + \delta S; \sigma + \delta\sigma, r + \delta r) - V(t, S; \sigma, r) \\
&\cong \frac{\partial V}{\partial S}\delta S + \frac{1}{2}\frac{\partial^2 V}{\partial S^2}(\delta S)^2 + \frac{\partial V}{\partial t}\delta t + \frac{\partial V}{\partial \sigma}\delta\sigma + \frac{\partial V}{\partial r}\delta r \\
&= \Delta_V\, \delta S + \frac{1}{2}\Gamma_V\, (\delta S)^2 + \theta_V\, \delta t + \mathcal{V}_V\, \delta\sigma + \rho_V\, \delta r.
\end{aligned}$$

To immunize the portfolio against small changes of a selected variable, we let the corresponding Greek be equal to zero. The resulting portfolio is said to be *neutral* relative to the Greek selected. In particular, a *delta-neutral* portfolio is protected against small changes of the underlying security price, and a *vega-neutral* portfolio is less sensitive to volatility movements.

As an example, let us consider the portfolio $(-1, x, y)$ composed of x stock shares with the current price S, a cash amount of y dollars invested without risk, and a short derivative security (i.e., its portfolio position equals -1) to be delta-hedged. Its time-t value is

$$\Pi(t, S) \equiv \Pi_t^{(-1,x,y)} = -V(t, S) + xS + y,$$

where $V(t, S)$ is the time-t price of the derivative for the spot value of $S(t) = S$. The delta of the portfolio is $\frac{\partial}{\partial S}\Pi(t, S) = -\Delta_V(t, S) + x$. If $\frac{\partial}{\partial S}\Pi(S) = 0$, then $x = \Delta_V(t, S)$, where $\Delta_V(t, S)$ is the delta of the derivative security. For any cash amount y, the delta of this portfolio is zero. It is convenient to choose the amount y so that the initial portfolio value is zero as well.

Example 3.2. Consider an investor armed with a log-normal pricing model, who writes and sells the call and put options, as described in Example 3.1.

(a) Construct a delta-neutral, zero-wealth portfolio that includes 100 short call options and a number of stock shares.

(b) Construct a delta-gamma-neutral portfolio that includes 100 short call options, stock shares, and three-month put options with the same strike price and maturity so that the initial portfolio value is zero.

Solution. (a) The time-t value of the portfolio with 100 short call options, x shares of stock worth xS dollars, and y dollars invested without risk is $\Pi(t, S) = -100 \cdot C(t, S) + x \cdot S + y$. The delta of this portfolio is $-100 \cdot \Delta_c(t, S) + x$. It is zero if $x = 100 \cdot \Delta_c(t, S)$. At time $t = 0$, the investor buys $x = 100 \cdot \Delta_c(0, 50) \cong 31.666$ shares of stock. The initial portfolio value is zero if

$$y = 100 \cdot C(0, 50) - x \cdot 50 \cong 142.220 - 1583.30 = -1441.08.$$

Since y is negative, the investor needs to borrow y dollars from the risk-free bank account. (b) Let us include z put options in the portfolio. Its time-t value is now given by

$$\Pi(t, S) = -100C(t, S) + x \cdot S + y + z \cdot P(t, S).$$

The delta and gamma of the portfolio are, respectively,

$$\Delta(t, S) = -100 \cdot \Delta_c(t, S) + x + z \cdot \Delta_p(t, S),$$
$$\Gamma(t, S) = -100 \cdot \Gamma_c(t, S) + z \cdot \Gamma_p(t, S).$$

At time $t = 0$, both Greeks and the portfolio value are zero:

$$\begin{cases} \Delta(0, 50) = -31.666 + x - 0.68334z = 0, \\ \Gamma(0, 50) = -4.747 + 0.04747z = 0, \\ \Pi(0, 50) = -142.220 + 50x + y + 53.73898z = 0. \end{cases}$$

Here, we used call and put values and their Greeks from Example 3.1. The solution to the above system of linear equations is

$$x = 100, \quad y = -5431.678, \quad z = 100.$$

So, the investor takes a long position on the stock and the put option, as well as a short position on the bank account. □

3.2.3 Europeans with Piecewise Linear Payoffs

Equations (3.38) and (3.39) are useful for pricing any European option having a *piecewise linear payoff* with possibly a finite (or countable) number of discontinuities. Actually, the standard call and put options are important special cases of such payoffs with no discontinuity and a discontinuity in their first derivatives. Let us now consider a simple example of a piecewise constant payoff with a single jump discontinuity.

Example 3.3. (Asset-or-Nothing Binary Call) Consider an option that pays the holder the value of the underlying share price of the stock if the stock price at expiry is above a given strike K and is otherwise worthless; i.e., the holder gets the asset (stock) or nothing with payoff function

$$\Lambda(S) = S\,\mathbb{I}_{\{S \geqslant K\}} = \begin{cases} S & \text{if } S \geqslant K, \\ 0 & \text{if } S < K. \end{cases}$$

Derive the risk-neutral pricing formula and the hedging position in the stock for the European option with this payoff.

Solution. Observe that the payoff is also the first term in the standard call option. The risk-neutral pricing formula immediately follows from (3.38):

$$V(t,S) = \mathrm{e}^{-r\tau}\,\widetilde{\mathrm{E}}_{t,S}\left[S(T)\,\mathbb{I}_{\{S(T)\geqslant K\}}\right] = S\,\mathcal{N}\left(d_+\big(\frac{S}{K},\tau\big)\right),$$

where $\tau = T - t$. The delta hedge is obtained by straightforward differentiation,

$$\Delta(t,S) = \mathcal{N}\left(d_+\big(\frac{S}{K},\tau\big)\right) + \frac{1}{\sigma\sqrt{\tau}}\,n\left(d_+\big(\frac{S}{K},\tau\big)\right).$$

Note that the time-t replicating portfolio is always long in the stock with a bank loan in the amount of $|V(t,S) - S\Delta(t,S)| = \frac{S}{\sigma\sqrt{\tau}}\,n\left(d_+\big(\frac{S}{K},\tau\big)\right)$. □

Recall that a cash-or-nothing binary call has payoff $\Lambda(S) = \mathbb{I}_{\{S\geqslant K\}}$. Hence, a standard call struck at K is equivalent to a portfolio consisting of K short positions in a cash-or-nothing call and one long position in an asset-or-nothing call, both struck at K. In total, we have four standard binary options. The risk-neutral pricing formulae for the other three options are provided below.

Asset-or-nothing put: $V(t,S) = \mathrm{e}^{-r\tau}\,\widetilde{\mathrm{E}}_{t,S}\left[S(T)\,\mathbb{I}_{\{S(T)<K\}}\right] = S\,\mathcal{N}\left(-d_+\big(\frac{S}{K},\tau\big)\right)$;

Cash-or-nothing call: $V(t,S) = \mathrm{e}^{-r\tau}\,\widetilde{\mathrm{E}}_{t,S}\left[\mathbb{I}_{\{S(T)\geqslant K\}}\right] = \mathrm{e}^{-r\tau}K\,\mathcal{N}\left(d_-\big(\frac{S}{K},\tau\big)\right)$;

Cash-or-nothing put: $V(t,S) = \mathrm{e}^{-r\tau}\,\widetilde{\mathrm{E}}_{t,S}\left[\mathbb{I}_{\{S(T)<K\}}\right] = \mathrm{e}^{-r\tau}K\,\mathcal{N}\left(-d_-\big(\frac{S}{K},\tau\big)\right)$.

Typical plots of binary options are given in Figures 3.5 and 3.6

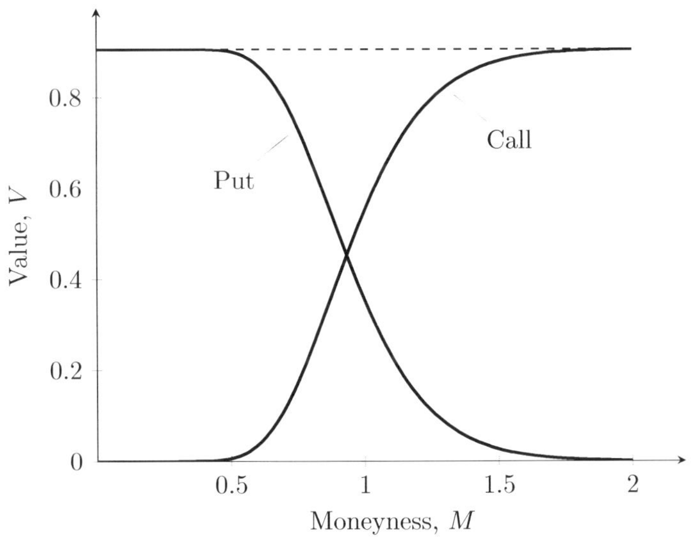

FIGURE 3.5: No-arbitrage values of cash-or-nothing call and put options as functions of moneyness $M := S/K$ under the (B,S) model with $\tau = 1$, $K = 100$, $r = 10\%$, and $\sigma = 25\%$. The dashed line represents the limiting value equal to $\mathrm{e}^{-r\tau} = \mathrm{e}^{-0.1} \cong 0.904837$.

We now consider a derivation of the risk-neutral pricing formula for a European option with arbitrary piecewise linear payoff, assuming the standard GBM model for the stock.

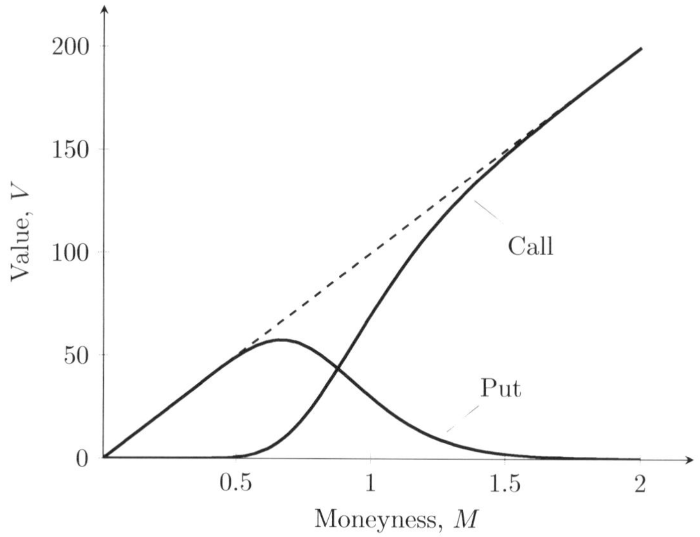

FIGURE 3.6: No-arbitrage values of asset-or-nothing call and put options as functions of moneyness $M := S/K$ under the (B, S) model with $\tau = 1$, $K = 100$, $r = 10\%$, and $\sigma = 25\%$. The dashed line has the equation $V = KM$.

As a general form for the payoff we consider the sum of linear functions restricted to any number $n \geqslant 1$ of nonoverlapping intervals:

$$\Lambda(S) = \sum_{i=1}^{n} (A_i S + B_i) \, \mathbb{I}_{\{a_i \leqslant S < b_i\}} \tag{3.50}$$

with any real constants A_i, B_i and where $0 \leqslant a_1 < b_1 \leqslant a_2 < b_2 \leqslant \ldots \leqslant a_n < b_n \leqslant \infty$. This function can account for a number of jump discontinuities (including no discontinuities), piecewise constant payoffs, and piecewise linear payoffs with any combination of positive and negative slopes.

As a concrete example, let $n = 1$. A call with $\Lambda(S) = (S - K)^+$ is then given by (3.50) with parameter choice $A_1 = 1$, $B_1 = -K$, $a_1 = K$, $b_1 = \infty$. A cash-or-nothing binary call with $\Lambda(S) = \mathbb{I}_{\{S \geqslant K\}}$ obtains with $A_1 = 0$, $B_1 = 1$, $a_1 = K$, $b_1 = \infty$. The payoff of an asset-or-nothing call, $\Lambda(S) = S \, \mathbb{I}_{\{S \geqslant K\}}$, corresponds to $A_1 = 1$, $B_1 = 0$, $a_1 = K$, $b_1 = \infty$. For $n = 2$, an example is the butterfly spread with strikes $K_1 < K_2 < K_3$, $K_2 = (K_1 + K_3)/2$, which is representable as a linear combination of piecewise linear functions with payoff $\Lambda(S) = (S - K_1) \, \mathbb{I}_{\{K_1 \leqslant S < K_2\}} + (K_3 - S) \, \mathbb{I}_{\{K_2 \leqslant S < K_3\}}$. This corresponds to setting the parameters in (3.50) to $A_1 = 1, B_1 = -K_1, a_1 = K_1, b_1 = K_2$ and $A_2 = -1, B_2 = K_3, a_2 = K_2, b_2 = K_3$.

In order to make use of the conditional expectation identities in (3.38) and (3.39), we write $\mathbb{I}_{\{a_i \leqslant S < b_i\}} = \mathbb{I}_{\{S \geqslant a_i\}} - \mathbb{I}_{\{S \geqslant b_i\}}$ and express the payoff in (3.50) as

$$\Lambda(S) = \sum_{i=1}^{n} \left\{ A_i S \big(\mathbb{I}_{\{S \geqslant a_i\}} - \mathbb{I}_{\{S \geqslant b_i\}} \big) + B_i \big(\mathbb{I}_{\{S \geqslant a_i\}} - \mathbb{I}_{\{S \geqslant b_i\}} \big) \right\}.$$

Applying the identities in (3.38) and (3.39) and the linearity property of expectations, we arrive at an analytical expression for the risk-neutral pricing formula of a European option with payoff in (3.50):

$$\begin{aligned} V(t,S) &= \mathrm{e}^{-r\tau}\,\widetilde{\mathrm{E}}_{t,S}\big[\Lambda(S(T))\big] \\ &= \mathrm{e}^{-r\tau}\sum_{i=1}^{n}\Big\{A_i\Big(\widetilde{\mathrm{E}}_{t,S}\big[S(T)\mathbb{I}_{\{S(T)\geqslant a_i\}}\big] - \widetilde{\mathrm{E}}_{t,S}\big[S(T)\mathbb{I}_{\{S(T)\geqslant b_i\}}\big]\Big) \\ &\qquad + B_i\Big(\widetilde{\mathbb{P}}_{t,S}\big(S(T)\geqslant a_i\big) - \widetilde{\mathbb{P}}_{t,S}\big(S(T)\geqslant b_i\big)\Big)\Big\} \\ &= \sum_{i=1}^{n}\Big\{A_i S\left[\mathcal{N}\Big(d_+(\frac{S}{a_i},\tau)\Big) - \mathcal{N}\left(d_+(\frac{S}{b_i},\tau)\right)\right] \\ &\qquad + \mathrm{e}^{-r\tau}B_i\left[\mathcal{N}\Big(d_-(\frac{S}{a_i},\tau)\Big) - \mathcal{N}\left(d_-(\frac{S}{b_i},\tau)\right)\right]\Big\} \end{aligned} \tag{3.51}$$

where $\tau = T - t$ and $d_\pm(x,\tau)$ functions defined in (3.37). We leave as an exercise the derivation of a general formula for the delta hedging position $\Delta(t,S)$ for this option. The pricing formula in (3.51) is applicable to several payoff forms that occur in practice and we leave some as assigned exercises at the end of this chapter.

Another approach to pricing derivatives with piecewise linear payoffs requires finding an equivalent portfolio that includes stock, cash, and standard call and put options with the same exercise date but possibly different strikes. If the total payoff of such a portfolio on the exercise date is the same as that of the target payoff function, then the portfolio replicates the derivative. By the law of one price, the no-arbitrage value of this portfolio equals the fair price of the replicated derivative.

In principle, any continuous piecewise payoff function can be manufactured from European call and put payoffs with different strikes. It can be an exact (e.g., for piecewise linear functions) or approximate representation. Practitioners often use standard European options as building blocks to create more sophisticated financial instruments. Let us review a few examples of such synthetic derivatives, which we call here *strategies*.

A *spread strategy* involves taking a position in two or more options of the same type with the same expiry date. For example, a bear (bull) spread is obtained by taking a position in one long and one short call (put) options. A *combination strategy* is an option portfolio that involves taking a position in both calls and puts on the same underlying security. Here are some examples of combinations.

Straddle combines one put and one call with the same strike and expiry date.

Strip combines one long call and two long puts with the same strike and expiry date.

Strap combines two long calls and one long put with the same strike and expiry date.

Strangle combines one put and one call with the same expiry date but different strikes.

Example 3.4.

Replicate a condor spread using (a) call options only; (b) put options only. The condor payoff is plotted in the figure on the right, where K_i with $i = 1,2,3,4$ are so that $0 < K_1 < K_2 < K_3 < K_4$, and $A > 0$ is the payoff value for $S \in [K_2, K_3]$.

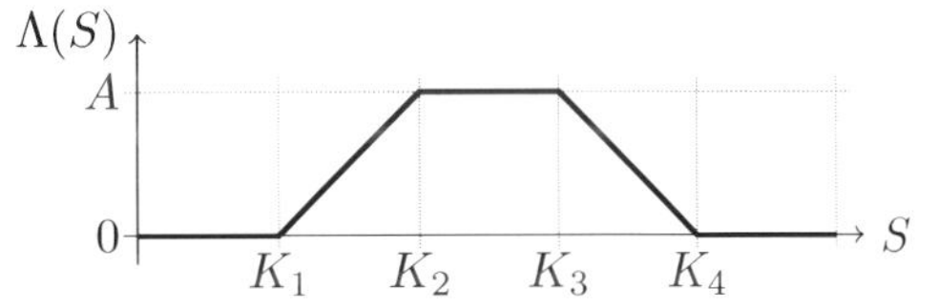

Solution. (a) Construct a portfolio of European calls that has the same payoff as the condor option step by step, starting from the origin, and moving from K_1 to K_4, and beyond.

Start with $\frac{A}{K_2-K_1}$ long calls struck at K_1. $\quad \frac{A}{K_2-K_1}(S-K_1)^+$

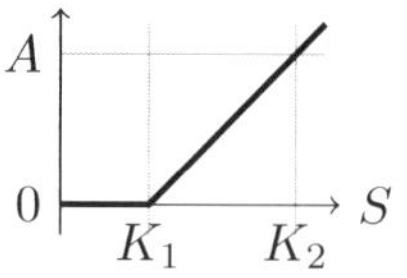

Add $\frac{A}{K_2-K_1}$ short calls struck at K_2. $\quad -\frac{A}{K_2-K_1}(S-K_2)^+$

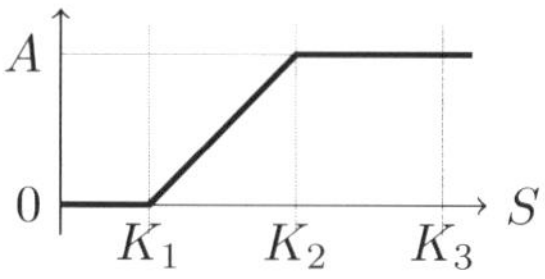

Add $\frac{A}{K_4-K_3}$ short calls struck at K_3. $\quad -\frac{A}{K_4-K_3}(S-K_3)^+$

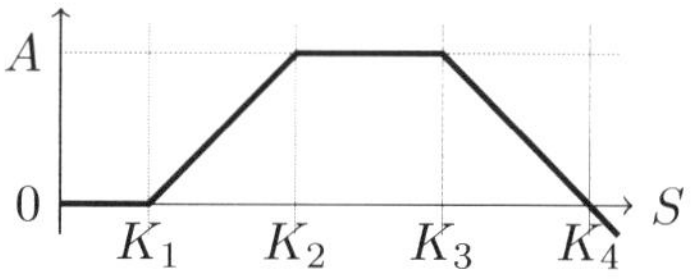

Add $\frac{A}{K_4-K_3}$ long calls struck at K_4. $\quad +\frac{A}{K_4-K_3}(S-K_4)^+$

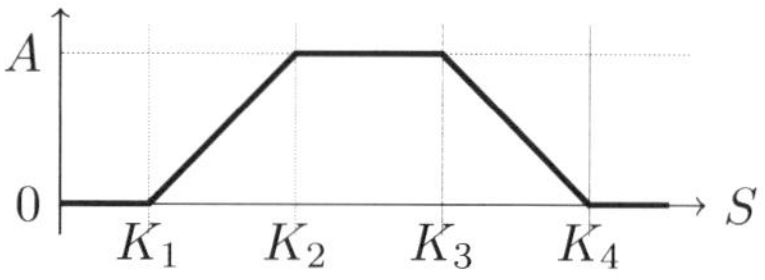

Thus, the no-arbitrage pricing formula for the spread with maturity time T is as follows:

$$V(t,S) = \frac{A}{K_2-K_1}\big(C(t,S;T,K_1) - C(t,S;T,K_2)\big) + \frac{A}{K_4-K_3}\big(C(t,S;T,K_4) - C(t,S;T,K_3)\big).$$

(b) Now construct an equivalent portfolio of European puts moving from the right of K_4 to K_1 and back to the origin.

Start with $\frac{A}{K_4-K_3}$ long puts struck at K_4. $\quad \frac{A}{K_4-K_3}(K_4-S)^+$

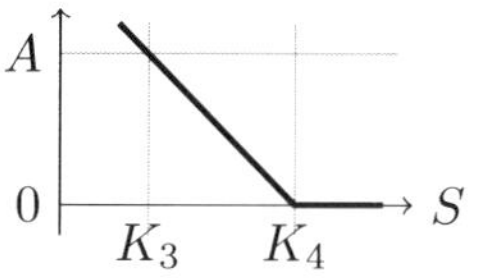

Add $\frac{A}{K_4-K_3}$ short puts struck at K_3. $\quad -\frac{A}{K_4-K_3}(K_3-S)^+$

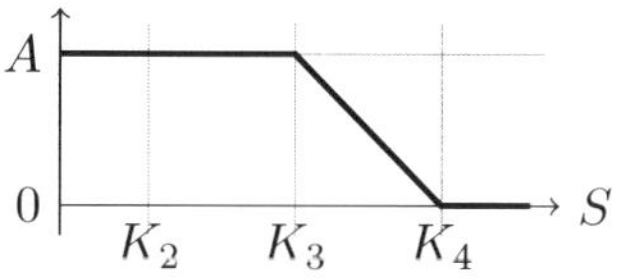

Add $\frac{A}{K_2-K_1}$ short puts struck at K_2. $\quad -\frac{A}{K_2-K_1}(K_2-S)^+$

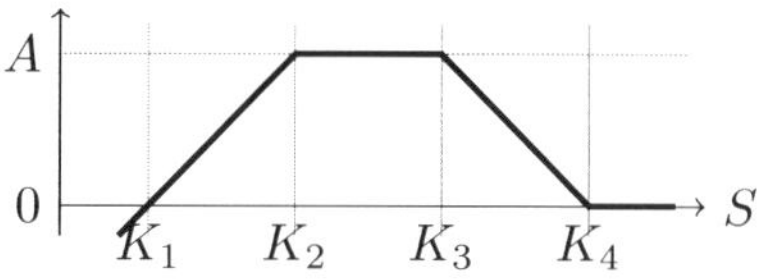

Add $\frac{A}{K_2-K_1}$ long puts struck at K_1. $\quad +\frac{A}{K_2-K_1}(K_1-S)^+$

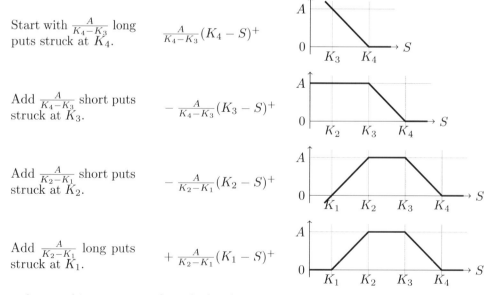

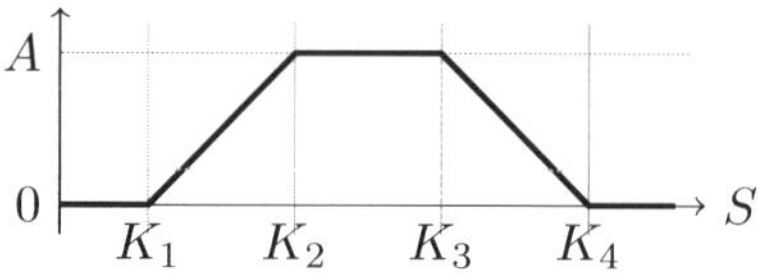

Thus, the no-arbitrage pricing formula for the spread with maturity time T is as follows:

$$V(t,S) = \frac{A}{K_2-K_1}\big(P(t,S;T,K_1) - P(t,S;T,K_2)\big) + \frac{A}{K_4-K_3}\big(P(t,S;T,K_4) - P(t,S;T,K_3)\big).$$

It is easy to show by employing the put-call parity relation that the pricing formulae derived in (a) and (b) are equivalent.

□

3.2.4 Power Options

Power options differ from vanilla European options in that the payoff function is not linear but raised to some power in the underlying spot. We now show how to analytically value such options under the GBM model. Typically, the payoff of a power option is a quadratic function of the stock price. The widest possible application of power options is for addressing the nonlinear risk of option sellers. There was proposed a class of soft-strike options which do not have a single fixed strike price but a continuous range of strikes spread over an interval. As was mentioned earlier, such options allow for addressing limitations of a standard delta hedging when the underlying asset is pinning the strike at the expiration of the option.

More generally, the payoff of a power option may involve the terminal stock price raised to some power, e.g., $S^\alpha(T) = (S(T))^\alpha$ with either positive or negative exponent $\alpha \neq 0$, as well as other terms involving $S^\alpha(T)$ times an indicator function restricting the value of $S(T)$ on some interval. Let us assume that the payoff is some linear combination of elemental payoffs having any of the three forms

$$S^\alpha(T) \quad \text{or} \quad S^\alpha(T)\,\mathbb{I}_{\{A_1<S(T)\leqslant A_2\}} \quad \text{or} \quad S^\alpha(T)\,\mathbb{I}_{\{S(T)>A\}}$$

where $A_1 < A_2$ and A are nonnegative constants. Hence, by the risk-neutral derivative pricing formulation the present value of a power option at time $t < T$ will involve expectations of the above payoffs under the risk-neutral measure $\widetilde{\mathbb{P}}$. We can handle all of the above three payoffs by simply deriving a formula for the conditional expectation of the payoff $S^\alpha(T)\,\mathbb{I}_{\{S(T)>A\}}$, for any $A \geqslant 0$, since

$$S^\alpha(T)\,\mathbb{I}_{\{A_1<S(T)\leqslant A_2\}} = S^\alpha(T)\,\mathbb{I}_{\{S(T)>A_1\}} - S^\alpha(T)\,\mathbb{I}_{\{S(T)>A_2\}}\,. \tag{3.52}$$

Note also that for $A_1 = 0$ and $A_2 = A$ we have

$$S^\alpha(T)\,\mathbb{I}_{\{0<S(T)\leqslant A\}} = S^\alpha(T) - S^\alpha(T)\,\mathbb{I}_{\{S(T)>A\}}$$

where $\mathbb{I}_{\{S(T)>0\}} = 1$ since the stock price is always positive. The expectation of $S^\alpha(T)$, conditional on a given spot value $S(t) = S > 0$, for any $t < T$, is easily calculated using (3.25) raised to the exponent α. Again we use the fact that $\widetilde{Z}$ in (3.26) is independent of $S(t)$ (under measure $\widetilde{\mathbb{P}}$), which allows us to remove the conditioning upon setting $S(t) = S$:

$$\begin{aligned}
\widetilde{\mathrm{E}}_{t,S}\big[S^\alpha(T)\big] \equiv \widetilde{\mathrm{E}}\big[S^\alpha(T)\,\big|\,S(t)=S\big] &= \mathrm{e}^{\alpha(r-\sigma^2/2)(T-t)}\,\widetilde{\mathrm{E}}\big[S^\alpha(t)\,\mathrm{e}^{\alpha\sigma\sqrt{T-t}\widetilde{Z}}\,\big|\,S(t)=S\big] \\
&= S^\alpha\,\mathrm{e}^{\alpha(r-\sigma^2/2)(T-t)}\,\widetilde{\mathrm{E}}\big[\mathrm{e}^{\alpha\sigma\sqrt{T-t}\widetilde{Z}}\big] \\
&= S^\alpha\,\mathrm{e}^{\alpha(r-\sigma^2/2)(T-t)}\,\mathrm{e}^{\frac{1}{2}\alpha^2\sigma^2(T-t)} \\
&= S^\alpha\,\mathrm{e}^{\alpha(r+\frac{1}{2}\sigma^2(\alpha-1))\tau}
\end{aligned} \tag{3.53}$$

where $\tau := T-t$. The only difference between the conditional expectation in (3.53) and that of $S^\alpha(T)\,\mathbb{I}_{\{S(T)>A\}}$ is the indicator function term. By the exact same step as in our derivation of the standard call and put options, the indicator random variable term simplifies,

$$\mathbb{I}_{\left\{\ln\frac{S(T)}{S(t)}>\ln\frac{A}{S}\right\}} = \mathbb{I}_{\{\widetilde{Z}>-d_-(\frac{S}{A},\tau)\}}\,,$$

with $d_\pm$ defined in (3.37), i.e., $d_\pm(\frac{S}{A},\tau) = \frac{\ln(S/A)+(r\pm\frac{1}{2}\sigma^2)\tau}{\sigma\sqrt{\tau}}$. Then, conditioning on $S(t) = S$ and using independence,

$$\begin{aligned}
\widetilde{\mathrm{E}}_{t,S}\left[S^\alpha(T)\,\mathbb{I}_{\{S(T)>A\}}\right] &= \mathrm{e}^{\alpha(r-\sigma^2/2)\tau}\,\widetilde{\mathrm{E}}\left[S^\alpha(t)\,\mathbb{I}_{\{\widetilde{Z}>-d_-(\frac{S(t)}{A},\tau)\}}\,\mathrm{e}^{\alpha\sigma\sqrt{\tau}\widetilde{Z}}\,\middle|\,S(t)=S\right] \\
&= S^\alpha\,\mathrm{e}^{\alpha(r-\sigma^2/2)\tau}\,\widetilde{\mathrm{E}}\big[\mathbb{I}_{\{\widetilde{Z}>-d_-(\frac{S}{A},\tau)\}}\,\mathrm{e}^{\alpha\sigma\sqrt{\tau}\widetilde{Z}}\big] \\
&= S^\alpha\,\mathrm{e}^{\alpha(r+\frac{1}{2}\sigma^2(\alpha-1))\tau}\,\mathcal{N}\left(d_+(\frac{S}{A},\tau) + (\alpha-1)\sigma\sqrt{\tau}\right).
\end{aligned} \tag{3.54}$$

The last expectation was computed using the identity in (B.1) of Appendix B and noting that $d_-(x,\tau) = d_+(x,\tau) - \sigma\sqrt{\tau}$. Note that this formula also recovers (3.53) in the limit $A \searrow 0$. This follows by monotone convergence of the expectations where $\mathbb{I}_{\{S(T)>A\}} \nearrow \mathbb{I}_{\{S(T)>0\}} = 1$, as $A \searrow 0$. Based on (3.52) and the linearity property of the expectation, using (3.54) for $A = A_1$ and for $A = A_2$ leads to the formula

$$\begin{aligned}
&\widetilde{\mathrm{E}}_{t,S}\left[S^\alpha(T)\,\mathbb{I}_{\{A_1<S(T)\leqslant A_2\}}\right] = \widetilde{\mathrm{E}}_{t,S}\left[S^\alpha(T)\,\mathbb{I}_{\{S(T)>A_1\}}\right] - \widetilde{\mathrm{E}}_{t,S}\left[S^\alpha(T)\,\mathbb{I}_{\{S(T)>A_2\}}\right] \\
&= S^\alpha\,\mathrm{e}^{\alpha(r+\frac{1}{2}\sigma^2(\alpha-1))\tau}\left[\mathcal{N}\left(d_-(\frac{S}{A_1},\tau)+\alpha\sigma\sqrt{\tau}\right) - \mathcal{N}\left(d_-(\frac{S}{A_2},\tau)+\alpha\sigma\sqrt{\tau}\right)\right]. \quad (3.55)
\end{aligned}$$

Let us apply equations (3.53) and (3.54) to pricing power options with quadratic payoffs.

Example 3.5. Derive the no-arbitrage pricing formula for each European-style option with the following payoffs:

(a) $\Lambda(S) = (K - S)^2$;

(b) $\Lambda(S) = (K^2 - S^2)^+$;

where $K > 0$ is a fixed strike.

Solution. (a) The payoff function can be written as $\Lambda(S) = K^2 - 2KS + S^2$. From the risk-neutral pricing formula for a European-style derivative, we have

$$\begin{aligned}
V(t,S) &= \mathrm{e}^{-r(T-t)}\widetilde{\mathrm{E}}_{t,S}\left[K^2 - 2KS(T) + S^2(T)\right] \\
&= \mathrm{e}^{-r(T-t)}\left[K^2 - 2K\widetilde{\mathrm{E}}_{t,S}[S(T)] + \widetilde{\mathrm{E}}_{t,S}[S^2(T)]\right] \\
&= \mathrm{e}^{-r(T-t)}\left[K^2 - 2K\mathrm{e}^{r(T-t)}S + \mathrm{e}^{(2r+\sigma^2)(T-t)}S^2\right] \\
&= \mathrm{e}^{-r\tau}K^2 - 2KS + \mathrm{e}^{(r+\sigma^2)\tau}S^2.
\end{aligned}$$

Here and below, $\tau = T - t$ denotes the time to maturity.
(b) First, write the payoff using an indicator function:

$$\Lambda(S) = (K^2 - S^2)\mathbb{I}_{\{S^2<K^2\}} = K^2\mathbb{I}_{\{S<K\}} - S^2\mathbb{I}_{\{S<K\}}.$$

Then, apply the risk-neutral pricing formula:

$$\begin{aligned}
V(t,S) &= \mathrm{e}^{-r(T-t)}\widetilde{\mathrm{E}}_{t,S}\left[K^2\mathbb{I}_{\{S(T)<K\}} - S^2(T)\mathbb{I}_{\{S(T)<K\}}\right] \\
&= \mathrm{e}^{-r(T-t)}K^2\widetilde{\mathbb{P}}_{t,S}(S(T)<K) - \mathrm{e}^{-r(T-t)}\widetilde{\mathrm{E}}_{t,S}[S^2(T)\mathbb{I}_{\{S(T)<K\}}] \\
&= \mathrm{e}^{-r\tau}K^2\mathcal{N}(-d_-(S/K,\tau)) \\
&\quad - \mathrm{e}^{-r\tau}S^2\mathrm{e}^{(2r+\sigma^2)\tau}\mathcal{N}(-d_+(S/K,\tau)-\sigma\sqrt{\tau}) \\
&= \mathrm{e}^{-r\tau}K^2\mathcal{N}(-d_-(S/K,\tau)) - \mathrm{e}^{(r+\sigma^2)\tau}S^2\mathcal{N}(-d_+(S/K,\tau)-\sigma\sqrt{\tau}).
\end{aligned}$$

Here, we used the identity

$$\begin{aligned}
\widetilde{\mathrm{E}}_{t,S}[S^2(T)\mathbb{I}_{\{S(T)<K\}}] &= \widetilde{\mathrm{E}}_{t,S}[S^2(T)(1-\mathbb{I}_{\{S(T)\geqslant K\}})] \\
&= \widetilde{\mathrm{E}}_{t,S}[S^2(T)] - \widetilde{\mathrm{E}}_{t,S}[S^2(T)\mathbb{I}_{\{S(T)\geqslant K\}}] \\
&= S^2\mathrm{e}^{(2r+\sigma^2)\tau} - S^2\mathrm{e}^{(2r+\sigma^2)\tau}\mathcal{N}(d_+(S/K,\tau)+\sigma\sqrt{\tau}) \\
&= S^2\mathrm{e}^{(2r+\sigma^2)\tau}\mathcal{N}(-d_+(S/K,\tau)-\sigma\sqrt{\tau}),
\end{aligned}$$

where $\tau = T - t$. □

We now use the formulae in (3.53)–(3.55) to price a soft-strike call option in the following example.

Example 3.6. (Soft-Strike Call Option) Consider the soft-strike European call option with payoff function

$$\Lambda_a(S) = \begin{cases} 0 & \text{if } S < K - a, \\ \frac{1}{4a}(S - K + a)^2 & \text{if } K - a \leqslant S \leqslant K + a, \\ S - K & \text{if } S > K + a, \end{cases} \tag{3.56}$$

where the constant $a \in [0, K]$ and K is a central strike value.

(a) Describe the main features of the graph of $\Lambda_a(S)$ for all $S > 0$.

(b) Assume the stock price process $\{S(t)\}_{t\geqslant 0}$ is a GBM with constant interest rate r and volatility σ. Let $S(t) = S$ be the spot price of the stock at current time $t < T$. Derive the no-arbitrage pricing formula for a European-style option with the above payoff function.

Solution.

We see in Figure 3.7 that $\Lambda_a(S) \geqslant (S - K)^+$, where $\Lambda_a(S) \searrow (S - K)^+$ as $a \searrow 0$; i.e., as a function of the parameter a, the soft-strike payoff decreases monotonically to the standard call payoff as $a \searrow 0$. In contrast to the standard call payoff $(S - K)^+$ whose derivative w.r.t. S has a unit jump discontinuity at $S = K$, the payoff $\Lambda_a(S)$ has a continuous derivative for all S. The left and right derivatives at $S = K - a$ are the same, $\Lambda_a'(K - a) = 0$, and the left and right derivatives at $S = K + a$ are the same, $\Lambda_a'(K + a) = 1$. Combining this with the derivative for $S \in (K - a, K + a)$, the payoff has a continuous derivative w.r.t. S given by

$$\Lambda_a'(S) = \begin{cases} 0 & \text{if } S < K - a, \\ \frac{1}{2a}(S - K + a) & \text{if } K - a \leqslant S \leqslant K + a, \\ 1 & \text{if } S > K + a. \end{cases} \tag{3.57}$$

Moreover, the payoff has a piecewise constant second derivative $\Lambda_a''(S)$. In fact, the payoff can be expressed as an integral over the standard call payoff function $(S - k)^+$ by employing a continuum of strikes $k \in (K - a, K + a)$ (see Exercise 3.20).

(b) Express $\Lambda_a(S(T))$ as a linear combination of payoffs involving the different powers of $S(T)$ multiplying indicator random variables:

$$\begin{aligned} \Lambda_a(S(T)) = {} & \frac{1}{4a}\left(S^2(T) - 2(K - a)S(T) + (K - a)^2\right) \mathbb{I}_{\{K - a \leqslant S(T) \leqslant K + a\}} \\ & + S(T)\, \mathbb{I}_{\{S(T) > K + a\}} - K\, \mathbb{I}_{\{S(T) > K + a\}}\,. \end{aligned} \tag{3.58}$$

The option value at current time t is then the discounted conditional expectation of this payoff. Let $V(t, S) \equiv C(t, S; K, a)$ denote the time-t value of the soft-strike call where we include the dependence on the parameters a, K defining the payoff. By linearity of expectations:

$$\begin{aligned} C(t, S; K, a) &= \mathrm{e}^{-r\tau}\, \widetilde{\mathrm{E}}_{t,S}\left[\Lambda_a(S(T))\right] \\ &= \mathrm{e}^{-r\tau}\left(\frac{1}{4a}\widetilde{\mathrm{E}}_{t,S}\left[S^2(T)\, \mathbb{I}_{\{K - a \leqslant S(T) \leqslant K + a\}}\right]\right. \\ &\quad - \frac{(K - a)}{2a}\widetilde{\mathrm{E}}_{t,S}\left[S(T)\, \mathbb{I}_{\{K - a \leqslant S(T) \leqslant K + a\}}\right] + \widetilde{\mathrm{E}}_{t,S}\left[S(T)\, \mathbb{I}_{\{S(T) > K + a\}}\right] \\ &\quad \left. + \frac{(K - a)^2}{4a}\widetilde{\mathbb{P}}_{t,S}\left(K - a \leqslant S(T) \leqslant K + a\right) - K\, \widetilde{\mathbb{P}}_{t,S}\left(S(T) > K + a\right)\right). \end{aligned}$$

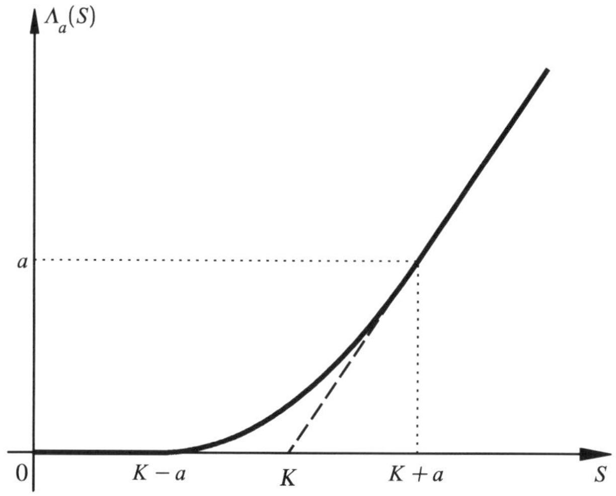

(a)

FIGURE 3.7: The payoff of a soft-strike call centred at strike K.

We now use the formulae given in (3.54) and (3.55) for respective powers of $\alpha = 1, 2$, as well as our previously derived formulae for the risk-neutral conditional probability that $S(T)$ lies above a strike level or within an interval. Combining all terms gives

$$\begin{aligned} C(t, S; K, a) = {} & S^2 \frac{e^{(r+\sigma^2)\tau}}{4a} \left[\mathcal{N}\left(d_+(\frac{S}{K-a}, \tau) + \sigma\sqrt{\tau}\right) - \mathcal{N}\left(d_+(\frac{S}{K+a}, \tau) + \sigma\sqrt{\tau}\right)\right] \\ & - \frac{(K-a)}{2a} S \left[\mathcal{N}\left(d_+(\frac{S}{K-a}, \tau)\right) - \mathcal{N}\left(d_+(\frac{S}{K+a}, \tau)\right)\right] \\ & + \frac{(K-a)^2}{4a} e^{-r\tau} \left[\mathcal{N}\left(d_-(\frac{S}{K-a}, \tau)\right) - \mathcal{N}\left(d_-(\frac{S}{K+a}, \tau)\right)\right] \\ & + S\mathcal{N}\left(d_+(\frac{S}{K+a}, \tau)\right) - K e^{-r\tau} \mathcal{N}\left(d_-(\frac{S}{K+a}, \tau)\right), \end{aligned} \tag{3.59}$$

where $\tau = T - t$ is the time to maturity and $d_\pm(x, \tau)$ are defined in (3.37). □

Based on the pricing formula in (3.59), we can compute the delta hedging position in the stock. Moreover, we can readily price the corresponding *soft-strike put* option with given central strike K and width a. We leave this as an exercise for the reader (see Exercise 3.21). There are also other power options that are readily priced with the use of the formulae in (3.53)–(3.55) and these are assigned as exercises at the end of this chapter.

3.2.5 Dividend Paying Stock

3.2.5.1 The Case of Continuous Dividend Paying Stock

Let us now consider the standard (B, S) model where the stock is a GBM which also pays a dividend with a constant continuous yield $q > 0$ per unit of time. During an infinitesimal time dt the stock holder receives a dividend payment of $qS(t)\,dt$ per share that is in proportion to the stock price at time t. We can see this from the return due only to the dividend, $\frac{S(t+\delta t)-S(t)}{S(t)} = e^{q\delta t} - 1 \approx q\delta t$ for small time interval $\delta t \approx 0$. By no arbitrage, this dividend payment to the stock holder must be exactly balanced by a decrease in the share price of the stock. Hence, in the physical $\mathbb{P}$-measure there is the additional negative drift term, $-qS(t)\,dt$, due to the dividend. The stock price process is a GBM with SDE

$$\mathrm{d}S(t) = (\mu - q)S(t)\,\mathrm{d}t + \sigma S(t)\,\mathrm{d}W(t)\,.$$

By the same unique risk-neutral measure $\widetilde{\mathbb{P}}$ as above, where $\widetilde{W}(t) := W(t) + \frac{(\mu-r)}{\sigma}t$ is a standard $\widetilde{\mathbb{P}}$-BM, the above SDE takes the form

$$\mathrm{d}S(t) = S(t)\big[(r - q)\,\mathrm{d}t + \sigma\,\mathrm{d}\widetilde{W}(t)\big]\,. \tag{3.60}$$

For a nonzero dividend, the risk-neutral drift of the stock price is $(r - q)$ in the place of r. Given an arbitrary initial stock value $S(0) > 0$, this SDE has a unique solution

$$S(t) = S(0)\,\mathrm{e}^{(r-q-\sigma^2/2)t+\sigma\widetilde{W}(t)}\,. \tag{3.61}$$

The time-T stock price is given in terms of the time-t price and a $\widetilde{\mathbb{P}}$-BM increment:

$$S(T) = S(t)\,\mathrm{e}^{(r-q-\sigma^2/2)(T-t)+\sigma(\widetilde{W}(T)-\widetilde{W}(t))}\,. \tag{3.62}$$

Clearly, in the presence of dividends the discounted stock price process $\overline{S}(t) := \mathrm{e}^{-rt}S(t)$ is not a $\widetilde{\mathbb{P}}$-martingale. Let us consider a portfolio that consists of one stock share at time 0. Suppose that every dividend payment is reinvested in the stock. The dividends are paid continuously, and the payment of $qS(t)\,dt$ received over time dt allows for buying $q\,dt$ additional stock shares for each share held at time t. So, the number of shares, $n(t)$, held at time t solves the first-order ordinary differential equation $\mathrm{d}n(t) = n(t)q\,\mathrm{d}t$ subject to the initial condition $n(0) = 1$. As a result, one share at time 0 grows to e^{qt} shares at time t. The time-t value of this portfolio is $\widehat{S}(t) := \mathrm{e}^{qt}S(t)$. Discounting the $\widehat{S}$ price process with the bank account gives $\widehat{S}(t)/B(t) \equiv \mathrm{e}^{-rt}\widehat{S}(t) \equiv \mathrm{e}^{-(r-q)t}S(t)$ as a $\widetilde{\mathbb{P}}$-martingale:

$$\begin{aligned}\widetilde{\mathrm{E}}\left[\mathrm{e}^{-rT}\widehat{S}(T) \mid \mathcal{F}_t\right] &= \mathrm{e}^{-(r-q)T}\widetilde{\mathrm{E}}\left[S(T) \mid \mathcal{F}_t\right] = \mathrm{e}^{-(r-q)T}\mathrm{e}^{(r-q)(T-t)}S(t)\\ &= \mathrm{e}^{-(r-q)t}S(t) = \mathrm{e}^{-rt}\widehat{S}(t).\end{aligned}$$

The portfolio value process $\{\widehat{S}(t)\}_{t\geqslant 0}$ satisfies the following SDE

$$\mathrm{d}\widehat{S}(t) = r\widehat{S}(t)\,\mathrm{d}t + \sigma\widehat{S}(t)\,\mathrm{d}\widetilde{W}(t).$$

The replicating portfolio is the same as given in (3.17). The portfolio is invested in the amount of $\Delta_t S(t)$ in the stock, so the dividend payment is $q\Delta_t S(t)\,dt$ over time dt. This term is now added to the differential change in the self-financing portfolio in (3.19), giving

$$\begin{aligned}\mathrm{d}\Pi_t &= r\beta_t B(t)\,\mathrm{d}t + \Delta_t\,\mathrm{d}S(t) + q\Delta_t S(t)\,\mathrm{d}t\\ &= \left[r\Pi_t - (r - q)\Delta_t S(t)\right]\mathrm{d}t + \Delta_t\,\mathrm{d}S(t)\\ &= r\Pi_t\,\mathrm{d}t + \sigma\Delta_t S(t)\,\mathrm{d}\widetilde{W}(t)\,.\end{aligned} \tag{3.63}$$

In the last line we used (3.60). This is of the same form as in (3.20) and hence (3.21) still holds where the discounted self-financing portfolio value process $\{\overline{\Pi}_t\}_{0\leqslant t\leqslant T}$ is a $\widetilde{\mathbb{P}}$-martingale. The risk-neutral pricing formula in (3.22) as well as (3.23) and (3.24) still hold. The conditions and arguments given in Section 3.2 that guarantee that a claim can be replicated (hedged) are the same as in the case of zero dividend on the stock.

The only difference is that the stock has drift parameter $(r - q)$ instead of r, i.e., $S(t)$ is given by (3.61) instead of (3.15) and (3.62) replaces (3.25). So the question is, in terms of pricing, what does this change? The general answer is actually very simple given the solution in the case that $q = 0$. Let us take a look at how (3.27)–(3.33) change. Using (3.62) in the place of (3.25), the expectations in (3.27) and (3.28) now become

$$V(t,S) = \mathrm{e}^{-r\tau}\int_{-\infty}^{\infty}\Lambda(S\,\mathrm{e}^{(r-q-\frac{1}{2}\sigma^2)\tau+\sigma\sqrt{\tau}z})\,n(z)\,\mathrm{d}z = \mathrm{e}^{-r\tau}\int_0^{\infty}\Lambda(y)\,\widetilde{p}(\tau;S,y)\,\mathrm{d}y\,, \quad (3.64)$$

where $\widetilde{p}(\tau; S, y)$ is now the risk-neutral transition PDF of the stock price GBM process with drift $r - q$, i.e., with r replaced by $r - q$ in (3.29). The generator for the stock price process with SDE in (3.60) is defined by $\mathcal{G}V := \frac{1}{2}\sigma^2 S^2 \frac{\partial^2}{\partial S^2}V + (r-q)S\frac{\partial}{\partial S}V$. Hence, the respective Black–Scholes partial differential equations in (3.30) and (3.31) are now

$$\frac{\partial v}{\partial \tau} = \frac{1}{2}\sigma^2 S^2\frac{\partial^2 v}{\partial S^2} + (r-q)S\frac{\partial v}{\partial S} - rv\,, \quad (3.65)$$

subject to $v(0, S) = \Lambda(S)$ and

$$\frac{\partial V}{\partial t} + \frac{1}{2}\sigma^2 S^2\frac{\partial^2 V}{\partial S^2} + (r-q)S\frac{\partial V}{\partial S} - rV = 0\,, \quad (3.66)$$

subject to $V(T, S) = \Lambda(S)$. Note that the dividend only changes the drift term where $rS\frac{\partial}{\partial S}$ has been replaced by $(r-q)S\frac{\partial}{\partial S}$. Everything else is the same, including the discount factor. Equations (3.32) and (3.33) are then also the same i.e., the delta hedge position is the same. Of course, the derivative value $V(t, S)$ for $q \neq 0$ differs from the value when $q = 0$.

We now show that the derivative pricing formula for $q \neq 0$ (nonzero stock dividend) obtains trivially from the corresponding pricing formula for $q = 0$ (zero dividend), given arbitrary payoff $V_T = \Lambda(S(T))$. To precisely describe this, let $V(t, S; r, q)$ be the time-t price of the derivative for given interest rate r and stock dividend q. This function has a dependence on parameters r and q, as well as other parameters that we simply suppress. The corresponding price when $q = 0$ is then $V(t, S; r, 0)$. The function $V(t, S; r, q)$ is given by (3.64) and $V(t, S; r, 0)$ is given by (3.27). Multiplying out the discount factor in both cases gives $\mathrm{e}^{r\tau}V(t,S;r,q) = [\mathrm{e}^{r\tau}V(t,S;r,0)]|_{r\to r-q}$, which is equivalent to

$$V(t,S;r,q) = \mathrm{e}^{-q(T-t)}V(t,S;r-q,0)\,. \quad (3.67)$$

From this relation we see that the pricing function (on the left) for $q \neq 0$ is given by the corresponding pricing function (on the right) for $q = 0$ after replacing the interest rate r by $r - q$ and multiplying the function by $\mathrm{e}^{-q\tau}$, $\tau = T - t$. The simple example below shows how (3.67) is very easily applied to a standard call and put.

Example 3.7. Apply the symmetry in (3.67) to (3.40) and obtain pricing formulae for the standard European call for a stock with continuous constant dividend yield q.

Solution. Let $C(t, S; r, q) \equiv C(\tau, S, K, \sigma, r, q)$ denote the pricing formula for the call with stock dividend q. For $q = 0$, $C(t, S; r, 0) \equiv C(\tau, S, K, \sigma, r, 0)$ is given by (3.40):

$$C(\tau,S,K,\sigma,r,0) = S\mathcal{N}\left(\frac{\ln\frac{S}{K} + (r+\frac{1}{2}\sigma^2)\tau}{\sigma\sqrt{\tau}}\right) - \mathrm{e}^{-r\tau}K\mathcal{N}\left(\frac{\ln\frac{S}{K} + (r-\frac{1}{2}\sigma^2)\tau}{\sigma\sqrt{\tau}}\right).$$

Applying (3.67), $C(t, S; r, q) = \mathrm{e}^{-q\tau}C(t, S; r - q, 0) = \mathrm{e}^{-q\tau}C(\tau, S, K, \sigma, r - q, 0)$:

$$\begin{aligned} C(t,S;r,q) &= \mathrm{e}^{-q\tau}S\mathcal{N}\left(\frac{\ln\frac{S}{K} + (r-q+\frac{1}{2}\sigma^2)\tau}{\sigma\sqrt{\tau}}\right) - \mathrm{e}^{-r\tau}K\mathcal{N}\left(\frac{\ln\frac{S}{K} + (r-q-\frac{1}{2}\sigma^2)\tau}{\sigma\sqrt{\tau}}\right) \\ &= \mathrm{e}^{-q\tau}S\mathcal{N}\left(d_+(\frac{\mathrm{e}^{-q\tau}S}{K},\tau)\right) - \mathrm{e}^{-r\tau}K\mathcal{N}\left(d_-(\frac{\mathrm{e}^{-q\tau}S}{K},\tau)\right) \end{aligned} \quad (3.68)$$

where $\tau = T - t$ and $d_\pm(x, \tau)$ are defined in (3.37). □

This example also points out another very useful simple symmetry where the standard option pricing function, for given dividend q, is given by the original (zero dividend) pricing function with spot value S replaced by the "effective spot" value $\mathrm{e}^{-q\tau}S$ (keeping everything else the same). This is an alternatively useful symmetry that we can apply to immediately obtain the pricing formula for $q \neq 0$ by substituting $\mathrm{e}^{-q\tau}S$ for S within the pricing formula for $q = 0$. We can express this additional symmetry as

$$V(t, S; r, q) = V(t, \mathrm{e}^{-q\tau}S; r, 0)\,. \tag{3.69}$$

We see quite trivially how this symmetry works in the above example of the call where setting $\mathrm{e}^{-q\tau}S$ for S in (3.40) gives (3.68). Applying either (3.67) or (3.69) to (3.43) gives the corresponding pricing formula for the put option on a dividend paying stock,

$$P(t, S; r, q) = \mathrm{e}^{-r\tau} K\mathcal{N}\left(-d_-(\frac{\mathrm{e}^{-q\tau}S}{K}, \tau)\right) - \mathrm{e}^{-q\tau}S\mathcal{N}\left(-d_+(\frac{\mathrm{e}^{-q\tau}S}{K}, \tau)\right). \tag{3.70}$$

The put-call parity relation in (3.42) now takes the more general form,

$$C(t, S) - P(t, S) = \mathrm{e}^{-q\tau}S - \mathrm{e}^{-r\tau}K \tag{3.71}$$

where we simply write $C(t, S) = C(t, S; r, q)$ and $P(t, S) = P(t, S; r, q)$.

Corresponding symmetry relations for the delta hedging position also follow. Let us denote $\Delta(t, S) \equiv \Delta(t, S; r, q) = \frac{\partial}{\partial S}V(t, S; r, q)$. Then,

$$\Delta(t, S; r, q) = \mathrm{e}^{-q\tau}\frac{\partial}{\partial S}V(t, S; r - q, 0) = \mathrm{e}^{-q\tau}\Delta(t, S; r - q, 0)\,. \tag{3.72}$$

By the chain rule, differentiating (3.69) gives us an alternative symmetry:

$$\Delta(t, S; r, q) = \mathrm{e}^{-q\tau}\frac{\partial}{\partial x}V(t, x; r, 0)\bigg|_{x=\mathrm{e}^{-q\tau}S} = \mathrm{e}^{-q\tau}\Delta(t, \mathrm{e}^{-q\tau}S; r, 0)\,. \tag{3.73}$$

For example, consider a call where (3.47) gives $\Delta_c(t, S; r, 0)$ and by either (3.72) or (3.73):

$$\Delta_c(t, S; r, q) = \mathrm{e}^{-q\tau}\mathcal{N}\left(d_+(\frac{\mathrm{e}^{-q\tau}S}{K}, \tau)\right) \equiv \mathrm{e}^{-q\tau}\mathcal{N}\left(\frac{\ln\frac{S}{K} + (r - q + \frac{1}{2}\sigma^2)\tau}{\sigma\sqrt{\tau}}\right). \tag{3.74}$$

We point out that this formula can also be obtained using our previous derivations, i.e., without use of the symmetry relation in (3.72).

If we assume no knowledge of the above symmetry and no prior pricing formula for $q = 0$, then we can of course derive the pricing formula in (3.68) from first principle, by employing the same steps that lead to (3.40) where we had $q = 0$. Namely, we substitute the expression for $S(T)$ in (3.62) into the discounted expectation in (3.36) and evaluate by using the identities in (3.38) and (3.39) where now the drift $r - q$ replaces r, i.e., (3.38) and (3.39) become

$$\widetilde{\mathrm{E}}_{t,S}\left[S(T)\,\mathbb{I}_{\{S(T)>K\}}\right] = \mathrm{e}^{(r-q)\tau}S\mathcal{N}\left(d_+(\frac{\mathrm{e}^{-q\tau}S}{K}, \tau)\right) \tag{3.75}$$

and

$$\widetilde{\mathbb{P}}_{t,S}(S(T) > K) = \mathcal{N}\left(d_-(\frac{\mathrm{e}^{-q\tau}S}{K}, \tau)\right). \tag{3.76}$$

Combining these into (3.36) gives the call pricing formula in (3.68). Using similar steps, or by put-call parity, we obtain the above put pricing formula. The pricing formulae for European derivatives on a dividend paying stock for all other types of payoffs, including those we considered in Sections 3.2.3 and 3.2.4, also follow by the above symmetry relation. Alternatively the formulae can be derived from first principles based on the identities in (3.75) and (3.76). In the case of power options on a dividend payoff stock, we have the identities in (3.53), (3.54), and (3.55) where r is replaced by $r - q$ (or equivalently do not replace r but replace S by $e^{-q\tau}S$).

Typical plots of pricing functions of call and put options written on a dividend paying stock are provided in Figures 3.8 and 3.9. We can see that the call option value decreases, whereas the put option value increases, when the dividend yield q gets larger.

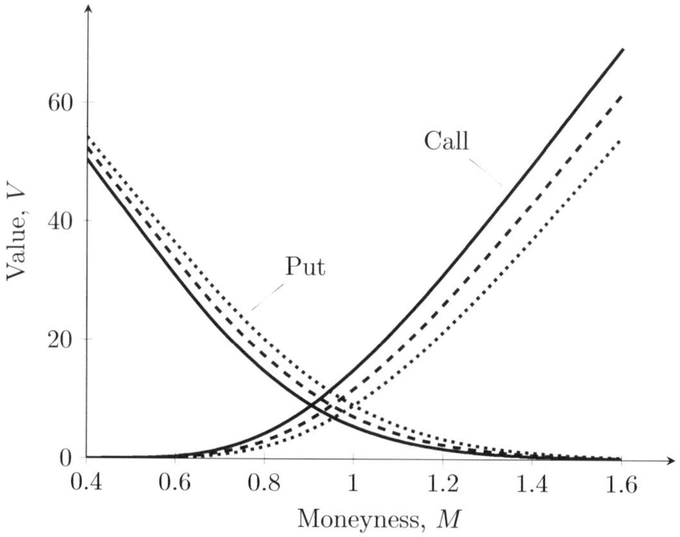

FIGURE 3.8: No-arbitrage values of standard European call and put options as functions of moneyness $M := S/K$ under the (B, S) model with $\tau = 1$, $K = 100$, $r = 10\%$, and $\sigma = 25\%$. The dividend yield q takes the following values: 0% (the solid line), 5% (the dashed line), and 10% (the dotted line).

In addition to put-call parity, we also have the *put-call symmetry relation*, which is written in the following two ways:

$$
\begin{aligned}
C(\tau, S, K, \sigma, r, q) &= P(\tau, K, S, \sigma, q, r), \\
P(\tau, S, K, \sigma, r, q) &= C(\tau, K, S, \sigma, q, r),
\end{aligned}
$$

where $C(t, S) \equiv C(\tau, S, K, \sigma, r, q)$ and $P(t, S) \equiv P(\tau, S, K, \sigma, r, q)$, respectively, denote the pricing functions for standard call and put options with time to maturity $\tau = T - t$, spot S, strike K, volatility σ, risk-free interest rate r, and stock dividend yield q. We prove the symmetry relation by exploiting the fact that the function $d_\pm$ transforms into $-d_\mp$ when we swap K with S and r with q, respectively. Indeed, we have

$$
\begin{aligned}
d_\pm(e^{-q\tau}S/K, \tau) &= \frac{\ln(S/K) + (r - q \pm \sigma^2/2)\tau}{\sigma\sqrt{\tau}} \xrightarrow[K\leftrightarrow S]{r\leftrightarrow q} \frac{\ln(K/S) + (q - r \pm \sigma^2/2)\tau}{\sigma\sqrt{\tau}} \\
&= -\frac{\ln(S/K) + (r - q \mp \sigma^2/2)\tau}{\sigma\sqrt{\tau}} = -d_\mp(e^{-q\tau}S/K, \tau).
\end{aligned}
$$

Therefore, the call pricing formula is transformed into the put pricing formula as follows:

$$C(\tau, S, K, \sigma, r, q) = \mathrm{e}^{-q\tau} S\mathcal{N}(d_+) - \mathrm{e}^{-r\tau} K\mathcal{N}(d_-) \underset{K \leftrightarrow S}{\overset{r \leftrightarrow q}{\longrightarrow}} \mathrm{e}^{-r\tau} K\mathcal{N}(-d_-) - \mathrm{e}^{-q\tau} S\mathcal{N}(-d_+)$$
$$= P(\tau, S, K, \sigma, r, q).$$

Thus, the above symmetry relations hold.

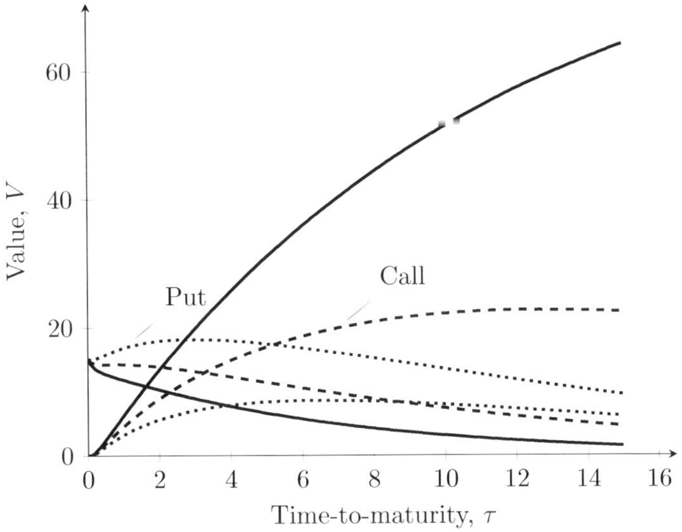

FIGURE 3.9: No-arbitrage values of standard European call and put options as functions of the time to maturity $\tau := T - t$ under the (B, S) model with $S = 85$, $K = 100$, $r = 10\%$, and $\sigma = 25\%$. The dividend yield q takes the following values: 0% (the solid line), 5% (the dashed line), and 10% (the dotted line)

[*Technical Remark*: We now give a more technical argument showing that the above symmetry relation in (3.67) holds as a particular case of a similar symmetry for more complex path-dependent payoffs. Consider an arbitrary European derivative where the payoff V_T has a path dependence on the stock, i.e., is a functional of the stock price process from time t to T. Examples of these derivatives are barrier options, Asian options, lookback options, etc. In general, the derivative price V_t is an $\mathcal{F}_t$-measurable random variable where (3.22) holds for any attainable claim V_T. More specifically, denote by $\{S^{(r-q)}(u) : t \leqslant u \leqslant T\}$ the path of the stock price process defined by the risk-neutral drift parameter $r - q$ (and volatility σ) and let $V_t = V_t^{(r,q)}$ be the corresponding derivative price at time t, for any constant interest rate r and constant dividend yield q. Now, let V_T depend on any segment of the stock price path, including the entire path history from time t to T. We write this as a functional, $V_T^{(r,q)} = F\left(\{S^{(r-q)}(u) : t \leqslant u \leqslant T\}\right)$. Note that $V_T^{(r-q,0)} = F\left(\{S^{(r-q-0)}(u) : t \leqslant u \leqslant T\}\right) = V_T^{(r,q)}$. Hence, by (3.22),

$$V_t^{(r,q)} = \mathrm{e}^{-r(T-t)} \widetilde{\mathrm{E}}\big[V_T^{(r,q)} \mid \mathcal{F}_t\big] = \mathrm{e}^{-q(T-t)} \cdot \mathrm{e}^{-(r-q)(T-t)} \widetilde{\mathrm{E}}\big[V_T^{(r-q,0)} \mid \mathcal{F}_t\big]$$
$$= \mathrm{e}^{-q(T-t)} V_t^{(r-q,0)}. \tag{3.77}$$

So (3.67) is recovered for standard non-path-dependent payoffs with spot $S(t) = S$ giving $V_t^{(r,q)} = V(t, S; r, q)$. However, we note that the relation in (3.69) does not generally hold for path-dependent derivative pricing.]

3.2.5.2 The Case of Discrete-Time Dividends

The continuous-time dividend payment model of a stock in the previous section led to analytical pricing formulae for European derivatives that have the same form as the formulae with zero dividend. We shall mostly adopt this model in further applications. In practice, however, stocks in the market pay dividends at discrete regular intervals of time. Let us suppose that at present time $t_0 < T$ we know the fixed future dividend dates, $T_i, i = 1, \ldots, N$, $t_0 < T_1 < \ldots < T_N \leqslant T$. At each time T_i the dividend payment $\text{div}(T_i)$ is a proportion of the stock value at time T_i, i.e., $\text{div}(T_i) = d_i S(T_i)$, where $0 \leqslant d_i \leqslant 1$ is the dividend percentage. When no dividend is paid at time T_i we have $d_i = 0$ and when the full share value of the stock is paid at time T_i then $d_i = 1$ and the stock becomes worthless for all time after T_i. Typically, we can assume $0 < d_i < 1$ for $i = 1, \ldots, N$. We remark that by writing $d_i = q_i (T_i - T_{i-1})$ then q_i is a dividend yield (rate) that is fixed within the time interval $(T_{i-1}, T_i]$.

The model for the stock is then as follows. Between the ith dividend date and just before the $(i+1)$th dividend date, i.e., within *any time interval between dividend payments*, the stock price evolves simply as a GBM according to the SDE in (3.14). Hence the stock price at time $t_0 \leqslant t < T_1$ is given in terms of the spot $S(t_0) \equiv S_0$ as

$$S(t) = S_0 \, \mathrm{e}^{(r - \frac{1}{2}\sigma^2)(t - t_0) + \sigma(\widetilde{W}(t) - \widetilde{W}(t_0))}.$$

At time $t = T_1^-$:

$$S(T_1^-) = S_0 \, \mathrm{e}^{(r - \frac{1}{2}\sigma^2)(T_1 - t_0) + \sigma(\widetilde{W}_1 - \widetilde{W}(t_0))},$$

where we use the notation $\widetilde{W}_i \equiv \widetilde{W}(T_i)$ for $i = 1, \ldots, N$. For times $t \in [T_{i-1}, T_i)$, $i = 2, \ldots, N$, we have

$$S(t) = S(T_{i-1}) \, \mathrm{e}^{(r - \frac{1}{2}\sigma^2)(t - T_{i-1}) + \sigma(\widetilde{W}(t) - \widetilde{W}_{i-1})},$$

and in particular for $t = T_i^-$, just before the ith dividend date,

$$S(T_i^-) = S(T_{i-1}) \, \mathrm{e}^{(r - \frac{1}{2}\sigma^2)(T_i - T_{i-1}) + \sigma(\widetilde{W}_i - \widetilde{W}_{i-1})}. \tag{3.78}$$

Finally, for the last interval $[T_N, T]$,

$$S(T) = S(T_N) \, \mathrm{e}^{(r - \frac{1}{2}\sigma^2)(T - T_N) + \sigma(\widetilde{W}(T) - \widetilde{W}_N)}.$$

Now, at each dividend payment date T_i, $i = 1, \ldots, N$, the stock price instantaneously decreases by a fraction d_i due to the dividend payment. This is expressed as

$$S(T_i) = (1 - d_i) S(T_i^-). \tag{3.79}$$

Hence, substituting the expression for $S(T_i^-)$ in (3.78) into this last equation gives the evolution of the stock price from one discrete dividend date to the next as

$$\frac{S(T_i)}{S(T_{i-1})} = (1 - d_i) \, \mathrm{e}^{(r - \frac{1}{2}\sigma^2)(T_i - T_{i-1}) + \sigma(\widetilde{W}_i - \widetilde{W}_{i-1})}, \quad i = 2, \ldots, N. \tag{3.80}$$

For time t_0 to just after the first dividend payment time T_1 we have

$$\frac{S(T_1)}{S(t_0)} = (1 - d_1) \, \mathrm{e}^{(r - \frac{1}{2}\sigma^2)(T_1 - t_0) + \sigma(\widetilde{W}_1 - \widetilde{W}(t_0))}. \tag{3.81}$$

Multiplying out the stock price ratios for all adjoining time intervals including the first and last intervals (note: this is a telescoping product) gives

$$\begin{aligned} S(T) &= S_0 \frac{S(T)}{S(t_0)} = S_0 \frac{S(T_1)}{S(t_0)} \left(\prod_{i=2}^{N} \frac{S(T_i)}{S(T_{i-1})} \right) \frac{S(T)}{S(T_N)} \\ &= S_0(1-d_1)\,\mathrm{e}^{(r-\frac{1}{2}\sigma^2)(T_1-t_0)+\sigma(\widetilde{W}_1-\widetilde{W}(t_0))} \qquad (3.82) \\ &\quad \times \prod_{i=2}^{N}(1-d_i)\,\mathrm{e}^{(r-\frac{1}{2}\sigma^2)(T_i-T_{i-1})+\sigma(\widetilde{W}_i-\widetilde{W}_{i-1})} \\ &\quad \times \mathrm{e}^{(r-\frac{1}{2}\sigma^2)(T-T_N)+\sigma(\widetilde{W}(T)-\widetilde{W}_N)} \\ &= S_0\Big[\prod_{i=1}^{N}(1-d_i)\Big]\mathrm{e}^{(r-\frac{1}{2}\sigma^2)(T-t_0)+\sigma(\widetilde{W}(T)-\widetilde{W}(t_0))} \\ &= \widetilde{S}_0\mathrm{e}^{(r-\frac{1}{2}\sigma^2)(T-t_0)+\sigma(\widetilde{W}(T)-\widetilde{W}(t_0))}, \qquad (3.83) \end{aligned}$$

where $\widetilde{S}_0 \equiv S_0 \prod_{i=1}^{N}(1-d_i) = S_0(1-d_1)\cdots(1-d_N)$. In the last line, we simplified the sum in the exponents where $\widetilde{W}_1-\widetilde{W}(t_0)+\widetilde{W}_2-\widetilde{W}_1+\cdots+\widetilde{W}_N-\widetilde{W}_{N-1}+\widetilde{W}(T)-\widetilde{W}_N = \widetilde{W}(T)-\widetilde{W}(t_0)$.

From (3.82) we see that the stock price $S(T)$ is a GBM random variable with drift r and volatility σ. The overall discount factor due to all the dividends is multiplying the actual initial stock price S_0 giving the initial value $\widetilde{S}_0$ *now acting as effective initial stock price* at time-t_0. We recall that in the case of a continuous dividend the quantity $\mathrm{e}^{-q\tau}S_0$ acts in the place of $\widetilde{S}_0$ when pricing a non-path-dependent European derivatives. Hence, all the time-t_0 pricing formulae for non-path-dependent (standard) European derivatives on the stock are obtained by using the initial value $\widetilde{S}_0$ in the place of S_0. This is clear by substituting the time-T stock price expression in (3.82) into the risk-neutral pricing (expectation) formula with (non-path-dependent) payoff $V_T = \Lambda(S(T))$. In particular, let $V(t_0, S; \mathbf{d})$ represent the European pricing formula for the above stock model with discrete dividends, within the time interval (t_0, T), grouped in a vector $\mathbf{d} = (d_1, \ldots, d_N)$. Accordingly, let $V(t_0, S) \equiv V(t_0, S; \mathbf{0})$ be the pricing function for zero dividends. Then, the analogue of (3.69) is the relation

$$V(t_0, S; \mathbf{d}) = V(t_0, \widetilde{S}_0), \quad \text{where } \widetilde{S}_0 \equiv S_0 \prod_{i=1}^{N}(1-d_i). \qquad (3.84)$$

The pricing functions for a standard call option, put option, power option, etc., follow immediately based on the formulae for zero dividends. For example, simply setting the spot value to $\widetilde{S}_0 \equiv S_0 \prod_{i=1}^{N}(1-d_i)$ in (3.40) gives the pricing formula for the time-t_0 call option with discrete dividends $d_1, \ldots, d_N$ within the time to maturity $\tau = T - t_0$:

$$C(t_0, S; \mathbf{d}) = \widetilde{S}_0 \mathcal{N}(\widetilde{d}_+) - \mathrm{e}^{-r\tau} K \mathcal{N}(\widetilde{d}_-)$$

where

$$\widetilde{d}_\pm := d_\pm\Big(\frac{\widetilde{S}_0}{K}, \tau\Big) = \frac{\ln\frac{S_0}{K} + \sum_{i=1}^{N}\ln(1-d_i) + (r \pm \frac{1}{2}\sigma^2)\tau}{\sigma\sqrt{\tau}}.$$

3.2.6 Option Pricing with the Stock Numéraire

The risk-neutral probability measure $\widetilde{\mathbb{P}}$ has the defining property that all discounted base asset price processes, $\frac{B(t)}{B(t)} \equiv 1$ and $\frac{S(t)}{B(t)}$ (or $\frac{\widehat{S}(t)}{B(t)}$ with $\widehat{S}(t) := \mathrm{e}^{qt}S(t)$ for a dividend paying stock), are $\widetilde{\mathbb{P}}$-martingales. When this property holds, we say that $\widetilde{\mathbb{P}}$ is an equivalent martingale measure (EMM) with the bank account B as a numéraire asset.

As in discrete-time market models discussed in Chapters 7 and 8 of Volume I, we can use different numéraires for discounting. The derivative pricing with general numéraire assets is discussed in the multi-asset setting in Section 4.2 of the next chapter. In this chapter, we deal with the (B, S) model that includes two base assets, each of which can be used as a numéraire. So far, we discussed the option pricing under the EMM $\widetilde{\mathbb{P}} \equiv \widetilde{\mathbb{P}}^{(B)}$ with the numéraire $g(t) = B(t) \equiv \mathrm{e}^{rt}$. We can also use the risky stock as a numéraire. Assuming that the stock pays dividends, we have $g(t) = \widehat{S}(t) \equiv \mathrm{e}^{qt} S(t)$. Let us find the EMM $\widehat{\mathbb{P}} \equiv \widetilde{\mathbb{P}}^{(\widehat{S})}$ defined so that the discounted process $\{B(t)/\widehat{S}(t)\}_{t \geqslant 0}$ is a $\widehat{\mathbb{P}}$-martingale. We note that $\widehat{S}(t)/\widehat{S}(t) \equiv 1$ is a trivial martingale.

As discussed in Chapter 2, an equivalent probability measure can be constructed with the use of a Radon–Nikodym derivative (RND). So, our first objective is to find an RND that allows for constructing an EMM from the real-world measure $\mathbb{P}$. We discuss two approaches to constructing EMMs, namely, Girsanov's Theorem and the change of numéraire. Secondly, we want to compare the SDEs for $S(t)$ and the corresponding strong solutions under the EMMs $\widetilde{\mathbb{P}}$ and $\widehat{\mathbb{P}}$. Lastly, we will see how the change of numéraire allows us to simplify the solution of option pricing problems.

Theorem 3.2 (Change of Numéraire). *Let $\widetilde{\mathbb{P}}$ be the EMM relative to the numéraire $g(t) = B(t) \equiv \mathrm{e}^{rt}$ so that $\frac{\widehat{S}(t)}{B(t)}$ is a $\widetilde{\mathbb{P}}$-martingale. Define the equivalent probability measure $\widehat{\mathbb{P}}$ by the Radon–Nikodym derivative*

$$\varrho_T \equiv \left(\frac{\mathrm{d}\widehat{\mathbb{P}}}{\mathrm{d}\widetilde{\mathbb{P}}} \right)_T = \frac{\widehat{S}(T)/\widehat{S}(0)}{B(T)/B(0)}.$$

Then the process $B(t)/\widehat{S}(t)$ is a $\widehat{\mathbb{P}}$-martingale. That is, $\widehat{\mathbb{P}} \equiv \widetilde{\mathbb{P}}^{(\widehat{S})}$ is an EMM relative to the numéraire $g(t) = \widehat{S}(t) \equiv \mathrm{e}^{qt} S(t)$. Moreover, the no-arbitrage time-t value of an attainable claim V_T is given by

$$V_t = B(t) \widetilde{\mathrm{E}} \left[\frac{V_T}{B(T)} \;\middle|\; \mathcal{F}_t \right] = \widehat{S}(t) \widehat{\mathrm{E}} \left[\frac{V_T}{\widehat{S}(T)} \;\middle|\; \mathcal{F}_t \right], \quad 0 \leqslant t \leqslant T, \tag{3.85}$$

where $\widetilde{\mathrm{E}}[\,]$ and $\widehat{\mathrm{E}}[\,]$ are expectations under the EMMs $\widetilde{\mathbb{P}}$ and $\widehat{\mathbb{P}}$, respectively.

Proof. The Radon–Nikodym derivative process $\varrho_t = \left(\frac{\mathrm{d}\widehat{\mathbb{P}}}{\mathrm{d}\widetilde{\mathbb{P}}} \right)_t := \frac{\widehat{S}(t)/\widehat{S}(0)}{B(t)/B(0)}$, $0 \leqslant t \leqslant T$, is a positive $\widetilde{\mathbb{P}}$-martingale with $\rho_0 = \widetilde{\mathrm{E}}[\varrho_T] = 1$. Using the identity in (2.118) of Proposition 2.15, with replacements $\mathbb{P} \to \widetilde{\mathbb{P}}$, $s \to t, t \to T$, for any $\widehat{\mathbb{P}}$-integrable random variable X, we have

$$\widehat{\mathrm{E}}[X \mid \mathcal{F}_t] = \widetilde{\mathrm{E}} \left[\frac{\varrho_T}{\varrho_t} X \mid \mathcal{F}_t \right] = \varrho_t^{-1} \widetilde{\mathrm{E}} \left[\varrho_T X \mid \mathcal{F}_t \right].$$

Note that by taking $X = \frac{1}{\varrho_T} = \frac{B(T)/B(0)}{\widehat{S}(T)/\widehat{S}(0)}$, we obtain that $\widehat{\mathrm{E}}[\frac{1}{\varrho_T} \mid \mathcal{F}_t] = \frac{1}{\varrho_t}$, and hence $\frac{1}{\varrho_t}$ is a $\widehat{\mathbb{P}}$-martingale. Thus, the process $B(t)/\widehat{S}(t)$ is a $\widehat{\mathbb{P}}$-martingale as well. Moreover, taking $X = \widehat{S}(t) \frac{V_T}{\widehat{S}(T)}$ within the above equation, where $\frac{\varrho_T}{\varrho_t} X = \frac{\widehat{S}(T)}{B(T)} \frac{B(t)}{\widehat{S}(t)} \widehat{S}(t) \frac{V_T}{\widehat{S}(T)} = B(t) \frac{V_T}{B(T)}$, and using the fact that $\widehat{S}(t)$ and $B(t)$ are $\mathcal{F}_t$–measurable, gives

$$\widehat{S}(t) \widehat{\mathrm{E}} \left[\frac{V_T}{\widehat{S}(T)} \;\middle|\; \mathcal{F}_t \right] = \widetilde{\mathrm{E}} \left[B(t) \frac{V_T}{B(T)} \;\middle|\; \mathcal{F}_t \right] = B(t) \widetilde{\mathrm{E}} \left[\frac{V_T}{B(T)} \;\middle|\; \mathcal{F}_t \right] = V_t. \qquad \square$$

Recall that under the real-world (physical) measure $\mathbb{P}$ the stock price process is governed by the SDE $\mathrm{d}S(t) = (\mu - q)S(t)\,\mathrm{d}t + \sigma S(t)\,\mathrm{d}W(t)$. Let us find the following three RND processes:

$$\text{(a) } \tilde{\varrho}_t \equiv \left(\frac{\mathrm{d}\widetilde{\mathbb{P}}}{\mathrm{d}\mathbb{P}}\right)_t, \quad \text{(b) } \hat{\varrho}_t \equiv \left(\frac{\mathrm{d}\widehat{\mathbb{P}}}{\mathrm{d}\mathbb{P}}\right)_t, \quad \text{(c) } \varrho_t \equiv \left(\frac{\mathrm{d}\widehat{\mathbb{P}}}{\mathrm{d}\widetilde{\mathbb{P}}}\right)_t.$$

(a) This problem was already solved in Example 2.25 for drift r instead of $(r - q)$ in the EMM $\widetilde{\mathbb{P}}$. Now we assume the stock price process under the EMM $\widetilde{\mathbb{P}}$ satisfies the SDE:

$$\mathrm{d}S(t) = (r - q)S(t)\,\mathrm{d}t + \sigma S(t)\,\mathrm{d}\widetilde{W}(t).$$

We follow the same steps as in the solution to Example 2.25. Here, we are starting with the SDE in the $\widetilde{\mathbb{P}}$- Brownian increment (both ways are equivalent!). In particular, we take

$$\tilde{\varrho}_t \equiv \varrho_t^{(\tilde{\gamma})} = \mathrm{e}^{-\frac{1}{2}\tilde{\gamma}^2 t + \tilde{\gamma} W(t)}$$

with constant $\tilde{\gamma}$. By Girsanov's theorem, $\mathrm{d}\widetilde{W}(t) = \mathrm{d}W(t) - \tilde{\gamma}\,\mathrm{d}t$. After plugging this into the above SDE, we find the value of $\tilde{\gamma}$ such that the SDE for the stock has drift rate $(\mu - q)$ in the $\mathbb{P}$-measure:

$$\begin{aligned}
\mathrm{d}S(t) &= (r - q)S(t)\,\mathrm{d}t + \sigma S(t)[\,\mathrm{d}W(t) - \tilde{\gamma}\,\mathrm{d}t] \\
&= (r - q - \tilde{\gamma}\sigma)S(t)\,\mathrm{d}t + \sigma S(t)\,\mathrm{d}W(t) \\
&= (\mu - q)S(t)\,\mathrm{d}t + \sigma S(t)\,\mathrm{d}W(t), \quad \text{if } \tilde{\gamma} = \frac{r - \mu}{\sigma}.
\end{aligned}$$

Therefore, $\tilde{\varrho}_t = \exp\left(-\frac{1}{2}\left(\frac{r-\mu}{\sigma}\right)^2 t + \left(\frac{r-\mu}{\sigma}\right) W(t)\right)$, i.e., the same expression as in (2.131). Note that in Example 2.25 we presented two methods to derive this.

(b) Recall that the numéraire $\widehat{S}(t)$ is the value of a portfolio that starts with one share and is managed so that all dividends are reinvested in the stock. As a result, one share grows to e^{qt} shares at time t. The EMM $\widehat{\mathbb{P}}$ is constructed so that the process $\left(\frac{B(t)}{\widehat{S}(t)}\right)_{t\geqslant 0}$ is a $\widehat{\mathbb{P}}$-martingale. By employing the Itô formula (product or quotient rule), we have the SDE for $\frac{B(t)}{\widehat{S}(t)}$:

$$\mathrm{d}\left(\frac{B(t)}{\widehat{S}(t)}\right) = (r - \mu + \sigma^2)\left(\frac{B(t)}{\widehat{S}(t)}\right)\mathrm{d}t - \sigma\left(\frac{B(t)}{\widehat{S}(t)}\right)\mathrm{d}W(t). \tag{3.86}$$

Let us construct the EMM $\widehat{\mathbb{P}}$ from $\mathbb{P}$ using the RND

$$\hat{\varrho}_t \equiv \varrho_t^{(\hat{\gamma})} = \mathrm{e}^{-\frac{1}{2}\hat{\gamma}^2 t + \hat{\gamma} W(t)}$$

so that the SDE for $\frac{B(t)}{\widehat{S}(t)}$ is driftless under $\widehat{\mathbb{P}}$ (and hence $B(t)/\widehat{S}(t)$ is a $\widehat{\mathbb{P}}$-martingale). We have $\widehat{W}(t) = W(t) - \hat{\gamma}t$ is a standard $\widehat{\mathbb{P}}$-BM. Using $\mathrm{d}W(t) = \mathrm{d}\widehat{W}(t) + \hat{\gamma}\,\mathrm{d}t$ in (3.86) gives

$$\mathrm{d}\left(\frac{B(t)}{\widehat{S}(t)}\right) = -\sigma\left(\frac{B(t)}{\widehat{S}(t)}\right)\mathrm{d}\widehat{W}(t), \quad \text{if } \hat{\gamma} = \frac{r - \mu + \sigma^2}{\sigma}.$$

From the original SDE for the stock price process in the $\mathbb{P}$-Brownian increment $\mathrm{d}W(t) = \mathrm{d}\widehat{W}(t) + \hat{\gamma}\,\mathrm{d}t$, we have (since $\sigma\hat{\gamma} = r - \mu + \sigma^2$)

$$\mathrm{d}S(t) = (r - q + \sigma^2)S(t)\,\mathrm{d}t + \sigma S(t)\,\mathrm{d}\widehat{W}(t).$$

Integrating the above SDE, we find the strong solution for the stock price in terms of the $\widehat{\mathbb{P}}$-BM:

$$S(t) = S(0)\mathrm{e}^{(r-q+\sigma^2/2)t+\sigma\widehat{W}(t)}.$$

This expression differs from the expression for $S(t)$ as a function of the $\widetilde{\mathbb{P}}$-BM:

$$S(t) = S(0)\mathrm{e}^{(r-q-\sigma^2/2)t+\sigma\widetilde{W}(t)}.$$

(c) There are multiple ways to solve this. We can find the RND $\varrho_t \equiv \left(\frac{\mathrm{d}\widehat{\mathbb{P}}}{\mathrm{d}\widetilde{\mathbb{P}}}\right)_t$ by relying on Girsanov's theorem or by using the change of numéraire method (it is discussed in Chapter 8 of Volume I, as well as in Section 4.2 in the next chapter). Let us apply both approaches. By Girsanov's theorem (used in parts (a) and (b)), we employed the fact that $\widetilde{W}(t) = W(t) - \tilde{\gamma}t$ and $\widehat{W}(t) = W(t) - \hat{\gamma}t$ are standard Brownian motions under the EMMs $\widetilde{\mathbb{P}}$ and $\widehat{\mathbb{P}}$, respectively. We can readily express one BM in terms of the other. Expressing $\widehat{W}(t)$ in terms of $\widetilde{W}(t)$,

$$\widehat{W}(t) = \widetilde{W}(t) - (\hat{\gamma} - \tilde{\gamma})t = \widetilde{W}(t) - \sigma t.$$

By Girsanov's theorem, we can conclude that $\widehat{\mathbb{P}}$ is constructed from $\widetilde{\mathbb{P}}$ using the RND

$$\varrho_t \equiv \varrho_t^{(\sigma)} = \mathrm{e}^{-\frac{1}{2}\sigma^2 t+\sigma\widetilde{W}(t)}.$$

Alternatively, the RND is given by the following ratio of numéraires:

$$\varrho_t \equiv \left(\frac{\mathrm{d}\widetilde{\mathbb{P}}^{(\widehat{S})}}{\mathrm{d}\widetilde{\mathbb{P}}^{(B)}}\right)_t = \frac{\widehat{S}(t)/\widehat{S}(0)}{B(t)/B(0)} = \frac{\mathrm{e}^{qt}S(t)/S(0)}{\mathrm{e}^{rt}} = \mathrm{e}^{(q-r)t}\frac{S(t)}{S(0)}.$$

Using the strong solution for $S(t)$ under $\widetilde{\mathbb{P}}$, we obtain

$$\varrho_t = \mathrm{e}^{(q-r)t}\mathrm{e}^{(r-q-\sigma^2/2)t+\sigma\widetilde{W}(t)} = \mathrm{e}^{-\frac{1}{2}\sigma^2 t+\sigma\widetilde{W}(t)}.$$

As we see, both approaches yield the same result.
Note: generally, we can obtain the other RND given we know two of the RNDs. Namely, we have $\left(\frac{\mathrm{d}\widetilde{\mathbb{P}}}{\mathrm{d}\mathbb{P}}\right)_t \cdot \left(\frac{\mathrm{d}\widehat{\mathbb{P}}}{\mathrm{d}\widetilde{\mathbb{P}}}\right)_t = \left(\frac{\mathrm{d}\widehat{\mathbb{P}}}{\mathrm{d}\mathbb{P}}\right)_t$, i.e., $\tilde{\varrho}_t \cdot \varrho_t = \widehat{\varrho}_t$. Hence,

$$\varrho_t = \frac{\widehat{\varrho}_t}{\tilde{\varrho}_t} = \mathrm{e}^{-\frac{1}{2}(\hat{\gamma}^2-\tilde{\gamma}^2)t+(\hat{\gamma}-\tilde{\gamma})W(t)} = \mathrm{e}^{-\frac{1}{2}[\hat{\gamma}^2-\tilde{\gamma}^2-2\tilde{\gamma}(\hat{\gamma}-\tilde{\gamma})]t+(\hat{\gamma}-\tilde{\gamma})\widetilde{W}(t)} = \mathrm{e}^{-\frac{1}{2}\sigma^2 t+\sigma\widetilde{W}(t)}.$$

The simplification was due to $\hat{\gamma}^2 - \tilde{\gamma}^2 - 2\tilde{\gamma}(\hat{\gamma} - \tilde{\gamma}) = \hat{\gamma}^2 - \tilde{\gamma}^2 - 2\tilde{\gamma}\sigma = \sigma^2$.

The no-arbitrage pricing in (3.85) allows us to choose either the bank account or the stock as numéraire; i.e., we can compute a conditional expectation of the payoff discounted by the numéraire asset price in either EMM $\widetilde{\mathbb{P}}$ or $\widehat{\mathbb{P}}$. Both give the same no-arbitrage price. Depending on the payoff, a given choice of numéraire may allow a simplification over another choice when deriving a no-arbitrage pricing formula. If the payoff is a function of the terminal stock price, $V_T = \Lambda(S(T))$, then the Markov property applied to (3.85) gives us the pricing formula, equivalent to (3.24):

$$V(t, S) = \mathrm{e}^{-q(T-t)}S\widehat{\mathrm{E}}_{t,S}\left[\frac{\Lambda(S(T))}{S(T)}\right] \equiv \mathrm{e}^{-q(T-t)}S\widehat{\mathrm{E}}\left[\frac{\Lambda(S(T))}{S(T)} \;\Big|\; S(t) = S\right]. \quad (3.87)$$

The following example gives a simple demonstration of the use of this pricing formula.

Example 3.8. Assume the standard (B, S) model with stock paying continuous dividends. Derive the no-arbitrage pricing function $V(t, S)$ for an asset-or-nothing derivative with the payoff function $\Lambda(S) = S\,\mathbb{I}_{\{K_1<S<K_2\}}$, where $0 \leqslant K_1 < K_2 \leqslant \infty$.

Solution. Let us derive $V(t,S)$ by choosing the stock as numéraire. From the payoff function, $\Lambda(S(T)) = S(T)\mathbb{I}_{\{K_1<S(T)<K_2\}}$, we have $\frac{\Lambda(S(T))}{S(T)} = \mathbb{I}_{\{K_1<S(T)<K_2\}}$. Hence, (3.87) gives

$$\begin{aligned} V(t,S) &= \mathrm{e}^{-q(T-t)} S\,\widehat{\mathrm{E}}_{t,S}\left[\mathbb{I}_{\{K_1<S(T)<K_2\}}\right] = \mathrm{e}^{-q(T-t)} S\,\widehat{\mathbb{P}}_{t,S}\big(K_1 < S(T) < K_2\big) \\ &= \mathrm{e}^{-q(T-t)} S\left[\widehat{\mathbb{P}}_{t,S}\big(S(T) < K_2\big) - \widehat{\mathbb{P}}_{t,S}\big(S(T) \leqslant K_1\big)\right]. \end{aligned}$$

To compute the conditional probability $\widehat{\mathbb{P}}_{t,S}(S(T) < K)$, we use the stock price in terms of $\widehat{W}$ ($\widehat{\mathbb{P}}$-BM): $S(t) = S(0)\mathrm{e}^{(r-q+\sigma^2/2)t+\sigma\widehat{W}(t)} \implies S(T) = S(t)\mathrm{e}^{(r-q+\sigma^2/2)(T-t)+\sigma(\widehat{W}(T)-\widehat{W}(t))}$ $\implies \ln\frac{S(T)}{S(t)} \stackrel{d}{=} (r-q+\sigma^2/2)(T-t) + \sigma\sqrt{T-t}\widehat{Z}$, where $\widehat{Z} := \frac{\widehat{W}(T)-\widehat{W}(t)}{\sqrt{T-t}} \sim Norm(0,1)$ under $\widehat{\mathbb{P}}$. Hence, by independence of $\widehat{Z}$ and $S(t)$,

$$\begin{aligned} \widehat{\mathbb{P}}_{t,S}(S(T) < K) = \widehat{\mathbb{P}}\left(\ln\frac{S(T)}{S(t)} < \ln\frac{K}{S}\right) &= \widehat{\mathbb{P}}\left(\widehat{Z} \leqslant -\frac{\ln(S/K) + (r-q+\sigma^2/2)(T-t)}{\sigma\sqrt{T-t}}\right) \\ &= \mathcal{N}\Big(-d_+\Big(\frac{\mathrm{e}^{-q(T-t)}S}{K}, T-t\Big)\Big). \end{aligned}$$

Therefore, the pricing formula takes the following form:

$$V(t,S) = \mathrm{e}^{-q(T-t)} S\left[\mathcal{N}\Big(-d_+\Big(\frac{\mathrm{e}^{-q(T-t)}S}{K_2}, T-t\Big)\Big) - \mathcal{N}\Big(-d_+\Big(\frac{\mathrm{e}^{-q(T-t)}S}{K_1}, T-t\Big)\Big)\right].$$

Since $\mathcal{N}(-x) = 1 - \mathcal{N}(x)$, we also have the equivalent expression:

$$V(t,S) = \mathrm{e}^{-q(T-t)} S\left[\mathcal{N}\Big(d_+\Big(\frac{\mathrm{e}^{-q(T-t)}S}{K_1}, T-t\Big)\Big) - \mathcal{N}\Big(d_+\Big(\frac{\mathrm{e}^{-q(T-t)}S}{K_2}, T-t\Big)\Big)\right]. \qquad \square$$

Now, we are going to apply the change of numéraire method to derive a general pricing for a standard call option. The pricing function is to be written in terms of two probabilities that the option is in the money (at maturity), which are calculated under the EMMs $\widetilde{\mathbb{P}}$ and $\widehat{\mathbb{P}}$, respectively.

Let us start with the standard risk-neutral pricing formula for a call option:

$$\begin{aligned} C(t,S) &= \mathrm{e}^{-r\tau}\widetilde{\mathrm{E}}_{t,S}\left[(S(T)-K)^+\right] = \mathrm{e}^{-r\tau}\widetilde{\mathrm{E}}_{t,S}\left[(S(T)-K)\mathbb{I}_{\{S(T)\geqslant K\}}\right] \\ &= \mathrm{e}^{-r\tau}\widetilde{\mathrm{E}}_{t,S}\left[S(T)\mathbb{I}_{\{S(T)\geqslant K\}}\right] - \mathrm{e}^{-r\tau}\widetilde{\mathrm{E}}_{t,S}\left[K\mathbb{I}_{\{S(T)\geqslant K\}}\right] \end{aligned}$$

with $\tau = T - t$. To calculate the first expectation, we can change the probability measure using the RND

$$\varrho_t \equiv \varrho_t^{\widehat{S}\to B} = \frac{B(t)/B(0)}{\widehat{S}(t)/\widehat{S}(0)} = \mathrm{e}^{(r-q)t}\frac{S(0)}{S(t)}.$$

As a result, we have

$$\begin{aligned} \mathrm{e}^{-r\tau}\widetilde{\mathrm{E}}_{t,S}\left[S(T)\mathbb{I}_{\{S(T)\geqslant K\}}\right] &= \mathrm{e}^{-r\tau}\widehat{\mathrm{E}}_{t,S}\left[\frac{\varrho_T}{\varrho_t}S(T)\mathbb{I}_{\{S(T)\geqslant K\}}\right] \\ &= \mathrm{e}^{-r\tau}\widehat{\mathrm{E}}_{t,S}\left[\frac{\mathrm{e}^{r\tau}}{\mathrm{e}^{q\tau}}\frac{S(t)}{S(T)}S(T)\mathbb{I}_{\{S(T)\geqslant K\}}\right] \\ &= \mathrm{e}^{-q\tau}S\,\widehat{\mathrm{E}}_{t,S}\left[\mathbb{I}_{\{S(T)\geqslant K\}}\right]. \end{aligned}$$

A more direct and simple way to arrive at this is to simply apply (3.87) to the first (asset-or-nothing) portion of the call with payoff $\Lambda_1(S(T)) := S(T)\mathbb{I}_{\{S(T)\geqslant K\}}$, i.e., $\frac{\Lambda_1(S(T))}{S(T)} = \mathbb{I}_{\{S(T)\geqslant K\}}$, giving the pricing function for this portion as we have above: $V_1(t,S) = \mathrm{e}^{-q\tau} S\,\widehat{\mathrm{E}}_{t,S}\left[\mathbb{I}_{\{S(T)\geqslant K\}}\right]$.

Therefore, the no-arbitrage call value is given by a difference of two conditional expectations of the same indicator function $\mathbb{I}_{\{S(T)\geqslant K\}}$ under two EMMs:

$$\begin{aligned} C(t,S) &= \mathrm{e}^{-q\tau} S\,\widehat{\mathrm{E}}_{t,S}\left[\mathbb{I}_{\{S(T)\geqslant K\}}\right] - \mathrm{e}^{-r\tau} K\,\widetilde{\mathrm{E}}_{t,S}\left[\mathbb{I}_{\{S(T)\geqslant K\}}\right] \\ &= \mathrm{e}^{-q\tau} S\,\widehat{\mathbb{P}}_{t,S}\left(S(T)\geqslant K\right) - \mathrm{e}^{-r\tau} K\,\widetilde{\mathbb{P}}_{t,S}\left(S(T)\geqslant K\right). \end{aligned}$$

This formula is quite general and works for any no-arbitrage asset price model including more advanced models with stochastic volatility and jumps. If the stock price follows the GBM process, we can calculate the two conditional expectations using the strong solutions (under $\widetilde{\mathbb{P}}$ and $\widehat{\mathbb{P}}$, respectively) and re-derive the well-known result for a standard European call option:

$$S(T) = S(t)\mathrm{e}^{(r-q-\frac{\sigma^2}{2})\tau+\sigma(\widetilde{W}(T)-\widetilde{W}(t))} \text{ under } \widetilde{\mathbb{P}} \implies \widetilde{\mathbb{P}}_{t,S}\left(S(T)\geqslant K\right) = \mathcal{N}(d_-(\mathrm{e}^{-q\tau}S/K,\tau)),$$
$$S(T) = S(t)\mathrm{e}^{(r-q+\frac{\sigma^2}{2})\tau+\sigma(\widehat{W}(T)-\widehat{W}(t))} \text{ under } \widehat{\mathbb{P}} \implies \widehat{\mathbb{P}}_{t,S}\left(S(T)\geqslant K\right) = \mathcal{N}(d_+(\mathrm{e}^{-q\tau}S/K,\tau)).$$

Using the put-call parity relation or, alternatively, repeating the steps outlined above, we can derive the following general pricing formula for a standard put option:

$$\begin{aligned} P(t,S) &= \mathrm{e}^{-r\tau} K\,\widetilde{\mathrm{E}}_{t,S}\left[\mathbb{I}_{\{S(T)<K\}}\right] - \mathrm{e}^{-q\tau} S\,\widehat{\mathrm{E}}_{t,S}\left[\mathbb{I}_{\{S(T)<K\}}\right] \\ &= \mathrm{e}^{-r\tau} K\,\widetilde{\mathbb{P}}_{t,S}\left(S(T)<K\right) - \mathrm{e}^{-q\tau} S\,\widehat{\mathbb{P}}_{t,S}\left(S(T)<K\right). \end{aligned}$$

We can generalize results of the above exercises and derive a simple formula for conditional expectations of the form $\widetilde{\mathrm{E}}_{t,S}[S(T)\mathbb{I}_A]$ with $0\leqslant t<T$, where the event A depends on stock prices:

$$\widetilde{\mathrm{E}}_{t,S}[S(T)\mathbb{I}_A] = \mathrm{e}^{(r-q)(T-t)}\widehat{\mathrm{E}}_{t,S}[S(t)\mathbb{I}_A] = \mathrm{e}^{(r-q)(T-t)} S\,\widehat{\mathbb{P}}_{t,S}(A). \tag{3.88}$$

3.3 Forward Starting, Chooser, and Compound Options

We are now equipped to readily develop pricing and hedging formulae for other classes of options besides the standard European options considered so far where we assumed a payoff $V_T = \Lambda(S(T))$. Rather than having a payoff that depends solely on the terminal stock price $S(T)$ at maturity T, there are European-style options that can have stipulations at a finite number of intermediate dates within the lifetime of the option. These types of derivatives are examples of what can be termed *multistage options*. As in most options, there are simpler and more complex versions of these contracts.

Examples of simpler contracts are *forward starting* options with one intermediate date $T_1 < T$. The holder enters the contract at a time $t < T_1$ such that at time T_1 the contract has the value of an option (say a European call) on an underlying stock which expires at T. Generally we can represent the option value at time T_1 as a function of $S(T_1)$: $V_{T_1} = V_{T_1}(T_1, S(T_1))$. The time-$T$ payoff, $V_T = \Lambda(S(T_1), S(T))$, of the option can depend upon $S(T_1)$ and $S(T)$. By risk-neutral pricing we have (discounting from T back to T_1)

$$V_{T_1} = \mathrm{e}^{-r(T-T_1)}\widetilde{\mathrm{E}}_{T_1,S(T_1)}[\,\Lambda(S(T_1),S(T))]\,. \tag{3.89}$$

Let $V(t,S) \equiv V(t,S;T_1,T)$ denote the time-t value of the forward starting option with given spot $S(t) = S$. By the Markov property of the stock price process, conditioning on $\mathcal{F}_t$ reduces to conditioning on $S(t)$. Applying again the risk-neutral pricing formula (discounting from time T_1 to t) and the tower property[4] finally gives us the time-t price as a single conditional expectation discounted from T to t:

$$\begin{aligned} V(t,S;T_1,T) &= \mathrm{e}^{-r(T_1-t)}\widetilde{\mathrm{E}}_{t,S}[\,V_{T_1}\,] \\ &= \mathrm{e}^{-r(T_1-t)}\cdot \mathrm{e}^{-r(T-T_1)}\widetilde{\mathrm{E}}_{t,S}\big[\widetilde{\mathrm{E}}_{T_1,S(T_1)}[\,\Lambda(S(T_1),S(T))]\big] \\ &= \mathrm{e}^{-r(T-t)}\widetilde{\mathrm{E}}_{t,S}\big[\,\Lambda(S(T_1),S(T))\,\big]. \end{aligned} \tag{3.90}$$

For instance, in a *forward starting call* the holder enters the contract at a time $t < T_1$, prior to an intermediate date T_1, whose value at time T_1 is a call on an underlying stock initiated at the "forward" time T_1. The call initiated at time T_1 expires at date $T > T_1$ and has some strike specification, which is generally a function of the stock price $S(T_1)$ (i.e., the strike is not a constant specified value). Specifically, the forward starting call can be specified as having strike $K_{T_1} = S(T_1)$, then $\Lambda(S(T_1),S(T)) = (S(T) - S(T_1))^+$. Hence, as viewed at present time t, the strike is a random variable corresponding to the price of the stock at time T_1. Inserting the payoff into (3.90) gives the time-t price of this forward starting call

$$C(t,S;T_1,T) = \mathrm{e}^{-r(T-t)}\widetilde{\mathrm{E}}_{t,S}\big[\,(S(T)-S(T_1))^+\,\big].$$

Let us assume the stock price is a GBM given by (3.15) with zero dividend. The addition of a continuous constant dividend yield q on the stock can be done trivially by applying the symmetry relation in (3.67) to the resulting pricing formula. The above expectation is readily evaluated by writing the payoff as

$$(S(T)-S(T_1))^+ = S(T_1)\left(\frac{S(T)}{S(T_1)} - 1\right)^+ = S(T_1)\,(Y-1)^+$$

where random variable $Y := \frac{S(T)}{S(T_1)} = \mathrm{e}^{(r-\frac{1}{2}\sigma^2)(T-T_1)+\sigma(\widetilde{W}(T)-\widetilde{W}(T_1))}$ is independent of $S(T_1)$. Hence, using the tower property in reverse gives a nested expectation with an inner expectation conditional on $S(T_1)$ as in (3.90):

$$\begin{aligned} C(t,S;T_1,T) &= \mathrm{e}^{-r(T-t)}\widetilde{\mathrm{E}}_{t,S}\big[\,\widetilde{\mathrm{E}}_{T_1,S(T_1)}[\,S(T_1)\,(Y-1)^+\,]\,\big] \\ &= \mathrm{e}^{-r(T-t)}\widetilde{\mathrm{E}}_{t,S}\big[\,S(T_1)\,\widetilde{\mathrm{E}}[(Y-1)^+\,]\,\big] \\ &= \mathrm{e}^{-r(T-T_1)}\widetilde{\mathrm{E}}[(Y-1)^+\,]\cdot \mathrm{e}^{-r(T_1-t)}\widetilde{\mathrm{E}}_{t,S}\big[\,S(T_1)\,\big]. \end{aligned} \tag{3.91}$$

Here, we pulled $S(T_1)$ out of the inner expectation as it is $\mathcal{F}_{T_1}$-measurable; then the inner condition is dropped since Y is independent of $S(T_1)$, and finally the unconditional expectation is a constant that is factored out of the expectation conditional on $S(t) = S$. Note that we have also factored the discount term into two parts. The first term on the right in (3.91) is recognized as the Black–Scholes price of a call with no dividend, time to maturity $T - T_1$, effective strike and spot of unity:

$$\mathrm{e}^{-r(T-T_1)}\widetilde{\mathrm{E}}[(Y-1)^+\,] = \mathcal{N}(d_+) - \mathrm{e}^{-r(T-T_1)}\mathcal{N}(d_-)\,, \tag{3.92}$$

---

[4] We remind the reader of the shorthand notation we have been adopting for conditional expectations. In particular, $\widetilde{\mathrm{E}}_{t,S}\,[(\,\cdot\,)] \equiv \widetilde{\mathrm{E}}[(\,\cdot\,) \mid S(t) = S]$ is a number where $S > 0$ is a spot value and $\widetilde{\mathrm{E}}_{T_1,S(T_1)}\,[(\,\cdot\,)] \equiv \widetilde{\mathrm{E}}[(\,\cdot\,) \mid S(T_1)]$ is a $\sigma(S(T_1))$-measurable (and hence $\mathcal{F}_{T_1}$-measurable) random variable where we are conditioning on the σ-algebra, $\sigma(S(T_1))$, generated by the stock at time T_1.

where

$$d_\pm \equiv d_\pm(1, T - T_1) = \frac{\ln(1) + (r \pm \frac{1}{2}\sigma^2)(T - T_1)}{\sigma\sqrt{T - T_1}} = \left(\frac{r}{\sigma} \pm \frac{1}{2}\sigma\right)\sqrt{T - T_1}\,.$$

The second term in (3.91) gives the spot S since $\widetilde{\mathrm{E}}_{t,S}\left[\, S(T_1)\,\right] = \mathrm{e}^{r(T_1-t)}S$. Multiplying the expression in (3.92) by S produces the pricing formula:

$$C(t, S; T_1, T) = S[\mathcal{N}(d_+) - \mathrm{e}^{-r(T-T_1)}\mathcal{N}(d_-)]\,. \tag{3.93}$$

We observe that this pricing function is linear in the spot S since the term in square brackets depends only on parameters r, σ, T_1, T. The delta hedge is then trivially given by the term in brackets, $\Delta(t, S) = \frac{\partial}{\partial S}C(t, S; T_1, T) = [\mathcal{N}(d_+) - \mathrm{e}^{-r(T-T_1)}\mathcal{N}(d_-)]$. This is a constant hedge position having no dependence on time t and spot S. What is then interesting is that this forward starting call can be hedged statically in time! Other related forward starting option problems are left as exercises for the reader (see Exercises 3.25 and 3.26). We only include one additional example with a quadratic payoff.

Example 3.9. Derive a no-arbitrage pricing formula for the forward starting option with the quadratic payoff $(S(T) - S(T_1))^2$ where $0 \leqslant T_1 < T$.

Solution. The no-arbitrage time-t value $V(t, S)$ with $t \in [0, T_1]$ and $S > 0$ is given by

$$V(t, S) = \mathrm{e}^{-r\,(T_1-t)}\widetilde{\mathrm{E}}_{t,S}\left[\mathrm{e}^{-r\,(T-T_1)}\,\widetilde{\mathrm{E}}_{T_1,S(T_1)}\left[(S(T) - S(T_1))^2\right]\right].$$

First, let us factor out $S^2(T_1)$ and use the fact that $S(T_1)$ is $\mathcal{F}_{T_1}$-measurable:

$$V(t, S) = \mathrm{e}^{-r\,(T_1-t)}\widetilde{\mathrm{E}}_{t,S}\left[\mathrm{e}^{-r\,(T-T_1)}\,S^2(T_1)\,\widetilde{\mathrm{E}}_{T_1,S(T_1)}\left[(S(T)/S(T_1) - 1)^2\right]\right].$$

Now, since $Y := \dfrac{S(T)}{S(T_1)} \stackrel{d}{=} \mathrm{e}^{(r-q-\sigma^2/2)(T-T_1)+\sigma\sqrt{T-T_1}Z}$, with $Z \sim Norm(0, 1)$, is independent of $\mathcal{F}_{T_1}$, we have

$$\begin{aligned} V(t, S) &= \mathrm{e}^{-r\,(T-t)}\widetilde{\mathrm{E}}_{t,S}\left[S^2(T_1)\right]\widetilde{\mathrm{E}}\left[\left(\frac{S(T)}{S(T_1)} - 1\right)^2\right] \\ &= \mathrm{e}^{-r\,(T-t)}\widetilde{\mathrm{E}}_{t,S}\left[S^2(T_1)\right]\left(\widetilde{\mathrm{E}}\left[Y^2\right] - 2\widetilde{\mathrm{E}}\left[Y\right] + 1\right). \end{aligned}$$

The above expectations can be easily computed using (3.53) with $\alpha = 2$ and the MGF of a normal distribution to yield

$$\begin{aligned} V(t, S) &= \mathrm{e}^{-r\,(T-t)}S^2\mathrm{e}^{2(r-q+\sigma^2/2)(T_1-t)}\left(\mathrm{e}^{2(r-q+\sigma^2/2)(T-T_1)} - 2\mathrm{e}^{(r-q)(T-T_1)} + 1\right) \\ &= \mathrm{e}^{(r-2q+\sigma^2)(T_1-t)}S^2\left(\mathrm{e}^{(r-2q+\sigma^2)(T-T_1)} - 2\mathrm{e}^{-q(T-T_1)} + \mathrm{e}^{-r(T-T_1)}\right). \end{aligned}$$ □

Another example of a multistage option is a so-called *chooser option*. An example of this is a contract that gives its holder the right to choose at a predetermined time T_1 (with $0 < T_1 < T$) whether the T-maturity option is a European call or a put. At time T_1, the value of the chooser option is

$$V(T_1, S(T_1)) = \max\left\{C(T_1, S(T_1)), P(T_1, S(T_1))\right\}$$

where $C(T_1, S) \equiv C(T_1, S; T, K)$ and $P(T_1, S) \equiv P(T_1, S; T, K)$ are, respectively, the time-T_1 values of the European call and put options with maturity T and strike K. Using the put-call parity relation $P(T_1, S) - C(T_1, S) = \mathrm{e}^{-r(T-T_1)}K - \mathrm{e}^{-q(T-T_1)}S$ gives

$$\begin{aligned} V(T_1, S(T_1)) &= \max\left\{C(T_1, S(T_1)), C(T_1, S(T_1)) + \mathrm{e}^{-r(T-T_1)}K - \mathrm{e}^{-q(T-T_1)}S(T_1)\right\} \\ &= C(T_1, S(T_1)) + \max\left\{0, \mathrm{e}^{-r(T-T_1)}K - \mathrm{e}^{-q(T-T_1)}S(T_1)\right\} \\ &= C(T_1, S(T_1)) + \mathrm{e}^{-q(T-T_1)}\left(\mathrm{e}^{-(r-q)(T-T_1)}K - S(T_1)\right)^+. \end{aligned} \tag{3.94}$$

At present time $t \in [0, T_1)$, the no-arbitrage value of the option is

$$V(t, S) = \mathrm{e}^{-r(T_1-t)}\, \widetilde{\mathrm{E}}_{t,S}\left[V(T_1, S(T_1))\right] \tag{3.95}$$

thanks to the martingale property. Let us plug (3.94) into (3.95) to obtain

$$\begin{aligned} V(t, S) &= \mathrm{e}^{-r(T_1-t)}\, \widetilde{\mathrm{E}}_{t,S}\left[C(T_1, S(T_1)) + \mathrm{e}^{-q(T-T_1)}\left(\mathrm{e}^{-(r-q)(T-T_1)}K - S(T_1)\right)^+\right] \\ &= \mathrm{e}^{-r(T_1-t)}\, \widetilde{\mathrm{E}}_{t,S}\left[C(T_1, S(T_1))\right] \\ &\quad + \mathrm{e}^{-q(T-T_1)}\, \mathrm{e}^{-r(T_1-t)}\, \widetilde{\mathrm{E}}_{t,S}\left[\left(\mathrm{e}^{-(r-q)(T-T_1)}K - S(T_1)\right)^+\right]. \end{aligned} \tag{3.96}$$

The martingale property of a discounted derivative value process allows us to simplify the first term in (3.96). The second term is a discounted value of a European put with strike $K_1 := \mathrm{e}^{-(r-q)(T-T_1)}K$ and maturity T_1. Therefore, the no-arbitrage price of the chooser option is as follows:

$$V(t, S) = C(t, S; T, K) + \mathrm{e}^{-q(T-T_1)}\, P(t, S; T_1, K_1). \tag{3.97}$$

Employing the known risk-neutral pricing formulae for standard European call and put options gives us the final expression for the no-arbitrage pricing formula.

The representation (3.97) allows us to find the delta of the chooser option as a sum of European call and put deltas:

$$\Delta_V(t, S) = \frac{\partial V}{\partial S}(t, S) = \Delta_c(t, S; T, K) + \mathrm{e}^{-q(T-T_1)}\, \Delta_p(t, S; T_1, K_1).$$

We now consider the problem of pricing a more complex class of options known as *compound options*. As the name implies, such contracts are options on options. Here, we shall assume the simplest types of compound options that involve an (outer) option to buy or sell another (inner or embedded) option at some future time. For standard European-style options, the payoff is simply a function of the underlying stock price at some expiry time $T > t$ where t is current calendar time. In a compound option, the essential difference is that its value at some future time, say $T_1 > t$, is a specified function of an option price on an underlying stock whereby the latter option is initiated at time T_1 and matures at a future time $T_2 > T_1$ with a specified payoff function. In a compound option, the role of an underlying asset is not played by the stock but rather *the embedded option on the stock plays the role of "underlying asset" for the (outer) option.*

Generally, a European compound option is defined by an outer payoff function $\phi^{(1)}$ and inner payoff function $\phi^{(2)}$. We are interested in pricing the compound option at time $t < T_1$. Given $t < T_1 < T_2$, let $V_t = V(t, S(t); T_1, T_2)$, $t \leqslant T_1$, be the *value process* of the compound option. Note that we have also denoted this as a function of given exercise times T_1, T_2. At time T_1, with stock price $S(T_1)$, the value of the compound option is given by $V_{T_1} = \phi^{(1)}(V^{(2)}_{T_1}(S(T_1), T_2))$, where $V^{(2)}_{T_1}(S(T_1), T_2)$ is the value of the underlying (inner) option at time T_1 with time to expiry $T_2 - T_1$ and payoff value $V^{(2)}_{T_2}(S(T_2), T_2) = \phi^{(2)}(S(T_2))$. Note that $V^{(2)}_{T_1}(S(T_1), T_2)$ is an ordinary function of the random variable $S(T_1)$ and can be viewed as a random asset value at time T_1. Given spot $S(t) = S$, then by the risk-neutral pricing formulation the arbitrage-free price of the compound option is given by the conditional expectation (under the risk-neutral measure $\widetilde{\mathbb{P}}$) of the discounted value, $\mathrm{e}^{-r(T_1-t)}V_{T_1}$, of the payoff of the outer option at time T_1:

$$V(t, S; T_1, T_2) = \mathrm{e}^{-r(T_1-t)}\widetilde{\mathrm{E}}_{t,S}\left[\phi^{(1)}(V^{(2)}_{T_1}(S(T_1), T_2))\right] \tag{3.98}$$

where

$$V^{(2)}_{T_1}(S(T_1), T_2) = \mathrm{e}^{-r(T_2-T_1)}\widetilde{\mathrm{E}}_{T_1, S(T_1)}\left[\phi^{(2)}(S(T_2))\right]. \tag{3.99}$$

The order of the steps for deriving the pricing function $V(t, S; T_1, T_2)$ by the above expectation approach is as follows.

1. Determine the time-T_1 price $V^{(2)}_{T_1}(S_1, T_2)$ of the embedded option on the stock having expiry $T_2 > T_1$ and spot variable S_1.

2. Set $S_1 = S(T_1)$ to obtain the payoff of the outer option at time T_1; as a random variable this payoff is $V_{T_1} = \phi^{(1)}(V^{(2)}_{T_1})$ where $V^{(2)}_{T_1} \equiv V^{(2)}_{T_1}(S(T_1), T_2)$.

3. Compute the discounted risk-neutral expectation in (3.98) i.e., the time-t price is $V(t, S) \equiv V(t, S; T_1, T_2) = \mathrm{e}^{-r(T_1-t)}\widetilde{\mathrm{E}}_{t,S}\left[V_{T_1}\right]$.

The most common examples of European compound options are a *call-on-a-call*, *put-on-a-call*, *put-on-a-put* and *call-on-a-put*. These four options are characterized by two expiration dates T_1 and T_2 and two strike values K_1 and K_2 and with respective payoff functions $\phi^{(1)}(x) = (x - K_1)^+$ and $\phi^{(2)}(x) = (x - K_2)^+$; $\phi^{(1)}(x) = (K_1 - x)^+$ and $\phi^{(2)}(x) = (x - K_2)^+$; $\phi^{(1)}(x) = (K_1 - x)^+$ and $\phi^{(2)}(x) = (K_2 - x)^+$; $\phi^{(1)}(x) = (x - K_1)^+$ and $\phi^{(2)}(x) = (K_2 - x)^+$. For example, the call-on-a-call contract gives the holder the right (but not the obligation) to buy an underlying call option for a fixed strike price K_1 at calendar time T_1 and where the underlying call is specified by strike K_2 and time to expiry $T_2 - T_1$.

As a concrete example, let us specifically value the call-on-a-call option by implementing (3.98) within the usual GBM process for the stock price process having dividend q in an economy with constant interest rate r. Denote the value of the underlying call at time T_1 by $C_{T_1} \equiv C_{T_1}(S(T_1), K_2, T_2)$. Hence, $V^{(2)}_{T_1}(S(T_1), T_2) = C_{T_1}(S(T_1), K_2, T_2)$ in (3.98) is given explicitly by the standard call price formula, i.e., for time-T_1 spot value $S(T_1) = S_1 > 0$ and time to maturity $T_2 - T_1$:

$$C_{T_1}(S_1, K_2, T_2) = \mathrm{e}^{-q(T_2-T_1)}S_1\mathcal{N}(d_+) - K_2\mathrm{e}^{-r(T_2-T_1)}\mathcal{N}(d_-), \tag{3.100}$$

where $d_\pm \equiv d_\pm(\frac{S_1}{K_2}, T_2 - T_1)$. Throughout this section, we define

$$d_\pm(x, \tau) := \frac{\ln x + (r - q \pm \frac{\sigma^2}{2})\tau}{\sigma\sqrt{\tau}}, \quad x, \tau > 0. \tag{3.101}$$

From (3.98), the call-on-a-call option value, denoted by $V^{cc}(t, S)$, is given by

$$V^{cc}(t, S) = \mathrm{e}^{-r(T_1-t)}\widetilde{\mathrm{E}}_{t,S}\left[(C_{T_1}(S(T_1), K_2, T_2) - K_1)^+\right]. \tag{3.102}$$

Note that the random variable within this expectation is nonzero only when $C_{T_1} > K_1$. Recall that the call pricing function $C_{T_1}(S_1, K_2, T_2)$ is a strictly increasing function of the spot variable S_1 where $C_{T_1}(S_1, K_2, T_2) \to 0$, as $S_1 \searrow 0$, and $C_{T_1}(S_1, K_2, T_2) \to \infty$, as $S_1 \to \infty$. The graph of $C_{T_1}(S_1, K_2, T_2)$ versus S_1 must therefore cross the level $K_1 > 0$ at exactly one (critical) point, i.e., at $S_1 = S_1^*$. This point is the root of the equation

$$C_{T_1}(S_1^*, K_2, T_2) = K_1.$$

Note that by (3.100) we see that this is a nonlinear algebraic equation so that S_1^*, being a function of K_1, K_2 and $T_2 - T_1$, is in practice obtained numerically. Given the point S_1^*, and since $C_{T_1}(S_1, K_2, T_2)$ is strictly increasing in S_1, we have the equivalence $\mathbb{I}_{\{C_{T_1}(S(T_1),K_2,T_2) > K_1\}} = \mathbb{I}_{\{S(T_1)>S_1^*\}}$, hence

$$(C_{T_1}(S(T_1), K_2, T_2) - K_1)^+ = (C_{T_1}(S(T_1), K_2, T_2) - K_1)\mathbb{I}_{\{S(T_1)>S_1^*\}}.$$

So (3.102) now reads

$$V^{cc}(t, S) = \mathrm{e}^{-r\tau_1}\widetilde{\mathrm{E}}_{t,S}\left[C_{T_1}(S(T_1), K_2, T_2)\mathbb{I}_{\{S(T_1)>S_1^*\}}\right] - K_1\mathrm{e}^{-r\tau_1}\widetilde{\mathrm{E}}_{t,S}\left[\mathbb{I}_{\{S(T_1)>S_1^*\}}\right] \quad (3.103)$$

where we define $\tau_1 := T_1 - t$ and $\tau_2 := T_2 - t$ in what follows.

The two conditional expectations in (3.103) are readily evaluated by using the strong solution representation of the stock price process. In particular, the second expectation is evaluated using

$$S(T_1) = S(t)\mathrm{e}^{(r-q-\sigma^2/2)\tau_1+\sigma(\widetilde{W}(T_1)-\widetilde{W}(t))} \quad (3.104)$$

and the fact that $\widetilde{W}(T_1) - \widetilde{W}(t)$ and $\widetilde{W}(t)$ are jointly independent. Hence, $\frac{S(T_1)}{S(t)}$ and $S(t)$ are jointly independent, giving

$$\begin{aligned}
\widetilde{\mathrm{E}}_{t,S}\left[\mathbb{I}_{\{S(T_1)>S_1^*\}}\right] &= \widetilde{\mathrm{E}}\left[\mathbb{I}_{\{\frac{S(T_1)}{S(t)} > \frac{S_1^*}{S(t)}\}} \mid S(t) = S\right] = \widetilde{\mathrm{E}}\left[\mathbb{I}_{\{\frac{S(T_1)}{S(t)} > \frac{S_1^*}{S}\}}\right] \\
&= \widetilde{\mathbb{P}}\left(\ln\frac{S(T_1)}{S(t)} > \ln\frac{S_1^*}{S}\right) \\
&= \widetilde{\mathbb{P}}\left(\frac{\widetilde{W}(T_1) - \widetilde{W}(t)}{\sqrt{\tau_1}} < a_-\right) = \mathcal{N}(a_-)
\end{aligned} \quad (3.105)$$

where we denote $a_\pm \equiv d_\pm(\frac{S}{S_1^*}, \tau_1)$. The last equality follows since $\frac{\widetilde{W}(T_1)-\widetilde{W}(t)}{\sqrt{\tau_1}} \sim Norm(0,1)$ under measure $\widetilde{\mathbb{P}}$.

The first conditional expectation in (3.103) is re-expressed as follows:

$$\begin{aligned}
&\widetilde{\mathrm{E}}_{t,S}\left[\mathbb{I}_{\{S(T_1)>S_1^*\}}C_{T_1}(S(T_1), K_2, T_2)\right] \\
&= \mathrm{e}^{-r(T_2-T_1)}\widetilde{\mathrm{E}}_{t,S}\left[\mathbb{I}_{\{S(T_1)>S_1^*\}}\widetilde{\mathrm{E}}_{T_1,S(T_1)}\left[(S(T_2) - K_2)^+\right]\right] \\
&= \mathrm{e}^{-r(\tau_2-\tau_1)}\widetilde{\mathrm{E}}_{t,S}\left[\widetilde{\mathrm{E}}_{T_1,S(T_1)}\left[\mathbb{I}_{\{S(T_1)>S_1^*,S(T_2)>K_2\}}(S(T_2) - K_2)\right]\right] \\
&= \mathrm{e}^{-r(\tau_2-\tau_1)}\widetilde{\mathrm{E}}_{t,S}\left[\mathbb{I}_{\{S(T_1)>S_1^*,\, S(T_2)>K_2\}}(S(T_2) - K_2)\right] \\
&= \mathrm{e}^{-r(\tau_2-\tau_1)}\widetilde{\mathrm{E}}_{t,S}\left[\mathbb{I}_{\{S(T_1)>S_1^*,\, S(T_2)>K_2\}}S(T_2)\right] \\
&\qquad - K_2\,\mathrm{e}^{-r(\tau_2-\tau_1)}\widetilde{\mathrm{E}}_{t,S}\left[\mathbb{I}_{\{S(T_1)>S_1^*,\, S(T_2)>K_2\}}\right].
\end{aligned} \quad (3.106)$$

Note that in the third line from the top we have moved the indicator random variable $\mathbb{I}_{\{S(T_1)>S_1^*\}}$ to the inside of the inner expectation since it is known at time T_1 (i.e., it is $\sigma(S(T_1))$-measurable). In the third line, we have altogether eliminated the inner conditional expectation (i.e., the conditioning on $S(T_1)$) simply by using iterated conditioning (i.e., the tower property). The two conditional expectations in the last equation line of (3.106) are evaluated as follows. The last expectation is a joint probability. Upon using the condition $S(t) = S$, the fact that $S(T_1)/S(t)$ and $S(T_2)/S(t)$ are both independent of $S(t)$, and using (3.104) and

$$S(T_2) = S(t)e^{(r-q-\sigma^2/2)\tau_2+\sigma(\widetilde{W}(T_2)-\widetilde{W}(t))} \tag{3.107}$$

we have

$$\begin{aligned}
\widetilde{\mathrm{E}}_{t,S}\left[\mathbb{I}_{\{S(T_1)>S_1^*,\, S(T_2)>K_2\}}\right] &= \widetilde{\mathrm{E}}\big[\mathbb{I}_{\left\{\frac{S(T_1)}{S(t)}>\frac{S_1^*}{S},\, \frac{S(T_2)}{S(t)}>\frac{K_2}{S}\right\}}\big] \\
&= \widetilde{\mathbb{P}}\left(\ln\frac{S(T_1)}{S(t)} > \ln\frac{S_1^*}{S},\, \ln\frac{S(T_2)}{S(t)} > \ln\frac{K_2}{S}\right) \\
&= \widetilde{\mathbb{P}}\left(\frac{\widetilde{W}(T_1)-\widetilde{W}(t)}{\sqrt{\tau_1}} > -a_-,\, \frac{\widetilde{W}(T_2)-\widetilde{W}(t)}{\sqrt{\tau_2}} > -b_-\right)
\end{aligned} \tag{3.108}$$

where $b_\pm \equiv d_\pm(\frac{S}{K_2}, \tau_2)$. Since the increments $\widetilde{W}(T_1) - \widetilde{W}(t)$ and $\widetilde{W}(T_2) - \widetilde{W}(t)$ are $Norm(0, \tau_1)$ and $Norm(0, \tau_2)$, respectively, the random variables $Z_1 := \frac{\widetilde{W}(T_1)-\widetilde{W}(t)}{\sqrt{\tau_1}}$ and $Z_2 := \frac{\widetilde{W}(T_2)-\widetilde{W}(t)}{\sqrt{\tau_2}}$ are both standard normals under the risk-neutral measure $\widetilde{\mathbb{P}}$. Moreover, using the independence of nonoverlapping Brownian increments, their covariance (in the $\widetilde{\mathbb{P}}$-measure) is given by

$$\begin{aligned}
\widetilde{\mathrm{Cov}}(Z_1, Z_2) &= \frac{1}{\sqrt{\tau_1\tau_2}}\widetilde{\mathrm{Cov}}(\widetilde{W}(T_1) - \widetilde{W}(t), \widetilde{W}(T_2) - \widetilde{W}(t)) \\
&= \frac{1}{\sqrt{\tau_1\tau_2}}\widetilde{\mathrm{Cov}}(\widetilde{W}(T_1) - \widetilde{W}(t), \widetilde{W}(T_2) - \widetilde{W}(T_1) + \widetilde{W}(T_1) - \widetilde{W}(t)) \\
&= \frac{1}{\sqrt{\tau_1\tau_2}}\widetilde{\mathrm{Var}}(\widetilde{W}(T_1) - \widetilde{W}(t)) = \frac{1}{\sqrt{\tau_1\tau_2}}\tau_1 = \sqrt{\frac{\tau_1}{\tau_2}} = \sqrt{\frac{T_1 - t}{T_2 - t}}\,.
\end{aligned}$$

Hence the vector (Z_1, Z_2) has standard normal bivariate distribution with correlation coefficient $\rho \equiv \sqrt{\tau_1/\tau_2}$. By symmetry, $(-Z_1, -Z_2)$ has the same bivariate distribution. Hence, (3.108) gives

$$\begin{aligned}
\widetilde{\mathrm{E}}_{t,S}\left[\mathbb{I}_{\{S(T_1)>S_1^*,\, S(T_2)>K_2\}}\right] &= \widetilde{\mathbb{P}}(Z_1 > -a_-, Z_2 > -b_-) \\
&= \widetilde{\mathbb{P}}(Z_1 < a_-, Z_2 < b_-) = \mathcal{N}_2(a_-, b_-; \rho)\,.
\end{aligned} \tag{3.109}$$

Following similar steps as led to (3.108) above, and inserting the exponential form in (3.107) for $S(T_2)$ where $S(t) = S$, the second to last conditional expectation in (3.106) is now conveniently rewritten in terms of Z_1 and Z_2 and evaluated:

$$\begin{aligned}
\widetilde{\mathrm{E}}_{t,S}\left[\mathbb{I}_{\{S(T_1)>S_1^*,\, S(T_2)>K_2\}}S(T_2)\right] &= S\,\widetilde{\mathrm{E}}\left[\mathbb{I}_{\{\ln\frac{S(T_1)}{S(t)}>\ln\frac{S_1^*}{S},\, \ln\frac{S(T_2)}{S(t)}>\ln\frac{K_2}{S}\}}S(T_2)/S(t)\right] \\
&= Se^{(r-q-\sigma^2/2)\tau_2}\,\widetilde{\mathrm{E}}\big[\mathbb{I}_{\{Z_1>-a_-,\, Z_2>-b_-\}}e^{\sigma\sqrt{\tau_2}Z_2}\big] \\
&= Se^{(r-q-\sigma^2/2)\tau_2}\,\widetilde{\mathrm{E}}\big[\mathbb{I}_{\{Z_1<a_-,\, Z_2<b_-\}}e^{-\sigma\sqrt{\tau_2}Z_2}\big] \\
&= Se^{(r-q-\sigma^2/2)\tau_2}\,e^{\frac{1}{2}\sigma^2\tau_2}\,\mathcal{N}_2(a_- + \rho\sigma\sqrt{\tau_2},\, b_- + \sigma\sqrt{\tau_2}\,; \rho) \\
&= Se^{(r-q)\tau_2}\,\mathcal{N}_2(a_- + \sigma\sqrt{\tau_1},\, b_- + \sigma\sqrt{\tau_2}\,; \rho) \\
&= Se^{(r-q)\tau_2}\,\mathcal{N}_2(a_+, b_+\,; \rho)\,.
\end{aligned} \tag{3.110}$$

We note that in evaluating the last expectation we used the identity in (B.25) of Appendix B. Alternatively, to compute the above expectation, we can use the stock as a numéraire asset and employ (3.88) with $S(t) = S(0)\mathrm{e}^{(r-q+\sigma^2/2)t+\sigma\widehat{W}(t)}$ under $\widehat{\mathbb{P}}$ to obtain

$$\begin{aligned}
\widetilde{\mathrm{E}}_{t,S}\left[\mathbb{I}_{\{S(T_1)>S_1^*,\,S(T_2)>K_2\}}S(T_2)\right] &= \mathrm{e}^{(r-q)(T_2-t)}S\,\widehat{\mathrm{E}}_{t,S}\left[\mathbb{I}_{\{S(T_1)>S_1^*,\,S(T_2)>K_2\}}\right] \\
&= \mathrm{e}^{(r-q)\tau_2}S\,\widehat{\mathbb{P}}_{t,S}\left(S(T_1)>S_1^*,\,S(T_2)>K_2\right) \\
&= \mathrm{e}^{(r-q)\tau_2}S\,\widehat{\mathbb{P}}\left(\ln\frac{S(T_1)}{S(t)}>\ln\frac{S_1^*}{S},\,\ln\frac{S(T_2)}{S(t)}>\ln\frac{K_2}{S}\right) \\
&= \mathrm{e}^{(r-q)\tau_2}S\,\widehat{\mathbb{P}}\left(\frac{\widehat{W}(T_1)-\widehat{W}(t)}{\sqrt{\tau_1}}>-a_+,\,\frac{\widehat{W}(T_2)-\widehat{W}(t)}{\sqrt{\tau_2}}>-b_+\right) \\
&= \mathrm{e}^{(r-q)\tau_2}S\,\mathcal{N}_2\left(a_+,\,b_+\,;\rho\right).
\end{aligned}$$

Finally, by combining the expressions in (3.110), (3.109), (3.106), and (3.105) into (3.103) gives the explicit formula for the compound call-on-a-call:

$$V^{cc}(t,S) = S\mathrm{e}^{-q\tau_2}\mathcal{N}_2\left(a_+,\,b_+\,;\rho\right) - K_2\mathrm{e}^{-r\tau_2}\mathcal{N}_2\left(a_-,\,b_-\,;\rho\right) - K_1\mathrm{e}^{-r\tau_1}\mathcal{N}(a_-) \quad (3.111)$$

where $a_\pm, b_\pm, \rho$ are defined above. Note that the option value is a function of the spot S, the two strike values K_1, K_2, and the two time to expiration values τ_1, τ_2.

The other types of compound options can be valued in similar fashion. For example, we leave the valuation of the put-on-a-put as an exercise. There also exists a form of put-call parity among some pairs of compound options. In particular, the call-on-a-call option value and the corresponding put-on-a-call option value $V^{pc}(t,S)$ are related by

$$V^{cc}(t,S) - V^{pc}(t,S) = C_t(S,K_2,T_2) - \mathrm{e}^{-r\tau_1}K_1. \quad (3.112)$$

Namely, the difference in the time-t value of the call-on-a-call and put-on-a-call (with spot $S(t)=S$ and given strike and maturity pairs K_1, T_1 and K_2, T_2) is simply the time-t value of a standard call (with spot $S(t)=S$, strike and maturity K_2, T_2) minus the discounted strike value $\mathrm{e}^{-r\tau_1}K_1$. This is shown as follows. According to (3.98), the put-on-a-call option has value

$$V^{pc}(t,S) = \mathrm{e}^{-r\tau_1}\widetilde{\mathrm{E}}_{t,S}\left[(K_1 - C_{T_1})^+\right]$$

where $C_{T_1} \equiv C_{T_1}(S(T_1),K_2,T_2)$ is the value of the (inner) call option initiated at T_1 and maturing at $T_2 > T_1$. Using the simple identity $(K_1 - C_{T_1})^+ = (C_{T_1} - K_1)^+ - C_{T_1} + K_1$ within the above expectation gives

$$V^{pc}(t,S) = \mathrm{e}^{-r\tau_1}\widetilde{\mathrm{E}}_{t,S}\left[(C_{T_1} - K_1)^+\right] - \mathrm{e}^{-r\tau_1}\widetilde{\mathrm{E}}_{t,S}\left[C_{T_1}\right] + \mathrm{e}^{-r\tau_1}K_1\,. \quad (3.113)$$

The first expectation is $V^{cc}(t,S)$. By the tower property, we now show that the second expectation reduces to the value $C_t(S,K_2,T_2)$ of a standard call with spot S, strike K_2, and maturity $T_2 > t$. Note that the call value C_{T_1} is a random variable expressed here as function of the time-T_1 spot random variable $S(T_1)$, and hence its value is given by the discounted expected value of the payoff $(S(T_2) - K_2)^+$ at time T_2, conditional on $S(T_1)$:

$$C_{T_1} \equiv C_{T_1}(S(T_1),K_2,T_2) = \widetilde{\mathrm{E}}_{T_1,S(T_1)}\left[\mathrm{e}^{-r(T_2-T_1)}(S(T_2)-K_2)^+\right].$$

Substituting this representation for C_{T_1} into the second expectation in (3.113) and invoking the tower property, while combining the discount factors, gives

$$\begin{aligned}
\mathrm{e}^{-r\tau_1}\widetilde{\mathrm{E}}_{t,S}\left[C_{T_1}\right] &= \mathrm{e}^{-r(T_1-t)}\widetilde{\mathrm{E}}_{t,S}\left[\widetilde{\mathrm{E}}_{T_1,S(T_1)}\left[\mathrm{e}^{-r(T_2-T_1)}(S(T_2)-K_2)^+\right]\right] \\
&= \mathrm{e}^{-r(T_2-t)}\widetilde{\mathrm{E}}_{t,S}\left[(S(T_2)-K_2)^+\right] = C_t(S,K_2,T_2). \quad (3.114)
\end{aligned}$$

Hence (3.112) is recovered from (3.113).

Another way to obtain (3.114) is simply to note that the discounted call price process $e^{-rt}C_t \equiv e^{-rt}C_t(S(t), K_2, T_2)$, for $t < T_2$ and fixed K_2, T_2, is a martingale under the risk-neutral measure $\widetilde{\mathbb{P}}$. Hence, combining the $\widetilde{\mathbb{P}}$-martingale and Markov properties:

$$e^{-rt}C_t = \widetilde{E}\left[e^{-rT_1}C_{T_1}|\mathcal{F}_t\right] = e^{-rT_1}\widetilde{E}_{t,S(t)}\left[C_{T_1}\right] = e^{-rT_1}\widetilde{E}_{t,S(t)}\left[C_{T_1}(S(T_1), K_2, T_2)\right].$$

Then, setting $S(t) = S$ gives (3.114). We remark that the above put-call parity type relation is valid for quite general models of the stock price process; i.e., it holds for GBM and other models where we assume the discounted stock price process is a martingale under the risk-neutral measure $\widetilde{\mathbb{P}}$ and the stock is not allowed to default. Of course, under more general models, the pricing formulas for the compound options will not involve univariate and bi-variate standard normal CDFs as we derived above for the GBM model. With the exception of some families of so-called solvable alternative models, one has to resort to numerical methods for pricing compound options under more complex stochastic models for the stock.

Let us evaluate conditional risk-neutral expectations similar to (3.109) and (3.110) but with other choices of indicator functions. We need such expectations for pricing other types of multistage options. Using the symmetry of the normal distribution, we obtain

$$\begin{aligned}
\widetilde{E}_{t,S}\left[\mathbb{I}_{\{S(T_1)>S_1^*,\, S(T)>K_2\}}\right] &= \widetilde{\mathbb{P}}(Z_1 > -a_-, Z_2 > -b_-) = \mathcal{N}_2\left(a_-, b_-; \rho\right), \\
\widetilde{E}_{t,S}\left[\mathbb{I}_{\{S(T_1)>S_1^*,\, S(T)<K_2\}}\right] &= \widetilde{\mathbb{P}}(Z_1 > -a_-, Z_2 < -b_-) = \mathcal{N}_2\left(a_-, -b_-; -\rho\right), \\
\widetilde{E}_{t,S}\left[\mathbb{I}_{\{S(T_1)<S_1^*,\, S(T)>K_2\}}\right] &= \widetilde{\mathbb{P}}(Z_1 < -a_-, Z_2 > -b_-) = \mathcal{N}_2\left(-a_-, b_-; -\rho\right), \\
\widetilde{E}_{t,S}\left[\mathbb{I}_{\{S(T_1)<S_1^*,\, S(T)<K_2\}}\right] &= \widetilde{\mathbb{P}}(Z_1 < -a_-, Z_2 < -b_-) = \mathcal{N}_2\left(-a_-, -b_-; \rho\right),
\end{aligned} \tag{3.115}$$

where $a_\pm \equiv d_\pm(\frac{S}{S_1^*}, \tau_1)$, $b_\pm \equiv d_\pm(\frac{S}{K_2}, \tau_2)$, and $\rho = \sqrt{\tau_1/\tau_2}$ with $\tau_i = T_i - t$ for $i = 1, 2$. Similarly, we have

$$\begin{aligned}
\widetilde{E}_{t,S}\left[\mathbb{I}_{\{S(T_1)>S_1^*,\, S(T)>K_2\}}S(T_2)\right] &= Se^{(r-q)\tau_2}\mathcal{N}_2\left(a_+, b_+; \rho\right), \\
\widetilde{E}_{t,S}\left[\mathbb{I}_{\{S(T_1)>S_1^*,\, S(T)<K_2\}}S(T_2)\right] &= Se^{(r-q)\tau_2}\mathcal{N}_2\left(a_+, -b_+; -\rho\right), \\
\widetilde{E}_{t,S}\left[\mathbb{I}_{\{S(T_1)<S_1^*,\, S(T)>K_2\}}S(T_2)\right] &= Se^{(r-q)\tau_2}\mathcal{N}_2\left(-a_+, b_+; -\rho\right), \\
\widetilde{E}_{t,S}\left[\mathbb{I}_{\{S(T_1)<S_1^*,\, S(T)<K_2\}}S(T_2)\right] &= Se^{(r-q)\tau_2}\mathcal{N}_2\left(-a_+, -b_+; \rho\right).
\end{aligned} \tag{3.116}$$

The next example demonstrates how the above formulae are used for pricing a complex chooser option.

Example 3.10. Consider a complex chooser binary option that gives its holder the right to choose at a predetermined time T between two cash-or-nothing options: a call with strike $K_c > 0$ and expiry $T_c > 0$ and a put with strike $K_p > 0$ and expiry $T_p > 0$. Here, we assume that $0 \leqslant T < \min\{T_c, T_p\}$. Derive a no-arbitrage pricing formula for this option.

Solution. Let $C(t, S) \equiv C(t, S; T_c, K_c)$ and $P(t, S) \equiv P(t, S; T_p, K_p)$ with $0 \leqslant t \leqslant T$ denote the pricing functions for the cash-or-nothing binary call and put options, respectively. The time-T value of the chooser option is

$$\begin{aligned}
V(T, S) &= \max\{C(T, S; T_c, K_c), P(T, S; T_p, K_p)\} \\
&= \max\left\{e^{-r(T_c-T)}\widetilde{E}_{T,S}\left[\mathbb{I}_{\{S(T_c)\geqslant K_c\}}\right], e^{-r(T_p-T)}\widetilde{E}_{T,S}\left[\mathbb{I}_{\{S(T_p)\leqslant K_p\}}\right]\right\}.
\end{aligned}$$

As seen in Figure 3.5, the pricing function of a cash-or-nothing put is a strictly decreasing function of S and where $\lim_{S\searrow 0} P(T, S) = e^{-r(T_p-T)}$ and $\lim_{S\to\infty} P(T, S) = 0$. On the other hand, the pricing function of a cash-or-nothing call is a strictly increasing function of S where $\lim_{S\searrow 0} C(T,S) = 0$ and $\lim_{S\to\infty} C(T,S) = \mathrm{e}^{-r(T_c-T)}$. Therefore, there exists S_1^* such that $P(T,S_1^*) = C(T,S_1^*)$, $P(T,S) > C(T,S)$ for $S < S_1^*$, and $P(T,S) < C(T,S)$ for $S > S_1^*$. Therefore, the time-T option value can be written without the max function as follows:

$$V(T,S) = C(T,S;T_c,K_c)\mathbb{I}_{\{S>S_1^*\}} + P(T,S;T_p,K_p)\mathbb{I}_{\{S\leqslant S_1^*\}}.$$

For $t \in [0,T]$, the time-t no-arbitrage value is

$$\begin{aligned} V(t,S) &= \mathrm{e}^{-r(T-t)}\widetilde{\mathrm{E}}_{t,S}[V(T,S(T))] \\ &= \mathrm{e}^{-r(T-t)}\widetilde{\mathrm{E}}_{t,S}\Big[\mathrm{e}^{-r(T_c-T)}\widetilde{\mathrm{E}}_{T,S(T)}\left[\mathbb{I}_{\{S(T_c)\geqslant K_c\}}\right]\mathbb{I}_{\{S(T)>S_1^*\}} \\ &\qquad + \mathrm{e}^{-r(T_p-T)}\widetilde{\mathrm{E}}_{T,S(T)}\left[\mathbb{I}_{\{S(T_p)\leqslant K_p\}}\right]\mathbb{I}_{\{S(T)\leqslant S_1^*\}}\Big]. \end{aligned}$$

Applying the tower property and linearity of conditional expectations, as well as the Markov property of $S(t)$, we can eliminate the inner conditional expectations and rewrite the above pricing formula as follows:

$$V(t,S) = \mathrm{e}^{-r(T_c-t)}\widetilde{\mathrm{E}}_{t,S}\left[\mathbb{I}_{\{S(T_c)\geqslant K_c,\, S(T)>S_1^*\}}\right] + \mathrm{e}^{-r(T_p-T)}\widetilde{\mathrm{E}}_{t,S}\left[\mathbb{I}_{\{S(T_p)\leqslant K_p,\, S(T)\leqslant S_1^*\}}\right].$$

In the right hand side, we recognize the conditional expectations from the collection in (3.115). Therefore, the no-arbitrage pricing formula takes the following final form:

$$\begin{aligned} V(t,S) &= \mathrm{e}^{-r(T_c-t)}\mathcal{N}_2\left(d_-(S/S_1^*,T-t), d_-(S/K_c,T_c-t); \sqrt{\frac{T-t}{T_c-t}}\right) \\ &\quad + \mathrm{e}^{-r(T_p-t)}\mathcal{N}_2\left(-d_-(S/S_1^*,T-t), -d_-(S/K_p,T_p-t); \sqrt{\frac{T-t}{T_p-t}}\right). \end{aligned}$$

□

3.4 Some European-Style Path-Dependent Derivatives

We now consider the application of risk-neutral pricing to path-dependent European options whose payoff is dependent on the underlying stock price history over the lifetime of the contract. We shall specialize to the pricing of two types of path-dependent options, namely, barrier options and lookback options. As seen in the examples below, these classes of options have a payoff that is a function of a combination of the stock price $S(T)$ at maturity $T > 0$ and the realized (sampled) maximum $M^S(T)$ or the realized minimum $m^S(T)$ of the stock price process $\{S(t)\}_{t\geqslant 0}$ where

$$M^S(t) := \sup_{0\leqslant u\leqslant t} S(u) \quad \text{and} \quad m^S(t) := \inf_{0\leqslant u\leqslant t} S(u) \tag{3.117}$$

for all $0 \leqslant t \leqslant T$. There are many variations of the payoff for these options. Some payoffs, such as in the case of a so-called double barrier option, are functions of the triplet $M^S(T), m^S(T), S(T)$. Here, we shall focus our attention on developing analytical pricing formulae for *single-barrier* options and lookback options whose payoff is *either* a function of the pair $M^S(T), S(T)$ *or* a function of the pair $m^S(T), S(T)$, separately. Given the joint distribution of either pair of random variables, there are two main types of payoffs for which we can in principle derive pricing formulae. In the first case, the payoff is assumed to be a (Borel) function, $\phi : R_+^2 \to \mathbb{R}$, of the terminal stock price and its realized maximum, and in the second case, it is a function of the terminal stock price and its realized minimum:

$$(i)\, V_T = \phi(M^S(T), S(T)) \quad \text{and} \quad (ii)\, V_T = \phi(m^S(T), S(T)) \,. \tag{3.118}$$

Let us first take a look at the payoffs that define some single-barrier option contracts. For such contracts the payoff simplifies into a product of a function of the terminal stock price $\Lambda(S(T))$ and an indicator function involving either the realized maximum $M^S(T)$ or minimum $m^S(T)$ of the stock price during the option's lifetime. There are two basic types of single-barrier options: (i) *knock-out* options that have a nonzero payoff only if a level $B > 0$ *is not attained* and (ii) *knock-in* options that have a nonzero payoff only if level B *is attained* during the option's lifetime. The different versions of these correspond to whether level B is a *lower barrier or an upper barrier*. Letting $\Lambda(S(T))$ be the effective payoff of a standard (non-path-dependent) European option, e.g., $\Lambda(x) = (x - K)^+$ for a call and $\Lambda(x) = (K - x)^+$ for a put struck at $K > 0$, we have the following four different payoffs for a single barrier at level B:

(a) Up-and-out: $V_T^{UO} = \Lambda(S(T))\,\mathbb{I}_{\{M^S(T)<B\}}$, where $\phi(M, S) := \Lambda(S)\,\mathbb{I}_{\{M<B\}}$;

(b) Down-and-out: $V_T^{DO} = \Lambda(S(T))\,\mathbb{I}_{\{m^S(T)>B\}}$, where $\phi(m, S) := \Lambda(S)\,\mathbb{I}_{\{m>B\}}$;

(c) Up-and-in: $V_T^{UI} = \Lambda(S(T))\,\mathbb{I}_{\{M^S(T)\geqslant B\}}$, where $\phi(M, S) := \Lambda(S)\,\mathbb{I}_{\{M\geqslant B\}}$;

(d) Down-and-in: $V_T^{DI} = \Lambda(S(T))\,\mathbb{I}_{\{m^S(T)\leqslant B\}}$, where $\phi(m, S) := \Lambda(S)\,\mathbb{I}_{\{m\leqslant B\}}$. (3.119)

For example, an up-and-out call with strike K is defined as having the payoff of a call, $(S(T) - K)^+$, if the realized maximum value of the underlying stock price stays below the barrier level B and has otherwise zero payoff if the stock price attains or goes above level B at any time until T. See Figure 3.10. We write this as

$$C_T^{UO} = (S(T) - K)^+\,\mathbb{I}_{\{M^S(T)<B\}}, \text{ where } \phi(M, S) = (S - K)^+\,\mathbb{I}_{\{M<B\}}.$$

The down-and-out call has payoff $C_T^{DO} = (S(T) - K)^+\,\mathbb{I}_{\{m^S(T)>B\}}$. In the case of a knock-out put, the up-and-out put has payoff $P_T^{UO} = (K - S(T))^+\,\mathbb{I}_{\{M^S(T)<B\}}$ and the down-and-out put has payoff $P_T^{DO} = (K - S(T))^+\,\mathbb{I}_{\{m^S(T)>B\}}$. See Figure 3.11.

On the other hand, an up-and-in call with strike K is defined as having a call payoff if the stock price has attained or has gone above B at any time until T and has otherwise zero payoff; i.e., the payoff is $C_T^{UI} = (S(T) - K)^+\,\mathbb{I}_{\{M^S(T)\geqslant B\}}$. Similarly, a down-and-in call has a payoff that is nonzero only if the stock price has fallen below or at level B, $C_T^{DI} = (S(T) - K)^+\,\mathbb{I}_{\{m^S(T)\leqslant B\}}$. For the up-and-in put and down-and-in put, with strike K, we have $P_T^{UI} = (K - S(T))^+\,\mathbb{I}_{\{M^S(T)\geqslant B\}}$ and $P_T^{DI} = (K - S(T))^+\,\mathbb{I}_{\{m^S(T)\leqslant B\}}$, respectively.

There is a very simple and useful symmetry relation between the knock-in and knock-out payoffs. Since we have the obvious relations

$$\mathbb{I}_{\{M^S(T)<B\}} + \mathbb{I}_{\{M^S(T)\geqslant B\}} = \mathbb{I}_{\{m^S(T)>B\}} + \mathbb{I}_{\{m^S(T)\leqslant B\}} = 1$$

then

$$V_T^{UO} + V_T^{UI} = V_T^{DO} + V_T^{DI} = \Lambda(S(T)) \,. \tag{3.120}$$

This is known as "knock-in-knock-out" symmetry. By computing the pricing formula for the knock-out (or knock-in) option then the pricing formula for the corresponding

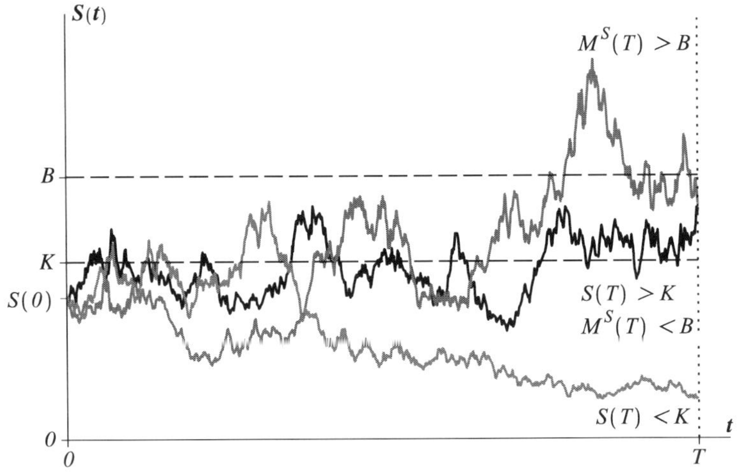

FIGURE 3.10: Three types of stock price paths starting at $S(0) < B$ are depicted for an up-and-out call with strike K. Only paths in the set $\{M^S(T) < B, S(T) > K\}$; i.e., paths that do not surpass level B and also end up above the strike at terminal time T give a positive payoff for an up-and-out call struck at K.

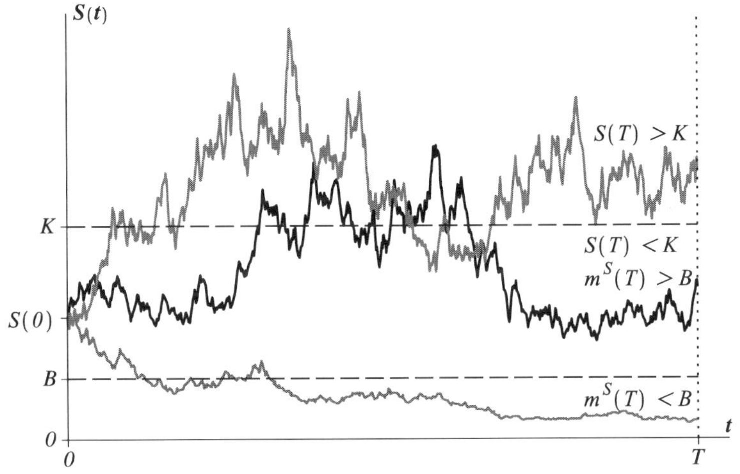

FIGURE 3.11: Three types of stock price paths starting at $S(0) > B$ are depicted for a down-and-out put with strike K. Only paths in the set $\{m^S(T) > B, S(T) < K\}$; i.e., paths that do not fall below level B and also end up below the strike at terminal time T give a positive payoff for a down-and-out put struck at K.

knock-in (or knock-out) follows simply by subtracting the former from the price of the standard European option having payoff $\Lambda(S(T))$. That is, letting $V_t^{UO}, V_t^{UI}, V_t^{DO}, V_t^{DI}$ represent the respective time-t barrier option prices, $t \leqslant T$, then by risk-neutral pricing we have

$$V_t^{UO} + V_t^{UI} = V_t^{DO} + V_t^{DI} = V_t\,, \tag{3.121}$$

where V_t is the time-t price of the standard European option with payoff $\Lambda(S(T))$.

We now turn to the definition of lookback options in the continuous time setting. We recall from the discrete-time setting (see Chapter 7 in Volume I) that there are two main kinds of lookback options, with either floating strike (LFS) or floating price (LFP). We list the four common lookback option payoffs:

$$\begin{aligned}
&\text{(a) Floating strike call (LFS call): } C_T^{LFS} = (S(T) - m^S(T))^+ = S(T) - m^S(T);\\
&\text{(b) Floating strike put (LFS put): } P_T^{LFS} = (M^S(T) - S(T))^+ = M^S(T) - S(T);\\
&\text{(c) Floating price call (LFP call): } C_T^{LFP} = (M^S(T) - K)^+;\\
&\text{(d) Floating price put (LFP put): } P_T^{LFP} = (K - m^S(T))^+.
\end{aligned} \tag{3.122}$$

Note that the LFS options are never out of the money since $m^S(T) \leqslant S(T) \leqslant M^S(T)$. For an LFS option, the strike price is floating as it is not preassigned but rather determined by the realized maximum or minimum value of the stock during the lifetime of the option. The payoff of an LFS option is the maximum difference between the stock's price at maturity and the floating strike. The LFS call gives its holder the right to buy at the lowest stock price realized during the option's lifetime, whereas the LFS put gives the right to sell at the highest realized stock price. For the LFP options, the payoffs are the maximum differences between the optimal (maximum or minimum) stock price and the fixed strike K. LFP options are designed so that the call (or put) has a payoff given by the stock price at its highest (or lowest) realized value during the option's lifetime.

3.4.1 Risk-Neutral Pricing under GBM

Before specializing and thereby simplifying the problem to the pricing of barrier options and lookback options, covered in Sections 3.4.2 and 3.4.3, we now present the risk-neutral pricing formulation for the two general types of payoffs in (3.118) above. We assume $\{S(t)\}_{t\geqslant 0}$ to be a GBM given by (3.15) if $q = 0$ or (3.61) if $q \neq 0$. The stock price in (3.61) is given by a strictly increasing exponential mapping

$$S(t) = S_0 \mathrm{e}^{\sigma X(t)},\ S(0) = S_0 \tag{3.123}$$

where the drifted $\widetilde{\mathbb{P}}$-BM process X is defined by (see (1.98) of Section 1.4.3)

$$X(t) := \widetilde{W}^{(\nu,1)}(t) \equiv \nu t + \widetilde{W}(t)\,,\ \ \nu := \frac{(r - q - \frac{1}{2}\sigma^2)}{\sigma}. \tag{3.124}$$

Note that $\nu := \frac{(r-\frac{1}{2}\sigma^2)}{\sigma}$ for the zero dividend case. Hence, the realized maximum and minimum of the stock price in (3.117) are related trivially to the maximum and minimum of the drifted BM:

$$M^S(t) = S_0 \mathrm{e}^{\sigma M^X(t)} \quad \text{and} \quad m^S(t) = S_0 \mathrm{e}^{\sigma m^X(t)} \tag{3.125}$$

where (see (1.100))

$$M^X(t) := \sup_{0\leqslant u\leqslant t} X(u) \quad \text{and} \quad m^X(t) := \inf_{0\leqslant u\leqslant t} X(u)\,. \tag{3.126}$$

In what follows we will be conditioning on $\mathcal{F}_t$ for any fixed current time t, $0 \leqslant t \leqslant T$. By the time homogeneity property of the stock price process, it is convenient to define $\tau := T - t$. Let us first consider expressing the joint random variables $M^S(T), S(T)$ as functions of the $\mathcal{F}_t$-measurable random variables $M^S(t), S(t)$. Using (3.123) the stock price at time T is

$$S(T) = S(t)\exp\left(\sigma[X(T) - X(t)]\right) = S(t)\exp\left(\sigma\mathcal{X}(\tau)\right) \tag{3.127}$$

where we define $\mathcal{X}(s) := X(s+t) - X(t) = \nu\, s + \widetilde{W}(s+t) - \widetilde{W}(t)$. Note that the process $\{\widetilde{W}(s+t) - \widetilde{W}(t)\}_{s \geqslant 0}$, for fixed t, is a standard $\widetilde{\mathbb{P}}$-BM $\{\widetilde{W}(s)\}_{s \geqslant 0}$; hence $\mathcal{X}(s)$ is the drifted BM, $\{\widetilde{W}^{(\nu,1)}(s)\}_{s \geqslant 0}$. The realized maximum of the stock price up to time T is the larger of the realized maximum up to time t and the realized maximum from time t to T:

$$M^S(T) = \max\left\{M^S(t), \sup_{t \leqslant u \leqslant T} S(u)\right\} = \max\left\{M^S(t), S(t)\, e^{\sigma M^{\mathcal{X}}(\tau)}\right\}, \tag{3.128}$$

where $M^{\mathcal{X}}(\tau) := \sup\limits_{0 \leqslant s \leqslant \tau} \mathcal{X}(s)$. To arrive at the last term, we employed the steps:

$$\begin{aligned}
\sup_{t \leqslant u \leqslant T} S(u) &= S(t) \cdot \exp\left(\sigma \sup_{t \leqslant u \leqslant T} [X(u) - X(t)]\right) \\
&= S(t) \cdot \exp\left(\sigma \sup_{0 \leqslant s \leqslant \tau} [X(s+t) - X(t)]\right) \\
&= S(t) \cdot \exp\left(\sigma \sup_{0 \leqslant s \leqslant \tau} \mathcal{X}(s)\right).
\end{aligned} \tag{3.129}$$

By similar steps, the sampled minimum of the stock price takes the form

$$m^S(T) = \min\left\{m^S(t), S(t)\, e^{\sigma m^{\mathcal{X}}(\tau)}\right\}, \tag{3.130}$$

where $m^{\mathcal{X}}(\tau) := \inf\limits_{0 \leqslant s \leqslant \tau} \mathcal{X}(s)$.

Based on (3.128) and (3.130), $\{(M^X(t), X(t))\}_{t \geqslant 0}$ and $\{(m^X(t), X(t))\}_{t \geqslant 0}$ are both (vector) Markov processes. Observe that both pairs of random variables $M^{\mathcal{X}}(\tau), \mathcal{X}(\tau)$ and $m^{\mathcal{X}}(\tau), \mathcal{X}(\tau)$ are $\mathcal{F}_t$-independent (and hence also independent of the random variables $S(t), m^S(t)$, and $M^S(t)$). Moreover, both pairs of random variables have the same joint distribution as $(M^X(\tau), X(\tau))$, and $(m^X(\tau), X(\tau))$, respectively. In particular, the joint PDF of $M^{\mathcal{X}}(\tau), \mathcal{X}(\tau)$ (in the risk-neutral measure $\widetilde{\mathbb{P}}$) is given by (sending $t \to \tau$ and $\mu \to \nu$ in (2.144) of Chapter 2)

$$\begin{aligned}
\widetilde{f}_{M^{\mathcal{X}}(\tau),\mathcal{X}(\tau)}(w,x) \equiv \widetilde{f}_{M^X(\tau),X(\tau)} &:= \frac{\partial^2}{\partial w \partial x}\widetilde{\mathbb{P}}(M^X(\tau) \leqslant w, X(\tau) \leqslant x) \\
&= \frac{2(2w - x)}{\tau\sqrt{2\pi\tau}} e^{-\frac{1}{2}\nu^2\tau + \nu x - (2w-x)^2/2\tau},
\end{aligned} \tag{3.131}$$

for $-\infty < x < w, w > 0$ and zero otherwise. The risk-neutral joint PDF of $m^{\mathcal{X}}(\tau), \mathcal{X}(\tau)$ is given by (sending $t \to \tau$ and $\mu \to \nu$ in (2.145) of Chapter 2)

$$\begin{aligned}
\widetilde{f}_{m^{\mathcal{X}}(\tau),\mathcal{X}(\tau)}(w,x) \equiv \widetilde{f}_{m^X(\tau),X(\tau)}(w,x) &:= \frac{\partial^2}{\partial w \partial x}\widetilde{\mathbb{P}}(m^X(\tau) \leqslant w, X(\tau) \leqslant x) \\
&= \frac{2(x - 2w)}{\tau\sqrt{2\pi\tau}} e^{-\frac{1}{2}\nu^2\tau + \nu x - (x-2w)^2/2\tau},
\end{aligned} \tag{3.132}$$

for $x > w, w < 0$ and zero otherwise.

By the joint Markov property and using (3.127) and (3.128), a European option with payoff (i) in (3.118) has time-t no-arbitrage price (expressed as an $\mathcal{F}_t$-measurable random variable) given by

$$\begin{aligned} V_t &= V(t, S(t), M^S(t)) \\ &= \mathrm{e}^{-r(T-t)}\widetilde{\mathrm{E}}[\,\phi\big(M^S(T), S(T)\big)\,|\mathcal{F}_t] \\ &= \mathrm{e}^{-r(T-t)}\widetilde{\mathrm{E}}[\,\phi\big(\max\{M^S(t), S(t)\,\mathrm{e}^{\sigma M^{\mathcal{X}}(\tau)}\}, S(t)\,\mathrm{e}^{\sigma\mathcal{X}(\tau)}\big)\,|S(t), M^S(t)]\,. \end{aligned} \tag{3.133}$$

For any positive real values $M^S(t) = M$, $S(t) = S > 0$, $M \geqslant S$; i.e., the spot values of the sampled maximum up to calendar time t and the stock price at calendar time t, the general pricing formula is obtained by computing this expectation while using the fact that $M^{\mathcal{X}}(\tau), \mathcal{X}(\tau)$ are $\mathcal{F}_t$-independent:

$$\begin{aligned} V(t,S,M) &= \mathrm{e}^{-r\tau}\widetilde{\mathrm{E}}[\,\phi\big(\max\{M^S(t), S(t)\,\mathrm{e}^{\sigma M^{\mathcal{X}}(\tau)}\}, S(t)\,\mathrm{e}^{\sigma\mathcal{X}(\tau)}\big)\,|S(t) = S, M^S(t) = M] \\ &= \mathrm{e}^{-r\tau}\widetilde{\mathrm{E}}[\,\phi\big(\max\{M, S\,\mathrm{e}^{\sigma M^{\mathcal{X}}(\tau)}\}, S\,\mathrm{e}^{\sigma\mathcal{X}(\tau)}\big)] \\ &= \mathrm{e}^{-r\tau}\int_0^\infty\int_{-\infty}^{w} \phi\big(\max\{M, S\,\mathrm{e}^{\sigma w}\}, S\,\mathrm{e}^{\sigma x}\big)\widetilde{f}_{M^X(\tau),X(\tau)}(w,x)\,\mathrm{d}x\,\mathrm{d}w\,, \end{aligned} \tag{3.134}$$

where $\tau = T - t$ is the time to maturity. This is a double integral of the joint density in (3.131) multiplied by the *effective payoff*, $h(w,x) := \phi\big(\max\{M, S\,\mathrm{e}^{\sigma w}\}, S\,\mathrm{e}^{\sigma x}\big)$, which is a function of w, x.

Applying the same steps by using (3.127) and (3.130), the time-t no-arbitrage pricing formula for the European option with payoff (ii) in (3.118) for given real positive spot values, $0 < m^S(t) = m \leqslant S = S(t)$, is given by

$$\begin{aligned} V(t,S,m) &= \mathrm{e}^{-r\tau}\widetilde{\mathrm{E}}[\,\phi\big(\min\{m^S(t), S(t)\,\mathrm{e}^{\sigma m^{\mathcal{X}}(\tau)}\}, S(t)\,\mathrm{e}^{\sigma\mathcal{X}(\tau)}\big)\,|S(t) = S, m^S(t) = m] \\ &= \mathrm{e}^{-r\tau}\widetilde{\mathrm{E}}[\,\phi\big(\min\{m, S\,\mathrm{e}^{\sigma m^{\mathcal{X}}(\tau)}\}, S\,\mathrm{e}^{\sigma\mathcal{X}(\tau)}\big)] \\ &= \mathrm{e}^{-r\tau}\int_{-\infty}^0\int_{w}^{\infty} \phi\big(\min\{m, S\,\mathrm{e}^{\sigma w}\}, S\,\mathrm{e}^{\sigma x}\big)\widetilde{f}_{m^X(\tau),X(\tau)}(w,x)\,\mathrm{d}x\,\mathrm{d}w\,, \end{aligned} \tag{3.135}$$

$\tau = T - t$. This is now a double integral involving the joint density in (3.132) and the effective payoff given by $g(w,x) := \phi\big(\min\{m, S\,\mathrm{e}^{\sigma w}\}, S\,\mathrm{e}^{\sigma x}\big)$.

[We remark that one can always generally write a stock price as in (3.123), using an exponential (monotonic) function of a Markov process X which is specified as a more complex process. Then, the pricing formulae in (3.134) and (3.135) can be used for more general stock price models as long as there exist joint densities $\widetilde{f}_{M^X(\tau),X(\tau)}$ and $\widetilde{f}_{m^X(\tau),X(\tau)}$ and also that the discounted stock price process is a $\widetilde{\mathbb{P}}$-martingale. Of course, for a more general stock price model that is not a GBM process, the process X is not specified simply as a drifted BM but as a more complex process. The joint densities for such processes will also be more complex than those for drifted BM given in (3.131) and (3.132).]

Note that both pricing formulae in (3.134) and (3.135) are functions of $\tau = T - t$. For example, we can write the price in (3.134) as a function $v(\tau, S, M)$ where $v(\tau, S, M) = V(t, S, M) = V(T - \tau, S, M)$ and similarly for the pricing function in (3.135). Note that the above pricing formulae are generally valid for any intermediate time and that the spot values $S(t) = S, M^S(t) = M, m^S(t) = m$ are known at intermediate time t. However, the payoff is generally a function of the realized maximum $M^S(T)$ (or minimum $m^S(T)$) involving the continuous sampling of the stock price *starting at a prior time* $t_0 = 0$. These are therefore referred to as "seasoned" contracts. This general situation is depicted in Figure 3.12.

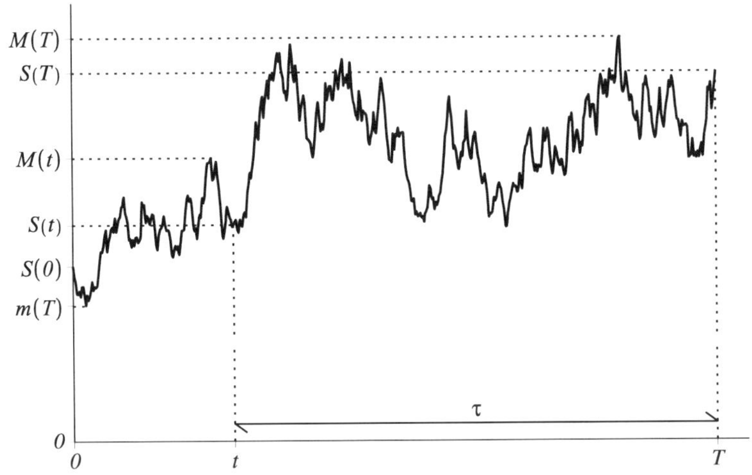

FIGURE 3.12: A sample stock price path is shown with its initial value, its value and realized maximum and minimum at both the intermediate (current) time t and at terminal time T.

Let us now specialize to the case where the realized maximum and minimum are computed *starting from current time* t. Then $S(t) = M^S(t) = m^S(t)$, i.e., with spot values $S = M = m$, where in the integrands of (3.134) and (3.135) we have, respectively,

$$\max\{M, S\,\mathrm{e}^{\sigma w}\} = \max\{S, S\,\mathrm{e}^{\sigma w}\} = S\,\mathrm{e}^{\sigma w}\,, \text{ since } w > 0\,,$$

$$\min\{m, S\,\mathrm{e}^{\sigma w}\} = \min\{S, S\,\mathrm{e}^{\sigma w}\} = S\,\mathrm{e}^{\sigma w}\,, \text{ since } w < 0\,.$$

Hence both payoff functions in (3.134) and (3.135) have the form $\phi(S\,\mathrm{e}^{\sigma w}, S\,\mathrm{e}^{\sigma x})$ and the option pricing formulae are functions of only the spot S and $\tau = T - t$. In particular, the pricing formula in (3.134) is reduced to $V(t, S, M) = V(t, S) = v(\tau, S)$:

$$v(\tau, S) = \mathrm{e}^{-r\tau} \int_0^\infty \int_{-\infty}^w \phi\big(S\,\mathrm{e}^{\sigma w}, S\,\mathrm{e}^{\sigma x}\big) \widetilde{f}_{M^X(\tau), X(\tau)}(w, x)\, \mathrm{d}x\, \mathrm{d}w\,. \tag{3.136}$$

Setting the current calendar time $t = 0$, $S = S(0) = S_0$, $\tau = T$ gives the price expressed as function of spot S_0, and the time to maturity, which is now represented by the variable T, $V(0, S_0) = v(T, S_0)$:

$$v(T, S_0) = \mathrm{e}^{-rT} \int_0^\infty \int_{-\infty}^w \phi\big(S_0\,\mathrm{e}^{\sigma w}, S_0\,\mathrm{e}^{\sigma x}\big) \widetilde{f}_{M^X(T), X(T)}(w, x)\, \mathrm{d}x\, \mathrm{d}w\,. \tag{3.137}$$

Of course, we need only compute one of these as (3.137) obtains trivially from (3.136) and vice versa. For options involving the realized minimum, (3.135) gives $V(t, S, m) = V(t, S) = v(\tau, S)$:

$$v(\tau, S) = \mathrm{e}^{-r\tau} \int_{-\infty}^0 \int_w^\infty \phi\big(S\,\mathrm{e}^{\sigma w}, S\,\mathrm{e}^{\sigma x}\big) \widetilde{f}_{m^X(\tau), X(\tau)}(w, x)\, \mathrm{d}x\, \mathrm{d}w \tag{3.138}$$

or expressed as a function of T and $S(0) = S_0$, where we simply make the variable replacements $S \to S_0$ and $\tau \to T$ in the derived pricing function $v(\tau, S)$.

3.4.2 Pricing Single Barrier Options

For barrier options, the contacts are specified such that the sampling of the maximum and minimum of the stock price starts at current time t. So we have the case discussed above where $S(t) = M^S(t) = m^S(t)$ (or $S_0 = S(0) = M^S(0) = m^S(0)$ for current time $t = 0$). Hence the pricing formulae in (3.136)–(3.138) are our general starting point. Given a spot value $S(t) = S$, we denote the respective time-t pricing functions for cases (a)–(d) defined in (3.119) by $V^{UO}(t, S; B)$, $V^{DO}(t, S; B)$, $V^{UI}(t, S; B)$, and $V^{DI}(t, S; B)$. As functions of time to maturity we write these pricing functions equally as $v^{UO}(\tau, S; B)$, $v^{DO}(\tau, S; B)$, $v^{UI}(\tau, S; B)$, and $v^{DI}(\tau, S; B)$. By knock-in-knock-out symmetry in (3.121), we need only derive a pricing formula for either knock-out or knock-in options as we can use the pricing formula for the standard (vanilla) option to obtain one pricing formula from the other:

$$V^{UO}(t, S; B) + V^{UI}(t, S; B) = V^{DO}(t, S; B) + V^{DI}(t, S; B) = V(t, S)\,, \tag{3.139}$$

where $V(t, S)$ is the time-t pricing formula for the standard European option with payoff function Λ.

For barrier options, we see that the overall payoff function ϕ in all cases (a)–(d) in (3.119) is a product of an indicator function in the first argument and the effective payoff function Λ in the second argument. Hence, in the integrand of (3.136)–(3.138) we have in the respective cases (a)–(d) in (3.119):

$$\begin{aligned}
&\text{(a) } \phi\big(S\,e^{\sigma w}, S\,e^{\sigma x}\big) = \mathbb{I}_{\{S\,e^{\sigma w} < B\}}\Lambda(S\,e^{\sigma x}) = \mathbb{I}_{\{w<b\}}\Lambda(S\,e^{\sigma x}), \ -\infty < x < w, w > 0;\\
&\text{(b) } \phi\big(S\,e^{\sigma w}, S\,e^{\sigma x}\big) = \mathbb{I}_{\{S\,e^{\sigma w} > B\}}\Lambda(S\,e^{\sigma x}) = \mathbb{I}_{\{w>b\}}\Lambda(S\,e^{\sigma x}), \ w < x < \infty, w < 0;\\
&\text{(c) } \phi\big(S\,e^{\sigma w}, S\,e^{\sigma x}\big) = \mathbb{I}_{\{S\,e^{\sigma w} \geqslant B\}}\Lambda(S\,e^{\sigma x}) = \mathbb{I}_{\{w\geqslant b\}}\Lambda(S\,e^{\sigma x}), \ -\infty < x < w, w > 0;\\
&\text{(d) } \phi\big(S\,e^{\sigma w}, S\,e^{\sigma x}\big) = \mathbb{I}_{\{S\,e^{\sigma w} \leqslant B\}}\Lambda(S\,e^{\sigma x}) = \mathbb{I}_{\{w\leqslant b\}}\Lambda(S\,e^{\sigma x}), \ w < x < \infty, w < 0;
\end{aligned} \tag{3.140}$$

where $b := \frac{1}{\sigma}\ln\frac{B}{S}$. These are all product functions in the integrand variables x and w. This leads to an important simplification in (3.136–(3.138) which reduce to single integrals, as given in the following result where the pricing formulae for knock-out barrier options are single integrals (in x) involving the effective payoff $\Lambda(S\,e^{\sigma x})$ and the risk-neutral probability density for the drifted BM in (3.124) that is killed at the effective barrier level b.

Proposition 3.3 (Pricing Formulae for Single-Barrier Knock-Out Options)**.** *Assume a constant interest rate r and constant continuous dividend yield q on a stock whose price process is a GBM with constant volatility σ. Let $B > 0$ be an arbitrary knock-out barrier level, $S(t) = S > 0$ be the stock spot price, and $\Lambda(\cdot)$ be the effective payoff function. Then, for $S < B$ the up-and-out option has value*

$$V^{UO}(t, S; B) = e^{-r\tau}\int_{-\infty}^{b} \Lambda(Se^{\sigma x})\,\widetilde{p}^{X_{(b)}}(\tau; 0, x)\,dx \tag{3.141}$$

and $V^{UO}(t, S; B) \equiv 0$ for $S \geqslant B$. For $S > B$, the down-and-out option has the value

$$V^{DO}(t, S; B) = e^{-r\tau}\int_{b}^{\infty} \Lambda(Se^{\sigma x})\,\widetilde{p}^{X_{(b)}}(\tau; 0, x)\,dx \tag{3.142}$$

and $V^{DO}(t, S; B) \equiv 0$ for $S \leqslant B$, where $\tau = T - t > 0$ is the time to maturity, $b := \frac{1}{\sigma}\ln\frac{B}{S}$, $\nu := \frac{(r-q-\frac{1}{2}\sigma^2)}{\sigma}$, and $\widetilde{p}^{X_{(b)}}$ is the (risk-neutral) density,

$$\begin{aligned}
\widetilde{p}^{X_{(b)}}(\tau; 0, x) &= p_0(\tau; x - \nu\tau) - e^{2\nu b}\,p_0(\tau; x - \nu\tau - 2b)\\
&\equiv \frac{1}{\sqrt{\tau}}n\left(\frac{x - \nu\tau}{\sqrt{\tau}}\right) - \left(\frac{B}{S}\right)^{\frac{2(r-q)}{\sigma^2}-1}\frac{1}{\sqrt{\tau}}n\left(\frac{x - \nu\tau - 2b}{\sqrt{\tau}}\right),
\end{aligned} \tag{3.143}$$

defined on the respective domains $(-\infty, b)$ and (b, ∞).

Proof. We prove (3.141), as (3.142) follows similarly. Using (a) in (3.140) within (3.136), changing the order of integration and evaluating the inner integral:

$$\begin{aligned} V^{UO}(t,S;B) &= \mathrm{e}^{-r\tau}\int_{-\infty}^{b}\Lambda(S\,\mathrm{e}^{\sigma x})\left(\int_0^b \widetilde{f}_{M^X(\tau),X(\tau)}(w,x)\,\mathrm{d}w\right)\mathrm{d}x \\ &= \mathrm{e}^{-r\tau}\int_{-\infty}^{b}\Lambda(S\,\mathrm{e}^{\sigma x})\,\frac{\partial}{\partial x}\widetilde{F}_{M^X(\tau),X(\tau)}(b,x)\,\mathrm{d}x \\ &= \mathrm{e}^{-r\tau}\int_{-\infty}^{b}\Lambda(S\,\mathrm{e}^{\sigma x})\,\widetilde{\mathbb{P}}(M^X(\tau)\leqslant b, X(\tau)\in\mathrm{d}x) \\ &= \mathrm{e}^{-r\tau}\int_{-\infty}^{b}\Lambda(S\,\mathrm{e}^{\sigma x})\,\widetilde{p}^{X_{(b)}}(\tau;0,x)\,\mathrm{d}x\,, \end{aligned}$$

for $b > 0$ and is identically zero for $b \leqslant 0$, i.e., $V^{UO}(t,S;B) \equiv 0$ for $S \geqslant B$. Here, we made use of (1.108) and (1.111) of Section 1.4.3 of Chapter 1, with the variable replacements for the drift $\mu \to \nu$ and level $m \to b$. Note that

$$2\nu b = \left(\frac{2(r-q)}{\sigma^2}-1\right)\ln\frac{B}{S} \implies \mathrm{e}^{2\nu b} = \mathrm{e}^{\left(\frac{2(r-q)}{\sigma^2}-1\right)\ln\frac{B}{S}} = \left(\frac{B}{S}\right)^{\frac{2(r-q)}{\sigma^2}-1}. \tag{3.144}$$

□

[Remark: The prices $v^{UO}(T,S_0;B) = V^{UO}(0,S_0;B)$ and $v^{DO}(T,S_0;B) = V^{DO}(0,S_0;B)$, expressing the current time-0 price with maturity T, follow in the obvious manner by setting $t = 0$, i.e., replacing $\tau \to T$ and $S \to S_0$ in the above formulae.]

Note that the density function in (3.143) is a linear combination of two normal densities. Hence, to apply (3.141) or (3.142) we need to compute an integral of the function $g(x) := \Lambda(S\,\mathrm{e}^{\sigma x})$, times $\mathbb{I}_{\{x<b\}}$ or $\mathbb{I}_{\{x>b\}}$, against a normal PDF in x. Let us now consider pricing an up-and-out call option where

$$\Lambda(S\,\mathrm{e}^{\sigma x}) = (S\,\mathrm{e}^{\sigma x}-K)\mathbb{I}_{\{S\,\mathrm{e}^{\sigma x}>K\}} = (S\,\mathrm{e}^{\sigma x}-K)\mathbb{I}_{\{x>\kappa\}} = S\,\mathrm{e}^{\sigma x}\mathbb{I}_{\{x>\kappa\}} - K\,\mathbb{I}_{\{x>\kappa\}}\,,$$

$\kappa := \frac{1}{\sigma}\ln\frac{K}{S}$ and we assume the nontrivial case with $S < B$. Substituting this expression into the integrand in (3.141) gives the price of the up-and-out call as a difference of two integrals:

$$C^{UO}(t,S,K;B) = \mathrm{e}^{-r\tau}S\int_{\kappa}^{b}\mathrm{e}^{\sigma x}\,\widetilde{p}^{X_{(b)}}(\tau;0,x)\,\mathrm{d}x - \mathrm{e}^{-r\tau}K\int_{\kappa}^{b}\widetilde{p}^{X_{(b)}}(\tau;0,x)\,\mathrm{d}x \tag{3.145}$$

if $\kappa < b \equiv \frac{1}{\sigma}\ln\frac{B}{S}$, i.e., $K < B$. Note that $C^{UO}(t,S,K;B) \equiv 0$ if $\kappa \geqslant b$ (i.e., $K \geqslant B$). It is also clear from Figure 3.10 that paths which are in the money (above the strike) are necessarily above or at level B. Since all paths give zero payoff, the price of the up-and-out call must be identically zero when $K \geqslant B$. For $K < B$ the price is given by computing the two integrals in (3.145) upon substituting the density in (3.143). The second integral in (3.145) is a combination of two integrals involving the standard normal PDF which are readily evaluated by changing variables or simply using either identity (B.1) or (B.2) in Appendix B:

$$\begin{aligned} &\int_{\kappa}^{b}\widetilde{p}^{X_{(b)}}(\tau;0,x)\,\mathrm{d}x \\ &= \int_{\kappa}^{b}\frac{\mathrm{e}^{-(x-\nu\tau)^2/2\tau}}{\sqrt{2\pi\tau}}\,\mathrm{d}x - \mathrm{e}^{2\nu b}\int_{\kappa}^{b}\frac{\mathrm{e}^{-(x-(\nu\tau+2b))^2/2\tau}}{\sqrt{2\pi\tau}}\,\mathrm{d}x \\ &= \mathcal{N}\left(\frac{b-\nu\tau}{\sqrt{\tau}}\right) - \mathcal{N}\left(\frac{\kappa-\nu\tau}{\sqrt{\tau}}\right) - \mathrm{e}^{2\nu b}\left[\mathcal{N}\left(-\frac{b+\nu\tau}{\sqrt{\tau}}\right) - \mathcal{N}\left(\frac{\kappa-2b-\nu\tau}{\sqrt{\tau}}\right)\right]. \end{aligned} \tag{3.146}$$

We can now express this in terms of the original parameters B, K, S, r, q, σ using (3.144) and the algebraic relations

$$\frac{b+\nu\tau}{\sqrt{\tau}} = d_-(\frac{B}{S},\tau)\,; \quad \frac{b-\nu\tau}{\sqrt{\tau}} = -d_-(\frac{S}{B},\tau); \quad \frac{\kappa-2b-\nu\tau}{\sqrt{\tau}} = -d_-(\frac{B^2}{KS},\tau)$$
$$\frac{\kappa+\nu\tau}{\sqrt{\tau}} = d_-(\frac{K}{S},\tau)\,; \quad \frac{\kappa-\nu\tau}{\sqrt{\tau}} = -d_-(\frac{S}{K},\tau)\,,$$

where we define $d_+(x,\tau) := \dfrac{\ln x + (r-q+\frac{1}{2}\sigma^2)\tau}{\sigma\sqrt{\tau}}$, $d_-(x,\tau) = d_+(x,\tau) - \sigma\sqrt{\tau}$. Substituting these expressions into (3.146) and using the identity $\mathcal{N}(x) + \mathcal{N}(-x) = 1$ gives the exact integral:

$$\begin{aligned}\int_{\kappa=\frac{1}{\sigma}\ln\frac{K}{S}}^{b=\frac{1}{\sigma}\ln\frac{B}{S}} \widetilde{p}^{X_{(b)}}(\tau;0,x)\,\mathrm{d}x &= \mathcal{N}\left(d_-(\frac{S}{K},\tau)\right) - \mathcal{N}\left(d_-(\frac{S}{B},\tau)\right)\\ &- \left(\frac{B}{S}\right)^{\frac{2(r-q)}{\sigma^2}-1}\left[\mathcal{N}\left(d_-(\frac{B^2}{KS},\tau)\right) - \mathcal{N}\left(d_-(\frac{B}{S},\tau)\right)\right].\end{aligned} \tag{3.147}$$

We leave it as an exercise for the reader to apply similar steps to show that the (discounted) first integral in (3.145) is given by

$$\begin{aligned}\mathrm{e}^{-(r-q)\tau}\int_{\kappa=\frac{1}{\sigma}\ln\frac{K}{S}}^{b=\frac{1}{\sigma}\ln\frac{B}{S}} \mathrm{e}^{\sigma x}\,\widetilde{p}^{X_{(b)}}(\tau;0,x)\,\mathrm{d}x &= \mathcal{N}\left(d_+(\frac{S}{K},\tau)\right) - \mathcal{N}\left(d_+(\frac{S}{B},\tau)\right)\\ &- \left(\frac{B}{S}\right)^{\frac{2(r-q)}{\sigma^2}+1}\left[\mathcal{N}\left(d_+(\frac{B^2}{KS},\tau)\right) - \mathcal{N}\left(d_+(\frac{B}{S},\tau)\right)\right].\end{aligned} \tag{3.148}$$

Substituting the integral expressions in (3.147) and (3.148) into (3.145) and combining terms gives the analytically exact pricing formula for the up-and-out call for $K < B$:

$$C^{UO}(t,S,K;B) = C(t,S,K) - C^{UI}(t,S,K;B) \tag{3.149}$$

where

$$C(t,S,K) = \mathrm{e}^{-q\tau} S\,\mathcal{N}\left(d_+(\frac{S}{K},\tau)\right) - \mathrm{e}^{-r\tau} K\mathcal{N}\left(d_-(\frac{S}{K},\tau)\right)$$

is the Black–Scholes pricing formula for a standard call on a dividend paying stock and C^{UI} is the *up-and-in call pricing formula* for $K < B$:

$$\begin{aligned}C^{UI}(t,S,K;B) &= e^{-q\tau} S\mathcal{N}\left(d_+(\frac{S}{B},\tau)\right) - e^{-r\tau} K\mathcal{N}\left(d_-(\frac{S}{B},\tau)\right)\\ &+ e^{-q\tau} S\left(\frac{B}{S}\right)^{\frac{2(r-q)}{\sigma^2}+1}\left[\mathcal{N}\left(d_+(\frac{B^2}{KS},\tau)\right) - \mathcal{N}\left(d_+(\frac{B}{S},\tau)\right)\right]\\ &- e^{-r\tau} K\left(\frac{B}{S}\right)^{\frac{2(r-q)}{\sigma^2}-1}\left[\mathcal{N}\left(d_-(\frac{B^2}{KS},\tau)\right) - \mathcal{N}\left(d_-(\frac{B}{S},\tau)\right)\right],\end{aligned} \tag{3.150}$$

where $\tau = T - t$ is time to maturity. Note that for $K \geqslant B$, $C^{UI}(t,S,K;B) = C(t,S,K)$.

Pricing formulae for other up-and-out (and up-and-in) options are readily derived using similar steps and by combining the above integral identities in (3.147) and (3.148) within (3.141). For down-and-out (and down-and-in), we use (3.142) and develop similar identities to (3.147) and (3.148) for evaluating the pricing integrals. The derivations of pricing formulae for down-and-out (and down-and-in) call and put options are left as exercises at the end of this chapter.

Example 3.11. (Up-and-Out Put Price) Derive the time-t, $t < T$, no-arbitrage pricing formula of an up-and-out put option with payoff

$$P_T^{UO} = (K - S(T))^+ \mathbb{I}_{\{M^S(T)<B\}}$$

where $B > 0$ is the knock-out barrier and $K > 0$ the strike. Assume the stock is a GBM with constant interest rate and continuous dividend yield q.

Solution. We take spot $S < B$. For an up-and-out option we use (3.141) with put payoff

$$\Lambda(S\,e^{\sigma x}) = K\,\mathbb{I}_{\{x<\kappa\}} - S\,e^{\sigma x}\mathbb{I}_{\{x<\kappa\}}\,, \quad \kappa \equiv \frac{1}{\sigma}\ln\frac{K}{S}\,.$$

The integral over the density is restricted to $x < b$, $b \equiv \frac{1}{\sigma}\ln\frac{B}{S}$. Since $\mathbb{I}_{\{x<\kappa\}}\mathbb{I}_{\{x<b\}} = \mathbb{I}_{\{x<b\wedge\kappa\}}$,

$$\Lambda(S\,e^{\sigma x})\mathbb{I}_{\{x<b\}} = K\,\mathbb{I}_{\{x<b\wedge\kappa\}} - S\,e^{\sigma x}\mathbb{I}_{\{x<b\wedge\kappa\}}$$

where $b\wedge\kappa \equiv \min(b,\kappa) = \frac{1}{\sigma}\ln\frac{K\wedge B}{S}$. The price of the up-and-out put is then given by

$$P^{UO}(t,S,K;B) = e^{-r\tau}K\int_{-\infty}^{b\wedge\kappa}\widetilde{p}^{X_{(b)}}(\tau;0,x)\,dx - e^{-r\tau}S\int_{-\infty}^{b\wedge\kappa}e^{\sigma x}\widetilde{p}^{X_{(b)}}(\tau;0,x)\,dx\,.$$

There are two cases: (i) $B < K$ or (ii) $B \geqslant K$. For $B < K$, $b\wedge\kappa = b$, and the price is

$$P^{UO}(t,S,K;B) = e^{-r\tau}K\int_{-\infty}^{b}\widetilde{p}^{X_{(b)}}(\tau;0,x)\,dx - e^{-r\tau}S\int_{-\infty}^{b}e^{\sigma x}\widetilde{p}^{X_{(b)}}(\tau;0,x)\,dx\,.$$

The two integrals can be computed using the same steps and identities used for the up-and-out call above. However, there is a shortcut based on (3.147) and (3.148) in the limit that the lower point of integration goes to $-\infty$. That is, the above two integrals correspond to taking the limit $K \searrow 0$, $\ln\frac{K}{S} \to -\infty$, in the expressions in (3.147) and (3.148). Since $d_\pm\left(\frac{B^2}{KS},\tau\right) \to \infty$, $d_\pm\left(\frac{S}{K},\tau\right) \to \infty$, all $\mathcal{N}(\cdot)$ terms with these arguments approach $\mathcal{N}(\infty) = 1$, as $K \searrow 0$. Upon using the symmetry $1 - \mathcal{N}(z) = \mathcal{N}(-z)$ in the resulting expressions, we obtain

$$\int_{-\infty}^{b=\frac{1}{\sigma}\ln\frac{B}{S}}\widetilde{p}^{X_{(b)}}(\tau;0,x)\,dx = \mathcal{N}\left(-d_-\left(\frac{S}{B},\tau\right)\right) - \left(\frac{B}{S}\right)^{\frac{2(r-q)}{\sigma^2}-1}\mathcal{N}\left(-d_-\left(\frac{B}{S},\tau\right)\right) \tag{3.151}$$

and

$$e^{-(r-q)\tau}\int_{-\infty}^{b=\frac{1}{\sigma}\ln\frac{B}{S}}e^{\sigma x}\,\widetilde{p}^{X_{(b)}}(\tau;0,x)\,dx = \mathcal{N}\left(-d_+\left(\frac{S}{B},\tau\right)\right) - \left(\frac{B}{S}\right)^{\frac{2(r-q)}{\sigma^2}+1}\mathcal{N}\left(-d_+\left(\frac{B}{S},\tau\right)\right). \tag{3.152}$$

Substituting these integrals gives the explicit pricing function for $B < K$:

$$\begin{aligned}P^{UO}(t,S,K;B) = e^{-r\tau}K&\left[\mathcal{N}\left(-d_-\left(\frac{S}{B},\tau\right)\right) - \left(\frac{B}{S}\right)^{\frac{2(r-q)}{\sigma^2}-1}\mathcal{N}\left(-d_-\left(\frac{B}{S},\tau\right)\right)\right]\\ -e^{-q\tau}S&\left[\mathcal{N}\left(-d_+\left(\frac{S}{B},\tau\right)\right) - \left(\frac{B}{S}\right)^{\frac{2(r-q)}{\sigma^2}+1}\mathcal{N}\left(-d_+\left(\frac{B}{S},\tau\right)\right)\right],\end{aligned} \tag{3.153}$$

$\tau = T - t$. For $B \geqslant K$, $\kappa \leqslant b$, $b \wedge \kappa = \kappa$, and the price is given by

$$P^{UO}(t, S, K; B) = \mathrm{e}^{-r\tau} K \int_{-\infty}^{\kappa} \widetilde{p}^{X_{(b)}}(\tau; 0, x)\, \mathrm{d}x - \mathrm{e}^{-r\tau} S \int_{-\infty}^{\kappa} \mathrm{e}^{\sigma x} \widetilde{p}^{X_{(b)}}(\tau; 0, x)\, \mathrm{d}x\,.$$

In this case, we express each integral on $(-\infty, \kappa)$ as the integral on $(-\infty, b)$ minus the integral on (κ, b). Then, we can use the difference of (3.151) and (3.147) to obtain the first integral on $(-\infty, \kappa)$ and the difference of (3.152) and (3.148). Combining terms and simplifying, we have the explicit pricing function for $B \geqslant K$:

$$\begin{aligned} P^{UO}(t, S, K; B) = P(t, S, K) &+ \mathrm{e}^{-q\tau} S \left(\frac{B}{S}\right)^{\frac{2(r-q)}{\sigma^2}+1} \mathcal{N}\left(-d_+\left(\frac{B^2}{KS}, \tau\right)\right) \\ &- \mathrm{e}^{-r\tau} K \left(\frac{B}{S}\right)^{\frac{2(r-q)}{\sigma^2}-1} \mathcal{N}\left(-d_-\left(\frac{B^2}{KS}, \tau\right)\right), \end{aligned} \tag{3.154}$$

$\tau = T - t$, where $P(t, S, K) = \mathrm{e}^{-r\tau} K \mathcal{N}\left(-d_-\left(\frac{S}{K}, \tau\right)\right) - \mathrm{e}^{-q\tau} S \mathcal{N}\left(-d_+\left(\frac{S}{K}, \tau\right)\right)$ is the Black–Scholes pricing formula for a standard put on a dividend paying stock. □

In closing this section, we show that barrier options can also be "delta hedged" and we also make the connection between the risk-neutral pricing approach and the corresponding Black–Scholes PDE (BSPDE) for pricing single barrier options. We focus our discussion on the up-and-out and down-and-out options. The analysis for knock-in barrier options follows from knock-in-knock-out symmetry. As we have shown above, the general pricing formulae for the knock-out options are given by (3.141) and (3.142) of Proposition 3.3. Assuming that the integrals in (3.141) and (3.142) exist, and that we can evaluate them, we have completely solved the pricing problem for single barrier options. Alternatively, we now show that the pricing function is a solution to a BSPDE subject to appropriate boundary conditions. We have already seen how the risk-neutral pricing formulation is related to the BSPDE for the case of a standard (no barrier) European option. In particular, the pricing function $V(t, S)$ in (3.28) is expressed as an (expectation) integral of the payoff against the risk-neutral transition PDF in (3.29) for the stock price process on the domain $(0, \infty)$. The discounted risk-neutral transition PDF, and therefore $V(t, S)$, solves the BSPDE (3.65) in the variables (S, τ) and (3.66) in the variables (t, S).

To see how the BSPDE arises for an up-and-out option, we apply a change of integration variables by letting $y = S\mathrm{e}^{\sigma x}$ ($x = \frac{1}{\sigma} \ln \frac{y}{S}$) in (3.141), which then takes the form

$$V^{UO}(t, S; B) = \mathrm{e}^{-r\tau} \int_0^B \Lambda(y)\, \widetilde{p}^{S_{(B)}}(\tau; S, y)\, \mathrm{d}y \tag{3.155}$$

where $\widetilde{p}^{S_{(B)}}(\tau; S, y)$ is defined for all S, y values on the interval $(0, B)$:

$$\begin{aligned} \widetilde{p}^{S_{(B)}}(\tau; S, y) \equiv\; & \frac{1}{\sigma y \sqrt{\tau}}\, n\left(\frac{\ln \frac{y}{S} - (r - q - \frac{1}{2}\sigma^2)\tau}{\sigma\sqrt{\tau}}\right) \\ & - \left(\frac{B}{S}\right)^{\frac{2(r-q)}{\sigma^2}-1} \frac{1}{\sigma y \sqrt{\tau}}\, n\left(\frac{\ln \frac{Sy}{B^2} - (r - q - \frac{1}{2}\sigma^2)\tau}{\sigma\sqrt{\tau}}\right). \end{aligned} \tag{3.156}$$

As shown in Exercise 3.35, this is the risk-neutral transition PDF for the stock price process killed at the first-hitting time to level B on *either* interval $(0, B)$ or (B, ∞). The latter interval is used for the down-and-out option where (3.142) takes the same form as in (3.155) but with (B, ∞) as the integration interval in place of $(0, B)$.

For any fixed y, the discounted transition PDF, $v(\tau, S, y) := \mathrm{e}^{-r\tau}\widetilde{p}^{S_{(B)}}(\tau; S, y)$, solves the time-homogeneous BSPDE in (3.65) subject to the initial condition $v(0+, S, y) = \delta(S - y)$ (with the Dirac delta function δ). In fact, it is a fundamental solution on the interval $(0, B)$ (as well as on the interval (B, ∞)) with zero boundary conditions at the barrier level B and at either endpoint $S \searrow 0$ or $S \to \infty$. Assuming the integral in (3.155) exists and the resulting pricing function is $C^{1,2}$ (continuously differentiable in t (or τ) and twice differentiable in S), we can apply the differential operator $\left(\frac{\partial}{\partial t} + \mathcal{L}^{BS}\right)$ (acting on variables t and S) on both sides of (3.155). Note that $\mathcal{L}^{BS}$ is the Black–Scholes operator as in (3.66). By interchanging the order of differentiation and integration (in the dummy variable y), and using the fact that $v(\tau, S, y)$ solves the BSPDE, gives

$$\frac{\partial}{\partial t}V^{UO} + \mathcal{L}^{BS}V^{UO} = \int_0^B \underbrace{\left(\frac{\partial}{\partial t}v(\tau, S, y) + \mathcal{L}^{BS}v(\tau, S, y)\right)}_{\equiv 0} \Lambda(y)\, \mathrm{d}y = 0\,. \tag{3.157}$$

Hence, $V^{UO} \equiv V^{UO}(t, S; B)$ is a solution to the BSPDE in (3.66) on the rectangular domain $0 < S < B, 0 \leqslant t < T$ or equivalently the BSPDE in (3.65) for $0 < S < B, \tau \in (0, T]$. The terminal condition (or initial condition $\tau \searrow 0$) is given by the payoff function where $V^{UO}(T, S; B) \equiv V^{UO}(T-, S; B) = \Lambda(S)$, for any continuous Λ and for $0 \leqslant S \leqslant B$. The boundary conditions at the endpoints of $(0, B)$ are given by (note: $S = 0$ is the limit $S \searrow 0$)

$$\begin{aligned} &V^{UO}(t, S = 0; B) = \mathrm{e}^{-r(T-t)}\Lambda(0)\,, && 0 \leqslant t \leqslant T\,, \\ &V^{UO}(t, S = B; B) = 0\,, && 0 \leqslant t < T\,. \end{aligned} \tag{3.158}$$

The boundary condition at $S = 0$ is due to the stock price staying at zero if it is set to zero and hence the payoff will be $\Lambda(0)$, which is discounted by $\mathrm{e}^{-r(T-t)}$ to obtain its time-t value. The second condition corresponds to the option being worthless if the spot is at the barrier level any time before maturity. As an example, for an up-and-out call its value at the lower boundary $S = 0$ is $C^{UO}(t, 0, K; B) = \mathrm{e}^{-r(T-t)}\Lambda(0) = \mathrm{e}^{-r(T-t)}(0 - K)^+ = 0$ and at $S = B$ we have $C^{UO}(t, B, K; B) = 0$. For an up-and-out put, $P^{UO}(t, 0, K; B) = \mathrm{e}^{-r(T-t)}(K - 0)^+ = \mathrm{e}^{-r(T-t)}K$ and $P^{UO}(t, B, K; B) = 0$ at $S = B$.

For a down-and-out option, the analysis is similar, leading to the same BSPDE in (3.66) for $V^{DO} \equiv V^{DO}(t, S; B)$ on the rectangular domain $B < S < \infty, 0 \leqslant t < T$ or equivalently the BSPDE in (3.65) for $B < S < \infty, \tau \in (0, T]$. The terminal (or initial) time condition is again the payoff function, $V^{DO}(T, S; B) = \Lambda(S)$, for $B \leqslant S < \infty$, and with boundary endpoint conditions:

$$\begin{aligned} &\lim_{S\to\infty} V^{DO}(t, S; B) = \lim_{S\to\infty} V(t, S)\,, && 0 \leqslant t \leqslant T\,, \\ &V^{DO}(t, S = B; B) = 0\,, && 0 \leqslant t < T\,. \end{aligned} \tag{3.159}$$

The first condition states that the value of the down-and-out option and the corresponding standard option value $V(t, S)$ should be the same in the limit of infinite stock value. This is due to the stock price staying close to infinity and not hitting the lower knock-out barrier (in finite time) if it starts close to infinity. For the GBM model, this is the case where the boundary at infinity is a natural boundary. The second boundary condition is again due to the option expiring worthless if the spot is at the barrier level before maturity. For example, a down-and-out call has value $C^{DO}(t, B, K; B) = 0$ at $S = B$ and $C^{DO}(t, S, K; B) \sim C(t, S, K) \sim \mathrm{e}^{-q(T-t)}S - \mathrm{e}^{-r(T-t)}K$, as $S \to \infty$. For a down-and-out put, $P^{DO}(t, B, K; B) = 0$ and $P^{DO}(t, S, K; B) \sim P(t, S, K) \sim 0$, as $S \to \infty$.

Let $V(t, S; B)$ denote either pricing function $V^{UO}(t, S; B)$ or $V^{DO}(t, S; B)$. For any given $B > 0$, we argued above that $V(t, S; B)$ is a $C^{1,2}$ function that solves the BSPDE in the (dummy) variables t, S. We can therefore apply Itô's formula to the discounted process defined via the function $V(t, S; B)$, i.e., $\{e^{-rt}V(t, S(t); B)\}_{t\geqslant 0}$. Taking the stochastic differential and using the fact that $V(t, S; B)$ solves the BSPDE gives

$$\mathrm{d}\left[e^{-rt}V(t, S(t); B)\right] = e^{-rt}\left(\frac{\partial}{\partial t} + \mathcal{L}^{BS}\right)V(t, S(t); B)\,\mathrm{d}t + \sigma\overline{S}(t)\frac{\partial}{\partial S}V(t, S(t); B)\,\mathrm{d}\widetilde{W}(t)\,,$$
$$= \sigma\overline{S}(t)\frac{\partial}{\partial S}V(t, S(t); B)\,\mathrm{d}\widetilde{W}(t)\,.$$

Note that this stochastic differential and that of the discounted price process for the knock-out barrier option are the same for all times before the stock price hits the barrier level B. Equating this with the expression in (3.20) gives the hedging position $\Delta_t = \frac{\partial}{\partial S}V(t, S(t); B)$. For a given realization of the stock price process, this is then the hedging position held in the stock for all times t up to the first hitting time to the (knock-out) level B, or otherwise up to maturity time T if the stock price does not attain the level B during the option's lifetime. In particular, for every spot value $S(t) = S < B$, the hedging formula for an up-and-out option is the delta of the pricing function, $\Delta^{UO}(t, S; B) = \frac{\partial}{\partial S}V^{UO}(t, S; B)$. Similarly, for $S(t) = S > B$, a down-and-out option is hedged using $\Delta^{DO}(t, S; B) = \frac{\partial}{\partial S}V^{DO}(t, S; B)$.

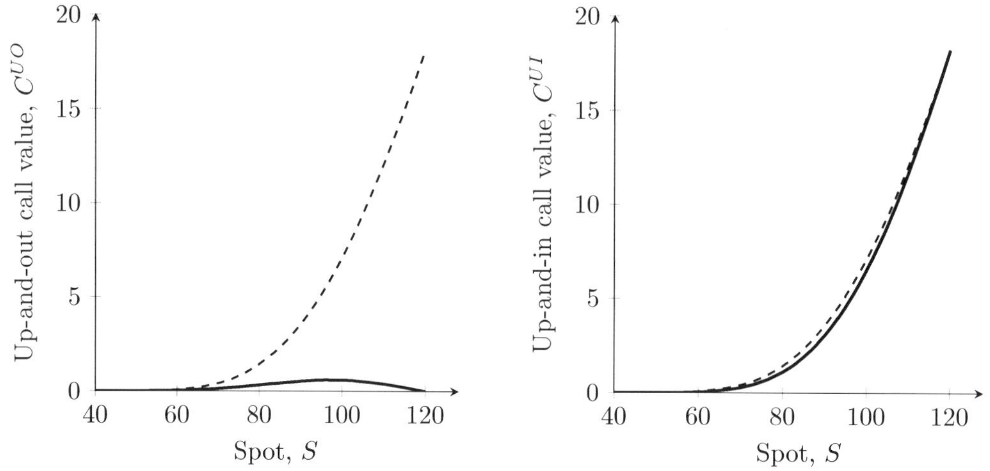

FIGURE 3.13: Plots of no-arbitrage values of the up-and-out and up-and-in European call options as functions of spot price S under the (B, S) model with $\tau = 1$, $r = 5\%$, $q = 10\%$, $\sigma = 25\%$, $K = 100$, and $B = 120$. The dashed line is a plot of the standard European call option.

3.4.3 Pricing Lookback Options

We can now proceed to derive pricing formulae for generally "seasoned" lookback options of types (a)–(d) with payoffs defined in (3.122) where conditioning is on knowledge of the sampled stock price maximum, $M^S(t) = M \geqslant S$, or minimum $m^S(t) = m \leqslant S$, i.e., we are entering the contract at time t where the realized maximum or minimum up to time t generally differs from the stock (spot) price $S(t) = S$. Our main pricing formulae are (3.134) and (3.135). We therefore need the effective payoff functions in the integrand of either case. For example, consider the floating strike (LFS) call with payoff $C_T^{LFS} = \phi(m^S(T), S(T)) = S(T) - m^S(T)$ in (a) of (3.122), i.e. $\phi(x, y) := y - x$. Hence, the effective payoff for this

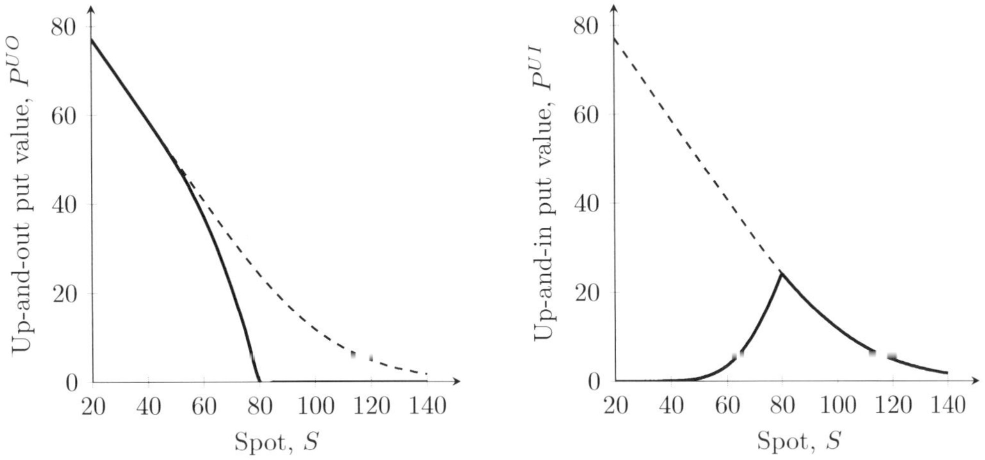

FIGURE 3.14: Plots of no-arbitrage values of the up-and-out and up-and-in European put options as functions of spot price S under the (B, S) model with $\tau = 1$, $r = 5\%$, $q = 10\%$, $\sigma = 25\%$, $K = 100$, and $B = 80$. The dashed line is a plot of the standard European put option.

option is the integrand function in (3.135) given by

$$\begin{aligned} g(w,x) &:= \phi\big(\min\{m, S\,\mathrm{e}^{\sigma w}\}, S\,\mathrm{e}^{\sigma x}\big) \\ &= S\,\mathrm{e}^{\sigma x} - \min\{m, S\,\mathrm{e}^{\sigma w}\} \\ &= S\,\mathrm{e}^{\sigma x} - \big[S\,\mathrm{e}^{\sigma w}\,\mathbb{I}_{\{S\,\mathrm{e}^{\sigma w} < m\}} + m\,\mathbb{I}_{\{S\,\mathrm{e}^{\sigma w} \geqslant m\}}\big] \\ &= S\,\mathrm{e}^{\sigma x} - S\,\mathrm{e}^{\sigma w}\,\mathbb{I}_{\{w < \widehat{m}\}} - m\,\mathbb{I}_{\{w \geqslant \widehat{m}\}} \end{aligned}$$

where $\widehat{m} := \frac{1}{\sigma}\ln\frac{m}{S} \leqslant 0$. For case (b) in (3.122), we have $\phi(M^S(T), S(T)) = M^S(T) - S(T)$, i.e., $\phi(x,y) := x - y$. Hence, the effective payoff in (3.134) is

$$\begin{aligned} h(w,x) := \phi\big(\max\{M, S\,\mathrm{e}^{\sigma w}\}, S\,\mathrm{e}^{\sigma x}\big) &= \max\{M, S\,\mathrm{e}^{\sigma w}\} - S\,\mathrm{e}^{\sigma x} \\ &= M\,\mathbb{I}_{\{w < \widehat{M}\}} + S\,\mathrm{e}^{\sigma w}\,\mathbb{I}_{\{w \geqslant \widehat{M}\}} - S\,\mathrm{e}^{\sigma x}\,, \end{aligned}$$

where $\widehat{M} := \frac{1}{\sigma}\ln\frac{M}{S} \geqslant 0$.

The reader can verify that the effective payoffs for cases (c) and (d) are as given below where we summarize the effective payoffs for lookbacks (a)–(d) in (3.122):

$$\text{(a) } g(w,x) = S\,\mathrm{e}^{\sigma x} - S\,\mathrm{e}^{\sigma w}\,\mathbb{I}_{\{w < \widehat{m}\}} - m\,\mathbb{I}_{\{w \geqslant \widehat{m}\}}; \tag{3.160}$$

$$\text{(b) } h(w,x) = M\,\mathbb{I}_{\{w < \widehat{M}\}} + S\,\mathrm{e}^{\sigma w}\,\mathbb{I}_{\{w \geqslant \widehat{M}\}} - S\,\mathrm{e}^{\sigma x}; \tag{3.161}$$

$$\text{(c) } h(w,x) = \begin{cases} (S\,\mathrm{e}^{\sigma w} - K)\,\mathbb{I}_{\{w > \kappa\}} & \text{for } M < K, \\ M\,\mathbb{I}_{\{w < \widehat{M}\}} + S\,\mathrm{e}^{\sigma w}\,\mathbb{I}_{\{w \geqslant \widehat{M}\}} - K & \text{for } M \geqslant K; \end{cases} \tag{3.162}$$

$$\text{(d) } g(w,x) = \begin{cases} K - m\,\mathbb{I}_{\{w \geqslant \widehat{m}\}} - S\,\mathrm{e}^{\sigma w}\,\mathbb{I}_{\{w < \widehat{m}\}} & \text{for } m < K, \\ (K - S\,\mathrm{e}^{\sigma w})\,\mathbb{I}_{\{w < \kappa\}} & \text{for } m \geqslant K, \end{cases} \tag{3.163}$$

where $\kappa := \frac{1}{\sigma}\ln\frac{K}{S}$.

It is important to note that the functions in (c) and (d) depend only on w (not x). Moreover, the functions in (a) and (b) are simply sums of functions that depend on only one of the variables, either x or w and not both. This therefore simplifies the pricing integrals in (3.134) and (3.135), which are then sums of single integrals involving either (risk-neutral) marginal density in $m^X(\tau)$ or $M^X(\tau)$. Recall that integrating a joint PDF in one of its arguments (over $\mathbb{R}$) produces the corresponding marginal PDF. This simplification is given explicitly below for the above cases (a)–(d) where the pricing formulae are reduced to single integrals involving the marginal CDF or PDF of $m^X(\tau)$ and $M^X(\tau)$ and other more trivial integrals for the expected value of the drifted BM.

We now state these CDFs and PDFs for further use below when computing expectation integrals within the $\widetilde{\mathbb{P}}$-measure. The CDFs of $m^X(\tau)$ and $M^X(\tau)$ were derived in Section 1.4.3 of Chapter 1. Under the risk-neutral measure $\widetilde{\mathbb{P}}$ we simply take the expressions in (1.113) and (1.120) where the process X now has drift $\nu := \frac{(r-q-\frac{1}{2}\sigma^2)}{\sigma}$ and the time variable is τ; i.e., replace $\mu \to \nu$, $t \to \tau$, and $m \to w$ in (1.113) and (1.120) to give

$$\widetilde{F}_{M^X(\tau)}(w) := \widetilde{\mathbb{P}}(M^X(\tau) \leqslant w) = \mathcal{N}\left(\frac{w-\nu\tau}{\sqrt{\tau}}\right) - \mathrm{e}^{2\nu w}\mathcal{N}\left(\frac{-w-\nu\tau}{\sqrt{\tau}}\right), \quad w > 0, \tag{3.164}$$

$\widetilde{F}_{M^X(\tau)}(w) \equiv 0$ for $w \leqslant 0$, and

$$\widetilde{F}_{m^X(\tau)}(w) := \widetilde{\mathbb{P}}(m^X(\tau) \leqslant w) = \mathcal{N}\left(\frac{w-\nu\tau}{\sqrt{\tau}}\right) + \mathrm{e}^{2\nu w}\mathcal{N}\left(\frac{w+\nu\tau}{\sqrt{\tau}}\right), \quad w < 0, \tag{3.165}$$

$\widetilde{F}_{m^X(\tau)}(w) \equiv 1$ for $w \geqslant 0$. Differentiating these CDFs gives the densities, i.e., $\mathrm{d}\widetilde{F}_{M^X(\tau)}(w) = \widetilde{f}_{M^X(\tau)}(w)\,\mathrm{d}w$ and $\mathrm{d}\widetilde{F}_{m^X(\tau)}(w) = \widetilde{f}_{m^X(\tau)}(w)\,\mathrm{d}w$, where

$$\widetilde{f}_{M^X(\tau)}(w) = \frac{1}{\sqrt{\tau}}n\left(\frac{w-\nu\tau}{\sqrt{\tau}}\right) + \frac{\mathrm{e}^{2\nu w}}{\sqrt{\tau}}\,n\left(\frac{w+\nu\tau}{\sqrt{\tau}}\right) - 2\nu\mathrm{e}^{2\nu w}\mathcal{N}\left(\frac{-w-\nu\tau}{\sqrt{\tau}}\right), \quad w > 0, \tag{3.166}$$

$$\widetilde{f}_{m^X(\tau)}(w) = \frac{1}{\sqrt{\tau}}n\left(\frac{w-\nu\tau}{\sqrt{\tau}}\right) + \frac{\mathrm{e}^{2\nu w}}{\sqrt{\tau}}\,n\left(\frac{w+\nu\tau}{\sqrt{\tau}}\right) + 2\nu\mathrm{e}^{2\nu w}\mathcal{N}\left(\frac{w+\nu\tau}{\sqrt{\tau}}\right), \quad w < 0. \tag{3.167}$$

Alternatively, the reader can verify that these same expressions are obtained by successively integrating the respective joint PDFs in (3.131) and (3.132).

Based on (3.134) and (3.135), we can now derive the main pricing formulae for the above four types of lookback options. Consider case (a), where we denote the time-t pricing function for the LFS call by $C^{LFS}(t, S, m)$ for all $0 < m \leqslant S < \infty$. Substituting (3.160) into (3.135) gives this pricing function as a sum of three integrals involving the joint PDF $\widetilde{f}(w, x) \equiv \widetilde{f}_{m^X(\tau), X(\tau)}(w, x)$. The integrals are respectively reduced to single integrals involving the marginal PDF of $X(\tau)$ and of $m^X(\tau)$ as follows:

$$\begin{aligned}
C^{LFS}(t, S, m) &= \mathrm{e}^{-r\tau}S\int_{\mathbb{R}}\left[\int_{\mathbb{R}}\widetilde{f}(w,x)\,\mathrm{d}w\right]\mathrm{e}^{\sigma x}\,\mathrm{d}x - \mathrm{e}^{-r\tau}S\int_{\mathbb{R}}\left[\int_{\mathbb{R}}\widetilde{f}(w,x)\,\mathrm{d}x\right]\mathrm{e}^{\sigma w}\mathbb{I}_{\{w<\widehat{m}\}}\,\mathrm{d}w \\
&\quad - \mathrm{e}^{-r\tau}m\int_{\mathbb{R}}\left[\int_{\mathbb{R}}\widetilde{f}(w,x)\,\mathrm{d}x\right]\mathbb{I}_{\{w>\widehat{m}\}}\,\mathrm{d}w \\
&= \mathrm{e}^{-r\tau}S\int_{\mathbb{R}}\widetilde{f}_{X(\tau)}(x)\,\mathrm{e}^{\sigma x}\,\mathrm{d}x - \mathrm{e}^{-r\tau}S\int_{-\infty}^{\widehat{m}}\widetilde{f}_{m^X(\tau)}(w)\,\mathrm{e}^{\sigma w}\,\mathrm{d}w \\
&\quad - \mathrm{e}^{-r\tau}m\int_{\widehat{m}}^{0}\widetilde{f}_{m^X(\tau)}(w)\,\mathrm{d}w\,.
\end{aligned} \tag{3.168}$$

The integrals are recognized as expectations, where the third integral is $\widetilde{\mathbb{P}}(m^X(\tau) > \widehat{m}) = 1 - \widetilde{\mathbb{P}}(m^X(\tau) \leqslant \widehat{m}) \equiv 1 - \widetilde{F}_{m^X(\tau)}(\widehat{m})$:

$$
\begin{aligned}
C^{LFS}(t, S, m) = \mathrm{e}^{-r\tau} S\, \widetilde{\mathrm{E}}[\mathrm{e}^{\sigma X(\tau)}] - \mathrm{e}^{-r\tau} S\, \widetilde{\mathrm{E}}[\mathrm{e}^{\sigma m^X(\tau)} \mathbb{I}_{\{m^X(\tau) < \widehat{m}\}}] \\
- \mathrm{e}^{-r\tau}\, m\, [1 - \widetilde{F}_{m^X(\tau)}(\widehat{m})]\,. \qquad (3.169)
\end{aligned}
$$

The first expectation is computed simply as $\widetilde{\mathrm{E}}[\mathrm{e}^{\sigma X(\tau)}] = \mathrm{e}^{\sigma\nu\tau} \widetilde{\mathrm{E}}[\mathrm{e}^{\sigma \widetilde{W}(\tau)}] = \mathrm{e}^{\sigma\nu\tau} \cdot \mathrm{e}^{\frac{1}{2}\sigma^2\tau} = \mathrm{e}^{(r-q)\tau}$. This holds true even for more complex models as long as the stock price $S(t)$ discounted by $\mathrm{e}^{-(r-q)t}$ is a $\widetilde{\mathbb{P}}$-martingale. Hence, the pricing formula for the LFS call is given equivalently by:

$$
\begin{aligned}
C^{LFS}(t, S, m) &= \mathrm{e}^{-q\tau} S - \mathrm{e}^{-r\tau} m\, [1 - \widetilde{F}_{m^X(\tau)}(\widehat{m})] - \mathrm{e}^{-r\tau} S \int_{-\infty}^{\widehat{m}} \widetilde{f}_{m^X(\tau)}(w)\, \mathrm{e}^{\sigma w}\, \mathrm{d}w \qquad (3.170) \\
&= \mathrm{e}^{-q\tau} S - \mathrm{e}^{-r\tau} m\, \widetilde{\mathbb{P}}(m^S(\tau) > m) - \mathrm{e}^{-r\tau} \widetilde{\mathrm{E}}\left[m^S(\tau)\, \mathbb{I}_{\{m^S(\tau) \leqslant m\}}\right] \\
&= \mathrm{e}^{-q\tau} S - \mathrm{e}^{-r\tau} m + \mathrm{e}^{-r\tau} \int_0^m \widetilde{\mathbb{P}}(m^S(\tau) \leqslant y)\, \mathrm{d}y
\end{aligned}
$$

where $\tau = T - t$. In the second equation line, we have the respective quantities expressed in terms of $m^S(\tau)$, $\widetilde{\mathrm{E}}[S\mathrm{e}^{\sigma m^X(\tau)} \mathbb{I}_{\{m^X(\tau) \leqslant \widehat{m}\}}] = \widetilde{\mathrm{E}}[m^S(\tau)\, \mathbb{I}_{\{m^S(\tau) \leqslant m\}}]$ and $\widetilde{\mathbb{P}}(m^X(\tau) > \widehat{m}) = \widetilde{\mathbb{P}}(m^S(\tau) > m)$ since the sampled minimum of the stock price (started at spot value S) for a time interval τ is $m^S(\tau) = S\mathrm{e}^{\sigma m^X(\tau)}$. The third line is obtained by re-expressing the expectation in the second line upon using an integration by parts,

$$
\begin{aligned}
\widetilde{\mathrm{E}}\left[m^S(\tau)\, \mathbb{I}_{\{m^S(\tau) \leqslant m\}}\right] = \int_0^m y\, \mathrm{d}\widetilde{F}_{m^S(\tau)}(y) &= m\widetilde{F}_{m^S(\tau)}(m) - \int_0^m \widetilde{F}_{m^S(\tau)}(y)\, \mathrm{d}y \\
&= m\widetilde{\mathbb{P}}(m^S(\tau) \leqslant m) - \int_0^m \widetilde{\mathbb{P}}(m^S(\tau) \leqslant y)\, \mathrm{d}y\,.
\end{aligned}
$$

This identity is valid for any number $m \geqslant 0$.

A similar derivation follows for the floating strike lookback (LFS) put option defined by the payoff in case (b) above where we denote the time-t pricing function for the LFS put by $P^{LFS}(t, S, M)$, for all $0 < S \leqslant M < \infty$. We now substitute (3.161) into (3.134) and this leads to a sum of three integrals involving the joint PDF of $\widetilde{f}_{M^X(\tau), X(\tau)}(w, x)$. Using similar steps as above, the reader can verify that the resulting pricing formula takes the equivalent expressions:

$$
\begin{aligned}
P^{LFS}(t, S, M) &= M\mathrm{e}^{-r\tau}\, \widetilde{F}_{M^X(\tau)}(\widehat{M}) - \mathrm{e}^{-q\tau} S + \mathrm{e}^{-r\tau} S \int_{\widehat{M}}^{\infty} \mathrm{e}^{\sigma w}\, \mathrm{d}\widetilde{F}_{M^X(\tau)}(w) \qquad (3.171) \\
&= M\mathrm{e}^{-r\tau} \widetilde{\mathbb{P}}(M^S(\tau) \leqslant M) - \mathrm{e}^{-q\tau} S + \mathrm{e}^{-r\tau} \widetilde{\mathrm{E}}\left[M^S(\tau)\, \mathbb{I}_{\{M^S(\tau) > M\}}\right] \\
&= \mathrm{e}^{-r\tau} M - \mathrm{e}^{-q\tau} S + \mathrm{e}^{-r\tau} \int_M^{\infty} \widetilde{\mathbb{P}}(M^S(\tau) > y)\, \mathrm{d}y\,,
\end{aligned}
$$

$\tau = T - t$. In the second equation line, we have the respective quantities expressed in terms of $M^S(\tau)$: $\widetilde{\mathrm{E}}[S\mathrm{e}^{\sigma M^X(\tau)} \mathbb{I}_{\{M^X(\tau) > \widehat{M}\}}] = \widetilde{\mathrm{E}}[M^S(\tau)\, \mathbb{I}_{\{M^S(\tau) > M\}}]$ and $\widetilde{\mathbb{P}}(M^X(\tau) > \widehat{M}) = \widetilde{\mathbb{P}}(M^S(\tau) > M)$ since the sampled maximum of the stock price is $M^S(\tau) = S\mathrm{e}^{\sigma M^X(\tau)}$. The third line is obtained by noting that $M^S(\tau)\, \mathbb{I}_{\{M^S(\tau) > M\}} = M^S(\tau) - M^S(\tau)\mathbb{I}_{\{M^S(\tau) \leqslant M\}}$, where $\mathbb{I}_{\{M^S(\tau) \leqslant M\}} = \mathbb{I}_{\{0 \leqslant M^S(\tau) \leqslant M\}}$. The expected value of the positive random variable $M^S(\tau)$ can be represented as an integral over its right tail (risk-neutral) probability:

$$
\widetilde{\mathrm{E}}[M^S(\tau)] = \int_0^{\infty} \widetilde{\mathbb{P}}(M^S(\tau) > y)\, \mathrm{d}y\,.
$$

The expected value of $M^S(\tau)\mathbb{I}_{\{0\leqslant M^S(\tau)\leqslant M\}}$ can be expressed by applying an integration by parts procedure as above,

$$\begin{aligned}\widetilde{\mathrm{E}}\left[M^S(\tau)\mathbb{I}_{\{0\leqslant M^S(\tau)\leqslant M\}}\right] &= \int_0^M y\,\mathrm{d}\widetilde{F}_{M^S(\tau)}(y) = M\widetilde{F}_{M^S(\tau)}(M) - \int_0^M \widetilde{F}_{M^S(\tau)}(y)\,\mathrm{d}y \\ &= M\widetilde{\mathbb{P}}(M^S(\tau)\leqslant M) - \int_0^M \widetilde{\mathbb{P}}(M^S(\tau)\leqslant y)\,\mathrm{d}y\,.\end{aligned}$$

Since $\widetilde{\mathbb{P}}(M^S(\tau)\leqslant y)+\widetilde{\mathbb{P}}(M^S(\tau)>y)=1$, for any $y\geqslant 0$, we can write the last integral as $\int_0^M \widetilde{\mathbb{P}}(M^S(\tau)\leqslant y)\,\mathrm{d}y = M - \int_0^M \widetilde{\mathbb{P}}(M^S(\tau)>y)\,\mathrm{d}y$, and then combine the above two expectations to establish the identity

$$\widetilde{\mathrm{E}}\left[M^S(\tau)\,\mathbb{I}_{\{M^S(\tau)>M\}}\right] = M\widetilde{\mathbb{P}}(M^S(\tau)>M) + \int_M^\infty \widetilde{\mathbb{P}}(M^S(\tau)>y)\,\mathrm{d}y$$

for any number $M\geqslant 0$. Substituting this into the second line of (3.171) gives the expression in the third line of (3.171). For the payoff in case (c) above we denote the time-t pricing function for the floating price lookback (LFP) call (on the maximum) with strike $K>0$ by $C^{LFP}(t,S,M;K)$, for all $0<S\leqslant M<\infty$. By using similar steps as in case (b) above, the reader can verify that the pricing formula takes on the equivalent expressions:

$$\begin{aligned}C^{LFP}(t,S,M;K) &= M\mathrm{e}^{-r\tau}\,\widetilde{F}_{M^X(\tau)}(\widehat{M}) - \mathrm{e}^{-r\tau}K + \mathrm{e}^{-r\tau}S\int_{\widehat{M}}^\infty \mathrm{e}^{\sigma w}\,\mathrm{d}\widetilde{F}_{M^X(\tau)}(w) \qquad (3.172)\\ &= M\mathrm{e}^{-r\tau}\widetilde{\mathbb{P}}(M^S(\tau)\leqslant M) - \mathrm{e}^{-r\tau}K + \mathrm{e}^{-r\tau}\widetilde{\mathrm{E}}\left[M^S(\tau)\,\mathbb{I}_{\{M^S(\tau)>M\}}\right]\\ &= \mathrm{e}^{-r\tau}\left[M-K+\int_M^\infty \widetilde{\mathbb{P}}(M^S(\tau)>y)\,\mathrm{d}y\right]\end{aligned}$$

for $M\geqslant K$, and

$$\begin{aligned}C^{LFP}(t,S,M;K) &= -\mathrm{e}^{-r\tau}K[1-\widetilde{F}_{M^X(\tau)}(\kappa)] + \mathrm{e}^{-r\tau}S\int_\kappa^\infty \mathrm{e}^{\sigma w}\,\mathrm{d}\widetilde{F}_{M^X(\tau)}(w) \qquad (3.173)\\ &= -\mathrm{e}^{-r\tau}K\,\widetilde{\mathbb{P}}(M^S(\tau)>K) + \mathrm{e}^{-r\tau}\widetilde{\mathrm{E}}\left[M^S(\tau)\,\mathbb{I}_{\{M^S(\tau)>K\}}\right]\\ &= \mathrm{e}^{-r\tau}\int_K^\infty \widetilde{\mathbb{P}}(M^S(\tau)>y)\,\mathrm{d}y\end{aligned}$$

for $M<K$, where $\tau=T-t$. Note that for $M<K$ the pricing function C^{LFP}, given by (3.173), is independent of the realized maximum M of the stock price at current time t.

In the last case (d) we denote the time-t pricing function for the floating price lookback (LFP) put (on the minimum) with strike $K>0$ by $P^{LFP}(t,S,m;K)$, for all $0<m\leqslant S<\infty$. By using similar steps as in case (a) above, the reader can verify that the pricing formula takes the equivalent forms:

$$\begin{aligned}P^{LFP}(t,S,m;K) &= K\mathrm{e}^{-r\tau}\,\widetilde{F}_{m^X(\tau)}(\kappa) - \mathrm{e}^{-r\tau}S\int_{-\infty}^\kappa \mathrm{e}^{\sigma w}\,\mathrm{d}\widetilde{F}_{m^X(\tau)}(w) \qquad (3.174)\\ &= K\mathrm{e}^{-r\tau}\,\widetilde{\mathbb{P}}(m^S(\tau)\leqslant K) - \mathrm{e}^{-r\tau}\widetilde{\mathrm{E}}\left[m^S(\tau)\,\mathbb{I}_{\{m^S(\tau)<K\}}\right]\\ &= \mathrm{e}^{-r\tau}\int_0^K \widetilde{\mathbb{P}}(m^S(\tau)\leqslant y)\,\mathrm{d}y\end{aligned}$$

for $m \geqslant K$, and

$$P^{LFP}(t,S,m;K) = K\mathrm{e}^{-r\tau} - m\,\mathrm{e}^{-r\tau}[1-\widetilde{F}_{m^X(\tau)}(\widehat{m})] - \mathrm{e}^{-r\tau}S\int_{-\infty}^{\widehat{m}} \mathrm{e}^{\sigma w}\,\mathrm{d}\widetilde{F}_{m^X(\tau)}(w) \tag{3.175}$$
$$= K\mathrm{e}^{-r\tau} - m\,\mathrm{e}^{-r\tau}\widetilde{\mathbb{P}}(m^S(\tau) > m) - \mathrm{e}^{-r\tau}\widetilde{\mathrm{E}}\left[m^S(\tau)\,\mathbb{I}_{\{m^S(\tau)\leqslant m\}}\right]$$
$$= \mathrm{e}^{-r\tau}\left[K - m + \int_0^m \widetilde{\mathbb{P}}(m^S(\tau) \leqslant y)\,\mathrm{d}y\right]$$

for $m < K$, where $\tau = T - t$. Note that for $m \geqslant K$ the pricing function P^{LFP} in (3.174) is independent of the realized minimum m of the stock price at current time t.

The relations in (3.170)–(3.175) can therefore be used to price all four main types of lookback options. These relations are valid for quite general (time-homogeneous Markov) models for the stock price with discounted process $\{\mathrm{e}^{-(r-q)t}S(t)\}_{t\geqslant 0}$ assumed to be a $\widetilde{\mathbb{P}}$-martingale. Of course, within the GBM model we have simple exact explicit formulae for all the necessary PDFs and CDFs that can now be used to derive analytically exact risk-neutral pricing formulae for all four types of lookback options. For instance, Example 3.12 below gives a derivation of $P^{LFS}(t,S,M)$ by implementing (3.171) within the GBM model for the stock price. Before presenting this example, we note that the pricing relations in (3.170), (3.171), (3.172), and (3.175) also further simplify in the case where the sampling of the maximum and minimum is started at the current time t: $S(t) = M^S(t) = m^S(t)$. This is seen by setting $M = S$ and $m = S$ and noting that $m^S(\tau) < S$ and $M^S(\tau) > S$ (a.s.), i.e., the probability $\widetilde{\mathbb{P}}(m^S(\tau) > m)$ becomes $\widetilde{\mathbb{P}}(m^S(\tau) > S) = 0$ and $\widetilde{\mathbb{P}}(M^S(\tau) > M)$ becomes $\widetilde{\mathbb{P}}(M^S(\tau) > S) = 1$. Moreover, the indicator functions simplify where $\mathbb{I}_{\{m^S(\tau)\leqslant m\}}$ becomes $\mathbb{I}_{\{m^S(\tau)\leqslant S\}} = 1$ and $\mathbb{I}_{\{M^S(\tau)>M\}}$ becomes $\mathbb{I}_{\{M^S(\tau)>S\}} = 1$. This is consistent with the fact that $\widehat{m} = \frac{1}{\sigma}\ln\frac{m}{S} = 0$ and $\widehat{M} = \frac{1}{\sigma}\ln\frac{M}{S} = 0$ when $M = m = S$. All the pricing formulae are then only functions of spot S and time t (or S and τ).

Example 3.12. (Floating Strike (LFS) Put Price) Derive the no-arbitrage pricing formula $P^{LFS}(t,S,M)$ for the lookback option with payoff (b) in (3.122). Assume the stock price is a GBM with a constant interest rate and a continuous dividend yield q.

Solution. It is convenient to obtain the pricing function $P^{LFS}(t,S,M)$ by using the first equation line in (3.171). The first term is evaluated explicitly by evaluating the CDF in (3.164) at $w = \widehat{M} \equiv \frac{1}{\sigma}\ln\frac{M}{S}$ and using the drift parameter $\nu = (r - q - \frac{1}{2}\sigma^2)/\sigma$,

$$\widetilde{F}_{M^X(\tau)}(\widehat{M}) = \mathcal{N}\left(\frac{\ln\frac{M}{S} - (r-q-\frac{1}{2}\sigma^2)\tau}{\sigma\sqrt{\tau}}\right) - \mathrm{e}^{\frac{2\nu}{\sigma}\ln\frac{M}{S}}\,\mathcal{N}\left(-\frac{\ln\frac{M}{S} + (r-q-\frac{1}{2}\sigma^2)\tau}{\sigma\sqrt{\tau}}\right)$$
$$= \mathcal{N}\left(-d_-\big(\tfrac{S}{M},\tau\big)\right) - \frac{S}{M}\left(\frac{M}{S}\right)^{\frac{2(r-q)}{\sigma^2}}\mathcal{N}\left(-d_-\big(\tfrac{M}{S},\tau\big)\right), \tag{3.176}$$

where we define $d_\pm(x,\tau) := \frac{\ln x + (r-q\pm\frac{1}{2}\sigma^2)\tau}{\sigma\sqrt{\tau}}$; $x > 0, \tau = T - t > 0$. Note that $\frac{2\nu}{\sigma} = \frac{2(r-q-\frac{1}{2}\sigma^2)}{\sigma^2} = \frac{2(r-q)}{\sigma^2} - 1$.

We now need to compute the (expectation) integral in (3.171). By substituting the density in (3.166), the integral is a sum of three integrals:

$$\int_{\widehat{M}}^{\infty} \mathrm{e}^{\sigma w}\widetilde{f}_{M^X(\tau)}(w)\,\mathrm{d}w = \int_{\widehat{M}}^{\infty} \mathrm{e}^{\sigma w}\frac{1}{\sqrt{\tau}}\,n\!\left(\frac{w-\nu\tau}{\sqrt{\tau}}\right)\mathrm{d}w + \int_{\widehat{M}}^{\infty} \mathrm{e}^{(\sigma+2\nu)w}\frac{1}{\sqrt{\tau}}\,n\!\left(\frac{w+\nu\tau}{\sqrt{\tau}}\right)\mathrm{d}w$$
$$- 2\nu\int_{\widehat{M}}^{\infty} \mathrm{e}^{(\sigma+2\nu)w}\mathcal{N}\left(\frac{-w-\nu\tau}{\sqrt{\tau}}\right)\mathrm{d}w\,. \tag{3.177}$$

The first two Gaussian integrals are readily evaluated by completing the square in the exponents, or simply by direct use of the integral identity (B.1) of Appendix B. It turns out that both integrals are given by

$$\int_{\widehat{M}}^{\infty} \mathrm{e}^{\sigma w} \frac{1}{\sqrt{\tau}} n\left(\frac{w - \nu\tau}{\sqrt{\tau}}\right) \mathrm{d}w = \int_{\widehat{M}}^{\infty} \mathrm{e}^{(\sigma+2\nu)w} \frac{1}{\sqrt{\tau}} n\left(\frac{w + \nu\tau}{\sqrt{\tau}}\right) \mathrm{d}w = \mathrm{e}^{(r-q)\tau} \mathcal{N}\left(d_+(\frac{S}{M}, \tau)\right). \tag{3.178}$$

Note that $\sigma + 2\nu = \frac{2(r-q)}{\sigma}$. So the third integral can be evaluated in two separate cases: (i) $r - q = 0$ and (ii) $r - q \neq 0$. We will treat the latter case since the pricing formula for case (i) can be obtained by taking the limit $(r - q) \to 0$ in the pricing formula for case (ii).

We now evaluate the third integral using a change of variables, $x = (\sigma + 2\nu)(w - \widehat{M})$, and write $\mathrm{e}^{(\sigma+2\nu)\widehat{M}} = \exp(\frac{2(r-q)}{\sigma^2} \ln \frac{M}{S}) = (\frac{M}{S})^{\frac{2(r-q)}{\sigma^2}}$, giving

$$\begin{aligned} 2\nu \int_{\widehat{M}}^{\infty} \mathrm{e}^{(\sigma+2\nu)w} \mathcal{N}\left(\frac{-w - \nu\tau}{\sqrt{\tau}}\right) \mathrm{d}w &= \frac{2\nu}{\sigma + 2\nu} \mathrm{e}^{(\sigma+2\nu)\widehat{M}} \int_0^{\infty} \mathrm{e}^x \mathcal{N}(Ax + B)\, \mathrm{d}x \\ &= \left[1 - \frac{\sigma^2}{2(r-q)}\right] \left(\frac{M}{S}\right)^{\frac{2(r-q)}{\sigma^2}} \int_0^{\infty} \mathrm{e}^x \mathcal{N}(Ax + B)\, \mathrm{d}x\,. \end{aligned} \tag{3.179}$$

Note that $\frac{2\nu}{\sigma+2\nu} = 1 - \frac{\sigma^2}{2(r-q)}$. Here, we define the constants $A \equiv -\frac{1}{(\sigma+2\nu)\sqrt{\tau}} = -\frac{\sigma}{2(r-q)\sqrt{\tau}}$ and $B \equiv -\frac{\widehat{M}+\nu\tau}{\sqrt{\tau}}$. We can assume that $r - q > 0$, i.e., $A < 0$, so that the integral identity in (B.5) of Appendix B can be directly applied. [We leave it to the reader to apply a change of variable and verify that the same result obtains by making use of an appropriate integral identity in Appendix B for the case that $r - q < 0$.] Applying (B.5) and simplifying the terms gives

$$\begin{aligned} \int_0^{\infty} \mathrm{e}^x \mathcal{N}(Ax + B)\, \mathrm{d}x &= -\mathcal{N}(B) + \mathrm{e}^{(1-2AB)/2A^2} \mathcal{N}\left(\frac{1 - AB}{|A|}\right) \\ &= -\mathcal{N}\left(-\frac{\widehat{M} + \nu\tau}{\sqrt{\tau}}\right) + \exp\left[\left(1 - 2\frac{(\widehat{M} + \nu\tau)/\sqrt{\tau}}{(\sigma + 2\nu)\sqrt{\tau}}\right) \frac{(\sigma + 2\nu)^2\tau}{2}\right] \\ &\quad \cdot \mathcal{N}\left(\left(1 - \frac{(\widehat{M} + \nu\tau)/\sqrt{\tau}}{(\sigma + 2\nu)\sqrt{\tau}}\right)(\sigma + 2\nu)\sqrt{\tau}\right) \\ &= -\mathcal{N}\left(-\frac{\widehat{M} + \nu\tau}{\sqrt{\tau}}\right) + \mathrm{e}^{\frac{\sigma}{2}(\sigma+2\nu)\tau - (\sigma+2\nu)\widehat{M}} \mathcal{N}\left(\frac{-\widehat{M} + (\sigma + \nu)\tau}{\sqrt{\tau}}\right) \\ &= -\mathcal{N}\left(-d_-\left(\frac{M}{S}, \tau\right)\right) + \mathrm{e}^{(r-q)\tau} \left(\frac{S}{M}\right)^{\frac{2(r-q)}{\sigma^2}} \mathcal{N}\left(d_+\left(\frac{S}{M}, \tau\right)\right). \end{aligned}$$

Substituting this expression into (3.179) and summing the resulting expression with the two equal expressions in (3.178) gives the left-hand side integral in (3.177):

$$\begin{aligned} \int_{\widehat{M}}^{\infty} \mathrm{e}^{\sigma w} \widetilde{f}_{M^X(\tau)}(w)\, \mathrm{d}w = 2\mathrm{e}^{(r-q)\tau} \mathcal{N}\left(d_+(\frac{S}{M}, \tau)\right) - \left[1 - \frac{\sigma^2}{2(r-q)}\right] \Bigg[\mathrm{e}^{(r-q)\tau} \mathcal{N}\left(d_+\left(\frac{S}{M}, \tau\right)\right) \\ - \left(\frac{M}{S}\right)^{\frac{2(r-q)}{\sigma^2}} \mathcal{N}\left(-d_-\left(\frac{M}{S}, \tau\right)\right)\Bigg]. \end{aligned} \tag{3.180}$$

Finally, by inserting this expression and the CDF in (3.176) into (3.171) and cancelling out two terms gives the pricing function for $r - q \neq 0$:

$$\begin{aligned}P^{LFS}(t,S,M) &= M\mathrm{e}^{-r\tau}\,\widetilde{F}_{M^X(\tau)}(\widehat{M}) - \mathrm{e}^{-q\tau}S + \mathrm{e}^{-r\tau}S\int_{\widehat{M}}^{\infty}\mathrm{e}^{\sigma w}\,\mathrm{d}\widetilde{F}_{M^X(\tau)}(w)\\ &= \mathrm{e}^{-q\tau}S\left[1+\frac{\sigma^2}{2(r-q)}\right]\mathcal{N}\left(d_+(\frac{S}{M},\tau)\right) + M\mathrm{e}^{-r\tau}\mathcal{N}\left(-d_-(\frac{S}{M},\tau)\right)\\ &\quad - \frac{\sigma^2}{2(r-q)}\mathrm{e}^{-r\tau}S\left(\frac{M}{S}\right)^{\frac{2(r-q)}{\sigma^2}}\mathcal{N}\left(-d_-(\frac{M}{S},\tau)\right) - \mathrm{e}^{-q\tau}S\,. \end{aligned} \tag{3.181}$$

The pricing formula for $r - q = 0$, i.e., when $r = q$, follows by taking the limit $r - q \to 0$. We leave it as a simple exercise in calculus (using L'Hôpital's Rule) to show that the sum of the two terms in $\frac{\sigma^2}{2(r-q)}$ cancel out when $r - q \to 0$. Then, using $r = q$, the final expression for the pricing function in case $r - q = 0$ simplifies to

$$\begin{aligned}P^{LFS}(t,S,M) &= \mathrm{e}^{-r\tau}\left[M\mathcal{N}\left(\frac{\ln\frac{M}{S}+\frac{1}{2}\sigma^2\tau}{\sigma\sqrt{\tau}}\right) - S\mathcal{N}\left(\frac{\ln\frac{M}{S}-\frac{1}{2}\sigma^2\tau}{\sigma\sqrt{\tau}}\right)\right]\\ &\quad + \sigma\sqrt{\tau}\mathrm{e}^{-r\tau}S\left[d_+(\frac{S}{M},\tau)\mathcal{N}\left(d_+(\frac{S}{M},\tau)\right) + n\left(d_+(\frac{S}{M},\tau)\right)\right], \end{aligned} \tag{3.182}$$

where $d_\pm(x,\tau) := \frac{\ln x \pm \frac{1}{2}\sigma^2\tau}{\sigma\sqrt{\tau}}$. □

In the above example, if we assume that the sampling of the maximum starts at current time t, i.e., $S = M$, then the pricing formulae in (3.181) and (3.182) simplify to

$$\begin{aligned}P^{LFS}(t,S) &= \mathrm{e}^{-q\tau}S\left[1+\frac{\sigma^2}{2(r-q)}\right]\mathcal{N}\left(\frac{(r-q+\frac{1}{2}\sigma^2)}{\sigma}\sqrt{\tau}\right) - \mathrm{e}^{-q\tau}S\\ &\quad + \mathrm{e}^{-r\tau}S\left[1-\frac{\sigma^2}{2(r-q)}\right]\mathcal{N}\left(-\frac{(r-q-\frac{1}{2}\sigma^2)}{\sigma}\sqrt{\tau}\right)\end{aligned} \tag{3.183}$$

for $r \neq q$ and

$$P^{LFS}(t,S) = \mathrm{e}^{-r\tau}S\left[(2+\sigma^2\tau/2)\,\mathcal{N}\left(\sigma\sqrt{\tau}/2\right) - 1 + \sigma\sqrt{\tau}n(\sigma\sqrt{\tau}/2)\right] \tag{3.184}$$

for $r = q$, where we simply write $P^{LFS}(t,S) \equiv P^{LFS}(t,S,M=S)$. The pricing functions in (3.183) and (3.184) are simply linear functions in S.

The derivations of explicit pricing functions for the other lookback options are left as exercises at the end of this chapter (see Exercise 3.39).

3.5 Structural Credit Risk Models

This section examines two classical models of credit risk known in the literature as the Merton (1974) and the Black & Cox (1976) models. Credit risk concerns the possibility of financial losses due to a borrower's failure to repay a loan or meet contractual obligations. Such failure is called a *default event*.

Here, we consider the structural approach, which assumes that the value of the firm's underlying assets follows a stochastic process, such as a GBM, and that the firm is considered in default when the value of its assets falls below a certain level compared to its liabilities. In the Merton model, a default may occur only at maturity. The Black–Cox model allows the firm to default prior to maturity.

3.5.1 The Merton Model

Let us assume that a firm is financed by bonds called *debt* and shares called *equity*. The time-t values of debt and equity are, respectively, denoted by $D(t)$ and $E(t)$. Let $A(t)$ be the total value of the firm's assets at time t. It is given by the balance sheet equation $A(t) = D(t) + E(t)$. Let the risk-neutral dynamics of $A(t)$ be described by a GBM:

$$\mathrm{d}A(t) = rA(t)\,\mathrm{d}t + \sigma A(t)\,\mathrm{d}\widetilde{W}(t) \text{ with } t \geqslant 0 \text{ and } A(0) > 0.$$

In addition, we assume that the firm's outstanding liability consists of a single zero-coupon bond with face value $F > 0$ and maturity $T > 0$.

If the total value of the firm's assets, $A(T)$, is not less than the firm's liabilities at maturity (i.e., if $A(T) \geqslant F$), the firm's debt will be paid in full, and the remainder will be distributed into the equity. If $A(T) < F$ holds, we say the firm defaults and the total value of the firm's assets at time T is distributed into the debt, whereas the equity receives nothing. Therefore, the debt and equity values at time T are, respectively,

$$D(T) = \min(F, A(T)) = F - (F - A(T))^+, \tag{3.185}$$

$$E(T) = \max(A(T) - F, 0) = (A(T) - F)^+. \tag{3.186}$$

As we see, the time-T debt and equity values can be expressed in terms of standard put and call payoffs written on the value $A(T)$.

The no-arbitrage time-t debt and equity values, $D(t) \equiv D(t, A(t))$ and $E(t) \equiv E(t, A(t))$, can be found by calculating discounted risk-neutral mathematical expectations of the values in (3.185) and (3.186), respectively:

$$\begin{aligned} D(t) &= \mathrm{e}^{-r(T-t)}\widetilde{\mathrm{E}}[D(T) \mid \mathcal{F}_t] = \mathrm{e}^{-r(T-t)}\widetilde{\mathrm{E}}\big[F - (F - A(T))^+ \mid \mathcal{F}_t\big] \\ &= \mathrm{e}^{-r(T-t)}F - P(t, A(t); T, F), \end{aligned} \tag{3.187}$$

$$\begin{aligned} E(t) &= \mathrm{e}^{-r(T-t)}\widetilde{\mathrm{E}}[E(T) \mid \mathcal{F}_t] = \mathrm{e}^{-r(T-t)}\widetilde{\mathrm{E}}\big[(A(T) - F)^+ \mid \mathcal{F}_t\big] \\ &= C(t, A(t); T, F). \end{aligned} \tag{3.188}$$

Here, $P(t, A(t); T, F)$ and $C(t, A(t); T, F)$, respectively, denote the Black–Scholes pricing functions for standard put and call options written on A with maturity T and strike price F. Using results from early sections, we can derive the following formulae for the firm's debt and equity values at time t under the Merton model:

$$D(t) = \mathrm{e}^{-r(T-t)}F\mathcal{N}(d_-) + A(t)\mathcal{N}(-d_+), \tag{3.189}$$

$$E(t) = A(t)\mathcal{N}(d_+) - \mathrm{e}^{-r(T-t)}F\mathcal{N}(d_-). \tag{3.190}$$

with $d_\pm \equiv d_\pm\left(\frac{A}{F}, T - t\right) = \frac{\ln(A/F) + (r \pm \sigma^2/2)(T - t)}{\sigma\sqrt{T - t}}$. Observe that $D(t) + E(t) = A(t)\big(\mathcal{N}(-d_+) + \mathcal{N}(d_+)\big) = A(t)$, as expected.

The Merton model allows for computing the probability of default (at time T):

$$\widetilde{\mathbb{P}}(\text{Default}) \equiv \widetilde{P}_D = \widetilde{\mathbb{P}}(A(T) < F) = \mathcal{N}(-d_-).$$

Assuming that the expected return of the assets is μ, we can also compute in Merton's model the probability of default in the real-world measure as follows:

$$\mathbb{P}(\text{Default}) \equiv P_D = \mathbb{P}(A(T) < F) = \mathcal{N}(-\hat{d}_-),$$

where $\hat{d}_- := \dfrac{\ln(A/F) + (\mu - \sigma^2/2)(T-t)}{\sigma\sqrt{T-t}}$. It is easy to show the following relationship between the risk-neutral and the real-world probabilities of default:

$$\widetilde{P}_D = \mathcal{N}\left(\mathcal{N}^{-1}(P_D) + \frac{\mu - r}{\sigma}\sqrt{T-t}\right).$$

Example 3.13. Consider a standard European call option on the firm's equity with maturity $T_1 \in (0, T)$ and strike price $K > 0$. Find the no-arbitrage price of this option, $C(t, A)$, as a function of time t, with $0 \leqslant t < T_1 < T$, and firm's time-t value $A = A(t)$.

Solution. Since $E(T_1) = C(T_1, A(T_1); T, F)$, the time-$T_1$ value of the call option on the firm's equity is

$$C(T_1, A(T_1)) = \left(C(T_1, A(T_1); T, F) - K\right)^+ = \left(\mathrm{e}^{-r(T-T_1)}\widetilde{\mathrm{E}}\left[(A(T) - F)^+ \mid \mathcal{F}_{T_1}\right] - K\right)^+.$$

The no-arbitrage value of the option at time t is therefore

$$\begin{aligned} C(t, A) &= \mathrm{e}^{-r(T_1-t)}\widetilde{\mathrm{E}}\left[(C(T_1, A(T_1); T, F) - K)^+ \mid A(t) = A\right] \\ &= \mathrm{e}^{-r(T_1-t)}\widetilde{\mathrm{E}}\left[\left(\mathrm{e}^{-r(T-T_1)}\widetilde{\mathrm{E}}\left[(A(T) - F)^+ \mid \mathcal{F}_{T_1}\right] - K\right)^+ \,\middle|\, A(t) = A\right]. \end{aligned}$$

Clearly, it is the price of a compound call-on-a-call option written on the firm's value $A(t)$ with the risk-neutral dynamics governed by the SDE $\mathrm{d}A(t) = rA(t)\,\mathrm{d}t + \sigma A(t)\,\mathrm{d}\widetilde{W}(t)$. By using (3.111), we can find the following explicit formula for the call on firm's equity:

$$C(t, A) = A\mathcal{N}_2\big(a_+, b_+; \rho\big) - F\mathrm{e}^{-r(T-t)}\mathcal{N}_2\big(a_-, b_-; \rho\big) - K\mathrm{e}^{-r(T_1-t)}\mathcal{N}(a_-), \tag{3.191}$$

where

$$\begin{aligned} a_\pm &= d_\pm\left(\frac{A}{A_1^*}, T_1 - t\right) = \frac{\ln(A/A_1^*) + (r \pm \sigma^2/2)(T_1 - t)}{\sigma\sqrt{T_1 - t}}, \\ b_\pm &= d_\pm\left(\frac{A}{F}, T - t\right) = \frac{\ln(A/F) + (r \pm \sigma^2/2)(T - t)}{\sigma\sqrt{T - t}}, \\ \rho &= \sqrt{\frac{T_1 - t}{T - t}}. \end{aligned}$$

The value A_1^* is a unique root of the equation $C(T_1, A_1; T, F) = K$. □

3.5.2 The Black–Cox Model

The Black–Cox model extends the Merton model in many respects. First, the Black–Cox model allows the default to occur prior to the maturity T of the debt. The creditors take over the borrowing firm when its value falls below a certain threshold, $L(t) = F\mathrm{e}^{-\gamma(T-t)}$ with $\gamma \geqslant r$, where $T - t$ is the time to maturity. Since the firm is in a good situation at time $t = 0$, we have that $A(0) > L(0)$ with $L(0) = F\mathrm{e}^{-\gamma T}$.

This approach makes the default time $\mathcal{T}_D$ uncertain. It is modeled as the first passage time:

$$\mathcal{T}_D = \inf\{t \in [0, T] \,:\, A(t) \leqslant L(t)\}.$$

We set $\mathcal{T}_D = \infty$ (no default occurs) if $\min_{0 \leqslant t \leqslant T}(A(t)/L(t)) > 1$ holds.

Let us find the risk-neutral probability of default before time $t \leqslant T$. Using the strong solution $A(t) = A(0)\mathrm{e}^{(r-\sigma^2/2)t+\sigma\widetilde{W}(t)}$, we can compute the probability that the firm is not in default by time t as follows:

$$\begin{aligned}
\widetilde{\mathbb{P}}(\mathcal{T}_D \geqslant t) &= \widetilde{\mathbb{P}}\big(A(u) \geqslant L(u) \text{ for all } u \in [0,t]\big) \\
&= \widetilde{\mathbb{P}}\left(\ln A(0) + (r-\sigma^2/2)u + \sigma\widetilde{W}(u) \geqslant \ln F - \gamma(T-u) \text{ for all } u \in [0,t]\right) \\
&= \widetilde{\mathbb{P}}\left((r-\gamma-\sigma^2/2)u + \sigma\widetilde{W}(u) \geqslant \ln F - \ln A(0) - \gamma T \text{ for all } u \in [0,t]\right) \\
&= \widetilde{\mathbb{P}}\big(m^X(t) \geqslant m\big),
\end{aligned}$$

where $X(t) \equiv W^{a,1}(t) := at + \widetilde{W}(t)$ with $a := \frac{r-\gamma-\sigma^2/2}{\sigma}$, $m := \frac{1}{\sigma}\ln\left(\frac{L(0)}{A(0)}\right) < 0$, and $m^X(t)$ denotes the sampled minimum of the process $\{X(u)\}_{0\leqslant u\leqslant t}$.

Using the joint distribution function for m^X and X given in (1.119) (or, equivalently, the CDF of the first hitting time down to level $m < 0$ given in (1.120)), we obtain

$$\begin{aligned}
\widetilde{\mathbb{P}}(\mathcal{T}_D < t) &= 1 - \widetilde{\mathbb{P}}(\mathcal{T}_D \geqslant t) = 1 - \widetilde{\mathbb{P}}(m^X(t) \geqslant m) = 1 - \widetilde{\mathbb{P}}(m^X(t) \geqslant m;\, X(t) \geqslant m) \\
&= \mathcal{N}\left(\frac{m-at}{\sqrt{t}}\right) + \mathrm{e}^{2am}\,\mathcal{N}\left(\frac{m+at}{\sqrt{t}}\right) \\
&= \mathcal{N}\left(\frac{\ln\frac{L(0)}{A(0)} - (r-\gamma-\sigma^2/2)t}{\sigma\sqrt{t}}\right) \\
&\quad + \left(\ln\frac{L(0)}{A(0)}\right)^{2\left(\frac{r-\gamma-\sigma^2/2}{\sigma^2}\right)} \mathcal{N}\left(\frac{\ln\frac{L(0)}{A(0)} + (r-\gamma-\sigma^2/2)t}{\sigma\sqrt{t}}\right).
\end{aligned} \tag{3.192}$$

The payoff to the debt holder at the maturity T is given by

$$\begin{aligned}
D(T) &= F\,\mathbb{I}_{\{\mathcal{T}_D \geqslant T\}} + \mathrm{e}^{r(T-\mathcal{T}_D)}A(\mathcal{T}_D)\mathbb{I}_{\{\mathcal{T}_D < T\}} \\
&= F\,\mathbb{I}_{\{\mathcal{T}_D \geqslant T\}} + F\mathrm{e}^{(r-\gamma)(T-\mathcal{T}_D)}\mathbb{I}_{\{\mathcal{T}_D < T\}} \\
&= F - F\left(1 - \mathrm{e}^{(r-\gamma)(T-\mathcal{T}_D)}\right)\mathbb{I}_{\{\mathcal{T}_D < T\}},
\end{aligned} \tag{3.193}$$

since $A(\mathcal{T}_D) = L(\mathcal{T}_D) = F\mathrm{e}^{-\gamma(T-\mathcal{T}_D)}$ and $\mathbb{I}_E = 1 - \mathbb{I}_{E^\complement}$. By using the balance sheet equation, we can also find the firm's equity value at maturity T

$$E(T) = A(T) - D(T) = A(T) - F\,\mathbb{I}_{\{\mathcal{T}_D \geqslant T\}} - F\mathrm{e}^{(r-\gamma)(T-\mathcal{T}_D)}\mathbb{I}_{\{\mathcal{T}_D < T\}}.$$

Note that the second term in (3.193), namely, $F\left(1 - \mathrm{e}^{(r-\gamma)(T-\mathcal{T}_D)}\right)\mathbb{I}_{\{\mathcal{T}_D < T\}}$, can be considered as a payoff of a down-and-in option that pays the amount of $F\left(1 - \mathrm{e}^{(r-\gamma)(T-\mathcal{T}_D)}\right)$ at maturity T if the firm defaults early before time T.

Let us find a pricing formula for the firm's debt $D(0)$ at time $t = 0$. To calculate the discounted risk-neutral expectation of $D(T)$ at time $t = 0$, we only need to know the PDF of $\mathcal{T}_D$ for times $t \in [0, T]$. Indeed, we have

$$\begin{aligned}
D(0) &= \mathrm{e}^{-rT}\widetilde{\mathrm{E}}[D(T)] = \mathrm{e}^{-rT}\widetilde{\mathrm{E}}\left[F - F\left(1 - \mathrm{e}^{(r-\gamma)(T-\mathcal{T}_D)}\right)\mathbb{I}_{\{\mathcal{T}_D < T\}}\right] \\
&= \mathrm{e}^{-rT}F - \mathrm{e}^{-rT}F\,\widetilde{\mathrm{E}}\left[\mathbb{I}_{\{\mathcal{T}_D < T\}}\right] + \mathrm{e}^{-\gamma T}F\,\widetilde{\mathrm{E}}\left[\mathrm{e}^{-(r-\gamma)\mathcal{T}_D}\mathbb{I}_{\{\mathcal{T}_D < T\}}\right] \\
&= \mathrm{e}^{-rT}F - \mathrm{e}^{-rT}F\,\widetilde{\mathbb{P}}(\mathcal{T}_D < T) + \mathrm{e}^{-\gamma T}F\int_0^T \mathrm{e}^{(\gamma-r)t}\,\frac{\mathrm{d}\widetilde{\mathbb{P}}(\mathcal{T}_D < t)}{\mathrm{d}t}\,\mathrm{d}t.
\end{aligned}$$

To compute the last integral in the above expression, we find the PDF of $\mathcal{T}_D$ by differentiating the CDF in (3.192):

$$\frac{\mathrm{d}\widetilde{\mathbb{P}}(\mathcal{T}_D < t)}{\mathrm{d}t} = n\left(\frac{\ln\frac{L(0)}{A(0)} - (r-\gamma-\sigma^2/2)t}{\sigma\sqrt{t}}\right)c_-(t) + \left(\ln\frac{L(0)}{A(0)}\right)^{2\left(\frac{r-\gamma-\sigma^2/2}{\sigma^2}\right)} n\left(\frac{\ln\frac{L(0)}{A(0)} + (r-\gamma-\sigma^2/2)t}{\sigma\sqrt{t}}\right)c_+(t),$$

where $c_\pm(t) := \frac{\ln\frac{L(0)}{A(0)} \pm (r-\gamma-\sigma^2/2)t}{2\sigma\sqrt{t^3}}$. The integral $\int_0^T e^{(\gamma-r)t}\frac{\mathrm{d}\widetilde{\mathbb{P}}(\mathcal{T}_D < t)}{\mathrm{d}t}\,dt$ can now be calculated numerically if $r \neq \gamma$. For the special case when $r = \gamma$, we have that

$$D(0) = e^{-rT}\widetilde{\mathrm{E}}[D(T)] = e^{-rT}F.$$

3.6 Exercises

Exercise 3.1. Assume the standard Black–Scholes model in an economy with constant continuously compounded interest rate r and with stock price process $\{S(t)\}_{t\geqslant 0}$ as a geometric Brownian motion (GBM) with constant volatility $\sigma > 0$ and constant continuous dividend yield q on the stock. Fix $S(0) = S > 0$ and time $T > 0$. Let the process $\{e^{(q-r)t}S(t)\}_{t\geqslant 0}$ be a $\widetilde{\mathbb{P}}$-martingale. Derive explicit analytical expressions for the risk-neutral probability of the following events:

(a) $\{S(T) < K\}$, with constant $K > 0$;

(b) $\{K_1 < S(T) < K_2\}$, with constants $K_2 > K_1 > 0$;

(c) $\{1/S^2(T) > K\}$, with constant $K > 0$;

(d) $\{S^\alpha(T_2) > S^\beta(T_1)\}$, with times $T_2 > T_1 > 0$ and constants $\alpha, \beta \neq 0$;

(e) $\{S(T_2) > S(T_1) > S(t)\}$, with times $T_2 > T_1 > t > 0$;

(f) $\{S(T_1) < K_1, S(T_2) > K_2\}$ for any $K_1, K_2 > 0$;

(g) $\{|S^\alpha(T) - K| \leqslant X\}$ with $0 < X < K$ and $\alpha \neq 0$.

Exercise 3.2. Assume the standard Black–Scholes model and stock price process as in Exercise 3.1. Derive the no-arbitrage pricing formula for the European-style option with the corresponding payoff functions (a)–(c), where $K > 0$ is a fixed strike and $a > 0$ is a constant. Express your answer in terms of the spot S and the time to maturity.

(a) $\Lambda(S) = a(S-K)^2$;

(b) $\Lambda(S) = (aS^2 - K)^+$;

(c) $\Lambda(S) = a|S^\alpha - K|$ with nonzero real constant α;

(d) $\Lambda(S) = a\left((S-K)^+\right)^2$.

Exercise 3.3. Assume the standard Black–Scholes model and stock price process as in Exercise 3.1. Consider a European option with payoff at expiry T:

$$\Lambda(S(T)) = \begin{cases} (S(T) - K_1)^+ & 0 \leqslant S(T) \leqslant X_1, \\ X_1 - K_1 & X_1 \leqslant S(T) \leqslant X_2, \\ (K_2 - S(T))^+ & S(T) \geqslant X_2. \end{cases}$$

where $K_1 < X_1 < X_2 < K_2$ and $X_2 = K_1 + K_2 - X_1$.

(a) Give a sketch of this payoff function and determine a replicating portfolio for $\Lambda(S(T))$ that consists of only calls or puts.

(b) Let $S(t) = S$ be the spot price of the stock at current time $t < T$. Derive the risk-neutral pricing formula for the time-t value of a European-style option with the above payoff function. Give an explicit answer in terms of all parameters in the model.

(c) Obtain a formula for the delta position in the stock at time $t < T$ that is required in a self-financing replicating strategy for the option.

Exercise 3.4. Assume the standard Black–Scholes model and stock price process as in Exercise 3.1. Let $S(t) = S > 0$ be the spot at time $t < T$, where T is the expiry date. Derive the corresponding arbitrage-free time-t pricing formula, $V(t, S)$, for a European option with the respective payoffs in (a) and (b) below.

(a) $\Lambda(S(T)) = \sum_{n=0}^{N} a_n S^n(T)$, $N \geqslant 1$, where a_n are real constant coefficients of the polynomial function.

(b) $\Lambda(S(T)) = \big(S^\alpha(T) - K\big)\mathbb{I}_{\{S(T)>K\}}$ where α is any nonzero real constant.

Exercise 3.5. Assume the standard Black–Scholes model and stock price process as in Exercise 3.1. A European *call spread* has payoff $\Lambda(S(T))$ equal to zero for $S(T) \leqslant K$, $S(T) - K$ for $K < S(T) < K + \epsilon$, and ϵ for $S(T) \geqslant K + \epsilon$, where K, ϵ are any positive values.

(a) Give a sketch of the payoff function.

(b) Derive a formula for the option's present value $V(t, S)$ and $\Delta(t, S) = \frac{\partial V}{\partial S}$. Express your answers in terms of spot S, time to maturity $T - t$, and parameters K,ϵ,r,q,σ.

(c) Find $V(t, S)$ in both limits $\epsilon \searrow 0$ and $\epsilon \to \infty$ and explain your results.

Exercise 3.6. Let $0 < K_1 < K_2 < K_3 < K_4 < K_5 < K_6$ and consider the payoff function:

$$\Lambda(S) = \begin{cases} (S - K_1)^+ & 0 \leqslant S \leqslant K_2, \\ K_2 - K_1 & K_2 \leqslant S \leqslant K_3, \\ K_2 - K_1 - (S - K_3) & K_3 \leqslant S \leqslant K_4, \\ K_2 - K_1 - (K_4 - K_3) & K_4 \leqslant S \leqslant K_5, \\ -(K_6 - S)^+ & S \geqslant K_5. \end{cases}$$

where we assume $K_4 - K_3 > K_2 - K_1$ and $K_2 - K_1 - (K_4 - K_3) = K_5 - K_6$.

(a) Give a sketch of this payoff function.

(b) Determine a replicating portfolio for $\Lambda(S)$ consisting of only calls (or puts, cash, and stock positions).

(c) Assuming a Black–Scholes economy as in Exercise 3.1, derive the no-arbitrage pricing formula for the European-style option with the above payoff.

Exercise 3.7. A so-called *range forward* European contract is specified as follows: at maturity T the holder must buy the underlying stock at price K_1 if $S(T) < K_1$, at price $S(T)$ if $K_1 \leqslant S(T) \leqslant K_2$, and at price K_2 for $S(T) > K_2$ where $K_1 < K_2$ are fixed strikes.

(a) Derive the explicit formula for the present value $V(t, S)$ of this contract with spot $S(t) = S$. Assume a Black–Scholes economy as in Exercise 3.1.

(b) Find the relationship between K_1 and K_2 such that the present value $V(t, S) = 0$.

Exercise 3.8. Consider a so-called *strangle* payoff function $\Lambda(S)$ defined by two strikes $0 < K_1 < K_2$:

$$\Lambda(S) = \begin{cases} K_2 - S & 0 \leqslant S \leqslant K_1, \\ K_2 - K_1 & K_1 \leqslant S \leqslant K_2, \\ S - K_1 & S \geqslant K_2. \end{cases}$$

(a) Give a sketch of $\Lambda(S)$.

(b) Determine a replicating portfolio for $\Lambda(S)$ in terms of positions in standard calls, stock, and cash.

(c) Assume the stock price $\{S(t)\}_{t \geqslant 0}$ is a GBM with constant volatility σ, constant continuous dividend q in an economy with constant interest rate r. Derive the arbitrage-free pricing formula for the present time $t < T$ value of a European-style derivative with the above payoff function.

(d) Provide an expression for the value, at calendar time $t < T$, of the position in the stock in the self-financed portfolio required to dynamically replicate the option value.

Exercise 3.9. Assume the standard Black–Scholes model and stock price process as in Exercise 3.1. Consider a spread with the payoff function as plotted below.

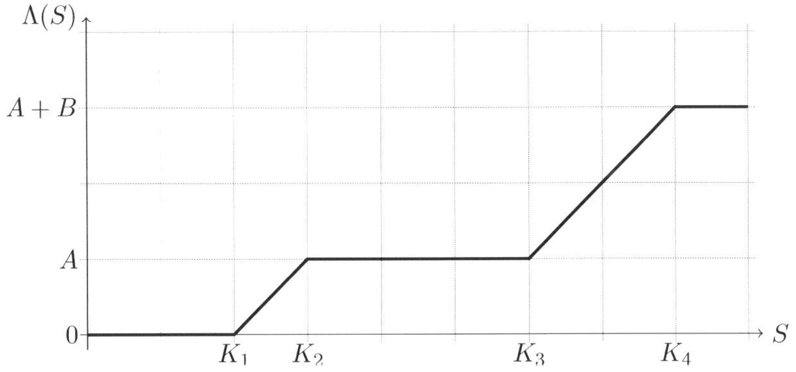

Here, $K_1, K_2, K_3, K_4 > 0$, $A = K_2 - K_1 > 0$, and $B = K_4 - K_3 > 0$.

(a) Write down the payoff $\Lambda(S)$ as a piece-wise function of spot S and determine a replicating portfolio for $\Lambda(S)$ that consists of only calls.

(b) Derive the pricing function $V(t, S)$ for time $t \in [0, T)$ and spot $S > 0$ of a European derivative with payoff $\Lambda(S(T))$ in terms of standard call values $C(t, S; T, K_i)$, $i = 1, 2, 3, 4$.

(c) Obtain a formula for the delta position $\Delta_V(t, S)$.

Exercise 3.10. Consider the spread from Exercise 3.9.

(a) Determine the initial no-arbitrage value and delta of the spread, if $K_1 = 95$, $K_2 = 100$, $K_3 = 105$, $K_4 = 110$, $S_0 = 100$, $r = 5\%$, $q = 0\%$, $\sigma = 20\%$, $T = 3/12$.

(b) Compute the approximate value of this spread at time $t = 1/12$ using the Δ-approximation if the stock price has raised up to $S(1/12) = 105$.

(c) At time 0, construct a delta-neutral portfolio that includes 10 short spreads and has the initial value of zero.

Exercise 3.11. Assume the standard Black–Scholes model and stock price process as in Exercise 3.1. Consider a European-style derivative with expiry $T > 0$ and payoff

$$\Lambda(S) = \begin{cases} K - S & \text{for } 0 < S < K - X, \\ X & \text{for } K - X \leqslant S < K, \\ K + X - S & \text{for } K \leqslant S < K + X, \\ 0 & \text{for } S \geqslant K + X, \end{cases}$$

where K and X are positive constants s.t. $X < K$.

(a) Give a sketch of the payoff function and determine a replicating portfolio for $\Lambda(S)$ that consists of only puts.

(b) Derive the pricing function $V(t, S)$ for time $t \in [0, T)$ and spot $S > 0$ of a European-style derivative with payoff $\Lambda(S(T))$ in terms of standard put values.

(c) Find the limiting value of $V(t, S)$ as $X \to 0$.

(d) Prove that for any t, S, K, X, and T, there exists M s.t. $V(t, S)$ equals the value $P(t, S; T, M)$ of a standard European put with strike M and maturity T.

Exercise 3.12. Assume the standard Black–Scholes model and stock price process as in Exercise 3.1. Consider European-style derivatives with payoffs sketched in Figure 3.15. For each derivative, do the following.

(i) Write down the payoff $\Lambda(S)$ as a piece-wise function of S and determine a replicating portfolio for $\Lambda(S)$ that consists of only calls (no puts), stock, and cash.

(ii) Let $S(t) = S$ be the spot price of the stock at current time $t \geqslant 0$. Derive a risk-neutral pricing formula for the time-t no-arbitrage value of the European derivative with payoff $\Lambda(S)$ and maturity $T > t$.

Exercise 3.13. A so-called *pay-later* European option costs the holder nothing (i.e., zero premium) to set up at present time $t = 0$. The payoff to the holder at maturity $T > 0$ is $(S(T) - K)^+$. Moreover, the holder must *pay out* X dollars to the writer in the case that $S(T) \geqslant K$. Derive an expression for the fair value of X. Determine the fair value for X in the limit of infinite volatility, $\sigma \to \infty$. Assume a Black–Scholes economy as in Exercise 3.1.

Exercise 3.14. Let $C(S, \tau)$ be the Black–Scholes pricing formula of a standard call option with spot S, strike K, fixed interest rate r, zero stock dividend, constant volatility σ, and time to maturity $\tau > 0$.

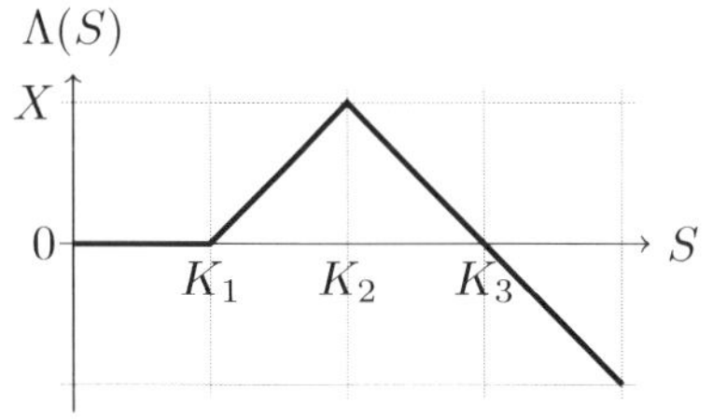

(a) A front spread payoff with $X = K_2 - K_1 = K_3 - K_2$.

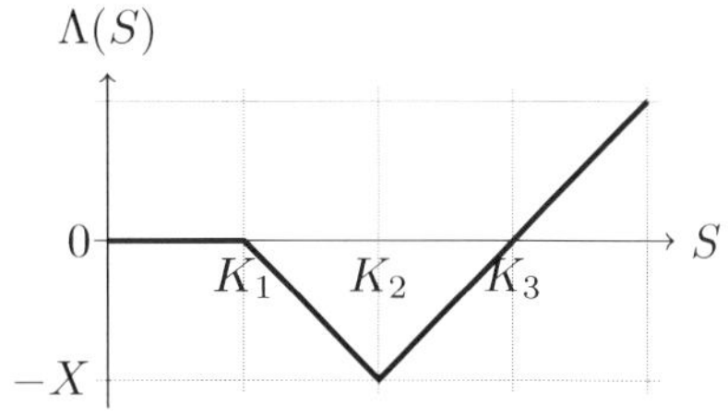

(b) A back spread payoff with $X = K_2 - K_1 = K_3 - K_2$.

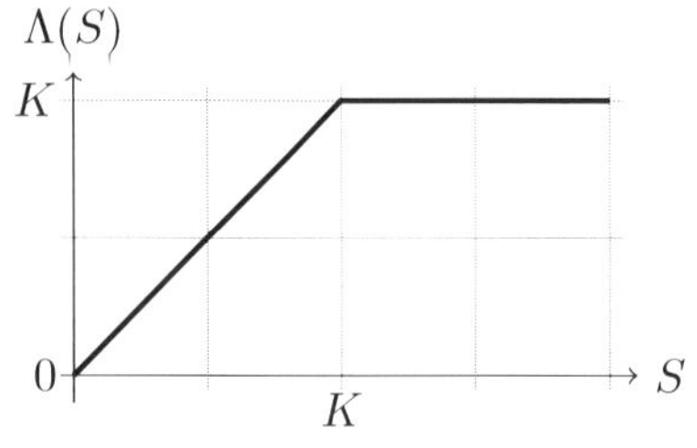

(c) A covered call payoff.

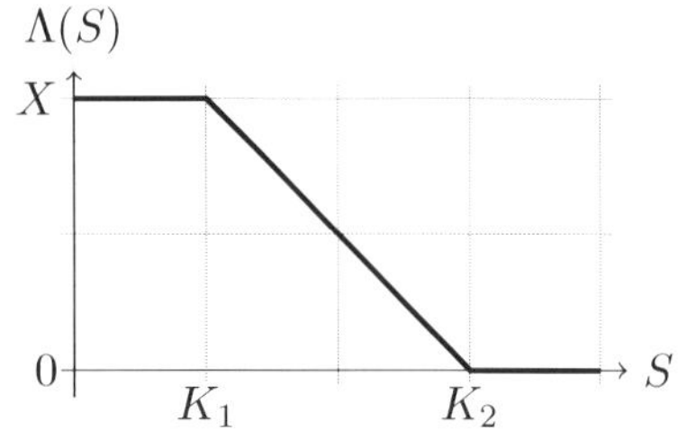

(d) A bear spread payoff with $X = K_2 - K_1$.

FIGURE 3.15: Payoff diagrams for Exercise 3.12.

(a) Show that the respective limiting values of the call price for vanishing and infinite volatility are given by

$$\lim_{\sigma \searrow 0} C(S,\tau) = \left(S - \mathrm{e}^{-r\tau} K\right)^+ \quad \text{and} \quad \lim_{\sigma \to \infty} C(S,\tau) = S.$$

(b) Give a financial interpretation of both limits. Note that the second limit is independent of the strike value K; give a financial intuition of this fact.

Exercise 3.15. Consider the value of a European call option written by an issuer who only has a fraction $0 \leqslant \alpha < 1$ of the underlying asset. That is, at expiration time T the payoff of this type of call is given by

$$V_T = (S(T) - K)^+ \,\mathbb{I}_{\{\alpha S(T) \geqslant S(T) - K\}} + \alpha S(T)\,\mathbb{I}_{\{\alpha S(T) < S(T) - K\}}\,.$$

Let $C_L(S,\tau;K,\alpha)$ denote the value of such a European call, where $\tau = T - t$ is the time to expiry, $K > 0$ is the strike, $S(t) = S$ is the spot of the underlying. Show that

$$C_L(S,\tau;K,\alpha) = C(S,K,\tau) - (1-\alpha)C\big(S, \frac{K}{1-\alpha}, \tau\big)$$

where $C\big(S, \frac{K}{1-\alpha}, \tau\big)$ is the price of a standard European call with strike $\frac{K}{1-\alpha}$, spot S, and time to expiry τ.
Note: you should not assume any model for the stock price process and therefore you need not provide any explicit formulae for any of the call price functions.

Exercise 3.16. Prove the following *invariance* property:

$$V_{BS}(t, aS; T, K) = aV_{BS}(t, S; T, K/a) \text{ with any } a > 0$$

where V_{BS} is the Black–Scholes price of a standard call or put option with expiry $T > t$, strike K, and sport S. You can prove the property by using the Black–Scholes pricing formula, the Black–Scholes PDE, or the expectation formula with the strong solution for $S(T)$. Explain what happens to the option value in a one-for-two stock split when a holder of one share before the split will hold two shares after the split.

Exercise 3.17. The price $V = V(t, S)$ of a standard European-style derivative satisfies the Black–Scholes PDE

$$\frac{\partial V}{\partial t} + \frac{1}{2}\sigma^2 S^2 \frac{\partial^2 V}{\partial S^2} + (r-q)S\frac{\partial V}{\partial S} - rV = 0, \quad 0 \leqslant t < T, \quad 0 < S < \infty,$$

subject to the terminal (payoff) condition at maturity

$$V(T, S) = \Lambda(S), \quad 0 < S < \infty.$$

(a) Show that the BSPDE reduces to a classical heat equation in spatial variable x and time variable τ,

$$\frac{\partial u(\tau, x)}{\partial \tau} = \frac{\partial^2 u(\tau, x)}{\partial x^2}, \quad 0 < \tau, \quad -\infty < x < \infty,$$

upon applying the following transformations:

$$\begin{aligned} S &= K\,\mathrm{e}^{x}, \quad t = T - \tfrac{\tau}{\sigma^2/2}, \\ V(t, S) &= K\,\mathrm{e}^{-\gamma x - (\beta^2 + \ell)\tau} u(\tau, x), \end{aligned}$$

where the constants γ, β, and ℓ are defined as follows:

$$\begin{aligned} \gamma &= \tfrac{1}{2}(\kappa - 1), \quad & \beta &= \tfrac{1}{2}(\kappa + 1), \\ \ell &= \tfrac{q}{\sigma^2/2}, \quad & \kappa &= \tfrac{r-q}{\sigma^2/2}. \end{aligned}$$

(b) Find the initial condition on the function $u(\tau, x)$.

Exercise 3.18. Suppose that the cost of carry on a commodity is b and assume a bank account with constant interest rate r. Let the price of the commodity follow a GBM model with constant volatility σ. Let $V = V(t, S)$ be the value of a European option on this commodity, where $S > 0$ is the spot value of the commodity at calendar time t.

(a) Show that V satisfies the Black–Scholes PDE:

$$\frac{\partial V}{\partial t} + \frac{\sigma^2}{2} S^2 \frac{\partial^2 V}{\partial S^2} + bS\frac{\partial V}{\partial S} - rV = 0\,.$$

(b) Based on (a), find the put-call parity relation between the put price $P(t, S)$ and call price $C(t, S)$, with common strike K and time to maturity $\tau = T - t$.

Exercise 3.19. Consider a portfolio Π with fixed positions θ_i in N securities each with price f_i, $i = 1, \ldots, N$, respectively. Assume the ith security has price f_i as a function of the same spot S at current time t and that each $f_i = f_i(S, T_i - t)$ satisfies the time-homogeneous Black–Scholes PDE with constant interest rate and volatility. The contract maturity dates T_i are allowed to be distinct. Find the algebraic relationship among the portfolio Greeks: $\Theta \equiv \frac{\partial \Pi}{\partial t}$, $\Delta \equiv \frac{\partial \Pi}{\partial S}$, and $\Gamma \equiv \frac{\partial^2 \Pi}{\partial S^2}$.

Exercise 3.20. Apply integration by parts twice to show that

$$\Lambda_a(S) = \frac{1}{2a}\int_{K-a}^{K+a} (S-k)^+ \, \mathrm{d}k$$

is an integral representation of the payoff in (3.56). Hence, this shows that the soft-strike call option with given central strike K and strike width $a > 0$ is replicated as a uniform superposition of standard call options of all strikes in the interval $(K-a, K+a)$.

Exercise 3.21. Recall the pricing of the soft-strike call in Example 3.6.

(a) Provide the corresponding definition of the payoff (as in (3.56)) of the soft-strike put option with central strike K and width a. Provide a plot of the payoff and describe its main features as done in Example 3.6.

(b) Derive a corresponding put-call parity relation for the soft-strike call and soft-strike put options having common central strike K and strike width a. Assuming the standard GBM model as in Example 3.6, derive the formula for $P(t, S; K, a)$, the no-arbitrage time-t price of the soft-strike put option.

Exercise 3.22. Assume the risk-neutral pricing formula in (3.24) holds. By using a general formula for the expectation of a random variable conditional on an event, in this case the event $\{S(T) > K\}$, derive the following general representations for the respective prices of a standard call and put option with strike K:

$$C(t,S) = \mathrm{e}^{-r(T-t)}\widetilde{\mathbb{P}}_{t,S}\big(S(T) > K\big) \cdot \Big(\widetilde{\mathrm{E}}_{t,S}\big[S(T) \mid S(T) > K\big] - K\Big),$$

$$P(t,S) = \mathrm{e}^{-r(T-t)}\widetilde{\mathbb{P}}_{t,S}\big(S(T) < K\big) \cdot \Big(K - \widetilde{\mathrm{E}}_{t,S}\big[S(T) \mid S(T) < K\big]\Big).$$

Give a probabilistic interpretation of these formulae.

Exercise 3.23. Assume a nondividend paying stock with price process $\{S(t)\}_{t\geqslant 0}$ as a GBM with constant volatility $\sigma > 0$ in the (B, S) economy with constant interest rate r. Let $C_t := C(t, S(t), K)$ be the price process at calendar time t, $0 \leqslant t < T$, with $C(t, S, K)$ as the pricing function of a standard European call option on the stock with given strike $K > 0$ and fixed maturity date T. It follows that C_t satisfies the SDE

$$\mathrm{d}C_t = \mu_c \, C_t \, \mathrm{d}t + \sigma_c \, C_t \, \mathrm{d}\widetilde{W}(t),$$

where $\{\widetilde{W}(t)\}_{t\geqslant 0}$ is a standard BM under the risk-neutral measure $\widetilde{\mathbb{P}}$.

(a) Find explicit expressions for μ_c and σ_c, i.e., the (log)-drift and (log)-volatility coefficient functions of the call price process.
Note: the (log)-volatility of the call price, σ_c, is a function of $S(t)$ (spot value) and parameters $K, \sigma, r, \tau = T - t$. Your expressions should be simplified as much as possible. Define all terms in your answer.

(b) Find the limiting expression for σ_c as $K \searrow 0$ (holding spot and other parameters fixed). Give a financial explanation of the resulting limit.

Exercise 3.24. Consider a stock price process as a GBM and assume *deterministic (non-random) time-dependent* volatility $\sigma(t)$, stock dividend yield $q(t)$, and interest rate $r(t)$. Assume that these are integrable functions of time $t \in [0, T]$.

(a) Derive the time-0 no-arbitrage pricing function $C(0, S_0, K)$ for the standard call option on this stock with $S(0) = S_0 > 0$ as spot, $K > 0$ as strike, and T as maturity.

(b) Derive the put-call parity relation between the time-0 values of the standard call and put prices $C(0, S_0, K)$ and $P(0, S_0, K)$.

(c) Now let the present time be any time $t \in [0, T)$ with spot $S(t) = S > 0$. Derive the time-t no-arbitrage pricing function for the standard call $C(t, S, K)$ and provide the corresponding put-call parity relation where $P(t, S, K)$ denotes the time-t pricing function for the standard put.

Exercise 3.25. A variant of the forward starting call option that we already considered in Section 3.3 is structured as follows. The holder receives at date $T_1 > t$ a call with strike $K_{T_1} = \alpha S(T_1)$ and maturity $T > T_1$. Here, α is a positive constant and $S(T_1)$ is the stock price realized at time T_1. Let the stock price be a GBM with constant interest rate r and dividend yield q.

(a) Let $S(t) = S$. Derive the time-t pricing formula $C(t, S; T_1, T)$ for this forward starting call and give a hedging strategy that applies up to time T_1. Is the strategy static or not?

(b) Show that the price of the forward starting call simplifies to that of a standard call struck at $K = \alpha S$ with time to maturity $T - t$ in the limiting case that $T_1 \to t$ (with t, T held fixed). On the other hand, show that in the limit $T \to T_1$ (with t, T_1 held fixed) the contract price is simply given by $S(1 - \alpha)^+$.

Exercise 3.26. Assume the stock price $\{S(t)\}_{t \geqslant 0}$ is a GBM with constant volatility σ and zero dividend in an economy with constant interest rate r. Let $t < T_1 < T$; i.e., T_1 is an arbitrary intermediate time before expiry time T, and consider a European-style option with payoff at time T:

$$V_T = \min\{S(T_1), S(T)\}.$$

(a) Show that the value V at any time $t \leqslant T_1$ of this option is given by

$$V = S[\mathcal{N}(-d_+) + \mathrm{e}^{-r(T-T_1)}\mathcal{N}(d_-)]$$

where $S(t) = S$ is the spot and $d_\pm := \dfrac{r(T - T_1) \pm \frac{1}{2}\sigma^2(T - T_1)}{\sigma\sqrt{T - T_1}} = \dfrac{(r \pm \frac{1}{2}\sigma^2)}{\sigma}\sqrt{T - T_1}$.

[Note: the option value is dependent on $T - T_1$ (where T and T_1 are fixed) and does not depend on t.]

(b) What is the position held in the stock at time $t \leqslant T_1$ in a self-financing replicating strategy? Is this position static over time? Justify whether or not a bank account is needed in the dynamic replication.

Exercise 3.27. Assume the standard Black–Scholes model and stock price process as in Exercise 3.1. Derive an arbitrage-free pricing formula for the forward starting option with the power payoff $V_T = |S(T) - S(T_1)|^n$ for $n = 2, 3, \ldots$ and $0 < T_1 < T$.

Exercise 3.28. Assume the standard Black–Scholes model and stock price process as in Exercise 3.1. Consider forward starting strip and strap written at time t with strike at $S(T_1)$ and expiry time T, where $0 \leqslant t < T_1 < T$. Their respective payoffs are given in Figure 3.16. For each derivative, do the following.

(a) Write the payoff as a piece-wise function of $S(T)$ and then express it using the positive part function $(x)^+$.

(b) First, characterize the derivative in terms of forward starting European call and put options, express the arbitrage-free value, $V(t,S)$, in the expectation form with standard call and put payoffs, and then find $V(t,S)$ in terms of the $\mathcal{N}$ function.

(c) What is the position, β_t, held in the bank account at time $t \in [0,T_1]$ in a self-financing replicating strategy?

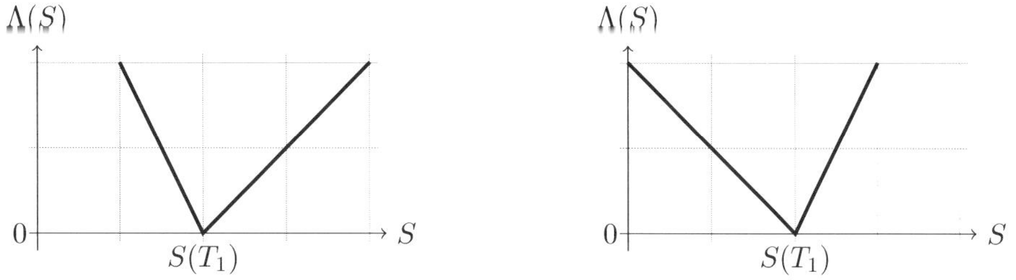

FIGURE 3.16: Payoffs of forward starting strip (left) and strap (right).

Exercise 3.29. Assume the standard Black–Scholes model and stock price process as in Exercise 3.1. Derive a put-call parity for the forward starting call and put options whose pricing functions are, respectively,

$$C(t,S) = \mathrm{e}^{-r(T-t)}\widetilde{\mathrm{E}}_{t,S}\big[(S(T)-\alpha S(T_1))^+\big] \quad \text{and} \quad P(t,S) = \mathrm{e}^{-r(T-t)}\widetilde{\mathrm{E}}_{t,S}\big[(\alpha S(T_1)-S(T))^+\big],$$

where $0 \leqslant t < T_1 < T$, $\alpha > 0$ and $S > 0$.

Exercise 3.30. Recall the derivation of the pricing function $V^{cc}(t,S)$ for the European call-on-a-call where $S(t) = S$ is the spot at calendar time $t < T_1 < T_2$. Assume the standard Black–Scholes model and stock price process as in Exercise 3.1.

(a) Using similar steps as in Section 3.3, derive the pricing formula for $V^{cp}(t,S)$, the arbitrage-free value of a *call-on-a-put*, with time-T_1 value $\big(P_{T_1} - K_1\big)^+$, where $P_{T_1} \equiv P_{T_1}(S(T_1), K_2, T_2)$ is the price of the (embedded) standard put with time to maturity $T_2 - T_1$ and strike K_2.

(b) Derive an expression for the delta position $\Delta_t = \Delta(t,S)$ in the stock at time $t < T_1$ needed to dynamically hedge the call-on-a-put option in (a).

Exercise 3.31. Assume the standard Black–Scholes model and stock price process as in Exercise 3.1. Consider a put-on-a-put option when one asset-or-nothing put written on another asset-or-nothing put. That is, at time T_1, the holder of this compound option receives the asset-or-nothing put on the underlying stock with strike K_2 and expiry T_2 (where $T_1 < T_2$) if the time-T_1 no-arbitrage value of this embedded put does not exceed K_1.

(a) Analyze the asymptotic behaviour of the time-t asset-or-nothing put value, $P(t,S)$, with maturity time T, strike K, and spot S, as $S \to 0$ or $S \to \infty$.

(b) Assume that K_1 is less than the maximum value of the embedded asset-or-nothing put with time to maturity $T_2 - T_1$ and strike K_2. Show that the arbitrage-free value, $V^{pp}(t, S)$, of this compound option at time t (where $0 \leqslant t < T_1$) is given by

$$V^{pp}(t,S) = \mathrm{e}^{-r(T_2-t)}\widetilde{\mathrm{E}}_{t,S}[S(T_2)\mathbb{I}_{\{S(T_2)\leqslant K_2\}}\left(\mathbb{I}_{\{S(T_1)\leqslant S_1^*\}} + \mathbb{I}_{\{S(T_1)\geqslant S_2^*\}}\right)]$$

for some S_1^* and S_2^* with $0 < S_1^* < S_2^*$. Explain how to compute S_1^* and S_2^*.

(c) Derive the pricing formula for $V^{pp}(t, S)$ in closed form.

Exercise 3.32. Assume the standard Black–Scholes model and stock price process as in Exercise 3.1. Consider asset-or-nothing binary call and put options with expiry T and strike K. Recall that their payoffs at expiry are $C_T = S(T)\mathbb{I}_{\{S(T)\geqslant K\}}$ and $P_T = S(T)\mathbb{I}_{\{S(T)\leqslant K\}}$, respectively. Let $C(t, S; T, K)$ and $P(t, S; T, K)$ denote their arbitrage-free values at time t. Consider a chooser option that gives its holder the right to choose at time T_1 with $T_1 < T$ between the call and put options. That it, at time T_1, the values of the chooser option is

$$V(T_1, S(T_1)) = \max\left\{C(T_1, S(T_1); T, K), P(T_1, S(T_1); T, K)\right\}.$$

(a) Derive the time-t no-arbitrage value, $V(t, S)$, of this chooser option for $0 \leqslant t < T_1$.

(b) Find the delta of the chooser option at time t.

Exercise 3.33. Assume the standard Black–Scholes model and stock price process as in Exercise 3.1. Consider asset-or-nothing call and put options with respective expiry times T_c and T_p and strike prices K_c and K_p. Let $C(t, S; T_c, K_c)$ and $P(t, S; T_p, K_p)$ denote their arbitrage-free values at time t for spot S. Consider a chooser option that gives its holder the right to choose at time T with $T < \min\{T_c, T_p\}$ between the call and put options. That it, at time T, the values of the chooser option is

$$V(T, S(T)) = \max\{C(T, S(T); T_c, K_c), P(T, S(T); T_p, K_p)\}.$$

Find the arbitrage-free value of this chooser option at time $t \in [0, T]$. Your solution can be given in terms of the bivariate normal distribution function $\mathcal{N}_2$.

Exercise 3.34. Assume the standard Black–Scholes model and stock price process as in Exercise 3.1. Derive a put-call parity for the call and put compound options written on

(a) an asset-or-nothing call option;

(b) an asset-or-nothing put option.

That is, the inner (embedded) option is the binary asset-or-nothing option on stock with expiry T_2 and strike K_2. The two outer options are standard European call and put with expiry T_1 and strike K_1 written on the underlying option.

Exercise 3.35. Consider the stock price process $\{S_{(B)}(t)\}_{t\geqslant 0}$ as a GBM that is killed at the first-hitting time $\mathcal{T}_B^S = \inf\{t \geqslant 0 : S(t) = B\}$ to level $B > 0$:

$$S_{(B)}(t) := \begin{cases} S(t) & \text{for } t < \mathcal{T}_B^S, \\ \partial^\dagger & \text{for } t \geqslant \mathcal{T}_B^S, \end{cases} \tag{3.194}$$

where $S(t)$ is given by (3.123) and $X(t)$ is given by (3.124). Note that the state space for the process is restricted to either interval $(0, B)$ or (B, ∞) corresponding to the two cases with $S(0) \in (0, B)$ or $S(0) \in (B, \infty)$, respectively.

(a) Show that the transition CDF for the process $S_{(B)}$ is given by

$$\widetilde{\mathbb{P}}(S_{(B)}(T) \leqslant y | S_{(B)}(t) = S) = \widetilde{\mathbb{P}}\left(X_{(b)}(\tau) \leqslant \frac{1}{\sigma} \ln \frac{y}{S}\right)$$

where $\tau = T - t > 0$ and $X_{(b)}$ is the BM process given by (3.124) with killing at the first-hitting time to level $b \equiv \frac{1}{\sigma} \ln \frac{B}{S}$ (see the definition in (1.106) of Chapter 1). By differentiating, obtain the (time-homogeneous) transition PDF:

$$\widetilde{p}^{S_{(B)}}(t, T; S, y) \equiv \widetilde{p}^{S_{(B)}}(\tau; S, y) = \frac{1}{\sigma y} \widetilde{p}^{X_{(b)}}(\tau; 0, x)$$

where $x = \frac{1}{\sigma} \ln \frac{y}{S}$ and $\widetilde{p}^{X_{(b)}}$ is the risk-neutral transition PDF for the killed BM process $X_{(b)}$. Then, using the density in (3.143), obtain the expression in (3.156).

(b) For a fixed y value, show that $v(\tau, S, y) := \mathrm{e}^{-r\tau} \widetilde{p}^{S_{(B)}}(\tau; S, y)$ solves the time-homogeneous BSPDE in (3.65) subject to the initial condition $v(0+, S, y) = \delta(S-y)$ and has zero boundary conditions $v(\tau, S = B, y) = 0$ and $v(\tau, S = 0+, y) = 0$ and $v(\tau, S = \infty, y) = 0$ for all $\tau > 0$ and all y.

Exercise 3.36. Using similar steps as in the derivation of the up-and-out call (and up-and-out put) pricing formula, derive explicit pricing functions and delta hedging positions for

(a) the down-and-out put and down-and-in put option;

(b) the down-and-out call and down-and-in call option.

Assume a barrier level $B > 0$, strike $K > 0$, time to maturity $\tau = T - t > 0$ and where the stock price is a GBM with constant interest rate r and stock dividend yield q.

Exercise 3.37. Assume the standard Black–Scholes model and stock price process as in Exercise 3.1. Find a *put-call parity relation* between the pricing functions for the following pairs of standard barrier options with common strike $K > 0$, barrier level $B > 0$, and time to maturity $\tau > 0$:

(a) *down-and-out* call and put options with respective pricing functions $C^{DO}(\tau, S, K; B)$ and $P^{DO}(\tau, S, K; B)$;

(b) *up-and-out* call and put options with respective pricing functions $C^{UO}(\tau, S, K; B)$ and $P^{UO}(\tau, S, K; B)$;

(c) *down-and-in* call and put options with respective pricing functions $C^{DI}(\tau, S, K; B)$ and $P^{DI}(\tau, S, K; B)$;

(d) *up-and-in* call and put options with respective pricing functions $C^{UI}(\tau, S, K; B)$ and $P^{UI}(\tau, S, K; B)$.

Exercise 3.38. Assume the standard Black–Scholes model and stock price process as in Exercise 3.1. Derive the no-arbitrage pricing formula $V(t, S) \equiv V(t, S; T_1, T_2)$ with spot $S(t) = S$, time t, and maturities T_1 and T_2 so that $t < T_1 \leqslant T_2$ for the following compound options:

(a) a call-on-a-forward compound option on the stock with outer and inner payoff functions defined respectively by $\phi^{(1)}(x) := (x - K_1)^+$ and $\phi^{(2)}(x) := x - K_2$ with $x > 0$;

(b) a binary-on-a-forward compound option on the stock with outer and inner payoff functions defined, respectively, by $\phi^{(1)}(x) := \mathbb{I}_{\{x>K_1\}}$ and $\phi^{(2)}(x) := x - K_2$ with $x > 0$.

Here, $K_1 > 0$ and $K_2 > 0$ are fixed strikes.

Exercise 3.39. Use similar steps as in Example 3.12 and make appropriate use of integral identities in Appendix B to derive explicit pricing functions for

$$\text{(i) } C^{LFS}(t, S, m), \text{ (ii) } C^{LFP}(t, S, M; K), \text{ and (iii) } P^{LFP}(t, S, m; K).$$

Assume the stock price is a GBM with constant interest rate r and dividend yield q.

Exercise 3.40. Assume the standard Black–Scholes model and stock price process as in Exercise 3.1. Consider the discrete geometric averaging of a stock price process at evenly distribute discrete times $t_j = t_0 + j\,\delta t$ with $j = 1, 2, \ldots, n$ and the time step $\delta t = (T - t_0)/n$. Define the discretely monitored geometric average by

$$G_k = \left[\prod_{j=1}^{k} S(t_j)\right]^{1/k} \quad \text{for } k = 1, 2, \ldots, n\,.$$

(a) Show that G_n is a log-normal random variable. Find the mean and variance of G_n (under the EMM $\widetilde{\mathbb{P}}$).

(b) Derive no-arbitrage time-t_0 prices of the fixed strike Asian call and put options with respective payoff functions $(G_n - K)^+$ and $(K - G_n)^+$, fixed strike $K > 0$, and maturity time T.

(c) Provide the corresponding put-call parity relation for the two fixed strike Asian options.

Exercise 3.41. Assume the standard Black–Scholes model and stock price process as in Exercise 3.1. Define the continuously monitored geometric average of $S(t)$ over a time period $[0, t]$ by

$$G(t) = \exp\left(\frac{1}{t}\int_0^t \ln S(u)\,\mathrm{d}u\right).$$

(a) Show that the process $\{\ln G(t)\}_{t\geqslant 0}$ is Gaussian.

(b) Show that $G(T)$ can be written as

$$G(T) = G(t)^{t/T} S(t)^{(T-t)/T} \exp(\bar{\mu} + \bar{\sigma}\widetilde{Z}) \text{ with } 0 \leqslant t < T$$

for some $\bar{\mu}, \bar{\sigma} \in \mathbb{R}$ and $\widetilde{Z} \sim Norm(0, 1)$ under $\widetilde{\mathbb{P}}$. Find the values of $\bar{\mu}$ and $\bar{\sigma}$. Calculate the mean and variance of $G(T)$ conditional on $G(t)$ and $S(t)$.

(c) Derive the no-arbitrage time-t pricing functions for the fixed strike Asian call and put options with respective payoff functions $(G(T) - K)^+$ and $(K - G(T))^+$ for $0 \leqslant t \leqslant T$. Express the pricing functions in terms of the spot values $G = G(t) > 0$, $S = S(t) > 0$, and times t and T.

(d) Establish the put-call parity relation for the fixed strike Asian call and put options.

Exercise 3.42. Assume the Merton model for the firm's value. Find SDEs describing the dynamics of the firm's equity $E(t)$ and debt $D(t)$ under the EMM $\widetilde{\mathbb{P}}$.

Exercise 3.43. Assume the Merton model for the firm's value. Consider a self-financing strategy consisting of the risk-free bank account $B(t)$ and the firm's debt $D(t)$. Let β_t and γ_t be the time-t positions in the bank account and the debt, respectively. Construct a self-financing strategy that hedges the firm's equity $E(t)$. That is, $\Pi_t := \beta_t B(t) + \gamma_t D(t) = E(t)$ and $\mathrm{d}\Pi_t = \beta_t\,\mathrm{d}B(t) + \gamma_t\,\mathrm{d}D(t)$ hold for all $t \geqslant 0$.

4

Risk-Neutral Pricing in a Multi-Asset Economy

In the previous chapter, we considered continuous-time risk-neutral derivative pricing in the simplest so-called (B, S) economy consisting of only one risky asset (stock) S and a risk-free bond or money market account B. We now extend the (B, S) economy to a continuous-time model of an economy consisting of an arbitrary number $n \geqslant 1$ of tradable risky base assets as well as a money market account. The simplest way to extend the classical (B, S) model to include multiple risky base assets is to assume that they are all independent of one another. For example, we can let each base asset price process be an Itô process, such as a geometric Brownian motion (GBM), driven by an independent Brownian motion (BM). Of course, this is a trivial and not very interesting extension. A more realistic extension is a model where the base assets are correlated stochastic processes. In particular, in this chapter, we shall remain within the framework of multidimensional correlated Itô processes. The interdependence among the base asset price processes arises by forming correlations among BMs that drive each price process. As was learned in Chapter 2, such correlations can be constructed by taking linear combinations of several, say d, independent BMs. The multidimensional GBM process is one such model for describing n stock price processes driven by d independent BMs. In all derivations of explicit pricing formulae for multi-asset derivatives considered in this chapter, we assume the "classical" multidimensional GBM model. This model offers fairly simple analytical tractability for many standard options written on multiple stocks. However, we shall first develop the pricing and hedging theory in a more general multidimensional continuous-time framework where all base asset price processes are quite general correlated Itô processes. We then simplify the framework to the classical multidimensional GBM model for the risky assets and also assume nonrandom interest rates and stock dividends. This allows us to price several standard (as well as some path dependent) European-style derivatives whose payoffs are dependent on the prices of multiply correlated assets. As in the case of a single asset, for non-path-dependent European-style derivatives the Markov property reduces the pricing problem to a conditional expectation where the multidimensional (discounted) Feynman–Kac formula leads to the Black–Scholes–Merton PDE for multi-asset contracts. Section 2.10 of Chapter 2 gives us this connection between the SDEs of Itô processes and the corresponding PDE problem.

As in the case of the (B, S) model covered in Chapter 3, we use a self-financing replicating portfolio strategy involving all base assets (risky stocks and bank account) and thereby obtain a general risk-neutral pricing formula expressing prices of European-style derivatives with attainable payoffs as a conditional expectation of the discounted payoff under the risk-neutral measure $\widetilde{\mathbb{P}}$ with the bank account as the numéraire asset. For the multidimensional GBM model, we have already shown (see the end of Section 2.10.3 of Chapter 2 where we applied Girsanov's Theorem) that the so-called *price of risk equations* must have a solution in order to guarantee the existence of a risk-neutral measure $\widetilde{\mathbb{P}}$. Essentially the same linear system of equations follows for our more general multi-asset model and hence similar conditions must be imposed on the model for the existence of a risk-neutral measure $\widetilde{\mathbb{P}}$. As shown in the discrete-time models, we will also see that the existence of a risk-neutral measure $\widetilde{\mathbb{P}}$ implies that the model is arbitrage-free. This is the statement of the so-called *First Fundamental Theorem of Asset Pricing*. Then, by an application of the multidimensional

DOI: 10.1201/9780429468889-4

Brownian Martingale Representation Theorem, we shall answer the question of completeness of the multi-asset market model; i.e., the question of whether derivative claims can be hedged (replicated) by a self-financing portfolio strategy. This will lead us to another important result, called the *Second Fundamental Theorem of Asset Pricing*, stating that the market model is complete if the risk-neutral measure $\widetilde{\mathbb{P}}$ exists and is unique. It then follows that the multidimensional GBM model with $n = d$, i.e., with same number of assets as independent BMs, having a nonsingular log-volatility matrix is a complete and arbitrage-free multi-asset market model. This is the model that we shall adopt in our examples of pricing derivatives.

The last topic covered in this chapter discusses the risk-neutral pricing framework in the more general context of *equivalent martingale measures (EMMs)* that correspond to different choices of numéraire assets for pricing. As was shown in the discrete-time models, there is a general numéraire invariant risk-neutral derivative pricing formula. A change of numéraire for pricing is just a special kind of change of measure that takes us from one EMM to another. A numéraire asset is any positive asset, such as a base asset or even a derivative asset, that is an Itô process driven by the same multidimensional BM. Hence, Girsanov's Theorem is again the tool that allows us to price derivatives under different choices of numéraires. We will present some examples of how changing numéraires facilitates the pricing of some multi-asset derivatives.

4.1 General Multi-Asset Market Model: Replication and Risk-Neutral Pricing

For a given time $T > 0$, let $\{\mathbf{W}(t) = (W_1(t), \ldots, W_d(t))\}_{0 \leqslant t \leqslant T}$ be a standard d-dimensional Brownian motion under the physical (real-world) measure $\mathbb{P}$ and fix a filtered probability space $(\Omega, \mathcal{F}, \mathbb{P}, \mathbb{F})$ with $\mathbb{F} = \mathbb{F}^{\mathbf{W}}$ as the natural filtration generated by the BM. One of our *base assets* is the bank account with value process $\{B(t)\}_{t \geqslant 0}$ where

$$B(t) = \exp\left(\int_0^t r(u)\,\mathrm{d}u\right) \implies \mathrm{d}B(t) = r(t)B(t)\,\mathrm{d}t\,, \quad B(0) = 1. \tag{4.1}$$

Hence, $B(t)$ is the value of one dollar invested in the bank account at time 0 and accruing interest at the continuously compounded rate $r(u)$ for all times $u \in [0, t]$. We also conveniently define the discount factor $D(t) := 1/B(t) = \mathrm{e}^{-\int_0^t r(u)\,\mathrm{d}u}$. The instantaneous interest rate process $\{r(t)\}_{t \geqslant 0}$ is assumed to be $\mathbb{F}$-adapted. The other base assets in our market model have prices assumed to be Itô processes $\{\mathbf{S}(t) = (S_1(t), \ldots, S_n(t))\}_{t \geqslant 0}$ solving the system of SDEs written equivalently as

$$\begin{aligned} \mathrm{d}S_i(t) &= S_i(t)\left[\mu_i(t)\,\mathrm{d}t + \sum_{j=1}^{d} \sigma_{ij}(t)\,\mathrm{d}W_j(t)\right] \\ &\equiv S_i(t)\left[\mu_i(t)\,\mathrm{d}t + \boldsymbol{\sigma}_i(t) \cdot \mathrm{d}\mathbf{W}(t)\right], \quad i = 1, \ldots, n. \end{aligned} \tag{4.2}$$

At the moment, we are assuming a zero dividend yield on all the risky base assets. Recalling our description of the multidimensional GBM process in Chapter 2, we note that (4.2) is of the same form as (2.226). However, the coefficients are now considered to be generally adapted processes. The multidimensional GBM process is recovered in the special case where we let all coefficients be constants.

The log-drift coefficients $\{\mu_i(t)\}_{t\geqslant 0}$ and the log-volatility coefficients $\{\sigma_{ij}\}_{t\geqslant 0}$ are assumed to be $\mathbb{F}$-adapted real-valued processes. The drift $\mu_i(t)$ gives the instantaneous physical rate of return (actual growth rate) of the ith asset. We denote the $n \times d$ matrix of log-volatilities by $\boldsymbol{\sigma}(t) := [\sigma_{ij}(t)]_{i=1,\ldots,n;\, j=1,\ldots,d}$. The ith row of $\boldsymbol{\sigma}(t)$ gives the log-volatility vector $\boldsymbol{\sigma}_i(t) = [\sigma_{i1}(t), \sigma_{i2}(t), \ldots, \sigma_{id}(t)]$ for the ith asset price process. The magnitude of each ith vector is denoted (without boldface) by

$$\sigma_i(t) \equiv \|\boldsymbol{\sigma}_i(t)\| = \left(\sigma_{i1}^2(t) + \ldots + \sigma_{id}^2(t)\right)^{1/2} \tag{4.3}$$

where we assume $\sigma_i(t) > 0$. Each component $\sigma_{ik}(t)$ is the log-volatility of the ith base asset (e.g., ith stock) w.r.t. the kth BM (kth source of randomness). We define the adapted log-diffusion $n \times n$ matrix by $\mathbf{C}(t) := \boldsymbol{\sigma}(t)[\boldsymbol{\sigma}(t)]^\top$ and the corresponding instantaneous log-correlation $n \times n$ matrix by $\boldsymbol{\rho}(t) = [\rho_{ij}(t)]_{i,j=1,\ldots,n}$, where

$$C_{ij}(t) = \boldsymbol{\sigma}_i(t) \cdot \boldsymbol{\sigma}_j(t) = \rho_{ij}(t)\sigma_i(t)\sigma_j(t) \tag{4.4}$$

and

$$\rho_{ij}(t) := \frac{\boldsymbol{\sigma}_i(t) \cdot \boldsymbol{\sigma}_j(t)}{\sigma_i(t)\sigma_j(t)} = \sum_{k=1}^{d} \frac{\sigma_{ik}(t)\sigma_{jk}(t)}{\sigma_i(t)\sigma_j(t)} . \tag{4.5}$$

Note that this is (2.195), but now these coefficients are *instantaneous log-correlations.* That is, multiplying the stochastic differentials of any pair of prices (relative to their prices at time t) using (4.2) gives

$$\frac{\mathrm{d}S_i(t)}{S_i(t)} \frac{\mathrm{d}S_j(t)}{S_j(t)} = \boldsymbol{\sigma}_i(t) \cdot \boldsymbol{\sigma}_j(t)\,\mathrm{d}t = C_{ij}(t)\,\mathrm{d}t = \rho_{ij}(t)\sigma_i(t)\sigma_j(t)\,\mathrm{d}t\,, \tag{4.6}$$

or in terms of the product of the price differentials,

$$\mathrm{d}S_i(t)\,\mathrm{d}S_j(t) = \rho_{ij}(t)\sigma_i(t)\sigma_j(t)S_i(t)S_j(t)\,\mathrm{d}t\,. \tag{4.7}$$

Recall from Chapter 2 that each SDE in (4.2) is of the form in (2.183). By the multi-factor Itô formula, each process has the representation in (2.184). Each ith stock price solving the SDE in (4.2) has a representation as its initial value, times a drift term and a martingale in the ith volatility vector:

$$\begin{aligned} S_i(t) &= S_i(0) \exp\left(\int_0^t (\mu_i(s) - \frac{1}{2}\|\boldsymbol{\sigma}_i(s)\|^2)\,\mathrm{d}s + \int_0^t \boldsymbol{\sigma}_i(s) \cdot \mathrm{d}\mathbf{W}(s)\right) \\ &= S_i(0)\,\mathrm{e}^{\int_0^t \mu_i(s)\,\mathrm{d}s}\,\mathcal{E}_t(\boldsymbol{\sigma}_i \cdot \mathbf{W})\,, \end{aligned} \tag{4.8}$$

where

$$\mathcal{E}_t(\boldsymbol{\sigma}_i \cdot \mathbf{W}) := \exp\left(-\frac{1}{2}\int_0^t \|\boldsymbol{\sigma}_i(s)\|^2\,\mathrm{d}s + \int_0^t \boldsymbol{\sigma}_i(s) \cdot \mathrm{d}\mathbf{W}(s)\right) . \tag{4.9}$$

Throughout we are assuming that all Itô integrals are martingales so the square-integrability condition is assumed to hold. This is assured by assuming the Novikov condition holds for each $i = 1, \ldots, n$,

$$\mathrm{E}\left[\exp\left(\frac{1}{2}\int_0^T \sigma_i^2(t)\,\mathrm{d}t\right)\right] < \infty\,, \tag{4.10}$$

where $\sigma_i^2(t) \equiv \|\boldsymbol{\sigma}_i(t)\|^2 = \sum_{j=1}^d \sigma_{ij}^2(t)$.

Let us take a small step back and briefly recall our application of Girsanov's Theorem in Section 2.10.3 of Chapter 2. There, we arrived at the conditions for the existence and uniqueness of a risk-neutral measure $\widetilde{\mathbb{P}}$ within the multidimensional GBM model. We showed that such a measure exists if and only if there is a solution to the linear system in (2.252) and then $\{e^{-rt}S_i(t)\}_{t\geqslant 0}$ are $\widetilde{\mathbb{P}}$-martingales, or equivalently all stock prices have the constant drift rate r as in (2.251). In the GBM model, the drift and volatility coefficients are all real constant parameters and we assumed a constant interest rate r.

Let us now again apply Girsanov's Theorem by using the same steps as in Section 2.10.3. The coefficients are now generally adapted processes so we need to use Theorem 2.25 for a generally adapted vector process $\{\boldsymbol{\gamma}(t) = (\gamma_1(t), \ldots, \gamma_d(t))\}_{0\leqslant t\leqslant T}$ satisfying the condition in (2.245). The measure change is defined more generally in terms of the Radon–Nikodym derivative having the form in (2.246). Repeating the same step by using the stochastic differential $d\widetilde{\mathbf{W}}(t) = d\mathbf{W}(t) - \boldsymbol{\gamma}(t)\,dt$, we again have (2.250), but now the coefficients are adapted (denoted by the time t argument):

$$dS_i(t) = S_i(t)\big[(\mu_i(t) + \boldsymbol{\sigma}_i(t)\cdot\boldsymbol{\gamma}(t))\,dt + \boldsymbol{\sigma}_i(t)\cdot d\widetilde{\mathbf{W}}(t)\big]. \tag{4.11}$$

The measure $\widetilde{\mathbb{P}}$ is risk-neutral if $\boldsymbol{\gamma}(t)$ is such that

$$dS_i(t) = S_i(t)\big[r(t)\,dt + \boldsymbol{\sigma}_i(t)\cdot d\widetilde{\mathbf{W}}(t)\big] \tag{4.12}$$

for all $i = 1, \ldots, n$. Note that this SDE is equivalent to stating that *all discounted stock price processes* $\{\overline{S}_i(t) := D(t)S_i(t)\}_{0\leqslant t\leqslant T}$ *are* $\widetilde{\mathbb{P}}$*-martingales.* This is easily seen by applying the Itô product rule to compute the differential

$$d[D(t)S_i(t)] \equiv d\overline{S}_i(t) = \overline{S}_i(t)\boldsymbol{\sigma}_i(t)\cdot d\widetilde{\mathbf{W}}(t), \tag{4.13}$$

which has zero drift and is hence a $\widetilde{\mathbb{P}}$-martingale. Alternatively, the $\widetilde{\mathbb{P}}$-martingale property of $\overline{S}_i(t)$ is shown by using the stochastic exponential representation of (4.12) and then multiplying by $D(t)$ and using $B(t) = e^{\int_0^t r(s)\,ds} = 1/D(t)$,

$$\overline{S}_i(t) \equiv D(t)S_i(t) = D(t)S_i(0)\,e^{\int_0^t r(s)\,ds}\,\mathcal{E}_t(\boldsymbol{\sigma}_i\cdot\widetilde{\mathbf{W}}) = S_i(0)\mathcal{E}_t(\boldsymbol{\sigma}_i\cdot\widetilde{\mathbf{W}}) \tag{4.14}$$

with exponential $\widetilde{\mathbb{P}}$-martingale

$$\mathcal{E}_t(\boldsymbol{\sigma}_i\cdot\widetilde{\mathbf{W}}) = \exp\left(-\frac{1}{2}\int_0^t \|\boldsymbol{\sigma}_i(s)\|^2\,ds + \int_0^t \boldsymbol{\sigma}_i(s)\cdot d\widetilde{\mathbf{W}}(s)\right). \tag{4.15}$$

The $\widetilde{\mathbb{P}}$-martingale property of the discounted prices hence follows simply by (4.14):

$$\widetilde{E}\left[\overline{S}_i(t)\mid\mathcal{F}_s\right] = \widetilde{E}\left[D(t)S_i(t)\mid\mathcal{F}_s\right] = S_i(0)\widetilde{E}\left[\mathcal{E}_t(\boldsymbol{\sigma}_i\cdot\widetilde{\mathbf{W}})\,\middle|\,\mathcal{F}_s\right] = S_i(0)\mathcal{E}_s(\boldsymbol{\sigma}_i\cdot\widetilde{\mathbf{W}}) = \overline{S}_i(s)$$

for all $0 \leqslant s \leqslant t \leqslant T$.

Hence, a risk-neutral measure exists if the log-drift coefficient in (4.11) equals $r(t)$, i.e., if and only if there exists a vector process $\boldsymbol{\gamma}(t)$ solving

$$\mu_i(t) + \boldsymbol{\sigma}_i(t)\cdot\boldsymbol{\gamma}(t) = r(t)\,,\quad i = 1, \ldots, n,$$

or in matrix form

$$\boldsymbol{\sigma}(t)\,\boldsymbol{\gamma}^\top(t) = \mathbf{b}(t), \tag{4.16}$$

with $n \times 1$ vectors $\mathbf{b}(t) := r(t)\mathbf{1} - \boldsymbol{\mu}(t)$, $\boldsymbol{\mu}(t) = [\mu_1(t), \ldots, \mu_n(t)]^\top$, $\mathbf{1} = [1, \ldots, 1]^\top$, and $d \times 1$ vector $\boldsymbol{\gamma}^\top(t) = [\gamma(t)]^\top$. This is the same linear system of n equations in d unknowns as in (2.252). This system is commonly referred to as the "price of risk equations" since a solution guarantees the existence of a risk-neutral measure $\widetilde{\mathbb{P}}$ and (as seen later) the existence of such a $\widetilde{\mathbb{P}}$ implies no-arbitrage in the above market model. Otherwise, if (4.16) has no solution (i.e., a $\widetilde{\mathbb{P}}$ does not exist) then it can be shown that an arbitrage portfolio strategy exists.

The discussion that follows (2.252) in Section 2.10.3 also pertains to the system (4.16) where the conditions must be satisfied for all values of $t \in [0, T]$ and (almost) all realizations of the multidimensional BM since the quantities in (4.16) are now generally adapted processes. For almost all outcomes ω and time $t \in [0, T]$, we must have a solution to the linear system $\boldsymbol{\sigma}(t,\omega)\,\boldsymbol{\gamma}^\top(t,\omega) = r(t,\omega)\mathbf{1} - \boldsymbol{\mu}(t,\omega)$. Hence, for almost all paths (t,ω), we require $\operatorname{rank}(\boldsymbol{\sigma}(t,\omega)) = n$. We simply state this as the condition that $\operatorname{rank}(\boldsymbol{\sigma}(t)) = n$. In the important case that $\operatorname{rank}(\boldsymbol{\sigma}(t)) = n = d$, i.e., when the *number of stocks equals the number of independent BMs and the $n \times n$ matrix $\boldsymbol{\sigma}(t)$ has an inverse $\boldsymbol{\sigma}^{-1}(t)$*, then the risk-neutral measure $\widetilde{\mathbb{P}}$ exists and is uniquely given by the adapted vector $\boldsymbol{\gamma}^\top(t) = \boldsymbol{\sigma}^{-1}(t)(r(t)\mathbf{1} - \boldsymbol{\mu}(t))$. We shall adopt this case when pricing derivatives later within the multidimensional GBM model.

We now assume that the above conditions are satisfied, i.e., that $\operatorname{rank}(\boldsymbol{\sigma}(t)) = n$, so a risk-neutral measure $\widetilde{\mathbb{P}}$ exists. As in Chapter 3, we consider the problem of no-arbitrage pricing of a derivative claim having value process $\{V_t\}_{0 \leqslant t \leqslant T}$. At maturity T, the derivative price equals the payoff V_T, which is an $\mathcal{F}_T$-measurable random variable. In our model, V_T is hence generally a functional of the multidimensional BM that is driving all of the $n+1$ base assets up to time T. In our multi-asset economy, a portfolio (trading) strategy is a continuous sequence of portfolios in the $n+1$ base assets:

$$(\beta_t, \boldsymbol{\Delta}_t) \equiv (\beta_t, \Delta_t^1, \ldots, \Delta_t^n)$$

where $\{\beta_t\}_{0 \leqslant t \leqslant T}$ and $\{\Delta_t^i\}_{0 \leqslant t \leqslant T}$, $i = 1, \ldots, n$, are adapted. As in the single stock economy, β_t represents the time-t position in the bank account. Each hedge position Δ_t^i denotes the number of shares held in the ith base asset (e.g., ith stock) at time t where $\Delta_t^i < 0$ corresponds to shorting the ith stock and $\Delta_t^i > 0$ is a long position. In continuous time, trading (i.e., portfolio re-balancing) is allowed at any moment in time. An investor begins with a given initial wealth Π_0 financing the initial portfolio with positions $(\beta_0, \Delta_0^1, \ldots, \Delta_0^n)$. A trading strategy over time $t \in [0, T]$ consists of holding Δ_t^i shares in each ith base asset and β_t units (as a loan or investment) in the bank account at time t; i.e., the portfolio value at time $t \in [0, T]$ is

$$\Pi_t = \boldsymbol{\Delta}_t \cdot \mathbf{S}(t) + \beta_t B(t). \tag{4.17}$$

As in the single risky asset model of Chapter 3, the only admissible portfolio replicating strategies are self-financing. That is, the differential change in portfolio value is only due to differential changes in the prices of all base assets while holding all the positions fixed. Formally, a self-financing portfolio strategy is a strategy $(\beta_t, \boldsymbol{\Delta}_t), 0 \leqslant t \leqslant T$, with cumulative gain in portfolio value given by (a.s.)

$$\begin{aligned}
\Pi_t &= \Pi_0 + \int_0^t \beta_u \, \mathrm{d}B(u) + \int_0^t \boldsymbol{\Delta}_u \cdot \mathrm{d}\mathbf{S}(u) \\
&= \Pi_0 + \int_0^t \beta_u r(u) B(u) \, \mathrm{d}u + \sum_{i=1}^n \int_0^t \Delta_u^i \, \mathrm{d}S_i(u)\,, \quad 0 \leqslant t \leqslant T.
\end{aligned} \tag{4.18}$$

In differential form,

$$\mathrm{d}\Pi_t = r(t)\beta_t B(t)\,\mathrm{d}t + \boldsymbol{\Delta}_t\cdot \mathrm{d}\mathbf{S}(t) \equiv r(t)\beta_t B(t)\,\mathrm{d}t + \sum_{i=1}^{n} \Delta_t^i\,\mathrm{d}S_i(t)\,.$$

Using (4.17) we express the bank account portion as $\beta_t B(t) = \Pi_t - \boldsymbol{\Delta}_t\cdot\mathbf{S}(t)$ and substituting this into the above differential gives

$$\mathrm{d}\Pi_t = r(t)(\Pi_t - \boldsymbol{\Delta}_t\cdot\mathbf{S}(t))\,\mathrm{d}t + \boldsymbol{\Delta}_t\cdot \mathrm{d}\mathbf{S}(t)\,. \tag{4.19}$$

This is the multidimensional extension of (3.19) in Chapter 3, giving the differential change in value of a self-financing portfolio in the $n+1$ base assets.

As in the single asset case, it is easy to show that the discounted self-financing portfolio value process, $\{\overline{\Pi}_t := D(t)\Pi_t\}_{0\leqslant t\leqslant T}$, is a $\widetilde{\mathbb{P}}$-martingale by computing its stochastic differential upon using (4.19) and simplifying (here, we use compact vector notation):

$$\begin{aligned}
\mathrm{d}\overline{\Pi}_t \equiv \mathrm{d}(D(t)\Pi_t) &= -r(t)D(t)\Pi_t\,\mathrm{d}t + D(t)\,\mathrm{d}\Pi_t \\
&= -r(t)D(t)\Pi_t\,\mathrm{d}t + D(t)\left[\,r(t)\Pi_t\,\mathrm{d}t - r(t)\boldsymbol{\Delta}_t\cdot\mathbf{S}(t)\,\mathrm{d}t + \boldsymbol{\Delta}_t\cdot\mathrm{d}\mathbf{S}(t)\,\right] \\
&= \boldsymbol{\Delta}_t\cdot\left[-r(t)D(t)\mathbf{S}(t)\,\mathrm{d}t + D(t)\,\mathrm{d}\mathbf{S}(t)\right] \\
&= \boldsymbol{\Delta}_t\cdot\,\mathrm{d}\left(D(t)\mathbf{S}(t)\right) \\
&\equiv \sum_{i=1}^{n}\Delta_t^i\,\mathrm{d}\overline{S}_i(t) = \sum_{i=1}^{n}\Delta_t^i\overline{S}_i(t)\,\boldsymbol{\sigma}_i(t)\cdot\mathrm{d}\widetilde{\mathbf{W}}(t)\,.
\end{aligned} \tag{4.20}$$

In the last line, we have two equivalent ways of expressing the $\widetilde{\mathbb{P}}$-martingale property, either as a linear combination of differentials in the discounted stock prices (which are all $\widetilde{\mathbb{P}}$-martingales) or, after substituting (4.13), as a linear combination of differentials in $\widetilde{\mathbf{W}}(t)$. Changes in the discounted (self-financing) portfolio are due only to changes in the discounted stock prices. The integral form of (4.20) expresses the cumulative change in the discounted portfolio value as

$$\overline{\Pi}_t = \overline{\Pi}_0 + \sum_{i=1}^{n}\int_0^t \Delta_u^i\,\mathrm{d}\overline{S}_i(u) = \Pi_0 + \sum_{i=1}^{n}\int_0^t \Delta_u^i\overline{S}_i(u)\,\boldsymbol{\sigma}_i(u)\cdot\mathrm{d}\widetilde{\mathbf{W}}(u)\,,\quad 0\leqslant t\leqslant T. \tag{4.21}$$

We are generally assuming that the Itô integral is square integrable and so the process in (4.21) is a $\widetilde{\mathbb{P}}$-martingale.

As in the (B,S) model of Chapter 3, no-arbitrage pricing of derivative claims relies on replicating the claim's payoff V_T by using a self-financing replicating strategy. In our continuous-time multi-asset model we have the following definition of such a strategy and market completeness.

Definition 4.1. A self-financing strategy $(\beta_t, \boldsymbol{\Delta}_t), 0\leqslant t\leqslant T$, is said to *replicate* the $\mathcal{F}_T$-measurable payoff V_T at maturity T if $\Pi_T = V_T$ (a.s.), i.e., $\mathbb{P}(\Pi_T = V_T) = 1$, and we say that the payoff V_T is *attainable*. A market model is *complete* if every $\mathcal{F}_T$-measurable payoff can be replicated; i.e., every derivative security can be hedged.

We now suppose that a self-financing replicating portfolio strategy for a given payoff V_T exists, i.e., that the payoff is attainable. Below we discuss the conditions for which this holds (i.e., market completeness) and the connection to the existence of risk-neutral measures. Repeating the same argument as in the (B,S) model of Chapter 3, the wealth Π_t at time $t\leqslant T$ needed to set up a self-financing replicating strategy up to time T must equal the time-t price of the derivative V_t. This is the fair price for an investor to hedge a short position in the derivative security with attainable payoff V_T at future time T, and hence avoid an obvious arbitrage. Therefore, $V_t = \Pi_t$, $0 \leqslant t \leqslant T$, for any attainable payoff V_T. Combining this with the fact that the value process of the discounted replicating portfolio is a $\widetilde{\mathbb{P}}$-martingale implies that the price process of the discounted derivative is a $\widetilde{\mathbb{P}}$-martingale. That is, $D(t)V_t = \widetilde{\mathrm{E}}[D(T)V_T \mid \mathcal{F}_t]$:

$$V_t = B(t)\,\widetilde{\mathrm{E}}\left[\frac{V_T}{B(T)}\,\middle|\,\mathcal{F}_t\right] = \widetilde{\mathrm{E}}\left[\mathrm{e}^{-\int_t^T r(u)\,\mathrm{d}u} V_T \mid \mathcal{F}_t\right], \quad 0 \leqslant t \leqslant T. \tag{4.22}$$

This is the risk-neutral pricing formula for the above multi-asset model with n base asset prices modelled as Itô processes obeying (4.11) and a bank account with a generally adapted (stochastic) interest rate process. Note that in the case of constant interest rate $r(t) = r$, this recovers (3.22).

Before moving on to the actual implementation of (4.22), we discuss the two main theorems of asset pricing in the context of the above multi-asset model. The first theorem tells us when the model can be used for risk-neutral pricing without arbitrage in the market and the second relates the completeness of the model and arbitrage. We begin by giving a simple definition of arbitrage in the above multi-asset continuous-time model. The definition mirrors that given in Chapter 8 of Volume I. The basic meaning is that the strategy requires no cost to execute and has a potential to create a profit at no risk. A market model that admits such a strategy is said to have an arbitrage.

Definition 4.2. A self-financing portfolio strategy $(\beta_t, \boldsymbol{\Delta}_t), t \geqslant 0$, is said to be an *arbitrage or arbitrage strategy* if it has zero initial value, $\Pi_0 = 0$, and satisfies the condition that $\Pi_T \geqslant 0$ (a.s), i.e., $\mathbb{P}(\Pi_T \geqslant 0) = 1$, and $\mathbb{P}(\Pi_T > 0) > 0$ for some $T > 0$.

Note that the probabilities in the definition are in the real-world (physical) measure. The above definition implies that if arbitrage exists then an investor can use a trading strategy to beat the bank account at no risk. To see that this is implied by the above definition of arbitrage, assume that an investor begins with some positive capital, say $\Pi_0' > 0$. The investor can then employ a strategy with value Π_t', $t \geqslant 0$, in which the initial capital Π_0' is invested in the bank account and executes a zero initial cost arbitrage strategy $(\beta_t, \boldsymbol{\Delta}_t)$ with value Π_t. At time $T > 0$ the investment in the bank account has value $B(T)\Pi_0'$ and the no-cost arbitrage strategy has value Π_T. The time-T value of the combined strategy is $\Pi_T' = \Pi_T + B(T)\Pi_0'$ and by the conditions in the above definition,

$$\mathbb{P}(\Pi_T \geqslant 0) = 1 \implies \mathbb{P}(\Pi_T' \geqslant B(T)\Pi_0') = 1$$

and

$$\mathbb{P}(\Pi_T > 0) > 0 \implies \mathbb{P}(\Pi_T' > B(T)\Pi_0') > 0\,.$$

The first condition states that (with certainty) the strategy has a return greater than or equal to that of the bank account investment. The second condition states that there is a positive probability that the strategy has a greater return than the bank account. We leave it as an exercise for the reader to prove the converse, i.e., that the existence of a strategy that beats the bank account at no risk implies the existence of an arbitrage portfolio strategy with zero initial cost.

Based on the above definition of arbitrage we can now state the so-called first fundamental theorem of asset (or derivative) pricing for the above continuous-time multi-asset market model as follows. The result is a simple consequence of the martingale property of any discounted self-financing portfolio strategy.

Theorem 4.1 (First Fundamental Theorem of Asset Pricing)**.** *If there exists a risk-neutral measure $\widetilde{\mathbb{P}}$ then there are no arbitrage strategies in the market.*

Proof. The discounted value process of any (admissible) self-financing portfolio is a $\widetilde{\mathbb{P}}$-martingale. Let $\{\Pi_t\}_{t\geqslant 0}$ be the value process of an assumed arbitrage portfolio strategy with $\Pi_0 = 0$. The $\widetilde{\mathbb{P}}$-martingale property gives

$$\widetilde{\mathrm{E}}[\overline{\Pi}_T] = \widetilde{\mathrm{E}}[D(T)\Pi_T] = \overline{\Pi}_0 = \Pi_0 = 0 \quad \text{for any } T \geqslant 0.$$

By the definition of an arbitrage portfolio, $\Pi_T \geqslant 0$ and hence $D(T)\Pi_T \geqslant 0$ (a.s.). So $D(T)\Pi_T$ is nonnegative (a.s.) and has zero expectation under measure $\widetilde{\mathbb{P}}$. This implies that $\widetilde{\mathbb{P}}(D(T)\Pi_T > 0) = 0$, which implies $\widetilde{\mathbb{P}}(\Pi_T > 0) = 0$ since $D(T) > 0$. Since the measure $\widetilde{\mathbb{P}}$ is equivalent to $\mathbb{P}$; i.e., zero probability events w.r.t. $\widetilde{\mathbb{P}}$ are zero probability events w.r.t. $\mathbb{P}$, then $\mathbb{P}(\Pi_T > 0) = 0$. We conclude that there are no zero-cost and zero-risk self-financing portfolio strategies such that $\mathbb{P}(\Pi_T > 0) > 0$, implying no arbitrage strategies are possible. □

This theorem provides a way of verifying if the market model can be used to calculate no-arbitrage prices of attainable derivative securities. That is, the above multi-asset model can be used for no-arbitrage pricing if the price of risk equations have a solution, i.e., if $\operatorname{rank}(\boldsymbol{\sigma}(t)) = n$. For example, there is no arbitrage in the model when the number of BMs is the same as the number of stocks, $d = n$, and the log-volatility matrix $\boldsymbol{\sigma}(t)$ is invertible. The multidimensional GBM model for n stocks driven by n BMs is an important example of a market model having no arbitrage. The (B, S) model of Chapter 3, where $d = n = 1$, is the simplest case of a GBM model with no arbitrage.

We now consider the question of completeness of the multi-asset model. As in Chapter 3, we consider all payoffs V_T that are square integrable. As well, all payoffs V_T are assumed to be $\mathcal{F}_T^{\mathbf{W}}$-measurable. The process $\{D(t)V_t\}_{0\leqslant t\leqslant T}$ is then a square-integrable $(\widetilde{\mathbb{P}}, \mathbb{F}^{\mathbf{W}})$-martingale. We can hence make use of the multidimensional Brownian Martingale Representation Theorem 2.26 where we identify $\widehat{P} \equiv \widetilde{\mathbb{P}}$, $\widehat{\mathbf{W}} \equiv \widetilde{\mathbf{W}}$, and $\widehat{M}(t) \equiv D(t)V_t$. In particular, there exists an adapted d-dimensional process, $\{\widetilde{\boldsymbol{\theta}}(t) = (\widetilde{\theta}_1(t), \ldots, \widetilde{\theta}_d(t))\}_{0\leqslant t\leqslant T}$, such that (a.s.)

$$D(t)V_t = V_0 + \int_0^t \widetilde{\boldsymbol{\theta}}(u) \cdot \mathrm{d}\widetilde{\mathbf{W}}(u)\,, \quad 0 \leqslant t \leqslant T. \tag{4.23}$$

In order to replicate the claim, i.e., to have $\Pi_t = V_t$ for all $t \in [0, T]$, we must have the equivalence of this expression with that in (4.21) (note $\overline{\Pi}_t = D(t)\Pi_t$). In other words, replication is possible if and only if both Itô integrands are equal,

$$\sum_{i=1}^{n} \Delta_t^i \overline{S}_i(t)\, \boldsymbol{\sigma}_i(t) = \widetilde{\boldsymbol{\theta}}(t)\,, \quad 0 \leqslant t \leqslant T. \tag{4.24}$$

Equating each vector component gives a linear system of d equations in n unknowns $\Delta_t^1, \ldots, \Delta_t^n$,

$$\sum_{i=1}^{n} \sigma_{ij}(t)\overline{S}_i(t)\Delta_t^i = \widetilde{\theta}_j(t)\,, \quad j = 1, \ldots, d.$$

In matrix form, this system is written compactly as

$$[\boldsymbol{\sigma}(t)]^\top \mathbf{v}(t) = \widetilde{\boldsymbol{\theta}}(t) \tag{4.25}$$

where $[\boldsymbol{\sigma}(t)]^\top$ is the $d \times n$ matrix transpose of $\boldsymbol{\sigma}(t)$; $\mathbf{v}(t) = \left[\overline{S}_1(t)\Delta_t^1, \ldots, \overline{S}_n(t)\Delta_t^n\right]^\top$ is an $n \times 1$ vector of the delta hedge position in each stock, scaled by each (strictly positive) discounted stock price; $\widetilde{\boldsymbol{\theta}}(t) = \left[\widetilde{\theta}_1(t), \ldots, \widetilde{\theta}_d(t)\right]^\top$ is a $d \times 1$ vector. The market model is complete (and can hedge any derivative) if there exists a solution to (4.25) for *any vector* $\widetilde{\boldsymbol{\theta}}(t) \in \mathbb{R}^d$. Hence, the market is complete if and only if $\text{rank}(\boldsymbol{\sigma}(t)) = d$. Note that the column and row rank of a matrix are the same, $\text{rank}(\boldsymbol{\sigma}(t)) = \text{rank}([\boldsymbol{\sigma}(t)]^\top)$. Given a solution $\mathbf{v}(t) = [v_1(t), \ldots, v_n(t)]^\top$ to the system in (4.25), the hedging positions in the stocks are given by $\Delta_t^1 = v_1(t)/\overline{S}_1(t), \ldots, \Delta_t^n = v_n(t)/\overline{S}_n(t)$. We note that this is not generally a practical method for obtaining the hedging positions. We are only focusing here on the existence of the hedging positions for replicating any payoff.

Let us assume that at least one risk-neutral measure $\widetilde{\mathbb{P}}$ exists, i.e., $\text{rank}(\boldsymbol{\sigma}(t)) = n$. If the market is complete, $\text{rank}(\boldsymbol{\sigma}(t)) = d = n$ and hence the solution to (4.16) must be unique and therefore $\widetilde{\mathbb{P}}$ is unique. Conversely, if $\widetilde{\mathbb{P}}$ is unique then $\text{rank}(\boldsymbol{\sigma}(t)) = n = d$ and therefore the market model is complete. We have therefore proven the following result, which is commonly stated as the "Second Fundamental Theorem of Asset Pricing." In a model having a risk-neutral measure (no arbitrage), this theorem ties together the uniqueness of a risk-neutral measure and market completeness.

Theorem 4.2 (Second Fundamental Theorem of Asset Pricing). *Assume the above multi-asset continuous-time market model has a risk-neutral probability measure $\widetilde{\mathbb{P}}$. Then, $\widetilde{\mathbb{P}}$ is unique if and only if the market model is complete.*

4.2 Equivalent Martingale Measures: Derivative Pricing with General Numéraire Assets

Consider an arbitrage-free multi-asset market model within a given domestic economy, as described in Section 4.1. Note that a model with two economies – a domestic one with assets denominated in domestic currency and a foreign one with assets denominated in foreign currency – is to be considered in Section 4.3.3. In particular, we assume that there exists a solution to the price of risk equations so that a risk-neutral measure $\widetilde{\mathbb{P}}$ exists. If $n = d$, where d is the number of BMs driving the base asset prices, then $\widetilde{\mathbb{P}}$ is unique and the market is complete. We shall keep the formulation more general with $n \leqslant d$ and such that a $\widetilde{\mathbb{P}}$ exists. Assuming no dividends on the assets, we recall that the measure $\widetilde{\mathbb{P}}$ has the defining property that all discounted base asset price processes $\{\frac{S_0(t)}{B(t)}, \frac{S_1(t)}{B(t)}, \ldots, \frac{S_n(t)}{B(t)}\}_{t\geqslant 0}$ are $\widetilde{\mathbb{P}}$-martingales. Here, $S_0(t) := B(t)$ and we note that $S_0(t)/B(t) \equiv 1$ is a trivial martingale. We recall that (4.12)–(4.15) hold. That is, for each base asset $i = 1, \ldots, n$,

$$\mathrm{d}\left(\frac{S_i(t)}{B(t)}\right) = \frac{S_i(t)}{B(t)}\boldsymbol{\sigma}_i(t)\cdot \mathrm{d}\widetilde{\mathbf{W}}(t). \tag{4.26}$$

Equivalently, each base asset price relative to the bank account value $B(t)$ is a $\widetilde{\mathbb{P}}$-martingale,

$$\frac{S_i(t)}{B(t)} = \frac{S_i(0)}{B(0)}\mathcal{E}_t(\boldsymbol{\sigma}_i \cdot \widetilde{\mathbf{W}}). \tag{4.27}$$

Dividing the asset price ratio at time t with the ratio at time 0 gives the exponential $\widetilde{\mathbb{P}}$-martingale:

$$\frac{S_i(t)/B(t)}{S_i(0)/B(0)} = \mathcal{E}_t(\boldsymbol{\sigma}_i \cdot \widetilde{\mathbf{W}}) := \exp\left(-\frac{1}{2}\int_0^t \|\boldsymbol{\sigma}_i(s)\|^2\,\mathrm{d}s + \int_0^t \boldsymbol{\sigma}_i(s)\cdot \mathrm{d}\widetilde{\mathbf{W}}(s)\right). \tag{4.28}$$

Hence, all base asset prices relative to $B(t)$, including $B(t)$, are $\widetilde{\mathbb{P}}$-martingales. When this property holds we say that $\widetilde{\mathbb{P}} \equiv \widetilde{\mathbb{P}}^{(B)}$ is an equivalent martingale measure (EMM) with bank account B as the numéraire (or numéraire asset). We shall denote a generic *positive asset price process by* $\{g(t)\}_{t\geqslant 0}$. This can be any one of the base assets, a portfolio in the base assets, or a derivative asset in the market model. Such an asset can be chosen as a *numéraire asset* and its price process is called the numéraire asset price process. We now show that this leads to an EMM for any given choice of numéraire. Just as in the discrete-time market models, we shall see that we are not restricted to using the bank account as the choice of numéraire for no-arbitrage derivative pricing.

Let $\{g(t)\}_{t\geqslant 0}$ be a (non-dividend-paying) numéraire asset price process with given adapted log-volatility vector $\boldsymbol{\sigma}^{(g)}(t) = [\sigma_1^{(g)}(t), \ldots, \sigma_d^{(g)}(t)]$. Since any numéraire is itself an asset, $\{\frac{g(t)}{B(t)}\}_{t\geqslant 0}$ is a $\widetilde{\mathbb{P}}$-martingale; i.e., we equivalently have

$$\mathrm{d}g(t) = g(t)\big[r(t)\,\mathrm{d}t + \boldsymbol{\sigma}^{(g)}(t)\cdot \mathrm{d}\widetilde{\mathbf{W}}(t)\big], \tag{4.29}$$

$$\mathrm{d}\left(\frac{g(t)}{B(t)}\right) = \frac{g(t)}{B(t)}\boldsymbol{\sigma}^{(g)}(t)\cdot \mathrm{d}\widetilde{\mathbf{W}}(t)\,, \tag{4.30}$$

and

$$\frac{g(t)}{B(t)} = \frac{g(0)}{B(0)}\,\mathcal{E}_t(\boldsymbol{\sigma}^{(g)}\cdot\widetilde{\mathbf{W}}) \implies \frac{g(t)/B(t)}{g(0)/B(0)} = \mathcal{E}_t(\boldsymbol{\sigma}^{(g)}\cdot\widetilde{\mathbf{W}})\,. \tag{4.31}$$

By dividing (4.31) into (4.28), we obtain a representation for the ratio process $\left\{\frac{S_i(t)}{g(t)}\right\}_{t\geqslant 0}$ for each $i = 0, 1, \ldots, n$:

$$\frac{S_i(t)/g(t)}{S_i(0)/g(0)} = \frac{S_i(t)/B(t)}{S_i(0)/B(0)}\left[\frac{g(t)/B(t)}{g(0)/B(0)}\right]^{-1} = \frac{\mathcal{E}_t(\boldsymbol{\sigma}_i\cdot\widetilde{\mathbf{W}})}{\mathcal{E}_t(\boldsymbol{\sigma}^{(g)}\cdot\widetilde{\mathbf{W}})}\,. \tag{4.32}$$

Note that the right-hand side ratio of exponential $\widetilde{\mathbb{P}}$-martingales is not a $\widetilde{\mathbb{P}}$-martingale. In particular, the ratio is *not* equal to $\mathcal{E}_t\big((\boldsymbol{\sigma}_i - \boldsymbol{\sigma}^{(g)})\cdot\widetilde{\mathbf{W}}\big)$.

We now apply Girsanov's Theorem 2.25 to change measures such that the right-hand side in (4.32) is a martingale under a new measure. This is accomplished by defining a new measure $\widetilde{\mathbb{P}}^{(g)}$ via the Radon–Nikodym derivative process

$$\varrho_t^{B\to g} \equiv \left(\frac{\mathrm{d}\widetilde{\mathbb{P}}^{(g)}}{\mathrm{d}\widetilde{\mathbb{P}}^{(B)}}\right)_t := \exp\left(-\frac{1}{2}\int_0^t \|\boldsymbol{\sigma}^{(g)}(s)\|^2\,\mathrm{d}s + \int_0^t \boldsymbol{\sigma}^{(g)}(s)\cdot \mathrm{d}\widetilde{\mathbf{W}}(s)\right) \tag{4.33}$$

for all $t \in [0, T]$. Note that $\widetilde{\mathbb{P}}^{(B)} \equiv \widetilde{\mathbb{P}}$, so $\widetilde{\mathbf{W}}(t) \equiv \widetilde{\mathbf{W}}^{(B)}(t)$ denotes the d-dimensional standard $\widetilde{\mathbb{P}}$-BM. Hence,

$$\widetilde{\mathbf{W}}^{(g)}(t) := \widetilde{\mathbf{W}}(t) - \int_0^t \boldsymbol{\sigma}^{(g)}(s)\,\mathrm{d}s \tag{4.34}$$

is a d-dimensional standard $\widetilde{\mathbb{P}}^{(g)}$-BM. In differential form, we have[1]

$$\mathrm{d}\widetilde{\mathbf{W}}^{(g)}(t) = \mathrm{d}\widetilde{\mathbf{W}}(t) - \boldsymbol{\sigma}^{(g)}(t)\,\mathrm{d}t \quad\text{or}\quad \mathrm{d}\widetilde{\mathbf{W}}(t) = \mathrm{d}\widetilde{\mathbf{W}}^{(g)}(t) + \boldsymbol{\sigma}^{(g)}(t)\,\mathrm{d}t\,. \tag{4.35}$$

---

[1] We remark that in the trivial case where $g(t) = B(t)$, $\boldsymbol{\sigma}^{(g)}(t) \equiv \boldsymbol{\sigma}^{(B)}(t) \equiv \mathbf{0}$ and $\varrho_t^{B\to g} \equiv 1$ and $\widetilde{\mathbf{W}}^{(g)}(t) \equiv \widetilde{\mathbf{W}}(t)$.

Expressing the ratio in (4.32) using (4.35), i.e. $\mathrm{d}\widetilde{\mathbf{W}}(s) = \mathrm{d}\widetilde{\mathbf{W}}^{(g)}(s) + \boldsymbol{\sigma}^{(g)}(s)\,\mathrm{d}s$,

$$\begin{aligned}
\frac{\mathcal{E}_t(\boldsymbol{\sigma}_i \cdot \widetilde{\mathbf{W}})}{\mathcal{E}_t(\boldsymbol{\sigma}^{(g)} \cdot \widetilde{\mathbf{W}})} &= \exp\left[-\frac{1}{2}\int_0^t \left(\|\boldsymbol{\sigma}_i(s)\|^2 - \|\boldsymbol{\sigma}^{(g)}(s)\|^2\right)\mathrm{d}s + \int_0^t \left(\boldsymbol{\sigma}_i(s) - \boldsymbol{\sigma}^{(g)}(s)\right)\cdot \mathrm{d}\widetilde{\mathbf{W}}(s)\right] \\
&= \exp\left[\int_0^t \left(-\frac{1}{2}\left(\|\boldsymbol{\sigma}_i(s)\|^2 - \|\boldsymbol{\sigma}^{(g)}(s)\|^2\right) + \left(\boldsymbol{\sigma}_i(s) - \boldsymbol{\sigma}^{(g)}(s)\right)\cdot\boldsymbol{\sigma}^{(g)}(s)\right)\mathrm{d}s \right.\\
&\qquad \left. + \int_0^t \left(\boldsymbol{\sigma}_i(s) - \boldsymbol{\sigma}^{(g)}(s)\right)\cdot \mathrm{d}\widetilde{\mathbf{W}}^{(g)}(s)\right] \\
&= \exp\left[-\frac{1}{2}\int_0^t \|\boldsymbol{\sigma}_i(s) - \boldsymbol{\sigma}^{(g)}(s)\|^2\,\mathrm{d}s + \int_0^t \left(\boldsymbol{\sigma}_i(s) - \boldsymbol{\sigma}^{(g)}(s)\right)\cdot \mathrm{d}\widetilde{\mathbf{W}}^{(g)}(s)\right] \\
&\equiv \mathcal{E}_t\left((\boldsymbol{\sigma}_i - \boldsymbol{\sigma}^{(g)}) \cdot \widetilde{\mathbf{W}}^{(g)}\right).
\end{aligned} \tag{4.36}$$

In the third line, we simplified the integrand by the vector identity $-\frac{1}{2}\left(\|\boldsymbol{\sigma}_i\|^2 - \|\boldsymbol{\sigma}^{(g)}\|^2\right) + \left(\boldsymbol{\sigma}_i - \boldsymbol{\sigma}^{(g)}\right)\cdot\boldsymbol{\sigma}^{(g)} = -\frac{1}{2}\|\boldsymbol{\sigma}_i - \boldsymbol{\sigma}^{(g)}\|^2$.

We have therefore shown that the *probability measure* $\widetilde{\mathbb{P}}^{(g)}$ *defined by (4.33) is an EMM where all base asset prices relative to the numéraire price* $g(t)$ *are* $\widetilde{\mathbb{P}}^{(g)}$*-martingales*:

$$\frac{S_i(t)}{g(t)} = \frac{S_i(0)}{g(0)}\,\mathcal{E}_t\left((\boldsymbol{\sigma}_i - \boldsymbol{\sigma}^{(g)}) \cdot \widetilde{\mathbf{W}}^{(g)}\right) \tag{4.37}$$

or as a stochastic differential expression

$$\mathrm{d}\left(\frac{S_i(t)}{g(t)}\right) = \frac{S_i(t)}{g(t)}\left(\boldsymbol{\sigma}_i - \boldsymbol{\sigma}^{(g)}\right)\cdot \mathrm{d}\widetilde{\mathbf{W}}^{(g)}(t) \tag{4.38}$$

for all $0 \leqslant t \leqslant T$.

Consider any *domestic non-dividend-paying asset* with a price process $\{A(t)\}_{t\geqslant 0}$ obeying the SDE

$$\mathrm{d}A(t) = A(t)\left[r(t)\,\mathrm{d}t + \boldsymbol{\sigma}^{(A)}(t)\cdot \mathrm{d}\widetilde{\mathbf{W}}(t)\right] \tag{4.39}$$

where $\boldsymbol{\sigma}^{(A)}(t)$ is an adapted log-volatility vector for asset A. For example, the asset price $A(t)$ can be any base asset price (or stock price) $S_i(t)$, any portfolio in the base assets, or any domestic derivative asset. In terms of the differential $\mathrm{d}\widetilde{\mathbf{W}}^{(g)}(t)$, the SDE in (4.39) gives

$$\mathrm{d}A(t) = A(t)\left[\left(r(t) + \boldsymbol{\sigma}^{(A)}(t)\cdot\boldsymbol{\sigma}^{(g)}(t)\right)\mathrm{d}t + \boldsymbol{\sigma}^{(A)}(t)\cdot \mathrm{d}\widetilde{\mathbf{W}}^{(g)}(t)\right]. \tag{4.40}$$

In particular, for the case where A is the ith domestic (non-dividend-paying) base asset (or stock), $A(t) = S_i(t)$:

$$\mathrm{d}S_i(t) = S_i(t)\left[\left(r(t) + \boldsymbol{\sigma}_i(t)\cdot\boldsymbol{\sigma}^{(g)}(t)\right)\mathrm{d}t + \boldsymbol{\sigma}_i(t)\cdot \mathrm{d}\widetilde{\mathbf{W}}^{(g)}(t)\right]. \tag{4.41}$$

Equation (4.40) is a useful SDE that gives the log-drift coefficient of an arbitrary (non-dividend-paying) domestic asset under the EMM $\widetilde{\mathbb{P}}^{(g)}$. We see that under the new measure $\widetilde{\mathbb{P}}^{(g)}$, the original ($\widetilde{\mathbb{P}}$-measure) risk-neutral drift $r(t)$ changes by an additional term given by the dot product of the log-volatility vector of asset A and that of the numéraire asset g. In particular, using (4.40) for $A(t) = g(t)$ gives the log-drift coefficient of the numéraire asset g under measure $\widetilde{\mathbb{P}}^{(g)}$,

$$\mathrm{d}g(t) = g(t)\left[\left(r(t) + \|\boldsymbol{\sigma}^{(g)}(t)\|^2\right)\mathrm{d}t + \boldsymbol{\sigma}^{(g)}(t)\cdot \mathrm{d}\widetilde{\mathbf{W}}^{(g)}(t)\right]. \tag{4.42}$$

Applying the Itô quotient rule and cancelling terms in the drift gives the driftless SDE

$$\mathrm{d}\left(\frac{A(t)}{g(t)}\right) = \frac{A(t)}{g(t)}\left(\boldsymbol{\sigma}^{(A)}(t) - \boldsymbol{\sigma}^{(g)}(t)\right) \cdot \mathrm{d}\widetilde{\mathbf{W}}^{(g)}(t)\,. \tag{4.43}$$

This is consistent with the fact that the ratio process $\{\frac{A(t)}{g(t)}\}_{t\geqslant 0}$ must be a $\widetilde{\mathbb{P}}^{(g)}$-martingale.

From (4.31) and the definition in (4.33) we see that the Radon–Nikodym derivative process at any time $0 \leqslant t \leqslant T$ is given by the ratio of the numéraire asset price relative to the bank account value at times t and initial time 0, which has the equivalent expressions:

$$\varrho_t^{B\to g} := \mathcal{E}_t(\boldsymbol{\sigma}^{(g)} \cdot \widetilde{\mathbf{W}}) = \frac{g(t)/B(t)}{g(0)/B(0)} = \frac{g(t)}{g(0)}\frac{B(0)}{B(t)}\,. \tag{4.44}$$

Using this expression for any time $t \leqslant T$, and for time T, gives us the ratio of the Radon–Nikodym derivative process at the two times,

$$\frac{\varrho_T^{B\to g}}{\varrho_t^{B\to g}} = \frac{g(T)/B(T)}{g(0)/B(0)}\frac{g(0)/B(0)}{g(t)/B(t)} = \frac{g(T)/B(T)}{g(t)/B(t)} = \frac{g(T)}{g(t)}\frac{B(t)}{B(T)}\,. \tag{4.45}$$

We also note that the change of measure in the opposing direction, $\widetilde{\mathbb{P}}^{(g)} \to \widetilde{\mathbb{P}}^{(B)}$, is specified by the Radon–Nikodym derivative process

$$\varrho_t^{g\to B} \equiv \left(\frac{\mathrm{d}\widetilde{\mathbb{P}}^{(B)}}{\mathrm{d}\widetilde{\mathbb{P}}^{(g)}}\right)_t = \left[\left(\frac{\mathrm{d}\widetilde{\mathbb{P}}^{(g)}}{\mathrm{d}\widetilde{\mathbb{P}}^{(B)}}\right)_t\right]^{-1} = \frac{1}{\varrho_t^{B\to g}} \tag{4.46}$$

and hence

$$\frac{\varrho_T^{g\to B}}{\varrho_t^{g\to B}} = \left(\frac{\varrho_T^{B\to g}}{\varrho_t^{B\to g}}\right)^{-1} = \frac{g(t)}{g(T)}\frac{B(T)}{B(t)}\,. \tag{4.47}$$

By applying the above measure change, $\widetilde{\mathbb{P}} \equiv \widetilde{\mathbb{P}}^{(B)} \to \widetilde{\mathbb{P}}^{(g)}$, to (4.22) while using the property in (2.120) with $\varrho_t \equiv \varrho_t^{g\to B}$, and noting that $\widetilde{\mathrm{E}}^{(B)}[\;] \equiv \widetilde{\mathrm{E}}[\;]$, we obtain

$$V_t = \widetilde{\mathrm{E}}^{(B)}\left[\frac{B(t)}{B(T)} V_T \,\middle|\, \mathcal{F}_t\right] = \widetilde{\mathrm{E}}^{(g)}\left[\frac{\varrho_T^{g\to B}}{\varrho_t^{g\to B}}\frac{B(t)}{B(T)} V_T \,\middle|\, \mathcal{F}_t\right] = \widetilde{\mathrm{E}}^{(g)}\left[\frac{g(t)}{g(T)} V_T \,\middle|\, \mathcal{F}_t\right]. \tag{4.48}$$

This is the *numéraire invariant form of the risk-neutral pricing formula for any attainable payoff V_T, given any choice of non-dividend-paying positive asset price process g as numéraire.* Since $g(t)$ is $\mathcal{F}_t$-measurable we can pull it out of the expectation, giving

$$V_t = g(t)\,\widetilde{\mathrm{E}}^{(g)}\left[\frac{V_T}{g(T)} \,\middle|\, \mathcal{F}_t\right]. \tag{4.49}$$

This general form of the asset pricing formula is also a statement of the fact that the derivative value relative to the numéraire asset price process, $\left\{\frac{V_t}{g(t)}\right\}_{0\leqslant t\leqslant T}$, is a $\widetilde{\mathbb{P}}^{(g)}$-martingale. The reader can also verify that the value of a self-financing replicating portfolio relative to the numéraire price, $\left\{\frac{\Pi_t}{g(t)}\right\}_{0\leqslant t\leqslant T}$, is a $\widetilde{\mathbb{P}}^{(g)}$-martingale. Then, since an attainable payoff can (by definition) be replicated, we have $\Pi_t = V_t$, which gives (4.49).

Note that the derivative price in (4.49) is expressed as an $\mathcal{F}_t$-measurable random variable. For most payoffs, as in standard and some path-dependent European derivatives, we can employ a (joint) Markov property, which then allows us to express V_t as a function of underlying random variables at time t. For example, in standard basket options with payoff $V_T = \Lambda(T, \mathbf{S}(T))$, we can use the Markov property of the vector stock price process $\{\mathbf{S}(t)\}_{t\geqslant 0}$. In particular, if the numéraire is also a function of the underlying stocks, i.e., $g(t) = g(t, \mathbf{S}(t))$, then the pricing formula gives $V_t = V(t, \mathbf{S}(t))$:

$$V(t, \mathbf{S}(t)) = g(t, \mathbf{S}(t))\, \widetilde{\mathrm{E}}^{(g)}\left[\frac{\Lambda(T, \mathbf{S}(T))}{g(T, \mathbf{S}(T))}\,\middle|\, \mathbf{S}(t)\right]. \tag{4.50}$$

Conditioning on $\mathcal{F}_t$ has been reduced to conditioning on the time-t stock price vector $\mathbf{S}(t)$. The pricing function, which is a function of time t and the spot variables $\mathbf{S} = (S_1, \ldots, S_n)$, is then given by setting $\mathbf{S}(t) = \mathbf{S}$,

$$V(t, \mathbf{S}) = g(t, \mathbf{S})\, \widetilde{\mathrm{E}}^{(g)}_{t,\mathbf{S}}\left[\frac{\Lambda(T, \mathbf{S}(T))}{g(T, \mathbf{S}(T))}\right]. \tag{4.51}$$

We now extend the above formulation to the case where the assets can pay dividends which are adapted processes. Any *dividend-paying asset* has a price process denoted by $\{A(t)\}_{t\geqslant 0}$ and obeys an SDE of the form

$$\mathrm{d}A(t) = A(t)\big[(r(t) - q_A(t))\, \mathrm{d}t + \boldsymbol{\sigma}^{(A)}(t)\cdot \mathrm{d}\widetilde{\mathbf{W}}(t)\big] \tag{4.52}$$

where $q_A(t)$ is an adapted dividend yield for asset A at time t. Note that this is equivalent to (4.39) when the dividend yield is zero. The asset price $A(t)$ can be any base stock price $S_i(t)$, any portfolio in the base assets, or any derivative asset. For the case with a general numéraire asset g, the SDE in (4.52) takes the form

$$\mathrm{d}A(t) = A(t)\big[(r(t) - q_A(t) + \boldsymbol{\sigma}^{(A)}(t)\cdot\boldsymbol{\sigma}^{(g)}(t))\, \mathrm{d}t + \boldsymbol{\sigma}^{(A)}(t)\cdot \mathrm{d}\widetilde{\mathbf{W}}^{(g)}(t)\big] \tag{4.53}$$

For a stock price process $\{S_i(t)\}_{t\geqslant 0}$ with a dividend yield process $\{q_i(t)\}_{t\geqslant 0}$ we have

$$\mathrm{d}S_i(t) = S_i(t)\big[\,\big(r(t) - q_i(t) + \boldsymbol{\sigma}_i(t)\cdot\boldsymbol{\sigma}^{(g)}(t)\big)\, \mathrm{d}t + \boldsymbol{\sigma}_i(t)\cdot \mathrm{d}\widetilde{\mathbf{W}}^{(g)}(t)\big], \tag{4.54}$$

and, in particular,

$$\mathrm{d}S_i(t) = S_i(t)\big[\,\big(r(t) - q_i(t)\big)\, \mathrm{d}t + \boldsymbol{\sigma}_i(t)\cdot \mathrm{d}\widetilde{\mathbf{W}}(t)\big], \tag{4.55}$$

It is important to note that we are still defining the same measure $\widetilde{\mathbb{P}}^{(g)}$ given by the Radon–Nikodym derivative in (4.33). However, we now also allow the numéraire (domestic) asset g to have an adapted dividend yield $q_g(t)$ where

$$\mathrm{d}g(t) = g(t)\big[(r(t) - q_g(t))\, \mathrm{d}t + \boldsymbol{\sigma}^{(g)}(t)\cdot \mathrm{d}\widetilde{\mathbf{W}}(t)\big]. \tag{4.56}$$

The dividend drift portions can be "eliminated" by defining the price processes

$$\widehat{S}_i(t) := \mathrm{e}^{\int_0^t q_i(s)\,\mathrm{d}s} S_i(t),\quad \widehat{A}(t) := \mathrm{e}^{\int_0^t q_A(s)\,\mathrm{d}s} A(t)\ \text{, and}\ \ \widehat{g}(t) := \mathrm{e}^{\int_0^t q_g(s)\,\mathrm{d}s} g(t)\,.$$

Recall that the process $\hat{A}(t)$ (where A can be replaced by S_i or g) equals the value of a portfolio that starts with one share and reinvests all dividends back into the asset. Note that $\widehat{A}(0) = A(0)$, $\widehat{g}(0) = g(0)$, $\widehat{S}_i(0) = S_i(0)$. It follows that the ratio processes $\left\{\frac{\widehat{A}(t)}{B(t)}\right\}_{t\geqslant 0}$, $\left\{\frac{\widehat{S}_i(t)}{B(t)}\right\}_{t\geqslant 0}$, for all i, and $\left\{\frac{\widehat{g}(t)}{B(t)}\right\}_{t\geqslant 0}$ are all $\widetilde{\mathbb{P}}$-martingales:

$$\frac{\widehat{A}(t)/B(t)}{A(0)/B(0)} = \mathcal{E}_t(\boldsymbol{\sigma}^{(A)}\cdot\widetilde{\mathbf{W}}),\quad \frac{\widehat{S}_i(t)/B(t)}{S_i(0)/B(0)} = \mathcal{E}_t(\boldsymbol{\sigma}_i\cdot\widetilde{\mathbf{W}}),\quad \frac{\widehat{g}(t)/B(t)}{g(0)/B(0)} = \mathcal{E}_t(\boldsymbol{\sigma}^{(g)}\cdot\widetilde{\mathbf{W}}).$$

The ratio process $\left\{\frac{\widehat{A}(t)}{\widehat{g}(t)}\right\}_{t\geqslant 0}$ and $\left\{\frac{\widehat{S}_i(t)}{\widehat{g}(t)}\right\}_{t\geqslant 0}$, for all i, are $\widetilde{\mathbb{P}}^{(g)}$-martingales, i.e.,

$$\frac{\widehat{A}(t)}{\widehat{g}(t)} = \frac{A(0)}{g(0)}\,\mathcal{E}_t((\boldsymbol{\sigma}^{(A)} - \boldsymbol{\sigma}^{(g)}) \cdot \widetilde{\mathbf{W}}^{(g)}) \tag{4.57}$$

and

$$\frac{\widehat{S}_i(t)}{\widehat{g}(t)} = \frac{S_i(0)}{g(0)}\,\mathcal{E}_t((\boldsymbol{\sigma}_i - \boldsymbol{\sigma}^{(g)}) \cdot \widetilde{\mathbf{W}}^{(g)})\,. \tag{4.58}$$

These reproduce our earlier formulae in case the dividend yields are zero.

The expressions in (4.44)–(4.47) are now given in terms of the process $\widehat{g}$:

$$\varrho_t^{B\to g} := \mathcal{E}_t(\boldsymbol{\sigma}^{(g)} \cdot \widetilde{\mathbf{W}}) = \frac{\widehat{g}(t)/B(t)}{g(0)/B(0)} = \frac{\widehat{g}(t)}{g(0)}\frac{B(0)}{B(t)}\,, \tag{4.59}$$

$$\frac{\varrho_T^{B\to g}}{\varrho_t^{B\to g}} = \frac{\widehat{g}(T)/B(T)}{\widehat{g}(t)/B(t)} = \frac{\widehat{g}(T)}{\widehat{g}(t)}\frac{B(t)}{B(T)} \tag{4.60}$$

and

$$\frac{\varrho_T^{g\to B}}{\varrho_t^{g\to B}} = \left(\frac{\varrho_T^{B\to g}}{\varrho_t^{B\to g}}\right)^{-1} = \frac{\widehat{g}(t)}{\widehat{g}(T)}\frac{B(T)}{B(t)}\,. \tag{4.61}$$

Finally, the *numéraire invariant form of the risk-neutral pricing formula for any attainable payoff V_T, given any choice of dividend-paying positive asset price process g as numéraire,* takes the equivalent form:

$$V_t = \widehat{g}(t)\,\widetilde{\mathrm{E}}^{(g)}\left[\frac{V_T}{\widehat{g}(T)}\,\middle|\,\mathcal{F}_t\right] = g(t)\,\widetilde{\mathrm{E}}^{(g)}\left[\mathrm{e}^{-\int_t^T q_g(s)\,\mathrm{d}s}\frac{V_T}{g(T)}\,\middle|\,\mathcal{F}_t\right]. \tag{4.62}$$

This general form of the asset pricing formula is also a statement of the fact that the derivative value relative to $\widehat{g}(t)$, $\left\{\frac{V_t}{\widehat{g}(t)}\right\}_{0\leqslant t\leqslant T}$, is a $\widetilde{\mathbb{P}}^{(g)}$-martingale. We remark that this formula recovers (4.49) in case the numéraire has no dividends. When $g = B$, $\widetilde{\mathbb{P}}^{(g)} = \widetilde{\mathbb{P}}$, the domestic bank account is the numéraire, and therefore, the dividend $q_g = q_B \equiv 0$, $g(t)/g(T) = B(t)/B(T)$ and (4.22) is recovered.

4.3 Black–Scholes PDE and Delta Hedging for Standard Multi-Asset Derivatives within a General Diffusion Model

As remarked for the case of the simplest (B, S) model of Chapter 3, the practical implementation of (4.22) depends on the complexity of the market model as well as the type of payoff. The general model considered in Section 4.1, where all the coefficients are adapted processes, is useful for a general theoretical discussion of the main concepts of pricing and replication. However, a general model is nearly impossible to calibrate to the market since the coefficients are too general, with possibly infinite numbers of parameters. In practice, we make additional assumptions on the model that simplify its use in both calibration and derivative pricing. Here, we simply formulate the problem of no-arbitrage pricing of *standard (non-path-dependent) options* within a market model where n risky base asset prices (e.g., n stock prices) are modelled as correlated positive diffusion processes and the interest rate $r(t)$ is assumed to be a nonrandom (ordinary) positive function of time. We also allow for a continuous dividend yield $q_i(t)$ on each stock i, where these are assumed to be ordinary functions of time. The general Black–Scholes PDE satisfied by the pricing function of such options then follows. The formula for computing the delta hedging positions in the multi-asset replicating strategy for a standard option is then derived using a simple application of the multidimensional Itô formula.

Let us consider $n \geqslant 1$ asset (stock) price processes $\{\mathbf{S}(t) = (S_1(t), \ldots, S_n(t))\}_{t \geqslant 0}$ solving

$$\mathrm{d}S_i(t) = S_i(t)\big[(\mu_i(t, \mathbf{S}(t)) - q_i(t))\,\mathrm{d}t + \boldsymbol{\sigma}_i(t, \mathbf{S}(t)) \cdot \mathrm{d}\mathbf{W}(t)\big], \quad i = 1, \ldots, n, \tag{4.63}$$

where $\mathbf{W}(t) = (W_1(t), \ldots, W_n(t))$; i.e., the number of BMs is the same as the number of base assets. Note that this system of SDEs is a special case of (4.2) where the adapted log-drift coefficients and log-volatility vectors are specified functions of the asset prices $\mathbf{S}(t)$ and time t: $\mu_i(t) \equiv \mu_i(t, \mathbf{S}(t))$ and $\boldsymbol{\sigma}_i(t) \equiv \boldsymbol{\sigma}_i(t, \mathbf{S}(t))$. We assume that the $n \times n$ log-diffusion matrix of elements $C_{ij}(t, \mathbf{S}(t)) := \boldsymbol{\sigma}_i(t, \mathbf{S}(t)) \cdot \boldsymbol{\sigma}_j(t, \mathbf{S}(t))$ is invertible (nonsingular). We also assume that all Itô integrals are square-integrable martingales and that the coefficient functions are such that a unique strong solution to (4.63) exists. Based on the analysis of the previous section, we hence have a complete market model with a unique risk-neutral probability measure $\widetilde{\mathbb{P}}$ where

$$\mathrm{d}S_i(t) = S_i(t)\big[(r(t) - q_i(t))\,\mathrm{d}t + \boldsymbol{\sigma}_i(t, \mathbf{S}(t)) \cdot \mathrm{d}\widetilde{\mathbf{W}}(t)\big], \quad i = 1, \ldots, n. \tag{4.64}$$

Our goal is to price a European option with a payoff being a function of the terminal random values of the asset prices

$$V_T = \Lambda(\mathbf{S}(T)) \equiv \Lambda(S_1(T), \ldots, S_n(T)). \tag{4.65}$$

This is referred to as a standard "basket option" on n assets or stocks. Later we derive explicit analytical pricing formulae for specific examples of such options within the multidimensional GBM model. By the (joint) Markov property of the time-T solution $\mathbf{S}(T)$ to (4.63), the conditioning on $\mathcal{F}_t$ in (4.22) is replaced with a conditioning on the time-t (random) prices $\mathbf{S}(t)$. The time-t derivative value is a function of the random vector $\mathbf{S}(t)$ and time t, $V_t = V(t, \mathbf{S}(t))$, where (4.22) now takes the form (note that $r(t)$ is nonrandom by assumption so it can be pulled out of the expectation):

$$V(t, \mathbf{S}(t)) = \mathrm{e}^{-\int_t^T r(u)\,\mathrm{d}u}\,\widetilde{\mathrm{E}}\big[V_T \mid \mathcal{F}_t\big] = \mathrm{e}^{-\int_t^T r(u)\,\mathrm{d}u}\,\widetilde{\mathrm{E}}\big[\Lambda(\mathbf{S}(T)) \mid \mathbf{S}(t)\big]. \tag{4.66}$$

The pricing function $V(t, \mathbf{S}) = V(t, S_1, \ldots, S_n)$ depends on calendar time t and the n spot values[2] (ordinary variables) $S_1 > 0, S_2 > 0, \ldots, S_n > 0$. This function is given by the discounted risk-neutral expectation of the payoff conditional on $\mathbf{S}(t) = \mathbf{S} \equiv (S_1, \ldots, S_n)$, i.e., with joint condition $S_1(t) = S_1, \ldots, S_n(t) = S_n$:

$$V(t, \mathbf{S}) = \mathrm{e}^{-\int_t^T r(u)\,\mathrm{d}u}\,\widetilde{\mathrm{E}}[\Lambda(\mathbf{S}(T)) \mid \mathbf{S}(t) = \mathbf{S}] \equiv \mathrm{e}^{-\int_t^T r(u)\,\mathrm{d}u}\,\widetilde{\mathrm{E}}_{t,\mathbf{S}}[\Lambda(\mathbf{S}(T))]. \tag{4.67}$$

By applying the discounted Feynman–Kac Theorem 2.24, we therefore have that the pricing function $V = V(t, S_1, \ldots, S_n)$ solves the *multi-asset Black–Scholes PDE (BSPDE)*

$$\frac{\partial V}{\partial t} + \frac{1}{2}\sum_{i=1}^{n}\sum_{j=1}^{n} C_{ij}(t, \mathbf{S}) S_i S_j \frac{\partial^2 V}{\partial S_i \partial S_j} + \sum_{i=1}^{n}(r(t) - q_i(t)) S_i \frac{\partial V}{\partial S_i} - r(t)V = 0 \tag{4.68}$$

[2]As in Chapter 3, in pricing formulae, we prefer to choose more appropriate letters for some of the ordinary variables. Here we denote the spot values by S_i, instead of using some other letter like x_i. The ordinary variables S_i should not be confused with the random variables such as $S_i(t)$ and $S_i(T)$.

subject to the terminal condition $V(T, \mathbf{S}) = \Lambda(\mathbf{S})$, where $\Lambda(\mathbf{S}) \equiv \Lambda(S_1, \ldots, S_n)$ is the payoff function. For continuous $\Lambda(\mathbf{S})$, we also have the $t \nearrow T$ limit $V(T-, \mathbf{S}) = V(T, \mathbf{S})$.

The BSPDE in (4.68) is also written compactly as $\frac{\partial V}{\partial t} + \mathcal{G}_{t,\mathbf{S}}V - r(t)V = 0$, where

$$\mathcal{G}_{t,\mathbf{S}}V := \frac{1}{2}\sum_{i=1}^{n}\sum_{j=1}^{n} C_{ij}(t, \mathbf{S})S_i S_j \frac{\partial^2 V}{\partial S_i \partial S_j} + \sum_{i=1}^{n}(r(t) - q_i(t))S_i \frac{\partial V}{\partial S_i} \tag{4.69}$$

is the generator for the system of SDEs in (4.64). The fundamental solution to (4.68) is the $\widetilde{\mathbb{P}}$-measure transition PDF, denoted by $\widetilde{p}(t, T; \mathbf{S}, \mathbf{y})$, multiplied by the discount factor $\mathrm{e}^{-\int_t^T r(u)\,\mathrm{d}u}$. In terms of this risk-neutral transition PDF, the price in (4.67) is given by an n-dimensional integral

$$V(t, \mathbf{S}) = \mathrm{e}^{-\int_t^T r(u)\,\mathrm{d}u} \int_{\mathbb{R}_+^n} \Lambda(\mathbf{y})\widetilde{p}(t, T; \mathbf{S}, \mathbf{y})\,\mathrm{d}\mathbf{y}\,. \tag{4.70}$$

We remark that if the log-volatility vectors are time independent, $\boldsymbol{\sigma}_i(t, \mathbf{S}(t)) \equiv \boldsymbol{\sigma}_i(\mathbf{S}(t))$, then $C_{ij}(t, \mathbf{S}) \equiv C_{ij}(\mathbf{S})$ is time independent. Moreover, if the interest rate $r(t) = r$ is constant and all dividend yields are constants, $q_i(t) = q_i$, then (4.64) is time homogeneous. The BSPDE is then time homogeneous and the pricing function can also be expressed as a function $v(\tau, \mathbf{S}) = V(t, \mathbf{S})$, where $\tau = T - t$ is time to maturity. The BSPDE in (4.68) is then equivalent to

$$\frac{\partial v}{\partial \tau} = \frac{1}{2}\sum_{i=1}^{n}\sum_{j=1}^{n} C_{ij}(\mathbf{S})S_i S_j \frac{\partial^2 v}{\partial S_i \partial S_j} + \sum_{i=1}^{n}(r - q_i)S_i \frac{\partial v}{\partial S_i} - rv \tag{4.71}$$

subject to the initial condition $v(0, \mathbf{S}) = \Lambda(\mathbf{S})$. The risk-neutral transition PDF is then time homogeneous, i.e. $\widetilde{p}(t, T; \mathbf{S}, \mathbf{y}) = \widetilde{p}(\tau; \mathbf{S}, \mathbf{y})$, and the pricing function is given by

$$v(\tau, \mathbf{S}) = \mathrm{e}^{-r\tau}\,\widetilde{\mathrm{E}}[\,\Lambda(\mathbf{S}(T)) \mid \mathbf{S}(t) = \mathbf{S}\,] = \mathrm{e}^{-r\tau} \int_{\mathbb{R}_+^n} \Lambda(\mathbf{y})\widetilde{p}(\tau; \mathbf{S}, \mathbf{y})\,\mathrm{d}\mathbf{y}\,. \tag{4.72}$$

Let us now turn to the derivation of the hedging positions required in a self-financing strategy that replicates the above standard European multi-asset option. What we will achieve shortly below is a more explicit form for the martingale representation in (4.23) since the option price process is a smooth (pricing) function of the stock price process. The first step is to apply the Itô formula and compute the stochastic differential $\mathrm{d}V_t = \mathrm{d}V(t, \mathbf{S}(t))$ of the option value process:

$$\begin{aligned}
\mathrm{d}V_t &= \left[\frac{\partial V}{\partial t}(t, \mathbf{S}(t)) + \frac{1}{2}\sum_{i=1}^{n}\sum_{j=1}^{n} C_{ij}(t, \mathbf{S})S_i(t)S_j(t)\frac{\partial^2 V}{\partial S_i \partial S_j}(t, \mathbf{S}(t))\right]\mathrm{d}t + \sum_{i=1}^{n}\frac{\partial V}{\partial S_i}(t, \mathbf{S}(t))\,\mathrm{d}S_i(t) \\
&= \left(\frac{\partial}{\partial t} + \mathcal{G}_{t,\mathbf{S}}\right)V(t, \mathbf{S}(t))\,\mathrm{d}t + \sum_{i=1}^{n} S_i(t)\frac{\partial V}{\partial S_i}(t, \mathbf{S}(t))\,\boldsymbol{\sigma}_i(t, \mathbf{S}(t))\cdot \mathrm{d}\widetilde{\mathbf{W}}(t) \\
&= r(t)V(t, \mathbf{S}(t))\,\mathrm{d}t + \sum_{i=1}^{n} S_i(t)\frac{\partial V}{\partial S_i}(t, \mathbf{S}(t))\,\boldsymbol{\sigma}_i(t, \mathbf{S}(t))\cdot \mathrm{d}\widetilde{\mathbf{W}}(t)\,.
\end{aligned} \tag{4.73}$$

In the last line, we used the fact that the pricing function $V(t, \mathbf{S})$ solves the BSPDE, i.e., $\frac{\partial V}{\partial t} + \mathcal{G}_{t,\mathbf{S}}V(t, \mathbf{S}) = r(t)V(t, \mathbf{S})$ for all $(t, \mathbf{S})$. The second step is to use the Itô product rule and obtain the stochastic differential of the discounted option value, which simplifies to a driftless expression upon using (4.73):

$$\begin{aligned} \mathrm{d}[D(t)V_t] &= D(t)\left[\mathrm{d}V(t,\mathbf{S}(t)) - r(t)V(t,\mathbf{S}(t))\,\mathrm{d}t\right] \\ &= \sum_{i=1}^{n} \overline{S}_i(t)\frac{\partial V}{\partial S_i}(t,\mathbf{S}(t))\,\boldsymbol{\sigma}_i(t,\mathbf{S}(t))\cdot \mathrm{d}\widetilde{\mathbf{W}}(t)\,. \end{aligned} \tag{4.74}$$

In integral form,

$$D(t)V_t \equiv D(t)V(t,\mathbf{S}(t)) = V(0,\mathbf{S}(0)) + \int_0^t \underbrace{\left(\sum_{i=1}^{n} \overline{S}_i(u)\frac{\partial V}{\partial S_i}(u,\mathbf{S}(u))\,\boldsymbol{\sigma}_i(u,\mathbf{S}(u))\right)}_{=\widetilde{\boldsymbol{\theta}}(u)} \cdot \mathrm{d}\widetilde{\mathbf{W}}(u)\,. \tag{4.75}$$

This is the representation in (4.23) where we identify the adapted vector process $\widetilde{\boldsymbol{\theta}}(t)$, for all $0 \leqslant t < T$. Finally, by equating this vector with that in (4.24), we obtain the *delta hedging position in each ith asset as the partial derivative of the option pricing formula w.r.t. the ith spot variable*:

$$\Delta_t^i = \frac{\partial V}{\partial S_i}(t,\mathbf{S}(t)) \equiv \frac{\partial V}{\partial S_i}(t,\mathbf{S})\bigg|_{\mathbf{S}=\mathbf{S}(t)}, \quad 0 \leqslant t < T. \tag{4.76}$$

This clearly generalizes the delta hedging formula in (3.33) in Chapter 3 for a standard European option on a single asset (stock). As in the single asset case, we define the delta hedging functions $\Delta^i(t,\mathbf{S}) \equiv \Delta^i(t,S_1,\ldots,S_n)$,

$$\Delta^i(t,\mathbf{S}) := \frac{\partial V}{\partial S_i}(t,\mathbf{S})\,, \quad i = 1,\ldots,n,$$

where $\Delta^i(t,\mathbf{S})$ gives the time-t position in the ith base asset given the spot values $\mathbf{S} = (S_1,\ldots,S_n)$. Given the pricing function $V(t,\mathbf{S})$, the partial derivatives $\Delta^i(t,\mathbf{S})$ can be computed and the self-financing replicating portfolio in (4.17) is specified by taking a (long or short) position $\Delta_t^1 = \Delta^1(t,\mathbf{S}(t))$ in the first asset with share price $S_1(t)$, $\Delta_t^2 = \Delta^2(t,\mathbf{S}(t))$ in the second asset with share price $S_2(t)$, ..., $\Delta_t^n = \Delta^n(t,\mathbf{S}(t))$ in the nth asset with share price $S_n(t)$ at time t. The time-t position in the bank account is then given by $\beta_t = D(t)[V(t,\mathbf{S}(t)) - \boldsymbol{\Delta}_t \cdot \mathbf{S}(t)]$; i.e., the value of the bank account portion of the portfolio at time t is $V(t,\mathbf{S}(t)) - \boldsymbol{\Delta}_t \cdot \mathbf{S}(t)$, where $\mathbf{S}(t)\cdot\boldsymbol{\Delta}_t = \sum_{i=1}^{n} \Delta_t^i S_i(t)$.

4.3.1 Standard European Option Pricing for Multi-Stock GBM

In order to be able to derive explicit pricing and hedging formulae for several standard options that occur in practice, we now specialize Section 4.3 to the classical multidimensional GBM market model where all log-volatility vectors, and hence all matrix elements of the covariance matrix of log-returns, are assumed to be *constants*. We shall assume a nonsingular $n \times n$ covariance matrix of log-returns and set $d = n$, i.e., the number of independent BMs and the number of risky assets to be equal. Hence, our multidimensional GBM market model has a unique risk-neutral measure $\widetilde{\mathbb{P}}$. By the fundamental theorems of asset pricing, the market model is therefore complete with no arbitrage opportunities.

We refer the reader to Section 2.10 of Chapter 2 where the multidimensional GBM process was introduced. Here, we also include a constant dividend yield q_i on each ith stock where the stock prices satisfy the system of SDEs

$$\mathrm{d}S_i(t) = S_i(t)\big[(r - q_i)\,\mathrm{d}t + \boldsymbol{\sigma}_i\cdot \mathrm{d}\widetilde{\mathbf{W}}(t)\big],\ i = 1,\ldots,n. \tag{4.77}$$

We recall our previous notation where the volatility vectors $\boldsymbol{\sigma}_i = [\sigma_{i1}, \ldots, \sigma_{in}]$ have square magnitude denoted by $\sigma_i^2 \equiv \|\boldsymbol{\sigma}_i\|^2$ and their inner product $\boldsymbol{\sigma}_i \cdot \boldsymbol{\sigma}_j = \rho_{ij}\sigma_i\sigma_j$. The correlation matrix $\boldsymbol{\rho}$ of constant elements $\rho_{ij} \in (-1, 1)$ gives the correlations among all pairwise log-returns, as shown in (2.236). The unique strong solution subject to initial prices $S_1(0), \ldots, S_n(0)$ is given by (see (2.229))

$$S_i(t) = S_i(0)\,\mathrm{e}^{(r-q_i-\frac{1}{2}\sigma_i^2)t+\boldsymbol{\sigma}_i\cdot\widetilde{\mathbf{W}}(t)} = S_i(0)\,\mathrm{e}^{(r-q_i)t}\mathcal{E}_t(\boldsymbol{\sigma}_i \cdot \widetilde{\mathbf{W}}) \tag{4.78}$$

where

$$\mathcal{E}_t(\boldsymbol{\sigma}_i \cdot \widetilde{\mathbf{W}}) = \exp\left[-\frac{1}{2}\sigma_i^2 t + \boldsymbol{\sigma}_i \cdot \widetilde{\mathbf{W}}(t)\right] = \exp\left[-\frac{1}{2}\sigma_i^2 t + \sum_{j=1}^{d}\sigma_{ij}\widetilde{W}_j(t)\right] \tag{4.79}$$

is an exponential $\widetilde{\mathbb{P}}$-martingale for each $i = 1, \ldots, n$. From (4.78) we have the stock price vector at time T expressed in terms of its value at time $t \leqslant T$:

$$\begin{aligned} S_i(T) &= S_i(t)\,\mathrm{e}^{(r-q_i-\frac{1}{2}\sigma_i^2)(T-t)+\boldsymbol{\sigma}_i\cdot(\widetilde{\mathbf{W}}(T)-\widetilde{\mathbf{W}}(t))} \\ &= S_i(t)\,\mathrm{e}^{(r-q_i-\frac{1}{2}\sigma_i^2)\tau+\sigma_i\sqrt{\tau}\widetilde{Z}_i} \end{aligned} \tag{4.80}$$

where $\tau = T - t$ is time to maturity and we define

$$\widetilde{Z}_i := \frac{\boldsymbol{\sigma}_i \cdot (\widetilde{\mathbf{W}}(T) - \widetilde{\mathbf{W}}(t))}{\sigma_i\sqrt{\tau}} = \frac{1}{\sigma_i\sqrt{\tau}}\sum_{j=1}^{d}\sigma_{ij}(\widetilde{W}_j(T) - \widetilde{W}_j(t)), \quad i = 1, \ldots, n.$$

Under measure $\widetilde{\mathbb{P}}$, these random variables are i.i.d. *Norm*$(0, 1)$ and their joint distribution is the multivariate standard normal with correlation matrix $\boldsymbol{\rho}$,

$$\widetilde{\mathbf{Z}} = [\widetilde{Z}_1, \ldots, \widetilde{Z}_n]^\top \sim Norm_n(\mathbf{0}, \boldsymbol{\rho})\,. \tag{4.81}$$

The pricing function $V(t, \mathbf{S}) = V(t, S_1, \ldots, S_n)$ is given as in (4.67) but now the interest rate is constant,

$$V(t, \mathbf{S}) = \mathrm{e}^{-r(T-t)}\,\widetilde{\mathrm{E}}[\,\Lambda(\mathbf{S}(T)) \mid \mathbf{S}(t) = \mathbf{S}\,] \equiv \mathrm{e}^{-r(T-t)}\,\widetilde{\mathrm{E}}_{t,\mathbf{S}}[\,\Lambda(\mathbf{S}(T))]\,. \tag{4.82}$$

By substituting (4.80) into (4.82), using the joint independence of $\mathbf{S}(t)$ and $\widetilde{\mathbf{Z}}$ and conditioning on the spot values $S_1(t) = S_1, \ldots, S_n(t) = S_n$, we obtain the pricing function as an n-dimensional integral in the standard Gaussian density:

$$\begin{aligned} V(t, \mathbf{S}) &= \mathrm{e}^{-r\tau}\,\widetilde{\mathrm{E}}[\,\Lambda(S_1\mathrm{e}^{(r-q_1-\frac{1}{2}\sigma_1^2)\tau+\sigma_1\sqrt{\tau}\widetilde{Z}_1}, \ldots, S_n\mathrm{e}^{(r-q_n-\frac{1}{2}\sigma_n^2)\tau+\sigma_n\sqrt{\tau}\widetilde{Z}_n})\,] \\ &= \mathrm{e}^{-r\tau}\int\!\!\ldots\!\int_{\mathbb{R}^n}\Lambda(S_1\mathrm{e}^{(r-q_1-\frac{1}{2}\sigma_1^2)\tau+\sigma_1\sqrt{\tau}z_1}, \ldots, S_n\mathrm{e}^{(r-q_n-\frac{1}{2}\sigma_n^2)\tau+\sigma_n\sqrt{\tau}z_n})\,n_n(\mathbf{z}; \boldsymbol{\rho})\,\mathrm{d}\mathbf{z} \end{aligned} \tag{4.83}$$

where $n_n(\mathbf{z}; \boldsymbol{\rho}) = n(z_1, \ldots, z_n; \boldsymbol{\rho})$ is the multivariate standard normal PDF in (2.238). This generalizes the pricing formula in (3.27) to the case of n stocks. By a similar change of integration variables, i.e., setting $y_i = S_i\,\mathrm{e}^{(r-q_i-\frac{1}{2}\sigma_i^2)\tau+\sigma_i\sqrt{\tau}z_i}$, we obtain the generalization of the pricing formula in (3.28),

$$V(t, \mathbf{S}) = \mathrm{e}^{-r\tau}\int_0^\infty \ldots \int_0^\infty \Lambda(\mathbf{y})\,\widetilde{p}(t, T; \mathbf{S}, \mathbf{y})\,\mathrm{d}\mathbf{y}\,. \tag{4.84}$$

The transition PDF in the risk-neutral measure, $\widetilde{p}(t,T;\mathbf{S},\mathbf{y})$, is time homogeneous; i.e., $\widetilde{p}(t,T;\mathbf{S},\mathbf{y}) \equiv \widetilde{p}(\tau;\mathbf{S},\mathbf{y})$ is given by the multivariate log-normal expression in (2.239) for all positive vectors $\mathbf{S},\mathbf{y}$, and where

$$a_i = \frac{\ln\frac{y_i}{S_i} - (r - q_i - \sigma_i^2/2)\tau}{\sigma_i\sqrt{\tau}}\,, \quad i = 1,\ldots,n.$$

The BSPDE in (4.68) simplifies to a parabolic PDE in the n spot variables:

$$\frac{\partial V}{\partial t} + \frac{1}{2}\sum_{i=1}^{n}\sum_{j=1}^{n}\rho_{ij}\sigma_i\sigma_j S_i S_j\frac{\partial^2 V}{\partial S_i\partial S_j} + \sum_{i=1}^{n}(r-q_i)S_i\frac{\partial V}{\partial S_i} - rV = 0 \tag{4.85}$$

subject to the terminal condition $V(T,\mathbf{S}) = \Lambda(\mathbf{S})$. In compact form this reads $\frac{\partial V}{\partial t} + \mathcal{G}_{\mathbf{S}}V - rV = 0$ where

$$\mathcal{G}_{\mathbf{S}}V := \frac{1}{2}\sum_{i=1}^{n}\sum_{j=1}^{n}\rho_{ij}\sigma_i\sigma_j S_i S_j\frac{\partial^2 V}{\partial S_i\partial S_j} + \sum_{i=1}^{n}(r-q_i)S_i\frac{\partial V}{\partial S_i} \tag{4.86}$$

is the (time-homogeneous) differential generator for the GBM price process solving (4.77). By time homogeneity, the pricing function is expressible as $v(\tau,\mathbf{S}) = V(t,\mathbf{S})$. The BSPDE in (4.85) is equivalent to

$$\frac{\partial v}{\partial \tau} = \frac{1}{2}\sum_{i=1}^{n}\sum_{j=1}^{n}\rho_{ij}\sigma_i\sigma_j S_i S_j\frac{\partial^2 v}{\partial S_i\partial S_j} + \sum_{i=1}^{n}(r-q_i)S_i\frac{\partial v}{\partial S_i} - rv \tag{4.87}$$

subject to the initial condition $v(0,\mathbf{S}) = \Lambda(\mathbf{S})$.

A common case is when $n = 2$ stocks. In this case, there are two log-volatility vectors $\boldsymbol{\sigma}_1 = [\sigma_1, 0]$ and $\boldsymbol{\sigma}_2 = [\sigma_2\rho, \sigma_2\sqrt{1-\rho^2}]$, with volatility parameters $\sigma_1,\sigma_2 > 0$ and only one correlation coefficient $\rho_{12} = \rho_{21} = \rho$ where $\boldsymbol{\sigma}_1\cdot\boldsymbol{\sigma}_2 = \rho\sigma_1\sigma_2$. The pricing function $V = V(t,S_1,S_2)$ solves the BSPDE

$$\begin{aligned}\frac{\partial V}{\partial t} + \frac{1}{2}\sigma_1^2 S_1^2\frac{\partial^2 V}{\partial S_1^2} + \frac{1}{2}\sigma_2^2 S_2^2\frac{\partial^2 V}{\partial S_2^2} + \rho\sigma_1\sigma_2 S_1 S_2\frac{\partial^2 V}{\partial S_1\partial S_2}&\\ +(r-q_1)S_1\frac{\partial V}{\partial S_1} + (r-q_2)S_2\frac{\partial V}{\partial S_2} - rV = 0\,,&\end{aligned} \tag{4.88}$$

subject to the terminal (payoff) condition $V(T,S_1,S_2) = \Lambda(S_1,S_2)$. The analogous BSPDE for $v = v(\tau,S_1,S_2)$ is given by (4.87) for $n = 2$.

We close this section by noting a useful relationship between the pricing function for zero stock dividends and that for dividends. In particular, we have the simple symmetry that extends (3.69) to the case of $n \geqslant 1$ stocks. From the pricing formula in (4.83) we observe that, for each i, the inclusion of a dividend q_i corresponds to changing the spot value $S_i \to S_i\mathrm{e}^{q_i\tau}$ in the pricing formula with $q_i = 0$. In particular, let $V(t,\mathbf{S};\mathbf{q}) \equiv V(t,S_1,\ldots,S_n;q_1,\ldots,q_n)$ be the pricing function for the case that the respective dividend yields on stocks $1,\ldots,n$ are $q_1,\ldots,q_n$ (which can have any real value including zero) and let $V(t,\mathbf{S};\mathbf{0}) \equiv V(t,\mathbf{S})$ be the pricing function in the case that all stock dividends are zero. Then,

$$V(t,\mathbf{S};\mathbf{q}) = V(t,\mathrm{e}^{-q_1(T-t)}S_1,\ldots,\mathrm{e}^{-q_n(T-t)}S_n) \tag{4.89}$$

or equivalently for the pricing functions expressed in terms of time to maturity τ, that is, $v(\tau,\mathbf{S};\mathbf{q}) = v(\tau,\mathrm{e}^{-q_1\tau}S_1,\ldots,\mathrm{e}^{-q_n\tau}S_n)$. Hence, when deriving the pricing function for a standard (non-path-dependent) multi-stock European option for the case of constant dividends, we can derive the pricing function for the case of all dividends being set to zero and then apply the above relation.

4.3.2 Explicit Pricing Formulae for the GBM Model

There are basically three analytical methods or approaches that we can combine or use separately in deriving pricing functions for multi-asset options. One main approach is to implement the risk-neutral pricing formula in (4.22) by computing the expectation by using nested conditioning. Another is the PDE approach, where we solve the BSPDE, which can involve some symmetry reduction technique whereby the original PDE is reduced to a lower dimensional PDE. The other is to use change of probability measures. The change of measure technique can be quite general and amounts to some judicious application of Girsanov's Theorem in order to simplify the evaluation of the expectation in the risk-neutral pricing formula. In Section 4.2, we will cover such an approach where measure changes correspond to a change of numéraires for risk-neutral pricing. In this section, we consider some examples of the first two methods for pricing some standard multi-stock options where the stock prices are correlated GBMs as specified in the previous section. In all our examples, we assume an economy with constant interest rate r and also allow for a constant continuous dividend yield q_i on each ith stock. The extension to the case of nonrandom time-dependent volatility vectors $\boldsymbol{\sigma}_i(t)$, interest rate $r(t)$, and dividend yields $q_i(t)$ is fairly straightforward as the price processes are still GBMs. We leave examples of such extensions as exercises for the reader.

4.3.2.1 Exchange and Other Related Options

A common type of standard European two-stock option is a so-called *exchange option* where the payoff is the positive part of the difference of the two asset prices at maturity. We now consider pricing this type of option with payoff

$$V_T = (\,S_2(T) - S_1(T)\,)^+ \tag{4.90}$$

where each random variable $S_i(T)$ is the terminal share price of stock $i = 1, 2$. In this example, we have $n = 2$ stocks where the payoff is nonzero only in the event that the terminal share price of the second stock is greater than that of the first stock, in which case the payoff is given by the difference of the two share prices. Let $\widetilde{\mathbb{P}}$ be the risk-neutral measure and let $\widetilde{\mathbf{W}}(t) = (\widetilde{W}_1(t), \widetilde{W}_2(t))$ be a standard two-dimensional vector Brownian motion under $\widetilde{\mathbb{P}}$, i.e., $\widetilde{W}_1(t)$ and $\widetilde{W}_2(t)$ are i.i.d. standard $\widetilde{\mathbb{P}}$-Brownian motions. The respective continuously compounded dividend yields on stocks 1 and 2 are denoted by q_1 and q_2. Defining $\tau := T - t > 0$, then by (4.80) we have

$$\begin{aligned} S_1(T) &= S_1(t)\,\mathrm{e}^{(r-q_1-\frac{1}{2}\sigma_1^2)\tau+\boldsymbol{\sigma}_1\cdot(\widetilde{\mathbf{W}}(T)-\widetilde{\mathbf{W}}(t))} \\ &\stackrel{d}{=} S_1(t)\,\mathrm{e}^{(r-q_1-\frac{1}{2}\sigma_1^2)\tau+\sigma_1\sqrt{\tau}\widetilde{Z}_1} \\ S_2(T) &= S_2(t)\,\mathrm{e}^{(r-q_2-\frac{1}{2}\sigma_2^2)\tau+\boldsymbol{\sigma}_2\cdot(\widetilde{\mathbf{W}}(T)-\widetilde{\mathbf{W}}(t))} \\ &\stackrel{d}{=} S_2(t)\,\mathrm{e}^{(r-q_2-\frac{1}{2}\sigma_2^2)\tau+\sigma_2\sqrt{\tau}(\rho\widetilde{Z}_1+\sqrt{1-\rho^2}\widetilde{Z}_2)} \end{aligned} \tag{4.91}$$

where

$$\widetilde{Z}_1 := \frac{\widetilde{W}_1(T) - \widetilde{W}_1(t)}{\sqrt{\tau}} \text{ and } \widetilde{Z}_2 := \frac{\widetilde{W}_2(T) - \widetilde{W}_2(t)}{\sqrt{\tau}}$$

are independent $Norm(0, 1)$ random variables in the measure $\widetilde{\mathbb{P}}$. Recall that $\boldsymbol{\sigma}_1 = [\sigma_1, 0]$ and $\boldsymbol{\sigma}_2 = [\sigma_2\rho, \sigma_2\sqrt{1-\rho^2}]$ are the respective volatility vectors of stocks 1 and 2 where ρ is the correlation coefficient of the log-returns of the two stocks, i.e., $\boldsymbol{\sigma}_1 \cdot \boldsymbol{\sigma}_2 = \rho\sigma_1\sigma_2$.

We denote the respective time-t spot prices of the two stocks by $S_1 > 0$ and $S_2 > 0$.

The option pricing function $v(\tau, S_1, S_2)$ is a function of time to maturity τ and the spot variables S_1, S_2. From the risk-neutral pricing formula in (4.82),

$$\begin{aligned} v(\tau, S_1, S_2) &\equiv \mathrm{e}^{-r\tau}\widetilde{\mathrm{E}}\big[\,(S_2(T) - S_1(T))^+ \mid S_1(t) = S_1, S_2(t) = S_2\big] \\ &= \mathrm{e}^{-r\tau}\widetilde{\mathrm{E}}\Big[\big(S_2\mathrm{e}^{(r-q_2-\frac{1}{2}\sigma_2^2)\tau+\sigma_2\sqrt{\tau}(\rho\widetilde{Z}_1+\sqrt{1-\rho^2}\widetilde{Z}_2)} \\ &\qquad - S_1\mathrm{e}^{(r-q_1-\frac{1}{2}\sigma_1^2)\tau+\sigma_1\sqrt{\tau}\widetilde{Z}_1}\big)^+\Big]. \end{aligned} \tag{4.92}$$

The second equation line is obtained by substituting the expressions in (4.91), conditioning on $S_1(t) = S_1, S_2(t) = S_2$, and using the independence between $\widetilde{Z}_1, \widetilde{Z}_2$ and the joint random variables $S_1(t), S_2(t)$. This expectation is now readily computed by applying nested conditioning. We first condition on $\widetilde{Z}_1$ and thereby isolate the exponential in $\widetilde{Z}_2$, giving

$$\begin{aligned} v(\tau, S_1, S_2) &= \mathrm{e}^{-r\tau}\widetilde{\mathrm{E}}\Big[\,\widetilde{\mathrm{E}}\Big[\big(S_2\mathrm{e}^{(r-q_2-\frac{1}{2}\sigma_2^2)\tau+\sigma_2\sqrt{\tau}(\rho\widetilde{Z}_1+\sqrt{1-\rho^2}\widetilde{Z}_2)} \\ &\qquad - S_1\mathrm{e}^{(r-q_1-\frac{1}{2}\sigma_1^2)\tau+\sigma_1\sqrt{\tau}\widetilde{Z}_1}\big)^+ \mid \widetilde{Z}_1\Big]\Big] \\ &= \mathrm{e}^{-r\tau}S_2\mathrm{e}^{(r-q_2-\frac{1}{2}\sigma_2^2)\tau}\widetilde{\mathrm{E}}\Big[\,\mathrm{e}^{\rho\sigma_2\sqrt{\tau}\widetilde{Z}_1}\,\widetilde{\mathrm{E}}\Big[\big(\,\mathrm{e}^{\sigma_2\sqrt{1-\rho^2}\sqrt{\tau}\widetilde{Z}_2} - X_1\,\big)^+ \mid \widetilde{Z}_1\Big]\Big] \end{aligned} \tag{4.93}$$

where $X_1 \equiv \frac{S_1}{S_2}\mathrm{e}^{(q_2-q_1+\frac{1}{2}(\sigma_2^2-\sigma_1^2))\tau+(\sigma_1-\rho\sigma_2)\sqrt{\tau}\widetilde{Z}_1}$ is a function of only random variable $\widetilde{Z}_1$ (not $\widetilde{Z}_2$) and hence has a fixed value within the inner expectation which is conditioned on $\widetilde{Z}_1$. Since $\widetilde{Z}_1$ and $\widetilde{Z}_2$ are independent, the inner conditional expectation is readily evaluated as an unconditional expectation, $\widetilde{\mathrm{E}}\big[\big(\,\mathrm{e}^{\sigma_2\sqrt{1-\rho^2}\sqrt{\tau}\widetilde{Z}_2} - X_1\,\big)^+ \mid \widetilde{Z}_1\big] = g(X_1)$ where

$$\begin{aligned} g(x_1) &:= \widetilde{\mathrm{E}}\big[\big(\,\mathrm{e}^{\sigma_2\sqrt{1-\rho^2}\sqrt{\tau}\widetilde{Z}_2} - x_1\,\big)^+\big] \\ &= \widetilde{\mathrm{E}}\left[\mathrm{e}^{\sigma_2\sqrt{1-\rho^2}\sqrt{\tau}\widetilde{Z}_2}\mathbb{I}_{\{\widetilde{Z}_2>\frac{\ln x_1}{\sigma_2\sqrt{1-\rho^2}\sqrt{\tau}}\}}\right] - x_1\widetilde{\mathrm{E}}\left[\mathbb{I}_{\{\widetilde{Z}_2>\frac{\ln x_1}{\sigma_2\sqrt{1-\rho^2}\sqrt{\tau}}\}}\right] \\ &= \mathrm{e}^{\frac{1}{2}\sigma_2^2(1-\rho^2)\tau}\mathcal{N}\left(\frac{-\ln x_1 + \sigma_2^2(1-\rho^2)\tau}{\sigma_2\sqrt{1-\rho^2}\sqrt{\tau}}\right) - x_1\mathcal{N}\left(\frac{-\ln x_1}{\sigma_2\sqrt{1-\rho^2}\sqrt{\tau}}\right). \end{aligned}$$

Note that in evaluating the above expectations we have applied the identity (B.1) in Appendix B, i.e., $\widetilde{\mathrm{E}}[\mathrm{e}^{b\widetilde{Z}_2}\mathbb{I}_{\{\widetilde{Z}_2>a\}}] = \mathrm{e}^{\frac{1}{2}b^2}\mathcal{N}(b-a)$, for any real constants a, b, where $\widetilde{Z}_2 \sim N(0,1)$ under $\widetilde{\mathbb{P}}$. Therefore, setting $x_1 = X_1$ into $g(x_1)$:

$$\begin{aligned} &\widetilde{\mathrm{E}}\big[\big(\,\mathrm{e}^{\sigma_2\sqrt{1-\rho^2}\sqrt{\tau}\widetilde{Z}_2} - X_1\,\big)^+ \,|\, \widetilde{Z}_1\big] \\ &= \mathrm{e}^{\frac{1}{2}\sigma_2^2(1-\rho^2)\tau}\mathcal{N}\left(\frac{-\ln X_1 + \sigma_2^2(1-\rho^2)\tau}{\sigma_2\sqrt{1-\rho^2}\sqrt{\tau}}\right) - X_1\mathcal{N}\left(\frac{-\ln X_1}{\sigma_2\sqrt{1-\rho^2}\sqrt{\tau}}\right) \\ &= \mathrm{e}^{\frac{1}{2}\sigma_2^2(1-\rho^2)\tau}\mathcal{N}(A\widetilde{Z}_1 + D) - X_1\mathcal{N}(A\widetilde{Z}_1 + C). \end{aligned} \tag{4.94}$$

Here, we conveniently define the constants

$$A := \frac{\rho\sigma_2 - \sigma_1}{\sigma_2\sqrt{1-\rho^2}}, \quad C := \frac{\ln(S_2/S_1) + [q_1 - q_2 + \frac{1}{2}(\sigma_1^2 - \sigma_2^2)]\tau}{\sigma_2\sqrt{1-\rho^2}\sqrt{\tau}}, \quad D := C + \sigma_2\sqrt{1-\rho^2}\sqrt{\tau}.$$

Substituting the random variable expression on the right-hand side of (4.94) into the outer expectation in (4.93), and writing X_1 in terms of $\widetilde{Z}_1$ and simplifying the exponents, gives

$$\begin{aligned}v(\tau,S_1,S_2) &= S_2 \mathrm{e}^{-(q_2+\frac{1}{2}\sigma_2^2)\tau}\Big(\mathrm{e}^{\frac{1}{2}\sigma_2^2(1-\rho^2)\tau}\widetilde{\mathrm{E}}\big[\,\mathrm{e}^{\rho\sigma_2\sqrt{\tau}\widetilde{Z}_1}\mathcal{N}(A\widetilde{Z}_1+D)\big] \\ &\qquad - \widetilde{\mathrm{E}}\big[\mathrm{e}^{\rho\sigma_2\sqrt{\tau}\widetilde{Z}_1}X_1\mathcal{N}(A\widetilde{Z}_1+C)\big]\Big) \\ &= S_2\mathrm{e}^{-(q_2+\frac{1}{2}\rho^2\sigma_2^2)\tau}\widetilde{\mathrm{E}}\big[\mathrm{e}^{\rho\sigma_2\sqrt{\tau}\widetilde{Z}_1}\mathcal{N}(A\widetilde{Z}_1+D)\big] \\ &\quad - S_1\mathrm{e}^{-(q_1+\frac{1}{2}\sigma_1^2)\tau}\widetilde{\mathrm{E}}\big[\,\mathrm{e}^{\sigma_1\sqrt{\tau}\widetilde{Z}_1}\mathcal{N}(A\widetilde{Z}_1+C)\big].\end{aligned} \tag{4.95}$$

Note that both expectations in (4.95) can be exactly evaluated by using the integral identity (B.4) in Appendix B, i.e., $\widetilde{\mathrm{E}}\big[\mathrm{e}^{b\widetilde{Z}_1}\mathcal{N}(a\widetilde{Z}_1+c)\big] = \mathrm{e}^{b^2/2}\mathcal{N}(\frac{ab+c}{\sqrt{1+a^2}})$, for any real constants a, b, c, and where $\widetilde{Z}_1 \sim N(0,1)$ under $\widetilde{\mathbb{P}}$. Using this identity twice, once for each expectation in (4.95), now gives

$$v(\tau,S_1,S_2) = S_2\mathrm{e}^{-q_2\tau}\mathcal{N}\left(\frac{\rho\sigma_2\sqrt{\tau}A+D}{\sqrt{1+A^2}}\right) - S_1\mathrm{e}^{-q_1\tau}\mathcal{N}\left(\frac{\sigma_1\sqrt{\tau}A+C}{\sqrt{1+A^2}}\right). \tag{4.96}$$

Using the definitions of A, B, C, D above, we now write the two arguments in the standard normal CDF function in terms of the original model parameters. In particular,

$$\sqrt{1+A^2} = \sqrt{1+\frac{(\rho\sigma_2-\sigma_1)^2}{\sigma_2^2(1-\rho^2)}} = \sqrt{\frac{\sigma_2^2(1-\rho^2)+(\rho\sigma_2-\sigma_1)^2}{\sigma_2^2(1-\rho^2)}} = \frac{\nu}{\sigma_2\sqrt{1-\rho^2}}$$

where $\nu := \sqrt{\sigma_1^2+\sigma_2^2-2\rho\sigma_1\sigma_2} = \|\boldsymbol{\sigma}_1-\boldsymbol{\sigma}_2\|$ is the magnitude of the difference of the volatility vectors of the two stocks. Then, the first argument is

$$\begin{aligned}\frac{\rho\sigma_2\sqrt{\tau}A+D}{\sqrt{1+A^2}} &= \frac{\sigma_2\sqrt{1-\rho^2}}{\nu}(\rho\sigma_2\sqrt{\tau}A+D) \\ &= \frac{\rho\sigma_2(\rho\sigma_2-\sigma_1)\tau+\ln(S_2/S_1)+[q_1-q_2+\frac{1}{2}(\sigma_1^2+\sigma_2^2-2\rho\sigma_2^2)]\tau}{\nu\sqrt{\tau}} \\ &= \frac{\ln(S_2/S_1)+[q_1-q_2+\frac{1}{2}(\sigma_1^2+\sigma_2^2-2\rho\sigma_1\sigma_2)]\tau}{\nu\sqrt{\tau}} \\ &= \frac{\ln(S_2/S_1)+(q_1-q_2+\frac{1}{2}\nu^2)\tau}{\nu\sqrt{\tau}}\end{aligned}$$

and the second argument is

$$\begin{aligned}\frac{\sigma_1\sqrt{\tau}A+C}{\sqrt{1+A^2}} &= \frac{\sigma_2\sqrt{1-\rho^2}}{\nu}\left(\frac{\sigma_1(\rho\sigma_2-\sigma_1)\tau+\ln(S_2/S_1)+[q_1-q_2+\frac{1}{2}(\sigma_1^2-\sigma_2^2)]\tau}{\sigma_2\sqrt{1-\rho^2}\sqrt{\tau}}\right) \\ &= \frac{\ln(S_2/S_1)+(q_1-q_2-\frac{1}{2}\nu^2)\tau}{\nu\sqrt{\tau}}.\end{aligned}$$

Inserting these expressions into (4.96) finally gives the option pricing formula

$$v(\tau,S_1,S_2) = S_2\mathrm{e}^{-q_2\tau}\mathcal{N}\left(d_+(\frac{S_2}{S_1},\tau)\right) - S_1\mathrm{e}^{-q_1\tau}\mathcal{N}\left(d_-(\frac{S_2}{S_1},\tau)\right) \tag{4.97}$$

where

$$d_\pm(x,\tau) := \frac{\ln x+(q_1-q_2\pm\frac{1}{2}\nu^2)\tau}{\nu\sqrt{\tau}}, \quad x,\tau>0. \tag{4.98}$$

The delta (hedging) positions in the two stocks can be computed by directly differentiating (4.97) w.r.t. S_1 and S_2 while adapting the algebraic identity in (3.48). We now give an alternate derivation of the hedging positions by exploiting the symmetry of the pricing function. In particular, note that the pricing function in (4.97) can be expressed as a product of S_1 and a function of $x := S_2/S_1$:

$$v(\tau, S_1, S_2) = S_1\left[\mathrm{e}^{-q_2\tau} x\mathcal{N}\big(d_+(x,\tau)\big) - \mathrm{e}^{-q_1\tau}\mathcal{N}\big(d_-(x,\tau)\big)\right] \equiv S_1 C(x, 1, \tau; q_1, q_2, \nu) \quad (4.99)$$

where we identify $C(x, 1, \tau; q_1, q_2, \nu)$ as the Black–Scholes pricing function for a standard European call option on a single underlying with effective "spot" x, "strike" 1, time to maturity τ, "interest rate" q_1, and "dividend" q_2. Hence, at calendar time $t = T - \tau$, the hedging position $\Delta^1(t, S_1, S_2)$ in the first stock is readily derived by taking the partial derivative w.r.t. S_1 while using the known relation for $\Delta_c = \frac{\partial C}{\partial x} = \mathrm{e}^{-q_2\tau}\mathcal{N}\big(d_+(x,\tau)\big)$ and $S_1\frac{\partial x}{\partial S_1} = -x$:

$$\begin{aligned}\Delta^1(t, S_1, S_2) = \frac{\partial v}{\partial S_1} = \frac{\partial}{\partial S_1}\left[S_1 C(x, 1, \tau; q_1, q_2, \nu)\right] &= C(x, 1, \tau; q_1, q_2, \nu) + S_1\frac{\partial x}{\partial S_1}\Delta_c \\ &= C(x, 1, \tau; q_1, q_2, \nu) - x\Delta_c \\ &= -\mathrm{e}^{-q_1\tau}\mathcal{N}\big(d_-(x,\tau)\big)\,. \end{aligned} \quad (4.100)$$

The hedging position in the second stock is computed by taking the partial derivative w.r.t. S_2 of (4.99) while using $S_1\frac{\partial x}{\partial S_2} = 1$,

$$\Delta^2(t, S_1, S_2) = \frac{\partial v}{\partial S_2} = \frac{\partial}{\partial S_2}\left[S_1 C(x, 1, \tau; q_1, q_2, \nu)\right] = S_1\frac{\partial x}{\partial S_2}\Delta_c = \mathrm{e}^{-q_2\tau}\mathcal{N}\big(d_+(x,\tau)\big)\,. \quad (4.101)$$

We note that for $t = T$ ($\tau = 0$) the respective positions are given by the limit $t \nearrow T$ ($\tau = 0+$) of the above expressions. Hence, from the positions (4.100) and (4.101), we see that the *stock portion of the self-financing replicating strategy completely replicates the exchange option.* That is, no bank account is needed (i.e., the position in the bank account $\beta(t) \equiv 0$) in order to replicate the option. The pricing function equals the value of the self-financing stock portfolio,

$$v(\tau, S_1, S_2) = S_1\Delta^1(t, S_1, S_2) + S_2\Delta^2(t, S_1, S_2). \quad (4.102)$$

As a stochastic price process we have $V_t = v(\tau, S_1(t), S_2(t)) = \Delta_t^1 S_1(t) + \Delta_t^2 S_2(t)$, where $\Delta_t^i = \Delta^i(t, S_1(t), S_2(t))$, $i = 1, 2$.

A European-style basket option whose payoff at some expiry date T is given by the maximum or minimum share price between two or more stocks can also be priced by the above methods. For example, in the case of the minimum price of two stocks at expiry T the payoff takes on the equivalent forms

$$\begin{aligned}V_T = \min\left(S_1(T), S_2(T)\right) &= S_1(T) - \left(S_1(T) - S_2(T)\right)^+ \\ &= S_2(T) - \left(S_2(T) - S_1(T)\right)^+. \end{aligned} \quad (4.103)$$

Similarly, for the case of the maximum price of two stocks, the payoff is

$$\begin{aligned}V_T = \max\left(S_1(T), S_2(T)\right) &= \left(S_2(T) - S_1(T)\right)^+ + S_1(T) \\ &= \left(S_1(T) - S_2(T)\right)^+ + S_2(T). \end{aligned} \quad (4.104)$$

The expressions in (4.103) and (4.104) follow from $\max(x, y) = (y-x)^+ + x = (x-y)^+ + y$ and $\min(x, y) = x - (x-y)^+ = y - (y-x)^+$. We also make note of a related useful identity:

$\min(x, y) + \max(x, y) = x + y$. As shown just below, the main point of (4.103) and (4.104) is that the problem of pricing a two-stock option with the payoff specified as either the minimum or maximum is immediately solved once we have priced the two-stock option with payoff in (4.90). The converse is also true. The pricing formula for the latter option was derived in the previous section with the explicit expression given in (4.97).

As in the previous section, let S_1, S_2 be the spot values, $\tau = T - t$ be the time to maturity, and let $v_{min}(\tau, S_1, S_2)$ denote the time-t price of the two-stock option on the minimum with payoff in (4.103). Then, by risk-neutral pricing:

$$\begin{aligned} v_{min}(\tau, S_1, S_2) &= \mathrm{e}^{-r\tau}\widetilde{\mathrm{E}}_{t,S_1,S_2}\big[\min\left(S_1(T), S_2(T)\right)\big] \\ &= \mathrm{e}^{-r\tau}\widetilde{\mathrm{E}}_{t,S_1,S_2}\big[S_2(T)\big] - \mathrm{e}^{-r\tau}\widetilde{\mathrm{E}}_{t,S_1,S_2}\big[\left(S_2(T) - S_1(T)\right)^+\big]. \end{aligned} \tag{4.105}$$

The second term in this last equation is simply the price $v(\tau, S_1, S_2)$ (see (4.92)) given by (4.97). The first term is readily evaluated by substituting the representation in (4.91) and using the independence of the Brownian increment $(\widetilde{\mathbf{W}}(T) - \widetilde{\mathbf{W}}(t))$ and the pair $(S_1(t), S_2(t))$, which are only functions of $\widetilde{\mathbf{W}}(t)$, i.e., $S_i(t) = S_i(0)\,\mathrm{e}^{(r-q_1-\frac{1}{2}\sigma_i^2)t+\boldsymbol{\sigma}_i\cdot\widetilde{\mathbf{W}}(t)}$, $i = 1, 2$. Hence,

$$\begin{aligned} \widetilde{\mathrm{E}}_{t,S_1,S_2}\big[S_2(T)\big] &\equiv \widetilde{\mathrm{E}}\big[S_2(T) \mid S_1(t) = S_1, S_2(t) = S_2\big] \\ &= S_2\,\mathrm{e}^{(r-q_2-\frac{1}{2}\sigma_2^2)\tau}\,\widetilde{\mathrm{E}}\big[\mathrm{e}^{\boldsymbol{\sigma}_2\cdot(\widetilde{\mathbf{W}}(T)-\widetilde{\mathbf{W}}(t))} \mid S_1(t) = S_1, S_2(t) = S_2\big] \\ &= S_2\,\mathrm{e}^{(r-q_2-\frac{1}{2}\sigma_2^2)\tau}\,\widetilde{\mathrm{E}}\big[\mathrm{e}^{\boldsymbol{\sigma}_2\cdot(\widetilde{\mathbf{W}}(T)-\widetilde{\mathbf{W}}(t))}\big] \text{ (by independence)} \\ &= S_2\,\mathrm{e}^{(r-q_2-\frac{1}{2}\sigma_2^2)\tau}\,\mathrm{e}^{\frac{1}{2}\sigma_2^2\tau} = S_2\,\mathrm{e}^{(r-q_2)\tau}. \end{aligned}$$

The last line follows by the m.g.f. of $\boldsymbol{\sigma}_2 \cdot (\widetilde{\mathbf{W}}(T) - \widetilde{\mathbf{W}}(t)) \sim Norm(0, \sigma_2^2\tau)$. We remark here that this conditional expectation also follows directly from the fact that the process $\{\mathrm{e}^{(q_2-r)t}S_2(t)\}_{t\geqslant 0}$ is a $\widetilde{\mathbb{P}}$-martingale, and combining this with the Markov property of the joint process $\{(S_1(t), S_2(t))\}_{t\geqslant 0}$ gives

$$\mathrm{e}^{(q_2-r)t}S_2(t) = \widetilde{\mathrm{E}}[\mathrm{e}^{(q_2-r)T}S_2(T) \mid \mathcal{F}_t] = \widetilde{\mathrm{E}}[\mathrm{e}^{(q_2-r)T}S_2(T) \mid S_1(t), S_2(t)].$$

Setting $(S_1(t), S_2(t)) = (S_1, S_2)$ and grouping the exponential terms gives the above result: $\widetilde{\mathrm{E}}\big[S_2(T) \mid S_1(t) = S_1, S_2(t) = S_2\big] = S_2\mathrm{e}^{(r-q_2)(T-t)}$. Hence, for each stock, we have $\widetilde{\mathrm{E}}_{t,S_1,S_2}\big[S_i(T)\big] = S_i\mathrm{e}^{(r-q_i)\tau}$, $i = 1, 2$.

Substituting this into (4.105) gives us the explicit pricing formula

$$\begin{aligned} v_{min}(\tau, S_1, S_2) &= \mathrm{e}^{-q_1\tau}S_1\,\mathcal{N}\Big(d_-\big(\frac{S_2}{S_1}, \tau\big)\Big) + \mathrm{e}^{-q_2\tau}S_2\left[1 - \mathcal{N}\Big(d_+\big(\frac{S_2}{S_1}, \tau\big)\Big)\right] \\ &= \mathrm{e}^{-q_1\tau}S_1\,\mathcal{N}\Big(d_-\big(\frac{S_2}{S_1}, \tau\big)\Big) + \mathrm{e}^{-q_2\tau}S_2\,\mathcal{N}\Big(-d_+\big(\frac{S_2}{S_1}, \tau\big)\Big) \end{aligned} \tag{4.106}$$

with $d_\pm(x, \tau)$ defined in (4.98). In the last equation line, we used the symmetry relation $\mathcal{N}(x) + \mathcal{N}(-x) = 1$.

Since $\max\left(S_1(T), S_2(T)\right) = S_1(T) + S_2(T) - \min\left(S_1(T), S_2(T)\right)$, the value of the option with payoff in (4.104) follows simply by (4.106):

$$\begin{aligned} v_{max}(\tau, S_1, S_2) &= \mathrm{e}^{-r\tau}\widetilde{\mathrm{E}}_{t,S_1,S_2}\left[\max\left(S_1(T), S_2(T)\right)\right] \\ &= \mathrm{e}^{-r\tau}\widetilde{\mathrm{E}}_{t,S_1,S_2}\left[S_1(T)\right] + \mathrm{e}^{-r\tau}\widetilde{\mathrm{E}}_{t,S_1,S_2}\left[S_2(T)\right] \\ &\quad - \mathrm{e}^{-r\tau}\widetilde{\mathrm{E}}_{S_1,S_2,t}\left[\min\left(S_1(T), S_2(T)\right)\right] \\ &= \mathrm{e}^{-q_1\tau}S_1 + \mathrm{e}^{-q_2\tau}S_2 - v_{min}(\tau, S_1, S_2) \\ &= \mathrm{e}^{-q_1\tau}S_1\,\mathcal{N}\Big(-d_-\big(\frac{S_2}{S_1}, \tau\big)\Big) + \mathrm{e}^{-q_2\tau}S_2\,\mathcal{N}\Big(d_+\big(\frac{S_2}{S_1}, \tau\big)\Big). \end{aligned} \tag{4.107}$$

Notice that the expressions in (4.106) and (4.107) are invariant with respect to interchanging all subscripts $1 \leftrightarrow 2$ on the dividends, volatilities, and spot values. This must be the case since the payoffs in (4.103) and (4.104) are invariant to the interchange $S_1(T) \leftrightarrow S_2(T)$. We leave it as an exercise for the reader (see Exercise 4.2) to derive expressions for the delta positions for the above two options on the maximum and minimum and to show that the relation in (4.102) holds; i.e., only the stocks are needed in the self-financing replicating strategies.

We now present a useful identity that can be used to more readily derive pricing functions for options whose payoff has a certain type of dependence on the terminal value of two correlated GBM processes. In particular, the payoffs in either (4.90), (4.103) or (4.104) can all be represented as a linear combination of *elemental payoffs*: $S_1(T)$, $S_2(T)$, $S_2(T)\mathbb{I}_{\{S_2(T)\geqslant S_1(T)\}}$, and $S_1(T)\mathbb{I}_{\{S_1(T)>S_2(T)\}}$ where $S_1(T)$ and $S_2(T)$ are correlated log-normal random variables. Other examples of payoffs that involve similar indicator functions in the terminal values of GBM processes occur when pricing foreign exchange options, as seen in Section 4.3.3.

Consider two log-normal random variables X and Y, represented by

$$X = x\mathrm{e}^{(\mu_X - \frac{1}{2}\sigma_X^2)\tau + \sigma_X\sqrt{\tau}Z_1}, \quad Y = y\mathrm{e}^{(\mu_Y - \frac{1}{2}\sigma_Y^2)\tau + \sigma_Y\sqrt{\tau}Z_2}$$

with positive parameters $x, y, \sigma_X, \sigma_Y, \tau$ and where Z_1, Z_2 are jointly bivariate standard normals (in a given measure $\mathbb{P}$) with $\mathrm{Corr}(Z_1, Z_2) = \rho$. Then, we have the following expectation formula (under measure $\mathbb{P}$):

$$\mathrm{E}\left[X\,\mathbb{I}_{\{X>Y\}}\right] = x\,\mathrm{e}^{\mu_X\tau}\,\mathcal{N}\left(\frac{\ln\frac{x}{y} + (\mu_X - \mu_Y + \frac{1}{2}\nu^2)\tau}{\nu\sqrt{\tau}}\right) \tag{4.108}$$

where $\nu^2 = \sigma_X^2 + \sigma_Y^2 - 2\rho\sigma_X\sigma_Y$. This identity is equivalent to considering any two correlated GBM processes

$$X(t) = X(0)\,\mathrm{e}^{(\mu_X - \frac{1}{2}\sigma_X^2)t + \boldsymbol{\sigma}^{(X)}\cdot\mathbf{W}(t)}, \quad Y(t) = Y(0)\,\mathrm{e}^{(\mu_Y - \frac{1}{2}\sigma_Y^2)t + \boldsymbol{\sigma}^{(Y)}\cdot\mathbf{W}(t)}, \quad t \geqslant 0,$$

where $\mathbf{W}(t)$ is a standard d-dimensional $\mathbb{P}$-BM, with $d \geqslant 2$, $\sigma_X = \|\boldsymbol{\sigma}^{(X)}\|$, $\sigma_Y = \|\boldsymbol{\sigma}^{(Y)}\|$, $\boldsymbol{\sigma}^{(X)}\cdot\boldsymbol{\sigma}^{(Y)} = \rho\sigma_X\sigma_Y$, and $\nu^2 = \|\boldsymbol{\sigma}^{(X)} - \boldsymbol{\sigma}^{(Y)}\|^2 = \sigma_X^2 + \sigma_Y^2 - 2\rho\sigma_X\sigma_Y$ with correlation coefficient ρ. Then, (4.108) is equivalent to the following conditional expectation formula:

$$\begin{aligned}\mathrm{E}_{t,x,y}\left[X(T)\,\mathbb{I}_{\{X(T)>Y(T)\}}\right] &\equiv \mathrm{E}\left[X(T)\,\mathbb{I}_{\{X(T)>Y(T)\}} \mid X(t) = x, Y(t) = y\right] \\ &= x\,\mathrm{e}^{\mu_X\tau}\,\mathcal{N}\left(\frac{\ln\frac{x}{y} + (\mu_X - \mu_Y + \frac{1}{2}\nu^2)\tau}{\nu\sqrt{\tau}}\right)\end{aligned} \tag{4.109}$$

where $\tau = T - t$.

Example 4.1. Use the identity in (4.109) to derive the pricing function $v_{max}(\tau, S_1, S_2)$ in (4.107).

Solution. The payoff has the form

$$V_T = \max\left(S_1(T), S_2(T)\right) = S_1(T)\mathbb{I}_{\{S_1(T)>S_2(T)\}} + S_2(T)\mathbb{I}_{\{S_2(T)\geqslant S_1(T)\}}.$$

We need only derive the price for the payoff $V_T^{(1)} \equiv S_1(T)\mathbb{I}_{\{S_1(T)>S_2(T)\}}$ since, by symmetry, the price for the second portion of the payoff obtains by interchanging the roles of the two stocks. That is, after deriving the expression for the pricing function $v^{(1)}(\tau, S_1, S_2)$ for payoff $V_T^{(1)}$ we interchange $S_1 \leftrightarrow S_2, \sigma_1 \leftrightarrow \sigma_2$, $q_1 \leftrightarrow q_2$ to obtain the pricing function $v^{(2)}(\tau, S_1, S_2)$

for payoff $V_T^{(2)} \equiv S_2(T)\mathbb{I}_{\{S_2(T)\geqslant S_1(T)\}}$. Finally, by linearity, we add the two pricing functions to obtain $v_{max}(\tau, S_1, S_2)$.

For payoff $V_T^{(1)}$ we have, by risk-neutral pricing,

$$v^{(1)}(\tau, S_1, S_2) = \mathrm{e}^{-r\tau}\widetilde{\mathrm{E}}_{t,S_1,S_2}\left[S_1(T)\mathbb{I}_{\{S_1(T)>S_2(T)\}}\right]$$

where the stock prices are two correlated GBM processes as in (4.78), i.e. as in (4.91). Hence, this conditional expectation is exactly of the form in (4.109), where now we are in the $\widetilde{\mathbb{P}}$ measure. We can therefore directly apply (4.109) once we identify the processes and the corresponding parameters. In this case, we identify the processes $X(t) = S_1(t), Y(t) = S_2(t)$, the spot variables $x = S_1, y = S_2$, the log-drift parameters $\mu_X = r - q_1, \mu_Y = r - q_2$, and log-volatility vectors $\boldsymbol{\sigma}^{(X)} = \boldsymbol{\sigma}_1 = [\sigma_1, 0], \boldsymbol{\sigma}^{(Y)} = \boldsymbol{\sigma}_2 = [\sigma_2\rho, \sigma_2\sqrt{1-\rho^2}]$, $\sigma_X = \sigma_1, \sigma_Y = \sigma_2$ where $\nu^2 = \|\boldsymbol{\sigma}_1 - \boldsymbol{\sigma}_2\|^2 = \sigma_1^2 + \sigma_2^2 - 2\rho\sigma_1\sigma_2$. Hence, (4.109) gives

$$\begin{aligned} v^{(1)}(\tau, S_1, S_2) &= \mathrm{e}^{-r\tau} S_1\,\mathrm{e}^{(r-q_1)\tau}\,\mathcal{N}\left(\frac{\ln\frac{S_1}{S_2} + ((r-q_1)-(r-q_2)+\frac{1}{2}\nu^2)\tau}{\nu\sqrt{\tau}}\right) \\ &= S_1\,\mathrm{e}^{-q_1\tau}\,\mathcal{N}\left(-d_-(\frac{S_2}{S_1}, \tau)\right) \end{aligned} \tag{4.110}$$

with $d_\pm(x, \tau)$ defined in (4.98). For the payoff $V_T^{(2)}$, the pricing function is given by applying the same identity or simply interchanging subscripts $1 \leftrightarrow 2$, giving (note that ν^2 remains the same)

$$v^{(2)}(\tau, S_1, S_2) = S_2\,\mathrm{e}^{-q_2\tau}\,\mathcal{N}\left(\frac{\ln\frac{S_2}{S_1} + (q_1 - q_2 + \frac{1}{2}\nu^2)\tau}{\nu\sqrt{\tau}}\right) = S_2\,\mathrm{e}^{-q_2\tau}\,\mathcal{N}\left(d_+(\frac{S_2}{S_1}, \tau)\right). \tag{4.111}$$

Adding the pricing functions for the two elemental payoffs gives the previously derived price in (4.107),

$$v_{max}(\tau, S_1, S_2) = v^{(1)}(\tau, S_1, S_2) + v^{(2)}(\tau, S_1, S_2).$$

□

The above exercise shows that the identity in (4.109) immediately gives us the pricing functions for the elemental payoffs. The payoff in (4.90) for the exchange option is also a linear combination of elemental payoffs, e.g.,

$$\begin{aligned} (\,S_2(T) - S_1(T)\,)^+ &= \max\left(S_1(T), S_2(T)\right) - S_1(T) \\ &= S_1(T)\mathbb{I}_{\{S_1(T)>S_2(T)\}} + S_2(T)\mathbb{I}_{\{S_2(T)\geqslant S_1(T)\}} - S_1(T). \end{aligned}$$

Hence, the pricing function in (4.97) for this payoff is obtained immediately by adding the pricing functions in (4.110) and (4.111) and subtracting $S_1\mathrm{e}^{-q_1\tau}$.

In the next example, we reconsider pricing the exchange option by solving the two-stock BSPDE problem. However, we do not solve the BSPDE by using the two-dimensional Feynman–Kac formula since this brings us right back to our previous methods. The key step in the method is to write the solution (i.e. the pricing function) in a form that reduces the original BSPDE in (4.88) to a *lower dimensional* PDE problem that is more readily solved. The functional form for the pricing function is dictated by the symmetry of the payoff function with respect to the spot variables. The lower dimensional PDE is then essentially like solving a simple pricing problem for a single asset with an effective payoff. This method is also useful in higher dimensions involving three or more stocks, assuming that the payoff has some simplifying (factoring) symmetry. The methodology is best demonstrated by example, as follows.

Example 4.2. (Symmetry Reduction in BSPDE) Derive the pricing formula for the exchange option with payoff function $\Lambda(S_1, S_2) = (S_2 - S_1)^+$ by solving the BSPDE in (4.88) while employing a symmetry reduction.

Solution. We note the symmetry of the payoff function, which can be written as a product of $S_2 > 0$ and a function of the ratio $x := S_1/S_2 > 0$, or as a product of S_1 and a function of the ratio S_2/S_1,

$$\Lambda(S_1, S_2) = (S_2 - S_1)^+ = S_2(1 - S_1/S_2)^+ = S_2(1 - x)^+ . \tag{4.112}$$

[Note: we can also write $(S_2 - S_1)^+ = S_1(S_2/S_1 - 1)^+$.] Hence, the pricing function for the terminal value of time $t = T$ is given by $V(T, S_1, S_2) = S_2(1 - x)^+$, i.e. for $t = T$ it is in fact a product of the spot variable S_2 and a function of variable $x = S_1/S_2$. We therefore make an "Ansatz" and *seek a solution for all* $t \leqslant T$ in the form of a product of the variable S_2 and some function $f(t, x)$:

$$V(t, S_1, S_2) = S_2 f(t, x) = S_2 f(t, S_1/S_2) . \tag{4.113}$$

By construction, for $t = T$ this relation holds by combining (4.113) and (4.112):

$$V(T, S_1, S_2) = S_2 f(T, x) = S_2(1 - x)^+ \implies f(T, x) = (1 - x)^+ \equiv \phi(x) . \tag{4.114}$$

The condition on the right is viewed as a terminal (effective payoff) condition $f(T, x) = \phi(x)$.

The main step is now to substitute the form in (4.113) into (4.88) and compute all partial derivative terms. The form for the solution is justified once the BSPDE is *simplified into a PDE in only the variables* t, x *and the function* $f(t, x)$. The S_2 variable should factor out completely; otherwise, either the form for the solution is not correct or some errors were made, such as in the calculation of the derivative terms. The partial derivatives in (4.88) are calculated by simply using the chain rule of ordinary calculus on the pricing function V in (4.113) and using $\frac{\partial x}{\partial S_2} = -\frac{S_1}{S_2^2} = -\frac{x}{S_2}$, $\frac{\partial x}{\partial S_1} = \frac{1}{S_2}$. For the first partial derivatives:

$$\frac{\partial V}{\partial S_2} = \frac{\partial}{\partial S_2}(S_2 f) = f + S_2 \frac{\partial x}{\partial S_2}\frac{\partial f}{\partial x} = f - x\frac{\partial f}{\partial x}, \quad \frac{\partial V}{\partial S_1} = S_2 \frac{\partial x}{\partial S_1}\frac{\partial f}{\partial x} = \frac{\partial f}{\partial x},$$

and $\frac{\partial V}{\partial t} = S_2 \frac{\partial f}{\partial t}$. For the second partial derivatives:

$$\begin{aligned}
\frac{\partial^2 V}{\partial S_1 \partial S_2} &= \frac{\partial}{\partial S_1}\left(\frac{\partial V}{\partial S_2}\right) = \frac{\partial x}{\partial S_1}\frac{\partial}{\partial x}\left(f - x\frac{\partial f}{\partial x}\right) = -\frac{x}{S_2}\frac{\partial^2 f}{\partial x^2} \\
\frac{\partial^2 V}{\partial S_1^2} &= \frac{\partial}{\partial S_1}\left(\frac{\partial V}{\partial S_1}\right) = \frac{\partial x}{\partial S_1}\frac{\partial^2 f}{\partial x^2} = \frac{1}{S_2}\frac{\partial^2 f}{\partial x^2} \\
\frac{\partial^2 V}{\partial S_2^2} &= \frac{\partial}{\partial S_2}\left(\frac{\partial V}{\partial S_2}\right) = \frac{\partial x}{\partial S_2}\frac{\partial}{\partial x}\left(f - x\frac{\partial f}{\partial x}\right) = \frac{x^2}{S_2}\frac{\partial^2 f}{\partial x^2}
\end{aligned}$$

Substituting all respective terms into the BSPDE in (4.88) leads to

$$\begin{aligned}
S_2\frac{\partial f}{\partial t} &+ \frac{1}{2}\sigma_1^2 S_1^2 \frac{1}{S_2}\frac{\partial^2 f}{\partial x^2} + \frac{1}{2}\sigma_2^2 S_2^2 \frac{x^2}{S_2}\frac{\partial^2 f}{\partial x^2} - \rho\sigma_1\sigma_2 S_1 S_2 \frac{x}{S_2}\frac{\partial^2 f}{\partial x^2} \\
&+ (r - q_1)S_1\frac{\partial f}{\partial x} + (r - q_2)S_2\left(f - x\frac{\partial f}{\partial x}\right) - rS_2 f = 0 .
\end{aligned}$$

We can finally simplify this equation by factoring out S_2 in all terms and substituting x for

$\frac{S_1}{S_2}$. Some terms cancel out and we can group together all terms in $x^2 \frac{\partial^2 f}{\partial x^2}$ and $x \frac{\partial f}{\partial x}$. What we obtain is in fact a PDE for f in the variables t, x:

$$\frac{\partial f}{\partial t} + \frac{1}{2}\nu^2 x^2 \frac{\partial^2 f}{\partial x^2} + (q_2 - q_1) x \frac{\partial f}{\partial x} - q_2 f = 0\,, \tag{4.115}$$

where $\nu^2 := \sigma_1^2 + \sigma_2^2 - 2\rho\sigma_1\sigma_2$, subject to the terminal condition in (4.114): $f(T, x) = (1-x)^+$.

The PDE in (4.115) is in only one spatial dimension instead of two. It can be solved by inspection at this point! This is because the PDE is a one-dimensional BSPDE for a GBM process corresponding to a single "asset or stock" with effective "spot variable" x, effective volatility parameter ν, effective "interest rate" q_2, and effective "stock dividend" q_1. The effective payoff function $\phi(x) = (1-x)^+$ is that of a put with strike $K = 1$. Hence, the function is given by the Black–Scholes pricing function for a put option on a dividend paying stock in (3.70) where we set $S = x$, $K = 1$, $\tau = T - t$, $r = q_2$, $\sigma = \nu$, and $q = q_1$:

$$f(t,x) = \mathrm{e}^{-q_2\tau}\mathcal{N}\left(-\frac{\ln x + (q_2 - q_1 - \frac{1}{2}\nu^2)\tau}{\nu\sqrt{\tau}}\right) - \mathrm{e}^{-q_1\tau} x \mathcal{N}\left(-\frac{\ln x + (q_2 - q_1 + \frac{1}{2}\nu^2)\tau}{\nu\sqrt{\tau}}\right). \tag{4.116}$$

[Alternatively this solution is found by using the one-dimensional discounted Feynman–Kac formula. We leave it to the reader to show this by using $f(t,x) = \mathrm{e}^{-q_2(T-t)}\mathrm{E}_{t,x}[\phi(X(T))]$ where $X(t)$ is the GBM process corresponding to the PDE in (4.115).]

The pricing function follows by using (4.116), with $x = S_1/S_2$, into (4.113):

$$V(t, S_1, S_2) = S_2 f(t, S_1/S_2) = S_2 \mathrm{e}^{-q_2\tau}\mathcal{N}\Big(d_+\big(\frac{S_2}{S_1}, \tau\big)\Big) - S_1 \mathrm{e}^{-q_1\tau}\mathcal{N}\Big(d_-\big(\frac{S_2}{S_1}, \tau\big)\Big) \tag{4.117}$$

Of course, this is the same expression as in (4.97) with $d_\pm$ defined in (4.98). □

4.3.2.2 Other Basket Options

The methods of the previous section can also be used to derive pricing functions for other standard European options where the payoff is a function of the terminal values of two or more stock prices. Within the multidimensional GBM model, we know that this generally involves a multivariate integral as given by (4.83). For the case of exchange-type options on two stocks and for other similar payoffs, we saw that the calculations are simplified to single dimensional integrals and the resulting pricing functions involve the standard normal CDF. This simplification is not possible for other types of payoffs on two stocks involving some fixed strike level. One example is the so-called *chooser max call* option on two stocks with payoff

$$V_T = \big(\max\{S_1(T), S_2(T)\} - K\big)^+. \tag{4.118}$$

A related option is a *chooser min put* option on two stocks with payoff

$$V_T = \big(K - \min\{S_1(T), S_2(T)\}\big)^+. \tag{4.119}$$

Both options are worth more than the corresponding call or put on either single underlying with strike $K > 0$. The valuation of these options leads to explicit expressions involving the bivariate standard normal CDF. We leave these as Exercises 4.13 and 4.14 at the end of this chapter. Another type of two-stock option is a so-called *spread option* where the payoff can be call-like or put-like on the difference of the terminal values of two assets. For a call spread we have

$$V_T = \big(S_2(T) - S_1(T) - K\big)^+ \tag{4.120}$$

for a given strike $K > 0$. The pricing function for this option does not reduce to a combination of any known functions such as standard normal CDFs. However, the pricing function can still be written as an integral. In the limit of small strike $K \searrow 0$, this option becomes the simpler exchange option with payoff in (4.90).

Certain types of options whose payoff depends on the terminal values of three or more stocks can also be priced analytically. The derivation depends on some simplifying symmetry of the payoff function. In some cases, the pricing function can be derived as an explicit expression involving the bivariate standard normal CDF. We refer the reader to some exercises at the end of this chapter.

4.3.3 Cross-Currency Option Valuation

We now consider the pricing of equity options whose payoff in domestic currency involves the prices of one or more assets denominated in a foreign currency as well as (possibly) the prices of domestic assets. These options are therefore subject to currency risk as well as foreign equity risk. For example, a *quanto option* refers generally to an option on some asset denominated in foreign currency and whose payoff is in domestic currency. There are several types of such options and their payoffs can be quite complex and path dependent. Four simple examples of call-like foreign exchange options (having obvious put-like analogues) include the following.

1. Foreign Equity Call struck in foreign currency: This is a call on a foreign asset (stock) S^f with a strike price K_f, both denominated in foreign currency, which is converted to domestic currency at the terminal value of the exchange rate $X(T)$ with payoff

$$C_T = X(T)\,(S^f(T) - K_f)^+ \,. \tag{4.121}$$

2. Foreign Equity Call struck in domestic currency: This is a call on a domestically converted foreign asset XS^f with a domestic strike price K with payoff

$$C_T = (X(T)S^f(T) - K)^+ \,. \tag{4.122}$$

3. Fixed Foreign Equity Rate Call: This is like the first call above, except that the foreign exchange rate is fixed to some preassigned value, say $\bar{X}$. The payoff is

$$C_T = \bar{X}\,(S^f(T) - K_f)^+ \,. \tag{4.123}$$

4. (Elf-X) Equity Linked Foreign Exchange Call: The holder has the right to purchase a foreign asset S^f by placing a lower value bound κ on the exchange rate for converting the asset value to domestic currency. The payoff is

$$C_T = S^f(T)(X(T) - \kappa)^+ \,. \tag{4.124}$$

Consider two markets or economies – a domestic one with assets denominated in domestic currency and a foreign one with assets denominated in a foreign currency. Examples of currencies are USD, CAD, EUR, GBP, JPY, etc. We therefore have two bank accounts, where one domestic dollar invested in the domestic account is worth $B(t) = \mathrm{e}^{\int_0^t r(u)\,\mathrm{d}u}$ in domestic currency and one foreign dollar invested in the foreign account has value $B^f(t) = \mathrm{e}^{\int_0^t r^f(u)\,\mathrm{d}u}$ in foreign currency. Note that we are still denoting the domestic interest rate process by $\{r(t)\}_{t\geqslant 0}$ while the foreign interest rate process is denoted by $\{r^f(t)\}_{t\geqslant 0}$. We first assume these processes are adapted and later simplify the model to make them constants or nonrandom functions of time so that we can derive simple pricing functions since we will work within the GBM model for all assets. Let $\{S^f(t)\}_{t\geqslant 0}$ be the price process for any foreign asset or stock and let $\{X^{f\to d}(t) \equiv X(t)\}_{t\geqslant 0}$ be the exchange rate process where $X(t)$ is the time-t exchange rate for converting an asset denominated in *foreign currency into domestic currency.* Hence, X has units of (domestic currency)/(foreign currency), e.g., CAD/USD, USD/CAD, USD/GBP, etc. In the above four examples, we only have one foreign asset. However, we can and do allow for option payoffs involving multiple assets in both market currencies.

We assume the existence of a risk-neutral measure $\widetilde{\mathbb{P}} \equiv \widetilde{\mathbb{P}}^{(B)}$ for the domestic market with the domestic bank account as the numéraire asset. The exchange rate process is assumed to satisfy the SDE

$$\mathrm{d}X(t) = X(t)\big[\widetilde{\mu}_X(t)\,\mathrm{d}t + \boldsymbol{\sigma}^{(X)}(t)\cdot \mathrm{d}\widetilde{\mathbf{W}}(t)\big], \tag{4.125}$$

where $\boldsymbol{\sigma}^{(X)}(t)$ is an adapted log-volatility vector and where $\widetilde{\mu}_X(t)$ is an adapted log-drift coefficient in the $\widetilde{\mathbb{P}}$-measure. Similarly, the foreign asset is assumed to satisfy the SDE

$$\mathrm{d}S^f(t) = S^f(t)\big[\widetilde{\mu}_S(t)\,\mathrm{d}t + \boldsymbol{\sigma}^{(S)}(t)\cdot \mathrm{d}\widetilde{\mathbf{W}}(t)\big], \tag{4.126}$$

where $\boldsymbol{\sigma}^{(S)}(t)$ is an adapted log-volatility vector and where $\widetilde{\mu}_S(t)$ is an adapted log-drift in the measure $\widetilde{\mathbb{P}}$. As in previous sections, $\widetilde{\mathbf{W}}(t)$ is a vector standard Brownian motion w.r.t. measure $\widetilde{\mathbb{P}}$. We determine the drifts $\widetilde{\mu}_X(t)$ and $\widetilde{\mu}_S(t)$ just below. First, let us recall that in the $\widetilde{\mathbb{P}}$-measure, all domestic non-dividend paying assets must have log-drift equal to the domestic interest rate $r(t)$ or otherwise an arbitrage exists. In other words, any domestic non-dividend paying asset, $A(t)$, discounted by the domestic bank account, $B(t) = \mathrm{e}^{\int_0^t r(u)\,du}$ should be a $\widetilde{\mathbb{P}}$-martingale or otherwise there is an arbitrage. If the asset pays a dividend, then the log-drift in the $\widetilde{\mathbb{P}}$-measure must equal the interest rate minus the dividend yield. In this case, as discussed in the previous chapter, the value of a self-financing portfolio strategy, where all dividend payments are reinvested into the stock, discounted by the domestic bank account is a $\widetilde{\mathbb{P}}$-martingale.

Note that $X(t)S^f(t)$ is the time-t price of a foreign asset converted to *domestic currency* and tradable in the domestic market. Hence, the process $\{X(t)S^f(t)\}_{t\geqslant 0}$ evolves as a domestic asset,

$$\mathrm{d}(X(t)S^f(t)) = X(t)S^f(t)\big[(r(t) - q_S(t))\,\mathrm{d}t + \boldsymbol{\sigma}^{(XS)}(t)\cdot \mathrm{d}\widetilde{\mathbf{W}}(t)\big], \tag{4.127}$$

where $\boldsymbol{\sigma}^{(XS)}(t) = \boldsymbol{\sigma}^{(X)}(t)+\boldsymbol{\sigma}^{(S)}(t)$ is the log-volatility vector of the product process XS. We recall from the Itô product rule that the log-volatility vector of a product of two processes is the sum of the log-volatility vectors of the two processes. We have also included a dividend yield $q_S(t)$ on the stock. Any additional domestic asset, say A, with adapted dividend yield $q_A(t)$ and log-volatility vector $\boldsymbol{\sigma}^{(A)}(t)$ has price process $\{A(t)\}_{t\geqslant 0}$ in domestic currency satisfying a similar SDE,

$$\mathrm{d}A(t) = A(t)\big[(r(t) - q_A(t))\,\mathrm{d}t + \boldsymbol{\sigma}^{(A)}(t)\cdot \mathrm{d}\widetilde{\mathbf{W}}(t)\big]. \tag{4.128}$$

One can easily show that the process $\left\{\mathrm{e}^{\int_0^t (q_A(u)-r(u))\,\mathrm{d}u} A(t)\right\}_{t\geqslant 0}$ is a $\widetilde{\mathbb{P}}$-martingale.

We now determine $\widetilde{\mu}_X(t)$ in (4.125) by using the fact that the foreign bank account investment converted to domestic currency, $X(t)B^f(t)$, must have log-drift $r(t)$. Applying the Itô product rule and the fact that $\mathrm{d}B^f(t) = r^f(t)B^f(t)\,\mathrm{d}t$ and $\mathrm{d}X(t)\,\mathrm{d}B^f(t) = 0$,

$$\begin{aligned}\frac{\mathrm{d}(X(t)B^f(t))}{X(t)B^f(t)} &= \frac{\mathrm{d}B^f(t)}{B^f(t)} + \frac{\mathrm{d}X(t)}{X(t)}\\ &= \big(r^f(t) + \widetilde{\mu}_X(t)\big)\,\mathrm{d}t + \boldsymbol{\sigma}^{(X)}(t)\cdot \mathrm{d}\widetilde{\mathbf{W}}(t)\,.\end{aligned} \tag{4.129}$$

Hence, $r^f(t) + \widetilde{\mu}_X(t) = r(t) \implies \widetilde{\mu}_X(t) = r(t) - r^f(t)$. We have hence determined the drift coefficient in (4.125), giving

$$\mathrm{d}X(t) = X(t)\big[(r(t) - r^f(t))\,\mathrm{d}t + \boldsymbol{\sigma}^{(X)}(t)\cdot \mathrm{d}\widetilde{\mathbf{W}}(t)\big]. \tag{4.130}$$

Next, we determine $\widetilde{\mu}_S(t)$ in (4.126) by first using the Itô product rule and combining (4.130) and (4.126) with the relation $\frac{\mathrm{d}X(t)}{X(t)}\frac{\mathrm{d}S^f(t)}{S^f(t)} = \boldsymbol{\sigma}^{(X)}(t)\cdot\boldsymbol{\sigma}^{(S)}(t)\,\mathrm{d}t$:

$$\begin{aligned}\frac{\mathrm{d}(X(t)S^f(t))}{X(t)S^f(t)} &= \frac{\mathrm{d}X(t)}{X(t)} + \frac{\mathrm{d}S^f(t)}{S^f(t)} + \frac{\mathrm{d}X(t)}{X(t)}\frac{\mathrm{d}S^f(t)}{S^f(t)} \\ &= \big(\widetilde{\mu}_S(t) + r(t) - r^f(t) + \boldsymbol{\sigma}^{(X)}(t)\cdot\boldsymbol{\sigma}^{(S)}(t)\big)\,\mathrm{d}t \\ &\quad + \big(\boldsymbol{\sigma}^{(X)}(t) + \boldsymbol{\sigma}^{(S)}(t)\big)\cdot \mathrm{d}\widetilde{\mathbf{W}}(t)\,.\end{aligned} \tag{4.131}$$

Equating this SDE with that in (4.127) gives $\widetilde{\mu}_S(t) = r^f(t) - q_S(t) - \boldsymbol{\sigma}^{(X)}(t)\cdot\boldsymbol{\sigma}^{(S)}(t)$, i.e., this determines the drift coefficient in (4.126):

$$\mathrm{d}S^f(t) = S^f(t)\left[\big(r^f(t) - q_S(t) - \boldsymbol{\sigma}^{(X)}(t)\cdot\boldsymbol{\sigma}^{(S)}(t)\big)\,\mathrm{d}t + \boldsymbol{\sigma}^{(S)}(t)\cdot \mathrm{d}\widetilde{\mathbf{W}}(t)\right]. \tag{4.132}$$

The solution to the system of SDEs in (4.130), (4.132), and (4.128) can be used for risk-neutral pricing of foreign exchange options with a payoff that is generally a function of the terminal values of the two assets and the foreign exchange rate, $V_T = \Lambda(S^f(T), A(T), X(T))$. Since the discounted domestic value process of the foreign exchange option (discounted by the domestic bank account) is a $\widetilde{\mathbb{P}}$-martingale, the no-arbitrage price of such an option is given by (4.22) where:

$$V_t = \widetilde{\mathrm{E}}\Big[\mathrm{e}^{-\int_t^T r(u)\,\mathrm{d}u}\,\Lambda(S^f(T), A(T), X(T)) \mid \mathcal{F}_t\Big], \quad 0 \leqslant t \leqslant T. \tag{4.133}$$

For quanto options with payoffs of the form $V_T = \Lambda(X(T), S^f(T))$, such as in (4.121)–(4.124), only (4.130) and (4.132) are needed. If the payoff is a function of the product $X(T)S^f(T)$, then (4.127) can be used directly.

For analytical tractability, we now adopt the GBM model where asset prices $\{A(t)\}_{t\geqslant 0}$ and $\{S^f(t)\}_{t\geqslant 0}$ and the exchange rate $\{X(t)\}_{t\geqslant 0}$ are GBM processes. We can therefore consider all coefficients to be nonrandom functions of time. Here, we simply assume constant interest rates $r(t) = r, r^f(t) = r^f$, constant dividend yields $q_S(t) = q_S, q_A(t) = q_A$, and constant log-volatility vectors $\boldsymbol{\sigma}^{(X)}(t) = \boldsymbol{\sigma}^{(X)}, \boldsymbol{\sigma}^{(S)}(t) = \boldsymbol{\sigma}^{(S)}, \boldsymbol{\sigma}^{(A)}(t) = \boldsymbol{\sigma}^{(A)}$. As in the case of basket options where the underlying assets are all domestic stocks modelled as GBM processes, the pricing formulation and pricing functions that follow also extend in the same fairly straightforward manner if we make these nonrandom functions of time rather than constants. The dot products of the log-volatility vectors are

$$\boldsymbol{\sigma}^{(X)}\cdot\boldsymbol{\sigma}^{(S)} = \rho_{XS}\sigma_X\sigma_S\,,\quad \boldsymbol{\sigma}^{(X)}\cdot\boldsymbol{\sigma}^{(A)} = \rho_{XA}\sigma_X\sigma_A\,,\quad \boldsymbol{\sigma}^{(A)}\cdot\boldsymbol{\sigma}^{(S)} = \rho_{AS}\sigma_A\sigma_S\,, \tag{4.134}$$

with constant correlation coefficients $\rho_{XS}, \rho_{XA}, \rho_{AS}$ and vector magnitudes $\sigma_X = \|\boldsymbol{\sigma}^{(X)}\|$, $\sigma_S = \|\boldsymbol{\sigma}^{(S)}\|$, $\sigma_A = \|\boldsymbol{\sigma}^{(A)}\|$. To compact notation, it is also useful to define the vector magnitudes

$$\sigma_{XS} \equiv \|\boldsymbol{\sigma}^{(X)} + \boldsymbol{\sigma}^{(S)}\| = \sqrt{\sigma_X^2 + \sigma_S^2 + 2\rho_{XS}\sigma_X\sigma_S}\,,$$

$$\sigma_{XA} \equiv \|\boldsymbol{\sigma}^{(X)} + \boldsymbol{\sigma}^{(A)}\| = \sqrt{\sigma_X^2 + \sigma_A^2 + 2\rho_{XA}\sigma_X\sigma_A}\,,$$

and $\sigma_{AS} \equiv \|\boldsymbol{\sigma}^{(A)} + \boldsymbol{\sigma}^{(S)}\| = \sqrt{\sigma_A^2 + \sigma_S^2 + 2\rho_{AS}\sigma_A\sigma_S}$.

As GBM processes, the above SDEs take the form

$$\frac{\mathrm{d}X(t)}{X(t)} = (r - r^f)\,\mathrm{d}t + \boldsymbol{\sigma}^{(X)} \cdot \mathrm{d}\widetilde{\mathbf{W}}(t)\,, \tag{4.135}$$

$$\frac{\mathrm{d}S^f(t)}{S^f(t)} = (r^f - q_S - \rho_{XS}\sigma_X\sigma_S)\,\mathrm{d}t + \boldsymbol{\sigma}^{(S)} \cdot \mathrm{d}\widetilde{\mathbf{W}}(t)\,, \tag{4.136}$$

$$\frac{\mathrm{d}\big[X(t)S^f(t)\big]}{X(t)S^f(t)} = (r - q_S)\,\mathrm{d}t + (\boldsymbol{\sigma}^{(X)} + \boldsymbol{\sigma}^{(S)}) \cdot \mathrm{d}\widetilde{\mathbf{W}}(t)\,, \tag{4.137}$$

$$\frac{\mathrm{d}A(t)}{A(t)} = (r - q_A)\,\mathrm{d}t + \boldsymbol{\sigma}^{(A)} \cdot \mathrm{d}\widetilde{\mathbf{W}}(t)\,. \tag{4.138}$$

From the unique solutions to the above linear SDEs, we have

$$X(T) = X(t)\,\mathrm{e}^{(r - r^f - \frac{1}{2}\sigma_X^2)(T-t) + \boldsymbol{\sigma}^{(X)}\cdot(\widetilde{\mathbf{W}}(T) - \widetilde{\mathbf{W}}(t))}\,, \tag{4.139}$$

$$S^f(T) = S^f(t)\,\mathrm{e}^{(r^f - q_S - \rho_{XS}\sigma_X\sigma_S - \frac{1}{2}\sigma_S^2)(T-t) + \boldsymbol{\sigma}^{(S)}\cdot(\widetilde{\mathbf{W}}(T) - \widetilde{\mathbf{W}}(t))}\,, \tag{4.140}$$

$$X(T)S^f(T) = X(t)S^f(t)\,\mathrm{e}^{(r - q_S - \frac{1}{2}\sigma_{XS}^2)(T-t) + (\boldsymbol{\sigma}^{(X)} + \boldsymbol{\sigma}^{(S)})\cdot(\widetilde{\mathbf{W}}(T) - \widetilde{\mathbf{W}}(t))}\,, \tag{4.141}$$

$$A(T) = A(t)\,\mathrm{e}^{(r - q_A - \frac{1}{2}\sigma_A^2)(T-t) + \boldsymbol{\sigma}^{(A)}\cdot(\widetilde{\mathbf{W}}(T) - \widetilde{\mathbf{W}}(t))}\,. \tag{4.142}$$

By the joint Markov property of the processes, and nonrandom interest rates, the conditioning on the filtration $\mathcal{F}_t$ in (4.133) is simply replaced by conditioning on the triplet $S^f(t), A(t), X(t)$, i.e. $V_t = V(t, S^f(t), A(t), X(t))$. Let $S^f(t) = S, A(t) = A, X(t) = x$ be the time-t *spot values of the foreign and domestic stocks and the exchange rate.* Assume that the domestic bank account $B(t) = \mathrm{e}^{rt}$ is served as the numéraire asset. The numéraire invariant form of the risk-neutral pricing formula provided in (4.62), (4.133) gives the pricing function $V(t, S, A, x)$ as the conditional expectation

$$V(t, S, A, x) = \mathrm{e}^{-r(T-t)}\,\widetilde{\mathrm{E}}\big[\Lambda(S^f(T), A(T), X(T)) \mid S^f(t) = S, A(t) = A, X(t) = x\big]. \tag{4.143}$$

Given a payoff function $\Lambda(S, A, x)$, we can therefore derive pricing functions by computing this expectation by whatever means. For a quanto option with payoff function $\Lambda(S, x)$; i.e., $V_T = \Lambda(S^f(T), X(T))$, the pricing function $V(t, S, x)$ is given by

$$V(t, S, x) = \mathrm{e}^{-r(T-t)}\,\widetilde{\mathrm{E}}\big[\Lambda(S^f(T), X(T)) \mid S^f(t) = S, X(t) = x\big]. \tag{4.144}$$

Let us consider a few examples of pricing quanto derivatives. We start with a forward contract for buying foreign currency. To find its no-arbitrage pricing formula, we only need to know the exchange rate dynamics.

Example 4.3. Consider a long forward contract for buying one unit of foreign currency for κ units of domestic currency at time T. Its payoff is $C_T = X(T) - \kappa$.

(a) Find a pricing formula for the time-t no-arbitrage value C_t of the contract.

(b) Find the fair forward exchange rate κ so that $C_t = 0$.

(c) Find the replicating portfolio strategy.

Solution. (a) The time-t no-arbitrage value is

$$C_t = \widetilde{\mathrm{E}}\left[\mathrm{e}^{-r(T-t)}\,C_T \mid X(t)\right] = \widetilde{\mathrm{E}}\left[\mathrm{e}^{-r(T-t)}\,(X(T) - \kappa) \mid X(t)\right].$$

Using the solution in (4.139), we calculate the conditional risk-neutral expectation of $X(T)$ given $X(t)$ as $\widetilde{\mathrm{E}}[X(T) \mid X(t)] = X(t)\mathrm{e}^{(r-r^f)(T-t)}$. Thus, the time-$t$ value of the contract is

$$C_t = \mathrm{e}^{-r^f(T-t)}X(t) - \mathrm{e}^{-r(T-t)}\kappa.$$

(b) Solve $C_t = 0$ for κ to obtain $\kappa = \mathrm{e}^{(r-r^f)(T-t)}X(t)$.
(c) The replicating portfolio strategy positions are easily derived from the following representation of C_t:

$$C_t = X(t)\mathrm{e}^{r^f t}\mathrm{e}^{-r^f T} - \mathrm{e}^{rt}\mathrm{e}^{-rT}\kappa = X(t)B^f(t)\underbrace{\mathrm{e}^{-r^f T}}_{=:\beta_f} + B(t)\underbrace{(-\mathrm{e}^{-rT}\kappa)}_{=:\beta_d}.$$

Thus, to replicate the forward contract, we need $\beta_f > 0$ units of foreign currency and $\beta_d < 0$ units of domestic currency. Clearly, this replication strategy is static. □

Another simple example of a quanto option is a standard call or put on the exchange rate $X(T)$. For example, a put option that allows for selling one unit of the foreign currency at maturity time T for κ units of domestic currency has the payoff $(\kappa - X(T))^+$. Its no-arbitrage value can be found by using the Black–Scholes pricing formula for a standard European put option on a stock with the effective dividend yield r^f and volatility σ_X:

$$\begin{aligned} P(t,x) &= \mathrm{e}^{-r(T-t)}\,\widetilde{\mathrm{E}}\big[(\kappa - X(T))^+ \mid X(t) = x\big] \\ &= \mathrm{e}^{-r(T-t)}\kappa\mathcal{N}(-d_-) - \mathrm{e}^{-r^f(T-t)}x\mathcal{N}(-d_+) \end{aligned}$$

where $d_\pm = \dfrac{\ln(x/\kappa) + (r - r^f \pm \sigma_X^2/2)(T-t)}{\sigma_X\sqrt{T-t}}$.

Example 4.4. Assume the foreign stock price is a GBM with constant log-volatility vector $\boldsymbol{\sigma}^{(S)}$ and having a dividend yield q_S and the exchange rate is a GBM with constant log-volatility vector $\boldsymbol{\sigma}^{(X)}$. Derive the pricing function $C(t,S,x)$ for

(a) the foreign equity call struck in domestic currency with payoff in (4.122);

(b) the foreign equity call struck in foreign currency with payoff in (4.121).

Solution. (a) This foreign equity call is essentially a standard European call on the asset $S^d(t) := X(t)S^f(t)$ governed by the SDE in (4.137). To find the function $C(t,S,x)$, we simply apply the Black–Scholes pricing formula for a call option on a GBM asset with the effective spot xS, dividend yield q_S, and volatility

$$\sigma_{XS} = \|\boldsymbol{\sigma}^{(X)} + \boldsymbol{\sigma}^{(S)}\| = \sqrt{\sigma_X^2 + 2\rho_{XS}\sigma_X\sigma_S + \sigma_S^2}.$$

As a result, we have

$$\begin{aligned} C(t,S,x) &= \mathrm{e}^{-r(T-t)}\widetilde{\mathrm{E}}\left[(X(T)S^f(T) - K_f)^+ \mid X(t)S^f(t) = xS\right] \\ &= xS\mathrm{e}^{-q_S\tau}\mathcal{N}(d_+(xS/K_f, T-t)) - K_f\mathrm{e}^{-r\tau}\mathcal{N}(d_-(xS/K_f, T-t)), \end{aligned}$$

where $d_\pm(xS/K_f, T-t) = \dfrac{\ln(xS/K_f) + (r - q_S \pm \sigma_{XS}^2/2)(T-t)}{\sigma_{XS}\sqrt{T-t}}$ for all $t < T$, $x > 0$, $S > 0$, and $K > 0$.

(b) The pricing function for this foreign exchange call option can be derived by evaluating the expectation in (4.144). The derivation is similar to that of the exchange option on two assets, where we write the payoff as follows:

$$C_T = X(T)\,(S^f(T) - K_f)^+ = (X(T)S^f(T) - K_f X(T))^+.$$

We identify the "second asset" price process as $\{X(t)S^f(t)\}_{t\geqslant 0}$ and $\{K_f X(t)\}_{t\geqslant 0}$ as the "first asset" price process. Their strong solutions are

$$\begin{aligned} K_f X(T) &= K_f X(t)\,\mathrm{e}^{(r-r^f-\frac{1}{2}\sigma_X^2)(T-t)+\boldsymbol{\sigma}^{(X)}\cdot(\widetilde{\mathbf{W}}(T)-\widetilde{\mathbf{W}}(t))}\,, \\ X(T)S^f(T) &= X(t)S^f(t)\,\mathrm{e}^{(r-q_S-\frac{1}{2}\sigma_{XS}^2)(T-t)+(\boldsymbol{\sigma}^{(X)}+\boldsymbol{\sigma}^{(S)})\cdot(\widetilde{\mathbf{W}}(T)-\widetilde{\mathbf{W}}(t))}\,, \end{aligned}$$

where $\sigma_{XS} = \|\boldsymbol{\sigma}^{(X)} + \boldsymbol{\sigma}^{(S)}\|$ is the volatility of $X(t)S^f(t)$, $\sigma_S = \|\boldsymbol{\sigma}^{(S)}\|$ is the volatility of $S^f(t)$, and $\sigma_X = \|\boldsymbol{\sigma}^{(X)}\|$ is the volatility of $K_f X(t)$. We can now simply identify the two effective dividends, volatilities, and spot values in (4.91) as: $q_1 \equiv r^f$, $q_2 \equiv q_S$, $\sigma_1^2 \equiv \sigma_X^2$, $\sigma_2^2 \equiv \sigma_{XS}^2 = \sigma_X^2 + \sigma_S^2 + 2\rho_{XS}\sigma_X\sigma_S$, $S_1 \equiv xK_f$, $S_2 \equiv xS$, where $X(t) = x$ and $S^f(t) = S$ are the actual spot values of the foreign exchange rate and stock price, respectively. The difference of the two log-volatility vectors $\boldsymbol{\sigma}^{(1)} \equiv \boldsymbol{\sigma}^{(X)}$ and $\boldsymbol{\sigma}^{(2)} \equiv \boldsymbol{\sigma}^{(XS)} \equiv \boldsymbol{\sigma}^{(X)} + \boldsymbol{\sigma}^{(S)}$ is simply $\boldsymbol{\sigma}^{(2)} - \boldsymbol{\sigma}^{(1)} = \boldsymbol{\sigma}^{(S)}$. Denote $\nu^2 := \|\boldsymbol{\sigma}^{(2)} - \boldsymbol{\sigma}^{(1)}\|^2 = \sigma_S^2$. Hence, a direct application of the formula in (4.97), with the above identifications of $S_1, S_2, q_1, q_2, \nu, \tau = T - t$, gives

$$\begin{aligned} C(t,S,x) &= \mathrm{e}^{-r(T-t)}\,\widetilde{\mathrm{E}}\big[(X(T)\,S^f(T) - K_f X(T))^+ \mid S^f(t) = S, X(t) = x\big] \\ &= xS\mathrm{e}^{-q_S(T-t)}\mathcal{N}(d_+) - xK_f\,\mathrm{e}^{-r^f(T-t)}\mathcal{N}(d_-) \\ &= x\left[S\mathrm{e}^{-q_S(T-t)}\mathcal{N}(d_+) - K_f\,\mathrm{e}^{-r^f(T-t)}\mathcal{N}(d_-)\right] \end{aligned}$$

where

$$d_\pm = \frac{\ln(S/K_f) + (r^f - q_S \pm \frac{1}{2}\sigma_S^2)(T-t)}{\sigma_S\sqrt{T-t}}$$

for all $t < T$, $x > 0, S > 0, K_f > 0$. □

As in the case of multi-asset pricing in a domestic economy where all assets are denominated in a single domestic market, we can now also apply the discounted Feynman–Kac Theorem 2.24 to obtain a corresponding BSPDE for the above option pricing functions. Consider the vector process $\{\mathbf{X}(t) \equiv (X_1(t), X_2(t)) := (S^f(t), X(t))\}_{t\geqslant 0}$ satisfying the system of two SDEs (4.135) and (4.136). We identify the constant log-drifts and log-volatility vectors of these two processes by $\mu_1 \equiv r^f - q_S - \rho_{XS}\sigma_X\sigma_S$, $\mu_2 \equiv r - r^f$, $\boldsymbol{\sigma}_1 \equiv \boldsymbol{\sigma}^{(S)}$ and $\boldsymbol{\sigma}_2 \equiv \boldsymbol{\sigma}^{(X)}$. The corresponding time-homogeneous generator for this pair of GBM processes in the spot variables $x_1 \equiv S, x_2 = x$ is then

$$\mathcal{G}_{(x_1,x_2)} := \frac{1}{2}\sum_{i=1}^{2}\sum_{j=1}^{2}\boldsymbol{\sigma}_i\cdot\boldsymbol{\sigma}_j\,x_i x_j\frac{\partial^2}{\partial x_i\partial x_j} + \sum_{i=1}^{2}\mu_i\frac{\partial}{\partial x_i}\,. \tag{4.145}$$

Writing out all the partial derivative terms explicitly and using the spot variable names S, x in the place of x_1, x_2, then, according to the Feynman–Kac Theorem 2.24, the pricing function $V = V(t,S,x)$ in (4.144) satisfies the PDE

$$\begin{aligned} &\frac{\partial V}{\partial t} + \frac{1}{2}\sigma_S^2 S^2\frac{\partial^2 V}{\partial S^2} + \frac{1}{2}\sigma_X^2 x^2\frac{\partial^2 V}{\partial x^2} + \rho_{XS}\sigma_S\sigma_X\,Sx\frac{\partial^2 V}{\partial S\partial x} \\ &+(r^f - q_S - \rho_{XS}\sigma_X\sigma_S)\,S\frac{\partial V}{\partial S} + (r - r^f)\,x\frac{\partial V}{\partial x} - rV = 0\,, \end{aligned} \tag{4.146}$$

subject to the terminal (payoff) condition $V(T, S, x) = \Lambda(S, x)$. This is the BSPDE for a quanto option where the payoff is a function of the share price of the foreign stock and the foreign exchange rate.

We leave it to the reader to show that by a similar application of the Feynman–Kac Theorem 2.24, the pricing function $V = V(t, S, A, x)$ in (4.143) satisfies the BSPDE

$$\begin{aligned}
&\frac{\partial V}{\partial t} + \frac{1}{2}\sigma_S^2 S^2 \frac{\partial^2 V}{\partial S^2} + \frac{1}{2}\sigma_A^2 A^2 \frac{\partial^2 V}{\partial A^2} + \frac{1}{2}\sigma_X^2 x^2 \frac{\partial^2 V}{\partial x^2} \\
&+ \rho_{XS}\sigma_S\sigma_X\, Sx \frac{\partial^2 V}{\partial S \partial x} + \rho_{AS}\sigma_A\sigma_S\, AS \frac{\partial^2 V}{\partial A \partial S} + \rho_{XA}\sigma_X\sigma_A\, Ax \frac{\partial^2 V}{\partial A \partial x} \\
&+ (r^f - q_S - \rho_{XS}\sigma_X\sigma_S)\, S \frac{\partial V}{\partial S} + (r - q_A)\, A \frac{\partial V}{\partial A} + (r - r^f)\, x \frac{\partial V}{\partial x} - rV = 0, \qquad (4.147)
\end{aligned}$$

subject to the terminal (payoff) condition $V(T, S, A, x) = \Lambda(S, A, x)$. The PDEs in (4.146) and (4.147) can be readily solved in cases where the payoff functions allow for a symmetry reduction to be employed. As an example, consider a quanto option having a payoff in the form of a product $V(T, S, x) = \Lambda(S, x) = xg(S)$. Then, the pricing function, solving (4.146) for all $t \leqslant T$, can be written in the form of a product, $V(t, S, x) = xf(t, S)$. By substituting this form into (4.146), computing all partial derivatives and factoring out the exchange spot variable x, the reader can verify that this leads to a reduced PDE for $f(t, S)$ in the variables (t, S) subject to the terminal condition $f(T, S) = g(S)$. The fundamental solution of the reduced PDE is readily obtained. Explicit examples of pricing quanto options by symmetry reduction of the above BSPDE in (4.146) are left as exercises at the end of this chapter.

4.3.4 Option Valuation with General Numéraire Assets

We finally consider some examples of how appropriate choices of numéraire can simplify derivative pricing problems. We leave several other applications of the change of numéraire technique for pricing multi-asset and foreign exchange options as exercises at the end of this chapter. The key idea is to choose a numéraire that simplifies the effective payoff $\frac{V_T}{g(T)}$ in the pricing problem.

Example 4.5. (Pricing an Exchange Option with a Stock as Numéraire) Reconsider the option pricing problem solved in Section 4.3.2.1 with the payoff in (4.90). Assume the stock prices are GBM processes driven by two independent BMs in a domestic economy with constant interest rate r, but now set the stock dividends to zero. Derive the European option pricing formula using the asset pricing formula with the *numéraire asset chosen as one of the stocks.*

Solution. As numéraire, let us choose $g(t) = S_1(t)$, $t \geqslant 0$. Then,

$$\frac{V_T}{g(T)} = \frac{(S_2(T) - S_1(T))^+}{S_1(T)} = \left(\frac{S_2(T)}{S_1(T)} - 1\right)^+ = (Y(T) - 1)^+$$

where we define the process $Y(t) := \frac{S_2(t)}{S_1(t)}, t \geqslant 0$. We can substitute this into either of the pricing formulas in (4.49)–(4.51). In particular, using (4.49) gives $V_t = V(t, S_1(t), S_2(t))$ as

$$V_t = S_1(t)\, \widetilde{\mathrm{E}}^{(S_1)}\left[(Y(T) - 1)^+ \mid \mathcal{F}_t\right].$$

To simplify notation, we shall simply denote $\widetilde{\mathbb{P}}^{(g)} = \widetilde{\mathbb{P}}^{(S_1)} \equiv \widehat{\mathbb{P}}$, $\widetilde{\mathrm{E}}^{(S_1)} \equiv \widehat{\mathrm{E}}$ and $\widetilde{\mathbf{W}}^{(g)}(t) = \widetilde{\mathbf{W}}^{(S_1)}(t) \equiv \widehat{\mathbf{W}}(t)$. Note that the conditioning on $\mathcal{F}_t$ can also be replaced by $Y(t)$.

The GBM process $\{Y(t)\}_{t\geqslant 0}$ is a $\widehat{\mathbb{P}}$-martingale with representation in (4.37) [where $g(t) = S_1(t)$, $\boldsymbol{\sigma}^{(g)} = \boldsymbol{\sigma}_1$, $S_i(t) = S_2(t)$]:

$$\frac{S_2(t)}{S_1(t)} \equiv Y(t) = Y(0)\,\mathcal{E}_t\big((\boldsymbol{\sigma}_2 - \boldsymbol{\sigma}_1)\cdot\widehat{\mathbf{W}}\big)\,, \quad Y(0) = \frac{S_2(0)}{S_1(0)}.$$

[We remark that this solution can also be arrived at by computing the stochastic differential $\mathrm{d}\left(\frac{S_2(t)}{S_1(t)}\right)$ by the Itô quotient rule and using (4.41), giving $\mathrm{d}Y(t) = Y(t)(\boldsymbol{\sigma}_2 - \boldsymbol{\sigma}_1)\cdot\mathrm{d}\widehat{\mathbf{W}}(t)$, which has the solution given by the above expression.] By the above solution, we have the random variable $X := \ln\frac{Y(T)}{Y(t)} = \ln\mathcal{E}_{T-t}\big((\boldsymbol{\sigma}_2 - \boldsymbol{\sigma}_1)\cdot\widehat{\mathbf{W}}\big)$, i.e.,

$$\begin{aligned} X &= -\frac{1}{2}\|\boldsymbol{\sigma}_2 - \boldsymbol{\sigma}_1\|^2(T-t) + (\boldsymbol{\sigma}_2 - \boldsymbol{\sigma}_1)\cdot(\widehat{\mathbf{W}}(T) - \widehat{\mathbf{W}}(t)) \\ &= -\frac{1}{2}\nu^2(T-t) + (\boldsymbol{\sigma}_2 - \boldsymbol{\sigma}_1)\cdot(\widehat{\mathbf{W}}(T) - \widehat{\mathbf{W}}(t)) \\ &= -\frac{1}{2}\nu^2(T-t) + \nu\sqrt{T-t}\widehat{Z} \end{aligned}$$

where $\nu^2 := \|\boldsymbol{\sigma}_2 - \boldsymbol{\sigma}_1\|^2 = \sigma_1^2 + \sigma_2^2 - 2\rho\sigma_1\sigma_2$ and $\widehat{Z} := \frac{(\boldsymbol{\sigma}_2 - \boldsymbol{\sigma}_1)\cdot(\widehat{\mathbf{W}}(T) - \widehat{\mathbf{W}}(t))}{\nu\sqrt{T-t}} \sim Norm(0,1)$ under measure $\widehat{\mathbb{P}}$. The Brownian increments are independent of $\mathcal{F}_t$ and therefore $\widehat{Z}$ (and X) is independent of $\mathcal{F}_t$. Since $Y(T) = Y(t)\mathrm{e}^X$, the above expectation is simplified to an unconditional one by independence, where $Y(t)$ is $\mathcal{F}_t$-measurable. The expectation below is very easily computed by applying the usual identities. However, observe that the expectation is just like that of a standard call option on a single stock "Y" with zero "interest rate," "zero dividend," "volatility" ν, "spot" $Y(t)$, "strike" of 1, and time to maturity $T-t$:

$$\begin{aligned} V_t &= S_1(t)\,\widehat{\mathrm{E}}\left[(Y(t)\mathrm{e}^X - 1)^+ \,|\, \mathcal{F}_t\right] \\ &= S_1(t)\,\left[Y(t)\mathcal{N}\big(d_+(Y(t), T-t\big) - 1\cdot\mathcal{N}(d_-(Y(t), T-t)\right] \\ &= S_1(t)\left[\frac{S_2(t)}{S_1(t)}\mathcal{N}\left(d_+\big(\frac{S_2(t)}{S_1(t)}, T-t\big)\right) - \mathcal{N}\left(d_-\big(\frac{S_2(t)}{S_1(t)}, T-t\big)\right)\right] \\ &= S_2(t)\mathcal{N}\left(d_+\big(\frac{S_2(t)}{S_1(t)}, T-t\big)\right) - S_1(t)\mathcal{N}\left(d_-\big(\frac{S_2(t)}{S_1(t)}, T-t\big)\right), \end{aligned}$$

where $d_\pm(x,\tau)$ are defined as in (4.98) with $q_1 = q_2 = 0$. This expresses the option value as a random variable $V_t = V(t, S_1(t), S_2(t))$. Setting $S_1(t) = S_1, S_2(t) = S_2$, gives the pricing formula as a function of calendar time $t < T$ and the spot values $S_1, S_2 > 0$:

$$V(t, S_1, S_2) = S_2\mathcal{N}\left(d_+\big(\frac{S_2}{S_1}, T-t\big)\right) - S_1\mathcal{N}\left(d_-\big(\frac{S_2}{S_1}, T-t\big)\right).$$

Note: this formula is exactly that in (4.97) for $q_1 = q_2 = 0$, $\tau = T - t$. □

To price quanto options, we used the domestic risk-neutral probability measure $\widetilde{\mathbb{P}} \equiv \widetilde{\mathbb{P}}^{(B)}$ with the domestic bank account $B(t) = \mathrm{e}^{rt}$ as the numéraire. Alternatively, we can use the foreign bank account denominated in the domestic currency as the numéraire asset to obtain another EMM, which we call the foreign risk-neutral probability measure.

Example 4.6. Consider the foreign bank account denominated in the domestic currency as the numéraire asset: $g(t) = X(t)B^f(t) = X(t)\mathrm{e}^{r^f t}$.

(a) Find the Radon–Nikodym derivative process $\varrho_t^{B\to g} \equiv \left(\frac{\mathrm{d}\widetilde{\mathbb{P}}^{(g)}}{\mathrm{d}\widetilde{\mathbb{P}}^{(B)}}\right)_t$.

(b) Find the SDEs for $X(t)$ and $S^f(t)$ under $\widehat{\mathbb{P}} \equiv \widetilde{\mathbb{P}}^{(g)}$.

(c) Let C_T^d and C_T^f be two claims denominated in the domestic and foreign currencies, respectively. Show the equivalence of no-arbitrage pricing formulae under $\widetilde{\mathbb{P}}$ and $\widehat{\mathbb{P}}$:

$$\mathrm{e}^{-r^f(T-t)}\,\widehat{\mathrm{E}}\left[\frac{X(t)}{X(T)}C_T^d \,\Big|\, \mathcal{F}_t\right] = \mathrm{e}^{-r(T-t)}\,\widetilde{\mathrm{E}}\left[C_T^d \mid \mathcal{F}_t\right] \text{ and}$$

$$\mathrm{e}^{-r^f(T-t)}\,\widehat{\mathrm{E}}\left[C_T^f \mid \mathcal{F}_t\right] = \mathrm{e}^{-r(T-t)}\,\widetilde{\mathrm{E}}\left[\frac{X(T)}{X(t)}C_T^f \,\Big|\, \mathcal{F}_t\right].$$

Solution. (a) We use the change of numéraire approach given in (4.45) to find the Radon–Nikodym derivative:

$$\varrho_t^{B\to g} = \frac{g(t)}{g(0)}\frac{B(0)}{B(t)} = \frac{\mathrm{e}^{r^f t}X(t)/X(0)}{\mathrm{e}^{rt}} = \mathrm{e}^{(r^f-r)t}\frac{X(t)}{X(0)}.$$

Using the strong solution (under $\widetilde{\mathbb{P}}$) given in (4.139), we can also write the Radon–Nikodym derivative in the exponential form used in Girsanov's theorem:

$$\varrho_t^{B\to g} = \mathrm{e}^{-\frac{1}{2}\sigma_X^2 t + \boldsymbol{\sigma}^{(X)}\cdot\widetilde{\mathbf{W}}(t)}.$$

(b) According to Girsanov's Theorem for Multidimensional Brownian Motion, the process $\widehat{\mathbf{W}}(t) := \widetilde{\mathbf{W}}(t) - \boldsymbol{\sigma}^{(X)}t$ is a $\widehat{\mathbb{P}}$-BM. After plugging $\widehat{\mathbf{W}}(t) + \boldsymbol{\sigma}^{(X)}t$ in place of $\widetilde{\mathbf{W}}(t)$ in the SDEs (4.135) and (4.136), we obtain the SDEs for $X(t)$ and $S^f(t)$ under $\widehat{\mathbb{P}}$:

$$\frac{\mathrm{d}X(t)}{X(t)} = (r - r^f + \sigma_X^2)\,\mathrm{d}t + \boldsymbol{\sigma}^{(X)}\cdot\mathrm{d}\widehat{\mathbf{W}}(t),$$

$$\frac{\mathrm{d}S^f(t)}{S^f(t)} = (r^f - q_S)\,\mathrm{d}t + \boldsymbol{\sigma}^{(S)}\cdot\mathrm{d}\widehat{\mathbf{W}}(t).$$

(c) Let us the prove the equivalence of pricing formulae for a domestic claim. The other identity is proved similarly. According to (4.49), the time-t no-arbitrage value of C_T^d is

$$C_t^d = B(t)\widetilde{\mathrm{E}}\left[\frac{C_T^d}{B(T)}\,\Big|\,\mathcal{F}_t\right] = \mathrm{e}^{-r(T-t)}\widetilde{\mathrm{E}}\left[C_T^d \mid \mathcal{F}_t\right].$$

Let us apply the change of measure, $\widetilde{\mathbb{P}} \equiv \widetilde{\mathbb{P}}^{(B)} \to \widetilde{\mathbb{P}}^{(g)} \equiv \widehat{\mathbb{P}}$, as done in (4.48), to obtain:

$$C_t^d = \widehat{\mathrm{E}}\left[\frac{\varrho_T^{g\to B}}{\varrho_t^{g\to B}}\frac{B(t)}{B(T)}C_T^d\,\Big|\,\mathcal{F}_t\right] = \widehat{\mathrm{E}}\left[\frac{X(t)\mathrm{e}^{r^f t}}{X(T)\mathrm{e}^{r^f T}}C_T^d\,\Big|\,\mathcal{F}_t\right] = \mathrm{e}^{-r^f(T-t)}\,\widehat{\mathrm{E}}\left[\frac{X(t)}{X(T)}C_T^d\,\Big|\,\mathcal{F}_t\right].$$

Note that this pricing formula can also be derived by exploiting the fact that the discounted price process $\left\{\frac{C_t^d}{X(t)\mathrm{e}^{r^f t}}\right\}_{t\geqslant 0}$ is a $\widehat{\mathbb{P}}$ martingale. □

Let us see how we can solve Example 4.4 using the foreign risk-neutral probability measure.

Example 4.7. Assume the foreign stock price is a GBM with constant log-volatility vector $\boldsymbol{\sigma}^{(S)}$ and having a dividend yield q_S and the exchange rate is a GBM with constant log-volatility vector $\boldsymbol{\sigma}^{(X)}$. Derive the pricing function $C(t, S, x)$ for the foreign equity call struck in foreign currency with payoff in (4.121) by using the change of numéraire method.

Solution. Let us use the numéraire asset $g(t) = X(t)B^f(t) = X(t)\mathrm{e}^{r^f t}$. As demonstrated in Example 4.6, under the EMM $\widehat{\mathbb{P}} \equiv \widetilde{\mathbb{P}}^{(g)}$, the foreign stock $S^f(t)$ is governed by the following SDE:

$$\frac{\mathrm{d}S^f(t)}{S^f(t)} = (r^f - q_S)\,\mathrm{d}t + \sigma_S\,\mathrm{d}\widehat{W}(t) \text{ with } \sigma_S = \|\boldsymbol{\sigma}^{(S)}\|\,.$$

The payoff C_T discounted by $g(T) = X(T)\mathrm{e}^{r^f T}$ takes the form

$$\frac{C_T}{g(T)} = \frac{X(T)(S^f(T) - K_f)^+}{X(T)\,\mathrm{e}^{r^f T}} = \mathrm{e}^{-r^f T}(S^f(T) - K_f)^+.$$

Thus, by the general pricing formula in (4.49), we have

$$\begin{aligned}
C(t,S,x) &= \widetilde{\mathrm{E}}^{(g)}\left[g(t)\frac{C_T}{g(T)} \;\Big|\; S^f(t) = S, X(t) = x\right] \\
&= \widetilde{\mathrm{E}}^{(g)}\left[X(t)\mathrm{e}^{r^f t}\mathrm{e}^{-r^f T}(S^f(T) - K_f)^+ \mid S^f(t) = S, X(t) = x\right] \\
&= x\mathrm{e}^{-r^f(T-t)}\widetilde{\mathrm{E}}^{(g)}\left[(S^f(T) - K_f)^+ \mid S^f(t) = S\right].
\end{aligned}$$

This expectation is readily evaluated by simply using the Black–Scholes pricing formula for a standard call option on a GBM asset with effective interest rate r^f, strike K_f, spot S, dividend yield q_S, volatility σ_S, and time-to-maturity $\tau := T - t$. Hence,

$$C(t,S,x) = \mathrm{e}^{-q_S\tau}xS\mathcal{N}(d_+) - \mathrm{e}^{-r^f\tau}xK_f\mathcal{N}(d_-)$$

with $d_\pm = \dfrac{\ln(S/K_f) + (r^f - q_S \pm \frac{1}{2}\sigma_S^2)\tau}{\sigma_S\sqrt{T-t}}$ for all $t < T$, $x > 0$, $S > 0$, and $K_f > 0$. □

Consider the case with a general domestic numéraire. For example, the numéraire asset is some function of the time and the joint process values, $g(t) = g(t, S^f(t), A(t), X(t))$. According to (4.53), the price process $\{A(t)\}_{t\geqslant 0}$ of a *domestic dividend-paying asset* (or a foreign stock price process converted to domestic currency, such as $S^f(t)X(t)$) obeys the following SDE:

$$\mathrm{d}A(t) = A(t)\left[\left(r(t) - q_A(t) + \boldsymbol{\sigma}^{(A)}(t)\cdot\boldsymbol{\sigma}^{(g)}(t)\right)\mathrm{d}t + \boldsymbol{\sigma}^{(A)}(t)\cdot\mathrm{d}\widetilde{\mathbf{W}}^{(g)}(t)\right]$$

where $q_A(t)$ is an adapted dividend yield for asset A at time t. For a foreign stock, we have (upon using $\mathrm{d}\mathbf{W}(t) = \mathrm{d}\widetilde{\mathbf{W}}^{(g)}(t) + \boldsymbol{\sigma}^{(g)}(t)\,\mathrm{d}t$ in (4.132))

$$\mathrm{d}S^f(t) = S^f(t)\left[\left(r^f(t) - q_S(t) - \boldsymbol{\sigma}^{(S)}(t)\cdot(\boldsymbol{\sigma}^{(X)}(t) - \boldsymbol{\sigma}^{(g)}(t))\right)\mathrm{d}t + \boldsymbol{\sigma}^{(S)}(t)\cdot\mathrm{d}\widetilde{\mathbf{W}}^{(g)}(t)\right], \tag{4.148}$$

and the foreign exchange rate process in (4.130) satisfies

$$\mathrm{d}X(t) = X(t)\left[\left(r(t) - r^f(t) + \boldsymbol{\sigma}^{(X)}(t)\cdot\boldsymbol{\sigma}^{(g)}(t)\right)\mathrm{d}t + \boldsymbol{\sigma}^{(X)}(t)\cdot\mathrm{d}\widetilde{\mathbf{W}}^{(g)}(t)\right]. \tag{4.149}$$

A relation similar to (4.51) holds for the foreign exchange derivatives, where the joint process $\{S^f(t), A(t), X(t)\}_{t\geqslant 0}$ is Markov and conditioning on the spot variables $S^f(t) = S, X(t) = X, A(t) = A$ gives

$$V(t,S,A,X) = g(t,S,A,X)\,\widetilde{\mathrm{E}}^{(g)}_{t,(S,A,X)}\left[\frac{\Lambda(T, S^f(T), A(T), X(T))}{g(t, S^f(T), A(T), X(T))}\right] \tag{4.150}$$

where we assume the numéraire is some function of the time and the joint process values, $g(t) = g(t, S^f(t), A(t), X(t))$.

Example 4.8. (Pricing an Equity-Linked Foreign Exchange Call) By using a different choice of numéraire than the domestic bank account, derive the pricing function for the equity-linked foreign exchange call with domestic payoff in (4.124). Assume the exchange process and the foreign stock price process are GBMs as in (4.135) and (4.136) and let the stock dividend $q_S = 0$.

Solution. The payoff $V_T \equiv C_T$ is given by (4.124), which we can write as

$$C_T = (X(T)S^f(T) - \kappa S^f(T))^+ .$$

Note that the domestic asset price process $\{X(t)S^f(t)\}_{t\geqslant 0}$ qualifies as a non-dividend-paying numéraire asset, as seen by the SDE in (4.137) where $q_S = 0$.

By choosing $g(t) = X(t)S^f(t)$, the payoff relative to this numéraire simplifies to

$$\frac{C_T}{g(T)} = \frac{(X(T)S^f(T) - \kappa S^f(T))^+}{X(T)S^f(T)} = (1 - \kappa X^{-1}(T))^+ \equiv (1 - \kappa Y(T))^+ = \kappa(\kappa^{-1} - Y(T))^+$$

where we define the reciprocal of the exchange process $Y(t) := X^{-1}(t)$ for all $t \geqslant 0$. The numéraire asset price is given by (4.141), where $q_S = 0$. Hence, the pricing formula in (4.49) gives the time-t price $V_t = C_t$,

$$C_t = g(t)\,\widetilde{\mathrm{E}}^{(g)}\left[\frac{C_T}{g(T)}\,\Big|\,\mathcal{F}_t\right] = \kappa\, X(t)S^f(t)\,\widehat{\mathrm{E}}\left[(\kappa^{-1} - Y(T))^+\,|\,\mathcal{F}_t\right] \tag{4.151}$$

where we abbreviate $\widetilde{\mathrm{E}}^{(g)} \equiv \widetilde{\mathrm{E}}^{(XS^f)} \equiv \widehat{\mathrm{E}}$ and $\widetilde{\mathbf{W}}^{(g)}(t) \equiv \widehat{\mathbf{W}}(t)$. Computing this expectation is essentially like pricing a standard put with effective strike κ^{-1}.

[Remark: C_t above is kept as an $\mathcal{F}_t$-measurable random variable. We can also perform all the calculations by computing the (ordinary) pricing function $C(t, S, x)$ for arbitrary spot variables $S^f(t) = S, X(t) = x$, $Y(t) \equiv X^{-1}(t) = x^{-1}$, giving

$$C(t, S, x) = \kappa\, xS\,\widehat{\mathrm{E}}\left[(\kappa^{-1} - Y(T))^+ \mid Y(t) = x^{-1}\right] . \tag{4.152}$$

This pricing function then also gives us $C_t = C(t, S^f(t), X(t))$. Alternatively, we could compute the expectation in (4.151), giving $C_t = C(t, S^f(t), X(t))$ and the pricing function then follows by setting $S^f(t) = S, X(t) = x$. The main point is that both calculations give us the price of the option, expressible as an ordinary function of the spot variables or as a random variable.]

We need to represent $Y(T)$ as a GBM random variable involving the $\widehat{\mathbb{P}}$-BM and then either of the above expectations is trivially computed. One way to do this[3] is to write the BM increment $\widetilde{\mathbf{W}}(T) - \widetilde{\mathbf{W}}(t)$, using (4.34) where $\boldsymbol{\sigma}^{(g)}$ is a constant vector, as

$$\widetilde{\mathbf{W}}^{(g)}(T) - \widetilde{\mathbf{W}}^{(g)}(t) + (T-t)\,\boldsymbol{\sigma}^{(g)} \equiv \widehat{\mathbf{W}}(T) - \widehat{\mathbf{W}}(t) + (T-t)\,\boldsymbol{\sigma}^{(XS^f)},$$

where $\boldsymbol{\sigma}^{(XS^f)} = \boldsymbol{\sigma}^{(X)} + \boldsymbol{\sigma}^{(S)}$. Substituting this BM increment into the representation (4.139) gives

$$X(T) = X(t)\,\mathrm{e}^{(\widehat{\mu}_X + \frac{1}{2}\sigma_X^2)(T-t) + \boldsymbol{\sigma}^{(X)}\cdot(\widehat{\mathbf{W}}(T) - \widehat{\mathbf{W}}(t))}$$

[3] Alternatively, the SDE in (4.149) may be directly used to compute the log-drift and log-volatility vector of the reciprocal process $Y = 1/X$ upon applying the Itô quotient rule, with $\boldsymbol{\sigma}^{(g)} = \boldsymbol{\sigma}^{(XS^f)}$, giving $\mathrm{d}Y(t) = Y(t)[\widehat{\mu}_Y\,\mathrm{d}t + \boldsymbol{\sigma}^{(Y)}\cdot \mathrm{d}\widehat{\mathbf{W}}(t)]$, where $\widehat{\mu}_Y = r^f - r - \boldsymbol{\sigma}^{(X)}\cdot\boldsymbol{\sigma}^{(S)}$ and $\boldsymbol{\sigma}^{(Y)} = -\boldsymbol{\sigma}^{(X)}$.

where $\widehat{\mu}_X = r - r^f + \boldsymbol{\sigma}^{(X)} \cdot \boldsymbol{\sigma}^{(S)} = r - r^f + \rho\sigma_X\sigma_S$, $\rho = \rho_{XS}$. Taking the reciprocal gives

$$Y(T) = Y(t)\, \mathrm{e}^{(\widehat{\mu}_Y - \frac{1}{2}\sigma_Y^2)(T-t) + \boldsymbol{\sigma}^{(Y)} \cdot (\widehat{\mathbf{W}}(T) - \widehat{\mathbf{W}}(t))}$$

where $\widehat{\mu}_Y = -\widehat{\mu}_X, \boldsymbol{\sigma}^{(Y)} = -\boldsymbol{\sigma}^{(X)}$, $\sigma_Y = \|\boldsymbol{\sigma}^{(Y)}\| = \|\boldsymbol{\sigma}^{(X)}\| = \sigma_X$. The random variable in the exponent

$$\boldsymbol{\sigma}^{(Y)} \cdot (\widehat{\mathbf{W}}(T) - \widehat{\mathbf{W}}(t)) \stackrel{d}{=} \sigma_Y \sqrt{T-t}\widehat{Z}\,, \quad \text{where } \widehat{Z} \sim Norm(0,1) \text{ under } \widehat{\mathbb{P}}$$

and $\widehat{Z}$ is $\mathcal{F}_t$-independent (independent of $Y(t)$) and $Y(t)$ is $\mathcal{F}_t$-measurable. Expressing $Y(T)$ in terms of $\widehat{Z}$, then by independence the expectation in either (4.151) or (4.152) reduces to an unconditional expectation. In particular, (4.152) gives

$$C(t,S,x) = \kappa\, x S\, \widehat{\mathrm{E}}\left[(\kappa^{-1} - x^{-1}\, \mathrm{e}^{(\widehat{\mu}_Y - \frac{1}{2}\sigma_Y^2)(T-t) + \sigma_Y\sqrt{T-t}\widehat{Z}})^+\right]. \tag{4.153}$$

This expectation is easily computed by applying the identity (B.2) of Appendix B or we can directly use the pricing function for a standard put. Let $P_{BS}(S, K, \tau; r, \sigma)$ be the Black–Scholes pricing function in (3.43) for a standard put with spot S, strike K, $\tau = T - t$, interest rate r, volatility σ, and zero dividend. Then, the above expectation equals $\mathrm{e}^{\widehat{\mu}_Y\tau} P_{BS}(x^{-1}, \kappa^{-1}, \tau; \widehat{\mu}_Y, \sigma_Y)$ and using the above relations for $\widehat{\mu}_Y$ and σ_Y in terms of the original parameters gives the pricing function

$$\begin{aligned} C(t,S,x) &= \kappa\, x S \mathrm{e}^{\widehat{\mu}_Y\tau} P_{BS}(x^{-1}, \kappa^{-1}, T-t; \widehat{\mu}_Y, \sigma_Y) \\ &= \kappa\, x S \left[\kappa^{-1}\mathcal{N}\left(-\frac{\ln(x^{-1}/\kappa^{-1}) + (\widehat{\mu}_Y - \frac{1}{2}\sigma_Y^2)(T-t)}{\sigma_Y\sqrt{T-t}}\right)\right. \\ &\qquad \left. - \mathrm{e}^{\widehat{\mu}_Y(T-t)} x^{-1}\mathcal{N}\left(-\frac{\ln(x^{-1}/\kappa^{-1}) + (\widehat{\mu}_Y + \frac{1}{2}\sigma_Y^2)(T-t)}{\sigma_Y\sqrt{T-t}}\right)\right] \\ &= S\left[x\mathcal{N}\left(d_+(\frac{x}{\kappa}, T-t)\right) - \mathrm{e}^{(r^f - r - \rho\sigma_X\sigma_S)(T-t)}\kappa\mathcal{N}\left(d_-(\frac{x}{\kappa}, T-t)\right)\right] \end{aligned} \tag{4.154}$$

where

$$d_\pm(\frac{x}{\kappa}, T-t) := \frac{\ln(x/\kappa) + (r - r^f + \rho\sigma_X\sigma_S \pm \frac{1}{2}\sigma_X^2)(T-t)}{\sigma_X\sqrt{T-t}}.$$

If the stock pays a constant dividend yield q_S, then the pricing function is given by (4.154) where the spot S is replaced by $S\mathrm{e}^{-q_S(T-t)}$. □

4.4 Exercises

Exercise 4.1. Consider three domestic stocks with prices satisfying correlated GBM:

$$\mathrm{d}S_i(t) = S_i(t)\big[(r - q_i)dt + \boldsymbol{\sigma}_i \cdot \mathrm{d}\widetilde{\mathbf{W}}(t)\big], \quad i = 1, 2, 3.$$

$\widetilde{\mathbf{W}}(t)$ is a standard three-dimensional standard BM under the risk-neutral measure $\widetilde{\mathbb{P}}$ with the bank account as the numéraire. The parameters $\|\boldsymbol{\sigma}_i\| = \sigma_i > 0$, and the correlation coefficients ρ_{ij}, where $\boldsymbol{\sigma}_i \cdot \boldsymbol{\sigma}_j = \rho_{ij}\sigma_i\sigma_j$, are all assumed constants. The interest rate r is constant and q_i are the respective constant dividend yields. Let $S_i(0) = S_i > 0$ be the initial prices for each respective stock $i = 1, 2, 3$.

(a) For $t \geqslant 0$, derive explicit expressions for:

(i) $\widetilde{\mathbb{P}}\left(S_1(t) < S_2(t) < S_3(t)\right)$,

(ii) $\widetilde{\mathbb{P}}\left(S_1(t) < K < S_2(t)\right)$, for any constant $K > 0$,

(iii) $\widetilde{\mathbb{P}}\left(S_3(t) > K\frac{S_2(t)}{S_1(t)}\right)$, for any constant $K > 0$,

(iv) $\widetilde{\mathbb{P}}\left(S_1^\alpha(t) < S_2^\beta(t)\right)$, for any constants $\alpha, \beta > 0$.

(b) Derive the time-t ($t < T$) no-arbitrage pricing formula for the value of the European basket option with payoff at maturity T given by

$$V_T = \max\{S_3(T), S_1(T)\} - \min\{S_2(T), S_1(T)\}.$$

(c) Find the Radon–Nykodim derivative $\left(\frac{\mathrm{d}\widehat{\mathbb{P}}}{\mathrm{d}\widetilde{\mathbb{P}}}\right)_t$ where $\widehat{\mathbb{P}}$ is the EMM with $g(t) := S_1(t)$ as the numéraire.

Exercise 4.2. Derive expressions for the hedge positions in the two stocks 1 and 2 for the options on the maximum and the minimum with pricing functions in (4.106) and (4.107). Show that the relation in (4.102) holds for both options.

Exercise 4.3. Derive the explicit time-t pricing formula $V(t, S_1, S_2)$ for the exchange option with payoff in (4.103) assuming the stock prices are GBM processes as in (4.77), but now let the interest rate $r(t)$, the dividend yields on the stocks $q_i(t)$, and the log-volatility vectors $\boldsymbol{\sigma}_i(t)$, $i = 1, 2$, be nonrandom integrable functions of time.

Exercise 4.4. A plain currency call option on a foreign exchange rate has payoff

$$C_T = (X(T) - \kappa)^+ .$$

(a) Derive the pricing function $C(t, x)$, $t < T$, for this call by evaluating the risk-neutral expectation formula in (4.144).

(b) Give the BSPDE for the pricing function $C(t, x)$.

Exercise 4.5. Assume that a foreign stock price process $\{S^f(t)\}_{t \geqslant 0}$ and the (foreign to domestic) exchange rate $\{X(t)\}_{t \geqslant 0}$ are correlated geometric Brownian motions with respective log-volatility vectors $\boldsymbol{\sigma}^{(S)} = [\sigma_S, 0]$ and $\boldsymbol{\sigma}^{(X)} = [\rho\sigma_X, \sqrt{1-\rho^2}\sigma_X]$. Assume the domestic and foreign interest rates r and r^f are constants and that the foreign stock pays no dividend.

(a) Derive a formula for the current time $t < T$ price C_t of a call option on foreign stock denominated in domestic currency with domestic payoff

$$C_T = \left(X(T)S^f(T) - K\right)^+.$$

(b) Similarly, derive a formula for the current price P_t of a put option with domestic payoff

$$P_T = \left(K - X(T)S^f(T)\right)^+.$$

(c) Derive a put-call parity formula relating the call and put prices C_t and P_t.

Exercise 4.6. Consider the fixed foreign equity rate call with payoff in (4.123). Assume the foreign stock price is a GBM with constant log-volatility vector $\boldsymbol{\sigma}^{(S)}$ and having a dividend yield q_S. Derive its pricing function $C(t,S)$ by

(a) evaluating the risk-neutral expectation formula in (4.144);

(b) solving the BSPDE in (4.146) by symmetry reduction.

Exercise 4.7. Consider the foreign equity call struck in foreign currency with payoff in (4.121). Assume the foreign stock price is a GBM with constant log-volatility vector $\boldsymbol{\sigma}^{(S)}$ and having a dividend yield q_S and the exchange rate is a GBM with constant log-volatility vector $\boldsymbol{\sigma}^{(X)}$. Derive its pricing function $C(t,S,x)$ by solving the BSPDE in (4.146) by symmetry reduction.

Exercise 4.8. Consider a self-financing portfolio process defined by

$$\Pi_t = \beta_t B(t) + \beta_t^f X(t) B^f(t) + \Delta_t^f X(t) S^f(t) + \Delta_t A(t) \tag{4.155}$$

where an agent in a domestic economy takes a time-t position β_t in the domestic bank account, β_t^f in the foreign bank account converted to domestic currency, a position Δ_t^f in the foreign asset (e.g., stock) converted to domestic currency, and a position Δ_t in the domestic asset A. Assume the domestic and foreign interest rates r and r^f are constant and that the processes are GBM satisfying (4.135)–(4.138). Let $V(t,S,A,x)$ be a smooth pricing function and $V_t = V(t, S^f(t), A(t), X(t)), t \geqslant 0$ be the price process of a standard (non-path-dependent) European foreign exchange derivative.

By applying a multidimensional Itô formula and assuming that the self-financing portfolio replicates the foreign exchange derivative, i.e. $\Pi_t = V_t$ and therefore $\mathrm{d}\Pi_t = \mathrm{d}V_t$, show that the dynamic *delta hedging positions* in (4.155) are given by

$$\Delta_t = \frac{\partial V}{\partial A}, \quad \Delta_t^f = \frac{1}{x}\frac{\partial V}{\partial S}, \tag{4.156}$$

and the *domestic value of the investment in the foreign bank account* is

$$X(t)\beta_t^f B^f(t) = x\frac{\partial V}{\partial x} - S\frac{\partial V}{\partial S}, \tag{4.157}$$

where S, A, x are the respective spot values of $S^f(t), A(t), X(t)$.

Exercise 4.9. Assume that a foreign stock price $\{S^f(t)\}_{t\geqslant 0}$, a foreign exchange rate process $\{X(t)\}_{t\geqslant 0}$, and a *domestic* asset price process $\{A(t)\}_{t\geqslant 0}$ (e.g., a stock denominated in domestic currency) are all geometric Brownian motions with respective constant log-volatility vectors $\boldsymbol{\sigma}^{(S)}$, $\boldsymbol{\sigma}^{(X)}$, and $\boldsymbol{\sigma}^{(A)}$:

$$\frac{dA(t)}{A(t)} = \mu_A dt + \boldsymbol{\sigma}^{(A)}\cdot d\mathbf{W}(t), \quad \frac{dS^f(t)}{S^f(t)} = \mu_S dt + \boldsymbol{\sigma}^{(S)}\cdot d\mathbf{W}(t), \quad \frac{dX(t)}{X(t)} = \mu_X dt + \boldsymbol{\sigma}^{(X)}\cdot d\mathbf{W}(t).$$

$\mathbf{W}(t)$ is a three-dimensional standard $\mathbb{P}$-BM in the physical measure $\mathbb{P}$ and the assets are correlated, where $\|\boldsymbol{\sigma}^{(S)}\| = \sigma_S$, $\|\boldsymbol{\sigma}^{(X)}\| = \sigma_X$, $\|\boldsymbol{\sigma}^{(A)}\| = \sigma_A$, $\boldsymbol{\sigma}^{(S)}\cdot\boldsymbol{\sigma}^{(A)} = \rho_{SA}\sigma_S\sigma_A$, $\boldsymbol{\sigma}^{(X)}\cdot\boldsymbol{\sigma}^{(A)} = \rho_{XA}\sigma_X\sigma_A$, $\boldsymbol{\sigma}^{(X)}\cdot\boldsymbol{\sigma}^{(S)} = \rho_{XS}\sigma_S\sigma_X$. Assume a domestic and a foreign economy with respective interest rates r and r^f as constants, zero dividends on all assets, and let $A(t) = A, S^f(t) = S, X(t) = X$ be the spot values.

(a) Derive the time $t < T$ pricing function for a domestic European option with payoff

$$V_T = \max\{X(T)S^f(T), A(T)\}.$$

(b) Based on part (a), provide formulas for the delta position in the domestic stock and the delta position in the foreign stock that are required in a dynamic hedging strategy for all $t < T$ (see Exercise 4.8).

(c) Derive the time $t < T$ pricing function for a domestic European-style option with payoff

$$V_T = X(T)S^f(T)\,\mathbb{I}_{\{X(T)\geqslant X_0\}} + A_T\,\mathbb{I}_{\{X(T)<X_0\}}$$

where $X(0) = X_0$ is a fixed positive initial exchange rate.

(d) Derive the time $t < T$ pricing function for a domestic European-style option with payoff

$$V_T = \big(aX(T)S^f(T) - bA(T)\big)^+, \quad \text{with positive constants } a, b.$$

Exercise 4.10. Consider a market with domestic and foreign bank accounts where $r(t)$ and $r^f(t)$ are deterministic time-dependent interest rates in the respective domestic and foreign currencies. Assume a domestic asset with price process $\{A(t)\}_{t\geqslant 0}$, a foreign asset with price process $\{A^f(t)\}_{t\geqslant 0}$, and a foreign exchange rate process $\{X(t)\}_{t\geqslant 0}$. Assume zero dividend yield on the assets and all processes are GBM with respective constant log-volatility vectors $\boldsymbol{\sigma}^{(A)}$, $\boldsymbol{\sigma}^{(A^f)}$, and $\boldsymbol{\sigma}^{(X)}$, where $\boldsymbol{\sigma}^{(X)}\cdot\boldsymbol{\sigma}^{(A)} = \rho_{XA}\sigma_X\sigma_A$, $\boldsymbol{\sigma}^{(X)}\cdot\boldsymbol{\sigma}^{(A^f)} = \rho_{XA^f}\sigma_X\sigma_{A^f}$, $\boldsymbol{\sigma}^{(A)}\cdot\boldsymbol{\sigma}^{(A^f)} = \rho_{AA^f}\sigma_A\sigma_{A^f}$.

(a) Find the log-drift of each of the following separate processes: (i) $X(t)$, (ii) $X(t)B^f(t)$, (iii) $A(t)$, and (iv) $A^f(t)$ under the measure $\widetilde{\mathbb{P}}^{(g)}$ with $g(t) = A(t)$ as the numéraire asset price. Express your answers explicitly in terms of all the relevant parameters.

(b) Derive the pricing function for a European contract at current time $t = 0$ whose payoff at maturity T is

$$V_T = \max\{A(T), X_0A^f(T)\}$$

where $X_0 = \frac{A(0)}{A^f(0)}$ is the current (known) exchange rate.

Hint: You can use the results of part (a).

Exercise 4.11. Assume $\{S^f(t)\}_{t\geqslant 0}$ and $\{(X(t)\}_{t\geqslant 0}$ are correlated GBM as in Exercise 4.5.

(a) By using any approach, derive the pricing function $P(t, S, X)$, $t < T$, for a put option on foreign stock denominated in domestic currency with domestic payoff:

$$P_T = X(T)\big(K_f - S^f(T)\big)^+, \quad K_f > 0.$$

(b) Determine the self-financed portfolio *delta position in the foreign stock* and the *domestic value of the investment in the foreign bank account* (as functions of the spot values and time) that an investor must hold at time $t < T$ in order to *dynamically replicate* the value P_t (see Exercise 4.8).

(c) Derive a put-call parity relation between the above put option value $P(t, S, X)$ and the corresponding call option price $C(t, S, K)$ with payoff $C_T = X(T)\big(S^f(T) - K_f\big)^+$.

Exercise 4.12. Assume a foreign stock S^f and exchange rate process X modelled as in Exercise 4.5 with time-0 (spot) values $S^f(0) = S$, $X(0) = X$. Derive the time-0 pricing formula, as a function of S, X, T, for a domestic European option having payoff at maturity $T > 0$ given by

$$V_T = \mathbb{I}_{\{M(T)<K\}}\,X(T)S^f(T)\,,$$

where $M(T)$ is the maximum realized value of the exchange rate up to time T:

$$M(T) := \max_{0 \leqslant t \leqslant T} X(t)\,.$$

Hint: Use an appropriate choice of numéraire asset that simplifies the derivative pricing. You will need to make use of the CDF of $M(T)$.

Exercise 4.13. Consider a domestic economy with constant interest rate r and two correlated GBM stock price processes given by (4.91). Let $S_1(t) = S_1, S_2(t) = S_2$, $0 \leqslant t \leqslant T$, be the stock spot values. Derive the pricing function $V(t, S_1, S_2)$ for a European chooser max call with payoff in (4.118).
Hint: You may rewrite the payoff using the identity $(\max\{a,b\} - c)^+ = \mathbb{I}_{\{a \geqslant b\}}(a-c)^+ + \mathbb{I}_{\{b>a\}}(b-c)^+$.

Exercise 4.14. Consider a domestic economy with two stocks as in Exercise 4.13. Derive the pricing function for a European chooser min put with payoff in (4.119).
Hint: You may use similar identities as in Exercise 4.13 to rewrite the payoff.

Exercise 4.15. Consider three stocks with GBM price dynamics as in Exercise 4.1 with constant interest rate r and constant dividend yield q_i on each stock $i = 1, 2, 3$.

(a) Derive the pricing function $V(t, S_1, S_2, S_3)$, $t < T$, for a European option with payoff

$$V_T = S_3(T)\, \mathbb{I}_{\{S_3(T) > S_1(T), S_3(T) > S_2(T)\}}.$$

(b) Derive the pricing function for a European option with payoff

$$V_T = \max\{S_1(T), S_2(T), S_3(T)\}.$$

Hint: You can use the result of part (a).

(c) Derive the pricing function for a European option with payoff

$$V_T = K\, \mathbb{I}_{\{S_1(T) < S_3(T),\, S_2(T) < S_3(T)\}}$$

for constant $K > 0$.

Exercise 4.16. Consider an exchange option on two stocks having a payoff

$$V_T = (aS_2(T) - bS_1(T)\,)^+$$

with positive constants a, b. Assume the stocks are GBM processes as in (4.91). Derive the time-t price V_t for this option by explicitly using one of the stocks as the numéraire asset and by implementing the risk-neutral pricing formula in (4.62). Note: the derivation is similar to that in Example 4.5.

Exercise 4.17. Consider a domestic economy with constant interest rate r and two domestic stock price processes as in Exercise 4.1. Derive the time-0 pricing function $V_0 = V(T, S_1, S_2)$, in the spot variables $S_1(0) = S_1$, $S_2(0) = S_2$, for a European path-dependent option with payoff at maturity T given by

$$V_T = S_1(T) \min_{0 \leqslant t \leqslant T} \frac{S_2(t)}{S_1(t)}.$$

Hint: Use an appropriate numéraire asset that simplifies the evaluation of the risk-neutral expectation. The CDF of the *minimum* of a GBM process can then be employed.

Exercise 4.18. Assume the stock prices are GBM processes as in (4.77). For $t \geqslant 0$, derive the explicit time-t pricing formula $V(t, S_1, S_2)$ for an exchange option that allows its holder to exchange a shares of S_1 for b shares of S_2 (or vice versa) with payoff

$$\Lambda(S_1(T), S_2(T)) = |aS_2(T) - bS_1(T)|\,,$$

where a and b are positive numbers.

Exercise 4.19. Consider a foreign exchange model with domestic bank account $B(t) = \mathrm{e}^{rt}$, foreign bank account $B^f(t) = \mathrm{e}^{r^f t}$, domestic asset $A(t)$, and exchange rate process $X(t)$ with the following real-world dynamics:

$$\frac{\mathrm{d}X(t)}{X(t)} = \mu_X\,\mathrm{d}t + \sigma_X\,\mathrm{d}W_1(t),$$
$$\frac{\mathrm{d}A(t)}{A(t)} = (\mu_A - q_A)\,\mathrm{d}t + \sigma_A\,\mathrm{d}\left(\rho W_1(t) + \sqrt{1-\rho^2}W_2(t)\right).$$

(a) Find the dynamics of $X^{-1}(t)$, $A(t)$, and $F(t) := X^{-1}(t)A(t)$ under the foreign risk-neutral EMM $\widetilde{\mathbb{P}} \equiv \widetilde{\mathbb{P}}^{(g)}$ with the numéraire asset $g(t) = X(t)B^f(t)$ so that $Y(t) := \dfrac{X^{-1}(t)B_d(t)}{B_f(t)}$ and $Z(t) := \dfrac{\mathrm{e}^{q_A t}F(t)}{B_f(t)}$ are $\widetilde{\mathbb{P}}$-martingales.

(b) Derive the no-arbitrage pricing formulae for quanto options with following payoffs:

(i) $V_T = \min\{F(T), K_f\}$, with $K_f > 0$;
(ii) $V_T = A(T)\,(\kappa - X^{-1}(T))^+$, with $\kappa > 0$.

Exercise 4.20. Assume $S^f(t)$ and $X(t)$ are correlated GBM processes as in Exercise 4.5. Assume the domestic and foreign interest rates r and r^f are constants and that the foreign stock pays no dividend. Let $S^f(t) = S$, $X(t) = x$ be the spot values at current time t with $0 \leqslant t < T$. Derive time-t no-arbitrage pricing formulae for each of the following payoffs:

(a) $V_T = \max\{X(T)S^f(T), K_d\}$;
(b) $V_T = X(T)\max\{S^f(T), K_f\}$;
(c) $V_T = \min\{X(T)S^f(T), K_d\}$;
(d) $V_T = X(T)\min\{S^f(T), K_f\}$;

where K_d and K_f are strike prices in the domestic and foreign currencies, respectively.

Exercise 4.21. Assume that a foreign stock price process $S^f(t)$ and an exchange rate process $X(t)$ are correlated geometric Brownian motions with respective constant log-volatility vectors $\boldsymbol{\sigma}^{(S)}$ and $\boldsymbol{\sigma}^{(X)}$, where $\boldsymbol{\sigma}^{(S)} \cdot \boldsymbol{\sigma}^{(X)} = \rho_{XS}\sigma_S\sigma_X$, $||\boldsymbol{\sigma}^{(S)}|| = \sigma_S$, $||\boldsymbol{\sigma}^{(X)}|| = \sigma_X$. Also assume that the stock pays constant continuous dividend yield q and the domestic and foreign interest rates r and r^f are constants. Let $S^f(t) = S$, $X(t) = x$ be the spot values at current time t with $t < T$. Derive the time-t no-arbitrage European pricing function $V(t, S, x)$ for each payoff:

(a) $V_T = S^f(T)X(T)\mathbb{I}_{\{S^f(T)X(T) \geqslant K\}}$ with constant domestic strike $K > 0$;

(b) $V_T = X(T)S^f(T)\,\mathbb{I}_{\{S^f(T) \geqslant K\}}$ with constant foreign strike $K > 0$;

(c) $V_T = X(T)\mathbb{I}_{\{S^f(T) \geqslant K\}}$ with constant foreign strike $K > 0$;

(d) $V_T = X(T)S^f(T)\,\mathbb{I}_{\{X(T) \geqslant \kappa\}}$ where $\kappa > 0$ is a constant exchange rate;

(e) $V_T = X(T)(K - S^f(T))^+$ with constant foreign strike $K > 0$;

(f) $V_T = (\kappa - X(T))^+ S^f(T)$ with constant exchange rate $\kappa > 0$.

Exercise 4.22. Consider a domestic asset price process $A(t)$ with constant continuous dividend yield q_A and constant log-volatility vector $\boldsymbol{\sigma}^{(A)}$, $||\boldsymbol{\sigma}^{(A)}|| = \sigma_A$. Assume a given spot value $A(t) = A$ and assume the foreign stock price and exchange rate are modelled as in Exercise 4.21 with constant interest rates. In addition, assume that this domestic asset is correlated with these processes where $\boldsymbol{\sigma}^{(A)} \cdot \boldsymbol{\sigma}^{(S)} = \rho_{AS}\sigma_A\sigma_S$ and $\boldsymbol{\sigma}^{(A)} \cdot \boldsymbol{\sigma}^{(X)} = \rho_{AX}\sigma_A\sigma_X$. Derive the no-arbitrage pricing function $V(t, S, A, x)$ (with time $t < T$ and spot values S, A, x) for each payoff:

(a) $(K - X(T)S^f(T))^+ \mathbb{I}_{\{X(T)S^f(T) \leqslant A(T)\}}$;

(b) $(K - A(T))^+ \mathbb{I}_{\{X(T)S^f(T) \geqslant A(T)\}}$;

(c) $\min\{A(T), K\} \mathbb{I}_{\{X(T)S^f(T) \leqslant A(T)\}}$;

(d) $\max\{X(T)S^f(T), K\} \mathbb{I}_{\{X(T)S^f(T) \leqslant A(T)\}}$;

with constant domestic strike $K > 0$.

Exercise 4.23. Consider the model with three domestic stocks as described in Exercise 4.1. Derive the no-arbitrage pricing function $V(t, S_1, S_2, S_3)$ (with time $t < T$ and spot values S_1, S_2, S_3) for the payoff $V_T = S_1(T)\mathbb{I}_{\{S_2(T) < KS_3(T)\}}$, with constant $K > 0$.

Exercise 4.24. Consider the model with three domestic stocks as described in Exercise 4.1. Determine an explicit put-call relation between the no-arbitrage pricing functions $C(t, S_1, S_2)$ and $P(t, S_1, S_2)$ for the respective payoffs:

(a) $C_T := (\max\{S_1(T), S_2(T)\} - K)^+$ and $P_T := (K - \max\{S_1(T), S_2(T)\})^+$;

(b) $C_T := (\min\{S_1(T), S_2(T)\} - K)^+$ and $P_T := (K - \min\{S_1(T), S_2(T)\})^+$;

(c) $C_T := ((S_2(T) - S_1(T))^+ - K)^+$ and $P_T := (K - (S_2(T) - S_1(T))^+)^+$;

with constant $K > 0$.

5

American Options

In this chapter we briefly present the theory for pricing early-exercise (American) options in continuous time. Recall that American options were first introduced in Chapter 4 of Volume I, and the discrete-time case was dealt with in Chapter 7 of Volume I. Let us recall some of the properties of early-exercise options. The key difference between European-style and American-style options is that the holder of an American option can exercise their rights at any time before the expiration date. This additional early exercise privilege should not be worthless. Thus, an American option is expected to be worth more than its European analogue. The extra premium paid on top of the price of the European option is called the *early-exercise premium.* We mainly focus our discussion on standard American call and put options with respective payoffs $(S-K)^+$ and $(K-S)^+$, although the theory can also be applied to other types of options. Throughout this chapter, we assume that we deal with only one underlying whose asset price process, $\{S(t)\}_{t\geqslant 0}$, follows the standard geometric Brownian motion model, the risk-neutral dynamics of which is given by

$$S(t) = S(0)\mathrm{e}^{(r-q-\sigma^2/2)t+\sigma\widetilde{W}(t)}, \quad t \geqslant 0. \tag{5.1}$$

Here, $r \geqslant 0$ is the risk-free interest rate, $q \geqslant 0$ is the continuous dividend yield, $\sigma > 0$ is the asset price volatility, and $\{\widetilde{W}(t)\}_{t\geqslant 0}$ is a Brownian motion considered under the risk-neutral probability measure $\widetilde{\mathbb{P}}$ with the bank account as the numéraire asset.

5.1 Basic Properties of Early-Exercise Options

Let $t_0 \geqslant 0$ be the contract inception time. American call and put options struck at K with expiration time $T > t_0$ are claims to payoffs $(S(t)-K)^+$ and $(K-S(t))^+$, respectively, that the holder can exercise at any intermediate time $t \in [t_0, T]$. As in previous chapters, the time-t value of an option with expiration date T is denoted by $V(t, S)$, where S is the current asset (spot) price and $t \in [t_0, T]$ is the calendar time. It is also convenient to express the option value as a function of the time to expiration $\tau = T - t$ when the asset price model is assumed to be a time-homogeneous stochastic process; the option value therefore depends on τ and, as in previous chapters, we also denote the pricing function as $v(\tau, S)$, where $v(T-t, S) = V(t, S)$.

An American option cannot be worth less than its corresponding intrinsic value, which is the payoff associated with immediate exercise. In contrast to the case of European calls and puts, whose no-arbitrage values satisfy a put-call parity, there exists a put-call parity estimate for American options (see Theorem 5.3), which gives lower and upper bounds on the difference of call and put values:

$$S\mathrm{e}^{-q\tau} - K \leqslant C(\tau, S) - P(\tau, S) \leqslant S - K\mathrm{e}^{-r\tau},$$

DOI: 10.1201/9780429468889-5

where C and P denote the prices of the American call and put options, respectively, with spot S and time to maturity τ. In the absence of dividends on the underlying asset, there is no advantage in exercising an American call prior to the expiry date. Hence, the American call being exercised at the expiration date is equivalent to its European counterpart (with the same strike and expiration). As a result, the American and European calls on a stock without dividends have the same value. If the stock pays dividends, the arguments above are not valid. The optimal exercise date depends on the dividend process. If dividends occur occasionally at certain dates, it may be optimal to exercise an American call option right before a dividend payment that is large enough. Therefore, in the presence of dividends, the American call can be worth more than the European call. Recall that this situation was studied for a binomial tree model in Chapter 7 of Volume I. Let us generalize our findings about relationships between American and European call/put options in the following propositions.

Proposition 5.1. *Let V^E and V, respectively, denote the values of European and American options having the same payoff $\Lambda(S)$ and expiration date T. The following two conditions are equivalent:*

(i) $V^E(t,S) \geqslant \Lambda(S)$ *for all* $S > 0$ *and all* $t \in [t_0, T]$;

(ii) $V(t,S) = V^E(t,S)$ *for all* $S > 0$ *and all* $t \in [t_0, T]$.

That is, if the corresponding European price is always above the intrinsic value during the contract lifetime, then it is never optimal to exercise the American option at any time earlier than expiry, i.e., there is no early-exercise premium and $V \equiv V^E$.

Proof. For any time t, the value $V^E(t,S)$ is the arbitrage-free price of a contract exercised at the expiry time T. Condition (i) implies that it is not optimal to exercise earlier for a lower value. For every time t, it is more beneficial to wait until expiration. The optimal exercise (stopping) time is therefore at expiry T. Hence, (i) implies (ii). To prove the converse, observe that the American option value is always above the intrinsic value, i.e., $V(t,S) \geqslant \Lambda(S)$ for all (t,S). Hence, condition (ii) implies (i). This result is essentially a statement of the fact that an early-exercise opportunity (and premium) arises if and only if the corresponding European option value falls below the intrinsic (payoff function) value. □

As a corollary of Proposition 5.1, we have the rather well-known result:

Proposition 5.2.

(1) An American call has a nonzero early-exercise premium if and only if $q > 0$.

(2) An American put has a nonzero early-exercise premium if and only if $r > 0$.

This result follows from an arbitrage-free argument. However, a simple and instructive proof goes as follows.

Proof. Recall the put-call parity relation for European call and put options with common expiry T and strike price K:

$$C^E(t,S) - P^E(t,S) = \mathrm{e}^{-q(T-t)}S - \mathrm{e}^{-r(T-t)}K.$$

Using the fact that $P^E(t,S) \geqslant 0$ gives

$$C^E(t,S) = P^E(t,S) + \mathrm{e}^{-q(T-t)}S - \mathrm{e}^{-r(T-t)}K \geqslant \mathrm{e}^{-q(T-t)}S - K. \tag{5.2}$$

Then, for $q = 0$, (5.2) gives $C^E(t,S) \geqslant S - K$. Hence, the European call value is always above its payoff function. From Proposition 5.1, we conclude that the European call value, $C^E(t,S)$, is equal to the American call value, $C(t,S)$, so that the early-exercise premium is zero. For the case $q > 0$, we use (5.2) and note that since the European put is a decreasing function of S, there exist large enough values of S such that $P^E(t,S) + \mathrm{e}^{-q(T-t)}S - \mathrm{e}^{-r(T-t)}K < 0$, i.e., $C^E(t,S) < S - K$ for some $S > K$. From the previous result we therefore have $C(t,S) > C^E(t,S)$ and hence conclude that the early-exercise premium is nonzero for $q > 0$. This proves (i), while statement (ii) is proved in a similar fashion by reversing the roles of S and q with K and r, respectively, and is left as an exercise. □

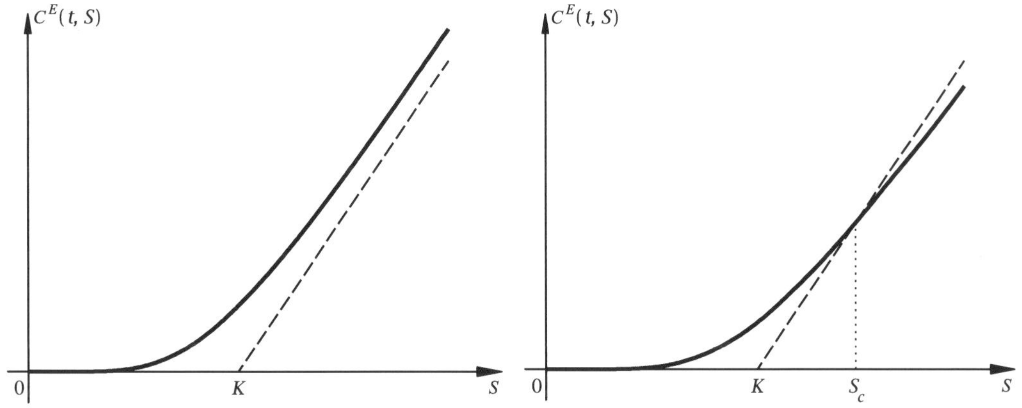

(a) The case with $q = 0$. (b) The case with $q > 0$.

FIGURE 5.1: The value of a European call option for two scenarios. The payoff is plotted as a dashed line. When $q > 0$, we observe that $C^E(t,S) < S - K$ for $S > S_c$. The critical value $S_c > 0$ solves the equation $C^E(t,S) = S - K$.

Statements (1) and (2) of Proposition 5.2 are illustrated in Figures 5.1 and 5.2, respectively. If $q = 0$ or $r = 0$, then $C^E(t,S) > (S-K)^+$ or, respectively, $P^E(t,S) > (K-S)^+$ for all $S > 0$. If $q > 0$ ($r > 0$), then the graph of the call (put) option value function intersects the payoff diagram at some critical point $S_c > 0$. For values of S greater (less) than S_c, the call (put) option price is strictly less than the payoff value. Note that $P(t,0) = \mathrm{e}^{-r(T-t)}K$, which equals K if $r = 0$ and is less that K if $r > 0$, for all $T - t > 0$.

When we considered the no-arbitrage valuation of options under the binomial tree model in Chapter 7 of Volume I, we analyzed the pricing of American options from both the holder's and writer's points of view. The writer sells an option in exchange for some initial capital, which can be used to hedge the short position in the derivative. This initial capital should provide the writer with sufficient funds to fulfil their obligations when the option is exercised. It is possible to find the best time for the holder (which is the worst time for the writer) to exercise the option. This time is called the *optimal exercise time*. The fair initial price of an American option is then defined as the smallest initial capital required to be hedged against exercise at the optimal time. Although the optimal exercise is not known in advance, the holder wishes to exercise the option as optimally as possible. Because the writer has no control over what exercise policy will be used by the buyer, the initial fair value of an American derivative with intrinsic value $\Lambda(S)$ is defined as the maximum of $\widetilde{\mathrm{E}}_0[\mathrm{e}^{-r\mathcal{T}}\Lambda(S(\mathcal{T}))]$ over all possible exercise polices $\mathcal{T} \in [t_0, T]$. It can be shown that both

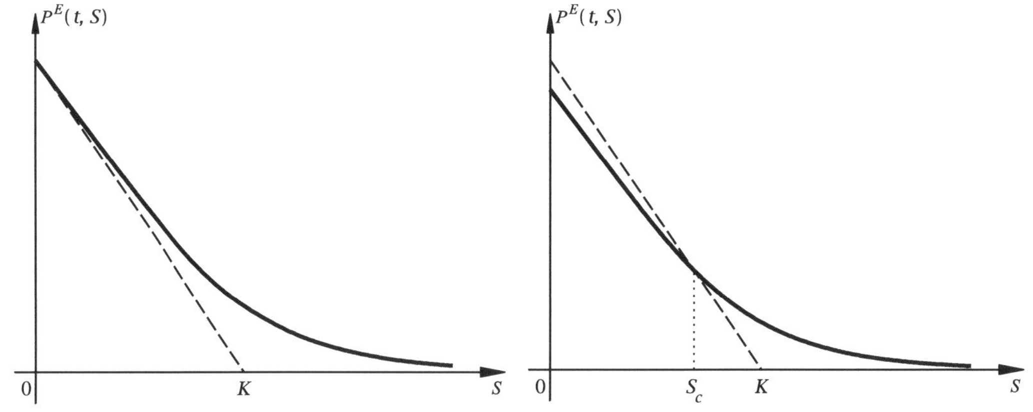

(a) The case with $r = 0$. (b) The case with $r > 0$.

FIGURE 5.2: The value of a European put option for two scenarios. The payoff is plotted as a dashed line. When $r > 0$, we observe that $P^E(t, S) < K - S$ for $S < S_c$. The critical value $S_c > 0$ solves the equation $P^E(t, S) = K - S$.

approaches give the same value. In the next section, we carry out this analysis in continuous time.

We conclude this section with two general results for American options. The proof of each of them is based on no-arbitrage argument, and hence the results are independent of the asset price model. The estimates provided below allow us to find no-arbitrage bounds on fair values of American call and put options.

Theorem 5.3 (Put-Call Parity Estimate). *Consider a continuous-time arbitrage-free pricing model for a risky stock with continuous dividend yield $q > 0$. The no-arbitrage argument implies the put-call parity estimate for American call and put options with common strike K and expiry T:*

$$e^{-q(T-t)}S - K \leqslant C(t, S) - P(t, S) \leqslant S - e^{-r(T-t)}K \text{ with } 0 \leqslant t \leqslant T.$$

Proof. Let us prove both inequalities by contradiction. Suppose that $C(t, S) - P(t, S) > S - e^{-r(T-t)}K$ holds at time t for spot price $S \equiv S(t)$. Write and sell the call option and buy one put option along with one stock share financing all transactions with a risk-free bank account. The time-t value of this portfolio is zero.

Scenario 1: The American call is not exercised at all. We sell the share for K by exercising the put option on the expiry date. The balance at time T is

$$K + (C(t, S) - P(t, S) - S)e^{r(T-t)} + \text{dividends} > 0.$$

Scenario 2: The American call is exercised at some time $u \in [t, T]$. After selling the stock share for K, the time-u value of our portfolio is

$$\begin{aligned} &K + (C(t, S) - P(t, S) - S)e^{r(u-t)} + P(u, S(u)) + \text{dividends} \\ &\geqslant (C(t, S) - P(t, S) - S - e^{-r(T-t)}K)e^{r(u-t)} > 0. \end{aligned}$$

Clearly, in both case we have an arbitrage opportunity, since the initial (at time t) value was zero. Our supposition contradicts to the fact that the market is arbitrage-free.

The proof of the second inequality is left as an exercise for the reader. □

Example 5.1. Consider a continuous-time arbitrage-free pricing model for a risky stock with continuous dividend yield $q > 0$. Using the no-arbitrage argument, show that

$$C^E(t,S) \leqslant C(t,S) \leqslant C^E(t,S) + S(1 - \mathrm{e}^{-q(T-t)})$$

holds for all $t \in [0,T]$ and all $S > 0$, where both European and American call options have the same strike K and expiry T.

Solution. The first inequality can be proved by contradiction. Suppose that $C(t,S) < C^E(t,S)$ holds at time t for spot price $S \equiv S(t)$. We form a portfolio with one long American call and one short European call. The proceeds $C^E(t,S) - C(t,S)$ are invested without risk at rate $r \geqslant 0$. The time t value of this portfolio is zero. At expiry, we exercise the American call if the European call is exercised. The guaranteed profit is $\mathrm{e}^{r(T-t)}(C^E(t,S) - C(t,S))$. It is an arbitrage, which leads to a contradiction. Thus, our supposition is incorrect.

Now, suppose that $C(t,S) > C^E(t,S) + S(1 - \mathrm{e}^{-q(T-t)})$ holds. At time t, we include the following in our portfolio with zero initial wealth: one short American call, one long European call, $1 - \mathrm{e}^{-q(T-t)}$ shares of stock. The proceeds are

$$V := C(t,S) - C^E(t,S) - S(1 - \mathrm{e}^{-q(T-t)}) > 0,$$

which are invested without risk. All dividend payments will be reinvested in the stock. As a result, $1 - \mathrm{e}^{-q(T-t)}$ shares of stock grows to $\mathrm{e}^{q\tau}(1 - \mathrm{e}^{-q(T-t)})$ shares over a period of length τ. Consider two possible scenarios.

Scenario 1: The American call is not exercised at all. The time-T value of our portfolio is

$$\mathrm{e}^{r(T-t)}V + S(T)(\mathrm{e}^{q(T-t)} - 1) + (S(T) - K)^+ > 0.$$

The last term represents the potential profit from exercising the European call at time T. Clearly, it is arbitrage, since the initial (at time t) value is zero. We have a contradiction.

Scenario 2: The American call is exercised at some time $u \in [t,T]$. To fulfil our obligation, we borrow one share of stock and sell it for K. The time-u value of the revised portfolio is

$$\mathrm{e}^{r(u-t)}V + C^E(u, S(u)) + S(u)(1 - \mathrm{e}^{-q(T-t)})\mathrm{e}^{q(u-t)} + K - S(u) \geqslant 0.$$

At time T, we exercise the European call option and buy the share for K. The terminal portfolio value is

$$\mathrm{e}^{r(T-t)}V + S(u)\left(((1 - \mathrm{e}^{-q(T-t)})\mathrm{e}^{q(u-t)} - 1)\mathrm{e}^{q(T-u)} - 1\right) + K(\mathrm{e}^{r(T-u)} - 1).$$

The stock position is $\mathrm{e}^{q(T-t)} - \mathrm{e}^{q(T-u)} \geqslant 0$. So, we have an arbitrage portfolio with a positive value. Again, our supposition contradicts to the fact that the market is arbitrage-free. □

5.2 Arbitrage-Free Pricing of American Options

5.2.1 Optimal Stopping Formulation and Early-Exercise Boundary

Consider an American option issued at time $t_0 \geqslant 0$ with payoff (intrinsic) value function Λ and expiration time $T > t_0$. Suppose that the option has not been exercised before time

$t \in [t_0, T]$. Additionally, suppose that the holder of the option follows some exercise policy $\mathcal{T}$, which is a stopping time taking values in the interval $[t, T]$ or taking the value ∞. Let $\mathcal{S}_{t,T}$ denotes the collection of all such stopping times. The no-arbitrage time-t value of the option, with underlying spot S, under a given exercise policy $\mathcal{T}$ is given by

$$\widetilde{\mathrm{E}}_{t,S}\left[\mathrm{e}^{-r(\mathcal{T}-t)}\,\Lambda(S(\mathcal{T}))\right].$$

Note that we are using our previous shorthand notation for the conditional expectation $\widetilde{\mathrm{E}}_{t,S}[\cdot] \equiv \widetilde{\mathrm{E}}[\,\cdot \mid S(t) = S]$. If $\mathcal{T}$ is infinite, we set $\mathrm{e}^{-r\infty}\,\Lambda(S(\infty))$ to be zero. Under uncertainty of what exercise rule will be used by the holder, the time-t value of an American option is defined to be

$$V(t, S) = \sup_{\mathcal{T} \in \mathcal{S}_{t,T}} \widetilde{\mathrm{E}}_{t,S}\left[\mathrm{e}^{-r(\mathcal{T}-t)}\,\Lambda(S(\mathcal{T}))\right]. \tag{5.3}$$

The optimal stopping time $\mathcal{T}^*$ maximizes the expectation on the right-hand side of (5.3). In other words, $\mathcal{T}^*$ is given implicitly by

$$V(t, S) = \widetilde{\mathrm{E}}_{t,S}\left[\mathrm{e}^{-r(\mathcal{T}^*-t)}\,\Lambda(S(\mathcal{T}^*))\right]. \tag{5.4}$$

By analogy with the case of the binomial tree model, the optimal exercise time $\mathcal{T}^*$ is the random variable corresponding to the first time when the option value equals the intrinsic value,

$$\mathcal{T}^* = \min\{u \;:\; t \leqslant u \leqslant T,\; V(u, S(u)) = \Lambda(S(u))\}. \tag{5.5}$$

That is, for maximal gain, along a stock price path $(u, S(u, \omega))$, $t \leqslant u \leqslant T$ with a fixed market scenario ω, the option should be exercised at the first time, say $t^* = \mathcal{T}^*(\omega) \in [t, T]$, when $V(t^*, S(t^*)) = \Lambda(S(t^*))$. Thus, letting $S(t) = S$, the (t, S)-plane consisting of the time and spot value points is separated into two sub-domains: a stopping domain where the option is exercised early,

$$\mathcal{D} = \{(t, S) \;:\; t_0 \leqslant t \leqslant T,\; V(t, S) = \Lambda(S)\}, \tag{5.6}$$

and a continuation domain for which the option is not exercised,

$$\mathcal{D}^c = \{(t, S) \;:\; t_0 \leqslant t \leqslant T,\; V(t, S) > \Lambda(S)\}. \tag{5.7}$$

The continuation domain is the complement of the stopping domain within the rectangle $[t_0, T] \times [0, \infty)$. As is seen from (5.6), the continuation domain is the set of all points (t, S) such that the option value $V(t, S)$ exceeds the value of the payoff function $\Lambda(S)$.

The geometry of the stopping domain may be quite complicated. It depends on the payoff function $\Lambda(S)$ and whether the underlying asset pays discrete or continuous dividends. However, for the case with a standard call/put payoff and continuous dividends, the stopping domain turns out to be simply connected. A typical shape of the stopping domain for a standard American put option is demonstrated in Figure 5.3. The *early-exercise boundary*, $\partial\mathcal{D}$, which separates the continuation and stopping domains, is a smooth curve on the (t, S)-plane and is of the form $\partial\mathcal{D} = \{(t, S) \;:\; 0 \leqslant t \leqslant T,\; S = S_t^*\}$, where the function S_t^* is given by

$$S_t^* = \min\{S > 0 \;:\; C(t, S) = (S - K)^+\} \tag{5.8}$$

for a call and

$$S_t^* = \max\{S > 0 \;:\; P(t, S) = (K - S)^+\} \tag{5.9}$$

for a put. Since the American option value is always nonnegative, the superscript plus signs in (5.8) and (5.9) are actually redundant. Because the option value is positive, the critical

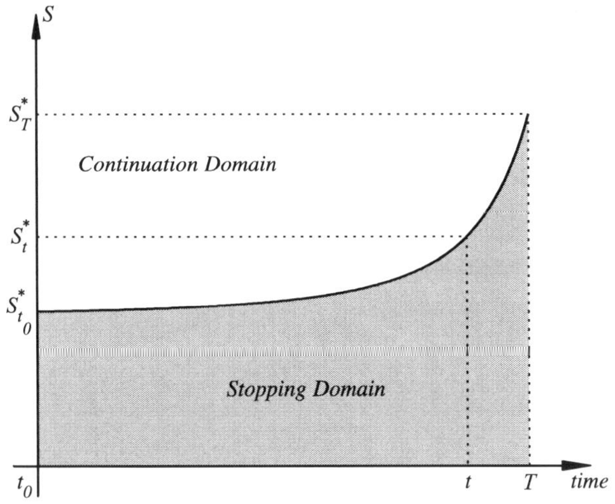

FIGURE 5.3: A typical stopping domain for an American put option. The option is exercised in the shaded area.

value S_t^* is larger than the strike K for an American call and it is less than K for an American put.

In general, the early-exercise curve S_t^* varies with time t. For a given calendar time $t \in [t_0, T]$, we define the continuation and stopping intervals in $[0, \infty)$ for spot price S, denoted by $\mathcal{D}_t^c$ and $\mathcal{D}_t$, respectively, so that $S \in \mathcal{D}_t$ iff $(t, S) \in \mathcal{D}$ and $\mathcal{D}_t^c := [0, \infty) \setminus \mathcal{D}_t$. For a given time t, the American option is exercised if $S \in \mathcal{D}_t$, and its value is equal to the payoff function in $\mathcal{D}_t$. For the American call, we have $\mathcal{D}_t^c = [0, S_t^*)$ and $\mathcal{D}_t = [S_t^*, \infty)$; while for the American put, we have $\mathcal{D}_t^c = (S_t^*, \infty)$ and $\mathcal{D}_t = [0, S_t^*]$. When S is in the stopping domain $\mathcal{D}_t$, the values of the American call and put are $C(t, S) = S - K$ and $P(t, S) = K - S$, respectively. In what follows, it will be convenient to express quantities in terms of time to maturity $\tau = T - t$. Consequently, we adopt the notations $S^*(\tau) \equiv S_t^*$, $\mathcal{D}(\tau) \equiv \mathcal{D}_t$, and $\mathcal{D}^c(\tau) \equiv \mathcal{D}_t^c$.

An obvious consequence of Proposition 5.2 is that (a) for an American call on a non-dividend-paying stock the exercise boundary is trivial (i.e., it is never optimal to exercise early where $\mathcal{D}^c(\tau) = [0, \infty)$) and (b) for an American put on a non-dividend-paying stock the exercise boundary is nontrivial (i.e., there is an optimal early-exercise time) if the interest rate is positive.

Note that for a general payoff function, the stopping domain may consist of several disconnected regions; therefore, the early-exercise boundary may be a union of separate curves. For example, in the case of an American strangle whose payoff is a sum of standard call and put payoffs, the stopping domain consists of upper and lower regions (assuming that both r and q are strictly positive).

As follows from the definition of the stopping domain, the optimal exercise time $\mathcal{T}^*$ defined in (5.5) is also the first passage time of the stopping domain $\mathcal{D}$. That is, the time $\mathcal{T}^*$ is the first time when the stock price process reaches the early-exercise boundary S^* (see Figure 5.4):

$$\mathcal{T}^* = \min\{t \in [t_0, T] \ : \ S(t) = S_t^*\}. \tag{5.10}$$

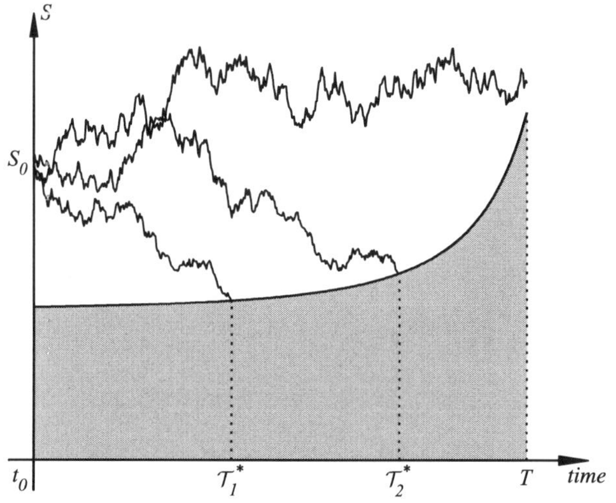

FIGURE 5.4: For two sample asset price paths corresponding to scenarios ω_1 and ω_2, an American put is exercised early at times $\mathcal{T}_1^* = \mathcal{T}^*(\omega_1)$ and $\mathcal{T}_2^* = \mathcal{T}^*(\omega_2)$, respectively. The option is not exercised early when a sample path does not cross the early-exercise boundary S^*.

5.2.2 The Smooth Pasting Condition

The American option should be exercised whenever the asset price S is in the stopping domain $\mathcal{D}$. This implies that the American option value $V(t,S)$ can be written as a piecewise continuous function of S which is equal to the payoff for all spot values $S \in \mathcal{D}_t$. For standard American call and put options, we have that $C(t,S) = (S-K)^+ = S-K$ if $S \geqslant S_t^*$ and $P(t,S) = (K-S)^+ = K-S$ if $S \leqslant S_t^*$, respectively. Clearly, the option value functions have to be continuous at the exercise boundary point $S = S_t^*$ or otherwise there exists an arbitrage opportunity. Moreover, as is demonstrated below, the optimal exercise condition requires the curve of $V(t,S)$ to be tangent to the payoff at the point S_t^*. In other words, the pricing function for an American option has a continuous derivative at the exercise boundary; that is, the delta of the option price is continuous at the early-exercise boundary. This property is known as the *smooth pasting condition*.

We need to prove that the derivative of the option pricing function is equal to the derivative of the payoff function at the point $S = S_t^*$ for any $t \in [t_0, T]$. Let us first consider the case with a put option. Note that the argument presented below also readily follows for any continuous and monotonic payoff function such as the standard call and put payoffs. Suppose that the American put has not been exercised before time t. Consider the collection $\mathcal{B}$ of all possible early-exercise boundaries defined by continuous functions $b\colon [t,T] \to \mathbb{R}_+$. For each $b \in \mathcal{B}$, there is an exercise policy $\mathcal{T}_b$, which is the first passage time for the boundary b. Let

$$P(t,S;b) = \widetilde{\mathrm{E}}_{t,S}\left[\mathrm{e}^{-r(\mathcal{T}_b - t)}\,(K - S(\mathcal{T}_b))^+\right]$$

be the put value under the exercise policy $\mathcal{T}_b$. Since every policy $\mathcal{T}_b$ is a stopping time in the set $\mathcal{S}_{[t,T]}$ and the function S_t^* of time t, defining the optimal exercise boundary, belongs to the collection $\mathcal{B}$, the value of the American put in (5.3) is given by $P(t,S) = \sup_{b\in\mathcal{B}} P(t,S;b)$. The optimal exercise boundary $b = S^*$ maximizes the function $b \mapsto P(t,S;b)$. So, this is a problem in calculus of variations. In general, if u is a function and $F(u)$ is some functional of u, then the calculus of variations is a technique that tries to find such a u that optimizes the value of F. If u^* is an optimal solution, the functional derivative $\frac{\mathrm{d}F(u)}{\mathrm{d}u}$ is zero at $u = u^*$. Since the optimal choice for the boundary b is the optimal early-exercise boundary S^*, the derivative of the put value P with respect to b is zero at $b = S^*$:

$$\left.\frac{\partial P(t,S;b)}{\partial b}\right|_{b=S^*} = 0.$$

Let us find the total derivative of the function P with respect to b along the boundary:

$$\frac{\mathrm{d}P}{\mathrm{d}b} = \left.\frac{\partial P(t,S;b)}{\partial S}\right|_{S=b} + \left.\frac{\partial P(t,S;b)}{\partial b}\right|_{S=b},$$

where we use the fact that $\frac{\partial S}{\partial b} = 1$ along the curve $S = b(t)$. On the one hand, when $b = S^*$, we have $\frac{\mathrm{d}P}{\mathrm{d}b} = \left.\frac{\partial P(t,S;b)}{\partial S}\right|_{S=S^*}$. On the other hand, the option value is equal to the payoff function when $S = b$. Therefore, $P(t,b;b) = (K - b)$ and then

$$\frac{\mathrm{d}P}{\mathrm{d}b}(t,b;b) = \frac{\mathrm{d}}{\mathrm{d}b}(K-b) = -1.$$

Putting the results together gives

$$\left.\frac{\partial P(t,S)}{\partial S}\right|_{S=S^*} = \left.\frac{\partial P(t,S;S^*)}{\partial S}\right|_{S=S^*} = -1 = \left.\frac{\mathrm{d}\Lambda(S)}{\mathrm{d}S}\right|_{S=S^*}. \tag{5.11}$$

where $\Lambda(S) = (K-S)^+$ is the put payoff and equals $K - S$ for all $S < K$. Note that $S^* < K$ for a put.

An alternative proof of the smooth pasting condition (5.11) is based on the no-arbitrage argument. If $S < S_t^*$, then $P(t,S) = K - S$. Therefore, the left-hand limit is $\lim_{S \nearrow S_t^*} \frac{\partial P(t,S)}{\partial S} = -1$. Now consider the right-hand limit. Since the American put value is always above the payoff values, we have $\lim_{S \searrow S_t^*} \frac{\partial P(t,S)}{\partial S} \geqslant -1$. Our objective is to show that this inequality is actually a strict equality. We prove this by showing that there exists an arbitrage if the limiting value is strictly greater than -1. Suppose that the asset price at calendar time t is at the boundary, i.e., $S \equiv S(t) = S_t^*$. Consider a portfolio of one long put option and one share of stock. The portfolio value is

$$\Pi(t) = S + P(t,S) = S + K - S = K.$$

After a sufficiently small time lapse δt, the asset price can move downward into the exercise domain or upward into the continuation domain. We denote the change in the asset price by $\delta S = S(t+\delta t) - S(t)$, so $S(t+\delta t) = S + \delta S$. If $\delta S < 0$, then $P(t+\delta t, S+\delta S) = K - (S+\delta S)$ and the change in the portfolio value is $\delta\Pi = \delta P + \delta S = 0$. If $\delta S > 0$, the small move δS creates a profit opportunity since $\delta S > 0$ and $\delta P < 0$ but in absolute value δP is smaller than δS. We thus have $\delta\Pi > 0$. Let us find the order of magnitude of $\delta\Pi$. Using the stochastic differential equation for the asset price process gives $\delta S = \mu S \delta t + \sigma S \delta W$, where $\delta W \sim \mathit{Norm}(0, \delta t)$. Therefore, we have

$$\begin{aligned}\delta\Pi &= \delta S + \delta P = \delta S + \left.\frac{\partial P}{\partial S}\right|_{S \searrow S_t^*} \delta S + \frac{1}{2}\left.\frac{\partial^2 P}{\partial S^2}\right|_{S \searrow S_t^*} (\delta S)^2 + o(\delta t) \\ &= \mathcal{O}(\delta t) + \left(\sigma S + \left.\frac{\partial P}{\partial S}\right|_{S \searrow S_t^*} \sigma S\right) \delta W = \mathcal{O}(\delta t) + \left(1 + \left.\frac{\partial P}{\partial S}\right|_{S \searrow S_t^*}\right) \sigma S \sqrt{\delta t}\,|Z|,\end{aligned}$$

where $Z \sim \mathit{Norm}(0,1)$, since $\delta W > 0$ for an upward movement of the stock price. Thus,

$$\mathrm{E}_t[\delta\Pi \mid \delta S > 0] = \mathcal{O}(\delta t) + \left(1 + \left.\frac{\partial P}{\partial S}\right|_{S \searrow S_t^*}\right) \sigma S \sqrt{\delta t}.$$

This represents the change in portfolio value conditional on the stock price having a positive change within an arbitrarily small time interval δt. Since the term in $\sqrt{\delta t}$ is arbitrarily larger than all other terms of order $\mathcal{O}(\delta t)$, the change in portfolio value is positive if $\frac{\partial P}{\partial S}|_{S \searrow S_t^*} > -1$, i.e., the upward return on the portfolio is positive and of order $\sqrt{\delta t}$. This is an arbitrage opportunity since the risk-free return should be a smaller quantity. Therefore, the condition (5.11) holds. In fact, the smooth pasting condition is applicable to all types of American options. The general result is presented below without a proof.

Proposition 5.4. *Consider an American option with a differentiable payoff function Λ at any point (t, S_t^*) on the early-exercise boundary. Then, the American option pricing function V satisfies the smooth pasting condition:*

$$\left.\frac{\partial V(t,S)}{\partial S}\right|_{S=S_t^*} = \Lambda'(S_t^*), \tag{5.12}$$

that is, the (spacial) derivative of V is continuous at the boundary of the stopping and continuation domains (as is illustrated in Figure 5.5). Additionally, the option value V satisfies the zero time-decay condition on the early-exercise domain,

$$\frac{\partial V(t,S)}{\partial t} = 0, \quad \textit{for all} \quad S \in \mathcal{D}_t. \tag{5.13}$$

Proof. The proof is left as an exercise for the reader. □

Remarks:

1. The condition in (5.12) is also obviously valid for all $S \in \mathcal{D}_t$ since $V(t,S) = \Lambda(S)$ on that region.

2. For a call (or put) the property (5.12) simply gives

$$\left.\frac{\partial V(t,S)}{\partial S}\right|_{S=S_t^*} = 1 \text{ (or } -1).$$

   This is illustrated in Figure 5.5.

3. The properties (5.12) and (5.13) are also valid under a general diffusion model.

5.2.3 Put-Call Symmetry Relation

An American option can be considered as providing the right to exchange one asset for another, namely, one share of the underlying stock worth S dollars (asset one) and K dollars of cash (asset two). Such an exchange can take place any time before expiration. So, a call option gives the right to exchange cash for one unit of stock (i.e., exchange asset two for asset one) while a put option gives the right to exchange one unit of stock for cash (i.e., exchange asset one for asset two). Assets one and two have dividend yields q and r, respectively. Given this similarity we might expect call and put prices to be equal when we interchange the role of the underlying stock and cash. Let $P(\tau, S; K, r, q)$ and $C(\tau, S; K, r, q)$ respectively denote the price of the American put and call options with asset price S, strike price K, time to expiration τ, dividend yield q, and risk-free interest rate r. After interchanging the role of the underlying asset and cash, the price of the modified American put is $P(\tau, K; S, q, r)$. Since the modified put is equivalent to the American call, we have

$$P(\tau, K; S, q, r) = C(\tau, S; K, r, q). \tag{5.14}$$

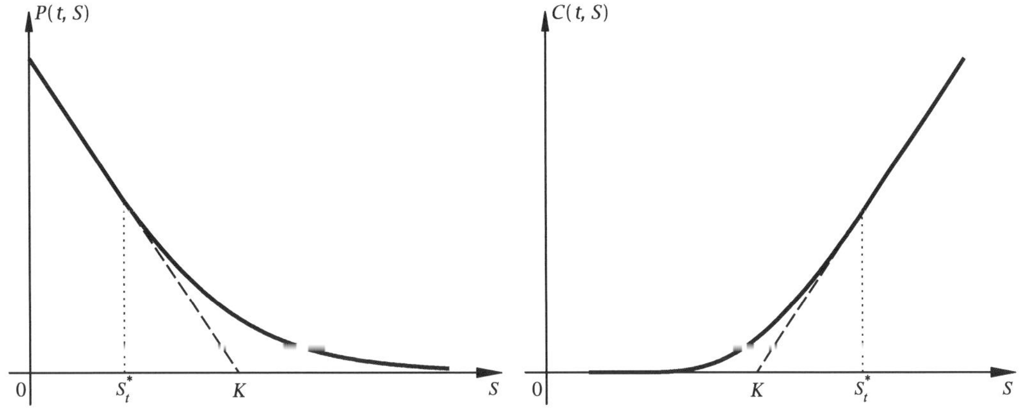

(a) American put. (b) American call.

FIGURE 5.5: The value functions for American put and call options satisfy the smooth pasting condition with slope equal to -1 and 1, respectively, at the optimal exercise boundary point S_t^*. The solid price curve touches the dashed payoff line tangentially at the point $(S_t^*, \Lambda(S_t^*))$, where $\Lambda(S)$ is the payoff function equal to $(S-K)^+$ and $(K-S)^+$ for the call and put options, respectively.

This symmetry between the value function of an American call and put is called the *put-call symmetry relation.* The result is true for both European and American styles of options. It is also true for both perpetual and finite-expiration contracts. The symmetry relation (5.14) implies that the prices of at-the-money (i.e., $S = K$) American call and put options are equal if $r = q$. A rigorous proof of this relation is left as an exercise for the reader (see Exercises 5.2 and 5.5).

5.2.4 Dynamic Programming Approach for Bermudan Options

Bermudans are contracts that essentially lie in between European and American derivatives. Exercise of a Bermudan option can only occur at a fixed set of dates. Let

$$\mathbf{T} := \{t_j \ : \ j = 1, 2, \ldots, M\}$$

be the set of allowable exercise dates, $t_0 \leqslant t_1 < t_2 < \cdots < t_M = T$. On the one hand, the value of a Bermudan option is defined as the maximum arbitrage-free price of the contract under all possible exercise policies. Every exercise policy is a stopping time taking its values in the set $\mathbf{T}$ or taking the value ∞. Let $\mathcal{S}_{\mathbf{T}}$ denote the collection of all stopping times taking values in $\mathbf{T} \cup \{\infty\}$. Assuming that the option has not been exercised before time t, its arbitrage-free value at time t is

$$V(t, S) = \sup_{\mathcal{T} \in \mathcal{S}_{\mathbf{T} \cap [t,T]}} \widetilde{\mathrm{E}}\left[e^{-r(\mathcal{T}-t)} \Lambda(S(\mathcal{T})) \mid S(t) = S\right], \tag{5.15}$$

where the underlying asset price process $\{S(t)\}_{t \geqslant 0}$ is assumed to be Markovian. We expect (5.15) to be a good approximation to (5.3) for small durations $\delta t_j = t_j - t_{j-1}$ between successive exercise dates. As $\max \delta t_j \to 0$ (and hence $M \to \infty$), the Bermudan option value in (5.15) converges to the (continuous-time exercise) American option value (5.3).

Since the continuous-time process $\{S(t)\}_{t \geqslant 0}$ is Markovian, $S(t_1), S(t_2), \ldots, S(t_M)$ form a Markov chain. It can be shown that the Bermudan option values at the discrete exercise dates satisfy the recurrence relation

$$V(t_j, S) = \max\left\{\Lambda(S), \widetilde{\mathrm{E}}_{t_j}\left[\mathrm{e}^{-r(t_{j+1}-t_j)}V(t_{j+1}, S(t_{j+1})) \mid S(t_j) = S\right]\right\} \tag{5.16}$$
$$=: \max\left\{\Lambda(S), V^{cont}(t_j, S)\right\} \quad \text{for } j = M-1, M-2, \ldots, 1,$$

where $V(T, S) = \Lambda(S)$ at the expiration time T. The conditional expectation in (5.16) denoted by $V^{cont}(t, S)$ represents the *continuation value* of the option at time t_i. It is the time-t value of the option that has not been exercised yet at time t.

Let the asset price process be a continuous-time stochastic process with assumed risk-neutral transition PDF $\tilde{p}(t, t'; S, S') := \frac{\widetilde{\mathbb{P}}(S(t') \in \mathrm{d}S' | S(t)=S)}{\mathrm{d}S'}$ with $S, S' > 0$ and $0 \leqslant t < t'$. Then, the continuation value in (5.16) is computed as

$$V^{cont}(t_j, S) = \mathrm{e}^{-r(t_{j+1}-t_j)} \int_0^\infty \tilde{p}(t_j, t_{j+1}; S, S')V(t_{j+1}, S')\,\mathrm{d}S'.$$

Recall that in the case of a time-homogeneous process, the transition PDF only depends on the duration $t' - t$ rather than on the individual time moments t and t', that is, we write

$$\tilde{p}(t, t'; S, S') = \tilde{p}(t' - t; S, S').$$

In particular, recall that for the geometric Brownian motion model in (5.1), $\tilde{p}$ is the log-normal density given by (3.29) with r replaced by $r - q$.

The dynamic-programming formulation (5.16) of the Bermudan option pricing problem does not require computation of the early-exercise boundary. However, the optimal exercise rule and early-exercise boundary can be obtained simultaneously while computing option values. In particular, the optimal exercise rule is

$$\mathcal{T}^* = \min\{t_j,\ j = 1, 2, \ldots, M \ :\ \Lambda(S(t_j)) = V(t_j, S(t_j))\}.$$

Essentially, the rule is defined in the same way as that for the binomial tree model in Chapter 6 of Volume I. Since the Bermudan option can only be exercised at times from the set $\mathbf{T}$, the stopping domain $\mathcal{D} \subset [0, T] \times \mathbb{R}_+$ is the union of lines

$$\mathcal{D} = \bigcup_{1 \leqslant j \leqslant M} \left\{(t_j, S) \ :\ \Lambda(S) \geqslant V^{cont}(t, S)\right\}.$$

The dynamic programming formulation of the American option pricing problem provides a basis for implementing a number of numerical methods for computing option prices using Monte Carlo simulations, finite-difference schemes, lattice methods, or a combination of such approaches.

5.3 Perpetual American Options

In this section, we consider a *perpetual option* with infinite time to expiration. That is, a perpetual option has no expiration date and can be exercised at any future time. Such options are not among traded securities. However, they are an instructive mathematical concept because they admit simple analytic solutions. Here, we derive closed-form pricing formulae for perpetual American calls and puts. Additionally, perpetual options represent an accurate approximation of finite-expiration American options when the time horizon of the expiry is very long.

Consider a perpetual derivative with intrinsic value $\Lambda(S)$. We are interested in its value as a function of the underlying asset price. The holder of a perpetual American option can exercise it at any time. Since the time to expiry is always infinite, the option value must be independent of time. That is, the value of a perpetual derivative, V, depends only on the asset price, i.e., we simply write $V(t, S) = V(S)$ for all $t \geqslant 0$. It is reasonable to expect that the early-exercise boundary that separates the continuation and stopping domains does not depend on the time variable t as well. The function S_t^* that describes the early-exercise boundary is constant, i.e., there is a fixed exercise level $S_t^* = S^*$ for all $t \geqslant 0$. The stopping domain (on the time and stock price plane) is $\mathcal{D} = \mathbb{R}_+ \times (0, S^*]$ for a perpetual put option and $\mathcal{D} = \mathbb{R}_+ \times [S^*, \infty)$ for a perpetual call option.

5.3.1 Pricing a Perpetual Put Option

Let us consider the case of a perpetual American put option with intrinsic value $(K - S)^+$. The holder will exercise the option when it is deep enough in the money. That is, there exists some *constant* boundary $S^* \leqslant K$ so that it is optimal to exercise as soon as the price $S(t)$ reaches the level S^*. From the definition of the optimal stopping time $\mathcal{T}^*$ as given in (5.4) and (5.10), we conclude that the value of a perpetual put, $P(S) \equiv P(S; K, r, q)$, is

$$P(S) = \widetilde{\mathrm{E}}_{0,S}\left[\mathrm{e}^{-r\mathcal{T}^*}\left(K - S(\mathcal{T}^*)\right)\right] \tag{5.17}$$

$$= (K - S^*)\widetilde{\mathrm{E}}_S\left[\mathrm{e}^{-r\mathcal{T}^*}\right] \quad \text{for } S > S^* \tag{5.18}$$

where $\widetilde{\mathrm{E}}_{0,S}[\,] \equiv \widetilde{\mathrm{E}}_S[\,]$ is the risk-neutral expectation conditional on $S(0) = S$, and $\mathcal{T}^*$ is the first passage time to the level S^*, i.e.,

$$\mathcal{T}^* = \inf\{t \geqslant 0 \;:\; S(t) = S^*\}.$$

Under the assumption that the asset price follows the geometric Brownian motion model, the time $\mathcal{T}^*$ is the first passage time of the drifted Brownian motion $X(t) = \mu t + \widetilde{W}(t)$, with drift parameter $\mu = \frac{r-q-\sigma^2/2}{\sigma}$, to the level $L^* := \frac{1}{\sigma}\ln\frac{S^*}{S}$, i.e.,

$$\mathcal{T}^* = \inf\{t \geqslant 0 \;:\; X(t) \leqslant L^*\}.$$

We recall from Section 1.4.2 of Chapter 1 that this corresponds to the first hitting time, $\tau_{L^*}^X$, of the drifted Brownian motion down to the level $L^* < 0$ in case $S > S^*$. The mathematical expectation in (5.18) is actually the Laplace transform of the PDF of $\mathcal{T}^*$, evaluated at r. This Laplace transform can be readily obtained analytically. However, it can also be computed via the Feynman–Kac formula for processes stopped at the first passage time. In this case, the stopping time $\mathcal{T}^*$ is the first passage time to the level S^*. According to the Feynman–Kac result, the function $v(S) = \widetilde{\mathrm{E}}_S\left[\mathrm{e}^{-r\mathcal{T}^*}\right]$ satisfies the ordinary differential equation (ODE) $\mathcal{G}v - rv = 0$ where $\mathcal{G}$ is the infinitesimal generator for the stock price process under the risk-neutral measure, i.e.,

$$\frac{1}{2}\sigma^2 S^2 \frac{\mathrm{d}^2 v}{\mathrm{d}S^2} + (r - q)S\frac{\mathrm{d}v}{\mathrm{d}S} - rv = 0, \quad S > S^*, \tag{5.19}$$

subject to the boundary conditions

$$\text{(i) } v(S^*) = 1, \quad \text{(ii) } \lim_{S\to\infty} v(S) = 0. \tag{5.20}$$

The level S^* is yet unknown but uniquely determined once $v(S)$ is obtained in terms of S^*, as described just below. The first boundary condition corresponds to $\mathcal{T}^*$ being zero ($\mathrm{e}^{-r\mathcal{T}^*} = 1$) if the process starts at $S = S^*$. The second condition in (5.20) arises since $\mathcal{T}^*$ goes to infinity ($\mathrm{e}^{-r\mathcal{T}^*} \to 0$) as the process is started at an arbitrarily large value S (i.e., the point at infinity is a natural boundary for the stock price process).

The value of the perpetual put given by $P(S) = (K - S^*)v(S)$ satisfies the ODE (5.19) subject to

$$\text{(i) } P(S^*) = K - S^*, \quad \text{(ii) } \lim_{S\to\infty} P(S) = 0. \tag{5.21}$$

Note that the ODE (5.19) is also obtained from the Black–Scholes partial differential equation (PDE) by setting the time derivative to zero. As is demonstrated below, the pricing function of a finite-expiration American option does indeed satisfy the Black–Scholes PDE (5.31) in the continuation domain. Since the perpetual option value V is time independent, the time derivative $\frac{\partial V}{\partial t}$ is identically zero. Hence, the Black–Scholes PDE (5.31) reduces to the ODE (5.19).

Equation (5.19) is of the Cauchy–Euler type, $ax^2y''(x) + bxy'(x) + cy(x) = 0$, with constants a, b, and c. The general solution has the form $y(x) = A_1x^{\lambda_1} + A_2x^{\lambda_2}$, where A_1 and A_2 are arbitrary constants. Putting the function $v(S) = S^\lambda$ into (5.19) gives the following auxiliary equation on the exponent λ:

$$\frac{\sigma^2}{2}\lambda(\lambda - 1) + (r - q)\lambda - r = 0.$$

Solving it gives

$$\lambda = \lambda_\pm := \frac{-(r - q - \sigma^2/2) \pm \sqrt{(r - q - \sigma^2/2)^2 + 2\sigma^2 r}}{\sigma^2}. \tag{5.22}$$

Assuming positive interest rate r, it is easy to show that $\lambda_- < 0$ and $\lambda_+ > 0$. So the general solution to (5.19) is $v(S) = a_+S^{\lambda_+} + a_-S^{\lambda_-}$. Now, we determine the coefficients $a_\pm$ in terms of S^*. To satisfy condition (5.20(ii)), we set

$$\lim_{S\to\infty} \left(a_+S^{\lambda_+} + a_-S^{\lambda_-}\right) = 0$$

and thus must have $a_+ = 0$. Then, a_- is determined from condition (5.20(i)):

$$v(S^*) = a_-(S^*)^{\lambda_-} = 1$$

for an arbitrary yet undetermined parameter S^*. Thus, the solution to the problem (5.19)–(5.20) is

$$v(S) = \left(\frac{S}{S^*}\right)^{\lambda_-}, \quad S > S^*.$$

In fact, the function $v(S)$ is the price of a so-called digital option that pays \$1 when the asset price breaches the lower level S^*. The perpetual put value is then

$$P(S) \equiv P(S; S^*) = \begin{cases} (K - S^*)\left(\frac{S}{S^*}\right)^{\lambda_-}, & S^* < S, \\ K - S, & 0 < S \leqslant S^*. \end{cases} \tag{5.23}$$

The last step is to determine S^*. Consider two methods: first, using the fact that S^* must be the optimal value that maximizes the option value $P(S)$ among all possible choices of $S^* > 0$; second, using the smooth pasting condition. Let us find the maximum of the function $S^* \mapsto P(S; S^*)$. Differentiating the option value with respect to S^* for $S^* < S$ gives

$$\frac{\partial P}{\partial S^*} = \frac{\partial}{\partial S^*}\left\{(K - S^*)\left(\frac{S}{S^*}\right)^{\lambda_-}\right\} = -\left(\frac{S}{S^*}\right)^{\lambda_-}\left(1 + \frac{K - S^*}{S^*}\lambda_-\right).$$

Set $\frac{\partial P}{\partial S^*} = 0$ to obtain the extremum

$$S^* = \frac{K\lambda_-}{\lambda_- - 1}. \tag{5.24}$$

Note that $\frac{\partial^2 P}{\partial (S^*)^2} = \frac{K\lambda_-}{(S^*)^2}\left(\frac{S}{S^*}\right)^{\lambda_-} < 0$ since $\lambda_- < 0$. Thus, S^* given by (5.24) is indeed a point of maximum. Finally, substituting (5.24) in (5.23) gives the perpetual put price:

$$P(S; K, r, q) = \begin{cases} \frac{K}{1-\lambda_-}\left(\frac{\lambda_- - 1}{\lambda_-}\right)^{\lambda_-}\left(\frac{S}{K}\right)^{\lambda_-} = -\frac{S^*}{\lambda_-}\left(\frac{S}{S^*}\right)^{\lambda_-}, & S^* < S, \\ K - S, & 0 < S \leqslant S^*. \end{cases} \tag{5.25}$$

The solution in (5.25) is also readily obtained by applying the smooth pasting condition to the function in (5.23). Namely, we differentiate the function for $S > S^*$ and set the derivative to -1 at $S = S^*$:

$$\left.\frac{\partial P}{\partial S}\right|_{S=S^*} = -\lambda_-\left(\frac{K - S^*}{S^*}\right) = -1.$$

Solving the above equation for S^* gives us the optimal value in (5.24).

Let us examine what happens to the exercise boundary in the limiting case when the interest rate r is zero. From (5.22), we obtain $\lambda_- = 0$. Thus, from (5.24) we see that $S^* = 0$. Hence, for a zero interest rate, the perpetual put is never exercised early. This observation is consistent with the property of a finite-expiration American put.

5.3.2 Pricing a Perpetual Call Option

Now we consider the perpetual American call struck at K. Let $C(S) \equiv C(S; K, r, q)$ denote its price when the stock price equals S. As in the case of the perpetual put option, the call pricing function $C(S)$ satisfies the ODE (5.19) for $S \in (0, S^*)$. In the stopping domain $[S^*, \infty)$, the call value is equal to the payoff value, i.e.,

$$C(S) = S - K \quad \text{for} \quad S \geqslant S^*.$$

Hence, the boundary conditions are

$$\text{(i)} \lim_{S \searrow 0} C(S) = 0, \quad \text{(ii)}\ C(S^*) = S^* - K. \tag{5.26}$$

The first condition states that the call is worthless if the stock price is zero since it will remain at zero indefinitely, i.e., the payoff is always zero. The second condition reflects the fact that $C(S)$ is continuous at $S = S^*$.

The general solution to (5.19) is $C(S) = a_+ S^{\lambda_+} + a_- S^{\lambda_-}$, where the exponents $\lambda_\pm$ are given by (5.22). Since $C(0+) = 0$ and $\lambda_- < 0$, we must set $a_- = 0$. Imposing the condition (5.26(ii)) gives $a_+ = (S^* - K)/(S^*)^{\lambda_+}$. Following the same procedure as above, we obtain the optimal value of S^*:

$$S^* = \frac{K\lambda_+}{\lambda_+ - 1}. \tag{5.27}$$

The perpetual call price is then

$$C(S;K,r,q)=\begin{cases}\frac{K}{\lambda_+-1}\left(\frac{\lambda_+-1}{\lambda_+}\right)^{\lambda_+}\left(\frac{S}{K}\right)^{\lambda_+}=\frac{S^*}{\lambda_+}\left(\frac{S}{S^*}\right)^{\lambda_+}, & 0<S<S^*,\\ S-K, & S^*\leqslant S.\end{cases} \tag{5.28}$$

A simple check by differentiation shows that this function satisfies the required smooth pasting condition for a call:

$$\left.\frac{\partial C}{\partial S}\right|_{S=S^*}=1.$$

Let us examine what happens to the exercise boundary in the limiting case when the dividend yield q is zero. From (5.22) and (5.27), we see that $\lambda_+\to 1$ and $S^*\to\infty$, as $q\searrow 0$. Hence, for a zero dividend yield, the perpetual call is never exercised early. Again, this is consistent with the property of a finite-expiration American call.

5.4 Finite-Expiration American Options

5.4.1 The PDE Formulation

Pricing an American option can be formulated as a boundary initial value problem for a partial differential equation with a time-dependent free boundary. The solution domain is divisible into a union of a stopping region, $\mathcal{D}$, where the American option is exercised, and a continuation domain, $\mathcal{D}^c$, where the option is not exercised. Inside the continuation domain, the option value function satisfies the Black–Scholes PDE. On the early-exercise boundary, the option value is equal to the payoff function $\Lambda(S)$. The early-exercise boundary is an unknown function of time which must also be determined as part of the solution. Here, we assume the payoff is time independent, although the formulation also extends to the case of a time-dependent payoff.

Delta hedging and continuous-time replication arguments apply to American options in the same way as they apply to European options. Therefore, within the continuation domain the option pricing function must satisfy the Black–Scholes PDE. The connection between the optimal stopping time formulation and the PDE approach can be shown using the following heuristic. Consider the recurrence relation (5.16) for a calendar time $t\in[t_0,T]$ and a small time step $\delta t>0$:

$$V(t,S)=\max\left\{\Lambda(S),\,\mathrm{e}^{-r\delta t}\widetilde{\mathrm{E}}\big[V(t+\delta t,S(t+\delta t))\mid S(t)=S\big]\right\}.$$

Assuming $V(t,S)$ is a sufficiently smooth function with continuous derivatives, we can expand $V(t+\delta t,S(t+\delta t))$ in a Taylor series while keeping terms up to $\mathcal{O}(\delta t)$. Assuming the underlying asset follows the GBM model in (5.1), we have:

$$\begin{aligned}V(t,S)&=\max\left\{\Lambda(S),\,(1-r\delta t)\widetilde{\mathrm{E}}_{t,S}\left[V(t,S(t))+\left(\frac{\partial V(t,S(t))}{\partial t}+(r-q)S(t)\frac{\partial V(t,S(t))}{\partial S}\right.\right.\right.\\&\quad\left.\left.\left.+\frac{1}{2}\sigma^2S^2(t)\frac{\partial^2V(t,S(t))}{\partial S^2}\right)\delta t+\sigma S(t)\frac{\partial V(t,S(t))}{\partial S}\delta\widetilde{W}_t\right]\right\}+\mathcal{O}\left(\delta t^2\right)\\&=\max\left\{\Lambda(S),\,V(t,S)+\left[\frac{\partial V(t,S)}{\partial t}+\mathcal{L}V(t,S)\right]\delta t\right\}+\mathcal{O}(\delta t^2).\end{aligned} \tag{5.29}$$

The second equation is obtained by evaluating the expectation conditional on $S(t)=S$ and then collecting terms up to $\mathcal{O}(\delta t)$. Note that the coefficient terms multiplying δt and $\delta\widetilde{W}_t$ are both $\mathcal{F}_t$-measurable with given spot value $S(t) = S$. Moreover, the conditional expectation term in $\delta\widetilde{W}_t$ vanishes by the independence of the Brownian increment $\delta\widetilde{W}_t = \widetilde{W}_{t+\delta t} - \widetilde{W}_t$ and $S(t)$, i.e.,

$$\widetilde{\mathrm{E}}_{t,S}\left[S(t)\frac{\partial V(t,S(t))}{\partial S}\delta\widetilde{W}_t\right] = S\frac{\partial V(t,S)}{\partial S}\widetilde{\mathrm{E}}_{t,S}[\delta\widetilde{W}_t] = 0$$

since $\widetilde{\mathrm{E}}_{t,S}[\delta\widetilde{W}_t] = \widetilde{\mathrm{E}}[\delta\widetilde{W}_t] = 0$. The expression in (5.29) has been written more compactly using the Black–Scholes differential operator $\mathcal{L} := \mathcal{G} - r$,

$$\mathcal{L}f := \frac{1}{2}\sigma^2 S^2\frac{\partial^2 f}{\partial S^2} + (r-q)S\frac{\partial f}{\partial S} - rf \tag{5.30}$$

For values of S in the continuation domain $\mathcal{D}_t^c$ the inequality $V(t,S) > \Lambda(S)$ is satisfied. In (5.29), we require the right-hand side to equal the left to small order $o(\delta t)$. Hence, the coefficient in δt must be zero, which gives the Black–Scholes PDE:

$$\frac{\partial V(t,S)}{\partial t} + \mathcal{L}V(t,S) = 0, \text{ for all } S \in \mathcal{D}_t^c \text{ and all } t \in [t_0, T].$$

Thanks to the time-homogeneous property of the solution, $V(t,S) = v(\tau,S)$, we have a PDE in terms of the time to maturity variable $\tau = T - t$ and spot S:

$$\frac{\partial v(\tau,S)}{\partial \tau} = \mathcal{L}v(\tau,S), \text{ for all } S \in \mathcal{D}^c(\tau) \text{ and all } \tau > 0. \tag{5.31}$$

The Black–Scholes PDE does not hold on the stopping domain, where the American option is given by the payoff function $V(S,\tau) = \Lambda(S)$. Since the payoff is time-independent, the derivative $\frac{\partial\Lambda(S)}{\partial\tau}$ is zero. Thus, the solution $v(\tau,S)$ on $\mathcal{D}(\tau)$ satisfies $\frac{\partial v}{\partial \tau} = 0$. Combining regions and assuming the payoff is a twice differentiable function gives a *nonhomogeneous* Black–Scholes PDE valid in the whole region:

$$\frac{\partial v(\tau,S)}{\partial \tau} = \mathcal{L}v(\tau,S) + f(\tau,S), \tag{5.32}$$

with function

$$f(\tau,S) = \begin{cases} 0, & S \in \mathcal{D}^c(\tau), \\ -\mathcal{L}\Lambda(S), & S \in \mathcal{D}(\tau). \end{cases} \tag{5.33}$$

Given the function $f(\tau,S)$ whose time dependence is determined in terms of the early-exercise boundary, the solution to (5.32), subject to the initial condition

$$v(0,S) = \Lambda(S) \tag{5.34}$$

and the boundary conditions

$$\lim_{S\searrow 0} v(\tau,S) = \lim_{S\searrow 0}\Lambda(S), \quad \lim_{S\to\infty} v(\tau,S) = \lim_{S\to\infty}\Lambda(S), \tag{5.35}$$

can be obtained in terms of the solution to the corresponding (homogeneous) Black–Scholes PDE, $\frac{\partial v(\tau,S)}{\partial\tau} = \mathcal{L}v(\tau,S)$.

For the American call and put, we recall the early-exercise curves $S = S^*(\tau)$ expressed as a function of time to maturity on the (S,τ)-plane. For the call, the set of spot values $S \in \mathcal{D}(\tau)$ is then equivalent to the set of values $S \geqslant S^*(\tau)$. For the put, the values $S \in \mathcal{D}(\tau)$

are the values $S \leqslant S^*(\tau)$. Hence, expressing the call and put option values as functions of τ and S, (5.32) and (5.33) specialize to

$$\frac{\partial C}{\partial \tau} - \frac{\sigma^2 S^2}{2}\frac{\partial^2 C}{\partial S^2} - (r-q)S\frac{\partial C}{\partial S} + rC = \begin{cases} 0 & S < S^*(\tau)\,, \\ qS - rK, & S \geqslant S^*(\tau)\,, \end{cases} \tag{5.36}$$

and

$$\frac{\partial P}{\partial \tau} - \frac{\sigma^2 S^2}{2}\frac{\partial^2 P}{\partial S^2} - (r-q)S\frac{\partial P}{\partial S} + rP = \begin{cases} rK - qS & S \leqslant S^*(\tau)\,, \\ 0 & S > S^*(\tau)\,, \end{cases} \tag{5.37}$$

respectively. For the call, we used the fact that $-\mathcal{L}\Lambda(S) = -\mathcal{L}(S-K) = -(rK - qS) = qS - rK$. For the put, $-\mathcal{L}\Lambda(S) = -\mathcal{L}(K-S) = rK - qS$. Here, we used $S^*(\tau)$ to denote the early-exercise boundary for the respective call and put with given strike K. The right-hand sides of these nonhomogeneous PDEs are nonzero only within the respective stopping regions.

We close this section by working out some other basic properties of the early-exercise boundary for a call and a put option. Let's consider the limit of infinitesimally small $\tau \searrow 0$. In particular, consider an American call struck at K with continuous dividend yield q. The pricing function $C(\tau, S; K)$ is an increasing function of τ, i.e., $C(\tau_2, S; K) \geqslant C(\tau_1, S; K)$ for $\tau_2 > \tau_1$. Also, the smooth pasting condition guarantees that the pricing functions join the intrinsic line at levels $S^*(\tau_1) - K$ and $S^*(\tau_2) - K$, respectively, giving $S^*(\tau_2) > S^*(\tau_1)$. We hence conclude that $S^*(\tau)$ is a continuously increasing function of $\tau > 0$ and attains the value of the early-exercise boundary of the corresponding perpetual call option, given by (5.27), in the limit $\tau \to \infty$. That is, an American call with greater time to maturity should be exercised deeper in the money to account for the loss of time value on the strike K. Since one would never prematurely exercise at a spot value that is below the strike level, the early-exercise boundary for an American call must satisfy the property $S^*(\tau) > K$ for all $\tau > 0$. To determine the boundary in the limit $\tau \searrow 0$, we note that the option value approaches the intrinsic value, i.e., at expiry the option value is given by the payoff $C(S, K, \tau = 0) = S - K$ for values on the exercise boundary. According to (5.36), we have

$$\frac{\partial C(S, K, 0+)}{\partial \tau} = rK - qS \tag{5.38}$$

for $S > K$. Since the condition $\partial C(S, K, 0+)/\partial \tau > 0$ ensures that the option is not yet exercised, the spot value S at which $\partial C(S, K, 0+)/\partial \tau$ becomes negative and hence for which the call is exercised at an instant just before expiry is given by $S = \frac{r}{q}K$. This is the case, however, if the value $\frac{r}{q}K$ is in the interval $S > K$; i.e., if $r > q > 0$. In this case just prior to expiry the call is not yet exercised if the spot is in the region $K < S < \frac{r}{q}K$, but would be exercised if $S \geqslant \frac{r}{q}K$. Hence, in the limiting case we have $S^*(0+) = \frac{r}{q}K$ for $r > q > 0$. For the other case we have $r \leqslant q$, so $\frac{r}{q}K \leqslant K$. Yet $S > K$, so that $S^*(0+) = K$ for $r > q$. Note that the condition $S^*(0+) > K$ is not possible in this case as this leads to a sub-optimal early exercise since the loss in dividends would have greater value than the interest earned over the infinitesimal time interval until expiry. Combining the above arguments, we arrive at the general limiting condition for the exercise boundary of an American call just prior to expiry:

$$\lim_{\tau \searrow 0} S^*(\tau) = \max\left(K, \frac{r}{q}K\right) = K \max\left(1, \frac{r}{q}\right). \tag{5.39}$$

From this property we see that $S^*(0+) \to \infty$ as $q \to 0$. So, for zero dividend yield the American call is never exercised early which is consistent with the fact that the American call has exactly the same worth as the European call. By using similar arguments as above, we readily work out $S^*(0+)$ for the American put struck at K with continuous dividend yield q. For the put, $S^*(\tau)$ is a continuously decreasing function of $\tau > 0$ and attains the value in (5.24) in the limit $\tau \to \infty$. We leave it as an exercise for the reader to show that

$$\lim_{\tau \searrow 0} S^*(\tau) = K \min\left(1, \frac{r}{q}\right). \tag{5.40}$$

5.4.2 The Integral Equation Formulation

Recall that the time-homogeneous risk-neutral transition PDF, $\tilde{p} = \tilde{p}(\tau; S, S')$, solves the forward Kolmogorov PDE in the S' variable and the backward Kolmogorov PDE in the spot variable S with zero boundary conditions at $S = 0$ and $S = \infty$ for all $\tau > 0$. As mentioned above, for the process (5.1) the PDF $\tilde{p}$ is just the log-normal density. We also know that the function $\mathrm{e}^{-r\tau}\tilde{p}$ solves the (homogeneous) Black–Scholes PDE. Combining these facts and applying Laplace transforms, one arrives at the well-known Duhamel solution to (5.32) in the form

$$\begin{aligned} v(\tau, S) &= \mathrm{e}^{-r\tau} \int_0^\infty \tilde{p}(\tau; S, S')\Lambda(S')\,\mathrm{d}S' + \int_0^\tau \mathrm{e}^{-r\tau'} \left[\int_0^\infty \tilde{p}(\tau'; S, S') f(\tau - \tau', S')\,\mathrm{d}S' \right] \mathrm{d}\tau' \\ &\equiv v^E(\tau, S) + v^e(\tau, S), \end{aligned} \tag{5.41}$$

where $f(\tau, S) = -\mathcal{L}\Lambda(S) \cdot \mathbb{I}_{\{S \in \mathcal{D}(\tau)\}}$ is defined in (5.33). An important aspect of this result is the fact that the American option value is expressible as a sum of two components. The first term is simply the standard European option value v^E, as given by the discounted risk-neutral expectation of the payoff. Hence the second term, denoted by v^e, must represent the early-exercise premium which the holder must pay to have the additional liberty of early exercise, i.e., $v^e(\tau, S) = v(\tau, S) - v^E(\tau, S)$.

For a call, $f(\tau, S) = (qS - rK)\mathbb{I}_{\{S \geqslant S^*(\tau)\}}$, i.e., $f(\tau - \tau', S') = (qS' - rK)\mathbb{I}_{\{S' \geqslant S^*(\tau - \tau')\}}$. For a given value of τ', this restricts the inner integral in (5.41) to the interval $[S^*(\tau - \tau'), \infty)$. Similarly, for a put, $f(\tau - \tau', S') = (rK - qS')\mathbb{I}_{\{0 < S' \leqslant S^*(\tau - \tau')\}}$, the inner integral in S' is restricted to the interval $(0, S^*(\tau - \tau')]$. Using (5.41), the solutions to (5.36) and (5.37) for the American call and put prices are given by

$$C(\tau, S; K) = C^E(\tau, S; K) + C^e(\tau, S; K)$$

and

$$P(\tau, S; K) = P^E(\tau, S; K) + P^e(\tau, S; K)$$

respectively, where the corresponding early-exercise premiums take on the integral forms

$$C^e(\tau, S; K) = \int_0^\tau \mathrm{e}^{-r\tau'} \left[\int_{S^*(\tau - \tau')}^\infty \tilde{p}(\tau'; S, S')(qS' - rK)\,\mathrm{d}S' \right] \mathrm{d}\tau' \tag{5.42}$$

and

$$P^e(\tau, S; K) = \int_0^\tau \mathrm{e}^{-r\tau'} \left[\int_0^{S^*(\tau - \tau')} \tilde{p}(\tau'; S, S')(rK - qS')\,\mathrm{d}S' \right] \mathrm{d}\tau'. \tag{5.43}$$

The above premiums can also be recast as

$$C^e(\tau, S; K) = \int_0^\tau \mathrm{e}^{-r\tau'} \widetilde{\mathrm{E}}_{0,S}\left[(qS(\tau') - rK)\, \mathbb{I}_{\{S(\tau') \geqslant S^*(\tau - \tau')\}} \right] \mathrm{d}\tau' \tag{5.44}$$

and

$$P^e(\tau, S; K) = \int_0^\tau \mathrm{e}^{-r\tau'} \widetilde{\mathrm{E}}_{0,S}\left[(rK - qS(\tau'))\,\mathbb{I}_{\{S(\tau')\leqslant S^*(\tau-\tau')\}}\right] \mathrm{d}\tau' \,, \tag{5.45}$$

where $\widetilde{\mathrm{E}}_{0,S}[\ ]$ denotes the time-0 risk-neutral expectation conditional on asset paths starting at $S(0) = S$. The time integral is over all intermediate times to maturity and the indicator functions ensure that all asset paths fall within the early-exercise region. The properties of the early-exercise boundaries established in the previous section guarantee that the early-exercise premiums are nonnegative. Using (5.39), for a dividend paying call we have $S(\tau') \geqslant S^*(\tau)$, i.e., $S(\tau') \geqslant K \max\left(1, \frac{r}{q}\right) \geqslant \frac{r}{q}K$. Hence, $qS(\tau') - rK \geqslant 0$ and $C^e \geqslant 0$. A similar analysis follows for the put premium where (5.40) gives $rK - qS(\tau') \geqslant 0$ and $P^e \geqslant 0$. The exercise premiums hence involve a continuous stream of discounted expected cash flows from contract inception until maturity.

For the geometric Brownian motion model (with constants r, q, and σ) the function $\tilde{p}$ is given by the log-normal density and the above double integrals readily simplify to single time integrals in terms of standard cumulative normal functions. Namely, the expectations in the integrands of (5.44) and (5.45) are easily computed by making use of the stock price representation

$$S(\tau') = S(0)\mathrm{e}^{(r-q-\sigma^2/2)\tau' + \sigma\widetilde{W}(\tau')} \stackrel{d}{=} S(0)\mathrm{e}^{(r-q-\sigma^2/2)\tau' + \sigma\sqrt{\tau'}\widetilde{Z}}$$

where $\widetilde{Z} \sim \mathit{Norm}(0,1)$ under measure $\widetilde{\mathbb{P}}$. Note that the quantity $S^*(\tau - \tau')$ is simply a constant, i.e., it is the value of the early-exercise boundary for the time to maturity $\tau - \tau'$. By employing the identity in (B.1) of Appendix B, and using the usual steps as in the derivation of the standard European call, the expectation in (5.44) is given by

$$\begin{aligned}
\widetilde{\mathrm{E}}_{0,S}\left[(qS(\tau') - rK)\,\mathbb{I}_{\{S(\tau')\geqslant S^*(\tau-\tau')\}}\right] &= qS\mathrm{e}^{(r-q-\sigma^2/2)\tau'}\, \widetilde{\mathrm{E}}\left[\mathrm{e}^{\sigma\sqrt{\tau'}\widetilde{Z}}\mathbb{I}_{\{\widetilde{Z}\geqslant -d_-^*(\tau')\}}\right] \\
&\quad - rK\widetilde{\mathbb{P}}(\widetilde{Z} \geqslant -d_-^*(\tau')) \\
&= qS\mathrm{e}^{(r-q)\tau'}\mathcal{N}(d_+^*(\tau')) - rK\mathcal{N}(d_-^*(\tau'))
\end{aligned}$$

where we define $d_\pm^*(\tau') := \dfrac{\ln\frac{S}{S^*(\tau-\tau')} + \left(r - q \pm \frac{1}{2}\sigma^2\right)\tau'}{\sigma\sqrt{\tau'}}$. By a similar calculation, the expectation in (5.45) is given by

$$\widetilde{\mathrm{E}}_{0,S}\left[(rK - qS(\tau'))\,\mathbb{I}_{\{S(\tau')\leqslant S^*(\tau-\tau')\}}\right] = rK\mathcal{N}(-d_-^*(\tau')) - qS\mathrm{e}^{(r-q)\tau'}\mathcal{N}(-d_+^*(\tau'))\,.$$

By inserting these expressions into (5.44) and (5.45), and using the pricing functions of the standard European call and put options, we have the explicit integral representations for the price of the American call and put:

$$\begin{aligned}
C(\tau, S; K) &= S\mathrm{e}^{-q\tau}\mathcal{N}(d_+) - K\mathrm{e}^{-r\tau}\mathcal{N}(d_-) \\
&\quad + \int_0^\tau \left[qS\mathrm{e}^{-q\tau'}\mathcal{N}(d_+^*(\tau')) - rK\mathrm{e}^{-r\tau'}\mathcal{N}(d_-^*(\tau'))\right] \mathrm{d}\tau',
\end{aligned} \tag{5.46}$$

$$\begin{aligned}
P(\tau, S; K) &= K\mathrm{e}^{-r\tau}\mathcal{N}(-d_-) - S\mathrm{e}^{-q\tau}\mathcal{N}(-d_+) \\
&\quad + \int_0^\tau \left[rK\mathrm{e}^{-r\tau'}\mathcal{N}(-d_-^*(\tau')) - qS\mathrm{e}^{-q\tau'}\mathcal{N}(-d_+^*(\tau'))\right] \mathrm{d}\tau',
\end{aligned} \tag{5.47}$$

where $d_\pm = \frac{\ln\frac{S}{K} + \left(r - q \pm \frac{1}{2}\sigma^2\right)\tau}{\sigma\sqrt{\tau}}$.

These integral representations are valid for $S \in (0, \infty)$, $\tau \geqslant 0$. By setting the spot equal to the boundary value, $S = S^*(\tau)$, and applying the respective boundary conditions, $C(\tau, S^*(\tau); K) = S^*(\tau) - K$ for the call and $P(\tau, S^*(\tau); K) = K - S^*(\tau)$ for the put, (5.46) and (5.47) give rise to integral equations for the early-exercise boundary. For the call

$$S^*(\tau) - K = S^*(\tau)\mathrm{e}^{-q\tau}\mathcal{N}(\hat{d}_+) - K\mathrm{e}^{-r\tau}\mathcal{N}(\hat{d}_-) \\ + \int_0^\tau \left[qS^*(\tau)\mathrm{e}^{-q\tau'}\mathcal{N}(\hat{d}^*_+(\tau')) - rK\mathrm{e}^{-r\tau'}\mathcal{N}(\hat{d}^*_-(\tau'))\right] \mathrm{d}\tau', \tag{5.48}$$

and separately for the put

$$K - S^*(\tau) = K\mathrm{e}^{-r\tau}\mathcal{N}(-\hat{d}_-) - S^*(\tau)\mathrm{e}^{-q\tau}\mathcal{N}(-\hat{d}_+) \\ + \int_0^\tau \left[rK\mathrm{e}^{-r\tau'}\mathcal{N}(-\hat{d}^*_-(\tau')) - qS^*(\tau)\mathrm{e}^{-q\tau'}\mathcal{N}(-\hat{d}^*_+(\tau'))\right] \mathrm{d}\tau', \tag{5.49}$$

where

$$\hat{d}_\pm = \frac{\ln \frac{S^*(\tau)}{K} + \left(r - q \pm \frac{1}{2}\sigma^2\right)\tau}{\sigma\sqrt{\tau}} \quad \text{and} \quad \hat{d}^*_\pm(\tau') = \frac{\ln \frac{S^*(\tau)}{S^*(\tau - \tau')} + \left(r - q \pm \frac{1}{2}\sigma^2\right)\tau'}{\sigma\sqrt{\tau'}}.$$

Note that (5.48) and (5.49) involve a variable upper integration limit and the integrands are nonlinear functions of $S^*(\tau)$, $S^*(\tau')$, τ, and τ'. From the theory of integral equations, (5.48) and (5.49) are known as *nonlinear Volterra integral equations.* Note that the solution $S^*(\tau)$, at time to maturity τ, is dependent on the solution $S^*(\tau - \tau')$ from zero time to maturity up to time τ. Although (5.48) and (5.49) are not analytically tractable, simple and efficient algorithms can be employed to solve for $S^*(\tau)$ numerically. A typical procedure involves the division of the solution domain into a regular mesh: $\tau_0 = 0$, $\tau_i = ih$, $i = 1, \ldots, n$, with n steps spaced as $h = \tau/n$. By approximating the time integral via a quadrature rule (e.g., the trapezoidal rule) one can obtain a system of algebraic equations in the values $S^*(\tau_i)$ which can be iteratively solved starting from the known value $S^*(\tau_0) = S^*(\tau = 0+)$ at zero time to maturity. Once the early-exercise boundary is determined, the integral in (5.46) or (5.47) for the respective call or put can be computed. In particular, a quadrature rule that makes use of the computed points $S^*(\tau_i)$ can be implemented. Accurate approximations to the early-exercise boundary are obtained by choosing the number n of points to be sufficiently large.

5.5 Exercises

Exercise 5.1. Prove Proposition 5.4 for an arbitrary American option with a differentiable payoff function Λ. In particular show the following.

(a) At any point (t, S^*_t) of the early-exercise boundary, the American option pricing function V satisfies the smooth pasting condition:

$$\left.\frac{\partial V(t, S)}{\partial S}\right|_{S = S^*_t} = \Lambda'(S^*_t).$$

(b) The option value V satisfies the zero time-decay condition on the early-exercise domain,

$$\frac{\partial V(t, S)}{\partial t} = 0, \quad \text{for all} \quad S \in \mathcal{D}_t.$$

Exercise 5.2. Let $P(S; K, r, q)$ and $C(S; K, r, q)$, respectively, denote the price functions of the perpetual American put and call options struck at K. The underlying asset price process follows geometric Brownian motion (5.1). Using the closed-form pricing formula (5.25) and (5.28), show that the option prices satisfy the put-call symmetry relation

$$P(K; S, q, r) = C(S; K, r, q).$$

Exercise 5.3. Using no-arbitrage argument, show for standard European and American put options that

$$P^E(t, S; T, K) \leqslant P(t, S; T, K) \leqslant P^E(t, S; T, K) + K(1 - \mathrm{e}^{-r(T-t)})$$

holds for all $t \leqslant T$ and all $S > 0$.

Exercise 5.4. Using no-arbitrage argument, prove for standard American call and put options with common strike K and expiry T that

$$\mathrm{e}^{-q(T-t)} S - K \leqslant C(t, S) - P(t, S)$$

holds for all $t \in [0, T]$ and all $S > 0$.

Exercise 5.5. Let $P(S, \tau; K, r, q)$ and $C(S, \tau; K, r, q)$, respectively, denote the price function of the American put and call options with strike price K and time to expiration τ. The underlying asset price process follows geometric Brownian motion (5.1). Show that $P(K, \tau; S, q, r)$ satisfies the Black–Scholes PDE along with the auxiliary conditions

$$P(K, 0; S, q, r) = (S - K)^+, \quad P(K, \tau; S, q, r) \geqslant (S - K)^+ \text{ for } \tau > 0.$$

Since the auxiliary conditions are identical to those of the American call option price, the put and call price functions satisfy the put-call symmetry relation

$$P(K, \tau; S, q, r) = C(S, \tau; K, r, q).$$

Exercise 5.6. Assume the standard Black–Scholes model and stock price process as in Exercise 5.8. Find the limiting values of the exercise boundary S^* and the pricing function for

(a) a perpetual American call, as the dividend yield q converges to 0;

(b) a perpetual American put, as the risk-free rate r converges to 0.

Exercise 5.7. Consider a call-like quadratic payoff $\Lambda(S) = a(S - K)^2 \mathbb{I}_{\{S \geqslant K\}}$, with fixed strike K and where a is some positive constant factor. Derive an analytical expression for the early-exercise boundary value S^* as well the perpetual American pricing function $V(S)$ for this payoff. Assume the geometric Brownian motion asset price model in (5.1). Express your answer in terms of the spot S and the parameters a, K, r, q, σ.

Exercise 5.8. Assume the standard Black–Scholes model in an economy with constant continuously compounded interest rate $r \geqslant 0$ and with stock price process $\{S(t)\}_{t \geqslant 0}$ as a GBM with constant volatility $\sigma > 0$ and continuous dividend yield $q \geqslant 0$. Derive the pricing function $V(S)$ for the perpetual American capped put option with the payoff function $\Lambda(S) = \min\{K - S, C\}$ with $0 < C \leqslant K$.

Exercise 5.9. Assume the standard Black–Scholes model and stock price process as in Exercise 5.8. Derive the pricing function $V(S)$ for a perpetual American power put option with the following payoff function:

(a) $\Lambda(S) = (K^n - S^n)^+$ with $n \geqslant 1$;

(b) $\Lambda(S) = ((K - S)^+)^n$ with $n \geqslant 1$;

(c) $\Lambda(S) = \min\{((K - S)^+)^n, C\}$ with $n \geqslant 1$ and $0 < C \leqslant K^n$.

where $K > 0$.

Exercise 5.10. Assume the standard Black–Scholes model and stock price process as in Exercise 5.8. Consider a perpetual American put option with payoff $\Lambda(S) = ((K - S)^+)^2 = (K - S)^2 \mathbb{I}_{\{S \leqslant K\}}$.

(a) Find the early-exercise boundary value S^* and the pricing function $V(S)$.

(b) Sketch the plot of $V(S)$. Also, include a graph of the payoff function and mark K and S^* on the S-axis.

(c) Find the limiting values of $V(S)$ and S^*, as $r \searrow 0$. Argue that the option is not optimal to exercise if $r = 0$.

Exercise 5.11. Consider a butterfly spread option with payoff centred at $K > 0$ and some positive width w where $K - w > 0$.

(a) Let spot S be the vertical axis and τ (time to maturity) the horizontal axis. Provide a sketch that depicts the early-exercise boundary curve(s), and clearly include labels for shaded regions corresponding to the continuation and early-exercise domains. Determine the asymptotes and asymptotic values for all the boundary values corresponding to $S^*(\tau = 0+)$ and $S^*(\tau = \infty)$.

(b) Provide a sketch of the American option value for this butterfly spread option as a function of spot S for a typical time to maturity $\tau > 0$. Include the payoff in your graph.

(c) Obtain an analytical pricing formula for this perpetual American butterfly option for all spot $S > 0$.

Exercise 5.12. Consider a strangle option with the payoff $\Lambda(S) = (K_1 - S)^+ + (S - K_2)^+$ where K_1 and K_2 are fixed strikes so that $0 < K_1 < K_2$. Derive an analytical expression for the early-exercise boundary value S^* as well the perpetual American pricing function $V(S)$ for this payoff. Assume the usual geometric Brownian motion asset price model in (5.1).

Exercise 5.13. Show that the pricing formulae for the American call and put in (5.46) and (5.47) satisfy the required boundary conditions at $S = 0$ and $S = \infty$.

Exercise 5.14. Using (5.46) and (5.47), derive integral representations for the delta, gamma, and vega sensitivities of the American call and put.

Exercise 5.15. Consider a Bermudan put option with strike K at maturity T with only a single intermediate early-exercise date $T_1 \in [0, T]$. Assume the underlying stock price process is a geometric Brownian motion within the risk-neutral measure, and let $P(S, T - t)$ denote the option value at calendar time t with spot S. Find an analytically closed-form expression for the present time $t = 0$ price $P(S_0, T)$. Hint: this problem is very closely related to the valuation of a compound option. In particular, proceed as follows. From backward recurrence show that

$$P(S_0, T) = \mathrm{e}^{-rT_1} \widetilde{\mathrm{E}}_0\big[P(S(T_1), T - T_1)\big], \tag{5.50}$$

with

$$P(S(T_1), T - T_1) = \begin{cases} P^E(S(T_1), T - T_1), & S(T_1) > S^*_{T_1}, \\ K - S(T_1), & S(T_1) \leqslant S^*_{T_1}, \end{cases} \tag{5.51}$$

where P^E is the price of the European put option and the critical value $S^*_{T_1}$ for the early-exercise boundary at calendar time T_1 solves

$$P_E(S^*_{T_1}, T - T_1) = K - S^*_{T_1}.$$

Compute the above expectation as a sum of two integrals: one over the domain $S_{T_1} > S^*_{T_1}$ and the other over $0 < S(T_1) \leqslant S^*_{T_1}$, to finally arrive at the expression for $P(S_0, T)$ in terms of univariate and bivariate cumulative normal functions. Show whether $S^*_{T_1}$ is a strictly increasing or decreasing function of the volatility σ and explain your answer. What is this functional dependency for the case of a Bermudan call? Explain.

6

Interest-Rate Modelling and Derivative Pricing

6.1 Basic Fixed Income Instruments

6.1.1 Bonds

The term *fixed income* refers to any type of investment that provides payments of a fixed amount on a fixed schedule. A most common type of fixed income investment is a bond. In Chapter 1 of Volume I, we introduced a zero-coupon bond (ZCB) that only returns the investor a redemption amount on the maturity date. Another example of a fixed income security is a coupon bond that provides regular payments (coupons) on a fixed schedule and a redemption value on the maturity date. Recall that $Z(t,T)$ denotes the time-t purchase price of a unit zero-coupon bond maturing at time T with $0 \leqslant t \leqslant T$, whose face value is equal to \$1. The following notations were also introduced in Chapter 1 of Volume I.

y is the continuously compounded yield rate (also called the spot rate); it is generally a function of calendar time t and maturity T for a given investment time period $[t,T]$, i.e., $y = y(t,T)$. When regarded as a function of the time to maturity, $\tau = T - t$, i.e., $y(\tau)$ is the yield rate earned by money invested for a period of τ years.

r is the short rate; it is a function of calendar time t, i.e., $r = r(t)$. The short rate is the rate on instantaneous borrowing or lending and is defined as $r(t) = y(t,t)$.

The continuously compounded yield rate y and zero-coupon bond price Z are simply related by

$$Z(t,T) = e^{-y(t,T)(T-t)}\,. \tag{6.1}$$

Any cash flow stream of multiple coupon payments can be replicated by means of a portfolio of zero-coupon bonds. In particular, the time-t value $V(t;\mathbf{c},\mathbf{T})$ of a cash flow stream $(\mathbf{c},\mathbf{T})$ with $\mathbf{c} = [c_1, c_2, \ldots, c_n]$ and $\mathbf{T} = [t_1, t_2, \ldots, t_n]$, where c_i is the payment at time t_i and $t \leqslant t_1 < t_2 < \cdots < t_n$ holds, is equal to the sum of discounted cash flows,

$$V(t;\mathbf{c},\mathbf{T}) = \sum_{i=1}^{n} c_i\, Z(t,t_i) = \sum_{i=1}^{n} c_i\, \mathrm{e}^{-y(t,t_i)(t_i-t)}. \tag{6.2}$$

The above formula allows for the pricing of coupon-paying bonds. On the other hand, if we are given the price of a coupon-paying bond for each maturity $t_1, t_2, \ldots, t_n$, then using (6.2) we can solve recursively for the prices $Z(t,t_1), Z(t,t_2), \ldots, Z(t,t_n)$ and therefore obtain the yield rates $y(t,t_1), y(t,t_2), \ldots, y(t,t_n)$. Indeed, let the coupon bond value $V_i(t) = c_1 Z(t,t_1) + c_2 Z(t,t_2) + \cdots + c_i Z(t,t_i)$ be known for all $i = 1,2,\ldots,n$. Then the ZCB values are

$$Z(t,t_1) = \frac{V_1(t)}{c_1} \quad \text{and} \quad Z(t,t_i) = \frac{V_i(t) - V_{i-1}(t)}{c_i} \quad \text{for } i = 2,3,\ldots,n.$$

This method of deducing zero-coupon bond values from coupon-paying bond prices is called *bootstrapping*.

DOI: 10.1201/9780429468889-6

6.1.2 Forward Rates

Suppose we wish to borrow an amount of A dollars for a period between times T and T' and we want to lock in the rate on the loan at time t with $t \leqslant T < T'$. To do this, at time t, we purchase A units of the zero-coupon (unit) bond maturing at time T and finance this purchase by selling short $A\frac{Z(t,T)}{Z(t,T')}$ units of a zero-coupon bond maturing at time T'. The cost of setting up this portfolio is zero. At time T, we receive \$$A$ from the long position in the T-maturity bonds. At time T', we are required to have $A\frac{Z(t,T)}{Z(t,T')}$ dollars to cover the short position in the T'-maturity bonds. Hence, to avoid arbitrage, A dollars invested during the time interval $[T, T']$ must yield an effective (continuously compounded) rate, denoted by $f(t; T, T')$, such that

$$A\mathrm{e}^{(T'-T)f(t;T,T')} = A\frac{Z(t,T)}{Z(t,T')} \implies f(t;T,T') = \frac{1}{T'-T}\ln\frac{Z(t,T)}{Z(t,T')}. \tag{6.3}$$

The rate $f(t; T, T')$ is called a *forward rate*. It is determined at time t for investing during the period $[T, T']$. Using (6.1), we express the forward rate in terms of yield rates as follows:

$$f(t;T,T') = \frac{1}{T'-T}\ln\left(\frac{\mathrm{e}^{-y(t,T)(T-t)}}{\mathrm{e}^{-y(t,T')(T'-t)}}\right) = y(t,T')\left(\frac{T'-t}{T'-T}\right) - y(t,T)\left(\frac{T-t}{T'-T}\right). \tag{6.4}$$

Note that, when $t = T$ the forward rate is determined in the beginning of the investment interval and we have

$$f(T;T,T') = y(T,T')\overbrace{\left(\frac{T'-T}{T'-T}\right)}^{1} - y(T,T)\overbrace{\left(\frac{T-T}{T'-T}\right)}^{0} = y(T,T').$$

That is, yield rates are forward rates for immediate delivery.

The *instantaneous forward rate* of maturity T is defined as

$$f(t,T) := \lim_{T'\to T} f(t;T,T'). \tag{6.5}$$

By applying the definition of the derivative of $\ln Z(t, T)$ w.r.t. T, for fixed t, (6.3) gives the instantaneous forward rate at time T as

$$f(t,T) := \lim_{T'\to T} f(t;T,T') = -\lim_{T'\to T}\frac{\ln Z(t,T') - \ln Z(t,T)}{T'-T} = -\frac{\partial}{\partial T}\ln Z(t,T). \tag{6.6}$$

Hence, integrating this equation, where $f(t,u) = -\frac{\partial}{\partial u}\ln Z(t,u)$, for $u \in [t,T]$, while using $Z(t,t) = 1$, gives the zero-coupon bond price in terms of instantaneous forward rates,

$$Z(t,T) = \exp\left(-\int_t^T f(t,u)\,\mathrm{d}u\right). \tag{6.7}$$

Taking the logarithm of both sides of (6.7) gives

$$\int_t^T f(t,u)\,\mathrm{d}u = -\ln Z(t,T). \tag{6.8}$$

Using (6.8), we can find the integral of the instantaneous forward rate of maturity changing from T to T' with $t \leqslant T \leqslant T'$:

$$\begin{aligned}\int_T^{T'} f(t,u)\,\mathrm{d}u &= \int_t^{T'} f(t,u)\,\mathrm{d}u - \int_t^T f(t,u)\,\mathrm{d}u = \ln Z(t,T) - \ln Z(t,T') \\ &= f(t;T,T')(T'-T).\end{aligned} \tag{6.9}$$

The forward rate is also related to the forward price for a unit zero-coupon bond maturing at time T' with settlement at time T. Recall that a forward contract is a contract under which one party is obliged to sell a specified asset for an agreed price to the other party at a designated future date. A *fixed income forward contract* is an agreement between two parties to pay a specified delivery price for a fixed income security (a zero-coupon bond, for instance) at a given delivery date. The *forward price* of the underlying security is the value of the delivery price that makes the forward contract have a no-arbitrage price of zero at initiation.

6.1.3 Arbitrage-Free Pricing

The risk-free interest rate has been assumed to be a constant or a deterministic function for most of the option pricing models applied in previous chapters. In this chapter, we will deal with stochastic models of interest rates. Assume a filtered probability space $(\Omega, \mathcal{F}, \mathbb{P}, \mathbb{F})$, where $\mathbb{F} = \{\mathcal{F}_t\}_{0 \leqslant t \leqslant T^*}$ is a filtration generated by the stochastic risk-free (short) rate process $\{r(t)\}_{0 \leqslant t \leqslant T^*}$ for some $T^* > 0$. Our general assumptions are as follows:

1. the short rate process is Markovian;
2. the zero-coupon bond price process $\{Z(t,T)\}_{0 \leqslant t \leqslant T}$ is adapted to $\mathbb{F}$ for every maturity T with $T \leqslant T^*$.

As in previous chapters, the bank account $\{B(t)\}_{t \geqslant 0}$ evolves according to

$$dB(t) = r(t)B(t)\,\mathrm{d}t, \quad B(0) = 1, \tag{6.10}$$

and is given by

$$B(t) = \exp\left(\int_0^t r(s)\,\mathrm{d}s\right).$$

Since $B(t)$ is a function of short rates $\{r(s)\}_{0 \leqslant s \leqslant t}$, the bank account is adapted to the filtration $\mathbb{F}$. Additionally, let us define the (stochastic) discount factor $D(t,T)$ from time t to time T with $0 \leqslant t \leqslant T \leqslant T^*$ given by

$$D(t,T) = \frac{B(t)}{B(T)} = \exp\left(-\int_t^T r(s)\,\mathrm{d}s\right).$$

The discount factor has the property $D(t,T)D(T,T') = D(t,T')$ for all $0 \leqslant t \leqslant T \leqslant T'$. Also, we recall from previous chapters that $D(t) := D(0,t) = B^{-1}(t)$ for $t \geqslant 0$.

One of the main problems discussed in this chapter is the no-arbitrage pricing of bonds, options on interest rate, and other fixed income derivatives. Here, bonds can be regarded as derivative assets since any bond is derived from the knowledge of the short (risk-free) rate, $r(t)$, which takes on the role of the underlying. The Fundamental Theorem of Asset Pricing (FTAP) is a cornerstone of the no-arbitrage pricing. Let us state the FTAP for the model of stochastic interest rates.

Theorem 6.1. *The market is arbitrage free if there exists a probability measure $\widetilde{\mathbb{P}}$ equivalent to the real-world measure $\mathbb{P}$, under which the discounted zero-coupon bond price*

$$\overline{Z}(t,T) := Z(t,T)D(t) = Z(t,T)/B(t),\ 0 \leqslant t \leqslant T,$$

is a martingale for each $T > 0$. Assuming the absence of arbitrage, the market is complete iff the equivalent martingale measure $\widetilde{\mathbb{P}}$ is unique.

The definition of a zero-coupon bond and definition of a martingale imply that[1]

$$\overline{Z}(t,T) = \widetilde{\mathrm{E}}_t\left[\overline{Z}(T,T)\right] = \widetilde{\mathrm{E}}_t\left[Z(T,T)D(T)\right] = \widetilde{\mathrm{E}}_t\left[D(T)\right].$$

Thus, the time-t price of the zero-coupon bond is

$$Z(t,T) = \widetilde{\mathrm{E}}_t[B(t)D(T)] = \widetilde{\mathrm{E}}_t[D(t,T)] = \widetilde{\mathrm{E}}_t\left[\exp\left(-\int_t^T r(s)\,\mathrm{d}s\right)\right]. \tag{6.11}$$

By using the results of the FTAP, we can also price derivatives. The time-t no-arbitrage value $V(t)$ of a payoff $V(T)$ payable at time T with $0 \leqslant t \leqslant T$ (note that $V(T)$ is $\mathcal{F}_T$-measurable) is

$$V(t) = \widetilde{\mathrm{E}}_t\left[D(t,T)V(T)\right] = \widetilde{\mathrm{E}}_t\left[\exp\left(-\int_t^T r(s)\,\mathrm{d}s\right)V(T)\right]. \tag{6.12}$$

As an example, consider a forward contract under which $\$K$ will be paid at time T in return for a repayment of \$1 at time T' with $T < T'$. Equivalently, this contract is arranged as if a zero-coupon bond maturing at time T' is delivered at time T in return for K dollars paid at the same time T. The time-T payoff is $Z(T,T') - K$. According to (6.12), the price of the forward contract at time t (with $t \leqslant T$) is

$$\begin{aligned} V(t) &= \widetilde{\mathrm{E}}_t\left[D(t,T)(Z(T,T') - K)\right] = \widetilde{\mathrm{E}}_t\left[D(t,T)\left(\widetilde{\mathrm{E}}_T[D(T,T')] - K\right)\right] \\ &= \widetilde{\mathrm{E}}_t\left[D(t,T')\right] - K\,\widetilde{\mathrm{E}}_t\left[D(t,T)\right] = Z(t,T') - K\,Z(t,T). \end{aligned}$$

Here, we combined the tower property with $Z(T,T') = \widetilde{\mathrm{E}}_T[D(T,T')]$ and the identity $D(t,T)D(T,T') = D(t,T')$. Choosing $K = Z(t,T')/Z(t,T)$ ensures that $V(t) = 0$. This value is called the *T-forward price* of the zero-coupon bond with maturity $T' > T$. The forward price is expressed in terms of the forward rate $f(t;T,T')$ as given in (6.3).

Consider now an asset X with price process $\{X(t)\}_{0 \leqslant t \leqslant T}$ and a forward contract that delivers one unit of X at time T in return for K. According to (6.12), the time-t price of this contract is

$$V(t) = \widetilde{\mathrm{E}}_t\left[D(t,T)(X(T) - K)\right] = \frac{1}{D(t)}\widetilde{\mathrm{E}}_t\left[D(T)X(T)\right] - K\widetilde{\mathrm{E}}_t\left[D(t,T)\right].$$

Since the discounted process $D(t)X(t)$ is a $\widetilde{\mathbb{P}}$-martingale and $\widetilde{\mathrm{E}}_t\left[D(t,T)\right] = Z(t,T)$, the above equation reduces to

$$V(t) = X(t) - KZ(t,T).$$

The present (time-t) value is zero iff $K = \frac{X(t)}{Z(t,T)}$. We call $\frac{X(t)}{Z(t,T)}$ the *T-forward price* at time $t \in [0,T]$ of the asset X. Note that $X(t) = Z(t,T')$ is a special case where the chosen asset is the time-T' maturity zero-coupon bond.

6.1.4 Fixed Income Derivatives

6.1.4.1 Options on Bonds

A European-style option on a zero-coupon bond is defined in the same way as the option on any other underlying security. A call option with maturity T and strike K on a bond

[1] We note the shorthand notation used throughout this chapter where $\widetilde{\mathrm{E}}_t[\,\cdot\,] \equiv \widetilde{\mathrm{E}}[\,\cdot \mid \mathcal{F}_t]$ is the $\widetilde{\mathbb{P}}$-measure expectation conditional on $\mathcal{F}_t$.

maturing at time $T' > T$ gives its holder the right but not the obligation to purchase the bond at time T for a fixed price K. The time-T payoff to the call option holder is $(Z(T,T') - K)^+$. Similarly, we define a put option with the time-T payoff $(K - Z(T,T'))^+$. If the joint (conditional) distribution of the discount factor $D(t,T)$ and bond price $Z(T,T')$ is known, then the no-arbitrage price at time t of a European option with payoff $V(T) = \Lambda(Z(T,T'))$ can be calculated using the risk-neutral pricing formula (6.12):

$$V(t) = \widetilde{\mathrm{E}}_t \left[D(t,T)\,\Lambda(Z(T,T')) \right], \quad 0 \leqslant t \leqslant T.$$

Note that many other common options on interest rates can be expressed as options on bonds. For example, a call option on the LIBOR rate considered in the last section of this chapter is equivalent to a put option on a zero-coupon bond.

We can also consider a European call option written on a coupon-bearing bond. The payoff of the option struck at exercise K, of maturity date T, can be written as $V(T) = (P(T) - K)^+$, where $P(T)$ is the value of the bond at maturity T:

$$P(T) = \sum_{j=1}^{n} c_j Z(T, T_j),$$

with cash flows $c_1, c_2, \ldots, c_n$ at times $T_1, T_2, \ldots, T_n$, respectively, with $T < T_1 < T_2 < \ldots < T_n$. Note that the sum involves all cash flows at future times past the maturity of the option, discounted back to time T.

6.1.4.2 Cap and Caplets

A *caplet* is a contract that gives its holder the right to pay the smaller of two simple interest rates: the floating rate f and fixed rate κ. The floating rate is typically the three- or six-month LIBOR. For the holder of a caplet over the interval $[T, T+\tau]$ with *tenor* τ, the rate of payment is capped at κ from T to $T+\tau$. So the simple interest paid on each dollar of the principal at time $T+\tau$ is the smaller of τf and $\tau\kappa$. Since without the caplet the interest payment would be τf, the caplet's worth to the holder is $\tau f - \min\{\tau f, \tau\kappa\} = (f-\kappa)^+\tau$. That is, the caplet pays $(f-\kappa)^+\tau$ to its holder at time $T+\tau$. In fact, we deal with a call payoff on the floating rate f with strike κ and maturity $T+\tau$. Under the equivalent martingale measure (EMM) $\widetilde{\mathbb{P}}$, the time-t value of the caplet is

$$\mathrm{Caplet}_{T+\tau}(t) = \tau \widetilde{\mathrm{E}}_t \left[D(t, T+\tau)(f-\kappa)^+ \right], \quad 0 \leqslant t \leqslant T. \tag{6.13}$$

As will be demonstrated below, the LIBOR rate is expressed in terms of zero-coupon bonds, and hence the application of the risk-neutral pricing formula in (6.13) is legitimate.

A *cap* is defined as a collection of caplets. Suppose that the payments are done at the times $T_1, T_2, \ldots, T_n$ with $T_{i+1} = T_i + \tau$. Let f_i denote the floating rate over $[T_{i-1}, T_i]$ for $i = 1, 2, \ldots, n$. A cap is defined as a stream of cash flows that pays to its holder $(f_i - \kappa)^+\tau$ at time T_i for all $i = 1, 2, \ldots, n$. The risk-neutral value of the cap with the tenor structure $\mathbf{T} := [T_1, T_2, \ldots, T_n]$ at time $t \leqslant T_0$ is

$$\mathrm{Cap}(t; \mathbf{T}) = \sum_{i=1}^{n} \mathrm{Caplet}_{T_i}(t) = \tau \widetilde{\mathrm{E}}_t \left[\sum_{i=1}^{n} D(t, T_i)(f_i - \kappa)^+ \right]. \tag{6.14}$$

6.1.4.3 Swap and Swaptions

Another basic fixed income instrument is an *interest rate swap*. It is an agreement between two parties in which one party makes fixed interest payments on some notional amount at regularly spaced dates in return for floating interest payments on the same principal at the same dates. The *payer swap* is a contract in which the floating rate f_i is swapped in arrears against a fixed rate κ at n intervals $[T_{i-1}, T_i]$ of length $\tau = T_i - T_{i-1}$ for all $i = 1, 2, \ldots, n$. The holder of the payer swap receives the cash flows $(f_1 - \kappa)\tau, \ldots, (f_n - \kappa)\tau$ at dates $T_1, \ldots, T_n$, respectively. The other party in the swap contract enters a *receiver swap*, in which a fixed rate is swapped against the floating rate. It suffices to only consider a payer swap which we simply call a swap. The time-t value of the swap is

$$\mathrm{Swap}(t; \mathbf{T}) = \tau \sum_{i=1}^{n} \widetilde{\mathrm{E}}_t \left[D(t, T_i)(f_i - \kappa) \right]. \tag{6.15}$$

The fixed rate κ can be defined so that the swap agreement costs zero at initiation. The *forward swap rate* is a fixed rate of interest that makes the swap contract worthless at current time $t \leqslant T_0$. In other words, the forward swap rate at time t is the rate κ that solves $\mathrm{Swap}(t; \mathbf{T}) = 0$.

Another contract called a *swaption* gives you the right but not the obligation to enter into a swap agreement with another party at time T_0. So a swaption delivers at time T_0 a swap when the swap value $\mathrm{Swap}(T_0; \mathbf{T})$ is positive. Thus, the time-t value of a swaption at time $t \leqslant T_0$ is

$$\mathrm{Swaption}(t; \mathbf{T}) = \widetilde{\mathrm{E}}_t \left[D(t, T_0) \big(\mathrm{Swap}(T_0; \mathbf{T}) \big)^+ \right]. \tag{6.16}$$

6.2 Single-Factor Models

A *single-factor model* is one that has a single, one-dimensional source of randomness affecting bond prices. Let a standard Brownian motion $\{W(t)\}_{t \geqslant 0}$ be such a source of randomness. Let the Brownian filtration $\mathbb{F}^W := \{\mathcal{F}_t^W\}_{0 \leqslant t \leqslant T^*}$ coincide with the filtration $\mathbb{F}$ generated by the short rate process $\{r(t)\}_{0 \leqslant t \leqslant T^*}$. This is the case when $r(t)$ is an Itô process governed by a nontrivial stochastic differential equation (SDE) of the form $\mathrm{d}r(t) = a(t)\,\mathrm{d}t + b(t)\,\mathrm{d}W(t)$ with $\mathbb{F}^W$-adapted processes a and b.

Let us derive an SDE for the zero-coupon bond $Z(t, T)$ under the EMM $\widetilde{\mathbb{P}}$. The discounted process $\overline{Z}(t, T) := D(t)Z(t, T)$ for $0 \leqslant t \leqslant T$, and a given fixed $T \in [0, T^*]$, is a $\widetilde{\mathbb{P}}$-martingale. By the Brownian Martingale Representation Theorem 2.18, there exists an $\mathbb{F}$-adapted process $\theta(t)$ such that $\overline{Z}(t, T) = \overline{Z}(0, T) + \int_0^t \theta(s)\,\mathrm{d}\widetilde{W}(s)$ (a.s.) for $0 \leqslant t \leqslant T$. Since $\overline{Z}(t, T)$ is strictly positive, we can define $\sigma_Z(t, T) = \theta(t)/\overline{Z}(t, T)$. Thus, we have

$$\mathrm{d}\overline{Z}(t, T) = \theta(t)\,\mathrm{d}\widetilde{W}(t) = \overline{Z}(t, T)\sigma_Z(t, T)\,\mathrm{d}\widetilde{W}(t).$$

By using the Itô product rule, we obtain the following expression for the stochastic differential of the discounted bond price:

$$\mathrm{d}\overline{Z}(t, T) = -\overline{Z}(t, T) r(t)\,\mathrm{d}t + D(t)\,\mathrm{d}Z(t, T).$$

As a result, we have the following SDE for $Z(t, T)$ under the EMM $\widetilde{\mathbb{P}}$:

$$\frac{\mathrm{d}Z(t, T)}{Z(t, T)} = r(t)\,\mathrm{d}t + \sigma_Z(t, T)\,\mathrm{d}\widetilde{W}(t). \tag{6.17}$$

This SDE has the equivalent integral form

$$Z(t, T) = Z(0, T) \mathrm{e}^{\int_0^t r(s)\,\mathrm{d}s - \frac{1}{2}\int_0^t \sigma_Z^2(s, T)\,\mathrm{d}s + \int_0^t \sigma_Z(s, T)\,\mathrm{d}\widetilde{W}(s)}. \tag{6.18}$$

By comparing SDEs (6.10) and (6.17), we notice that the bond Z is riskier than the bank account B since the SDE in (6.17) contains an extra Brownian term.

To find the SDE for $Z(t,T)$ under the original (physical) $\mathbb{P}$-measure, we use the equivalence of the measures $\mathbb{P}$ and $\widetilde{\mathbb{P}}$. By Girsanov's Theorem 2.17, there exists a change of measure generated by an $\mathbb{F}$-adapted process $\gamma(t)$ such that

$$\widetilde{W}(t) = W(t) + \int_0^t \gamma(s)\,\mathrm{d}s \tag{6.19}$$

is a standard $\widetilde{\mathbb{P}}$-BM, given that $W(t)$ is a standard $\mathbb{P}$-BM, for all $t \in [0, T^*]$. Substituting $\mathrm{d}W(t) + \gamma(t)\,\mathrm{d}t$ in place of $\mathrm{d}\widetilde{W}(t)$ in (6.17) gives the following SDE under $\mathbb{P}$:

$$\frac{\mathrm{d}Z(t,T)}{Z(t,T)} = (r(t) + \gamma(t)\sigma_Z(t,T))\,\mathrm{d}t + \sigma_Z(t,T)\,\mathrm{d}W(t)\,, \tag{6.20}$$

for all $t \in [0,T]$, $T \leqslant T^*$. The quantity $\gamma(t)$ is the excess rate of return over the risk-free rate of return $r(t)$ per one unit of volatility; it is known as the *market price of risk* or *risk premium*. This term represents the extra reward we receive for investing in the risky bond rather than in the risk-free bank account. Since $\gamma(t)$ is adapted to the σ-algebra $\mathcal{F}_t$ generated by the short rate process $\{r(t)\}_{t\geqslant 0}$ and since $\{r(t)\}_{t\geqslant 0}$ is a Markov process, the risk premium is a function of t and $r(t)$, i.e., $\gamma = \gamma(t, r(t))$. The market price of risk can be estimated from observable values of rate and bond prices. However, it is a common practice to assume that γ is constant.

6.2.1 Diffusion Models for the Short Rate Process

Suppose the short rate process $\{r(t)\}_{t\geqslant 0}$ is a diffusion described by an SDE

$$\mathrm{d}r(t) = a(t, r(t))\,\mathrm{d}t + b(t, r(t))\,\mathrm{d}W(t)\,. \tag{6.21}$$

The coefficients a and b are smooth functions that meet standard conditions required to ensure the existence of solutions to (6.21). Here is a list of popular models of interest rates falling in the class of diffusions as in (6.21):

$$\mathrm{d}r(t) = \alpha\,\mathrm{d}t + \sigma\,\mathrm{d}W(t) \qquad \text{(the *Merton model*)} \tag{6.22}$$

$$\mathrm{d}r(t) = \beta r(t)\,\mathrm{d}t + \sigma r(t)\,\mathrm{d}W(t) \qquad \text{(the *Dothan model*)} \tag{6.23}$$

$$\mathrm{d}r(t) = (\alpha - \beta r(t))\,\mathrm{d}t + \sigma\,\mathrm{d}W(t) \qquad \text{(the *Vasiček model*)} \tag{6.24}$$

$$\mathrm{d}r(t) = (\alpha - \beta r(t))\,\mathrm{d}t + \sigma r(t)\,\mathrm{d}W(t) \qquad \text{(the *Brennan–Schwartz model*)} \tag{6.25}$$

$$\mathrm{d}r(t) = (\alpha - \beta r(t))\,\mathrm{d}t + \sigma\sqrt{r(t)}\,\mathrm{d}W(t) \qquad \text{(the *Cox–Ingersoll–Ross model*)} \tag{6.26}$$

$$\mathrm{d}r(t) = \alpha(t)\,\mathrm{d}t + \sigma\,\mathrm{d}W(t) \qquad \text{(the *Ho–Lee model*)} \tag{6.27}$$

$$\mathrm{d}r(t) = \alpha(t)r(t)\,\mathrm{d}t + \sigma(t)\,\mathrm{d}W(t) \qquad \text{(the *Black–Derman–Toy model*)} \tag{6.28}$$

$$\mathrm{d}r(t) = (\alpha(t) - \beta(t)r(t))\,\mathrm{d}t + \sigma(t)\,\mathrm{d}W(t) \qquad \text{(the *Hull–White model*)} \tag{6.29}$$

$$\mathrm{d}r(t) = r(t)(\alpha(t) - \beta(t)\ln r(t))\,\mathrm{d}t + \sigma(t)r(t)\,\mathrm{d}W(t) \qquad \text{(the *Black–Karasinski model*)} \tag{6.30}$$

Here, α, β, and σ are constants; $\alpha(t)$, $\beta(t)$, and $\sigma(t)$ are nonrandom continuous functions of time. Assuming that the coefficients a and b in (6.21) are independent of t, i.e., $a = a(r(t))$

and $b = b(r(t))$, we obtain a time-homogeneous model for short rates. The models (6.22)–(6.26) are time homogeneous, whereas the models (6.27)–(6.30) are time-inhomogeneous.

Each model has some motivation behind it. There are three main characteristics to be taken into account when a model for interest rates is designed.

- Positiveness of interest rates. For instance, the Merton model and the Vasiček model do not guarantee that $r(t)$ remains positive. However, these models can still be used for short intervals of time.

- The rate process $r(t)$ is mean reverting, meaning that $r(t)$ fluctuates near a fixed long-term mean level given by $\lim_{t\to\infty} \mathrm{E}[r(t)]$. So the process cannot wander off toward $+\infty$ (or to zero) since a negative drift (or a positive drift) will eventually pull the path back to the long-term level. For instance, the Vasiček model has the long-term mean level equal to α/β. The drift rate of the diffusion in (6.24) is positive for $r(t) < \alpha/\beta$ and is negative for $r(t) > \alpha/\beta$.

- The model has to be tractable. Ideally, formulae for bond prices and for prices of some derivatives can be derived in closed form.

Note that any model is only an approximation of reality, and a financial model is worthy of consideration if it gives a good approximation of what is observed in the market. The coefficient functions a and b involve parameters that need to be estimated. For instance, they can be chosen so that model values of bonds and other derivatives are as close to the respective market values as possible.

6.2.2 PDE for the Zero-Coupon Bond Value

A model for the term structure of interest rates and bond pricing can be developed from a model for the short rate process. Let us derive the governing differential equation for the zero-coupon bond price. Our first approach employs the (discounted) Feynman–Kac formula. The second method presented here is based on the no-arbitrage argument.

The modelling equation (6.21) for $r(t)$ is specified under the real-world measure. The price $Z(t,T)$ of a zero-coupon bond satisfies (6.11), where the expectation is taken under the EMM $\widetilde{\mathbb{P}}$. So we need to know the risk-neutral dynamics of the short rate process. Using (6.19), we change measure in (6.21) to obtain the SDE

$$\mathrm{d}r(t) = \big(a(t,r(t)) - \gamma(t,r(t))b(t,r(t))\big)\,\mathrm{d}t + b(t,r(t))\,\mathrm{d}\widetilde{W}(t)\,. \tag{6.31}$$

The short rate process $\{r(t)\}_{t\geqslant 0}$ is hence Markovian. The expectation conditional on $\mathcal{F}_t$ in the right-hand side of (6.11) is then simply a conditional expectation given the value of $r(t)$. That is, as a random variable the zero-coupon bond price is given as a function of the random variable $r(t)$, $Z(t,T) \equiv Z(t,T,r(t))$, where

$$Z(t,T,r(t)) = \widetilde{\mathrm{E}}[\mathrm{e}^{-\int_t^T r(s)\,\mathrm{d}s} \mid \mathcal{F}_t] = \widetilde{\mathrm{E}}[\mathrm{e}^{-\int_t^T r(s)\,\mathrm{d}s} \,\big|\, r(t)]\,.$$

Conditioning on the known (spot) value of the short rate, $r(t) = r$, gives the pricing function $Z(t,T,r)$ for a zero-coupon bond. Thus, the time-t value of a zero-coupon bond is an ordinary function (i.e., $f(t,r) = Z(t,T,r)$) of the current short rate r and time t (for fixed T) where

$$Z(t,T,r) = \widetilde{\mathrm{E}}_{t,r}[\mathrm{e}^{-\int_t^T r(s)\,\mathrm{d}s}] \equiv \widetilde{\mathrm{E}}[\mathrm{e}^{-\int_t^T r(s)\,\mathrm{d}s} \,\big|\, r(t) = r]\,. \tag{6.32}$$

By associating this expectation with that in (2.97), the (discounted) Feynman–Kac Theorem 2.13 shows that the pricing function $Z = Z(t,T,r)$ satisfies the following PDE[2] (see (2.98)):

$$\frac{\partial Z}{\partial t} + \frac{b^2(t,r)}{2}\frac{\partial^2 Z}{\partial r^2} + \big(a(t,r) - \gamma(t,r)b(t,r)\big)\frac{\partial Z}{\partial r} - rZ = 0 \tag{6.33}$$

for all $t < T$ and all r, subject to the terminal condition

$$Z(T,T,r) = 1 \text{ for all } r. \tag{6.34}$$

Once the short rate model and the market price of risk γ are specified, the bond price can be determined by solving (6.33) subject to (6.34).

An alternative derivation of the PDE (6.33) is based on hedging and no-arbitrage arguments. Applying the Itô formula to the bond price process $Z(t,T) \equiv Z(t,T,r(t))$, which is a function of the stochastic short rate $r(t)$ and time t, gives the following SDE for the bond price process:

$$\mathrm{d}Z(t,T) = \left(\frac{\partial Z}{\partial t} + a\frac{\partial Z}{\partial r} + \frac{b^2}{2}\frac{\partial^2 Z}{\partial r^2}\right)\mathrm{d}t + b\frac{\partial Z}{\partial r}\,\mathrm{d}W(t)$$

or, equivalently, in log-normal form

$$\frac{\mathrm{d}Z(t,T)}{Z(t,T)} = \mu\,\mathrm{d}t + \sigma\,\mathrm{d}W(t)\,,$$

where the log-drift rate μ and the log-diffusion coefficient σ are, respectively,

$$\mu(t,T,r) = \frac{1}{Z(t,T,r)}\left[\frac{\partial Z(t,T,r)}{\partial t} + a(t,r)\frac{\partial Z(t,T,r)}{\partial r} + \frac{b^2(t,r)}{2}\frac{\partial^2 Z(t,T,r)}{\partial r^2}\right], \tag{6.35}$$

$$\sigma(t,T,r) = \frac{b(t,r)}{Z(t,T,r)}\frac{\partial Z(t,T,r)}{\partial r}\,. \tag{6.36}$$

The underlying short rate process is not a traded security and it cannot therefore be used for hedging bonds. Instead, we try to hedge one bond with another one of a different maturity. Consider two zero-coupon bonds maturing at times T_1 and T_2 with $T_1 < T_2$, respectively. At time t, we buy T_1-bonds of value $V_1(t)$ (a long position) and sell T_2-bonds of value $V_2(t)$ (a short position). The total value of this portfolio at time t is $\Pi(t) = V_1(t) - V_2(t)$. The change in portfolio value from t to $t + \mathrm{d}t$ is

$$\begin{aligned}
\mathrm{d}\Pi(t) &= \frac{V_1(t)}{Z(t,T_1)}\,\mathrm{d}Z(t,T_1) - \frac{V_2(t)}{Z(t,T_2)}\,\mathrm{d}Z(t,T_2)\\
&= V_1(t)(\mu_1\,\mathrm{d}t + \sigma_1\,\mathrm{d}W(t)) - V_2(t)(\mu_2\,\mathrm{d}t + \sigma_2\,\mathrm{d}W(t))\\
&= (V_1(t)\mu_1 - V_2(t)\mu_2)\,\mathrm{d}t + (V_1(t)\sigma_1 - V_2(t)\sigma_2)\,\mathrm{d}W(t)\,,
\end{aligned}$$

where, to compact notations, we denote

$$\mu_i = \mu(t,T_i,r(t)) \text{ and } \sigma_i = \sigma(t,T_i,r(t)) \text{ for } i = 1,2.$$

[2]We remark that the role of the process $X(t)$ with SDE in (2.49) is now played by $r(t)$ with SDE in (6.31) and the dummy variable x in (2.98) is now called r. The variable r is not to be confused with the function denoted by $r(t,x)$ in Theorem 2.13, which is now given by $r(t,x) \equiv x$, as seen by identifying (2.97) with (6.32) and where $\phi(x) \equiv 1$.

Suppose that V_1 and V_2 are chosen such that

$$\frac{V_1(t)}{V_2(t)} = \frac{\sigma_2}{\sigma_1} \equiv \frac{\sigma(t,T_2,r(t))}{\sigma(t,T_1,r(t))}$$

holds for all t. Then $V_1\sigma_1 - V_2\sigma_2 \equiv 0$ and hence the $\mathrm{d}W(t)$ term in the stochastic differential $\mathrm{d}\Pi$ vanishes. As a result, the equation for $\mathrm{d}\Pi$ becomes

$$\frac{\mathrm{d}\Pi(t)}{\Pi(t)} = \frac{\mu_1\sigma_2 - \mu_2\sigma_1}{\sigma_2 - \sigma_1} \equiv \frac{\mu(t,T_1,r(t))\sigma(t,T_2,r(t)) - \mu(t,T_2,r(t))\sigma(t,T_1,r(t))}{\sigma(t,T_2,r(t)) - \sigma(t,T_1,r(t))}\,\mathrm{d}t$$

Thus, we have a risk-free, self-financing portfolio strategy. To avoid arbitrage, the rate of return has to be equal to the risk-free rate $r(t)$; that is,

$$\frac{\mu_1\sigma_2 - \mu_2\sigma_1}{\sigma_2 - \sigma_1} = r \iff \frac{\mu_1 - r}{\sigma_1} = \frac{\mu_2 - r}{\sigma_2}.$$

The above relation is valid for arbitrary maturity dates T_1 and T_2, so the ratio $\frac{\mu(t,T,r)-r}{\sigma(t,T,r)}$ is independent of T for all $T > t$. We can hence define

$$\gamma(t,r) = \frac{\mu(t,T,r) - r}{\sigma(t,T,r)} \tag{6.37}$$

so that the drift rate of the bond price process is

$$\mu(t,T,r) = r + \gamma(t,r)\sigma(t,T,r)\,. \tag{6.38}$$

Comparing the above equation with (6.20), we conclude that γ is nothing but the market price of risk for the short rate process. Equating the formulae (6.35) and (6.38) gives the governing PDE (6.33) for the price of a zero-coupon bond.

6.2.3 Affine Term Structure Models

A short-rate model that produces the bond pricing function of the form

$$Z(t,T,r) = \mathrm{e}^{A(t,T)-C(t,T)r}, \tag{6.39}$$

where r is the short rate at time t, and the functions $A(t,T)$ and $C(t,T)$ are independent of r, is called an *affine term structure* model. Let the short rate process follow the SDE of the form in (6.31):

$$\mathrm{d}r(t) = \tilde{a}(t,r(t))\,\mathrm{d}t + b(t,r(t))\,\mathrm{d}\widetilde{W}(t)\,, \tag{6.40}$$

where $\tilde{a}(t,r) := a(t,r) - \gamma(t,r)b(t,r)$ and $\widetilde{W}(t)$ is a standard BM under the EMM $\widetilde{\mathbb{P}}$.

Certain conditions must be set on the short rate process $r(t)$ in order that the zero-coupon bond price admits the form (6.39). Applying the Itô formula to the bond price $Z(t,T,r(t)) = \mathrm{e}^{A(t,T)-C(t,T)r(t)}$, where $r(t)$ follows (6.40), gives

$$\frac{\mathrm{d}Z(t,T,r(t))}{Z(t,T,r(t))} = \left[\frac{\partial A(t,T)}{\partial t} - \frac{\partial C(t,T)}{\partial t}r(t) - C(t,T)\tilde{a}(t,r(t)) + \frac{1}{2}C^2(t,T)b^2(t,r(t))\right]\mathrm{d}t$$
$$- C(t,T)b(t,r(t))\,\mathrm{d}\widetilde{W}(t)\,.$$

We also know that the risk-neutral dynamics of the bond price is given by (6.17). That is, the drift rate in the above SDE has to be equal to the risk-neutral rate $r(t)$. It follows that the ordinary (nonrandom) function $g(t,r)$ defined by

$$g(t,r) = \frac{\partial A(t,T)}{\partial t} - \frac{\partial C(t,T)}{\partial t}r - C(t,T)\tilde{a}(t,r) + \frac{1}{2}C^2(t,T)b^2(t,r) - r$$

is identically zero for all t and r. Since A and C are independent of r, then $g(t,r) \equiv 0$ holds only if $\tilde{a}(t,r)$ and $b^2(t,r)$ are both linear functions of r. That is, for the bond pricing formula to be of the affine form (6.39), it is necessary that the risk-neutral drift and the square of the diffusion coefficient both be affine (i.e., linear) functions of r:

$$\tilde{a}(t,r) = a_0(t) + a_1(t)r \quad \text{and} \quad b^2(t,r) = b_0(t) + b_1(t)r, \tag{6.41}$$

where $a_0(t)$, $a_1(t)$, $b_0(t)$, and $b_1(t)$ are only functions of time.

The zero-coupon bond pricing function $Z(t,T,r)$ solves the PDE (6.33) with terminal conditions $Z(T,T,r) = 1$. Substituting the solution in (6.39) into (6.33) yields

$$\frac{\partial A(t,T)}{\partial t} - \left(1 + \frac{\partial C(t,T)}{\partial t}\right) r + \frac{b^2(t,r)}{2} C^2(t,T) - \tilde{a}(t,r)C(t,T) = 0 \text{ for } t < T, \tag{6.42}$$

with terminal conditions $A(T,T) = 0$ and $C(T,T) = 0$.

Substituting (6.41) into (6.42) gives

$$\begin{aligned} &\frac{\partial A(t,T)}{\partial t} - a_0(t)C(t,T) + \frac{b_0(t)}{2}C^2(t,T) \\ &- \left(\frac{\partial C(t,T)}{\partial t} + a_1(t)C(t,T) - \frac{b_1(t)}{2}C^2(t,T) + 1\right) r = 0\,. \end{aligned}$$

Since the left-hand side of the above equation is identically zero for *all* values of the rate r, the functions A and C must solve the following pair of differential equations:

$$\frac{\partial C(t,T)}{\partial t} + a_1(t)C(t,T) - \frac{b_1(t)}{2}C^2(t,T) + 1 = 0, \tag{6.43}$$

$$\frac{\partial A(t,T)}{\partial t} - a_0(t)C(t,T) + \frac{b_0(t)}{2}C^2(t,T) = 0, \tag{6.44}$$

for $t < T$, subject to the respective boundary conditions at $t = T$: $A(T,T) = 0$ and $C(T,T) = 0$. The equation in (6.43) is a first order nonlinear ODE and is known as the *Ricatti equation.* For some special cases of a_1 and b_1, it is possible to solve equation (6.43) in closed form. Once an analytic solution for C is available, the solution for A is obtained by integrating (6.44) with respect to t. In the next three subsections, we consider three short-rate models that admit the bond pricing formula as in (6.39) where A and C are given in analytically closed form.

6.2.4 The Ho–Lee Model

The short rate in the Ho–Lee model satisfies (6.27). We note that the Merton model is a special case of the Ho–Lee model with constant drift rate α. For pricing bonds, we need the short rate dynamics under the EMM $\widetilde{\mathbb{P}}$. Assume that the market price of risk γ is independent of r, then by (6.31) the SDE for $r(t)$ under $\widetilde{\mathbb{P}}$ is

$$\mathrm{d}r(t) = \alpha(t)\,\mathrm{d}t + \sigma\,\mathrm{d}\widetilde{W}(t)\,,$$

where $\tilde{\alpha}(t) = \alpha(t) - \gamma(t)\sigma$. The strong solution to this linear SDE is a drifted and scaled Brownian motion:

$$r(s) = r(t) + \int_t^s \tilde{\alpha}(u)\,\mathrm{d}u + \sigma(\widetilde{W}(s) - \widetilde{W}(t)) \tag{6.45}$$

for $0 \leqslant t \leqslant s$. The rate $r(s)$ conditional on $r(t) = r$ is normally distributed with mean $r + \int_t^s \tilde{\alpha}(u)\,\mathrm{d}u$ and variance $\sigma^2(s-t)$. Since $\widehat{W}(s-t) := \widetilde{W}(s) - \widetilde{W}(t)$ is independent of $\mathcal{F}_t$

for every $s > t$, the random variable $\int_t^T \widehat{W}(s-t)\,\mathrm{d}s \sim Norm(0, (T-t)^3/3)$ is independent of $\mathcal{F}_t$, i.e., it is independent of $r(t)$. Hence, the conditional expectation simplifies to an unconditional expectation, and we obtain the no-arbitrage pricing function:

$$\begin{aligned} Z(t,T,r) &= \widetilde{\mathrm{E}}_{t,r}\left[\mathrm{e}^{-\int_t^T r(s)\,\mathrm{d}s}\right] = \widetilde{\mathrm{E}}\left[\mathrm{e}^{-\int_t^T \left(r+\int_t^s \tilde{\alpha}(u)\,\mathrm{d}u+\sigma\widehat{W}(s-t)\right)\mathrm{d}s}\right] \\ &= \mathrm{e}^{-r(T-t)-\hat{\alpha}(t,T)(T-t)^2/2}\widetilde{\mathrm{E}}\left[\mathrm{e}^{-\sigma\int_t^T \widehat{W}(s-t)\,\mathrm{d}s}\right] \\ &= \mathrm{e}^{-r(T-t)-\hat{\alpha}(t,T)(T-t)^2/2+\sigma^2(T-t)^3/6}, \end{aligned} \tag{6.46}$$

where $\hat{\alpha}(t,T) := \frac{2}{(T-t)^2}\int_t^T\int_t^s \tilde{\alpha}(u)\,\mathrm{d}u\,\mathrm{d}s$. In the case of the Merton model with constant $\tilde{\alpha}$, we have $\hat{\alpha}(t,T) = \tilde{\alpha}$ and

$$Z(t,T,r) = \mathrm{e}^{-r(T-t)-\tilde{\alpha}(T-t)^2/2+\sigma^2(T-t)^3/6}. \tag{6.47}$$

The yield rate is an affine function of r,

$$y(t,T,r) = -\frac{\ln Z(t,T,r)}{T-t} = r + \hat{\alpha}(t,T)\frac{(T-t)}{2} - \sigma^2\frac{(T-t)^2}{6}.$$

Since the distribution of the short rate $r(t)$ is normal, it then follows that $y(t,T,r(t))$ is also normally distributed and the distribution of $Z(t,T,r(t))$ is log-normal.

The Ho–Lee model has several serious shortcomings:

1. the short rate can become negative (with nonzero probability);
2. $Z(t,T,r) \to \infty$ and $y(t,T,r) \to -\infty$ as $T \to \infty$;
3. the yield rates $y(t,T_1,r(t))$ and $y(t,T_2,r(t))$ for different maturities T_1 and T_2 are both a linear function of the short rate $r(t)$, and they are hence perfectly correlated.

The Ho–Lee model is an affine model with $a_0(t) = \tilde{\alpha}(t)$, $a_1(t) \equiv 0$, $b_0(t) = \sigma^2$, and $b_1(t) \equiv 0$ in (6.41). Therefore, the bond pricing function in (6.47) can also be found by solving (6.43)–(6.44) for $A(t,T)$ and $C(t,T)$, which take the following simpler form:

$$\begin{aligned} &\frac{\partial C(t,T)}{\partial t} + 1 = 0, \\ &\frac{\partial A(t,T)}{\partial t} - \tilde{\alpha}(t)C(t,T) + \frac{\sigma^2}{2}C^2(t,T) = 0, \end{aligned}$$

subject to $A(T,T) = C(T,T) = 0$. The first equation is trivially integrated to give $C(t,T) = T-t$. Substituting this into the second equation, integrating and applying the condition $A(T,T) = 0$, gives

$$A(t,T) = -\int_t^T \tilde{\alpha}(u)(T-u)\,\mathrm{d}u + \frac{\sigma^2}{2}\int_t^T (T-u)^2\,\mathrm{d}u = -\hat{\alpha}(t,T)\frac{(T-t)^2}{2} + \sigma^2\frac{(T-t)^3}{6}.$$

This recovers the above same formula for the yield and zero-coupon bond price in (6.47). For the case of constant $\tilde{\alpha}(t) \equiv \tilde{\alpha}$, the formulae simplify where $\hat{\alpha}(t,T) = \tilde{\alpha}$.

6.2.5 The Vasiček Model

The short rate process in the Vasiček model satisfies SDE (6.24). Recall that the solution to (6.24) is also known as the Ornstein–Uhlenbeck process. Assuming the market price of risk γ is constant, we obtain the following risk-neutral dynamics of $r(t)$:

$$\mathrm{d}r(t) = (\tilde{\alpha} - \beta r(t))\,\mathrm{d}t + \sigma\,\mathrm{d}\widetilde{W}(t), \tag{6.48}$$

where $\tilde{\alpha} = \alpha - \gamma\sigma$, β, and σ are positive parameters. The diffusion solving the above SDE is called a mean-reverting process since the instantaneous drift $\tilde{\alpha} - \beta r(t) = \beta(\tilde{\alpha}/\beta - r(t))$ pulls the process toward the constant mean level $\tilde{\alpha}/\beta$ with magnitude proportional to the deviation of the process from the mean.

Integrating the above SDE (see Examples 2.12 and 2.23) gives

$$r(T) = \mathrm{e}^{-\beta(T-t)} r(t) + \frac{\tilde{\alpha}}{\beta}\left(1 - \mathrm{e}^{-\beta(T-t)}\right) + \sigma \int_t^T \mathrm{e}^{-\beta(T-s)}\, \mathrm{d}\widetilde{W}(s)\,. \tag{6.49}$$

The probability distribution of $r(T)$ conditional on $r(t)$ is normal with mean

$$\widetilde{\mathrm{E}}[r(T) \mid r(t)] = \mathrm{e}^{-\beta(T-t)} r(t) + \frac{\alpha}{\beta}(1 - \mathrm{e}^{-\beta(T-t)})$$

and variance

$$\widetilde{\mathrm{Var}}(r(T) \mid r(t)) = \sigma^2 \frac{1 - \mathrm{e}^{-2\beta(T-t)}}{2\beta}\,.$$

Note that the long-term mean and variance (as $T - t \to \infty$) are given by the constants:

$$\lim_{T\to\infty} \widetilde{\mathrm{E}}[r(T) \mid r(t)] = \frac{\tilde{\alpha}}{\beta} \quad \text{and} \quad \lim_{T\to\infty} \widetilde{\mathrm{Var}}(r(T) \mid r(t)) = \frac{\sigma^2}{2\beta}\,.$$

We now proceed to the calculation of bond prices. The drift and diffusion coefficients in (6.48) correspond to $a_0 = \tilde{\alpha}$, $a_1 = -\beta$, $b_0 = \sigma^2$, and $b_1 = 0$ in (6.41). We have the following pair of ODEs for $A(t,T)$ and $C(t,T)$:

$$\frac{\partial C(t,T)}{\partial t} - \beta C(t,T) + 1 = 0, \tag{6.50}$$

$$\frac{\partial A(t,T)}{\partial t} - \tilde{\alpha} C(t,T) + \frac{\sigma^2}{2} C^2(t,T) = 0, \tag{6.51}$$

for $t < T$, subject to $A(T,T) = C(T,T) = 0$. Solving the above system, we obtain

$$C(t,T) = \frac{1}{\beta}\left(1 - \mathrm{e}^{-\beta(T-t)}\right) \text{ and } A(t,T) = (C(t,T) - (T-t))y_\infty - \frac{\sigma^2}{4\beta} C^2(t,T), \tag{6.52}$$

where $y_\infty := \frac{\tilde{\alpha}}{\beta} - \frac{\sigma^2}{2\beta^2}$. Thus, according to (6.39), the bond pricing formula for the Vašiček model is

$$Z(t,T,r) = \exp\left[\left(\frac{1 - \mathrm{e}^{-\beta(T-t)}}{\beta}\right)(y_\infty - r) - y_\infty(T-t) - \frac{\sigma^2}{4\beta}\left(\frac{1 - \mathrm{e}^{-\beta(T-t)}}{\beta}\right)^2\right]. \tag{6.53}$$

The yield rate is found to be

$$y(t,T,r) = y_\infty - \frac{1}{\beta}(1 - \mathrm{e}^{-\beta(T-t)})\frac{y_\infty - r}{T-t} + \frac{\sigma^2}{4\beta(T-t)}\left(\frac{1 - \mathrm{e}^{-\beta(T-t)}}{\beta}\right)^2.$$

By taking $T \to \infty$, the last two terms of the above equation vanish so that the long-term yield rate is equal to y_∞.

6.2.6 The Cox–Ingersoll–Ross Model

One common drawback of the Merton model and the Vasiček model is that the short rate can be negative due to its normal distribution. The first tractable model for the short rate process $r(t)$ that keeps rates positive was proposed by Cox, Ingersoll, and Ross (the CIR model). The short-rate model follows the square root diffusion described by the SDE in (6.26). Assuming the market price of risk is equal to $\gamma\sqrt{r}$ with a constant γ, we obtain the following risk-neutral dynamics:

$$\mathrm{d}r(t) = (\alpha - \tilde{\beta} r(t))\,\mathrm{d}t + \sigma\sqrt{r(t)}\,\mathrm{d}\widetilde{W}(t),$$

where $\tilde{\beta} = \beta + \gamma\sigma$. With a nonnegative initial interest rate, $r(t)$ stays nonnegative. Moreover, the CIR process is mean-reverting with the long-run mean level $\alpha/\tilde{\beta}$.

The CIR process $r(t)$ is reduced to the squared Bessel (SQB) process $X(t)$, which solves the SDE (7.30) by means of a scale and time transformation,

$$r(t) = e^{-\tilde{\beta} t}\frac{\sigma^2}{4} X(\tau_t),$$

where the (strictly increasing) time transformation τ_t is defined to be

$$\tau_t := \begin{cases} t & \text{if } \tilde{\beta} = 0, \\ \frac{e^{\tilde{\beta} t} - 1}{\tilde{\beta}} & \text{if } \tilde{\beta} \neq 0. \end{cases} \tag{6.54}$$

The index of the SQB process in (7.30) is $\mu = \frac{2\alpha}{\sigma^2} - 1$. In what follows, we assume that $\mu \geqslant 0$ (or, equivalently, $2\alpha \geqslant \sigma^2$), and hence the left-hand endpoint 0 is an entrance boundary for the short-rate CIR process.

The transition PDF for the CIR process relates to that of the SQB process. Under the risk-neutral measure $\widetilde{\mathbb{P}}$ it is given by

$$\tilde{p}(t; r_0, r) = c_t e^{\tilde{\beta} t}\left(\frac{r e^{\tilde{\beta} t}}{r_0}\right)^{\frac{\mu}{2}} \exp\left(-c_t(r e^{\tilde{\beta} t} + r_0)\right) I_\mu\left(2c_t\sqrt{r r_0 e^{\tilde{\beta} t}}\right), \tag{6.55}$$

where $c_t := \frac{2}{\sigma^2 \tau_t}$.

By expressing the above SDE in integral form, with $r(0) = r_0$, and taking expectations (under measure $\widetilde{\mathbb{P}}$) on both sides while denoting the mean by $\widetilde{\mathrm{E}}[r(t)] \equiv m(t)$, we have

$$m(t) = r_0 + \int_0^t (\alpha - \tilde{\beta} m(u))\,\mathrm{d}u\,.$$

Here, we assume that the Itô integral $\int_0^t \sqrt{r(u)}\,\mathrm{d}\widetilde{W}(u)$ has zero expectation (i.e., it is a $\widetilde{\mathbb{P}}$-martingale). Differentiating gives a linear first order ODE

$$\frac{\mathrm{d}}{\mathrm{d}t} m(t) = \alpha - \tilde{\beta} m(t)\,, \quad t \geqslant 0\,,$$

subject to $m(0) = r_0$. Solving gives

$$m(t) \equiv \widetilde{\mathrm{E}}[r(t)] = \frac{\alpha}{\tilde{\beta}}(1 - e^{-\tilde{\beta} t}) + r_0 e^{-\tilde{\beta} t}\,.$$

Here, we assume that $\tilde{\beta} \neq 0$. Note that the mean of the process converges to the long-term level $\alpha/\tilde{\beta}$, as $t \to \infty$.

The CIR model is within the class of affine term structure models with a bond pricing formula of the form in (6.39). The respective ODEs in (6.43) and (6.44) for $A(t,T)$ and $C(t,T)$ are

$$\frac{\partial C(t,T)}{\partial t} - \tilde{\beta} C(t,T) - \frac{\sigma^2}{2} C^2(t,T) + 1 = 0, \tag{6.56}$$

$$\frac{\partial A(t,T)}{\partial t} - \alpha C(t,T) = 0, \tag{6.57}$$

subject to $A(T,T) = C(T,T) = 0$. The first equation is a first order ODE with a quadratic nonlinear term. The trick in solving this equation is to turn it into a linear second order ODE by invoking the transformation

$$\psi(t) := \exp\left(\frac{\sigma^2}{2} \int_t^T C(s,T)\,\mathrm{d}s\right).$$

Taking derivatives gives (denoting $C'(t,T) \equiv \partial C(t,T)/\partial t$):

$$\frac{\psi'(t)}{\psi(t)} \equiv \frac{\mathrm{d}}{\mathrm{d}t} \ln \psi(t) = -\frac{\sigma^2}{2} C(t,T) \implies C(t,T) = -\frac{2}{\sigma^2} \frac{\psi'(t)}{\psi(t)},$$

$$\frac{\psi''(t)}{\psi(t)} = -\frac{\sigma^2}{2} C'(t,T) - \frac{\sigma^2}{2} C(t,T) \frac{\psi'(t)}{\psi(t)} \implies C'(t,T) = -\frac{2}{\sigma^2} \frac{\psi''(t)}{\psi(t)} + \frac{\sigma^2}{2} C^2(t,T).$$

Substituting the above two expressions for $C(t,T)$ and $C'(t,T)$ into (6.56), and simplifying, gives

$$\psi''(t) - \tilde{\beta}\psi'(t) - \frac{\sigma^2}{2}\psi(t) = 0. \tag{6.58}$$

This is a second order linear ODE with constant coefficients. Its solution is found by standard methods. In particular, this ODE has the general solution

$$\psi(t) = c_1 \mathrm{e}^{\frac{1}{2}(\tilde{\beta}+\vartheta)t} + c_2 \mathrm{e}^{\frac{1}{2}(\tilde{\beta}-\vartheta)t}$$

with derivative

$$\psi'(t) = \frac{c_1}{2}(\tilde{\beta}+\vartheta)\mathrm{e}^{\frac{1}{2}(\tilde{\beta}+\vartheta)t} + \frac{c_2}{2}(\tilde{\beta}-\vartheta)\mathrm{e}^{\frac{1}{2}(\tilde{\beta}-\vartheta)t},$$

where $\vartheta := \sqrt{\tilde{\beta}^2 + 2\sigma^2}$. The constants $c_{1,2}$ are determined by applying the boundary conditions, $\psi(T) = \mathrm{e}^0 = 1$ and $\psi'(T) = -\frac{\sigma^2}{2} C(T,T)\psi(T) = 0$, giving a 2×2 linear system in c_1 and c_2:

$$\mathrm{e}^{\frac{1}{2}(\tilde{\beta}+\vartheta)T} c_1 + \mathrm{e}^{\frac{1}{2}(\tilde{\beta}-\vartheta)T} c_2 = 1,$$

$$\frac{1}{2}(\tilde{\beta}+\vartheta)\mathrm{e}^{\frac{1}{2}(\tilde{\beta}+\vartheta)T} c_1 + \frac{1}{2}(\tilde{\beta}-\vartheta)\mathrm{e}^{\frac{1}{2}(\tilde{\beta}-\vartheta)T} c_2 = 0.$$

Solving gives

$$c_1 = \left(\frac{1}{2} - \frac{\tilde{\beta}}{2\vartheta}\right)\mathrm{e}^{-\frac{1}{2}(\tilde{\beta}+\vartheta)T} \quad \text{and} \quad c_2 = \left(\frac{1}{2} + \frac{\tilde{\beta}}{2\vartheta}\right)\mathrm{e}^{-\frac{1}{2}(\tilde{\beta}-\vartheta)T}.$$

Using these coefficients gives the unique explicit expression for $\psi(t)$, which can be equivalently written as

$$\begin{aligned}\psi(t) &= \left(\frac{1}{2} - \frac{\tilde{\beta}}{2\vartheta}\right)\mathrm{e}^{-\frac{1}{2}(\tilde{\beta}+\vartheta)(T-t)} + \left(\frac{1}{2} + \frac{\tilde{\beta}}{2\vartheta}\right)\mathrm{e}^{-\frac{1}{2}(\tilde{\beta}-\vartheta)(T-t)} \\ &= \mathrm{e}^{-\frac{1}{2}\tilde{\beta}(T-t)}\left[\cosh\frac{\vartheta(T-t)}{2} + \frac{\tilde{\beta}}{\vartheta}\sinh\frac{\vartheta(T-t)}{2}\right].\end{aligned}$$

Differentiating this expression, and using the fact that $C(t,T) = -\frac{2}{\sigma^2}\frac{\psi'(t)}{\psi(t)}$, finally gives

$$C(t,T) = \frac{2\sinh\frac{\vartheta(T-t)}{2}}{\vartheta\cosh\frac{\vartheta(T-t)}{2} + \tilde{\beta}\sinh\frac{\vartheta(T-t)}{2}} = \frac{2\,\mathrm{e}^{\vartheta(T-t)} - 2}{(\tilde{\beta}+\vartheta)\left(\mathrm{e}^{\vartheta(T-t)} - 1\right) + 2\,\vartheta}\,. \tag{6.59}$$

Having solved for $C(t,T)$, the function $A(t,T)$ is obtained by integrating (6.57), with $\int_t^T \frac{\partial A(u,T)}{\partial u}\,\mathrm{d}u = A(T,T) - A(t,T) = -A(t,T)$, giving

$$A(t,T) = -\alpha\int_t^T C(u,T)\,\mathrm{d}u\,.$$

We can compute this integral using the fact that $-C(t,T) = \frac{2}{\sigma^2}\frac{\mathrm{d}}{\mathrm{d}t}\ln\psi(t)$, i.e.,

$$A(t,T) = \frac{2\alpha}{\sigma^2}\int_t^T \frac{\mathrm{d}}{\mathrm{d}u}\ln\psi(u)\,\mathrm{d}u = \frac{2\alpha}{\sigma^2}\ln\frac{\psi(T)}{\psi(t)} = \frac{2\alpha}{\sigma^2}\ln\frac{1}{\psi(t)}\,.$$

Note that $\psi(T) = 1$. Using the above expression for ψ gives the explicit form for $A(t,T)$, which we can write equivalently as

$$\begin{aligned} A(t,T) &= \frac{2\alpha}{\sigma^2}\ln\left(\frac{\vartheta\,\mathrm{e}^{\frac{1}{2}\tilde{\beta}(T-t)}}{\vartheta\cosh\frac{\vartheta(T-t)}{2} + \tilde{\beta}\sinh\frac{\vartheta(T-t)}{2}}\right) \\ &= \frac{2\alpha}{\sigma^2}\ln\left(\frac{2\vartheta\,\mathrm{e}^{\frac{1}{2}(\tilde{\beta}+\vartheta)(T-t)}}{(\tilde{\beta}+\vartheta)(\mathrm{e}^{\vartheta(T-t)} - 1) + 2\vartheta}\right). \end{aligned} \tag{6.60}$$

Inserting the coefficients in (6.60) and (6.59) into (6.39) gives us the closed form analytical expression for the price of a zero-coupon bond under the CIR model.

6.3 Heath–Jarrow–Morton Formulation

When we use a short-rate model such as one of those considered in previous subsections, the bond prices, yield rates, and forward rates are the output of the model. We usually find that the model bond prices $Z(t,T)$ do not perfectly match the market (observed) bond prices $Z_{\text{obs}}(t,T)$. The usual approach is to calibrate the short-rate model parameters to achieve the best possible agreement between the model prices and the market prices. For example, we can use the least squares method and find the optimal model parameters by minimizing $\sum_i [Z(t,T_i) - Z_{\text{obs}}(t,T_i)]^2$, the sum of squares of the differences between the model and observed bond prices across a given set of maturities $T_1, T_2, \ldots$. However, since the number of bonds with different maturities exceeds the number of model parameters, we will find that the observed prices are closer to the prices produced by the calibrated model but still differ to some extent. This issue can be dealt with by using one of the following two approaches. One way is to construct a time-inhomogeneous model with multiple random factors. Such a model will be more flexible than a single-factor model with constant parameters and can be calibrated to historical data with a greater degree of accuracy. Multifactor models are briefly discussed further in Section 6.4. Several time-inhomogeneous short-rate models, including the Ho–Lee model, the Black–Derman-Toy model, and the Hull–White model, are presented in the beginning of the previous section.

The other approach is to construct a no-arbitrage term-structure model where observed bond prices, yield rates, or forward rates are taken as input variables. The ideal result would be if the model prices $Z(t,T)$ precisely match the market prices $Z_{\text{obs}}(t,T)$ at the time of calibration t. The Heath–Jarrow–Morton (HJM) framework attempts to construct a model for the family of forward rate curves $\{f(t,T)\}_{0\leqslant t\leqslant T}$ with $0 \leqslant T \leqslant T^*$. Under the HJM model, the forward rate $f(t,T)$ (for *arbitrarily fixed* $T \in [0,T^*]$) follows the SDE

$$\mathrm{d}f(t,T) = \alpha_F(t,T)\,\mathrm{d}t + \sigma_F(t,T)\,\mathrm{d}W(t)\,, \quad 0 \leqslant t \leqslant T\,, \tag{6.61}$$

where $\{W(t)\}_{t\geqslant 0}$ is a standard Brownian motion under the physical measure $\mathbb{P}$, and the processes $\alpha_F(t,T)$ and $\sigma_F(t,T)$ are adapted to its natural filtration $\mathbb{F}^W$. We note that the differential is taken w.r.t. the calendar time variable t and T acts as a fixed parameter. To simplify the analysis in what follows, we shall assume a single Brownian motion where the forward rate process starts with the initial rate $f(0,T)$. The framework readily extends to the case where the forward rates are driven by a multidimensional Brownian motion.

Let us assume that there exists an EMM (risk-neutral measure) $\widetilde{\mathbb{P}}$ such that the discounted bond price process $\{\overline{Z}(t,T)\}_{0\leqslant t\leqslant T}$ is a $\widetilde{\mathbb{P}}$-martingale. As is demonstrated below, the EMM assumption implies that the drift coefficient α_F is determined by the diffusion coefficient σ_F when the SDE (6.61) for forward rates is considered under $\widetilde{\mathbb{P}}$.

6.3.1 HJM under Risk-Neutral Measure

Let us work out the risk-neutral dynamics of the bond prices. The zero-coupon bond price can be expressed in terms of forward rates, as given in (6.7). The discounted bond price is

$$\overline{Z}(t,T) \equiv D(t)Z(t,T) = \exp\left(-\int_0^t r(u)\,\mathrm{d}u - \int_t^T f(t,u)\,\mathrm{d}u\right),$$

for $0 \leqslant t \leqslant T \leqslant T^*$. By the Itô formula, we have

$$\mathrm{d}\overline{Z}(t,T) = \overline{Z}(t,T)\left(\mathrm{d}X(t) + \frac{1}{2}\,\mathrm{d}[X,X](t) - r(t)\,\mathrm{d}t\right), \tag{6.62}$$

where $X(t)$ denotes the log-price of the bond and is given by

$$X(t) := \ln Z(t,T) = -\int_t^T f(t,u)\,\mathrm{d}u\,. \tag{6.63}$$

In (6.62) we used the Itô formula $\mathrm{d}e^{Y(t)} = e^{Y(t)}\left(\mathrm{d}Y(t) + \frac{1}{2}\,\mathrm{d}[Y,Y](t)\right)$, where $Y(t) = X(t) - \int_0^t r(u)\,\mathrm{d}u$, $\mathrm{d}Y(t) = \mathrm{d}X(t) - r(t)\,\mathrm{d}t$, and $\mathrm{d}[Y,Y](t) = \mathrm{d}[X,X](t) = \mathrm{d}X(t)\,\mathrm{d}X(t)$.

The next step is to derive an SDE for the discounted bond price in terms of the drift and diffusion functions that are driving the forward rate in (6.61). We need to compute the stochastic differential of $X(t)$ defined by (6.63), which is a Riemann integral of the process $f(t,u)$ w.r.t. the parameter $u \in [t,T]$. It can be shown that[3]

$$\mathrm{d}X(t) \equiv \mathrm{d}\left(-\int_t^T f(t,u)\,\mathrm{d}u\right) = f(t,t)\,\mathrm{d}t - \int_t^T \mathrm{d}f(t,u)\,\mathrm{d}u\,. \tag{6.64}$$

[3]If $f : \mathbb{R}^2 \to \mathbb{R}$ is a nonrandom (ordinary) function, then ordinary calculus gives the ordinary derivative w.r.t. t as $\frac{\mathrm{d}}{\mathrm{d}t}\left(-\int_t^T f(t,u)\,\mathrm{d}u\right) = f(t,t) - \int_t^T \frac{\partial f}{\partial t}(t,u)\,\mathrm{d}u$, i.e., multiplying both sides by $\mathrm{d}t$, with $\frac{\partial f}{\partial t}(t,u)\,\mathrm{d}t = \mathrm{d}f(t,u)$ for fixed parameter u, gives the ordinary differential $\mathrm{d}\left(-\int_t^T f(t,u)\,\mathrm{d}u\right) = f(t,t)\,\mathrm{d}t - \int_t^T \mathrm{d}f(t,u)\,\mathrm{d}u$. If $f(t,u) : \Omega \to \mathbb{R}$ is an $\mathcal{F}_t$-measurable Itô process (for all parameter values u), then this result applies where $\mathrm{d}f(t,u)$ is a stochastic differential w.r.t. t.

The first term is $r(t)\,\mathrm{d}t$ since the forward rate corresponds to the instantaneous short rate at time t when $T = t$, i.e., $f(t,t) = r(t)$ is the observed rate on any risk-free investment at current time t. The second term can now be written as a stochastic differential of an Itô process upon substituting the form in (6.61), using the parameter u in the place of T, within the integral and interchanging the order of the differentials (as follows by applying Fubini's Theorem twice):

$$\int_t^T \mathrm{d}f(t,u)\,\mathrm{d}u = \int_t^T (\alpha_F(t,u)\,\mathrm{d}t + \sigma_F(t,u)\,\mathrm{d}W(t))\,\mathrm{d}u$$
$$= \left(\int_t^T \alpha_F(t,u)\,\mathrm{d}u\right)\mathrm{d}t + \left(\int_t^T \sigma_F(t,u)\,\mathrm{d}u\right)\mathrm{d}W(t)\,.$$

Defining the new drift and volatility functions (which are proportional the instantaneous drift and volatility functions of the forward rate process averaged over all maturity times between t and T),

$$A_F(t,T) := \int_t^T \alpha_F(t,u)\,\mathrm{d}u \quad \text{and} \quad \Sigma_F(t,T) := \int_t^T \sigma_F(t,u)\,\mathrm{d}u\,, \tag{6.65}$$

gives, according to (6.64),

$$\mathrm{d}X(t) = [r(t) - A_F(t,T)]\,\mathrm{d}t - \Sigma_F(t,T)\,\mathrm{d}W(t)\,. \tag{6.66}$$

We note that, by differentiating the integrals in (6.65) w.r.t. the maturity variable T, we have

$$\frac{\partial}{\partial T}A_F(t,T) = \alpha_F(t,T) \quad \text{and} \quad \frac{\partial}{\partial T}\Sigma_F(t,T) = \sigma_F(t,T)\,. \tag{6.67}$$

Therefore, using (6.66) gives $\mathrm{d}[X,X](t) = \Sigma_F^2(t,T)\,\mathrm{d}t$ and the SDE in (6.62) for the discounted bond price takes the form

$$\frac{\mathrm{d}\overline{Z}(t,T)}{\overline{Z}(t,T)} = \left(\frac{1}{2}\Sigma_F^2(t,T) - A_F(t,T)\right)\mathrm{d}t - \Sigma_F(t,T)\,\mathrm{d}W(t)\,. \tag{6.68}$$

Note that this SDE is in terms of the Brownian increment in the physical measure $\mathbb{P}$.

The HJM model includes a zero-coupon bond for each maturity $T \in [0, T^*]$. Hence, to avoid arbitrage when trading in any of these bonds we make recourse to the First Fundamental Theorem of Chapter 3. Namely, if there exists of a risk-neutral measure $\widetilde{\mathbb{P}}$ under which all discounted bond price processes are $\widetilde{\mathbb{P}}$-martingales, then there are no arbitrage strategies in the model. The risk-neutral measure $\widetilde{\mathbb{P}}$ is the measure under which $\{\overline{Z}(t,T)\}_{0\leqslant t\leqslant T}$ are martingales for all choices of $T \in [0, T^*]$. By Girsanov's Theorem 2.17, we can change measures from $\mathbb{P}$ to $\widetilde{\mathbb{P}}$ where

$$\widetilde{W}(t) := W(t) + \int_0^t \theta(s)\,\mathrm{d}s\ ,\ t \geqslant 0,$$

is a standard $\widetilde{\mathbb{P}}$-BM with $\{\theta(t)\}_{0\leqslant t\leqslant T}$ as an adapted process. Using the differential form $\mathrm{d}W(t) = \mathrm{d}\widetilde{W}(t) - \theta(t)\,\mathrm{d}t$ into (6.68) gives

$$\frac{\mathrm{d}\overline{Z}(t,T)}{\overline{Z}(t,T)} = \left(\frac{1}{2}\Sigma_F^2(t,T) - A_F(t,T) + \Sigma_F(t,T)\,\theta(t)\right)\mathrm{d}t - \Sigma_F(t,T)\,\mathrm{d}\widetilde{W}(t)\,. \tag{6.69}$$

Hence, the measure $\widetilde{\mathbb{P}}$ defines the risk-neutral measure if the drift term in this expression is *identically zero*, i.e., if $\theta(t)$ satisfies the equation

$$\frac{1}{2}\Sigma_F^2(t,T) - A_F(t,T) + \Sigma_F(t,T)\,\theta(t) = 0, \tag{6.70}$$

for all $0 \leqslant t \leqslant T$, and for each $T \in [0, T^*]$. As we saw in Chapter 2, this corresponds to a so-called market price of risk equation where the unknown $\theta(t)$ is the market price of risk. Here, we note that $\theta(t)$ must solve the above equation for each maturity value $T \in [0, T^*]$. By taking partial derivatives w.r.t. the maturity T on both sides of (6.70), and using the derivatives defined in (6.67), gives:

$$\alpha_F(t,T) = \sigma_F(t,T)\left[\Sigma_F(t,T) + \theta(t)\right]. \tag{6.71}$$

In order to have no arbitrage in the HJM model, the drift and volatility functions and the averaged volatility $\Sigma_F(t,T)$ for the forward price process must necessarily satisfy a relation of this form for every maturity value T. In fact, if the relation in (6.71) holds, then the HJM model has no arbitrage, and assuming a nonzero volatility function $\sigma_F(t,T)$, the risk-neutral measure is given uniquely by solving for $\theta(t)$ in (6.71):

$$\theta(t) = \frac{\alpha_F(t,T)}{\sigma_F(t,T)} - \Sigma_F(t,T) \tag{6.72}$$

for all $t \in [0, T]$. This is the statement of the Heath–Jarrow–Morton Theorem, which states that the above HJM model driven by a single Brownian motion admits no arbitrage if there exists an adapted process $\theta(t)$ solving (6.71) for all values of time t and maturities T.

The Heath–Jarrow–Morton Theorem is now readily proven by showing that (6.71), for all $0 \leqslant t \leqslant T \leqslant T^*$, implies (6.70), i.e., that $\widetilde{\mathbb{P}}$ exists. Moreover, if $\sigma_F(t,T) \neq 0$, then the risk-neutral measure $\widetilde{\mathbb{P}}$ is uniquely specified by (6.72). Indeed, by assumption, (6.71) holds if we replace the maturity T by the (dummy) variable s, $0 \leqslant s \leqslant T \leqslant T^*$, i.e.,

$$\alpha_F(t,s) = \sigma_F(t,s)\left[\Sigma_F(t,s) + \theta(t)\right].$$

Integrate both sides of this equation w.r.t. s, from $s = t$ to $s = T > t$, while fixing t:

$$\begin{aligned}
\int_t^T \alpha_F(t,s)\,\mathrm{d}s &= \int_t^T \sigma_F(t,s)\Sigma_F(t,s)\,\mathrm{d}s + \theta(t)\int_t^T \sigma_F(t,s)\,\mathrm{d}s \\
&= \int_t^T \frac{1}{2}\frac{\partial}{\partial s}(\Sigma_F^2(t,s))\,\mathrm{d}s + \theta(t)\int_t^T \frac{\partial}{\partial s}\Sigma_F(t,s)\,\mathrm{d}s \\
&= \frac{1}{2}\left[\Sigma_F^2(t,T) - \Sigma_F^2(t,t)\right] + \theta(t)\left[\Sigma_F(t,T) - \Sigma_F(t,t)\right] \\
&= \frac{1}{2}\Sigma_F^2(t,T) + \theta(t)\Sigma_F(t,T),
\end{aligned}$$

where $\Sigma_F(t,t) \equiv 0$. The integral on the left-hand side is, by definition, $A_F(t,T)$, and hence we recover the relation in (6.70).

Hence, assuming no arbitrage in the HJM model, i.e., assuming (6.71) holds, the SDE in (6.69) is driftless, i.e.,

$$\frac{\mathrm{d}\overline{Z}(t,T)}{\overline{Z}(t,T)} = -\Sigma_F(t,T)\,\mathrm{d}\widetilde{W}(t), \tag{6.73}$$

where discounted zero-coupon bond price processes of all maturities T are $\widetilde{\mathbb{P}}$-martingales. In particular, the ratio $\frac{\overline{Z}(t,T)}{\overline{Z}(0,T)}$ is the stochastic exponential of the process $\{-\Sigma_F(t,T)\}_{0\leqslant t\leqslant T}$ w.r.t. the Brownian motion $\widetilde{W}$ on the time interval $[0,t]$:

$$\begin{aligned}
\overline{Z}(t,T) &= \overline{Z}(0,T)\mathcal{E}_t(-\Sigma_F \cdot \widetilde{W}) \\
&= Z(0,T)\exp\left[-\frac{1}{2}\int_0^t \Sigma_F^2(s,T)\,\mathrm{d}s - \int_0^t \Sigma_F(s,T)\,\mathrm{d}\widetilde{W}(s)\right].
\end{aligned} \tag{6.74}$$

Note that $\overline{Z}(0,T) = Z(0,T)$ since $B(0) = 1$. We recall that the unit bank account has value $B(t) = \exp\left(\int_0^t r(s)\,\mathrm{d}s\right)$. Hence, the risk-neutral value process of a zero-coupon bond takes the form

$$Z(t,T) = Z(0,T)\exp\left[-\frac{1}{2}\int_0^t \Sigma_F^2(s,T)\,\mathrm{d}s - \int_0^t \Sigma_F(s,T)\,\mathrm{d}\widetilde{W}(s) + \int_0^t r(s)\,\mathrm{d}s\right]. \tag{6.75}$$

The SDE for $Z(t,T)$ under $\widetilde{\mathbb{P}}$ has the form

$$\frac{\mathrm{d}Z(t,T)}{Z(t,T)} = r(t)\,\mathrm{d}t - \Sigma_F(t,T)\,\mathrm{d}\widetilde{W}(t)\,. \tag{6.76}$$

We recall from Chapter 3 that a (domestic) nondividend paying asset has (log-)drift equal to $r(t)$ under the risk-neutral measure. Clearly, a zero-coupon bond is an example of a nondividend paying asset.

Finally, upon using (6.71), the SDE in (6.61) for the forward rate process takes the form (w.r.t. the $\widetilde{\mathbb{P}}$-BM):

$$\begin{aligned}\mathrm{d}f(t,T) &= [\alpha_F(t,T) - \theta(t)\sigma_F(t,T)]\,\mathrm{d}t + \sigma_F(t,T)\,\mathrm{d}\widetilde{W}(t)\\ &= \sigma_F(t,T)\Sigma_F(t,T)\,\mathrm{d}t + \sigma_F(t,T)\,\mathrm{d}\widetilde{W}(t)\,.\end{aligned} \tag{6.77}$$

This SDE shows us that the instantaneous (time-t) risk-neutral drift of the forward rate process (for given maturity T) is determined by its instantaneous volatility and its (integrated) volatility across all times up to the maturity T. That is, the forward rate has SDE of the form

$$\mathrm{d}f(t,T) = \widetilde{\alpha}_F(t,T)\,\mathrm{d}t + \sigma_F(t,T)\,\mathrm{d}\widetilde{W}(t)\,, \tag{6.78}$$

with risk-neutral drift $\widetilde{\alpha}_F(t,T) = \sigma_F(t,T)\int_t^T \sigma_F(t,u)\,\mathrm{d}u$. This is necessarily the form for the drift of the forward rate process under the risk-neutral measure $\widetilde{\mathbb{P}}$.

6.3.2 Relationship between HJM and Affine Yield Models

We can formulate any one-factor short-rate model within the HJM framework. Consider an affine term structure model. The short rate process follows the SDE (6.40) under $\widetilde{\mathbb{P}}$ with coefficients given by (6.41):

$$\mathrm{d}r(t) = \big(a_0(t) + a_1(t)r(t)\big)\,\mathrm{d}t + \sqrt{b_0(t) + b_1(t)r(t)}\,\mathrm{d}\widetilde{W}(t)\,.$$

Note that $b(t,r) \equiv \sqrt{b_0(t) + b_1(t)r}$ defines the diffusion function of the short rate process, where $b^2(t,r(t)) = b_0(t) + b_1(t)r(t)$. The bond price is in the affine form (6.39):

$$Z(t,T) = \mathrm{e}^{A(t,T) - r(t)C(t,T)},$$

where the functions C and A solve the ODEs (6.43) and (6.44), respectively. According to (6.6), the forward rates are

$$f(t,T) = -\frac{\partial}{\partial T}\ln Z(t,T) = -\frac{\partial A(t,T)}{\partial T} + r(t)\frac{\partial C(t,T)}{\partial T}.$$

Applying the Itô formula, where $A(t,T)$ and $C(t,T)$ are nonrandom functions of time t, we obtain the stochastic differential of the forward rate in the form

$$\begin{aligned} \mathrm{d}f(t,T) &= \frac{\partial C(t,T)}{\partial T}\,\mathrm{d}r(t) + r(t)\frac{\partial^2 C(t,T)}{\partial t \partial T}\,\mathrm{d}t - \frac{\partial^2 A(t,T)}{\partial t \partial T}\,\mathrm{d}t \\ &= \left[\frac{\partial C(t,T)}{\partial T}\big(a_0(t) + a_1(t)r(t)\big) + r(t)\frac{\partial^2 C(t,T)}{\partial t \partial T} - \frac{\partial^2 A(t,T)}{\partial t \partial T}\right]\mathrm{d}t \\ &\quad + \frac{\partial C(t,T)}{\partial T} b(t, r(t))\,\mathrm{d}\widetilde{W}(t)\,. \end{aligned}$$

Equating the diffusion term in the above SDE with (6.77) gives

$$\sigma_F(t,T) = \frac{\partial C(t,T)}{\partial T} b(t, r(t)) = \frac{\partial C(t,T)}{\partial T}\sqrt{b_0(t) + b_1(t)r(t)}\,. \tag{6.79}$$

Equating the drift term with $\widetilde{\alpha}_F(t,T)$ in (6.77) gives

$$\begin{aligned} &\frac{\partial C(t,T)}{\partial T}\big(a_0(t) + a_1(t)r(t)\big) + r(t)\frac{\partial^2 C(t,T)}{\partial t \partial T} - \frac{\partial^2 A(t,T)}{\partial t \partial T} \\ &= \frac{\partial C(t,T)}{\partial T} b^2(t, r(t)) \int_t^T \frac{\partial C(t,u)}{\partial u}\,\mathrm{d}u \\ &= b^2(t, r(t))\frac{\partial C(t,T)}{\partial T}\big(C(t,T) - C(t,t)\big) \\ &= \big(b_0(t) + b_1(t)r(t)\big)\frac{\partial C(t,T)}{\partial T} C(t,T). \end{aligned} \tag{6.80}$$

This is essentially the no-arbitrage condition that must be satisfied by any affine yield model for all times $t \leqslant T$.

6.3.2.1 The Ho–Lee Model in the HJM Framework

Suppose that the diffusion coefficient $\sigma_F(t,T)$ in the SDE (6.61) for forward rates is constant and equal to σ. The function Σ_F is then given by

$$\Sigma_F(t,T) = \int_t^T \sigma\,\mathrm{d}u = \sigma(T-t)\,.$$

The SDE (6.77) takes the form

$$\mathrm{d}f(t,T) = \sigma^2(T-t)\,\mathrm{d}t + \sigma\,\mathrm{d}\widetilde{W}(t)\,.$$

Integrating the above equation w.r.t. time t gives

$$f(t,T) = f(0,T) + \sigma^2 t(T - t/2) + \sigma\widetilde{W}(t)\,. \tag{6.81}$$

Setting $T = t$ gives the short rate (since $r(t) = f(t,t)$)

$$r(t) = f(0,t) + \sigma^2 t^2/2 + \sigma\widetilde{W}(t)\,.$$

In the above equation, we recognize the Ho–Lee model (6.45). Isolating the Brownian term in (6.81) in terms of the forward rates and substitution into the last equation gives

$$f(t,T) = r(t) + f(0,T) - f(0,t) + \sigma^2 t(T-t)\,. \tag{6.82}$$

As is seen from the above equation, the spot rate $r(t)$ and the forward rate $f(t,T)$ are linearly dependent and hence perfectly correlated.

Using (6.7) and (6.81) for $T=u$, we find the bond price:

$$Z(t,T) = \exp\left[-\int_t^T \left(f(0,u) + \sigma^2 t(u - t/2) + \sigma\widetilde{W}(t)\right) du\right]$$
$$= \exp\left[-\int_t^T f(0,u)\,du - \frac{1}{2}\sigma^2 tT(T-t) - \sigma(T-t)\widetilde{W}(t)\right].$$

From (6.9), we obtain $-\int_t^T f(0,u)\,du = \ln\frac{Z(0,T)}{Z(0,t)}$ and then

$$Z(t,T) = \frac{Z(0,T)}{Z(0,t)} \exp\left[-\frac{1}{2}\sigma^2 tT(T-t) - \sigma(T-t)\widetilde{W}(t)\right]. \tag{6.83}$$

Since $\frac{Z(0,T)}{Z(0,t)} = \mathrm{e}^{-f(0;t,T)(T-t)}$, the yield rate is

$$y(t,T) = -\frac{\ln Z(t,T)}{T-t} = f(0;t,T) + \frac{\sigma^2 tT}{2} + \sigma\widetilde{W}(t).$$

By eliminating the Brownian term using the equation just below (6.81), we express the yield rate in terms of forward rates (where $r(t) = f(t,t)$):

$$y(t,T) = f(t,t) - f(0,t) + f(0;t,T) + \frac{1}{2}\sigma^2 t(T-t). \tag{6.84}$$

6.3.2.2 The Vasiček Model in the HJM Framework

Let us derive the forward rate SDE and verify the no-arbitrage condition in (6.80) for the Vasiček model. The short-rate process (under $\widetilde{\mathbb{P}}$) is driven by the SDE (6.48) which we repeat here:

$$\mathrm{d}r(t) = (\tilde{\alpha} - \beta r(t))\,\mathrm{d}t + \sigma\,\mathrm{d}\widetilde{W}(t).$$

The functions A and C for the bond price are given in (6.52). Using (6.79) and the expression for $C(t,T)$ in (6.52), we obtain the diffusion coefficient of the forward rate:

$$\sigma_F(t,T) = \sigma\frac{\partial C(t,T)}{\partial T} = \sigma\frac{\partial}{\partial T}\left(\frac{1}{\beta}\left(1 - \mathrm{e}^{-\beta(T-t)}\right)\right) = \sigma\mathrm{e}^{-\beta(T-t)}. \tag{6.85}$$

Let us verify that (6.80) holds. Using the expression for $C(t,T)$ and (6.51), we have

$$\frac{\partial^2 C(t,T)}{\partial t\partial T} = \beta\mathrm{e}^{-\beta(T-t)},$$
$$\frac{\partial^2 A(t,T)}{\partial t\partial T} = \left(\tilde{\alpha} - \frac{\sigma^2}{\beta}\right)\mathrm{e}^{-\beta(T-t)} + \frac{\sigma^2}{\beta}\mathrm{e}^{-2\beta(T-t)}.$$

For any given $r(t) = r$, the left-hand side of (6.80) is hence given by

$$\mathrm{e}^{-\beta(T-t)}(\tilde{\alpha} - \beta r) + r\beta\mathrm{e}^{-\beta(T-t)} - \tilde{\alpha}\mathrm{e}^{-\beta(T-t)} + \frac{\sigma^2}{\beta}\mathrm{e}^{-\beta(T-t)} - \frac{\sigma^2}{\beta}\mathrm{e}^{-2\beta(T-t)}$$
$$= \frac{\sigma^2}{\beta}\left(\mathrm{e}^{-\beta(T-t)} - \mathrm{e}^{-2\beta(T-t)}\right), \tag{6.86}$$

and the right-hand side of (6.80) is

$$\frac{\sigma^2}{\beta}\mathrm{e}^{-\beta(T-t)}\left(1-\mathrm{e}^{-\beta(T-t)}\right). \tag{6.87}$$

Hence the expressions in (6.86) and (6.87) are equal, i.e., the Vasiček short-rate model is of the HJM type for which the no-arbitrage condition holds. Using (6.85), the risk-neutral drift of the forward rate is given by

$$\widetilde{\alpha}_F(t,T) \equiv \sigma_F(t,T)\Sigma_F(t,T) = \sigma\mathrm{e}^{-\beta(T-t)}\int_t^T \sigma\mathrm{e}^{-\beta(u-t)}\,\mathrm{d}u = \frac{\sigma^2}{\beta}\mathrm{e}^{-\beta(T-t)}\left(1-\mathrm{e}^{-\beta(T-t)}\right).$$

Thus, for the Vasiček model, the forward rates follow the SDE

$$\mathrm{d}f(t,T) = \frac{\sigma^2}{\beta}\left(\mathrm{e}^{-\beta(T-t)} - \mathrm{e}^{-2\beta(T-t)}\right)\mathrm{d}t + \sigma\mathrm{e}^{-\beta(T-t)}\,\mathrm{d}\widetilde{W}(t) \tag{6.88}$$

under the risk-neutral measure $\widetilde{\mathbb{P}}$.

6.4 Multifactor Affine Term Structure Models

In the previous two sections, we discussed one-factor models where Brownian motion is the only source of randomness. One-factor short-rate models offer good analytical tractability. For many models, a closed-form solution for the bond price can be found. However, one-factor models of interest rates have many drawbacks, including the fact that yield rates for different maturities are perfectly correlated. So the term structure of interest rates for a one-factor model is oversimplified. One of the possible solutions is to consider multifactor interest rate models that involve the short rate along with other random parameters. Consider the following example. Let the short rate follow

$$\mathrm{d}r(t) = \alpha\big(\bar{r} - \beta r(t)\big)\,\mathrm{d}t + \sigma\,\mathrm{d}W_1(t).$$

The *stochastic volatility model* assumes that the volatility σ of the short rate process is stochastic. For example, its square (or variance) is governed by a square-root model,

$$\mathrm{d}\sigma^2(t) = \big(\gamma - \delta\sigma^2(t)\big)\,\mathrm{d}t + \xi\sigma\,\mathrm{d}W_2(t),$$

where W_1 and W_2 are correlated Brownian motions, and α, γ, δ, ξ are constant parameters. So the interest rate volatility $\sigma(t)$ is included as the second random variable. This approach can be extended by including another additional factor: the stochastic mean level of the short rate $\bar{r}$. As a result, one can construct a three-factor model with three state variables: the short rate $r(t)$, the volatility $\sigma(t)$, and the mean level $\bar{r}(t)$.

The multifactor approach provides a greater flexibility in modelling the stochastic term structure of interest rates. However, increasing the number of random factors, reduces the analytical tractability of a model. Valuation of fixed income derivatives often relies on efficient numerical methods. Calibration of a multifactor model can also be a challenge. In this section we discuss a general multifactor affine term structure model and provide several examples of two- and three-factor models. These models are popular due to their good analytical tractability.

Consider a stochastic interest rate model with n state variables $X_1(t), X_2(t), \ldots, X_n(t)$ (or, as an n-by-1 vector, $\mathbf{X}(t) = [X_1(t), X_2(t), \ldots, X_n(t)]^\top$). The model is said to be *affine* if the zero-coupon bond prices admit the following exponential form:

$$Z(t,T) = \exp\left[A(t,T) + \sum_{j=1}^{n} c_j(t,T)X_j(t)\right] = \exp\left[A(t,T) + \mathbf{C}(t,T)^\top \mathbf{X}(t)\right], \qquad (6.89)$$

where $\mathbf{C}(t,T)^\top := [c_1(t,T), c_2(t,T), \ldots, c_n(t,T)]$. The model is time homogeneous if the state variables $X_j(t)$ are all time-homogeneous processes and A and $\mathbf{C}$ are functions of the time to maturity $\tau = T - t$ only. In this case the bond price function takes the form

$$Z(t, t+\tau) = \exp\left[A(\tau) + \mathbf{C}(\tau)^\top \mathbf{X}(t)\right]. \qquad (6.90)$$

Hence, the yield rate is

$$y(t, t+\tau) = -\frac{1}{\tau}\left(A(\tau) + \mathbf{C}(\tau)^\top \mathbf{X}(t)\right).$$

By taking the limit $\tau \searrow 0$, we obtain the short-rate process

$$r(t) = y(t,t) = -\frac{\mathrm{d}A}{\mathrm{d}\tau}(0) - \frac{\mathrm{d}\mathbf{C}^\top}{\mathrm{d}\tau}(0)\mathbf{X}(t).$$

As was demonstrated in Section 6.2, a single-factor time-homogeneous model admits a bond pricing function in the affine form, if the short-rate process follows an SDE with a linear drift and a linear squared diffusion coefficient,

$$\mathrm{d}r(t) = \left(\alpha + \beta r(t)\right)\mathrm{d}t + \sqrt{\lambda + \mu r(t)}\, \mathrm{d}\widetilde{W}(t).$$

Clearly, certain conditions have to be set on the dynamics of the state vector $\mathbf{X}(t)$ in order that $Z(t, t+\tau)$ has the form (6.90). Duffie and Kan proved that the SDE for $\mathbf{X}(t)$ under $\widetilde{\mathbb{P}}$ has to be of the form

$$\mathrm{d}\mathbf{X}(t) = (\boldsymbol{\alpha} + \boldsymbol{B}\mathbf{X}(t))\,\mathrm{d}t + \mathbf{\Sigma}\mathbf{D}(\mathbf{X}(t))\,\mathrm{d}\widetilde{\mathbf{W}}(t), \quad t \geqslant 0, \qquad (6.91)$$

where $\mathbf{D}$ is a diagonal matrix with

$$\sqrt{\lambda_1 + \boldsymbol{\mu}_1^\top \mathbf{X}(t)}, \sqrt{\lambda_2 + \boldsymbol{\mu}_2^\top \mathbf{X}(t)}, \ldots, \sqrt{\lambda_n + \boldsymbol{\mu}_n^\top \mathbf{X}(t)}$$

on the main diagonal, $\boldsymbol{\alpha}$ and $\boldsymbol{\mu}_i$, $i = 1, 2, \ldots, n$, are constant n-dimensional vectors, $\boldsymbol{B} = [\beta_{ij}]_{i,j=1}^n$ and $\mathbf{\Sigma} = [\sigma_{ij}]_{i,j=1}^n$ are constant n-by-n matrices, and $\{\widetilde{\mathbf{W}}(t)\}_{t\geqslant 0}$ is an n-dimensional Brownian motion under $\widetilde{\mathbb{P}}$. Certain conditions on the model parameters are required to ensure that each of the variance processes $\lambda_i + \boldsymbol{\mu}_i^\top \mathbf{X}(t)$, $i = 1, 2, \ldots, n$, remains positive.

6.4.1 Gaussian Multifactor Models

Let us set the coefficient vectors $\boldsymbol{\mu}_1, \boldsymbol{\mu}_2, \ldots, \boldsymbol{\mu}_n$ to be zero. The SDE (6.91) reduces to the Gaussian form

$$\mathrm{d}\mathbf{X}(t) = (\boldsymbol{\alpha} + \boldsymbol{B}\mathbf{X}(t))\,\mathrm{d}t + \mathbf{\Sigma}\,\mathrm{d}\widetilde{\mathbf{W}}(t), \qquad (6.92)$$

where all parameters and matrices are constant. It can be shown that the distribution of $\mathbf{X}(t)$ is multivariate normal.

6.4.2 Equivalent Classes of Affine Models

The number of possible affine models as defined by (6.91) can be quite large. However, by a transformation of variables, a model can be represented in different ways. Dai and Singleton claimed that the models are considered equivalent if they generate identical prices for all contingent claims. Affine models can be classified according to the number of factors and the number of state variables appearing in the volatility matrix $\mathbf{D}$. There are only two equivalent classes of one-factor models: the Vasiček model and the CIR model. When the number of factors is two, we have three equivalent classes, listed below in their canonical forms.

1. The two-factor Vasiček model

$$\begin{aligned} \mathrm{d}X_1(t) &= -\beta_{11}X_1(t)\,\mathrm{d}t + \mathrm{d}\widetilde{W}_1(t), \\ \mathrm{d}X_2(t) &= \big(-\beta_{21}X_1(t) - \beta_{22}X_2(t)\big)\,\mathrm{d}t + \mathrm{d}\widetilde{W}_2(t). \end{aligned} \tag{6.93}$$

2. The two-factor CIR model (for example, the Longstaff–Schwartz model)

$$\begin{aligned} \mathrm{d}X_1(t) &= \big(\mu_1 - \beta_{11}X_1(t) - \beta_{12}X_2(t)\big)\,\mathrm{d}t + \sqrt{X_1(t)}\,\mathrm{d}\widetilde{W}_1(t), \\ \mathrm{d}X_2(t) &= \big(\mu_2 - \beta_{21}X_1(t) - \beta_{22}X_2(t)\big)\,\mathrm{d}t + \sqrt{X_2(t)}\,\mathrm{d}\widetilde{W}_2(t). \end{aligned} \tag{6.94}$$

3. The two-factor stochastic volatility model (for example, the Fong–Vasiček model)

$$\begin{aligned} \mathrm{d}X_1(t) &= (\mu_1 - \beta_{11}X_1(t))\,\mathrm{d}t + \sqrt{X_1(t)}\,\mathrm{d}\widetilde{W}_1(t), \\ \mathrm{d}X_2(t) &= \big(\mu_2 - \beta_{21}X_1(t) - \beta_{22}X_2(t)\big)\,\mathrm{d}t + \big(1 + \delta_{21}\sqrt{X_1(t)}\big)\,\mathrm{d}\widetilde{W}_2(t). \end{aligned} \tag{6.95}$$

In each of the above models, $(\widetilde{W}_1(t), \widetilde{W}_2(t))$ is a standard two-dimensional Brownian motion, under the risk-neutral measure $\widetilde{\mathbb{P}}$ with bank account as numéraire. The short rate is assumed to be an affine (linear) function of the two factors:

$$r(t) = a_0 + a_1X_1(t) + a_2X_2(t)\,. \tag{6.96}$$

The coefficients a_0, a_1, a_2 are typically assumed to be constants, although they can also be chosen as nonrandom functions of time t. These parameters, along with those arising from the SDE for each model, can be used to calibrate the affine model to the spot yield curve.

In the two-factor Vasiček model, the parameters in (6.93) and (6.96) are assumed to take on real values where $\beta_{11} > 0, \beta_{22} > 0$. It is readily shown that the factors $X_1(t), X_2(t)$ are jointly normal random variables and hence the short rate is a normal random variable as long as a_1, a_2 are not both zero. Hence, there is a positive probability that $r(t) < 0$ for any time $t > 0$. In the two-factor CIR model the parameters in (6.94) are chosen such that $\mu_1 \geqslant 0, \mu_2 \geqslant 0, \beta_{11} > 0, \beta_{22} > 0, \beta_{12} \leqslant 0, \beta_{21} \leqslant 0$. Under these conditions it can be shown that the factors are nonnegative processes. That is, if the factors start with nonnegative values $X_1(0) \geqslant 0, X_2(0) \geqslant 0$, then $X_1(t) \geqslant 0, X_2(t) \geqslant 0$ for all time $t \geqslant 0$ (almost surely). Assuming a nonnegative initial short rate $r(0) \geqslant 0$, together with the conditions that $a_0 \geqslant 0, a_1 > 0, a_2 > 0$ in (6.96), guarantees that $r(t) \geqslant 0$ for all time $t \geqslant 0$.

In the interest of space, we do not present the details for pricing bonds under these two-factor affine models. Rather, we summarize the basic steps that are similar to those given for the single-factor affine models and leave the rest of the details as exercises at the end of this chapter. For any of the above two-factor affine models we have prices of zero-coupon bonds that are driven by a two-dimensional system of SDEs for the vector $\mathbf{X}(t) = [X_1(t), X_2(t)]^\top$. Hence, the corresponding bond pricing function can be expressed as a function of the calendar time t and the time-t value of the two factors, i.e., the zero-coupon bond price process $Z(t,T) = V(t, X_1(t), X_2(t))$, where $V = V(t, x_1, x_2)$ is a smooth differentiable function of time t and twice differentiable function of the spot values $X_1(t) = x_1, X_2(t) = x_2$. Recalling our analysis in Chapter 2, we see that the pricing function can be determined by applying the (two-dimensional) discounted Feynman–Kac Theorem 2.24. We can simply use Theorem 2.24 to express $V(t, x_1, x_2)$ as a solution to a PDE for $t < T$, subject to the terminal condition $V(T, x_1, x_2) = Z(T,T) = 1$. For an affine model, the solution necessarily has the form given by (6.89) for $n = 2$, $X_1(t) = x_1, X_2(t) = x_2$:

$$V(t, x_1, x_2) = \mathrm{e}^{A(t,T)+c_1(t,T)x_1+c_2(t,T)x_2}. \tag{6.97}$$

Note that the exponent is linear in the factor variables x_1 and x_2. This is an extension of the form of the solution in (6.39) to include two factors rather than just the single short rate variable r. For a time-homogeneous model we have $V(t, x_1, x_2) = v(\tau, x_1, x_2)$, where the coefficients are functions of $\tau = T - t$: $A(t,T) = A(\tau)$, $c_1(t,T) = c_1(\tau)$, $c_2(t,T) = c_2(\tau)$. Substitution of the above exponential form for V into the corresponding bond-pricing PDE (for the particular model) leads to a system of first order ordinary differential equations in the functions $A(\tau)$, $c_1(\tau)$, and $c_2(\tau)$. This system can then be solved subject to the initial conditions: $A(0) = c_1(0) = c_2(0) = 0$. Given the solutions for these coefficient functions, the bond pricing function is then given by (6.97) (see Exercise 6.13).

6.5 Pricing Derivatives under Forward Measures

6.5.1 Forward Measures

Taking the zero-coupon bond Z rather than the bank account B as a numéraire asset allows us to simplify the derivative pricing formula (6.12). Let $\widehat{\mathbb{P}}_T^{(Z)}$ denote the EMM relative to the numéraire $g(t) = Z(t,T)$. The probability measure $\widehat{\mathbb{P}}_T^{(Z)}$ is called the *T-forward measure.* It is defined so that the process $\{B(t)/Z(t,T)\}_{0\leqslant t\leqslant T}$ is a $\widehat{\mathbb{P}}_T^{(Z)}$-martingale. The Radon–Nikodym derivative of $\widetilde{\mathbb{P}}^{(B)} \equiv \widetilde{\mathbb{P}}$ w.r.t. $\widetilde{\mathbb{P}}^{(g)} \equiv \widehat{\mathbb{P}}_T^{(Z)} \equiv \widehat{\mathbb{P}}$ is

$$\varrho \equiv \frac{\mathrm{d}\widetilde{\mathbb{P}}}{\mathrm{d}\widehat{\mathbb{P}}} = \frac{B(T)/B(0)}{Z(T,T)/Z(0,T)} = Z(0,T)B(T). \tag{6.98}$$

Note that $B(0) = Z(T,T) = 1$. The respective Radon–Nikodym derivative process is

$$\varrho_t \equiv \left(\frac{\mathrm{d}\widetilde{\mathbb{P}}}{\mathrm{d}\widehat{\mathbb{P}}}\right)_t = \frac{B(t)/B(0)}{Z(t,T)/Z(0,T)} = \frac{Z(0,T)B(t)}{Z(t,T)}, \quad 0 \leqslant t \leqslant T, \tag{6.99}$$

where $\varrho_T = \varrho$. We recall, from (4.44) in Chapter 4, that $\varrho_t = \varrho_t^{g\to B}$, where $\varrho_t^{B\to g} = 1/\varrho_t^{g\to B} = 1/\varrho_t$ corresponds to changing numéraires from the bank account to the zero-coupon bond in this case. Let payoff $V(T)$ with maturity time T be $\mathcal{F}_T$-measurable and integrable w.r.t. $\widetilde{\mathbb{P}}$. Applying the change of numéraire theorem, i.e., as follows immediately by the risk-neutral pricing formula (4.49) with $g(t) = Z(t,T)$ as numéraire asset price, the time-t price of the claim is given by the conditional expectation under the T-forward measure $\widehat{\mathbb{P}}$:

$$V(t) = g(t)\,\widehat{\mathrm{E}}_t\left[\frac{V(T)}{g(T)}\right] = Z(t,T)\widehat{\mathrm{E}}_t\left[V(T)\right]. \tag{6.100}$$

The expected value $\widehat{\mathrm{E}}_t[V(T)] = V(t)/Z(t,T)$ is called the forward price at time t of the payoff $V(T)$ with maturity time T. Recall that the forward price makes the time-t value of forward delivery of $V(T)$ at time T zero.

Example 6.1. Compute the expectation of the future short rate $r(T)$, conditional on $\mathcal{F}_t$, under the forward measure $\widehat{\mathbb{P}} \equiv \widehat{\mathbb{P}}_T^{(Z)}$.

Solution. We note that the Radon–Nikodym derivative process of $\widehat{\mathbb{P}} \equiv \widehat{\mathbb{P}}_T^{(Z)}$ w.r.t. $\widetilde{\mathbb{P}}$ is $\left(\frac{\mathrm{d}\widehat{\mathbb{P}}}{\mathrm{d}\widetilde{\mathbb{P}}}\right)_t = \frac{1}{\varrho_t}$. Hence, by (6.99),

$$\frac{\left(\frac{\mathrm{d}\widehat{\mathbb{P}}}{\mathrm{d}\widetilde{\mathbb{P}}}\right)_T}{\left(\frac{\mathrm{d}\widehat{\mathbb{P}}}{\mathrm{d}\widetilde{\mathbb{P}}}\right)_t} = \frac{1/\varrho_T}{1/\varrho_t} = \frac{\varrho_t}{\varrho_T} = \frac{1}{Z(t,T)}\frac{B(t)}{B(T)}.$$

Thus, we have (by the property(2.118) where $r(T)$ is $\mathcal{F}_T$-measurable),

$$\begin{aligned}\widehat{\mathrm{E}}_t[r(T)] &= \widetilde{\mathrm{E}}_t\left[\frac{1}{Z(t,T)}\frac{B(t)}{B(T)}r(T)\right] = \frac{1}{Z(t,T)}\widetilde{\mathrm{E}}_t\left[\exp\left(-\int_t^T r(u)\,\mathrm{d}u\right)r(T)\right]\\ &= \frac{1}{Z(t,T)}\widetilde{\mathrm{E}}_t\left[-\frac{\partial}{\partial T}\exp\left(-\int_t^T r(u)\,\mathrm{d}u\right)\right]\end{aligned}$$

(assume that we can interchange the operations of differentiation and integration)

$$= -\frac{1}{Z(t,T)}\frac{\partial}{\partial T}\left\{\widetilde{\mathrm{E}}_t\left[\exp\left(-\int_t^T r(u)\,\mathrm{d}u\right)\right]\right\} = -\frac{1}{Z(t,T)}\frac{\partial Z(t,T)}{\partial T},$$

where the bond price formula (6.11) is used. Using (6.6), we have

$$-\frac{1}{Z(t,T)}\frac{\partial Z(t,T)}{\partial T} = -\frac{\partial \ln Z(t,T)}{\partial T} = f(t,T).$$

Hence, the instantaneous forward rate $f(t,T)$ is equal to the mathematical expectation of the short rate $r(T) \equiv f(T,T)$ conditional on $\mathcal{F}_t$ under the T-forward measure:

$$\widehat{\mathrm{E}}_t[r(T)] = \widehat{\mathrm{E}}_t[f(T,T)] = f(t,T)$$

Thus, we conclude that the forward rate process $\{f(t,T)\}_{0\leqslant t\leqslant T}$ is a $\widehat{\mathbb{P}}_T^{(Z)}$-martingale. □

Let the dynamics of the T-maturity bond price be governed by the SDE (6.17) which we repeat here:

$$\frac{\mathrm{d}Z(t,T)}{Z(t,T)} = r(t)\,\mathrm{d}t + \sigma_Z(t,T)\,\mathrm{d}\widetilde{W}(t),$$

where $\widetilde{W}$ is a $\widetilde{\mathbb{P}}$-Brownian motion. Under the HJM framework with forward rates following the SDE (6.61), we have the SDE (6.76). Hence, equating the volatility terms gives

$$\sigma_Z(t,T) = -\Sigma_F(t,T) = -\int_t^T \sigma_F(t,u)\,\mathrm{d}u. \tag{6.101}$$

The bond price is given by (6.18). We rewrite it as follows:

$$Z(t,T) = Z(0,T)B(t)\exp\left(-\frac{1}{2}\int_0^t \sigma_Z^2(s,T)\,\mathrm{d}s + \int_0^t \sigma_Z(s,T)\,\mathrm{d}\widetilde{W}(s)\right). \tag{6.102}$$

The measure $\widehat{\mathbb{P}} \equiv \widehat{\mathbb{P}}_T^{(Z)}$ is defined by the Radon–Nikodym derivative $\frac{\mathrm{d}\widehat{\mathbb{P}}}{\mathrm{d}\widetilde{\mathbb{P}}} = \widehat{\varrho}_T$, where

$$\widehat{\varrho}_t := \frac{1}{\varrho_t} = \frac{Z(t,T)}{B(t)Z(0,T)}, \quad 0 \leqslant t \leqslant T$$

with ϱ_t given in (6.99). Using the bond price solution (6.102), we have

$$\widehat{\varrho}_t \equiv \left(\frac{\mathrm{d}\widehat{\mathbb{P}}}{\mathrm{d}\widetilde{\mathbb{P}}}\right)_t = \exp\left(-\frac{1}{2}\int_0^t \sigma_Z^2(s,T)\,\mathrm{d}s + \int_0^t \sigma_Z(s,T)\,\mathrm{d}\widetilde{W}(s)\right).$$

Note that this is precisely the definition of the change of measure as given by the exponential $\widetilde{\mathbb{P}}$-martingale in (4.33) since $\widehat{\varrho}_t = \varrho_t^{B\to g}$ with $g(t) = Z(t,T)$. Since we are assuming that asset prices are driven by a single Brownian component, the volatility vector of the numéraire asset is now simply a scalar, $\sigma^{(g)}(t) \equiv \sigma_Z(t,T)$ for fixed maturity T. By Girsanov's Theorem 2.17, the process

$$\widehat{W}(t) := \widetilde{W}(t) - \int_0^t \sigma_Z(s,T)\,\mathrm{d}s, \quad 0 \leqslant t \leqslant T$$

is a standard $\widehat{\mathbb{P}}$-Brownian motion. Using

$$\mathrm{d}\widetilde{W}(t) = \mathrm{d}\widehat{W}(t) + \sigma_Z(t,T)\,\mathrm{d}t\,, \tag{6.103}$$

we have the following SDE for $Z(t,T)$ under the T-forward measure $\widehat{\mathbb{P}}$:

$$\frac{\mathrm{d}Z(t,T)}{Z(t,T)} = \left(r(t) + \sigma_Z^2(t,T)\right)\mathrm{d}t + \sigma_Z(t,T)\,\mathrm{d}\widehat{W}(t)\,.$$

Within the HJM framework, we hence obtain

$$\begin{aligned}
\mathrm{d}f(t,T) &= \sigma_F(t,T)\Sigma_F(t,T)\,\mathrm{d}t + \sigma_F(t,T)\,\mathrm{d}\widetilde{W}(t) \\
&= \sigma_F(t,T)\Sigma_F(t,T)\,\mathrm{d}t + \sigma_F(t,T)\left(\mathrm{d}\widehat{W}(t) - \Sigma_F(t,T)\,\mathrm{d}t\right) \\
&= \sigma_F(t,T)\,\mathrm{d}\widehat{W}(t)\,.
\end{aligned} \tag{6.104}$$

Thus, the forward rate satisfies the driftless SDE $\mathrm{d}f(t,T) = \sigma_F(t,T)\,\mathrm{d}\widehat{W}(t)$ and is hence a $\widehat{\mathbb{P}}$-martingale. We arrived at the same conclusion in Example 6.1.

6.5.2 Pricing Stock Options under Stochastic Interest Rates

In this section we present a generalized Black–Scholes formula for option prices under an asset price model with stochastic interest rates. Consider a risky asset such as a stock without dividends. Let the price dynamics of a nondividend paying stock $S(t)$ and the bond price $Z(t,T)$ at time t be governed by

$$\frac{\mathrm{d}S(t)}{S(t)} = r(t)\,\mathrm{d}t + \sigma_S(t)\,\mathrm{d}\widetilde{W}(t), \tag{6.105}$$

$$\frac{\mathrm{d}Z(t,T)}{Z(t,T)} = r(t)\,\mathrm{d}t + \sigma_Z(t,T)\,\mathrm{d}\widetilde{W}(t), \tag{6.106}$$

with respective log-volatilities $\sigma_S(t)$ and $\sigma_Z(t,T)$ and where $\widetilde{W}$ is a standard (one-dimensional) $\widetilde{\mathbb{P}}$-Brownian motion. For a given maturity T, we define

$$F_S(t) = \frac{S(t)}{Z(t,T)}, \quad 0 \leqslant t \leqslant T,$$

which is the time-t price of the T-maturity forward of the stock. By the definition of the T-forward measure $\widehat{\mathbb{P}} \equiv \widehat{\mathbb{P}}_T^{(Z)}$, all (domestic) *nondividend paying asset prices divided by the bond price $Z(t,T)$ are $\widehat{\mathbb{P}}$-martingales.* That is, the forward price of the stock is a $\widehat{\mathbb{P}}$-martingale (see (4.43) where $A(t) = S(t), g(t) = Z(t,T)$). This is also easily shown by computing the stochastic differential of $F_S(t)$ expressed in terms of the Brownian increment $\mathrm{d}\widehat{W}(t)$ (see Exercise 6.11):

$$\frac{\mathrm{d}F_S(t)}{F_S(t)} = (\sigma_S(t) - \sigma_Z(t,T))\,\mathrm{d}\widehat{W}(t) = \sigma_{F_S}(t)\,\mathrm{d}\widehat{W}(t)\,, \tag{6.107}$$

where $\sigma_{F_S}(t) \equiv \sigma_S(t) - \sigma_Z(t,T)$ is the log-volatility of $F_S(t)$ and $\widehat{W}$ is a standard $\widehat{\mathbb{P}}$-Brownian motion. For the sake of analytical tractability, we now assume $\sigma_S(t)$ and $\sigma_Z(t,T)$ to be nonrandom (deterministic) continuous functions of time. Using the strong solution to the above SDE, i.e., $F_S(t) = F_S(0)\mathcal{E}_t(\sigma_{F_S}\cdot\widehat{W})$, we have

$$\begin{aligned} F_S(T) &= F_S(t)\exp\left[-\frac{1}{2}\int_t^T(\sigma_S(u)-\sigma_Z(u,T))^2\,\mathrm{d}u + \int_t^T(\sigma_S(u)-\sigma_Z(u,T))\,\mathrm{d}\widehat{W}(u)\right] \\ &\stackrel{d}{=} F_S(t)\exp\left[-\frac{1}{2}\bar{\sigma}_{t,T}^2(T-t) + \bar{\sigma}_{t,T}\sqrt{T-t}\widehat{Z}\right], \end{aligned} \tag{6.108}$$

where $\bar{\sigma}_{t,T}^2 := \frac{1}{T-t}\int_t^T\big(\sigma_S(u)-\sigma_Z(u,T)\big)^2\,\mathrm{d}u$, and $\widehat{Z} \sim Norm(0,1)$ (under measure $\widehat{\mathbb{P}}$) is independent of $F_S(t)$.

According to (6.100), the time-t price of a standard European call on the stock with maturity T and strike K is

$$C(t) = Z(t,T)\widehat{\mathrm{E}}_t\left[(S(T)-K)^+\right].$$

Since $S(T) = S(T)/Z(T,T) = F_S(T)$, we have

$$\begin{aligned} C(t) &= Z(t,T)\,\widehat{\mathrm{E}}_t\left[(F_S(T)-K)^+\right] \\ &= Z(t,T)\,\widehat{\mathrm{E}}\left[\left(F_S(t)\mathrm{e}^{-\frac{1}{2}\bar{\sigma}_{t,T}^2(T-t)+\bar{\sigma}_{t,T}\sqrt{T-t}\widehat{Z}} - K\right)^+ \,\Big|\, F_S(t)\right] \end{aligned} \tag{6.109}$$

where the expectation is conditional on the time-t forward price $F_S(t) = S(t)/Z(t,T)$. Note that, since $F_S(t)$ is independent of $\widehat{Z}$, this expectation is computed by making use of Independence Proposition A.6. In particular, the expectation is the same as that of a standard call with strike K, zero "effective interest rate and dividend on the underlying", spot value $F_S(t)$, volatility $\bar{\sigma}_{t,T}$, and time to maturity $T-t$. Hence, we obtain the well-known *Black formula* for the value of a call option on a stock:

$$\begin{aligned} C(t) &= Z(t,T)\left[F_S(t)\mathcal{N}(d_+(t)) - K\mathcal{N}(d_-(t))\right] \\ &= S(t)\mathcal{N}(d_+(t)) - KZ(t,T)\mathcal{N}(d_-(t)) \end{aligned} \tag{6.110}$$

where

$$d_\pm(t) = \frac{1}{\bar{\sigma}_{t,T}\sqrt{T-t}}\left[\ln\left(\frac{S(t)}{K\,Z(t,T)}\right) \pm \frac{1}{2}\bar{\sigma}_{t,T}^2(T-t)\right].$$

Note that here we have expressed the time-t call price in terms of the time-t stock price and zero-coupon bond price $S(t)$ and $Z(t,T)$. Plugging in the time-t spot values for the stock and the bond gives the pricing function for the call. The above formulation is readily extended to the case with multiple stocks driven by a vector Brownian motion where the stocks are correlated with each other as well as being correlated with the bond price process (see Exercise 6.12).

6.5.3 Pricing Options on Zero-Coupon Bonds

The risk-neutral pricing formulae (6.12) and (6.100) allow for computing no-arbitrage prices of any attainable claim. We assume that an EMM (risk-neutral measure) exists, implying the absence of arbitrage. Consider a European-style claim with maturity T on a zero-coupon bond maturing at time $T' > T$. The payoff function $V(T)$ of such a claim has the form $\Lambda(Z(T,T'))$. The no-arbitrage value $V(t)$, $t \leqslant T$, of the claim is given by

$$V(t) = \widetilde{\mathrm{E}}_t\left[D(t,T)\Lambda\big(Z(T,T')\big)\right] = \widetilde{\mathrm{E}}_t\left[\exp\left(-\int_t^T r(u)\,\mathrm{d}u\right)\Lambda\big(Z(T,T')\big)\right] \tag{6.111}$$

under the risk-neutral measure $\widetilde{\mathbb{P}} \equiv \widetilde{\mathbb{P}}^{(B)}$, or

$$V(t) = Z(t,T)\widehat{\mathrm{E}}_t\left[\Lambda\big(Z(T,T')\big)\right] \tag{6.112}$$

when the equivalent T-forward measure $\widehat{\mathbb{P}} \equiv \widehat{\mathbb{P}}_T^{(Z)}$ is used. The main drawback of the pricing formula (6.111) is that it is necessary to find the joint distribution of $\int_t^T r(u)\,\mathrm{d}u$ and $Z(T,T')$ under $\widetilde{\mathbb{P}}$ to find the expectation. When using (6.112), we only need to find the dynamics of the bond price $Z(t,T')$ under the T-forward measure.

Let $F_Z(t) \equiv F_Z(t;T,T')$ denote the T-forward price of the T'-maturity bond at time t:

$$F_Z(t) = \frac{Z(t,T')}{Z(t,T)}, \quad 0 \leqslant t \leqslant T \leqslant T'.$$

Note that the bond price $Z(t,T')$ is a domestic (nondividend paying) asset. Hence, under the risk-neutral measure $\widetilde{\mathbb{P}}$, its price follows the SDE in (6.106) (now with maturity T' replacing T),

$$\mathrm{d}Z(t,T') = Z(t,T')\big[r(t)\,\mathrm{d}t + \sigma_Z(t,T')\,\mathrm{d}\widetilde{W}(t)\big]\,.$$

In the T-forward measure, with $g(t) = Z(t,T)$ as numéraire asset price, the forward price $F_Z(t)$ is a $\widehat{\mathbb{P}}$-martingale satisfying a driftless SDE,

$$\frac{\mathrm{d}F_Z(t)}{F_Z(t)} = (\sigma_Z(t,T') - \sigma_Z(t,T))\,\mathrm{d}\widehat{W}(t)\,. \tag{6.113}$$

The reader will recognize this as an application of (4.43) in Chapter 4, where $A(t) = Z(t,T'), g(t) = Z(t,T)$. Since the forward price $F_Z(t)$ is a $\widehat{\mathbb{P}}_T^{(Z)}$-martingale, we have $\widehat{\mathrm{E}}_t\left[F_Z(T)\right] = F_Z(t)$. By the above definition we also have $F_Z(T) = \frac{Z(T,T')}{Z(T,T)} = Z(T,T')$, which gives

$$\widehat{\mathrm{E}}_t\left[Z(T,T')\right] = F_Z(t)\,. \tag{6.114}$$

In other words, $F_Z(t)$ is the time-t forward price for delivery of $Z(T,T')$ at time T.

Alternatively, to find the dynamics of $F_Z(t)$, we use the solution (6.18) for deducing a relation between bond prices with maturities T and T':

$$F_Z(t) = \frac{Z(t,T')}{Z(t,T)} = \frac{Z(0,T')}{Z(0,T)}\mathrm{e}^{-\frac{1}{2}\int_0^t\left(\sigma_Z^2(s,T')-\sigma_Z^2(s,T)\right)\mathrm{d}s+\int_0^t\left(\sigma_Z(s,T')-\sigma_Z(s,T)\right)\mathrm{d}\widetilde{W}(s)} \tag{6.115}$$

(now we use (6.103) to change the probability measure)

$$\begin{aligned} &= \frac{Z(0,T')}{Z(0,T)}\mathrm{e}^{-\frac{1}{2}\int_0^t\left(\sigma_Z^2(s,T')-2\sigma_Z(s,T')\sigma_Z(s,T)+\sigma_Z^2(s,T)\right)\mathrm{d}s+\int_0^t\left(\sigma_Z(s,T')-\sigma_Z(s,T)\right)\mathrm{d}\widehat{W}(s)} \\ &= \frac{Z(0,T')}{Z(0,T)}\mathrm{e}^{-\frac{1}{2}\int_0^t\left(\sigma_Z(s,T')-\sigma_Z(s,T)\right)^2\mathrm{d}s+\int_0^t\left(\sigma_Z(s,T')-\sigma_Z(s,T)\right)\mathrm{d}\widehat{W}(s)}. \end{aligned} \tag{6.116}$$

On the right hand side of (6.116) we recognize a stochastic exponential. Hence, (6.113) follows. In particular, we can relate the forward price of the T'-maturity bond at time t to that at time T by simply dividing (6.116) into the same expression for $t = T$:

$$\frac{F_Z(T)}{F_Z(t)} = \frac{Z(T,T')}{F_Z(t)} = \mathrm{e}^{-\frac{1}{2}\int_t^T \left(\sigma_Z(s,T')-\sigma_Z(s,T)\right)^2 \mathrm{d}s + \int_t^T \left(\sigma_Z(s,T')-\sigma_Z(s,T)\right) \mathrm{d}\widehat{W}(s)}. \tag{6.117}$$

The reader will note that this also follows by directly integrating (6.113).

Let us consider pricing standard European call and put options on the bond. For example, the call option with maturity T and strike K on the bond $Z(T,T')$ gives the right to buy the bond at time T for K. For some term structure models, such as the Gaussian HJM model, one is able to derive the pricing formulae for standard European options in closed form.

Suppose that the bond volatility function $\sigma_Z(t,T)$ is a deterministic (nonrandom) function. Under the HJM framework we have the relation (6.101), which implies that the diffusion coefficient of the forward rate in the SDE (6.61) is also a deterministic function of time t. It follows that the drift and diffusion coefficients in both (6.77) and (6.104) are nonrandom functions of t. Hence, the forward rate is a Gaussian process in either measure $\widetilde{\mathbb{P}}$ or $\widehat{\mathbb{P}}$. We see from (6.117) that $\frac{F_Z(T)}{F_Z(t)}$ is a log-normal random variable. In particular, taking logarithms on both sides of (6.117) gives

$$\begin{aligned}\ln Z(T,T') &= \ln F_Z(t) - \frac{1}{2}\int_t^T \left(\sigma_Z(s,T') - \sigma_Z(s,T)\right)^2 \mathrm{d}s + \int_t^T \left(\sigma_Z(s,T') - \sigma_Z(s,T)\right) \mathrm{d}\widehat{W}(s) \\ &\stackrel{d}{=} \ln F_Z(t) - \frac{1}{2}\bar{\sigma}^2(T-t) + \bar{\sigma}\sqrt{T-t}\widehat{Z} \end{aligned} \tag{6.118}$$

where $\widehat{Z} \sim \mathit{Norm}(0,1)$ (under $\widehat{\mathbb{P}}$) and we conveniently define the constant (for given t, T, T')

$$\bar{\sigma}^2 \equiv \bar{\sigma}^2_{t,T,T'} := \frac{1}{T-t}\int_t^T \left(\sigma_Z(s,T') - \sigma_Z(s,T)\right)^2 \mathrm{d}s\,.$$

This corresponds to the time-averaged square difference of the volatilities of the bond price with respective maturities of T' and T. Note that the last line in (6.118) is an equality in distribution where we used the fact that the Itô integral is a normal random variable with zero mean and variance $\bar{\sigma}^2(T-t)$. The reader can readily verify that this follows by Itô isometry since the integrand is a nonrandom square-integrable function of s (for fixed T, T'). Moreover, the above Itô integral is independent of $\mathcal{F}_t$, i.e., $\widehat{Z}$ is independent of $\mathcal{F}_t$ (and, of course, also independent of $F_Z(t)$).

Based on the above properties, we can now readily price a call option with maturity T and strike K, which is written on the underlying bond with value $Z(T,T')$. The price of the call at time t is given by computing the conditional expectation in (6.112). The steps are exactly as those for computing the price of a standard European call. We condition on $\mathcal{F}_t$, where $F_Z(t)$ is $\mathcal{F}_t$-measurable and independent of $\widehat{Z}$. According to (6.118), we substitute $Z(T,T') = F_Z(t)\mathrm{e}^{-\frac{1}{2}\bar{\sigma}^2(T-t)+\bar{\sigma}\sqrt{T-t}\widehat{Z}}$ into the conditional expectation, and we use the independence proposition to reduce the calculation to an unconditional expectation. The latter is then computed by the usual expectation identities in Appendix B (or by simply recognizing the call pricing function at hand in the second line below):

$$\begin{aligned}C(t) &= Z(t,T)\,\widehat{\mathrm{E}}\big[(Z(T,T') - K)^+ \mid \mathcal{F}_t\big]\\&= Z(t,T)\,\widehat{\mathrm{E}}\Big[\Big(F_Z(t)\mathrm{e}^{-\frac{1}{2}\bar{\sigma}^2(T-t)+\bar{\sigma}\sqrt{T-t}\widehat{Z}} - K\Big)^+ \Big| \mathcal{F}_t\Big]\\&= Z(t,T)\,\widehat{\mathrm{E}}\Big[\Big(x\mathrm{e}^{-\frac{1}{2}\bar{\sigma}^2(T-t)+\bar{\sigma}\sqrt{T-t}\widehat{Z}} - K\Big)^+\Big]\Big|_{x=F_Z(t)}\\&= Z(t,T)\,[F_Z(t)\mathcal{N}(d_+(t)) - K\,\mathcal{N}(d_-(t))]\\&= Z(t,T')\mathcal{N}(d_+(t)) - K\,Z(t,T)\mathcal{N}(d_-(t))\,,\end{aligned} \tag{6.119}$$

where

$$d_\pm(t) := \frac{1}{\bar{\sigma}\sqrt{T-t}}\left[\ln\left(\frac{Z(t,T')}{K\,Z(t,T)}\right) \pm \frac{1}{2}\bar{\sigma}^2(T-t)\right].$$

Note: $Z(t,T)F_Z(t) = Z(t,T')$ and $\frac{F_Z(t)}{K} = \frac{Z(t,T')}{KZ(t,T)}$.

Example 6.2. Consider a one-factor model for a non-dividend paying stock $S(t)$ governed by the SDE in (6.105), where the short rate $r(t)$, $t \geqslant 0$ follows the Ho–Lee model with the same BM as in (6.105).

(a) Derive the SDE for the forward price $F_S(t)$, $0 \leqslant t \leqslant T$ and the probability distribution of $F_S(t)$ under $\widehat{\mathbb{P}}$.

(b) Obtain the Black pricing formula for a call on a unit ZCB with maturity $T > 0$.

Solution.

(a) For the Ho–Lee model, the log-volatility of the unit ZCB $Z(t,T)$ is $\sigma_Z(t,T) = -\sigma(T-t)$. Therefore, the SDE in (6.107) takes the following form:

$$\mathrm{d}F_S(t) = F_S(t)\,[\sigma_S(t) + \sigma(T-t)]\,\mathrm{d}\widehat{W}(t)$$

with $\widehat{\mathbb{P}}$-BM $\widehat{W}(t)$. The strong solution is then

$$F_S(T) = F_S(t)\exp\left(-\frac{1}{2}\bar{\sigma}_{t,T}^2(T-t) + \int_t^T (\sigma_S(u) + \sigma(T-u))\,\mathrm{d}\widehat{W}(u)\right),$$

where $\bar{\sigma}_{t,T}^2 := \frac{1}{T-t}\int_t^T (\sigma_S(u) + \sigma(T-u))\,\mathrm{d}u$. Therefore, the log-return on the forward stock price is a normal random variable:

$$\ln\left(\frac{F_S(T)}{F_S(t)}\right) \sim Norm\left(-\frac{1}{2}\bar{\sigma}_{t,T}^2(T-t),\, \bar{\sigma}_{t,T}^2(T-t)\right).$$

(b) Consider a call option on $Z(t,T')$ with maturity T s.t. $T < T'$. The time-t no-arbitrage value $C(t)$ with $0 \leqslant t \leqslant T$ is given by the Black pricing formula in (6.119). To apply it, we only need to find the average log-volatility of the forward price:

$$\bar{\sigma}_{t,T}^2 = \frac{1}{T-t}\int_t^T (-\sigma(T'-u) + \sigma(T-u))^2\,\mathrm{d}u = \sigma^2(T'-T)^2.$$

Thus, we have $C(t) = Z(t,T')\mathcal{N}(d_+(t)) - K\,Z(t,T)\mathcal{N}(d_-(t))$, where

$$d_\pm(t) = \frac{1}{\sigma(T'-T)\sqrt{T-t}}\left[\ln\left(\frac{Z(t,T')}{K\,Z(t,T)}\right) \pm \frac{1}{2}\sigma^2(T'-T)^2(T-t)\right]$$

and the ZCB price is given in (6.46). □

Example 6.3. Using the change of numéraire approach, derive the following general pricing formula for a call on $Z(t,T')$ with strike $K>0$, maturity $T>0$ so that $0\leqslant t\leqslant T<T'$:

$$C(t)=Z(t,T')\,\widehat{\mathbb{P}}^{(T')}(Z(T,T')>K\mid\mathcal{F}_t)-K\,Z(t,T)\,\widehat{\mathbb{P}}^{(T)}(Z(T,T')>K\mid\mathcal{F}_t)$$

where $\widehat{\mathbb{P}}^{(T)}\equiv\widehat{\mathbb{P}}^{(Z)}_T$ and $\widehat{\mathbb{P}}^{(T')}\equiv\widehat{\mathbb{P}}^{(Z)}_{T'}$ denote the T- and T'-forward probability measures, respectively.

Solution. Using the risk-neutral probability measure $\widetilde{\mathbb{P}}$ with the numéraire $B(t)$ in the standard pricing formula gives us

$$\begin{aligned}C(t)&=\widetilde{\mathrm{E}}\left[\frac{B(t)}{B(T)}(Z(T,T')-K)^+\;\Big|\;\mathcal{F}_t\right]\\&=\widetilde{\mathrm{E}}\left[\frac{B(t)}{B(T)}Z(T,T')\mathbb{I}_{\{Z(T,T')\geqslant K\}}\;\Big|\;\mathcal{F}_t\right]-\widetilde{\mathrm{E}}\left[\frac{B(t)}{B(T)}K\,\mathbb{I}_{\{Z(T,T')\geqslant K\}}\;\Big|\;\mathcal{F}_t\right].\end{aligned}$$

Now we change from the risk-neutral probability measure $\widetilde{\mathbb{P}}$ to the forward probabilities measures $\widehat{\mathbb{P}}^{(Z)}_{T'}$ and $\widehat{\mathbb{P}}^{(Z)}_T$ in the first and second expectations, respectively. In doing so, we use the following Radon–Nikodym derivatives:

$$\left(\frac{\mathrm{d}\widetilde{\mathbb{P}}}{\mathrm{d}\widehat{\mathbb{P}}^{(Z)}_{T'}}\right)_t=\frac{B(t)/B(0)}{Z(t,T')/Z(0,T')}\quad\text{and}\quad\left(\frac{\mathrm{d}\widetilde{\mathbb{P}}}{\mathrm{d}\widehat{\mathbb{P}}^{(Z)}_{T}}\right)_t=\frac{B(t)/B(0)}{Z(t,T)/Z(0,T)}\quad\text{for } 0\leqslant t\leqslant T.$$

Thus, we obtain

$$\begin{aligned}C(t)&=\widehat{\mathrm{E}}^{(T')}\left[\frac{B(T)/B(t)}{Z(T,T')/Z(t,T')}\frac{B(t)}{B(T)}Z(T,T')\mathbb{I}_{\{Z(T,T')\geqslant K\}}\;\Big|\;\mathcal{F}_t\right]\\&\quad-\widehat{\mathrm{E}}^{(T)}\left[\frac{B(T)/B(t)}{Z(T,T)/Z(t,T)}\frac{B(t)}{B(T)}K\,\mathbb{I}_{\{Z(T,T')\geqslant K\}}\;\Big|\;\mathcal{F}_t\right]\\&=Z(t,T')\widehat{\mathrm{E}}^{(T')}\left[\mathbb{I}_{\{Z(T,T')\geqslant K\}}\mid\mathcal{F}_t\right]-Z(t,T)\widehat{\mathrm{E}}^{(T)}\left[\mathbb{I}_{\{Z(T,T')\geqslant K\}}\mid\mathcal{F}_t\right]\\&=Z(t,T')\widehat{\mathbb{P}}^{(T')}(Z(T,T')\geqslant K\mid\mathcal{F}_t)-Z(t,T)\widehat{\mathbb{P}}^{(T)}(Z(T,T')\geqslant K\mid\mathcal{F}_t).\end{aligned}$$

□

6.6 LIBOR Model

6.6.1 LIBOR Rates

LIBOR, which stands for London InterBank Offer Rate, refers to the market interest rate. Let $L(t;T,T+\tau)$ denote an annual simple rate of interest that is locked at time t for borrowing from time T to time $T+\tau$, where $0\leqslant t\leqslant T\leqslant T+\tau$. That is, \$1 invested at time T will grow to $1+\tau L(t;T,T+\tau)$ dollars at time $T+\tau$. We call $L(t;T,T+\tau)$ the *forward LIBOR*.

In Section 6.1.2, we discussed how the rate on a loan for a period between times T and $T'=T+\tau$ can be locked in at time t. We purchase one zero-coupon bond maturing at time T and finance this purchase by selling $\frac{Z(t,T)}{Z(t,T+\tau)}$ units of the bond maturing at time $T+\tau$. The cost of setting up this portfolio is zero. The forward LIBOR rate, which is the effective simple rate of interest applied over the interval $[T,T+\tau]$, is calculated as follows:

$$1+\tau L(t;T,T+\tau)=\frac{Z(t,T)}{Z(t,T+\tau)}\implies L(t;T,T+\tau)=\frac{1}{\tau}\left[\frac{Z(t,T)}{Z(t,T+\tau)}-1\right].\qquad(6.120)$$

The interest period τ is often referred as the *tenor*, and it is typically equal to 0.25 for a three-month LIBOR or 0.5 for a six-month LIBOR. If $t = T$, then we call $L(T; T, T + \tau)$ the *spot LIBOR*. It is given by

$$L(T; T, T + \tau) = \frac{1}{\tau}\left[\frac{1}{Z(T, T + \tau)} - 1\right].$$

6.6.2 The Brace–Gatarek–Musiela Model of LIBOR Rates

Pricing interest rate derivatives such as caps and swaps requires a tractable model of floating rates. To adapt the Black–Scholes formula for stock options to the case with fixed-income products, it is desirable to assume that the risk-neutral dynamics of floating rates is log-normal. Suppose that caps and swaps are written on forward rates $f(t, T)$. It can be shown that if the diffusion coefficient in the SDE (6.61) is proportional to the forward rate, i.e., if $\sigma_F(t, T) = \sigma(t, T)f(t, T)$ with $\sigma(t, T)$ assumed to be a nonrandom function, then the forward rates governed by the risk-neutral SDE (6.77) can explode in finite time. Such behaviour of forward rates is caused by the risk-neutral drift term $\widetilde{\alpha}(t, T)$ in (6.77), which takes the form

$$\widetilde{\alpha}(t, T) = \sigma(t, T)f(t, T)\int_t^T \sigma(t, u)f(t, u)\, du\,.$$

To overcome this problem, Brace, Gatarek, and Musiela (BGM) have suggested using the LIBOR forward rates $L(t; T, T+\tau)$, which are simple rates of interest, instead of continuously compounding forward rates.

Let us consider the case with a single maturity $T \leqslant T^*$ and tenor $\tau > 0$. For notational simplicity we write $L(t) \equiv L(t; T, T + \tau)$. Here, we present the theory for LIBOR forward rates driven by a single Brownian factor, although the extension to multiple Brownian factors follows readily. Hence, suppose that the bond price process is governed by the SDE in the form of (6.17). As follows from (6.113), the $(T+\tau)$-forward price of the bond, $\frac{Z(t,T)}{Z(t,T+\tau)}$, is a $\widehat{\mathbb{P}}$-martingale, for all $0 \leqslant t \leqslant T$, where $\widehat{\mathbb{P}} \equiv \widehat{\mathbb{P}}^{(Z)}_{T+\tau}$ is the $(T + \tau)$-forward measure. From (6.120), we have that the LIBOR rate $L(t)$ is a strictly positive $\widehat{\mathbb{P}}$-martingale as well. According to the Brownian martingale representation theorem 2.18, there exists an $\mathbb{F}$-adapted process $\upsilon(t, T)$ such that

$$\mathrm{d}L(t) = \upsilon(t, T)L(t)\, \mathrm{d}\widehat{W}(t), \quad 0 \leqslant t \leqslant T, \tag{6.121}$$

where $\widehat{W}$ is a $\widehat{\mathbb{P}}$-Brownian motion. The process $\upsilon(t, T)$ relates to the zero-coupon volatilities as follows:

$$\upsilon(t, T) = \left(\frac{1 + \tau L(t)}{\tau L(t)}\right)\left[\sigma_Z(t, T) - \sigma_Z(t, T + \tau)\right]. \tag{6.122}$$

The proof of (6.122) is left as an exercise at the end of this chapter. Notice that $\upsilon(t, T) \equiv \sigma_L^{(\tau)}(t, T)$ is the log-volatility of the process $L(t; T, T + \tau)$ for $0 \leqslant t \leqslant T$.

The Brace–Gatarek–Musiela (BGM) model for forward LIBOR rates is constructed such that $\upsilon(t, T)$ is a deterministic (nonrandom) function of time for $0 \leqslant t \leqslant T$. As a result, the forward LIBOR rate $L(T)$ conditional on $L(t)$ has a log-normal distribution under the measure $\widehat{\mathbb{P}}$ with the following conditional mean and variance:

$$\widehat{\mathrm{E}}_t[\ln L(T)] = \ln L(t) - \frac{1}{2}(T - t)\bar{\upsilon}^2_{t,T}, \tag{6.123}$$

$$\widehat{\mathrm{Var}}_t(\ln L(T)) = \bar{\upsilon}^2_{t,T}(T - t) \tag{6.124}$$

with time-averaged variance $\bar{v}_{t,T}^2 := \frac{1}{T-t}\int_t^T v^2(s,T)\,\mathrm{d}s$. These expressions follow readily by simply using the solution to the SDE (6.121) as an exponential $\widehat{\mathbb{P}}$-martingale for $L(t)$ at time t and $L(T)$ at time T,

$$L(T) = L(t)\mathrm{e}^{-\frac{1}{2}\int_t^T v^2(s,T)\,\mathrm{d}s+\int_t^T v(s,T)\,\mathrm{d}\widehat{W}(s)} \stackrel{d}{=} L(t)\mathrm{e}^{-\frac{1}{2}(T-t)\bar{v}_{t,T}^2+\bar{v}_{t,T}\sqrt{T-t}\,\widehat{Z}}. \tag{6.125}$$

The second equality is in distribution where $\int_t^T v(s,T)\,\mathrm{d}\widehat{W}(s) \sim \mathit{Norm}(0, \bar{v}_{t,T}\sqrt{T-t})$ under $\widehat{\mathbb{P}}$, and $\widehat{Z}$ denotes a standard normal random variable under $\widehat{\mathbb{P}}$. The log-normality of the forward LIBOR rate, i.e., the representation in (6.125), allows for the construction of pricing formulae for caps and swaps in closed form. One of the main results is the Black caplet formula presented in the next subsection.

6.6.3 Pricing Caplets, Caps, and Swaps

A cap is defined as a portfolio of caplets. Hence, to find the no-arbitrage price of a cap, it suffices to find the price of a single caplet, Caplet(t), with $0 \leqslant t \leqslant T$. Under the BGM model, a caplet is a European call on the spot LIBOR rate. Its payoff at maturity $T+\tau$ is $(L(T)-\kappa)^+\tau$, where $L(T) \equiv L(T;T,T+\tau)$ and $\kappa > 0$ is a strike rate. By (6.13), the time-t price of the caplet is

$$\begin{aligned}
\text{Caplet}(t) &= \tau\widetilde{\mathrm{E}}_t\left[D(t,T+\tau)(L(T)-\kappa)^+\right] \\
&= \widetilde{\mathrm{E}}_t\left[\frac{B(t)}{B(T+\tau)}\left(\frac{1}{Z(T,T+\tau)}-1-\kappa\tau\right)^+\right].
\end{aligned}$$

From the bond price formula (6.11), we have

$$Z(T,T+\tau) = \widetilde{\mathrm{E}}_T\left[\frac{B(T)}{B(T+\tau)}\right].$$

Using the tower property and $\mathcal{F}_T$-measurability of $Z(T,T+\tau)$ gives

$$\begin{aligned}
\text{Caplet}(t) &= \widetilde{\mathrm{E}}_t\left[\widetilde{\mathrm{E}}_T\left[\frac{B(t)}{B(T+\tau)}\left(\frac{1}{Z(T,T+\tau)}-1-\kappa\tau\right)^+\right]\right] \\
&= \widetilde{\mathrm{E}}_t\left[\frac{B(t)}{B(T)}\left(\frac{1}{Z(T,T+\tau)}-1-\kappa\tau\right)^+\widetilde{\mathrm{E}}_T\left[\frac{B(T)}{B(T+\tau)}\right]\right] \\
&= \widetilde{\mathrm{E}}_t\left[\frac{B(t)}{B(T)}\left(\frac{1}{Z(T,T+\tau)}-1-\kappa\tau\right)^+ Z(T,T+\tau)\right] \\
&= (1+\kappa\tau)\,\widetilde{\mathrm{E}}_t\left[\frac{B(t)}{B(T)}\left(\frac{1}{1+\kappa\tau}-Z(T,T+\tau)\right)^+\right].
\end{aligned}$$

That is, the caplet is equivalent to a put option on the zero-coupon bond $Z(T,T+\tau)$ with strike $\frac{1}{1+\kappa\tau}$ and maturity T.

The price of a caplet is easier to evaluate by using a forward measure. As was demonstrated in the previous subsection, the forward LIBOR rate $L(T)$ conditional on $L(t)$ has a log-normal distribution under the forward measure $\widehat{\mathbb{P}}_{T+\tau}^{(Z)}$. The price of the caplet at time t is

$$\text{Caplet}(t) = \tau Z(t,T+\tau)\widehat{\mathrm{E}}_t\left[(L(T)-\kappa)^+\right], \tag{6.126}$$

where $\widehat{\mathrm{E}}[\,\cdot\,]$ denotes the expectation under $\widehat{\mathbb{P}}^{(Z)}_{T+\tau}$. We leave it to the reader to show (by using similar steps as pricing a standard call with the use of (6.125)) that Black's caplet formula obtains:

$$\text{Caplet}(t) = \tau Z(t, T+\tau)\big[L(t)\,\mathcal{N}(d_+(t)) - \kappa\,\mathcal{N}(d_-(t))\big], \tag{6.127}$$

where

$$d_\pm(t) = \frac{1}{\bar{\upsilon}_{t,T}\sqrt{T-t}}\left[\ln\frac{L(t)}{\kappa} \pm \frac{1}{2}\bar{\upsilon}^2_{t,T}(T-t)\right].$$

A cap is a series of caplets that pays $\tau\big(L(T_{i-1}; T_{i-1}, T_i) - \kappa\big)^+$ at time $T_i = T_0 + i\tau$ for all $i = 1, 2, \ldots, n$. The total value of the cap at time $t \leqslant T_0$ is equal to the sum of all caplet values. Each ith caplet is valued by considering the expectation $\widehat{\mathrm{E}}^{(T_i)}_t[\,\cdot\,]$ conditional on $\mathcal{F}_t$ under the T_i-forward measure $\widehat{\mathbb{P}}^{(Z)}_{T_i}$. Summing each caplet value gives the price of the cap:

$$\begin{aligned}
\text{Cap}(t) &= \sum_{i=1}^{n} \tau Z(t, T_i)\,\widehat{\mathrm{E}}^{(T_i)}_t\left[(L(T_{i-1}; T_{i-1}, T_i) - \kappa)^+\right] \\
&= \tau \sum_{i=1}^{n} Z(t, T_i)\big[L(t; T_{i-1}, T_i)\,\mathcal{N}(d^{(i-1)}_+(t)) - \kappa\,\mathcal{N}(d^{(i-1)}_-(t))\big]
\end{aligned} \tag{6.128}$$

where

$$d^{(i-1)}_\pm(t) = \frac{1}{\bar{\upsilon}_{t,T_{i-1}}\sqrt{T_{i-1}-t}}\left[\ln\frac{L(t; T_{i-1}, T_i)}{\kappa} \pm \frac{1}{2}\bar{\upsilon}^2_{t,T_{i-1}}(T_{i-1}-t)\right],$$

and $\upsilon^2_{t,T_{i-1}} = \frac{1}{T_{i-1}-t}\int_t^{T_{i-1}} \upsilon^2(s, T_{i-1})\,\mathrm{d}s$.

The price of a payer swap at time t is also easy to evaluate by using forward measures. The holder of the swap receives $\tau\big(L(T_{i-1}; T_{i-1}, T_i) - \kappa\big)$ at time $T_i = T_0 + i\tau$ for all $i = 1, 2, \ldots, n$. The no-arbitrage price at time t is now readily computed by choosing the T_i-forward measure $\widehat{\mathbb{P}}^{(Z)}_{T_i}$ for each ith payoff term:

$$\text{Swap}(t) = \sum_{i=1}^{n} \tau Z(t, T_i)\,\widehat{\mathrm{E}}^{(T_i)}_t\big[L(T_{i-1}; T_{i-1}, T_i) - \kappa\big]$$

(using the fact that the LIBOR rate $L(t; T_{i-1}, T_i)$ is a $\widehat{\mathbb{P}}^{(Z)}_{T_i}$-martingale)

$$\begin{aligned}
&= \tau \sum_{i=1}^{n} Z(t, T_i)\big[L(t; T_{i-1}, T_i) - \kappa\big] = \sum_{i=1}^{n} \tau Z(t, T_i)\left[\frac{Z(t, T_{i-1}) - Z(t, T_i)}{\tau Z(t, T_i)} - \kappa\right] \\
&= \sum_{i=1}^{n}\big[Z(t, T_{i-1}) - Z(t, T_i)\big] - \tau\kappa \sum_{i=1}^{n} Z(t, T_i) &(6.129) \\
&= \big[Z(t, T_0) - Z(t, T_n)\big] - \tau\kappa \sum_{i=1}^{n} Z(t, T_i)\,. &(6.130)
\end{aligned}$$

In obtaining (6.129) we used the definition in (6.120), i.e., $L(t; T_{i-1}, T_i) = \frac{Z(t,T_{i-1}) - Z(t,T_i)}{\tau Z(t,T_i)}$.

The forward swap rate that makes the swap contract have a no-arbitrage price of zero at time t is hence

$$\kappa(t; T_0, T_n) = \frac{Z(t, T_0) - Z(t, T_n)}{\tau \sum_{i=1}^{n} Z(t, T_i)}\,. \tag{6.131}$$

Substituting the expression $Z(t,T_0) - Z(t,T_n)$, in terms of this forward swap rate, into (6.130) gives

$$\text{Swap}(t) = \tau\big[\kappa(t;T_0,T_n) - \kappa\big]\sum_{i=1}^{n} Z(t,T_i)\,. \tag{6.132}$$

6.7 Exercises

Exercise 6.1. Suppose that the (continuously compounded) spot rates for the next three years are

T	1	2	3
$y(0,T)$	3%	3.25%	3.5%

Find the forward rates $f(0;1,2)$, $f(0;1,3)$, and $f(0;2,3)$.

Exercise 6.2. Consider a derivative which has the payoff $(y(T,T+1) - y(T,T+5))^+$. Why might it be inappropriate to use a one-factor interest rate model to value this derivative?

Exercise 6.3. Consider the Hull–White model (6.29). Show that the short rate $r(t)$ is a Gaussian process. Find the mean and variance of $r(T)$ conditional on $r(t)$ for $0 \leqslant t \leqslant T$.

Exercise 6.4. The short-rate prices under the Pearson–Sun (PS) model follows the SDE

$$\mathrm{d}r(t) = \big(\alpha - \beta r(t)\big)\,\mathrm{d}t + \sigma\sqrt{r(t) - \lambda}\,\mathrm{d}W(t).$$

Make use of the CIR bond pricing formula to derive bond prices for the PS model.

Exercise 6.5. Find the forward rates $f(t,T)$ in the Vasiček model.

Exercise 6.6. Consider an extended CIR model where the short rate $r(t)$ follows the SDE

$$\mathrm{d}r(t) = \big(\alpha(t) - \beta(t)r(t)\big)\,\mathrm{d}t + \sigma(t)\sqrt{r(t)}\,\mathrm{d}\widetilde{W}(t)$$

under the risk-neutral measure $\widetilde{\mathbb{P}}$. Here, $\alpha(t)$, $\beta(t)$, and $\sigma(t) > 0$ are continuous ordinary (nonrandom) functions. Show that the price of a zero-coupon bond is

$$Z(t,T) = \exp\left(-\int_t^T C(s,T)\alpha(s)\,\mathrm{d}s - C(t,T)r(t)\right), \quad t \leqslant T,$$

where $C(t,T)$ solves the first order differential equation

$$C'(t,T) = \beta(t)C(t,T) + \frac{\sigma^2(t)}{2}C^2(t,T) - 1$$

for $t < T$, subject to $C(T,T) = 0$.

Exercise 6.7. Recall that under $\widetilde{\mathbb{P}}$ the ZCB price satisfies the SDE

$$\mathrm{d}Z(t,T) = Z(t,T)\big[r(t)\,\mathrm{d}t + \sigma_Z(t,T)\,\mathrm{d}\widetilde{W}(t)\big]\,.$$

For an affine-term model with $Z(t,T,r(t)) = \mathrm{e}^{A(t,T) - C(t,T)\,r(t)}$, the coefficient σ_Z in the above SDE is given by $\sigma_Z(t,T) = -C(t,T)b(t,r(t))$ where $b(t,r(t))$ is the diffusion coefficient of the SDE for $r(t)$.

(a) Consider the yield rate process $y(t,T) = -\dfrac{\ln Z(t,T)}{T-t}$ with $0 \leqslant t \leqslant T$. Use Itô's formula to show that

$$\mathrm{d}y(t,T) = \frac{1}{T-t}\big[(y(t,T) + \frac{1}{2}\sigma_Z^2(t,T) - r(t))\,\mathrm{d}t - \sigma_Z(t,T)\,\mathrm{d}\widetilde{W}(t)\big].$$

(b) Find the SDE for $Z(t,T)$ under the Vasiček model.

(c) Using the results of (a) and (b), derive the SDE for $y(t,T)$ under the Vasiček model.

Exercise 6.8. Consider the Vasiček model for the short rate process (under the risk-neutral probability measure $\widetilde{\mathbb{P}}$): $\mathrm{d}r(t) = (\tilde{\alpha} - \beta r(t))\,\mathrm{d}t + \sigma\,\mathrm{d}\widetilde{W}(t)$. Additionally, consider a stock with the price $S(t)$ following the SDE

$$\frac{\mathrm{d}S(t)}{S(t)} = r(t)\,\mathrm{d}t + \sigma_S\,\mathrm{d}\widetilde{W}(t),$$

with constant stock volatility $\sigma_S > 0$.

(a) What SDE does the T-forward price process $\{F_S(t) = S(t)/Z(t,T)\}_{0\leqslant t\leqslant T}$ follow under the equivalent T-forward probability measure $\widehat{\mathbb{P}}_T^{(Z)}$?

(b) What SDE does the stock price process $S(t)$ with $0 \leqslant t \leqslant T$ follow under $\widehat{\mathbb{P}}_T^{(Z)}$?

(c) Derive a formula for the time-t no-arbitrage value of a put option with maturity T and strike K, which is written on the stock S. Particularly, find an explicit expression for the effective squared volatility $\bar{\sigma}^2$ in terms of model parameters.

Exercise 6.9. Consider the model for the short rate $r(t), t \geqslant 0$, with SDE

$$dr(t) = (\tilde{\alpha} - \beta r(t))dt + e^{\beta t}d\widetilde{W}(t),$$

with nonzero constants $\tilde{\alpha}, \beta$ and $\widetilde{W}(t), t \geqslant 0$, as a standard BM in the risk-neutral measure $\widetilde{\mathbb{P}} \equiv \widetilde{\mathbb{P}}^{(B)}$.

(a) Determine $\widetilde{\mathrm{E}}[r(T)|r(t) = r]$ and $\widetilde{\mathrm{Var}}\big(r(T)|r(t) = r\big)$, for all $T > t$, within the measure $\widetilde{\mathbb{P}}$.

(b) Derive an explicit formula for the zero-coupon bond price $Z(t,T,r)$ and the corresponding yield $y(t,T,r)$, for given spot rate $r(t) = r$.

(c) Determine explicitly the SDE (w.r.t. standard BM $\widetilde{W}(t)$) satisfied by the instantaneous forward rate process $f(t,T)$, $t \leqslant T$.

Exercise 6.10. A two-factor HJM model is given by the SDE

$$\mathrm{d}f(t,T) = \alpha(t,T)\,\mathrm{d}t + \sigma_1(t,T)\,\mathrm{d}W_1(t) + \sigma_2(t,T)\,\mathrm{d}W_2(t),$$

where W_1 and W_2 are independent Brownian motions.

(a) Find the SDE for the discounted zero-coupon bond price process $\overline{Z}(t,T) = D(t)Z(t,T)$ and show that

$$\overline{Z}(t,T) = \overline{Z}(0,T)\exp\left(-\int_0^t A(s,T)\,\mathrm{d}s - \int_0^t \Sigma_1(s,T)\,\mathrm{d}W_1(s) - \int_0^t \Sigma_2(s,T)\,\mathrm{d}W_2(s)\right),$$

where $\Sigma_i(t,T) = \int_t^T \sigma_i(t,u)\,\mathrm{d}u$ for $i = 1,2$ and $A(t,T) = \int_t^T \alpha(t,u)\,\mathrm{d}u$.

(b) Show that the no-arbitrage condition is given by

$$\widetilde{\alpha}(t,T) = \sigma_1(t,T)\int_t^T \sigma_1(t,s)\,\mathrm{d}s + \sigma_2(t,T)\int_t^T \sigma_2(t,s)\,\mathrm{d}s\,.$$

(c) Provide the multidimensional extension to the formulas in part (a) and (b) for the more general case of a d-factor HJM model for any $d \geqslant 1$.

Exercise 6.11. Let the dynamics of the stock price $S(t)$ and the bond price $Z(t,T)$ be respectively governed by (6.105) and (6.106). Show that the dynamics of the forward price $F(t) = \frac{S(t)}{Z(t,T)}$ under the T-forward measure $\widehat{\mathbb{P}} \equiv \widehat{\mathbb{P}}_T^{(Z)}$ is governed by (6.107).

Exercise 6.12. Consider the multidimensional extension of (6.105)–(6.106) where each stock price process $\{S_i(t)\}_{0\leqslant t\leqslant T}$, $i = 1,\ldots,n$, and the zero-coupon bond price $Z(t,T)$ are all driven by a vector Brownian motion $\mathbf{W}(t) = (W_1(t),\ldots,W_d(t))$. Assume the existence of a risk-neutral measure $\widetilde{\mathbb{P}} \equiv \widetilde{\mathbb{P}}^{(B)}$, i.e.,

$$\frac{\mathrm{d}S_i(t)}{S_i(t)} = r(t)\,\mathrm{d}t + \boldsymbol{\sigma}_i(t)\cdot \mathrm{d}\widetilde{\mathbf{W}}(t)\,,$$

$$\frac{\mathrm{d}Z(t,T)}{Z(t,T)} = r(t)\,\mathrm{d}t + \boldsymbol{\sigma}_Z(t)\cdot \mathrm{d}\widetilde{\mathbf{W}}(t)\,,$$

where $\boldsymbol{\sigma}_i(t)$ and $\boldsymbol{\sigma}_Z(t)$ are the respective log-volatility vectors of each stock and bond price. The forward price of each stock is defined by $F_i(t) = \frac{S_i(t)}{Z(t,T)}$.

(a) Obtain the analogue to the SDE in (6.107) for each forward price $F_i(t)$, $i = 1,\ldots,n$.

(b) Assume a given stock $S_1(t) \equiv S(t)$ has constant log-volatility vector $\boldsymbol{\sigma}_S(t) = \boldsymbol{\sigma}_S$ with magnitude $||\boldsymbol{\sigma}_S|| = \sigma_S$ and assume a constant vector $\boldsymbol{\sigma}_Z(t) = \boldsymbol{\sigma}_Z$ with magnitude $||\boldsymbol{\sigma}_Z|| = \sigma_Z$, where $\boldsymbol{\sigma}_S \cdot \boldsymbol{\sigma}_Z = \rho\,\sigma_S\sigma_Z$, $\rho \in (-1,1)$. By employing the measure $\widehat{\mathbb{P}} \equiv \widehat{\mathbb{P}}_T^{(Z)}$, derive a pricing formula for a European call with payoff $(S(T) - K)^+$.

(c) By employing the measure $\widehat{\mathbb{P}}$, obtain a pricing formula for an exchange option with payoff $(S_2(T) - S_1(T))^+$, assuming constant log-volatility vectors with $||\boldsymbol{\sigma}_i|| = \sigma_i$, $\boldsymbol{\sigma}_1 \cdot \boldsymbol{\sigma}_2 = \rho\sigma_1\sigma_2$, $||\boldsymbol{\sigma}_Z|| = \sigma_Z$, and $\boldsymbol{\sigma}_Z \cdot \boldsymbol{\sigma}_i = \rho_i\sigma_i\sigma_Z$, $\rho \in (-1,1)$, $\rho_i \in (-1,1)$, $i = 1,2$.

Exercise 6.13. Consider the two-factor Vasiček model defined by the two-dimensional system of SDEs in (6.93).

(a) Provide the time-homogeneous PDE for the pricing function $V(t,x_1,x_2)$ of a zero-coupon bond under this model by directly applying Theorem 2.24 to the conditional expectation

$$V(t,x_1,x_2) = \widetilde{\mathrm{E}}_{t,x_1,x_2}\left[\mathrm{e}^{-\int_t^T r(u,\mathbf{X}(u))\,du}\right].$$

Note that the short rate is given by (6.96). Namely, as a process we have $r(t) = r(t,\mathbf{X}(t)) = a_0 + a_1X_1(t) + a_2X_2(t)$ with spot value $r(t,x_1,x_2) = a_0 + a_1x_1 + a_2x_2$.

(b) Let $V(t,x_1,x_2) = v(\tau,x_1,x_2)$, $\tau = T - t$, be the solution to the PDE in part (a) where $V(t,x_1,x_2)$ has the affine form in (6.97), i.e., let

$$v(\tau,x_1,x_2) = \mathrm{e}^{A(\tau)+c_1(\tau)x_1+c_2(\tau)x_2}$$

for $0 \leqslant \tau \leqslant T$. By substituting this form into the PDE of part (a), show that the coefficient functions must satisfy the first order system of ordinary differential equations:

$$\begin{aligned}
A'(\tau) &= \frac{1}{2}\left(c_1^2(\tau) + c_2^2(\tau)\right) - a_0\,,\\
c_1'(\tau) &= -\beta_{11}c_1(\tau) - \beta_{21}c_2(\tau) - a_1\,,\\
c_2'(\tau) &= -\beta_{22}c_2(\tau) - a_2\,,
\end{aligned}$$

subject to the initial conditions $A(0) = c_1(0) = c_2(0) = 0$.

(c) Solve the system of equations in part (b) and hence obtain the pricing function for the zero-coupon bond.

Exercise 6.14. Consider the two-factor CIR model defined by the two-dimensional system of SDEs in (6.94). Find the corresponding first order system of ODEs on $A(\tau)$, $c_1(\tau)$, and $c_2(\tau)$.

Exercise 6.15. Prove that the forward LIBOR rate volatility $v(t,T)$ and the T- and $(T+\tau)$-maturity bond volatilities $\sigma_Z(t,T)$ and $\sigma_Z(t,T+\tau)$ are related by

$$v(t,T) = \left(\frac{1+\tau L(t;T,T+\tau)}{\tau L(t;T,T+\tau)}\right)\left[\sigma_Z(t,T) - \sigma_Z(t,T+\tau)\right].$$

Exercise 6.16. A *floorlet* is similar to a caplet except that the floating rate is bounded from below. The effective payoff of a floorlet at time T is $\tau\left(\kappa - L(T;T,T+\tau)\right)^+$. Derive the Black pricing formula for a floorlet. Is there a relationship between a floorlet and a caplet?

Exercise 6.17. *Caps* and *floors* are collections of caplets and floorlets, respectively, applied to periods $[T_j, T_j+\tau]$ with $j = 1, 2, \ldots, n$. Show that a model-independent relationship *cap* = *floor* + *swap* exists.

Exercise 6.18. Under the BGM model, find the time-t no-arbitrage value of:

(a) a floorlet with payoff $(\kappa - L(T))^+\tau$ paid at time $(T+\tau)$, where $L(t) = L(t;T,T+\tau)$;

(b) a floor contract with the tenor structure $\mathbf{T} = [T_1, T_2, \ldots, T_n]$ where $0 \leqslant t \leqslant T_0 < T_1 < \ldots < T_n$ and $T_i = T_{i-1} + \tau$.

Exercise 6.19. What SDE does $X(t) := \tau L(t;T,T+\tau) + 1$ follow under the EMM $\widehat{\mathbb{P}}^{(Z)}_{T+\tau}$? Use it to find the no-arbitrage prices of a caplet and a floorlet.

7

Alternative Models of Asset Price Dynamics

In the study of stochastic processes and their applications to finance, geometric Brownian motion (GBM) is the simplest model used for continuous-time asset pricing. The volatility in the GBM model is constant, i.e., the diffusion coefficient is a linear function of the underlying asset price, as is the drift coefficient. For a long time, the GBM model has stood as one of the few known analytically tractable, continuous-time stochastic models for asset pricing. Such models admit transition probability density functions and pricing formulae for various standard, barrier, and lookback European-style options in closed form. However, despite its simplicity, it is commonly recognized that the GBM model only partially captures the complexity of financial markets. The log-normal assumption of asset price returns disagrees with most empirical evidence. For example, let us consider a series of options and then plot the value of implied volatility against strike and time to expiry. A notable defect of the GBM model is that, by construction, the implied volatility surface is supposed to be completely flat. In contrast, the market observed implied volatility surfaces of stock index options exhibit pronounced smile and skew patterns. This phenomenon is commonly called "the volatility smile." These and other important market observations have spawned the development of more realistic pricing models based on alternative stochastic processes. This chapter demonstrates how the standard Black–Scholes model can be extended to make it consistent with the volatility smile. In the mathematical finance literature, more sophisticated models have been proposed for asset prices, interest rates, and other financial variables. Examples of alternative models include processes with jumps, stochastic volatility models, and local (i.e., state-dependent) volatility diffusions. This chapter focuses on several alternative models of asset price dynamics, namely, the local volatility model, the constant elasticity of variance (CEV) diffusion model, the Heston stochastic volatility model, jump diffusion processes, and the variance gamma model. We start with a bit of a technical section on characteristic functions, which are helpful in pricing standard European options under non-standard asset price models.

7.1 Characteristic Functions

7.1.1 Definition and Properties

The characteristic function (ChF) of an absolutely continuous random variable is defined as the Fourier transform of the probability density function (PDF). Characteristic functions have many valuable properties. For example, by differentiating a ChF, one can compute the moments of the random variable. Many probability distributions with the PDF unknown in closed-form can be analyzed through their ChF.

The method of characteristic functions has proved to be a handy tool in option pricing. Particularly, in the no-arbitrage valuation of standard European call and put options, one can use the Gil–Pelaez formula for calculating cumulative probabilities for asset prices at

DOI: 10.1201/9780429468889-7

maturity or employ the Carr–Madan method. The latter is based on the Fourier transform of the option pricing function w.r.t. the log strike price. Both ways rely on the inverse Fourier transform of some expressions involving the characteristic function of the log asset price.

The *Fourier transform* of a function $f\colon \mathbb{R} \to \mathbb{R}$, denoted $\hat{f}$, is defined as

$$\hat{f}(u) := \int_{-\infty}^{\infty} \mathrm{e}^{iux} f(x)\,\mathrm{d}x = \int_{-\infty}^{\infty} (\cos(ux) + i\sin(ux)) f(x)\,\mathrm{d}x\,, \quad \text{for } u \in \mathbb{R}. \tag{7.1}$$

Here, we assume that f is an integrable function such that

$$\int_{-\infty}^{\infty} |f(x)|\,\mathrm{d}x < \infty\,.$$

Having the Fourier transform $\hat{f}$ available, we can recover the original function f via the *inverse Fourier transform* as follows:

$$f(x) = \frac{1}{2\pi}\int_{-\infty}^{\infty} \mathrm{e}^{-iux} \hat{f}(u)\,\mathrm{d}u = \frac{1}{2\pi}\int_{-\infty}^{\infty} (\cos(ux) - i\sin(ux)) \hat{f}(u)\,\mathrm{d}u\,. \tag{7.2}$$

Let f_X be the PDF of some real random variable X. The Fourier transform of f_X, denoted φ_X, is given by

$$\varphi_X(u) = \mathrm{E}[\mathrm{e}^{iuX}] = \int_{-\infty}^{\infty} \mathrm{e}^{iux} f_X(x)\,\mathrm{d}x\,. \tag{7.3}$$

The function φ_X is called the *characteristic function* (ChF) of X. It is easy to show that a ChF of random variable X is defined for all $u \in \mathbb{R}$ thanks to the Jensen inequality:

$$|\varphi_X(u)| = \left|\mathrm{E}\left[\mathrm{e}^{iuX}\right]\right| \leqslant \mathrm{E}\left[\left|\mathrm{e}^{iuX}\right|\right] = \mathrm{E}[1] = 1\,.$$

The moment generating function (MGF) M_X of X, defined by $M_X(u) := \mathrm{E}[\mathrm{e}^{uX}]$, can be recovered from the ChF φ_X, and vice versa the ChF can be obtained from the MGF:

$$\begin{aligned} M_X(u) &= \mathrm{E}[\mathrm{e}^{uX}] = \mathrm{E}[\mathrm{e}^{i(-iu)X}] = \varphi_X(-iu)\,, \\ \varphi_X(u) &= \mathrm{E}[\mathrm{e}^{iuX}] = M_X(iu)\,. \end{aligned}$$

An important property is a one-to-one correspondence between characteristic functions and probability distributions. One can show that random variables X and Y are identically distributed, denoted $X \overset{d}{=} Y$, iff $\varphi_X(u) = \varphi_Y(u)$ for all $u \in \mathbb{R}$. In other words, we can identify the probability distribution of a random variable by analyzing its characteristic function. Some other useful properties of characteristic functions are listed below.

1. Let $X_1, X_2, \ldots, X_n$ be independent random variables with characteristic functions $\varphi_1, \varphi_2, \ldots, \varphi_n$, respectively. Then the characteristic function of the sum

$$S_n := X_1 + X_2 + \cdots + X_n$$

is given by the product of characteristic functions φ_k, $k = 1, 2, \ldots, n$:

$$\varphi_{S_n}(u) = \mathrm{E}[\mathrm{e}^{iuS_n}] = \mathrm{E}\left[\mathrm{e}^{iu(X_1+X_2+\cdots+X_n)}\right] = \mathrm{E}\left[\prod_{k=1}^{n} \mathrm{e}^{iuX_k}\right] = \prod_{k=1}^{n} \varphi_k(u)\,.$$

Additionally, if X_k, $k = 1, 2, \ldots, n$ are identically distributed (i.e., we deal with i.i.d. random variables), then $\varphi_k(u) = \varphi_1(u)$ for all k and all u, and thus $\varphi_{S_n}(u) = [\varphi_1(u)]^n$ holds.

2. Let $\{X_k\}_{k\geqslant 1}$ be a sequence of independent random variables, and φ_k is the characteristic function of X_k for $k \geqslant 1$. Suppose that N is another positive-integer-valued random variable that is independent of the X sequence. For example, N has a Poisson probability distribution. Consider the random sum $S := X_1 + X_2 + \cdots + X_N$. Its characteristic function can be calculated by conditioning on N:

$$\varphi_S(u) = \mathrm{E}\left[\mathrm{E}\left[\mathrm{e}^{iu(X_1+X_2+\cdots+X_N)} \mid N\right]\right] = \mathrm{E}\left[\prod_{k=1}^{N} \varphi_k(u)\right].$$

For the i.i.d. case, we have

$$\begin{aligned}\varphi_S(u) &= \mathrm{E}\left[(\varphi_1(u))^N\right] = \mathrm{E}\left[\mathrm{e}^{N \ln \varphi_1(u)}\right] \\ &= \varphi_N(-i \ln \varphi_1(u)) = M_N(\ln \varphi_1(u)),\end{aligned}$$

where φ_N and M_N are, respectively, the ChF and the MGF of the random variable N.

3. The complex conjugate of ChF φ_X is given by

$$\overline{\varphi_X(u)} = \overline{\mathrm{E}[\mathrm{e}^{iuX}]} = \mathrm{E}[\overline{\mathrm{e}^{iuX}}] = \mathrm{E}[\mathrm{e}^{-iuX}] = \varphi_X(-u).$$

Therefore, if $X \overset{d}{=} -X$, i.e., X has a symmetric probability distribution, then its ChF φ_X is a real-valued function:

$$\varphi_X(u) = \varphi_{-X}(u) = \mathrm{E}[\mathrm{e}^{-iuX}] = \varphi_X(-u) = \overline{\varphi_X(u)}.$$

The converse is also true. That is, the ChF φ_X of random variable X is real-valued iff $X \overset{d}{=} -X$. For example, as shown below, the ChF of a standard normal random variable is $\varphi(u) = \mathrm{e}^{-u^2/2}$, which is a real-valued function.

4. Suppose the nth derivative of ChF φ_X exists. Switching the order of integration and differentiation gives

$$\varphi_X^{(n)}(u) = \mathrm{E}\left[\frac{\mathrm{d}^n}{\mathrm{d}u^n}\left(\mathrm{e}^{iuX}\right)\right] = \mathrm{E}\left[(iX)^n \mathrm{e}^{iuX}\right] \implies \varphi_X^{(n)}(0) = \mathrm{E}\left[(iX)^n\right] = i^n \mathrm{E}[X^n].$$

Therefore, $\mathrm{E}[X^n] = i^{-n}\varphi_X^{(n)}(0)$. An important special case is $\mathrm{E}[X] = -i\varphi'_X(0)$. The moment generating property of characteristic functions can also be represented as a series:

$$\varphi_X(u) = 1 + \sum_{k=1}^{n} \mathrm{E}[X^k] \cdot \frac{(iu)^k}{k!} + o(|u|^n).$$

In particular, setting $n = 2$ in the above formula gives

$$\varphi_X(u) = 1 + iu\mathrm{E}[X] - \frac{u^2}{2}\mathrm{E}[X^2] + o(u^2).$$

5. Let X be a random variable with ChF φ_X. Then, the ChF of the random variable $aX+b$ with constant a and b is

$$\varphi_{aX+b}(u) = \mathrm{E}\left[\mathrm{e}^{i(aX+b)u}\right] = \mathrm{e}^{ibu}\mathrm{E}\left[\mathrm{e}^{iauX}\right] = \mathrm{e}^{ibu}\varphi_X(au).$$

Thus, the ChF of a demeaned random variable $Y = X - \mathrm{E}[X]$ is

$$\varphi_Y(u) = \mathrm{e}^{-iu\mathrm{E}[X]}\varphi_X(u) = \mathrm{e}^{-\varphi'_X(0)u}\varphi_X(u).$$

The PDF f of a probability distribution can be recovered from the ChF φ by using the inverse Fourier transform as given in (7.2). The formula takes the following form:

$$f(x) = \frac{1}{2\pi}\int_{-\infty}^{\infty} \mathrm{e}^{-iux}\varphi(u)\,\mathrm{d}u\,. \tag{7.4}$$

Using the symmetry property $\varphi(-u) = \overline{\varphi(u)}$, the formula for the real part of a complex number $\Re z := \frac{z+\bar{z}}{2}$, as well as properties of the complex conjugate

$$\overline{z_1 \cdot z_2} = \overline{z_1}\cdot\overline{z_2} \quad \text{and} \quad \overline{\mathrm{e}^{i\theta}} = \mathrm{e}^{-i\theta},$$

we can simplify (7.4) and express the PDF f as an integral of a real-valued function:

$$\begin{aligned} f(x) &= \frac{1}{2\pi}\left[\int_{-\infty}^{0} \mathrm{e}^{-iux}\varphi(u)\,\mathrm{d}u + \int_{0}^{\infty} \mathrm{e}^{-iux}\varphi(u)\,\mathrm{d}u\right] = \frac{1}{2\pi}\int_0^{\infty}\left(\mathrm{e}^{iux}\varphi(-u) + \mathrm{e}^{-iux}\varphi(u)\right)\mathrm{d}u \\ &= \frac{1}{2\pi}\int_0^{\infty}\left(\overline{\mathrm{e}^{-iux}\varphi(u)} + \mathrm{e}^{-iux}\varphi(u)\right)\mathrm{d}u = \frac{1}{2\pi}\int_0^{\infty} 2\Re\left(\mathrm{e}^{-iux}\varphi(u)\right)\mathrm{d}u \\ &= \frac{1}{\pi}\int_0^{\infty}\Re\left(\mathrm{e}^{-iux}\varphi(u)\right)\mathrm{d}u\,. \end{aligned} \tag{7.5}$$

Example 7.1. Find the characteristic function for

(a) the normal distribution $\mathit{Norm}(\mu, \sigma^2)$;

(b) the Poisson distribution $\mathit{Pois}(\lambda)$ with $\lambda > 0$;

(c) the drifted and scaled Brownian motion $\{X(t) = \mu t + \sigma W(t)\}_{t\geqslant 0}$;

(d) the Poisson process $\{N(t)\}_{t\geqslant 0}$ with intensity $\lambda > 0$.

Solution.

(a) The ChF of a standard normal random variable Z is

$$\begin{aligned} \varphi_Z(u) = \mathrm{E}\left[\mathrm{e}^{iuZ}\right] &= \int_{-\infty}^{\infty}\frac{1}{\sqrt{2\pi}}\mathrm{e}^{iuz-z^2/2}\,\mathrm{d}z \\ &= \int_{-\infty}^{\infty}\frac{1}{\sqrt{2\pi}}\mathrm{e}^{-(z-iu)^2/2}\mathrm{e}^{-u^2/2}\,\mathrm{d}z = \mathrm{e}^{-u^2/2}. \end{aligned}$$

Thus, for $X \sim \mathit{Norm}(\mu, \sigma^2)$, which can be written as $X = \mu + \sigma Z$ with $Z \sim \mathit{Norm}(0, 1)$, we have

$$\varphi_X(u) = \varphi_{\mu+\sigma Z}(u) = \mathrm{e}^{i\mu u}\varphi_Z(\sigma u) = \mathrm{e}^{i\mu u - \sigma^2 u^2/2}.$$

(b) Let $N \sim \mathit{Pois}(\lambda)$. Then

$$\varphi_N(u) = \sum_{k=0}^{\infty}\mathrm{e}^{iuk}\frac{\lambda^k}{k!}\mathrm{e}^{-\lambda} = \sum_{k=0}^{\infty}\frac{(\lambda\mathrm{e}^{iu})^k}{k!}\mathrm{e}^{-\lambda} = \mathrm{e}^{-\lambda}\mathrm{e}^{\lambda\mathrm{e}^{iu}} = \mathrm{e}^{\lambda(\mathrm{e}^{iu}-1)}.$$

(c) Since $X(t) \sim \mathit{Norm}(\mu t, \sigma^2 t)$, the ChF of $X(t)$ is $\varphi_{X(t)}(u) = \mathrm{e}^{i\mu t u - \sigma^2 u^2 t/2}$. As we can see, the ChF depends on the model parameters, μ and σ, and the time variable t. We can also observe that $\ln\varphi_{X(t)}(u) = t\Psi(u)$ for some function $\Psi(u)$ that does not depend on t.

(d) Note that the Poisson process is discussed in Section 7.4. For the solution, we only need the probability distribution of the process. The time-t value $N(t)$ of the Poisson process is a $\mathit{Pois}(\lambda t)$-distributed random variable. Thus, from part (b), we obtain

$$\varphi_{N(t)}(u) = \mathrm{e}^{\lambda t(\mathrm{e}^{iu}-1)}.$$

□

7.1.2 Recovering the Distribution Function

If the ChF φ of a continuous probability distribution is known, the PDF f can be recovered by the inverse Fourier transform as given in (7.5). The cumulative distribution function (CDF) F can be obtained by integrating the density function. So, if the PDF can only be found using the Fourier inversion, then it seems that we need to calculate a double integral to recover the CDF:

$$F(x) = \int_{-\infty}^{x} f(y)\,\mathrm{d}y = \frac{1}{2\pi}\int_{-\infty}^{x}\int_{-\infty}^{\infty} \mathrm{e}^{-iuy}\varphi(u)\,\mathrm{d}u\,\mathrm{d}y\,.$$

However, it is possible to avoid the double integration and retrieve the CDF F directly from the ChF φ using a single integration as shown below.

First, let us find the Fourier transform of $\mathrm{e}^{-\alpha x}F(x)$, where $\mathrm{e}^{-\alpha x}$ with $\alpha > 0$ is a dumping factor, which ensures that the "dumped" CDF converges to zero, as $x \to \infty$. We have

$$\begin{aligned}
\int_{-\infty}^{\infty} \mathrm{e}^{iux}\mathrm{e}^{-\alpha x}F(x)\,\mathrm{d}x &= \int_{-\infty}^{\infty} \mathrm{e}^{-(\alpha-iu)x}F(x)\,\mathrm{d}x = -\frac{1}{\alpha-iu}\int_{-\infty}^{\infty} F(x)\,\mathrm{d}\left(\mathrm{e}^{-(\alpha-iu)x}\right) \\
&= -\frac{1}{\alpha-iu}F(x)\mathrm{e}^{-(\alpha-iu)x}\Big|_{x=-\infty}^{x=\infty} + \frac{1}{\alpha-iu}\int_{-\infty}^{\infty}\mathrm{e}^{-(\alpha-iu)x}\,\mathrm{d}F(x) \\
&= \frac{1}{\alpha-iu}\int_{-\infty}^{\infty}\mathrm{e}^{i(u+i\alpha)x}f(x)\,\mathrm{d}x \\
&= \frac{1}{\alpha-iu}\varphi(u+i\alpha)\,.
\end{aligned}$$

Using the Fourier inversion gives

$$\begin{aligned}
\mathrm{e}^{-\alpha x}F(x) &= \frac{1}{2\pi}\int_{-\infty}^{\infty}\mathrm{e}^{-iux}\frac{1}{\alpha-iu}\varphi(u+i\alpha)\,\mathrm{d}u \implies \\
F(x) &= \frac{1}{2\pi}\mathrm{e}^{\alpha x}\int_{-\infty}^{\infty}\mathrm{e}^{-iux}\frac{1}{\alpha-iu}\varphi(u+i\alpha)\,\mathrm{d}u\,.
\end{aligned}$$

Alternatively, the Gil–Pelaez formula can also be used to compute the CDF:

$$F(x) = \frac{1}{2} + \frac{1}{2\pi}\int_{0}^{\infty}\frac{\mathrm{e}^{iux}\varphi(-u) - \mathrm{e}^{-iux}\varphi(u)}{iu}\,\mathrm{d}u\,. \tag{7.6}$$

Using the properties of the complex conjugate, we have

$$F(x) = \frac{1}{2} - \frac{1}{\pi}\int_{0}^{\infty}\frac{\Im\left(\mathrm{e}^{-iux}\varphi(u)\right)}{u}\,\mathrm{d}u\,. \tag{7.7}$$

The complementary CDF is calculated as follows:

$$F^c(x) = 1 - F(x) = \frac{1}{2} + \frac{1}{\pi}\int_{0}^{\infty}\frac{\Im\left(\mathrm{e}^{-iux}\varphi(u)\right)}{u}\,\mathrm{d}u\,. \tag{7.8}$$

Here, $\Im z := \frac{z-\bar{z}}{2}$ denotes the imaginary part of a complex number z.

7.1.3 Pricing Standard European Options

Here, we consider two approaches to pricing standard European put and call options. The first approach is based on the fundamental pricing formulae derived in earlier chapters. We only need to compute the distribution function of stock prices at maturity (or their log values) to implement it. The second approach is based on the Carr–Madan solution, which involves taking a Fourier transform of the option pricing function w.r.t. the log strike price $\kappa := \ln K$.

Recall that the initial no-arbitrage prices of call and put options can be written using the distribution function of the stock price $S(T)$ at maturity as follows:

$$\begin{aligned} C(0, S_0; T, K) &= \mathrm{e}^{-qT} S_0 \widehat{\mathbb{P}}(S(T) \geqslant K \mid S(0) = S_0) - \mathrm{e}^{-rT} K \widetilde{\mathbb{P}}(S(T) \geqslant K \mid S(0) = S_0) \\ &= \mathrm{e}^{-qT} S_0 \big(1 - \widehat{F}_{S(T)}(K)\big) - \mathrm{e}^{-rT} K \big(1 - \widetilde{F}_{S(T)}(K)\big), \qquad (7.9) \\ P(0, S_0; T, K) &= \mathrm{e}^{-rT} K \widetilde{\mathbb{P}}(S(T) \leqslant K \mid S(0) = S_0) - \mathrm{e}^{-qT} S_0 \widehat{\mathbb{P}}(S(T) \leqslant K \mid S(0) = S_0) \\ &= \mathrm{e}^{-rT} K \widetilde{F}_{S(T)}(K) - \mathrm{e}^{-qT} S_0 \widehat{F}_{S(T)}(K), \qquad (7.10) \end{aligned}$$

where $\widetilde{\mathbb{P}}$ and $\widehat{\mathbb{P}}$ denote the equivalent martingale measures (EMMs) with the bank account $B(t) = \mathrm{e}^{rt}$ and the stock $\widehat{S}(t) \equiv \mathrm{e}^{qt} S(t)$ as numéraires, respectively. The functions $\widetilde{F}_{S(T)}$ and $\widehat{F}_{S(T)}$ denote the respective CDFs of $S(T)$ conditional on $S(0) = S_0$. As usual, r is the risk-free rate of interest compounded continuously, q is the continuous dividend yield, T is the maturity time, K is the strike price, and S_0 is the initial asset value.

As shown in the previous subsection, a CDF can be recovered from the characteristic function. It is more convenient to work with the log prices $X(t) := \ln S(t)$, $t \in [0, T]$ and the log strike $\kappa := \ln K$. The equations (7.9) and (7.10) can be rewritten as follows:

$$C(0, S_0; T, K) = \mathrm{e}^{-qT} S_0 \big(1 - \widehat{F}_{X(T)}(\kappa)\big) - \mathrm{e}^{-rT} K \big(1 - \widetilde{F}_{X(T)}(\kappa)\big), \qquad (7.11)$$

$$P(0, S_0; T, K) = \mathrm{e}^{-rT} K \widetilde{F}_{X(T)}(\kappa) - \mathrm{e}^{-qT} S_0 \widehat{F}_{X(T)}(\kappa), \qquad (7.12)$$

where $\widetilde{F}_{X(T)}$ and $\widehat{F}_{X(T)}$ denote the CDFs of $X(T)$ conditional on $X(0) = \ln S_0$ under $\widetilde{\mathbb{P}}$ and $\widehat{\mathbb{P}}$, respectively.

At the first glance, as follows from (7.11) and (7.12), we need to find both characteristic functions $\widetilde{\varphi}$ and $\widehat{\varphi}$ of the log price $X(T)$, under $\widetilde{\mathbb{P}}$ and $\widehat{\mathbb{P}}$, respectively. To do so, we need to know the probability distributions of $X(T)$ under each EMMs. However, as demonstrated below, the function $\widehat{\varphi}$ can be derived from $\widetilde{\varphi}$ using the change of numéraire method. Indeed, the Radon–Nikodym derivative $\varrho_T \equiv \left(\frac{\mathrm{d}\widehat{\mathbb{P}}}{\mathrm{d}\widetilde{\mathbb{P}}}\right)_T$ is given by $\varrho_T = \dfrac{\widehat{S}(T)/\widehat{S}(0)}{B(T)/B(0)}$. Using the fact that $S(T) = \mathrm{e}^{X(T)}$ and the martingale property

$$\widetilde{\mathrm{E}}[\widehat{S}(T)/B(T)] = \widehat{S}(0)/B(0) \implies B(T)/B(0) = \widetilde{\mathrm{E}}[\widehat{S}(T)/\widehat{S}(0)],$$

we have

$$\varrho_T = \frac{\widehat{S}(T)/\widehat{S}(0)}{\widetilde{\mathrm{E}}[\widehat{S}(T)/\widehat{S}(0)]} = \frac{S(T)/S(0)}{\widetilde{\mathrm{E}}[S(T)/S(0)]} = \frac{S(T)}{\widetilde{\mathrm{E}}[S(T)]} = \frac{\mathrm{e}^{X(T)}}{\widetilde{\mathrm{E}}[\mathrm{e}^{X(T)}]}. \qquad (7.13)$$

Thus, the characteristic function $\widehat{\varphi}$ of $X(T)$ conditional on $X(0) = \ln S_0$ under the EMM $\widehat{\mathbb{P}}$ is expressed in terms of the ChF $\widetilde{\varphi}$ as follows:

$$\begin{aligned} \widehat{\varphi}(u) &= \widehat{\mathrm{E}}\left[\mathrm{e}^{iuX(T)}\right] = \widetilde{\mathrm{E}}\left[\varrho_T\, \mathrm{e}^{iuX(T)}\right] = \widetilde{\mathrm{E}}\left[\frac{\mathrm{e}^{X(T)}}{\widetilde{\mathrm{E}}[\mathrm{e}^{X(T)}]}\, \mathrm{e}^{iuX(T)}\right] \\ &= \frac{\widetilde{\mathrm{E}}\left[\mathrm{e}^{X(T)+iuX(T)}\right]}{\widetilde{\mathrm{E}}[\mathrm{e}^{X(T)}]} = \frac{\widetilde{\mathrm{E}}\left[\mathrm{e}^{i(u-i)X(T)}\right]}{\widetilde{\mathrm{E}}[\mathrm{e}^{i(-i)X(T)}]} = \frac{\widetilde{\varphi}(u-i)}{\widetilde{\varphi}(-i)}. \qquad (7.14) \end{aligned}$$

Here, $\widetilde{\mathrm{E}}[\;]$ and $\widehat{\mathrm{E}}[\;]$ denote mathematical expectations under the respective EMMs.

Combining equations (7.8), (7.11), and (7.14), we obtain the final pricing formula for a standard call option (and a similar formula can be easily derived for a European put option):

$$C(0, S_0; T, K) = S_0 \mathrm{e}^{-qT} \left(\frac{1}{2} + \frac{1}{\pi} \int_0^\infty \Im \left[\frac{\mathrm{e}^{-iu\kappa} \widetilde{\varphi}(u-i)}{u\widetilde{\varphi}(-i)} \right] du \right) - K\mathrm{e}^{-rT} \left(\frac{1}{2} + \frac{1}{\pi} \int_0^\infty \Im \left[\frac{\mathrm{e}^{-iu\kappa} \widetilde{\varphi}(u)}{u} \right] du \right), \tag{7.15}$$

where $\kappa = \ln K$ is the log strike price. Clearly, the integrals in (7.15) can be combined together to speed up numerical computations.

Example 7.2. Find the characteristic functions $\widetilde{\varphi}$ and $\widehat{\varphi}$ for the log-normal asset price model with the risk-neutral dynamics given by

$$S(t) = S_0 \, \mathrm{e}^{(r-q-\sigma^2/2)t + \sigma \widetilde{W}(t)}, \quad t \geqslant 0 \, .$$

Solution. The log price $X(T) = \ln S_0 + (r-q-\sigma^2/2)T + \sigma\widetilde{W}(T)$ has the normal distribution $Norm(\ln S_0 + (r-q-\sigma^2/2)T, \sigma^2 T)$. Thus, according to the result of Example 7.1, the ChF $\widetilde{\varphi}(u) := \widetilde{\mathrm{E}}[\mathrm{e}^{iuX(T)}]$ is

$$\widetilde{\varphi}(u) = \mathrm{e}^{i(\ln S_0 + (r-q-\sigma^2/2)T)u - \sigma^2 u^2 T/2}.$$

According to (7.14), the ChF $\widehat{\varphi}(u) := \widehat{\mathrm{E}}[\mathrm{e}^{iuX(T)}]$ is then

$$\begin{aligned} \widehat{\varphi}(u) &= \frac{\mathrm{e}^{i(\ln S_0 + (r-q-\sigma^2/2)T)(u-i) - \sigma^2 (u-i)^2 T/2}}{\mathrm{e}^{i(\ln S_0 + (r-q-\sigma^2/2)T)(-i) - \sigma^2 (-i)^2 T/2}} \\ &= \mathrm{e}^{i(\ln S_0 + (r-q-\sigma^2/2)T)u - \sigma^2 [(u-i)^2 + 1] T/2} \\ &= \mathrm{e}^{i(\ln S_0 + (r-q+\sigma^2/2)T)u - \sigma^2 u^2 T/2}. \end{aligned}$$

Note that the same result can be derived by using the strong solution for $S(t)$ under $\widehat{\mathbb{P}}$:

$$S(t) = S_0 \, \mathrm{e}^{(r-q+\sigma^2/2)t + \sigma \widehat{W}(t)}, \quad t \geqslant 0 \, .$$

Indeed, we have

$$\begin{aligned} \widehat{\varphi}(u) &= \widehat{\mathrm{E}}[\mathrm{e}^{iuX(T)}] = \widehat{\mathrm{E}}[\mathrm{e}^{iu(\ln S_0 + (r-q+\sigma^2/2)T + \sigma \widehat{W}(T))}] \\ &= \mathrm{e}^{i(\ln S_0 + (r-q+\sigma^2/2)T)u - \sigma^2 u^2 T/2}. \end{aligned}$$

□

Example 7.3. Consider the following asset price model (under $\widetilde{\mathbb{P}}$):

$$S(t) = S_0 \, \mathrm{e}^{(r-q)t - \omega t + X(t)}, \quad t \geqslant 0,$$

with risk-free rate r, dividend yield q, constant ω, and some random process $\{X(t)\}_{t\geqslant 0}$ s.t. $X_0 = 0$ and $\mathrm{E}[\mathrm{e}^{X(t)}] < \infty$ for all $t \geqslant 0$. Let $\varphi_{X(t)}(u)$ denote the characteristic function of $X(t)$ under $\widetilde{\mathbb{P}}$. Assume that $\ln \varphi_{X(t)}(u) = t\Psi(u)$ holds, where Ψ is some function that does not depend on t. Find the value of ω such that $\widetilde{\mathrm{E}}[S(t) \mid S(0) = S_0] = \mathrm{e}^{(r-q)t} S_0$.

Solution. The characteristic function of $\ln S(t)$ is

$$\varphi_{\ln S(t)}(u) = \widetilde{\mathrm{E}}[\mathrm{e}^{iu \ln S(t)}] = \mathrm{e}^{iu[\ln S_0 + (r-q)t - \omega t]} \widetilde{\mathrm{E}}[\mathrm{e}^{iuX(t)}] = \mathrm{e}^{iu[\ln S_0 + (r-q)t - \omega t]} \varphi_{X(t)}(u).$$

Substituting $-i$ for u yields

$$\widetilde{\mathrm{E}}[S(t)] = \widetilde{\mathrm{E}}[\mathrm{e}^{i(-i)\ln S(t)}] = \mathrm{e}^{\ln S_0 + (r-q)t - \omega t}\varphi_{X(t)}(-i) = S_0 \mathrm{e}^{(r-q)t}\mathrm{e}^{-\omega t}\varphi_{X(t)}(-i).$$

So, $\widetilde{\mathrm{E}}[S(t)] = \mathrm{e}^{(r-q)t}S_0$ holds iff $\mathrm{e}^{-\omega t}\varphi_{X(t)}(-i) = 1$, which gives

$$\omega = \frac{1}{t}\ln\varphi_{X(t)}(-i) = \Psi(-i). \tag{7.16}$$

For example, if $S(t)$ is GBM, then $X(t) = \sigma W(t)$ and $\ln\varphi_{X(t)}(u) = -\sigma^2 u^2 t/2$. Thus, $\omega = \sigma^2/2$. □

Remark. It is not difficult to show that if the process $X(t)$ with ω given in (7.16) has stationary and independent increments then the process $\mathrm{e}^{-(r-q)t}S(t)$, $t \geqslant 0$ is a $\widetilde{\mathbb{P}}$-martingale. For example, $X(t)$ is a jump diffusion process, which we study in Section 7.4 of this chapter.

7.1.4 The Carr–Madan Method for Pricing Vanilla Options

Consider a standard European call whose initial no-arbitrage price is given as an integral of the discounted payoff multiplied by the risk-neutral density of the log price:

$$C(0, S_0; T, K) = \mathrm{e}^{-rT}\int_{\kappa}^{\infty}(\mathrm{e}^x - \mathrm{e}^{\kappa})\,\tilde{f}(x)\,\mathrm{d}x,$$

where $\kappa = \ln K$ is the log strike price, and $\tilde{f}(x)$ is the risk-neutral PDF of $X(T) := \ln S(T)$ conditional on $X(0) = \ln S_0$. Let, as before, $\widetilde{\varphi}$ denote the ChF of $X(T)$ under $\widetilde{\mathbb{P}}$.

The no-arbitrage price $C(0, S_0; T, K)$ can be viewed as a function of the log strike $\kappa \in (-\infty, \infty)$. Thus, we can take the Fourier transform of the pricing function w.r.t. κ. As demonstrated below, the Fourier transform of C is written in terms of the characteristic function $\widetilde{\varphi}$. Thus, the no-arbitrage price can be recovered by a single integration. The pricing function needs to be absolutely integrable in κ over the entire line. However, $C(0, S_0; T, \mathrm{e}^{\kappa}) \to \mathrm{e}^{-qT}S_0$ as $\kappa \to -\infty$, and hence the integral of $C(0, S_0; T, \mathrm{e}^{\kappa})$ over $\kappa \in (-\infty, \infty)$ diverges. To resolve this issue, we introduce a multiplicative dumping factor $\mathrm{e}^{\alpha\kappa}$ with some $\alpha > 0$.

Let us define $c(\kappa) := \mathrm{e}^{\alpha\kappa}C(0, S_0; T, \mathrm{e}^{\kappa})$. The Fourier transform of $c(\kappa)$ is

$$\psi(v) = \int_{-\infty}^{\infty}\mathrm{e}^{iv\kappa}c(\kappa)\,\mathrm{d}\kappa\,.$$

Firstly, we derive an analytic expression for ψ in terms the ChF $\widetilde{\varphi}$ and then obtain the call price numerically using the inverse transform:

$$C(0, S_0; T, K) = \frac{\mathrm{e}^{-\alpha\kappa}}{2\pi}\int_{-\infty}^{\infty}\mathrm{e}^{-iv\kappa}\psi(v)\,\mathrm{d}v\,.$$

Since $C(0, S_0; T, K)$ is real, the function $\mathrm{e}^{-iv\kappa}\psi(v)$ is odd in its imaginary part and even in its real part. Thus, we can calculate the option price by integrating the real part of $\mathrm{e}^{-iv\kappa}\psi(v)$:

$$C(0, S_0; T, K) = \frac{\mathrm{e}^{-\alpha\kappa}}{\pi}\int_0^{\infty}\Re\left[\mathrm{e}^{-iv\kappa}\psi(v)\right]\,\mathrm{d}v\,.$$

Now, let us derive $\psi(v)$:

$$
\begin{aligned}
\psi(v) &= \int_{-\infty}^{\infty} \mathrm{e}^{iv\kappa}\mathrm{e}^{\alpha\kappa} C(0, S_0; T, \mathrm{e}^{\kappa})\,\mathrm{d}\kappa \\
&= \int_{-\infty}^{\infty} \mathrm{e}^{iv\kappa}\mathrm{e}^{\alpha\kappa}\mathrm{e}^{-rT} \int_{\kappa}^{\infty} (\mathrm{e}^{x} - \mathrm{e}^{\kappa})\,\tilde{f}(x)\,\mathrm{d}x\,\mathrm{d}\kappa \\
&= \int_{-\infty}^{\infty} \mathrm{e}^{-rT}\tilde{f}(x) \int_{-\infty}^{x} \left(\mathrm{e}^{iv\kappa+\alpha\kappa+x} - \mathrm{e}^{iv\kappa+\alpha\kappa+\kappa}\right)\mathrm{d}\kappa\,\mathrm{d}x \\
&= \int_{-\infty}^{\infty} \mathrm{e}^{-rT}\tilde{f}(x) \left(\frac{\mathrm{e}^{(iv+\alpha+1)x}}{iv+\alpha} - \frac{\mathrm{e}^{(iv+\alpha+1)x}}{iv+\alpha+1}\right)\mathrm{d}x \\
&= \frac{\mathrm{e}^{-rT}}{iv+\alpha}\int_{-\infty}^{\infty} \mathrm{e}^{(iv+\alpha+1)x}\tilde{f}(x)\,\mathrm{d}x - \frac{\mathrm{e}^{-rT}}{iv+\alpha+1}\int_{-\infty}^{\infty} \mathrm{e}^{(iv+\alpha+1)x}\tilde{f}(x)\,\mathrm{d}x \\
&= \frac{\mathrm{e}^{-rT}}{iv+\alpha}\int_{-\infty}^{\infty} \mathrm{e}^{i(v-i(\alpha+1))x}\tilde{f}(x)\,\mathrm{d}x - \frac{\mathrm{e}^{-rT}}{iv+\alpha+1}\int_{-\infty}^{\infty} \mathrm{e}^{i(v-i(\alpha+1))x}\tilde{f}(x)\,\mathrm{d}x \\
&= \frac{\mathrm{e}^{-rT}}{iv+\alpha}\widetilde{\varphi}(v-i(\alpha+1)) - \frac{\mathrm{e}^{-rT}}{iv+\alpha+1}\widetilde{\varphi}(v-i(\alpha+1)) \\
&= \frac{\mathrm{e}^{-rT}\widetilde{\varphi}(v-i(\alpha+1))}{(\alpha+iv)(\alpha+1+iv)}.
\end{aligned}
$$

The no-arbitrage value of a put option, $P(0, S_0; T, K)$, can be found in the same way. Let us introduce the function $p(\kappa) := \mathrm{e}^{\alpha\kappa}P(0, S_0; T, \mathrm{e}^{\kappa})$ with a dumping factor $\alpha < 0$. The dumping component forces the convergence as $\kappa \to \infty$. First, we can find the Fourier transform of $p(\kappa)$. Following the same steps as for the case with a call option, we derive:

$$
\psi(v) = \int_{-\infty}^{\infty} \mathrm{e}^{iv\kappa} p(\kappa)\,\mathrm{d}\kappa = \frac{\mathrm{e}^{-rT}\widetilde{\varphi}(v-i(\alpha+1))}{(\alpha+iv)(\alpha+1+iv)}.
$$

Second, we can calculate the option price:

$$
P(0, S_0; T, K) = \frac{\mathrm{e}^{-\alpha\kappa}}{\pi}\int_{0}^{\infty} \Re\left[\mathrm{e}^{-iv\kappa}\psi(v)\right]\,\mathrm{d}v\,.
$$

Notice that the functions $c(\kappa)$ and $p(\kappa)$ have exactly the same Fourier transform. What option value, the call or the put, is obtained by inverting the Fourier transform $\psi(v)$ depends on the sign of the dumping parameter α.

7.2 Stochastic Volatility Diffusion Models

7.2.1 Local Volatility Models

A local volatility model is one which considers the volatility σ as a deterministic function of the current calendar time and the current asset value, i.e., $\sigma = \sigma(t, S)$. The volatility function can be chosen to match the model option values to the market counterparts precisely. However, we cannot generally find closed-form pricing formulae for European options except for some special cases of σ.

Let the asset price process, considered under the risk-neutral probability measure $\widetilde{\mathbb{P}}$, follow the stochastic differential equation (SDE)

$$
\frac{\mathrm{d}S(t)}{S(t)} = (r-q)\,\mathrm{d}t + \sigma(t, S(t))\,\mathrm{d}\widetilde{W}(t)\,,\ t \geqslant 0\,,\quad S(0) = S_0 > 0, \tag{7.17}
$$

where $r \geqslant 0$ is a constant interest rate, $q \geqslant 0$ is a constant dividend yield and $\{\widetilde{W}(t)\}_{t\geqslant 0}$ is Brownian motion under the measure $\widetilde{\mathbb{P}}$. The *time- and state-dependent volatility* $\sigma(t,S)$ is sometimes called the *local volatility function*. In general, σ is a nonnegative continuous function. We are also assuming that the process $\{\mathrm{e}^{-(r-q)t}S(t)\}_{t\geqslant 0}$ is a $\widetilde{\mathbb{P}}$-martingale. For some special cases of σ, the corresponding distribution of the process with SDE (7.17) can be obtained analytically. The most known example of a solvable state-dependent volatility model is the constant elasticity of variance (CEV) diffusion model, which is discussed in the next subsection. As is shown below, the volatility surface $\sigma(t,S)$ can be calibrated to empirical data so that the model successfully produces option values consistent with all market prices across different strikes and maturities.

Let $\widetilde{p} = \widetilde{p}(t,t';S,S')$ with $t < t'$ and $S,S' \in \mathbb{R}_+$ denote a risk-neutral transition probability function associated with (7.17). Under quite general conditions on σ, we recall that the function $\widetilde{p}$ satisfies both the backward and forward Kolmogorov equations. The backward partial differential equation (PDE) reads

$$\frac{\partial \widetilde{p}}{\partial t} + \frac{1}{2}\sigma^2(t,S)S^2\frac{\partial^2 \widetilde{p}}{\partial S^2} + (r-q)S\frac{\partial \widetilde{p}}{\partial S} = 0\,, \tag{7.18}$$

and the corresponding forward PDE is

$$\frac{\partial \widetilde{p}}{\partial t'} = \frac{1}{2}\frac{\partial^2}{\partial S'^2}\left(\sigma^2(t',S')S'^2\,\widetilde{p}\right) - (r-q)\frac{\partial}{\partial S'}\left(S'\,\widetilde{p}\right), \tag{7.19}$$

with the initial (or final) time condition

$$\widetilde{p}(t,t;S,S') = \widetilde{p}(t',t';S,S') = \delta(S-S')\,.$$

Consider a European-style derivative written at time $t_0 \geqslant 0$ on the underlying asset S. The no-arbitrage derivative value $V(t,S)$, which is a function of the current calendar time $t \in [t_0,T]$ and the current asset price $S = S(t)$, is given by the risk-neutral pricing formula

$$V(t,S) = \mathrm{e}^{-r(T-t)}\widetilde{\mathrm{E}}_{t,S}[V(T,S(T))]\,. \tag{7.20}$$

By the (discounted) Feynman–Kac Theorem, the pricing function $V(t,S)$ satisfies the Black–Scholes partial differential equation (BSPDE)

$$\frac{\partial V(t,S)}{\partial t} + \frac{1}{2}\sigma^2(t,S)S^2\frac{\partial^2 V(t,S)}{\partial S^2} + (r-q)S\frac{\partial V(t,S)}{\partial S} - rV(t,S) - 0\,. \tag{7.21}$$

Suppose that the European claim of interest is a standard call option with strike price K and expiration time T. The pricing function of a European call option, $C(t,S;T,K)$, satisfies the BSPDE in (7.21). Surprisingly, the function $C(t,S;T,K)$, regarded explicitly as a function of the strike and maturity time arguments (T,K) (instead of functions of the arguments (t,S) which are held fixed), also satisfies a PDE known as Dupire's equation.

Theorem 7.1 (Dupire's Equation). *The pricing function, $c(T,K) := C(t,S;T,K)$, of a European call option satisfies the PDE*

$$\frac{\partial c(T,K)}{\partial T} = \frac{1}{2}\sigma^2(T,K)K^2\frac{\partial^2 c(T,K)}{\partial K^2} + (q-r)K\frac{\partial c(T,K)}{\partial K} - qc(T,K). \tag{7.22}$$

We note that this is a PDE in the so-called dual variables (T,K), where $K > 0$ and $T > t$, and it is also sometimes called the dual BSPDE. The validity of this theorem also rests upon certain technical assumptions which are stated in the proof.

Proof. The conditional expectation in (7.20) is an integral of the product of the payoff function and the risk-neutral transition PDF $\widetilde{p}(t, T; S, S')$, i.e., the call option value is given by

$$c(T, K) = \mathrm{e}^{-r(T-t)}\widetilde{\mathrm{E}}_{t,S}[(S(T) - K)^+] = \mathrm{e}^{-r(T-t)} \int_K^\infty \widetilde{p}(t, T; S, S')(S' - K)\,\mathrm{d}S'. \quad (7.23)$$

The first and second derivatives of (7.23) with respect to K give

$$\frac{\partial c(T, K)}{\partial K} = -\mathrm{e}^{-r(T-t)} \int_K^\infty \widetilde{p}(t, T; S, S')\,\mathrm{d}S' \quad (7.24)$$

and

$$\frac{\partial^2 c(T, K)}{\partial K^2} = \mathrm{e}^{-r(T-t)}\widetilde{p}(t, T; S, K). \quad (7.25)$$

The derivative of the option price with respect to expiration T is

$$\begin{aligned}\frac{\partial c(T, K)}{\partial T} &= -r\mathrm{e}^{-r(T-t)} \int_K^\infty \widetilde{p}(t, T; S, S')(S' - K)\,\mathrm{d}S' \\ &\quad + \mathrm{e}^{-r(T-t)} \int_K^\infty \frac{\partial \widetilde{p}(t, T; S, S')}{\partial T}(S' - K)\,\mathrm{d}S'.\end{aligned} \quad (7.26)$$

Using the forward Kolmogorov equation (7.19) for $\widetilde{p}$ with $t' = T$ gives

$$\begin{aligned}\frac{\partial c(T, K)}{\partial T} &= -rc(T, K) + \mathrm{e}^{-r(T-t)} \int_K^\infty \Big[-(r - q)\frac{\partial}{\partial S'}(S'\widetilde{p}(t, T; S, S')) \\ &\quad + \frac{1}{2}\frac{\partial^2}{\partial S'^2}\left(\sigma^2(T, S')S'^2\widetilde{p}(t, T; S, S')\right)\Big](S' - K)\,\mathrm{d}S'.\end{aligned} \quad (7.27)$$

The integral containing the first derivative with respect to S' can be evaluated by parts as follows:

$$\begin{aligned}\int_K^\infty (S' - K)\frac{\partial}{\partial S'}(S'\widetilde{p})\,\mathrm{d}S' &= \widetilde{p}S'(S' - K)\Big|_{S'=K}^{S'=\infty} - \int_K^\infty S'\widetilde{p}\,\mathrm{d}S' \\ &= -\int_K^\infty (S' - K)\widetilde{p}\,\mathrm{d}S' - K\int_K^\infty \widetilde{p}\,\mathrm{d}S' \\ &= \left[-c + K\frac{\partial c}{\partial K}\right]\mathrm{e}^{r(T-t)},\end{aligned}$$

where $\widetilde{p}S'(S' - K)|_{S'=K} = 0$, and we simply wrote $S' = (S' - K) + K$ in the second equation line and used (7.23) and (7.24). Its important to note here that we are assuming $\widetilde{p}S'(S' - K) \to 0$, as $S' \to \infty$. That is, the process is assumed to have a transition PDF that decays faster than $(S')^{-2}$, as $S' \to \infty$, i.e., we are assuming $\lim_{S'\to\infty}(S')^2\widetilde{p}(t, T; S, S') = 0$. These limits are implied by the existence of the integral in (7.27), where it is also assumed that $S'\frac{\partial^2}{\partial S'^2}(\sigma^2(T, S')S'^2\widetilde{p})$ decays faster than $1/S'$, as $S' \to \infty$. The latter also implies that $\frac{\partial}{\partial S'}(\sigma^2(T, S')S'^2\widetilde{p}) \to 0$, as $S' \to \infty$. Hence, evaluating the second integral term in (7.27) by parts gives

$$\begin{aligned}\int_K^\infty (S' - K)\frac{\partial^2}{\partial S'^2}(\sigma^2(T, S')S'^2\widetilde{p})\,\mathrm{d}S' &= -\int_K^\infty \frac{\partial}{\partial S'}(\sigma^2(T, S')S'^2\widetilde{p})\,\mathrm{d}S' \\ &= \sigma^2(T, K)K^2\widetilde{p}(t, T; S, K) \\ &= \mathrm{e}^{r(T-t)}\sigma^2(T, K)K^2\frac{\partial^2 c}{\partial K^2}.\end{aligned}$$

Collecting all the intermediate results obtained above into (7.27) gives the following dual Black–Scholes PDE:

$$\begin{aligned}\frac{\partial c}{\partial T} &= -rc + (r-q)c - (r-q)K\frac{\partial c}{\partial K} + \frac{1}{2}\sigma^2(T,K)K^2\frac{\partial^2 c}{\partial K^2} \\ &= -qc + (q-r)K\frac{\partial c}{\partial K} + \frac{1}{2}\sigma^2(T,K)K^2\frac{\partial^2 c}{\partial K^2}\,. \end{aligned}$$ □

The following formula for the local volatility may be used in practice to calibrate a local volatility surface $\sigma(t,S)$ using market European call option prices across a range of maturities and strikes.

Theorem 7.2 (The Derman–Kani–Dupire Formula). *Let $q = 0$. The local volatility function is expressed in analytically closed form as follows in terms of call option prices:*

$$\sigma^2(T,K) = \frac{2}{K^2}\frac{\frac{\partial C}{\partial T} + rK\frac{\partial C}{\partial K}}{\frac{\partial^2 C}{\partial K^2}}. \tag{7.28}$$

7.2.2 Constant Elasticity of Variance Model

In the realm of state-dependent volatility models, the *constant elasticity of variance* (CEV) model has provided an introduction to nonlinear diffusion models that exhibit an implied volatility (half) smile as a function of the strike. The CEV diffusion model has a power-law volatility function with two adjustable parameters. The model admits closed-form pricing formulae for standard European, barrier, and lookback options. Spectral expansions and Laplace transform techniques are very useful in deriving analytical pricing formulae and transition densities for the CEV process and other time-homogeneous models with more complex nonlinear local volatility functions.

7.2.2.1 Definition and Basic Properties

The constant elasticity of variance diffusion model assumes that the asset price is a time-homogeneous diffusion process $\{S(t)\}_{t\geqslant 0}$ that obeys the stochastic differential equation (SDE)

$$\mathrm{d}S(t) = \nu S(t)\,\mathrm{d}t + \alpha(S(t))^{\beta+1}\,\mathrm{d}W(t)\,,\ t\geqslant 0\,,\quad S(0) = S_0 > 0\,, \tag{7.29}$$

where ν, β, α are real parameters (with $\alpha > 0$) and $\{W(t)\}_{t\geqslant 0}$ is a standard Brownian motion under a given probability measure $\mathbb{P}$. The SDE (7.29) has a linear drift coefficient and power-type nonlinear diffusion coefficient $\alpha S^{\beta+1}$. If $\beta = 0$, then the CEV diffusion reduces to a geometric Brownian motion governed by the SDE $\mathrm{d}S(t) = \nu S(t)\,\mathrm{d}t + \alpha S(t)\,\mathrm{d}W(t)$. Thus the CEV model can be viewed as a generalization of the standard Black–Scholes model for asset prices, where the log-volatility σ is a deterministic nonlinear function of the asset price S given by $\sigma(S) := \alpha S^{\beta}$. We will refer to $\sigma(S)$ as the *local volatility function*. The parameter β can be interpreted as the *elasticity*[1] *of the local volatility function* with the property $\sigma'(S) = \beta\sigma(S)/S$, and α is a scale parameter fixing the instantaneous volatility to equal a constant σ_0 at an initial spot value S_0, i.e., $\sigma_0 = \sigma(S_0) = \alpha S_0^{\beta}$. Typical values of the CEV elasticity implicit in equity index option markets are negative and can be as low as $\beta = -4$. For this and other reasons discussed later, β is assumed to take on negative values in what follows.

[1] In physics, elasticity is the ability of a deformed material body to return to its original shape and size when the forces causing the deformation are removed.

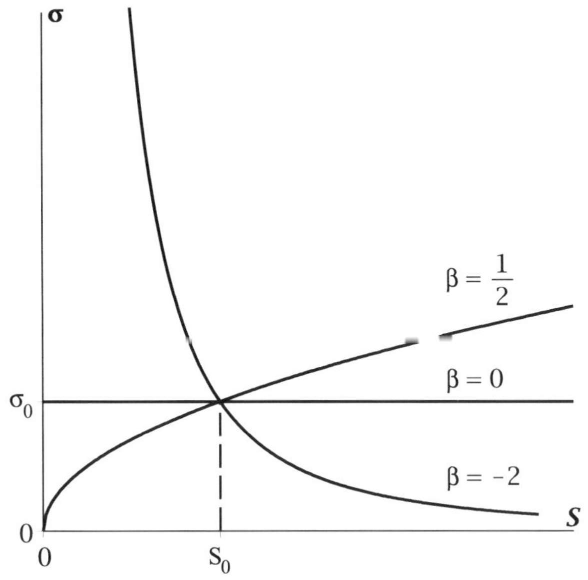

FIGURE 7.1: The local volatility function $\sigma(S) = \alpha S^{\beta}$ of the CEV model. The parameter α is chosen such that $\sigma(S_0) = \sigma_0$ is fixed.

The CEV process is a nonnegative process. To discuss its behaviour at the boundary points 0 and ∞, several definitions are required. For a diffusion process, $\{S(t)\}_{t \geqslant 0}$, defined on an interval (a, b), we say that $x = a$ and $x = b$ are boundary points of the *state space* of the process. We say that point $x \in \{a, b\}$ is

(a) an *entrance* boundary if the process S can enter the state space starting from x but cannot reach x in finite time starting from the interior of the state space;

(b) an *exit* boundary if the process S can reach x in finite time starting from the interior of the state space but cannot enter the state space starting from x;

(c) a *regular* boundary if it is both an exit and an entrance boundary;

(d) a *natural* boundary if it is neither an exit nor an entrance boundary.

We note that "entrance" really means entrance and not exit, and "exit" means exit and not entrance. It is possible to impose a boundary condition for the diffusion S at a regular boundary. A regular boundary is called a *killing* (or *reflecting*) boundary when an instantaneously killing (or reflecting) boundary condition is imposed.

Given the CEV process satisfying (7.29), we can classify the boundary points 0 and ∞ according to the value of β. For $\beta < 0$, infinity is a natural boundary point. For $-\frac{1}{2} \leqslant \beta < 0$, the origin is an exit boundary point. For $\beta < -\frac{1}{2}$, the origin is a regular boundary point. For $\beta = 0$, i.e., for GBM, both boundary points 0 and ∞ are natural boundaries. For $\beta > 0$, the origin is a natural boundary and infinity is an entrance boundary point.

7.2.2.2 Transition Probability Law

A combination of a change of variables and a scale and time transformation allows us to represent the CEV diffusion process $\{S(t)\}_{t \geqslant 0}$ as a function of a squared Bessel (SQB) process $\{X(t)\}_{t\geqslant 0}$, which has generator $\mathcal{G}f(x) := (2\mu+2)f'(x)+2xf''(x)$ and hence satisfies the SDE

$$\mathrm{d}X(t) = (2\mu+2)\,\mathrm{d}t + 2\sqrt{X(t)}\,\mathrm{d}W(t), \tag{7.30}$$

where μ is the so-called index of the process. The left-hand endpoint 0 is an entrance boundary if $\mu \geqslant 0$, a regular boundary if $-1 < \mu < 0$, or an exit boundary if $\mu \leqslant -1$. The right-hand endpoint ∞ is a natural boundary. The state space of the SQB process is the interval $[0,\infty)$ or $(0,\infty)$ depending on the value of μ and the boundary condition at zero. The SQB diffusion is a time-homogeneous Markov process. The transition probability density function (PDF) is given by

$$p_\mu(t;x_0,x) \equiv \frac{\mathbb{P}(X(t)\in\mathrm{d}x \,|\, X(0)=x_0)}{\mathrm{d}x} = \left(\frac{x}{x_0}\right)^{\frac{\mu}{2}} \frac{e^{-(x+x_0)/(2t)}}{2t} I_{\tilde{\mu}}\left(\frac{\sqrt{xx_0}}{t}\right), \tag{7.31}$$

where $\tilde{\mu} = \mu$ if $\mu \geqslant 0$ or if $\mu \in (-1,0)$ and 0 is specified as a regular reflecting boundary, and $\tilde{\mu} = |\mu|$, $\mu < 0$, if 0 is an exit or a regular killing boundary. Here, $I_\mu(z)$ denotes the modified Bessel function of the first kind, of order μ and argument z.

Let us find the transition PDF of the CEV process (with $\beta \neq 0$) by reducing it to the SQB process. Firstly, a CEV process with a nonzero drift parameter ν is obtained from a driftless CEV process denoted $\{F(t)\}_{t\geqslant 0}$ by employing a scale and time change:

$$S(t) = \mathrm{e}^{\nu t} F(\tau_t), \text{ where } \tau_t = \begin{cases} \frac{1}{2\nu\beta}\left(\mathrm{e}^{2\nu\beta t} - 1\right) & \nu \neq 0, \\ t & \nu = 0. \end{cases} \tag{7.32}$$

Secondly, the change of variables via the monotonic mapping

$$\mathsf{X}\colon F \to F^{-2\beta}/(\alpha^2\beta^2) \tag{7.33}$$

gives us the SQB process $X(t) := \mathsf{X}(F(t)) = F(t)^{-2\beta}/(\alpha^2\beta^2)$ with the index $\mu = \frac{1}{2\beta}$. Note that this mapping is strictly increasing for $\beta < 0$. Thus, the CEV diffusion can be expressed in terms of the SQB process as follows:

$$S(t) = \mathrm{e}^{\nu t}\left(\alpha^2\beta^2 X(\tau_t)\right)^{-\frac{1}{2\beta}}, \tag{7.34}$$

where the process $\{X(t)\}_{t\geqslant 0}$ solves the SDE (7.30) with $\mu = \frac{1}{2\beta}$. Although the detailed mathematical proof of the representation (7.34) is left as an exercise for the reader, we briefly discuss the transformation. Two techniques are applied. First, Itô's formula is applied when transforming $F(t) \to X(t)$. Using the change of variables $x = \mathsf{X}(S)$ in the PDF (7.31) gives us the transition PDF of the driftless CEV process:

$$p^{(0)}(t;S_0,S) = p_\mu(t;\mathsf{X}(S_0),\mathsf{X}(S)) \cdot \mathsf{X}'(S). \tag{7.35}$$

Second, the removal of drift is done with the use of the Kolmogorov PDE for the transition PDF. Denote the transition PDF of the CEV process with the drift parameter ν by $p^{(\nu)}(t;S_0,S)$. The function $p^{(\nu)}$ solves the corresponding forward Kolmogorov PDE,

$$\frac{\partial p^{(\nu)}}{\partial t} = \frac{1}{2}\frac{\partial^2}{\partial S^2}\left(\sigma^2(S)\,p^{(\nu)}\right) - \frac{\partial}{\partial S}\left(\nu S\,p^{(\nu)}\right), \tag{7.36}$$

subject to the Dirac delta initial condition, $p^{(\nu)}(0;S_0,S) = \delta(S-S_0)$, and imposed homogeneous boundary conditions at the endpoints 0 and ∞. According to the transformation (7.32), the PDFs $p^{(0)}$ and $p^{(\nu)}$ (with $\nu \neq 0$) relate to each other as follows:

$$p^{(\nu)}(t;S_0,S) = \mathrm{e}^{-\nu t} p^{(0)}(\tau_t;S_0,\mathrm{e}^{-\nu t}S). \tag{7.37}$$

The reader may check that $p^{(\nu)}$ satisfies (7.36), assuming that $p^{(0)}$ solves (7.36) with $\nu = 0$. This proves that the process $\{S(t)\}_{t\geqslant 0}$ in (7.32) satisfies the SDE (7.29) for the CEV process with drift. As a result, the PDF $p^{(\nu)}$ is expressed in terms of the transition PDF p_μ of the SQB process:

$$p^{(\nu)}(t; S_0, S) = \mathrm{e}^{-\nu t} p_\mu(\tau_t; \mathsf{X}(S_0), \mathsf{X}(\mathrm{e}^{-\nu t} S)) \cdot \mathsf{X}'(\mathrm{e}^{-\nu t} S). \tag{7.38}$$

Assume that the parameter β is negative. Here, we consider the case when the endpoint $S = 0$ is a regular killing boundary (if $\beta < -0.5$) or exit (if $-0.5 \leqslant \beta < 0$). The transition PDF $p^{(0)}(t; S_0, S)$ with $S_0, S > 0$ and $t > 0$ for the driftless CEV process $F(t)$ takes the form

$$p^{(0)}(t; S_0, S) = \frac{S^{-2\beta-\frac{3}{2}} S_0^{\frac{1}{2}}}{\delta^2 |\beta| t} \exp\left(-\frac{S^{-2\beta} + S_0^{-2\beta}}{2\delta^2 \beta^2 t} \right) I_{\frac{1}{2|\beta|}} \left(\frac{S^{-\beta} S_0^{-\beta}}{\delta^2 \beta^2 t} \right). \tag{7.39}$$

The density $p^{(0)}(t; S_0, S)$ does not integrate (with respect to S) to 1, since $S = 0$ is an absorbing point. However, the driftless CEV diffusion process satisfies the martingale property (w.r.t. the natural filtration $(\mathcal{F}_t)_{t\geqslant 0}$):

$$\mathrm{E}[F(t+u) \mid \mathcal{F}_t] = F(t) \quad \text{for all} \quad t, u \geqslant 0.$$

For $\beta < 0$, the probability of absorption at zero (bankruptcy) before time $t > 0$ is

$$\mathbb{P}(S(t) = 0) = 1 - \mathbb{P}(S(t) > 0) = 1 - \int_0^\infty p^{(\nu)}(t; S_0, S)\,\mathrm{d}S = G\left(\frac{1}{2|\beta|}, \frac{\mathsf{X}(S_0)}{2\tau_t} \right), \tag{7.40}$$

where $G(\mu, a)$ is the complementary gamma function given by

$$G(\mu, a) = \frac{1}{\Gamma(\mu)} \int_a^\infty \mathrm{e}^{-t}\, t^{\mu-1}\,\mathrm{d}t.$$

Here, $\Gamma(x)$ is the gamma function defined by $\Gamma(x) = \int_0^\infty t^{x-1} \mathrm{e}^{-t}\,\mathrm{d}t$.

Note that if the reflecting boundary condition is imposed at 0, then there is no absorption at the endpoint. The corresponding transition density (for the case with $\beta < -0.5$) is given by (7.39) with the replacement $I_{\frac{1}{2|\beta|}} \to I_{\frac{1}{2\beta}}$. The driftless process $\{F(t)\}_{t\geqslant 0}$ becomes a strict submartingale.

7.2.2.3 Pricing European Options

We assume $\beta < 0$, so the driftless CEV process (with $\nu = 0$) obeys the martingale property. Generally, we have

$$\mathrm{E}[S(t+u) \mid \mathcal{F}_t] = \mathrm{e}^{\nu(t+u)}\, \mathrm{E}[F(\tau_{t+u}) \mid \mathcal{F}_{\tau_t}] = \mathrm{e}^{\nu u} \mathrm{e}^{\nu t} F(\tau_t) = \mathrm{e}^{\nu u} S(t),$$

for all $t, u \geqslant 0$. Therefore, the forward price process $\{\mathrm{e}^{-\nu t} S(t)\}_{t\geqslant 0}$ is a $\mathbb{P}$-martingale. Under the risk-neutral probability measure $\widetilde{\mathbb{P}}$ we set $\nu = r - q$. The no-arbitrage price of a European call option (with expiry T and strike K) under the CEV process with the transition density in (7.35)–(7.39) is then given by

$$\begin{aligned} C(0, S_0; T, K) &= \mathrm{e}^{-rT}\, \widetilde{\mathrm{E}}_{0,S_0}\left[(S(T) - K)^+ \right] = \mathrm{e}^{-rT} \int_K^\infty (S - K)\, \widetilde{p}^{(\nu)}(T; S_0, S)\,\mathrm{d}S \\ &= \mathrm{e}^{-qT} S_0 \int_m^\infty \frac{1}{2} \left(\frac{y_0}{y} \right)^{\frac{1}{4\beta}} \exp\left(-\frac{y + y_0}{2} \right) I_{\frac{1}{2|\beta|}}(\sqrt{y y_0})\,\mathrm{d}y \\ &\quad - \mathrm{e}^{-rT} K \int_m^\infty \frac{1}{2} \left(\frac{y}{y_0} \right)^{\frac{1}{4\beta}} \exp\left(-\frac{y + y_0}{2} \right) I_{\frac{1}{2|\beta|}}(\sqrt{y y_0})\,\mathrm{d}y, \end{aligned} \tag{7.41}$$

where

$$m = \frac{2\nu K^{-2\beta}}{\alpha^2\beta\left(1 - e^{-2\nu\beta T}\right)}, \quad y_0 = \frac{2\nu S_0^{-2\beta}}{\alpha^2\beta\left(\mathrm{e}^{2\nu\beta T} - 1\right)}$$

for the case when $\nu = r - q \neq 0$. If $\nu = 0$, then $m = \frac{K^{-2\beta}}{\alpha^2\beta^2 T}$ and $y_0 = \frac{S_0^{-2\beta}}{\alpha^2\beta^2 T}$.

Proof. Using (7.39), we express the price of a European call under the CEV model as follows:

$$\begin{aligned} C_0 \equiv C(0, S_0; T, K) &= \mathrm{e}^{-rT} \int_K^\infty (S - K)\, \widetilde{p}^{(\nu)}(T; S_0, S)\, \mathrm{d}S \\ &= \mathrm{e}^{-rT} \int_K^\infty (S - K)\, \mathrm{e}^{-\nu T} p^{(0)}(\tau_T; S_0, \mathrm{e}^{-\nu T} S)\, \mathrm{d}S\,. \end{aligned}$$

By changing variables defined by $S' = \mathrm{e}^{-\nu T} S$, and then renaming the dummy variable S' as S,

$$\begin{aligned} C_0 &= \mathrm{e}^{-rT} \int_{Ke^{-\nu T}}^\infty \left(\mathrm{e}^{\nu T} S - K\right)\, p^{(0)}(\tau; S_0, S)\, \mathrm{d}S \\ &= \mathrm{e}^{-qT} \int_{K\mathrm{e}^{-\nu T}}^\infty S\, p^{(0)}(\tau; S_0, S)\, \mathrm{d}S - \mathrm{e}^{-rT} K \int_{K\mathrm{e}^{-\nu T}}^\infty p^{(0)}(\tau; S_0, S)\, \mathrm{d}S\,. \end{aligned}$$

Using (7.35) and applying the change of variables $x = \mathsf{X}(S) := \frac{S^{-2\beta}}{\alpha^2\beta^2}$ gives

$$\begin{aligned} C_0 &= \mathrm{e}^{-qT} \int_{K\mathrm{e}^{-\nu T}}^\infty S\, p_\mu(\tau; \mathsf{X}(S_0), \mathsf{X}(S)) \cdot \mathsf{X}'(S)\, \mathrm{d}S \\ &\quad - \mathrm{e}^{-rT} K \int_{K\mathrm{e}^{-\nu T}}^\infty p_\mu(\tau; \mathsf{X}(S_0), \mathsf{X}(S)) \cdot \mathsf{X}'(S)\, \mathrm{d}S \\ &= \mathrm{e}^{-qT} \int_k^\infty (\alpha^2\beta^2 x)^{-\frac{1}{2\beta}} p_\mu(\tau; x_0, x)\, \mathrm{d}x - \mathrm{e}^{-rT} K \int_k^\infty p_\mu(\tau; x_0, x)\, \mathrm{d}x \\ &= \mathrm{e}^{-qT} S_0 \int_k^\infty \left(\frac{x}{x_0}\right)^{-\frac{1}{2\beta}} p_\mu(\tau; x_0, x)\, \mathrm{d}x - \mathrm{e}^{-rT} K \int_k^\infty p_\mu(\tau; x_0, x)\, \mathrm{d}x\,, \end{aligned}$$

where $\tau \equiv \tau_T$ is given by (7.32), $k = \mathsf{X}(\mathrm{e}^{-\nu T} K) = \frac{(\mathrm{e}^{-\nu T} K)^{-2\beta}}{\alpha^2\beta^2}$, $x_0 = \mathsf{X}(S_0) = \frac{S_0^{-2\beta}}{\alpha^2\beta^2}$, and $\mu = \frac{1}{2\beta}$. The formula (7.41) is deduced by applying (7.31) and changing variables in the above two integrals: $y = \frac{x}{\tau}$, $y_0 = \frac{x_0}{\tau}$, and $m = \frac{k}{\tau}$. □

The two integrals in (7.41) can be computed numerically with the use of some quadrature rule. Alternatively, the call option pricing formula can be expressed in terms of the complementary noncentral chi-square distribution function. The latter can be represented as a series of elementary functions. When the call option price is calculated, the put option value can be obtained by using put-call parity.

The *noncentral chi-square distribution* denoted $\chi_\upsilon^2(\lambda)$ is a continuous probability distribution with parameters $\upsilon \in (0, \infty)$ and $\lambda \in [0, \infty)$ and with the PDF

$$f(x; \upsilon, \lambda) = \frac{1}{2}\left(\frac{x}{\lambda}\right)^{(\upsilon-2)/4} I_{(\upsilon-2)/2}(\sqrt{\lambda x}) \exp\left(-\frac{\lambda + x}{2}\right), \quad x > 0. \tag{7.42}$$

For integer υ, this distribution arises as a probability distribution of a sum of squared normal variables. Let $Z_1, Z_2, \ldots, Z_\upsilon$ be independent standard normal random variables, and $a_1, a_2, \ldots, a_\upsilon$ be some real constants, then $Y = \sum_{j=1}^{\upsilon}(Z_j + a_j)^2$ is said to have the noncentral chi-square distribution with υ degrees of freedom and noncentrality parameter

$\lambda = \sum_{j=1}^{\upsilon} a_j^2$. If all $a_j = 0$, then Y has the central chi-square distribution with υ degrees of freedom, which is denoted as usual by χ_υ^2.

Since the integrands in (7.41) are both the noncentral chi-square PDFs of the form (7.42), the integrals in (7.41) can be expressed in terms of the complementary distribution function for $\chi_\upsilon^2(\lambda)$:

$$Q(x; \upsilon, \lambda) = \int_x^\infty f(y; \upsilon, \lambda)\,\mathrm{d}y. \tag{7.43}$$

Recall that the cumulative and complementary distribution functions, respectively denoted by F and Q, relate to each other as $F(x) + Q(x) = 1$.

The first integrand in (7.41) is the noncentral chi-square PDF $f\left(y; 2 + \frac{1}{|\beta|}, y_0\right)$ with $2 + \frac{1}{|\beta|}$ degrees of freedom and noncentrality parameter y_0. Integrating with respect to y from m to infinity gives the corresponding complementary distribution function $Q\left(m; 2 + \frac{1}{|\beta|}, y_0\right)$. The second integrand in (7.41) equals $f\left(y_0; 2 + \frac{1}{|\beta|}, y\right)$, where y is now the noncentrality parameter. The second integral is therefore equal to $1 - Q\left(y_0; \frac{1}{|\beta|}, m\right)$ (see Exercise 7.6). Thus, the call option value is

$$C(0, S_0; T, K) = \mathrm{e}^{-qT} S_0 Q\big(m; 2 + \frac{1}{|\beta|}, y_0\big) - \mathrm{e}^{-rT} K\left(1 - Q\big(y_0; \frac{1}{|\beta|}, m\big)\right). \tag{7.44}$$

The complementary noncentral chi-square distribution function $Q(x; \upsilon, \lambda)$ can be computed using the following formula:

$$Q(2x; 2\upsilon, 2\lambda) = 1 - \sum_{n=1}^{\infty} g\,(n + \upsilon, x) \sum_{j=1}^{n} g\,(j, \lambda)\,, \tag{7.45}$$

where $g(m, x)$ is a gamma PDF given by

$$g(m, x) = \frac{\mathrm{e}^{-x}\, x^{m-1}}{\Gamma(m)}.$$

7.3 The Heston model

While in local volatility models, more realistic behaviour of asset prices is achieved by introducing a nonlinear, state-dependent volatility function, in the stochastic volatility framework, volatility changes over time according to another random process correlated with the underlying price process. Stochastic volatility models are quite popular among practitioners since such models can produce pronounced volatility smiles and skews. The volatility process is usually described by another stochastic differential equation. Therefore, we deal with a two-dimensional SDE for asset price modelling.

Consider an asset price model, where the underlying asset price $S(t)$ is geometric Brownian motion but with a stochastic volatility that follows the Ornstein–Uhlenbeck process:

$$\begin{cases} \mathrm{d}S(t) = \mu S(t)\,\mathrm{d}t + \sigma(t) S(t)\,\mathrm{d}W_1(t), \\ \mathrm{d}\sigma(t) = -\beta b(t)\,\mathrm{d}t + \delta\,\mathrm{d}W_2(t), \end{cases} \tag{7.46}$$

where $W_1(t)$ are $W_2(t)$ are dependent standard Brownian motions with correlation coefficient $\rho \in (-1, 1)$. In practice, the value of ρ is typically negative and can be close to -1. If $\rho < 0$, then volatility increases as the asset price decreases. The nonlinear volatility function of the CEV model in (7.29) with $\beta < 0$ demonstrates the same type of behaviour.

Applying Itô's formula gives that the instantaneous variance, equal to the squared volatility, $\mathrm{v}(t) = \sigma^2(t)$ follows the SDE

$$\mathrm{dv}(t) = (\delta^2 - 2\beta v(t))\,\mathrm{d}t + 2\delta\sqrt{\mathrm{v}(t)}\,\mathrm{d}W_2(t).$$

The above equation describes a square-root, mean-reverting diffusion process. Define the parameters $\kappa := 2\beta$, $\theta := \delta^2/(2\beta)$, and $\xi := 2\delta$ to obtain the well-known *Heston model*:

$$\begin{cases} \mathrm{d}S(t) = \mu S(t)\,\mathrm{d}t + \sqrt{\mathrm{v}(t)}S(t)\,\mathrm{d}W_1(t), \\ \mathrm{dv}(t) = \kappa(\theta - \mathrm{v}(t))\,\mathrm{d}t + \xi\sqrt{\mathrm{v}(t)}\,\mathrm{d}W_2(t). \end{cases} \tag{7.47}$$

Applying Itô's formula to $S(t)$ gives us the SDE for the log price $X(t) := \ln S(t)$:

$$\mathrm{d}X(t) = \left(\mu - \frac{\mathrm{v}(t)}{2}\right)\mathrm{d}t + \sqrt{\mathrm{v}(t)}\,\mathrm{d}W_1(t).$$

Since equations in (7.47) have two sources of uncertainty, namely, the Brownian motions $W_1(t)$ and $W_2(t)$ controlling movements of the stock price $S(t)$ and instantaneous variance $\mathrm{v}(t)$, respectively, the Heston model is an incomplete market model.

The parameters in the model (7.47) have the following interpretation:

μ is the rate of return on the underlying asset.

θ is the mean reversion level of the variance process $\mathrm{v}(t)$; that is, θ is the long-run average variance, since the expected value of $\mathrm{v}(t)$ goes to θ, as $t \to \infty$.

κ is the mean reversion speed at which the variance process $\mathrm{v}(t)$ reverts to θ.

ξ is the volatility of the volatility; as the name suggests, it determines the variance of $\mathrm{v}(t)$.

ρ is the coefficient of correlation between the log price and the instantaneous volatility.

$\mathrm{v}(0)$ is the initial value of the variance; it can be treated as another parameter.

To guarantee the positiveness of the instantaneous variance $\mathrm{v}(t)$, the model parameters should satisfy

$$2\kappa\theta \geqslant \xi^2.$$

As mentioned above, the variance process is *mean-reverting*. Indeed, the drift term $\kappa(\theta - v)$ is negative for $v > \theta$ and positive for $v < \theta$. In both cases, the drift term pushes the process toward the level of θ. Notice that the expected value and variance of $\mathrm{v}(t)$ are

$$\mathrm{E}[\mathrm{v}(t)] = \theta + (\mathrm{v}(0) - \theta)\mathrm{e}^{-\kappa t},$$
$$\mathrm{Var}[\mathrm{v}(t)] = \frac{\mathrm{v}(0)\xi^2\mathrm{e}^{-\kappa t}}{\kappa}(1 - \mathrm{e}^{-\kappa t}) + \frac{\theta\xi^2}{2\kappa}(1 - \mathrm{e}^{-\kappa t})^2,$$

respectively. Thus, the expected value $\mathrm{E}[\mathrm{v}(t)]$ converges to the long-run average variance θ, as $t \to \infty$ or $\kappa \to \infty$, and $\mathrm{Var}[\mathrm{v}(t)] \to 0$, as $\kappa \to \infty$. The larger κ is, the faster $\mathrm{v}(t)$ converges to θ.

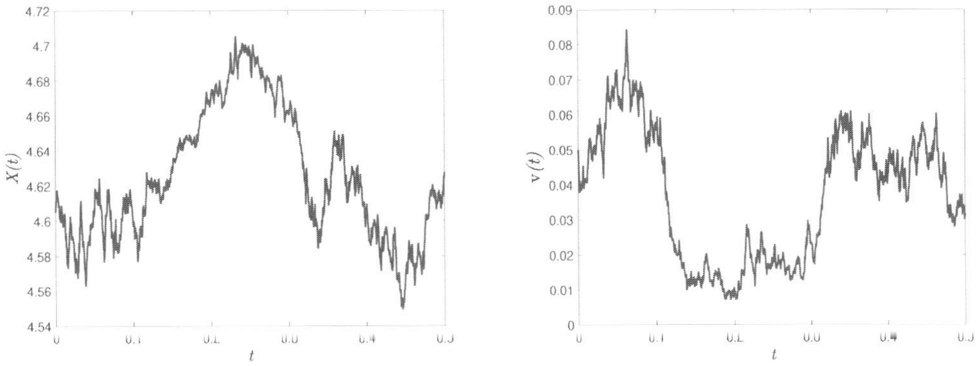

(a) A sample path of the log price process $X(t)$ for the Heston Model.

(b) A sample path of the volatility process $\mathrm{v}(t)$ for the Heston Model.

FIGURE 7.2: Sample paths of the log price $X(t)$ and the variance $\mathrm{v}(t)$ for the Heston model with parameters $X(0) = \ln 100$, $\mathrm{v}(0) = 0.5$, $r = 0.04$, $\tilde{\kappa} = 5$, $\tilde{\theta} = 0.05$, $\xi = 0.5$, and $\rho = -0.8$.

The risk-neutral probability measure is obtained by applying Girsanov's theorem with

$$\widetilde{W}_1(t) = W_1(t) + \frac{\mu - r}{\sqrt{\mathrm{v}(t)}} \cdot t \quad\text{and}\quad \widetilde{W}_2(t) = W_2(t) + \frac{\lambda(S(t), \mathrm{v}(t), t)}{\xi\sqrt{\mathrm{v}(t)}} \cdot t.$$

Here, $\lambda(S(t), \mathrm{v}(t), t)$ is a so-called volatility risk premium. A typical assumption is that λ is proportional to the instantaneous variance, i.e., $\lambda(S(t), \mathrm{v}(t), t) = \lambda_0 \mathrm{v}(t)$ for some constant λ_0. Thus, under the risk-neutral probability measure $\widetilde{\mathbb{P}}$, the joint dynamics of $S(t)$ and $\mathrm{v}(t)$ is governed by

$$\begin{cases} \mathrm{d}S(t) = rS(t)\,\mathrm{d}t + \sqrt{\mathrm{v}(t)}S(t)\,\mathrm{d}\widetilde{W}_1(t), \\ \mathrm{d}\mathrm{v}(t) = \tilde{\kappa}(\tilde{\theta} - \mathrm{v}(t))\,\mathrm{d}t + \xi\sqrt{\mathrm{v}(t)}\,\mathrm{d}\widetilde{W}_2(t), \end{cases} \tag{7.48}$$

where $\tilde{\kappa} := \kappa + \lambda_0$ and $\tilde{\theta} := \kappa\theta/(\kappa + \lambda_0) = \kappa\theta/\tilde{\kappa}$. Applying Itô's formula, we find the risk-neutral dynamics of the log price $X(t) \equiv \ln S(t)$:

$$\mathrm{d}X(t) = \left(r - \frac{\mathrm{v}(t)}{2}\right)\mathrm{d}t + \sqrt{\mathrm{v}(t)}\,\mathrm{d}\widetilde{W}_1(t).$$

Sample paths of the log price $X(t)$ and the variance $\mathrm{v}(t)$ for the Heston model are provided in Figure 7.2. We can notice the mean-reverting dynamics of the variance process with the mean reversion level $\tilde{\theta} = 0.05$. When $t \in [0.15, 0.3]$, the variance $\mathrm{v}(t)$ is close to zero, and hence the path of log price $X(t)$ exhibits less variability in comparison with its values for $t < 0.1$ and $t > 0.3$. Also, on the interval $t \in [0.1, 0.3]$, we can observe that the paths of $X(t)$ and $\mathrm{v}(t)$ are moving in the opposite directions, since the processes are negatively correlated ($\rho = -80\%$).

Pricing derivatives under the Heston model is not a straightforward procedure. First, the joint distribution of $S(t)$ and $\mathrm{v}(t)$ is not available in closed form. Therefore, the risk-neutral expectation of a payoff function is hard to compute when $\rho \neq 0$. If the price process and the volatility process are uncorrelated (i.e., $\rho = 0$), then the price of a standard European option can be written as an integral over Black–Scholes prices. In the general case, one can obtain derivative prices by using Monte Carlo simulations, the PDE technique, or the

Fourier transform method. In what follows, we consider one approach to pricing standard European calls and puts, namely, the transform method.

Recall that the time-t no-arbitrage value of a European call option is given by:

$$C^E(t,S,v;T,K) = \mathrm{e}^{-q\tau} S\,\widehat{\mathbb{P}}_{t,S,v}(S(T) \geqslant K) - \mathrm{e}^{-r\tau} K\,\widetilde{\mathbb{P}}_{t,S,v}(S(T) \geqslant K),$$

where $\tau = T - t$ is the time to maturity, and S and v are spot values of the asset price and the variance, respectively. Here, $\widehat{\mathbb{P}}$ and $\widetilde{\mathbb{P}}$ denote the EMMs with the stock and the risk-free bank account as numéraires, respectively. The subscripts t, S, v indicate that the above probabilities are conditional on the event $\{S(t) = S,\ \mathrm{v}(t) = v\}$:

$$\widehat{\mathbb{P}}_{t,S,v}(S(T) \geqslant K) \equiv \widehat{\mathbb{P}}(S(T) \geqslant K \mid S(t) = S, \mathrm{v}(t) = v) \text{ and}$$
$$\widetilde{\mathbb{P}}_{t,S,v}(S(T) \geqslant K) \equiv \widetilde{\mathbb{P}}(S(T) \geqslant K \mid S(t) = S, \mathrm{v}(t) = v).$$

The above probabilities can be calculated by inverting respective characteristic functions. Let $\varphi(u;\tau,x,v)$ denote the characteristic function of the log price $X(T)$ under the probability measure $\widetilde{\mathbb{P}}$ conditional on $S(t) = \mathrm{e}^x$ and $\mathrm{v}(t) = v$. It is given by

$$\varphi(u;\tau,x,v) := \widetilde{\mathrm{E}}\left[\mathrm{e}^{iuX(T)} \mid X(t) = x, \mathrm{v}(t) = v\right], \text{ where } \tau = T - t.$$

Thus, φ, considered as a function of τ, x, and v, satisfies the following PDE problem, thanks to the Feynman–Kac theorem:

$$-\frac{\partial\varphi}{\partial\tau} + \frac{1}{2}\left(v\frac{\partial^2\varphi}{\partial x^2} + 2\xi\rho v\frac{\partial^2\varphi}{\partial x\partial v} + \xi^2 v\frac{\partial^2\varphi}{\partial v^2}\right)$$
$$+\left(r - \frac{v}{2}\right)\frac{\partial\varphi}{\partial x} + (\tilde{\theta} - v)\tilde{\kappa}\frac{\partial\varphi}{\partial v} = 0, \quad \tau > 0, \tag{7.49}$$
$$\varphi(\tau = 0) = \mathrm{e}^{iux}. \tag{7.50}$$

Let us use the following *anzats* as a solution to the problem (7.49)–(7.50):

$$\varphi(u;\tau,x,v) = \mathrm{e}^{iux + C(\tau;u) + D(\tau;u)v}. \tag{7.51}$$

Since φ satisfies the initial condition in (7.50), we have

$$C(0;u) = 0 \text{ and } D(0;u) = 0 \text{ for all } u \in \mathbb{R}. \tag{7.52}$$

The partial derivatives of φ in (7.51) are

$$\frac{\partial\varphi}{\partial\tau} = (C + Dv)\varphi,$$
$$\frac{\partial\varphi}{\partial x} = iu\varphi, \qquad \frac{\partial\varphi}{\partial v} = D\varphi,$$
$$\frac{\partial^2\varphi}{\partial x^2} = -u^2\varphi, \qquad \frac{\partial^2\varphi}{\partial v^2} = D^2\varphi, \quad \frac{\partial^2\varphi}{\partial x\partial v} = iuD\varphi.$$

Plug them in (7.49) and collect powers of v:

$$\left(-\frac{\partial C}{\partial\tau} + iur + \theta\kappa D\right) + \left(-\frac{\partial D}{\partial\tau} + \frac{\xi^2}{2}D^2 + -\kappa D i\xi\rho u D - \frac{iu}{2} - \frac{u^2}{2}\right)v = 0.$$

Clearly, both terms in parentheses do not depend on the variable v. Since the above equation is valid for all values of v, both expressions in parentheses equal zero. Therefore, the function C and D satisfies the following system of first-order ordinary differential equations (ODEs):

$$\begin{cases} \dfrac{\partial D}{\partial \tau} = \dfrac{\xi^2}{2} D^2 - \tilde{\kappa} D + i\xi\rho u D - \dfrac{iu}{2} - \dfrac{u^2}{2}, \\ \dfrac{\partial C}{\partial \tau} = iur + \tilde{\theta}\tilde{\kappa} D, \end{cases} \tag{7.53}$$

subject to the initial condition in (7.52). The first equation in (7.53) is known as the Riccati equation. Once, D is found, the function C can be obtained by a straightforward integration of the second equation in (7.53). We then plug both C and D in (7.51) to obtain φ. To find the characteristic function $\hat{\varphi}$ of the log price $X(T)$ under the EMM $\widehat{\mathbb{P}}$, we use (7.14):

$$\hat{\varphi}(u; \tau, x, v) = \frac{\varphi(u - i; \tau, x, v)}{\varphi(-i; \tau, x, v)}.$$

After that, we can recover the CDF of the log price under both EMMs $\widetilde{\mathbb{P}}$ and $\widehat{\mathbb{P}}$ by employing the Fourier inversion as described in Section 7.1.2 and compute prices of standard European put and call options. Alternatively, we can use the Carr–Madan method described in Subsection 7.1.4.

7.3.1 Solution to the Ricatti Equation

In what follows, we provide a solution to the first equation in (7.53). First, we introduce the Riccati equation and explain how its solution is derived. The reader can find other details in many standard textbooks on ordinary differential equations.

The Riccati equation for function $y(t)$ with a continuous first-order derivative is

$$y'(t) = P + Qy(t) + Ry(t)^2. \tag{7.54}$$

The solution $y(t)$ can be expressed in terms of function $z(t)$ that solves

$$z''(t) - Qz'(t) + PRz(t) = 0 \tag{7.55}$$

as follows: $y(t) = -\frac{1}{R}\frac{z'(t)}{z(t)}$. Indeed, let us differentiate this expression to obtain

$$\begin{aligned} y'(t) &= \frac{\mathrm{d}}{\mathrm{d}t}\left(-\frac{1}{R}\frac{z'(t)}{z(t)}\right) = \frac{1}{R}\left(\frac{z'(t)^2}{z(t)^2} - \frac{z''(t)}{z(t)}\right) = \frac{1}{R}\frac{z'(t)^2}{z(t)^2} - \frac{1}{R}\left(\frac{Qz'(t) - PRz(t)}{z(t)}\right) \\ &= R\left(\frac{1}{R}\frac{z'(t)}{z(t)}\right)^2 - \frac{Q}{R}\frac{Qz'(t)}{z(t)} + \frac{PR}{R}\frac{z(t)}{z(t)} = Ry(t)^2 + Qy(t) + P. \end{aligned}$$

The second-order ODE in (7.55) admits a solution of the form

$$z(t) = K_1 \mathrm{e}^{\lambda_1 t} + K_2 \mathrm{e}^{\lambda_2 t},$$

where K_1 and K_2 are two arbitrary constants, and λ_1 and λ_2 are solutions to the characteristic equation $\lambda^2 - Q\lambda + PR = 0$ given by

$$\lambda_1 = \frac{Q + \sqrt{Q^2 - 4PR}}{2} \quad \text{and} \quad \lambda_2 = \frac{Q - \sqrt{Q^2 - 4PR}}{2}$$

with $Q^2 - 4PR > 0$. The general solution to (7.54) is then

$$y(t) = -\frac{1}{R}\frac{K_1\lambda_1 e^{\lambda_1 t} + K_2\lambda_2 e^{\lambda_2 t}}{K_1 e^{\lambda_1 t} + K_2 e^{\lambda_2 t}} = -\frac{1}{R}\frac{K\lambda_1 e^{\lambda_1 t} + \lambda_2 e^{\lambda_2 t}}{K e^{\lambda_1 t} + e^{\lambda_2 t}},$$

where $K = K_1/K_2$ is an arbitrary constant.

The equation for $D(t)$ in (7.53) can be written as

$$\frac{\mathrm{d}D}{\mathrm{d}\tau} = RD^2 - QD + P,$$

where $R = \xi^2/2$, $Q = \tilde{\kappa} - i\xi\rho u$, and $P = -u^2/2 - iu/2$. Therefore, we obtain

$$D(\tau) = -\frac{1}{R}\frac{K\lambda_1 e^{\lambda_1 t} + \lambda_2 e^{\lambda_2 t}}{K e^{\lambda_1 t} + e^{\lambda_2 t}}, \tag{7.56}$$

where λ_1 and λ_2 solves the auxiliary equation $\lambda^2 + (\tilde{\kappa} - i\xi\rho u)\lambda - (u^2 + iu)\xi^2/4$. The roots are

$$\lambda_{1,2} = \frac{-(\tilde{\kappa} - i\xi\rho u) \pm \sqrt{(\tilde{\kappa} - i\xi\rho u)^2 + (u^2 + iu)\xi^2}}{2}.$$

To find the constant K, we use the initial condition $D(0; u) = 0$. When $\tau = 0$ is plugged in (7.56), we obtain $D(0) = -\frac{1}{R}\frac{K\lambda_1 + \lambda_2}{K+1}$. Thus, $K\lambda_1 + \lambda_2 = 0$, from which $K = -\lambda_2/\lambda_1$. Therefore, the solution for D can be written as

$$\begin{aligned} D(\tau; u) &= -\frac{1}{R}\left(\frac{-\lambda_2 e^{\lambda_1\tau} + \lambda_2 e^{\lambda_2\tau}}{-\lambda_2/\lambda_1 e^{\lambda_1\tau} + e^{\lambda_2\tau}}\right) = -\frac{\lambda_2}{R}\left(\frac{1 - e^{(\lambda_1 - \lambda_2)\tau}}{1 - (\lambda_2/\lambda_1)e^{(\lambda_1-\lambda_2)\tau}}\right) \\ &= \frac{(\tilde{\kappa} - i\xi\rho u) + g(u)}{\xi^2}\frac{1 - e^{g(u)\tau}}{1 - h(u)e^{g(u)\tau}}, \end{aligned}$$

where

$$\begin{aligned} g(u) &:= \lambda_1 - \lambda_2 = \sqrt{(\tilde{\kappa} - i\xi\rho u)^2 + (u^2 + iu)\xi^2}, \\ h(u) &:= \lambda_2/\lambda_1 = -(g(u) + (\tilde{\kappa} - i\xi\rho u))/(g(u) - (\tilde{\kappa} - i\xi\rho u)). \end{aligned}$$

The solution for C is found by integrating the second differential equation in (7.53) and using the initial condition $C(0; u) = 0$:

$$\begin{aligned} C(\tau; u) &= \int_0^\tau (iur + \tilde{\theta}\tilde{\kappa}D(s))\,\mathrm{d}s = iur\tau + \tilde{\theta}\tilde{\kappa}\frac{(\tilde{\kappa} - i\xi\rho u) + g(u)}{\xi^2}\int_0^\tau \frac{1 - e^{g(u)s}}{1 - h(u)e^{g(u)s}}\,\mathrm{d}s \\ &= iur\tau + \tilde{\theta}\tilde{\kappa}\frac{(\tilde{\kappa} - i\xi\rho u) + g(u)}{\xi^2}\left(\tau + \left(\frac{1 - h(u)}{g(u)h(u)}\right)\ln\left(\frac{1 - h(u)e^{g(u)\tau}}{1 - h(u)}\right)\right). \end{aligned}$$

We can now plug the above expressions for the functions C and D in the formula (7.51) for the characteristic function φ. After simplifying the resulting equation, we obtain

$$\varphi(u; \tau, x, v) = \frac{\exp\left(iux + iru\tau + \dfrac{\tilde{\kappa}\tilde{\theta}\beta(u)\tau}{\xi^2} - \dfrac{(u^2 + iu)v}{g(u)\coth\left(\frac{g(u)\tau}{2}\right) + \beta(u)}\right)}{\cosh\left(\frac{g(u)\tau}{2}\right) + \left(\frac{\beta(u)}{g(u)}\right)\sinh\left(\frac{g(u)\tau}{2}\right)}, \tag{7.57}$$

where $\beta(u) := \tilde{\kappa} - i\xi\rho u$. Here, we use hyperbolic functions

$$\cosh(x) = \frac{e^x + e^{-x}}{2}, \quad \sinh(x) = \frac{e^x - e^{-x}}{2}, \quad \text{and} \quad \coth(x) = \frac{\cosh(x)}{\sinh(x)} = \frac{e^x + e^{-x}}{e^x - e^{-x}}.$$

A typical plot of the characteristic function $\varphi(u)$ for the Heston model is given in Figure 7.3. Since $\varphi(u)$ is a complex-valued function, we plot its real, $\Re\varphi(u)$, and imaginary, $\Im\varphi(u)$, parts separately. We can notice the oscillatory behaviour of both functions. Also, the amplitude of each of $\Re\varphi(u)$ and $\Im\varphi(u)$ is decreasing and converges to zero, as $u \to \infty$.

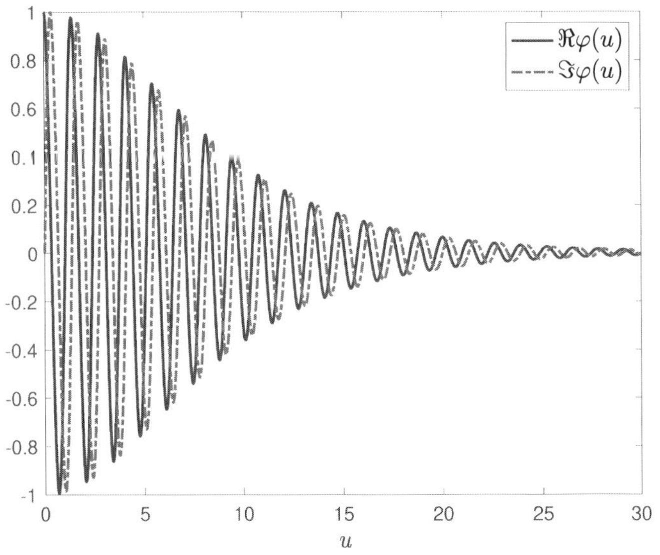

FIGURE 7.3: A typical plot of the characteristic function for the Heston model. The real part, $\Re\varphi(u)$, and the imaginary part, $\Im\varphi(u)$, are plotted for the following choice of parameters: $\tau = 0.5$, $x = \ln 100$, $v = 0.5$, $r = 0.04$, $\tilde{\kappa} = 5$, $\tilde{\theta} = 0.05$, $\xi = 0.5$, and $\rho = -0.8$.

7.3.2 Implied Volatility for the Heston Model

One of the important properties of the Heston model is its ability to produce non-flat shapes of the *implied Black–Scholes volatility*. Let $c_{\mathrm{BS}}(\tau, S; \sigma)$ denote the no-arbitrage value of a standard European call option for time-to-maturity τ and spot price S produced by the Black–Scholes model with volatility σ. We fix other parameters such as the strike price K, the dividend yield q, and the risk-free interest rate r. The implied Black–Scholes volatility, denoted σ_{BS}, is defined as a solution to the equation

$$c_{\mathrm{BS}}(\tau, S; \sigma) = c_{\mathrm{market}},$$

where we equate the Black–Scholes price to the market value of the option. Since c_{BS} is an increasing function of σ, the solution σ_{BS} exists and is unique, as long as the market price c_{market} admits no arbitrage. It is known that the Black–Scholes volatility implied by the market prices of European call and put options is not a constant function of the moneyness $M := S/K$. Instead, σ_{BS} being plotted as a function of M exhibits a skewed "smile" shape. Two options on the same underlying with the same maturity but different strikes can imply different volatilities. The pattern of the implied BS volatility differs across various markets. This fact indicates deficiencies in the standard Black–Scholes option pricing model from the market's viewpoint. We show some volatility smiles in Figure 7.4b.

The Black–Scholes model assumes constant volatility and normal distributions of underlying asset's log returns. To address imperfections of the model, practitioners use advanced asset price models with stochastic and state-dependent volatilities, such as the Heston model and the CEV model, as well as the jump diffusion processes considered in the next section. The implied BS volatility is also a useful visual tool for studying option price models. In Figure 7.4, we plot the no-arbitrage call value, $c(\tau, S)$, and the respective implied Black–Scholes volatility, $\sigma_{\rm BS}$, as functions of moneyness M for the Heston model, where the correlation coefficient ρ equals -50%, 0%, or 50%. The shape of the implied BS vol is a symmetric smile when $\rho = 0$. It means that options whose strike price K differs substantially from the spot price S have higher prices (and thus larger implied volatilities) than what is given by the standard Black–Scholes option pricing model.

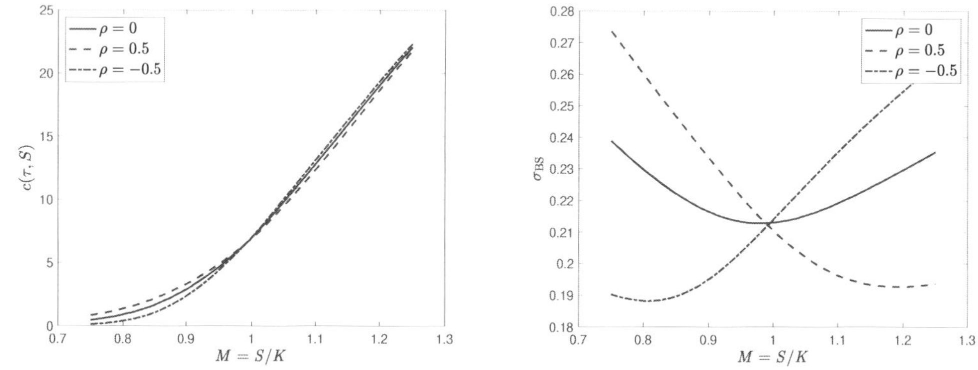

(a) No-arbitrage values of the standard call option under the Heston model.

(b) Implied Black–Scholes volatility under the Heston model.

FIGURE 7.4: Plots of the no-arbitrage call value, $c(\tau, S)$, and the respective implied Black–Scholes volatility, $\sigma_{\rm BS}$, as functions of moneyness $M := S/K$ under the Heston model with $S = 100$, $\tau = 0.5$, $\mathrm{v}(0) = 0.5$, $r = 0.04$, $\tilde{\kappa} = 2$, $\tilde{\theta} = 0.05$, $\xi = 0.5$, and $\rho \in \{-0.5, 0, 0.5\}$.

7.4 Models with Jumps

7.4.1 The Poisson Process

Suppose that events of interest are occurring at random time moments. Let $N(t)$ be the number of occurrences from time 0 to time t. By varying $t \in \mathbb{R}_+$, we obtain a random process $\{N(t)\}_{t\geqslant 0}$. Initially, we set $N(0) = 0$. Assume that only an individual event may occur at a time. Therefore, each possible realization (a sample path) of $N(t)$ is a nondecreasing step function that only changes by jumps of size one. As a result, we obtain a fundamental pure jump process known as the *Poisson process.*

Definition 7.1. A Poisson process is an integer-valued stochastic process $\{N(t)\}_{t\geqslant 0}$ with $N(0) = 0$ that satisfies the following properties.

(a) The numbers of events that occur in disjoint time intervals are independent. That is, for all s_1, s_2, t_1, t_2 with $0 \leqslant s_1 < s_2 \leqslant t_1 < t_2$ the random variables $N(s_2) - N(s_1)$ and $N(t_2) - N(t_1)$ are independent.

(b) The distribution of the number of events that occur on a given time interval only depends on the length and does not depend on the location of the interval. That is, for all s and

t with $0 \leqslant s < t$, the probability distribution of $N(t) - N(s)$ depends on $t - s$ and does not depend on t and s considered individually.

(c) There exists a constant $\lambda > 0$ such that for any small interval of length δt, the probability of having one event occurring in a given small time interval, $[t, t + \delta t]$, is approximately $\lambda \delta t$ plus an error of small order $o(\delta t)$. That is,

$$\lim_{\delta t \searrow 0} \frac{\mathbb{P}(N(t + \delta t) - N(t) = 1)}{\delta t} = \lambda,$$

whereas the probability to have two or more events occurring in $[t, t + \delta t]$ is $o(\delta t)$. That is,

$$\lim_{\delta t \searrow 0} \frac{\mathbb{P}(N(t + \delta t) - N(t) \geqslant 2)}{\delta t} = 0.$$

Properties (a) and (b) mean that $\{N(t)\}_{t \geqslant 0}$ is a process with independent and stationary increments. Additionally, as is shown just below, the above properties lead to the fact that all increments of the process $\{N(t)\}_{t \geqslant 0}$ are Poisson distributed.

Proposition 7.3. *For all s and t such that $0 \leqslant t < s$,*

$$N(s) - N(t) \sim \mathit{Pois}(\lambda(s - t)).$$

Proof. The proof is left as an exercise for the reader. □

Since $N(0) = 0$, Proposition 7.3 implies that for any $t > 0$ the value $N(t)$ has the Poisson probability distribution $\mathit{Pois}(\lambda t)$. Therefore, the mean and the variance of $N(t)$ are both equal to

$$\mathrm{E}[N(t)] = \mathrm{Var}(N(t)) = \lambda t.$$

In summary, the Poisson process is a process with independent, stationary, and Poisson-distributed increments. The number $\lambda > 0$ is called the *rate* or *intensity*. For the Poisson process, the rate equals the average number of jumps per one unit of time, i.e.,

$$\lambda = \frac{\mathrm{E}[N(s) - N(t)]}{s - t} \quad \text{for} \quad 0 \leqslant t < s.$$

The Poisson process with rate λ is sometimes denoted by $N_\lambda(t)$. Note that although we assume here that λ is constant, in general, the intensity can be a function of time t defined by

$$\lambda(t) = \lim_{s \to t} \frac{\mathrm{E}[N(s) - N(t)]}{s - t}.$$

Let T_k be the occurrence time of the kth event or, equivalently, the moment when the Poisson process makes the kth jump for $k = 1, 2, \ldots$. These random times $T_1, T_2, \ldots$ form an increasing sequence of positive reals. Let us find the probability distribution of T_1. The probability that $T_1 > t$ for some $t > 0$ is the same as the probability that no jumps occur in the interval $[0, t]$. So, we have

$$\mathbb{P}(T_1 > t) = \mathbb{P}(N(t) = 0) = \mathrm{e}^{-\lambda t}.$$

Therefore, the PDF of T_1 is

$$f_{T_1}(T) = \frac{\partial \mathbb{P}(T_1 \leqslant t)}{\partial t} = \frac{\partial (1 - \mathrm{e}^{-\lambda t})}{\partial t} = \lambda \mathrm{e}^{-\lambda t},$$

for all $t > 0$, and zero otherwise. It is an exponential density, and hence $T_1 \sim \mathit{Exp}(\lambda)$.

To find the probability distribution of T_k for any $k \geqslant 1$, we use the equivalence of the events $\{T_k > t\}$ and $\{N(t) \leqslant k-1\}$ for any $t > 0$ and $k \geqslant 1$. In other words, a Poisson process has its kth jump after time t iff its time-t value does not exceed $k-1$. So, we have $\mathbb{P}(T_k > t) = \mathbb{P}(N(t) \leqslant k-1)$. This equivalence allows us to find the probability distribution of T_k. As shown below, it coincides with the gamma distribution $Gamma(k, \lambda)$, also known as the Erlang distribution. Using the fact that $N(t) \sim Pois(\lambda t)$, we can compute the probability of the event $\{N(t) \leqslant k-1\}$ and hence find the CDF of T_k as follows:

$$F_{T_k}(t) = 1 - \mathbb{P}(T_k > t) = 1 - \mathbb{P}(N(t) \leqslant k-1) = 1 - \sum_{\ell=0}^{k-1} \frac{(\lambda t)^\ell \mathrm{e}^{-\lambda t}}{\ell!}.$$

Now, we can find the PDF of T_k by differentiating the CDF

$$\begin{aligned} f_{T_k}(t) = F'_{T_k}(t) &= \lambda \mathrm{e}^{-\lambda t} \sum_{\ell=0}^{k-1} \frac{(\lambda t)^\ell}{\ell!} - \mathrm{e}^{-\lambda t} \sum_{\ell=1}^{k-1} \frac{\ell \lambda (\lambda t)^{\ell-1}}{\ell!} \\ &= \lambda \mathrm{e}^{-\lambda t} \left(\sum_{\ell=0}^{k-1} \frac{(\lambda t)^\ell}{\ell!} - \sum_{\ell=0}^{k-2} \frac{(\lambda t)^\ell}{\ell!} \right) = \lambda \mathrm{e}^{-\lambda t} \frac{(\lambda t)^{k-1}}{(k-1)!} = \frac{\lambda^k}{\Gamma(k)} t^{k-1} \mathrm{e}^{-\lambda t}. \end{aligned}$$

In the last line, we can now recognize the PDF of the gamma probability distribution $Gamma(\alpha, \lambda)$ with shape parameter α equal to k and rate parameter λ.

In addition to the above result, one can show that the increments

$$\tau_j := T_j - T_{j-1} \text{ for } j = 1, 2, \ldots$$

with $T_0 = 0$ form a sequence of i.i.d. random variables having the exponential distribution with parameter $\lambda > 0$. In other words, the joint complementary cumulative distribution function of $\tau_1, \tau_2, \ldots, \tau_n$ is given by

$$\mathbb{P}(\tau_1 > t_1, \tau_2 > t_2, \ldots, \tau_n > t_n) = \mathrm{e}^{-\lambda(t_1 + t_2 + \cdots + t_n)}$$

for all $n \geqslant 1$ and all $t_1, t_2, \ldots, t_n \geqslant 0$. Thus, the time T_k of the kth jump is a sum of k i.i.d. $Exp(\lambda)$-distributed random variables:

$$T_k = T_1 - T_0 + T_2 - T_1 + \cdots + T_k - T_{k-1} = \tau_1 + \tau_2 + \cdots + \tau_k, \quad k \geqslant 1.$$

Since $\mathrm{E}[\tau_j] = {}^1/_\lambda$ for any $j \geqslant 1$, the mean of T_k is $\mathrm{E}[T_k] = {}^k/_\lambda$.

This observation leads us to the second definition of the Poisson process presented below. It is more constructive than the first one since it also gives us a simple algorithm for generating sample paths of a Poisson process.

Definition 7.2. Consider a sequence of i.i.d. exponentially distributed random variables, $\{\tau_j\}_{j \geqslant 1}$, having common mean $\lambda^{-1} > 0$. The *Poisson process* $\{N(t)\}_{t \geqslant 0}$ is defined as

$$N(t) := \max\{k \,:\, T_k \leqslant t\} = \sum_{k \geqslant 1} \mathbb{I}_{\{T_k \leqslant t\}}, \tag{7.58}$$

where $T_k := \sum_{j=1}^k \tau_j$ for $k \geqslant 1$.

We can show the equivalence of our two definitions of the Poisson process. First of all, the process defined in (7.58) starts at zero. It makes a jump of size one at each time T_k, i.e., T_k is the moment of the kth jump. Each τ_j is a time lag between two successive jumps of the process. Let us find the distribution of $N(t)$ for $t > 0$. First, notice that, since T_k is

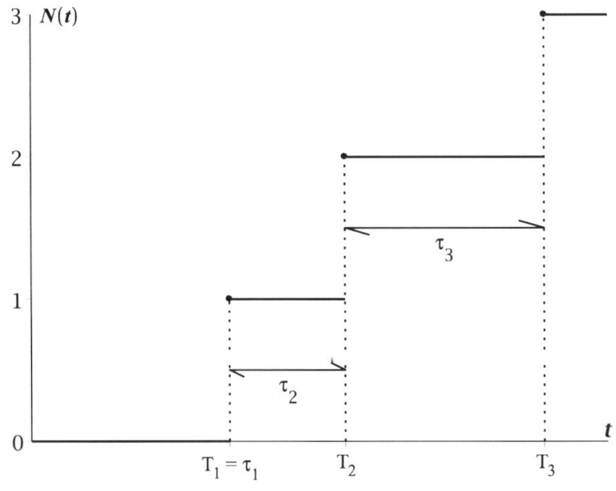

FIGURE 7.5: A sample path of a Poisson process.

a sum of k i.i.d. $Exp(\lambda)$-distributed random variables for any $k \geqslant 1$, the occurrence time T_k has the gamma distribution $Gamma(k, \lambda)$.The probability of having at least k jumps by time t is

$$\mathbb{P}(N(t) \geqslant k) = \mathbb{P}(T_k \leqslant t) = \int_0^t f_{T_k}(x)\,\mathrm{d}x = \int_0^t \frac{\lambda^k x^{k-1}}{(k-1)!} \mathrm{e}^{-\lambda x}\,\mathrm{d}x.$$

Integration by parts gives

$$\int_0^t \frac{\lambda^{k+1} x^k}{k!} \mathrm{e}^{-\lambda x}\,\mathrm{d}x = \int_0^t \frac{\lambda^k x^{k-1}}{(k-1)!} \mathrm{e}^{-\lambda x}\,\mathrm{d}x - \frac{(\lambda t)^k}{k!}\mathrm{e}^{-\lambda t}.$$

Therefore, the probability of having exactly k jumps by time t is

$$\begin{aligned}\mathbb{P}(N(t) = k) &= \mathbb{P}(N(t) \geqslant k) - \mathbb{P}(N(t) \geqslant k+1)\\ &= \int_0^t \frac{\lambda^k x^{k-1}}{(k-1)!} \mathrm{e}^{-\lambda x}\,\mathrm{d}x - \int_0^t \frac{\lambda^{k+1} x^k}{k!} \mathrm{e}^{-\lambda x}\,\mathrm{d}x = \frac{(\lambda t)^k}{k!}\mathrm{e}^{-\lambda t}\end{aligned}$$

for any $k = 0, 1, 2, \ldots$. That is, $N(t) \sim Pois(\lambda t)$. Lastly, one can show that $N(t) - N(s) \sim Pois(\lambda(t-s))$ and that $N(t) - N(s)$ and $N(s)$ are independent for all t and s with $0 \leqslant s < t$. The proof is a bit technical, and hence we omit it here.

The Poisson process only changes its value at the time moments $T_1, T_2, T_3, \ldots$ forming a strictly increasing sequence $0 < T_1 < T_2 < T_3 < \ldots$. The process is constant in between these jump times. According to the definition in (7.58), the sample paths are right-continuous, nondecreasing, piecewise-constant functions of time. A typical sample path is shown in Figure 7.5. The mean function of the Poisson process is a linear function of time: $\mathrm{E}[N(t)] = \lambda t$. Lastly, we define the *compensated Poisson process* $\{X(t) := N(t) - \lambda t\}_{t \geqslant 0}$, which has zero mean function

$$\mathrm{E}[X(t)] = \mathrm{E}[N(t)] - \lambda t = \lambda t - \lambda t = 0.$$

Moreover, the compensated Poisson process is a martingale. The proof of this fact is left as an exercise for the reader.

7.4.2 Jump Diffusion Models with a Compound Poisson Component

A standard Poisson process has deterministic jumps of size one. A *compound Poisson process* is defined as a Poisson process with jumps of random sizes. Such a process is helpful in insurance to model claims that arrive at times generated by a Poisson process, and the amounts claimed are positive random variables. It is also useful for modelling an asset price process with jumps of random sizes.

Let $\{N(t)\}_{t\geqslant 0}$ be the Poisson process with intensity $\lambda > 0$; let $Y_1, Y_2, \ldots$ be a sequence of i.i.d. random variables with finite common mean $\mathrm{E}[Y]$. Suppose that they are all independent of the Poisson process. The compound Poisson process, denoted $Q(t)$, is defined as

$$Q(t) = \sum_{k=1}^{N(t)} Y_k, \quad t \geqslant 0.$$

That is, the time-t value of the process $Q(t)$ is the cumulative value of all jumps occurred on the interval $[0, t]$. In the example with claims made by policyholders to an insurance company, $\{Y_k\}_{k\geqslant 1}$ are the amounts claimed, $N(t)$ is the number of claims made by time t, and $Q(t)$ gives the total amount claimed by time t. The jumps in $\{Q(t)\}_{t\geqslant 0}$ occur at the same time as the jumps in $\{N(t)\}_{t\geqslant 0}$, but whereas the jumps in $\{N(t)\}_{t\geqslant 0}$ are always of size one, the jumps in $\{Q(t)\}_{t\geqslant 0}$ are of random size. The kth jump that occurs at time T_k is of size Y_k. The standard Poisson process is a special case of the compound Poisson process with $Y_k \equiv 1$ for all $k \geqslant 1$.

Properties of the compound Poisson process are similar to those of the standard Poisson process. The mean function of the compound Poisson process is linear in time as well:

$$\mathrm{E}[Q(t)] = \mathrm{E}\left[\sum_{k=1}^{N(t)} Y_k\right] = \mathrm{E}[N(t)] \cdot \mathrm{E}[Y] = \lambda t\, \mathrm{E}[Y] .$$

Sample paths of the compound Poisson process are right-continuous, piecewise-constant functions of time. If the jumps Y_k are nonnegative (with probability one), then the sample paths are nondecreasing functions. Typical sample paths of a standard Poisson process and a compound Poisson process with jumps uniformly distributed in $(0, 1)$ are given in Figures 7.6a and 7.6b, respectively.

The probability distribution of $Q(t)$ can be derived by conditioning on the number of occurrences. Set $X_n = \sum_{k=1}^{n} Y_k$ for $n \geqslant 1$. Then,

$$\begin{aligned}\mathbb{P}(Q(t) \leqslant x) &= \mathrm{E}\big[\,\mathbb{P}(Q(t) \leqslant x \mid N(t))\,\big] = \sum_{n=0}^{\infty} \mathbb{P}(X_n \leqslant x) \cdot \mathbb{P}(N(t) = n) \\ &= \sum_{n=0}^{\infty} \mathbb{P}(X_n \leqslant x) \cdot \mathrm{e}^{-\lambda t} \frac{(\lambda t)^n}{n!}, \quad \text{for } x \in \mathbb{R}.\end{aligned}$$

For example, it is not difficult to show that if all $Y_k \sim Bin(1, p)$, $0 < p \leqslant 1$, then $\{Q(t)\}_{t\geqslant 0}$ is a Poisson process with intensity λp.

A *jump diffusion process* is defined as the sum of a scaled Brownian motion with drift $\mu t + \sigma W(t)$ and a compound Poisson process $Q(t)$. Assume that all stochastic factors are jointly independent. If the jump diffusion process $X(t)$ starts at X_0, we can write the solution for $X(t)$ as follows:

$$X(t) = X_0 + \mu t + \sigma W(t) + \sum_{k=1}^{N(t)} Y_k, \quad t \geqslant 0. \tag{7.59}$$

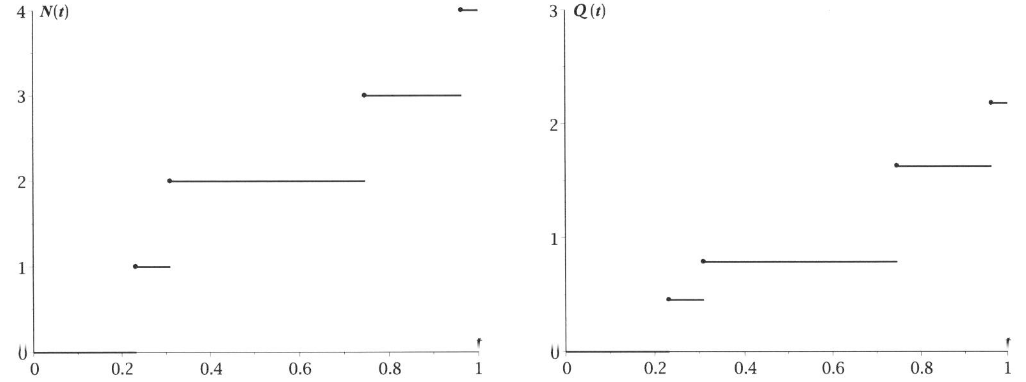

(a) A standard Poisson process with jumps of size one.

(b) A compound Poisson process with uniformly distributed jump sizes.

FIGURE 7.6: Sample paths of standard and compound Poisson processes with $\lambda = 4$.

The jump diffusion process follows Brownian motion (with continuous paths) between the moments of jumps. At the moment of a jump, the value of the process changes by the jump size. That is, the value of at time t right after a jump of size Y occurs is given by $X(t) = X(t^-) + Y$, where $X(t^-)$ is the value right before time t. Thus, a sample path of $X(t)$ is a right-continuous function of time. In Figure 7.7, we can see a typical sample path of a jump diffusion process with normally distributed jump sizes, $Y_k \sim Norm(\alpha, \beta^2)$. Note that a jump diffusion process belongs to a large family of Lévy processes. The scaled Brownian motion with drift, the Poisson process, and the variance gamma process, which is considered in the very end of this chapter, are also members of that family.

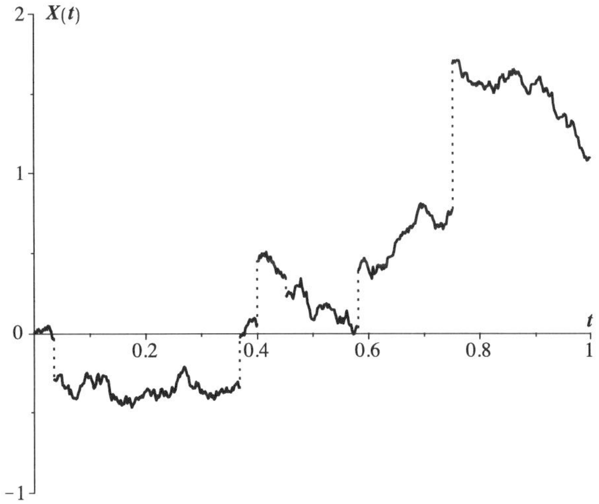

FIGURE 7.7: A sample path of a jump diffusion process with normally distributed jump sizes. The path has six jumps on the interval $[0, 1]$. The parameters are: $X_0 = 0$, $\mu = 0.1$, $\sigma = 0.5$, $\lambda = 4$, $\alpha = 0$, and $\beta = 0.5$.

Let us consider the following jump diffusion model for asset prices;

$$\ln S(t) = \ln S(0) + \mu t + \sigma W(t) + \sum_{k=1}^{N(t)} Y_k\,,\ t \geqslant 0\,, \quad S(0) > 0\,, \tag{7.60}$$

where the standard Brownian motion $\{W(t)\}_{t\geqslant 0}$ is independent of the compound Poisson component. Taking the exponential of both parts of (7.60) gives the asset price solution:

$$S(t) = S(0)\exp\Bigg(\mu t + \sigma W(t) + \underbrace{\sum_{k=1}^{N(t)} Y_k}_{\equiv Q(t)}\Bigg) = S(0)\mathrm{e}^{\mu t+\sigma W(t)+Q(t)}. \tag{7.61}$$

Since the exponential function takes strictly positive values, the price $S(t)$ is strictly positive at every time t. As seen from (7.61), the asset price process follows geometric Brownian motion (with continuous paths) between the moments of jumps. At the moment of a jump, the asset price is multiplied by the exponential of the jump size, $J \equiv \mathrm{e}^Y$. That is, the asset value at time t right after a jump of size J is given by $S(t) = S(t^-) \cdot J$, where $S(t^-)$ is the price right before time t.

To write the asset price process (7.61) as a solution to some SDE, we first define the stochastic integral of a stochastic process $\{\varphi_t\}_{0\leqslant t\leqslant T}$ w.r.t. the compound Poisson process $\{Q(t)\}_{0\leqslant t\leqslant T}$ by

$$\int_0^T \varphi_t \,\mathrm{d}Q(t) = \int_0^T \varphi_t Y_{N(t)} \,\mathrm{d}N(t) := \sum_{k=1}^{N(T)} \varphi_{T_k} Y_k. \tag{7.62}$$

Here, the process $\{\varphi_t\}_{0\leqslant t\leqslant T}$ is adapted to the natural filtration generated by $\{Q(t)\}_{0\leqslant t\leqslant T}$; T_k, $k \geqslant 1$ are jump times of the Poisson process $\{N(t)\}_{0\leqslant t\leqslant T}$. The infinitesimal increment $\mathrm{d}N(t)$ equals 1 if $t = T_k$ for some k and 0 otherwise. The integral in (7.62) can be interpreted as the time-T value of a portfolio containing φ_t shares of a risk asset at time t, whose value follows the compound Poisson process $\{Q(t)\}_{0\leqslant t\leqslant T}$.

The jump diffusion process in (7.59) admits the stochastic integral representation

$$X(t) = X_0 + \int_0^t \mu \,\mathrm{d}u + \int_0^t \sigma \,\mathrm{d}W(u) + \int_0^t Y_{N(u)} \,\mathrm{d}N(u).$$

Thus, the log price $\ln S(t)$ in (7.60) can be written as a sum of the stochastic integral w.r.t. Brownian motion $W(t)$ and the compound Poisson component

$$\ln S(t) = \ln S(0) + \int_0^t \mu \,\mathrm{d}u + \int_0^t \sigma \,\mathrm{d}W(u) + \int_0^t Y_{N(u)} \,\mathrm{d}N(u).$$

This integral representation of the log price process can also be written as the SDE

$$\mathrm{d}\ln S(t) = \mu \,\mathrm{d}t + \sigma \,\mathrm{d}W(t) + Y_{N(t)} \,\mathrm{d}N(t). \tag{7.63}$$

As is seen, if no jump occurs, then the log price simply evolves as scaled Brownian motion with drift. Thus, we can apply Itô's formula between the jump moments to obtain

$$\frac{\mathrm{d}S(t)}{S(t)} = (\mu + \sigma^2/2)\,\mathrm{d}t + \sigma \,\mathrm{d}W(t) \text{ for all } t \in [T_{k-1}, T_k),\ k \geqslant 1,$$

where $T_0 \equiv 0$. If a jump occurs at time t, then the log jump size, $Y \equiv \ln J$, is added to the log price $\ln S(t^-)$, or, equivalently, the asset price value $S(t^-)$ is multiplied by J. Using the relations $\mathrm{d}S(t) := S(t) - S(t^-)$ and $\mathrm{d}N(t) = 1$, we write $\mathrm{d}S(t) = (J-1)S(t^-)\,\mathrm{d}N(t)$. That is, the proportional change in the stock price at time t when a jump occurs is given by $\frac{\mathrm{d}S(t)}{S(t^-)} = (J_{N(t)} - 1)\,\mathrm{d}N(t)$, $k \geqslant 1$. As a result, the asset price process $\{S(t)\}_{t\geqslant 0}$ follows the SDE

$$\frac{\mathrm{d}S(t)}{S(t^-)} = (\mu + \sigma^2/2)\,\mathrm{d}t + \sigma\,\mathrm{d}W(t) + (J_{N(t)} - 1)\,\mathrm{d}N(t),\ t \geqslant 0. \tag{7.64}$$

This SDE can also be written in the integral form

$$S(t) = S(0) + \int_0^t (\mu + \sigma^2/2)S(u)\,\mathrm{d}u + \int_0^t \sigma S(u)\,\mathrm{d}W(u) + \int_0^t S(u^-)(J_{N(u)} - 1)\,\mathrm{d}N(u).$$

As we know, the asset price model is arbitrage free, if there exists an equivalent martingale measure (EMM). Under the EMM $\widetilde{\mathbb{P}}$, the coefficient μ in (7.61) has to be selected such that $\widetilde{\mathrm{E}}_{t,S}[S(u+t)] = \mathrm{e}^{(r-q)t}S$ holds for all $u, t \geqslant 0$. Using the facts that the processes $\{W(t)\}_{t\geqslant 0}$ and $\{N(t)\}_{t\geqslant 0}$ have independent and time stationary increments and that $\{W(t)\}_{t\geqslant 0}$, $\{N(t)\}_{t\geqslant 0}$, and $\{Y_k\}_{k\geqslant 1}$ are mutually independent, we have

$$\begin{aligned}
\mathrm{E}_{t,S}[S(u+t)] &= S\mathrm{e}^{\mu t}\mathrm{E}\left[\mathrm{e}^{\sigma(W(u+t)-W(u))}\right]\mathrm{E}\left[\mathrm{E}[\mathrm{e}^{Y}]^{N(u+t)-N(u)}\right] \\
&= S\mathrm{e}^{\mu t}\mathrm{E}\left[\mathrm{e}^{\sigma W(t)}\right]\mathrm{E}\left[\mathrm{E}[\mathrm{e}^{Y}]^{N(t)}\right] \\
&= S\mathrm{e}^{\mu t}\mathrm{e}^{t\sigma^2/2}\mathrm{e}^{\lambda t(a-1)} = S\mathrm{e}^{(\mu+\sigma^2/2+\lambda(a-1))t},
\end{aligned}$$

where $a := \mathrm{E}[\mathrm{e}^Y]$, and $Y \stackrel{d}{=} Y_k$, $k \geqslant 1$. Therefore, under the EMM $\widetilde{\mathbb{P}}$, the drift coefficient μ is given by

$$\mu = r - q - \sigma^2/2 + \lambda(1-a). \tag{7.65}$$

The SDE in (7.64) takes the following form under $\widetilde{\mathbb{P}}$:

$$\frac{\mathrm{d}S(t)}{S(t^-)} = (r - q + \lambda(1-a))\,\mathrm{d}t + \sigma\,\mathrm{d}\widetilde{W}(t) + (J_{N(t)} - 1)\,\mathrm{d}N(t). \tag{7.66}$$

The solution for $S(t)$ given in (7.61) is then

$$S(t) = S_0 \exp\left((r - q - \sigma^2/2 + \lambda(1-a))t + \sigma\widetilde{W}(t) + \sum_{k=1}^{N(t)} Y_k\right). \tag{7.67}$$

Note that $\widetilde{\mathbb{P}}$ is one of several martingale measures (for the numéraire asset $B(t) = \mathrm{e}^{rt}$) that are equivalent to the real-world measure $\mathbb{P}$. Since the asset price solution $S(t)$ has two sources of randomness, namely, $W(t)$ and $Q(t)$, a jump diffusion asset pricing model is an incomplete market model that admits infinitely many EMMs. In the construction of the EMM $\widetilde{\mathbb{P}}$ we assumed that the intensity λ and the jump-size probability distribution remain the same. Other EMMs can be constructed by changing the distribution of the compound Poisson process $Q(t)$. The general structure of a Radon–Nikodym derivative for a jump diffusion process is discussed in Subsection 7.4.5.

The jump size $J \equiv \mathrm{e}^Y$ can be taken to be a nonrandom number or a random variable. By assuming that J has a log-normal probability distribution (and hence the log jump Y is normal), we obtain the *Merton model*. In the case of normal log jumps, we can derive a closed-form pricing formula for standard European options by conditioning on the number of jumps.

7.4.3 The Merton Jump Diffusion Model

Let all log jumps Y_j, $j \geqslant 1$ be independent normally distributed random variables with common mean α and variance β^2. When the value of $N(t)$ in (7.60) is fixed, the log price $\ln S(t)$ becomes a sum of normal random variables and is therefore a normal variable as well. The distribution of the log price $\ln S(t)$ conditional on the value of the Poisson process $N(t)$ has the mean

$$\mathrm{E}[\ln S(t) \mid N(t)] = \ln S_0 + \mu t + N(t)\mathrm{E}[Y] = \ln S_0 + \mu t + \alpha N(t)$$

and the variance

$$\mathrm{Var}(\ln S(t) \mid N(t)) = \sigma^2 t + N(t)\,\mathrm{Var}(Y) = \sigma^2 t + \beta^2 N(t).$$

The log price $\ln S(t)$ conditional on $N(t)$ is expressed in terms of a standard normal variate $Z \sim \mathit{Norm}(0,1)$ as follows:

$$(\ln S(t) \mid N(t)) \stackrel{d}{=} \big(\ln S_0 + \mu t + \alpha N(t)\big) + \sqrt{\sigma^2 + \beta^2 N(t)/t}\,\sqrt{t}Z,$$

The conditional CDF of $\ln S(t)$ is

$$F_{\ln S(t)|N(t)}(x|n) = \mathbb{P}(\ln S(t) \leqslant x \mid N(t) = n) = \mathcal{N}\left(\frac{x - (\ln S_0 + \mu t + \alpha n)}{\sqrt{\sigma^2 t + \beta^2 n}}\right).$$

To obtain the unconditional CDF of the log price, we use the law of total probability. The unconditional CDF of $\ln S(t)$ is then

$$\begin{aligned} F_{\ln S(t)}(x) &= \mathbb{P}(\ln S(t) \leqslant x) = \mathrm{E}\big[\mathbb{P}(\ln S(t) \leqslant x \mid N(t))\big] \\ &= \sum_{n=0}^{\infty} \mathbb{P}(\ln S(t) \leqslant x \mid N(t) = n) \cdot \mathbb{P}(N(t) = n) \\ &= \sum_{n=0}^{\infty} \mathcal{N}\left(\frac{x - (\ln S_0 + \mu t + \alpha n)}{\sqrt{\sigma^2 t + \beta^2 n}}\right) \cdot \mathrm{e}^{-\lambda t}\frac{(\lambda t)^n}{n!}. \end{aligned}$$

The PDF $f_{\ln S(t)}(x)$ is produced by taking a derivative of the above CDF:

$$f_{\ln S(t)}(x) = F'_{\ln S(t)}(x) = \sum_{n=0}^{\infty} \frac{1}{\sqrt{\sigma^2 t + \beta^2 n}} n\left(\frac{x - (\ln S_0 + \mu t + \alpha n)}{\sqrt{\sigma^2 t + \beta^2 n}}\right) \cdot \mathrm{e}^{-\lambda t}\frac{(\lambda t)^n}{n!}.$$

Thus, the probability distribution of the log price $\ln S(t)$ is a mixture of normal distributions with Poisson weights.

In the conclusion of this subsection, we derive a closed-form formula of the no-arbitrage price of a European-style option under the Merton jump diffusion model. Let T be the maturity time, and let Λ denote a payoff function. The pricing is performed under the risk-neutral probability measure $\widetilde{\mathbb{P}}$ so that the stock price is given by (7.67) with $a = \mathrm{e}^{\alpha+\beta^2/2}$. Let us fix the Poisson process value $N(T) = n$ and rewrite the formula for $S(T)$ in the GBM form:

$$S(T) = \hat{S}_n \exp\left((r - q - \hat{\sigma}_n^2/2)T + \hat{\sigma}_n \widetilde{W}(T)\right), \tag{7.68}$$

where

$$\hat{S}_n = S_0 \mathrm{e}^{(\lambda(1-a)+(\hat{\sigma}_n^2-\sigma^2)/2)T+\alpha n} = S_0 \mathrm{e}^{\lambda(1-a)T+(\alpha+\beta^2/2)n}, \quad \hat{\sigma}_n^2 = \sigma^2 + n\beta^2/T, \tag{7.69}$$

and $a \equiv \mathrm{E}[\mathrm{e}^{Y}] = \mathrm{e}^{\alpha+\beta^2/2}$ since $Y \sim Norm(\alpha, \beta^2)$. Hence, according to (7.68), the Merton asset price model (with fixed $N(T) = n$) can be reduced to the Black–Scholes model where the underlying asset is modelled as a GBM process with the effective spot value $\hat{S}_n$ and the asset volatility $\hat{\sigma}_n$.

The no-arbitrage value of a standard European option such as a call and a put with strike K and expiry T under the Merton model with fixed $N(T)$ is given by a regular formula. Let us denote the Black–Scholes (BS) no-arbitrage option value by $V_{BS}(\hat{S}_n; \hat{\sigma}_n)$. This is the pricing function for an underlying stock price following a GBM with spot value $\hat{S}_n$, volatility parameter $\hat{\sigma}_n$, interest rate r, and stock dividend yield q. Now the option pricing function under the Merton jump diffusion model, denoted V_M, is obtained by taking the expectation of the BS option value with respect to $N(T)$:

$$
\begin{aligned}
V_M(S_0) &= \widetilde{\mathrm{E}}\big[\mathrm{e}^{-rT}\Lambda(S(T))\big] = \widetilde{\mathrm{E}}\left[\widetilde{\mathrm{E}}\big[\mathrm{e}^{-rT}\Lambda(S(T)) \mid N(T)\big]\right] = \widetilde{\mathrm{E}}\left[V_{BS}(\hat{S}_{N(T)}; \hat{\sigma}_{N(T)})\right] \\
&= \sum_{n=0}^{\infty} V_{BS}(\hat{S}_n; \hat{\sigma}_n)\widetilde{\mathbb{P}}(N(T) = n) = \sum_{n=0}^{\infty} \mathrm{e}^{-\lambda T}\frac{(\lambda T)^n}{n!} V_{BS}(\hat{S}_n; \hat{\sigma}_n),
\end{aligned}
$$

where $\hat{S}_n$ and $\hat{\sigma}_n$ are given by (7.69). For a standard European call option, the pricing formula takes the form

$$
C_M(0, S_0; T, K) = \sum_{n=0}^{\infty} \mathrm{e}^{-\lambda T}\frac{(\lambda T)^n}{n!}\left(\hat{S}_n \mathrm{e}^{-qT}\mathcal{N}(d_+(\hat{S}_n, \hat{\sigma}_n)) - K\mathrm{e}^{-rT}\mathcal{N}(d_-(\hat{S}_n, \hat{\sigma}_n))\right),
$$

where $d_\pm(\hat{S}, \hat{\sigma}) = \frac{\ln(\hat{S}/K)+(r-q\pm\hat{\sigma}^2/2)T}{\hat{\sigma}\sqrt{T}}$. In practice, we truncate the above series when calculating the option price. The number of terms left in the summation is not large, although it depends on the option type, maturity T, and intensity λ.

7.4.4 Characteristic Function for a Jump Diffusion Process

Recall a general solution for a jump diffusion process:

$$
X(t) = X(0) + \mu t + \sigma W(t) + \sum_{k=1}^{N(t)} Y_k, \quad t \geqslant 0.
$$

Here, the standard Brownian motion $\{W(t)\}_{t\geqslant 0}$, the Poisson process $\{N(t)\}_{t\geqslant 0}$ with intensity $\lambda > 0$, and the i.i.d. jump sizes $\{Y_k\}_{k\geqslant 1}$ are jointly independent. Thus, the ChF of $X(t)$ is a product of the ChF of the drifted and scaled Brownian motion starting at $X(0)$, $B(t) := X(0)+\mu t+\sigma W(t)$, and the ChF of the compound Poisson process $Q(t) := \sum_{k=1}^{N(t)} Y_k$. Since the distribution of $B(t)$ is normal, we use results of Example 7.1 to obtain

$$
\varphi_{B(t)}(u) = \exp\left(iuX(0) + iu\mu t - \sigma^2 u^2 t/2\right). \tag{7.70}
$$

The ChF of $Q(t)$ can be found by conditioning on the value of $N(t)$:

$$
\begin{aligned}
\varphi_{Q(t)}(u) &= \mathrm{E}\big[\exp(iuY(t))\big] = \mathrm{E}\left[\mathrm{E}\left[\exp\left(iu\sum_{k=1}^{N} Y_k\right) \,\Big|\, N(t) = N\right]\right] \\
&= \mathrm{E}\left[\mathrm{E}\left[\exp(iuY_1)^N \mid N(t) = N\right]\right] = \mathrm{E}\left[\mathrm{E}\left[\exp(iuY_1)\right]^{N(t)}\right] \\
&= \mathrm{E}\left[\varphi_Y(u)^{N(t)}\right] = \mathrm{E}\left[\mathrm{e}^{(\ln\varphi_Y(u))N(t)}\right] = M_{N(t)}(\ln\varphi_Y(u)).
\end{aligned}
$$

Here, $M_{N(t)}$ denotes the MGF of $N(t)$, and φ_Y is the ChF of the jump size distribution. Using again results of Example 7.1, we obtain

$$\varphi_{Q(t)}(u) = \exp\left(\lambda t\left(\varphi_Y(u) - 1\right)\right). \tag{7.71}$$

Setting $u = -iv$ in the above formula gives us the MGF of the compound Poisson process

$$M_{Q(t)}(v) = \exp\left(\lambda t\left(M_Y(v) - 1\right)\right), \tag{7.72}$$

where $M_Y(v)$ denotes the MGF of Y.

Combining (7.70) and (7.71) gives

$$\begin{aligned}\varphi_{X(t)}(u) &= \varphi_{B(t)}(u)\varphi_{Q(t)}(u)\\ &= \exp\left(iuX(0) + iu\mu t - \sigma^2u^2t/2 + \lambda t\left(\varphi_Y(u) - 1\right)\right).\end{aligned} \tag{7.73}$$

If $X(0) = 0$, then the ChF $\varphi_{X(t)}(u)$ can be represented in the form

$$\varphi_{X(t)}(u) = \mathrm{e}^{t\Psi(u)}, \text{ where } \Psi(u) := iu\mu - \sigma^2u^2/2 + \lambda\left(\varphi_Y(u) - 1\right)$$

for all $t \geqslant 0$ and all $u \in \mathbb{R}$. The function $\Psi(u)$ is called the *characteristic exponent* of X.

We can consider following choices for the probability distribution of jump sizes Y_k:

The normal distribution *Norm*(α, β^2). Its ChF is $\varphi(u) = \mathrm{e}^{i\alpha u - \beta^2u^2/2}$.

The asymetric double exponential distribution with the PDF

$$f(x) = p\eta_1\mathrm{e}^{-\eta_1 x}\mathbb{I}_{\{x\geqslant 0\}} + (1-p)\eta_2\mathrm{e}^{-\eta_2|x|}\mathbb{I}_{\{x<0\}},$$

where $\eta_1 > 1$, $\eta_2 > 0$, and $p \in (0, 1)$. Here, η_1 and η_2 are the rates of positive and negative jumps, respectively; p is the probability of a positive jump (hence, $1 - p$ is the probability of a negative jump). The ChF is

$$\varphi(u) = \frac{p\eta_1}{\eta_1 - iu} + \frac{(1-p)\eta_2}{\eta_2 + iu}.$$

The Laplace distribution with the PDF $f(x) = \frac{1}{b}\exp\left(-\frac{|x-a|}{b}\right)$ with $b > 0$. Its ChF is given by

$$\varphi(u) = \frac{\mathrm{e}^{iau}}{1 + b^2u^2}.$$

Example 7.4. Consider the jump diffusion asset price model with normally distributed jump sizes. The risk-neutral dynamics (under $\widetilde{\mathbb{P}}$) of asset prices $S(t)$ is described by the solution

$$S(t) = S(0)\,\mathrm{e}^{(r-q)t - \omega t + \sigma\widetilde{W}(t) + \widetilde{Q}(t)}, \quad t \geqslant 0,$$

with risk-free rate $r \geqslant 0$, dividend yield $q \geqslant 0$, volatility $\sigma > 0$, standard Brownian motion $\widetilde{W}(t)$, and compound Poisson process $\widetilde{Q}(t) = \sum_{k=1}^{N_\lambda(t)} Y_k$ with $Y_k \sim$ *Norm*(α, β^2).

(a) Find the characteristic function $\widetilde{\varphi}(u)$ of the log price $X(t) := \ln S(t)$.

(b) For what value of ω, the process $\mathrm{e}^{-(r-q)t}S(t)$ is a $\widetilde{\mathbb{P}}$-martingale?

Solution. The log price $X(t)$ is a jump diffusion process given by

$$X(t) = \ln S(0) + (r - q - \omega)t + \sigma\widetilde{W}(t) + \sum_{k=1}^{N_\lambda(t)} Y_k, \quad t \geqslant 0.$$

Since the jump sizes Y_k are $Norm(\alpha, \beta^2)$-distributed, the ChF of Y_k is $\varphi_Y(u) = \mathrm{e}^{i\alpha u - \beta^2 u^2/2}$. Using Equation (7.73) gives us the ChF

$$\widetilde{\varphi}(u) = \exp\left(iu \ln S(0) + iu(r - q - \omega)t - \sigma^2 u^2/2 + \lambda t\left(\mathrm{e}^{i\alpha u - \beta^2 u^2/2} - 1\right)\right).$$

The risk-neutral value of ω can be found using the result of Example 7.3 or the strong solution for $S(t)$ as given in Equation (7.67). Indeed, from Equation (7.16) we obtain that

$$\begin{aligned}\omega &= \frac{1}{t}\ln \varphi_{\sigma\widetilde{W}(t)+\widetilde{Q}(t)}(-i) \\ &= \frac{1}{t}\left(-\sigma^2(-i)^2 t/2 + \lambda t\left(\mathrm{e}^{-\alpha i^2 - \beta^2(-i)^2/2} - 1\right)\right) \\ &= \sigma^2/2 + \lambda(\mathrm{e}^{\alpha+\beta^2/2} - 1).\end{aligned}$$

Alternatively, Equation (7.67) gives us the same expression for ω:

$$\omega = \sigma^2/2 + \lambda(a - 1) = \sigma^2/2 + \lambda(\mathrm{e}^{\alpha+\beta^2/2} - 1),$$

where $a = \mathrm{E}[\mathrm{e}^Y] = \mathrm{e}^{\alpha+\beta^2/2}$ for $Y \sim Norm(\alpha, \beta^2)$. □

In Figure 7.8, we plot the implied Black–Scholes (BS) volatility surface, σ_{BS}, as a function of moneyness $M = S/K$ and time-to-maturity τ for the Merton model. Recall that the volatility is constant and independent of M and τ for the standard Black–Scholes option pricing model. Under the Merton model, the implied BS volatility surface has a pronounced smile for τ close to zero. The surface becomes flat with the growth of τ.

7.4.5 Change of Measure for Jump Diffusion Processes

Consider a jump diffusion process $X(t) = X_0 + \mu t + \sigma W(t) + \sum_{k=1}^{N(t)} Y_k$, $0 \leqslant t \leqslant T$ with $T > 0$, defined on a filtered probability space $\left(\Omega, \mathcal{F}, \mathbb{P}, \mathbb{F} = (\mathcal{F}_t)_{0\leqslant t\leqslant T}\right)$. Assume that the standard Brownian motion $\{W(t)\}_{0\leqslant t\leqslant T}$, the Poisson process $\{N(t)\}_{0\leqslant t\leqslant T}$ with intensity $\lambda > 0$, and random jump sizes $\{Y_k\}_{k\geqslant 1}$ are mutually independent. As we know, probability measures can be introduced via a Radon–Nikodym derivative. Girsanov's theorem, which is discussed in Chapter 2, allows for constructing an equivalent probability measure that only affects the drift rate of the diffusion part. Indeed, define $\varrho_t = \exp\left(-\frac{1}{2}\gamma^2 t + \gamma W(t)\right)$ with $0 \leqslant t \leqslant T$ and the probability measure $\widehat{\mathbb{P}}$ by the Radon–Nikodym derivative $\left(\frac{\mathrm{d}\widehat{\mathbb{P}}}{\mathrm{d}\mathbb{P}}\right)_T \equiv \varrho_T$. Then, the process $\{\widehat{W}(t)\}_{0\leqslant t\leqslant T}$ given by $\widehat{W}(t) = W(t) - \gamma t$ is a standard $\widehat{\mathbb{P}}$-BM w.r.t. filtration $\mathbb{F}$. The solution for the jump diffusion $X(t)$ under $\widehat{\mathbb{P}}$ is $X(t) = X_0 + (\mu + \gamma\sigma)t + \sigma\widehat{W}(t) + \sum_{k=1}^{N(t)} Y_k$. This method of constructing equivalent probability measures is not the only one for processes with jumps. We can also work with the compound Poisson process and introduce equivalent probability measures that change the intensity λ and the probability distribution of jump sizes. Our objective is to derive a more general Radon–Nikodym derivative and show that there are multiple equivalent probabilities measures. In particular, we can construct multiple EMMs, and hence the asset price model based on a jump diffusion model is incomplete.

Let us begin with an analogue of Girsanov's theorem for a standard Poisson process.

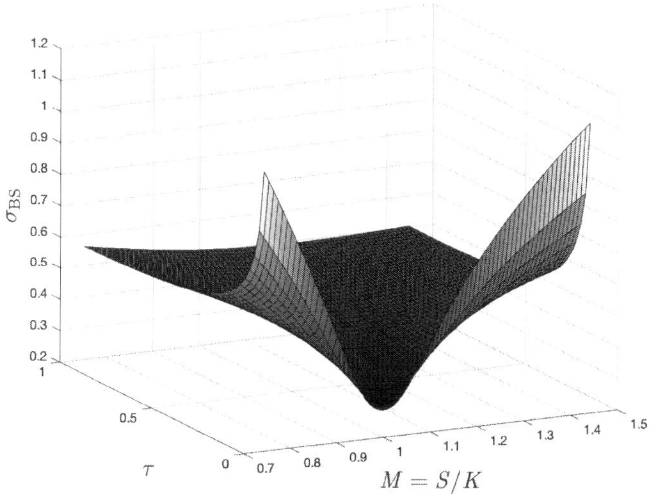

FIGURE 7.8: The implied Black–Scholes volatility σ_{BS} is plotted as a function of moneyness $M := S/K$ and time-to-maturity τ for the Merton jump diffusion model with $S = 100$, $r = 0.04$, $q = 0$, $\sigma = 0.2$, $\lambda = 1$, $\alpha = 0$, $\beta = 0.5$.

Proposition 7.4. *Let $N(t)$, $0 \leqslant t \leqslant T$, be a Poisson process with intensity $\lambda > 0$ under the probability measure $\mathbb{P}_\lambda$. For $\widetilde{\lambda} > 0$, define the equivalent probability measure $\mathbb{P}_{\widetilde{\lambda}}$ by the Radon–Nikodym derivative*

$$\varrho_T \equiv \left(\frac{\mathrm{d}\mathbb{P}_{\widetilde{\lambda}}}{\mathrm{d}\mathbb{P}_\lambda}\right)_T = \mathrm{e}^{(\lambda-\widetilde{\lambda})T}\left(\frac{\widetilde{\lambda}}{\lambda}\right)^{N(T)}.$$

Then, under $\mathbb{P}_{\widetilde{\lambda}}$, $\{N(t)\}_{0\leqslant t\leqslant T}$ is a Poisson process with rate $\widetilde{\lambda}$.

Proof. It is easy to see that the process

$$\rho_t := \mathrm{e}^{(\lambda-\widetilde{\lambda})t}\left(\frac{\widetilde{\lambda}}{\lambda}\right)^{N(t)} = \mathrm{e}^{(\lambda-\widetilde{\lambda})t+N(t)\ln(\widetilde{\lambda}/\lambda)}, \quad 0 \leqslant t \leqslant T$$

is a $\mathbb{P}_\lambda$-martingale. Indeed, using the independence and stationarity of increments of $N(t)$, for any s and t with $0 \leqslant s < t \leqslant T$, we have

$$\begin{aligned}
\mathrm{E}^{\mathbb{P}_\lambda}[\varrho_t \mid \mathcal{F}_s] &= \mathrm{e}^{(\lambda-\widetilde{\lambda})t}\mathrm{E}^{\mathbb{P}_\lambda}\left[\mathrm{e}^{N(t)\ln(\widetilde{\lambda}/\lambda)} \mid \mathcal{F}_s\right] \\
&= \mathrm{e}^{(\lambda-\widetilde{\lambda})t+N(s)\ln(\widetilde{\lambda}/\lambda)}\mathrm{E}^{\mathbb{P}_\lambda}\left[\mathrm{e}^{(N(t)-N(s))\ln(\widetilde{\lambda}/\lambda)} \mid \mathcal{F}_s\right] \\
&= \mathrm{e}^{(\lambda-\widetilde{\lambda})t+N(s)\ln(\widetilde{\lambda}/\lambda)}\mathrm{E}^{\mathbb{P}_\lambda}\left[\mathrm{e}^{N(t-s)\ln(\widetilde{\lambda}/\lambda)}\right] \\
&= \mathrm{e}^{(\lambda-\widetilde{\lambda})t+N(s)\ln(\widetilde{\lambda}/\lambda)}\exp(\lambda(t-s)(\mathrm{e}^{\ln(\widetilde{\lambda}/\lambda)}-1)) \\
&= \mathrm{e}^{(\lambda-\widetilde{\lambda})t+N(s)\ln(\widetilde{\lambda}/\lambda)}\mathrm{e}^{(\widetilde{\lambda}-\lambda)(t-s)} \\
&= \mathrm{e}^{(\lambda-\widetilde{\lambda})s}\left(\frac{\widetilde{\lambda}}{\lambda}\right)^{N(s)} = \varrho_s.
\end{aligned}$$

Here, $\mathrm{E}^{\mathbb{P}_\lambda}[\,]$ denotes the mathematical expectation w.r.t. $\mathbb{P}_\lambda$. A similar notation is adopted for other probability measures. At time $t = 0$, we have $\varrho_0 = 1$. So, ϱ_T is indeed a Radon–Nikodym derivative. Clearly, under $\mathbb{P}_{\widetilde{\lambda}}$, the probability distribution of $N(t)$ is *Pois*$(\widetilde{\lambda}t)$ for any $t \in (0, T]$:

$$\mathbb{P}_{\widetilde{\lambda}}(N(t) = k) = \mathrm{E}^{\mathbb{P}_\lambda}\left[\varrho_t \mathbb{I}_{\{N(t)=k\}}\right] = \mathrm{e}^{(\lambda-\widetilde{\lambda})t}\left(\frac{\widetilde{\lambda}}{\lambda}\right)^k \mathbb{P}_\lambda(N(t) = k)$$

$$= \mathrm{e}^{(\lambda-\widetilde{\lambda})t}\left(\frac{\widetilde{\lambda}}{\lambda}\right)^k \mathrm{e}^{-\lambda t}\frac{(\lambda t)^k}{k!} = \mathrm{e}^{-\widetilde{\lambda}t}\frac{(\widetilde{\lambda}t)^k}{k!}.$$

Lastly, to show that under $\mathbb{P}_{\widetilde{\lambda}}$ the increment $N(t) - N(s)$ is independent of $\mathcal{F}_s$ and have the *Pois*$(\lambda(t-s))$ distribution for any s and t with $0 \leqslant s < t \leqslant T$, we compute the conditional MGF as follows:

$$\begin{aligned}
\mathrm{E}^{\mathbb{P}_{\widetilde{\lambda}}}\left[\mathrm{e}^{u(N(t)-N(s))} \,\middle|\, \mathcal{F}_s\right] &= \mathrm{E}^{\mathbb{P}_\lambda}\left[\frac{\varrho_t}{\varrho_s}\mathrm{e}^{u(N(t)-N(s))} \,\middle|\, \mathcal{F}_s\right] \\
&= \mathrm{e}^{(\lambda-\widetilde{\lambda})(t-s)}\mathrm{E}^{\mathbb{P}_\lambda}\left[\mathrm{e}^{(u+\ln(\widetilde{\lambda}/\lambda))(N(t)-N(s))} \,\middle|\, \mathcal{F}_s\right] \\
&= \mathrm{e}^{(\lambda-\widetilde{\lambda})(t-s)}\mathrm{E}^{\mathbb{P}_\lambda}\left[\mathrm{e}^{(u+\ln(\widetilde{\lambda}/\lambda))N(t-s)}\right] \\
&= \mathrm{e}^{(\lambda-\widetilde{\lambda})(t-s)}\exp(\lambda(t-s)(\mathrm{e}^{u+\ln(\widetilde{\lambda}/\lambda)} - 1)) \\
&= \mathrm{e}^{\widetilde{\lambda}(t-s)(\mathrm{e}^u-1)}.
\end{aligned}$$

We can recognize the unconditional MGF of the *Pois*$(\widetilde{\lambda}(t-s))$ distribution in the final line of the above expression. □

In the case of compound Poisson processes, we can obtain a more general result, where the change measure is extended to variations in jump sizes in addition to intensity variations.

Proposition 7.5. *Let $\{Q(t)\}_{0\leqslant t\leqslant T}$ be a compound Poisson process with intensity $\lambda > 0$ and jump size probability measure $\nu(x)$ under the measure $\mathbb{P}_{\lambda,\nu}$. For $\widetilde{\lambda} > 0$ and another jump size probability measure $\widetilde{\nu}(x)$, equivalent to $\nu(x)$, define the equivalent measure $\mathbb{P}_{\widetilde{\lambda},\widetilde{\nu}}$ by the Radon–Nikodym derivative*

$$\varrho_T \equiv \left(\frac{\mathrm{d}\mathbb{P}_{\widetilde{\lambda},\widetilde{\nu}}}{\mathrm{d}\mathbb{P}_{\lambda,\nu}}\right)_T = \mathrm{e}^{(\lambda-\widetilde{\lambda})T}\left(\frac{\widetilde{\lambda}}{\lambda}\right)^{N(T)}\prod_{j=1}^{N(T)} h(Y_j), \tag{7.74}$$

where $h(x) := \frac{\mathrm{d}\widetilde{\nu}}{\mathrm{d}\nu}(x)$ is the likelihood ratio. Then, under $\mathbb{P}_{\widetilde{\lambda},\widetilde{\nu}}$, the process $Q(t) = \sum_{j=1}^{N(t)} Y_j$, $0 \leqslant t \leqslant T$ is a compound Poisson process with intensity $\widetilde{\lambda}$ and jump size probability measure $\widetilde{\nu}(x)$.

Remarks.

1. Recall that probability measures $\widetilde{\nu}$ and $\widetilde{\nu}(x)$ are said to be *equivalent* if all sets of ν-measure zero are $\widetilde{\nu}$-measure zero and vice versa. That is, for any Borel set A, $\widetilde{\nu}(A) = 0$ iff $\nu(A) = 0$.
2. The function $h(Y)$ with $Y \sim \nu$ is the Radon–Nikodym derivative $\frac{\mathrm{d}\widetilde{\nu}}{\mathrm{d}\nu}$, and for any

measurable and integrable function g of $Y \sim \widetilde{\nu}$, we have

$$\mathrm{E}^{\mathbb{P}_{\widetilde{\nu}}}\left[g(Y)\right] \equiv \int_{-\infty}^{\infty} g(x)\,\mathrm{d}\widetilde{\nu}(x) = \int_{-\infty}^{\infty} g(x)\frac{\mathrm{d}\widetilde{\nu}(x)}{\mathrm{d}\nu(x)}\,\mathrm{d}\nu(x)$$
$$= \int_{-\infty}^{\infty} h(x)g(x)\,\mathrm{d}\nu(x) \equiv \mathrm{E}^{\mathbb{P}_{\nu}}\left[h(Y)g(Y)\right].$$

3. When both jump size probability measures ν and $\widetilde{\nu}$ are absolutely continuous with respective densities $f_\nu(x)$ and $f_{\widetilde{\nu}}(x)$ defined on the same support, the likelihood ratio $h(x) := \frac{\mathrm{d}\widetilde{\nu}}{\mathrm{d}\nu}(x)$ is given by the ratio of PDFs:

$$h(x) = \frac{f_{\widetilde{\nu}}(x)}{f_\nu(x)}.$$

For example, let both ν and $\widetilde{\nu}$ be normal probability measures having the distributions $Norm(a, b^2)$ and $Norm(\widetilde{a}, \widetilde{b}^2)$, respectively. Then, the likelihood ratio is

$$h(x) = \frac{b}{\widetilde{b}}\exp\left(\frac{(x-a)^2}{2b^2} - \frac{(x-\widetilde{a})^2}{2\widetilde{b}^2}\right).$$

When $b = \widetilde{b}$ holds, the above expressions is simplified: $h(x) = \exp\left(\frac{a^2-\widetilde{a}^2-2(a-\widetilde{a})x}{2b^2}\right)$.

Note that if $f_\nu(x)$ and $f_{\widetilde{\nu}}(x)$ have different supports, then there exists an interval $\mathcal{I} := [u, v]$, $u < v$ so that one of the integrals $\nu(\mathcal{I}) = \int_u^v f_\nu(x)\,\mathrm{d}x$ and $\widetilde{\nu}(\mathcal{I}) = \int_u^v f_{\widetilde{\nu}}(x)\,\mathrm{d}x$ is zero, whereas the other is not. In this case, the probability measures $\nu(x)$ and $\widetilde{\nu}(x)$ are not equivalent.

4. When both jump size probability measures $\nu(x)$ and $\widetilde{\nu}(x)$ are discrete with respective PMFs $p_\nu(x)$ and $p_{\widetilde{\nu}}(x)$ defined on the same countable support $D = \{x_1, x_2, \ldots, x_m, \ldots\}$, the likelihood ratio $h(x)$ equals the ratio of PMFs:

$$h(x) = \sum_{k\geqslant 1}\frac{p_{\widetilde{\nu}}(x_k)}{p_\nu(x_k)}\mathbb{I}_{\{x=x_k\}}.$$

Clearly, if there exists a point x so that $p_\nu(x) > 0$ and $p_{\widetilde{\nu}}(x) = 0$ or vice versa, the probability measures $\nu(x)$ and $\widetilde{\nu}(x)$ are not equivalent.

Proof. It is not difficult to prove that the process

$$\varrho_t := \mathrm{e}^{(\lambda-\widetilde{\lambda})t}\left(\frac{\widetilde{\lambda}}{\lambda}\right)^{N(t)}\prod_{j=1}^{N(t)} h(Y_j) = \mathrm{e}^{(\lambda-\widetilde{\lambda})t+\sum_{j=1}^{N(t)}(\ln h(Y_j)+\ln(\widetilde{\lambda}/\lambda))}, \quad 0 \leqslant t \leqslant T \qquad (7.75)$$

is a $\mathbb{P}_\lambda$-martingale that starts at 1 (i.e., $\varrho_0 = 1$). Thus, ϱ_T is indeed a Radon–Nikodym derivative. The proof is left as an exercise for the reader. To show that under $\mathbb{P}_{\widetilde{\lambda},\widetilde{\nu}}$ the increment $Q(t) - Q(s)$ is independent of $\mathcal{F}_s$ and have the correct probability distribution for any s and t with $0 \leqslant s < t \leqslant T$, we compute the conditional MGF as follows:

$$\mathrm{E}^{\mathbb{P}_{\widetilde{\lambda},\widetilde{\nu}}}\left[\mathrm{e}^{u(Q(t)-Q(s))}\;\middle|\;\mathcal{F}_s\right] = \mathrm{E}^{\mathbb{P}_{\lambda,\nu}}\left[\frac{\varrho_t}{\varrho_s}\mathrm{e}^{u(Q(t)-Q(s))}\;\middle|\;\mathcal{F}_s\right]$$
$$= \mathrm{E}^{\mathbb{P}_{\lambda,\nu}}\left[\mathrm{e}^{(\lambda-\widetilde{\lambda})(t-s)+\sum_{j=N(s)+1}^{N(t)}(\ln h(Y_j)+\ln(\widetilde{\lambda}/\lambda))}\mathrm{e}^{u\sum_{j=N(s)+1}^{N(t)}Y_j}\;\middle|\;\mathcal{F}_s\right]$$
$$= \mathrm{e}^{(\lambda-\widetilde{\lambda})(t-s)}\mathrm{E}^{\mathbb{P}_{\lambda,\nu}}\left[\mathrm{e}^{\sum_{j=1}^{N(t-s)}(uY_j+\ln h(Y_j)+\ln(\widetilde{\lambda}/\lambda))}\right],$$

where we use the fact that the increment $Q(t) - Q(s)$ is independent of $\mathcal{F}_s$ and that under $\mathbb{P}_{\lambda,\nu}$ we have $Q(t) - Q(s) \stackrel{d}{=} Q(t-s) = \sum_{j=1}^{N(t-s)} Y_j$. Let us calculate the above expectation by conditioning on $N(t-s)$ and using the independence of jump sizes Y_j:

$$\begin{aligned}
&\mathrm{e}^{(\lambda-\widetilde{\lambda})(t-s)}\mathrm{E}^{\mathbb{P}_{\lambda,\nu}}\left[\mathrm{e}^{\sum_{j=1}^{N(t-s)}(uY_j+\ln h(Y_j)+\ln(\widetilde{\lambda}/\lambda))}\right] \\
&= \sum_{k=1}^{\infty} \mathrm{e}^{(\lambda-\widetilde{\lambda})(t-s)}\left(\frac{\widetilde{\lambda}}{\lambda}\right)^k \mathrm{e}^{-\lambda(t-s)}\frac{(\lambda(t-s))^k}{k!}\mathrm{E}^{\mathbb{P}_{\lambda,\nu}}\left[\exp\left(\sum_{j=1}^{k}(Y_j+\ln h(Y_j))\right)\right] \\
&= \sum_{k=1}^{\infty} \mathrm{e}^{-\widetilde{\lambda}(t-s)}\frac{(\widetilde{\lambda}(t-s))^k}{k!}\prod_{j=1}^{k}\mathrm{E}^{\mathbb{P}_{\lambda,\nu}}\left[h(Y_j)\mathrm{e}^{uY_j}\right] \\
&= \sum_{k=1}^{\infty} \mathrm{e}^{-\widetilde{\lambda}(t-s)}\frac{(\widetilde{\lambda}(t-s))^k}{k!}\prod_{j=1}^{k}\widetilde{M}(u)^k = \exp\left(\widetilde{\lambda}(t-s)(\widetilde{M}(u)-1)\right), \qquad (7.76)
\end{aligned}$$

where we use that for any $j \geqslant 1$

$$\mathrm{E}^{\mathbb{P}_{\lambda,\nu}}\left[h(Y_j)\mathrm{e}^{uY_j}\right] = \int_{-\infty}^{\infty} h(x)\mathrm{e}^{ux}\,\mathrm{d}\nu(x) = \int_{-\infty}^{\infty}\mathrm{e}^{ux}\,\mathrm{d}\widetilde{\nu}(x) = \mathrm{E}^{\mathbb{P}_{\widetilde{\lambda},\widetilde{\nu}}}\left[\mathrm{e}^{uY_j}\right] =: \widetilde{M}(u).$$

In (7.76), we can recognize the MGF of the compound Poisson process as given in (7.72) with time $t-s$, intensity $\widetilde{\lambda}$, and the MGF of jump sizes $\widetilde{M}(u)$ under $\mathbb{P}_{\widetilde{\lambda},\widetilde{\nu}}$. □

7.4.6 The Variance Gamma Model

The *variance gamma* (VG) process is a three-parameter generalization of the Brownian motion model for the dynamics of the logarithm of the asset price. The VG process is obtained by evaluating the Brownian motion with drift at a random time given by a gamma process. The resulting process is a pure jump process.

Let $\{B(t;\theta,\sigma) \equiv W^{(\theta,\sigma)}(t) = \theta t + \sigma W(t)\}_{t\geqslant 0}$ denote a scaled Brownian motion with drift rate θ and scale parameter σ, as defined in (1.32) of Section 1.3.1. Hence, this is a normal random variable,

$$B(t;\theta,\sigma) \sim Norm(\theta t, \sigma^2 t).$$

The gamma process $\{G(t) \equiv G(t;\mu,\upsilon)\}_{t\geqslant 0}$ with mean rate μ and variance rate υ is a random process starting at zero and with independent gamma distributed increments over nonoverlapping intervals of time. The increment $\Delta G = G(t+h) - G(t)$ over time interval $[t, t+h]$ with $0 \leqslant t < t+h$ has the gamma distribution with mean μh and variance υh. Its probability density is

$$f_{\Delta G}(g;\mu,\upsilon,h) = \left(\frac{\mu}{\upsilon}\right)^{\mu^2 h/\upsilon}\frac{g^{\mu^2 h/\upsilon - 1}\exp\left(-\frac{\mu}{\upsilon}g\right)}{\Gamma\left(\frac{\mu^2 h}{\upsilon}\right)}, \quad g > 0. \qquad (7.77)$$

Here, $\Gamma(x)$ is the gamma function defined below Equation (7.40).

The VG process $\{X(t) \equiv X(t;\sigma,\upsilon,\theta)\}_{t\geqslant 0}$ is obtained by evaluating the Brownian motion at a random time given by the gamma process with mean rate 1 and variance rate υ,

$$X(t;\sigma,\upsilon,\theta) := B(G(t;1,\upsilon);\theta,\sigma) = \theta G(t;1,\upsilon) + \sigma W(G(t;1,\upsilon)).$$

The process starts at zero: $X(0) = 0$. The PDF of the VG process at time t can be expressed by conditioning on the realization of the gamma time change G as a normal density function:

$$f_{X(t)|G(t)}(x \mid g) = \frac{1}{\sigma\sqrt{2\pi g}}\exp\left(-\frac{(x-\theta g)^2}{2\sigma^2 g}\right).$$

The unconditional probability density $f_{X(t)}$ of $X(t)$ may then be obtained by employing the density (7.77), with $h = t$, for the time change g and integrating out g. This gives us the PDF in the following integral form:

$$f_{X(t)}(x) = \int_0^\infty f_{X(t)|G(t)}(x \mid g) f_{G(t)}(g)\,\mathrm{d}g$$
$$= \int_0^\infty \frac{1}{\sigma\sqrt{2\pi g}} \exp\left(-\frac{(x-\theta g)^2}{2\sigma^2 g}\right) \frac{g^{\frac{t}{v}-1}\exp(-\frac{g}{v})}{v^{\frac{t}{v}}\Gamma(\frac{t}{v})}\,\mathrm{d}g\,. \tag{7.78}$$

The new specification for the stock price dynamics is obtained by replacing the role of Brownian motion in the original Black–Scholes geometric Brownian motion model by the variance gamma process. The stock price process is given by the geometric VG law with parameters σ, v, and θ. The risk-neutral process for the asset price is given by

$$S(t) := S_0 \exp((r-q-\omega)t + X(t;\sigma,v,\theta)), \tag{7.79}$$

where r and q have the usual meaning, and the constant $\omega = \ln(1-\theta v - \sigma^2 v/2)/v$ is chosen so that the discounted asset price process $\{\mathrm{e}^{-(r-q)t}S(t)\}_{t\geqslant 0}$ is a martingale.

The no-arbitrage price of a European-style derivative, $V(S_0;T)$ with spot S_0, time to maturity T, and payoff Λ, is given by the usual risk-neutral pricing formula, $V(S_0;T) = \mathrm{e}^{-rT}\widetilde{\mathrm{E}}_{0,S_0}[\Lambda(S(T))]$, where the expectation is taken under the risk-neutral measure $\widetilde{\mathbb{P}}$. The evaluation of the derivative price proceeds by conditioning on the random time change $G(T)$, which is independent of the Brownian motion. Conditional on the value of $G(T)$, the VG process $X(T)$ is normally distributed. Thus, the asset price $S(T)$ conditional on time $G(T) = g$ has a log-normal probability distribution:

$$(S(T) \mid G(T) = g) \stackrel{d}{=} S_0\mathrm{e}^{(r-q-\omega)T+\theta g+\sigma W(g)} = \hat{S}_g \mathrm{e}^{(r-q-\sigma^2/2)g+\sigma W(g)},$$

where $\hat{S}_g := S_0\mathrm{e}^{(r-q-\omega)T-(r-q-\sigma^2/2)g}$. The Black–Scholes (BS) formula gives the option value with the asset price S at the expiry time g. Let V_{BS} denote the BS derivative pricing function. The derivative price with expiry T under the VG risk-neutral dynamics, denoted V_{VG}, is then obtained by integrating this conditional BS price with respect to the gamma time change:

$$V_{\mathrm{VG}}(S_0;T) = \mathrm{E}\big[V_{\mathrm{BS}}(\hat{S}_{G(T)}; G(T))\big] = \int_0^\infty V_{\mathrm{BS}}(\hat{S}_g; g) f_{G(T)}(g)\,\mathrm{d}g.$$

The latter integral can be evaluated numerically.

One can also use the method of characteristic functions for pricing European-style options. The derivation of the ChF $\varphi_{\ln S(t)}(u)$ of the log price $\ln S(t)$ is left as an exercise for the reader.

7.5 Exercises

Exercise 7.1. Consider the standard GBM pricing model for a nondividend paying stock with price $S(t)$. Derive the Black–Scholes pricing formula for a standard put options using the characteristic function method. First, using the strong solution, find the conditional characteristic function of $X(T) = \ln S(T)$ given that $X(t) = \ln S$ for some $S > 0$ under each of the EMMs $\widetilde{\mathbb{P}}^{(B)}$ and $\widetilde{\mathbb{P}}^{(S)}$. After that, use the general pricing formula

$$P(t,S) = e^{-r(T-t)}\, K\,\widetilde{\mathbb{P}}^{(B)}(X(T) < \ln K \mid X(t) = \ln S) \\ -\, S\,\widetilde{\mathbb{P}}^{(S)}(X(T) < \ln K \mid X(t) = \ln S)$$

to obtain the result. The following two facts may be assumed in your solution:

1. By definition, the *error function* is
$$\mathrm{Erf}(x) = \frac{2}{\sqrt{\pi}} \int_0^x e^{-t^2}\, dt.$$
We have that
$$\int_0^\infty \frac{1}{u} e^{-au^2} \sin(bu)\, du = \frac{\pi}{2}\mathrm{Erf}\left(\frac{b}{2\sqrt{a}}\right) \quad \text{for } a > 0.$$
2. The standard normal CDF can be expressed in terms of the error function as follows:
$$\mathcal{N}(x) = \frac{1 + \mathrm{Erf}(x/\sqrt{2})}{2}.$$

Exercise 7.2. Consider a Gaussian model for risk-free interest rates under the EMM $\widetilde{\mathbb{P}}$:

$$r(t) = r_0 + \mu t + \sigma \widetilde{W}(t),\ t \geqslant 0,$$

where μ and $\sigma > 0$ are real constants, and $\widetilde{W}(t)$ is a standard Brownian motion.

(a) Show that the probability distribution of the integrated short rate process
$$R(t,T) := \int_t^T r(u)\, du \text{ with } 0 \leqslant t < T$$
is normal. Find its mean and variance given that $r(t) = r$ for some $r > 0$. Recall that the stochastic integral $\int_0^s W(u)\, du$ has the normal distribution with mean 0 and variance $s^3/3$ for any $s > 0$.

(b) Find the characteristic function $\varphi_{t,T,r}(u) = \widetilde{\mathrm{E}}\left[e^{iuR(t,T)} \mid r(t) = r\right]$ of $R(t,T)$.

(c) Derive the no-arbitrage price of a zero-coupon bond $Z(t,T,r) = \widetilde{\mathrm{E}}\left[e^{-R(t,T)} \mid r(t) = r\right]$ from $\varphi_{t,T,r}(u)$.

Exercise 7.3. Let the process $\{F(t)\}_{t\geqslant 0}$ solve the SDE

$$dF(t) = \alpha (F(t))^{\beta+1}\, dW(t).$$

Show that the process $\{X(t)\}_{t\geqslant 0}$ defined by $X(t) := (F(t))^{-2\beta}/(\alpha^2\beta^2)$ is an SQB process with index $\mu = \frac{1}{2\beta}$, i.e., it satisfies the SDE (7.30).

Exercise 7.4. Suppose that $p^{(0)}(t; S_0, S)$ solves the PDE

$$\frac{\partial p^{(0)}}{\partial t} = \frac{1}{2}\frac{\partial^2}{\partial S^2}\left(\sigma^2(S)\, p^{(0)}\right)$$

with $\sigma(S) = \alpha S^{\beta+1}$. Show that the function $p^{(\nu)}$ defined by

$$p^{(\nu)}(t; S_0, S) = e^{-\nu t} p^{(0)}(\tau_t; S_0, e^{-\nu t} S)$$

with $\tau_t = \frac{1}{2\nu\beta}\left(e^{2\nu\beta t} - 1\right)$ and $\nu \neq 0$ satisfies the PDE

$$\frac{\partial p^{(\nu)}}{\partial t} = \frac{1}{2}\frac{\partial^2}{\partial S^2}\left(\sigma^2(S)\, p^{(\nu)}\right) - \frac{\partial}{\partial S}\left(\nu S\, p^{(\nu)}\right).$$

Exercise 7.5. Let $\{F(t)\}_{t\geqslant 0}$ be a forward price process solving the SDE

$$\mathrm{d}F(t) = \sigma(F(t))\,\mathrm{d}W(t)$$

with some nonlinear $\sigma(F)$. Consider a transformed process $S(t) = \mathrm{e}^{rt}F(\tau_t)$ with some strictly increasing smooth function τ_t such that

$$\tau_0 = 0 \quad \text{and} \quad \tau_t'|_{t=0} = 1. \tag{7.80}$$

The function τ is a time change. The transition PDF $p^{(r)}$ for the process $\{S(t)\}_{t\geqslant 0}$ is given by

$$p^{(r)}(t; S_0, S) = \mathrm{e}^{-rt}p^{(0)}(\tau_t; S_0, \mathrm{e}^{-rt}S),$$

where $p^{(0)}$ is a transition PDF of the forward price F. Show the following.

(a) The PDF $p^{(r)}$ satisfies the forward Kolmogorov PDE

$$\frac{\partial p}{\partial t} = \frac{1}{2}\frac{\partial^2}{\partial S^2}\left(\tilde{\sigma}^2(t,S)\,p\right) - r\frac{\partial}{\partial S}\left(Sp\right),$$

where $\tilde{\sigma}(t,S) = \mathrm{e}^{rt}\sqrt{\tau'(t)}\sigma(\mathrm{e}^{-rt}S)$ is a time- and state-dependent function. The process $\{S(t)\}_{t\geqslant 0}$ hence solves the time-inhomogeneous SDE

$$\mathrm{d}S(t) = rS(t)\,\mathrm{d}t + \tilde{\sigma}(t, S(t))\,\mathrm{d}W(t).$$

(b) Show that $\frac{\partial\tilde{\sigma}}{\partial t} = 0$ holds (hence $\tilde{\sigma}$ is time homogeneous) only if τ_t solves the ordinary differential equation (ODE)

$$\tau''(t) + 2r\left(1 - \frac{\mathrm{e}^{-rt}S\sigma'(\mathrm{e}^{-rt}S)}{\sigma(\mathrm{e}^{-rt}S)}\right)\tau'(t) = 0 \tag{7.81}$$

subject to the initial conditions (7.80).

(c) The solution to (7.81) is independent of S only if the function $\sigma(F)$ satisfies

$$\left(\frac{F\sigma'(F)}{\sigma(F)}\right)' = 0.$$

The only solution to this ODE is a power function: $\sigma(F) = \alpha F^{\beta+1}$, where α and β are constants. The respective solution to (7.81) is $\tau_t = \frac{\mathrm{e}^{2r\beta t}-1}{2r\beta}$.

Thus, the CEV process is the only model, where the above scale and time change give a time-homogeneous diffusion with linear drift. Moreover, the transformation in this case is volatility preserving, i.e., $\tilde{\sigma}(S,t) = \sigma(S)$.

Exercise 7.6. Prove the following property of the noncentral chi-square PDF:

$$\int_a^\infty f(x; \upsilon, y)\,\mathrm{d}y = 1 - Q(x, \upsilon - 2, a).$$

Hint: Use the following identities:

$$(a)\ I_\nu(z) = \left(\frac{z}{2}\right)^\nu \sum_{j=0}^\infty \frac{(z^2/4)^j}{j!\Gamma(\nu+j+1)}, \quad (b)\ \int_a^\infty g(n,y)\,\mathrm{d}y = \sum_{j=1}^n g(j,a),$$

where $g(m,x) = \frac{\mathrm{e}^{-x}x^{m-1}}{\Gamma(m)}$.

Exercise 7.7. Prove that the compensated Poisson process, $\{N(t)-\lambda t\}_{t\geqslant 0}$, is a martingale with respect to its natural filtration.

Exercise 7.8. Consider the exponential process $\{\exp(at+N(t))\}_{t\geqslant 0}$. For what value of a is this process a martingale with respect to its natural filtration?

Exercise 7.9. Let $N(t)$, $t \geqslant 0$ be a Poisson process with intensity $\lambda > 0$. Let $\sigma > -1$ be a given constant. Show that the process $Y(t) = e^{N(t)\ln(\sigma+1)-\lambda\sigma t}$, $t \geqslant 0$ is a martingale. [Hint: The MGF of Poisson random variable X with intensity μ is $\mathrm{E}[\mathrm{e}^{uX}] = \exp(\mu(\mathrm{e}^u - 1))$.]

Exercise 7.10. Consider a jump diffusion process of the form

$$X(t) = \mu t + \sigma W(t) + \sum_{j=1}^{N(t)} Y_j,$$

where μ and σ are constants, $W(t)$ is Brownian motion, $N(t)$ is a Poisson process with intensity λ, and Y_j, $j \geqslant 1$ are i.i.d. jump sizes having the symmetric double-exponential distribution with probability density $f(y) = \frac{a}{2}\mathrm{e}^{-a|y|}$ with $a > 0$.

(a) Find the characteristic function $\varphi_{X(t)}(u) = \mathrm{E}[\mathrm{e}^{iuX(t)}]$.

(b) Let $r > 0$ be a risk-free rate. For what value of μ, we have $\mathrm{E}\left[\mathrm{e}^{X(t)}\right] = \mathrm{e}^{rt}$?

Exercise 7.11. Show that ϱ_T defined in Equation (7.74) from Proposition 7.5 is the Radon–Nikodym derivative $\left(\frac{\mathrm{d}\mathbb{P}_{\tilde{\lambda},\tilde{\nu}}}{\mathrm{d}\mathbb{P}_{\lambda,\nu}}\right)_T$. That is, $\varrho_0 = 1$, and the process ϱ_t, $0 \leqslant t \leqslant T$, defined in (7.75), is a $\mathbb{P}_{\lambda,\nu}$-martingale.

Exercise 7.12. Consider the Heston model (7.46) with uncorrelated Brownian motions $W_1(t)$ and $W_2(t)$.

(a) Show that the instantaneous variance $\mathrm{v}(t)$ in (7.46) can be expressed in terms of the SQB process by means of a suitable scale and time change.

(b) Find the transition PDF of the variance process $\mathrm{v}(t)$.

(c) Write the price of a standard European option with payoff Λ as an integral over the Black–Scholes prices.

Exercise 7.13. Apply the following integral representation of the modified Bessel function of the second kind (of order ν), denoted K_ν, to find a closed-form nonintegral expression for the transition PDF of the VG process:

$$K_\nu(z) = \frac{1}{2}\left(\frac{z}{2}\right)^\nu \int_0^\infty \exp\left(-t - \frac{z^2}{4t}\right)\frac{\mathrm{d}t}{t^{\nu+1}}.$$

Exercise 7.14. Consider a variance gamma (VG) process $\{X(t) \equiv X(t;\sigma,\upsilon,\theta)\}_{t\geqslant 0}$, which is obtained by evaluating the scaled Brownian motion with drift, $W^{(\theta,\sigma)}(t) \equiv \theta t + \sigma W(t)$, at a random time given by the gamma process $G(t;1,\upsilon)$ with mean rate 1 and variance rate υ:

$$X(t;\sigma,\upsilon,\theta) := W_{\theta,\sigma}(G(t;1,\upsilon)) = \theta G(t;1,\upsilon) + \sigma W(G(t;1,\upsilon)).$$

The gamma process $\{G(t) \equiv G(t;\mu,\upsilon)\}_{t\geqslant 0}$ with mean rate μ and variance rate υ is a random process starting at zero and having independent, gamma-distributed increments. For any t and s so that $0 \leqslant s < t$, the increment $G(t)-G(s)$ over the time interval $[s,t]$ has the gamma distribution with mean $(t-s)\mu$ and variance $(t-s)\upsilon$.

(a) Find the characteristic function of a gamma-distributed random variable with shape parameter α and rate parameter β. The PDF is given by

$$f_X(x) = \frac{\beta^\alpha}{\Gamma(\alpha)} x^{\alpha-1} \mathrm{e}^{-\beta x}, \quad x > 0.$$

[Note: if $X \sim \text{Gamma}(\alpha, \beta)$, then $\mathrm{E}[X] = \frac{\alpha}{\beta}$ and $\mathrm{Var}(X) = \frac{\alpha}{\beta^2}$.]

(b) Find the conditional characteristic function of a VG process given the gamma time: $\mathrm{E}[\mathrm{e}^{iuX(t)} \mid G(t) = g]$.

(c) Now, calculate the characteristic function of a VG process by integrating over g and using the result of (b):

$$\begin{aligned}\varphi_{X(t)}(u) &= \mathrm{E}[\mathrm{e}^{iuX(t)}] = \mathrm{E}\big[\mathrm{E}[\mathrm{e}^{iuX(t)} \mid G(t)]\big] \\ &= \int_0^\infty \mathrm{E}[\mathrm{e}^{iuX(t)} \mid G(t) = g] \, \frac{g^{\frac{t}{v}-1} \exp(-\frac{g}{v})}{v^{\frac{t}{v}} \Gamma(\frac{t}{v})} \, dg \,.\end{aligned}$$

Note that you should calculate the integral in closed form.

(d) Obtain the characteristic function $\varphi(u)$ of the log asset value $\ln S(t)$, where the risk-neutral dynamics of $S(t)$ is given by (7.79). Then, derive the characteristic function $\psi(u)$ used in the Carr–Madan method for calculating no-arbitrage prices of standard European options.

A

Essentials of General Probability Theory

In this chapter, we present some measure-theoretic foundations of probability theory and random variables. This then leads us into the main tools and formulas for computing expectations and conditional expectations of random variables (more importantly, continuous random variables) under different probability measures. The main formulas that are provided here are used in further chapters for the understanding and quantitative modelling of continuous-time stochastic financial models.

A.1 Random Variables and Lebesgue Integration

Within the foundation of general probability theory, the mathematical expectation of any real-valued random variable X defined on a probability space $(\Omega, \mathcal{F}, \mathbb{P})$ is a so-called *Lebesgue integral w.r.t. a given probability measure* $\mathbb{P}$ over the sample space Ω. In order to define an integral of a random variable (i.e., a measurable set function) w.r.t. some given measure, we need a measurable space and a measure. The measurable space here is the pair $(\Omega, \mathcal{F})$, where $\mathcal{F}$ is a σ-algebra of events in Ω, and the measure is the probability measure function $\mathbb{P}\colon \Omega \to [0,1]$. Before providing a precise definition of such an integral, we make a couple of remarks. Namely, Ω is any abstract set having either a finite, infinitely countable, or uncountable number of elements. So here we are dealing with a general probability space that includes the finite probability spaces studied in previous chapters as special cases. As usual, we denote every element in Ω by ω, where $\{\omega\}$ is a singleton set in Ω. Any random variable X has a *positive part* $X^+ := \max\{X, 0\} \geqslant 0$ and a *negative part* $X^- := \max\{-X, 0\} \geqslant 0$, where $X = X^+ - X^-$ and $|X| = X^+ + X^-$. Note that both random variables X^+ and X^- are *nonnegative*. The expectation of X w.r.t. the measure $\mathbb{P}$ is defined as the difference of two nonnegative expectations:

$$\mathrm{E}[X] = \mathrm{E}[X^+] - \mathrm{E}[X^-] .$$

We write this, term by term, in the notation of a Lebesgue integral w.r.t. the measure $\mathbb{P}$ as follows:

$$\underbrace{\int_\Omega X(\omega)\,\mathrm{d}\mathbb{P}(\omega)}_{\mathrm{E}[X]} = \underbrace{\int_\Omega X^+(\omega)\,\mathrm{d}\mathbb{P}(\omega)}_{\mathrm{E}[X^+]} - \underbrace{\int_\Omega X^-(\omega)\,\mathrm{d}\mathbb{P}(\omega)}_{\mathrm{E}[X^-]} . \tag{A.1}$$

The absolute value of X has expectation $\mathrm{E}[\,|X|\,] = \mathrm{E}[X^+]+\mathrm{E}[X^-]$. X is said to be integrable w.r.t. measure $\mathbb{P}$ iff $\mathrm{E}[\,|X|\,] < \infty$; i.e., $\mathrm{E}[X^+] < \infty$ and $\mathrm{E}[X^-] < \infty$, hence $\mathrm{E}[X] < \infty$. A common notation used to state this is to write $X \in L^1(\Omega, \mathcal{F}, \mathbb{P})$. If $\mathrm{E}[X^+] = \infty$ and $\mathrm{E}[X^-] < \infty$, then we set $\mathrm{E}[X] = \infty$; if $\mathrm{E}[X^+] < \infty$ and $\mathrm{E}[X^-] = \infty$, then we set $\mathrm{E}[X] = -\infty$; if $\mathrm{E}[X^+] = \mathrm{E}[X^-] = \infty$, then $\mathrm{E}[X]$ is not defined. Note that for strictly positive or nonnegative X, $X^+ \equiv X$ ($X^- \equiv 0$). For strictly negative or nonpositive X, we have

DOI: 10.1201/9780429468889-A

$X^- \equiv -X$ ($X^+ \equiv 0$). The Lebesgue integral of a random variable X over any subset $A \in \mathcal{F}$ is defined as the Lebesgue integral (over Ω) of the random variable $\mathbb{I}_A X$:

$$\int_A X(\omega)\,\mathrm{d}\mathbb{P}(\omega) \equiv \int_\Omega \mathbb{I}_A(\omega)X(\omega)\,\mathrm{d}\mathbb{P}(\omega) \equiv \mathrm{E}[\mathbb{I}_A X], \tag{A.2}$$

where $\mathbb{I}_A(\omega) = 1$ if $\omega \in A$ and $\mathbb{I}_A(\omega) = 0$ if $\omega \notin A$. Moreover, putting $X \equiv 1$ gives the probability under measure $\mathbb{P}$ of any event $A \in \mathcal{F}$ as a Lebesgue integral w.r.t. $\mathbb{P}$ over A:

$$\mathrm{E}[\mathbb{I}_A] = \mathbb{P}(A) = \int_A \mathrm{d}\mathbb{P}(\omega) = \int_\Omega \mathbb{I}_A(\omega)\,\mathrm{d}\mathbb{P}(\omega)\,. \tag{A.3}$$

Since $\mathbb{P}$ is an assumed probability measure, we must have $\mathbb{P}(\Omega) = \mathrm{E}[\mathbb{I}_\Omega] = 1$.

We now develop the Lebesgue integral of X w.r.t. measure $\mathbb{P}$ by first considering its definition for any ($\mathbb{P}$-a.s.) nonnegative real-valued[1] random variable $X\colon \Omega \to [0,\infty)$. The symbol $\mathbb{P}$-a.s. stands for almost surely w.r.t. to probability measure $\mathbb{P}$, i.e., a relation holds $\mathbb{P}$-a.s. means that it holds with probability one. So $X \geqslant 0$ ($\mathbb{P}$-a.s.) means that the set for which $X \geqslant 0$ has probability one: $\mathbb{P}(\{\omega \in \Omega : X(\omega) \geqslant 0\}) = 1$. We form a partition of the positive real line: $0 = y_0 < y_1 < y_2 < \ldots < y_k < y_{k+1} < \ldots < y_n$, where $y_n \to \infty$ as $n \to \infty$. A partition is defined such that the maximum sub-interval spacing approaches zero, i.e., $\lim_{n\to\infty} \max_{0\leqslant k\leqslant n-1} (y_{k+1} - y_k) = 0$ is implied. Then, the sets in $\mathcal{F}$ defined by $A_k := X^{-1}([y_k, y_{k+1})) \equiv \{\omega \in \Omega : y_k \leqslant X(\omega) < y_{k+1}\}, k = 0, 1, \ldots, n-1$, and $A_n := X^{-1}([y_n, \infty)) = \{\omega \in \Omega : X(\omega) \geqslant y_n\}$ are mutually exclusive and form a partition of the sample space Ω for every $n \geqslant 0$. The Lebesgue integral is defined as the limit of a partial sum:

$$\int_\Omega X(\omega)\,\mathrm{d}\mathbb{P}(\omega) := \lim_{n\to\infty} \sum_{k=0}^{n} y_k \mathbb{P}(A_k) \equiv \sum_{k=0}^{\infty} y_k \mathbb{P}(A_k)\,. \tag{A.4}$$

Assuming it exists, this sum gives a unique value (including cases where it may be infinite). This definition uses the lower value y_k of X on every set A_k. An equivalent definition replaces the value y_k with the upper value y_{k+1} for every $k \geqslant 1$. For any real $X = X^+ - X^-$, we simply use (A.4) for each respective nonnegative Lebesgue integral of X^+ and X^- and subtract to obtain the Lebesgue integral of X according to (A.1).

Assume $\{A_i\}$ is a *countable collection of disjoint sets* in $\mathcal{F}$. Then, applying the identity $\mathbb{I}_{\cup_i A_i} = \sum_i \mathbb{I}_{A_i}$ to (A.2) and using the linearity property of the Lebesgue integral gives

$$\int_{\cup_i A_i} X(\omega)\,\mathrm{d}\mathbb{P}(\omega) = \sum_i \int_\Omega X(\omega)\mathbb{I}_{A_i}(\omega)\,\mathrm{d}\mathbb{P}(\omega) = \sum_i \int_{A_i} X(\omega)\,\mathrm{d}\mathbb{P}(\omega)\,, \tag{A.5}$$

i.e., $\mathrm{E}[X\,\mathbb{I}_{\cup_i A_i}] = \sum_i \mathrm{E}[X\,\mathbb{I}_{A_i}]$. For $X \equiv 1$ we have $\mathrm{E}[\mathbb{I}_{\cup_i A_i}] = \sum_i \mathrm{E}[\mathbb{I}_{A_i}]$:

$$\mathbb{P}\Big(\bigcup_i A_i\Big) = \sum_i \mathbb{P}(A_i).$$

This recovers the countable additivity property of $\mathbb{P}$, which must hold for $\mathbb{P}$ to be a measure.

To see how the Lebesgue integral in (A.4) works, consider the class of random variables having the form of a finite sum of indicator random variables $X = \sum_{j=0}^N x_j \mathbb{I}_{C_j}$ with each $x_j \in [0,\infty)$, $C_j = \{\omega \in \Omega : X(\omega) = x_j\} \subset \mathcal{F}$, and where $\{C_j\}_{j=0}^N$ forms a partition of Ω.

[1]The definition also extends to the case $X\colon \Omega \to [0,\infty]$ where X can equal ∞, i.e., $X = X \cdot \mathbb{I}_{\{0\leqslant X<\infty\}} + \infty \cdot \mathbb{I}_{\{X=\infty\}}$ where the usual convention $0 \cdot \infty = 0$ is adopted. If $\mathbb{P}(X = \infty) = 0$, then we simply take $X = X \cdot \mathbb{I}_{\{0\leqslant X<\infty\}}$. If $\mathbb{P}(X = \infty) > 0$ i.e., the set $\{\omega \in \Omega\colon X(\omega) = \infty\}$ has positive $\mathbb{P}$-measure, then $\mathrm{E}[X] = \infty$.

This is called a *simple function* or *simple random variable.* Any random variable that can take only finitely many different values, $x_0, \ldots, x_N$, (including possibly a zero value) has this form. Note also that any simple random variable can be represented as the difference of two *nonnegative simple random variables.* For each sufficiently small interval $[y_k, y_{k+1})$ we have $A_k = X^{-1}([y_k, y_{k+1})) = X^{-1}(x_j) = C_j$ if there is a value $x_j \in [y_k, y_{k+1})$ for some $0 \leqslant j \leqslant N$, or otherwise $A_k = X^{-1}([y_k, y_{k+1})) = \emptyset$. For sufficiently large n, each x_j value will be contained in exactly one sub-interval and the nth partial sum in (A.4) does not change value as n is increased indefinitely since only N intervals will contain an x_j value (with probability measure $\mathbb{P}(C_j)$) and the rest of the intervals yield $A_k = \emptyset$ (i.e., $\mathbb{P}(A_k) = 0$). Hence, the Lebesgue integral in (A.4) is given as the finite sum

$$\int_\Omega X(\omega)\,\mathrm{d}\mathbb{P}(\omega) = \sum_{j=0}^{N} x_j \mathbb{P}(C_j) = \sum_{j=0}^{N} x_j \mathbb{P}(X = x_j)\,. \tag{A.6}$$

For a discrete random variable X taking on a countably infinite number of values $\{x_0, x_1, \ldots\}$, with infinitely countable or uncountable Ω, it readily follows that the Lebesgue integral recovers the familiar expectation formula:

$$\mathrm{E}[X] = \int_\Omega X(\omega)\,\mathrm{d}\mathbb{P}(\omega) = \sum_{j=0}^{\infty} x_j \mathbb{P}(X = x_j),$$

where the probabilities $p_j \equiv \mathbb{P}(X = x_j)$ define the probability mass function (PMF) of X in the measure $\mathbb{P}$.

Simple random variables are particularly useful and provide an alternative equivalent way of defining the Lebesgue integral. In standard measure and integration theory, one begins by taking (A.6) as the definition for the Lebesgue integral of any simple random variable. Then, for any $X \geqslant 0$ (a.s.) the Lebesgue integral in (A.4) is equivalently defined as the supremum of Lebesgue integral values over the set of all nonnegative simple random variables having value not greater than X:

$$\mathrm{E}[X] = \int_\Omega X(\omega)\,\mathrm{d}\mathbb{P}(\omega) := \sup\left\{\int_\Omega X^*(\omega)\,\mathrm{d}\mathbb{P}(\omega) : X^* \text{ is simple } 0 \leqslant X^* \leqslant X\right\}. \tag{A.7}$$

This last definition is rather abstract but it gives rise to other more explicit representations for the Lebesgue integral upon using the fact that a nonnegative X can be expressed (a.s.) as a limiting sequence of nonnegative simple random variables. Then, the Lebesgue integral of any nonnegative X can be computed as the limit of the Lebesgue integrals corresponding to the sequence of nonnegative simple random variables that converges to X. To see how this arises, we need to first recall the concept of pointwise convergence of a sequence of random variables. We recall that a sequence of random variables, $X_n, n = 1, 2, \ldots$, is said to converge pointwise almost surely to some random variable X when $X_n \to X$ (a.s.) as $n \to \infty$, i.e., $\mathbb{P}\left(\left\{\omega \in \Omega : \lim_{n\to\infty} X_n(\omega) = X(\omega)\right\}\right) = 1$. A sequence of nonnegative random variables $X_n, n = 1, 2, \ldots$, is said to *converge pointwise monotonically* to X (a.s.) if $X_n \to X$ and $0 \leqslant X_1 \leqslant X_2 \leqslant \ldots \leqslant X$ (a.s.), i.e., $\mathbb{P}(\{\omega \in \Omega : X_n(\omega) \nearrow X(\omega)\}) = 1$. If we have such a sequence, then each successive random variable in the sequence will approximate X better than the previous one and the approximation becomes exact in the limit $n \to \infty$. It stands to reason that the corresponding sequence of expected values (Lebesgue integrals for each X_n) will be increasingly better approximations and will give the expected value (Lebesgue integral) of X in the limit $n \to \infty$. This is summarized (without proof) in the following well-known theorem, appropriately called the Monotone Convergence Theorem (MCT) for random variables.

Theorem A.1 (Monotone Convergence for random variables). *If $X_n, n = 1, 2, \ldots,$ is a sequence of nonnegative random variables converging pointwise monotonically to X (a.s.), then*

$$\lim_{n\to\infty} \mathrm{E}[X_n] = \mathrm{E}[X]\,, \quad i.e., \quad \lim_{n\to\infty} \int_\Omega X_n(\omega)\,\mathrm{d}\mathbb{P}(\omega) = \int_\Omega X(\omega)\,\mathrm{d}\mathbb{P}(\omega)\,.$$

In fact, we have monotonic convergence, i.e., $\mathrm{E}[X_n] \nearrow \mathrm{E}[X]$ *as* $n \to \infty$.

The MCT has many applications. For instance, we recover the known continuity properties of a probability measure $\mathbb{P}$. That is, let $A_1, A_2, \ldots, A_n, \ldots$ be subsets (events) in Ω. If $A_1 \subset A_2 \subset \cdots$, then $\lim_{n\to\infty} A_n = \bigcup_{n=1}^{\infty} A_n$ and we have (monotone continuity from below):

$$\mathbb{P}\left(\bigcup_{n=1}^{\infty} A_n\right) = \lim_{n\to\infty} \mathbb{P}(A_n)\,.$$

If $A_1 \supset A_2 \supset \cdots$, then $\lim_{n\to\infty} A_n = \bigcap_{n=1}^{\infty} A_n$ and we have (monotone continuity from above):

$$\mathbb{P}\left(\bigcap_{n=1}^{\infty} A_n\right) = \lim_{n\to\infty} \mathbb{P}(A_n).$$

These follow as a corollary of Theorem A.1. For example, monotone continuity from above is obtained by defining the sequence of monotonically increasing random variables $X_n = 1 - \mathbb{I}_{A_n}$, where $A_{n+1} \subset A_n$ for all $n \geqslant 1$. Hence, $X_n \nearrow X \equiv 1 - \mathbb{I}_A$, $A = \bigcap_{n=1}^{\infty} A_n$. By MCT we have $\lim_{n\to\infty} \mathrm{E}[X_n] = \mathrm{E}[X]$, i.e.,

$$\lim_{n\to\infty} \mathrm{E}[1 - \mathbb{I}_{A_n}] = \mathrm{E}[1 - \mathbb{I}_A] \implies \lim_{n\to\infty} \mathbb{P}(A_n) = \mathbb{P}(A) = \mathbb{P}\left(\bigcap_{n=1}^{\infty} A_n\right).$$

We now apply MCT to obtain a more explicit formula for $\mathrm{E}[X]$ when $X \geqslant 0$ (a.s.) by producing a monotonically increasing sequence of nonnegative simple random variables. There are many ways to do so. An explicit example is the sequence defined by

$$X_n(\omega) := \begin{cases} k/2^n & \text{if } \omega \in A_n^{(k)} \\ 0 & \text{otherwise} \end{cases} \tag{A.8}$$

with sets $A_n^{(k)} := \{\omega \in \Omega : k/2^n \leqslant X(\omega) < (k+1)/2^n\}$, $k = 0, \ldots, 2^{2n}$. For every $n \geqslant 1$, X_n is a simple random variable:

$$X_n(\omega) = \sum_{k=0}^{2^{2n}} \frac{k}{2^n}\,\mathbb{I}_{A_n^{(k)}}(\omega) = \sum_{k=0}^{2^{2n}} \frac{k}{2^n}\,\mathbb{I}_{\{X \in [\frac{k}{2^n}, \frac{k+1}{2^n})\}}(\omega)\,. \tag{A.9}$$

We leave it as an exercise for the reader to verify that $X_n(\omega) \nearrow X(\omega)$ for any $X \geqslant 0$. Therefore, by using (A.6) as the expected value for every simple X_n and passing to the limit with the use of MCT, the Lebesgue integral (expectation) of any nonnegative random variable X is given by

$$\mathrm{E}[X] = \lim_{n\to\infty} \mathrm{E}[X_n] = \lim_{n\to\infty} \sum_{k=0}^{2^{2n}} \frac{k}{2^n}\,\mathbb{P}\left(\frac{k}{2^n} \leqslant X < \frac{k+1}{2^n}\right). \tag{A.10}$$

This is equivalent to (A.4) but here the partitions are chosen such that the convergence is monotonic, whereas the series in (A.4) is generally not monotonic. For arbitrary $X = X^+ - X^-$ we use the above construction for the two separate nonnegative random variables X^+ and X^- and subtract the two expectations, giving $\mathrm{E}[X]$, assuming we don't have $\infty - \infty$.

The Lebesgue integral above was presented in the general context of an abstract state space Ω with generally uncountable numbers of abstract elements ω and random variables X (i.e., measurable set functions) on a probability space $(\Omega, \mathcal{F}, \mathbb{P})$. It is important to consider the case where $\Omega \subset \mathbb{R}$; i.e., every ω is a real number. Recall the Borel σ-algebra $\mathcal{B}(\mathbb{R})$. This is the collection of subsets of $\mathbb{R}$ generated by all finite intervals (open or closed) or equivalently all semi-infinite intervals, i.e., all open sets in $\mathbb{R}$. The pair $(\mathbb{R}, \mathcal{B}(\mathbb{R}))$, i.e., $\Omega = \mathbb{R}$, $\mathcal{F} = \mathcal{B}(\mathbb{R})$, is a measurable space. The *Lebesgue measure*[2] on $\mathbb{R}$, which we denote by m, is a measure that assigns a nonnegative real value or infinity to each Borel set $B \in \mathcal{B}(\mathbb{R})$, i.e., $m : B \to [0, \infty) \cup \infty$, such that all intervals $[a, b], a \leqslant b$, have measure equal to their length: $m([a, b]) = b - a$. All semi-infinite intervals $(-\infty, a], (-\infty, a), [b, \infty)$, or (b, ∞) have infinite Lebesgue measure. A single point has Lebesgue measure zero, $m(\{x\}) = 0$ for any point $x \in \mathbb{R}$. The empty set also has zero measure. Any set B that is a countable union of points also has zero measure since m is countably additive, i.e.,

$$m\Big(\bigcup_{n=1}^{\infty} B_n\Big) = \sum_{n=1}^{\infty} m(B_n)$$

for all disjoint Borel sets $B_n, n \geqslant 1$. This also implies finite additivity holds where $m(\cup_{n=1}^{N} B_n) = \sum_{n=1}^{N} m(B_n)$ for any $N \geqslant 1$. The set of rational numbers $\mathbb{Q}$ is countable with Lebesgue measure zero. The irrationals $\mathbb{R} \cap \mathbb{Q}^{\mathsf{C}} = \mathbb{R} \backslash \mathbb{Q}$ are uncountable. Since $(\mathbb{R} \backslash \mathbb{Q}) \cup \mathbb{Q} = \mathbb{R}$, the Lebesgue measure of any Borel set B is unchanged if we remove all the rational numbers from it, i.e., as a disjoint union, $B = (B \backslash \mathbb{Q}) \cup (B \cap \mathbb{Q})$, hence $m(B) = m(B \backslash \mathbb{Q}) + m(B \cap \mathbb{Q}) = m(B \backslash \mathbb{Q})$. Note that $B \cap \mathbb{Q} \subset \mathbb{Q}$ so $m(B \cap \mathbb{Q}) \leqslant m(\mathbb{Q}) = 0 \implies m(B \cap \mathbb{Q}) = 0$.

There are also other more peculiar Borel sets that are uncountable but yet have Lebesgue measure zero. The well-known Cantor ternary set on $[0, 1]$, which is discussed in detail in many textbooks on real analysis, is such an example. In the interest of space, we don't discuss this set in any detail here. It suffices to note that the points in this set are too sparsely distributed over the unit interval $[0, 1]$ and hence do not accumulate any length measure. On the other hand, the points in the Cantor set cannot be counted (i.e., listed in a sequence in one-to-one with the integers). The Cantor set is constructed by starting with $[0, 1]$ and removing the middle third interval $(\frac{1}{3}, \frac{2}{3})$ and then repeating this process of removing the middle third for all remaining intervals in succession. One can then make a simple argument to show that the set is not countable, in essentially the same manner that the total number of branches in a binomial tree having an infinite number of time steps cannot be counted either.

By the above discussion, we see that the triplet $(\Omega, \mathcal{F}, \mathbb{P}) := ([0, 1], \mathcal{B}([0, 1]), m)$ serves as an example of a probability space, where m acts as a *uniform probability measure* on the set $\Omega = [0, 1] \equiv \{x \in \mathbb{R} : 0 \leqslant x \leqslant 1\}$. By our previous notation, we may also write this as $\Omega = \{\omega \in \mathbb{R} : 0 \leqslant \omega \leqslant 1\}$. Any real-valued random variable $X : [0, 1] \to \mathbb{R}$ is a Borel (measurable) function, where $X^{-1}(B) \in \mathcal{B}([0, 1])$ for every $B \in \mathcal{B}(\mathbb{R})$. Its expected value is

[2]The Lebesgue measure m of an interval $I_n = [a_n, b_n]$, or $(a_n, b_n]$, or $[a_n, b_n)$, or (a_n, b_n), is its length, $m(I_n) = \ell(I_n) := b_n - a_n, a_n \leqslant b_n$. The measure m of a Lebesgue-measurable set (for our purposes a Borel set) B is defined precisely as the smallest total length among all countable unions of intervals in $\mathbb{R}$ that (cover) contain B, i.e., for any $B \in \mathcal{B}(\mathbb{R})$, $m(B) := \inf\{\sum_{n=1}^{\infty} \ell(I_n) : B \subset \cup_{n=1}^{\infty} I_n\}$.

then the Lebesgue integral w.r.t. the Lebesgue (uniform probability) measure m:

$$\mathrm{E}[X] = \int_{[0,1]} X(\omega)\, \mathrm{d}m(\omega)\,. \tag{A.11}$$

For example, the outcome of the experiment of picking a real number in $[0, 1]$ uniformly at random is captured by the value of the random variable $X(\omega) = \omega$. The probability that a number is chosen within some arbitrary subinterval $[a, b] \subset [0, 1]$ must therefore be the length of the interval. In this case the event is represented as the set $\{X \in [a, b]\} \equiv \{a \leqslant \omega \leqslant b\}$. Its probability is the Lebesgue integral of the indicator random variable $\mathbb{I}_{\{X\in[a,b]\}}(\omega) = \mathbb{I}_{\{a\leqslant\omega\leqslant b\}}$:

$$\mathbb{P}(X \in [a, b]) = \mathrm{E}[\mathbb{I}_{\{X\in[a,b]\}}] = \int_{[0,1]} \mathbb{I}_{[a,b]}(\omega)\, \mathrm{d}m(\omega) \equiv \int_{[a,b]} \mathrm{d}m(\omega) = m([a, b]) = b - a\,.$$

Note that $\mathbb{P}(\Omega) = \mathbb{P}(X \in [0, 1]) = m([0, 1]) = 1$. Combining this with the countable additivity property of m and the fact that $m : B \to [0, 1]$ for any $B \in \mathcal{B}([0, 1]) = \mathcal{B}(\mathbb{R}) \cap [0, 1]$ shows that the Lebesgue measure m restricted to the unit interval $[0, 1]$ is a proper probability measure. Note that we also get the probability of picking any finite or countable set of numbers in $[0, 1]$ is zero since the Lebesgue measure of such a set is zero. In fact, the probability of picking a rational number is zero:

$$\mathbb{P}(X \in [0, 1] \cap \mathbb{Q}) = \int_{[0,1]\cap\mathbb{Q}} dm(\omega) = m([0, 1] \cap \mathbb{Q}) = 0\,.$$

The probability of picking an irrational number is 1, $\mathbb{P}(X \in [0, 1]\backslash\mathbb{Q}) = m([0, 1]\backslash\mathbb{Q}) = 1$. As well, the probability that we pick a number to be in the Cantor set $\mathcal{C}$ is zero since $\mathbb{P}(X \in \mathcal{C}) = m(\mathcal{C}) = 0$.

If we have a Lebesgue-measurable set $\Omega \subset \mathbb{R}$ with finite nonzero measure, $0 < m(\Omega) < \infty$, then the Lebesgue measure restricted to Ω, denoted by m_Ω, gives $m_\Omega(B) = m(B)$ for any set $B \in \mathcal{B}(\Omega)$. [Note: we assume that Ω is a generated since any Lebesgue-measurable set is either a Borel set or close to it in the sense that all sets that are Lebesgue-measurable and not Borel are null sets of Lebesgue measure zero.] The set Ω has "nonzero finite length." Typically, Ω is an interval or a combination of intervals, but need not be. In the above example, $\Omega = [0, 1]$ (i.e., the unit interval) and $m_{[0,1]}(B) = m(B)$ for any $B \in \mathcal{B}([0, 1])$. Then, $m_\Omega : B \to [0, c]$, $c \equiv m(\Omega)$, for any $B \in \mathcal{B}(\Omega)$. So m_Ω is a uniform measure on the space $(\Omega, \mathcal{B}(\Omega))$. We simply normalize this measure to obtain a *uniform probability measure* defined by $\mathbb{P}(B) := \frac{1}{c} \cdot m_\Omega(B)$, for all $B \in \mathcal{B}(\Omega)$, with $\mathbb{P}(\Omega) = \frac{1}{c} \cdot m_\Omega(\Omega) = \frac{c}{c} = 1$. Hence, $(\Omega, \mathcal{B}(\Omega), \mathbb{P})$ is a probability space, where $\mathcal{B}(\Omega)$ contains all the events.

At this point we recall (from real analysis) the Lebesgue integral over $\mathbb{R}$ w.r.t. the Lebesgue measure m, which is defined regardless of any association to a probability space. The Lebesgue integral, w.r.t. m over $\mathbb{R}$, is defined for any Lebesgue-measurable function f, i.e., if $f^{-1}(I)$ is a Lebesgue-measurable set for any interval $I \in \mathbb{R}$. For our purposes, it suffices to assume that f is any Borel-measurable real-valued function, i.e., the set $f^{-1}(B) \equiv \{x \in \mathbb{R} : f(x) \in B\} \in \mathcal{B}(\mathbb{R})$ for any $B \in \mathcal{B}(\mathbb{R})$. The Lebesgue integral (w.r.t. m) over $\mathbb{R}$ is first defined for a nonnegative Borel function f. By this we mean f is nonnegative *almost everywhere* (abbreviated a.e. or m-a.e.), i.e., the set of points where $f < 0$ has Lebesgue measure zero: $m(\{x \in \mathbb{R} : f(x) < 0\}) = 0$. Essentially, by putting $\Omega = \mathbb{R}$, $\mathcal{F} = \mathcal{B}(\mathbb{R})$, and replacing $\mathbb{P}(\omega)$ by $m(x)$ and $X(\omega)$ by $f(x)$, equivalent definitions follow as those displayed above for the probability (expectation) Lebesgue integrals on the generally abstract measurable sets $(\Omega, \mathcal{F})$ with measure $\mathbb{P}$. In the Lebesgue integral the measurable space is $(\mathbb{R}, \mathcal{B}(\mathbb{R}))$ and the measure is the Lebesgue measure m. The Lebesgue integral of a nonnegative Borel function f (w.r.t. m over $\mathbb{R}$) can be defined by

$$\int_{\mathbb{R}} f(x)\, dm(x) := \lim_{n\to\infty} \sum_{k=0}^{n} y_k\, m(B_k)\,, \tag{A.12}$$

assuming the sum converges in $\mathbb{R}$ or equals ∞. This is the analogue of the definition in (A.4). Here, the partitions $y_k, k \geqslant 0$ are defined above (A.4) and $B_k := f^{-1}([y_k, y_{k+1})) \equiv \{x \in \mathbb{R} : y_k \leqslant f(x) < y_{k+1}\}, k = 0, 1, \ldots, n-1$, $B_n := f^{-1}([y_n, \infty)) = \{x \in \mathbb{R} : f(x) \geqslant y_n\}$.

An equivalent (and more standard definition) is to first define the integral for any simple function $\varphi(x) = \sum_{k=1}^{n} a_k \mathbb{I}_{A_k}(x)$, with a_k's as real numbers and A_k's as Lebesgue-measurable sets (for our purposes these are Borel sets in $\mathbb{R}$):

$$\int_{\mathbb{R}} \varphi(x)\, dm(x) := \sum_{k=1}^{n} a_k\, m(A_k)\,. \tag{A.13}$$

[Note that the usual convention $0 \cdot \infty = 0$ is used throughout.] Then, we define the Lebesgue integral for any nonnegative f as the supremum over the set of Lebesgue integral values of all nonnegative simple functions that are less than or equal to f:

$$\int_{\mathbb{R}} f(x)\, dm(x) := \sup\left\{ \int_{\mathbb{R}} \varphi(x)\, dm(x) : \varphi \text{ is a simple function, } 0 \leqslant \varphi \leqslant f \right\}. \tag{A.14}$$

The Lebesgue integral, w.r.t. m over $\mathbb{R}$, of any Lebesgue-measurable (or Borel function) $f = f^+ - f^-$, with $f^\pm \geqslant 0$, is then

$$\int_{\mathbb{R}} f(x)\, dm(x) = \int_{\mathbb{R}} f^+(x)\, dm(x) - \int_{\mathbb{R}} f^-(x)\, dm(x)\,. \tag{A.15}$$

As in the above case of the expectation of a random variable, f is integrable iff $\int_{\mathbb{R}} f^+\, dm < \infty$ and $\int_{\mathbb{R}} f^-\, dm < \infty$, i.e., $|f| = f^+ + f^-$ has a finite integral. If $\int_{\mathbb{R}} f^+\, dm = \infty$ and $\int_{\mathbb{R}} f^-\, dm < \infty$, then $\int_{\mathbb{R}} f\, dm = \infty$; if $\int_{\mathbb{R}} f^+\, dm < \infty$ and $\int_{\mathbb{R}} f^-\, dm = \infty$, then $\int_{\mathbb{R}} f\, dm = -\infty$; if $\int_{\mathbb{R}} f^+\, dm = \int_{\mathbb{R}} f^-\, dm = \infty$, then $\int_{\mathbb{R}} f\, dm = \infty - \infty$ is not defined. Note that the Lebesgue integral of f over any measurable (Borel) set $A \in \mathcal{B}(\mathbb{R})$ is the integral of the (Borel) measurable function $\mathbb{I}_A f$ over $\mathbb{R}$:

$$\int_A f(x)\, dm(x) = \int_{\mathbb{R}} \mathbb{I}_A(x) f(x)\, dm(x)\,. \tag{A.16}$$

[Note: f is assumed (Borel) measurable so that the function $\mathbb{I}_A f$ is also measurable for any measurable set A.] The positive and negative parts of $\mathbb{I}_A f$ are $(\mathbb{I}_A f)^\pm = \mathbb{I}_A f^\pm$, giving

$$\begin{aligned} \int_A f(x)\, dm(x) &= \int_{\mathbb{R}} \mathbb{I}_A(x) f^+(x)\, dm(x) - \int_{\mathbb{R}} \mathbb{I}_A(x) f^-(x)\, dm(x) \\ &= \int_A f^+(x)\, dm(x) - \int_A f^-(x)\, dm(x)\,. \end{aligned} \tag{A.17}$$

The MCT for random variables (Theorem A.1) has an obvious analogue for sequences of Lebesgue-measurable functions, which we now state for Borel-measurable functions. Recall that a sequence of functions, $f_n, n = 1, 2, \ldots$, converges pointwise almost everywhere (a.e.) to a function f when $f_n(x) \to f(x)$ (a.e.) as $n \to \infty$. That is, the set for which convergence does not hold is a null set with Lebesgue measure zero: $m\big(\{x \in \mathbb{R} : \lim_{n\to\infty} f_n(x) \neq f(x)\}\big) = 0$. A sequence of nonnegative functions $f_n, n = 1, 2, \ldots$, is said to *converge pointwise monotonically* to f (a.e.) if $f_n(x) \to f(x)$ and $0 \leqslant f_1(x) \leqslant f_2(x) \leqslant \ldots \leqslant f(x)$ (a.e.), i.e., the set of values of $x \in \mathbb{R}$ for which these relations do not hold has Lebesgue measure zero. If we have such a sequence, then each successive function will better approximate f and the approximation becomes exact in the limit $n \to \infty$. Moreover, if the sequence of functions are Borel functions, then their corresponding Lebesgue integrals converge to the Lebesgue integral of the limiting function f. This is summarized in the MCT (Monotone Convergence Theorem) for functions. Its proof is given in any standard textbook on real analysis.

Theorem A.2 (Monotone Convergence for functions). *If $f_n(x), n = 1, 2, \ldots, x \in \mathbb{R}$, is a sequence of nonnegative Borel functions converging pointwise monotonically (a.e.) to some function f, then*

$$\lim_{n\to\infty} \int_{\mathbb{R}} f_n(x)\,\mathrm{d}m(x) = \int_{\mathbb{R}} f(x)\,\mathrm{d}m(x).$$

In fact, we have monotonic convergence, i.e., $\int_{\mathbb{R}} f_n(x)\,\mathrm{d}m(x) \nearrow \int_{\mathbb{R}} f(x)\,\mathrm{d}m(x)$, as $n \to \infty$.

For every nonnegative measurable (or Borel) function f, there exists a sequence of nonnegative simple functions, $f_n, n \geqslant 1$, converging monotonically to f, i.e., $f_n(x) \nearrow f(x)$ (a.e.) as $n \to \infty$. In fact, the analogue of the sequence in (A.9) is the sequence of simple functions:

$$f_n(x) = \sum_{k=0}^{2^{2n}} \frac{k}{2^n} \mathbb{I}_{\{f^{-1}\left([\frac{k}{2^n}, \frac{k+1}{2^n})\right)\}}(x) \equiv \sum_{k=0}^{2^{2n}} \frac{k}{2^n} \mathbb{I}_{\{f(x)\in[\frac{k}{2^n}, \frac{k+1}{2^n})\}}. \tag{A.18}$$

Using MCT (Theorem A.2) and the simple formula in (A.13) gives us the following representation for the Lebesgue integral, w.r.t. m over $\mathbb{R}$, of any nonnegative function f:

$$\int_{\mathbb{R}} f(x)\,\mathrm{d}m(x) = \lim_{n\to\infty} \int_{\mathbb{R}} f_n(x)\,\mathrm{d}m(x) = \lim_{n\to\infty} \sum_{k=0}^{2^{2n}} \frac{k}{2^n} m\big(A_k^n\big), \tag{A.19}$$

where $A_k^n := f^{-1}\big([\frac{k}{2^n}, \frac{k+1}{2^n})\big) \equiv \{x \in \mathbb{R} : \frac{k}{2^n} \leqslant f(x) < \frac{k+1}{2^n}\}$. Such a series can be obtained for both f^+ and f^- and then the Lebesgue integral of $f = f^+ - f^-$ w.r.t. m over $\mathbb{R}$ is given by the difference, provided the result is defined (not $\infty - \infty$). Note also that the Lebesgue integral of a nonnegative f over any measurable (Borel) set B, using (A.16), has the representation

$$\int_B f(x)\,\mathrm{d}m(x) = \lim_{n\to\infty} \int_{\mathbb{R}} \mathbb{I}_B(x) f_n(x)\,\mathrm{d}m(x) = \lim_{n\to\infty} \sum_{k=0}^{2^{2n}} \frac{k}{2^n} m\big(A_k^n \cap B\big). \tag{A.20}$$

The theory of Lebesgue integration w.r.t. m provides a framework for integrating a general class of real-valued functions. Lebesgue integration also forms the foundation for probability theory. The actual computation of many specific integrals is, in practice, a difficult task, as there are no known techniques for integrating particular functions. However, most Lebesgue integrals that we encounter are equivalent to a corresponding Riemann integral. This then allows us to use all the known powerful techniques of elementary calculus to compute Riemann integrals. Consider a continuous function $f : [a, b] \to \mathbb{R}$. Then, f is integrable and the function $F(x) := \int_{[a,x]} f(y)\,\mathrm{d}m(y) \equiv \int_a^x f(y)\,\mathrm{d}m(y)$ is differentiable for $x \in (a, b)$, i.e., $F'(x) = f(x)$. This is the fundamental theorem of calculus. The theorem just below relates the Lebesgue integral w.r.t. m and the Riemann integral of a bounded function over a finite interval. This theorem is proven in most standard textbooks on real analysis. Note that the statement "f is continuous (a.e.)" means that the set of points for which f is not continuous has Lebesgue measure zero.

Theorem A.3 (Lebesgue versus Riemann Integration). *Let $f : [a, b] \to \mathbb{R}$ be bounded. Then:*

(i) f is Riemann-integrable, i.e., $\int_a^b f(x)\,\mathrm{d}x$ is defined, if and only if f is continuous (a.e.).

(ii) If f is Riemann-integrable, then the Lebesgue integral over $[a, b]$ is also defined and the two integrals are the same, i.e., $\int_a^b f(x)\,\mathrm{d}x = \int_{[a,b]} f(x)\,\mathrm{d}m(x)$.

Hence, when computing the Lebesgue integral of a function over an interval with a well-defined Riemann integral, we can simply equate the Lebesgue integral with the corresponding Riemann integral. For example, we write $\int_a^b f(x)\,\mathrm{d}x \equiv \int_{[a,b]} f(x)\,\mathrm{d}m(x)$. As well, assuming the existence of Riemann integrals for semi-infinite or infinite intervals, we write $\int_a^\infty f(x)\,\mathrm{d}x \equiv \int_{[a,\infty)} f(x)\,\mathrm{d}m(x)$, $\int_{-\infty}^b f(x)\,\mathrm{d}x \equiv \int_{(-\infty,b]} f(x)\,\mathrm{d}m(x)$, $\int_{-\infty}^\infty f(x)\,\mathrm{d}x \equiv \int_{\mathbb{R}} f(x)\,\mathrm{d}m(x)$. The same goes for other improper well-defined Riemann integrals. More generally, if the integral is over some arbitrary Borel set B, it is also customary to use shorthand notation for the Lebesgue integral w.r.t. m over B as $\int_B f(x)\,\mathrm{d}x \equiv \int_B f(x)\,\mathrm{d}m(x)$. If the integrand function f (or $f \cdot \mathbb{I}_B$) is not continuous (a.e.), then the Riemann integral is not defined and the integral is understood to be the corresponding Lebesgue integral, assuming it exists.

The expected value of a general real-valued random variable X defined on a probability space $(\Omega, \mathcal{F}, \mathbb{P})$ is a Lebesgue integral w.r.t. $\mathbb{P}$ over Ω. This is a construction that is general, yet not always practical when dealing with a generally abstract sample space Ω. For discrete random variables, the expectation reduces to a sum involving the probability mass function. We also saw that, for a continuous uniform random variable, its expectation (Lebesgue integral w.r.t. $\mathbb{P}$) reduces to a Lebesgue integral w.r.t. m and hence the latter can also be expressed as a Riemann integral. The transformation from a Lebesgue integral w.r.t. measure $\mathbb{P}$ over Ω into a Lebesgue integral (or Riemann) over $\mathbb{R}$ makes the theory more practical. This allows $\mathrm{E}[X]$ to be expressed in terms of integrals over $\mathbb{R}$, rather than over Ω, as follows.

We begin by recalling what a distribution measure is for a random variable X. Let B be any Borel set in $\mathbb{R}$. The *distribution measure* of X w.r.t. a probability measure $\mathbb{P}$ is defined by the set function

$$\mu_X(B) := \mathbb{P}(X^{-1}(B)) \equiv \mathbb{P}(X \in B). \tag{A.21}$$

It is important to note that μ_X measures subsets in $\mathbb{R}$, whereas $\mathbb{P}$ measures subsets in Ω. Namely, the probability of the event $\{X \in B\}$ is computed as a μ_X-measure of B. The cumulative distribution function (CDF), F_X, of the random variable X, w.r.t. $\mathbb{P}$, is then given in terms of this measure by

$$F_X(x) := \mu_X((-\infty, x]) = \mathbb{P}(X \in (-\infty, x]) \equiv \mathbb{P}(X \leqslant x),\ x \in \mathbb{R}. \tag{A.22}$$

Recall that any CDF is generally a right-continuous monotone nondecreasing function with limiting values $\lim_{x\to-\infty} F_X(x) \equiv F_X(-\infty) = 0$ and $\lim_{x\to\infty} F_X(x) \equiv F_X(\infty) = 1$.

Since the measure $\mathbb{P}$ is countably additive, μ_X is countably additive. Indeed, for any countable collection of pairwise disjoint Borel sets $\{B_i\}$ the corresponding pre-images $\{X^{-1}(B_i)\}$ are pairwise disjoint sets in Ω. Hence,

$$\mu_X\Big(\bigcup_i B_i\Big) = \mathbb{P}\Big(X^{-1}\Big(\bigcup_i B_i\Big)\Big) = \mathbb{P}\Big(\bigcup_i X^{-1}(B_i)\Big) = \sum_i \mathbb{P}(X^{-1}(B_i)) = \sum_i \mu_X(B_i).$$

Moreover, the measure is normalized, $\mu_X(\mathbb{R}) = \mathbb{P}(X \in \mathbb{R}) = 1$, so $(\mathbb{R}, \mathcal{B}, \mu_X)$ is in fact a probability space.

Since $(\mathbb{R}, \mathcal{B}, \mu_X)$ is a measure space, we can define a Lebesgue integral of a Borel function $f(x)$ w.r.t. $\mu_X(x)$ over $\mathbb{R}$ in a similar manner as the Lebesgue integral w.r.t. the measure m. For a simple function $\varphi(x) = \sum_{k=1}^{n} a_k \mathbb{I}_{A_k}(x)$, with A_k's as Borel sets in $\mathbb{R}$:

$$\int_{\mathbb{R}} \varphi(x)\,\mathrm{d}\mu_X(x) := \sum_{k=1}^{n} a_k\,\mu_X(A_k)\,. \tag{A.23}$$

This is the analogue of (A.13). Then, the Lebesgue integral w.r.t. $\mu_X(x)$ over $\mathbb{R}$ for any nonnegative f is defined as the supremum over Lebesgue integral values of all nonnegative simple functions that are less than or equal to f:

$$\int_{\mathbb{R}} f(x)\,\mathrm{d}\mu_X(x) := \sup\left\{\int_{\mathbb{R}} \varphi(x)\,\mathrm{d}\mu_X(x) : \varphi \text{ is a simple function, } 0 \leqslant \varphi \leqslant f\right\}. \tag{A.24}$$

The Lebesgue integral, w.r.t. μ_X over $\mathbb{R}$, of any Borel function $f = f^+ - f^-$, is then given by the difference of the two nonnegative Lebesgue integrals:

$$\int_{\mathbb{R}} f(x)\,\mathrm{d}\mu_X(x) = \int_{\mathbb{R}} f^+(x)\,\mathrm{d}\mu_X(x) - \int_{\mathbb{R}} f^-(x)\,\mathrm{d}\mu_X(x)\,, \tag{A.25}$$

and for any Borel set B in $\mathbb{R}$ we have

$$\int_{B} f(x)\,\mathrm{d}\mu_X(x) = \int_{\mathbb{R}} \mathbb{I}_B(x) f(x)\,\mathrm{d}\mu_X(x) = \int_{\mathbb{R}} \mathbb{I}_B(x) f^+(x)\,\mathrm{d}\mu_X(x) - \int_{\mathbb{R}} \mathbb{I}_B(x) f^-(x)\,\mathrm{d}\mu_X(x)\,. \tag{A.26}$$

Using MCT and the sequence of simple functions in (A.18), the Lebesgue integral of any nonnegative function f, w.r.t. μ_X over $\mathbb{R}$, is given by

$$\int_{\mathbb{R}} f(x)\,\mathrm{d}\mu_X(x) = \lim_{n\to\infty} \int_{\mathbb{R}} f_n(x)\,\mathrm{d}\mu_X(x) = \lim_{n\to\infty} \sum_{k=0}^{2^{2n}} \frac{k}{2^n}\,\mu_X\left(A_k^n\right), \tag{A.27}$$

where $A_k^n := f^{-1}\left(\left[\frac{k}{2^n}, \frac{k+1}{2^n}\right)\right)$.

Based on the above construction, we have the following result, which gives the expected value of a random variable $g(X)$ as a Lebesgue integral of the (ordinary) function $g(x)$ w.r.t. the distribution measure of X over $\mathbb{R}$. Here, we assume that $g(X)$ is integrable, i.e., $\mathrm{E}[\,|g(X)|\,] < \infty$.

Theorem A.4. *Given a random variable X on $(\Omega, \mathcal{F}, \mathbb{P})$ and a Borel function $g : \mathbb{R} \to \mathbb{R}$,*

$$\mathrm{E}[g(X)] \equiv \int_{\Omega} g(X(\omega))\,\mathrm{d}\mathbb{P}(\omega) = \int_{\mathbb{R}} g(x)\,\mathrm{d}\mu_X(x)\,. \tag{A.28}$$

Proof. In many analysis textbooks we find a standard way to prove this by first showing that (A.28) follows trivially for the simplest case of a Boolean indicator function $g(x) = \mathbb{I}_A(x)$ and then by the linearity property of integrals the result is shown to hold for any simple function $g(x) = \sum_{k=1}^{n} a_k \mathbb{I}_{A_k}(x)$. Equation (A.28) is then shown to hold for any nonnegative function by using MCT and finally it follows for any $g(x) = g^+(x) - g^-(x)$. As an alternate proof, it is now instructive to see how (A.28) follows directly using (A.27) for nonnegative

g with the sequence g_n defined as in (A.18) with f replaced by g:

$$\begin{aligned}\int_{\mathbb{R}} g(x)\,\mathrm{d}\mu_X(x) = \lim_{n\to\infty}\int_{\mathbb{R}} g_n(x)\,\mathrm{d}\mu_X(x) &= \lim_{n\to\infty}\sum_{k=0}^{2^{2n}}\frac{k}{2^n}\,\mu_X(A_k^n)\\ &= \lim_{n\to\infty}\sum_{k=0}^{2^{2n}}\frac{k}{2^n}\,\mathbb{P}(X^{-1}(A_k^n))\\ &= \lim_{n\to\infty}\sum_{k=0}^{2^{2n}}\frac{k}{2^n}\,\mathbb{P}\Big(g(X)\in[\frac{k}{2^n},\frac{k+1}{2^n})\Big)\\ &\equiv \int_{\Omega} g(X(\omega))\,\mathrm{d}\mathbb{P}(\omega)\,.\end{aligned}$$

Here, we used the definition in (A.21) for each set $A_k^n = g^{-1}([\frac{k}{2^n},\frac{k+1}{2^n}))$ and manipulated the set $X^{-1}(A_k^n) \equiv \{\omega\in\Omega : X(\omega)\in A_k^n\} = \{\omega\in\Omega : g(X(\omega))\in g(A_k^n)\} = \{\omega\in\Omega : g(X(\omega))\in[\frac{k}{2^n},\frac{k+1}{2^n})\}$. Hence, (A.28) holds for both nonnegative parts g^+ and g^- of g, i.e., it must hold for any Borel function g. □

The above expectation formula is useful if the integral on the right-hand side of (A.28) can be computed more explicitly. This is still in the form of a Lebesgue integral w.r.t. the measure μ_X. However, we can reduce this to more familiar forms depending on the type of random variable X.

In the simplest case of a constant random variable $X\equiv a$, the distribution measure is the *Dirac measure*, $\mu_X(\,\cdot\,) = \delta_a(\,\cdot\,)$, i.e., for any Borel set B,

$$\delta_a(B) := \begin{cases}1 & \text{if } a\in B\\ 0 & \text{if } a\notin B.\end{cases} \tag{A.29}$$

In particular, $\delta_a(\{a\}) = 1$ and $\delta_a(\{x\}) = 0$ for $x\neq a$. By (A.22), the CDF is simply $F_X(x) := \mu_X((-\infty,x]) = \delta_a((-\infty,x]) = \mathbb{I}_{\{x\geqslant a\}}$. The expected value as a Lebesgue integral w.r.t. $\mathbb{P}$ is trivially given since $\mathrm{E}[g(X)] = g(a)\mathbb{P}(\Omega) = g(a)$. According to (A.28), this value must equal

$$\mathrm{E}[g(X)] = \int_{\mathbb{R}} g(x)\,\mathrm{d}\delta_a(x) = g(a). \tag{A.30}$$

This gives us the formula for computing an integral w.r.t. the Dirac measure and is known as the sifting property since the Dirac measure picks out only the integrand value for $x=a$. Extending this to any purely discrete random variable that can take on distinct values a_i with probability $p_i>0, i=1,2,\ldots, \sum_i p_i = 1$, the distribution measure is a linear combination of Dirac measures at each point in the range of X: $\mu_X(B) = \sum_i p_i\delta_{a_i}(B)$. In particular, $\mu_X(\{a_i\}) = p_i$. Then, using (A.30) and the fact that the integral w.r.t. a linear combination of measures is the linear combination of integrals w.r.t. each measure,

$$\mathrm{E}[g(X)] = \int_{\mathbb{R}} g(x)\,\mathrm{d}\mu_X(x) = \sum_i p_i\int_{\mathbb{R}} g(x)\,\mathrm{d}\delta_{a_i}(x) = \sum_i p_i g(a_i). \tag{A.31}$$

Using (A.22), the CDF is given as the piecewise constant (staircase function),

$$F_X(x) = \sum_i p_i\delta_{a_i}((-\infty,x]) = \sum_i p_i\mathbb{I}_{\{x\geqslant a_i\}}\,, \tag{A.32}$$

with jump discontinuities only at points $x = a_i$, i.e., $F_X(a_i) - F_X(a_i-) = p_i$ and $F_X(x) - F_X(x-) = 0$ for all x values not equal to any a_i. This is the familiar form for the CDF of a purely discrete random variable. Observe that, if g is continuous at the points a_i, the expectation of $g(X)$ is equal to the Riemann–Stieltjes integral of g with F_X as integrator:

$$\mathrm{E}[g(X)] = \int_{\mathbb{R}} g(x)\,\mathrm{d}F_X(x) = \sum_i g(a_i)(F_X(a_i) - F_X(a_i-)) = \sum_i g(a_i)p_i. \tag{A.33}$$

We can also define a distribution measure for a random variable X given as a so-called mixture of random variables with each having its own distribution measure, i.e., $\mu_X(B) = \sum_i p_i\mu_i(B)$ where $p_i \geqslant 0, i = 1, 2, \ldots, \sum_i p_i = 1$, and each μ_i is a distribution measure on $(\mathbb{R}, \mathcal{B})$. Then, the expected value of $g(X)$ is given as a linear combination of Lebesgue integrals w.r.t. each measure μ_i:

$$\mathrm{E}[g(X)] = \int_{\mathbb{R}} g(x)\,\mathrm{d}\mu_X(x) = \sum_i p_i \int_{\mathbb{R}} g(x)\,\mathrm{d}\mu_i(x)\,. \tag{A.34}$$

Let us now consider the application of Theorem A.4 to the most common important case of a continuous random variable. This is the case where the random variable has a probability density function (PDF) $f_X(x)$. In this case there is a nonnegative integrable Borel function f_X such that

$$\mu_X(B) = \int_B f_X(x)\,\mathrm{d}m(x)\,, \text{ for all Borel sets } B, \tag{A.35}$$

with CDF

$$F_X(b) := \int_{(-\infty,b]} f_X(x)\,\mathrm{d}m(x) \tag{A.36}$$

for all $b \in \mathbb{R}$. In this case, the CDF F_X is continuous and hence has no jumps, i.e., $F_X(x) = F_X(x-) = F_X(x+)$ for all x. Moreover, F_X is not just continuous but in fact an *absolutely continuous function*. The reader may wish to consult a textbook on real analysis to learn more about this technical detail. It suffices here to point out that F_X is differentiable and its derivative is the PDF: $F_X' \equiv f_X$. The distribution measure is said to be *absolutely continuous w.r.t. the Lebesgue measure m*. In this case, X is an absolutely continuous random variable but we simply say that it is a continuous random variable.[3] Based on (A.35), it is easy to prove, using similar steps as in the above proof of (A.28) combined with the linearity property of the Lebesgue integral w.r.t. m and where $\mu_X(A_k^n) = \int_{A_k^n} f_X(x)\,\mathrm{d}m(x)$, that the expectation is a Lebesgue integral of gf_X w.r.t. m:

$$\mathrm{E}[g(X)] = \int_{-\infty}^{\infty} g(x)\,f_X(x)\,\mathrm{d}m(x)\,. \tag{A.37}$$

In most (and for our purposes essentially all) applications, the density f_X (when it exists) is bounded and continuous (a.e.) on $\mathbb{R}$; i.e., the CDF is the Riemann integral of the PDF,

$$F_X(b) := \int_{-\infty}^{b} f_X(x)\,\mathrm{d}x\,,\ b \in \mathbb{R}. \tag{A.38}$$

[3]There also exist very special types of random variables X, where F_X is not an absolutely continuous function, yet it has zero derivative $F_X'(x) \equiv 0$ (a.e.). In this case X is said to be *singularly continuous*, where the CDF is a nondecreasing monotone continuous function with zero derivative (a.e.) and hence there does not exist a PDF f_X, i.e., (A.36) (and (A.38)) does not hold. The expectation of $g(X)$ is still defined as a Lebesgue integral in (A.28). The so-called Cantor function on $[0, 1]$ is a well-known textbook example of a CDF of a singularly continuous random variable that is uniformly distributed on the Cantor set $\mathcal{C}$. The Cantor CDF is constant on the complement of the Cantor set, i.e., $F_X'(x) \equiv 0$ for $x \in [0, 1]\backslash\mathcal{C}$. There are other known interesting properties of such random variables. However, throughout this text we will have no need for such singular cases so that all continuous random variables are also absolutely continuous.

Moreover, if g is continuous (a.e.) on $\mathbb{R}$, then (A.37) reduces to the familiar well-known formula for the expected value of $g(X)$ as a Riemann integral:

$$\mathrm{E}[g(X)] = \int_{-\infty}^{\infty} g(x)\, f_X(x)\, \mathrm{d}x\,, \tag{A.39}$$

assuming both $g^{\pm}$ are Riemann-integrable, $\mathrm{E}[\,|g(X)|\,] \equiv \int_{-\infty}^{\infty} |g(x)|\, f_X(x)\, \mathrm{d}x < \infty$, i.e., $\mathrm{E}[g^+] < \infty$ and $\mathrm{E}[g^-] < \infty$.

The standard normal $X \sim Norm(0,1)$ is an important example of a continuous random variable having positive Gaussian density $f_X(x) = n(x) := \frac{1}{\sqrt{2\pi}} \mathrm{e}^{-x^2/2}$, for all real x. The distribution μ_X is absolutely continuous w.r.t. the Lebesgue measure. In fact, f_X is bounded and continuous on $\mathbb{R}$ with the CDF given by the Riemann integral:

$$F_X(b) := \mathbb{P}(X \leqslant b) = \int_{-\infty}^{b} n(x)\, \mathrm{d}x \equiv \mathcal{N}(b)\,, \;\; b \in \mathbb{R}.$$

Clearly, $\mathcal{N}'(x) = n(x)$ where $F_X(x) = \mathcal{N}(x)$ is a proper CDF since it is monotonically increasing from $\mathcal{N}(-\infty) = 0$ to $\mathcal{N}(\infty) = 1$.

We now wish to make one last connection of the expectation in (A.28) to the so-called *Lebesgue–Stieltjes integral* encountered in real analysis. For any type of random variable X, we then realize that $\mathrm{E}[g(X)]$, when g is continuous (a.e.), is simply a Riemann–Stieltjes integral of g with CDF F_X as integrator. Beginning with (A.22), observe that for any *semi-open interval* $(a,b]$ the probability $\mathbb{P}(X \in (a,b])$ is given equivalently by the distribution measure $\mu_X((a,b])$ or the difference of the CDF values at the interval endpoints:

$$\mu_X((a,b]) = \mu_X((-\infty,b]) - \mu_X((-\infty,a]) = F_X(b) - F_X(a)\,. \tag{A.40}$$

For any semi-open interval, $\ell_{F_X}((a,b]) := F_X(b) - F_X(a)$ defines its "length relative to F_X." Since $\ell_{F_X}((a,c]) = \ell_{F_X}((a,b]) + \ell_{F_X}((b,c])$ for $a < b < c$, this length is additive (cumulative) for adjoining intervals. In the special case that $F_X(x) = x$, we recover the usual length $b - a$. In contrast to the usual length, an infinitesimal interval does not necessarily have length zero relative to F_X since F_X is a CDF that may have jump discontinuities. In fact, a singleton set has length equal to the size of the jump discontinuity of the CDF:

$$\ell_{F_X}(\{x\}) = \lim_{\epsilon \to 0} \ell_{F_X}\big((x-\epsilon, x]\big) = F_X(x) - \lim_{\epsilon \to 0} F_X(x-\epsilon) = F_X(x) - F_X(x-)\,.$$

For a purely continuous random variable X, there are no jumps so all points have zero such length, but for a random variable having a discrete part there is a nonzero length given by the PMF values $p_i = F_X(x_i) - F_X(x_i-)$ at the points corresponding to the countable set of discrete values $\{x_i\}$ of X. For the other types of intervals, we have $\ell_{F_X}([a,b]) = F_X(b) - F_X(a-), \ell_{F_X}((a,b)) = F_X(b-) - F_X(a), \ell_{F_X}([a,b)) = F_X(b-) - F_X(a-)$.

Based on the above definition of the length ℓ_F, for a given CDF F, the definition of the Lebesgue measure m is generalized to the *Lebesgue–Stieltjes measure generated by* F. This measure is denoted by m_F. The measure $m_F : \mathbb{R} \to [0,1]$, defined for any Borel set $B \in \mathcal{B}(\mathbb{R})$, is the smallest total length relative to F of all countable unions of semi-open intervals in $\mathbb{R}$ that contain B:

$$m_F(B) := \inf \left\{ \sum_{n=1}^{\infty} \ell_F(I_n) : I_n = (a_n, b_n],\, a_n \leqslant b_n,\, B \subset \bigcup_{n=1}^{\infty} I_n \right\}. \tag{A.41}$$

Hence, m_F is a measure that assigns a value in $[0,1]$ for each Borel set B, such that all semi-open intervals $I_n = (a_n, b_n]$ have measure $m_F((a_n,b_n]) = \ell_F((a_n,b_n]) = F(b_n) - F(a_n)$.

All intervals, including semi-infinite intervals, have finite measure; in particular, $m_F(\mathbb{R}) = F(\infty) - F(-\infty) = 1 - 0 = 1$. This measure is countably additive on $\mathcal{B}(\mathbb{R})$:

$$m_F\bigg(\bigcup_{n=1}^{\infty} B_n\bigg) = \sum_{n=1}^{\infty} m_F(B_n)$$

for all disjoint Borel sets $B_n, n \geqslant 1$. Because of the equivalence relation in (A.40) and the fact that the σ-algebra generated by all semi-open intervals in $\mathbb{R}$ is $\mathcal{B}(\mathbb{R})$, the distribution measure is the same as the Lebesgue–Stieltjes measure generated by the CDF, i.e., $\mu_X(B) = m_{F_X}(B)$ for every Borel set B. For example, $X \equiv a$ has CDF $F_X(x) = \mathbb{I}_{\{x \geqslant a\}}$. So, the measure $m_{F_X} = \delta_a$ is the Dirac measure concentrated at a. For any purely discrete random variable, with CDF in (A.32), the measure $m_{F_X} = \sum_i p_i \delta_{a_i}$ is a weighted sum of the Dirac measures for each point with its corresponding PMF value, i.e., $m_{F_X}(B) = \sum_i p_i \delta_{a_i}(B)$.

The Lebesgue–Stieltjes integral of a function w.r.t. m_{F_X} is defined in the same manner as in (A.23), (A.24), (A.25), and (A.26) with the notation that μ_X is replaced by m_{F_X}. In particular, the expectation in (A.28) is now recognized as the Lebesgue-Stieltjes integral of g w.r.t. m_{F_X} over $\mathbb{R}$, denoted by $\int_{\mathbb{R}} g(x)\,\mathrm{d}m_{F_X}(x)$, where we write equivalently

$$\mathrm{E}[g(X)] = \int_{\mathbb{R}} g(x)\,\mathrm{d}\mu_X(x) \equiv \int_{\mathbb{R}} g(x)\,\mathrm{d}m_{F_X}(x)\,. \tag{A.42}$$

If g is continuous (a.e.) then it can be shown that the Lebesgue-Stieltjes integral is the same as the corresponding Riemann-Stieltjes integral of g with CDF F_X as integrator function:

$$\mathrm{E}[g(X)] = \int_{\mathbb{R}} g(x)\,\mathrm{d}m_{F_X}(x) = \int_{\mathbb{R}} g(x)\,\mathrm{d}F_X(x)\,. \tag{A.43}$$

Since any CDF is a monotone (nondecreasing) bounded function, then it is of bounded variation and hence can always be used as integrator. If X is absolutely continuous, then $\mathrm{d}F_X(x) = F_X'(x)\,\mathrm{d}x = f_X(x)\,\mathrm{d}x$ and the Riemann-Stieltjes integral is the same as the usual Riemann integral in (A.39). For a purely discrete random variable with CDF in (A.32), the Riemann-Stieltjes integral in (A.43) is given by (A.33). The Riemann-Stieltjes integral in (A.43) also gives $\mathrm{E}[g(X)]$ for all other types of mixture random variables, as shown in Section 2.2.1, for example. Such random variables can have a discrete and continuous part (the continuous part being either absolutely continuous and/or singularly continuous).

Based on the above expectation formulas, one can also proceed to compute various quantities such as the moments $\mathrm{E}[X^n], n \geqslant 1$, of a real-valued random variable X (assuming $\mathrm{E}[|X|^n] < \infty$); the moment generating function $M_X(t) := \mathrm{E}[\mathrm{e}^{tX}]$, which is either infinite or a function of the real parameter t on some interval of convergence about $t = 0$; the characteristic function $\phi_X(t) := \mathrm{E}[\mathrm{e}^{itX}]$, $i \equiv \sqrt{-1}$, which is bounded for all $t \in \mathbb{R}$, i.e., $|\phi_X(t)| \leqslant \mathrm{E}[\,|\mathrm{e}^{itX}|\,] = 1$. The characteristic function (or the moment generating function) is useful for computing the mean and variance as well as various moments of a random variable. The relevant formulas and theorems related to these functions are part of standard material that is covered in most textbooks on probability theory and are hence (in the interest of space) simply omitted here.

In closing this section, we mention one other general result, which is the *change of variable formula for an expectation* given in (A.44) just below. This allows us to compute the expectation $\mathrm{E}[g(X)]$ as an integral w.r.t. the distribution measure μ_Y of the random variable defined by $Y := g(X)$. Note that, since X is a random variable on $(\Omega, \mathcal{F}, \mathbb{P})$, then Y is also a random variable on $(\Omega, \mathcal{F}, \mathbb{P})$. That is, g is a Borel function, $g^{-1}(B) \in \mathcal{B}$, giving $Y^{-1}(B) = X^{-1}(g^{-1}(B)) \in \mathcal{F}$ for every $B \in \mathcal{B}$. Assuming Y is integrable, i.e., $\mathrm{E}[\,|Y|\,] < \infty$, then

$$\mathrm{E}[Y] := \int_{\Omega} Y(\omega)\,\mathrm{d}\mathbb{P}(\omega) = \int_{\mathbb{R}} g(x)\,\mathrm{d}\mu_X(x) = \int_{\mathbb{R}} y\,\mathrm{d}\mu_Y(y)\,, \tag{A.44}$$

where $\mu_Y(B) := \mathbb{P}(Y^{-1}(B)) \equiv \mathbb{P}(Y \in B)$, for every $B \in \mathcal{B}$. The proof of this formula follows readily from the relation between the two distribution measures: $\mu_Y(B) := \mathbb{P}(Y^{-1}(B)) = \mathbb{P}(X^{-1}(g^{-1}(B))) = \mu_X(g^{-1}(B))$. Hence, in the first equation line in the proof of Theorem A.4 we have $\mu_X(A_k^n) = \mu_X(g^{-1}([\frac{k}{2^n}, \frac{k+1}{2^n}))) = \mu_Y([\frac{k}{2^n}, \frac{k+1}{2^n}))$, i.e., for any nonnegative g:

$$\int_{\mathbb{R}} g(x)\,\mathrm{d}\mu_X(x) = \lim_{n\to\infty} \sum_{k=0}^{2^{2n}} \frac{k}{2^n}\,\mu_X(A_k^n) = \lim_{n\to\infty} \sum_{k=0}^{2^{2n}} \frac{k}{2^n}\,\mu_Y\big(\big[\frac{k}{2^n}, \frac{k+1}{2^n}\big)\big). = \int_{\mathbb{R}} y\,\mathrm{d}\mu_Y(y)$$

This proves the formula for nonnegative g and similar steps can be used to prove the result for any $g = g^+ - g^-$. In summary, we see that $\mathrm{E}[g(X)]$ can be evaluated in three different ways: (i) by integrating the random variable $Y \equiv g(X)$ w.r.t. $\mathbb{P}$ over Ω, (ii) by integrating the function $g(x)$ w.r.t. distribution measure $\mu_X(x)$ over $\mathbb{R}$, or (iii) by integrating the function $f(y) = y$ w.r.t. distribution measure $\mu_Y(y)$ over $\mathbb{R}$.

A.2 Multidimensional Lebesgue Integration

The above integration theory for Borel functions of a single variable (and random variables defined as functions of a single random variable) extends into the general multi-dimensional case. The construction of Lebesgue integrals mirrors the above single-variable case. We recall the Borel sets in $\mathbb{R}^n$, $\mathcal{B}_n \equiv \mathcal{B}(\mathbb{R}^n)$ and Borel functions defined over $\mathbb{R}^n$. In $\mathbb{R}^2$, we denote the Lebesgue measure by $m_2 \colon \mathcal{B}_2 \to [0, \infty]$. It can be defined formally as an extension of the definition given above for the Lebesgue measure m. Given any two intervals I_1, I_2, then m_2 measures the area of the rectangle $I_1 \times I_2$: $m_2(I_1 \times I_2) = \ell(I_1)\ell(I_2)$. In terms of the Lebesgue measure in one dimension we have the *product measure* $m_2(I_1 \times I_2) = m(I_1)m(I_2)$. Note that the null sets (having zero measure w.r.t. m_2) include any countable union of points in $\mathbb{R}^2$ as well as some uncountable sets of the form $A \times \{b\}$, $A \subset \mathbb{R}$, $b \in \mathbb{R}$ or $\{a\} \times B$, $a \in \mathbb{R}$, $B \subset \mathbb{R}$. Also, any graph or curve in $\mathbb{R}^2$ has zero m_2 measure. The Lebesgue integral of a Borel function over $\mathbb{R}^2$ w.r.t. measure m_2 is defined in similar fashion to what we have above for the single variable case. A simple Borel function $\varphi(x, y) = \sum_{k=1}^n a_k \mathbb{I}_{A_k}(x, y)$, $a_k \in \mathbb{R}$, with all $A_k \in \mathcal{B}_2$, has Lebesgue integral

$$\int_{\mathbb{R}^2} \varphi(x, y)\,\mathrm{d}m_2(x, y) := \sum_{k=1}^{n} a_k\, m_2(A_k)\,. \tag{A.45}$$

The Lebesgue integral of any nonnegative Borel function f is defined by

$$\int_{\mathbb{R}^2} f(x, y)\,\mathrm{d}m_2(x, y) := \sup\left\{\int_{\mathbb{R}^2} \varphi(x, y)\,\mathrm{d}m_2(x, y) : \varphi \text{ is a simple function, } 0 \leqslant \varphi \leqslant f\right\}. \tag{A.46}$$

The Lebesgue integral, w.r.t. m_2 over $\mathbb{R}^2$, of any Borel function $f = f^+ - f^-$, with $f^{\pm} \geqslant 0$, is then

$$\int_{\mathbb{R}^2} f\,\mathrm{d}m_2 \equiv \int_{\mathbb{R}^2} f(x, y)\,\mathrm{d}m_2(x, y) = \int_{\mathbb{R}^2} f^+(x, y)\,\mathrm{d}m_2(x, y) - \int_{\mathbb{R}^2} f^-(x, y)\,\mathrm{d}m_2(x, y)\,. \tag{A.47}$$

f is integrable iff $\int_{\mathbb{R}^2} |f|\,\mathrm{d}m_2 < \infty$. We denote this by writing $f \in L^1(\mathbb{R}^2, \mathcal{B}_2, m_2)$. The Lebesgue integral of f over any Borel set $B \subset \mathbb{R}^2$ is the Lebesgue integral of $\mathbb{I}_B f$ over $\mathbb{R}^2$:

$$\int_B f\,\mathrm{d}m_2 = \int_{\mathbb{R}^2} \mathbb{I}_B(x, y) f(x, y)\,\mathrm{d}m_2(x, y)\,. \tag{A.48}$$

Assuming an integrable function, $f \in L^1(\mathbb{R}^2, \mathcal{B}_2, m_2)$, then Fubini's Theorem can be applied for interchanging the order of integration:

$$\int_{\mathbb{R}^2} f \, \mathrm{d}m_2 = \int_{\mathbb{R}} \left(\int_{\mathbb{R}} f(x,y) \, \mathrm{d}m(x) \right) \mathrm{d}m(y) = \int_{\mathbb{R}} \left(\int_{\mathbb{R}} f(x,y) \, \mathrm{d}m(y) \right) \mathrm{d}m(x) . \tag{A.49}$$

What is important for us is when f is continuous (m_2–a.e.) in (A.47) or $\mathbb{I}_B f$ is continuous in (A.48). The Lebesgue integral in (A.47) is then equal to the Riemann (double) integral over $\mathbb{R}^2$:

$$\int_{\mathbb{R}^2} f \, \mathrm{d}m_2 = \iint_{\mathbb{R}^2} f(x,y) \, \mathrm{d}x \, \mathrm{d}y = \int_{\mathbb{R}} \left(\int_{\mathbb{R}} f(x,y) \, \mathrm{d}x \right) \mathrm{d}y = \int_{\mathbb{R}} \left(\int_{\mathbb{R}} f(x,y) \, \mathrm{d}y \right) \mathrm{d}x \tag{A.50}$$

where we assume $f \in L^1(\mathbb{R}^2, \mathcal{B}_2, m_2)$, i.e., the function is integrable. In all our applications this will be the case where, for fixed $x \in \mathbb{R}$, f is continuous in y and for fixed $y \in \mathbb{R}$, f is continuous in x. The set B in (A.48) is usually a rectangular region $[a, b] \times [c, d]$, or of type $a \leqslant x \leqslant b, h_1(x) \leqslant y \leqslant h_2(x)$, or $g_1(y) \leqslant x \leqslant g_2(y), c \leqslant y \leqslant d$, etc. Given a set $B = B_1 \times B_2 = \{(x, y) \in \mathbb{R}^2 \colon x \in B_1, y \in B_2\}$, and assuming $\mathbb{I}_B f$ is continuous and integrable on $\mathbb{R}^2$, the Lebesgue integral in (A.48) is then

$$\int_{B_1 \times B_2} f \, \mathrm{d}m_2 = \int_{B_2} \left(\int_{B_1} f(x,y) \, \mathrm{d}x \right) \mathrm{d}y = \int_{B_1} \left(\int_{B_2} f(x,y) \, \mathrm{d}y \right) \mathrm{d}x . \tag{A.51}$$

We note that when the integrals only have meaning as Lebesgue integrals, we interpret the Riemann integrals as convenient shorthand notation for the corresponding Lebesgue integrals.

In $\mathbb{R}^3$, the Lebesgue measure, $m_3 \colon \mathcal{B}_3 \to [0, \infty]$, measures the volume of $I_1 \times I_2 \times I_3$, where $m_3(I_1 \times I_2 \times I_3) = \ell(I_1)\ell(I_2)\ell(I_3)$. For any $B = B_1 \times B_2 \times B_3 \in \mathcal{B}_3$, we have the product measure $m_3(B) = m(B_1)m(B_2)m(B_3)$. The null sets of m_3 include any countable union of points in $\mathbb{R}^3$ as well as uncountable sets of the form $A \times B \times \{c\}$, $A, B \subset \mathbb{R}$, $c \in \mathbb{R}$, or $A \times \{b\} \times C$, $A, C \subset \mathbb{R}$, $b \in \mathbb{R}$, or $\{a\} \times B \times C$, $B, C \subset \mathbb{R}$, $a \in \mathbb{R}$, all surfaces and lines, etc.. The Lebesgue integral of a Borel function $f \colon \mathbb{R}^3 \to \mathbb{R}$ w.r.t. measure m_3 over $\mathbb{R}^3$ is defined in analogy with the above construction in $\mathbb{R}^2$. If $f(x_1, x_2, x_3)$ is integrable w.r.t. m_3 over $\mathbb{R}^3$ (denoted as $f \in L^1(\mathbb{R}^3, \mathcal{B}_3, m_3)$) and is furthermore a continuous function (m_3–a.e.) of the three variables, then its Lebesgue integral is a Riemann (triple) integral over $\mathbb{R}^3$:

$$\int_{\mathbb{R}^3} f \, \mathrm{d}m_3 = \int_{\mathbb{R}} \int_{\mathbb{R}} \int_{\mathbb{R}} f(x_1, x_2, x_3) \, \mathrm{d}x_1 \, \mathrm{d}x_2 \, \mathrm{d}x_3 \, , \tag{A.52}$$

where we can also change the order of integration by successive application of Fubini's Theorem. For a Borel set $B = B_1 \times B_2 \times B_3 = \{(x_1, x_2, x_3) \in \mathbb{R}^3 \colon x_1 \in B_1, x_2 \in B_2, x_3 \in B_3\}$ we have

$$\begin{aligned} \int_{B_1 \times B_2 \times B_3} f \, \mathrm{d}m_3 &= \int_{B_3} \int_{B_2} \int_{B_1} f(x_1, x_2, x_3) \, \mathrm{d}m(x_1) \, \mathrm{d}m(x_2) \, \mathrm{d}m(x_3) \\ &= \int_{B_3} \int_{B_2} \int_{B_1} f(x_1, x_2, x_3) \, \mathrm{d}x_1 \, \mathrm{d}x_2 \, \mathrm{d}x_3 \\ &= \int_{\mathbb{R}^3} f(x_1, x_2, x_3) \, \mathbb{I}_B(x_1, x_2, x_3) \, \mathrm{d}x_1 \, \mathrm{d}x_2 \, \mathrm{d}x_3 \, , \end{aligned} \tag{A.53}$$

where the Riemann integral is used as shorthand for the Lebesgue integral and is equivalent to it when $f(x_1, x_2, x_3) \, \mathbb{I}_B(x_1, x_2, x_3)$ is a continuous function of x_1, x_2, x_3.

More generally, in $\mathbb{R}^n$ the Lebesgue measure, $m_n\colon \mathcal{B}_n \to [0,\infty]$, gives the n-dimensional volume of any n-dimensional cube: $m_n(I_1\times\ldots\times I_n) = \ell(I_1)\times\ldots\times\ell(I_n)$. For every cartesian n-tuple $B = B_1\times\ldots\times B_n \in \mathcal{B}_n$, $m_n(B) = m(B_1)m(B_2)\cdots m(B_n)$ is a product measure. Null sets of m_n are sets having zero n-dimensional volume and these include any countable union of points in $\mathbb{R}^n$, hyperplanes, lines, etc. The Lebesgue integral of a Borel function $f\colon \mathbb{R}^n \to \mathbb{R}$ w.r.t. m_n over $\mathbb{R}^n$ for all $n \geqslant 2$ is constructed as in the above case of $n = 2$. If $f \in L^1(\mathbb{R}^n, \mathcal{B}_n, m_n)$, i.e., $f(\mathbf{x}) \equiv f(x_1,\ldots,x_n)$ is integrable w.r.t. m_n over $\mathbb{R}^n$, and is furthermore a continuous function (m_n–a.e.) of the n variables $\mathbf{x}$, then its Lebesgue integral is equal to its Riemann integral over $\mathbb{R}^n$:

$$\int_{\mathbb{R}^n} f\,\mathrm{d}m_n = \int_{\mathbb{R}}\cdots\int_{\mathbb{R}} f(x_1,\ldots,x_n)\,\mathrm{d}x_1\ldots\,\mathrm{d}x_n\,. \tag{A.54}$$

For a Borel set $B = B_1\times\ldots\times B_n = \{\mathbf{x}\in\mathbb{R}^n\colon x_1\in B_1,\ldots,x_n\in B_n\}$, we have

$$\begin{aligned}\int_{B_1\times\ldots\times B_n} f\,\mathrm{d}m_n &= \int_{B_n}\cdots\int_{B_1} f(x_1,\ldots,x_n)\,\mathrm{d}m(x_1)\ldots\,\mathrm{d}m(x_n)\\ &= \int_{B_n}\cdots\int_{B_1} f(x_1,\ldots,x_n)\,\mathrm{d}x_1\ldots\,\mathrm{d}x_n\\ &= \int_{\mathbb{R}^n} f(\mathbf{x})\,\mathbb{I}_B(\mathbf{x})\,\mathrm{d}^n\mathbf{x}\,,\end{aligned} \tag{A.55}$$

where the Riemann integral is used as shorthand for the Lebesgue integral and is equivalent to it when $f(\mathbf{x})\,\mathbb{I}_B(\mathbf{x})$ is continuous on $\mathbb{R}^n$.

A.3 Multiple Random Variables and Joint Distributions

Let us now see how distributions and expectations are formulated for multiple random variables (i.e., random vectors) by first considering a pair of random variables $(X,Y)\colon\Omega\to\mathbb{R}^2$ defined on the same probability space $(\Omega,\mathcal{F},\mathbb{P})$. The *joint distribution measure*

$$\mu_{X,Y}\colon \mathcal{B}_2 \to [0,1]$$

is the measure induced by the pair (X,Y) and defined by

$$\mu_{X,Y}(B) := \mathbb{P}((X,Y)\in B),\ \ B\in\mathcal{B}_2. \tag{A.56}$$

This measure is countably additive and assigns a number in $[0,1]$ to a Borel set B in $\mathbb{R}^2$ which corresponds to the probability of the event $\{(X,Y)\in B\} \equiv \{\omega\in\Omega\colon (X(\omega),Y(\omega))\in B\}$. This measure is normalized so that $\mu_{X,Y}(\mathbb{R}^2) = \mathbb{P}((X,Y)\in\mathbb{R}^2) = 1$ and so the measure space $(\mathbb{R}^2,\mathcal{B}_2,\mu_{X,Y})$ is also a probability space. Writing $B = B_1\times B_2$, $B_1,B_2\in\mathcal{B}$, then we see that

$$\mu_{X,Y}(B_1\times B_2) = \mathbb{P}(X^{-1}(B_1)\cap Y^{-1}(B_2)) = \mathbb{P}(X\in B_1, Y\in B_2) \tag{A.57}$$

gives the probability of the joint event $\{X\in B_1\}\cap\{Y\in B_2\} \equiv \{X\in B_1, Y\in B_2\}$. This joint measure determines the univariate (marginal distribution) measures of X and Y by letting $B_1 = \mathbb{R}$ or $B_2 = \mathbb{R}$:

$$\mu_{X,Y}(B\times\mathbb{R}) = \mathbb{P}(X\in B, Y\in\mathbb{R}) = \mathbb{P}(X\in B) = \mu_X(B), \tag{A.58}$$
$$\mu_{X,Y}(\mathbb{R}\times B) = \mathbb{P}(X\in\mathbb{R}, Y\in B) = \mathbb{P}(Y\in B) = \mu_Y(B), \tag{A.59}$$

for all $B\in\mathcal{B}$.

Letting $B_1 = (-\infty, x], B_2 = (-\infty, y]$ in (A.57) gives the joint CDF of (X, Y):

$$F_{X,Y}(x,y) := \mu_{X,Y}((-\infty,x] \times (-\infty,y]) = \mathbb{P}(X \leqslant x, Y \leqslant y), \ x, y \in \mathbb{R}. \tag{A.60}$$

This CDF is right-continuous on $\mathbb{R}^2$, monotone in both x and y, and recovers the univariate (marginal) CDF of X or Y in the respective limits:

$$\lim_{y\to\infty} F_{X,Y}(x,y) \equiv F_{X,Y}(x,\infty) = \mathbb{P}(X \leqslant x) = \mu_X((-\infty,x]) = F_X(x), \ x \in \mathbb{R}, \tag{A.61}$$

$$\lim_{x\to\infty} F_{X,Y}(x,y) \equiv F_{X,Y}(\infty,y) = \mathbb{P}(Y \leqslant y) = \mu_Y((-\infty,y]) = F_Y(y), \ y \in \mathbb{R}. \tag{A.62}$$

Taking the limit of infinite argument in the marginal CDF (in either case) gives

$$\lim_{x\to\infty, y\to\infty} F_{X,Y}(x,y) \equiv F_{X,Y}(\infty,\infty) = F_X(\infty) = F_Y(\infty) = 1.$$

Taking a decreasing sequence of numbers $x_n \searrow -\infty$ gives a decreasing sequence of sets approaching the empty set: $(-\infty, x_n] \times (-\infty, y] \searrow \emptyset$. By monotone continuity (from above) of the measure, $\mu_{X,Y}((-\infty,x_n] \times (-\infty,y]) \searrow \mu_{X,Y}(\emptyset) = 0$, i.e., for any $y \in \mathbb{R}$,

$$F_{X,Y}(-\infty,y) \equiv \lim_{x\to-\infty} F_{X,Y}(x,y) = \lim_{x_n\searrow-\infty} \mu_{X,Y}((-\infty,x_n] \times (-\infty,y]) = \mu_{X,Y}(\emptyset) = 0\,.$$

Similarly, $\lim_{y\to-\infty} F_{X,Y}(x,y) \equiv F_{X,Y}(x,-\infty) = 0\,,\ x \in \mathbb{R}$. These two relations must clearly hold since X and Y are in $\mathbb{R}$ and so $\mathbb{P}(X < -\infty, Y \leqslant y) = 0$ and $\mathbb{P}(X \leqslant x, Y < -\infty) = 0$.

Let x_1, x_2, y_1, y_2 be real numbers such that $x_1 < x_2$ and $y_1 < y_2$; then the joint measure of the semi-open rectangle $(x_1, x_2] \times (y_1, y_2]$ is given by

$$\begin{aligned}\mu_{X,Y}((x_1,x_2] \times (y_1,y_2]) &= \mathbb{P}(x_1 < X \leqslant x_2,\ y_1 < Y \leqslant y_2)\\ &= F_{X,Y}(x_2,y_2) - F_{X,Y}(x_1,y_2) - F_{X,Y}(x_2,y_1) + F_{X,Y}(x_1,y_1)\,.\end{aligned} \tag{A.63}$$

Based on this relation, the definition of a Lebesgue–Stieltjes measure for a single random variable, defined in (A.41), can be extended to a Lebesgue–Stieltjes measure generated by the joint CDF $F_{X,Y}$ of (X, Y). The quantity in (A.63) can be viewed as a measure of an "area relative to the joint CDF" $F_{X,Y}$ for any semi-open rectangle in $\mathbb{R}^2$. The Lebesgue–Stieltjes measure generated by $F_{X,Y}$, which we denote by $m_F^{X,Y}$, is the measure function $m_F^{X,Y} : \mathbb{R}^2 \to [0,1]$, defined for any Borel set $B = B_1 \times B_2 \in \mathcal{B}_2$, that assigns the smallest total area relative to $F_{X,Y}$ (using (A.63)) of all countable unions of semi-open rectangles $I_k \times J_l \equiv (a_k, b_k] \times (c_l, d_l]$, $a_k \leqslant b_k$, $c_l \leqslant d_l$ in $\mathbb{R}^2$ that contain B:

$$m_F^{X,Y}(B) := \inf\left\{\sum_{k=1}^{\infty}\sum_{l=1}^{\infty} \mu_{X,Y}(I_k \times J_l) : B_1 \subset \bigcup_{k=1}^{\infty} I_k,\ B_2 \subset \bigcup_{l=1}^{\infty} J_l\right\}. \tag{A.64}$$

This measure is equivalent to the joint distribution measure, i.e., $m_F^{X,Y}(B) = \mu_{X,Y}(B)$.

Since $(\mathbb{R}^2, \mathcal{B}_2, \mu_{X,Y})$ is a measure, we can define the Lebesgue integral w.r.t. $\mu_{X,Y}$ over $\mathbb{R}^2$ (i.e., the Lebesgue–Stieltjes integral w.r.t. $\mu_{X,Y}$ or equivalently w.r.t. $m_F^{X,Y}$) in very similar manner as was done above for the Lebesgue–Stieltjes integral w.r.t. μ_X in (A.23) - (A.68). For any simple function

$$\varphi(x,y) = \sum_{k=1}^{K}\sum_{l=1}^{L} a_{k,l}\mathbb{I}_{B_1^k \times B_2^l}(x)$$

with $B_1^k \times B_2^l \in \mathcal{B}_2$, $a_{k,l} \in \mathbb{R}$, its Lebesgue integral w.r.t. $\mu_{X,Y}$ over $\mathbb{R}^2$ is defined by

$$\int_{\mathbb{R}^2} \varphi(x,y)\,\mathrm{d}\mu_{X,Y}(x,y) := \sum_{k=1}^{K}\sum_{l=1}^{L} a_{k,l}\,\mu_{X,Y}(B_1^k \times B_2^l)\,. \tag{A.65}$$

Based on this definition, the Lebesgue–Stieltjes integral w.r.t. $\mu_{X,Y}$ over $\mathbb{R}^2$ for any nonnegative Borel function $f\colon \mathbb{R}^2 \to \mathbb{R}$ is defined as the supremum over integral values of all nonnegative simple functions $\varphi \leqslant f$:

$$\int_{\mathbb{R}^2} f(x,y)\,\mathrm{d}\mu_{X,Y}(x,y) := \sup\left\{\int_{\mathbb{R}^2} \varphi(x,y)\,\mathrm{d}\mu_{X,Y}(x,y) : \varphi \text{ is simple},\ 0 \leqslant \varphi \leqslant f\right\}. \tag{A.66}$$

For any Borel function $f = f^+ - f^-$, the Lebesgue–Stieltjes integral is given by the difference of the two nonnegative integrals:

$$\int_{\mathbb{R}^2} f(x,y)\,\mathrm{d}\mu_{X,Y}(x,y) = \int_{\mathbb{R}^2} f^+(x,y)\,\mathrm{d}\mu_{X,Y}(x,y) - \int_{\mathbb{R}^2} f^-(x,y)\,\mathrm{d}\mu_{X,Y}(x,y)\,, \tag{A.67}$$

and for any Borel set B in $\mathbb{R}^2$ we have

$$\int_{B} f(x,y)\,\mathrm{d}\mu_{X,Y}(x,y) = \int_{\mathbb{R}^2} \mathbb{I}_B(x,y) f(x,y)\,\mathrm{d}\mu_{X,Y}(x,y)\,. \tag{A.68}$$

We note that MCT is a general property that also applies to all Lebesgue–Stieltjes integrals.

Based on the above construction, we have the following result for the expected value of $h(X,Y)(\omega) \equiv h(X(\omega), Y(\omega))$, defined as a Borel function of two random variables (X,Y). Here, we assume that $h(X,Y)$ is integrable, i.e., $\mathrm{E}[\,|h(X,Y)|\,] < \infty$.

Theorem A.5. *Given a pair of random variables (X,Y) on $(\Omega, \mathcal{F}, \mathbb{P})$ and a Borel function $h : \mathbb{R}^2 \to \mathbb{R}$,*

$$\mathrm{E}[h(X,Y)] \equiv \int_{\Omega} h(X(\omega), Y(\omega))\,\mathrm{d}\mathbb{P}(\omega) = \int_{\mathbb{R}^2} h(x,y)\,\mathrm{d}\mu_{X,Y}(x,y)\,. \tag{A.69}$$

The proof of (A.69) is very similar to the proof of (A.28). In the special case, $h(x,y) = g(x)$, (A.69) recovers (A.28). The Lebesgue–Stieltjes integral in (A.69) is a very general representation for $\mathrm{E}[h(X,Y)]$, where h is a Borel function. That is, X or Y can be any type of random variable, i.e., any combination of discrete, absolutely continuous, or singularly continuous. The expectation in (A.69) reduces to various useful and familiar formulas for $\mathrm{E}[h(X,Y)]$ that a student learns in a standard course in probability theory. For our purpose, there are two main cases: discrete or continuous (we simply say continuous to mean absolutely continuous).

Assume that h is a continuous function (m_2–a.e.). This is virtually always the case in practice and certainly the case for all our applications in this text. The Lebesgue–Stieltjes is then a Riemann–Stieltjes integral over $\mathbb{R}^2$. Let us consider the simple case, where both X and Y are discrete random variables and $h(x,y)$ is continuous at all values $(x,y) = (x_i, y_j)$ in the range of (X,Y); then the Riemann–Stieltjes integral simply recovers the summation formula in the joint PMF $p_{X,Y}(x,y) \equiv \mathbb{P}(X = x, Y = y)$ of (X,Y) at the support values:

$$\mathrm{E}[h(X,Y)] = \sum_{\text{all } x_i}\sum_{\text{all } y_j} p_{X,Y}(x_i,y_j)\,h(x_i,y_j)$$

assuming $\mathrm{E}[|h(X,Y)|] < \infty$ (i.e., summation converging for both negative and positive parts of h). Letting $h(X,Y) = \mathbb{I}_{\{X \leqslant x, Y \leqslant y\}}$, for any fixed real values (x,y), recovers the joint CDF as a (two-dimensional piecewise constant) staircase function in (x,y):

$$F_{X,Y}(x,y) = \mathbb{P}(X \leqslant x, Y \leqslant y) = \mathrm{E}[\mathbb{I}_{\{X \leqslant x, Y \leqslant y\}}] = \sum_{x_i \leqslant x} \sum_{y_j \leqslant y} p_{X,Y}(x_i, y_j)$$

with jump discontinuities at only the support values $(x,y) = (x_i, y_j)$ of the PMF. This recovers the formula in (A.32) when $y \to \infty$.

Let us now consider the case, where (X,Y) are continuous with joint density denoted by $f_{X,Y}$. In this case, every Borel set $B \in \mathcal{B}_2$ has joint measure given by the Lebesgue integral of a nonnegative integrable Borel function $f_{X,Y} : \mathbb{R}^2 \to \mathbb{R}$ (namely, the joint PDF) over B:

$$\mu_{X,Y}(B) = \int_B f_{X,Y}(x,y)\,\mathrm{d}m_2(x,y) \equiv \int_{-\infty}^{\infty}\int_{-\infty}^{\infty} \mathbb{I}_B(x,y) f_{X,Y}(x,y)\,\mathrm{d}x\,\mathrm{d}y\,. \tag{A.70}$$

Recall that we sometimes simply write the Lebesgue integral as a Riemann integral using the convention we adopted in the previous section. Of course, the two are equal if $\mathbb{I}_B f_{X,Y}$ is a continuous function in (x,y). Since (A.70) holds for all $B \in \mathcal{B}_2$, then it holds for all sets of the form $B = (-\infty, a] \times (-\infty, b]$. Using $\mathbb{I}_B(x,y) = \mathbb{I}_{\{x \leqslant a, y \leqslant b\}}$ and the definition (A.60), we have the joint CDF

$$F_{X,Y}(a,b) = \int_{-\infty}^{b}\int_{-\infty}^{a} f_{X,Y}(x,y)\,\mathrm{d}x\,\mathrm{d}y\,, \quad a,b \in \mathbb{R}. \tag{A.71}$$

Since $F_{X,Y}(\infty,\infty) = 1$, then $f_{X,Y}$ integrates to unity on all of $\mathbb{R}^2$. In fact, (A.71) holds iff (A.70) holds for all Borel sets $B \in \mathcal{B}_2$. The joint CDF is continuous on $\mathbb{R}^2$ and related to the joint PDF by differentiating (A.71),

$$f_{X,Y}(x,y) = \frac{\partial^2}{\partial x \partial y} F_{X,Y}(x,y)\,, \quad x,y \in \mathbb{R}.$$

The marginal CDF of X and Y are given by (A.61)–(A.62) and taking either limit $a \to \infty$ or $b \to \infty$ in (A.71) gives

$$F_X(a) = \int_{-\infty}^{a}\left(\int_{-\infty}^{\infty} f_{X,Y}(x,y)\,\mathrm{d}y\right)\mathrm{d}x \text{ and } F_Y(b) = \int_{-\infty}^{b}\left(\int_{-\infty}^{\infty} f_{X,Y}(x,y)\,\mathrm{d}x\right)\mathrm{d}y,$$

for all $a,b \in \mathbb{R}$. Hence, the existence of the joint PDF $f_{X,Y}$ implies the existence of the respective marginal densities of X and Y:

$$f_X(x) = \int_{-\infty}^{\infty} f_{X,Y}(x,y)\,\mathrm{d}y \quad \text{and} \quad f_Y(y) = \int_{-\infty}^{\infty} f_{X,Y}(x,y)\,\mathrm{d}x. \tag{A.72}$$

We note that the converse is generally not true. Recall from our discussion of a single random variable, the marginal densities are nonnegative Borel functions that exist whenever (see (A.35) and (A.36))

$$\mu_X(B) = \int_B f_X(x)\,\mathrm{d}x \quad \text{and} \quad \mu_Y(B) = \int_B f_Y(y)\,\mathrm{d}y$$

for all Borel sets $B \subset \mathbb{R}$, or equivalently whenever

$$F_X(a) = \int_{-\infty}^{a} f_X(x)\,\mathrm{d}x \quad \text{and} \quad F_Y(b) = \int_{-\infty}^{b} f_Y(y)\,\mathrm{d}y$$

for all $a, b \in \mathbb{R}$. The expectations $\mathrm{E}[h_1(X)]$ and $\mathrm{E}[h_2(Y)]$, for single-variable Borel functions h_1 and h_2, are therefore given by the respective Riemann (Lebesgue) integrals over $\mathbb{R}$:

$$\mathrm{E}[h_1(X)] = \int_{\mathbb{R}} h_1(x) f_X(x)\,\mathrm{d}x \quad \text{and} \quad \mathrm{E}[h_2(Y)] = \int_{\mathbb{R}} h_2(y) f_Y(y)\,\mathrm{d}y\,. \tag{A.73}$$

For jointly continuous (X, Y), (A.70) holds, and it is readily proven (in the same manner that (A.37) or (A.39) is proven) that the expected value in (A.69) is given by the integral over $\mathbb{R}^2$ of the joint PDF multiplied by h, i.e.,

$$\mathrm{E}[h(X,Y)] = \int_{-\infty}^{\infty}\int_{-\infty}^{\infty} h(x,y) f_{X,Y}(x,y)\,\mathrm{d}x\,\mathrm{d}y\,. \tag{A.74}$$

Recall that X and Y are mutually independent iff

$$\mathbb{P}(X \in B_1, Y \in B_2) = \mathbb{P}(X \in B_1)\,\mathbb{P}(Y \in B_2) \tag{A.75}$$

for all $B_1, B_2 \in \mathcal{B}(\mathbb{R})$. That is, for a Borel rectangle $B = B_1 \times B_2$ the joint distribution measure given by (A.57) is now a product of the marginal distribution measures:

$$\mu_{X,Y}(B_1 \times B_2) = \mathbb{P}(X \in B_1)\,\mathbb{P}(Y \in B_2) = \mu_X(B_1)\,\mu_Y(B_2) := \mu_{X\times Y}(B_1 \times B_2)\,. \tag{A.76}$$

Hence, (A.75) and (A.76) are equivalent. From (A.60), we also have that independence is equivalent to

$$F_{X,Y}(x,y) = F_X(x)\,F_Y(y), \quad x, y \in \mathbb{R}. \tag{A.77}$$

Moreover, two continuous random variables (X, Y) are independent if and only if their joint PDF is the product of the marginal PDFs,

$$f_{X,Y}(x,y) = f_X(x)\,f_Y(y), \quad x, y \in \mathbb{R}. \tag{A.78}$$

This is easily proven. In particular, assuming (X, Y) are independent, then (A.71) gives

$$\begin{aligned}\int_{-\infty}^{b}\int_{-\infty}^{a} f_{X,Y}(x,y)\,\mathrm{d}x\,\mathrm{d}y &= F_{X,Y}(a,b) = F_X(a)\,F_Y(b)\\ &= \int_{-\infty}^{a} f_X(x)\,\mathrm{d}x \int_{-\infty}^{b} f_Y(y)\,\mathrm{d}y = \int_{-\infty}^{b}\int_{-\infty}^{a} f_X(x) f_Y(y)\,\mathrm{d}x\,\mathrm{d}y\end{aligned}$$

for all $a, b \in \mathbb{R}$. This implies $f_{X,Y}(x,y) = f_X(x) f_Y(y)$. We leave the proof of the converse as an exercise for the reader. When (X, Y) are independent, the general expectation formula for all types of random variables, as given by (A.69), is now a Lebesgue–Stieltjes integral w.r.t. the above product measure $\mu_{X\times Y}$:

$$\mathrm{E}[h(X,Y)] = \int_{\mathbb{R}^2} h(x,y)\,\mathrm{d}\mu_{X\times Y}(x,y) = \int_{\mathbb{R}}\int_{\mathbb{R}} h(x,y)\,\mathrm{d}\mu_X(x)\,\mathrm{d}\mu_Y(y)\,, \tag{A.79}$$

where the order of integration in μ_X and μ_Y is interchangeable according to Fubini's Theorem. In the case that $h(X, Y) = h_1(X) h_2(Y)$,

$$\mathrm{E}[h(X,Y)] = \left(\int_{\mathbb{R}} h_1(x)\,\mathrm{d}\mu_X(x)\right)\left(\int_{\mathbb{R}} h_2(y)\,\mathrm{d}\mu_Y(y)\right) = \mathrm{E}[h_1(X)]\,\mathrm{E}[h_2(Y)]\,. \tag{A.80}$$

Of course, for continuous (X, Y) this product of expectations is given by (A.73). Taking $h_1(x) = x, h_2(y) = y, h(x,y) = xy$ shows that two mutually independent random variables have zero covariance

$$\mathrm{Cov}(X,Y) \equiv \mathrm{E}[XY] - \mathrm{E}[X]\,\mathrm{E}[Y] = 0\,. \tag{A.81}$$

The converse is generally not true.

An important example of a jointly continuous random vector (X, Y) is the standard bivariate normal distribution where $\mathrm{E}[X] = \mathrm{E}[Y] = 0$, $\mathrm{Cov}(X, Y) = \rho$, $|\rho| < 1$. The well-known joint PDF is

$$f_{X,Y}(x, y) = n_2(x, y; \rho) := \frac{1}{2\pi\sqrt{1-\rho^2}} \exp\left(-\frac{x^2 + y^2 - 2\rho xy}{2(1-\rho^2)}\right), \quad x, y \in \mathbb{R}. \tag{A.82}$$

The joint distribution measure $\mu_{X,Y}$ is absolutely continuous w.r.t. Lebesgue measure m_2; i.e., for all $B \in \mathcal{B}_2$ we have

$$\mu_{X,Y}(B) = \int_B n_2(x, y; \rho)\, \mathrm{d}m_2(x, y) \equiv \int_{-\infty}^{\infty}\int_{-\infty}^{\infty} \mathbb{I}_B(x, y) n_2(x, y; \rho)\, \mathrm{d}x\, \mathrm{d}y. \tag{A.83}$$

The joint CDF is

$$F_{X,Y}(a, b) = \mathcal{N}_2(x, y; \rho) := \int_{-\infty}^{b}\int_{-\infty}^{a} n_2(x, y; \rho)\, \mathrm{d}x\, \mathrm{d}y, \quad a, b \in \mathbb{R}. \tag{A.84}$$

The functions n_2 and $\mathcal{N}_2$ denote the standard bivariate normal PDF and CDF, respectively, where $n_2(x, y; \rho) = \frac{\partial^2}{\partial x \partial y}\mathcal{N}_2(x, y; \rho)$. We note also the symmetry: $n_2(x, y, \rho) = n_2(y, x, \rho)$ and $\mathcal{N}_2(x, y, \rho) = \mathcal{N}_2(y, x, \rho)$. The marginal CDFs of X and Y are the standard normal CDF and follow simply from the limiting values of the joint CDF (see (A.61)–(A.62)) :

$$F_X(x) = F_{X,Y}(x, \infty) = \mathcal{N}_2(x, \infty; \rho) = \mathcal{N}(x), \quad F_Y(y) = F_{X,Y}(\infty, y) = \mathcal{N}_2(\infty, y; \rho) = \mathcal{N}(y).$$

Here, we used the integral definition of $\mathcal{N}_2$ in (A.84). Hence, X and Y are identically distributed $Norm(0, 1)$ random variables with standard normal (marginal) PDF

$$f_X(x) = \mathcal{N}'(x) = n(x), \quad f_Y(y) = \mathcal{N}'(y) = n(y),$$

$n(z) := \frac{1}{\sqrt{2\pi}}\mathrm{e}^{-z^2/2}$, $z \in \mathbb{R}$. [Note that these marginal PDFs also follow by integrating the joint PDF according to (A.72).] We observe that the pair (X, Y) is mutually independent if and only if the correlation coefficient $\rho = 0$, i.e.,

$$f_{X,Y}(x, y) = n_2(x, y; 0) = \frac{1}{2\pi}\mathrm{e}^{-(x^2+y^2)/2} = n(x)n(y) = f_X(x)\, f_Y(y), \quad x, y \in \mathbb{R},$$

and the integral in (A.84) factors into

$$F_{X,Y}(a, b) = \mathcal{N}_2(a, b; 0) = \mathcal{N}(a)\,\mathcal{N}(b) = F_X(a)\, F_Y(b), \quad a, b \in \mathbb{R}.$$

Substituting the above joint PDF into (A.74) gives the expectation of a Borel function of the normal pair (X, Y) as

$$\mathrm{E}[h(X, Y)] = \int_{-\infty}^{\infty}\int_{-\infty}^{\infty} h(x, y) n_2(x, y; \rho)\, \mathrm{d}x\, \mathrm{d}y. \tag{A.85}$$

For $\rho = 0$, $n_2(x, y; 0) = n(x)n(y)$, and for $h(x, y) = h_1(x)h_2(y)$ this expectation reduces to (A.80), where

$$\mathrm{E}[h_1(X)] = \int_{\mathbb{R}} h_1(x)\, \mathrm{d}\mu_X(x) = \int_{\mathbb{R}} h_1(x)n(x)\, \mathrm{d}x,$$

$$\mathrm{E}[h_2(Y)] = \int_{\mathbb{R}} h_2(x)\, \mathrm{d}\mu_Y(y) = \int_{\mathbb{R}} h_2(y)n(y)\, \mathrm{d}y.$$

The above formulation extends to the more general case of an *n-dimensional real-valued random vector* $\mathbf{X} = (X_1, \ldots, X_n) \in \mathbb{R}^n$, for all integers $n \geqslant 1$. Each X_i is a random variable on $(\Omega, \mathcal{F}, \mathbb{P})$ where $X_i^{-1}(B_i) \in \mathcal{F}$ for every $B_i \in \mathcal{B}(\mathbb{R})$, $i = 1, \ldots, n$. As a random vector $\mathbf{X}\colon \Omega \to \mathbb{R}^n$, for every Borel set $B = B_1 \times \ldots \times B_n \in \mathcal{B}_n$,

$$\mathbf{X}^{-1}(B) \equiv \{\mathbf{X} \in B\} \equiv \{X_1 \in B_1, \ldots, X_n \in B_n\} \in \mathcal{F}.$$

The joint distribution measure of $\mathbf{X} = (X_1, \ldots, X_n)$, which generalizes (A.56), is defined by

$$\mu_{\mathbf{X}}(B) \equiv \mu_{X_1,\ldots,X_n}(B) := \mathbb{P}(\mathbf{X} \in B) \equiv \mathbb{P}(X_1 \in B_1, \ldots, X_n \in B_n). \tag{A.86}$$

This measure assigns a probability to a Borel set B in $\mathbb{R}^n$ which corresponds to the probability of the joint event $\{\omega \in \Omega\colon X_1(\omega) \in B_1, \ldots, X_n(\omega) \in B_n\} = \{X_1 \in B_1\} \cap \ldots \cap \{X_n \in B_n\}$. It is normalized, $\mu_{\mathbf{X}}(\mathbb{R}^n) = \mathbb{P}(\mathbf{X} \in \mathbb{R}^n) = 1$, so $(\mathbb{R}^n, \mathcal{B}_n, \mu_{\mathbf{X}})$ is a probability space.

The joint (n-dimensional) measure $\mu_{\mathbf{X}}$ determines all the univariate, bivariate, trivariate, etc., distribution measures for all single random variables X_i, pairs (X_i, X_j), triples (X_i, X_j, X_k), etc. This follows by setting some of the appropriate sets among $B_1, \ldots, B_n$ equal to $\mathbb{R}$. For example, setting all sets $B_j = \mathbb{R}$ for all $j \neq i$, and $B_i = A \in \mathcal{B}$, gives the univariate marginal distribution measures

$$\mu_{\mathbf{X}}(B_1 \times \ldots \times B_n) = \mathbb{P}(X_i \in A) = \mu_{X_i}(A), \ \ i = 1, \ldots, n.$$

The bivariate (marginal) distribution measure of a pair (X_i, X_j), $i < j$, is obtained by setting all sets $B_k = \mathbb{R}$ for all $k \neq i, k \neq j$, and $B_i = A \in \mathcal{B}, B_j = B \in \mathcal{B}$:

$$\mu_{\mathbf{X}}(B_1 \times \ldots \times B_n) = \mathbb{P}(X_i \in A, X_j \in B) = \mu_{X_i,X_j}(A \times B), \ \ i < j = 1, \ldots, n.$$

Letting $B_1 = (-\infty, x_1], B_2 = (-\infty, x_2], \ldots, B_n = (-\infty, x_n]$ in (A.86) gives the multivariate joint CDF of $(X_1, \ldots, X_n)$:

$$F_{\mathbf{X}}(\mathbf{x}) \equiv F_{(X_1,\ldots,X_n)}(x_1, \ldots, x_n) = \mathbb{P}(X_1 \leqslant x_1, \ldots, X_n \leqslant x_n), \tag{A.87}$$

$\mathbf{x} = (x_1 \ldots, x_n) \in \mathbb{R}^n$. This CDF is right-continuous on $\mathbb{R}^n$ and is a nondecreasing monotonic function in all variables $x_1, \ldots, x_n$. The marginal CDFs of each X_i are recovered in the limit that $x_j \to \infty$, for all $j \neq i$ in (A.87):

$$F_{X_i}(x_i) = \mathbb{P}(X_i \leqslant x_i) = F_{(X_1,\ldots,X_n)}(\infty, \ldots, \infty, x_i, \infty, \ldots, \infty), \ x_i \in \mathbb{R}. \tag{A.88}$$

Similarly, the (marginal) joint CDF for each random vector pair (X_i, X_j), $i < j$, is obtained by letting $x_k \to \infty$ for all $k \neq i, k \neq j$:

$$F_{X_i,X_j}(x_i, x_j) = \mathbb{P}(X_i \leqslant x_i, X_j \leqslant x_j) = \lim_{\textit{all } x_k \to \infty; k \neq i, k \neq j} F_{(X_1,\ldots,X_n)}(x_1, \ldots, x_n).$$

All other (marginal) joint CDFs are obtained in the appropriate limits. For example, we can consider any k-dimensional random vector such as $(X_1, \ldots, X_k)$, for any $1 \leqslant k \leqslant n$, having joint CDF

$$F_{X_1,\ldots,X_k}(x_1, \ldots, x_k) = F_{(X_1,\ldots,X_n)}(x_1, \ldots, x_k, \infty, \ldots, \infty).$$

More generally, the joint CDF $F_{X_{i_1},\ldots,X_{i_k}}(y_1, \ldots, y_k)$, where $(y_1, \ldots, y_k) \in \mathbb{R}^k$, of any k-dimensional random vector $(X_{i_1}, \ldots, X_{i_k})$, $1 \leqslant i_1 < \ldots < i_k \leqslant n$, $1 \leqslant k \leqslant n$, taken from $(X_1, \ldots, X_n)$ is obtained by setting $x_j = \infty$ for all $j \notin \{i_1, i_2, \ldots, i_k\}$ in the n-dimensional joint CDF $F_{(X_1,\ldots,X_n)}(x_1, \ldots, x_n)$. This corresponds to the probability

$$F_{X_{i_1},\ldots,X_{i_k}}(y_1, \ldots, y_k) = \mathbb{P}(X_{i_1} \leqslant y_1, \ldots, X_{i_k} \leqslant y_k). \tag{A.89}$$

As in the two-dimensional case, the joint CDF evaluates to zero when setting any one of its arguments to $-\infty$. Setting all $x_i = \infty$ gives unity: $F_{\mathbf{X}}(\infty, \ldots, \infty) = \mathbb{P}(\mathbf{X} \in \mathbb{R}^n) = 1$.

The random vector $\mathbf{Y} = (Y_1, \ldots, Y_k) := (X_{i_1}, \ldots, X_{i_k})$, for any $1 \leqslant k \geqslant n$, has a joint distribution measure defined by

$$\mu_{\mathbf{Y}}(B_1 \times \ldots \times B_k) = \mathbb{P}(X_{i_1} \in B_1, \ldots, X_{i_k} \in B_k)$$

for any Borel set $B_1 \times \ldots \times B_k \in \mathcal{B}_k \equiv \mathcal{B}(\mathbb{R}^k)$. Hence, $(\mathbb{R}^k, \mathcal{B}_k, \mu_{\mathbf{Y}})$ is a probability space for each $k = 1, \ldots, n$.

The relation in (A.63) can be extended to any n-dimensional semi-open rectangle with the use of the n-dimensional joint CDF $F_{\mathbf{X}}(\mathbf{x})$. Moreover, the Lebesgue–Stieltjes measure defined in (A.64) can be extended into n dimensions accordingly, as generated by $F_{\mathbf{X}}$. In fact, the n-dimensional joint distribution measure $\mu_{\mathbf{X}}$ in (A.86) is the same Lebesgue–Stieltjes measure on $\mathbb{R}^n$. The above construction of the Lebesgue–Stieltjes integral w.r.t. $\mu_{X,Y}$ (for dimension $n = 2$), provided by (A.65),(A.66), (A.67), and (A.68), extends in the obvious manner into dimension $n \geqslant 2$. We write the Lebesgue–Stieltjes integral of a Borel function $f \colon \mathbb{R}^n \to \mathbb{R}$ w.r.t. the joint distribution measure $\mu_{\mathbf{X}}$ as the difference of two nonnegative integrals

$$\int_{\mathbb{R}^n} f(\mathbf{x})\, \mathrm{d}\mu_{\mathbf{X}}(\mathbf{x}) = \int_{\mathbb{R}^n} f^+(\mathbf{x})\, \mathrm{d}\mu_{\mathbf{X}}(\mathbf{x}) - \int_{\mathbb{R}^n} f^-(\mathbf{x})\, \mathrm{d}\mu_{\mathbf{X}}(\mathbf{x})\,, \tag{A.90}$$

and for any Borel set B in $\mathbb{R}^n$ we have

$$\int_B f(\mathbf{x})\, \mathrm{d}\mu_{\mathbf{X}}(\mathbf{x}) = \int_{\mathbb{R}^n} \mathbb{I}_B(\mathbf{x}) f(\mathbf{x})\, \mathrm{d}\mu_{\mathbf{X}}(\mathbf{x})\,. \tag{A.91}$$

Here, $f(\mathbf{x}) \equiv f(x_1, \ldots, x_n)$, $\mathrm{d}\mu_{\mathbf{X}}(\mathbf{x}) \equiv \mathrm{d}\mu_{X_1,\ldots,X_n}(x_1, \ldots, x_n)$ is shorthand vector notation.

Given a Borel function, $h : \mathbb{R}^n \to \mathbb{R}$, of a random vector $\mathbf{X} = (X_1, \ldots, X_n)$ on a probability space $(\Omega, \mathcal{F}, \mathbb{P})$, Theorem A.5 is generalized to give the expected value of the random variable $h(\mathbf{X}) \equiv h(X_1, \ldots, X_n)$ as an integral of $h(\mathbf{x}) = h(x_1, \ldots, x_n)$ w.r.t. the joint distribution measure $\mu_{\mathbf{X}}$ over $\mathbb{R}^n$:

$$\mathrm{E}[h(\mathbf{X})] \equiv \int_\Omega h(\mathbf{X}(\omega))\, \mathrm{d}\mathbb{P}(\omega) = \int_{\mathbb{R}^n} h(\mathbf{x})\, \mathrm{d}\mu_{\mathbf{X}}(\mathbf{x})\,. \tag{A.92}$$

This formula can be proven in the same manner as the proof of (A.69). This Lebesgue–Stieltjes integral is a general representation for the expected value $\mathrm{E}[h(\mathbf{X})]$, where h is a Borel function on $\mathbb{R}^n$. So, each component of the vector $(X_1, \ldots, X_n)$ can be any type of random variable, i.e., any combination of discrete, absolutely continuous, or singularly continuous random variables. The two main types of random variables of interest to us are either discrete or continuous (i.e., absolutely continuous).

The case where all X_i are discrete (as in the binomial and multinomial financial models considered in previous chapters) simply generalizes the above double summation formulas in the case of two variables to multiple (n-fold) summation formulas involving the joint PMF $p_{X_1,\ldots,X_n}(x_1, \ldots, x_n) \equiv \mathbb{P}(X_1 = x_1, \ldots, X_n = x_n)$ at the support values:

$$\mathrm{E}[h(X_1, \ldots, X_n)] = \sum_{\text{all } x_1} \cdots \sum_{\text{all } x_n} p_{X_1,\ldots,X_n}(x_1, \ldots, x_n)\, h(x_1, \ldots, x_n)\,.$$

Here, we assume $\mathrm{E}[\,|h(X_1, \ldots, X_n)|\,] < \infty$ (i.e., we assume the sums converge for both negative and positive parts of h). Choosing the indicator function $h(\mathbf{X}) = \mathbb{I}_{\{X_1 \leqslant a_1, \ldots, X_n \leqslant a_n\}}$ recovers the joint CDF:

$$F_{(X_1,\ldots,X_n)}(a_1, \ldots, a_n) = \mathrm{E}[\mathbb{I}_{\{X_1 \leqslant a_1, \ldots, X_n \leqslant a_n\}}] = \sum_{x_1 \leqslant a_1} \cdots \sum_{x_n \leqslant a_n} p_{X_1,\ldots,X_n}(x_1, \ldots, x_n)\,.$$

This is a (n-dimensional piecewise constant) staircase function in the variables $(a_1, \ldots, a_n)$ with jump discontinuities occurring at only the support values of the PMF.

The most important case for continuous random variables is when $(X_1, \ldots, X_n)$ are continuous with joint density $f_{X_1,\ldots,X_n}(x_1, \ldots, x_n) \equiv f_{\mathbf{X}}(\mathbf{x})$ as a nonnegative integrable Borel function $f_{\mathbf{X}}: \mathbb{R}^n \to \mathbb{R}$, i.e., when every Borel set $B \in \mathcal{B}_n$ has joint measure given by the Lebesgue integral of the joint density over B:

$$\mu_{\mathbf{X}}(B) = \int_B f_{\mathbf{X}} \, \mathrm{d}m_n \equiv \int_{\mathbb{R}^n} \mathbb{I}_B(\mathbf{x}) f_{\mathbf{X}}(\mathbf{x}) \, \mathrm{d}^n\mathbf{x} . \tag{A.93}$$

This is the generalization of (A.70), where the Lebesgue integral is written as a Riemann integral on $\mathbb{R}^n$. The Lebesgue and Riemann integrals are equal if $\mathbb{I}_B f_{\mathbf{X}}$ is (m_n a.e.) continuous on $\mathbb{R}^n$. The joint CDF is obtained by setting $B = (-\infty, x_1] \times \ldots \times (-\infty, x_n]$, where $\mathbb{I}_B(y_1, \ldots, y_n) = \mathbb{I}_{\{y_1 \leqslant x_1, y_n \leqslant x_n\}}$:

$$F_{X_1,\ldots,X_n}(x_1, \ldots, x_n) = \int_{-\infty}^{x_n} \cdots \int_{-\infty}^{x_1} f_{X_1,\ldots,X_n}(y_1, \ldots, y_n) \, \mathrm{d}y_1 \ldots \mathrm{d}y_n , \tag{A.94}$$

for all $x_1, \ldots, x_n \in \mathbb{R}$. Note that the joint PDF $f_{\mathbf{X}}$ integrates to unity since $\mu_{\mathbf{X}}(\mathbb{R}^n) = \mathbb{P}(\mathbf{X} \in \mathbb{R}^n) = 1$. As we proved for $n = 2$, the relation in (A.94) is equivalent to (A.93). The joint CDF is continuous on $\mathbb{R}^n$ and related to the joint PDF by differentiation,

$$f_{X_1,\ldots,X_n}(x_1, \ldots, x_n) = \frac{\partial^n}{\partial x_1 \ldots \partial x_n} F_{X_1,\ldots,X_n}(x_1, \ldots, x_n), \quad x_1, \ldots, x_n \in \mathbb{R}. \tag{A.95}$$

Note that (A.95) implies the existence of all marginal PDFs (densities) for all univariate X_i, bivariate (X_i, X_j), etc. In particular, all k-dimensional random vectors $(X_{i_1}, \ldots, X_{i_k})$, $1 \leqslant i_1 < \ldots < i_k \leqslant n$, $1 \leqslant k \leqslant n$, have CDF as in (A.89). Using this relation in (A.94) gives us the joint (marginal) PDF of $(X_{i_1}, \ldots, X_{i_k})$ as an $(n-k)$-dimensional integral of the joint PDF over $\mathbb{R}^{n-k}$ in the integration variables y_j, for all $j \notin \{i_1, i_2, \ldots, i_k\}$. For example, the CDF of the random vector consisting of the first k variables, $(X_1, \ldots, X_k)$, is

$$F_{X_1,\ldots,X_k}(x_1, \ldots, x_k) = F_{X_1,\ldots,X_k,X_{k+1},\ldots,X_n}(x_1, \ldots, x_k, \infty, \ldots, \infty) \tag{A.96}$$

$$= \int_{-\infty}^{x_k} \cdots \int_{-\infty}^{x_1} \left(\int_{-\infty}^{\infty} \cdots \int_{-\infty}^{\infty} f_{X_1,\ldots,X_n}(y_1, \ldots, y_n) \, \mathrm{d}y_{k+1} \ldots \mathrm{d}y_n \right) \mathrm{d}y_1 \ldots \mathrm{d}y_k . \tag{A.97}$$

The $(n-k)$-dimensional (inner) integral is the joint PDF of $(X_1, \ldots, X_k)$. This is an integrable Borel function $f_{X_1,\ldots,X_k}: \mathbb{R}^k \to R$, given by

$$f_{X_1,\ldots,X_k}(x_1, \ldots, x_k) = \int_{-\infty}^{\infty} \cdots \int_{-\infty}^{\infty} f_{X_1,\ldots,X_n}(x_1, \ldots, x_k, x_{k+1}, \ldots, x_n) \, \mathrm{d}x_{k+1} \ldots \mathrm{d}x_n .$$

Hence, for every $k = 1, \ldots, n$, we have the marginal CDF and PDF relations:

$$F_{X_1,\ldots,X_k}(x_1, \ldots, x_k) = \int_{-\infty}^{x_k} \cdots \int_{-\infty}^{x_1} f_{X_1,\ldots,X_k}(y_1 \ldots y_k) \, \mathrm{d}y_1 \ldots \mathrm{d}y_k , \tag{A.98}$$

and

$$f_{X_1,\ldots,X_k}(x_1, \ldots, x_k) = \frac{\partial^k}{\partial x_1 \ldots \partial x_k} F_{X_1,\ldots,X_k}(x_1, \ldots, x_k), \quad x_1, \ldots, x_k \in \mathbb{R}. \tag{A.99}$$

Based on (A.93), it can be proven that the expectation formula in (A.92) takes the form of an integral over $\mathbb{R}^n$ involving the joint PDF:

$$\begin{aligned} \mathrm{E}[h(\mathbf{X})] &= \int_{\mathbb{R}^n} h(\mathbf{x}) f_{\mathbf{X}}(\mathbf{x})\, \mathrm{d}^n\mathbf{x} \\ &\equiv \int_{-\infty}^{\infty} \cdots \int_{-\infty}^{\infty} h(x_1, \ldots, x_n)\, f_{X_1,\ldots,X_n}(x_1, \ldots, x_n)\, \mathrm{d}x_1 \ldots \mathrm{d}x_n \,. \end{aligned} \tag{A.100}$$

Note that (A.74) is a special case of this formula for $n = 2$ dimensions. All marginal CDFs in (A.89) are also conveniently expressed as expectations of indicator functions where

$$F_{X_{i_1},\ldots,X_{i_k}}(y_1, \ldots, y_k) = \mathbb{P}(X_{i_1} \leqslant y_1, \ldots, X_{i_k} \leqslant y_k) = \mathrm{E}[\mathbb{I}_{\{X_{i_1} \leqslant y_1, \ldots, X_{i_k} \leqslant y_k\}}] \,.$$

It is convenient to define $\mathbf{Y} = (Y_1, \ldots, Y_k) := (X_{i_1}, \ldots, X_{i_k})$. Now, differentiating all k arguments of the (marginal) joint CDF gives the (marginal) joint PDF for a continuous random vector $\mathbf{Y}$,

$$f_{Y_1,\ldots,Y_k}(y_1, \ldots, y_k) = \frac{\partial^k}{\partial y_1 \ldots \partial y_k} F_{Y_1,\ldots,Y_k}(y_1, \ldots, y_k) \,.$$

Hence, if $h\colon \mathbb{R}^k \to \mathbb{R}$ is a Borel function of only $(X_{i_1}, \ldots, X_{i_k}) \equiv (Y_1, \ldots, Y_k)$, with $1 \leqslant k \leqslant n$ components from $\mathbf{X}$, (A.100) reduces to a k-dimensional integral involving the (marginal) joint PDF of $\mathbf{Y}$:

$$\mathrm{E}[h(Y_1, \ldots, Y_k)] = \int_{-\infty}^{\infty} \cdots \int_{-\infty}^{\infty} h(y_1, \ldots, y_k)\, f_{Y_1,\ldots,Y_k}(y_1, \ldots, y_k)\, \mathrm{d}y_1 \ldots \mathrm{d}y_k \,. \tag{A.101}$$

Based on (A.101), and choosing appropriate functions for h, we can in principle compute several quantities of interest, such as moments, product moments, joint moment generating functions, joint characteristic functions, etc., as long as the integrals exist. In particular, the covariance between any two continuous random variables in $\mathbf{X}$, say X_i and X_j, is computed by making use of the joint PDF of the pair (X_i, X_j) and the marginal densities of X_i and X_j:

$$\begin{aligned} \mathrm{Cov}(X_i, X_j) &:= \mathrm{E}[X_i X_j] - \mathrm{E}[X_i]\, \mathrm{E}[X_j] \\ &= \int_{\mathbb{R}} \int_{\mathbb{R}} x y f_{X_i,X_j}(x, y)\, \mathrm{d}x\, \mathrm{d}y - \left(\int_{\mathbb{R}} x f_{X_i}(x)\, \mathrm{d}x \right) \left(\int_{\mathbb{R}} y f_{X_j}(y)\, \mathrm{d}y \right) . \end{aligned} \tag{A.102}$$

Let us now consider the case where $X_1, \ldots, X_n$ are independent. It follows that the joint distribution measure in (A.86) is now a product measure on $\mathbb{R}^n$:

$$\mu_{X_1,\ldots,X_n}(B) = \prod_{i=1}^{n} \mathbb{P}(X_i \in B_i) = \prod_{i=1}^{n} \mu_{X_i}(B_i) := \mu_{X_1 \times \ldots \times X_n}(B) \tag{A.103}$$

for all Borel sets $B = B_1 \times \ldots \times B_n$ in $\mathbb{R}^n$. The joint CDF is then the product of marginal CDFs,

$$F_{X_1,\ldots,X_n}(x_1, \ldots, x_n) = \prod_{i=1}^{n} \mathbb{P}(X_i \leqslant x_i) = \prod_{i=1}^{n} F_{X_i}(x_i) \,. \tag{A.104}$$

For continuous random variables then, by differentiating (A.104) according to (A.95), the joint PDF is the product of marginal densities

$$f_{X_1,\ldots,X_n}(x_1, \ldots, x_n) = \prod_{i=1}^{n} f_{X_i}(x_i) \,. \tag{A.105}$$

In fact, it can be shown that (A.105) and (A.104) are equivalent in the case of continuous random variables.

The expectation formula in (A.79) extends to n dimensions,

$$\mathrm{E}[h(\mathbf{X})] = \int_{\mathbb{R}} \cdots \int_{\mathbb{R}} h(\mathbf{x})\,\mathrm{d}\mu_{X_1}(x_1)\ldots \mathrm{d}\mu_{X_n}(x_n)\,, \tag{A.106}$$

where the order of integration is interchangeable according to Fubini's Theorem. Similarly, in the case where $h(\mathbf{X}) = h_1(X_1)h_2(X_2)\cdots h(X_n)$, (A.80) extends to a product of n expectations:

$$\mathrm{E}[h_1(X_1)h_2(X_2)\cdots h(X_n)] = \prod_{i=1}^{n} \int_{\mathbb{R}} h_i(x_i)\,\mathrm{d}\mu_{X_i}(x_i) = \prod_{i=1}^{n} \mathrm{E}[h_i(X_i)]\,, \tag{A.107}$$

where we assume that all product functions are integrable, $\mathrm{E}[\,|h_i(X_i)|\,] < \infty$, $i = 1, \ldots, n$. For continuous random variables we have the usual formula for the expectation involving the marginal densities,

$$\mathrm{E}[h_1(X_1)h_2(X_2)\cdots h(X_n)] = \prod_{i=1}^{n} \mathrm{E}[h_i(X_i)] = \prod_{i=1}^{n} \int_{-\infty}^{\infty} h_i(x) f_{X_i}(x)\,\mathrm{d}x\,. \tag{A.108}$$

The above formulas in the case of independence have analogues for any sub-collection of random variables, i.e., for any random vector $\mathbf{Y} = (Y_1, \ldots, Y_k) := (X_{i_1}, \ldots, X_{i_k})$, $1 \leqslant i_1 < i_2 < \ldots < i_k \leqslant n$, $1 \leqslant k \leqslant n$, as discussed above. If all components are independent, then the joint distribution measure of $\mathbf{Y}$ is simply the product measure on $\mathbb{R}^k$:

$$\mu_{Y_1,\ldots,Y_k}(B) = \prod_{i=1}^{k} \mu_{Y_i}(B_i) := \mu_{Y_1 \times \ldots \times Y_k}(B)$$

for all Borel sets $B = B_1 \times \ldots \times B_k$ in $\mathbb{R}^k$. The joint CDF of $\mathbf{Y}$ is the product of the marginal CDFs

$$F_{Y_1,\ldots,Y_k}(y_1, \ldots, y_n) = \prod_{i=1}^{k} F_{Y_i}(y_i)\,,$$

with joint PDF (for the case of a continuous random vector) as a product of the marginal densities

$$f_{Y_1,\ldots,Y_k}(y_1, \ldots, y_n) = \prod_{i=1}^{k} f_{Y_i}(y_i)\,.$$

Note that in the case that X_i and X_j are independent, $f_{X_i,X_j}(x, y) = f_{X_i}(x) f_{X_j}(y)$, $\mathrm{E}[X_i X_j] = \mathrm{E}[X_i]\mathrm{E}[X_j]$, so (A.102) gives zero covariance, as required.

A.4 Conditioning

Now that we are equipped with general probability theory, we revisit the subject of conditioning and conditional expectations of random variables. A thorough discussion on conditioning w.r.t. a σ-algebra can also be found in Chapter 6 of Volume I. We recall the formal definition as follows, for any X as a random variable on $(\Omega, \mathcal{F}, \mathbb{P})$, with σ-algebra $\mathcal{G} \subset \mathcal{F}$.

Definition A.1. A random variable Y is called the *conditional expectation* of X given $\mathcal{G}$, denoted by $Y \equiv \mathrm{E}[X \mid \mathcal{G}]$, if

(i) Y is a $\mathcal{G}$-measurable random variable, i.e., $\sigma(Y) \subset \mathcal{G}$;

(ii) $\mathrm{E}[X\, \mathbb{I}_B] = \mathrm{E}[Y\, \mathbb{I}_B]$ for every event $B \in \mathcal{G}$.

Property (i) is satisfied if the σ-algebra generated by the random variable Y, denoted by $\sigma(Y)$, is contained in $\mathcal{G}$. In the case where $\mathcal{G} = \sigma(\mathcal{P})$, with $\mathcal{P}$ as a partition, property (i) states that $\mathrm{E}[X \mid \mathcal{G}]$ is a random variable that is constant on every atom of $\mathcal{P}$. Property (ii) is the so-called *partial averaging property.* That is, the expected value of X restricted to any given event B in $\mathcal{G}$ is the same as the expected value of $\mathrm{E}[X \mid \mathcal{G}]$ restricted to the same event B in $\mathcal{G}$. Note that for $B = \Omega$, property (ii) implies the nested expectation identity $\mathrm{E}[X] = \mathrm{E}[\mathrm{E}[X \mid \mathcal{G}]]$; i.e., the expected value of X is the same as the expected value of X that has been conditioned on any information set $\mathcal{G} \subset \mathcal{F}$.

We recall that for $\mathcal{P} = \{A_i\}_{i\geqslant 1}$ as a countable partition of Ω with $\mathcal{G} = \sigma(\mathcal{P})$, we have (see Chapter 6 in Volume I):

$$\mathrm{E}[X \mid \mathcal{G}] \equiv \mathrm{E}[X \mid \sigma(\mathcal{P})] = \sum_{i\geqslant 1} \mathrm{E}[X \mid A_i]\, \mathbb{I}_{A_i}, \tag{A.109}$$

where the expected value of X conditional on an event (atom) A_i is given by

$$\mathrm{E}[X \mid A_i] = \frac{\mathrm{E}[X\, \mathbb{I}_{A_i}]}{\mathbb{P}(A_i)}. \tag{A.110}$$

We observe that $\mathrm{E}[X \mid \mathcal{G}](\omega) = \mathrm{E}[X \mid A_i]$, for every $\omega \in A_i$, for given $i \geqslant 1$; i.e., the random variable $\mathrm{E}[X \mid \mathcal{G}]$ is constant on each atom A_i. The atoms A_i can correspond to events generated by some discrete random variables. Note that the formula in (A.109) is only valid for $\mathcal{G}$ generated by a partition (i.e., countable atoms). More generally, if $\mathcal{G}$ is not generated by a partition (e.g., if $\mathcal{G}$ is a σ-algebra generated by a single or multiple continuous random variable(s)) then (A.109) is nonsensical. In the general case, one needs to show that properties (i) and (ii) hold in the above definition. If both properties hold, then $Y \equiv \mathrm{E}[X \mid \mathcal{G}]$ is unique (a.s.).

In the case where $\mathcal{G} = \sigma(\mathcal{P})$, we see immediately that the sum on the r.h.s. of (A.109), written as $Y := \sum_{i\geqslant 1} y_i\, \mathbb{I}_{A_i}$, $y_i := \mathrm{E}[X \mid A_i]$, is a $\mathcal{G}$-measurable random variable since each $\mathbb{I}_{A_i}$ is $\mathcal{G}$-measurable, i.e., $A_i \subset \mathcal{G} \equiv \sigma(\{A_i\}_{i\geqslant 1})$. For property (ii), it suffices to show that it holds for every atom $B = A_k$, $k \geqslant 1$. Indeed, the reader can readily verify that $\mathrm{E}[X\, \mathbb{I}_{A_k}] = \mathrm{E}[Y\, \mathbb{I}_{A_k}]$, for all $k \geqslant 1$. Explicit steps showing this are given in Chapter 6 of Volume I.

Since any expectation is in fact a Lebesgue integral w.r.t. a given probability measure $\mathbb{P}$, we can also state Definition A.1 in the equivalent manner using Lebesgue integral notation. In particular, property (ii) in Definition A.1 reads

$$\int_B X(\omega)\, \mathrm{d}\mathbb{P}(\omega) = \int_B Y(\omega)\, \mathrm{d}\mathbb{P}(\omega)$$

for every $B \in \mathcal{G}$, and the expectation conditional an event A_i in (A.110) reads

$$\mathrm{E}[X \mid A_i] = \frac{1}{\mathbb{P}(A_i)} \int_{A_i} X(\omega)\, \mathrm{d}\mathbb{P}(\omega). \tag{A.111}$$

Let X and Y be two random variables on $(\Omega, \mathcal{F}, \mathbb{P})$. Formally, the conditional expectation of X given Y is equal to the conditional expectation of X given the σ-algebra generated by Y:

$$\mathrm{E}[X \mid Y] := \mathrm{E}[X \mid \sigma(Y)].$$

For every $\omega \in \Omega$, $\mathrm{E}[X \mid Y](\omega) := \mathrm{E}[X \mid Y = Y(\omega)]$. Note that in this case the sub-σ-algebra, $\mathcal{G} = \sigma(Y) \subset \mathcal{F}$, contains all the information about Y only. The random variable $\mathrm{E}[X \mid Y]$ is $\sigma(Y)$-measurable. If the joint distribution of the pair (X, Y) is known, then one can obtain the conditional distribution of X given the value of Y. Then $\mathrm{E}[X \mid Y = y]$ is the expected value of X relative to the conditional distribution of X given $Y = y$. In any standard text on probability theory, the reader can find formulas of such conditional expectations for the cases with discrete or continuous random variables.

Many calculations involve expectations of functions of two or more random variables conditional on a σ-algebra $\mathcal{G}$, where one of more of the random variables is $\mathcal{G}$-measurable and the rest are independent of $\mathcal{G}$. It is hence important to recall (see also Chapter 6 of Volume I) the so-called *Independence Proposition* which is used throughout this volume. We re-state this proposition (for two and any number of random variables) as follows. Note that in all cases we are assuming that h is integrable.

Proposition A.6 (Independence Proposition for Two Random Variables). *Let X and Y be random variables on $(\Omega, \mathcal{F}, \mathbb{P})$. Suppose that X is $\mathcal{G}$-measurable and that Y is independent of the σ-algebra $\mathcal{G} \subset \mathcal{F}$. Let $h : \mathbb{R}^2 \to \mathbb{R}$ be a Borel function. Then*

$$\mathrm{E}[h(X, Y) \mid \mathcal{G}] = g(X),$$

with function $g : \mathbb{R} \to \mathbb{R}$ given by the unconditional expectation $g(x) := \mathrm{E}[h(x, Y)]$.

Proof. Let us first consider the case when X is a $\mathcal{G}$-measurable simple random variable: $X = \sum_{k=1}^{M} x_k \mathbb{I}_{A_k}$, where $A_k = \{X = x_k\} \subset \mathcal{G}$ are atoms in the partition $\{A_k\}_{k \geqslant 1}$, i.e., X has range (support) $\{x_1, \ldots, x_M\}$ for some $M \geqslant 1$. Property (i) in A.1 holds since $g(x) := \mathrm{E}[h(x, Y)]$ is a Borel function for $x \in \mathbb{R}$, which implies $\sigma(g(X)) \subset \sigma(X) \subset \mathcal{G}$. We need to show that property (ii) in A.1 holds, which in this case reads:

$$\mathrm{E}[h(X, Y)] \cdot \mathbb{I}_B] = \mathrm{E}[g(X) \cdot \mathbb{I}_B], \ \ \forall B \in \mathcal{G}.$$

Indeed, for each $\omega \in A_k$, $h(X, Y)(\omega) = h(X(\omega), Y(\omega)) = h(x_k, Y(\omega))$, i.e., $h(X, Y) = \sum_{k=1}^{M} h(x_k, Y)\mathbb{I}_{A_k}$. Hence, by using the independence of Y and $\mathcal{G}$ (i.e., $h(x_k, Y)$ is independent of $\mathbb{I}_{A_k \cap B}$ where $A_k \cap B \in \mathcal{G}$, $\mathbb{I}_{A_k}\mathbb{I}_B = \mathbb{I}_{A_k \cap B}$):

$$\mathrm{E}[h(X, Y) \cdot \mathbb{I}_B] = \sum_{k=1}^{M} \mathrm{E}[h(x_k, Y)\mathbb{I}_{A_k \cap B}] = \sum_{k=1}^{M} \mathrm{E}[h(x_k, Y)]\mathbb{P}(A_k \cap B) = \mathrm{E}[g(X) \cdot \mathbb{I}_B].$$

The last equality follows again by independence, where $g(X) = \sum_{k=1}^{M} \mathrm{E}[h(x_k, Y)]\mathbb{I}_{A_k}$.

The proof for a general random variable X follows by considering a sequence of $\mathcal{G}$-measurable simple random variables $\{X_n\}_{n \geqslant 1}$ defined by (A.9), i.e., $X_n = \sum_{k=0}^{4^n} x_n^{(k)} \mathbb{I}_{A_n^{(k)}}$, $x_n^{(k)} = \frac{k}{2^n}$, where $A_n^{(k)} = \{X \in [\frac{k}{2^n}, \frac{k+1}{2^n})\} \in \mathcal{G}$ for each $k = 0, \ldots, 4^n$, $n = 1, \ldots$. For simplicity we are assuming a nonnegative random variable X. The proof for any X follows by considering both the nonnegative and nonpositive parts of $X = X^+ - X^-$. For each $n \geqslant 1$, we have $g(X_n) = \sum_{k=0}^{4^n} \mathrm{E}[h(x_n^{(k)}, Y)] \cdot \mathbb{I}_{A_n^{(k)}}$ and $h(X_n, Y) = \sum_{k=0}^{4^n} h(x_n^{(k)}, Y) \cdot \mathbb{I}_{A_n^{(k)}}$. Again, property (i) holds since $g(x) := \mathrm{E}[h(x, Y)]$ is a Borel function. For every $n \geqslant 1$, each $\mathbb{I}_{A_n^{(k)}}$ is independent of Y and it readily follows that $\mathrm{E}[h(X_n, Y) \cdot \mathbb{I}_B] = \mathrm{E}[g(X_n) \cdot \mathbb{I}_B]$. In the limit $n \to \infty$, we recover property (ii) above, where $g(X_n) \to g(X)$ and $h(X_n, Y) \to h(X, Y)$ (a.s.). ☐

We recall the way in which Proposition A.6 is used in practice and its essence. Firstly, we fix the random variable X to some ordinary variable (parameter) x in $h(X,Y)$ and compute the function $g(x)$ as the unconditional expectation, $g(x) = \mathrm{E}[h(x,Y)]$. After having determined $g(x)$, we put back the random variable X in place of x, giving the random variable $g(X)$, which is the same as $\mathrm{E}[h(X,Y) \mid \mathcal{G}]$. That is, conditioning on the information in $\mathcal{G}$ allows us to hold X as constant (as a given value x) and then the conditional expectation of $h(x,Y)$ given the information in $\mathcal{G}$ becomes an unconditional expectation since Y, and hence $h(x,Y)$, is independent of $\mathcal{G}$. Note that when $\mathcal{G} = \sigma(X)$ we recover the well-known formula for computing the expectation of a function of two random variables, say $h(X,Y)$, conditional on one of the random variables, say X, where Y is independent of X (i.e., Y is independent of $\sigma(X)$):

$$\mathrm{E}[h(X,Y) \mid \sigma(X)] \equiv \mathrm{E}[h(X,Y) \mid X] = g(X),$$

where $g(x) = \mathrm{E}[h(X,Y) \mid X = x] = \mathrm{E}[h(x,Y)]$.

Proposition A.7 (Independence Proposition for Several Random Variables). *Consider two random vectors* $\mathbf{X} = (X_1, \ldots, X_m)$ *and* $\mathbf{Y} = (Y_1, \ldots, Y_n)$*, where all components of* $\mathbf{X}$ *are assumed* $\mathcal{G}$*-measurable and all components of* $\mathbf{Y}$ *are assumed independent of* $\mathcal{G}$*. Let* $h : \mathbb{R}^{m+n} \to \mathbb{R}$ *be a Borel function. Then,* $\mathrm{E}[h(\mathbf{X},\mathbf{Y}) \mid \mathcal{G}] = g(\mathbf{X})$ *where* $g : \mathbb{R}^m \to \mathbb{R}$ *is defined by the unconditional expectation* $g(\mathbf{x}) := \mathrm{E}[h(\mathbf{x},\mathbf{Y})]$*, i.e.,*

$$\mathrm{E}[h(X_1, \ldots, X_m, Y_1, \ldots, Y_n) \mid \mathcal{G}] = g(X_1, \ldots, X_m)$$

where $g(x_1, \ldots, x_m) := \mathrm{E}[h(x_1, \ldots, x_m, Y_1, \ldots, Y_n)]$.

We note that the proof of this result follows similar steps as in the above proof of Proposition A.6 where vectors of random variables are employed.

It is instructive to see how Proposition A.6 follows in the case of continuous random variables. Let X and Y be jointly (absolutely) continuous and independent random variables possessing a joint PDF $f_{X,Y}(x,y) = f_X(x)f_Y(y)$ with marginal PDFs $f_X(x)$ and $f_Y(y)$. The conditional PDF of Y given $X = x$ is hence the marginal PDF of Y: $f_{Y|X}(y|x) = f_{X,Y}(x,y)/f_X(x) = f_Y(y)$. Evaluating the conditional expectation in the usual manner gives

$$\begin{aligned}\mathrm{E}[h(X,Y) \mid X = x] &= \mathrm{E}[h(x,Y) \mid X = x]\\ &= \int_{\mathbb{R}} h(x,y) f_{Y|X}(y|x)\,\mathrm{d}y = \int_{\mathbb{R}} h(x,y) f_Y(y)\,\mathrm{d}y = \mathrm{E}[h(x,Y)]\,.\end{aligned}$$

Hence, as a random variable we have $\mathrm{E}[h(X,Y) \mid X] = \int_{\mathbb{R}} h(X,y) f_Y(y)\,\mathrm{d}y := g(X)$, i.e., $\mathrm{E}[h(X,Y) \mid X](\omega) = \int_{\mathbb{R}} h(X(\omega),y) f_Y(y)\,\mathrm{d}y := g(X(\omega))$, for each $\omega \in \Omega$.

Now, to formally show that the above expectation formula is in fact the correct one we need to verify properties (i) and (ii) of Definition A.1 with $\mathcal{G} = \sigma(X)$. Property (i) holds since g defined by the above integral is a Borel function and hence $\sigma(g(X)) \subset \sigma(X)$, i.e., $g(X)$ is $\sigma(X)$-measurable. For property (ii), we need to show that

$$\mathrm{E}\big[\mathbb{I}_{X \in B} \cdot \mathrm{E}[h(X,Y) \mid X]\big] \equiv \mathrm{E}\big[\mathbb{I}_{X \in B} \cdot g(X)\big] = \mathrm{E}\big[\mathbb{I}_{X \in B} \cdot h(X,Y)\big]$$

for every Borel set B in $\sigma(X)$; i.e., B is any Borel set in the range of X and so we take any $B \in \mathcal{B}(\mathbb{R})$. Expressing these expectations using the joint PDF gives

$$\mathrm{E}\big[\mathbb{I}_{X \in B} \cdot g(X)\big] = \int_B g(x) \left(\int_{\mathbb{R}} f_{X,Y}(x,y)\,\mathrm{d}y \right) \mathrm{d}x = \int_B g(x) f_X(x)\,\mathrm{d}x$$

and

$$\mathrm{E}\big[\mathbb{I}_{X\in B}\cdot h(X,Y)\big] = \int_B\left(\int_{\mathbb{R}} h(x,y)f_{X,Y}(x,y)\,\mathrm{d}y\right)\mathrm{d}x\,.$$

For these two quantities to be equal, for every Borel set B, we necessarily must have the equivalence of the x-integrands, i.e.,

$$g(x) = \int_{\mathbb{R}} h(x,y)\frac{f_{X,Y}(x,y)}{f_X(x)}\,\mathrm{d}y = \int_{\mathbb{R}} h(x,y)f_{Y|X}(y|x)\,\mathrm{d}y = \int_{\mathbb{R}} h(x,y)f_Y(y)\,\mathrm{d}y\,.$$

This proves the above expression for $g(x)$ and hence for $\mathrm{E}[h(X,Y)\mid X]$.

If $\mathbf{X}$ and $\mathbf{Y}$ are jointly continuous and independent random vectors, then their joint PDF is $f_{\mathbf{X},\mathbf{Y}}(\mathbf{x},\mathbf{y}) = f_{\mathbf{X}}(\mathbf{x})f_{\mathbf{Y}}(\mathbf{y})$ with marginal PDFs $f_{\mathbf{X}}(\mathbf{x})$ and $f_{\mathbf{Y}}(\mathbf{y})$. The conditional PDF of $\mathbf{Y}$ given $\mathbf{X}=\mathbf{x}$ is the marginal PDF of $\mathbf{Y}$: $f_{\mathbf{Y}|\mathbf{X}}(\mathbf{y}|\mathbf{x}) = f_{\mathbf{Y}}(\mathbf{y})$. By basic probability theory, the conditional expectation is given by the n-dimensional integral:

$$\mathrm{E}[h(\mathbf{X},\mathbf{Y})\mid\mathbf{X}=\mathbf{x}] = \int_{\mathbb{R}^n} h(\mathbf{x},\mathbf{y})f_{\mathbf{Y}|\mathbf{X}}(\mathbf{y}|\mathbf{x})\,\mathrm{d}^n\mathbf{y} = \int_{\mathbb{R}^n} h(\mathbf{x},\mathbf{y})f_{\mathbf{Y}}(\mathbf{y})\,\mathrm{d}^n\mathbf{y} = \mathrm{E}[h(\mathbf{x},\mathbf{Y})]\,.$$

A similar analysis as given above (for the case of two random variables) formally shows that this is the correct formula satisfying properties (i) and (ii) of Definition A.1 with $\mathcal{G}=\sigma(\mathbf{X})$. In this case the Borel sets $B\subset\mathbb{R}^n$.

Many of the important properties of conditioning are covered in detail in Chapter 6 of Volume I. It is useful to summarize these in the following theorem as they pertain to the general theory of random variables. Many of the proofs are rather straightforward and are standard in real analysis so we don't repeat them here. We note that some of the properties are proven in Chapter 6 of Volume I.

Theorem A.8. *Let X and Y be random variables on $(\Omega,\mathcal{F},\mathbb{P})$ and assume appropriate integrability of all random variables. Then, the following hold (a.s.).*

1. *(Linearity) For any real constants a,b:*

$$\mathrm{E}[a\,X+b\,Y\mid\mathcal{G}] = a\,\mathrm{E}[X\mid\mathcal{G}] + b\,\mathrm{E}[Y\mid\mathcal{G}]\,.$$

2. *(Nested Expectation)*

$$\mathrm{E}[\,\mathrm{E}[X\mid\mathcal{G}]\,] = \mathrm{E}[X]\,.$$

3. *(Tower Property) For σ-algebras $\mathcal{H}\subset\mathcal{G}\subset\mathcal{F}$,*

$$\mathrm{E}[\mathrm{E}[X\mid\mathcal{G}]\mid\mathcal{H}] = \mathrm{E}[X\mid\mathcal{H}].$$

We note that (in reverse order of conditioning): $\mathrm{E}[\mathrm{E}[X\mid\mathcal{H}]\mid\mathcal{G}] = \mathrm{E}[X\mid\mathcal{H}]$. *This follows by property #5 below, since* $\mathrm{E}[X\mid\mathcal{H}]$ *is $\mathcal{H}$-measurable, and hence, it is $\mathcal{G}$-measurable.*

4. *(Independence) If X is independent of $\mathcal{G}$, then*

$$\mathrm{E}[X\mid\mathcal{G}] = \mathrm{E}[X]\,.$$

5. *(Measurability) If X is $\mathcal{G}$-measurable, then*

$$\mathrm{E}[X\mid\mathcal{G}] = X\,.$$

6. *(Positivity) If $X\geqslant 0$, then* $\mathrm{E}[X\mid\mathcal{G}]\geqslant 0$.

7. *(Monotone Convergence) If $X_n, n \geqslant 1$, is a nonnegative sequence of random variables on $(\Omega, \mathcal{F}, \mathbb{P})$ and increases (a.s.) to X, then the sequence $\mathrm{E}[X_n \mid \mathcal{G}], n \geqslant 1$, increases to $\mathrm{E}[X \mid \mathcal{G}]$.*

8. *(Pulling out what is known) If X is $\mathcal{G}$-measurable, then*

$$\mathrm{E}[XY \mid \mathcal{G}] = X\mathrm{E}[Y \mid \mathcal{G}] .$$

9. *(Conditional Jensen's Inequality) If $\phi : \mathbb{R} \to \mathbb{R}$ is a convex function, then*

$$\mathrm{E}[\phi(X) \mid \mathcal{G}] \geqslant \phi\left(\mathrm{E}[X \mid \mathcal{G}]\right) .$$

Note that setting $\mathcal{G} = \mathcal{F}_0 \equiv \{\emptyset, \Omega\}$ recovers the known unconditional Jensen's Inequality: $\mathrm{E}[\phi(X)] \geqslant \phi\left(\mathrm{E}[X]\right)$. Recall: $\mathrm{E}[\cdot \mid \mathcal{F}_0] \equiv \mathrm{E}[\cdot]$.

A.5 Changing Probability Measures

Here, we shall keep the discussion very succinct. Let us begin by defining a new probability measure $\widehat{\mathbb{P}} \equiv \widehat{\mathbb{P}}^{(\varrho)}$ by

$$\widehat{\mathbb{P}}(A) \equiv \int_A \mathrm{d}\widehat{\mathbb{P}}(\omega) := \int_A \varrho(\omega)\,\mathrm{d}\mathbb{P}(\omega) , \tag{A.112}$$

for all $A \in \mathcal{F}$, i.e., $\widehat{\mathbb{P}}(A) \equiv \widehat{\mathrm{E}}[\mathbb{I}_A] := \mathrm{E}[\varrho\,\mathbb{I}_A]$, such that ϱ is chosen to be a nonnegative (a.s.) random variable on $(\Omega, \mathcal{F})$ having unit expectation under measure $\mathbb{P}$:

$$\mathrm{E}[\varrho] \equiv \int_\Omega \varrho(\omega)\,\mathrm{d}\mathbb{P}(\omega) = 1.$$

Note that in order for $\widehat{\mathbb{P}}$ to be a probability measure we necessarily have $1 = \widehat{\mathbb{P}}(\Omega) = \mathrm{E}[\varrho\mathbb{I}_\Omega] = \mathrm{E}[\varrho]$. The measure $\widehat{\mathbb{P}}$ is also countably additive from the countable additivity property of the Lebesgue integral w.r.t. $\mathbb{P}$, i.e., for any countable collection of pairwise disjoint sets $\{A_i\} \in \mathcal{F}$ we have, setting $A \equiv \cup_i A_i$ in (A.112),

$$\widehat{\mathbb{P}}(\cup_i A_i) = \mathrm{E}[\varrho\,\mathbb{I}_{\cup_i A_i}] = \mathrm{E}[\varrho \sum_i \mathbb{I}_{A_i}] = \sum_i \mathrm{E}[\varrho\,\mathbb{I}_{A_i}] = \sum_i \widehat{\mathbb{P}}(A_i).$$

Hence, $\widehat{\mathbb{P}}$ is a probability measure and $(\Omega, \mathcal{F}, \widehat{\mathbb{P}})$ is a probability space.

The random variable ϱ corresponds to the so-called *Radon–Nikodym derivative* of $\widehat{\mathbb{P}}$ w.r.t. $\mathbb{P}$. Note that $\mathrm{d}\widehat{\mathbb{P}}(\omega) = \varrho(\omega)\,\mathrm{d}\mathbb{P}(\omega)$. It is customary notation to denote ϱ by $\frac{\mathrm{d}\widehat{\mathbb{P}}}{\mathrm{d}\mathbb{P}}$, i.e., $\mathrm{d}\widehat{\mathbb{P}}(\omega) = \frac{\mathrm{d}\widehat{\mathbb{P}}}{\mathrm{d}\mathbb{P}}(\omega)\,\mathrm{d}\mathbb{P}(\omega)$. The notation arises naturally when we compute the expectation of a random variable in the two different measures. The expectation of a random variable X in the original $\mathbb{P}$-measure is denoted by $\mathrm{E}[X]$ and we let $\widehat{\mathrm{E}}[X]$ denote the expectation in the new $\widehat{\mathbb{P}}$-measure. Using the definition in (A.112), the expectation of X under measure $\widehat{\mathbb{P}}$ equals the expectation of $X\varrho \equiv X\frac{\mathrm{d}\widehat{\mathbb{P}}}{\mathrm{d}\mathbb{P}}$ under measure $\mathbb{P}$:

$$\widehat{\mathrm{E}}[X] = \int_\Omega X(\omega)\,\mathrm{d}\widehat{\mathbb{P}}(\omega) = \int_\Omega X(\omega)\varrho(\omega)\,\mathrm{d}\mathbb{P}(\omega) = \mathrm{E}[X\varrho] \equiv \mathrm{E}\left[X\frac{\mathrm{d}\widehat{\mathbb{P}}}{\mathrm{d}\mathbb{P}}\right] . \tag{A.113}$$

Moreover, if ϱ is strictly positive (a.s.) and Y is integrable under measure $\mathbb{P}$, then its expectation under measure $\mathbb{P}$ equals the expectation of $\frac{Y}{\varrho} \equiv Y\frac{\mathrm{d}\mathbb{P}}{\mathrm{d}\widehat{\mathbb{P}}}$ under measure $\widehat{\mathbb{P}}$:

$$\mathrm{E}[Y] = \int_\Omega Y(\omega)\,\mathrm{d}\mathbb{P}(\omega) = \int_\Omega Y(\omega)\frac{1}{\varrho(\omega)}\,\mathrm{d}\widehat{\mathbb{P}}(\omega) = \widehat{\mathrm{E}}\left[\frac{Y}{\varrho}\right] = \widehat{\mathrm{E}}\left[Y\frac{\mathrm{d}\mathbb{P}}{\mathrm{d}\widehat{\mathbb{P}}}\right]. \tag{A.114}$$

Note that $\frac{\mathrm{d}\mathbb{P}}{\mathrm{d}\widehat{\mathbb{P}}} = \left(\frac{\mathrm{d}\widehat{\mathbb{P}}}{\mathrm{d}\mathbb{P}}\right)^{-1} = \frac{1}{\varrho}$ for any strictly positive ϱ.

Remark: here, we don't state the formal Radon–Nikodym theorem. There are different versions of it in measure theory, where measures can be more general than probability measures. One version of the Radon–Nikodym theorem goes as follows. Given two finite measures μ and ν on a space $(\Omega, \mathcal{F})$, where ν is absolutely continuous w.r.t. μ (i.e., all sets of μ-measure zero are ν-measure zero), then there is a nonnegative $\mathcal{F}$-measurable function $h \equiv \frac{\mathrm{d}\nu}{\mathrm{d}\mu}$ such that the ν-measure of any set $A \in \mathcal{F}$ is given by a Lebesgue integral w.r.t. μ: $\nu(A) = \int_A h\,\mathrm{d}\mu$. Moreover, this function is unique w.r.t. measure μ. For our purposes, the Radon–Nikodym theorem guarantees that, given two equivalent probability measures $\widehat{\mathbb{P}}$ and $\mathbb{P}$, there is nonnegative random variable ϱ satisfying the above relations. Moreover, ϱ is unique (a.s.). The Radon–Nikodym theorem also applies to distribution measures of random variables and joint random variables.

We have already seen how measure changes are applied for discrete random variables. Let us briefly see how measure changes can be applied in the case of absolutely continuous random variables. Assume $X \in \mathbb{R}$ has a PDF $f(x)$ under measure $\mathbb{P}$ and a PDF $\widehat{f}(x)$ under measure $\widehat{\mathbb{P}}$. Although we can further generalize, we shall also assume that these densities are (a.e.) positive on $\mathbb{R}$. Then, for any $b \in \mathbb{R}$ we have the CDF of X under measure $\widehat{\mathbb{P}}$, $\widehat{F}_X(b) \equiv \widehat{\mu}_X((-\infty, b])$ as

$$\widehat{F}_X(b) = \widehat{\mathrm{E}}[\mathbb{I}_{\{X\leqslant b\}}] \equiv \int_{-\infty}^b \widehat{f}(x)\,\mathrm{d}x = \int_{-\infty}^b \frac{\widehat{f}(x)}{f(x)}f(x)\,\mathrm{d}x \equiv \mathrm{E}\left[\frac{\widehat{f}(X)}{f(X)}\mathbb{I}_{\{X\leqslant b\}}\right]. \tag{A.115}$$

By the Radon–Nikodym derivative, we see that the ratio of densities gives the Radon–Nikodym derivative for changing distribution measures of X. In this case, $\frac{\mathrm{d}\widehat{\mu}_X}{\mathrm{d}\mu_X}(x) = \frac{\widehat{f}(x)}{f(x)}$. This is known as a likelihood ratio. As a random variable we have the Radon–Nikodym derivative $\varrho = \varrho(X) = \frac{\widehat{f}(X)}{f(X)}$. This also generalizes to the multidimensional case in the obvious manner as the ratio of the joint PDFs.

An example of the likelihood ratio in measure changes is to consider a normal random variable. Say that $X \sim Norm(a, \sigma^2)$, i.e., that it has mean a and variance σ^2 under measure $\mathbb{P}$. Then, defining the change of measure $\mathbb{P} \to \widehat{\mathbb{P}}$, i.e., change of distribution measure $\mu_X \to \widehat{\mu}_X$, by the Radon–Nikodym random variable

$$\varrho = \frac{\mathrm{d}\widehat{\mathbb{P}}}{\mathrm{d}\mathbb{P}} = \frac{\sigma}{\widehat{\sigma}}\exp\left[\frac{(X-a)^2}{2\sigma^2} - \frac{(X-\widehat{a})^2}{2\widehat{\sigma}^2}\right] \equiv h(X)$$

we have that $X \sim Norm(\widehat{a}, \widehat{\sigma}^2)$ under measure $\widehat{\mathbb{P}}$. This follows from the above likelihood ratio of densities. The PDF of X under measure $\widehat{\mathbb{P}}$ equals its PDF under measure $\mathbb{P}$ times the likelihood ratio $\frac{\mathrm{d}\widehat{\mu}_X}{\mathrm{d}\mu_X}(x) = h(x) = \frac{\widehat{f}(x)}{f(x)}$. Hence,

$$h(X) = \frac{\widehat{f}(X)}{f(X)} = \frac{\frac{1}{\widehat{\sigma}\sqrt{2\pi}}\mathrm{e}^{\frac{-(X-\widehat{a})^2}{2\widehat{\sigma}^2}}}{\frac{1}{\sigma\sqrt{2\pi}}\mathrm{e}^{\frac{-(X-a)^2}{2\sigma^2}}}$$

which gives the above result. We note that this change of measure changes both the mean and variance of a normal random variable.

The next (and last) result of this chapter gives us a formula for computing the expectation (under a given measure $\mathbb{P}$) of a random variable conditional on any sub-σ-algebra $\mathcal{G} \subset \mathcal{F}$ via the corresponding conditional expectation of ϱX under another equivalent measure $\widehat{\mathbb{P}}$. The new measure $\widehat{\mathbb{P}}$ is defined in (A.112) with (Radon–Nikodym derivative) random variable ϱ. The theorem is useful when considering measure changes while calculating conditional expectations involving stochastic processes such as those driven by Brownian motions. The sub-σ-algebras are part of a filtration for Brownian motion. The conditioning on the filtration simplifies even further when we are dealing with practical applications involving Markov processes.

Theorem A.9 (General Bayes Formula). *Let $(\Omega, \mathcal{F}, \mathbb{P})$ and $(\Omega, \mathcal{F}, \widehat{\mathbb{P}})$ be two probability spaces with $\mathbb{P} \sim \widehat{\mathbb{P}}$ (i.e., $\mathbb{P}$ and $\widehat{\mathbb{P}}$ are equivalent probability measures). Let the random variable X be integrable w.r.t. $\widehat{\mathbb{P}}$ and set $\varrho \equiv \frac{d\widehat{\mathbb{P}}}{d\mathbb{P}}$. Then, the random variable ϱX is integrable w.r.t. $\mathbb{P}$ and its expectation under measure $\widehat{\mathbb{P}}$ conditional on a sub-σ-algebra $\mathcal{G} \subset \mathcal{F}$ is given by (a.s.)*

$$\widehat{\mathrm{E}}\big[X \mid \mathcal{G}\big] = \frac{\mathrm{E}[\varrho X \mid \mathcal{G}]}{\mathrm{E}[\varrho \mid \mathcal{G}]}\,. \tag{A.116}$$

Proof. All we need to verify is that the right-hand side of (A.116), $Y := \frac{\mathrm{E}[\varrho X|\mathcal{G}]}{\mathrm{E}[\varrho|\mathcal{G}]}$, is (almost surely) the random variable $\widehat{\mathrm{E}}\big[X \mid \mathcal{G}\big]$, i.e., the expectation of X under $\widehat{\mathbb{P}}$ conditional on $\mathcal{G}$. Note that Y must satisfy the two properties in Definition A.1 with $\widehat{\mathbb{P}}$ as measure. Hence, we need to show: (i) Y is $\mathcal{G}$-measurable; (ii) $\widehat{\mathrm{E}}\big[\mathbb{I}_A Y\big] = \widehat{\mathrm{E}}\big[\mathbb{I}_A X\big]$, for every event $A \in \mathcal{G}$. Property (i) is obviously satisfied since Y is a ratio of two $\mathcal{G}$-measurable random variables and hence is $\mathcal{G}$-measurable. Property (ii) is shown by first applying the change of measure $\widehat{\mathbb{P}} \to \mathbb{P}$ for an unconditional expectation via (A.113), then making use of the tower property $\mathrm{E}[\mathrm{E}[\,\cdot \mid \mathcal{G}]] = \mathrm{E}[\,\cdot\,]$ in reverse order, pulling out the $\mathcal{G}$-measurable random variable $\mathbb{I}_A Y$ in the inner conditional expectation, cancelling out the $\mathrm{E}[\varrho \mid \mathcal{G}]$ term, re-applying the tower property and changing back measures $\mathbb{P} \to \widehat{\mathbb{P}}$ in the final expectation:

$$\begin{aligned}
\widehat{\mathrm{E}}\big[\mathbb{I}_A Y\big] = \mathrm{E}[\varrho\, \mathbb{I}_A Y] = \mathrm{E}[\mathrm{E}[\varrho\, \mathbb{I}_A Y \mid \mathcal{G}]] = \mathrm{E}\left[\mathbb{I}_A Y\, \mathrm{E}[\varrho \mid \mathcal{G}]\right] &= \mathrm{E}\left[\mathbb{I}_A\, \mathrm{E}[\varrho X \mid \mathcal{G}]\right] \\
&= \mathrm{E}\left[\mathrm{E}[\varrho \mathbb{I}_A X \mid \mathcal{G}]\right] \\
&= \mathrm{E}\left[\varrho\, \mathbb{I}_A X\right] = \widehat{\mathrm{E}}\left[\mathbb{I}_A X\right].
\end{aligned}$$

Finally, the assumption that X is integrable w.r.t. $\widehat{\mathbb{P}}$, i.e., $\widehat{\mathrm{E}}[\,|X|\,] < \infty$, implies that ϱX is integrable w.r.t. $\mathbb{P}$ by changing measures:

$$\widehat{\mathrm{E}}[\,|X|\,] = \mathrm{E}[\varrho\,|X|\,] = \mathrm{E}[\,|\varrho\, X|\,] < \infty\,.$$

□

B

Some Useful Integral (Expectation) Identities and Symmetry Properties of Normal Random Variables

Throughout all the formulas below, a, b, c, A, B, C are any real constants and X is a normal random variable with mean $\mu \in \mathbb{R}$ and standard deviation $\sigma > 0$, i.e., $X \sim Norm(\mu, \sigma^2)$ with PDF $\varphi_{\mu,\sigma}(x) \equiv \frac{e^{-(x-\mu)^2/2\sigma^2}}{\sigma\sqrt{2\pi}}, x \in \mathbb{R}$. The functions $\mathcal{N}(x)$ and $\mathcal{N}_2(x, y; \rho)$ are the standard normal univariate and bivariate cumulative distribution functions, respectively.

$$\mathrm{E}\left[e^{BX}\mathbb{I}_{\{X>A\}}\right] \equiv \int_A^\infty e^{Bx}\varphi_{\mu,\sigma}(x)\,dx = e^{\mu B + \frac{1}{2}\sigma^2 B^2}\mathcal{N}\left(\sigma B + \frac{\mu - A}{\sigma}\right) \tag{B.1}$$

$$\mathrm{E}\left[e^{BX}\mathbb{I}_{\{X<A\}}\right] \equiv \int_{-\infty}^A e^{Bx}\varphi_{\mu,\sigma}(x)\,dx = e^{\mu B + \frac{1}{2}\sigma^2 B^2}\mathcal{N}\left(-\sigma B + \frac{A - \mu}{\sigma}\right) \tag{B.2}$$

$$\mathrm{E}\left[\mathcal{N}(AX + C)\right] \equiv \int_{-\infty}^\infty \mathcal{N}(Ax + C)\varphi_{\mu,\sigma}(x)\,dx = \mathcal{N}\left(\frac{\mu A + C}{\sqrt{1 + \sigma^2 A^2}}\right) \tag{B.3}$$

$$\begin{aligned}\mathrm{E}\left[e^{BX}\mathcal{N}(AX + C)\right] &\equiv \int_{-\infty}^\infty e^{Bx}\mathcal{N}(Ax + C)\varphi_{\mu,\sigma}(x)\,dx \\ &= e^{\mu B + \frac{1}{2}\sigma^2 B^2}\mathcal{N}\left(\frac{\mu A + C + \sigma^2 AB}{\sqrt{1 + \sigma^2 A^2}}\right)\end{aligned} \tag{B.4}$$

$$\int_0^\infty \mathcal{N}(Ax + B)e^x\,dx = -\mathcal{N}(B) + e^{(1-2AB)/2A^2}\mathcal{N}\left(\frac{1 - AB}{|A|}\right) \quad (\text{for } A < 0) \tag{B.5}$$

$$\int_0^\infty \mathcal{N}(Ax + B)e^{-x}\,dx = \mathcal{N}(B) + \operatorname{sgn}(A)\,e^{(1+2AB)/2A^2}\mathcal{N}\left(-\frac{1 + AB}{|A|}\right) \tag{B.6}$$

$$\begin{aligned}\int_{-\infty}^\infty \mathcal{N}(Ax + B)e^x\,dx &= e^{(1-2AB)/2A^2}\mathcal{N}\left(\frac{1 - AB}{|A|}\right) \\ &\quad - e^{(1+2AB)/2A^2}\mathcal{N}\left(-\frac{1 + AB}{|A|}\right) \quad (\text{for } A < 0)\end{aligned} \tag{B.7}$$

$$\int_0^\infty \mathcal{N}(Ax + B)\,dx = \frac{1}{|A|}\left[B\mathcal{N}(B) + \frac{e^{-B^2/2}}{\sqrt{2\pi}}\right] \quad (\text{for } A < 0) \tag{B.8}$$

DOI: 10.1201/9780429468889-B

$$\mathrm{E}\left[\mathcal{N}(AX+C)\mathbb{I}_{\{X<B\}}\right] \equiv \int_{-\infty}^{B} \mathcal{N}(Ax+C)\varphi_{\mu,\sigma}(x)\,\mathrm{d}x$$
$$= \mathcal{N}_2\left(\frac{B-\mu}{\sigma}, \frac{\mu A+C}{\sqrt{1+\sigma^2A^2}}; \frac{-\sigma A}{\sqrt{1+\sigma^2A^2}}\right) \tag{B.9}$$

$$\mathrm{E}\left[\mathcal{N}(AX+C)\mathbb{I}_{\{X>B\}}\right] \equiv \int_{B}^{\infty} \mathcal{N}(Ax+C)\varphi_{\mu,\sigma}(x)\,\mathrm{d}x$$
$$= \mathcal{N}_2\left(\frac{\mu-B}{\sigma}, \frac{\mu A+C}{\sqrt{1+\sigma^2A^2}}; \frac{\sigma A}{\sqrt{1+\sigma^2A^2}}\right) \tag{B.10}$$

Note that formulas (B.1)–(B.10) simplify in the special case when $X \stackrel{d}{=} Z \sim Norm(0,1)$ by setting $\mu = 0, \sigma = 1$; e.g., $\mathrm{E}[\mathrm{e}^{BZ}\mathbb{I}_{\{Z>A\}}] = \mathrm{e}^{B^2/2}\mathcal{N}(B-A)$.

Other useful expectation identities also follow by differentiating the above expressions w.r.t. a parameter. In particular, by differentiating $n \geqslant 0$ times w.r.t. parameter B on both sides of (B.1) gives:

$$\mathrm{E}\left[X^n\mathrm{e}^{BX}\mathbb{I}_{\{X>A\}}\right] = \frac{\partial^n}{\partial B^n}\left[\mathrm{e}^{\mu B+\frac{1}{2}\sigma^2B^2}\mathcal{N}\left(\sigma B + \frac{\mu-A}{\sigma}\right)\right], \tag{B.11}$$

$$\mathrm{E}\left[X^n\mathbb{I}_{\{X>A\}}\right] = \frac{\partial^n}{\partial B^n}\left[\mathrm{e}^{\mu B+\frac{1}{2}\sigma^2B^2}\mathcal{N}\left(\sigma B + \frac{\mu-A}{\sigma}\right)\right]\Bigg|_{B=0}. \tag{B.12}$$

For example, for $n = 1$ we have:

$$\mathrm{E}\left[X\mathrm{e}^{BX}\mathbb{I}_{\{X>A\}}\right] = \mathrm{e}^{\mu B+\frac{1}{2}\sigma^2B^2}\left[(\mu+\sigma^2B)\mathcal{N}\left(\sigma B + \frac{\mu-A}{\sigma}\right) + \sigma n\left(\sigma B + \frac{\mu-A}{\sigma}\right)\right], \tag{B.13}$$

$$\mathrm{E}\left[X\,\mathbb{I}_{\{X>A\}}\right] = \mu\mathcal{N}\left(\frac{\mu-A}{\sigma}\right) + \sigma n\left(\frac{\mu-A}{\sigma}\right), \tag{B.14}$$

where $n(z) \equiv \mathcal{N}'(z) = \frac{\mathrm{e}^{-z^2/2}}{\sqrt{2\pi}}$. For $X \equiv Z$ ($\mu = 0, \sigma = 1$), (B.13)-(B.14) simplify to

$$\mathrm{E}\left[Z\mathrm{e}^{BZ}\mathbb{I}_{\{Z>A\}}\right] = \mathrm{e}^{B^2/2}[B\mathcal{N}(B-A) + n(B-A)], \tag{B.15}$$

$$\mathrm{E}\left[Z\,\mathbb{I}_{\{Z>A\}}\right] = n(-A) = n(A). \tag{B.16}$$

Note that analogous idenities to (B.11)–(B.16) are generated by differentiating (B.2) w.r.t. B.

Differentiating $n \geqslant 0$ times w.r.t. parameter B on both sides of (B.4) produces yet other useful expectation formulas:

$$\mathrm{E}\left[X^n\mathrm{e}^{BX}\mathcal{N}(AX+C)\right] = \frac{\partial^n}{\partial B^n}\left[\mathrm{e}^{\mu B+\frac{1}{2}\sigma^2B^2}\mathcal{N}\left(\frac{\mu A+C+\sigma^2AB}{\sqrt{1+\sigma^2A^2}}\right)\right], \tag{B.17}$$

$$\mathrm{E}\left[X^n\mathcal{N}(AX+C)\right] = \frac{\partial^n}{\partial B^n}\left[\mathrm{e}^{\mu B+\frac{1}{2}\sigma^2B^2}\mathcal{N}\left(\frac{\mu A+C+\sigma^2AB}{\sqrt{1+\sigma^2A^2}}\right)\right]\Bigg|_{B=0}. \tag{B.18}$$

In particular, for $n = 1$:

$$\mathrm{E}\left[X\mathrm{e}^{BX}\mathcal{N}(AX+C)\right] = \mathrm{e}^{\mu B+\frac{1}{2}\sigma^2B^2}\left[(\mu+\sigma^2B)\mathcal{N}\left(\frac{\mu A+C+\sigma^2AB}{\sqrt{1+\sigma^2A^2}}\right) + \frac{\sigma^2A}{\sqrt{1+\sigma^2A^2}}n\left(\frac{\mu A+C+\sigma^2AB}{\sqrt{1+\sigma^2A^2}}\right)\right], \tag{B.19}$$

$$\mathrm{E}\left[X\mathcal{N}(AX+C)\right] = \mu\mathcal{N}\left(\frac{\mu A+C}{\sqrt{1+\sigma^2A^2}}\right) + \frac{\sigma^2A}{\sqrt{1+\sigma^2A^2}}n\left(\frac{\mu A+C}{\sqrt{1+\sigma^2A^2}}\right). \tag{B.20}$$

For standard normal $X = Z$, the above formulas simplify by setting $\mu = 0, \sigma = 1$.

The bivariate normal CDF has useful symmetry relations such as

$$\mathcal{N}_2(a,b;\rho) = \mathcal{N}_2(b,a;\rho)\,, \tag{B.21}$$

$$\mathcal{N}_2(a,b;\rho) + \mathcal{N}_2(a,-b;-\rho) = \mathcal{N}(a)\,. \tag{B.22}$$

If X, Y are any two absolutely continuous random variables, then their joint CDF has the equivalent representation

$$F_{X,Y}(a,b) = \int_{-\infty}^{a} f_X(x)F_{Y|X}(b|x)\,\mathrm{d}x = \int_{-\infty}^{b} f_Y(y)F_{X|Y}(a|y)\,\mathrm{d}y,$$

where $F_{Y|X}$ and $F_{X|Y}$ are the respective conditional CDFs, of Y given X and X given Y; f_X and f_Y are the respective PDFs of X and Y. Differentiating gives $\frac{\partial}{\partial a}F_{X,Y}(a,b) = f_X(a)F_{Y|X}(b|a)$ and $\frac{\partial}{\partial b}F_{X,Y}(a,b) = f_Y(b)F_{X|Y}(a|b)$. Applying these relations to standard normal random variables $X = Z_1, Y = Z_2$ with $\mathrm{Cov}(Z_1, Z_2) = \rho$, where $Y|\{X = x\} \stackrel{d}{=} Norm(\rho x, 1-\rho^2)$ and $X|\{Y = y\} \stackrel{d}{=} Norm(\rho y, 1-\rho^2)$, i.e., $F_{Y|X}(b|x) = \mathcal{N}\left(\frac{b-\rho x}{\sqrt{1-\rho^2}}\right)$ and $F_{X|Y}(a|y) = \mathcal{N}\left(\frac{a-\rho y}{\sqrt{1-\rho^2}}\right)$, gives

$$\mathcal{N}_2(a,b;\rho) = \int_{-\infty}^{a} n(x)\mathcal{N}\left(\frac{b-\rho x}{\sqrt{1-\rho^2}}\right)\mathrm{d}x = \int_{-\infty}^{b} n(y)\mathcal{N}\left(\frac{a-\rho y}{\sqrt{1-\rho^2}}\right)\mathrm{d}y\,. \tag{B.23}$$

Hence,

$$\frac{\partial}{\partial a}\mathcal{N}_2(a,b;\rho) = n(a)\mathcal{N}\left(\frac{b-\rho a}{\sqrt{1-\rho^2}}\right), \quad \frac{\partial}{\partial b}\mathcal{N}_2(a,b;\rho) = n(b)\mathcal{N}\left(\frac{a-\rho b}{\sqrt{1-\rho^2}}\right). \tag{B.24}$$

In the expectation formulas below, Z_1, Z_2 are i.i.d. standard normal random variables with covariance $\mathrm{Cov}(Z_1, Z_2) = \rho$, $|\rho| < 1$, i.e., with joint PDF $n_2(x,y;\rho) \equiv \frac{\partial^2}{\partial x\partial y}\mathcal{N}_2(x,y;\rho) = \frac{1}{2\pi\sqrt{1-\rho^2}}\mathrm{e}^{-(x^2+y^2-2\rho xy)/2(1-\rho^2)}$.

$$\mathrm{E}\left[\mathrm{e}^{-BZ_2}\,\mathbb{I}_{\{Z_1<a,Z_2<b\}}\right] \equiv \iint_{\mathbb{R}^2} \mathrm{e}^{-By} n_2(x,y;\rho)\mathbb{I}_{\{x<a,y<b\}}\,\mathrm{d}x\,\mathrm{d}y$$
$$= \mathrm{e}^{\frac{1}{2}B^2}\mathcal{N}_2\left(a+\rho B, b+B;\rho\right) \tag{B.25}$$

$$\mathrm{E}\left[\mathrm{e}^{-BZ_2}\,\mathbb{I}_{\{Z_1>a,Z_2<b\}}\right] = \mathrm{e}^{\frac{1}{2}B^2}\mathcal{N}_2\left(-(a+\rho B), b+B;-\rho\right) \tag{B.26}$$

$$\mathrm{E}\left[\mathrm{e}^{-BZ_2}\,\mathbb{I}_{\{Z_1<a,Z_2>b\}}\right] = \mathrm{e}^{\frac{1}{2}B^2}\mathcal{N}_2\left(a+\rho B, -(b+B);-\rho\right) \tag{B.27}$$

$$\mathrm{E}\left[\mathrm{e}^{-BZ_2}\,\mathbb{I}_{\{Z_1>a,Z_2>b\}}\right] = \mathrm{e}^{\frac{1}{2}B^2}\mathcal{N}_2\left(-(a+\rho B), -(b+B);\rho\right) \tag{B.28}$$

Note: interchanging $a \leftrightarrow b$ in (B.25)–(B.28) gives, respectively, equivalent formulas for the expectations $\mathrm{E}\left[\mathrm{e}^{-BZ_1}\,\mathbb{I}_{\{Z_1<a,Z_2<b\}}\right]$, $\mathrm{E}\left[\mathrm{e}^{-BZ_1}\,\mathbb{I}_{\{Z_1<a,Z_2>b\}}\right]$, $\mathrm{E}\left[\mathrm{e}^{-BZ_1}\,\mathbb{I}_{\{Z_1>a,Z_2<b\}}\right]$, and $\mathrm{E}\left[\mathrm{e}^{-BZ_1}\,\mathbb{I}_{\{Z_1>a,Z_2>b\}}\right]$. We remark that the limit $b \to \infty$ in (B.25)–(B.26) produces related identities:

$$\mathrm{E}\left[\mathrm{e}^{-BZ_2}\,\mathbb{I}_{\{Z_1<a\}}\right] = \mathrm{e}^{\frac{1}{2}B^2}\mathcal{N}_2\left(a+\rho B, \infty;\rho\right) = \mathrm{e}^{\frac{1}{2}B^2}\mathcal{N}(a+\rho B), \tag{B.29}$$
$$\mathrm{E}\left[\mathrm{e}^{-BZ_2}\,\mathbb{I}_{\{Z_1>a\}}\right] = \mathrm{e}^{\frac{1}{2}B^2}\mathcal{N}_2\left(-(a+\rho B), \infty;-\rho\right) = \mathrm{e}^{\frac{1}{2}B^2}\mathcal{N}(-(a+\rho B)). \tag{B.30}$$

Differentiating the expressions in (B.25)–(B.28) w.r.t. parameter B leads to other identities. In particular, differentiating the expression in (B.25), $n \geqslant 0$ times w.r.t. B, gives

$$\mathrm{E}\left[Z_2^n\mathrm{e}^{-BZ_2}\,\mathbb{I}_{\{Z_1<a,Z_2<b\}}\right] = (-1)^n\frac{\partial^n}{\partial B^n}\left[\mathrm{e}^{\frac{1}{2}B^2}\mathcal{N}_2\left(a+\rho B, b+B;\rho\right)\right], \tag{B.31}$$

$$\mathrm{E}\left[Z_2^n\,\mathbb{I}_{\{Z_1<a,Z_2<b\}}\right] = (-1)^n\frac{\partial^n}{\partial B^n}\left[\mathrm{e}^{\frac{1}{2}B^2}\mathcal{N}_2\left(a+\rho B, b+B;\rho\right)\right]\Bigg|_{B=0}. \tag{B.32}$$

The partial derivatives w.r.t. B are computed using (B.24) and the chain rule. For $n = 1$:

$$\mathrm{E}\left[Z_2\mathrm{e}^{-BZ_2}\,\mathbb{I}_{\{Z_1<a,Z_2<b\}}\right] = -\mathrm{e}^{\frac{1}{2}B^2}\left[B\mathcal{N}_2\left(a+\rho B, b+B;\rho\right) + n(b+B)\mathcal{N}\left(\frac{a-\rho b}{\sqrt{1-\rho^2}}\right)\right.$$
$$\left.+\,\rho n(a+\rho B)\mathcal{N}\left(\frac{b-\rho a+(1-\rho^2)B}{\sqrt{1-\rho^2}}\right)\right], \tag{B.33}$$

$$\mathrm{E}\left[Z_2\,\mathbb{I}_{\{Z_1<a,Z_2<b\}}\right] = -\rho n(a)\mathcal{N}\left(\frac{b-\rho a}{\sqrt{1-\rho^2}}\right) - n(b)\mathcal{N}\left(\frac{a-\rho b}{\sqrt{1-\rho^2}}\right). \tag{B.34}$$

Analogous identities follow readily by differentiating the expressions in (B.26)–(B.28).

C

Answers and Hints to Exercises

C.1 Chapter 1

1.1. (a) $\begin{bmatrix} X \\ Y \end{bmatrix} \sim Norm_2\left(\begin{bmatrix} \mu_X \\ \mu_Y \end{bmatrix}, \begin{bmatrix} \Sigma_{XX} & \Sigma_{XY} \\ \Sigma_{XY} & \Sigma_{YY} \end{bmatrix}\right)$, where $\mu_X = \mu_1 + \mu_2$, $\mu_Y = \mu_1 - \mu_2$, $\Sigma_{XX} = \mathrm{Var}(X) = \sigma_1^2 + \sigma_2^2 + 2\rho\sigma_1\sigma_2$, $\Sigma_{YY} = \mathrm{Var}(Y) = \sigma_1^2 + \sigma_2^2 - 2\rho\sigma_1\sigma_2$, $\Sigma_{XY} = \mathrm{Cov}(X, Y) = \rho\sigma_1\sigma_2$.

(b) $X|\{Y = y\} \stackrel{d}{=} Norm\left(\mu_X + \frac{\Sigma_{XY}}{\Sigma_{YY}}(y - \mu_Y), \Sigma_{XX} - \frac{\Sigma_{XY}^2}{\Sigma_{YY}}\right)$,

$Y|\{X = x\} \stackrel{d}{=} Norm\left(\mu_Y + \frac{\Sigma_{XY}}{\Sigma_{XX}}(x - \mu_X), \Sigma_{YY} - \frac{\Sigma_{XY}^2}{\Sigma_{XX}}\right)$.

1.2. (a) $\mathrm{E}[\mathbb{I}_{\{S>K\}}] = \mathcal{N}\left(\frac{a - \ln K}{|b|}\right)$ for all $b \neq 0$.

(b) $\mathrm{E}[S\,\mathbb{I}_{\{S>K\}}] = \mathrm{e}^{a+b^2/2}\mathcal{N}\left(|b| + \frac{a-\ln K}{|b|}\right)$ for all $b \neq 0$.

(c) $\mathrm{E}[\max(S, K)] = \mathrm{e}^{a+b^2/2}\mathcal{N}\left(|b| + \frac{a-\ln K}{|b|}\right) + K\,\mathcal{N}\left(\frac{\ln K - a}{|b|}\right)$ for all $b \neq 0$.

1.3. (a) $Y \sim Norm(0, 11)$. (b) $\begin{bmatrix} Y \\ Z \end{bmatrix} \sim Norm_2\left(\begin{bmatrix} 1 \\ 0 \end{bmatrix}, \begin{bmatrix} 11 & -2 \\ -2 & 8 \end{bmatrix}\right)$.

(c) $\mathrm{Cov}(aY + bZ, X_1) = 4a$. Hence, W is independent of X_1 only when it is a multiple of Z (i.e., $a = 0$ and b is arbitrary).

1.4. (a) Hint: Each X_k has a normal distribution, since it is a linear combination of k i.i.d. standard normals. (b) $\mathrm{Cov}(X_k, X_j) = \min(j, k)$. (c) $C_{ij} = \min(j, k)$ for $1 \leqslant j, k \leqslant n$. (d) The joint PDF of the normal vector $\mathbf{X}$ is given by

$$f_{\mathbf{X}}(\mathbf{x}) = \frac{1}{(2\pi)^{n/2}} \exp\left(-\frac{1}{2}\sum_{k=1}^{n}(x_k - x_{k-1})^2\right) = \prod_{k=1}^{n} n(x_k - x_{k-1}),$$

where $x_0 \equiv 0$ and $n(z) := \frac{\mathrm{e}^{-z^2/2}}{\sqrt{2\pi}}$.

1.6. Hint: Assuming a given filtration for BM, in each case show that the process (i) starts at zero, (ii) has continuous paths, (iii) has independent non-overlapping increments, and (iv) has normally distributed increments, $X(t) - X(s) \stackrel{d}{=} Norm(0, t-s)$, for $s < t$.

1.7. $\mathrm{Corr}(B(t), X(t)) = \frac{1}{\sqrt{2}} \frac{\mathrm{Var}(B(t))}{\sqrt{\mathrm{Var}(B(t))\,\mathrm{Var}(X(t))}} = \frac{1}{\sqrt{2}}$.

1.8. $X := W(1) + W(2) + \cdots + W(n) \stackrel{d}{=} Norm\left(0, \frac{n(n+1)(2n+1)}{6}\right)$. Hint: X is a normal random variable so it suffices to compute its mean and variance. Consider successive Brownian increments and write $W(k) = Z_1 + Z_2 + \cdots + Z_k$ with i.i.d. $Norm(0, 1)$ random variables Z_j, $j \geqslant 1$.

DOI: 10.1201/9780429468889-C

1.9. $X_n := \sum_{k=1}^{n} a_k W(t_k)$ is a linear combination of normal random variables and is hence a normal random variable. Since each $W(t_k)$ has zero mean, $\mathrm{E}[X_n] = 0$. Define $Z_k := W(t_k) - W(t_{k-1}) \sim Norm(0, t_k - t_{k-1})$, for $1 \leqslant k \leqslant n$, $t_0 \equiv 0$. Note that $Z_k, k \geqslant 1$, are mutually independent. Each $W(t_k)$ can be expressed in terms of Z_k: $W(t_k) = Z_1 + Z_2 + \cdots + Z_k$. Thus, $X_n = \sum_{k=1}^{n} c_k Z_k$, where $c_k := \sum_{j=k}^{n} a_j$. Hence, we have $X_n \sim Norm\left(0, \sum_{k=1}^{n}\left(\sum_{j=k}^{n} a_j\right)^2 (t_k - t_{k-1})\right)$.

1.10. Hint: You may write $W(t) = Y + W(s)$, where $Y := W(t) - W(s)$ is independent of $W(s)$ and $\mathcal{F}_s$.

(a) $W^3(s) + 3(t-s)W(s)$; (b) $W^2(s) - 2tW(s) + t^2 + t - s$; (c) 0;

(d) $t - s$; (e) $2s^2$; (f) $\mathrm{e}^{x+\frac{1}{2}(t-s)}\mathcal{N}\left(\frac{b-x}{\sqrt{t-s}} - \sqrt{t-s}\right)$.

1.12. $\mathrm{Cov}(X(t), Y(s)) = \sigma_x\sigma_y \,\mathrm{Cov}(W(t), W(s)) = \sigma_x\sigma_y \min(s,t)$.

1.13. (a) $\mathrm{Cov}(X(s), X(t)) = s\sigma(s)\sigma(t)$; $P(s,t;x,y) = \mathcal{N}\left(\frac{y'-x'}{\sqrt{t-s}}\right)$ and $p(s,t;x,y) = \frac{1}{\sigma(t)\sqrt{t-s}} n\left(\frac{y'-x'}{\sqrt{t-s}}\right)$ where x', y' are given in part (b) of the exercise question.

(b) Same expression for $P(s,t;x,y)$ as in (a).

1.14. Hint: The identity in (B.1) in Appendix B can be useful.

1.16. (a) $P(s,t;x,y) = \mathcal{N}\left(\frac{\ln(y/x)-\alpha\mu(t-s)}{\alpha\sigma\sqrt{t-s}}\right)$; $p(s,t;x,y) = \frac{1}{y\alpha\sigma\sqrt{t-s}} n\left(\frac{\ln(y/x)-\alpha\mu(t-s)}{\alpha\sigma\sqrt{t-s}}\right)$.

(b) $P(s,t;x,K_1) + 1 - P(s,t;x,K_2) = \mathcal{N}\left(\frac{\ln(K_1/x)-\alpha\mu(t-s)}{\alpha\sigma\sqrt{t-s}}\right) + \mathcal{N}\left(\frac{\ln(x/K_2)+\alpha\mu(t-s)}{\alpha\sigma\sqrt{t-s}}\right)$.

(c) $\mathbb{P}(Y(t) < b | Y(T) = y) = \mathcal{N}\left(\frac{\ln[(b/y_0)(y_0/y)^{t/T}]}{\alpha\sigma\sqrt{t(T-t)/T}}\right)$; $f_{Y(t)|Y(T)}(x|y) = \frac{p(0,t;y_0,x)p(t,T;x,y)}{p(0,T;y_0,y)}$, with p given in part (a); or equivalently, $f_{Y(t)|Y(T)}(x|y) = \frac{\partial}{\partial x}\mathbb{P}(Y(t) < x|Y(T) = y) = \frac{1}{x\alpha\sigma\sqrt{t(T-t)/T}} n\left(\frac{\ln[(x/y_0)(y_0/y)^{t/T}]}{\alpha\sigma\sqrt{t(T-t)/T}}\right)$.

1.17. The transition CDF is

$$P(s,t;x,y) := \mathbb{P}(S^n(t) \leqslant y \mid S^n(s) = x) = \mathcal{N}\left(\frac{\ln\frac{y}{x} - n\mu(t-s)}{n\sigma\sqrt{t-s}}\right).$$

Thus, the transition PDF is

$$p(s,t;x,y) = \frac{\partial}{\partial y}P(s,t;x,y) = \frac{1}{yn\sigma\sqrt{2\pi(t-s)}}\exp\left(-\frac{\left[\ln\frac{y}{x} - n\mu(t-s)\right]^2}{2(n\sigma)^2(t-s)}\right)$$

$x, y > 0$ and $t > s \geqslant 0$.

1.18. The transition CDF is

$$P(s,t;x,y) := \mathbb{P}(|W(t)| \leqslant y \mid |W(s)| = x) = \mathcal{N}\left(\frac{y-x}{\sqrt{t-s}}\right) - \mathcal{N}\left(\frac{-(y+x)}{\sqrt{t-s}}\right).$$

Thus, the transition PDF is

$$p(s,t;x,y) = \frac{\partial}{\partial y}P(s,t;x,y) = \frac{1}{\sqrt{2\pi(t-s)}}\left[\mathrm{e}^{-(y-x)^2/2(t-s)} + \mathrm{e}^{-(y+x)^2/2(t-s)}\right]$$

for all $x, y \geqslant 0$ and $t > s \geqslant 0$.

1.19. Hint: Define the mutually independent Brownian increments $Z_k := W(t_k) - W(t_{k-1})$, for $1 \leqslant k \leqslant n$, $t_0 \equiv 0$, $W(t_0) \equiv W(0) = 0$. The Brownian vector $\mathbf{X} := [W(t_1), W(t_2), \ldots, W(t_n)]^\top$ and the vector of Brownian increments, $\mathbf{Z} := [Z_1, Z_2, \ldots, Z_n]^\top$, are then related by a linear transformation, $\mathbf{X} = \mathbf{BZ}$, where $\mathbf{B}$ is a lower triangular $n \times n$ matrix of 1's s.t. $B_{ij} = 0$ if $i < j$ and $B_{ij} = 1$ if $i \geqslant j$. Thus, $\mathbf{X}$ is a multivariate normal with mean zero and covariance matrix $\mathbf{C} = \mathbf{BC}^Z\mathbf{B}^\top$, where $\mathbf{C}^Z$ denotes the covariance matrix of the $\mathbf{Z}$.

1.20. First, show that $\text{Var}(X(0)) = 0$, and hence $X(0) = m(0) = 0$. Second, show that $\text{Cov}(X(v) - X(u), X(t) - X(s)) = 0$ for $u, v, s, t \geqslant 0$ s.t. $u < v \leqslant s < t$ and that $X(t) - X(s) \sim Norm(0, |t - s|)$ for all $s, t \geqslant 0$.

1.21. $m(t) = a + (b - a)\frac{t-t_0}{T-t_0}$; $c(s,t) = \text{Cov}(B^{(0,0)}_{[t_0,T]}(s), B^{(0,0)}_{[t_0,T]}(t)) = s \wedge t - t_0 - \frac{(s-t_0)(t-t_0)}{T-t_0}$.

1.22. $m(t) = S_0 \mathrm{e}^{(\mu + \frac{1}{2}\sigma^2)t}$ and $c(s,t) = S_0^2 \mathrm{e}^{(\mu+\frac{1}{2}\sigma^2)(s+t)}\left[\mathrm{e}^{\sigma^2(s\wedge t)} - 1\right]$ for all $s, t \geqslant 0$.

1.23. (a) $F_{M^S(t),S(t)}(y,x) = \mathcal{N}\left(\frac{\ln \frac{x}{S_0} - \mu t}{\sigma\sqrt{t}}\right) - \left(\frac{y}{S_0}\right)^{\frac{2\mu}{\sigma^2}} \mathcal{N}\left(\frac{\ln \frac{xS_0}{y^2} - \mu t}{\sigma\sqrt{t}}\right)$;

(b) $F_{M^S(t)}(y) = \mathcal{N}\left(\frac{\ln \frac{y}{S_0} - \mu t}{\sigma\sqrt{t}}\right) - \left(\frac{y}{S_0}\right)^{\frac{2\mu}{\sigma^2}} \mathcal{N}\left(\frac{\ln \frac{S_0}{y} - \mu t}{\sigma\sqrt{t}}\right)$;

(c) $F_{\mathcal{T}_B^S}(t) = \mathcal{N}\left(\frac{\ln \frac{S_0}{B} + \mu t}{\sigma\sqrt{t}}\right) + \left(\frac{B}{S_0}\right)^{\frac{2\mu}{\sigma^2}} \mathcal{N}\left(\frac{\ln \frac{S_0}{B} - \mu t}{\sigma\sqrt{t}}\right)$;

(d) $\mathbb{P}(M^S(t) \leqslant y, S(t) \in \mathrm{d}x) = \frac{1}{x\sigma\sqrt{t}}\left[n\left(\frac{\ln \frac{x}{S_0} - \mu t}{\sigma\sqrt{t}}\right) - \left(\frac{y}{S_0}\right)^{\frac{2\mu}{\sigma^2}} n\left(\frac{\ln \frac{xS_0}{y^2} - \mu t}{\sigma\sqrt{t}}\right)\right] \mathrm{d}x$;

(e) $\mathbb{P}(m^S(t) \geqslant y, S(t) \geqslant x) = \mathcal{N}\left(\frac{\ln \frac{S_0}{x} + \mu t}{\sigma\sqrt{t}}\right) - \left(\frac{y}{S_0}\right)^{\frac{2\mu}{\sigma^2}} \mathcal{N}\left(\frac{\ln \frac{y^2}{xS_0} + \mu t}{\sigma\sqrt{t}}\right)$;

(f) $F_{\mathcal{T}_B^S}(t) = \mathcal{N}\left(\frac{\ln \frac{B}{S_0} - \mu t}{\sigma\sqrt{t}}\right) + \left(\frac{B}{S_0}\right)^{\frac{2\mu}{\sigma^2}} \mathcal{N}\left(\frac{\ln \frac{B}{S_0} + \mu t}{\sigma\sqrt{t}}\right)$;

(g) $\mathbb{P}(m^S(t) \geqslant y, S(t) \in \mathrm{d}x) = \frac{1}{x\sigma\sqrt{t}}\left[n\left(\frac{\ln \frac{S_0}{x} + \mu t}{\sigma\sqrt{t}}\right) - \left(\frac{y}{S_0}\right)^{\frac{2\mu}{\sigma^2}} n\left(\frac{\ln \frac{y^2}{xS_0} + \mu t}{\sigma\sqrt{t}}\right)\right] \mathrm{d}x$.

1.24. Hint: make appropriate use of the identity (valid for any two absolutely continuous random variables X and Y): $\mathbb{P}(Y \leqslant b | X = x) = \frac{\mathbb{P}(Y \leqslant b, X \in \mathrm{d}x)}{\mathbb{P}(X \in \mathrm{d}x)} \equiv \frac{1}{f_X(x)}\frac{\partial}{\partial x}F_{Y,X}(b,x)$.

1.25. $\mathbb{P}(M^S(t) \leqslant B\,|S(T) = y) = 1 - \left(\frac{B}{S_0}\right)^{\frac{2\mu}{\sigma^2}} \frac{n\left(\frac{\ln(yS_0/B^2) - \mu T}{\sigma\sqrt{T}}\right)}{n\left(\frac{\ln(y/S_0) - \mu T}{\sigma\sqrt{T}}\right)}$, where $n(z) := \frac{\mathrm{e}^{-z^2/2}}{\sqrt{2\pi}}$.

1.26. (a) $F_{M^X(t)}(m) - F_{M^X(t),X(t)}(m,x)$;
(b) $F_{M^X(t),X(t)}(m,b) - F_{M^X(t),X(t)}(m,a)$;
(c) $F_{X(t)}(b) - F_{X(t)}(a) + F_{M^X(t),X(t)}(m,a) - F_{M^X(t),X(t)}(m,b)$;
(d) $F_{m^X(t),X(t)}(m,b) - F_{m^X(t),X(t)}(m,a)$;
(e) $1 - F_{X(t)}(x) - F_{m^X(t)}(m) + F_{m^X(t),X(t)}(m,x)$.

1.27. For $S_0 > B$:

$$\mathbb{P}(\mathcal{T}_B^S < \infty) = \begin{cases} 1 & \text{if } \mu \leqslant 0, \\ \left(\frac{B}{S_0}\right)^{\frac{2\mu}{\sigma^2}} & \text{if } \mu > 0. \end{cases}$$

For $S_0 < B$:

$$\mathbb{P}(\mathcal{T}_B^S < \infty) = \begin{cases} 1 & \text{if } \mu \geqslant 0, \\ \left(\frac{B}{S_0}\right)^{\frac{2\mu}{\sigma^2}} & \text{if } \mu < 0. \end{cases}$$

Note: the probability that the process never attains level B in a finite time is given by $\mathbb{P}(\mathcal{T}_B^S = \infty) = 1 - \mathbb{P}(\mathcal{T}_B^S < \infty)$.

1.28. (a) Show that both processes are normally distributed with the same mean and covariance functions.
(b) $B^{(\mu)}(t) | \{B^{(\mu)}(s) = x\} \stackrel{d}{=} Norm\left(m(t) + \frac{T-t}{T-s}(x - m(s)), \frac{t}{T}(T-t) - \frac{s}{T}\frac{(T-t)^2}{T-s}\right)$, $m(t) \equiv a + (b-a)\frac{t}{T}$. Hence,

$$p^{B^{(\mu)}}(s,t;x,y) = \frac{n\left(\frac{y-[m(t)+\frac{T-t}{T-s}(x-m(s))]}{\sqrt{\frac{t}{T}(T-t)-\frac{s}{T}\frac{(T-t)^2}{T-s}}}\right)}{\sqrt{\frac{t}{T}(T-t) - \frac{s}{T}\frac{(T-t)^2}{T-s}}}, \; s < t, \; x, y \in \mathbb{R}.$$

1.29. (a) Make use of (1.47) and (1.58).

(b) $f_{S^{(a,b)}_{[t_0,T]}(t)}(x) = \frac{1}{x\sigma\sqrt{(t-t_0)(T-t)/(T-t_0)}} n\left(\frac{\ln[(x/a)(a/b)^{(t-t_0)/(T-t_0)}]}{\sigma\sqrt{(t-t_0)(T-t)/(T-t_0)}}\right), x \in (0, \infty)$.

C.2 Chapter 2

2.1. In each case, check for measurability with respect to any filtration $\mathbb{F} \equiv \{\mathcal{F}_t\}_{t\geqslant 0}$ for Brownian motion and check for square-integrability of the integrand.

(a) The Itô integral is well-defined.
(b) The stochastic integral is not well-defined upon showing that $\int_0^1 \mathrm{E}[X^2(t)]\,\mathrm{d}t = \infty$.
(c) The integrand $X(t) = W(\frac{1}{t})$ is not $\mathcal{F}_t$–measurable for $t \in (0,1)$, since $\frac{1}{t} > t$ for all $0 < t < 1$. Hence, the stochastic integral is not well-defined.
(d) The stochastic integral is well-defined.
(e) The stochastic integral is well-defined iff $a > -1/2$.

2.2. (a) The integrand $X(t) = W(2t)$ is not $\mathcal{F}_t$–measurable. Hence, the stochastic integral is not well-defined.
(b) The stochastic integral is well-defined.
(c) Although $X(t) := W(\frac{1}{t})$ is $\mathcal{F}_t$–measurable for $t \geqslant 1$, the stochastic integral is not well-defined upon showing that $\int_1^\infty \mathrm{E}[X^2(t)]\,\mathrm{d}t = \infty$.
(d) The stochastic integral is well-defined.

2.3. $\int_0^T W(t) \diamond \mathrm{d}_\alpha W(t) = (2\alpha - 1)\int_0^T W(t)\,\mathrm{d}W(t) + 2(1-\alpha)\int_0^T W(t) \circ \mathrm{d}W(t)$.

2.4. $\frac{3}{2}W^2(T) - \frac{T}{2}$.

2.5. $\frac{1}{2}(W^2(t) - t)$.

2.6. (a) $\frac{2}{3}\sqrt{\frac{2}{\pi}}t^{3/2}$; (b) $t^3 + \frac{3}{2}t^4 + \frac{t^5}{5}$; (c) $7t^4$; (d) $\frac{1}{2}(\mathrm{e}^{2t} - 1)$; (e) $(\frac{b}{2} - \frac{1}{4})\mathrm{e}^{2b} - (\frac{a}{2} - \frac{1}{4})\mathrm{e}^{2a}$; (f) 0.

2.7. (a) $\mathrm{d}(\mathrm{e}^{W(t)}) = \frac{1}{2}\mathrm{e}^{W(t)}\,\mathrm{d}t + \mathrm{e}^{W(t)}\,\mathrm{d}W(t)$;
(b) $\mathrm{d}(W^k(t)) = \frac{1}{2}k(k-1)W^{k-2}(t)\,\mathrm{d}t + kW^{k-1}(t)\,\mathrm{d}W(t)$;
(c) $\mathrm{d}\cos(tW(t)) = -\big(W(t)\sin(tW(t)) + \frac{t^2}{2}\cos(tW(t))\big)\,\mathrm{d}t - t\sin(tW(t))\,\mathrm{d}W(t)$;
(d) $\mathrm{d}\big(\mathrm{e}^{W^2(t)}\big) = \mathrm{d}f(W(t)) = (1+2W^2(t))\mathrm{e}^{W^2(t)}\,\mathrm{d}t + 2W(t)\mathrm{e}^{W^2(t)}\,\mathrm{d}W(t)$;
(e) $\mathrm{d}\tan^{-1}(t+W(t)) = \frac{1+(t+W(t))(t+W(t)-1)}{(1+(t+W(t))^2)^2}\,\mathrm{d}t + \frac{1}{1+(t+W(t))^2}\,\mathrm{d}W(t)$.

2.8. $g(t) = C\mathrm{e}^{-\alpha^2 t/2}$, where $C = g(0)$ is an arbitrary constant.

2.10. For $Z \sim Norm(0,1)$, we have $\mathrm{E}[Z^{2n}] = (2n-1)(2n-3)\cdots 3\cdot 1 \equiv (2n-1)!!$ and $\mathrm{E}[Z^{2n-1}] = 0$ for all $n \geqslant 1$.

2.12. (a) $\tan^{-1}(W(T)) + \int_0^T \frac{W(t)}{(1+W^2(t))^2}\,\mathrm{d}t$;
(b) $\frac{1}{2}\sin\big(W^2(T)+T\big) - \int_0^T \big[\cos\big(W^2(t)+t\big) - W^2(t)\sin\big(W^2(t)+t\big)\big]\,\mathrm{d}t$;
(c) $\big(W^2(t) - 2W(t) + 2\big)\mathrm{e}^{W(t)} - 2 - \frac{1}{2}\int_0^t W(u)(W(u)+2)\mathrm{e}^{W(u)}\,\mathrm{d}u$;
(d) $(W(t)-1)\mathrm{e}^{W(t)} + 1 - \frac{1}{2}\int_0^t (W(u)+1)\mathrm{e}^{W(u)}\,\mathrm{d}u$;
(e) $(W(T)-1)\mathrm{e}^{W(T)} + \frac{T}{2}W^2(T) + 1 - \frac{1}{2}\int_0^T \big(W^2(t) + (1+W(t))\mathrm{e}^{W(t)} + t\big)\,\mathrm{d}t$.

2.13. (a) $\int_0^t \mathrm{e}^{W(s)}\,\mathrm{d}W(s)$; (b) $\mathrm{E}[X(t)] = 0$, $\mathrm{Var}(X(t)) = \frac{1}{2}(\mathrm{e}^{2t}-1)$.

2.14. $Z(t)$ satisfies $\mathrm{d}Z(t) = Z(t)\left[\frac{\sigma^2}{2}\,\mathrm{d}t + \sigma\,\mathrm{d}W(t)\right]$. Thus, $m(t)$ solves $\frac{\mathrm{d}}{\mathrm{d}t}m(t) = \frac{\sigma^2}{2}m(t)$ with initial condition $m(0) = 1$.

2.15. The drift and diffusion coefficients are $\mu(t) \equiv 0$ and $\sigma(t) = \frac{1}{\sqrt{T-t}}n\left(\frac{W(t)}{\sqrt{T-t}}\right)$. The limiting value is $X(T-) = \begin{cases} 1, & \text{if } W(T) > 0, \\ 0, & \text{if } W(T) < 0. \end{cases}$ The state space of the process is $(0,1)$.

2.16. $\frac{\mathrm{d}Z(t)}{Z(t)} = \big(\mu_Y - \mu_X + \sigma_X(\sigma_X - \sigma_Y)\big)\,\mathrm{d}t + \big(\sigma_Y - \sigma_X\big)\,\mathrm{d}W(t)$, i.e., $\mu_Z = \mu_Y - \mu_X + \sigma_X(\sigma_X - \sigma_Y)$ and $\sigma_Z = \sigma_Y - \sigma_X$.
For correlated case, $\frac{\mathrm{d}Z(t)}{Z(t)} = (\mu_Y - \mu_X + \sigma_X^2 - \rho\sigma_X\sigma_Y)\,\mathrm{d}t + \sigma_Y\,\mathrm{d}W^Y(t) - \sigma_X\,\mathrm{d}W^X(t)$.

2.17. $\mathcal{G}^Y f(y) = \left(\frac{3}{2}y - \frac{1}{y}\right)f'(y) + \frac{1}{2}f''(y)$, $y > 0$ or $y < 0$.

2.18. $\mathrm{d}Z(t) = \big(W^2(t) + t\big(1 - 2W(t) + \frac{1}{2}W^2(t)\big)\big)\mathrm{e}^{-W(t)}\,\mathrm{d}t + t(2W(t) - W^2(t))\mathrm{e}^{-W(t)}\mathrm{d}W(t)$. $\mathrm{E}[Z(t)] = t^2(1+t)\mathrm{e}^{t/2}$ and $\mathrm{Var}(Z(t)) = t^4\big(3 + 24t + 16t^2\big)\mathrm{e}^{2t} - t^4(1+t)^2\mathrm{e}^t$.

2.19. (a) $Y(t) = x^2 + 2\sigma\int_0^t X(s)\,\mathrm{d}W(s)$. (b) $\mathrm{Var}(Y(t)) = 4\sigma^2\left[\left(x^2 + \frac{\sigma^2}{2c}\right)\frac{\mathrm{e}^{2ct}-1}{2c} - \frac{\sigma^2}{2c}t\right]$ and hence the square-integrability condition is satisfied. The mean follows by the martingale property, $\mathrm{E}[Y(t)] = Y(0) = x^2$.

2.20. Defining $\hat{\beta}(t) := \int_0^t \beta(s)\,\mathrm{d}s$, we have

$$X(t) = \mathrm{e}^{\hat{\beta}(t)}\left(x_0 + \int_0^t \alpha(s)\mathrm{e}^{-\hat{\beta}(s)}\,\mathrm{d}s\right) + \mathrm{e}^{\hat{\beta}(t)}\int_0^t \gamma(s)\mathrm{e}^{-\hat{\beta}(s)}\,\mathrm{d}W(s)\,.$$

Hence, $X(t)$ is a normal random variable with mean and covariance functions: $m_X(t) = \mathrm{e}^{\hat{\beta}(t)}\left(x_0 + \int_0^t \alpha(s)\mathrm{e}^{-\hat{\beta}(s)}\,\mathrm{d}s\right)$ and $c_X(t,s) = \mathrm{e}^{\hat{\beta}(t)+\hat{\beta}(s)}\int_0^{s\wedge t}\gamma^2(u)\mathrm{e}^{-2\hat{\beta}(u)}\,\mathrm{d}u$.

2.21. (a) $\mathrm{d}Y(t) = \sigma \mathrm{e}^{\sigma W(t) - \frac{1}{2}\sigma^2 t}\,\mathrm{d}W(t)$; (b) $\mathrm{d}Z(t) = W(t)f'(t)\,\mathrm{d}t + f(t)\,\mathrm{d}W(t)$.

2.22. $\mathrm{d}Y(t) = \left(\frac{1}{1+\mathrm{e}^{-Y(t)}} - \frac{1}{2}\right)\mathrm{d}t + \mathrm{d}W(t)$.

2.23. $\mathrm{d}X(t) = -\frac{X(t)}{(1-t)}\,\mathrm{d}t + \mathrm{d}W(t)$.

2.24. (a) $X(t) = \mathrm{e}^{\int_0^t \left(W(s) - \frac{1}{2}W^2(s)\right)\mathrm{d}s + \frac{1}{2}(W^2(t) - t)}$;
(b) $X(t) = \mathrm{e}^{-(\alpha + \frac{1}{2}\sigma^2)t + \sigma W(t)}\left(x + \alpha\theta \int_0^t \mathrm{e}^{(\alpha+\frac{1}{2}\sigma^2)s - \sigma W(s)}\,\mathrm{d}s\right)$;
(c) $X(t) = x\mathrm{e}^{\int_0^t a(s)\,\mathrm{d}s - \frac{1}{2}\sigma^2 t + \sigma W(t)}$;
(d) $X(t) = \mathrm{e}^{\frac{1}{2}t + W(t)}$;
(e) $\mathrm{e}^{-t} + 2(1 - \mathrm{e}^{-t}) + \int_0^t \mathrm{e}^{-(t-s)} W(s)\,\mathrm{d}W(s)$;
(f) $1 + \int_0^t W(s)\,\mathrm{d}s + \frac{1}{2}(W^2(t) - t)$;
(g) $\left(1 - \frac{1}{2b}\right)\mathrm{e}^{-bt} + \frac{1}{2b}\mathrm{e}^{bt} + \int_0^t s\mathrm{e}^{-b(t-s)}\,\mathrm{d}W(s)$.

2.25. Hint: use an appropriate Itô formula.

2.26. Hint: given the SDE, $\mathrm{d}X(t) = \mu(X(t))\,\mathrm{d}t + \sigma(X(t))\,\mathrm{d}W(t)$, first identify $g(y) \equiv \sigma(y)$ and then check that $\mu(y) = \frac{1}{2}g(y)g'(y)$. Having g, solve for $f(x)$; i.e., solve for $y = y(x) \equiv f(x)$ as a solution (within an integration constant C) to the separable first-order ODE: $\mathrm{d}y = g(y)\,\mathrm{d}x$. Finally, $X(t) = f(W(t))$ solves the SDE with $x_0 = X(0) = f(0)$ determining the constant C.
(a) $X(t) = \left(\frac{\sigma}{2}W(t) + \sqrt{x_0}\right)^2$, $0 \leqslant t < \infty$;
(b) $X(t) \equiv 0$ for $x_0 = 0$ and $X(t) = \frac{1}{x_0^{-1} - W(t)}$ for $x_0 \neq 0$, $0 \leqslant t < \mathcal{T}_{x_0^{-1}}$;
(c) $X(t) = \ln \frac{1}{\mathrm{e}^{-x_0} - W(t)}$, $0 \leqslant t < \mathcal{T}_{\mathrm{e}^{-x_0}}$, where $\mathcal{T}_b := \inf\{t \geqslant 0 : W(t) = b\}$.

2.27. Hint: since $g(t,x) = \left[\frac{\partial^n}{\partial u^n}\mathrm{e}^{ux - u^2 t/2}\right]_{u=0}$, for all $n = 1, 2, 3, \ldots$, solves the backward PDE for Brownian motion, $\left(\frac{\partial}{\partial t} + \frac{1}{2}\frac{\partial^2}{\partial x^2}\right) g(t,x) = 0$, the process $Y(t) := g(t, W(t))$ satisfies a driftless SDE.

2.28. (a) $f(x) = C_0 + C_1 \int^x \mathfrak{s}(y)\,\mathrm{d}y$, where $\mathfrak{s}(y) := \exp\left(-2\int \frac{\mu(y)}{\sigma^2(y)}\,\mathrm{d}y\right)$, and C_0, C_1 are arbitrary real constants. The condition is $\int_0^T \mathrm{E}\left[\sigma^2(X(u))\mathfrak{s}^2(X(u))\right]\mathrm{d}u < \infty$.

(b) $f(x) = C_0 + C_1 \mathrm{e}^{-\frac{2\mu}{\sigma^2}x}$, where C_0, C_1 are arbitrary real constants. The square-integrability condition in this case holds for all $t \geqslant 0$ since

$$\int_0^t \mathrm{E}\left[\left(f'(X(u))\right)^2\right]\mathrm{d}u < \infty \iff \int_0^t \mathrm{e}^{\frac{8\mu^2}{\sigma^4}u}\,\mathrm{d}u < \infty\,.$$

2.29. $\mathrm{d}X(t) = -\frac{1}{2}X(t)\,\mathrm{d}t - Y(t)\,\mathrm{d}W(t)$ and $\mathrm{d}Y(t) = -\frac{1}{2}Y(t)\,\mathrm{d}t + X(t)\,\mathrm{d}W(t)$.

2.30. (i) $\mu(x) = \sqrt{1+x^2} + \frac{x}{2}$ and $\sigma(x) = \sqrt{1+x^2}$.
(ii) $\frac{\partial}{\partial t}p + \frac{1}{2}(1+x^2)\frac{\partial^2}{\partial x^2}p + \left(\sqrt{1+x^2} + \frac{x}{2}\right)\frac{\partial}{\partial x}p = 0$ subject to the terminal condition $p(T-, T; x, y) = \delta(x-y)$.
(iii) The CDF and PDF are, respectively, $P(s,t;x,y) = \mathcal{N}\left(\frac{1}{\sqrt{t-s}}\ln\frac{y+\sqrt{1+y^2}}{x+\sqrt{1+x^2}} - \sqrt{t-s}\right)$ and $p(s,t;x,y) = \frac{\partial}{\partial y}P(s,t;x,y) = \frac{1}{\sqrt{t-s}\sqrt{1+y^2}}\,n\left(\frac{1}{\sqrt{t-s}}\ln\frac{y+\sqrt{1+y^2}}{x+\sqrt{1+x^2}} - \sqrt{t-s}\right)$ for all times $0 \leqslant s < t$ and $x, y \in \mathbb{R}$.

2.31. $f(t,x) = \mathrm{E}_{t,x}[X^2(T)] = x^2 e^{(T-t)}$ for $t \leqslant T$ and $x \in \mathbb{R}$, where the process $\{X(t)\}_{t\geqslant 0}$ solves $dX(t) = X(t)\,dW(t)$.

2.32. Hint: the solution is valid only if the integral is well-defined; i.e., the integrand must be integrable w.r.t. y on $(-\infty, \infty)$.

2.33. $V(t,x) = \int_{-\infty}^{\infty} \frac{f(y)}{\sqrt{2\pi(1-t)}} \exp\left(-\frac{(y-x-a(1-t))^2}{2(1-t)}\right) dy$, for all $t < 1$.

2.34. $f(t,x) = \mathcal{N}\left(\frac{\ln\frac{x}{K_1}+(\mu-\frac{1}{2}\sigma^2)(T-t)}{\sigma\sqrt{T-t}}\right) - \mathcal{N}\left(\frac{\ln\frac{x}{K_2}+(\mu-\frac{1}{2}\sigma^2)(T-t)}{\sigma\sqrt{T-t}}\right)$.

2.35. We have $f(t,x) = \mathrm{E}[\phi(x+\mu(T-t)+\bar\sigma\sqrt{T-t}Z)]$, where $\bar\sigma\sqrt{T-t} = \sigma\sqrt{\frac{T^2-t^2}{2}}$, i.e., $\bar\sigma = \sigma\sqrt{(T+t)/2}$.

(a) $f(t,x) = (x+\mu(T-t))\mathcal{N}\left(\frac{x-y+\mu(T-t)}{\bar\sigma\sqrt{T-t}}\right) + \bar\sigma\sqrt{T-t}\, n\left(\frac{x-y+\mu(T-t)}{\bar\sigma\sqrt{T-t}}\right)$;

(b) $f(t,x) = \mathcal{N}\left(\frac{y-x-\mu(T-t)}{\bar\sigma\sqrt{T-t}}\right)$.

2.37. The Radon–Nikodym derivative is $\rho_t = e^{-\frac{1}{2}\sigma^2 t+\sigma\widehat{W}(t)}$. The SDE is $dS(t) = S(t)\left[(r+\sigma^2)\,dt + \sigma\, d\widehat{W}(t)\right]$.

2.38. $\mathrm{E}[S(t)] = S_0 e^{\mu t}$ and $\varrho_t = \exp\left[-\frac{1}{2}\left(\frac{r-\mu}{\sigma}\right)^2 \int_0^t S^{-2\beta}(u)\,du + \frac{r-\mu}{\sigma}\int_0^t S^{-\beta}(u)\,dW(u)\right]$.

C.3 Chapter 3

Throughout this section, we define $d_\pm(x,t) := \frac{\ln x+(r-q\pm\frac{\sigma^2}{2})t}{\sigma\sqrt{t}}$.

3.1. (a) $1-\mathcal{N}(d_-(S/K,T))$.
(b) $\mathcal{N}(d_-(S/K_1,T)) - \mathcal{N}(d_-(S/K_2,T))$. (c) $\mathcal{N}(-d_-(S\sqrt{K},T))$.
(d) $\mathcal{N}\left(\frac{(\alpha-\beta)\ln S+(r-q-\sigma^2/2)(\alpha T_2-\beta T_1)}{\sigma\sqrt{\alpha^2 T_2+\beta^2 T_1-2\alpha\beta T_1}}\right)$. (e) $\mathcal{N}(d_-(1,T_2-T_1))\cdot\mathcal{N}(d_-(1,T_1-t))$.
(f) $\mathcal{N}_2(-d_-(S/K_1,T_1), d_-(S/K_2,T_1); -\rho)$ where $\rho = \frac{\min\{T_1,T_2\}}{\sqrt{T_1T_2}}$.
(g) $\mathcal{N}\left(d_-\left(\frac{S}{(K+X)^{1/\alpha}},T\right)\right) - \mathcal{N}\left(d_-\left(\frac{S}{(K-X)^{1/\alpha}},T\right)\right)$ if $\alpha<0$;
$\mathcal{N}\left(d_-\left(\frac{S}{(K-X)^{1/\alpha}},T\right)\right) - \mathcal{N}\left(d_-\left(\frac{S}{(K+X)^{1/\alpha}},T\right)\right)$ if $\alpha>0$.

3.2. (a) $V(t,S) = ae^{-r\tau}\left(S^2 e^{2(r-q)\tau+\sigma^2\tau} - 2KSe^{(r-q)\tau} + K^2\right)$.
(b) $V(t,S) = \widehat{S}e^{-q\tau}\mathcal{N}(\hat d_+(\widehat{S}/K,\tau)) - Ke^{-r\tau}\mathcal{N}(\hat d_-(\widehat{S}/K,\tau))$ where $\widehat{S} = aS^2 e^{(r-q+\sigma^2)\tau}$ and $\hat d_\pm(\widehat{S}/K,\tau) = \frac{\ln(\widehat{S}/K)+(r-q\pm 2\sigma^2)\tau}{2\sigma\sqrt{\tau}}$.
(c) $V(t,S) = aS_\alpha e^{-q\tau}\left(2\mathcal{N}(\hat d_+(S_\alpha/K,\tau))-1\right) - aKe^{-r\tau}\left(2\mathcal{N}(\hat d_-(S_\alpha/K,\tau))-1\right)$ where $\hat d_\pm(x,\tau) := \frac{\ln x+(r-q\pm\sigma_\alpha^2/2)\tau}{\sigma_\alpha\sqrt{\tau}}$, $S_\alpha = S^\alpha e^{(\alpha-1)(r-q+\alpha\sigma^2/2)\tau}$ and $\sigma_\alpha = |\alpha|\sigma$.
(d) $V(t,S) = ae^{(r-2q)\tau+\sigma^2\tau}S^2\mathcal{N}\left(d_+(S/K,\tau)+\sigma\sqrt{\tau}\right) - 2ae^{-q\tau}SK\mathcal{N}\left(d_+(S/K,\tau)\right) + ae^{-r\tau}K^2\mathcal{N}\left(d_+(S/K,\tau)-\sigma\sqrt{\tau}\right)$.

3.3. (a) It can be replicated by purchasing a European call with strike K_1, writing a European call with strike X_1, purchasing a European call with strike K_2, and writing a European call with strike X_2. All options have the same time to maturity.
(b) $V(\tau, S) = Se^{-q\tau}\big(\mathcal{N}(d_+(S/K_1,\tau)) - \mathcal{N}(d_+(S/X_1,\tau)) + \mathcal{N}(d_+(S/K_2,\tau)) - \mathcal{N}(d_+(S/X_2,\tau))\big) - e^{-r\tau}\big(K_1\mathcal{N}(d_-(S/K_1,\tau)) - X_1\mathcal{N}(d_-(S/X_1,\tau)) + K_2\mathcal{N}(d_-(S/K_2,\tau)) - X_2\mathcal{N}(d_-(S/X_2,\tau))\big)$.
(c) $\Delta(\tau, S) = e^{-q\tau}\big(\mathcal{N}(d_+(S/K_1,\tau)) - \mathcal{N}(d_+(S/X_1,\tau)) + \mathcal{N}(d_+(S/K_2,\tau)) - \mathcal{N}(d_+(S/X_2,\tau))\big)$.

3.4. (a) $V(t, S) = e^{-r\tau}\left(a_0 + \sum_{n=1}^N a_n S^n e^{n(r-q+\sigma^2(n-1)/2)\tau}\right)$.
(b) $V(t,S) = e^{-q\tau} S_\alpha \mathcal{N}\left(d_-\left(\frac{S}{K},\tau\right) + \alpha\sigma\sqrt{\tau}\right) - e^{-r\tau} K \mathcal{N}\left(d_-\left(\frac{S}{K},\tau\right)\right)$

3.5. (b) $V(t,S) = e^{-q\tau}S\big[\mathcal{N}(d_+(S/K,\tau)) - \mathcal{N}(d_+(S/(K+\varepsilon),\tau))\big] - e^{-r\tau}\big[K\mathcal{N}(d_-(S/K,\tau)) - (K+\varepsilon)\mathcal{N}(d_-(S/(K+\varepsilon),\tau))\big]$ where $\tau = T - t$.
(c) $\Delta(t,S) = e^{-q\tau}\big[\mathcal{N}(d_+(S/K,\tau)) - \mathcal{N}(d_+(S/(K+\varepsilon),\tau))\big]$.

3.6. (b) It can be replicated by purchasing the following European options (with the same time to maturity): a long call struck at K_1, a short call struck at K_2, a short call struck at K_3, a long call struck at K_4, a long call struck at K_5, and a short call struck at K_6.
(c) $V(t,S) = C(t,S;T,K_1) - C(t,S;T,K_2) - C(t,S;T,K_3) + C(t,S;T,K_4) + C(t,S;T,K_5) - C(t,S;T,K_6)$.

3.7. (a) $V(t,S) = e^{-q\tau}S\big[\mathcal{N}(d_+(S/K_2,\tau)) + \mathcal{N}(-d_+(S/K_1,\tau))\big] - e^{-r\tau}\big[K_2\mathcal{N}(d_-(S/K_2,\tau)) + K_1\mathcal{N}(-d_-(S/K_1,\tau))\big]$ with $\tau = T - t$. (b) $K = Se^{(r-q)\tau}$.

3.8. (b) The strangle option payoff is replicated by a portfolio with one short share of stock, one long standard call with strike K_2, one long standard call with strike K_1, and K_2 units of cash.
(c) $V(t,S) = e^{-q\tau}S\big[\mathcal{N}(d_+(S/K_1,\tau)) - \mathcal{N}(-d_+(S/K_2,\tau))\big] - e^{-r\tau}\big[K_1\mathcal{N}(d_-(S/K_1,\tau)) - K_2\mathcal{N}(-d_-(S/K_2,\tau))\big]$.
(d) $\Delta(t,S) = e^{-q\tau}\big[\mathcal{N}(d_+(S/K_1,\tau)) - \mathcal{N}(-d_+(S/K_2,\tau))\big]$.

3.9. (a) $\Lambda(S) = 0$ if $S < K_1$, $\Lambda(S) = S - K_1$ if $K_1 \leqslant S < K_2$, $\Lambda(S) = A$ if $K_2 \leqslant S < K_3$, $\Lambda(S) = S - K_3 + A$ if $K_3 \leqslant S < K_4$, $\Lambda(S) = A + B$ if $K_4 \leqslant S$.
The replicating portfolio consists of a long call struck at K_1, a short call struck at K_2, a long call stuck at K_3, and a short call struck at K_4.
(b) $V(t,S) = C(t,S;T,K_1) - C(t,S;T,K_2) + C(t,S;T,K_3) - C(t,S;T,K_4)$.
(c) $\Delta_V(t,S) = \Delta_C(t,S;T,K_1) - \Delta_C(t,S;T,K_2) + \Delta_C(t,S;T,K_3) - \Delta_C(t,S;T,K_4) = e^{-q\tau}\mathcal{N}(d_+(S/K_1,\tau)) - e^{-q\tau}\mathcal{N}(d_+(S/K_2,\tau)) + e^{-q\tau}\mathcal{N}(d_+(S/K_3,\tau)) - e^{-q\tau}\mathcal{N}(d_+(S/K_4,\tau))$ where $\tau = T - t$.

3.10. (a) $V(0,100) \cong \$4.3862$; $\Delta_V(0,100) \cong 0.3437$. (b) $V(1/12,105) \approx \$6.1047$.
(c) The portfolio consists of $z = -10$ derivatives (spreads), $x \cong 3.437$ stock shares. and $y \cong -\$299.84$ in the risk-free bank account.

3.11. (a) $\Lambda(S) = (K + X - S)^+ - (K - S)^+ + (K - X - S)^+$. The replicating portfolio consists of a long put struck at $K + X$, a short put struck at K, and a long put stuck at $K - X$.
(b) $V(t,S) = P(t,S;T,K+X) - P(t,S;T,K) + P(t,S;T,K-X)$. (c) $P(t,S;T,K)$.
(d) Use the fact that $(K-S)^+ \leqslant \Lambda(S) \leqslant (K+X-S)^+$ for any $X \geqslant 0$, the Law of One Price, and properties of the no-arbitrage pricing function for a standard European put option.

3.12. (a) $\Lambda(S) = (S-K_1)^+ - 2(S-K_2)^+$. The replicating portfolio consists of one long call struck at K_1 and two short calls struck at K_2; $V(t,S) = \mathrm{e}^{-q\tau}S\big(\mathcal{N}(d_+(S/K_1,\tau)) - 2\mathcal{N}(d_+(S/K_2,\tau))\big) - \mathrm{e}^{-r\tau}\big(K_1\mathcal{N}(d_-(S/K_1,\tau)) - 2K_2\mathcal{N}(d_-(S/K_2,\tau))\big)$ where $\tau = T-t$.
(b) $\Lambda(S) = -(S-K_1)^+ + 2(S-K_2)^+$. The replicating portfolio consists of one short call struck at K_1 and two long calls struck at K_2; $V(t,S) = \mathrm{e}^{-q\tau}S\big(2\mathcal{N}(d_+(S/K_2,\tau)) - \mathcal{N}(d_+(S/K_1,\tau))\big) - \mathrm{e}^{-r\tau}\big(2K_2\mathcal{N}(d_-(S/K_2,\tau)) - K_1\mathcal{N}(d_-(S/K_1,\tau))\big)$.
(c) $\Lambda(S) = S-(S-K)^+$. The replicating portfolio consists of one long stock share and one short call struck at K; $V(t,S) = \mathrm{e}^{-q\tau}S\mathcal{N}(-d_+(S/K,\tau)) + \mathrm{e}^{-r\tau}K\mathcal{N}(d_-(S/K,\tau))$.
(d) $\Lambda(S) = X - (S-K_1)^+ + (S-K_2)^+$. The replicating portfolio consists of X dollars, one short call struck at K_1 and one long call struck at K_2. $V(t,S) = \mathrm{e}^{-r\tau}X + \mathrm{e}^{-q\tau}S\big(\mathcal{N}(d_+(S/K_2,\tau)) - \mathcal{N}(d_+(S/K_1,\tau))\big) - \mathrm{e}^{-r\tau}\big(K_2\mathcal{N}(d_-(S/K_2,\tau)) - K_1\mathcal{N}(d_-(S/K_1,\tau))\big)$.

3.13. $V(t,S) = \mathrm{e}^{-q\tau}S\mathcal{N}(d_+(S/K,\tau)) - \mathrm{e}^{-r\tau}(K+X)\mathcal{N}(d_-(S/K,\tau))$. We have that $V(t,S)=0$ iff $X = \frac{\mathrm{e}^{(r-q)\tau}S\cdot\mathcal{N}(d_+(S/K,\tau))}{\mathcal{N}(d_-(S/K,\tau))} - K$. As $\sigma\to\infty$, $X\to\infty$.

3.14. (a) Show that, as $\sigma \searrow 0$, $\mathcal{N}(d_\pm(S/K,\tau)) \to \mathcal{N}(\infty) = 1$ if $S/K > \mathrm{e}^{-r\tau}$, $\mathcal{N}(d_\pm(S/K,\tau)) \to \mathcal{N}(-\infty) = 0$ if $S/K < \mathrm{e}^{-r\tau}$, and $\mathcal{N}(d_\pm(S/K,\tau)) \to \mathcal{N}(0) = 1/2$ if $S/K = \mathrm{e}^{-r\tau}$. As $\sigma\to\infty$, $\mathcal{N}(d_+(S/K,\tau)) \to \mathcal{N}(\infty) = 1$ and $\mathcal{N}(d_-(S/K,\tau)) \to \mathcal{N}(-\infty) = 0$.

3.15. Show that $\Lambda(S(T)) = (S(T)-K)^+ - (1-\alpha)\big(S(T)-(1-\alpha)^{-1}K\big)^+$.

3.19. $\Theta + \frac{1}{2}\sigma^2S^2\Gamma + rS\Delta - r\Pi = 0$.

3.21. (a) The soft-strike put option has payoff

$$\Lambda_a(S) = \begin{cases} K-S & \text{if } S < K-a, \\ \frac{1}{4a}(K+a-S)^2 & \text{if } K-a \leqslant S \leqslant K+a, \\ 0 & \text{if } S > K+a. \end{cases}$$

It is greater than the payoff of a standard put option with strike K: $\Lambda_a(S) \geqslant (K-S)^+$. We have $\Lambda_a(S) \searrow (K-S)^+$, as $a \searrow 0$. The derivative of $\Lambda_a(S)$ is a continuous function of S.
(b) $\Lambda_a^{\text{call}}(S) - \Lambda_a^{\text{put}}(S) = S-K$ for all $S>0$. we have the following put-call parity: $C(t,S;K,a) - P(t,S;K,a) = \mathrm{e}^{-q\tau}S - \mathrm{e}^{-r\tau}K$ where $\tau = T-t$.

3.22. Use the expectation identity $\mathbb{P}(A)\mathrm{E}[X|A] = \mathrm{E}[X\mathbb{I}_A]$ for any measure $\mathbb{P}$, event A, and random variable X.

3.23. (a) We have $\mu_c = r$ and

$$\sigma_c(t,S) = \sigma S\frac{\Delta_c(t,S)}{C(t,S)} = \frac{\sigma S\mathcal{N}(d_+(S/K,\tau))}{S\mathcal{N}(d_+(S/K,\tau)) - \mathrm{e}^{-r\tau}K\mathcal{N}(d_-(S/K,\tau))}.$$

(b) $\lim_{K\searrow 0}\sigma_c = \sigma$.

3.24. (a) $C(0,S_0,K) = \mathrm{e}^{-\bar{q}T}S_0\mathcal{N}\big(d_+\big(\frac{S_0}{K},T\big)\big) - \mathrm{e}^{-\bar{r}T}K\mathcal{N}\big(d_-\big(\frac{S_0}{K},T\big)\big)$ where $d_\pm\big(\frac{S_0}{K},T\big) := \frac{\ln\frac{S_0}{K} + (\bar{r}-\bar{q}\pm\frac{1}{2}\bar{\sigma}^2)T}{\bar{\sigma}\sqrt{T}}$, $\bar{r} = \frac{1}{T}\int_0^T r(s)\,\mathrm{d}s$, $\bar{q} = \frac{1}{T}\int_0^T q(s)\,\mathrm{d}s$, and $\bar{\sigma}^2 = \frac{1}{T}\int_0^T \sigma^2(s)\,\mathrm{d}s$.
(b) $C(0,S_0,K) - P(0,S_0,K) = \mathrm{e}^{-\bar{q}T}S_0 - \mathrm{e}^{-\bar{r}T}K$.

3.25. (a) $C(t,S;T_1,T) = S\big[\mathrm{e}^{-q(T-t)}\mathcal{N}(d_+(1/\alpha,T-T_1)) - \alpha\mathrm{e}^{-r(T-T_1)-q(T_1-t)}\mathcal{N}(d_-(1/\alpha,T-T_1))\big]$ where $d_\pm(1/\alpha,T-T_1) = \frac{\ln\frac{1}{\alpha}+[r-q\pm\frac{1}{2}\sigma^2](T-T_1)}{\sigma\sqrt{T-T_1}}$.
For all $t \leqslant T_1$, $\Delta(t,S) = \mathrm{e}^{-q(T-t)}\mathcal{N}(d_+(1/\alpha,T-T_1)) - \alpha\mathrm{e}^{-r(T-T_1)-q(T_1-t)}\mathcal{N}(d_-(1/\alpha,T-T_1))$.
(b) $\lim_{T_1\searrow t} C(t,S;T_1,T) = S\mathrm{e}^{-q(T-t)}\mathcal{N}(d_+(1/\alpha,T-t)) - \alpha S\mathrm{e}^{-r(T-t)}\mathcal{N}(d_-(1/\alpha,T-t))$.
$\lim_{T_1\nearrow T} C(t,S;T_1,T) = \mathrm{e}^{-q(T_1-t)}S\cdot(1-\alpha)^+$.

3.26. (a) $V(t,S) = S[\mathcal{N}(-d_+) + \mathrm{e}^{-r(T-T_1)}\mathcal{N}(d_-)]$ where $d_\pm = \frac{1}{\sigma}(r \pm \frac{1}{2}\sigma^2)\sqrt{T-T_1}$.
(b) For all $t \leqslant T_1$, we have $\Delta(t,S) = \mathcal{N}(-d_+) + \mathrm{e}^{-r(T-T_1)}\mathcal{N}(d_-)$. Hence, until time T_1 the option can be replicated using a static portfolio in Δ shares of stock and no involvement of the bank account.

3.27. $V(t,S) = \mathrm{e}^{-r(T-t)}S^n\mathrm{e}^{n(r-q+(n-1)\sigma^2/2)(T_1-t)}\sum_{k=0}^{n}\binom{n}{k}\mathrm{e}^{k(r-q+(k-1)\sigma^2/2)(T-T_1)}$
$\Big[(-1)^{n-k}\mathcal{N}\big(d_-(1,T-T_1)+k\sigma\sqrt{T-T_1}\big) + (-1)^k\mathcal{N}\big(-d_-(1,T-T_1)-k\sigma\sqrt{T-T_1}\big)\Big]$.

3.28. For the forward starting strip, we have:
(a) $\Lambda(S(T)) = (S(T)-S(T_1))^+ + 2(S(T_1)-S(T))^+$.
(b) $V(t,S) = \mathrm{e}^{-q\tau_1}S\big[\mathrm{e}^{-q\tau}(1-3\mathcal{N}(-d_+(1,\tau))) + \mathrm{e}^{-r\tau}(3\mathcal{N}(-d_-(1,\tau))-1)\big]$ where $\tau = T-T_1$ and $\tau_1 = T_1 - t$. (c) $\beta_t = 0$.
For the forward starting strap, we have:
(a) $\Lambda(S(T)) = 2(S(T)-S(T_1))^+ + (S(T_1)-S(T))^+$.
(b) $V(t,S) = \mathrm{e}^{-q\tau_1}S\big[\mathrm{e}^{-q\tau}(3\mathcal{N}(d_+(1,\tau))-1) + \mathrm{e}^{-r\tau}(1-3\mathcal{N}(d_-(1,\tau)))\big]$ where $\tau = T-T_1$ and $\tau_1 = T_1-t$. (c) $\beta_t = 0$.

3.29. $C(t,S) - P(t,S) = \mathrm{e}^{-q\tau}S - \mathrm{e}^{-r\tau}K$ where $\tau = T-t$ and $K = \alpha\mathrm{e}^{(r-q)(T_1-t)}S$.

3.30. (a) $V^{cp}(t,S) = \mathrm{e}^{-r(T_2-t)}K_2\mathcal{N}_2(-a_-,-b_-;\rho) - S\mathrm{e}^{-q(T_2-t)}\mathcal{N}_2(-a_+,-b_+;\rho)$ $-\mathrm{e}^{-r(T_1-t)}K_1\mathcal{N}(-a_-)$ where $a_\pm = d_\pm(S/S_1^*,T_1-t)$, $b_\pm = d_\pm(S/K_2,T_2-t)$, and $\rho = \sqrt{\frac{T_1-t}{T_2-t}}$.
(b) $\Delta(t,S) = -\mathrm{e}^{-q(T_2-t)}\mathcal{N}_2\left(-d_+(\frac{S}{S_1^*},T_1-t), -d_+(\frac{S}{K_2},T_2-t); \sqrt{\frac{T_1-t}{T_2-t}}\right)$.

3.31. (a) $P(t,S) \sim \mathrm{e}^{-q(T-t)}S$, as $S\to 0$; $P(t,S)\to 0$, as $S\to\infty$.
(b) $\{P(T_1,S_1;T_2,K_2) \leqslant K_1\} - \{S_1 \leqslant S_1^*\}\cup\{S_1 \geqslant S_2^*\}$, where S_1^* and S_2^* are solutions to $P(T_1,S_1;T_2,K_2) = K_1$.
(c) $V(t,S) = \mathrm{e}^{-q\tau_2}S\Big(\mathcal{N}_2\left(-d_+^{(1)},-d_+^{(3)};\sqrt{\frac{\tau_1}{\tau_2}}\right) + \mathcal{N}_2\left(d_+^{(2)},-d_+^{(3)};-\sqrt{\frac{\tau_1}{\tau_2}}\right)\Big)$
where $d_+^{(1,2)} = \frac{\ln(S/S_{1,2}^*)+(r-q+\sigma^2/2)\tau_1}{\sigma\sqrt{\tau_1}}$, $d_+^{(3)} = \frac{\ln(S/K_2)+(r-q+\sigma^2/2)\tau_2}{\sigma\sqrt{\tau_2}}$, $\tau_1 = T_1-t$, and $\tau_2 = T_2 - t$.

3.32. (a) $V(t,S) = \mathrm{e}^{-q(T-t)}S\Big[\mathcal{N}_2\left(d_+^{(1)},d_+^{(2)};\sqrt{\frac{\tau_1}{\tau_2}}\right) + \mathcal{N}_2\left(-d_+^{(1)},-d_+^{(2)};\sqrt{\frac{\tau_1}{\tau_2}}\right)\Big]$
where $d_+^{(1)} = d_+\left(\frac{S}{S_1^*},\tau_1\right)$, $d_+^{(2)} = d_+\left(\frac{S}{K},\tau_2\right)$, $\tau_1 = T_1-t$, $\tau_2 = T-t$, and S_1^* is a unique solution to $C(T_1,S_1;T,K) = P(T_1,S_1;T,K)$.

(b) The delta is given by

$$\Delta_V(t,S) = \mathrm{e}^{-q\tau_2}\left[\mathcal{N}\left(d_+^{(1)}, d_+^{(2)}; \sqrt{\frac{\tau_1}{\tau_2}}\right) + \frac{n(d_+^{(1)})}{\sigma\sqrt{\tau_1}}\left(2\mathcal{N}\left(\frac{\sqrt{\tau_2}d_+^{(2)} - \sqrt{\tau_1}d_+^{(1)}}{\sqrt{\tau_2-\tau_1}}\right) - 1\right)\right.$$
$$\left. + \mathcal{N}\left(-d_+^{(1)}, -d_+^{(2)}; \sqrt{\frac{\tau_1}{\tau_2}}\right) + \frac{n(d_+^{(2)})}{\sigma\sqrt{\tau_2}}\left(2\mathcal{N}\left(\frac{\sqrt{\tau_2}d_+^{(1)} - \sqrt{\tau_1}d_+^{(2)}}{\sqrt{\tau_2-\tau_1}}\right) - 1\right)\right].$$

3.33. The pricing function is

$$V(t,S) = \mathrm{e}^{-q(T_c-t)}S\mathcal{N}_2\left(d_+\left(\frac{S}{S_1^*}, T-t\right), d_+\left(\frac{S}{K_c}, T_c - t\right); \sqrt{\frac{T-t}{T_c-t}}\right)$$
$$+ \mathrm{e}^{-q(T_p-t)}S\mathcal{N}_2\left(-d_+\left(\frac{S}{S_1^*}, T-t\right), -d_+\left(\frac{S}{K_p}, T_p - t\right); \sqrt{\frac{T-t}{T_p-t}}\right),$$

where S_1^* is a unique solution to $C(T, S_1; T_c, K_c) = P(T, S_1; T_p, K_p)$.

3.34. (a) $V^{cc}(t,S) - V^{pc}(t,S) = \mathrm{e}^{-q(T_2-t)}S\mathcal{N}(d_+(S/K_2, T_2-t)) - \mathrm{e}^{-r(T_1-t)}K_1$.
(b) $V^{cp}(t,S) - V^{pp}(t,S) = \mathrm{e}^{-q(T_2-t)}S\mathcal{N}(-d_+(S/K_2, T_2-t)) - \mathrm{e}^{-r(T_1-t)}K_1$.

3.36. (a) If $S > B$ and $K > B$, then

$$P^{DO}(t,S,K;B) = P(t,S,K) - P(t,S,B) - \left(\frac{B}{S}\right)^{\mu}\left[P\left(t, \frac{B^2}{S}, K\right) - P\left(t, \frac{B^2}{S}, B\right)\right]$$
$$+ \mathrm{e}^{-r\tau}(K-B)\left[\left(\frac{B}{S}\right)^{\mu}\mathcal{N}\left(-d_-(\tfrac{B}{S}, \tau)\right) - \mathcal{N}\left(-d_-(\tfrac{S}{B}, \tau)\right)\right],$$

where $P(t,S,K) = \mathrm{e}^{-r\tau}K\mathcal{N}\left(-d_-(\frac{S}{K}, \tau)\right) - \mathrm{e}^{-q\tau}S\mathcal{N}\left(-d_+(\frac{S}{K}, \tau)\right)$ is the pricing function for a standard put with spot S and strike K. By symmetry, we have

$$\begin{aligned} P^{DI}(t,S,K;B) &= P(t,S,K;B) - P^{DO}(t,S,K;B) \\ &= P(t,S,B) + \left(\frac{B}{S}\right)^{\mu}\left[P\left(t, \frac{B^2}{S}, K\right) - P\left(t, \frac{B^2}{S}, B\right)\right] \\ &\quad + \mathrm{e}^{-r\tau}(B-K)\left[\left(\frac{B}{S}\right)^{\mu}\mathcal{N}\left(-d_-(\tfrac{B}{S}, \tau)\right) - \mathcal{N}\left(-d_-(\tfrac{S}{B}, \tau)\right)\right], \end{aligned}$$

for $S > B$ and $K > B$. If $S \leqslant B$ or $K \leqslant B$, then $P^{DO}(t,S,K;B) = 0$ and $P^{DI}(t,S,K;B) = P(t,S,K;B)$.
As for the delta hedging positions, we have

$$\Delta^{DO}(t,S,K) = \begin{cases} \frac{\partial P^{DO}}{\partial S}(t,S,K) & \text{for } S > B, \\ 0 & \text{for } S < B \end{cases} \quad \text{and}$$

$$\Delta^{DI}(t,S,K) = \begin{cases} \Delta(t,S,K) - \Delta^{DO}(t,S,K) & \text{for } S > B, \\ \Delta(t,S,K) & \text{for } S < B \end{cases}$$

where $\Delta(t,S,K) := \frac{\partial P(t,S,K)}{\partial S} = -\mathrm{e}^{-q\tau}\mathcal{N}\big(-d_+(\frac{S}{K},\tau)\big)$, and

$$\begin{aligned}\Delta^{DO} &= \Delta(t,S,K) - \Delta(t,S,B) + \left(\frac{B}{S}\right)^{\mu+2}\left[\Delta\left(t,\frac{B^2}{S},K\right) - \Delta\left(t,\frac{B^2}{S},B\right)\right]\\ &+ \frac{\mu}{S}\left(\frac{B}{S}\right)^{\mu}\left[P\left(t,\frac{B^2}{S},K\right) - P\left(t,\frac{B^2}{S},B\right) + \mathrm{e}^{-r\tau}(B-K)\mathcal{N}\left(-d_-(\frac{B}{S},\tau)\right)\right]\\ &+ \frac{2\,\mathrm{e}^{-r\tau}(K-B)}{S\sigma\sqrt{\tau}}\,n\left(d_-(\frac{S}{B},\tau)\right).\end{aligned}$$

(b) If $S > B$ and $K \geqslant B$, then

$$C^{DO}(t,S,K) = C(t,S,K) - \left(\frac{B}{S}\right)^{\frac{2(r-q)}{\sigma^2}-1} C\left(t,\frac{B^2}{S},K\right),$$

where $C(t,S,K) \equiv \mathrm{e}^{-q\tau}S\mathcal{N}\left(d_+(\frac{S}{K},\tau)\right) - \mathrm{e}^{-r\tau}K\mathcal{N}\left(d_-(\frac{S}{K},\tau)\right)$ is the pricing function for a standard call with spot S and strike K. If $S > B$ and $K < B$, then

$$\begin{aligned}C^{DO}(t,S,B) &= C(t,S,B) - \left(\frac{B}{S}\right)^{\frac{2(r-q)}{\sigma^2}-1} C\left(t,\frac{B^2}{S},B\right)\\ &+ \mathrm{e}^{-r\tau}(B-K)\left[\mathcal{N}\left(d_-(\frac{S}{B},\tau)\right) - \left(\frac{B}{S}\right)^{\frac{2(r-q)}{\sigma^2}-1}\mathcal{N}\left(d_-(\frac{B}{S},\tau)\right)\right].\end{aligned}$$

For the trivial case when $S \leqslant B$, we have $C^{DO} \equiv 0$. By knock-in-knock-out symmetry, we have $C^{DI}(t,S,K) = C(t,S,K) - C^{DO}(t,S,K)$. Hence, for $S > B$, the pricing function for the down-and-in call is

$$C^{DI}(t,S,K) = \left(\frac{B}{S}\right)^{\frac{2(r-q)}{\sigma^2}-1} C\left(t,\frac{B^2}{S},K\right)$$

for $K \geqslant B$, and

$$\begin{aligned}C^{DI}(t,S,K) &= C(t,S,K) - C(t,S,B) + \left(\frac{B}{S}\right)^{\frac{2(r-q)}{\sigma^2}-1} C\left(t,\frac{B^2}{S},B\right)\\ &+ \mathrm{e}^{-r\tau}(K-B)\left[\mathcal{N}\left(d_-(\frac{S}{B},\tau)\right) - \left(\frac{B}{S}\right)^{\frac{2(r-q)}{\sigma^2}-1}\mathcal{N}\left(d_-(\frac{B}{S},\tau)\right)\right]\end{aligned}$$

for $K < B$. For the trivial case when $S \leqslant B$, we have $C^{DI}(t,S,K) = C(t,S,K)(t,S,K)$, since $C^{DO}(t,S,K) \equiv 0$. The delta hedging positions for the down-and-out and down-and-in call options are

$$\Delta^{DO}(t,S,K) = \begin{cases}\frac{\partial C^{DO}(t,S,K)}{\partial S} & \text{for } S > B,\\ 0 & \text{for } S < B,\end{cases} \quad \text{and}$$

$$\Delta^{DI}(t,S,K) = \begin{cases}\Delta(t,S,K) - \Delta^{DO}(t,S,K) & \text{for } S > B,\\ \Delta(t,S,K) & \text{for } S < B,\end{cases}$$

where $\Delta(t,S,K) := \frac{\partial C(t,S,K)}{\partial S} = \mathrm{e}^{-q\tau}\mathcal{N}\big(d_+(\frac{S}{K},\tau)\big)$ is the delta hedging position for the standard call. Consider the nontrivial case where $S > B$. For $K > B$, we have

$$\begin{aligned}\Delta^{DO}(t,S,K) &= \mathrm{e}^{-q\tau}\mathcal{N}\big(d_+(\frac{S}{K},\tau)\big)\\ &+ \left(\frac{B}{S}\right)^{\mu}\left[(\mu+1)\left(\frac{B}{S}\right)^{2}\mathrm{e}^{-q\tau}\mathcal{N}\big(d_+(\frac{B^2}{KS},\tau)\big) - \mu\frac{K}{S}\mathrm{e}^{-r\tau}\mathcal{N}\big(d_-(\frac{B^2}{KS},\tau)\big)\right].\end{aligned}$$

For $K < B$, we have

$$\Delta^{DO}(t,S,K) = \mathrm{e}^{-q\tau}\mathcal{N}\left(d_+(\frac{S}{B},\tau)\right) + \frac{2(B-K)}{S\sigma\sqrt{\tau}}\mathrm{e}^{-r\tau}\mathfrak{n}\left(d_-(\frac{S}{B},\tau)\right)$$
$$+\left(\frac{B}{S}\right)^{\mu}\left[(\mu+1)\left(\frac{B}{S}\right)^2\mathrm{e}^{-q\tau}\mathcal{N}(d_+(\frac{B}{S},\tau)) - \mu\frac{K}{S}\mathrm{e}^{-r\tau}\mathcal{N}\left(d_-(\frac{B}{S},\tau)\right)\right].$$

3.37. (a) $C^{DO}(\tau,S,K;B) - P^{DO}(\tau,S,K;B) = V^{DO}(\tau,S,K;B)$ where $V^{DO} = 0$ if $S \leqslant B$ and $V^{DO} = \mathrm{e}^{-q\tau}S\big[\mathcal{N}(d_+(S/B,\tau)) - (B/S)^{\frac{2(r-q)}{\sigma^2}+1}\mathcal{N}(d_+(B/S,\tau))\big] - \mathrm{e}^{-r\tau}K\big[\mathcal{N}(d_-(S/B,\tau)) - (B/S)^{\frac{2(r-q)}{\sigma^2}-1}\mathcal{N}(d_-(B/S,\tau))\big]$ if $S > B$.
(b) $C^{UO}(\tau,S,K;B) - P^{UO}(\tau,S,K;B) = V^{UO}(\tau,S,K;B)$ where $V^{UO} = 0$ if $S \geqslant B$ and $V^{UO} = \mathrm{e}^{-q\tau}S\big[\mathcal{N}(-d_+(S/B,\tau)) - (B/S)^{\frac{2(r-q)}{\sigma^2}+1}\mathcal{N}(-d_+(B/S,\tau))\big] - \mathrm{e}^{-r\tau}K\big[\mathcal{N}(-d_-(S/B,\tau)) - (B/S)^{\frac{2(r-q)}{\sigma^2}-1}\mathcal{N}(-d_-(B/S,\tau))\big]$ if $S < B$.
(c) $C^{DI}(\tau,S,K;B) - P^{DI}(\tau,S,K;B) = \mathrm{e}^{-q\tau}S - \mathrm{e}^{-r\tau}K - V^{DO}(\tau,S,K;B)$.
(d) $C^{UI}(\tau,S,K;B) - P^{UI}(\tau,S,K;B) = \mathrm{e}^{-q\tau}S - \mathrm{e}^{-r\tau}K - V^{UO}(\tau,S,K;B)$.

3.38. (a) $V(t,S) = \mathrm{e}^{-q(T_2-T_1)}\big[\mathrm{e}^{-q(T_1-t)}S\mathcal{N}(d_+(S/\widetilde{K},T_1-1)) - \mathrm{e}^{-r(T_1-t)}\widetilde{K}\mathcal{N}(d_-(S/\widetilde{K},T_1-t))\big]$ where $\widetilde{K} = \big(K_1+\mathrm{e}^{-r(T_2-T_1)}K_2\big)\mathrm{e}^{q(T_2-T_1)}$. (b) $V(t,S) = \mathrm{e}^{-r(T_1-t)}\mathcal{N}(-d_-(S/\widetilde{K},T_1-t))$ where $\widetilde{K}$ is defined in (a).

3.39. (i) Let $C(t,S,m) = \mathrm{e}^{-r\tau}S\mathcal{N}\left(d_+(\frac{S}{m},\tau)\right) - \mathrm{e}^{-r\tau}m\mathcal{N}\left(d_-(\frac{S}{m},\tau)\right)$ is the pricing formula for a standard call option with spot S, strike m, and time to maturity $\tau = T-t$.

$$C^{LFS}(t,S,m) = C(t,S,m) + \frac{\sigma^2\mathrm{e}^{-q\tau}S}{2(r-q)}\left[\mathrm{e}^{-(r-q)\tau}\left(\frac{m}{S}\right)^{\frac{2(r-q)}{\sigma^2}}\mathcal{N}\left(d_-(\frac{m}{S},\tau)\right) - \mathcal{N}\left(-d_+(\frac{S}{m},\tau)\right)\right].$$

For the case where $r = q$,

$$C^{LFS}(t,S,m) = C(t,S,m) + \sigma\sqrt{\tau}\mathrm{e}^{-r\tau}S\left[\mathfrak{n}\left(d_+(\frac{S}{m},\tau)\right) - d_+(\frac{S}{m},\tau)\mathcal{N}\left(-d_+(\frac{S}{m},\tau)\right)\right].$$

(ii) For the case where $K \leqslant M$, $r \neq q$, $S \leqslant M < \infty$, $\tau = T-t > 0$, we have

$$C^{LFP}(t,S,m) = \mathrm{e}^{-r\tau}(M-K) + C(t,S,M)$$
$$+\frac{\sigma^2\mathrm{e}^{-q\tau}S}{2(r-q)}\left[\mathcal{N}\left(d_+\left(\frac{S}{M},\tau\right)\right) - \mathrm{e}^{-(r-q)\tau}\left(\frac{M}{S}\right)^{\frac{2(r-q)}{\sigma^2}}\mathcal{N}\left(-d_-\left(\frac{M}{S},\tau\right)\right)\right]$$

For the case where $M < K$, $r \neq q$, $S \leqslant M < \infty$, $\tau = T-t > 0$, we have

$$C^{LFP}(t,S,m) = C(t,S,K)$$
$$+\frac{\sigma^2\mathrm{e}^{-q\tau}S}{2(r-q)}\left[\mathcal{N}\left(d_+(\frac{S}{K},\tau)\right) - \mathrm{e}^{-(r-q)\tau}\left(\frac{K}{S}\right)^{\frac{2(r-q)}{\sigma^2}}\mathcal{N}\left(-d_-(\frac{K}{S},\tau)\right)\right]$$

For $r = q$, $M \geqslant K$, we have

$$C^{LFP} = \mathrm{e}^{-r\tau}(M-K) + C(t,S,M)$$
$$+\mathrm{e}^{-r\tau}S\sigma\sqrt{\tau}\left[d_+(\frac{S}{M},\tau)\mathcal{N}\left(d_+(\frac{S}{M},\tau)\right) + \mathfrak{n}\left(d_+(\frac{S}{M},\tau)\right)\right].$$

For $r = q$, $M < K$, we have

$$C^{LFP} = C(t,S,K) + \mathrm{e}^{-r\tau} S\sigma\sqrt{\tau}\left[d_+(\frac{S}{K},\tau)\mathcal{N}\left(d_+(\frac{S}{K},\tau)\right) + n\left(d_+(\frac{S}{K},\tau)\right)\right].$$

(iii) For $m \leqslant S < \infty$, $r \neq q$, $m < K$:

$$\begin{aligned} P^{LFP}(t,S,m;K) &= \mathrm{e}^{-r\tau}(K-m) + P(t,S,m) \\ &+ \frac{\sigma^2 \mathrm{e}^{-q\tau} S}{2(r-q)}\left[\mathrm{e}^{-(r-q)\tau}\left(\frac{m}{S}\right)^{\frac{2(r-q)}{\sigma^2}} \mathcal{N}\left(d_-(\frac{m}{S},\tau)\right) - \mathcal{N}\left(-d_+(\frac{S}{m},\tau)\right)\right] \end{aligned}$$

where $P(t,S,m) = \mathrm{e}^{-r\tau} m\mathcal{N}\left(-d_-(\frac{S}{m},\tau)\right) - \mathrm{e}^{-q\tau} S\mathcal{N}\left(-d_+(\frac{S}{m},\tau)\right)$ is the price of a standard put option with spot S, strike m, and time to maturity $\tau = T - t$. The corresponding pricing formula (for $m < K$) in case $r = q$ is

$$\begin{aligned} P^{LFP}(t,S,m;K) &= \mathrm{e}^{-r\tau}(K-m) + P(t,S,m) \\ &+ \mathrm{e}^{-r\tau} S\sigma\sqrt{\tau}\left[n\left(d_+(\frac{S}{m},\tau)\right) - d_+(\frac{S}{m},\tau)\,\mathcal{N}\left(-d_+(\frac{S}{m},\tau)\right)\right]. \end{aligned}$$

For $m \geqslant K$, $m \leqslant S < \infty$, $r \neq q$:

$$\begin{aligned} P^{LFP}(t,S,m;K) &= P(t,S,K) \\ &+ \frac{\sigma^2 \mathrm{e}^{-q\tau} S}{2(r-q)}\left[\mathrm{e}^{-(r-q)\tau}\left(\frac{K}{S}\right)^{\frac{2(r-q)}{\sigma^2}} \mathcal{N}\left(d_-(\frac{K}{S},\tau)\right) - \mathcal{N}\left(-d_+(\frac{S}{K},\tau)\right)\right] \end{aligned}$$

The corresponding pricing formula (for $m \geqslant K$) in case $r = q$ is

$$\begin{aligned} P^{LFP} &= P(t,S,K) \\ &+ \sigma\sqrt{\tau}\mathrm{e}^{-r\tau} S\left[n\left(d_+(\frac{S}{K},\tau)\right) - d_+(\frac{S}{K},\tau)\,\mathcal{N}\left(-d_+(\frac{S}{K},\tau)\right)\right]. \end{aligned}$$

3.40. (a) The distribution of $\ln G_n$ conditional on $S_0 \equiv S(t_0)$ is normal:

$$\ln G_n \sim Norm\left(\ln S_0 + \frac{n+1}{2}(r - q - \sigma^2/2)\delta t, \frac{(n+1)(2n+1)}{6n}\sigma^2\delta t\right).$$

The probability distribution of G_n is log-normal:

$$G_n \stackrel{d}{=} S_0 \exp\left(\frac{n+1}{2n}(r-q-\sigma^2/2)(T-t_0) + \sqrt{\frac{(n+1)(2n+1)}{6n^2}(T-t_0)}\sigma Z\right)$$

where $Z \sim Norm(0,1)$. The mean and variance of G_n are, respectively, given by

$$\mathrm{E}[G_n] = \exp\left(\mathrm{E}[\ln G_n] + \frac{\mathrm{Var}(\ln G_n)}{2}\right) \quad \text{and}$$
$$\mathrm{Var}(G_n) = \exp\left(2\mathrm{E}[\ln G_n] + \mathrm{Var}(\ln G_n)\right)\left(\mathrm{e}^{\mathrm{Var}(\ln G_n)} - 1\right)$$

where

$$\mathrm{E}[\ln G_n] = \ln S_0 + \frac{n+1}{2}(r-q-\sigma^2/2)\delta t \text{ and } \mathrm{Var}(\ln G_n) = \frac{(n+1)(2n+1)}{6n}\sigma^2\delta t.$$

(b) The pricing functions for the fixed strike Asian call and put options are, respectively,

$$C(t_0, S_0) = \mathrm{e}^{-r(T-t_0)} \left[\hat{G}_0 \mathcal{N}(d_+) - K\mathcal{N}(d_-)\right] \text{ and}$$
$$P(t_0, S_0) = \mathrm{e}^{-r(T-t_0)} \left[K\mathcal{N}(-d_-) - \hat{G}_0 \mathcal{N}(-d_+)\right]$$

where

$$\hat{G}_0 = S_0 \exp\left[\frac{n+1}{2n}\left(r - q - \frac{n^2-1}{6n}\sigma^2\right)(T-t_0)\right],$$
$$\hat{\sigma}^2 = \frac{(n+1)(2n+1)}{6n^2}\sigma^2,$$
$$d_\pm = \frac{\ln(\hat{G}_0/K) \pm \hat{\sigma}^2(T-t_0)/2}{\hat{\sigma}\sqrt{T-t_0}}.$$

(c) $C(t_0, S_0) - P(t_0, S_0) = \mathrm{e}^{-r(T-t_0)} \left[\hat{G}_0 - K\right]$.

3.41. (a) The probability distribution of $\ln G(t)$ is normal:

$$\ln G(t) \sim Norm\left(\ln S(0) + (\mu - \frac{1}{2}\sigma^2)\frac{t}{2}, \frac{1}{3}\sigma^2 t\right),$$

where $\mu = r - q$ under $\widetilde{\mathbb{P}}$. (b) We have:

$$\mathrm{E}[G(T) \mid G(t), S(t)] = G(t)^{t/T} S(t)^{(T-t)/T} \mathrm{e}^{\bar{\mu}+\bar{\sigma}^2/2} \text{ and}$$
$$\mathrm{Var}(G(T) \mid G(t), S(t)) = G(t)^{2t/T} S(t)^{2(T-t)/T} \mathrm{e}^{2\bar{\mu}+\bar{\sigma}^2}(e^{\bar{\sigma}^2} - 1),$$

where $\bar{\mu} = (r - q - \frac{\sigma^2}{2})\frac{(T-t)^2}{2T}$ and $\bar{\sigma} = \frac{\sigma}{T}\sqrt{\frac{1}{3}(T-t)^3}$.
(c) The time-t no-arbitrage prices of the fixed strike Asian call and put options are, respectively,

$$C(t, S, G) = \mathrm{e}^{-r(T-t)} \left[\mathcal{N}(d_+) - K\mathcal{N}(d_-)\right] \text{ and}$$
$$P(t, S, G) = \mathrm{e}^{-r(T-t)} \left[K\mathcal{N}(-d_-) - A\mathcal{N}(-d_+)\right].$$

where $A = G^{t/T} S^{(T-t)/T} \mathrm{e}^{\bar{\mu}+\frac{1}{2}\bar{\sigma}^2}$ and

$$d_+ = \frac{\ln(A/K)}{\bar{\sigma}} + \frac{1}{2}\bar{\sigma} = \frac{\ln\left(G^{t/T} S^{(T-t)/T}/K\right) + \bar{\mu} + \bar{\sigma}^2}{\bar{\sigma}} \quad \text{and} \quad d_- = d_+ - \bar{\sigma}.$$

(d) The put-call parity is

$$C(t, S, G) - P(t, S, G) = \mathrm{e}^{-r(T-t)+\bar{\mu}+\bar{\sigma}^2/2} G^{t/T} S^{(T-t)/T} - \mathrm{e}^{-r(T-t)} K.$$

3.42. $\mathrm{d}E(t) = rE(t) + \sigma A(t)\mathcal{N}(d_+(t))\widetilde{W}(t)$ and $\mathrm{d}D(t) = rD(t) + \sigma A(t)\mathcal{N}(-d_+(t))\widetilde{W}(t)$, where $d_+(t) := \frac{\ln(A(t)/F)+(r+\sigma^2/2)(T-t)}{\sigma\sqrt{T-t}}$.

3.43. $\gamma_t = \frac{\mathcal{N}(d_+)}{\mathcal{N}(-d_+)}$ and $\beta_t = -\frac{\mathrm{e}^{-r(T-t)}F}{B(t)}\frac{\mathcal{N}(d_-)}{\mathcal{N}(-d_+)}$, where $d_\pm = \frac{\ln(A(t)/F)+(r\pm\sigma^2/2)(T-t)}{\sigma\sqrt{T-t}}$.

C.4 Chapter 4

4.1. (a) (i) $\mathcal{N}_2(d_1, d_2; \rho)$ where

$$d_1 = \frac{\ln\frac{S_2}{S_1} + \left(q_1 - q_2 + \frac{1}{2}(\sigma_1^2 - \sigma_2^2)\right)t}{\sqrt{\sigma_1^2 + \sigma_2^2 - 2\rho_{12}\sigma_1\sigma_2}\sqrt{t}},$$

$$d_2 = \frac{\ln\frac{S_3}{S_2} + \left(q_2 - q_3 + \frac{1}{2}(\sigma_2^2 - \sigma_3^2)\right)t}{\sqrt{\sigma_2^2 + \sigma_3^2 - 2\rho_{23}\sigma_2\sigma_3}\sqrt{t}},$$

$$\rho = \frac{\rho_{12}\sigma_1\sigma_2 + \rho_{23}\sigma_2\sigma_3 - \rho_{13}\sigma_1\sigma_3 - \sigma_2^2}{\sqrt{\sigma_1^2 + \sigma_2^2 - 2\rho_{12}\sigma_1\sigma_2}\sqrt{\sigma_2^2 + \sigma_3^2 - 2\rho_{23}\sigma_2\sigma_3}}.$$

(ii) $\mathcal{N}_2(-d_1, d_2; \rho)$ where $\rho = -\rho_{12}$ and

$$d_i = \frac{\ln\frac{S_i}{K} + \left(r - q_i - \frac{1}{2}\sigma_i^2\right)t}{\sigma_i\sqrt{t}}, \quad i = 1, 2.$$

(iii) $\mathcal{N}\left(-\dfrac{\ln\frac{S_1S_3}{KS_2} + \left[r + q_2 - q_1 - q_3 + \frac{1}{2}(\sigma_2^2 - \sigma_1^2 - \sigma_3^2)\right]t}{\sigma\sqrt{t}}\right)$ where

$$\sigma = \sqrt{\sigma_1^2 + \sigma_2^2 + \sigma_3^2 + 2(\rho_{13}\sigma_1\sigma_3 - \rho_{12}\sigma_1\sigma_2 - \rho_{23}\sigma_2\sigma_3)}.$$

(iv) $\mathcal{N}\left(\dfrac{\beta\ln S_2(0) - \alpha\ln S_1(0) + \beta(r - q_2 - \sigma_2^2/2)t - \alpha(r - q_1 - \sigma_1^2/2)t}{\sqrt{\alpha^2\sigma_1^2 + \beta^2\sigma_2^2 - 2\alpha\beta\sigma_1\sigma_2\rho}\sqrt{t}}\right).$

(b) The no-arbitrage pricing function is

$$v(\tau, S_1, S_2) = S_1e^{-q_1\tau}\mathcal{N}\left(d_+^{(1,2)}\left(\frac{S_1}{S_2}, \tau\right)\right) - S_2e^{-q_2\tau}\mathcal{N}\left(d_-^{(1,2)}\left(\frac{S_1}{S_2}, \tau\right)\right)$$
$$+ S_3e^{-q_3\tau}\mathcal{N}\left(d_+^{(3,1)}\left(\frac{S_3}{S_1}, \tau\right)\right) - S_1e^{-q_1\tau}\mathcal{N}\left(d_-^{(3,1)}\left(\frac{S_3}{S_1}, \tau\right)\right)$$

where

$$d_\pm^{(1,2)}(x, \tau) = \frac{\ln x + (q_2 - q_1 \pm \frac{1}{2}\nu_{1,2}^2)\tau}{\nu_{1,2}\sqrt{\tau}}, \quad \nu_{1,2} = \sqrt{\sigma_1^2 + \sigma_2^2 - 2\rho_{12}\sigma_1\sigma_2},$$

$$d_\pm^{(3,1)}(x, \tau) = \frac{\ln x + (q_1 - q_3 \pm \frac{1}{2}\nu_{1,3}^2)\tau}{\nu_{1,3}\sqrt{\tau}}, \quad \nu_{1,3} = \sqrt{\sigma_1^2 + \sigma_3^2 - 2\rho_{13}\sigma_1\sigma_3}.$$

for $x > 0$ and $\tau = T - t > 0$.

(c) $\left(\dfrac{d\widehat{\mathbb{P}}}{d\widetilde{\mathbb{P}}}\right)_t = \exp\left(-\dfrac{\|\boldsymbol{\sigma}_1\|^2}{2}t + \boldsymbol{\sigma}_1 \cdot \widetilde{\mathbf{W}}(t)\right).$

4.2. For the option on the minimum, we have

$$\Delta^1(t, S_1, S_2) = e^{-q_1\tau}\mathcal{N}\left(d_-\left(\frac{S_2}{S_1}, \tau\right)\right) \text{ and } \Delta^2(t, S_1, S_2) = e^{-q_2\tau}\mathcal{N}\left(-d_+\left(\frac{S_2}{S_1}, \tau\right)\right).$$

For the option on the maximum, we have

$$\Delta^1(t, S_1, S_2) = e^{-q_1\tau}\mathcal{N}\left(-d_-\left(\frac{S_2}{S_1}, \tau\right)\right) \text{ and } \Delta^2(t, S_1, S_2) = e^{-q_2\tau}\mathcal{N}\left(d_+\left(\frac{S_2}{S_1}, \tau\right)\right).$$

4.3. The no-arbitrage pricing function is

$$v_{min}(\tau, S_1, S_2) = \mathrm{e}^{-\bar{q}_1\tau} S_1 \mathcal{N}\Big(d_-\big(\frac{S_2}{S_1},\tau\big)\Big) + \mathrm{e}^{-\bar{q}_2\tau} S_2 \mathcal{N}\Big(-d_+\big(\frac{S_2}{S_1},\tau\big)\Big)$$

where

$$d_\pm(x,\tau) := \frac{\ln x + (\bar{q}_1 - \bar{q}_2 \pm \frac{1}{2}\bar{\nu}^2)\tau}{\bar{\nu}\sqrt{\tau}}, \quad x, \tau > 0,$$

and $\bar{\nu} = \sqrt{\bar{\sigma}_1^2 + \bar{\sigma}_2^2 - 2\bar{\rho}\bar{\sigma}_1\bar{\sigma}_2}$. Here $\bar{q}_i$ and $\bar{\sigma}_i^2$ denote time-averaged (over $[t,T]$) dividend yield and squared volatility for each $i = 1, 2$, respectively, given by

$$\bar{q}_i = \frac{1}{\tau}\int_t^T q_i(u)\,\mathrm{d}u \text{ and } \bar{\sigma}_i^2 = \frac{1}{T-t}\int_t^T \sigma_i^2(u)\,\mathrm{d}u\,.$$

The correlation coefficient is $\bar{\rho} = \dfrac{\int_t^T \boldsymbol{\sigma}_1(u)\cdot\boldsymbol{\sigma}_2(u)\,\mathrm{d}u}{\sqrt{\int_t^T\sigma_1^2(u)\,\mathrm{d}u\int_t^T\sigma_2^2(u)\,\mathrm{d}u}}$.

4.4. (a) The no-arbitrage pricing function is

$$C(t,x) = \mathrm{e}^{-r^f(T-t)}x\mathcal{N}(d_+) - \mathrm{e}^{-r(T-t)}\kappa\mathcal{N}(d_-)$$

where $d_\pm = \dfrac{\ln(x/\kappa) + (r - r^f \pm \frac{1}{2}\sigma_X^2)(T-t)}{\sigma_X\sqrt{T-t}}$, $x > 0$, $\kappa > 0$, $t < T$.
(b) The BSPDE takes the form

$$\frac{\partial V}{\partial t} + \frac{1}{2}\sigma_X^2 x^2\frac{\partial^2 V}{\partial x^2} + (r - r^f)\,x\frac{\partial V}{\partial x} - rV = 0\,,$$

subject to the terminal condition $C(T,x) = (x-\kappa)^+$, for all $x > 0$.

4.5. (a) The no-arbitrage pricing function for the call with spot values $S^f(t) = S > 0, X(t) = x > 0$, maturity time T, and strike $K > 0$ is

$$C(t,S,x) = xS\mathcal{N}(d_+) - \mathrm{e}^{-r(T-t)}K\,\mathcal{N}(d_-)$$

where $d_\pm = \dfrac{\ln(xS/K) + (r \pm \frac{1}{2}\sigma_{XS}^2)(T-t)}{\sigma_{XS}\sqrt{T-t}}$, $t < T$, and $\sigma_{XS} = \sqrt{\sigma_X^2 + \sigma_S^2 + 2\rho\,\sigma_X\sigma_S}$.
(b) For the put, we have

$$P(t,S,x) = \mathrm{e}^{-r(T-t)}K\,\mathcal{N}(-d_-) - xS\mathcal{N}(-d_+).$$

(c) $C_t - P_t = \mathrm{e}^{-r(T-t)}X(t)S^f(t) - \mathrm{e}^{-r(T-t)}\,K$.

4.6. The no-arbitrage pricing function is

$$C(t,S) = \mathrm{e}^{-(r-r^f+q_S+\rho_{XS}\sigma_X\sigma_S)(T-t)}\bar{X}S\mathcal{N}(d_+) - \mathrm{e}^{-r(T-t)}\bar{X}K_f\,\mathcal{N}(d_-)$$

where $d_\pm = \dfrac{\ln(S/K_f) + (r^f - q_S - \rho_{XS}\sigma_X\sigma_S \pm \frac{\sigma_S^2}{2})(T-t)}{\sigma_S\sqrt{T-t}}$ for all $t < T$ and $S, K_f > 0$.

4.7. The no-arbitrage pricing function is

$$C(t,S,x) = x\left[S\mathrm{e}^{-q_S(T-t)}\mathcal{N}(d_+) - K_f\,\mathrm{e}^{-r^f(T-t)}\mathcal{N}(d_-)\right]$$

where $d_\pm = \dfrac{\ln(S/K_f) + (r^f - q_S \pm \frac{\sigma_S^2}{2})(T-t)}{\sigma_S\sqrt{T-t}}$ for all $t < T$ and $x, S, K_f > 0$.

4.9. (a) The no-arbitrage pricing function is

$$V(t,S,A,x) = xS\mathcal{N}\Big(d_+(\frac{xS}{A}, T-t)\Big) + A\mathcal{N}\Big(-d_-(\frac{xS}{A}, T-t)\Big)$$

where $d_\pm(x,\tau) := \frac{\ln x \pm \frac{1}{2}\sigma_Y^2(T-t)}{\sigma_Y\sqrt{T-t}}$, for all $x, \tau > 0$, and

$$\sigma_Y = \sqrt{\sigma_X^2 + \sigma_S^2 + \sigma_A^2 + 2\rho_{XS}\sigma_X\sigma_S - 2\rho_{AS}\sigma_A\sigma_S - 2\rho_{XA}\sigma_X\sigma_A}\,.$$

(b) $\Delta_t = \mathcal{N}(-d_-(y,\tau)) = \mathcal{N}\Big(-d_-(\frac{xS}{A}, T-t)\Big)$ and $\Delta_t^f = \mathcal{N}\Big(d_+(\frac{xS}{A}, T-t)\Big)$.

(c) The no-arbitrage pricing function is

$$V(t,S,A,x) = xS\mathcal{N}\left(\frac{\ln(x/X_0) + (r - r^f + \rho_{XS}\sigma_X\sigma_S + \frac{1}{2}\sigma_X^2)(T-t)}{\sigma_X\sqrt{T-t}}\right)$$
$$+ A\mathcal{N}\left(-\frac{\ln(x/X_0) + (r - r^f + \rho_{XA}\sigma_X\sigma_A - \frac{1}{2}\sigma_X^2)(T-t)}{\sigma_X\sqrt{T-t}}\right).$$

(d) The no-arbitrage pricing function is

$$V(t,S,A,x) = axS\mathcal{N}\Big(d_+(\frac{axS}{bA}, T-t)\Big) - bA\mathcal{N}\Big(d_-(\frac{axS}{bA}, T-t)\Big)$$

with functions $d_\pm$ defined in part (a).

4.10. (a) We have the log-drift in each case (i)-(iv) given by the coefficient in $\mathrm{d}t$:

$$\frac{\mathrm{d}X(t)}{X(t)} = (r(t) - r^f(t) + \rho_{XA}\sigma_X\sigma_A)\,\mathrm{d}t + \boldsymbol{\sigma}^{(X)}\cdot\mathrm{d}\widehat{\mathbf{W}}(t),$$
$$\frac{\mathrm{d}A^f(t)}{A^f(t)} = (r^f(t) - \rho_{XA^f}\sigma_X\sigma_{A^f} + \rho_{AA^f}\sigma_A\sigma_{A^f})\,\mathrm{d}t + \boldsymbol{\sigma}^{(A^f)}\cdot\mathrm{d}\widehat{\mathbf{W}}(t),$$
$$\frac{\mathrm{d}\big[X(t)B^f(t)\big]}{X(t)B^f(t)} = (r(t) + \rho_{XA}\sigma_X\sigma_A)\,\mathrm{d}t + \boldsymbol{\sigma}^{(X)}\cdot\mathrm{d}\widehat{\mathbf{W}}(t),$$
$$\frac{\mathrm{d}A(t)}{A(t)} = (r(t) + \sigma_A^2)\,\mathrm{d}t + \boldsymbol{\sigma}^{(A)}\cdot\mathrm{d}\widehat{\mathbf{W}}(t).$$

(b) The no-arbitrage pricing function is

$$V_0 = A(0)\left[\mathrm{e}^{\bar{\mu}_Y T}\mathcal{N}(d_+(1,T)) + \mathcal{N}(-d_-(1,T))\right] \text{ where } d_\pm(1,T) = \frac{(\bar{\mu}_Y \pm \frac{1}{2}\sigma_Y^2)\sqrt{T}}{\sigma_Y},$$

$\sigma_Y = \sqrt{\sigma_A^2 + \sigma_{A^f}^2 - 2\rho_{AA^f}\sigma_A\sigma_{A^f}}$, and $\bar{\mu}_Y = \bar{r}^f - \bar{r} - \sigma_{A^f}(\rho_{XA^f}\sigma_X + \rho_{AA^f}\sigma_A)$.

4.11. (a) $P(t,S,x) = x\left[\mathrm{e}^{-r^f(T-t)}K_f\mathcal{N}(-d_-) - S\mathcal{N}(-d_+)\right]$ where $d_\pm \equiv \frac{\ln\frac{S}{K_f}+(r^f\pm\frac{1}{2}\sigma_S^2)(T-t)}{\sigma_S\sqrt{T-t}}$ and $t<T$.
(b) $\Delta_t^f = -\mathcal{N}(-d_+)$, $X(t)\beta_t^f B^f(t) = \mathrm{e}^{-r^f(T-t)}xK_f\mathcal{N}(-d_-)$ with $d_\pm$ defined in part (a) and where S and x are the respective spot values of $S^f(t)$ and $X(t)$.
(c) $C(t,S,x) - P(t,S,x) = xS - \mathrm{e}^{-r^f(T-t)}x\,K_f$.

4.12. The no-arbitrage initial price for spot values $X(0)=X$ and $S^f(0)=S$ is

$$V_0 = XS\left[\mathcal{N}\left(-d_+(\frac{X}{K},T)\right) - \left(\frac{K}{X}\right)^{\frac{2(r-r^f+\rho\sigma_X\sigma_S)}{\sigma_X^2}+1}\mathcal{N}\left(-d_+(\frac{K}{X},T)\right)\right]$$

where $d_+(x,T) := \dfrac{\ln x + (r - r^f + \rho\sigma_X\sigma_S + \frac{1}{2}\sigma_X^2)T}{\sigma_X\sqrt{T}}$.

4.13. The pricing formula is as follows:

$$\begin{aligned}
&V(t,S_1,S_2)\\
&= \mathrm{e}^{-q_2\tau}S_2\mathcal{N}_2\left(\frac{\ln\frac{S_2}{S_1}+(q_1-q_2+\frac{1}{2}\nu^2)\tau}{\nu\sqrt{\tau}}, \frac{\ln\frac{S_2}{K}+(r-q_2+\frac{1}{2}\sigma_2^2)\tau}{\sigma_2\sqrt{\tau}}; \frac{\sigma_2-\rho\sigma_1}{\nu}\right)\\
&- \mathrm{e}^{-r\tau}K\mathcal{N}_2\left(\frac{\ln\frac{S_2}{S_1}+(q_1-q_2+\frac{1}{2}(\sigma_1^2-\sigma_2^2))\tau}{\nu\sqrt{\tau}}, \frac{\ln\frac{S_2}{K}+(r-q_2-\frac{1}{2}\sigma_2^2)\tau}{\sigma_2\sqrt{\tau}}; \frac{\sigma_2-\rho\sigma_1}{\nu}\right)\\
&+ \mathrm{e}^{-q_1\tau}S_1\mathcal{N}_2\left(\frac{\ln\frac{S_1}{S_2}+(q_2-q_1+\frac{1}{2}\nu^2)\tau}{\nu\sqrt{\tau}}, \frac{\ln\frac{S_1}{K}+(r-q_1+\frac{1}{2}\sigma_1^2)\tau}{\sigma_1\sqrt{\tau}}; \frac{\sigma_1-\rho\sigma_2}{\nu}\right)\\
&- \mathrm{e}^{-r\tau}K\mathcal{N}_2\left(\frac{\ln\frac{S_1}{S_2}+(q_2-q_1+\frac{1}{2}(\sigma_2^2-\sigma_1^2))\tau}{\nu\sqrt{\tau}}, \frac{\ln\frac{S_1}{K}+(r-q_1-\frac{1}{2}\sigma_1^2)\tau}{\sigma_1\sqrt{\tau}}; \frac{\sigma_1-\rho\sigma_2}{\nu}\right).
\end{aligned}$$

4.14. The pricing formula is as follows:

$$\begin{aligned}
&V(t,S_1,S_2)\\
&= \mathrm{e}^{-r\tau}K\,\mathcal{N}_2\left(\frac{\ln\frac{S_2}{S_1}+(q_1-q_2+\frac{1}{2}(\sigma_1^2-\sigma_2^2))\tau}{\nu\sqrt{\tau}}, -\frac{\ln\frac{S_1}{K}+(r-q_1-\frac{1}{2}\sigma_1^2)\tau}{\sigma_1\sqrt{\tau}}; \frac{\sigma_1-\rho\sigma_2}{\nu}\right)\\
&- \mathrm{e}^{-q_1\tau}S_1\mathcal{N}_2\left(\frac{\ln\frac{S_2}{S_1}+(q_1-q_2-\frac{1}{2}\nu^2)\tau}{\nu\sqrt{\tau}}, -\frac{\ln\frac{S_1}{K}+(r-q_1+\frac{1}{2}\sigma_1^2)\tau}{\sigma_1\sqrt{\tau}}; \frac{\sigma_1-\rho\sigma_2}{\nu}\right)\\
&+ \mathrm{e}^{-r\tau}K\,\mathcal{N}_2\left(\frac{\ln\frac{S_1}{S_2}+(q_2-q_1+\frac{1}{2}(\sigma_2^2-\sigma_1^2))\tau}{\nu\sqrt{\tau}}, -\frac{\ln\frac{S_2}{K}+(r-q_2-\frac{1}{2}\sigma_2^2)\tau}{\sigma_2\sqrt{\tau}}; \frac{\sigma_2-\rho\sigma_1}{\nu}\right)\\
&- \mathrm{e}^{-q_2\tau}S_2\mathcal{N}_2\left(\frac{\ln\frac{S_1}{S_2}+(q_2-q_1-\frac{1}{2}\nu^2)\tau}{\nu\sqrt{\tau}}, -\frac{\ln\frac{S_2}{K}+(r-q_2+\frac{1}{2}\sigma_2^2)\tau}{\sigma_2\sqrt{\tau}}; \frac{\sigma_2-\rho\sigma_1}{\nu}\right).
\end{aligned}$$

4.15. (a) The pricing formula is

$$V(t,S_1,S_2,S_3) = \mathrm{e}^{-q_3\tau}S_3\mathcal{N}_2\left(\frac{\ln\frac{S_3}{S_2}+(q_2-q_3+\frac{1}{2}\nu_1^2)\tau}{\nu_1\sqrt{\tau}}, \frac{\ln\frac{S_3}{S_1}+(q_1-q_3+\frac{1}{2}\nu_2^2)\tau}{\nu_2\sqrt{\tau}}; \beta\right)$$

for all $\tau = T - t > 0$, where $\beta = \dfrac{\sigma_3(\sigma_3 - \rho_{13}\sigma_1 - \rho_{23}\sigma_2) + \rho_{12}\sigma_1\sigma_2}{\nu_1\nu_2}$.

(b) We have $V(t, S_1, S_2, S_3) = V^{(1)}(t, S_1, S_2, S_3) + V^{(2)}(t, S_1, S_2, S_3) + V^{(3)}(t, S_1, S_2, S_3)$ where

$$V^{(1)} = \mathrm{e}^{-q_1\tau} S_1 \mathcal{N}_2\left(\frac{\ln\frac{S_1}{S_3} + (q_3 - q_1 + \frac{1}{2}\nu_2^2)\tau}{\nu_2\sqrt{\tau}}, \frac{\ln\frac{S_1}{S_2} + (q_2 - q_1 + \frac{1}{2}\nu_3^2)\tau}{\nu_3\sqrt{\tau}}; \beta_1\right),$$

$$V^{(2)} = \mathrm{e}^{-q_2\tau} S_2 \mathcal{N}_2\left(\frac{\ln\frac{S_2}{S_3} + (q_3 - q_2 + \frac{1}{2}\nu_1^2)\tau}{\nu_1\sqrt{\tau}}, \frac{\ln\frac{S_2}{S_1} + (q_1 - q_2 + \frac{1}{2}\nu_3^2)\tau}{\nu_3\sqrt{\tau}}; \beta_2\right),$$

$$V^{(3)} = \mathrm{e}^{-q_3\tau} S_3 \mathcal{N}_2\left(\frac{\ln\frac{S_3}{S_2} + (q_2 - q_3 + \frac{1}{2}\nu_1^2)\tau}{\nu_1\sqrt{\tau}}, \frac{\ln\frac{S_3}{S_1} + (q_1 - q_3 + \frac{1}{2}\nu_2^2)\tau}{\nu_2\sqrt{\tau}}; \beta_3\right),$$

$$\beta_1 := \frac{\sigma_1(\sigma_1 - \rho_{12}\sigma_2 - \rho_{13}\sigma_3) + \rho_{23}\sigma_2\sigma_3}{\nu_2\nu_3},$$

$$\beta_2 := \frac{\sigma_2(\sigma_2 - \rho_{12}\sigma_1 - \rho_{23}\sigma_3) + \rho_{13}\sigma_1\sigma_3}{\nu_1\nu_3},$$

$$\beta_3 := \frac{\sigma_3(\sigma_3 - \rho_{13}\sigma_1 - \rho_{23}\sigma_2) + \rho_{12}\sigma_1\sigma_2}{\nu_1\nu_2}.$$

(c) The pricing function is

$$V(t, S_1, S_2, S_3) = \mathrm{e}^{-r\tau} K \mathcal{N}_2\left(\frac{\ln\frac{S_3}{S_2} + (q_2 - q_3 + \frac{1}{2}(\sigma_2^2 - \sigma_3^2))\tau}{\nu_1\sqrt{\tau}}, \frac{\ln\frac{S_3}{S_1} + (q_1 - q_3 + \frac{1}{2}(\sigma_1^2 - \sigma_3^2))\tau}{\nu_2\sqrt{\tau}}; \beta\right)$$

where β is the same as in part (a).

4.16. The pricing function is

$$V(t, S_1, S_2) = aS_2\mathrm{e}^{-q_2\tau}\mathcal{N}\left(d_+\left(\frac{aS_2}{bS_1}, \tau\right)\right) - bS_1\mathrm{e}^{-q_1\tau}\mathcal{N}\left(d_-\left(\frac{aS_2}{bS_1}, \tau\right)\right)$$

where $d_\pm(x, \tau) := \dfrac{\ln x + (q_1 - q_2 \pm \frac{1}{2}\nu^2)\tau}{\nu\sqrt{\tau}}$ and $\nu = \sqrt{\sigma_1^2 + \sigma_2^2 - 2\rho_{12}\sigma_1\sigma_2}$.

4.17. We have

$$V(T, S_1, S_2) = S_2\mathrm{e}^{-q_2T}\left[1 + \frac{\sigma^2}{2(q_1 - q_2)}\right]\mathcal{N}\left(-\left(\frac{q_1 - q_2}{\sigma} + \frac{\sigma}{2}\right)\sqrt{T}\right) + S_2\mathrm{e}^{-q_1T}\left[1 - \frac{\sigma^2}{2(q_1 - q_2)}\right]\mathcal{N}\left(\left(\frac{q_1 - q_2}{\sigma} - \frac{\sigma}{2}\right)\sqrt{T}\right)$$

if $q_1 - q_2 \neq 0$, and

$$V(T, S_1, S_2) = \mathrm{e}^{-qT}S_2\left[\left(2 + \frac{\sigma^2}{2}T\right)\mathcal{N}\left(-\frac{\sigma}{2}\sqrt{T}\right) - \sigma\sqrt{\frac{T}{2\pi}}\mathrm{e}^{-\frac{\sigma^2}{8}T}\right]$$

if $q_1 = q_2 \equiv q$.

4.18. The pricing function is

$$V(t, S_1, S_2) = aS_2\mathrm{e}^{-q_2\tau}\left[2\mathcal{N}\left(d_+\left(\frac{aS_2}{bS_1}, \tau\right)\right) - 1\right] - bS_1\mathrm{e}^{-q_1\tau}\left[2\mathcal{N}\left(d_-\left(\frac{aS_2}{bS_1}, \tau\right)\right) - 1\right]$$

where $d_\pm(x,\tau) := \dfrac{\ln x + (q_1 - q_2 \pm \frac{1}{2}\nu^2)\tau}{\nu\sqrt{\tau}}$ and $\nu = \sqrt{\sigma_1^2 + \sigma_2^2 - 2\rho_{12}\sigma_1\sigma_2}$.

4.19. (a) The SDEs are

$$\begin{aligned}
\frac{\mathrm{d}X^{-1}(t)}{X^{-1}(t)} &= (r_f - r_d)\,\mathrm{d}t - \sigma_X\,\mathrm{d}\widetilde{W}_1(t),\\
\frac{\mathrm{d}A(t)}{A(t)} &= (r_d - q_A + \sigma_X\sigma_A\rho)\,\mathrm{d}t + \sigma_A\left(\rho\,\mathrm{d}\widetilde{W}_1(t) + \sqrt{1-\rho^2}\,\mathrm{d}\widetilde{W}_2(t)\right),\\
\frac{\mathrm{d}F(t)}{F(t)} &= (r_f - q_A)\,\mathrm{d}t + (\rho\sigma_A - \sigma_X)\,\mathrm{d}\widetilde{W}_1(t) + \sigma_A\sqrt{1-\rho^2}\,\mathrm{d}\widetilde{W}_2(t).
\end{aligned}$$

(b) For spot values $A \equiv A(t) > 0$ and $x \equiv X(t) > 0$, we have the following pricing functions:
(i) $V(t,A,x) = \mathrm{e}^{-q_A\tau}(A/x)\mathcal{N}(-d_+) + \mathrm{e}^{-r_f\tau}K_f\mathcal{N}(d_-)$ where $\tau = T - t > 0$,
$d_\pm = \dfrac{\ln(A/(xK_f)) + (r_f - q_A \pm \sigma_F^2/2)\tau}{\sigma_F\sqrt{\tau}}$, and $\sigma_F^2 = \sigma_A^2 - 2\rho\sigma_A\sigma_X + \sigma_X^2$.
(ii) $V(t,A,x) = \mathrm{e}^{-q_\kappa\tau}\kappa A\mathcal{N}(d_+) - \mathrm{e}^{-q_A\tau}(A/x)\mathcal{N}(d_-)$ where $\tau = T - t > 0$,
$d_\pm = \dfrac{\ln(\kappa x) + (q_A - q_\kappa \pm \sigma_X^2/2)\tau}{\sigma_X\sqrt{\tau}}$, and $q_\kappa := r_f - r_d + q_A - \sigma_X\sigma_A\rho$.

4.20. (a) $V(t,S,x) = \mathrm{e}^{-q_S\tau}xS\mathcal{N}(a_+) + \mathrm{e}^{-r\tau}K_d\mathcal{N}(-a_-)$
where $\tau = T - t$ and $a_\pm = \dfrac{\ln(xS/K_d) + (r - q_S \pm v^2/2)\tau}{v\sqrt{\tau}}$.
(b) $V(t,S,x) = \mathrm{e}^{-q_S\tau}xS\mathcal{N}(b_+) + \mathrm{e}^{-r^f\tau}xK_f\mathcal{N}(-b_-)$
where $\tau = T - t$ and $b_\pm = \dfrac{\ln(S/K_f) + (r^f - q_S \pm \sigma_S^2/2)\tau}{\sigma_S\sqrt{\tau}}$.
(c) $V(t,S,x) = \mathrm{e}^{-q_S\tau}xS\mathcal{N}(-a_+) + \mathrm{e}^{-r\tau}K_d\mathcal{N}(a_-)$ with τ and $a_\pm$ defined in (a).
(d) $V(t,S,x) = \mathrm{e}^{-q_S\tau}xS\mathcal{N}(-b_+) + \mathrm{e}^{-r^f\tau}xK_f\mathcal{N}(b_-)$ with τ and $b_\pm$ defined in (b).

4.21. (a) $V(t,S,x) = \mathrm{e}^{-q_S\tau}xS\mathcal{N}(d_+)$ where $d_+ = \frac{\ln(xS/K)+(r-q_S+\sigma_{XS}^2/2)\tau}{\sigma_{XS}\sqrt{\tau}}$
and $\sigma_{XS}^2 = \sigma_X^2 + 2\rho\sigma_X\sigma_S + \sigma_S^2$.
(b) $V(t,S,x) = \mathrm{e}^{-q_S\tau}xS\mathcal{N}(d_+)$ where $d_+ = \frac{\ln(S/K)+(r^f-q_S+\sigma_S^2/2)\tau}{\sigma_S\sqrt{\tau}}$.
(c) $V(t,S,x) = \mathrm{e}^{-r^f\tau}xK\mathcal{N}(d_-)$ where $d_- = \frac{\ln(S/K)+(r^f-q_S-\sigma_S^2/2)\tau}{\sigma_S\sqrt{\tau}}$.
(d) $V(t,S,x) = \mathrm{e}^{-q_S\tau}xS\mathcal{N}(d_-)$ where $d_- = \frac{\ln(x/\kappa)+(r-r^f+\rho\sigma_X\sigma_S+\sigma_X^2/2)\tau}{\sigma_X\sqrt{\tau}}$.

(e) $V(t,S,x) = \mathrm{e}^{-r^f\tau}xK\mathcal{N}(-d_-) - \mathrm{e}^{-q_S\tau}xS\mathcal{N}(-d_+)$ where $d_\pm = \frac{\ln(S/K)+(r^f-q_S\pm\sigma_S^2/2)\tau}{\sigma_S\sqrt{\tau}}$.
(f) $V(t,S,x) = \mathrm{e}^{-(r-r^f+q_S+\rho\sigma_X\sigma_S)\tau}\kappa S\mathcal{N}(d_-) - \mathrm{e}^{-q_S\tau}xS\mathcal{N}(d_+)$
where $d_\pm = \frac{\ln(x/\kappa)+(r^f-r-\rho\sigma_X\sigma_S\pm\sigma_X^2/2)\tau}{\sigma_X\sqrt{\tau}}$.

4.22. Define the following:
$\sigma_Y^2 = \sigma_X^2 + \sigma_S^2 + \sigma_A^2 + 2(\rho_{XS}\sigma_X\sigma_S - \rho_{AX}\sigma_A\sigma_X - \rho_{AS}\sigma_A\sigma_S)$,
$\sigma_{XS}^2 = \sigma_X^2 + 2\sigma_X\sigma_S\rho_{XS} + \sigma_S^2$,
$a = -\dfrac{\ln(xS/K) + (r - q_S - \sigma_{XS}^2/2)\tau}{\sigma_{XS}\sqrt{\tau}}$,
$b = -\dfrac{\ln(xS/A) + (q_A - q_S + (\sigma_A^2 - \sigma_{XS}^2)/2)\tau}{\sigma_Y\sqrt{\tau}}$,

$c = -\dfrac{\ln(A/K) + (r - q_A - \sigma_A^2/2)\tau}{\sigma_A\sqrt{\tau}}$,

$\rho_1 = \dfrac{\sigma_{XS}^2 - \rho_{AS}\sigma_A\sigma_S - \rho_{AX}\sigma_A\sigma_X}{\sigma_{XS}\sigma_Y}$, $\rho_2 = \dfrac{\sigma_A - \rho_{AX}\sigma_X - \rho_{AS}\sigma_S}{\sigma_Y}$, and $\tau = T - t$.
(a) $V(t,S,A,x) = \mathrm{e}^{-r\tau}K\mathcal{N}_2(a, b; \rho_1) - \mathrm{e}^{-q_S\tau}xS\mathcal{N}_2(a - \sigma_{XS}\sqrt{\tau}, b - \rho_1\sigma_{XS}\sqrt{\tau}; \rho_1)$;
(b) $V(t,S,A,x) = \mathrm{e}^{-r\tau}K\mathcal{N}_2(-b, c; \rho_2) - \mathrm{e}^{-q_A\tau}A\mathcal{N}_2(-b - \rho_2\sigma_A\sqrt{\tau}, c - \sigma_A\sqrt{\tau}; \rho_2)$;
(c) $V(t,S,A,x) = \mathrm{e}^{-q_A\tau}A\mathcal{N}_2(b + \rho_2\sigma_A\sqrt{\tau}, c - \sigma_A\sqrt{\tau}; -\rho_2) + \mathrm{e}^{-r\tau}K\mathcal{N}(b) - \mathrm{e}^{-r\tau}K\mathcal{N}_2(b, c; -\rho_2)$;
(d) $V(t,S,A,x) = \mathrm{e}^{-q_S\tau}xS\mathcal{N}_2(-a+\sigma_{XS}\sqrt{\tau}, b-\rho_1\sigma_{XS}\sqrt{\tau}; -\rho_1) + \mathrm{e}^{-r\tau}K\mathcal{N}_2(a, b; \rho_1)$.

4.23. $V(t, S_1, S_2, S_3) = \mathrm{e}^{-q_1\tau}S_1\mathcal{N}\left(\frac{\ln(KS_3/S_2)+(q_2-q_3+(\sigma_2^2-\sigma_3^2-2\rho_{12}\sigma_1\sigma_2+2\rho_{13}\sigma_1\sigma_3)/2)\tau}{\sqrt{\sigma_2^2+\sigma_3^2-2\rho_{23}\sigma_2\sigma_3}\sqrt{\tau}}\right)$.

4.24. For $C-P \equiv C(t, S_1, S_2) - P(t, S_1, S_2)$, we have: (a) $C - P = \mathrm{e}^{-q_1\tau}S_1\mathcal{N}(-d_-(S_2/S_1, \tau)) + \mathrm{e}^{-q_2\tau}S_2\mathcal{N}(d_+(S_2/S_1, \tau)) - \mathrm{e}^{-r\tau}K$;
(b) $C - P = \mathrm{e}^{-q_1\tau}S_1\mathcal{N}(d_-(S_2/S_1, \tau)) + \mathrm{e}^{-q_2\tau}S_2\mathcal{N}(-d_+(S_2/S_1, \tau)) - \mathrm{e}^{-r\tau}K$;
(c) $C - P = \mathrm{e}^{-q_2\tau}S_2\mathcal{N}(d_+(S_2/S_1, \tau)) - \mathrm{e}^{-q_1\tau}S_1\mathcal{N}(d_-(S_2/S_1, \tau)) - \mathrm{e}^{-r\tau}K$;
where $d_\pm(x, \tau) = \frac{\ln x + (q_1 - q_2 \pm \nu^2/2)\tau}{\nu\sqrt{\tau}}$, $\nu^2 = \sigma_1^2 + \sigma_2^2 - 2\rho_{12}\sigma_1\sigma_2$, and $\tau = T - t$.

C.5 Chapter 5

5.1 (a) Hint: fix $t > 0$. Let $S(t) = S_t^*$. Form a portfolio with one option and $|a|$ shares of stock, where $a \equiv \Lambda'(S_t^*)$. Show that after a small time lapse δt, the change in the portfolio value is $\delta\Pi = \delta V + |a|\,\delta S = \left(\left.\frac{\partial V}{\partial S}\right|_{S\to S_t^*} + |a|\right)\sigma S\,\delta W + \mathcal{O}(\delta t)$. Use this fact to prove the existence of arbitrage if $\left.\frac{\partial V(t,S)}{\partial S}\right|_{S\nearrow S_t^*} < a$ or $\left.\frac{\partial V(t,S)}{\partial S}\right|_{S\searrow S_t^*} > a$.

5.2 Hint: use the no arbitrage pricing formulae for the standard perpetual call and put options and the fact that $\lambda_\pm(q, r) = 1 - \lambda_\mp(r, q)$.

5.6 (a) $S^* \to \infty$ and $C(S) \to S$, as $q \searrow 0$.
(b) $S^* \to 0$ and $P(S) \to K$, as $r \searrow 0$.

5.7 The pricing function is given by

$$V(S) = \begin{cases} a_+\, S^{\lambda_+} & \text{if } 0 < S < S^*, \\ a\,(S-K)^2 & \text{if } S \geqslant S^*, \end{cases}$$

where $\lambda_\pm = \dfrac{-(r - q - \sigma^2/2) \pm \sqrt{(r - q - \sigma^2/2)^2 + 2\sigma^2 r}}{\sigma^2}$, $a_+ = a\,(S^* - K)^2 (S^*)^{-\lambda_+}$, and $S^* = \dfrac{\lambda_+ K}{\lambda_+ - 2}$.

5.8 The pricing function is

$$V(S) = \begin{cases} (K - S^*)(S/S^*)^{\lambda_-} & \text{if } S > S^*, \\ \min\{K - S, C\} & \text{if } 0 < S \leqslant S^* \end{cases}$$

where $S^* = \max\left\{K - C, \frac{K\lambda_-}{\lambda_- - 1}\right\}$.

5.9 (a) The pricing function is

$$V(S) = \begin{cases} (K^n - (S^*)^n)(S/S^*)^{\lambda_-} & \text{if } S > S^*, \\ K^n - S^n & \text{if } 0 < S \leqslant S^*, \end{cases}$$

where $S^* = K\left(\frac{\lambda_-}{\lambda_- - n}\right)^{1/n}$ with λ_- defined in the answer to Exercise 5.7.
(b) The pricing function is

$$V(S) = \begin{cases} (K - S^*)^n (S/S^*)^{\lambda_-} & \text{if } S > S^*, \\ (K - S)^n & \text{if } 0 < S \leqslant S^*, \end{cases}$$

where $S^* = K\left(\frac{\lambda_-}{\lambda_- - n}\right)$.
(c) The pricing function is

$$V(S) = \begin{cases} (K - S^*)^n (S/S^*)^{\lambda_-} & \text{if } S > S^*, \\ \min\{(K - S)^n, C\} & \text{if } 0 < S \leqslant S^*, \end{cases}$$

where $S^* = \max\left\{K - C^{1/n}, K\left(\frac{\lambda_-}{\lambda_- - n}\right)\right\}$.

5.10 (a) The pricing function is defined in the answer to Exercise 5.9(b) for $n = 2$.
(c) $S^* \to 0$ and $V(S) \to K^2$, as $r \searrow 0$.

5.11 (a) The early exercise boundary curve is then the union of two early exercise curves: one for the call and the other for the put. This gives two sets of asymptotic values as $\tau \searrow 0$ and $\tau \to \infty$. For the call portion, we have $S_c^*(0^+) = (K - w)\max\left\{1, \frac{r}{q}\right\}$ and $S_c^*(\infty) = \frac{(K-w)\lambda_+}{\lambda_+ - 1}$ For the put portion, we have $S_p^*(0^+) = (K + w)\min\left\{1, \frac{r}{q}\right\}$ and $S_p^*(\infty) = \frac{(K+w)\lambda_-}{\lambda_- - 1}$ with $\lambda_\pm \equiv \lambda_\pm(r, q)$ defined in the answer to Exercise 5.7.
(c) The pricing function is

$$V(S) = \begin{cases} \frac{S_c^* - (K-w)}{(S_c^*)^{\lambda_+}} S^{\lambda_+} & \text{if } 0 < S \leqslant S_c^*, \\ S - (K - w) & \text{if } S_c^* \leqslant S \leqslant K, \\ S - K + w & \text{if } S_c^* \leqslant S \leqslant K, \\ \frac{(K+w) - S_p^*}{(S_p^*)^{\lambda_-}} S^{\lambda_-} & \text{if } S_p^* \leqslant S < \infty, \end{cases}$$

where $S_c^* = \frac{(K-w)\lambda_+}{\lambda_+ - 1}$ and $S_p^* = \frac{(K+w)\lambda_-}{\lambda_- - 1}$.

5.12 The pricing function is

$$V(S) = \begin{cases} K - S & \text{if } 0 < S \leqslant S_1^*, \\ C_- S^{\lambda_-} + C_+ S^{\lambda_+} & \text{if } S_1^* < S < S_2^*, \\ S - K & \text{if } S \geqslant S_2^* \end{cases}$$

where S_1^*, S_2^*, and $C_\pm$ can be found by solving numerically the following:

$$\begin{cases} \lambda_- C_- (S_1^*)^{\lambda_- - 1} + \lambda_+ C_+ (S_1^*)^{\lambda_+ - 1} = -1, \\ \lambda_- C_- (S_2^*)^{\lambda_- - 1} + \lambda_+ C_+ (S_2^*)^{\lambda_+ - 1} = 1 \end{cases} \implies \begin{cases} C_- = \frac{(K_1 - S_1^*)(S_2^*)^{\lambda_+} - (S_2^* - K_2)(S_1^*)^{\lambda_+}}{(S_1^*)^{\lambda_-}(S_2^*)^{\lambda_+} - (S_1^*)^{\lambda_+}(S_2^*)^{\lambda_-}}, \\ C_+ = \frac{(S_2^* - K_2)(S_1^*)^{\lambda_-} - (K_1 - S_1^*)(S_2^*)^{\lambda_-}}{(S_1^*)^{\lambda_-}(S_2^*)^{\lambda_+} - (S_1^*)^{\lambda_+}(S_2^*)^{\lambda_-}} \end{cases}$$

with $\lambda_\pm$ defined in the answer to Exercise 5.7.

5.15 The time-0 no-arbitrage value of the Bermudan put option is

$$P_0 = K\mathrm{e}^{-rT}\mathcal{N}_2\left(d_-\left(\frac{S_0}{S^*_{T_1}}, T_1\right), -d_-\left(\frac{S_0}{K}, T\right); -\sqrt{\frac{T_1}{T}}\right)$$
$$- S_0\mathrm{e}^{-qT}\mathcal{N}_2\left(d_+\left(\frac{S_0}{S^*_{T_1}}, T_1\right), -d_+\left(\frac{S_0}{K}, T\right); -\sqrt{\frac{T_1}{T}}\right)$$
$$+ K\mathrm{e}^{-rT_1}\mathcal{N}\left(-d_-\left(\frac{S_0}{S^*_{T_1}}, T_1\right)\right) - S_0\mathrm{e}^{-qT_1}\mathcal{N}\left(-d_+\left(\frac{S_0}{S^*_{T_1}}, T_1\right)\right),$$

where the critical value $S = S^*_{T_1}$ is the spot value at the intersection of the European price curve $P^E(S, K, T - T_1)$ and the payoff function $K - S$, as functions of S.

C.6 Chapter 6

6.1. $f(0;1,2) = 3.5\%$, $f(0;1,3) = 3.75\%$, $f(0;2,3) = 4\%$.

6.2. For a one-factor affine model, the yield rate is a linear function of the short rate. Thus, the payoff is a linear function of $r(T)$.

6.3. $\mathrm{E}[r(T) \mid r(t)] = r(t)\,\mathrm{e}^{-\int_t^T \beta(u)\,\mathrm{d}u} + \int_t^T \alpha(s)\mathrm{e}^{-\int_s^T \beta(u)\,\mathrm{d}u}\,\mathrm{d}s$ and $\mathrm{Var}(r(T) \mid r(t)) = \int_t^T \sigma^2(s)\mathrm{e}^{-2\int_s^T \beta(u)\,\mathrm{d}u}\,\mathrm{d}s$.

6.4. Let $Z(t,T,x)$ be the risk-neutral bond pricing function for the CIR process with the drift coefficient $(\alpha - \lambda\beta) - \beta x$ and diffusion coefficient $\sigma\sqrt{x}$. Then, $Z(r,T,r-\lambda)$ is the bond pricing function for the PS model.

6.5. $f(t,T) = f(0,T) + \mathrm{e}^{-\beta(T-t)}\left(r(t) - f(0,t)\right) + \frac{\sigma^2\mathrm{e}^{-\beta(T-t)}}{2\beta^2}\left(1 - \mathrm{e}^{-2\beta t}\right)\left(1 - \mathrm{e}^{-\beta(T-t)}\right)$.

6.7. (b) $\frac{\mathrm{d}Z(t,T)}{Z(t,T)} = \left(r(t)\,\mathrm{d}t - \frac{\sigma}{\beta}(1 - \mathrm{e}^{-\beta(T-t)})\,\mathrm{d}\widetilde{W}(t)\right)$;
(c) $\mathrm{d}y(t,T) = \frac{1}{T-t}\left[\left(y(t,T) - r(t) + \frac{\sigma^2}{2\beta^2}(1 - \mathrm{e}^{-\beta(T-t)})^2\right)\mathrm{d}t + \frac{\sigma}{\beta}(1 - \mathrm{e}^{-\beta(T-t)})\,\mathrm{d}\widetilde{W}(t)\right]$.

6.8. (a) $\frac{\mathrm{d}F_S(t)}{F_S(t)} = \left(\sigma_0 + \frac{\sigma}{\beta}(1 - \mathrm{e}^{-\beta(T-t)})\right)\mathrm{d}\widehat{W}(t)$;
(b) $\frac{\mathrm{d}S(t)}{S(t)} = (r(t) + \sigma_S(t)\sigma_Z(t,T))\,\mathrm{d}t + \sigma_S(t)\,\mathrm{d}\widehat{W}(t)$;
(c) $P(t) = Z(t,T)K\mathcal{N}(-d_-) - S(t)\mathcal{N}(-d_+)$ where $d_\pm = \dfrac{\ln(F_S(t)/K) \pm \bar{\sigma}^2_{t,T}(T-t)/2}{\bar{\sigma}_{t,T}\sqrt{T-t}}$
and $\bar{\sigma}^2_{t,T} = \frac{1}{T-t}\int_t^T (\sigma_S(u) - \sigma_Z(u,T))^2\,\mathrm{d}u$.

6.9. (a) $\widetilde{\mathrm{E}}[r(T) \mid r(t) = r] = r\mathrm{e}^{-\beta(T-t)} + \frac{\tilde{\alpha}}{\beta}(1 - \mathrm{e}^{-\beta(T-t)})$ and $\widetilde{\mathrm{Var}}(r(T) \mid r(t) = r) = \frac{1}{4\beta}\left(\mathrm{e}^{2\beta T} - \mathrm{e}^{4\beta t - 2\beta T}\right)$;
(b) $Z(t,T,r) = \mathrm{e}^{A(t,T) - C(t,T)r}$ and $y(t,T,r) = -\frac{A(t,T) - C(t,T)r}{T-t}$,

where

$$C(t,T) = \frac{1}{\beta}(1 - e^{-\beta(T-t)}) \text{ and}$$

$$A(t,T) = \frac{1}{2\beta^2}\left[\frac{e^{2\beta T} - e^{2\beta t}}{2\beta} - \frac{2e^{-\beta T}}{3\beta}(e^{3\beta T} - e^{3\beta t}) + \frac{e^{-2\beta T}}{4\beta}(e^{4\beta T} - e^{4\beta t})\right] - \frac{\tilde{\alpha}}{\beta}\left[T - t - \frac{1}{\beta}(1 - e^{-\beta(T-t)})\right];$$

(c) $\mathrm{d}f(t,T) = \tilde{\alpha}_F(t,T)\,\mathrm{d}r + \sigma_F(t,T)\,\mathrm{d}\widetilde{W}(t)$ where $\tilde{\alpha}_F(t,T) = \frac{e^{2\beta t}}{\beta}(e^{-\beta(T-t)} - e^{-2\beta(T-t)})$ and $\sigma_F(t,T) = e^{2\beta t - \beta T}$.

6.11. $\mathrm{d}F_S(t) = (\sigma_S(t) - \sigma_Z(t,T))F_S(t)\mathrm{d}\widehat{W}(t)$.

6.12. (a) $\mathrm{d}\frac{F_i(t)}{F_i(t)} = (\boldsymbol{\sigma}_i(t) - \boldsymbol{\sigma}_Z(t))\cdot \mathrm{d}\widehat{\mathbf{W}}(t)$.

(b) $C_t = S(t)\mathcal{N}\left(d_+\left(\frac{S(t)}{KZ(t,T)}, T-t\right)\right) - KZ(t,T)\mathcal{N}\left(d_-\left(\frac{S(t)}{KZ(t,T)}, T-t\right)\right)$

where $d_\pm(x,\tau) := \frac{\ln x \pm \frac{1}{2}\sigma_{F_S}^2\tau}{\sigma_{F_S}\sqrt{\tau}}$ and $\sigma_{F_S} = \|\boldsymbol{\sigma}_S - \boldsymbol{\sigma}_Z\|$.

(c) $V_t = S_2(t)\mathcal{N}\left(d_+\left(\frac{S_2(t)}{S_1(t)}, T-t\right)\right) - S_1(t)\mathcal{N}\left(d_-\left(\frac{S_2(t)}{S_1(t)}, T-t\right)\right)$

where $d_\pm(x,\tau) := \frac{\ln x \pm \frac{1}{2}\nu^2\tau}{\nu\sqrt{\tau}}$ and $\nu := \|\boldsymbol{\sigma}_{F_2} - \boldsymbol{\sigma}_{F_1}\| = \sqrt{\sigma_1^2 + \sigma_2^2 - 2\rho\sigma_1\sigma_2}$.

6.13. (a) $\frac{\partial V}{\partial t} + \frac{1}{2}\frac{\partial^2 V}{\partial x_1^2} + \frac{1}{2}\frac{\partial^2 V}{\partial x_2^2} - \beta_{11}x_1\frac{\partial V}{\partial x_1} - (\beta_{21}x_1 + \beta_{22}x_2)\frac{\partial V}{\partial x_2} - (a_0 + a_1x_1 + a_2x_2)V = 0$ with $V = V(t, x_1, x_2)$.

(c) For $\beta_{11} \neq \beta_{22}$, we have

$$c_1(\tau) = c_{10} + c_{11}e^{-\beta_{11}\tau} + c_{12}e^{-\beta_{22}\tau},$$

$$A(\tau) = \left[\frac{1}{2}\left(\frac{a_2^2}{\beta_{22}^2} + c_{10}^2\right) - a_0\right]\tau + c_{10}c_{11}\left(\frac{1 - e^{-\beta_{11}\tau}}{\beta_{11}}\right) + \left(c_{10}c_{12} - \frac{a_2^2}{\beta_{22}^2}\right)\left(\frac{1 - e^{-\beta_{22}\tau}}{\beta_{22}}\right) + \frac{1}{4}c_{11}^2\left(\frac{1 - e^{-2\beta_{11}\tau}}{\beta_{11}}\right) + \frac{1}{4}\left(\frac{a_2^2}{\beta_{22}^2} + c_{12}^2\right)\left(\frac{1 - e^{-2\beta_{22}\tau}}{\beta_{22}}\right) + c_{11}c_{12}\left(\frac{1 - e^{-(\beta_{11}+\beta_{22})\tau}}{\beta_{11} + \beta_{22}}\right)$$

with constants defined by

$$c_{10} := \frac{a_2\beta_{21} - a_1\beta_{22}}{\beta_{11}\beta_{22}}, \quad c_{11} := \frac{a_2\beta_{21}}{\beta_{11}(\beta_{11} - \beta_{22})} + \frac{a_1}{\beta_{11}}, \quad c_{12} := -\frac{a_2\beta_{21}}{\beta_{22}(\beta_{11} - \beta_{22})}.$$

The bond pricing formula is $v(\tau, x_1, x_2) = e^{A(\tau) + c_1(\tau)x_1 + c_2(\tau)x_2}$.

6.14. Functions $A(\tau)$, $c_1(\tau)$, and $c_2(\tau)$ satisfy

$$A'(\tau) = \mu_1 c_1(\tau) + \mu_2 c_2(\tau) - a_0\,,$$
$$c_1'(\tau) = \frac{1}{2}c_1^2(\tau) - \beta_{11}c_1(\tau) - \beta_{21}c_2(\tau) - a_1\,,$$
$$c_2'(\tau) = \frac{1}{2}c_2^2(\tau) - \beta_{12}c_1(\tau) - \beta_{22}c_2(\tau) - a_2\,.$$

6.16. $\text{Caplet}(t) - \text{Floorlet}(t) = \tau Z(t, T+\tau)L(t) - \tau\kappa Z(t, T+\tau)$.

$\text{Floorlet}(t) = \tau\kappa Z(t, T+\tau)\mathcal{N}(-a_-(t, L(t))) - \tau Z(t, T+\tau)L(t)\mathcal{N}(-a_+(t, L(t)))$

where $a_\pm(t, L) = \frac{1}{\bar{v}_{t,T}\sqrt{T-t}}\left[\ln\frac{L}{\kappa} \pm \frac{1}{2}\bar{v}_{t,T}^2(T-t)\right]$ and $\bar{v}_{t,T}^2 = \frac{1}{T-t}\int_t^T v^2(u,T)\,\mathrm{d}u$.

6.18. (a) $\text{Floorlet}_{T+\tau}(t) = \tau Z(t, T+\tau)\left[\kappa\mathcal{N}(-a_-(T, L(t))) - L(t)\mathcal{N}(-a_+(T, L(t)))\right]$

(b) $\text{Floor}(t) = \tau\sum_{i=1}^n Z(t, T_i)\left[\kappa\mathcal{N}(-a_-(T_{i-1}, L_{i-1}(t))) - L_{i-1}(t)\mathcal{N}(-a_+(T_{i-1}, L_{i-1}(t)))\right]$;

where $a_\pm(t, L) := \frac{1}{\bar{v}_{t,T}\sqrt{T-t}}\left[\ln\frac{L}{\kappa} \pm \frac{1}{2}\bar{v}_{t,T}^2(T-t)\right]$ and $\bar{v}_{t,T}^2 := \frac{1}{T-t}\int_t^T v^2(u,T)\,\mathrm{d}u$.

C.7 Chapter 7

7.2. (a) $Norm(r(T-t)+\mu(T-t)^2/2, \sigma^2(T-t)^3/3)$.
(b) $\varphi(u) = \exp(iu(r(T-t)+\mu(T-t)^2/2) - \sigma^2 u^2 (T-t)^3/3)$.
(c) $Z(t,T,r) = \exp(-r(T-t) - \mu(T-t)^2/2 - \sigma^2(T-t)^3/3)$.

7.3. $X(t)$ satisfies the SDE $\mathrm{d}X(t) = (2\mu+2)\,\mathrm{d}t + 2\sqrt{X(t)}\,\mathrm{d}\widehat{W}(t)$ where $\mu = 1/(2\beta)$ and $\widehat{W}(t) = -\mathrm{sgn}(\beta)W(t)$.

7.8. $a = \lambda(1-\mathrm{e})$.

7.10. (a) $\varphi_{X(t)}(u) = \mathrm{e}^{iu\mu t - u^2\sigma^2 t/2 - \lambda t u^2/(a^2+u^2)}$.
(b) $\mu = r - \sigma^2/2 - \lambda/(a^2-1)$.

7.12. (a) $\mathrm{v}(t) = \mathrm{e}^{-\kappa t}\left(\frac{\xi^2}{4}\right)X(\tau_t)$, where $m = \frac{2\kappa\theta}{\xi^2} - 1$ and $\tau_t = (\mathrm{e}^{\kappa t}-1)/\kappa$.
(b) $f_{\mathrm{v}(t)}(v) = p_X\left(\tau_t; v_0\left(\frac{4}{\xi^2}\right), v\mathrm{e}^{\kappa t}\left(\frac{4}{\xi^2}\right)\right)\mathrm{e}^{\kappa t}\left(\frac{4}{\xi^2}\right)$.

7.13. $$f_{X(t)}(x) = \frac{2\exp\left(\frac{\theta x}{\sigma^2}\right)}{\sigma\sqrt{2\pi v}\,\Gamma(\frac{t}{v})}\left(\frac{b}{a}\right)^{c/2} K_c(2\sqrt{ab}).$$

7.14. (a) $\varphi_X(u) = \left(\frac{\beta}{\beta - iu}\right)^{\alpha}$. (b) $\varphi_{X(t)|G(t)}(u|g) = \mathrm{e}^{i\theta g u - \sigma^2 g u^2/2}$.
(c) $\varphi_{X(t)}(u) = (1 - i\theta u v + \sigma^2 u^2 v/2)^{-t/v}$.
(d) $\varphi_{\ln S(t)}(u) = \mathrm{e}^{iu(S_0 + (r-q-\omega)t)}\varphi_{X(t)}(u)$ and $\psi(v) = \frac{\mathrm{e}^{-rT}\varphi_{\ln S(T)}(v - i(\alpha+1))}{(\alpha+iv)(\alpha+1+iv)}$.

D

Glossary of Symbols and Abbreviations

$B(t)$ or B_t	value of a risk-free security (e.g. a bond or a bank account) at time t
$b(x; n, p)$	probability mass function for the binomial law $Bin(n, p)$
$\mathcal{B}(x; n, p)$	(cumulative) probability distribution function for the binomial law $Bin(n, p)$
$Bin(n, p)$	binomial probability distribution with number of trials n and success probability p
C^A	American call
C^E	European call
CDF	cumulative (probability) distribution function
$\mathrm{Corr}(X, Y)$	correlation coefficient of X and Y
$\mathrm{Cov}(X, Y)$	covariance of X and Y
CRR	Cox–Ross–Rubinstein
D	downward move in a binomial tree
div	dividend
$d(t, T)$	discount factor from time t to time T
$d(t)$	discounting function over a time period of length t
$\frac{\mathrm{d}\mathbb{P}}{\mathrm{d}\mathbb{Q}}$	Radon–Nikodym derivative of $\mathbb{P}$ w.r.t. $\mathbb{Q}$
$\left(\frac{\mathrm{d}\mathbb{P}}{\mathrm{d}\mathbb{Q}}\right)_t$	Radon–Nikodym derivative process at time t
EMM	equivalent martingale measure
$Exp(\lambda)$	exponential probability distribution with rate λ
$\mathrm{E}[X]$	mathematical expectation of X
$\widetilde{\mathrm{E}}[X]$	risk-neutral mathematical expectation of X
$\mathrm{E}[X \mid \mathcal{F}]$	mathematical expectation of X conditional on a σ-algebra $\mathcal{F}$
$\mathrm{E}_t[X]$	mathematical expectation of X conditional on $\mathcal{F}_t$
$\widetilde{\mathrm{E}}^{(g)}[X]$	mathematical expectation of X w.r.t. the probability measure $\mathbb{P}^{(g)}$
$\widetilde{\mathrm{E}}_t^{(g)}[X]$	mathematical expectation of X conditional on $\mathcal{F}_t$ w.r.t. the probability measure $\mathbb{P}^{(g)}$

DOI: 10.1201/9780429468889-D

$\mathrm{E}_{t,x}[X]$	mathematical expectation of X conditional on an underlying process having value x at time t
$\mathrm{E}_{t,\mathbf{x}}[X]$	mathematical expectation of X conditional on a underlying vector process having value $\mathbf{x}$ at time t
$\mathcal{E}_t(\gamma \cdot W)$	exponential martingale process of an adapted process γ w.r.t. Brownian motion W
$\mathcal{E}_t(\boldsymbol{\gamma} \cdot \mathbf{W})$	exponential martingale process of an adapted vector process $\boldsymbol{\gamma}$ w.r.t. vector Brownian motion W
$\mathcal{F}_t$	σ-algebra generated by information available at time t
$\mathbb{F}$	filtration
$f(t;T,T')$	forward rate at time t for interval $[T,T']$
$f(t,T)$	instantaneous forward rate at time t for maturity T
$F(t,T)$	forward price at time t for maturity T
f_X, f_D	probability density function (of random variable X or probability distribution D)
F_X, F_D	(cumulative) distribution function (of random variable X or probability distribution D)
$Gamma(\kappa,\lambda)$	gamma probability distribution with shape parameter κ and rate parameter λ
GBM	geometric Brownian motion
$\mathbb{I}_A$	indicator of event (or set) A
X,Y,Z	random variables
iff	if and only if
i.i.d.	independent and identically distributed
K	strike price
$\Lambda(\cdot)$	payoff function
m_t^X	minimum over $[0,t]$ of the process X
M_t^X	maximum over $[0,t]$ of the process X
$n(x)$	probability density function for a standard normal law
$n_2(x,y;\rho)$	joint probability density function for two standard normal random variables with correlation coefficient ρ
$n_n(x_1,\ldots,x_n;\boldsymbol{\rho})$	joint probability density function for n standard normal random variables with correlation matrix $\boldsymbol{\rho}$
$\mathcal{N}(x)$	(cumulative) probability distribution function for a standard normal law

$\mathcal{N}_2(x, y; \rho)$	joint probability distribution function for two standard normal random variables with correlation coefficient ρ
$\mathcal{N}_n(x_1, \ldots, x_n; \boldsymbol{\rho})$	joint probability distribution function for n standard normal random variables with correlation matrix $\boldsymbol{\rho}$
$Norm(\mu, \sigma^2)$	normal probability distribution with mean μ and variance σ^2
$Norm_n(\mathbf{m}, \boldsymbol{\Sigma})$	n-variate normal probability distribution with mean vector $\mathbf{m}$ and covariance matrix $\boldsymbol{\Sigma}$
ω	scenario (element of a state space)
Ω	state space
$\mathbb{P}$	probability measure
$\widetilde{\mathbb{P}} \equiv \widetilde{\mathbb{P}}^{(B)}$	risk-neutral probability measure with bank account as numéraire
$\widetilde{\mathbb{P}}^{(g)}$	risk-neutral probability measure (or EMM) with asset g as numéraire
$\mathbb{P} \sim \mathbb{Q}$	equivalent probability measures
$\mathbb{P}(A)$	probability of event A
$\mathbb{P}(A \mid B)$	probability of event A conditional on event B
$\mathcal{P}_t$	partition of Ω generated by information available at time t
$\mathcal{P}(X)$	partition of Ω generated by random variable X
$\mathcal{P}(X_1, \ldots, X_n)$	partition of Ω generated by random variables $X_1, \ldots, X_n$
P^A	American put
P^E	European put
$\Pi(t)$ or Π_t	portfolio value at time t
$\overline{\Pi}(t)$ or $\overline{\Pi}_t$	discounted portfolio value at time t
$p(s, t; x, y)$	transition PDF for a one-dimensional diffusion
$p(t; x, y)$	time-homogeneous transition PDF for a one-dimensional diffusion
$p(s, t; \mathbf{x}, \mathbf{y})$	transition PDF for a multidimensional diffusion
$p(t; \mathbf{x}, \mathbf{y})$	time-homogeneous transition PDF for a multidimensional diffusion
PDF	probability density function
$Pois(\lambda)$	Poisson probability distribution with rate λ
ϱ	Radon–Nikodym derivative
ϱ_t	Radon–Nikodym derivative process at time t
ρ	correlation coefficient
r	interest rate

$r(t)$	instantaneous interest rate at time t
$\mathbb{R}$	set of real numbers $(-\infty, \infty)$
$\mathbb{R}_+$	set of nonnegative real numbers $[0, \infty)$
$\sigma(X)$	σ-algebra generated by random variable X
$\sigma(\{X_\lambda\})$	σ-algebra generated by a collection $\{X_\lambda\}$
$S(t)$ or S_t	price of a risky asset (e.g. a stock) at time t
$S_i(t)$ or S_t^i	price of the ith risky asset at time t
SDE	stochastic differential equation
SQB	squared Bessel process
T	maturity time; expiry time; exercise time
$\mathcal{T}_b^X$	first hitting time of X at level b
$\mathcal{T}_{(a,b)}^X$	first exit time of X from the interval (a, b)
U	upward move in a binomial tree
Unif(a, b)	uniform probability distribution on an interval (a, b)
$V(t)$	(accumulated) value function at time t
$v(\tau, S)$	derivative pricing function of time to maturity τ and spot S
$V(t, S)$ or $V_t(S)$	derivative pricing function of calendar time t and spot S
$\overline{V}(t, S)$ or $\overline{V}_t(S)$	discounted derivative pricing function
$\mathrm{Var}(X)$	variance of X
VaR	Value at Risk
w.r.t.	with respect to
W	Brownian motion
$W^{(\mu,\sigma)}$	scaled Brownian motion with a linear drift
$y(\tau)$	yield rate for time to maturity τ
$y(t, T)$	yield rate at time t for maturity T
$Z(t, T)$	zero-coupon bond price at time t for maturity T
ZCB	zero-coupon bond

Greek Alphabet

A, α alpha

B, β beta

Γ, γ gamma

Δ, δ delta

E, ϵ, ε epsilon

Z, ζ zeta

H, η eta

Θ, θ, ϑ theta

I, ι iota

K, κ kappa

Λ, λ lambda

M, μ mu

N, ν nu

Ξ, ξ xi

O, o omicron

Π, π pi

R, ρ, ϱ rho

Σ, σ sigma

T, τ tau

Υ, υ upsilon

Φ, ϕ, φ phi

X, χ chi

Ψ, ψ psi

Ω, ω omega

References

Theory of Probability and Stochastic Processes

D. Applebaum. *Lévy Processes and Stochastic Calculus.* Cambridge University Press, 2009.

K.B. Athreya and S.N. Lahiri. *Measure Theory and Probability Theory.* Springer Texts in Statistics. Springer-Verlag, 2006.

Z. Brzeźniak and T. Zastawniak. *Basic Stochastic Processes: A Course Through Exercices.* Springer-Verlag, 1999.

M. Capinski and P.E. Kopp. *Measure, Integral and Probability.* Springer Undergraduate Mathematics Series. Springer-Verlag, 2004.

R. Cont and P. Tankov. *Financial Modelling with Jump Processes.* Chapman & Hall/CRC Financial Mathematics Series. Taylor & Francis, 2004.

W. Feller. *An Introduction to Probability Theory and Its Applications*, volume 1. John Wiley & Sons, 1971a.

W. Feller. *An Introduction to Probability Theory and Its Applications*, volume 2. John Wiley & Sons, 1971b.

A. Gut. *Probability: a Graduate Course.* Springer-Verlag, 2005.

A. Gut. *An Intermediate Course in Probability.* Springer-Verlag, 2009.

R.V. Hogg and E.A. Tanis. *Probability and Statistical Inference.* Pearson/Prentice Hall, 8th edition, 2010.

M. Jeanblanc, M. Yor, and M. Chesney. *Mathematical Methods for Financial Markets.* Springer Finance. Springer-Verlag, 2009.

I. Karatzas and S.E. Shreve. *Brownian Motion and Stochastic Calculus.* Graduate Texts in Mathematics. Springer New York, 1991.

S. Karlin and H.M. Taylor. *A First Course in Stochastic Processes.* Academic Press, 1975.

S. Karlin and H.M. Taylor. *A Second Course in Stochastic Processes.* Academic Press, 1981.

F.C. Klebaner. *Introduction to Stochastic Calculus with applications.* Imperial College Press, 2005.

H.H. Kuo. *Introduction to Stochastic Integration.* Springer-Verlag, 2006.

B.K. Øksendal. *Stochastic Differential Equations: An Introduction with Applications.* Springer-Verlag, 6th edition, 2003.

M.M. Rao and R.J. Swift. *Probability Theory with Applications.* Mathematics and Its Applications. Springer-Verlag, 2006.

A.N. Shiryaev. *Probability.* Graduate Texts in Mathematics. Springer-Verlag, 1996.

Introduction to Mathematics of Finance

R. Brown, S. Kopp, and P. Zima. *Mathematics of Finance.* McGraw-Hill Ryerson Limited, 7th edition, 2011.

J.R. Buchanan. *An Undergraduate Introduction to Financial Mathematics.* World Scientific, 2008.

M. Capiński and T. Zastawniak. *Mathematics for Finance: An Introduction to Financial Engineering.* Springer Undergraduate Mathematics Series. Springer-Verlag, 2003.

M. Davis, L. Bachelier, A. Etheridge, and P.A. Samuelson. *Louis Bachelier's Theory of Speculation: The Origins of Modern Finance.* Princeton University Press, 2011.

D. Lovelock, M. Mendel, and A.L. Wright. *An Introduction to the Mathematics of Money: Saving and Investing.* Springer-Verlag, 2007.

T. Mikosch. *Elementary Stochastic Calculus: with Finance in View.* World Scientific, 1998.

S. Roman. *Introduction to the Mathematics of Finance: From Risk Management to Options Pricing.* Undergraduate Texts in Mathematics. Springer-Verlag, 2004.

S.M. Ross. *An Elementary Introduction to Mathematical Finance.* Cambridge University Press, 2011.

P. Wilmott, S. Howison, and J. Dewynne. *The Mathematics of Financial Derivatives: A Student Introduction.* Cambridge University Press, 1995.

Mathematics of Finance (Discrete-Time)

M. Capiński and E. Kopp. *Discrete Models of Financial Markets.* Mastering Mathematical Finance. Cambridge University Press, 2012.

H. Föllmer and A. Schied. *Stochastic Finance: An Introduction in Discrete Time.* De Gruyter Textbook Series. De Gruyter, 2011.

P.K. Medina and S. Merino. *Mathematical Finance and Probability. A Discrete Introduction.* Birkhauser, 2004.

S.R. Pliska. *Introduction to Mathematical Finance: Discrete Time Models.* John Wiley & Sons, 1997.

S.E. Shreve. *Stochastic Calculus for Finance I: The Binomial Asset Pricing Model.* Springer Finance. Springer-Verlag, 2012.

Mathematics of Finance (Continuous-Time)

C. Albanese and G. Campolieti. *Advanced Derivatives Pricing and Risk Management: Theory, Tools and Hands-on Programming Application.* Academic Press advanced finance series. Elsevier Academic Press, 2006.

M. Avellaneda and P. Laurence. *Quantitative Modeling of Derivative Securities: From Theory To Practice.* Chapman & Hall/CRC, 1999.

K. Back. *A Course in Derivative Securities: Introduction to Theory and Computation.* Springer Finance. Springer-Verlag, 2005.

M. Baxter and A. Rennie. *Financial Calculus: An Introduction to Derivative Pricing.* Cambridge University Press, 1996.

D. Brigo and F. Mercurio. *Interest Rate Models - Theory and Practice: With Smile, Inflation and Credit.* Springer Finance. Springer-Verlag, 2007.

A.J.G. Cairns. *Interest Rate Models: An Introduction.* Princeton University Press, 2004.

R.-A. Dana and M. Jeanblanc. *Financial Markets in Continuous Time.* Springer Finance. Springer-Verlag, 2003.

R.J. Elliott and P.E. Kopp. *Mathematics of Financial Markets.* Springer-Verlag, 2005.

A. Etheridge. *A Course in Financial Calculus.* Cambridge University Press, 2002.

J.-P. Fouque, G. Papanicolaou, and K.R. Sircar. *Derivatives in Financial Markets with Stochastic Volatility.* Cambridge University Press, 2000.

Y.K. Kwok. *Mathematical Models of Financial Derivatives.* Springer Finance. Springer-Verlag, 2008.

A.L. Lewis. *Option Valuation under Stochastic Volatility: With Mathematica Code.* Finance Press, 2000.

A.N. Shiryaev. *Essentials of Stochastic Finance: Facts, Models, Theory.* Advanced series on statistical science & applied probability. World Scientific, 1999.

S.E. Shreve. *Stochastic Calculus for Finance II: Continuous-Time Models.* Springer Finance. Springer-Verlag, 2010.

P. Wilmott. *Derivatives: the Theory and Practice of Financial Engineering.* Wiley Frontiers in Finance Series. John Wiley & Sons, 1998.

Computational Methods

G. Fusai and A. Roncoroni. *Implementing Models in Quantitative Finance: Methods and Cases: Methods and Cases.* Springer Finance. Springer-Verlag, 2007.

P. Glasserman. *Monte Carlo Methods in Financial Engineering.* Applications of mathematics: stochastic modelling and applied probability. Springer-Verlag, 2004.

A. Hirsa. *Computational Methods in Finance.* Chapman & Hall/CRC Financial Mathematics Series. Taylor & Francis, 2012.

P.E. Kloeden and E. Platen. *Numerical Solution of Stochastic Differential Equations.* Applications of mathematics: stochastic modelling and applied probability. Springer-Verlag, 1992.

R. Korn, E. Korn, and G. Kroisandt. *Monte Carlo Methods and Models in Finance and Insurance.* Chapman & Hall/CRC Financial Mathematics Series. Taylor & Francis, 2010.

D.L. McLeish. *Monte Carlo Simulation and Finance.* Wiley Finance. John Wiley & Sons, 2011.

E. Platen and N. Bruti-Liberati. *Numerical Solution of Stochastic Differential Equations with Jumps in Finance.* Stochastic modelling and applied probability. Springer-Verlag, 2010.

E. Platen and D. Heath. *A Benchmark Approach to Quantitative Finance.* Springer Finance. Springer-Verlag, 2006.

D. Tavella. *Quantitative Methods in Derivatives Pricing: An Introduction to Computational Finance.* Wiley Finance. John Wiley & Sons, 2003.

Financial Economics

S.L. Allen. *Financial Risk Management: A Practitioner's Guide to Managing Market and Credit Risk.* Wiley Finance. John Wiley & Sons, 2012.

J.P. Danthine and J.B. Donaldson. *Intermediate Financial Theory.* Academic Press advanced finance series. Elsevier Academic Press, 2005.

J. Gatheral. *The Volatility Surface: A Practitioner's Guide.* John Wiley & Sons, 2011.

J.C. Hull. *Options, Futures, and Other Derivatives.* Pearson Education, 8th edition, 2011.

D.G. Luenberger. *Investment Science.* Oxford University Press, 1997.

R.L. McDonald. *Derivatives Markets.* Addison-Wesley series in finance. Addison Wesley, 2003.

H.H. Panjer and P.P. Boyle. *Financial Economics: With Applications to Investments, Insurance, and Pensions.* Actuarial Foundation, 1998.

Index

Printed and bound by CPI Group (UK) Ltd, Croydon, CR0 4YY
28/10/2024
01780248-0001